COVER ILLUSTRATIONS

Section 1 Cover:

This Voyager 2 image of Saturn was obtained August 11, 1981 from a range of 14.7 million kilometers (9.1 million miles). North is at the upper right edge of the disc. Seen above the planet are the satellites Dione (right) and Enceladus. This false-color print shows a green spot at the south edge of a yellow band; in true color, the spot would appear brown and the band white. A bright yellow spot slightly above and to the left in this image moves eastward relative to the green spot at a rate that allows it to pass the green feature in about 50 days. The convective clouds that appear between the two spots are typical of the region. Here, the smallest visible structures measure about 270 km (170 mi.). The Voyager project is managed for NASA by the Jet Propulsion Laboratory, Pasadena, California. JPL Photo P-23914C.

Section 2 Cover:

This synthesized picture illustrates the brightness contours that represent the fine structure observed in Saturn's F-ring using a stellar occulation profile compiled by the Voyager 2 photopolarimeter. Because of the instrument's high sensitivity and the small apparent size of the star used, Delta Scorpii, structure was observed on a much finer scale—hundreds of meters—that would otherwise be possible. The highest-resolution of the Voyager imaging system is seldom better than 10 kilometers. The F-ring was represented here by sweeping the "trace" in longitude with much the same overall form as seen at several points by the imaging system, with a dense outer core and a diffuse set of inner strands. A great deal of radial structure at the subkilometer scale, however, is evident in these data. It must be remembered that the azimuthal symmetry and lack of kinks or knots here are only artifacts of the representation, which takes a single narrow slice of the rings. JPL Photo P-23960C.

GEOCHIMICA ET COSMOCHIMICA ACTA

Supplement 16

PROCEEDINGS OF THE TWELFTH LUNAR AND PLANETARY SCIENCE CONFERENCE

Proceedings of Lunar and Planetary Science

Volume 12, Part B

Houston, Texas, March 16–20, 1981

LUNAR AND PLANETARY INSTITUTE PROCEEDINGS PUBLICATIONS

Numbers to left refer to *Geochimica et Cosmochimica Acta* Supplement number.
1 *Proceedings of the Apollo 11 Lunar Science Conference* (1970) A. A. Levinson, ed., Pergamon Press, N.Y. 3 v., 2492 pp.
2 *Proceedings of the Second Lunar Science Conference* (1971) A. A. Levinson, ed., MIT Press (distributed by Pergamon Press, N.Y.). 3 v., 2818 pp.
3 *Proceedings of the Third Lunar Science Conference* (1972) E. A. King, D. Heymann, and D. R. Criswell, eds., MIT Press (distributed by Pergamon Press, N.Y.). 3 v., 3263 pp.
4 *Proceedings of the Fourth Lunar Science Conference* (1973) W. A. Gose, ed., Pergamon Press, N.Y. 3 v., 3290 pp.
5 *Proceedings of the Fifth Lunar Science Conference* (1974) W. A. Gose, ed., Pergamon Press, N.Y. 3 v., 3134 pp.
6 *Proceedings of the Sixth Lunar Science Conference* (1975) R. B. Merrill, ed., Pergamon Press, N.Y. 3 v., 3637 pp.
7 *Proceedings of the Seventh Lunar Science Conference* (1976) R. B. Merrill, ed., Pergamon Press, N.Y. 3 v., 3651 pp.
8 *Proceedings of the Eighth Lunar Science Conference* (1977) R. B. Merrill, ed., Pergamon Press, N.Y. 3 v., 3965 pp.
9 *Mare Crisium: The View from Luna 24.* Proceedings of the Conference on Luna 24, December 1977, Houston, Texas (1978) R. B. Merrill and J. J. Papike, eds., Pergamon Press, N.Y. 719 pp.
10 *Proceedings of the Ninth Lunar and Planetary Science Conference* (1978) R. B. Merrill, ed., Pergamon Press, N.Y. 3 v., 3973 pp.
11 *Proceedings of the Tenth Lunar and Planetary Science Conference* (1979) R. B. Merrill, ed., Pergamon Press, N.Y. 3 v., 3077 pp.
12 *Proceedings of the Conference on the Lunar Highlands Crust.* November 1979, Houston, Texas (1980) J. J. Papike and R. B. Merrill, eds., Pergamon Press, N.Y. 505 pp.
13 *Proceedings of the Conference on the Ancient Sun.* October 1979, Boulder, Colorado (1980) R. O. Pepin, J. A. Eddy, and R. B. Merrill, eds., Pergamon Press, N.Y. 581 pp.
14 *Proceedings of the Eleventh Lunar and Planetary Science Conference* (1980) R. B. Merrill, ed., Pergamon Press, N.Y. 3 v., 2502 pp.
15 *Proceedings of the Conference on Multi-ring Basins: Formation and Evolution* (1981) Proceedings of Lunar and Planetary Science, vol. 12A. P. H. Schultz and R. B. Merrill, eds., Pergamon Press, N.Y. 295 pp.
16 *Proceedings of the Twelfth Lunar and Planetary Science Conference* (1981) Proceedings of Lunar and Planetary Science, vol. 12B. R. Ridings and R. B. Merrill, eds. Pergamon Press, N.Y. 2 sections, 1833 pp.

Other volumes compiled by the Lunar and Planetary Institute are:

Impact and Explosion Cratering Proceedings of the Symposium on Planetary Cratering Mechanics, September 1976, Flagstaff, Arizona (1977) D. J. Roddy, R. O. Pepin, and R. B. Merrill, eds., Pergamon Press, N.Y. 1301 pp.

Index to the Proceedings of the Lunar and Planetary Science Conferences 1970–1978. Index to Geochimica et Cosmochimica Acta Supplements 1–8, and 10 (1979) Compiled by A. R. Masterson; R. B. Merrill, ed., Pergamon Press, N.Y. 261 pp. (Includes table of contents for each publication indexed.)

GEOCHIMICA ET COSMOCHIMICA ACTA
Journal of The Geochemical Society and The Meteoritical Society

Supplement 16

PROCEEDINGS OF THE TWELFTH LUNAR AND PLANETARY SCIENCE CONFERENCE
Houston, Texas, March 16–20, 1981

Compiled by the
Lunar and Planetary Institute
Houston, Texas

Section 2
Planets, Asteroids and Satellites

Proceedings of Lunar and Planetary Science
Volume 12, Part B

Pergamon Press

New York • Oxford • Toronto • Sydney • Frankfurt • Paris

Pergamon Press Offices:

U.S.A.	Pergamon Press Inc., Maxwell House, Fairview Park, Elmsford, New York 10523, U.S.A.
U.K.	Pergamon Press Ltd., Headington Hill Hall, Oxford OX30BW, England
CANADA	Pergamon of Canada Ltd., 150 Consumers Road, Willowdale, Ontario M2J 1P9, Canada
AUSTRALIA	Pergamon Press (Aust) Pty. Ltd., P.O. Box 544 Potts Point, NSW 2011, Australia
FRANCE	Pergamon Press SARL, 24 rue des Ecoles 75240 Paris, Cedex 05, France
FEDERAL REPUBLIC OF GERMANY	Pergamon Press GmbH, 6242 Kronberg/Taunus, Pferdstrasse 1, Federal Republic of Germany

Material in this volume may be copied without restraint for library, abstract service, educational or personal research purposes, however, republication of any paper or portion thereof requires the written permission of the author(s) as well as appropriate acknowledgement of this source.

First edition 1982

Type set by European Printing Corporation Ltd.,
printed by Publishers Production International,
and bound by Arnold's Bindery
in the United States of America

Library of Congress Cataloging in Publication Data

Lunar and Planetary Science Conference (12th :
    1981 : Houston, Tex.)
    Proceedings of the Twelfth Lunar and Planetary
Science Conference, Houston, Texas, March 16-20,
1981.

    (Geochimica et cosmochimica acta. Supplement ;
no. 16)
    Bibliography: p.
    Includes index.
    Contents: v. 1. The moon -- v. 2. Planets
asteroids, and satellites.
    1. Lunar geology--Congresses. 2. Planets--
Geology--Congresses. 3. Meteorites--Congresses.
I. Lunar and Planetary Institute. II. Title
III. Series.
QB592.L84   1981   559.9'1   81-17703
ISBN 0-08-028074-9   AACR2

MANAGING EDITORS

R. B. Merrill, Lunar and Planetary Institute, Houston, Texas 77058
R. Ridings, Lunar and Planetary Institute, Houston, Texas 77058

EDITORIAL BOARD

F. Hörz, Geology Branch, NASA Johnson Space Center, Houston, Texas 77058
W. Mendell, Geology Branch, NASA Johnson Space Center, Houston, Texas 77058
D. Phinney, Physics Department, Wheaton College, Wheaton, Illinois, 60187, and Lockheed Engineering and Management Services Company, Incorporated, Houston, Texas 77058
W. C. Phinney, Geology Branch, NASA Johnson Space Center, Houston, Texas 77058

BOARD OF ASSOCIATE EDITORS

A. Basu, Department of Geology, Indiana University, Bloomington, Indiana 47405
M. J. Cintala, Department of Geological Sciences, Brown University, Providence, Rhode Island 02912
S. K. Croft, Lunar and Planetary Institute, Houston, Texas 77058
W. D. Daily, Lawrence Livermore Laboratory, Livermore, California 94550
J. S. Delaney, Department of Mineral Science, American Museum of Natural History, New York, New York 10024
J. W. Delano, Department of Earth and Space Sciences, State University of New York, Stony Brook, New York 11794
E. K. Gibson, Geochemistry Branch, NASA Johnson Space Center, Houston, Texas 77058
R. A. F. Grieve, Department of Geological Sciences, Brown University, Providence, Rhode Island 02912
A. W. Harris, Jet Propulsion Laboratory, Pasadena, California 91109
G. F. Herzog, Department of Earth and Space Sciences, State University of New York, Stony Brook, New York 11794
R. H. Hewins, Department of Geological Sciences, Rutgers University, New Brunswick, New Jersey 08903
R. M. Housley, Rockwell Science Center, Thousand Oaks, California 93160
I. D. Hutcheon, Enrico Fermi Institute, University of Chicago, Chicago, Illinois 60637
J. F. Kerridge, Institute of Geophysics, University of California at Los Angeles, Los Angeles, California 90024
M. M. Lindstrom, Department of Earth and Planetary Science, Washington University, St. Louis, Missouri 63130
G. E. Lofgren, Geology Branch, NASA Johnson Space Center, Houston, Texas 77058
M.-S. Ma, Radiation Center, Oregon State University, Corvallis, Oregon 97331

M. C. Malin, Department of Geology, Arizona State University, Tempe, Arizona 85281

H. J. Melosh, Department of Earth and Space Sciences, State University of New York, Stony Brook, New York 11794

R. V. Morris, Geochemistry Branch, NASA Johnson Space Center, Houston, Texas 77058

L. E. Nyquist, Geochemistry Branch, NASA Johnson Space Center, Houston, Texas 77058

C. Pieters, Department of Geological Sciences, Brown University, Providence, Rhode Island 02912

F. A. Podosek, Department of Earth and Planetary Sciences, Washington University, St. Louis, Missouri 63130

G. Ryder, Northrop Services, Incorporated, Houston, Texas 77034

P. D. Spudis, U.S. Geological Survey, Flagstaff, Arizona 86001

E. M. Stolper, Division of Geological and Planetary Sciences, California Institute of Technology, Pasadena, California 91125

L. A. Taylor, Department of Geological Sciences, University of Tennessee, Knoxville, Tennessee 37916

C. A. Wood, Geology Branch, NASA Johnson Space Center, Houston, Texas 77058

H. A. Zook, Geology Branch, NASA Johnson Space Center, Houston, Texas 77058

TECHNICAL EDITOR

K. Hrametz, Lunar and Planetary Institute, Houston, Texas 77058

EDITORIAL STAFF

L. Gaddy
R. Edwards
T. McCasey
J. Samuels
Lunar and Planetary Institute, Houston, Texas 77058

Global maps of Mars' properties

Available information about Mars has been greatly increased by spacecraft and Earth-based observations during the last decade. As the resolution of observations has improved, the quantity of data has increased to the point where computer processing has become an essential analytical tool. In order to coordinate the efforts of scientists attempting to use these data—which vary widely in type and resolution—the Mars Consortium was formed. This organization grew out of discipline groups within the Viking program; its first formal meeting was held in Flagstaff, Arizona, in October, 1980. A major effort of the Consortium is to transform diverse data sets into a single format that facilitates display of individual data sets and comparisons among them. The Consortium also coordinates the work of many scientists studying Mars.

Spatial resolution of martian observations ranges from tens of meters for low-altitude Viking imaging to hundreds of kilometers for gravity and Earth-based photometry. Data sets such as geologic maps that are derived from primary observations typically have a resolution of tens of kilometers. The standard representation chosen is a compromise between high resolution and modest storage requirements, and between ease of access and a statistically unbiased representation. The format chosen is equivalent to a simple cylindrical projection, each element being 1/4° latitude by 1/4° longitude, and covers the entire planet. The internal storage arrangement is chosen so that direct output on conventional graphics devices is in the Mars mapping convention (north at top, 180° longitude at the left and right borders). Data can have resolution differing by powers of 2; in addition, some data sets have been constructed with resolution of 10° or integral divisions of 10°.

Data can be submitted for Consortium processing as maps in several projections, as vector-format data representing multi-parameter observations at single points, or as digital raster arrays in any standard mapping projection. The data are archived as received and converted into the Consortium's standard format. If in analog form (maps), the data are digitized in their original projections. The data are converted into image or vector form as appropriate, retaining all original spatial and intensity resolution. The computer files may be 8-, 16-, or 32-bit, as appropriate. Smoothing and interpolation techniques, as appropriate and as warranted by the data, are used to generate as complete a global image as is faithful to the original observations. High speed mapping transforms and sorting routines are used to place the information into simple cylindrical projections. Sinusoidal equal-area projections are made for the purpose of statistical analyses. Images in conformal projections (Mercator, Lambert, and polar stereographic) are produced as desired.

All data are scaled into 8-bit integers, as most martian data sets can be adequately represented with a resolution of one part in 256. However, most of the image-processing routines for filtering, correlation, and statistics are available in 8-bit and 16-bit integer and 32-bit floating-point versions.

The format of the Consortium is similar to that developed for lunar data (see Frontispiece, *Proceedings of the Eighth Lunar Science Conference*, 1977); many of the computer routines are descendant from those developed for the lunar data.

Some of the Consortium data sets are displayed here (Plates 1–15). The development of these and other sets is described in more detail by Kieffer, Davis, and Soderblom in these Proceedings. Major additions of cratering information and Earth-based radar data are anticipated in the near future. Data processing is by K. Edwards, C. E. Isbell, E. M. Sanchez, D. Casebier, and P. J. Helm.

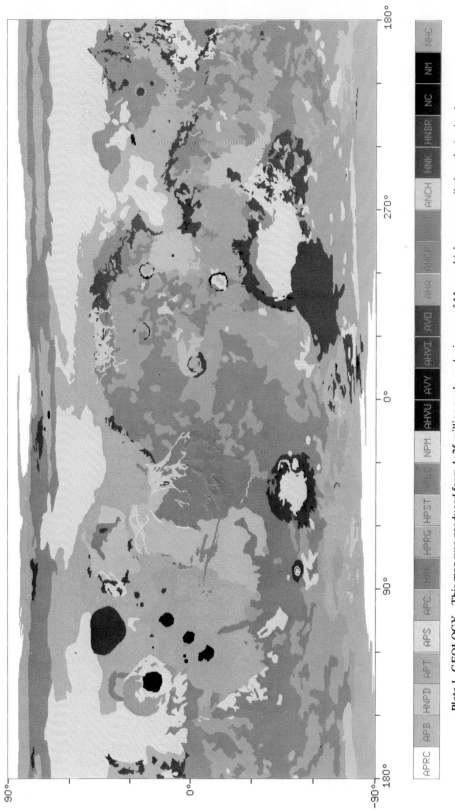

Plate 1. GEOLOGY—This map was produced from 1:25 million-scale geologic map of Mars, which was compiled on the basis of Mariner 9 images. The unit designations are identical and color encoding similar to the earlier map. (See Scott, D. H., and Carr, M. H., 1978, Geologic Map of Mars: U.S. Geological Survey Misc. Inv. Ser. Map I-1083.)

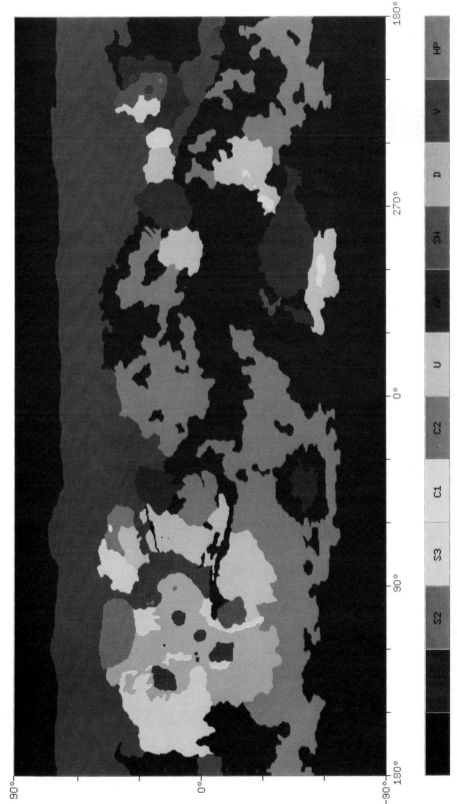

Plate 2. VOLCANIC UNITS—The principal types and ages of volcanic units were mapped using Viking and Mariner images. This map was digitized in Mercator and polar stereographic projections, which were combined into a global cylindrical projection. The volcanic units shown are: Simple flows (plateau plains, lower Hesperian, upper Hesperian); complex flows (Elysium age, Tharsis age); undifferentiated northern plains flows; central volcanoes (Alba Patera, highland paterae, shields, domes); questionable volcanic material. (See Greeley, R., and Spudis, P. D., 1981, *Reviews of Geophys. and Space Physics,* v. 19, p. 13–41.)

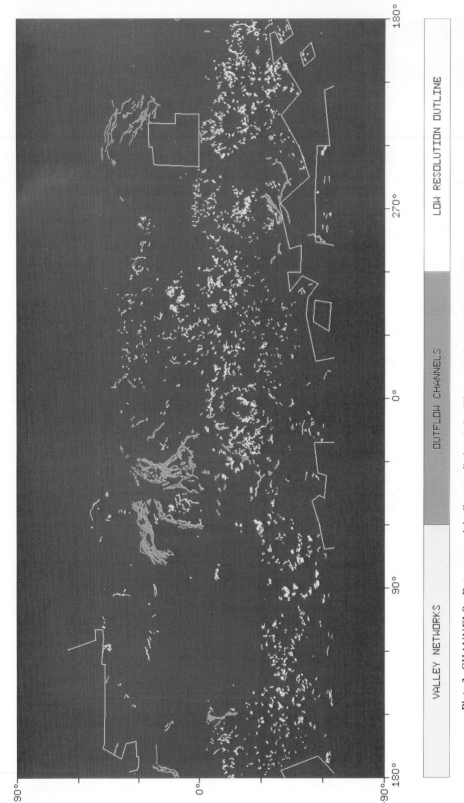

Plate 3. CHANNELS—Data were originally compiled at 1:5 million-scale from individual Viking images. Escarpments bounding outflow channels and flow-line features are combined in this representation; areas with no data are outlined in white. (See Carr, M. H., and Clow, G. D., 1981, *Icarus*, in press.)

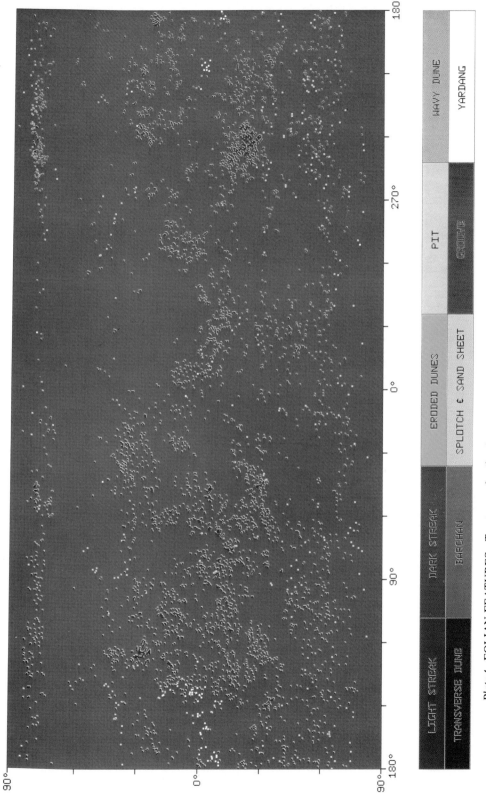

Plate 4. EOLIAN FEATURES—Ten types of eolian features were identified in Viking images and mapped at 1:5 million-scale. Their location, implied wind direction, and season of observations were digitized and converted into a global cylindrical projection. Pixels have been enlarged to 3/4° square for visibility and represent groups of features, not individual landforms. Geology by A. W. Ward, N. Witbeck, M. Weisman, and K. B. Doyle, U.S. Geological Survey, Flagstaff, Arizona.

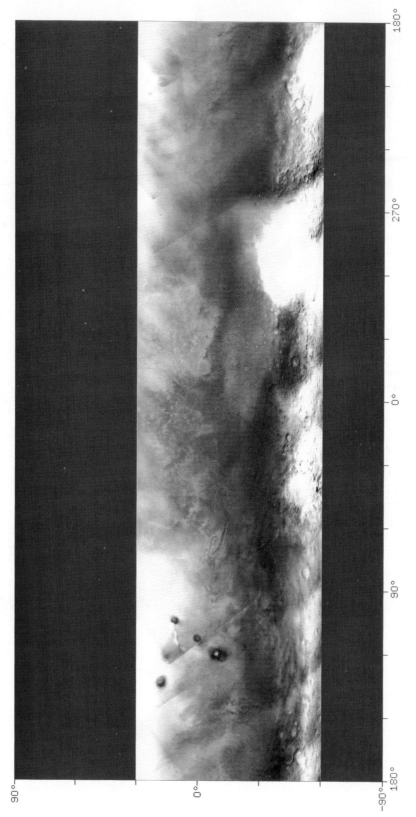

Plates 5 and 6. VIKING 2 APPROACH COLOR—Images in three colors acquired with resolutions between 10 and 20 km/line pair were processed to remove a Minnaert phase function, mosaicked in Mercator projection, and a seam removal technique applied. Regions near both latitude limits are subject to some atmospheric obscuration; some equatorial areas were observed only near the terminator. Central wavelengths of the violet, green, and red images were 0.45, 0.53, and 0.59 μm, respectively. A "natural" color

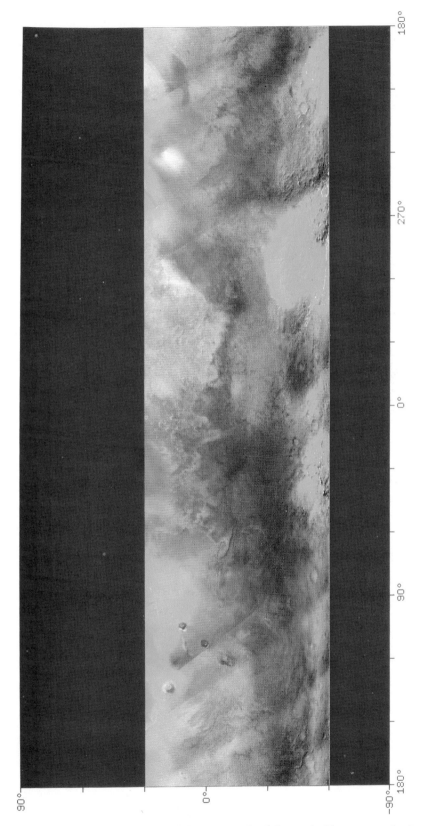

composite (Plate 5) is the result of violet, green, and red data as the blue, green and red components. A "hybrid" composite print (Plate 6) is the result of the 0.45/0.53 μm ratio as the blue component, the albedo at 0.59 μm as the green component, and the 0.59/0.45 μm ratio as the red component. (See Soderblom et al., 1978, Icarus, v. 34, p. 446; Mercator projection published therein.)

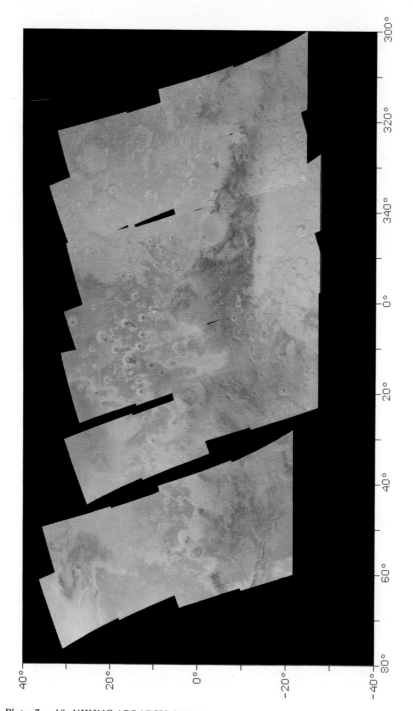

Plates 7 and 8. VIKING APOAPSIS COLOR—Images in three colors with a resolution of approximately 1 km/line pair have been obtained for nearly all the martian equatorial area. Strips of eight images that had been geometrically corrected and reduced to a common photometric function were digitally mosaicked in Mercator projection with approximately this resolution. Thirty-two images in each color were converted to cylindrical projection

and digitally mosaicked. The natural-color rendition (Plate 7) is the result of using the violet, green, and red data as the blue, green, and red color components. The color enhancement "hybrid" (Plate 8) was processed in the same manner as Plate 6. The color contouring in Plate 6 is an artifact caused by the coarse digitization (underexposure) of the original 0.53 μm data.

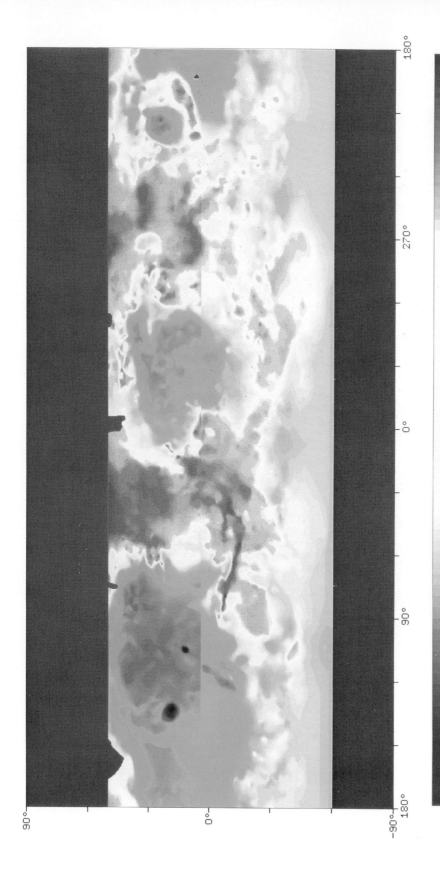

Plate 9. PREDAWN RESIDUAL TEMPERATURE—20-μm brightness temperatures measured by the Viking Infrared Thermal Mapper (IRTM) are referenced to the IRTM thermal model. Observations were obtained between midnight and dawn. Data south of latitude 4° N were obtained by Viking Orbiter 1 in September, 1976; those north of this seam line were obtained by Viking Orbiter 2 during November, 1977, through March, 1978. The contour lines at 2 K intervals were digitized from Mercator maps, interpolated, and combined in cylindrical projection. (See Kieffer et al., 1977, *J. Geophys. Res.*, v. 82, p. 4249; and Zimbelman, J.R., and Kieffer, H. H., 1979, *J. Geophys. Res.*, v. 84, p. 8239; Mercator version published therein.)

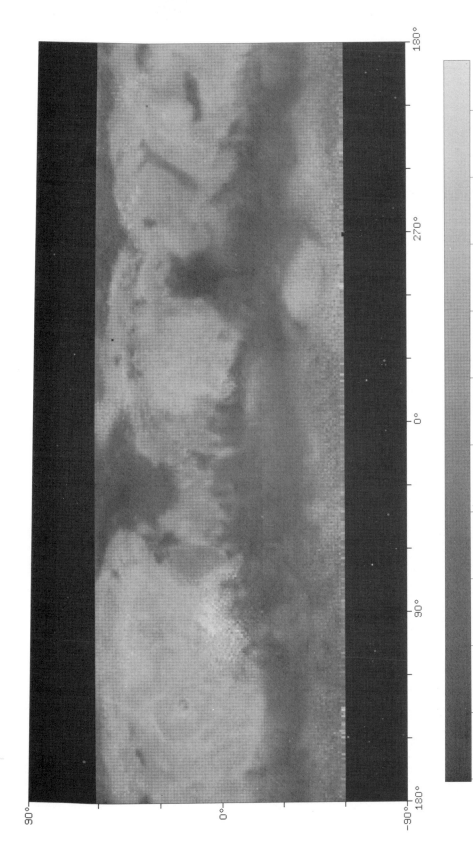

Plate 10. SOLAR ALBEDO—Broadband brightness (0.3 to 3.0 μm) measured by the Viking IRTM was reduced to Lambert albedo for four martian seasons largely free of atmospheric dust or local brightening. These data were averaged into resolution elements of 1° latitude by 1° longitude. Absolute albedos range from 0.10 to 0.33. (See Pleskot, L., and Miner, E. D., 1981, *Icarus*, in press.)

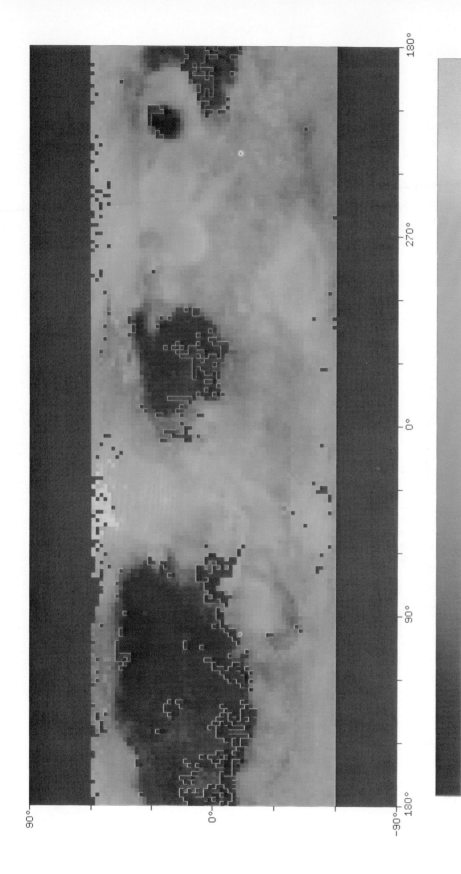

Plate 11. THERMAL INERTIA—Viking IRTM observations of 20-μm brightness temperatures between October 5, 1977, and August 9, 1978, were sorted into bins of 2° latitude by 2° longitude. The diurnal temperature variation, excluding a mid-afternoon period of unexpectedly rapid cooling, was fitted for thermal inertia and radiometric albedo simultaneously. Thermal inertia is in units of 10^{-3} cal cm^{-2} s$^{-1/2}$ K. (See Palluconi, F. D., and Kieffer, H. H., 1981, *Icarus*, v. 45, p. 415.)

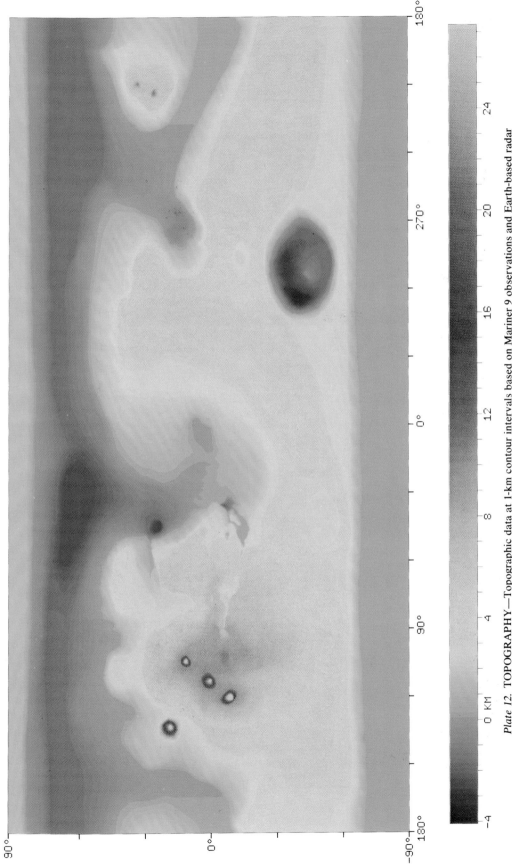

Plate 12. TOPOGRAPHY—Topographic data at 1-km contour intervals based on Mariner 9 observations and Earth-based radar observations through 1971 were digitized at the scale and projections of three parts of the 1:25 million-scale topographic map of Mars. The areas between contour lines were filled by successive generation of intermediate contours. (See Topographic Map of Mars, U.S. Geological Survey Misc. Inv. Ser. Map I-961.)

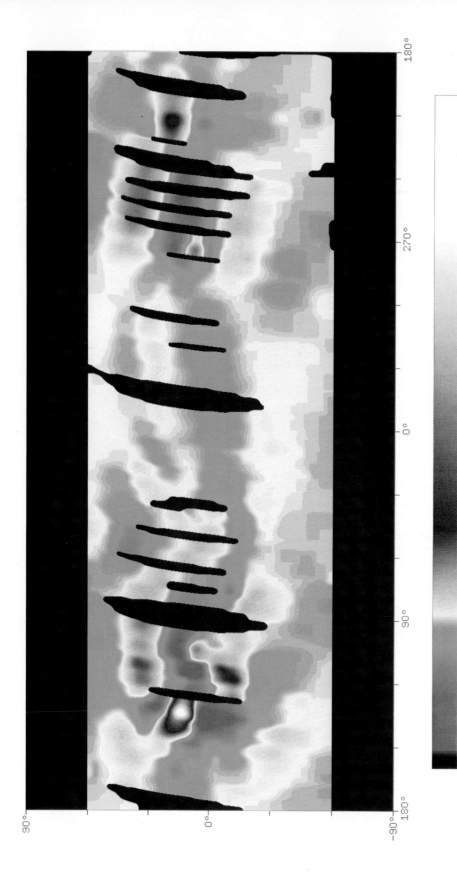

Plate 13. GRAVITY—Line-of-sight acceleration residuals from a fourth degree and order field at individual points along the Viking spacecraft track were separated into five sets based on, with altitude boundaries at 250, 350, 550, 850, 1500 and 1700 km. The gravity data in each set were smoothed with a digital filter of width equal to the minimum altitude in that set. The five sets were then combined with weighting functions inversely proportional to the square of the minimum altitude in each set. The unfiltered data range from −80 mgal to +344 mgal. (See Sjogren, W. L., 1979, *Science*, v. 203, p. 1006.)

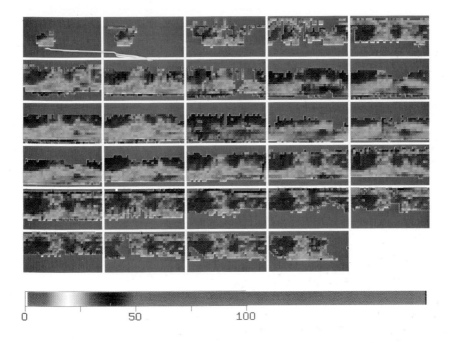

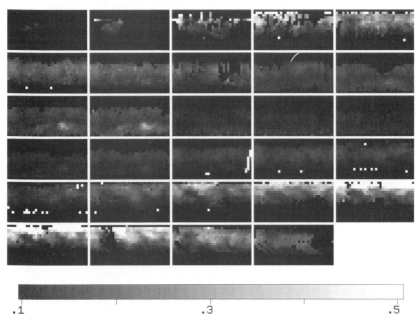

Plates 14 and 15. WATER-VAPOR ABUNDANCE AND 1.4-μm ALBEDO—Data from the Viking Mars Atmospheric Water Detector (MAWD) were sorted into 10° bins for seasonal increments of about 15° of L_s, for a total of 29 seasons. Each map covers all of Mars. Water-vapor abundance (Plate 14) varies from 1 precipitable μm to >60 precipitable μm and is highly season dependent. The 1.4-μm albedo maps (Plate 15) assume a Lambert phase function and vary little except for periods of martial dust storms. The beginning seasons (L_S) for each map as arranged above on both plates are:

85	95	110	123	137	169	186	201	216	232
248	276	289	305	317	335	350	6	22	
37	56	65	80	95	110	126	140	157	

(See Farmer et al., 1977, J. Geophys. Res., p. 4225; and Farmer, C. B., and Doms, P. E., 1979, J. Geophys. Res., v. 84, p. 2881.)

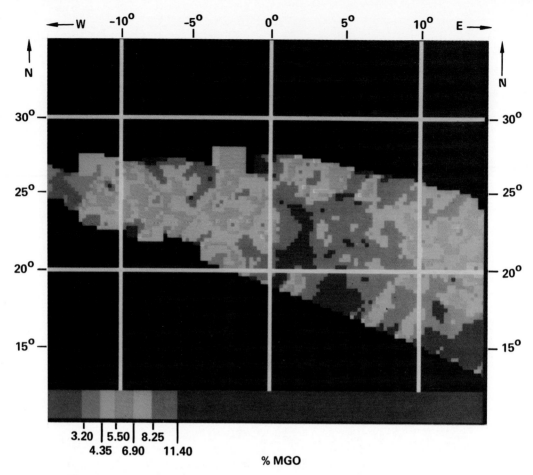

Plate 16. X-ray fluorescence (XRF) Mg/Si intensity ratios in the Hadley-Apennine region derived from measurements made by an X-ray processor assembly flown on the orbiting Service Module during the Apollo 15 mission. The instrument used to collect the XRF measurements and techniques used to reduce these data are described in a paper in these *Proceedings* (Clark and Hawke), as well as in earlier works referenced there. Ranges of % MgO to which colors correspond are shown beneath the color band at the bottom of the picture. Colors were picked deliberately to show the maximum amount of variation in the predominantly low magnesium region. Longitude is marked across the top, and latitude along the sides of the picture. The Apennine Bench and Palus Putredinis appear as distinct low Mg/Si features, as would be expected for high KREEP compositions. Yellow areas are more mafic, possibly showing some pyroclastic contamination. Mg/Si values increase further (red) on the Imbrium border to the south, approaching Serenitatis to the east, and in Sulpicius Gallus, where pyroclastics would be expected. The Alpes/Appeninus facies boundary on the Appenine Front can be clearly seen. Magnesium values are slightly higher for the Alpes facies to the north and east. This increase in magnesium would be expected for low-K Fra Mauro basalt. There are distinct west to east (KREEP component) and north to south (LKFM component) trends of Mg/Si variation.

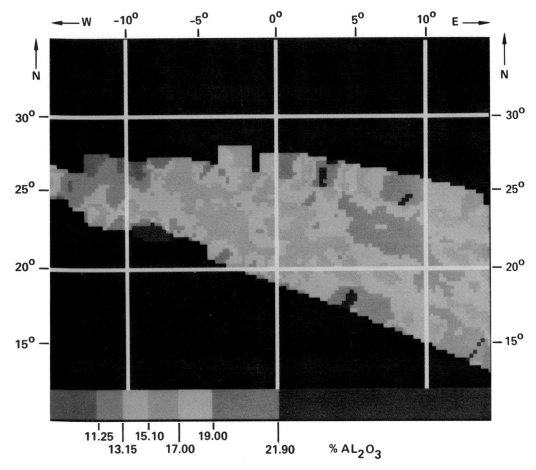

Plate 17. X-ray fluorescence (XRF) Al/Si intensity ratios in the Hadley-Appenine region derived from measurements made by an X-ray processor assembly flown on the orbiting Service Module during the Apollo 15 mission. The instrument used to collect the XRF measurements and techniques used to reduce these data are described in a paper in these *Proceedings* (Clark and Hawke), as well as in earlier works referenced there. Ranges of % Al_2O_3 to which colors correspond are shown at the bottom of the picture. Colors were picked deliberately to show the maximum amount of variation in aluminum in this region. Longitude and latitude are marked as in Plate 16. The Apennine Bench is quite distinct, as are the Apennine crest and the backslope units. Both the Apennine Bench and Palus Putredinis are low in Al/Si. The Apennine crest is slightly lower than the Imbrium backslope area to the east, possibly the result of pyroclastic deposits on the northern Apennine crest. There is an obvious west to east increase in Al/Si on this map. This trend, which correlates with a similar trend in % Mg/Si on Plate 16, also could be due to a KREEP compositional variation. Generally, Al/Si and Mg/Si values are well correlated in this region.

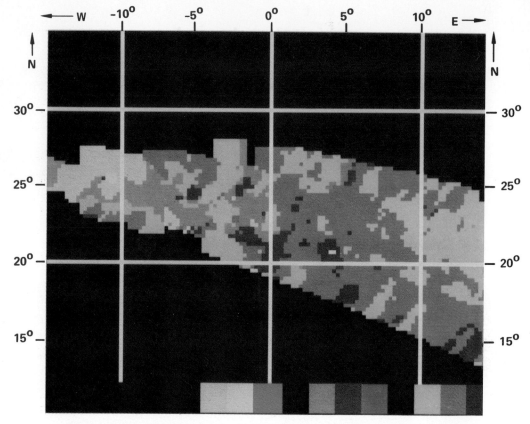

Plate 18. Color correlation of X-ray fluorescence Mg/Si and Al/Si intensity ratios in the Hadley-Apennine region derived from measurements made by an X-ray processor assembly flown on the orbiting Service Module during the Apollo 15 mission. The instrument used to collect the XRF measurements and techniques used to reduce these data are described in a paper in these *Proceedings* (Clark and Hawke), as well as in earlier works referenced there. In this plate color is assigned as a mnemonic device to indicate the degree of correlation between the two types of intensity ratios, which are mapped separately in the two preceding plates. The predominance of red shades indicates a positive correlation. Color assignments are:

		Al/Si		
		0.40 to 0.89	0.89 to 1.13	1.13 to 2.60
	0.45 to 0.83	magenta	lavender	blue
Mg/Si	0.83 to 1.00	orange	red	purple
	1.00 to 2.80	yellow	tan	maroon

Note the generally positive trend in the correlation. The Apennine Bench is low in both elements (magenta), indicative of KREEP basalt; and the Apennine Front is more moderate in both values (red and orange). The more mafic areas, with relatively high Mg/Si and relatively low Al/Si values, stand out as yellow.

CONTENTS

Section Two—Planets, Asteroids and Satellites

METEORITES

Carbon-14 ages of Allan Hills meteorites and ice E. L. Fireman and T. Norris 1019

Microchrons: The ^{87}Rb–^{87}Sr dating of microscopic samples D. A. Papanastassiou and G. J. Wasserburg 1027

Chondrites

A unique type 3 ordinary chondrite containing graphite-magnetite aggregates—Allan Hills A77011 S. G. McKinley, E. R. D. Scott, G. J. Taylor, and K. Keil 1039

The Elga meteorite: Silicate inclusions and shock metamorphism Eu. G. Osadchii, G. V. Baryshnikova, and G. V. Novikov 1049

SEM, optical and Mössbauer studies of submicrometer chromite in Allende R. M. Housley 1069

Origin of rims on coarse-grained inclusions in the Allende meteorite G. J. MacPherson, L. Grossman, J. M. Allen, and J. R. Beckett 1079

Petrogenesis of light and dark portions of the Leighton gas-rich chondritic breccia H. Y. McSween, Jr. S. Biswas, and M. E. Lipschutz 1093

Mineralogical aspects of terrestrial weathering effects in chondrites from Allan Hills, Antarctica J. L. Gooding 1105

Conditions of formation of pyroxene excentroradial chondrules R. H. Hewins, L. C. Klein, and B. V. Fasano 1123

A revision of metallographic cooling rate curves for chondrites J. Willis and J. I. Goldstein 1135

Ordinary chondrite parent body: An internal heating model M. Miyamoto, N. Fujii, and H. Takeda 1145

Electron microscopy of carbonaceous matter in Allende acid residues G. R. Lumpkin 1153

Carbon in the Allende meteorite: Evidence for poorly graphitized carbon rather than carbyne P. P. K. Smith and P. R. Buseck 1167

Noble gas trapping by laboratory carbon condensates S. Niemeyer and K. Marti 1177

Silicon and oxygen isotopes in selected Allende inclusions R. H. Becker and S. Epstein 1189

An estimate of atmospheric contamination of Allende coarse-grained inclusion 3529Z R. Warasila, O. A. Schaeffer, and K. Frank 1199

Irradiation records of Acapulco and other small meteorites derived from ^{53}Mn and rare gas measurements P. Englert, U. Herpers, and W. Herr 1209

Stable NRM and mineralogy in Allende: Chondrules P. Wasilewski and C. Saralkar 1217

The composition of natural remanent magnetization of an Antarctic chondrite, ALHA 76009 (L$_6$) T. Nagata and M. Funaki 1229

The magnetic properties of the Abee meteorite: Evidence for a strong magnetic field in the early solar system. N. Sugiura and D. W. Strangway 1243

Achondrites and Mesosiderites

Howardites and polymict eucrites: Regolith samples from the eucrite parent
body. Petrology of Bholgati, Bununu, Kapoeta, and ALHA76005
    M. Fuhrman and J.J. Papike 1257

Ion probe analysis of plagioclase in three howardites and three eucrites
    I. M. Steele and J. V. Smith 1281

Thermal and impact histories of pyroxenes in lunar eucrite-like gabbros and
eucrites   H. Takeda, H. Mori, T. Ishii, and M. Miyamoto 1297

Metamorphism in mesosiderites   J. S. Delaney, C. E. Nehru, M. Prinz, and
G. E. Harlow 1315

Roaldite, a new nitride in iron meteorites   H. P. Nielsen and V. F. Buchwald   1343

Complementary rare earth element patterns in unique achondrites, such as
ALHA 77005 and shergottites, and in the earth   M.-S. Ma, J. C. Laul,
and R. A. Schmitt 1349

SNC Meteorites: Igneous rocks from Mars?   C. A. Wood and L. D. Ashwal   1359

MARS AND VENUS

Natural radioactivity of the moon and planets   Yu. A. Surkov 1377

A possible common origin for the rare gases on Venus, Earth, and Mars
    C. J. Hostetler 1387

Regional Geology

Mars global properties: Maps and applications   H. H. Kieffer, P. A. Davis,
and L. A. Soderblom 1395

High resolution visual, thermal, and radar observations in the northern Syrtis
Major region of Mars   J. R. Zimbelman and R. Greeley 1419

Late-stage summit activity of Martian shield volcanoes   P. J. Mouginis-Mark   1431

Mars: A large highland volcanic province revealed by Viking images
    D. H. Scott and K. L. Tanaka 1449

A secondary origin for the central plateau of Hebes Chasma   C. Peterson . . 1459

Surface Properties

Spectral reflectance of weathered terrestrial and martian surfaces
    D. L. Evans, T. G. Farr, and J. B. Adams 1473

Mars weathering analogs: Secondary mineralization in Antarctic basalts
    J. L. Berkley 1481

Landing induced dust clouds on Venus and Mars   J. B. Garvin 1493

Composition of Venus

Density constraints on the composition of Venus   K. A. Goettel,
J. A. Shields, and D. A. Decker 1507

Metal chloride and elemental sulfur condensates in the Venusian troposphere:
Are they possible?   V. L. Barsukov, I. L. Khodakovsky, V. P. Volkov,
Yu. I. Sidorov, V. A. Dorofeeva, and N. E. Andreeva 1517

GALILEAN SATELLITES

An Io thermal model with intermittent volcanism   G. J. Consolmagno 1533

Microstructure and particulate properties of the surfaces of Io and Ganymede: Comparison with other solar system bodies   K. D. Pang, K. Lumme, and E. Bowell 1543

Structures on Europa   B. K. Lucchitta, L. A. Soderblom, and H. M. Ferguson 1555

The sputter-generation of planetary coronae: Galilean satellites of Jupiter   C. C. Watson 1569

Tectonic deformation of Galileo Regio and limits to the planetary expansion of Ganymede   W. B. McKinnon 1585

Dark-ray craters on Ganymede   J. Conca 1599

EXPERIMENTAL AND THEORETICAL STUDIES

Impact Cratering

A method for estimating the initial impact conditions of terrestrial cratering events, exemplified by its application to Brent Crater, Ontario   R. A. F. Grieve and M. J. Cintala 1607

Structural study of Cactus Crater   J. Vizgirda and T. J. Ahrens 1623

Impact accretion experiments   V. Werle, H. Fechtig, and E. Schneider 1641

Impact cratering experiments in Bingham materials and the morphology of craters on Mars and Ganymede   J. H. Fink, R. Greeley, and D. E. Gault . 1649

Fragmentation of ice by low velocity impact   M. A. Lange and T. J. Ahrens . 1667

Initial energy partitioning and some excavation stage phenomenology in laboratory-scale cratering calculations in clay   M. G. Austin, J. M. Thomsen, and S. F. Ruhl 1689

Secondary cratering effects on lunar microterrain: Implications for the micrometeoroid flux   R. J. Allison and J. A. M. McDonnell 1703

The stochastic variability of asteroidal regolith depths   K. R. Housen 1717

Physical Properties

Xenon diffusion following ion implantation into feldspar: Dependence on implantation dose   C. L. Melcher, D. S. Burnett, and T. A. Tombrello ... 1725

A brief note on the effect of interface bonding on seismic dissipation   B. R. Tittmann, M. Abdel-Gawad, C. Salvado, J. Bulau, L. Ahlberg, and T. W. Spencer 1737

On the estimation of lunar paleointensities—Studies of synthetic analogues of stably magnetized samples   J. R. Dunn, M. Fuller, and D. Clauter 1747

Flow behavior of ten iron-containing silicate compositions   L. C. Klein, B. V. Fasano, and J. M. Wu 1759

Effects of body shape on disk-integrated spectral reflectance   J. Gradie and J. Veverka 1769

Some key issues in isotopic anomalies: Astrophysical history and aggregation   D. D. Clayton 1781

Comments on "Xe^{129} and the origin of CCF xenon in meteorites" by R. S. Lewis and E. Anders   *D. Heymann* 1803

Cosmic-ray-produced stable nuclides: Various production rates and their implications   *R. C. Reedy* 1809

Errata ....

Lunar Sample Index i

Heavenly Body Index ix

Subject Index xiii

Author Index xix

Carbon-14 ages of Allan Hills meteorites and ice

E. L. Fireman and T. Norris

Smithsonian Astrophysical Observatory, Cambridge, Massachusetts 02138

Abstract—^{14}C and ^{39}Ar measurements for ALHA meteorites 77282, 77294, and 77297 are presented. ^{39}Ar is undetectable in all three, indicating terrestrial ages greater than 1.2×10^3 years. The small amounts of ^{14}C observed in 77282 and 77294 indicate that their terrestrial ages are $\sim 30 \times 10^3$ years; no ^{14}C is detected in ALHA 77297, giving it a lower limit age of 35×10^3 years. The trapped air compositions and the specific ^{14}C activities of the CO_2 in two surface and two subsurface samples of ALHA ice are measured and compared to those in Byrd core ice. The CO_2 abundances and the specific activities of the surface samples are very high, which indicates that melting and refreezing had occurred and the specific ^{14}C activities were enriched by nuclear debris. The CO_2 abundances of the subsurface samples are toward the top of the range observed for Byrd core samples and their specific activities are approximately that for contemporary air. If no melting and refreezing occurred in the subsurface samples, then the ALHA ice is young, less than 3×10^3 years for one sample and less than 6×10^3 years for another.

1. INTRODUCTION

Allan Hills is a blue ice region of approximately 100 km^2 area in Antarctica where many meteorites have been found exposed on the ice (Cassidy et al., 1977; Cassidy, 1978). More than 1000 meteorites have been recovered from the Allan Hills collection site.

The terrestrial ages of the Allan Hills meteorites, which are obtained from their cosmogenic nuclide abundances, are important time markers that can reflect the history of ice movement to the site. ^{39}Ar (270-year half-life), ^{14}C (5730-year half-life), ^{36}Cl (3.1×10^5-year half-life), ^{26}Al (7.3×10^5-year half-life), and ^{53}Mn (3.7×10^6-year half-life) are cosmogenic nuclides that have been used for terrestrial age determinations of meteorites. Measurements of these nuclides in a number of Allan Hills and Yamato meteorites have been reported (Evans et al., 1979; Fireman et al., 1979; Fireman, 1979, 1980a,b; Nishiizumi et al., 1979a,b; Nishiizumi and Arnold, 1980). The radioactivity studies indicate that most of the Allan Hills meteorites fell between 20×10^3 and 200×10^3 years ago. Rare-gas studies (Weber and Schultz, 1980) indicate that multiple shower fragments are rare. ^{14}C measurements (Fireman, 1979, 1980a,b) gave terrestrial ages greater than 30×10^3 years for seven ALHA meteorites and ages of less than 25×10^3 years for two. ^{26}Al and ^{36}Cl measurements (Evans et al., 1979; Nishiizumi and Arnold, 1980) gave terrestrial ages of less than 200×10^3 years for all of ~ 50 meteorites except two, which were determined to be $\sim 600 \times 10^3$ years old. Terrestrial age determinations for many meteorites of known recovery locations will eventually be necessary to reveal the detailed history of the ice movement. The age of the ice underlying the meteorites is unknown.

The terrestrial age is the sum of a travel time and a surface residence time. The travel time is for the meteorite, solidly encased in ice, moving with the ice to the Allan Hills. The surface residence time occurs after the Allan Hills barrier has caused the ice to stagnate and winds have ablated the ice causing the meteorite to be exposed on the surface. The age of the ice underlying a meteorite gives its travel time.

Our principal purpose in studying the terrestrial ages of ALHA meteorites is to locate samples of ancient ice and analyze their trapped gas contents; the terrestrial ages of the ALHA meteorites and their times of travel to Allan Hills are also of interest.

2. ^{14}C AND ^{39}Ar TERRESTRIAL AGES OF ALHA METEORITES

Our experimental procedure for ^{14}C and ^{39}Ar determinations in meteorites has been described and ^{14}C and ^{39}Ar measurements for nine ALHA meteorites have been presented (Fireman, 1979, 1980a,b). We measured the ^{14}C and ^{39}Ar activities for three additional ALHA meteorites, 77294, 77297, and 77282, and made improved measurements for 77004. The amounts of CO_2 obtained in stepwise heatings and the ^{14}C and ^{39}Ar activities are given in Table 1. Since the argon carrier used in the first extraction was reused for the other extractions, the ^{39}Ar activity is the total for all extractions. No ^{39}Ar activity was observed in any of the ALHA samples. The ^{39}Ar activity limits combined with the 10 ± 1 dpm/kg of ^{39}Ar observed for the Bruderheim standard (a recent fall) give terrestrial ages greater than 1200 years.

Small amounts of ^{14}C activity are observed in the melts of 77294 and 77282 but not in those of 77004 and 77297. A 500°C extraction, which gives information on the weathering time, was done for ALHA 77004 and 77294 but not for 77297 and 77282. The comparison of the ^{14}C activities with that in Bruderheim gives the ^{14}C terrestrial ages. The terrestrial ages of thes four meteorites together with the eight previously reported are given in Table 2.

3. ^{14}C AGES AND TRAPPED GAS COMPOSITIONS IN ICE SAMPLES

The ice movement at Allan Hills is being studied (Annexstad and Nishio, 1980); the ^{36}Cl activity in Allan Hills ice is being measured (Finkel *et al.*, 1980); and preliminary ^{14}C results on two Allan Hills samples have been reported (Fireman, 1980b).

Table 1. Amounts of CO_2, ^{14}C, and ^{39}Ar in ALHA meteorites.

ALHA # Type (Wgt.) (Weathering class.)	Extraction Temp. (°C)	CO_2 (cm^3 STP)*	^{14}C (dpm/kg)**	^{39}Ar (dpm/kg)
77004 H4 (10.0 g) (C)	500	8.6	<0.8	
	1000	5.4	<0.7	
	Melt	4.5	<0.7	<0.3
	Remelt	3.5	<0.5	
77294 H5 (10.5 g) (A)	500	1.34	<0.2	
	1000	1.04	0.6 ± 0.2	<0.4
	Melt	0.28	1.0 ± 0.1	
	Remelt	0.10	<0.2	
77297 L6 (5.16 g) (A/B)	1000	0.93	counting in prog.	
	Melt	1.58	<0.4	<0.3
	Remelt	0.72	—	—
77282 L6 (10.2 g) (B)	1000	3.2	counting in prog.	
	Melt	4.18	1.3 ± 0.3	0.3 ± 0.2
	Remelt	1.23	—	
Bruderheim L6 (7.0 g)	500	1.80	<1.0	
	1000	2.08	23 ± 2	
	Melt	1.29	38 ± 2	10 ± 1
	Remelt	0.11	<0.2	

*Amount of CO_2 in excess of the estimated carrier.
**Error contains the uncertainty in the counting efficiencies and 1σ statistical counting error.

Table 2. The ^{14}C terrestrial ages of twelve ALHA meteorites.

Meteorite	Class	Weathering Category	^{14}C (dpm/kg)	Terrestrial ^{14}C Age (10^3 yr)
76005	Eucrite	—	<1.0	>34
76006	H6	B	<1.7	>32
76007	L6	C	<1.2	>34
76008	H6	C	<1.7	>32
77003	L3	A	4.6 ± 1.0	21^{+4}_{-3}
77004	H4	C	<1.5	>33
77214	L	C	<3.0	>25
77256	Diogenite	A	16.0 ± 1.5	11.1 ± 1.0
77272	L6	B	<0.5	>38
77282	L6	B	$(2.0 \pm 0.5)^*$	$(\sim 30)^*$
77294	H5	A	1.6 ± 0.3	30 ± 2
77297	L6	A/B	$(<1.0)^*$	$(>35)^*$

*Counting in progress.

The Allan Hills site is a region of ice ablation. There being essentially no previous glaciological or isotopic studies on ice ablation regions, we compare measurements on Allan Hills ice with those on Byrd core ice. The Byrd core was drilled to 2200 m depth in a snow-accumulation region (Gow et al., 1968). The Byrd core ice has been studied extensively and provides a continuous stratigraphic record for the past ~100,000 years (Epstein et al., 1970; Johnson et al., 1972). Age depth relationships on the basis of glaciological evidence and $^{18}O/^{16}O$ measurements on melt water have been given (Epstein et al., 1970; Johnson et al., 1972).

Our gas-extraction procedure for the ice, together with a description of procedures for the CO_2 separation from the gas and the counting of the ^{14}C activity in the CO_2, are given in Appendix I.

Table 3 gives the amounts of gas, the percentage of CO_2 in the gas, the ^{14}C specific activity of the CO_2 and the ^{14}C ages of four Allan Hills ice samples, a frozen distilled water sample, and seven Byrd core samples. Maps of the Allan Hills site given by Annexstad and Nishio (1979, 1980) show the meteorite concentrations, the locations of the numbered stakes, and the strain flower region, marked by four stakes (ABCD) about 500 m north of Stake 11. Stake 11 is at the center of the highest meteorite concentration region.

The gas in the frozen distilled water, which should typify dissolved air, has the highest CO_2 abundance, 1.35%. The gas in Allan Hills surface ice samples (from Stakes 12 and 18) have the next highest CO_2 abundances, 0.187 and 0.129%, respectively, which are attributed to dissolved air because melting and refreezing occur on the surface of the ice at Allan Hills. The CO_2 abundances (0.054 and 0.049%) in the gas from the Allan Hills subsurface samples (5 to 25 cm depth from a location between Stakes 11 and 12 and 10 to 20 cm depth from the strain flower area) are essentially the same as that in the 362 to 363 m Byrd sample, 0.051%. This measurement indicates that the amounts of air dissolved in these subsurface samples are small.

The ^{14}C specific activities in the CO_2 from the Allan Hills surface ice samples (27 ± 3) and $(25 \pm 6) \times 10^{-3}$ dpm/cm^3, are much higher than in contemporary air and must be attributed to isotopic exchange with nuclear debris. The specific activities in the subsurface Allan Hills samples are essentially the same as contemporary air, and give the young ^{14}C limit ages of less than 6×10^3 year for the ice between Stakes 11 and 12 and 3×10^3 year for the ice at the strain flower location. The possibility exists, however, that a small amount of surface melt water percolated in the subsurface ice and lowered the ^{14}C age. Lower specific ^{14}C activities are observed in the Byrd core, the lowest measured specific activity, $<3.0 \times 10^{-3}$ dpm per cm^3 STP of CO_2, is in the Byrd sample from 1071 m depth, giving it a ^{14}C age greater than 8×10^3 years.

Table 3. Amounts of gas, percentage of CO_2 in the gas, specific activity of the CO_2 and ^{14}C ages.

Sample (loc., depth, wgt.)	Extraction (temp, pH)	Gas (cm³/kg)	CO_2 %	$^{14}C/CO_2$ (10^{-3} dpm/cm³)	^{14}C age (10^3 yr)
Allan Hills					
Stake 12, sur. 31 kg	24°C, 1; 55°C, 1	47.0 / 0.9	0.187	27 ± 3	Nuclear debris
Stake 18, sur. 16.6 kg	24°C, 5.5; 55°C, 1	20.1 / 0.4	0.129	25 ± 6	Nuclear debris
Stake 11–12, 5–25 cm 12.9 kg	24°C, 5.5; 55°C, 1	46.0 / 1.5	0.054	7.0 ± 3.0	≤6
Strain flower, 10–20 cm 10.3 kg	24°C, 5.5; 24°C, 1	52.5 / 0.15	0.049	11.0 ± 3.0	≤2
Frozen water 7.8 kg	24°C, 5.5; 55°C, 1	20.4 / 0.06	1.35	9.0 ± 0.5	0 ± 0.7
Byrd core					
271 m 7.6 kg	24°C, 5.5; 55°C, 1	66.8 / 0.1	0.0336	6.8 ± 1.0	1.0 ± 1.0
272 m 8.8 kg	24°C, 5.5; 55°C, 1	57.5 / 0.1			
362 m 6.2 kg	24°C, 1; 55°C, 1	74.3 / 0.5	0.051	6.0 ± 0.5	2.0 ± 0.7
363 m 8.0 kg	24°C, 1; 55°C, 1	69.3 / 1.3			
1068 m 9.2 kg	24°C, 5.5; 55°C, 1	117.8 / 0.2	0.0216	—	—
1071 m 6.2 kg	24°C, 5.5; 55°C, 1	101.2 / 7.5	0.0356	≤3.0	≥8.0
1469 m 9.5 kg	24°C, 5.5; 55°C, 1	115.1 / 0.4	0.0237	—	—

The gas in two Byrd core samples from the very large depths of 1068 m and 1469 m have very low CO_2 abundances, 0.0216 and 0.0237%, respectively. This result is consistent with the Byrd core results of Berner et al. (1980).

Another source of information about the history of the ice is the composition of the trapped gas. The composition of the gas is interesting because it provides information on the melting history of the ice as well as the composition of the ancient atmosphere.

The oxygen is removed from the gas by combustion with a carbon filament and the remainder is analyzed by means of a mass spectrometer. The procedures used for the oxygen removal and the mass spectrometric analyses are described in Appendix 2.

Table 4 gives the nitrogen and argon abundances, the $\delta\,^{15}N$ relative to room air nitrogen, and the $^{40}Ar/^{36}Ar$ ratios of the gas extracted from the ice samples. The N_2 abundances in the trapped gas are the same as in normal air (within 2σ) for all ice samples except for the frozen distilled water. The frozen water has a low N_2 abundance and a high Ar abundance as expected on the basis of the solubilities of N_2 and Ar in water. The Ar abundances in the trapped air of all the Antarctic ice samples are the same as in normal air

Table 4. N_2 and Ar abundances and isotopic compositions.

Gas sample	N_2(%) ($\sigma = \pm 1.0$)	Ar(%) ($\sigma = \pm 0.05$)	$\delta\ ^{15}N$(‰) ($\sigma = \pm 0.30$)	$^{40}Ar/^{36}Ar$ ($\sigma = \pm 2$)
Allan Hills				
Stake 12 (surface)	77.4	0.89	−0.26	296
Stake 18 (surface)	77.6	0.92	0.51	294
Stake 10–11 (5–25 cm)	79.7	0.94	1.25	296
Strain flower (10–20 cm)	78.0	0.93	−2.72	296
Frozen water	71.8	1.30	—	—
Air	78.0	0.93	0.0	296
Byrd core				
271 m	79.3	0.93	1.76	295
272 m	77.9	0.93	—	—
362 m	79.3	0.95	−1.38	295
363 m	77.7	0.82	−1.71	294
1068 m	77.6	0.82	1.08	—
1071 m	78.6	0.92	3.70	296
1468 m	78.8	0.71	−0.07	292

$$\delta\ ^{15}N = \left[\frac{(^{15}N/^{14}N)_{sample}}{(^{15}N/^{14}N)_{air}} - 1\right] \times 1000$$

(within 2σ) except for the deepest Byrd sample (1468 m), which has the low abundance of 0.71%. The $^{40}Ar/^{36}Ar$ ratios in all the gas samples are the same as in normal air.

The $\delta\ ^{15}N$ was measured with higher precision, $\sigma = \pm 0.30$‰. Both positive and negative $\delta\ ^{15}N$ values outside the 2σ error are observed. The reason for these variations is unknown.

Our results on the gas contents and compositions can be summarized in the following way. The amounts of air trapped in the Allan Hills ice samples are lower than in the Byrd core samples and the CO_2 abundance of the Allan Hills ice samples are higher. On the other hand, the N_2 and Ar abundances and isotopic compositions are quite similar at the two sites.

4. CONCLUSIONS

The conclusion that follows from the ^{14}C terrestrial ages combined with the ^{36}Cl and ^{26}Al results of others (Evans et al., 1979; Nishiizumi and Arnold, 1980) is that most ALHA meteorites fell between 20,000 and 200,000 years ago; however, there are a few younger and a few older meteorites. The conclusions from the studies of the ^{14}C activities and trapped gas compositions of the Allan Hills ice samples are as follows: (1) Age determinations can not be made on surface ice samples because of melting and refreezing with the incorporation of nuclear debris; (2) If percolation of surface melt water does not occur to the depths of the ice collected between Stakes 11 and 12 and of the strain flower ice, then the ice at these sites is very young, less than 6×10^3 years and 3×10^3 years, respectively. The travel times of meteorites collected at these sites would then be short compared to their likely terrestrial ages.

Acknowledgments—We thank J. C. DeFelice for his help in all phases of this work, W. A. Cassidy and J. O. Annexstad for the Allan Hills ice, and C. C. Langway, Jr. for the Byrd core ice. This research was supported by NSF Grant DDP 78-05730 and NASA Grant NGR 09-015-145.

REFERENCES

Annexstad J. O. and Nishio F. (1979) Glaciological studies in Allan Hills 1978–79. *Antarct. J. U.S.* **14**, 87–88.
Annexstad J. O. and Nishio F. (1980) Glaciological studies in Allan Hills 1979–80. *Antarct. J. U.S.* **15**, 65–66.
Berner W., Oeschger H., and Stauffer B. (1980) Information of the CO_2 cycle from ice core studies. *Radiocarbon* **22**, 227–235.
Cassidy W. A. (1978) Antarctic search for meteorites during the 1977–1978 field season. *Antarct. J. U.S.* **13**, 39–40.
Cassidy W. A., Olsen E., and Yanai K. (1977) Antarctica: A deep-freeze storehouse for meteorites. *Science* **198**, 727–731.
Dienes P. (1970) Mass spectrometric correction factors for the determination of small isotopic composition variations of carbon and oxygen. *Intl. J. Mass Spec. Ion Physics* **4**, 283–295.
Epstein S., Sharp R. P., and Gow A. J. (1970) Antarctic ice sheet: Stable isotope analyses of Byrd station cores and interhemispheric climate implications. *Science* **168**, 1570–1572.
Evans J. C., Rancitelli L. A., and Reeves J. H. (1979) ^{26}Al content of Antarctic meteorites: Implications for terrestrial ages and bombardment history. *Proc. Lunar Planet. Sci. Conf. 10th*, p. 1061–1072.
Finkel R. C., Nishiizumi K., Elmore D., Ferraro R. D., and Gove H. E. (1980) ^{36}Cl in polar ice, rainwater, and seawater. *Geophys. Res. Lett.* **7**, 983–986.
Fireman E. L. (1979) ^{14}C and ^{39}Ar abundances in Allan Hills meteorites. *Proc. Lunar Planet. Sci. Conf. 10th*, p. 1053–1060.
Fireman E. L. (1980a) Carbon-14 dating of Antarctic meteorites and Antarctic ice (abstract). In *Lunar and Planetary Science XI*, p. 288–290. Lunar and Planetary Institute, Houston.
Fireman E. L. (1980b) Carbon-14 and argon-39 in ALHA meteorites. *Proc. Lunar Planet. Sci. Conf. 11th*, p. 1215–1221.
Fireman E. L., Rancitelli L. A., and Kirsten T. (1979) Terrestrial ages of four Allan Hills meteorites: Consequences for Antarctic ice. *Science* **203**, 453–455.
Gow A. J., Ueda H. T., and Garfield D. E. (1968) Antarctic ice sheets: Preliminary results of first core hole to bedrock. *Science* **161**, 1011–1013.
Johnson S. J., Dansgaard W., Clausen H. B., and Langway C. C. (1972) Oxygen isotope profiles through the Antarctic and Greenland ice sheets. *Nature* **235**, 429–434.
Langway C. C., Herron M., and Cregin J. H. (1974) Chemical profile of the Ross ice shelf. *J. Glaciology* **13**, 431–437.
Mook W. G. and Grootes P. M. (1973) The measuring procedure and corrections for the high-precision mass-spectrometric analysis of isotopic abundance ratios, especially referring to carbon, oxygen and nitrogen. *Intl. J. Mass Spec. Ion Physics* **12**, 273–298.
Nishiizumi K. and Arnold J. R. (1980) Ages of Antarctic meteorites (abstract). In *Lunar and Planetary Science XI*, p. 815–817. Lunar and Planetary Institute, Houston.
Nishiizumi K., Arnold J. R., Elmore D., Ferraro R. D., Gove H. E., Finkel R. C., Beukens R. P., Chang K. H., and Kilius L. R. (1979b) Measurements of ^{36}Cl in Antarctic meteorites and Antarctic ice using a Van de Graff accelerator. *Earth Planet. Sci. Lett.* **45**, 275–285.
Nishiizumi K., Imamura M., and Honda M. (1979a) Cosmic-ray produced radio-nuclides in Antarctic meteorites. *Memoirs of the National Institute of Polar Research, Special Issue No. 12* (T. Nagata, ed.), p. 161–177. Nat'l. Inst. Polar Res., Tokyo.
Weber H. W. and Schultz L. (1980) Noble gases in ten stone meteorites from Antarctica. *Z. Naturforsch.* **35a**, 44–49.

APPENDIX I: EXTRACTION OF GAS FROM THE ICE

The ice samples are cleaned by an established procedure (Langway *et al.*, 1974) before insertion into the gas extraction system shown in Fig. A1. The 6-in.-diameter glass cylinder containing the ice is sealed and the large cold finger above the cylinder is cooled to $-76°C$. The air around the ice is removed by evacuations and helium purges. During these cleansing purges, the surface ice melts and the surface melt water is stripped of gaseous contaminants. Approximately 5% of the ice is lost by this gas cleaning procedure. The glass cylinder is then filled with He to 1.3 atm. pressure and the remaining ice left to melt overnight..The next day, the melt water is purged with helium. The helium flows past the cold finger, through a $Mg(ClO_4)_2$ drier, then through a fine spiral glass CO_2 trap kept at $-196°C$ and a molecular sieve (#5A) gas trap also kept at $-196°C$, and then exits. The CO_2 is removed from its trap and measured volumetrically. The gas minus the CO_2 is then removed from its trap and also measured volumetrically. The water is then acidified to a pH of 1 using either sulfuric or nitric acid, heated to 55°C, and purged again to ensure that CO_2 and air are completely removed from the water. The CO_2 is counted for ^{14}C in the smallest of the low-level proportional minicounters used for meteorites (Fireman, 1979).

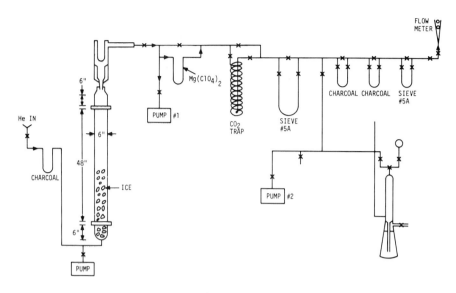

Fig. A1. Gas-extraction system.

APPENDIX II: MASS SPECTROMETRIC ANALYSIS

A small aliquot of the gas minus CO_2 is measured volumetrically and placed into a glass reaction vessel to remove O_2. The vessel consists of two electrical feedthroughs, which are connected to platinum rods with a carbon filament connecting the platinum rods. The vessel, containing the sample, is placed in a liquid nitrogen bath and the carbon filament is heated to a dull red glow. The O_2 combines with C at the filament forming CO_2, which freezes out on the cold walls. The reaction takes about one hour with a 95% conversion of O_2 to CO_2. The unreacted gas is purified by reacting it with copper oxide at 300°C to remove any residual CO and with copper at 300°C to remove any traces of oxygen.

The purified gas (N_2 + Ar) is placed in a mass spectrometer. The spectrometer is a 6-in. Nuclide Corp. instrument that was converted from a noble gas instrument into a dual-inlet and dual-collector isotope ratio-mass spectrometer. An Airco purified N_2 gas bulb is used as the internal working standard for the nitrogen isotope analysis. The $^{15}N/^{14}N$ data are corrected following standard procedures (Dienes, 1970; Mook and Grootes, 1973). The $^{40}Ar/^{28}N_2$ and the $^{40}Ar/^{36}Ar$ ratios are obtained using the machine as a single collector mass spectrometer. These data are corrected for isotopic fractionation and ionization efficiencies by comparisons of the sample to a room air standard.

Microchrons: The ^{87}Rb–^{87}Sr dating of microscopic samples

D. A. Papanastassiou and G. J. Wasserburg

The Lunatic Asylum of the Charles Arms Laboratory,
Division of Geological and Planetary Sciences, California Institute of Technology,
Pasadena, California 91125

Abstract—We have developed analytical techniques for sample handling and for the measurement of Rb and Sr isotopic compositions and Rb/Sr atom ratios in microsamples containing 10^{11}–10^{12} atoms of Sr. Using a modified direct loading technique, contamination for these procedures can be kept at the level of 10^9 atoms of Rb and $(2–8) \times 10^9$ atoms of Sr. The Sr ionization efficiency for these measurements is 2–5% and permits the determination of ^{87}Sr/^{86}Sr to a precision of 1–2‰ (2σ) and of ^{84}Sr/^{88}Sr to 1% (2σ). The ^{84}Sr measurements are obtained on $\sim 10^9$ atoms per analysis. Using these techniques we have determined an isochron for the KREEPUTh-rich basalt 14276 on a total sample of 6×10^{-6} g. The age and initial ^{87}Sr/^{86}Sr obtained from the microsamples agree with data previously obtained on macrosamples of this basalt. Data on meteoritic samples and on two tektites are also presented and are consistent with previously published results on similar materials. The data presented demonstrate this approach to be sound and readily feasible, so that these techniques can now be applied to the study of microsamples (interplanetary dust particles, lunar regolith grains) as well as for the study of element migration during metamorphism on a 100 μm scale.

INTRODUCTION

We report the results of an effort to extend the Rb–Sr dating technique to microscopic samples of individual mineral grains as small as ~ 30 μm in size. A successful method of this type would permit the study of interplanetary dust grains and lunar regolith particles and would further allow study of meteoritic and terrestrial crystals on a scale commensurate with chemical element redistribution. For the past several years, extensive efforts have been directed toward the measurement of isotopic and chemical abundances on microscopic samples using a variety of techniques. All approaches which use mass spectrometry must seek to optimize the ionization efficiency for the atomic species under investigation and eliminate the presence of interfering isobaric species. Chemical element separation has been necessary in most instances for the elimination of isobaric interferences and for increased ionization efficiency. For cases where isobaric interferences are absent, and by using a thermal ionization source, it has been possible to carry out analyses for Mg, Ca, and Pb on small samples containing 10^{11}–10^{12} atoms of the atomic species of interest at a precision of $\sim 1‰$ without chemical separations. Data on these elements have been obtained on lunar spherules, grains from meteorites, and interplanetary dust particles [Tera and Wasserburg, 1975; Wasserburg, Lee, and Papanastassiou, 1977; Lee, Papanastassiou, and Wasserburg, 1977; Esat, Brownlee, Papanastassiou, and Wasserburg, 1979]. For elements which have been purified by chemical separation procedures, it has been possible in special circumstances to achieve a much higher sensitivity. However, one of the major difficulties encountered in chemical separations has been laboratory contamination which often seriously affects the analyses. The smallest amounts of contamination we have reported for microchemical separation procedures of silicates are 2×10^{10} and 1.4×10^{11} atoms (3×10^{-12} and 2×10^{-11} g) for Rb and Sr respectively [Papanastassiou and Wasserburg, 1973]. Measurement of the isotopic composition and the relative number of atoms of Rb and Sr has presented difficulties because of the need to separate Rb from Sr due to the isobaric interference at mass 87. This separation is critical since the changes in ^{87}Sr/^{86}Sr are directly related to the age of the sample.

The approach used here is a modified version of the direct loading technique [Wasserburg et al., 1977; Lee et al., 1977]. Results using this approach were reported at the 12th Lunar and Planetary Science Conference [Papanastassiou and Wasserburg, 1981]. A sample (30–300 μm in diameter) was dissolved in small amounts of reagents (HF, HNO$_3$, and HCl) and isotopic tracers were added to aliquots of the sample solution. The mixture of sample and tracers was then loaded directly on a filament. The filament was heated in the ion source and the isotopic ratios of the selected elements were measured at the temperatures which favor each element being analyzed. The selection of the temperatures was such as to strongly suppress all sources of emission which would cause interference at the masses of interest. In the following we shall exhibit tests of this procedure for Rb and Sr and show some age determinations on selected samples as a demonstration of the validity of this approach.

EXPERIMENTAL PROCEDURES

Mass spectrometry

Samples were loaded on heavily oxidized, short (0.3 cm long), V-shaped, zone-refined Ta filaments. Prior to use, these filaments are heated for 2 hours at ~2100°C, after which they emit essentially no Rb or Sr. We have determined a loading blank for these filaments of 2×10^7 and 3×10^7 atoms of Rb and Sr respectively, which is negligible relative to the sample size used and the contamination introduced by the sample dissolution procedure (Table 1). The Sr isotopic composition of this blank was also determined. During the experiments involving microsamples, it has been necessary to strictly control the use of the mass-spectrometer. The ion source of the mass-spectrometer was removed and cleaned prior to a series of experiments on microsamples. After installation of the source, a clean blank filament was introduced in the source and heated to ~1500°C in order to reduce possible hydrocarbon interferences. A check of the Rb–Sr mass region using this blank filament also provided a confirmation of the absence of interfering species. No macrosamples were introduced in the mass-spectrometers when a series of microsample experiments was in progress. To eliminate memory effects, high Rb/Sr microsamples were analyzed and then the mass-spectrometer source was cleaned before analyzing low

Table 1. Rb–Sr contamination.

Blanks	Rb 10^9 atoms	Sr 10^9 atoms	$^{87}Sr/^{86}Sr$
Filament Loading	0.02	0.03	
Sample Dissolution	1.4	2.8	0.712
	1.0	2.1	0.713
	0.9	2.1	0.722
Total Procedure	1.1	6.4	
	—	8.2	

Reagents	10^9 atoms/ml	10^9 atoms/ml	
HF	21	20	
HNO$_3$, conc	1.9	17	
HNO$_3$, 1N	0.4	2.5	
HCl, 2.5N	0.14	4.1	
H$_2$O	0.14	0.9	

Citric Acid (satur.)	0.3 ppb	34 ppb	
Sucrose (satur.)	0.9 ppb	8 ppb	
Orange Juice	0.5 ppb	12 ppb	

Rb/Sr samples. The latter procedure was adopted after an analysis of an aliquot of the solution of a plagioclase crystal from basalt 14276 (plagioclase-1, Table 4) showed the presence of Rb interference with ^{85}Rb/^{87}Rb attributable to that in the spiked quintessence sample from this basalt analyzed in the mass-spectrometer immediately preceding the plagioclase analysis. No significant spectral interferences other than ^{87}Rb were observed. To minimize hydrocarbon interferences a "cold finger," near the sample filament, was filled with liquid nitrogen during the analyses.

All data on the microsamples were obtained using a 17-stage electron multiplier operated as an analog amplifier with a gain of $\sim 10^4$. As a check of the multiplier response, data were obtained during a Sr run over a range in ^{88}Sr beam intensity from 6×10^{-14} A to 3×10^{-13} A. The Sr isotopic ratios showed no systematic drift as a function of ^{88}Sr intensity. These observations coupled with the agreement of multiplier and Faraday cup collector data for ^{87}Sr/^{86}Sr and for ^{84}Sr/^{88}Sr (see Table 2) show that the multiplier response is linear for the ion currents used in this work to within the levels of precision. A typical mass spectrum of the Rb–Sr region is shown in Fig. 1; the intensity of ^{88}Sr is 2×10^{-13} A. Different isotopes are shown at appropriate sensitivities, while the regions between spectral lines are scanned at the highest sensitivity. Note the absence of a signal at mass 85 $[I(^{85}\text{Rb}) < 10^{-17}$ A]. Background measurements are obtained at the positions indicated by arrows and chosen so as to avoid minor hydrocarbon contributions as can be seen on the high mass side of ^{86}Sr at the level of $\sim 3 \times 10^{-17}$ A. ^{88}Sr beams at the level of $1-2 \times 10^{-13}$ A were obtained typically for several hours for samples of 6×10^{11} Sr atoms loaded on the filament. This corresponds to an ionization efficiency of 2–5% (ions detected/atoms loaded on filament). Achieving this high ion yield was critical in carrying out these experiments. As a further check on the ionization efficiency, we analyzed an aliquot of the ^{84}Sr tracer solution containing 2.6×10^9 atoms of ^{84}Sr. We obtained a stable beam of 0.5×10^{-9} A on the electron multiplier (gain 8×10^3) for 12 minutes; this corresponds to a ^{84}Sr primary beam of 6×10^{-14} A and an ionization efficiency slightly

Table 2. Comparison of macro- and direct-loading techniques[1].

	$(^{85}\text{Rb}/^{87}\text{Rb})_m$	Sr 10^{11} atoms loaded	$(^{84}\text{Sr}/^{88}\text{Sr})_c^2 \times 10^3$	$(^{87}\text{Sr}/^{86}\text{Sr})_c^3$	$(^{87}\text{Rb}/^{86}\text{Sr})^4$	T_{BABI}^5 AE
OLIVENZA (0.622 g)						
Spiked—CHEM	0.659	2×10^4	63.40	0.74386 ± 2	0.6936	4.51 ± 0.02
Spiked—DLT	0.658	4	63.6	0.7452 ± 7	0.6910	4.66 ± 0.12
Unspiked—DLT	2.60	7	6.73 ± 3	0.7439 ± 7	—	4.53 ± 0.11
	2.58	2	6.71 ± 8	0.7462 ± 21	—	4.76 ± 0.23
	2.60	2	6.72 ± 5	0.7439 ± 14	—	4.53 ± 0.17
MURCHISON (0.248 g)						
Spiked—CHEM	0.902	2×10^4	70.80	0.72990 ± 3	0.5330	4.06 ± 0.02
Spiked—DLT	0.895	4	70.8	0.7298 ± 12	0.5263	4.09 ± 0.18
	0.905	5	70.8	0.7306 ± 8	0.5361	4.12 ± 0.13
Unspiked—DLT	2.61	7	6.74 ± 2	0.7311 ± 9	—	4.22 ± 0.13

[1]CHEM = Full macroscopic chemistry; DLT = modified direct loading technique of spiked or unspiked sample; m = ratio as measured; c = corrected ratios (for Rb interference and for Sr isotopic fractionation using ^{86}Sr/^{88}Sr = 0.1194).
[2]The normal ^{84}Sr/^{88}Sr is $(6.745 \pm 0.001) \times 10^{-3}$ [Papanastassiou and Wasserburg, 1978]. Ratios much higher than this listed in this column reflect addition of the ^{84}Sr tracer for concentration measurements.
[3]Uncertainties correspond to the last significant figures and are $2\sigma_{\text{mean}}$.
[4]Errors in ^{87}Rb/^{86}Sr are less than 2%, where the dominant uncertainty is due to instrumental isotope fractionation.
[5]Model age is calculated relative to $(^{87}\text{Sr}/^{86}\text{Sr})_{\text{BABI}} = 0.69898$ and using $\lambda(^{87}\text{Rb}) = 0.0139$ AE^{-1}. Uncertainties were calculated by using the quoted uncertainty in $(^{87}\text{Sr}/^{86}\text{Sr})_c$ and the estimated maximum uncertainty in ^{87}Rb/^{86}Sr of 2%.

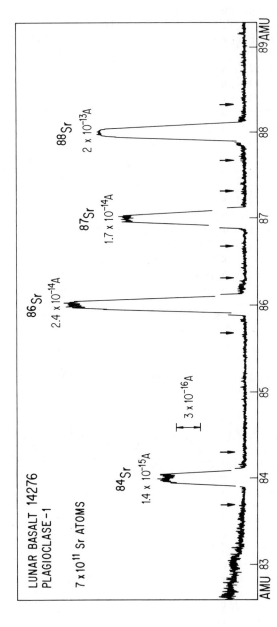

Fig. 1. Mass spectrum of Rb–Sr region. Note that isotopes are shown at different sensitivities and that the regions between integral mass numbers are scanned at high sensitivity. The sensitivity shown for 3×10^{-16} A applies to the regions between the integral masses. The arrows point to the mass positions (on both sides of each isotope) where background measurements were obtained. Note the absence of a ^{85}Rb signal. The sloping background below mass 83 is due to the reflection of the ^{88}Sr beam off the vacuum envelope. The noise on top of the Sr ion beams (especially for ^{84}Sr) as well as for the reflected ions below mass 84 reflects the ion counting statistics.

greater than 10%. We believe that the main reason for the high ionization efficiency is the direct contact of most of the sample atoms with the ionizing surface. The heavy oxidation of the filament, resulting in heavily pitted areas, which are clearly visible using a binocular microscope, may also contribute by providing an effective increase in the available surface area and a "porous" structure in which the sample is loaded. We note that during this work with microsamples we observed similar Sr ionization efficiencies for directly loaded sample solutions and for aliquots of Sr salt solutions, so that the increase in Sr ionization efficiency at this level of sample size is not correlated with the chemical purity of the Sr.

The observed Sr isotope fractionation during a run ranged from $+0.5‰$ per atomic mass unit to $\sim 0‰$ amu^{-1} at the end of the run. This is consistent with the range of fractionation obtained for macrosamples using the simple Faraday cup collector and with the expected fractionation effects due to the use of the electron multiplier.

During a sample analysis, Rb was measured in the early part of the run and yielded data of good precision. In order to minimize mass dependent isotope fractionation, Rb data for each run were obtained in a systematic way over a wide range of ion beam intensities [$I(^{87}Rb)$ from 10^{-14} A to 3×10^{-13} A]. We note that this technique yields Rb isotopic data which are subject to a wider variation in mass dependent isotope fractionation than the technique using silica gel developed and described by Papanastassiou and Wasserburg (1971). The latter technique involves loading a fixed amount of Rb (~ 2 ng) on a Re V-shaped filament with silica gel and yields variations in $^{85}Rb/^{87}Rb$ during a run and between runs of the same sample which are typically less than 0.3% and always less than 0.5%. This precision in the Rb isotopic ratios results in uncertainties in the Rb concentration and in the calculated $^{87}Rb/^{86}Sr$ ratios for the sample which are less than 0.5%. For the experiments on microsamples we rejected the use of the silica gel technique because the improved emission of Rb results in significant mass interference and contamination, even though the silica gel improves also the Sr ionization efficiency, to a value several times higher than indicated above. The use of silica gel may warrant further investigation. From the reproducibility of the Rb isotopic composition using the direct loading technique on microsamples we believe that uncertainties in $^{85}Rb/^{87}Rb$ and in the calculated $^{87}Rb/^{86}Sr$ from spiked samples are less than 2%, with this uncertainty being due to uncertainties in isotope fractionation effects.

After Rb data acquisition, the temperature was increased, Rb mostly burned off, and Sr data acquired. ^{85}Rb was measured carefully in all Sr analyses. The ratio of the ion beams at mass 87 to mass 86 as a function of the $^{85}Rb^+/^{88}Sr^+$ ratio throughout the run permitted the determination of the ratio $^{87}Rb/^{85}Rb$ emitted during Sr data acquisition. An extreme example of this for a high Rb/Sr sample is shown in Fig. 2a. A more typical case also for a high Rb/Sr sample is shown in Fig. 2b. For most cases the isobaric interference due to ^{87}Rb was small (in the range 0.01% to 0.2% of ^{87}Sr) during a significant part of the Sr data acquisition interval. We found that the ratio of $^{87}Rb/^{85}Rb$ obtained during Sr emission was the same as that obtained during the Rb part of the run. This provides strong support for the emission of Rb from the sample on the filament with a constant and well-defined $^{87}Rb/^{85}Rb$ during the run and thus permits correction for isobaric interference at mass 87 in a consistent, quantitative manner. Due to optimal spiking of the main aliquots of the sample solutions, $^{87}Rb/^{85}Rb$ of the spiked samples were typically in the narrow range of 4–1 so that the interference at mass 87 did not become inordinately high due to overspiking of the Rb. For the analyses of microsamples (Tables 3 and 4) we have listed the range of the Rb interference observed in each mass-spectrometer analysis. For this, we have listed the ratio of the ion beam intensity (I) at mass 85 relative to the intensity of mass 88. For each analysis, a large fraction of the Sr data was obtained at the lower limit of $I(85)/I(88)$ given in the tables. For a given $I(85)/I(88)$ value in the tables, the approximate level of the Rb interference correction can easily be reconstructed for unspiked as well as for spiked samples since the $^{87}Rb/^{85}Rb$ for spiked samples were always in the range 1 to 4.

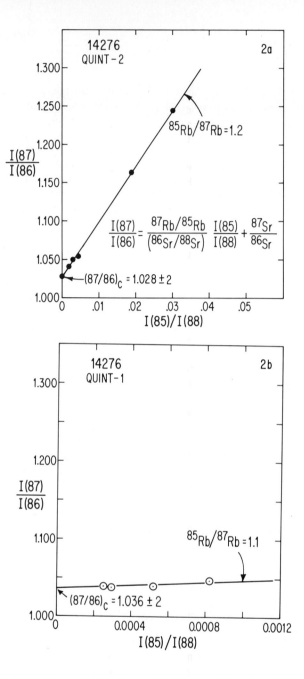

Fig. 2. Spectral interference effects at mass 87. We show the ratios of the measured beam intensities (I) at masses 85, 86, 87, and 88 during the time of Sr data acquisition. As Rb is volatilized much faster than Sr, if the species involved are the isotopes of Rb and Sr with well-defined isotopic compositions, the data points will determine a mixing line described by:

$$I(87)/I(86) = [(^{87}Rb/^{85}Rb)/(^{86}Sr/^{88}Sr)]I(85)/I(88) + ^{87}Sr/^{86}Sr,$$

where the isotopic ratios are those of the sample being analyzed. The slope of the line is proportional to $^{87}Rb/^{85}Rb$ and the intercept with the y-axis yields the $^{87}Sr/^{86}Sr$ in the sample being analyzed. The data in Figures a, b define straight lines whose slopes yield $^{85}Rb/^{87}Rb$ ratios which are consistent with the same $^{85}Rb/^{87}Rb$ ratios measured in the Rb part of the runs. Note that Fig. 2a shows an extreme case with substantial correction for ^{87}Rb interference while Fig. 2b shows a more typical case with much smaller effects. The difference between the two analyses is dominantly the amount of time allowed for Rb volatilization at a temperature slightly below the Sr running temperature.

Contamination

We list in Table 1 the determinations of the level of contamination for the procedures used. We note the extremely low contamination from the filament. The contamination for sample dissolution is relatively constant at $1-1.4 \times 10^9$ atoms Rb and $2-3 \times 10^9$ atoms Sr, and is consistent with the measured contamination in the reagents used (see below). The total procedural blanks including sample handling are $\sim 3 \times 10^9$ atoms Rb and $6-8 \times 10^{10}$ atoms Sr respectively. These higher blanks, especially for Sr, are variable and attributable to the amounts of the mixture of citric acid and sucrose ("orange juice") used for picking up and transferring samples. While the citric acid and sucrose used for the orange juice are analytical grade reagents and have relatively low Rb and Sr levels, it is clear that at this level they provide significant contamination. This problem will require the identification of even cleaner reagents or the further reduction in the amounts used in the sample handling procedures.

Chemistry

Samples were dissolved in HF and HNO_3. The sample dissolution was observed under a binocular microscope. To prevent loss due to charge build-up, a crystal was transferred into a teflon beaker containing a drop ($9 \mu l$) of water; $9 \mu l$ of concentrated HF were added, followed by $4.5 \mu l$ of concentrated HNO_3 after the HF had reacted with the sample. The solution was dried and the HF and HNO_3 treatment repeated. The sample was then treated with $4.5 \mu l$ of concentrated HNO_3 a third time, and then picked up with $14 \mu l$ of dilute HCl; aliquots were then obtained for mass-spectrometric analyses. During the chemical procedures samples were handled on a very restricted area in the teflon beakers to minimize contamination. For each sample a small aliquot ($<5\%$) was spiked to determine the Rb/Sr ratio and the approximate numbers of atoms in the solution. A fraction of the main solution was then spiked optimally for Rb and Sr. This method of spiking a small aliquot of a sample solution so that the major aliquot could then be spiked optimally, especially when the concentrations can vary widely, has been a standard procedure for careful isotope dilution work for more than two decades (cf. Burnett and Wasserburg, 1967).

Mixed tracers

For all this work, mixed $^{87}Rb-^{84}Sr$ isotope tracers were prepared and calibrated. Four sets of tracers were prepared gravimetrically with a range in both the concentrations and in the ratio of $^{87}Rb/^{84}Sr$ to cover three orders of magnitude so as to provide for optimal spiking of a variety of samples with a wide range in Rb/Sr (e.g., sanidine, quintessence, plagioclase, etc.). The mixed tracers were calibrated by isotope dilution with normal standards; the concentrations of the mixed tracers were found in all cases to be consistent with the gravimetry to within 0.05% for Sr and 0.06% for Rb. The isotopic compositions of the mixed tracers were also measured and found to be essentially constant, in agreement with the low levels of contamination in the reagents used for dilution. Sample solution aliquoting and tracer addition required use of micro-syringes (Hamilton) equipped with intra-medic polyethylene tubing ("PE-60" and "PE-20"). These techniques permitted the addition of the optimal amounts of tracers to the samples. The sample aliquoting and the addition of the mixed tracer to the sample aliquots were done volumetrically based on the calibrated volume of the polyethylene tubing. The volumes of solutions used were measured approximately with a clean ruler placed near and along the polyethylene tubing containing the solution. The use of mixed tracers results in precisely measurable Rb/Sr atom ratios while the individual Rb and Sr concentrations may be known only to 20%.

Procedural tests

As a test of this approach we dissolved a large sample of the Olivenza meteorite. Two-thirds of the sample solution was spiked with ^{87}Rb and ^{84}Sr tracers. A fraction of the spiked solution was passed through the standard macro-chemistry and the separated Rb and Sr were analyzed on a mass-spectrometer. The results are given in Table 2 as spiked-CHEM. A small drop of the spiked solution of the sample (containing 4.5×10^{11} atoms of Sr) which had not gone through any chemical separation was directly loaded on a heavily oxidized Ta filament and analyzed for Rb and Sr. The results are given in Table 2 as spiked-DLT. Inspection of the data shows the DLT results and those obtained by standard procedures to be in good agreement for both ^{85}Rb/^{87}Rb and the Sr isotopic composition in the spiked sample within 2‰. The atom ratios of Rb/Sr calculated from the measured isotopic compositions are also in good agreement. Similarly, the model ages calculated from these data are in excellent agreement. One third of the original sample solution was not spiked. Three drops of this solution containing from 7×10^{11} to 2×10^{11} atoms (1.1 to 0.3 picomoles) of Sr were also analyzed by the DLT. The results show good agreement for the ^{87}Sr/^{86}Sr in the sample with precision from 1‰ to 3‰. The ^{84}Sr/^{88}Sr ratio was also determined and found to be in excellent agreement with the normal value to within 0.5% with uncertainties for each analysis of up to 1%. The results on ^{84}Sr were obtained on as few as 1.2×10^9 atoms.

A similar experiment was performed using a solution of a large sample (0.25 g) of the Murchison meteorite. Comparison of the DLT data on small samples and those obtained by standard procedures on macroscopic samples shows them to be in good agreement. The model age for Murchison was found to be distinctly young and indicates relatively young Rb–Sr migration as expected also from data by earlier workers [Mittlefehldt and Wetherill, 1979].

These experiments show that it is feasible to obtain Rb–Sr mass-spectrometric data on picomole and subpicomole quantities of Sr (10^{12}–10^{11} atoms) with 1–2‰ precision (2σ) in

Table 3. Results on diverse microsamples[1].

	Sr 10^{11} atoms loaded	I(85)/I(88)[2] $\times 10^4$	(^{84}Sr/^{88}Sr)$_c$ $\times 10^3$	(^{87}Sr/^{86}Sr)$_c$	^{87}Rb/^{86}Sr	T$_{BABI}$ AE
Murchison chondrule (300 μm; F = 0.4[3])						
Aliquot	0.6	—	125	—	1.08	—
Unspiked	2	0.3–0.7	6.68 ± 4	0.7220 ± 9	—	1.57 ± 0.07
Spiked	6	0.5–1.5	17.3	0.7234 ± 24	1.044	1.66 ± 0.17
Spiked	6	0.2–4.0	17.3	0.7215 ± 10	1.045	1.53 ± 0.07
Olivenza chondrule ($690 \times 190 \times 70$ μm^3; F = 0.1)						
Aliquot	6	—	28	—	0.76	—
Spiked	4	0.2–1.2	173	0.7450 ± 11	0.717	4.48 ± 0.14
Colomera						
Sanidine-4	4	4–18	105	6.357 ± 9	88.7	4.45 ± 0.09
$380 \times 250 \times 80$ μm^3; F = 0.02						
Sanidine-6	24	3–4	50	7.016 ± 12	102.0	4.32 ± 0.09
$250 \times 175 \times 100$ μm^3; F = 0.1						
Diopside-2	4	0.6–6.0	67	0.7573 ± 14	0.545	7.32 ± 0.23
$380 \times 300 \times 100$ μm^3; F = 0.25						
Microtektites [Indian Ocean Core, 39°23'S; 104°22'E]						
BG1 (240 μm; F = 0.04)	8	0.5–12	205	0.7168 ± 7	1.528	0.83 ± 0.04
BG2 (210 μm; F = 0.05)	8	0.1–0.7	176	0.7187 ± 7	0.628	2.22 ± 0.09

[1] See Table 1 footnotes.
[2] Range of ion beam intensity ratios throughout the Sr analysis.
[3] F = fraction of the solution of the sample used in all experiments.

^{87}Sr/^{86}Sr. The corresponding numbers of ^{87}Sr and ^{86}Sr atoms, on which this data precision depends, are a factor of ten lower (10^{11}–10^{10} atoms). Data on the low abundance isotope ^{84}Sr can also be obtained, down to 10^9 atoms, with 1% (2σ) precision. We note that the number of radiogenic ^{87}Sr* atoms in a sample (e.g., for chondritic matter, ~4.5 AE old) is approximately equal to the number of ^{84}Sr atoms. Therefore, the ability to measure the ^{84}Sr isotopic abundance with 1% (2σ) precision provides direct support for the ability to determine Rb–Sr ages at the 1% level.

Microsamples

A series of experiments was carried out on some "large" meteorite samples, which included single chondrules from Murchison and Olivenza, and sanidine and diopside crystals from the Colomera iron meteorite. The Murchison chondrule (300 μ in diameter) was dissolved and a small aliquot taken and spiked to determine the number of Rb and Sr atoms in the solution. The remainder was divided into four drops and one drop was analyzed for the Sr isotopic composition without spiking; two of the other drops were spiked. Repeat analyses for the Murchison chondrule (Table 3) show excellent agreement between all runs for the ^{87}Sr/^{86}Sr ratios at the level of ± 1.3‰. The unspiked sample also yielded a value for ^{84}Sr/^{88}Sr within 10‰ of normal composition ($2\sigma_m = \pm 6$‰ for 10^9 atoms of ^{84}Sr). The model age for this chondrule is very young in accord with the results on the macroscopic sample of Murchison (Table 2). The data on Murchison indicate (a) the existence of a very recent disturbance, and (b) that it may be possible to place very strong upper limits to the time of disturbance by analyzing microsamples as opposed to bulk or whole meteorite samples. The Olivenza chondrule was dissolved and small aliquots were spiked for analysis. The results yield a model age of 4.48 ± 0.14 AE, which is consistent

Table 4. Microchron of basalt 14276[1].

	Sr 10^{11} atoms loaded	I(85)/I(88)[2] $\times 10^4$	(^{84}Sr/^{88}Sr)$_c$ $\times 10^3$	(^{87}Sr/^{86}Sr)$_c$	^{87}Rb/^{86}Sr
Quintessence-1 ($100 \times 50 \times 50$ μm; F = 1)					
Aliquot	0.3	—	111	—	5.96
Spiked	3	3–9	53.7	1.036 ± 2	6.11
Unspiked	3	3–9	—	1.040 ± 3	—
Quintessence-2 ($50 \times 50 \times 50$ μm; F = 1)					
Tot. Spiked	6	20–40	33.3	1.028 ± 2	6.13
Quintessence C-2 ($110 \times 95 \times 60$ μm; F = 0.50)					
Aliquot	0.6	2–4	401	0.842 ± 2	2.89
Spiked	7	4–6	49.1	0.8742 ± 22	3.23
Plagioclase-1 ($130 \times 100 \times 100$ μm; F = 0.45)					
Aliquot	6	0.4–0.6	90.8	0.6985 ± 9	
Spiked[3]	7	0.1–0.6	25.5	0.7062 ± 12	0.0087
Unspiked	7	0.1–0.7	6.74 ± 5	0.7010 ± 15	
Plagioclase C-3 ($120 \times 110 \times 60$ μm; F = 0.40)					
Aliquot	2	0.2–0.6	47.8	0.7005 ± 5	0.0028
Spiked	5	0.1–0.5	17.3	0.7015 ± 12	0.0047
Unspiked	7	0.1–0.6	6.71 ± 3	0.7015 ± 9	—

[1]Subscript c denotes ratios corrected for isotope fractionation and for interference at mass 87; for each sample we give the total fraction (F) of the solution which was used for all the experiments listed (see also Table 2 footnotes).
[2]See Table 3, footnote 2.
[3]The ratio of (^{87}Sr/^{86}Sr)$_c$ for this analysis is too high due to memory of radiogenic Sr in the ion-source (see text).

with the internal isochron age of this meteorite of 4.63 ± 0.16 AE [Sanz and Wasserburg, 1969].

Two microscopic samples of a coarse-grained sanidine crystal taken from the Colomera meteorite were analyzed. These are extremely radiogenic and have values of Rb/Sr and $^{87}Sr/^{86}Sr$ comparable to the macrosamples measured before [Wasserburg, Sanz, and Bence, 1968; Sanz, Burnett, and Wasserburg, 1970]. The two crystals give model ages of 4.45 and 4.32 AE. These ages are 3.5 and 6.3% lower than the age of 4.61 ± 0.04 AE obtained previously for Colomera inclusions [Sanz et al., 1970]. The sanidine crystals were obtained from an inclusion on the exterior of the meteorite and were stained with rust. These materials show minor but clear alteration. This presumably is reflected in the dispersion of the Rb-Sr data and model age relations. We note that for these extremely radiogenic crystals only a small fraction of the solution (as little as 2%) was needed for a measurement, so that samples as small as 50 μm would be fully adequate for analysis. A diopside crystal was also analyzed which gave a reasonable Sr isotopic composition but an extremely high model age. This is almost certainly due to the precipitation of K and Rb during dissolution since $HClO_4$ was also added to aid in dissolving this more resistant crystal. Alkali precipitation occurred because the sample solution was partly evaporated after the addition of $HClO_4$, before tracer addition.

For diversity, two microtektites obtained from an Indian Ocean drilled core (Table 3) were kindly provided for us by Dr. B. Glass. The Rb-Sr data of one of these microtektites (BG-1) fall well within the field of data for Australasian tektites [Schnetzler and Pinson, 1964; Compston and Chapman, 1969; Shaw and Wasserburg, 1981]. Following the analysis of the tektite data by Shaw and Wasserburg [1981], BG-1 is, therefore, consistent with an inferred time of weathering and sedimentation of the source material of the tektite of ~ 250 my. The second micro-tektite (BG-2) shows $^{87}Sr/^{86}Sr$ comparable to BG-1, but has an atypically low Rb/Sr and falls far off the data for Australasian tektites. A distinct process, e.g., loss of Rb by volatilization, may be responsible for this behavior.

Lunar KREEPUTh basalt 14276

An attempt was made to determine an isochron on small lunar samples. The sample chosen for investigation was Rb-rich lunar basalt 14276, which we had previously studied using conventional techniques [Wasserburg and Papanastassiou, 1971]. Materials were picked from previously obtained bulk mineral separates. In addition a small amount of material was crushed for microscopic examination. In this crushed material we identified micro-rocks composed of several crystals of different minerals. These micro-rocks were used to extract pairs of adjacent quintessence and plagioclase crystals for analysis.

A single grain of quintessence (#1, $100 \times 50 \times 50$ μm^3) was chosen, it dissolved readily and an aliquot of the solution was used to determine the number of Rb and Sr atoms. The remaining solution was split into two drops, one of which was spiked. The Rb-Sr determinations on the aliquots to which the ^{87}Rb and ^{84}Sr tracer was added are of good precision and give $^{87}Rb/^{86}Sr$ which are in excellent agreement. The $^{87}Sr/^{86}Sr$ data determined on spiked and unspiked aliquots are also in excellent agreement. A second high purity quintessence grain (#2) was selected and analyzed. This yielded results almost indistinguishable from #1. These samples of quintessence are a factor of two more radiogenic than any macroscopic samples obtained by conventional mineral separation procedures [Wasserburg and Papanastassiou, 1971]. This indicates that the crystals analyzed here may be the pure endmember. A plagioclase crystal (#1) was selected and analyzed. It showed a relatively primitive $^{87}Sr/^{86}Sr$ value and a low $^{87}Rb/^{86}Sr$ ratio; however, the ^{87}Sr abundance in the spiked aliquot was too high as a result of emission of radiogenic Sr which remained in the ion source from a preceding quintessence analysis. The ion source was cleaned and another unspiked fraction of the same solution analyzed, yielding superior results. A micro-rock (labeled C) was then processed under the microscope; it consisted of a quintessence grain (C-2) and a plagioclase grain (C-3). These grains were separated from each other using a micro-chisel and processed. The C-2

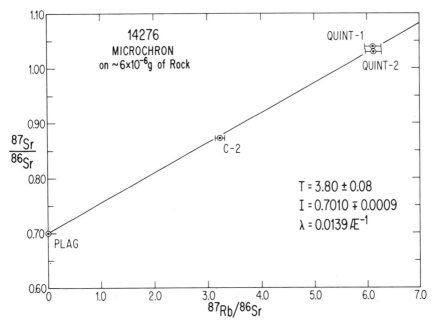

Fig. 3. Rb–Sr evolution diagram for KREEPUTh-rich basalt 14276, using 6 μg of sample. Note the radiogenic nature and reproducibility of the Quintessence-1 and -2 crystals indicating that they may represent the endmember composition. Quintessence C-2 contained substantial amounts of plagioclase. There are several data points in the circle indicated as plagioclase. The data define a straight line which we interpret as a microchron.

quintessence was dissolved and a small aliquot (6×10^{10} atoms of Sr) was adequate to determine the approximate number of atoms in the sample. Using this information, the total solution was spiked and analyzed. This quintessence crystal gave a lower enrichment of ^{87}Sr as compared to the first two crystals analyzed; this is consistent with the significant amount of adhering plagioclase in the C-2 quintessence grain. The plagioclase (C-3) obtained from micro-rock C, was then analyzed three times and gave good data which confirmed the observations on plagioclase #1 obtained after the ion source cleaning. The results on basalt 14276 are exhibited in Fig. 3. The data form a good linear array, which is interpreted to be an isochron. The age calculated from these data is 3.80 ± 0.08 AE in good agreement with the age of 3.88 ± 0.01 AE obtained previously on macroscopic samples [Wasserburg and Papanastassiou, 1971]. Similarly, the initial ^{87}Sr/^{86}Sr obtained from the microsamples is consistent with data obtained by the "macro" techniques. These results, which included many replicate analyses, were obtained on a total of 6×10^{-6} g of lunar sample.

CONCLUSIONS

We have demonstrated by analyzing small aliquots of sample solutions of known Rb/Sr and Rb and Sr isotopic compositions that we can measure the Rb and Sr isotopic compositions and Rb/Sr atom ratios in microsamples containing 10^{11}–10^{12} atoms of Sr. We have shown that good quality isochrons may now be obtained by using individual mineral grains. Micro-rocks can be separated into their constituent crystals and these analyzed individually. Grains can also be extracted from a thin-section after examination by optical microscopy and after electron microprobe analyses. The level of precision in ^{87}Sr/^{86}Sr obtainable is comparable (1–2‰) with that achieved in the mid-1960s by the then conventional techniques on macrosamples. These analyses may be carried out quickly and simply. We call the isochrons determined on truly microscopic samples "microchrons." The levels of contamination can be reduced to acceptable levels for these analyses. The only technical details which require attention are (1) establishing that the sample is dissolved and fully equilibrated with the tracer, and (2) developing efficient and clean micromanipulation techniques for removing and transferring individual crystals.

The approach given here will permit selection of pure mineral grains which are desired for isotopic age studies within individual chondrules or inclusions in iron meteorites. These techniques will also permit investigation of the distances and times of element migration in adjacent crystals resulting from metamorphism. Rb–Sr model ages may prove to be of considerable importance in identifying and studying interplanetary dust particles possibly of cometary origin. Extensive efforts are now underway to apply these techniques to interplanetary dust grains and deep-sea spherules of presumed extraterrestrial origin.

Acknowledgments—We thank Dr. Billy P. Glass for generously providing us with the micro-tektites. We thank J. Brown for his careful sample manipulation and T. Wen for diligently honed chemical and mass-spectrometric skills. We appreciated constructive comments on the manuscript by D. M. Unruh, L. E. Nyquist, and an anonymous reviewer. This work was done with steady hands and controlled by NASA grant NGL-05-002-188. Div. Contr. No. 3563(384).

REFERENCES

Burnett D. S. and Wasserburg G. J. (1967) ^{87}Rb–^{87}Sr ages of silicate inclusions in iron meteorites. *Earth Planet. Sci., Lett.* **2**, 397–408.

Compston W. and Chapman D. R. (1969) Sr isotope patterns within the Southeast Australasian strewn field. *Geochim. Cosmochim. Acta* **33**, 1023–1036.

Esat T. M., Brownlee D. E., Papanastassiou D. A., and Wasserburg G. J. (1979) Mg isotopic composition of interplanetary dust particles. *Science* **206**, 190–197.

Lee T., Papanastassiou D. A., and Wasserburg G. J. (1977) Mg and Ca isotopic study of individual microscopic crystals from the Allende meteorite by the direct loading technique. *Geochim. Cosmochim. Acta* **41**, 1473–1485.

Mittlefehldt D. W. and Wetherill G. W. (1979) Rb–Sr studies of CI and CM chondrites. *Geochim. Cosmochim. Acta* **43**, 201–206.

Papanastassiou D. A. and Wasserburg G. J. (1971) Rb–Sr ages of igneous rocks from the Apollo 14 mission and the age of the Fra Mauro formation. *Earth Planet. Sci. Lett.* **12**, 36–48.

Papanastassiou D. A. and Wasserburg G. J. (1973) Rb–Sr ages and initial Sr in basalts from Apollo 15. *Earth Planet. Sci. Lett.* **17**, 324–337.

Papanastassiou D. A. and Wasserburg G. J. (1978) Sr isotopic anomalies in the Allende meteorite. *Geophys. Res. Lett.* **5**, 595–598.

Papanastassiou D. A. and Wasserburg G. J. (1981) Microchrons: The ^{87}Rb–^{87}Sr dating of microscopic samples (abstract). In *Lunar and Planetary Science XII*, p. 802–804. Lunar and Planetary Institute, Houston.

Sanz H. G., Burnett D. S., and Wasserburg G. J. (1970) A precise ^{87}Rb/^{87}Sr age and initial ^{87}Sr/^{86}Sr for the Colomera iron meteorite. *Geochim. Cosmochim. Acta* **34**, 1227–1239.

Sanz H. G. and Wasserburg G. J. (1969) Determination of an internal ^{87}Rb–^{87}Sr isochron for the Olivenza chondrite. *Earth Planet. Sci. Lett.* **5**, 335–345.

Schnetzler C. C. and Pinson W. H., Jr. (1964) Variation of strontium isotopes in tektites. *Geochim. Cosmochim. Acta* **28**, 953–969.

Shaw H. F. and Wasserburg G. J. (1981) Sm–Nd and Rb–Sr isotopic systematics of Australasian tektites (abstract). In *Lunar and Planetary Science XII*, p. 967–969. Lunar and Planetary Institute, Houston.

Tera F. and Wasserburg G. J. (1975) Precise isotopic analysis of Pb in picomole and subpicomole quantities. *Anal. Chem.* **47**, 2214–2220.

Wasserburg G. J., Lee T., and Papanastassiou D. A. (1977) Mg and Ca isotopic study of individual microscopic crystals from the Allende meteorite by the direct loading technique (abstract). In *Lunar Science VIII* p. 991–993, The Lunar Science Institute, Houston.

Wasserburg G. J. and Papanastassiou D. A. (1971) Age of an Apollo 15 mare basalt; lunar crust and mantle evolution. *Earth Planet. Sci. Lett.* **13**, 97–104.

Wasserburg G. J., Sanz H. G., and Bence A. E. (1968) Potassium feldspar phenocrysts in the surface of Colomera, an iron meteorite. *Science* **161**, 684–687.

A unique type 3 ordinary chondrite containing graphite-magnetite aggregates—Allan Hills A77011

Susan G. McKinley, Edward R. D. Scott, G. Jeffrey Taylor, and Klaus Keil

Department of Geology and Institute of Meteoritics, University of New Mexico, Albuquerque, New Mexico 87131

Abstract—Allan Hills A77011 has all the petrographic properties of type 3 ordinary chondrites: it contains sharply-defined chondrules with glassy groundmasses, heterogeneous olivine (Fa_{1-39}, PMD 39), pyroxene (Fs_{1-40}, PMD 56), and metallic Fe,Ni (PMD of Co in kamacite is 7.3), significant concentrations of Cr (≤ 0.9 wt%) and Si (≤ 0.3 wt%) in metal, and a fine-grained, silicate matrix ("Huss matrix"). The Huss matrix makes up ~ 15 vol% of the meteorite and consists of equal amounts of opaque and recrystallized material; its $FeO/(FeO + MgO)$ ratio is 1.1 times that of the bulk silicate portion of the rock. These properties, along with high C (1 wt%) and $^{36}Ar_P$ (100×10^{-8} cc STP/g) contents indicate that ALHA77011 is a petrologic type 3.2 ± 0.2. Metal concentrations and published chemical analyses are typical of L chondrites.

ALHA77011 is unique because it is the only L3 chondrite we know that contains a few vol% of aggregates of graphite and magnetite crystals, which are generally micron or submicron in size. These aggregates, which range in size from <5 to $200 \mu m$, are intimately associated with metallic Fe,Ni. Graphite-magnetite has widely varying bulk Fe and C concentrations; however, these elements covary and all others total <3 wt%. The unique occurrence of graphite-magnetite allows us to pair unambiguously 34 L3 meteorite specimens from the 1977–1979 Allan Hills collections. ALHA77011 and Sharps, an H3 with 11.3 vol% of graphite-magnetite, are intermediate in their properties between normal type 3 ordinary chondrites and a new kind of type 3 ordinary chondrite that forms clasts in ordinary-chondrite regolith breccias and contains no Huss matrix, only graphite-magnetite and chondrules. We suggest that the ordinary chondrites formed from at least four separate components: Mg-rich chondrules, Fe-rich Huss matrix, metal-troilite, and graphite-magnetite.

INTRODUCTION

We are studying a collection of 145 undescribed meteorites from the 1977 collection from Allan Hills, Antarctica, that weigh ≤ 150 g. The purpose of this study is to increase the chances of discovering new types of meteorites and of recovering additional samples of rare types of meteorites. Preliminary investigation of the 145 specimens indicates that 122 of them are equilibrated ordinary chondrites. The remaining 23 specimens include a CO3 carbonaceous chondrite (A77029), two enstatite chondrites which may represent the same meteorite (A77156 and A77295), a type 3 ordinary chondrite (A77176), and 19 specimens of the unique type 3 ordinary chondrite named ALHA77011. This paper reports the results of a detailed petrologic study of ALHA77011.

Type 3 ordinary chondrites contain a matrix composed of fine-grained (grain size typically $\leq 1 \mu m$), Fe-rich silicates (Huss *et al.*, 1981). In thin sections, this matrix is opaque in the least metamorphosed of the unequilibrated ordinary chondrites (those in category I of Huss *et al.*, 1981 and classes 3.0–3.2 of Sears *et al.*, 1980), but is generally more transparent, hence, presumably recrystallized, in more metamorphosed type 3 ordinary chondrites. We refer to the opaque and recrystallized silicate matrix as "Huss matrix". We prefer this term to the term "matrix", which is commonly used for all material interstitial to chondrules, or the term "silicate matrix", which may include chondrule and lithic fragments. (We recognize that other workers such as Wood (1962), Kurat (1970), Christophe Michel-Lévy (1976) and Ashworth (1977) described this material. However, Huss *et al.* (1981) were the first to provide a detailed characterization of the matrix in a whole suite of unequilibrated ordinary chondrites).

ALHA77011 is distinctive in that it contains, besides Huss matrix, numerous aggregates consisting of micron- and submicron-sized graphite and magnetite (Scott et al., 1981a, b, c). We have not observed abundant graphite-magnetite aggregates in other type 3 ordinary chondrites, except for Sharps. This property, therefore, makes ALHA77011 distinctive and has allowed us to pair the 19 specimens from the 1977 collection of undescribed specimens with 15 other specimens that were collected in the Allan Hills, Antarctica during the 1977, 1978 and 1979 field seasons (King et al., 1980; Score et al., 1981). Applying the procedures of the Meteoritical Society Nomenclature Committee (Graham, 1980), we have designated the lowest number, 77011, as the name of the paired specimens. [In preliminary reports (Rubin et al. 1981b; Scott et al. 1981b) written before our study was complete, this meteorite was described as ALHA77043.]

METHODS OF STUDY

Each meteorite specimen was examined and described macroscopically and then portions were made into polished thin sections. These were examined in both reflected and transmitted light. Modal analysis was performed on two representative sections of ALHA77011 with an automated stage counting 2000 points.

Mineral compositions were determined using an automated ARL EMX-SM electron microprobe. Analyses were performed with crystal spectrometers at an accelerating voltage of 15 kV and sample current of $\sim 0.015\ \mu A$. Quantitative analyses of olivine and pyroxene within and between chondrules were made for Fe, Mg, Ca and Si.

The composition of opaque Huss matrix was determined using the broad beam (25 μm) microprobe technique. Corrections were made with a normative procedure described by Huss et al. (1981). Graphite-magnetite inclusions were also analyzed with a beam 25 μm in diameter, but because of uncertainties in the correction factors for C, analyses were not corrected, except for background and drift.

Metallic Fe,Ni was analyzed using pure elements as standards for Fe, Ni, Co, and Cr, a synthetic Fe$_3$Si for Si, and apatite for P. Analyses were corrected for differential matrix effects by the standard ZAF method. Co analyses were also corrected for the overlap of the Co K by the Fe K peak; a typical correction factor is 0.2% of the amount of Fe present. Detection limits for minor elements are as follows: Cr, 0.07 wt.%, Ca, 0.03 wt.%; Mg, Si and P, 0.02 wt.%.

RESULTS

Modal analysis (Table 1) demonstrates that ALHA77011 contains ~ 3 vol% graphite-magnetite aggregates. This property sets it apart from all other known type 3 ordinary

Table 1. Modal analyses (in vol %) of Allan Hills A77011 and Bishunpur (L3).

Phase	A77011*	Bishunpur†
Graphite-magnetite	3.2	n.o.
Opaque Huss matrix	7.0	10.7
Recrystallized Huss matrix	7.7	3.0
Silicates‡	74.1	77.1
Metallic Fe,Ni	2.2	3.6
Weathered metallic Fe,Ni	2.1	–
Troilite	3.7	5.5
	100.0	100.0

n.o. not observed.
*2000 points were counted on specimen numbers ALHA77043 and ALHA77178.
†Data from Huss (1979).
‡Silicates in chondrules and chondrule, lithic and mineral fragments.

chondrites, which contain ≪ 1% (compare with mode of Bishunpur, Table 1), except for the Sharps H3 chondrite (Rubin et al., 1981b). Consequently, it is easy to recognize in thin section specimens of this meteorite. Our study of the 145 meteorites in the 1977 Allan Hills collection reveals 19 that are definitely the same meteorite. We also examined sections of all type 3 ordinary chondrites collected at Allan Hills from 1976 to 1979, that are listed in Score et al. (1981). Fifteen of these large specimens contain a few percent of graphite-magnetite inclusions and can, therefore, be paired unambiguously with the 19 samples studied by us. Cassidy (1980) previously suggested that 9 of the 1977 Allan Hills L3 chondrites were paired. Table 2 lists the numbers and weights of the paired specimens. Other paired specimens may be present among the 1978 Allan Hills specimens weighing ≤ 150 g which were not described by Score et al. (1981) and to which we had no access.

Petrographic descriptions of several specimens of ALHA77011 appear in King et al. (1980). ALHA77011 contains a wide variety of sharply-defined chondrules (Fig. 1) that range from 0.2 to 4 mm in diameter. As in other unequilibrated ordinary chondrites, porphyritic chondrules are the most abundant; barred olivine, radiating pyroxene and aphanitic chondrules are also present. Many chondrules contain glass, which is predominantly turbid or partially devitrified, and less commonly pink-brown and clear. Figure 2 shows results of random analyses of olivine and low-Ca pyroxene grains in specimens ALHA77043 and ALHA77178; both minerals are highly heterogeneous. Olivine ranges from Fa_1 to Fa_{37} (average Fa_{17}) and has a standard deviation of 8.1 mole% Fa, percent mean deviation (PMD) is 39%. (We advocate the use of standard deviation rather than the traditional PMD because it is a more useful statistical parameter.) Low-Ca pyroxene is mostly monoclinic and frequently polysynthetically twinned. Its composition ranges from Fs_1 to Fs_{40} (average Fs_{12}) and has a standard deviation of 8.3 mole% Fs (PMD = 56%) (Table 3).

Roughly equal amounts of opaque and recrystallized Huss matrix are present. These materials occur in areas between chondrules, sometimes partially rimming chondrules. The composition of the opaque Huss matrix (Table 4) is generally similar to that in other type 3 ordinary chondrites (Huss et al., 1981), characterized by a higher FeO concen-

Table 2. Paired L3 specimens from the 1977–79 Allan Hills collection and their original weights.

ALHA number	Weight grams	ALHA number	Weight grams
77011	291.5*	77167	611.2*
77015	411.1*	77170	12.20
77031	0.48	77175	23.3
77033	9.3*	77178	5.73
77034	1.77	77185	28.0
77036	8.45	77211	26.7
77043	11.45	77214	2111.0*
77047	20.4	77241	144.1
77049	7.29	77244	39.5
77050	84.2	77249	503.6*
77052	112.2	77260	744.3*
77115	154.4	77303	78.6
77140	78.6*		
77160	70.4*	78038	363.0*
77163	24.3	78188	0.87*
77164	38.1*		
77165	30.5*	79001	32.3*
77166	139.8	79045	115.4*

*Weights from Score et al. (1981). Weights of other specimens from the 1977 collection, which are previously undescribed, are from R. Score (pers. comm.).

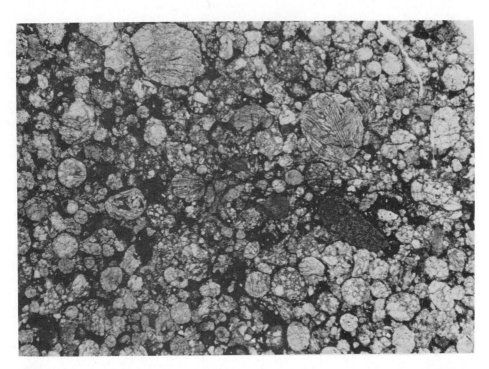

Fig. 1. Photograph of a thin section of the L3 chondrite, Allan Hills A77011, in transmitted light showing abundant well-defined chondrules characteristic of type 3 ordinary chondrites. Max. length: 17 mm (Specimen number A77036).

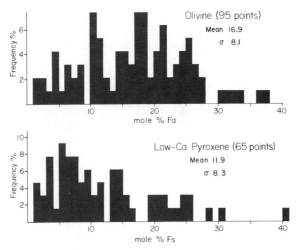

Fig. 2. Histograms showing random electron-microprobe analyses of olivines in mole % fayalite (Fe_2SiO_4) and low-Ca pyroxenes in mole % ferrosilite ($FeSiO_3$) in Allan Hills A77011. The extreme heterogeneity is typical of the most unequilibrated type 3 ordinary chondrites.

tration than the bulk silicate portion of chondrites. However, concentrations of Al_2O_3, Na_2O and K_2O are somewhat low; interestingly, the opaque Huss matrix in Sharps shares these characteristics. Graphite-magnetite aggregates range in size from <5 to 200 μm and are exceedingly widespread. Individual graphite and magnetite grains within these aggregates are typically submicron in size, though some graphite crystals are large enough to

Table 3. Composition of olivines and low-Ca pyroxenes in Allan Hills A77011*.

	Olivine mole % Fa	Low-Ca pyroxene mole % Fs
Range	1–37	1–40
Mean	16.9	11.9
σ	8.1	8.3
PMD (%)	39	56

*Specimen numbers A77043 and A77178.

display the characteristic optical anisotropy. Graphite-magnetite aggregates are commonly associated with Fe,Ni metal (Fig. 3). Commonly, graphite-magnetite is mixed with metallic Fe,Ni, but less often it rims metallic Fe,Ni grains. It also occurs on chondrule rims or as patches between chondrules. Graphite-magnetite is highly heterogeneous in composition. Table 4 shows the composition of graphite-magnetite assuming that S is present in the form of FeS and other elements except C are present as oxides. After allowance for FeS, Fe is listed as Fe_3O_4, and C calculated by difference. On a plot of the apparent concentrations of Fe and C (Scott et al., 1981b,c), analyses of graphite-magnetite in ALHA77011, like those in other meteorites, lie along a curve which is interpreted as a magnetite-graphite mixing line. Concentrations of elements other than Fe and C are highly variable but total < 3 wt%.

Grains of metallic Fe,Ni commonly show minor terrestrial weathering effects. Most metal occurs as irregular to spherical grains that reside between chondrules. Some metal occurs as patches on chondrule rims, and a small amount occurs as spheres within

Table 4. Mean compositions (wt%) of graphite-magnetite aggregates and opaque Huss matrix in Allan Hills A77011.

Constituent	Opaque Huss matrix		Graphite-magnetite	
	(a)	(b)	Range	Mean
SiO_2	32.6	39.0	0.7–2.5	1.5
TiO_2	0.09	0.11		<0.06
Al_2O_3	1.6	1.9	<0.05–0.24	0.13
Cr_2O_3	0.54	0.65		<0.07
FeO	23.8	28.5	—	—
Fe_3O_4	—	—	8–23	16
MnO	0.46	0.55		n.m.
MgO	22.6	27.0	0.05–1.4	0.6
CaO	1.04	1.24	0.11–0.3	0.2
Na_2O	0.40	0.48	<0.03–0.13	0.05
K_2O	0.12	0.14		<0.03
P_2O_5	0.33	0.39	<0.05–0.17	0.1
FeS	0.84	—	0.3–1.8	1.1
NiO	—	—	0.4–3.8	1.3
Fe–Ni	7.4	—	—	—
C(calc.)	n.m.	—	(75–90)	(79)
	91.8	100.0		100

n.m. not measured.
(a) Probe analyses corrected by normative procedures of Huss et al. (1981).
(b) Analysis in (a) recalculated to total 100% neglecting troilite and metallic Fe,Ni.

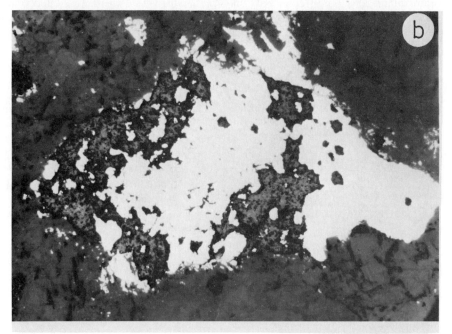

Fig. 3. Reflected light photomicrographs of Allan Hills A77011 (specimen number ALHA77260,16) showing metallic Fe,Ni (white), troilite (light gray), and graphite-magnetite aggregates (dark gray) surrounded by silicates (medium gray). (a) Graphite-magnetite forms over 50 aggregates <5 to 50 μm in size between chondrules; larger aggregates are associated with metal. A few aggregates are marked by arrows. Width: 1.7 mm. (b) A higher magnification photograph of the graphite-magnetite-metal assemblage in the upper right of (a) showing numerous white blobs of metal in graphite-magnetite and dark gray blobs of graphite-magnetite in metal. Width: 440 μm.

chondrules. Except for the metal inside chondrules, virtually all of it is associated with graphite-magnetite. Kamacite ranges quite widely in both Ni (2.9 to 5.9 wt%) and Co (0.3 to 1.1 wt%); the PMD of Co in kamacite is 7.3. Most kamacite outside chondrules has Cr, Si and P below detection limits, although a few grains contain significant amounts of Cr (≤ 0.9 wt%) and Si (≤ 0.3 wt%). In contrast, most kamacite inside chondrules contains detectable quantities of both Cr (≤ 0.7 wt%) and Si (≤ 0.2 wt%); no P was detected. Single taenite grains within the matrix contain no detectable Cr, Si or P. Taenite associated with kamacite contains minor amounts of Cr (≤ 0.6 wt.%) and Si (≤ 0.2 wt.%). Taenite within chondrules contains some measurable Si (≤ 0.6 wt%). Taenite grains are typically $< 10\,\mu$m in diameter and have central Ni contents of 35 to 45 wt%.

Troilite occurs intergrown with Huss matrix, as irregular grains between chondrules and associated with metallic Fe,Ni spherules. Probe analyses show stoichiometric FeS. However, troilite associated with metallic Fe,Ni commonly contains dusty inclusions; analyses of this troilite reveal minor amounts of Mg (≤ 0.23 wt.%), Cr (≤ 0.16 wt.%), Ca (≤ 0.13 wt.%), P (≤ 0.11 wt.%), Si (≤ 0.19 wt.%) probably from these inclusions, as Fujimaki *et al.* (1981) reported.

DISCUSSION

Classification

Chemical analyses of ALHA77011 by Jarosewich (1980) and Rhodes and Fulton (1981) (both analyses on specimen ALHA77214) yield atomic Al/Si, Ca/Si and Mg/Si ratios typical of ordinary chondrites, and the Fe/Si atomic ratio of 0.59 lies within the L group range (Wasson, 1974). Similarly the Ni/Mg, S/Mg and Fe/Mg ratios fall within the L field of Jarosewich and Dodd (1981). Classification as an L-group chondrite was originally suggested by King *et al.* (1980) on the basis of petrographic observations of large specimens of ALHA77011.

The presence of distinct chondrules with glassy groundmasses, the heterogeneity of olivine and pyroxene compositions (Fig. 2), and the abundance of monoclinic low-Ca pyroxene demonstrate that ALHA77011 is a petrologic type 3 ordinary chondrite. Sears *et al.* (1980) suggested a classification scheme to subdivide the type 3 chondrites into categories 3.0 through 3.9. The classification is based on the parameters listed in Table 5. (Sears *et al.* also use thermoluminescence sensitivity as a parameter, but such data are not yet available for ALHA77011.) The individual parameters indicate classifications ranging from 3.0 to 3.7. Although this range is larger than that exhibited by most type 3 ordinary chondrites, all type 3 chondrites display some range in subtype classification. For

Table 5. Petrologic subtype of Allan Hills A77011 according to criteria of Sears *et al.* (1980).

Parameter	ALHA77011 Value	Subtype
PMD of Fe in olivine	39	3.2
Carbon (wt.%)*	1.0	3.0
^{36}Ar$_P$(10^{-8} cm^3 STP g^{-1})†	100	3.0
% matrix recrystallization	50	3.5
(FeO)/(FeO + MgO) in Huss matrix relative to whole rock	1.1	3.7
PMD Co in kamacite	7.3	3.5

*Specimen numbers ALHA77214 (Jarosewich, 1980) and ALHA77214,28 (Gibson and Andrawes, 1980).
†Primordial ^{36}Ar calculated from analyses of ALHA77015, ALHA77167 and ALHA77260 by Nagao *et al.* (1981).

example, Chainpur ranges from 3.1 to 3.5, Tieschitz from 3.3 to 3.8, and Bishunpur from 3.0 to 3.4 (Sears, pers. comm., 1981). We note that use of a single parameter can lead to large errors in classification (≤ 0.7). Bearing in mind the heterogeneities among and within type 3 ordinary chondrites we classify ALHA77011 as an L group chondrite of petrologic type 3.2 ± 0.2, or category II in the scheme of Huss et al. (1981).

The composition of metallic Fe,Ni supports the classification of ALHA77011 as one of the least-metamorphosed type 3 meteorites. The Ni contents of kamacite grains range widely, which is characteristic of the least metamorphosed chondrites (Afiattalab and Wasson, 1980). The high Cr and Si concentrations in some grains are also a property common to type 3.0–3.4 chondrites. Similar high contents of these normally lithophile elements have been reported in Bishunpur (Rambaldi et al., 1980), Krymka (Rambaldi, 1981), Hallingeberg (Woolum et al., 1981), Semarkona and ALHA77278 (Taylor et al, 1981), and Sharps (our unpublished data). [An exception is Bovedy, an L4 chondrite that contains ~ 0.1 wt.% Si in its metal phases, no detectable Cr in kamacite and 0.08 wt.% in taenite (Rubin et al., 1981a)].

Relation to other type 3 ordinary chondrites

Most types 3 ordinary chondrites contain ~ 15 vol% Huss matrix (Huss et al., 1981) and ≤ 0.1 vol% graphite-magnetite. We have found these small amounts of graphite-magnetite in similar associations to those observed in ALHA77011 in the following type 3 ordinary chondrites: ALHA77299, Bremervörde and Dhajala (H chondrites), Mezö-Madaras (L) and ALHA77278 (LL). The only other meteorite known to contain *abundant* graphite-magnetite besides ALHA77011 is Sharps (H), which has 11 vol% graphite-magnetite and 9 vol% Huss matrix (Table 6). Graphite-magnetite is also abundant in four type 3 chondrite clasts found in ordinary-chondrite regolith breccias (Rubin et al., 1981b; Scott et al., 1981c), which contain 14–36 vol% graphite-magnetite and no Huss matrix.

The relative abundances of Huss matrix and graphite-magnetite in these meteorites suggest that there are three kinds of type 3 ordinary chondrites (Table 5): those with little or no graphite-magnetite, those with little or no Huss matrix, and those like Allan Hills A77011 with both Huss matrix and graphite-magnetite. We cannot rule out the possibility that the assignment into three categories is due to biased sampling of the chondrite population; all gradations from no graphite-magnetite to no Huss matrix may exist.

Association of graphite-magnetite and metallic Fe,Ni

Except for spherical grains within chondrules, all the metallic Fe,Ni in ALHA77011 is intimately associated with graphite-magnetite (Fig. 3). We have observed this in Sharps as

Table 6. Proportions of graphite-magnetite aggregates and Huss matrix in vol % in the three kinds of type 3 ordinary chondrites.

	Graphite-magnetite	Huss matrix
Normal type 3*	$\ll 0.5$	14–16
Intermediate type 3		
ALHA77011	3.2	14.7
Sharps	11.3	8.6
Graphite-magnetite-rich type 3†	14–36	<0.05

*Data for four category I type 3 chondrites (type 3.0–3.4) from Huss (1979).
†Data for four clasts (Scott et al., 1981b).

well. In some cases the metal occurs within graphite-magnetite inclusions. This suggests that some metal may have formed by reduction by C from magnetite. We cannot determine whether this reduction took place before or after the graphite-magnetite was incorporated into the meteorites. In most cases there is clearly too much metal present in a given metal-graphite-magnetite inclusion for the metal to have formed directly from the graphite and magnetite. These inclusions must have formed by the accretion of metal and graphite-magnetite into composite aggregates before ALHA77011 and Sharps were assembled. The common occurrence of rims of graphite-magnetite around metal is consistent with this idea. The low abundance of silicates in these aggregates suggests that the solids present in the region of the nebula where the metal-graphite-magnetite aggregates formed were (at least for a time) largely composed of Fe and C. How metal, graphite and magnetite were fractionated from silicates is not clear.

Major components of chondrites

The discovery of graphite-magnetite aggregates in type 3 ordinary chondrites suggests that this material may have been an important component in the formation of ordinary chondrites. It appears that chondrites formed from variable amounts of at least four distinct components: Mg-rich chondrules, Fe-rich Huss matrix, metallic Fe, Ni-troilite, and graphite-magnetite. These components clearly formed in different environments and/or places in the solar nebula.

Acknowledgments—We thank the NSF Polar Programs Office for supporting the recovery of Antarctic meteorites and the NSF-NASA Antarctic Meteorite Working Group for making the meteorites available for study. R. A. Score provided invaluable assistance in helping us locate sections of L3 chondrites from Allan Hills. We thank R. Hewins, E. King and J. Wasson for lending us thin sections, and A. Rubin for assistance. This work was supported in part by NASA Grant No. NGL 32-004-064.

REFERENCES

Afiattalab F. and Wasson J. T. (1980) Composition of the metal phases in ordinary chondrites: implications regarding classification and metamorphism. *Geochim. Cosmochim. Acta* **44**, 431–446.
Ashworth J. R. (1977) Matrix textures in unequilibrated ordinary chondrites. *Earth Planet. Sci. Lett.* **35**, 24–34.
Cassidy W. A. (1980) Catalog of Antarctic meteorites 1977–1978; Discussion. *Smithsonian Contrib. Earth Sci.* **23**, 42–44.
Christophe Michel-Levy M. (1976) La matrice noire et blanche de la chondrite de Tieschitz (H3). *Earth Planet. Sci. Lett.* **30**, 143–150.
Fujimaki H., Aoki K., Sunagawa I. and Matsu-ura M. (1981) Study on some minerals in ALHA 77015 (L3) chondrite (abstract). *Sixth Symposium on Antarctic Meteorites*, p. 28–29, Japan.
Gibson E. K. Jr. and Andrawes F. F. (1980) The Antarctic environment and its effects upon the total carbon and sulfur abundances in recovered meteorites. *Proc. Lunar Planet. Sci. Conf. 11th*, p. 1223–1234.
Graham A. L. (1980) Meteoritical Bulletin Number 57. *Meteoritics* **15**, 93–103.
Huss G. R. (1979) The matrix of unequilibrated ordinary chondrites: Implications for the origin and history of chondrites. Masters Thesis, Univ. New Mexico, Albuquerque, p. 1–139.
Huss G. R., Keil K., and Taylor G. J. (1981) The matrices of unequilibrated ordinary chondrites: implications for the origin and history of chondrites. *Geochim. Cosmochim. Acta* **45**, 33–51.
Jarosewich E. (1980) Catalog of Antarctic meteorites 1977–1978; Chemical analyses of some Allan Hills meteorites. *Smithsonian Contrib. Earth Sci.* **23**, 48.
Jarosewich E. and Dodd R. T. (1981) Chemical variations among L-group chondrites, II. Chemical distinctions between L3 and LL3 chondrites. *Meteoritics* **16**, 83–91.
King T. V. V., Score R., Gabel E. M., and Mason B. (1980) Catalog of Antarctic meteorites 1977–1978; meteorite descriptions. *Smithsonian Contrib. Earth Sci.* **23**, 12–44.
Kurat G. (1970) Zur Genese des kohligen Materials im Meteoriten von Tieschitz. *Earth Planet. Sci. Lett.* **7**, 317–324.
Nagao K., Sarto K., Ohba Y., and Takaoka N. (1981) Rare gas studies of the Antarctic meteorites (abstract). *Sixth Symposium on Antarctic Meteorites*, p. 60–61, Japan.
Rambaldi E. R. (1981) The nebula record in Krymka, a highly unequilibrated LL chondrite (abstract). In *Lunar and Planetary Science XII*, p. 908–910. Lunar and Planetary Institute, Houston.

Rambaldi E. R., Sears D. W., and Wasson J. T. (1980) Si-rich Fe–Ni grains in highly unequilibrated chondrites. *Nature* **287**, 817–820.

Rhodes J. M. and Fulton C. R. (1981) Chemistry of some Antarctic meteorites (abstract). In *Lunar and Planetary Science XII*, p. 880–882. Lunar and Planetary Institute, Houston.

Rubin A. E., Keil K., Taylor G. J., Ma M.-S. Schmitt R. A., and Bogard D. D. (1981a) Derivation of a heterogeneous lithic fragment in the Bovedy L-group chondrite from impact-melted porphyritic chondrules. *Geochim. Cosmochim. Acta.* In press.

Rubin A. E., McKinley S. G., Scott E. R. D., Taylor G. J., and Keil K. (1981b) A new kind of unequilibrated ordinary chondrite with graphite-magnetite matrix (abstract). In *Lunar and Planetary Science XII*, p. 908–910. Lunar and Planetary Institute, Houston.

Score R., Schwarz C. M., King T. V. V., Mason B., Bogard D. D., and Gabel E. M. (1981) Antarctic meteorite descriptions 1976–1977–1978–1979. Publ. 54, Lunar Curatorial Facility, NASA Johnson Space Center, Houston. 144 pp.

Scott E. R. D., Rubin A. E., Taylor G. J., and Keil K. (1981a) New kind of type 3 chondrite with a graphite-magnetite matrix. *Earth Planet. Sci. Lett.* In press.

Scott E. R. D., Taylor G. J., Rubin A. E., Okada A., and Keil K. (1981b) Graphite-magnetite inclusions in ordinary chondrites: an important constituent in the early solar system (abstract). In *Lunar and Planetary Science XII*, p. 955–957. Lunar and Planetary Institute, Houston.

Scott E. R. D., Taylor G. J., Rubin A. E., Okada A., and Keil K. (1981c) Graphite-magnetite aggregates in ordinary chondritic meteorites. *Nature* **291**, 544–546.

Sears D. W., Grossman J. N., Melcher C. L., Ross L. M., and Mills A. A. (1980) Measuring metamorphic history of unequilibrated ordinary chondrites. *Nature* **287**, 791–795.

Taylor G. J., Okada A., Scott E. R. D., Rubin A. E., Huss G. R., and Keil K. (1981) The occurrence and implications of carbide-magnetite assemblages in unequilibrated ordinary chondrites (abstract). In *Lunar and Planetary Science XII*, p. 1076–1078. Lunar and Planetary Institute, Houston.

Wasson J. T. (1974) *Meteorites-Classification and Properties.* Springer-Verlag, N.Y. 360 pp.

Wood J. A. (1962) Metamorphism in chondrites. *Geochim. Cosmochim. Acta* **26**, 739–749.

Woolum D. S. (1981) Bi, Pb, Cr and Si-rich metals in Hallingeberg (L3) (abstract). In *Lunar and Planetary Science XII*, p. 1209–1211. Lunar and Planetary Institute, Houston.

The Elga meteorite: Silicate inclusions and shock metamorphism

Eu. G. Osadchii[1], G. V. Baryshnikova[2], and G. V. Novikov[1]

[1]Institute of Experimental Mineralogy, USSR Academy of Science, Chernogolovka, Moscow Region, USSR [2]V. I. Vernadsky Institute of Geochemistry and Analytical Chemistry, USSR Academy of Science, Moscow, USSR

Abstract—The Elga type IIE meteorite is an octahedrite with silicate inclusions (Wasson, 1970). Detailed analyses under the optical microscope and by the electron microprobe have revealed five petrologic types of silicate inclusions dissimilar in mineralogy and structure. In addition to Fe–Ni phases, the matrix was found to contain K–Na feldspar and feldspar glass with excess silica, chromian augite, bronzite, olivine, fluorapatite (the first report for meteorites), whitlockite (?), magnetite (?), chromite, schreibersite, Fe–Ni–P and Fe–Ni–P–S alloys, presumably heterosite ($FePO_4$), tridymite, troilite, rutile (?), and phosphate glasses of complex composition.

Structural characteristics of silicate inclusions, the occurrence of injection and liquid immiscibility structures, relict minerals (phosphates and olivine) as well as structural relationships between the silicate inclusions and metal matrix strongly suggest that the melting of primary inclusions was caused by high dynamic pressures (impact melting). The original mineral assemblage, which consisted of phosphate (whitlockite?), Fe-rich olivine and K–Na feldspar, underwent decomposition by the scheme:

$$\text{K–Na feldspar} + \text{olivine} + \text{phosphate} + \text{Fe–Ni phases} \xrightarrow{\text{melting and decomposition}} \text{alkali glass with normative K–Na feldspar and } SiO_2 +$$

monoclinic pyroxene + schreibersite + Wüstite (magnetite).

The melting of silicate inclusions *in situ* did not considerably heat the metal matrix. The silicate melt must have crystallized rapidly and at low (after-shock) pressures to produce augite and K–Na feldspar. Large contents of trivalent iron (to 50%) in clinopyroxene indicate that the oxygen activity in this process had been high.

The cosmic history of the Elga meteorite must have had at least two impact events of different intensity. In the first one, the silicate inclusions fused practically completely and the melt migrated between the inclusions to form structures which resemble liquid immiscibility. The second impact was not so strong and produced partial maskelynitization of K–Fa feldspar, basal twins, clinopyroxene cleavage and, possibly, partial fusion of the silicate.

It is suggested from the silicate inclusion compositions that the material of the elga meteorite is the product from the differentiation that occurred early in the history of the parent body.

INTRODUCTION

The Elga meteorite was found in 1959 in Yakutia (Vronsky, 1962; Plyashkevich, 1962). It weighs 28.8 kg. The first description of the find was given by Plyashkevich (1962) on the basis of optical observations which established the following minerals in the silicate inclusions: potassium feldspar (anorthoclase), monoclinic pyroxene (pigeonite), maskelynite, merrillite, hortonolite, glass, chromite (?), lawrencite (?) and two or three unidentified minerals. It is believed that spiral-like and crumpled structures as well as displacement of the Neumann lines, and jointing in schreibersite have resulted from deformation and secondary melting. The Ni, Ga and Ge contents in the metal matrix (Wasson, 1970) place the Elga meteorite in IIE group. In subsequent studies (Kvasha *et al.*, 1974), the mineralogy of silicate inclusions and shock metamorphism were defined more exactly. The microprobe analysis indicated K–Na feldspar, chromediopside and bronzite. The Elga meteorite was found to be similar to other iron meteorites in which silicate inclusions contain K–Na feldspar (Kodaikanal; Bence and Burnett, 1969).

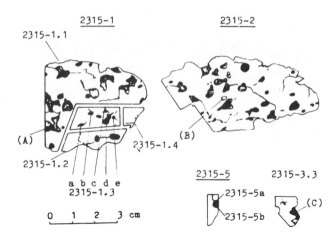

Fig. 1. The general view of the samples of the Elga meteorite. Silicate inclusions and veinlets are shaded. Inclusion boundaries are sharp and clear. (A)-specimens (30 mg) for Mössbauer investigations. (B)-troilite inclusion. (C)-silicate inclusion abundant in Fe–Ni–P alloy segregations of micrographic structure (Fig. 3). 2315-5a-the green part of the silicate inclusion with the liquid immiscibility structure, 2315-5b-the yellow part. Combined numbers designate electron microprobed inclusions.

The present study aims to define the mineralogy of silicate inclusions, their petrologic types and structural characteristics of the meteorite and to provide the genetic interpretation thereof. Detailed characterization of the silicate inclusions will be done in our later work.

The methods used in this study were optical microscopy, electron microprobe, X-ray powder diffraction and Mössbauer spectroscopy. The samples used in the study included specimen 2315 (from the USSR Acad. Sci. Collection), composed of two plates ~5 mm thick and ~25 cm^2 total area, several smaller specimens of 0.5–1 cm^2 each (Fig. 1), and 28 transparent polished thin sections prepared from intact or fragmented inclusions.

PETROLOGIC TYPES OF SILICATE INCLUSIONS
Structural characteristics

At the saw-cut surface the Elga meteorite looks like a raisin bun. The silicate inclusions vary in size from several tenths of a millimeter to 15 mm (Fig. 1). Inclusions, although dispersed uniformly in polished specimen 2315, are non-uniform over the entire meteorite. Specimen 2315 was cut out from the inclusion-rich portion. Macroscopically, the silicate inclusions tend to be irregular or elongated in shape with smoothed, rounded edges, and are only occasionally spherical. Virtually all of the inclusions branch off as veinlets ~0.5 mm thick filled with a silicate; they often form connections between the inclusions. Irregular inclusions are greenish-gray on the polished surfaces of specimens (Fig. 1) while spherical or nearly spherical inclusions are light ranging from milky-white to grayish-white. Microscopical observations have shown the former to be polymineralic and the latter to be monomineralic or of various types of glass.

Detailed microscopic investigations of separate silicate inclusions in the Elga meteorite indicated five distinct types with respect to their structure, mineralogy and petrology.

I. Inclusions are essentially of micropoikilitic and occasionally intersertal texture. Prismatic crystals of clinopyroxenes occur as intergrowths in elongated spherulitic crystals of K–Na feldspar (Fig. 2). Clinopyroxene crystals are commonly oriented and occur as radiating bundles from the narrow part of inclusions. In a wider part, the border is often composed entirely of K–Na feldspar. These two major minerals occur in greatly varying amounts. Orthopyroxene is a minor mineral and phosphate (?), chromite, olivine

Fig. 2. The general view of the structure of a type-I inclusion. Augite prismatic crystals are mantled by the groundmass of K–Na feldspar spherocrystals. (Transmitted polarized light, crossed nicols.) Scale bar is 0.4 mm.

and rutile (?) are accessories. Schreibersite is located at the border of inclusions and often forms continuous rims. In addition to schreibersite, the rims contain a Fe–Ni–P alloy as angular grains with a micrographic structure of unclear origin (Fig. 3). The Fe–Ni–P alloy and, partly, schreibersite form reaction rims around inclusions.

Type I inclusions predominate in meteorite and appear to be similar to the known (i)-type inclusions in Kodaikanal meteorite (Bunch and Olsen, 1968).

II. Inclusions which show a microporphyritic structure consisting of prismatic clinopyroxene crystals set in a glass groundmass (Fig. 4). Normally, monoclinic pyroxene occurs as radiating crystals, though random and very rare graphic-like arrangements are observed. The relative amounts of phenocrysts and groundmass vary greatly. The end zones of this type of inclusion often consist entirely of glass. Orthopyroxene is a minor mineral, and rutile (?), chromite, phosphate (whitlockite ?), and tridymite are accessories. Schreibersite is commonly located at the borders.

In Elga, type-II inclusions are less abundant than type-I inclusions. Earlier Kvasha *et al.* (1974) described briefly these silicate inclusions without differentiating various types.

III. Inclusions with a fully crystalline, hypidiomorphicgranular structure (Fig. 5). They are composed of clinopyroxene euhedral crystals twinned polysynthetically (twin plane 001), fully anhedral olivine grains and K–Na feldspar. These inclusions are rarely found.

IV. These are monomineralic sphere-shaped inclusions. The studied specimens contained one unbroken inclusion and several fragments. The intact inclusion is composed of two parts (Fig. 6). The round core at the center is a spherulite composed of K–Na feldspar needles radiating from the center surrounded by several feldspar crystals of a tabular habit. Along the periphery is a brown rim of weakly polarizing needle-shaped, radiating aggregates, also of the feldspar composition. No other minerals have been detected. So far, no inclusions of this type have been reported in meteorites with silicate inclusions.

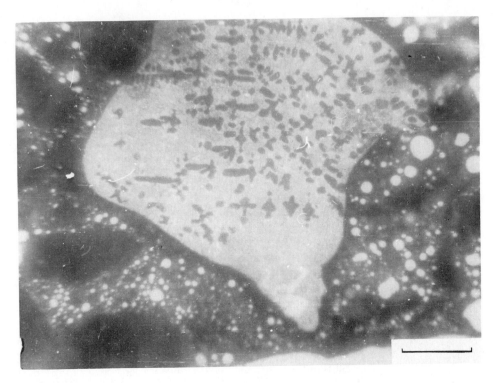

Fig. 3. Micrographic structure of a Fe–Ni–P alloy located within 2315-1.3 inclusion; graphic inclusions (dark-grey) contain Si and Ca. Reflected light, in oil immersion. Scale bar is 10 μm.

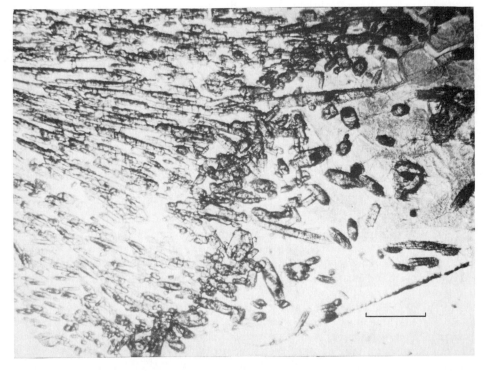

Fig. 4. The general view of a type-II inclusion. Elongated prismatic crystals of augite are mantled by transparent K–Na feldspar glass. (Transmitted polarized light, no analyser.) Scale bar is 0.2 mm.

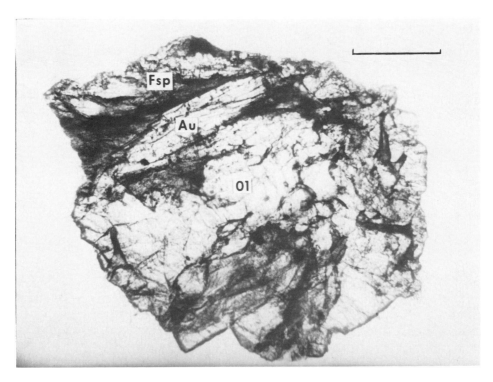

Fig. 5. A type-III silicate inclusion with the holocrystalline hypidiomorphic-granular texture; Ol–olivine, Au–augite, Fsp–feldspar. (Transmitted polarized light, no analyser.) Scale bar is 0.4 mm.

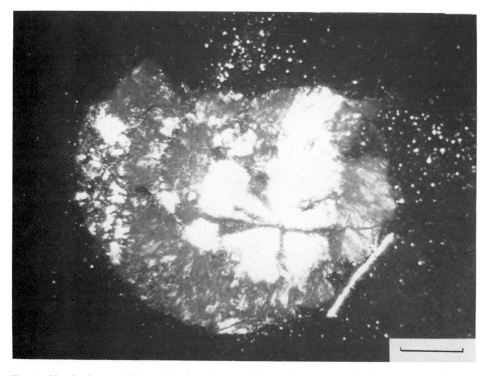

Fig. 6. Nearly intact silicate inclusion of type-IV. In the center a K–Na feldspar spherulite surrounded by K–Na feldspar crystals and spherulites. (Transmitted polarized light, crossed nicols.) Scale bar is 0.2 mm.

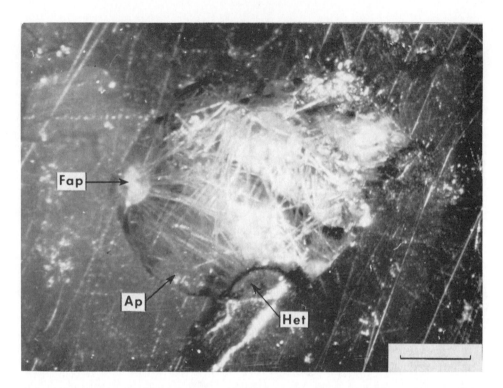

Fig. 7. An intact type-V silicate inclusion. Absolutely transparent K–Na feldspar glass containing needle-shaped bronzite crystals with crystallization centers at the border of the inclusion. Het–heterosite(?); Fap–fluorapatite; Ap–phosphate glass containing K and Na. (Incident reflected light.) Scale bar is 0.3 mm.

Earlier studies on the Elga (Kvasha et al., 1974) and Kodaikanal meteorites (Bence and Burnett, 1969), have found spherical inclusions to consist of glass containing rutile (?) needles. No such inclusions were detected in the present study.

V. To this type belongs only one inclusion which was found in a polished section and which has unusual structure and composition. It consists of transparent glass of the K–Na feldspar composition + 33.8% normative silica with bundles of orthopyroxene needles radiating from centers of crystallization at the inclusion boundaries (Fig. 7). The relict minerals, fluorapatite and potassium- and sodium-bearing apatite are centers of crystallization. Also present may be heterosite (?).

MINERALOGY

Most of the silicate inclusions in the Elga meteorite contain nearly equal amounts of clinopyroxene and K–Na feldspar, including alkali feldspar glasses; orthopyroxene is a minor mineral, and olivine, whitlockite (?), fluorapatite, rutile (?), phosphate glass, tridymite (?) are accessories. Opaque minerals include, in addition to the Fe–Ni phase in the matrix, schreibersite, Fe–Ni–P and Fe–Ni–P–S alloys, troilite, chromite and magnetite.

Transparent minerals

K–NA FELDSPAR forms the groundmass in type-I inclusions, and occurs as elongated, feather-like, brownish-pink crystals; it also is one of the major minerals in the III and IV-type inclusions (Figs. 2, 4, 5). In several type-I inclusions it occurs as elongated

Table 1. Electron microprobe analyses of feldspathic glass (1–4) and feldspar (5–8) from the Elga meteorite (in weight percent).

	Feldspathic glass				Feldspar				Average
	(1) 2315-1.3a	(2) 2315-1.3a	(3) 2315-1.3d	(4) 2315-1.3c	(5) 2315-1.3c	(6) 2315-1.3c	(7) 2315-5a	(8) 2315-5b	
SiO$_2$	70.28	71.50	77.66	75.61	72.93	73.31	68.58	68.37	72.28
TiO$_2$	0.38	0.24	0.39	0.75	0.23	0.43	0.51	0.46	0.42
Al$_2$O$_3$	15.59	15.56	14.39	13.19	15.80	15.26	16.62	16.66	15.38
Cr$_2$O$_3$	0.00	0.00	0.00	0.00	0.00	0.00	0.02	0.02	0.01
FeO	0.87	0.65	0.98	0.63	0.98	1.29	0.52	0.48	0.80
MnO	0.00	0.00	0.00	0.01	0.00	0.00	0.03	0.03	0.01
MgO	0.00	0.00	0.00	0.16	0.00	0.00	0.01	0.09	0.03
CaO	0.00	0.00	0.00	0.14	0.00	0.00	0.11	0.08	0.04
Na$_2$O	3.80	3.45	2.23	3.97	7.93	7.93	5.27	5.16	4.97
K$_2$O	6.89	6.76	6.97	7.03	2.38	3.61	6.86	6.86	5.87
Sum	97.78	98.17	102.61	101.49	100.25	101.81	98.53	98.21	99.81

Structural formulae

Si	3.168	3.194	3.290	3.270	3.167	3.164	3.092	3.091	3.180
Ti	0.012	0.008	0.012	0.024	0.008	0.014	0.017	0.016	0.014
Al	0.825	0.819	0.718	0.672	0.809	0.776	0.883	0.888	0.799
Cr	0	0	0	0	0	0	0.001	0.001	0
Fe	0.033	0.024	0.035	0.023	0.036	0.046	0.020	0.018	0.029
Mn	0	0	0	0	0	0	0.001	0.001	0
Mg	0	0	0	0.010	0	0	0.001	0.006	0.001
Ca	0	0	0	0.006	0	0	0.005	0.004	0.002
Na	0.332	0.299	0.183	0.333	0.668	0.664	0.461	0.452	0.424
K	0.396	0.385	0.377	0.388	0.132	0.199	0.394	0.396	0.333
O	8	8	8	8	8	8	8	8	8
Ab	45.6	43.7	32.7	45.8	83.5	76.9	53.6	53.0	55.9
Or	54.4	56.3	67.3	53.4	16.5	23.1	45.8	46.5	43.9
An	0	0	0	0.8	0	0	0.6	0.5	0.2
Ab	31.4	28.1	16.7	30.6	63.3	62.9	44.7	43.9	40.0
Or	37.5	36.1	34.3	35.6	12.5	18.9	38.2	38.4	31.4
Q	31.1	35.8	49.0	33.8	24.2	18.2	17.1	17.7	28.6

crystals toothed at the ends, with occasional microclinic twins. The K-Na feldspar spherulitic crystals characteristically have a system of fractures across the long axis. The mineral grain size varies. The optical constants are $Np = 1.5144 \pm 0.0015$; $Ng = 1.5215 \pm 0.0015$, 2V—not clear; cleavage extinction angle 16–22°.

The alkali feldspar in Elga was analysed earlier (Kvasha et al., 1974), but only an average analysis is available. Electron microprobe analyses in this study have shown this feldspar to range in composition from $Ab_{53}Or_{46.5}An_{0.5}$ to $Ab_{83.5}Or_{16.3}$ with SiO_2 in considerable excess—17–24 mol.% of the normative silica (Table 1, analyses 5 and 8, Figs. 8, 9). The variations in SiO_2 content within a single inclusion are not large.

Both morphologically and optically, the alkali feldspar is similar to that in Kodaikanal meteorite (Bunch and Olsen, 1968; Bence and Burnett, 1969), though it contains on the average more SiO_2, less Al_2O_3 and Na_2O, and no CaO. It is possible that the alkali feldspar in Elga is a submicroscopic mixture of two alkali feldspars—cryptoantiperthite, the more so that the microprobe analyses correspond to two types of feldspar—essentially Na and K-Na feldspars (Table 1, analyses 5 and 7, Fig. 8).

ALKALI GLASSES constitute the entire groundmass in type-II and type-V inclusions. They occur as transparent, colorless (in places, brownish—pink or lilac) glasses which are characteristically not perfectly isotropic. In some type-I inclusions, particularly on borders where there are no augite insets, the glass is most abundant in minute microlites—rutile (?) needles. Sometimes (Bence and Burnett, 1969; Kvasha et al., 1974) this glass is called "rutilated". The glass refractive index is 1.504 ± 0.0006.

Analyses indicate that the glasses are K-Na feldspathic, ranging in composition from $Ab_{32.7}Or_{67.3}$ to $Ab_{45.6}Or_{54.4}$ (Table 1, analyses 1 and 3, Fig. 9), with even larger SiO_2 excess compared to K-Na feldspar (from 31.3 to 49% normative silica. The glasses are higher in potassium but lower in sodium than K-Na feldspar). These features are common to all Weekeroo Station-type meteorites (Bunch et al., 1970). Although Elga is similar to

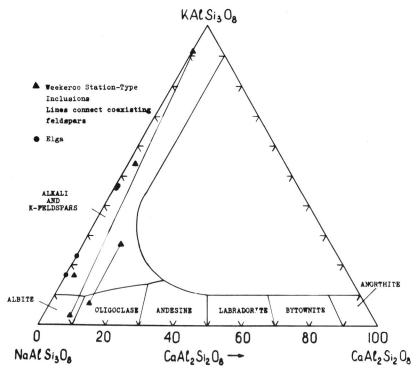

Fig. 8. The Ab–Or–An diagram for compositions of feldspar glasses and K–Na feldspar from inclusions of various petrologic types (Table 1).

Kodaikanal (Bence and Burnett, 1969), in most cases the Elga glasses are not K–Na maskelynite and are believed to have formed from the earlier K–Na feldspar after shock melting.

CLINOPYROXENE occurs commonly in the first three types of silicate inclusions as euhedral, strongly elongated prismatic crystals; small prismatic crystals are rare. The crystals are not more than 0.5 mm long and 0.2–0.3 mm wide. Most crystals are twined polysynthetically on (001), with well-developed cleavage along the same direction. Winchell et al. (1953) argued that such basal twins are produced by shift strain and are characteristic of augites. The optical constants determined in this study are close to those of Plyashkevich (1962) and Kvasha et al. (1974): $2V$—positive $\sim 60°$; $C:Ng = 40$–$48°$ (for in-glass crystals); weak undulatory extinction; $Ng > 1.690$, $Np = 1.6755 \pm 0.0025$. Some crystals are zoned, as is indicated by variations in extinction angle between central (larger) and marginal (smaller) parts.

Plyashkevich (1962) identified clinopyroxene as pigeonite while electron microprobe studies by Kvasha et al. (1974) revealed that it is a chromediopside.

Clinopyroxene crystals from the type-I and II inclusions were analysed with the electron microprobe. It should be noted that the minerals in the two types are similar in their major element contents and dissimilar in Na and K—the clinopyroxene in the glass groundmass contains 0.66–0.68 wt.% Na_2O and 0.10 wt.% K_2O (Table 2, analyses 4 and 5), while in type-I pyroxenes these elements are absent. The clinopyroxenes from Kodaikanal and Colomera meteorites (Bunch et al., 1970) show the same trend. Although we identified the clinopyroxene as chromian augite, and occasionally endiopside (Table 2, analyses 1–6, Fig. 10), its chemical composition truly does not correspond to either variety and no analogues have been found.

Mössbauer spectroscopy was applied in an attempt to characterize the clinopyroxene more fully. A part of type-I inclusion (A) from specimen 2315-1.1 (Fig. 1) was chosen for

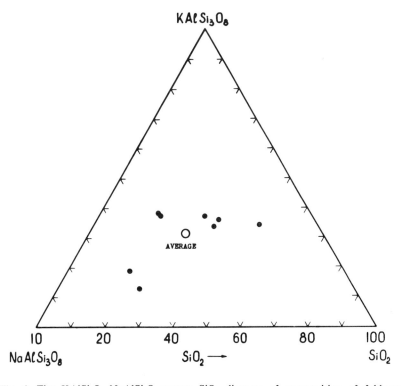

Fig. 9. The $KAlSi_3O_8$–$NaAlSi_3O_8$–excess SiO_2 diagram of composition of feldspar glasses and K–Na feldspar from inclusions of various petrologic types (Table 1).

Table 2. Electron microprobe analyses of clinopyroxene, orthopyroxene and olivine from the Elga meteorite (in weight percent).

	Augite						Orthopyroxene		Olivine
	(1) 2315-1.3a	(2) 2315-1.3a	(3) 2315-1.3b	(4) 2315-5a	(5) 2315-5b	(6) 2315-5b	(7) 2315-5b	(8) 2315-1.3e	(9)
SiO_2	53.00	53.76	55.01	52.19	51.81	53.19	56.08	56.03	37.80
TiO_2	0.90	0.74	0.00	0.49	0.48	0.49	0.22	0.52	0.26
Al_2O_3	0.00	0.00	0.00	0.97	0.00	–	–	0.30	0.11
Cr_2O_3	1.49	1.42	1.70	1.06	1.09	1.36	0.60	0.45	0.00
Fe_2O_3	2.92	3.52	3.53	3.76	3.97	3.51	–	–	–
FeO	2.62	3.17	3.18	3.38	3.57	3.16	10.80	9.67	21.04
MnO	0.00	0.00	0.00	0.36	0.26	0.25	0.04	0.24	0.50
MgO	16.15	17.20	17.30	16.12	16.43	16.94	30.76	31.08	40.71
CaO	21.18	20.34	20.11	19.80	19.92	20.70	1.42	0.22	0.00
Na_2O	0.00	0.00	0.00	0.68	0.66	–	–	0.11	0.00
K_2O	0.00	0.00	0.00	0.10	0.09	–	–	0.07	0.00
Sum	98.26	100.15	100.93	98.91	98.28	99.60	99.93	98.69	100.42
	Structural formulae								
Si	1.970	1.962	1.988	1.940	1.944	1.938	1.983	1.996	0.991
Ti	0.025	0.020	0	0.014	0.014	0.013	0.006	0.014	0.005
Al	0	0	0	0.042	0	–	–	0.012	0.003
Cr	0.041	0.038	0.046	0.029	0.030	0.037	0.016	0.010	0
Fe^{3+}	0.082	0.097	0.096	0.105	0.112	0.096	–	–	–
Fe^{2+}	0.082	0.097	0.096	0.105	0.112	0.096	0.319	0.288	0.460
Mn	0	0	0	0.009	0.008	0.008	0.001	0.007	0.011
Mg	0.895	0.936	0.932	0.893	0.919	0.920	1.622	1.650	1.601
Ca	0.844	0.795	0.779	0.757	0.789	0.808	0.054	0.008	0
Na	0	0	0	0.049	0.048	–	–	0.008	0
K	0	0	0	0.05	0.004	–	–	0.003	0
O	6	6	6	6	6	6	6	6	4
Fs	8.6	10.1	10.1	11.3	11.6	10.4	16.0	14.8	22.3 (Fa)
En	47.0	48.6	49.0	48.0	47.6	47.7	81.3	84.8	77.7 (Fo)
Wo	44.4	41.3	40.9	40.7	40.8	41.9	2.7	0.4	

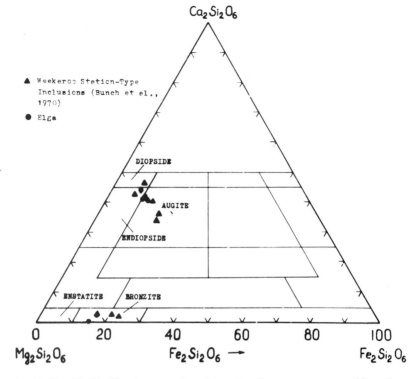

Fig. 10. The En–Fs–Wo diagram of augite and orthopyroxene compositions from inclusions of various petrologic types (Table 2).

the analysis. The NGR method was considered most suitable in this case as it allows a reliable distinction to be drawn between augite and diopside. The Elga material is especially suited to this kind of analyses since the sample contains no other iron-bearing phases.

Mössbauer spectra of specimen 2315-1.1 (A) at two temperatures (300 K and 82 K) are given in Fig. 11. The sample contains both ferrous and ferric ions. The ferrous ions can be assigned conventionally to two different crystallographic sites, because in addition to an intense doublet A there is distinct though less intense doublet B, with a slightly larger quadrupole splitting and smaller half-widths of the components. Doublet A appears to be complex due to Fe^{2+} ions occupying the same site but having different adjacent neighbours. Doublet C corresponds to Fe^{3+} ions in the structure. Decomposition of the spectrum into components showed that the relative amount of Fe^{3+} is ~50%. At low temperatures, decomposition of components improves considerably (Fig. 11b) to yield more reliable quantitative results.

Comparing hyperfine-structure parameters (HFS) of the test sample NGR spectra with those for natural clinopyroxenes in the diopside-hedenbergite series of different composition at 300 and 82 K (Table 3) shows that the Elga clinopyroxene cannot belong to the diopside-hedenbergite series. The HFS parameters of clinopyroxene coincide virtually with those for natural augite* (Table 3) used as a reference. The structure of this augite was given elsewhere (Arganova, pers. comm.). The inference that the Elga clinopyroxene has the augite composition is further confirmed by the fact that the spectra of oriented augite and clinopyroxene diopside-hedenbergite series have different polarisation ratios in

---

*Composition: SiO_2 50.09, TiO_2 1.15, Al_2O_3 5.34, Fe_2O_3 4.23, FeO 6.13, MnO 0.15, MgO 10.49, CaO 19.17, Na_2O 2.78, K_2O 0.04, H_2O^+ 0.07, H_2O^- 0.09, total 99.79 (Arganova, pers. comm.).

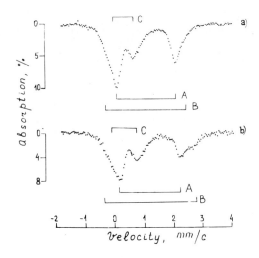

Fig. 11. Mössbauer spectra of augite HFS from a type-I inclusion, specimen 2315-1.1(A) (see Fig. 1). a) 300°K; b) 82°K. A and B-two Fe^{2+} sites, C–Fe^{3+}.

their HFS. These results were obtained on samples prepared by pressing powder particles into paraffin pellets. In this procedure, the particles show anisotropy of their external shapes and the flat surfaces become oriented parallel to the pellet plane. For this mechanical texture, the relative intensity of the doublet's right components decreases in the augite spectra, and increases in the diopside spectra. Consequently, the dependence of the mechanical properties on the anisotropy as the size of the sample is reduced, is the most convincing evidence of two essentially different structures. This last criterion may be useful in discriminating between closely similar structures with the help of Mössbauer measurements at room temperature.

Deer et al. (1963) reported that augite is commonly richer in Fe^{3+} at relatively low total iron (to 7 wt.% FeO) than diopside.

Table 3. Parameters of augite and clinopyroxene HFS spectra.

Parameters	Quadrupole splitting, mm/c			$\dfrac{Fe^{3+}}{Fe_{tot.}}$	T°, K
	Fe^{2+}		Fe^{3+}		
Sample	A	B	C		
2315-1.1(A)*	2.01	2.74	0.65	0.44	300
	2.09	3.04	0.874	0.44	82
Augite[1]	2.17	2.58	0.58	0.233	300
	2.12	2.84	0.71	0.217	82
Diopside[2]	1.96				300
	2.60				82
Hedenbergite[3]	2.24				300
	2.75				82

*See Fig. 1.
1. Arganova's sample
2. Diopside, $X_{Fe} = 0.2$
3. Hedenbergite, $X_{Fe} = 0.8$
A and B—two sites Fe^{2+}, C—Fe^{3+} (see Fig. 11)

In addition to the described augite, another clinoproxene with distinct morphological and optical properties was observed on the border of type-I inclusion in one of the polished sections. It either occurs as distinct needle-like dark-brown crystals 0.5 mm long, or forms the core of the largest augite crystals; the assemblage is similar in appearance to zoned crystals observed in lunar KREEP—basalts (Longhi et al., 1972). This dark-brown pyroxene has $C:Ng = 20-30°$, i.e., much lower than for augite, and $Np = 1.6953 \pm 0.0033$, $Ng > 1.700$—higher than for augite. It can be identified by these constants as pigeonite (?).

ORTHOPYROXENE is a minor mineral in type-I and II silicate inclusions and constitutes by visual inspection only 5% of the total inclusion volume. It occurs as relatively large, prismatic, colorless crystals or short prismatic crystals (0.7 mm) that differ from augite in lower birefringence and straight extinction. Characteristically, the basal cleavage exhibits no twins. Certain crystals are hollow or skeleton in shape. In type-V inclusions orthopyroxene occurs as needle-like crystals in the glass matrix. This orthopyroxene is chemically similar to bronzite (Table 2, analyses 7 and 8, Fig. 10).

Optically and chemically, this orthopyroxene is very similar to the one found earlier in the silicate inclusions in this octahedrite (Kvasha et al., 1974) and in Kodaikanal (Bence and Burnett, 1969), but has very low CaO content (0.22 wt.% vs. 1.02 and 1.43 wt.%, respectively).

OLIVINE grains were found in inclusions of types I and II. Only in one case did the olivine occur as an euhedral dipyramidal-prismatic crystal; commonly, its anhedral grains show diffuse, indistinct reaction-like boundaries with associated augite. In all probability, in the last case olivine is a relict mineral. Electron microprobe analyses of an euhedral grain gave the composition Fa 22.3 mol.% (Table 2, analysis 9).

According to Plyashkevich's (1962) data, olivine on the border of one of the inclusions in Elga octahedrite has a horthonolite (?) composition.

WHITLOCKITE—Several euhedral dipyramidal-prismatic crystals with hexagonal habit at the cut were observed in the glass matrix of type-II inclusions. The color is brownish-violet to yellow. The crosscuts of the crystals are isotropic. The mineral is uniaxial, negative, with low birefringence. Crystals are small (20 μm).

FLUORAPATITE has not been known to occur in meteorites. Only a single, reliably identified grain (Table 4, analysis 1, Fig. 7) occurs as a semi-sphere adjacent to the wall of a type-V inclusion composed of feldspar glass. Fluorapatite is a nucleus of crystallization of bronzite needle-shaped crystals. The electron microprobe analysis recalculated per a theoretical formula confirms the crystalline state of the grain.

PHOSPHATE GLASSES are common in inclusions with a high silica content in feldspar

Table 4. Electron microprobe analyses of fluorapatite and phosphate glasses from the Elga meteorite (in weight percent). Sample 2315-1.3e.

	(1)	(2)	(3)	(4)
SiO_2	0.10	1.62	0.97	0.85
TiO_2	0.09	1.33	1.14	0.23
Al_2O_3	0.18	0.00	0.51	0.00
FeO	1.21	18.20	18.57	9.52
MnO	0.08	0.40	0.39	0.55
MgO	0.82	19.46	20.42	26.37
CaO	51.33	3.43	3.28	3.11
Na_2O	–	6.53	5.71	6.66
K_2O	–	1.21	0.00	0.00
P_2O_5	41.05	44.31	43.29	47.95
F	5.02	–	–	–
Sum	99.59	96.49	94.28	95.24

glass (type-V). The melting of glasses under the microprobe electron beam is a problem in the analysis. The composition of glasses are presented in Table 4 and show high sodium (to 6.66% Na_2O) and titanium (to 1.33% TiO_2) contents.

TRIDYMITE (?)—transparent polished thin sections contain minute tablets (to 20 μm) of pseudohexagonal habit, presumably tridymite, which occur as clusters of several crystals in interstices between the clinopyroxene crystals enclosed in the transparent glass (type-II inclusions). Birefringence is low, though somewhat higher than in the host glass.

RUTILE (?)—The glass matrix of certain type-II inclusions, especially on the border, are full of the tiniest needles-microlites of a highly-refractive mineral, possibly rutile (?). The needles are yellowish, with straight extinction, to 100 μm length. Such "rutilized" glass has already been described for Elga (Kvasha et al., 1974) and Kodaikanal (Bence and Burnett, 1969).

Opaque minerals and alloys

SCHREIBERSITE is the most widespread opaque mineral (Table 5, analysis 1). It accumulates mainly around silicate inclusions, commonly in continuous rims and is dispersed uniformly as small inclusions throughout the metal matrix. Schreibersite associated with silicate inclusions is, for the most part, fractured normally to the contact.

FE–NI–P ALLOY, like schreibersite, forms rims around the silicate inclusions and injection structures (see next section). Phosphorus content is lower than schreibersite (Table 5, analyses 4–7). The decomposition structures of the Fe–Ni–P alloy—Fe–Ni phases and submicroscopic schreibersite make the identification of composition somewhat difficult. Grains of this alloy are irregular in shape and also occur within the inclusions. They are very similar to troilite, and oxidize readily on the polished surface exhibiting micrographic structures (Fig. 3). Electron microprobe analysis showed Ca and Si in the dark areas; no Mg, Cr and S were found. It should be noted that grains within silicate inclusions in this alloy exhibit micrographic structure. These grains have higher Ni content than the alloy in contact with silicate inclusions and globules (Table 5, analyses 5 and 6).

FE–NI–P–S ALLOY commonly occurs as spheres within silicate injections and contains up to 1.81 wt.% Si (Table 5, analyses 2 and 3). Although these spheres appear homogeneous at ×1600 magnification, they may well be submicroscopic growths or decomposition structures composed of the Fe–Ni phases with troilite and schreibersite.

TROILITE is less common than schreibersite. Troilite occurs as spherical inclusions, 1.5 mm in diameter (Fig. 1) in a metal matrix, or as bands, globules and small rounded particles (~30 μm) in silicate veinlets which connect the inclusions. The structure is fine-grained, with most grains of nearly the same size but differently oriented. This structure is characteristic of shock—metamorphosed troilite (Heymann et al., 1966). Consequently the troilite in Elga exhibits two structural characteristics induced by shock metamorphism. These are either spherical formations in the silicate, which result from remelting, or a uniformly granular structure (Heymann et al., 1966).

Table 5. Electron microprobe analyses (PPD) of schreibersite and alloys from the Elga meteorite (in weight percent). Sample 2315-1.3c.

	(1)	(2)	(3)	(4)	(5)	(6)	(7)
Si	<0.05	1.81	0.92	1.35	<0.05	1.81	<0.05
S	<0.05	4.05	1.40	<0.05	<0.05	<0.05	<0.05
P	17.94	3.95	5.75	3.28	10.58	9.34	6.01
Fe	66.40	87.26	81.32	87.21	76.46	76.39	83.82
Ni	17.66	7.91	11.23	10.19	15.32	14.59	12.76
Sum	102.00	104.98	100.62	102.03	102.36	102.13	102.59

MAGNETITE occurs only in veinlets connecting silicate inclusions as narrow bands and as lenticular bodies on the silicate (or troilite) metal contact. It has been strongly weathered since it fell on earth.

CHROMITE is a rare mineral in the Elga meteorite. Only several small (~10 μm) grains have been found on contacts of silicate inclusions with metal matrix. Chemical analysis of two grains is given in Table 6, analysis 1. Chromite corresponds to ulvöspinel in the Ti content.

STRUCTURAL CHARACTERSITICS RELATED TO IMPACT MELTING

The Fe–Ni matrix exhibits some deformation which shows up as displacement of the Neumann lines in kamacite bands, crumpled structure, twisting of separate kamacite crystals and jointing in schreibersite (Axon, 1968; Lipschutz, 1968; Heymann et al., 1966). This as well as inhomogeneous distribution of deformed areas throughout the meteorite was noted in earlier studies (Plyashkevich, 1962; Kvasha et al., 1974).

One of the most distinct and common structural features is the occurrence of schreibersite near the contacts of silicate inclusions with the matrix metal. Occasionally, Fe–Ni–P(S) alloys are present rather than schreibersite. Schreibersite shows characteristic jointing normal to contacts (Plyashkevich, 1962); its origin has been related to shock deformations (Lipschutz, 1968).

Nearly all inclusions feature injection structures which occur as 1) silicate-into-silicate, 2) silicate-into-metal, 3) metal-into-silicate, and 4) Fe–Ni–P(S) alloy-into-metal veinlet-like injections. Some injections are up to 1 mm long and from fractions to tens of microns wide. The injections have been detected under the microscope in reflected light in oil immersion. The injection structures are located on contacts of silicate inclusions with the matrix or near the peripheries of metal segregates or on alloys of complex composition located within the silicate inclusions. In the last case, injections resemble liquid immiscibility and must have formed involving metal segregation. Fluid immiscibility is most distinctly exhibited at the branching points of the silicate veinlets.

The silicate injections contain metal spheres with dissolved phosphorus or phosphorus and sulfur (Fig. 12). Certain spheres are deformed to imply that they had formed before the injection took place.

Metal-into-silicate injections are less distinct. They contain (Fig. 13) both glassy (Table 6) and devitrified silicate spheres. The latter consist of K–Na feldspar and augite. The quantitative relationship of K–Na feldspar to augite varies greatly.

Table 6. Electron microprobe analyses of chromite and silicate spheres from the Elga meteorite (in weight percent).

	(1)	(2) 2315-1.3c	(3) 2315-5a
SiO_2	2.01	56.26	53.77
TiO_2	3.77	0.64	0.54
Al_2O_3	2.27	6.65	11.03
Cr_2O_3	55.93	0.00	0.11
FeO	30.49	15.25	16.78
MnO	0.63	0.00	0.06
MgO	3.80	9.90	5.54
CaO	0.00	10.78	5.82
Na_2O	0.00	0.00	3.35
K_2O	0.00	1.57	5.07
Sum	98.90	101.89	102.07

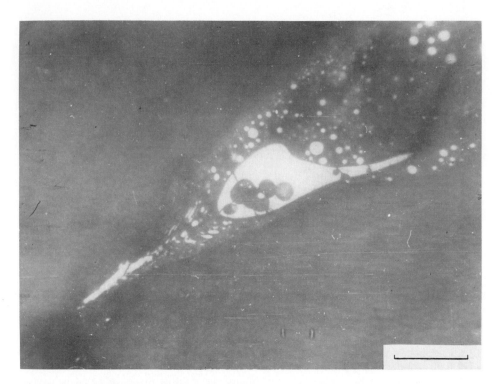

Fig. 12. A silicate-into-silicate injection with the globules of Fe–Ni–P(S) alloy. (Reflected light, in oil immersion.) Scale bar is 5 m.

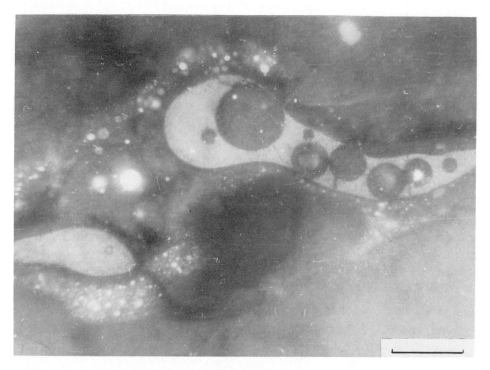

Fig. 13. A silicate-into-silicate injection with elongated formations of Fe–Ni–P(S) alloy of eutectic structure. At the periphery there are metal (white) and Fe–Ni–P(S) alloy (grey) globules; a large, elongated segregation of Fe–Ni–P(S) alloy incorporates silicate augite-bearing globules (glass and devitrified). (Reflected light, in oil immersion.) Scale bar is 10 m.

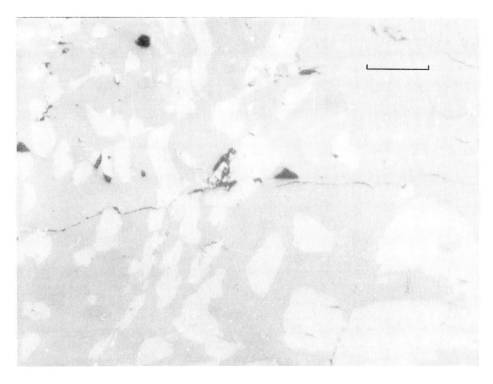

Fig. 14. Liquid immiscibility structure in a silicate inclusion (2315-5). Upward from the crack—the green part of inclusion (2315-5a); downward—the yellow one (2315-5b) (Fig. 1, Tables 1 and 2). In the center—an augite crystal growing across the boundary between the green and yellow parts of the inclusion. (Reflected light.) Scale bar is 0.1 mm.

Metal-into-metal injections represent Fe–Ni–P(S) intrusions into the Fe–Ni matrix. In contrast to silicate inclusions, they are large and elongated in shape. In certain cases, Fe–Ni–P alloy decomposed to schreibersite and submicroscopic Fe–Ni phases. Silicate spheres in these injections are larger than in other injections and occasionally are deformed.

Typically, metal globules are located inside silicate globules and vice versa. For example, there are silicate globules that accomodate more than ten metal globules. Globule-in-glouble structures were also observed—e.g., a metal globule located inside the silicate one, a silicate inside the metal, and so on, up to six alternations.

Liquid immiscibility structures are of particular interest. One of the inclusions inspected under the binocular microscope was found to consist of two differently-colored glasses separated by a narrow crack (Fig. 14). Elongated crystals of augite penetrate through this boundary causing no displacement along the crack. The compositions in the yellow and green portions of the inclusion, as well as the augite compositions in these parts, are very similar to each other (Table 1, analyses 7 and 8). However these augites are somewhat more alkali-rich than those found in other types of inclusions (<0.1 wt.%).

DISCUSSION

The mineralogy and structure of the Elga meteorite indicate that it must have had at least two impact events of different intensity early in its history. The first, most vigorous impact induced the melting of primary silicate inclusions or "whole rock" melt (Schaal and Hörz, 1977; 1980) and determined the major structural-mineralogical characteristics of the meteorite. The second event, much less intense, only resulted in schreibersite jointing and, possibly, partial maskelynitization of the K–Na feldspar.

The first impact produced new minerals as is evidenced by the relict minerals occurring in the reaction relationship with the secondary minerals (Fig. 5). The relict minerals are Fe-rich olivine (hortonolite) (Plyashkevich, 1962), alkali-bearing phosphates of complex composition, and fluorapatite. Therefore, the primary pre-shock silicate inclusions appear to have contained K–Na feldspar with no excess silica, hortonolite and phosphate.

The impact melting of the primary inclusions caused the decomposition of olivine and phosphate with the subsequent formation of augite, schreibersite and Fe–Ni–P(S) alloys, Si-rich feldspar glass and wüstite or magnetite. This process may be represented schematically as the reaction:

$$\text{K–Na feldspar} + \text{olivine} + \text{phosphate} + \text{Fe–Ni phases} \xrightarrow{\text{melting and decomposition}} \text{alkali glass with normative K–Na feldspar and } SiO_2 + \text{augite} + \text{schreibersite} + \text{wüstite (magnetite)} \quad (1)$$

This reaction accounts most satisfactorily for many of the observed mineralogical and structural characteristics of the Elga octahedrite. The whitlockite decomposes to phosphorus and oxygen. The phosphorus reacts with the metal to form denser schreibersite and Fe–Ni–P(S) alloy which commonly occurs as rims around the silicate inclusions. The oxygen is extended to form Fe oxides. The formation of wüstite or magnetite in the presence of metallic Fe depends largely on temperature and pressure (Olsen and Fredriksson, 1966; Robie et al., 1978). The temperature gradients may promote the formation of both high-temperature wüstite and magnetite.

There seems to be two plausible explanations for the 'excess' SiO_2 in feldspar glasses. 1) If the parent inclusions were more abundant in olivine (hortonolite) than in calcium phosphate, the impact melt from which the available augite had crystallized should retain 'excess' SiO_2. 2) If the olivine composition in the parent inclusions was, for instance, Fa 50 mol.%, the 'excess' SiO_2 in the glasses could develop if silica had been present in the parent inclusions as a separate phase. It should be noted that in the first case, too, we should not rule out the possibility for SiO_2 to occur in the parent inclusions as a separate phase. The first explanation seems more attractive because, while the Elga octahedrite contains hortonolite relics (Plyashkevich, 1962), crystalline varieties of silica have not yet been found in great abundance in iron meteorites.

The Mössbauer analyses showed that the augite which crystallized from the melt contained to 50% Fe^{3+}. This corresponds to $Fe–Fe_3O_4$ rather than $Fe–FeO$ equilibrium and suggests a very high oxygen activity in the melt. However, this could only take place at temperatures low enough for silicate inclusions to melt, and at pressures to several kilobars, although the observed injection and other structures indicate that the silicate melt had low viscosity.

High temperatures are evidenced by injection and liquid immiscibility structures, metal globules in the silicate and silicate globules in the metal, confined to the boundary of the inclusions. It appears that injection structures result from the migration of a silicate melt between inclusions through a system of connecting veinlets rather than from the equilibrium coexistence of two immiscible liquids (Visser et al., 1979). With the melts having practically identical compositions, the boundary between them could persist, irrespective of considerable temperature differences between the melts brought in contact and subsequent rapid cooling of the whole inclusion. The oxygen fugacity also varied in these melts and may have contributed to the persistence of the boundary as is evidenced by different colors of the melts.

Consequently, the silicate inclusions retain indications of the past exposure to high temperatures which is impossible to evaluate quantitatively. The maximum pressure in the shock metamorphism could not be below 1000 kb, as indicated by the troilite structures (Heymann et al., 1966).

The Fe–Ni matrix of the Elga meteorite is deformed. This shows up as displacement of the Neumann lines in kamacite crystals, crumpled structure, and elastic twisting of kamacite crystals. However, the Wimanstätten pattern persists, which would have been impossible above 800 K and subsequent relatively slow cooling (Wood, 1967). The observed inconsistencies, such as the melting of silicate inclusions in relatively cold matrix, may be attributed to high dynamic pressures during high-speed collisions of meteorite bodies. As shock waves propagate through a medium that is inhomogeneous in terms of density, the energy preferentially releases (dissipates) in areas with reduced density (Zeldovich and Riser, 1966; Adadurov et al., 1966). In the Elga meteorite the areas with reduced density are composed of silicate inclusions. The propagation of a shock wave of an adequate intensity could make the silicate inclusions melt completely in a "cold" Fe–Ni matrix. After the shock wave had passed, the melt crystallized at low pressure, as is evidenced by low alumina content of the augite, by analogy with monoclinic pyroxenes (Wells, 1977).

Indeed, it still remains a puzzle (Levin, 1977) how the silicate inclusions in the Elga octahedrite could retain the tracks of solar cosmic rays (Kashkarov et al., 1975) despite the fact that the silicate substance had melted and crystallized inside the Fe–Ni matrix, which shielded the silicate from the solar radiation.

It appears from the composition of pre-shock inclusions that they are products of the earliest low-temperature melting.

Acknowledgments—The authors thank Dr. I. M. Romanenko for assistance in microprobe analyses, Drs. A. Ya. Skrypnik and N. I. Zaslavskaya for generously providing the Elga meteorite specimens, and G. B. Lakoza for translating the text into English.

REFERENCES

Adadurov G. A., Breusov O. N., Dremin A. N., and Lasarev A. I. (1966) To the problem of conservation of inorganic matter under the action of shock waves. *Phisika gorenya i vzryva*, No. 4, 130–135.
Axon H. J. (1968) The metallographic structure of the Kodaikanal meteorite. *Mineral. Mag.* 36, 687–690.
Bence A. E. and Burnett D. S. (1969) Chemistry and mineralogy of the silicates and metal of the Kodaikanal meteorite. *Geochim. Cosmochim. Acta* 33, 387–408.
Bunch T. E., Keil K., and Olsen E. (1970) Mineralogy and petrology of silicate inclusions in iron meteorites. *Contrib. Mineral. Petrol.* 25, 297–340.
Bunch T. E. and Olsen E. (1968) Potassium feldspar in Weekeroo Station, Kodaikanal, and Colomera iron meteorites. *Science* 160, 1223–1225.
Deer W. A., Howie R. A., and Zussman J. (1963) *Rock Forming Minerals*. Longmans, Green & Co Ltd. London. 371 pp.
Heymann D., Lipschutz M. E., Nilson B., and Anders E. (1966) Canyon Diablo meteorite: Metallographic and mass spectrometric study of 56 fragments. *J. Geophys. Res.* 71, 619–641.
Kashkarov L. L., Korotkova N. N., and Lavrukhina A. K. (1975) Relict irradiation by low-energy heavy nuclei of cosmic rays the iron meteorite matter. *Dokl. Akad. Nauk. SSSR.* 221, 198–200.
Kvasha L. G., Lavrentjev Ya. G., and Sobolev N. V. (1974) On silicate inclusions and shock metamorphism features in the Elga meteorite. *Meteoritika* 33, 143–147.
Levin B. Ya. (1977) On the origin of meteorites. *Meteoritika* 36, 3–23.
Lipschutz M. E. (1968) Shock effects in iron meteorites: A review. In *Shock Metamorphism of Natural Materials* (B. M. French and N. M. Short, eds.), p. 571–583. Mono, Baltimore.
Longhi J., Walker D., and Hays J. F. (1972) Petrography and crystallization history of basalts 14310 and 14072. *Proc. Lunar. Sci. Conf. 3rd*, p. 131–139.
Olsen E. and Fredriksson K. (1966) Phosphates in iron and pallasite meteorites. *Geochim. Cosmochim. Acta* 30, 459–470.
Plyashkevich L. N. (1962) Some data on composition and structure of the Elga iron meteorite. *Meteoritika* 22, 51–60.
Robie R. A., Hemingway B. S., and Fisher J. R. (1978) Thermodynamic properties of minerals and related substances at 298 K and 1 bar (10^5 pascals) pressure and at higher temperature. *U.S. Geol. Surv. Bull.* 1452, 456 pp.
Schaal R. B. and Hörz F. (1977) Shock metamorphism of lunar and terrestrial basalts. *Proc. Lunar. Sci. Conf. 8th*, p. 1697–1729.
Schaal R.B. and Hörz F. (1980) Experimental shock metamorphism of lunar soil. *Proc. Lunar. Planet. Sci. Conf. 11th*, p. 1679–1695.
Visser W., Angust F., and Koster Van Groos. (1979) Effect of pressure on liquid immiscibility in the system K_2O–FeO–Al_2O_3–SiO_2–P_2O_5. *Amer. J. Sci.* 279, 1160–1175.
Vronsky B. I. (1962) On the discovery of the Elga Iron meteorite. *Meteoritika* 22, 47–50.

Wasson J. T. (1970) Ni, Ga, Ge and Ir in the metal of iron meteorites with silicate inclusions. *Geochim. Cosmochim. Acta* **34**, 957–964.
Wells P. R. A. (1977) Pyroxene thermometry in simple and complex system. *Contrib. Mineral. Petrol.* **62**, 129–139.
Winchell A. N. and Winchell V. (1953) *Optical Mineralogy*. Foreign Languages Publishing House, Moscow. 561 pp.
Wood J. A. (1967) The cooling rates and parent planes of several iron meteorites, their thermal history and parent bodies. *Geochim. Cosmochim. Acta* **31**, 1733–1770.
Zeldovich Ya. B. and Riser Yu. P. (1966) Physics of shock waves and high-temperature hydrodynamic phenomena. Nauka, Moscow. 632 pp.

SEM, optical, and Mössbauer studies of submicrometer chromite in Allende

R. M. Housley

Rockwell International Science Center, Thousand Oaks, California 91360

Abstract—We present new scanning electron and optical microscope results showing that submicrometer chromite is abundant along healed cracks and grain boundaries in Allende chondrule olivine. Some wider healed cracks also contain pentlandite and euhedral Ni$_3$Fe grains. We also report Mössbauer measurements on Allende HF-HCl residues confirming a high Fe^{+++}/Fe^{++} ratio.

INTRODUCTION

The presence of submicrometer chromite in Allende was discovered by Lewis et al. (1975) in the course of a search for the host phase of an anomalous Xe component. They found that a residue insoluble in successive HF and HCl treatments of the meteorite contained a large fraction of the total inert gas inventory including the anomalous component. This residue (constituting about 0.5 percent of the initial mass) consisted predominantly of carbon, spinel, and sub-micrometer chromite. Lewis et al. (1979) indicated that their recovery of fine-grained chromite was 0.153 percent of the initial mass. Using the Cr$_2$O$_3$ content of Allende reported by Clarke et al. (1970) this implies that at least about 20 percent of the total Cr in the meteorite is present as submicrometer chromite.

We recently reported some results of a transmission electron microscope study of Allende HF-HCl residues (Housley and Clarke, 1980). This included energy dispersive X-ray analyses of 20 individual submicrometer chromite grains (Table 3). Based on our experience with analogous synthetic samples we believe that the reported Cr/Fe ratios are accurate within about 5%. As pointed out in that report, many of these compositions are clearly anomalous.

The spinel structure contains 32 oxygens per unit cell and has 16 octahedral cation sites plus eight tetrahedral ones (Deer et al., 1962). Trivalent chromium Cr^{+++} has a very large octahedral site preference energy (Burns, 1975). Therefore, both chromite FeCr$_2$O$_4$ and magnesiochromite MgCr$_2$O$_4$ are expected to be normal spinels with Cr^{+++} entirely on the octahedral sites and the divalent cations on the tetrahedral sites. Hercynite FeAl$_2$O$_4$ is observed to be a normal spinel (Deer et al., 1962). On the other hand, magnetite FeFe$_2$O$_4$ and magnesioferrite MgFe$_2$O$_4$ are inverse spinels with Fe^{+++} on the tetrahedral sites.

Of the compositions listed in Table 3 of Housley and Clarke (1980), only three—AF-2, All-N8, and All-N13—can be simple solid solutions of magnetite in chromite. The remainder contain excess Cr + Al. Divalent chromium Cr^{++} does not seem likely in view of the P$_{O_2}$ inferred from the oxide-sulfide assemblages in chondrules (McMahon and Haggerty, 1980). In addition, spinels of such a composition have not been observed.

If we rule out Cr^{++} and require that Cr^{+++} occupy only octahedral sites, then seven of the grains analyzed must contain both major tetrahedral Fe^{+++} and tetrahedral cation vacancies. These are AF-1, All-3, 4, 5, 6, 7, and 15. They appear to require formation under unusual and highly oxidizing conditions. If we assume that Al can occupy both octahedral and tetrahedral sites, then the remaining 10 grains must contain a modest number of cation vacancies, but need not contain Fe^{+++}. Therefore, they need not have formed under strongly oxidizing conditions.

Here we report new Mössbauer observations bearing on the state of Fe in these residue chromites. We also present scanning electron microscope and optical microscope observations of the *in situ* occurrence of submicrometer chromite in Allende chondrules.

MÖSSBAUER RESULTS

A sample for Mössbauer study was prepared from Allende HF-HCl residue obtained from Walker's group. This residue in turn was prepared according to the procedures of Lewis *et al.* (1975) and is further characterized in Fraundorf *et al.* (1977). About 10 mg of this material which we designate by the label AF was put in suspension as described in Housley and Clarke (1980). All coarse grained components, mainly spinel and hercynite, were removed by centrifuging and decanting. The suspension was then coagulated with 3N HCl, centrifuged, and dryed.

Two small flakes of this coagulated residue were examined in a scanning electron microscope (SEM). No remaining mineral grains as large as 1 μm could be found. The elemental composition was obtained by energy dispersive X-ray analysis while scanning over a large area and at many individual points. The composition showed little variation from point to point and was consistent with the expectation that chromite was the dominant mineral phase.

A room temperature Mössbauer spectrum of this material is shown in Fig. 1. In comparison to the synthetic ferrichromites studies by Robbins *et al.* (1971) the lines are broad and poorly resolved. The Fe^{++} line in particular appears to be very broad and to partially underlie the Fe^{+++} part of the spectrum. The Fe^{+++}/Fe^{++} ratio appears to be about 1, but cannot be determined with much precision. A similar conclusion was reached by Virgo (pers. comm.; Lewis *et al.*, 1975). Part of the Fe^{+++} intensity appears to be centered to the left of the indicated positions for octahedral Fe^{+++}. This is consistent with

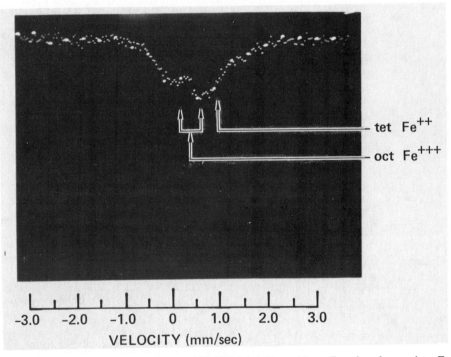

Fig. 1. Mössbauer spectrum of Allende HF-HCl insoluble residues. Zero is referenced to Fe metal at room temperature. Indicated line positions were observed in synthetic ferrichromite (Robbins *et al.*, 1971).

the suggestion made on the basis of our interpretation of the STEM results that a substantial fraction of the Fe^{+++} is in tetrahedral sites.

In a preliminary report (Housley, 1981) we presented Mössbauer and X-ray results on a sample separated from lightly crushed Allende material with a magnetic needle. This material was believed to consist largely of chondrules and chondrule fragments. However, an effort to confirm the results using hand-sorted chondrule material was unsuccessful. Subsequently, we have found that hand-sorted fusion crust gives a Mössbauer spectrum nearly identical to that shown in Housley (1981). Fusion crust is much more magnetic than the chondrule material and we now believe it dominated the original results.

MICROSCOPIC OBSERVATIONS

All the observations we report were made on one standard polished thin sections of Allende. A very thin layer of Au was sputtered onto the polished surface. Sputtering conditions were empirically determined such that the Au was thin enough not to significantly interfere with optical microscope studies including reflected light work, yet was thick enough to prevent significant charging in the SEM. In the SEM, using the normal secondary electron mode, contrast differences could be observed between silicates and oxides plus sulfides, and in turn between them and metal. No useful contrast between silicates except sodalite and other high alkali phases was observed. Also, little contrast was observed between troilite, pentlandite, magnetite, and chromite. These phases were identified either from the optical images or by energy dispersive X-ray analysis. In some cases the back-scattered electron mode provided better contrast in images and was used.

The facet of our microscopic work, which will be discussed here, was intended to contribute to understanding the significance of submicrometer chromite in Allende. Part of this effort involved searching for submicrometer chromite in various environments, another part involved a general assessment of the chemical behavior of Cr in Allende.

Several searches of matrix material for chromite proved tantalizing but frustrating. Scans over large areas of matrix always showed significant average Cr contents consistent with the value of 0.38 wt.% reported by McSween and Richardson (1977). Using the spot analysis mode, local areas that had Cr contents several times higher than this average could easily be found. However, nothing definitive has yet been recognized in these analyses or the corresponding images to prove that chromite is present, or if it is to shed light on its mode of formation. No significant Cr content was found in large area scans of two millimeter sized sulfide nodules in the matrix.

Recalling the report of Cr in Allende metal by Fuchs and Olsen (1973), we examined the metal associated with several small sulfide-oxide nodules in the chondrules. In most cases apparent Cr concentrations of 0.4 to 0.6%, close to those reported, were observed. However, these nodules also contained oxides with Cr concentrations of a few percent similar to those described by Haggerty and McMahon (1979). Therefore, since these metal grains were small, we suspect that the apparent Cr contents were largely due to secondary fluorescence. A few small chondrule metal grains free from visible oxide association were also essentially free of Cr. We suspect that chondrule metal in Allende generally does not contain significant Cr.

A Cr K_α X-ray map of one of the oxide-containing nodules showed some enrichment of Cr near the border with the surrounding forsterite and a faint streak enriched in Cr running out into the forsterite. This put us on an interesting track. We soon found we could recognize that streak and similar ones on the basis of contrast variations in the SEM images. Based on their irregular curvature we identified them as healed cracks. Some classic examples from a different grain are illustrated in Figs. 2 and 3. Figure 4 illustrates similar streaks along grain boundaries between olivine crystals in yet another chondrule.

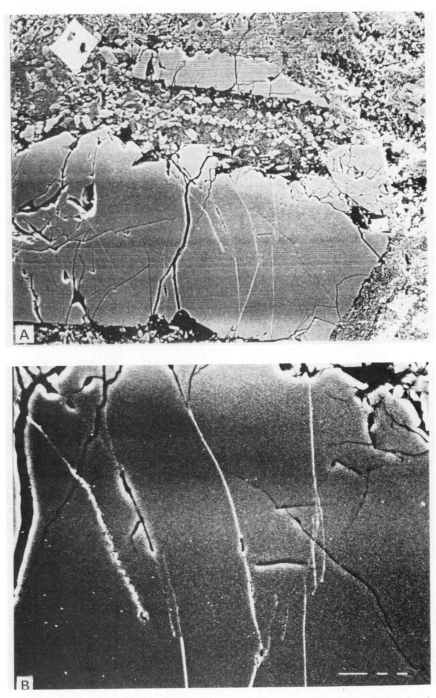

Fig. 2. SEM backscattered electron images of a large olivine grain in Allende. The left hand scale bar in both images is 10 μm. In A, large chromite grains are visible in the upper left and middle right. Fine grained sulfides are abundant along the far right. Bright lines are cracks healed largely with submicrometer chromite.

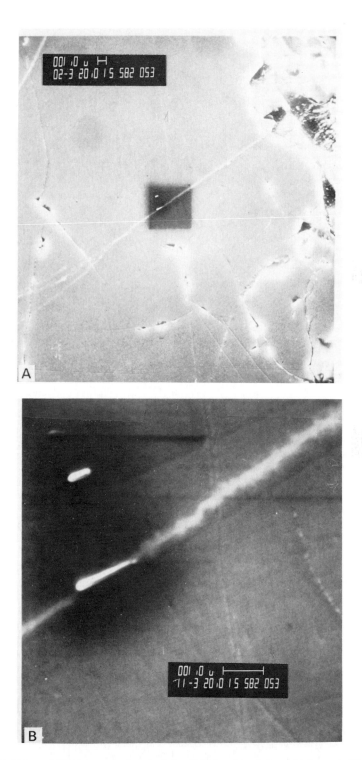

Fig. 3. Enlarged SEM images of part of the right hand healed crack shown in Fig. 2. Note sample orientation has been changed. Image B is an enlargement of the dark area in the center of A. The bright streaks in B indicate areas analyzed. Note individual chromite grains must be much smaller than 1 μm.

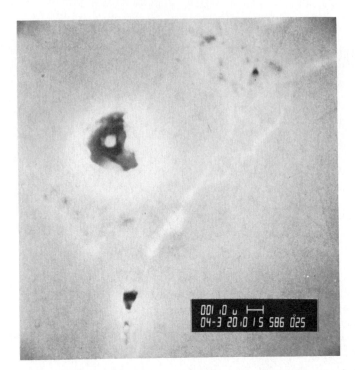

Fig. 4. SEM image showing submicrometer chromite concentrations along grain boundaries between olivines in a chondrule.

We observed X-ray spectra of approximately one hundred of these bright streaks, in about ten different chondrules. Invariably, Cr and Fe are strongly enriched. In some cases there are subordinant enrichments in Al and V. When a careful effort was made to subtract the large contributions from the surrounding olivine, the difference spectra were dominated by Cr and Fe in about a 2:1 ratio plus Mg, Al, and V in lesser amounts. The bright streaks thus seem to be chromite.

Our observations may be summarized as follows: Most Allende olivine chondrules contain abundant submicrometer chromite along healed cracks and grain boundaries. The narrower cracks seem to contain only chromite while the wider ones may also contain pentlandite and occasional euhedral Ni_3Fe grains. The left hand crack in Fig. 2B is an example containing all 3 phases.

We alternatively studied the sample in the SEM and with an optical microscope. In otherwise clear areas, healed cracks can be most easily recognized in optical transmission by their dark brown color. Long healed cracks can also be easily seen as bright streaks in optical reflection. Typical examples are shown in Fig. 5. All such optical features proved to be chromite, excluding of course the wider sulfide filled cracks which can be recognized directly by their optical reflectivity.

During our SEM searches of chondrules we found 3 grains rich in Cr, but containing little Fe. These grains are about 1 μm in size and are not clearly associated with cracks or grain boundaries. They occur as isolated crystals in very forsteritic olivine. Two principally contain Cr while one contains about twice as much Al as Cr. In our previous STEM study (Housley and Clarke, 1980) we found, but did not report, one grain which was largely Cr_2O_3 with very little Fe. We suggest that these isolated Cr rich grains may be eskolaite.

SEM examination of the fusion crust showed it to contain abundant octahedral crystals about 1 μm in diameter in a glassy matrix. These consisted largely of Fe oxide containing minor amounts of Cr.

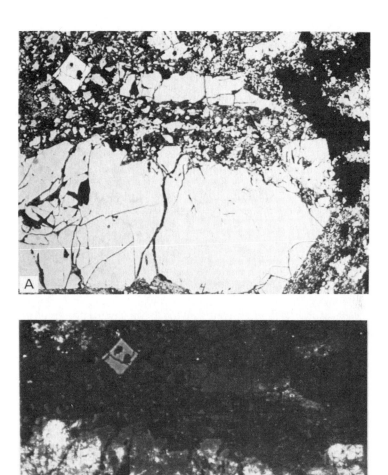

Fig. 5. Optical microscope images of about the same area as shown in Fig. 2A. Image A in reflected light shows large chromites and chromite along healed cracks clearly. Image B with transmitted light also shows the dark appearance of the healed cracks in transmission.

DISCUSSION

The Mössbauer data show that the average Fe^{+++}/Fe^{++} ratio in Allende submicrometer chromite is in the order of 1. The width of the lines makes more detailed interpretation difficult but suggests a range of local environments more varied than in the synthetic ferrichromites of similar average Fe^{+++}/Fe^{+++} ratio studied by Robbins *et al.* (1971). Both the Fe^{+++}/Fe^{++} ratio and the width of the component lines lend credibility to our interpretation of the STEM analyses of individual grains.

A simple interpretation of the microscopic observations at the descriptive level seems possible. It appears that excess concentrations of Cr^{+++} were incorporated in rapidly crystallizing olivines during the initial stages of chondrule formation. At a later time when the chondrules were cooled sufficiently differential contraction caused cracks to form and the excess Cr^{+++} was precipitated as chromite along them. The mobilization of chromite into the cracks from an outside source seems less likely since neither a convenient source nor a plausible mobilization mechanism are evident.

The possibility that Cr entered the olivine as Cr^{++} and later precipitated as Cr^{+++} in chromite under more oxidizing conditions has also been considered. It seems unlikely in view of fairly oxidized state of the Fe associated with the chondrules.

It seems improbable that chromite precipitated from chondrule olivine as hypothesized above would have the unusually oxidized compositions required for the seven Cr rich grains mentioned in the introduction. We therefore believe that they must have originated from an as yet undetermined environment in the matrix.

Of the remaining 13 grains we believe based on the systematic trends observed in the partitioning of Mg between olivine and chromite (Bunch and Olsen, 1975, and references therein) that the four Mg rich grains AF-3, All-N1, All-N1, and All-N9 (Housley and Clarke, 1980, Table 3) probably originated in forsteritic chondrules. Thus, although providing only a rough constraint, the present data are consistent with the possibility that chondrules and matrix make comparable contributions to the total submicrometer chromite in Allende.

Although differential leaching and/or oxidation cannot be completely ruled out in these samples, in view of the way they were obtained, we doubt that it is important. A simulation experiment with a sample in the ulvöspinel-chromite series did not show differential leaching although a very small amount of surface material was etched away.

The data in Housley (1981) along with the microscopic observations of fusion crust and new Mössbauer observations mentioned here suggest, contrary to expectation, that maghemite, rather than magnetite, is the dominant magnetic oxide in Allende fusion crust.

SUMMARY

New Mössbauer data support the anomalous oxidation state of some Allende submicrometer chromite inferred from previous STEM analyses. Abundant submicrometer chromite was observed by optical and SEM techniques in Allende chondrule olivines. Indirect evidence suggests that comparable amounts reside in unidentified sites in the matrix.

Acknowledgments—The generous help of Mitzi Housley with the SEM work is appreciated. The sample for Mössbauer study was provided by R. M. Walker. This work was supported by NASA Contract NAS9-11539.

REFERENCES

Bunch T. E. and Olsen E. (1975) Distribution and significance of chromium in meteorites. *Geochim. Cosmochim. Acta* **39**, 911–927.

Burns R. G. (1975) Crystal field effects in chromium and its partitioning in the mantle. *Geochim. Cosmochim. Acta* **39**, 857–864.

Clarke R. S., Jarosewich E., Mason B., Nelen J., Gomez M., and Hyde J. R. (1970) The Allende, Mexico, meteorite shower. *Smithson. Contrib. Earth Sci.* **5**, 53 pp.

Deer W. A., Howie R. A., and Zussman J. (1962) *Rock-forming Minerals*, vol. 5. Wiley, N.Y. 372 pp.

Fraundorf P., Flynn G. J., Shrirck J. R., and Walker R. M. (1977) Search for fission tracks from superheavy elements in Allende. *Earth Planet. Sci. Lett.* **37**, 285–295.

Fuchs L. H. and Olsen E. (1973) Composition of metal in type III carbonaceous chondrites and its relevance to the source-assignment of lunar metal. *Earth Planet. Sci. Lett.* **18**, 379–384.

Haggerty S. E. and McMahon B. M. (1979) Magnetite-sulfide-metal complexes in the Allende meteorite. *Proc. Lunar Planet. Sci. Conf. 10th*, p. 851–870.

Housley R. M. (1981) Oxidation of Allende parent material (abstract). In *Lunar and Planetary Science XII*, p. 477–478. Lunar and Planetary Institute, Houston.

Housley R. M. and Clarke D. R. (1980) XPS and STEM studies of Allende acid insoluble rsidues. *Proc. Lunar Planet. Sci. Conf. 11th*, p. 945–958.

Lewis R. S., Alaerts L., and Anders E. (1979) Ferrichromite: A major host phase of isotopically anomalous noble gases in primitive meteorites (abstract). In *Lunar and Planetary Science X*, p. 725–727.

Lewis R. S., Srinivasan B., and Anders E. (1975) Host phase of a strange xenon component in Allende. *Science* **190**, 1251–1262.

McMahon B. M. and Haggerty S. E. (1980) Experimental studies bearing on the magnetite-alloy-sulfide association in the Allende meteorite: Constraints on the conditions of chondrule formation. *Proc. Lunar Planet. Sci. Conf. 11th*, p. 1003–1025.

McSween H. Y. Jr. and Richardson S. M. (1977) The composition of carbonaceous chondrite matrix. *Geochim. Cosmochim. Acta* **41**, 1145–1161.

Robbins M., Wertheim G. K., Sherwood R. C., and Buchanan D. N. E. (1971) Magnetic properties and site distributions in the system $FeCr_2O_4 - Fe_3O_4(Fe^{2+}Cr_{2-x}Fe_x^{3+}O_4)$. *J. Phys. Chem. Solids* **32**, 717–729.

Origin of rims on coarse-grained inclusions in the Allende meteorite

Glenn J. MacPherson[1], Lawrence Grossman[1,2], John M. Allen[3] and John R. Beckett[1]

[1]Dept. of Geophysical Sciences, University of Chicago, 5734 S. Ellis Avenue, Chicago, Illinois 60637
[2]Enrico Fermi Institute, University of Chicago, 5734 S. Ellis Avenue, Chicago, Illinois 60637
[3]Dept. of Geology, University of Toronto, Toronto, Ontario, Canada M5S 1A1

Abstract—Rim sequences on coarse-grained inclusions in Allende are of two types. Poorly-developed rims consist mostly of feldspathoids and olivine and are found on inclusions having a low volume fraction of secondary alteration products. Well-developed rims have abundant diopside, fassaite, hedenbergite, andradite and feldspathoids, and are found on highly altered inclusions. Rims were formed as by-products of the alteration process, prior to accretion of the inclusions into the meteorite. Inward diffusion of silica and alkalies from the surrounding, nebular gas caused breakdown of coarse-grained melilite to anorthite, grossular and nepheline, releasing excess calcium which diffused outward. Rim layers formed at inclusion/gas interfaces in response to steep chemical potential gradients of diffusing components. The rim layers are thus steady-state, diffusion-controlled features whose origin is analogous to that of terrestrial banded calc-silicate deposits at contacts between marbles and pelitic schists. The different types of rim sequences are a consequence of the different degrees of alteration of host inclusions because the amount of calcium available at the inclusion surface is a function of the amount of melilite that is altered.

INTRODUCTION

Coarse-grained, calcium-rich inclusions in the Allende meteorite are enclosed by sequences of very thin, mineralogically distinct rim layers. Wark and Lovering (1977) gave detailed descriptions of these rims and stated that different types of refractory inclusions have different types of rim sequences. Rims on Type A inclusions (Grossman, 1975) were described as being composed of the following layers, from innermost to outermost: FeO-rich spinel + perovskite, anorthite + feldspathoids + olivine + grossular, Ti-Al-rich clinopyroxene which grades outward in composition to aluminous diopside and, finally, hedenbergite + andradite + wollastonite. Rim sequences on Type B inclusions were described as being similar to those on Type A's, but lacking the Ti-Al-pyroxene and diopside bands so prominent in the rims on those inclusions. Sometimes, we observe only a single layer of feldspathoids + olivine + minor hedenbergite outside of the spinel-perovskite band in those inclusions whose rims are poor in diopside. For reasons which will become evident later in this paper, we prefer to call the diopside-rich rim sequences "well-developed" and the others "poorly-developed". It is not only the abundance of diopside which distinguishes the rim sequences defined in this way, but the abundances of *all* calcium-rich phases. Well-developed rims are also rich in hedenbergite and andradite, while the poorly-developed ones contain little of these phases and are composed mostly of feldspathoids and olivine.

Three problems arise from the presence of rims. First, from their mode of occurrence, it is probable that they formed in gas-condensed phase reactions in the solar nebula; yet, some of their constituents (nepheline, grossular, andradite, wollastonite) are not known to be in equilibrium with a gas of solar composition at any temperature. Did they form under non-equilibrium conditions, or in a gas of unusual composition? Second, did the rim minerals originally condense elsewhere and then accrete as sedimentary material around the inclusions or did they nucleate and crystallize directly on inclusion margins? The only evidence bearing on this is the lack of granularity in some layers, suggesting, if anything, that they are not sedimentary. Third, why do Type A inclusions have different rim sequences from Type B's, as suggested by Wark and Lovering (1977)? Is it because rims formed by reaction of inclusion margins with a gas phase and Type A's presented reacting surfaces to the gas which were of different composition from those of Type B's? Or did Types A and B inclusions react with gases of different composition in different nebular regions? In this paper, we present new observations bearing on all these questions, and offer a model for the formation of rims.

ANALYTICAL TECHNIQUES

All of the samples were examined optically and with a JEOL JSM-35 scanning electron microscope equipped with a KEVEX Si (Li) X-ray detector. All electron microscope photographs are backscattered electron images. Fine-grained phases were identified by visually comparing X-ray spectra of unknowns to spectra of phases analyzed previously by electron microprobe in other inclusions. Coarse-grained phases were analyzed quantitatively on the Chicago microprobe, using operating conditions and data reduction described previously (Allen *et al.*, 1978).

OBSERVATIONS

A zoned alteration vein

Coarse-grained inclusions are altered, some heavily so. A fine-grained assemblage of nepheline, grossular, anorthite and wollastonite sometimes lines cavities formed by corrosion of the margins of grains of primary mineral phases, mostly melilite, and fills veins which cross-cut the grains. The fine-grained minerals are clearly an assemblage of secondary alteration products of the coarse-grained phases. In an SEM study of the alteration products, Allen *et al.* (1978) found that they often occur as euhedral crystals that project into void space, suggesting growth from a vapor phase. This implies that the primary minerals reacted with a vapor phase to form the alteration products. Although the same minerals found as alteration products also occur in rims, the genetic relationship, if any, between the alteration and rimming processes is poorly understood. We have discovered a unique alteration vein that cross-cuts a Type B1 (Wark and Lovering, 1977) inclusion in our collection, TS34F1 (Fig. 4 of Clayton *et al.*, 1977), and sheds considerable light on this relationship.

TS34F1 is a spheroidal inclusion, ~1 cm in diameter, which consists of an outer mantle of coarse-grained melilite enclosing an inner core of melilite + anorthite + fassaite + spinel. The inclusion is mantled by a thin, poorly-developed rim sequence consisting of, from inside to outside, spinel + perovskite, feldspathoids, olivine + rare grains of hedenbergite.

The vein extends inward from the edge of the inclusion, cutting across both the rim layers and the coarse-grained melilite mantle of the inclusion. It is over 500 μm long, up to 75 μm wide and symmetrically zoned about its middle (Fig. 1). The zone closest to the axis of the vein consists of hedenbergite + andradite, followed by successive layers of aluminous diopside and anorthite + grossular towards the melilite. The anorthite and grossular crystals embay and corrode coarse-grained melilite with which they are in contact. Except for the absence of sodium-bearing phases from the vein and of a spinel-perovskite band between the anorthite-bearing layer and the fresh melilite, the vein's mineral zonation is virtually identical to that in well-developed rim sequences and must have formed in the same process. The implications of this vein for the origin of rim sequences is discussed later.

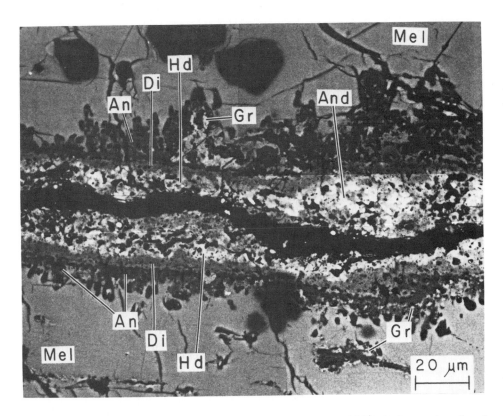

Fig. 1. Back-scattered electron image of the zoned alteration vein in TS34F1. Dark central portion of vein is epoxy from thin section preparation. And: andradite; An: anorthite; Di: aluminous diopside; Gr: grossular; Hd: hedenbergite; Mel: melilite.

A sinuous inclusion

Davis et al. (1980) noted an inclusion consisting almost entirely of a well-developed rim sequence juxtaposed with a poorly-developed one. The characteristics of this object, called the sinuous inclusion, also place important constraints on the origin of rims.

In polished thin section, the inclusion is U-shaped (Fig. 2) and has a total length of 18 mm and a width of only 0.25–0.50 mm. It is highly contorted and is broken in places, so the entire object occupies an area of only 3 × 6 mm. The shape of the inclusion in three dimensions is unknown. Either it is a two-dimensional chain of crystals whose long axis fortuitously lies in the plane of the section or, more probably, it is a complexly folded three-dimensional sheet which is seen in cross-section in the thin section.

The axis of the inclusion is a prominent layer of coarse pink spinel and perovskite, on one side of which are numerous irregular patches of melilite. This spinel-perovskite-melilite band, which we call the SP layer, and broken fragments of it are mantled on both sides by layers of fine-grained material, the precise mineralogy and zoning of which depend on whether they overlie the melilite-rich side of the SP layer or the side opposite the melilite (Fig. 3). Visible in Fig. 2 are three outer rim layers which enclose the entire sinuous inclusion and all of the previously mentioned layers. Each of these outer layers has a mineralogy and loosely-packed texture that closely resembles the matrix of Allende. These are the "clastic" rims of MacPherson and Grossman (1981) and will not be discussed further here. The inner rim layers will each be described in detail.

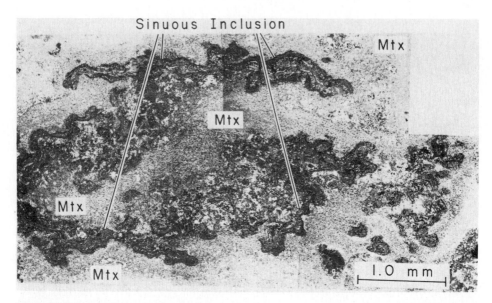

Fig. 2. Back-scattered electron image of the entire sinuous inclusion and the matrix (Mtx) surrounding it. Part of the matrix in this case includes "clastic rim" material as described by MacPherson and Grossman (1981).

SP layer

One side of the SP layer consists of a continuous ribbon of spinel, along the center of which is a chain of perovskites. The other side of the SP layer is composed of melilite, in some places as a continuous band and in others as isolated islands within spinel. Within the melilite are often found, grossular- and anorthite-lined voids. Ti-Fe-oxides and hibonite are very minor accessories in the SP layer.

Spinel contains 1.72–8.30% FeO, within the composition range found in spinels in rims on coarse-grained inclusions (Allen *et al.*, 1978; Haggerty and Merkel, 1976; Wark *et al.*, 1979). There is no evidence for systematic variation of FeO across the spinel layer, such as that noted by Wark and Lovering (1977) in rim spinels on Type A inclusions. Spinel also contains 0–0.58% TiO_2 and 0.23–0.50% V_2O_3, comparable to values in rim spinels from the Type A inclusion studied by Allen *et al.* (1978). Perovskite is nearly pure $CaTiO_3$, as no other elements having atomic number in the range 11 to 30 were detected with the solid state analyzer. Hibonite contains 5.05% MgO and 9.08% TiO_2, more of these oxides than reported so far in meteoritic hibonite (Grossman, 1975; Allen *et al.*, 1978; Keil and Fuchs, 1971; Macdougall, 1979; Allen *et al.*, 1980). The hibonite also contains 0.25% FeO. Grossular contains 4.84% Fe_2O_3 and 1.36% MgO, corresponding to small proportions of the andradite and pyrope components. Melilite shows no magnesium peak in its X-ray spectra and is therefore close to pure gehlenite in composition.

Layers overlying the melilite-rich side of SP

The layer directly overlying the melilite is continuous and 18–30 μm thick. Nepheline and anorthite in subequal amounts, plus lesser sodalite, are the only minerals positively identified in it. An unknown Ca-Al-Si phase with an X-ray spectrum similar to zoisite was also found in amounts of up to 10% of the layer in places. Anorthite and sodalite occur in monomineralic areas up to 20 μm in size. Voids make up about 5% of this layer. Regular boundaries to the voids suggest crystallographic control of their shapes. Anorthite crystals, up to 5 μm in length, locally project into these voids, suggesting that the anorthite condensed from a vapor. This layer has a coarsely scalloped boundary with the SP layer and a more regular, but very finely bumpy, contact with the next layer out.

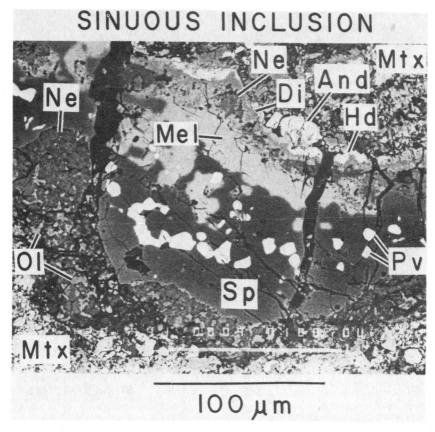

Fig. 3. Back-scattered electron image of part of the sinuous inclusion, showing the central spinel-perovskite layer with melilite on one side only, and with a well-developed rim sequence overlying the melilite and a poorly-developed rim overlying the spinel. Abbreviations same as in previous figures, except Ne: nepheline; Ol: olivine; Pv: perovskite; Sp: spinel.

This next layer is continuous, 4–12 µm thick and consists almost entirely of diopside and fassaite, with minor perovskite and no voids. The innermost diopside contains about 5% Al_2O_3 and up to 1% TiO_2, but both components decrease in amount towards the next layer out. The perovskite occurs locally as trains of tiny crystals in the most aluminous diopside.

The outermost layer is discontinuous. Where present, it consists mostly of variable proportions of hedenbergite and andradite, with lesser nepheline and sodalite. In places where a sequence can be recognized, hedenbergite occurs closest to the diopside of the next layer in, while andradite forms large ($\leqslant 10$ µm) euhedral crystals overlying the hedenbergite. Nepheline and sodalite fill interstices between the hedenbergite and andradite crystals. The contact between this layer and the diopside layer is smooth and regular, whereas the outer margin of this layer is jagged due to projecting blocky crystals of andradite and hedenbergite.

In short, the series of layers overlying the melilite-rich side of the SP layer is identical to well-developed rim sequences on coarse-grained inclusions.

Layer overlying the spinel-rich side of SP

A single layer, 20–40 µm thick, mantles the spinel-rich side of the SP layer, opposite the melilite-rich side. The contact of this layer with SP is quite irregular and scalloped, as if the underlying spinel had been partially dissolved away before or during deposition. The layer consists of porous and fractured nepheline and sodalite in subequal amounts in which are embedded rounded grains of olivine and rare clinopyroxene. Scattered masses of spinel may be original pieces of SP isolated during its dissolution. Clinopyroxene ranges in composition from aluminous diopside ($Al_2O_3 \sim 4.5\%$, $TiO_2 \sim 0.2\%$) to hedenbergite. There is a tendency for clinopyroxene to be concentrated in the outer parts of the layer. Olivine is magnesium-rich, and the small amounts of aluminum seen in its X-ray spectra are apparently due to tiny inclusions of nepheline.

In short, this layer overlying the spinel-rich side of the SP layer is identical to many poorly-developed rims on coarse-grained inclusions.

Rim composition and the degree of alteration in coarse-grained inclusions

As mentioned earlier, Wark and Lovering (1977) found that what we call well-developed rims are restricted to Type A inclusions and poorly-developed rims are restricted to Type B's. Contrary to those observations, however, we find that this correlation is not strictly true. To be sure, well-developed rims are present on all 6 fluffy Type A inclusions (MacPherson and Grossman, 1979) which we have examined and poorly-developed rims are present on all three Type B1's in our collection. We also find, however, that two out of three of the compact Type A's that we have examined have poorly-developed rim sequences and that three out of four Type B2's have well-developed rim sequences. The one Type I inclusion (Grossman, 1975; 1980) has no rim at all.

What we have found is a general correlation between rim type and the degree of secondary alteration of coarse-grained phases, mainly melilite, in the interior of the inclusions. Point-count modes (Beckett *et al.*, 1980) show 60–75 volume % alteration products in fluffy Type A's, 20–45% in B2's, 15–35% in compact A's, 7–20% in B1's, and 10% in the Type I, which contains no melilite. The most altered inclusions all have well-developed rims, the least altered inclusions have poorly-developed rims and inclusions with intermediate degrees of alteration can have either well-developed or poorly-developed rims. In all cases, melilite is the principal coarse-grained phase that is altered, suggesting that melilite breakdown is critical to rim formation. Significantly, the one inclusion with no melilite has no rim.

DISCUSSION

Origin of rims

Each of the observations described above gives important clues to the origin of rim sequences on coarse-grained inclusions.

Because the zoned vein in TS34F1 cross-cuts the coarse-grained melilite, formation of the vein post-dates melilite formation. Because the vein extends deep into the interior of the inclusion and is zoned symmetrically about its long axis, the minerals filling it cannot have been mechanically deposited. Rather, they must have been chemically precipitated from a fluid that entered the inclusion along a crack. The secondary nature of the vein, the fine grain size of the phases within it and the abundance of grossular and anorthite suggest that this vein is merely an unusually large example of the veins of secondary alteration products which are found elsewhere in this and other coarse-grained inclusions and which formed by fluid phase alteration of the melilite. In fact, the highly embayed melilite at its interface with the vein and the filling of those embayments with anorthite and grossular from the vein show that the minerals in this particular vein indeed formed at the expense of the melilite, presumably by reaction of melilite with the fluid. The sequence of mineral zones in the vein is almost identical to that in well-developed rim sequences, implying that rims are a product of the secondary alteration process. As Blander and Fuchs (1975) and Wark and Lovering (1977) showed by means of textural arguments, rimming occurred before accretion, rather than in the parent body regolith as suggested by Bunch *et al.* (1980). This implies that secondary alteration of the inclusions was also pre-accretionary because of the connection we have drawn between rimming and secondary alteration. This suggests very strongly that the fluid responsible for these processes was the solar nebular gas, rather than some hydrothermal fluid in the parent body as suggested by Bunch and Chang (1979) and McSween (1979).

An important difference between the vein and rim sequences is that there is no spinel-perovskite band between the primary melilite and the anorthite-bearing layer in the vein, while such a band is ubiquitous in rim sequences on the outsides of coarse-grained inclusions. From this and the fact that the vein cross-cuts the spinel-perovskite band around the outside of TS34F1, we conclude that alteration post-dated formation of the spinel-perovskite band and that the latter was formed in a way unrelated to this alteration process.

The existence of two types of rim sequences on coarse-grained inclusions raises the possibility that some inclusions were rimmed in one gaseous reservoir under one set of conditions, while other inclusions were rimmed in another reservoir under different conditions. The structure of the sinuous inclusion indicates that such a model is not feasible. The sinuous inclusion has a well-developed rim sequence on the melilite-rich side of the SP layer and a poorly-developed rim sequence on the spinel-rich side. Although one could imagine rimming each side of SP at different times and in different nebular reservoirs, severe problems are encountered in sheltering one side while the other is being rimmed and in preventing the first-rimmed side from becoming re-rimmed during deposition of rims on the second side. Rather, the structure of the sinuous inclusion leads to the conclusion that the two different rim sequences resulted from reaction of a single gas with two different materials, melilite on one side of SP and spinel on the other. Extension of this argument to TS34F1 provides a ready explanation for the fact that the vein strongly resembles a well-developed rim sequence, while the rim on the same inclusion is poorly-developed. We propose that the spinel band on the outside of the inclusion reacted with the nebular gas to form a poorly-developed rim, while the melilite bordering the vein reacted with the same gas to produce a sequence of zones resembling a well-developed rim sequence.

The question remains as to why some coarse-grained inclusions have well-developed rim sequences and others have poorly-developed ones, even though all rims are deposited on a spinel-perovskite band. The critical evidence is that well-developed, calcium-rich rims occur on heavily altered inclusions whereas poorly-developed, calcium-poor rims occur on inclusions that are relatively fresh. If the gas was unable to attack the interior of an inclusion to a significant degree, such as was the case in the dense Type B1 spheroids, only a small amount of melilite would be altered and little calcium would be available for rim formation. In this case, the predominant reaction might have been

$$2MgAl_2O_{4(c)} + 5SiO_{(g)} + 4Na_{(g)} + 7H_2O_{(g)} \rightarrow 4NaAlSiO_{4(c)} + Mg_2SiO_{4(c)} + 7H_{2(g)}. \qquad (1)$$
spinel $\qquad\qquad\qquad\qquad\qquad\qquad\qquad$ nepheline $\qquad$ forsterite

If, however, the gas had freer access to the inclusion interior, as it apparently did in the case of the fluffy Type A inclusions, extensive alteration of melilite would make relatively large amounts of calcium available for rim formation. In this case, the predominant reaction may have been

$$5Ca_2Al_{1.6}Mg_{0.2}Si_{1.2}O_{7(c)} + 4SiO_{(g)} + Na_{(g)} + H_{2(g)} \rightarrow$$
melilite, Åk 20

$$2CaAl_2Si_2O_{8(c)} + Ca_3Al_2Si_3O_{12(c)} + NaAlSiO_{4(c)} +$$
anorthite $\qquad$ grossular $\qquad$ nepheline

$$CaMgSi_2O_{6(c)} + 4Ca_{(g)} + Al_{(g)} + H_2O(g). \qquad (2)$$
diopside

Gaseous calcium and aluminum are written as products in reaction (2) for the purposes of simplification. In reality, however, we envision that these elements, once liberated in reaction (2), will be consumed immediately to form such phases as aluminous pyroxene, hedenbergite and andradite in these and the other rim layers by reaction with other incoming components. Also, we have assumed that the typical composition of melilite which undergoes alteration is Åk 20. We emphasize that reactions (1) and (2) probably occur in all inclusions, but that reaction (1) predominates in little-altered inclusions while (2) is dominant in heavily altered ones. Oxidized iron appears to have also played a significant role in rim formation. Its presence in the gas phase was apparently responsible for the FeO contents of olivine and spinel in poorly-developed rims and for formation of hedenbergite and andradite in well-developed rims.

We envision the mechanism of formation of multi-layered rims on coarse-grained inclusions as analogous to that proposed for terrestrial banded calc-silicate deposits, such as those which develop between marbles and pelitic schists in metamorphic rocks (e.g., Thompson, 1975). In the present case, however, the reaction is between the crystalline phases in the interiors of the inclusions and the surrounding nebular gas, rather than between the two different solids in the terrestrial analogy. In both cases, it is a diffusion-controlled process in which all the layers form simultaneously, rather than sequentially, when elements migrate along composition gradients across the contact between two materials of contrasting composition. Calcium is the dominant mobile element in the terrestrial cases and seems to figure prominently in inclusion rims as well. In the case of the inclusions, alteration proceeds by inward diffusion of sodium, silicon and oxidized iron from the nebular gas through the spinel layer, aided by grain boundaries and fractures. Calcium, in turn, diffuses outward where it forms compounds with incoming components. Wark (1981) also suggests that calcium migrated outward and silicon, iron and alkalies inward during alteration. Using the terrestrial analogy, the rim layers are steady-state features formed simultaneously in response to chemical potential gradients across the inclusion/gas interface. Each layer formed in local equilibrium, given the chemical potentials of the components at its particular location. The banded structure is thus an intermediate stage between the initial and final equilibrium states of the inclusion. The phases that form in this way are not necessarily the same as those that would form at

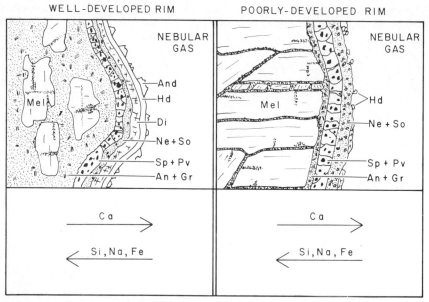

Fig. 4. Schematic diagram illustrating the steady-state diffusion model for rim formation. A highly altered inclusion on the left forms a well-developed rim. A slightly altered inclusion on the right forms a poorly-developed rim. At bottom are shown the directions of diffusion of major components. Abbreviations same as in previous figures.

equilibrium if the entire inclusion reacted completely with the same gas phase. This may explain why many phases observed in the rims are not predicted in thermodynamic models to be equilibrium condensates from a gas of solar composition. We note that element concentrations in the rims do not vary monotonically with distance from the inclusion edge. Although monotonic variations in chemical potentials of diffusing components is the usual case, models of metasomatic zoning do not require monotonic variations of concentrations in the solid assemblage. For an example of this, see Frantz and Mao (1979). Our model is illustrated schematically in Fig. 4.

Origin of the SP layer in the sinuous inclusion

In their study of a fluffy Type A inclusion, Allen *et al.* (1978) showed that the spinel-perovskite band passes through coarse melilite crystals in places and does not simply overlie the melilites as do the other rim layers. This suggests that the spinel-perovskite band formed while melilite crystals were still growing. One possibility is that this band formed during a molten stage in the history of the inclusions and represents the first-crystallizing phases. MacPherson and Grossman (1979) showed that textural features in fluffy Type A inclusions are difficult to reconcile with a liquid origin. Thus, the above explanation encounters problems, as spinel-perovskite bands are found on each individual fragment of some of these irregularly-shaped inclusions, faithfully following even the most convoluted contours in their surfaces. Although we do not yet have an alternative explanation for the origin of these spinel-perovskite bands, it appears that they formed by reaction of the surface of the inclusion with the gas prior to and at higher temperature than the secondary alteration process discussed above.

Models for formation of the SP layer in the sinuous inclusion are strongly constrained by the asymmetrical distribution of melilite within it. Formation by deposition or reaction during condensation is difficult, as such a model would involve protecting one side of a spinel-perovskite sheet while the other side was exposed to the gas. One possibility is that the SP layer began as a spinel-perovskite band on a coarse-grained inclusion and that it was removed from the inclusion prior to deposition of the rim layers. In this case, the melilite would have originated as portions of the coarse-grained interior which adhered to the spinel-perovskite band during removal of the latter. One problem with this idea is that any process energetic enough to remove the spinel rim from an inclusion probably would have shattered and dispersed the rim fragments as well. The length of the sinuous inclusion argues that this did not occur, suggesting that a more gentle process was involved. Removal of the spinel rim would have been much easier had the coarse-grained inclusion been heavily altered, as the alteration is often most extensive immediately beneath the spinel layer. This model thus requires that alteration, and therefore rimming, predated removal of the spinel-perovskite layer. The well-developed rim sequence that now overlies the melilite-rich side of the SP layer in the sinuous inclusion would have had to form *inside* of the original inclusion, beneath the spinel rim. Because no such rim structures have ever been observed to the *inside* of the spinel-perovskite rim on a coarse-grained inclusion, we conclude that this is an unlikely model for the origin of the SP sheet.

The model which we propose for the origin of the SP layer is based on some observations we have made on coarse-grained inclusions, such as those illustrated in Fig. 5. The inclusion in Fig. 5a contains a large number of large, rounded cavities. All of these cavities are lined with delicate needles of wollastonite, implying that the cavities were not produced artificially during sawing of the slab surface visible in the photo. The smooth borders of many of them and the nearly perfectly circular cross-sections of some of them imply that they are not corrosion cavities produced during alteration, but, instead, are vesicles which formed while the inclusion was molten. We thus interpret the large circular re-entrant in the lower right of the figure as a vapor bubble which was frozen in as it broke the surface of the inclusion. In Fig. 5b, we see an example of an inclusion in which many small bubbles presumably coalesced into a single, large, central one. The melt then

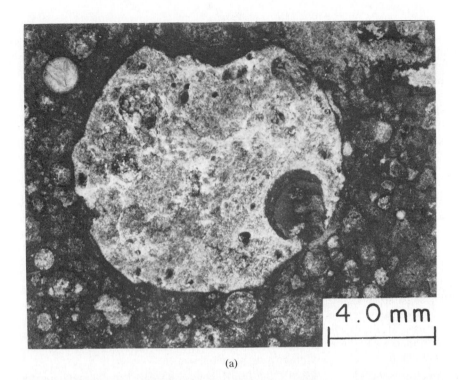

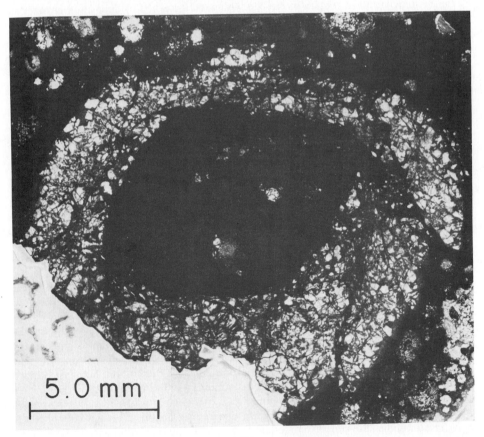

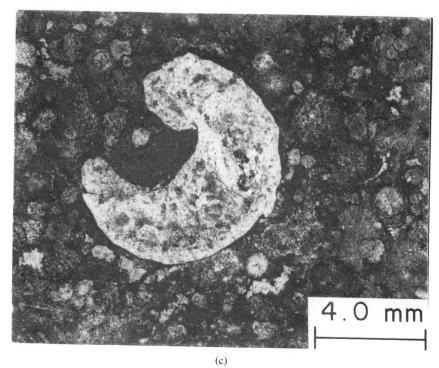

(c)

Fig. 5. (a) Slab surface photograph of a coarse-grained inclusion having many rounded cavities, most of which are lined with wollastonite needles (not visible in photo). The large embayment at lower right is filled with Allende matrix material. (b) Transmitted light photograph (plane polarized light) of a toroidal coarse-grained inclusion. The dark center of the inclusion is filled with Allende matrix. This a Type B2 inclusion containing melilite, fassaite, spinel and anorthite. (c) Slab surface photograph of a crescent-shaped, coarse-grained inclusion.

solidified into a solid spherical shell around it. In Fig. 5c, we see an inclusion which solidified just as a large bubble was escaping, leaving a crescent-shaped inclusion behind.

In our model for SP, the sinuous inclusion was originally a hollow shell such as that in Fig. 5b, but much thinner. The inclusion became rimmed by spinel on its outer surface only, because the nebular gas did not have access to the interior surface. In support of this idea, we note that the shell in that figure has a spinel band on its outer surface only. Fragmentation of the inclusion, perhaps due to collision with another object, exposed both an inner, melilite-rich surface and an outer, spinel-lined surface to later alteration and rimming by the gas. The well-developed rim sequence formed on the melilite-rich side and the poorly-developed one on the spinel-rich side.

Relation of rims to fine-grained inclusions

Fine-grained inclusions are white- to pink-colored objects that are aggregates of 30–50 μm sized bodies (Wark and Lovering, 1977). Each of these tiny bodies is concentrically layered, with cores of spinel + perovskite or, more rarely, hibonite, mantled by successive rims of feldspathoids, aluminous diopside and finally hedenbergite + andradite. Wark and Lovering (1977) pointed out that the zonation of these bodies is the same as the sequence of rim layers on coarse-grained inclusions. Any model for the origin of rims must therefore be able to explain these structures in fine-grained inclusions.

The irregular shapes of many fine-grained inclusions and the loosely-packed textures of their component particles suggest that these inclusions are mechanical aggregates of the tiny spinel-centered bodies. We suggest that each of these bodies originally consisted of a core of spinel or hibonite and a mantle of melilite. Such bodies could have been precursors to the fluffy Type A inclusions. Gas phase alteration of the melilite mantles on these bodies could have produced well-developed rim sequences around the spinel cores. Because no melilite has ever been observed in fine-grained inclusions, it must be postulated that the alteration process resulted in complete obliteration of melilite. Considering both the small sizes of the spinel-centered bodies and the intensity of alteration we have seen in the interiors of large, coarse-grained inclusions, this postulate does not seem unreasonable. An alternative explanation is that fine-grained inclusions are vestiges of nearly completely altered coarse-grained inclusions. Since nothing resembling such spinel-centered bodies has ever been reported, however, even in very heavily altered inclusions, we consider this explanation unlikely.

CONCLUSIONS

We have proposed that rims on coarse-grained inclusions formed in a non-equilibrium process involving those reactions between coarse-grained melilite and spinel and a nebular gas phase that also produced the secondary alteration products in the interiors of these inclusions. Because it was a non-equilibrium process, it is difficult to infer whether the reactions took place in a system of solar composition or under what physical conditions they occurred. It is hoped that information bearing on these questions as well as on the time scale for alteration can be extracted from diffusion models similar to those that have been used to model the terrestrial analogs. Such work is in progress in this laboratory.

Acknowledgments—We thank J. B. Brady, R. N. Clayton, A. M. Davis, R. C. Newton and S. Huebner for helpful discussions. This work was supported by funds from NASA grant NGR 14-001-249 (L.G.), NSF grant EAR-7823420 (L.G.), and NSERC of Canada (J. M. A.).

REFERENCES

Allen J. M., Grossman L., Davis A. M., and Hutcheon I. D. (1978) Mineralogy, textures and mode of formation of a hibonite-bearing Allende inclusion. *Proc. Lunar Planet. Sci. Conf. 9th*, p. 1209–1233.

Allen J. M., Grossman L., Lee T., and Wasserburg G. J. (1980) Mineralogy and petrography of HAL, an isotopically-unusual Allende inclusion. *Geochim. Cosmochim. Acta* **44**, 685–699.

Beckett J. R., MacPherson G. J., and Grossman L. (1980) Major element compositions of coarse-grained Allende inclusions (abstract). In *Meteoritics* **15**, 263.

Blander M. and Fuchs L. H. (1975) Calcium-aluminum-rich inclusions in the Allende meteorite: Evidence for a liquid origin. *Geochim. Cosmochim. Acta* **39**, 1605–1619.

Bunch T. E. and Chang S. (1979) Thermal metamorphism (shock?) and hydrothermal alteration in C3V meteorites (abstract). In *Lunar and Planetary Science X*, p. 164–166. Lunar and Planetary Institute, Houston.

Bunch T. E., Chang S., and Ott U. (1980) Regolith origin for Allende meteorite (abstract). In *Lunar and Planetary Science XI*, p. 119–121. Lunar and Planetary Institute, Houston.

Clayton R. N., Onuma N., Grossman L., and Mayeda T. K. (1977) Distribution of the pre-solar component in Allende and other carbonaceous chondrites. *Earth Planet. Sci. Lett.* **34**, 209–224.

Davis A. M., Allen J. M., Tanaka T., Grossman L., and MacPherson G. J. (1980) A sinuous inclusion from Allende: Trace element analysis of a rim (abstract). In *Meteoritics* **15**, 279–280.

Frantz J. D. and Mao H. K. (1979) Bimetasomatism resulting from intergranular diffusion: II. Prediction of multimineralic zone sequences. *Amer. J. Sci.* **279**, 302–323.

Grossman L. (1975) Petrography and mineral chemistry of Ca-rich inclusions in the Allende meteorite. *Geochim. Cosmochim. Acta* **39**, 433–454.

Grossman L. (1980) Refractory inclusions in the Allende meteorite. *Ann. Rev. Earth Planet. Sci.* **8**, 559–608.

Haggerty S. E. and Merkel G. A. (1976) Primordial condensation in the early solar system: Allende mineral chemistry (abstract). In *Lunar Science VII*, p. 342–344. The Lunar Science Institute, Houston.

Keil K. and Fuchs L. H. (1971) Hibonite [$Ca_2(Al,Ti)_{24}O_{38}$] from the Leoville and Allende chondritic meteorites. *Earth Planet. Sci. Lett.* **12**, 184–190.

Macdougall J. D. (1979) Refractory-element-rich inclusions in CM meteorites. *Earth Planet. Sci. Lett.* **42**, 1–6.

MacPherson G. J. and Grossman L. (1979) Melted and non-melted coarse-grained Ca-, Al-rich inclusions in Allende (abstract). In *Meteoritics* **14**, 479–480.

MacPherson G. J. and Grossman L. (1981) Clastic rims on inclusions: Clues to the accretion of the Allende parent body (abstract). In *Lunar and Planetary Science XII*, p. 646–647. Lunar and Planetary Institute, Houston.

McSween H. Y. Jr. (1979) Are carbonaceous chondrites primitive or processed? A review. *Rev. Geophys. Space Phys.* **17**, 1059–1078.

Thompson A. B. (1975) Calc-silicate diffusion zones between marble and pelitic schist. *J. Petrol.* **16**, 314–346.

Wark D. A. (1981) The pre-alteration compositions of Allende Ca-Al-rich condensates (abstract). In *Lunar and Planetary Science XII*, p. 1148–1150. Lunar and Planetary Institute, Houston.

Wark D. A. and Lovering J. F. (1977) Marker events in the early evolution of the solar system: Evidence from rims on Ca-Al-rich inclusions in carbonaceous chondrites. *Proc. Lunar Sci. Conf. 8th*, p. 95–112.

Wark D. A., Wasserburg G. J., and Lovering J. F. (1979) Structural features of some Allende coarse-grained Ca-Al-rich inclusions: Chondrules within chondrules? (abstract). In *Lunar and Planetary Science X*, p. 1292–1294. Lunar and Planetary Institute, Houston.

Petrogenesis of light and dark portions of the Leighton gas-rich chondritic breccia

Harry Y. McSween, Jr.[1], S. Biswas[2,3], and Michael E. Lipschutz[2]

[1]Department of Geological Sciences, University of Tennessee, Knoxville, Tennessee 37916 [2]Department of Chemistry, Purdue University, West Lafayette, Indiana 47907 [3]Duke Power Co., Charlotte, North Carolina 28242

Abstract—Individual dark fractions of the Leighton chondritic breccia are all enriched in volatile/mobile trace elements. Concentrations of the most highly volatile/mobile elements correlate with degree of equilibration, as expressed by compositional variability for silicates. Individual light clasts in the same meteorite are equilibrated and generally volatile-poor, except for enrichments in Rb and Cs. These alkali elements may have been redistributed by feldspar formation during metamorphism. Enrichments of other trace elements in dark portions may have been caused by mobilization during metamorphism in the parent body interior and deposition in the cooler surface layers. An alternative model involves accretional mixing of a diverse compositional population of primitive H group materials to form dark materials. Subsequent integration of light and dark materials was accomplished by impacts into a regolith.

INTRODUCTION

A significant number of ordinary chondrites are regolith breccias composed of light and dark colored portions with different trace element compositions. In an earlier paper (McSween and Lipschutz, 1980), we studied the petrographic and chemical characteristics of a typical pair of light and dark samples from each of a number of brecciated H chondrites. This paper will explore the range of petrographic and chemical variations among a larger number of light and dark fractions of the *same* meteorite, Leighton. These data will be used to test models for the petrogenesis of chondrites with light/dark structures.

SAMPLING AND ANALYSIS PROCEDURES

Light and dark portions of Leighton were selected for study in the main mass (507g) obtained from the Field Museum of Natural History (sample no. Me 768). This specimen has a cut surface (here called the front face) of approximately 40 cm^2 surface area from which most specimens were taken (Fig. 1). The back face is a fractured surface. The cut surface was faced with a carbide tool, and individual portions of dark and light material were chipped out. Samples were identified by letters F or B (indicating whether the portion was extracted from the front or back face of the specimen), followed by L or D (noting the light or dark coloration of that portion), plus an identifying number, as in sample FD3. The light and dark portions of Leighton studied in our previous survey (McSween and Lipschutz, 1980) were numbered FD0 and FL0 and are included here for comparison.

Polished thin sections were prepared from each light and dark portion of Leighton removed from the main mass. Many sections were small, with surface areas ranging from 10 to 40 mm^2. Several sections were examined for abundant amounts of feldspar using a Nuclide luminoscope. Compositions of olivine, pyroxene, and metal were determined using a MAC Model 400S automated electron microprobe with appropriate synthetic and natural standards, and data were reduced using empirical corrections by Ziebold and Ogilvie (1964) for metal and Bence and Albee (1968) and Albee and Ray (1970) for silicates. All metal grains larger than about 10 μ in the sections were analyzed. Olivine and pyroxene grains were selected for analysis using a grid pattern. Mean compositions and standard deviations for metals and silicates are presented in Table 1. The more commonly used percent mean deviations (PMD) for silicates, as defined by Dodd et al. (1967), are shown in Fig. 2.

Neutron activation analyses of 12 volatile/mobile trace elements in aliquots of the same light and dark portions of Leighton were performed using procedures outlined by Bart and Lipschutz (1979).

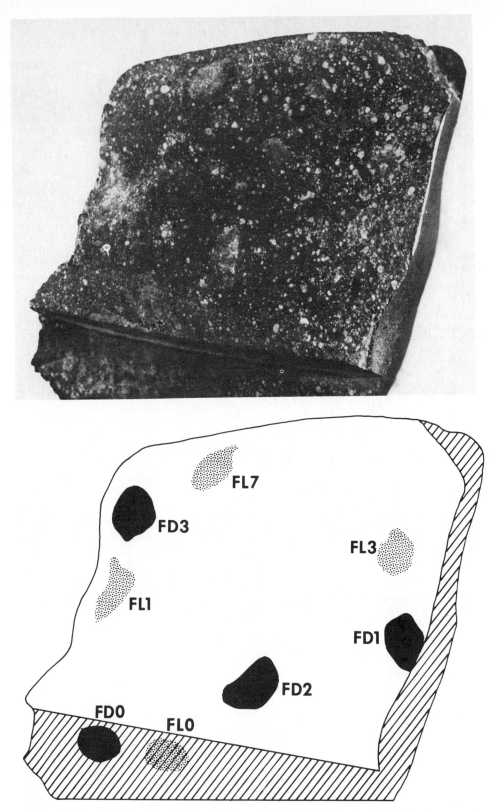

Fig. 1. Photograph of the cut surface (front face) of the Leighton chondritic breccia, measuring 7.0 × 5.6 cm. Sketch illustrates locations of light and dark portions of this meteorite removed for petrographic and chemical studies. The locations of FD0 and FL0 are not exact, as they were taken earlier from a wedge now cut from the front face.

Here, we are concerned primarily with possible correlations between petrographic and chemical parameters, and we present a subset of the chemical data only in graphical form. The complete chemical data and their further significance will be reported in another paper (in preparation).

Table 1. Mean compositions and variability for olivine, pyroxene, and metals in the Leighton chondrite. (Numbers in parentheses indicate standard deviations.) [Numbers in brackets indicate how many analyses were averaged.]

Dark portions	FD0	FD1	BD1	FD2	FD3
olivine $\overline{Fa}$	16.8(3.85)[89]	19.3(.97)[32]	15.9(4.70)[70]	19.9(6.65)[52]	16.5(7.95)[60]
pyroxene $\overline{Fs}$	14.9(4.12)[56]	16.7(2.53)[30]	13.2(5.22)[42]	12.9(6.54)[46]	13.7(6.46)[55]
kamcite $\overline{Ni}$	6.24(1.26)[31]	5.88(1.10)[27]	5.09(1.28)[40]	5.10(1.13)[21]	4.70(1.65)[29]
kamacite $\overline{Co}$	0.48(.11)	0.45(.09)	0.49(.05)	0.52(.13)	0.48(.11)
taenite $\overline{Ni}$	34.9(11.0)[16]	29.1(8.9)[14]	37.1(2.7)[16]	33.0(9.6)[10]	27.7(7.6)[16]
taenite $\overline{Co}$	0.16(.12)	0.09(.07)	0.03(.02)	0.11(.07)	0.10(.06)

Light portions	FL0	BL1	FL1	FL3	FL7
olivine $\overline{Fa}$	18.1(.52)[64]	16.4(1.66)[68]	18.8(.52)[43]	18.8(.66)[32]	19.0(.41)[26]
pyroxene $\overline{Fs}$	16.9(.98)[72]	16.4(2.25)[20]	16.2(.54)[42]	17.1(.64)[20]	17.3(.49)[24]
kamcite $\overline{Ni}$	7.37(.28)[25]	7.30(.30)[12]	7.41(.31)[20]	7.37(.24)[15]	7.39(.32)[10]
kamacite $\overline{Co}$	0.48(.09)	0.43(.08)	0.45(.09)	0.50(.09)	0.46(.11)
taenite $\overline{Ni}$	34.2(2.8)[14]	33.9(2.7)[10]	34.1(2.0)[14]	34.0(2.7)[10]	34.4(3.0)[7]
taenite $\overline{Co}$	0.11(.02)	0.10(.02)	0.08(.02)	0.07(.02)	0.06(.02)

PETROGRAPHY OF THE DARK PORTIONS

Previous descriptions of light/dark chondrites have generally alluded to light clasts set in a dark matrix (Noonan and Nelen, 1976; McSween and Lipschutz, 1980; Keil and Fodor, 1980). Although this is the most common textural relationship, there are clasts of dark material as well as dark areas with indistinct boundaries that could be either clasts or matrix concentrations. The dark materials studied here are mostly of the last type, having obscure contact relations with the surrounding material. The dark sample of Leighton from our previous survey of chondrite breccias, numbered FD0, is a sample of dark matrix. The term "matrix" in this sense refers to any fine-grained material enclosing clasts, and it does not appear to be the same aphanitic matrix that comprises up to 15 vol% of normal type 3 ordinary chondrites (Huss et al., 1981).

All dark portions contain clearly defined chondrules and abraded lithic fragments without evidence of textural integration. Glass is common in many of the chondrules. These composite objects, along with large mineral fragments, are enclosed in a crystalline groundmass whose fine grain-size produces the dark appearance. Important minerals in the dark fractions are olivine, low-calcium pyroxene, metal, and troilite, with minor high-calcium pyroxene, feldspar, apatite, and chromite.

The compositional ranges of olivine, pyroxene, and metal overlap those of unequilibrated H chondrites (McSween and Lipschutz, 1980), indicating possible affinity of the dark portions with this meteorite group. Most silicate grains are compositionally homogeneous, but different grains in the same sample may vary in composition. Compositional scatter among olivines and pyroxenes (as measured by PMD, Table 1) ranges from appreciable, as in FD3, to slight, as in FD1 (Fig. 2). Individual dark portions of Leighton may most likely be classified as petrologic types 3 and 4. We previously noted (McSween and Lipschutz, 1980) that silicate composition histograms for dark *matrix* portions could be interpreted as peaks of equilibrated silicate compositions superimposed on unequilibrated (scattered) distributions, as in FD0 (Fig. 2). This pattern might arise from admixture of pulverized light material with dark; however, the other dark portions

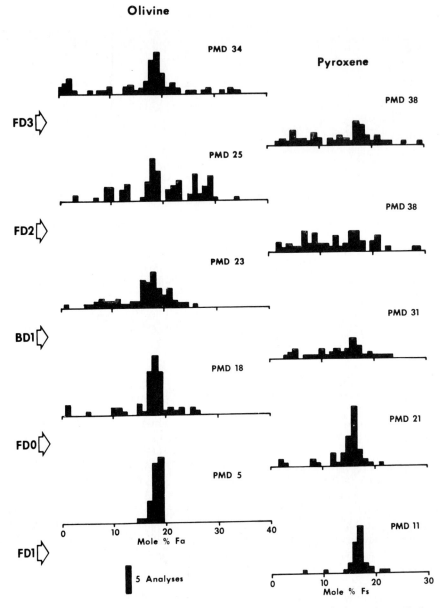

Fig. 2. Histograms of olivine and pyroxene compositions in 5 dark portions of the Leighton chondrite. These are ordered from top to bottom by degree of equilibration.

in this study do not exhibit this feature so clearly and may be less contaminated than dark matrix.

All dark portions contain relatively small grains of kamacite, taenite, and tetrataenite. Compositions of metal phases in terms of Co and Ni are scattered (Fig. 3), and standard deviations do not correlate with those of silicates (Table 1). The ranges of metal compositions presumably were established when individual metal grains cooled prior to compaction in a regolith, as has been determined for several other chondritic breccias (Scott and Rajan, 1980). That post-compaction reheating did not occur is also suggested by the presence of solar flare tracks and solar wind gases in Leighton (Wilkening, 1976; Schultz and Kruse, 1978). These features presumably were produced in a regolith and would have been erased or lost during any subsequent heating.

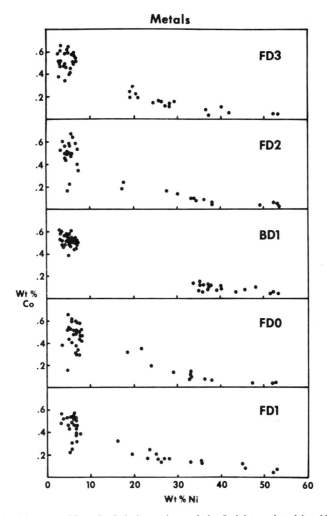

Fig. 3. Metal compositions in 5 dark portions of the Leighton chondrite. Ni and Co contents of kamacite and taenite are variable; metal compositions with slightly greater than 50 wt. % Ni are tetrataenite.

PETROGRAPHY OF THE LIGHT PORTIONS

Samples of light material occur as clearly visible, subangular clasts up to 17 mm apparent diameter within the Leighton meteorite. Most of these show coarse recrystallized textures in which chondrule outlines have been largely obliterated. However, one clast studied (FL7) has a finer clastic texture produced by severe brecciation, and several other clasts show similar effects though to lesser degrees. The mineralogy of light clasts is similar to dark portions except for the growth of feldspar at the expense of chondrule glass. Metal grains are significantly larger in light fractions than in dark, presumably a result of metamorphic recrystalization.

Silicate and metal compositions are the same as those of other equilibrated H group chondrites (McSween and Lipschutz, 1980). Olivines and pyroxenes in all light clasts show almost no compositional variability, and standard deviations of metal compositions are small (Table 1). Graphical representations of the compositions of silicates and metals in the light clasts are very similar to those already published for the light fraction FL0 of Leighton (Figs. 1 and 3 of McSween and Lipschutz, 1980). Light clasts are classified as H5-6 chondrite material.

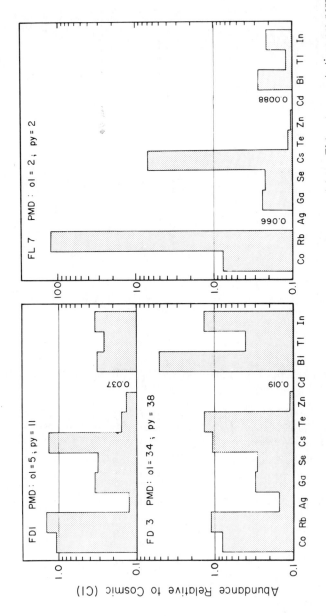

Fig. 4. Typical volatile element patterns in dark (FD1 and FD3) and light (FL7) portions of Leighton. Element concentrations are normalized to C1 (cosmic) values and are plotted from left to right in order of increasing nebular volatility.

CHEMICAL COMPARISON OF DARK AND LIGHT PORTIONS

One of the unsolved problems in chondritic breccias is understanding the origin of the anomalous concentrations of volatile elements contained in the various light and dark fractions. We have analyzed a suite of volatile trace elements in the separated portions of Leighton in order to determine if variations in these elements are related to petrographic characteristics.

Three typical volatile element patterns are shown in Fig. 4. Abundances normalized to C1 chondrite (cosmic) values are plotted in order of increasing nebular volatility. The figure includes data from the most equilibrated (FD1) and least equilibrated (FD3) dark portions and a light portion (FL7). The dark portions are compositionally variable and volatile-rich; the light portions also vary in composition but are generally volatile-poor. However, in both dark and light portions certain volatile elements may exceed cosmic

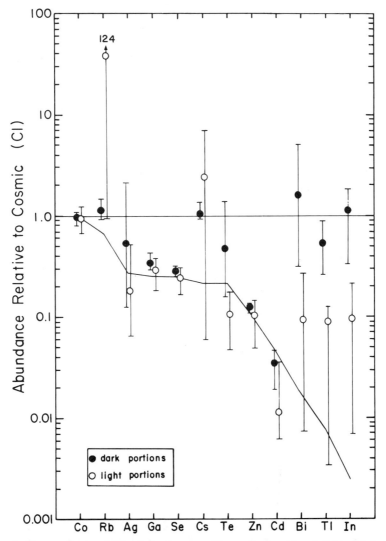

Fig. 5. Cl-normalized volatile element compositions of all analyzed light and dark portions of Leighton. Average values for light and dark portions are circles, and vertical bars indicate ranges. The jagged diagonal line represents the abundances in unequilibrated H group chondrites (Binz et al., 1976).

values by significant amounts, e.g., Rb and Cs in FL7 and Bi, In, and Te in FD3. The ranges and average values for concentrations of these elements in five light and five dark portions of Leighton are illustrated in Fig. 5. The diagonal line represents the average values in H3 chondrites (Binz et al., 1976).

Several patterns emerge from these chemical data on light and dark portions. Light portions are consistently enriched over dark in the alkali elements Rb and Cs. Dark portions are consistently enriched over light in the highly volatile elements Bi, In, Te, and Tl. These same four volatile elements are highly correlated with degree of equilibration, as expressed by PMD values for olivine and pyroxene. In all cases these correlations are at the 98 percent confidence level or greater. Plots of Bi, Tl, and In values versus PMD for olivine and pyroxene are illustrated in Fig. 6.

PETROGENESIS OF DARK AND LIGHT PORTIONS

Petrographic data indicate that both dark and light fractions of Leighton are of H chondrite parentage. However, volatile trace element concentrations in dark and light fractions of this meteorite (and other H chondrite breccias) are clearly not the same as those of other unequilibrated and equilibrated H chondrites. It has been recognized for some time that dark portions of chondritic breccias may be enriched in certain volatile elements (e.g., Rieder and Wänke, 1969); this study has demonstrated that light portions may also be enriched in certain elements to supercosmic levels (Fig. 4). Previous suggestions that volatile element concentrations in dark portions reflect admixture of carbonaceous chondrite (Suess et al., 1964; Müller and Zähringer, 1966; Wilkening, 1976) appear to be untenable because of the mismatch between volatile element enrichment patterns in dark portions and in carbonaceous chondrites (Rieder and Wänke, 1969; Bart and Lipschutz, 1979). In fact, chemical variations among dark portions of Leighton found

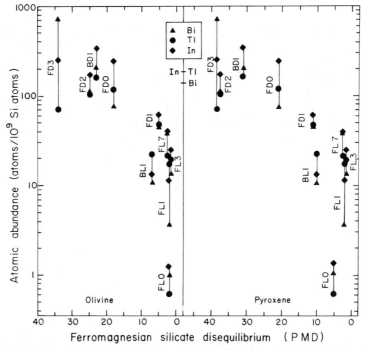

Fig. 6. Bi, In, and Ti abundances versus PMD of olivine and of pyroxene in 10 Leighton samples. Some abundances exceed Cl values indicated on the vertical center line.

in this study preclude an origin by mixing of a small number of components with *any* fixed compositions.

The marked enrichments of Rb and Cs in light clasts suggest that the distribution of these elements may have been affected by thermal metamorphism. One of the characteristics of chondrite equilibration is the progressive development of feldspar, primarily from chondrule glasses. For H group chondrites, these are alkali-rich feldspars with compositons near $Ab_{82}An_{12}Or_6$ (Van Schmus and Ribbe, 1968). Other alkali metals such as Rb and Cs are almost certainly sited in feldspars; thus, anomalous values for these elements in equilibrated light clasts could reflect different proportions of feldspar. Presumably feldspars were distributed inhomogeneously during recrystallization, so that some small clasts may have enrichments in alkalis and others may have depletions. Unfortunately, no clear increase in modal abundance of feldspar can be seen optically in those light clasts with the highest Rb and Cs contents. One end of the FL7 sample (with extreme Rb and Cs values) exhibits a blue cathodoluminescence that might be attributable to finely granulated feldspars. However, estimates of feldspar abundance in chondrites are notoriously inaccurate due to the fine grain size (Van Schmus and Ribbe, 1968), and the problem is exacerbated in light clasts (particularly FL7) because of overprinting of brecciation effects.

It is conceivable that the volatile element patterns in the dark portions of Leighton may also reflect a metamorphic process. Dreibus and Wänke (1980) suggested that certain mobile elements could have emanated from the chondrite parent body interior during metamorphism, ultimately deposited in the cooler exterior region of the parent body. Materials from these exterior regions would then be compacted into the material now recovered as dark portions of gas-rich meteorites.

To test this model we have examined patterns of the most mobile elements in these chondrites. It is important in this connection to distinguish between elemental *volatility* which governs vapor-to-solid deposition during nebular condensation and *mobility* which governs vaporization from the solid during thermal episodes in the parent body (cf. Ngo and Lipschutz, 1980). Relative mobilities are estimated from trace element retention in samples heated at some fixed temperature relative to unheated material; however, the sequence is affected by choice of the reference temperature and parent material. If we assume 1000°C as the temperature of the putative source region of the H-group parent from which the trace elements evolved and a thermal response similar to that of the H3 chondrite Tieschitz, the twelve elements studied here would be listed in order of increasing mobility as Co, Ga, Se, Rb, Cs, Zn, Ag, Te, In Tl, Bi, Cd (Ikramuddin *et al.*, 1977). (Ikramuddin *et al.* (1977) did not determine Rb or Cd but other studies (cf. Ngo and Lipschutz, 1980) indicate Rb and Cs to be of about equal mobility and Cd to be the most mobile trace element in representative chondrites.) In the general case, we prefer a somewhat different order—Co, Se, Ga, Rb, Cs, Te, Bi, In, Ag, Zn, Tl, Cd—based upon mean retentivities in five chondrites at 1000°C (Walsh and Lipschutz, in preparation), but since we are considering specifically the H chondrite parent body, the sequence derived from the Tieschitz H3 chondrite may be more appropriate.

If the model of Wänke and Dreibus (1980) is valid, the greater an element's mobility, the greater its probability for enrichment in the dark portions. In fact, four of the most mobile elements—Te, In, Tl, and Bi—are significantly enriched in the dark portions relative to the light. Furthermore, if silicate PMD values were established during metamorphism and reflect depth in the parent body, a correlation between these values and elemental concentrations might be expected; Te, In, Tl, and Bi exhibit such correlations. The failure of Cd to follow either trend can be explained in an *ad hoc* manner by postulating that temperatures in the sink regions were too high to allow highly mobile Cd to condense. This cannot explain the cosmic Cs and Rb levels in the dark portions, their high and variable abundances in the light portions of Leighton and, indeed, all other gas-rich chondrites (cf. Bart and Lipschutz, 1979), or the H-chondrite-like abundances of mobile Ag, Zn and Cd. Furthermore, we would not expect the observed Bi, In, or Tl abundances in light portions generally exceeding those of H3 chondrites (Fig. 5).

Chou et al. (1981) have suggested that high volatile element contents of the dark matrix portions of chondritic breccias were produced by localized impacts in a regolith. Redistribution of elements by this mechanism would be erratic for impacts of low intensity; high intensity impacts devolatilize chondritic matter (Walsh and Lipschutz, in preparation). Under the time-temperature conditions necessary to alter olivine and pyroxene compositions, chondritic matter would not retain mobile elements at levels present in Leighton (Ngo and Lipschutz, 1980). Furthermore, a correlation between highly mobile elements and silicate PMD would not be expected to arise from shock heating.

Bart and Lipschutz (1979) have previously suggested another alternative, that both light and dark portions represent primitive H-group parent material more compositionally diverse than that sampled by existing H3-6 chondrites. According to this model, dark material condensed from the nebula and accreted at generally lower temperatures than those for light. Except for this generalization (and its consequences, i.e. enrichment of the most volatile elements—Bi, In and Tl—and their correlation with PMD in dark material), this suggestion is predictively unsatisfying. It requires components of no fixed composition (or the equivalent, a very large number of components of fixed composition), and it is a very difficult model to test. Since trace element patterns would be primary, one would expect abundances of elements less volatile than Bi to accord with those of H-chondrites, and most do (Fig. 5). High (and variable) abundances of Cs and Rb are not accounted for by this suggestion but they can be accomodated by an *ad hoc* assumption of localized alkali metal enrichments.

CONCLUSIONS

Chondritic breccias consist primarily of equilibrated light and unequilibrated dark fractions of the same chemical group. Enrichments of alkali metals in light clasts of brecciated chondrites may result from inhomogeneous redistribution of feldspar components during thermal metamorphism in the parent body interior. Enrichments of other volatile trace elements in dark portions cannot result from mixing of a small number of petrologic components of fixed compositions. A more complex mixing model involving sampling of a diverse compositional population is possible. Another possible model involves metamorphic mobilization and deposition of these elements in the upper layers of the parent body. Additional studies may allow the evaluation of these two models in more detail.

Acknowledgments—We are grateful to Dr. E. Olsen for providing samples of Leighton for our study. Dr. O. C. Kopp kindly provided access to the luminoscope. Constructive reviews were provided by Dr. G. J. Taylor and Dr. D. W. Sears. This work was supported by the National Aeronautics and Space Administration under grants NSG 7413 to HYM and NGL 15-005-140 to MEL.

REFERENCES

Albee A. and Ray L. (1970) Correction factors of electron probe microanalysis of silicates, oxides, carbonates, phosphates and sulfates. *Anal. Chem.* **42**, 1408–1414.
Bart G. and Lipschutz M. E. (1979) On volatile element trends in gas rich meteorites. *Geochim. Cosmochim. Acta* **43**, 1499–1504.
Bence A. E. and Albee A. (1968) Empirical correction factors of the electron microanalysis of silicates and oxides. *J. Geol.* **76**, 382–403.
Binz C. M., Ikramuddin M., Rey P., and Lipschutz M. E. (1976) Trace elements and interelement relationships in unequilibrated ordinary chondrites. *Geochim. Cosmochim. Acta* **40**, 59–71.
Chou C. L., Sears D. W., and Wasson J. T. (1981) Composition and classification of clasts in the St. Mesmin LL chondrite breccia. *Earth Planet. Sci. Lett.* In press.
Dodd R. T., Van Schmus W. R., and Koffman D. M. (1967) A survey of the unequilibrated ordinary chondrites. *Geochim. Cosmochim. Acta* **31**, 921–951.
Dreibus, G. and Wänke H. (1980) On the origin of the excess of volatile trace elements in the dark portion of gas-rich chondrites (abstract). *Meteoritics* **15**, 284–285.
Huss G. R., Keil K., and Taylor G. J. (1981) The matrices of unequilibrated ordinary chondrites: implications for the origin and history of chondrites. *Geochim. Cosmochim. Acta* **45**, 33–52.

Ikramuddin M., Matza S., and Lipschutz M. E. (1977) Thermal metamorphism of primitive meteorites—V. Ten trace elements in Tieschitz H3 chondrite heated at 400–1000°C. *Geochim. Cosmochim. Acta* **41**, 1247–1256.

Keil K. and Fodor R. B. (1980) Origin and history of the polymict-brecciated Tysnes Island chondrite and its carbonaceous and noncarbonaceous lithic fragments. *Chem. Erde* **39**, 1–26.

McSween H. Y. and Lipschutz M. E. (1980) Origin of volatile-rich H chodrites with light/dark structures. *Proc. Lunar Planet. Sci. Conf. 11th*, p. 853–864.

Müller O. and Zähringer J. (1966) Chemische Unterschiede bei Uredelgashaltigen Steinmeteoriten. *Earth Planet. Sci. Lett.* **1**, 25–29.

Ngo H. T. and Lipschutz M. E. (1980) Thermal metamorphism of primitive meteorites—X. Additional trace elements in Allende (C3V) heated to 1400°C. *Geochim. Cosmochim. Acta* **44**, 731–739.

Noonan A. F. and Nelen J. A. (1976) A petrographic and mineral chemistry study of the Weston, Connecticut, chondrite (abstract). *Meteoritics* **11**, 111–130.

Rieder R. and Wänke H. (1969) Study of trace element abundances in meteorites by neutron activation. In *Meteorite Research* (P. M. Millman, ed), p. 75–86. Reidel. Dordrecht, Netherlands.

Schultz L. and Kruse H. (1978) Light noble gases in stony meteorites—a compilation. *Nucl. Track Detection* **2**, 65–103.

Scott E. R. D. and Rajan R. S. (1980) Thermal history of some xenolithic ordinary chondrites (abstract). In *Lunar and Planetary Science XI*, p. 1015–1017. Lunar and Planetary Institute, Houston.

Suess H. E., Wänke H., and Wlotzka F. (1964) On the origin of gas-rich meteorites. *Geochim. Cosmochim. Acta* **28**, 595–607.

Van Schmus W. R. and Ribbe P. H. (1968) The composition and structural state of feldspar from chondritic meteorites. *Geochim. Cosmochim. Acta* **32**, 1327–1342.

Wilkening L. L. (1976) Carbonaceous chondritic xenoliths and planetary-type noble gases in gas-rich meteorites. *Proc. Lunar Sci. Conf. 7th.*, p. 3549–3559.

Ziebold T. O. and Ogilvie R. E. (1964) An empirical method for electron microanalysis. *Anal. Chem.* **36**, 322–327.

Mineralogical aspects of terrestrial weathering effects in chondrites from Allan Hills, Antarctica

James L. Gooding*

Earth and Space Sciences Division, Jet Propulsion Laboratory, California Institute of Technology, Pasadena, California 91109

Abstract—Mineralogical effects of terrestrial weathering in two H6 and four L6 chondrites from Allan Hills, Antarctica are similar to those in weathered specimens of the Holbrook, Arizona (L6) chondrite and imply formation in the presence of liquid water. The dominant weathering products were formed by alteration of Ni–Fe metal and sulfide and consist of complex, multiple-phase, hydrous ferric oxides which contain ~75–82 wt. % Fe_2O_3, ~1–10% NiO, ~0.5–8% SO_3, ~0.2–4% Cl, traces of Na, Mg, and Ca, and an inferred total complement of ~10–15% H_2O (and CO_2), as determined by electron microprobe analyses. In the most intensely weathered chondrites such as ALHA76008, Fe-oxides may comprise ~15–20% of some samples although combined data from x-ray diffractometry and differential thermal analysis strongly suggest that ≤5% goethite [or other FeO(OH) polymorph] is present. Thus, ~2/3 of the Fe-oxides may be amorphous or otherwise not identifiable as well-crystallized minerals. Severity of weathering shows no obvious correlation with terrestrial residence age among the ALHA samples.

INTRODUCTION

Field and laboratory observations of meteorites recovered from Antarctica have shown that many of the specimens are significantly weathered (Olsen *et al.*, 1978; King *et al.*, 1980). Consequently, detailed knowledge of the processes and products of Antarctic meteorite weathering is required in order to analytically reconstruct meteorite "finds" and to quantitatively assess the degree of preservation of rare or unusual specimens which might later become subjects for intense study of their pre-terrestrial properties. Furthermore, study of weathered Antarctic stony meteorites might provide insight into possible processes by which mafic and ultramafic extraterrestrial igneous rocks decompose in cold and relatively dry environments which may be analogous to those on the surface of Mars. Accordingly, a study of six weathered ordinary chondrites from Allan Hills, Antarctica was performed with emphasis placed on a survey of bulk mineralogical features which might be correlated with previously observed (Biswas *et al.*, 1980; Gibson and Andrawes, 1980) geochemical effects of Antarctic weathering.

Previous mineralogical studies of weathering effects in meteorites other than those from Antarctica (e.g., Buddhue, 1957; Ramdohr, 1963, 1973) have emphasized identification of individual minerals which comprise the weathering products. Ramdohr (1963, 1973), in particular, has successfully applied microscopic techniques (especially reflected-light microscopy of polished sections) in identifying a variety of secondary (weathering product) minerals, some of which occur only in trace amounts. However, the problem of quantitatively assessing relative degrees of weathering by application of instrumental methods of mineralogical analysis to bulk samples of meteorites has not been as actively pursued. Furthermore, the direct mineralogical comparison of weathering effects in Antarctic meteorites with those in petrologically similar meteorites recovered from other environments has not been made.

The work reported here represents a first attempt at comparing the mineralogical aspects of weathering effects in selected Antarctic chondrites with those present in a

---

*Current address: Code SN2, NASA Johnson Space Center, Houston, Texas 77058.

Table 1. Allan Hills, Antarctica chondrites studied in the present investigation.

Sample No.	Chemical-Petrologic Type[a]	Mass (g) of Parent Stone[b]	Weathering Index[a]	Fracture Index[a]	Terrestrial Residence Age (10^3 y)
ALHA76006, 3	H6	1137	C	B	>32[c], <200[d,e]
ALHA76008, 6	H6	1150	B/C	B	>32[c], 1540[e]
ALHA77155, 13	L6	305	A/B	A	<400[d]
ALHA77270, 15	L6	589	A/B	B	—
ALHA77272, 31	L6	674	B/C	B	>34[c], <700[d]
ALHA77296, 9	L6	963	A/B	A	—

[a] Score et al. (1981)
[b] King et al. (1980)
[c] based on ^{14}C, Fireman (1979)
[d] based on ^{26}Al, Evans and Rancitelli (1980)
[e] based on ^{26}Al, Fireman et al. (1979)

petrologically similar chondrite which was weathered in the Arizona desert. The methods of analysis employed include some (x-ray diffractometry, differential thermal analysis, reflectance spectrophotometry) of well-known utility (e.g., Zussman, 1977) in the examination of limited quantities of polymineralic bulk samples. Principal findings include the remarkable similarity of bulk weathering effects in the Antarctic and Arizonan samples, despite ostensibly major differences in their respective weathering environments, and evidence for the apparently dominant role of liquid water in weathering of the Antarctic samples. However, these results are not presented as a comprehensive treatment of weathering effects in Antarctic or other meteorites. Instead, they are offered as a survey of effects which deserve further specialized study.

SAMPLE SELECTION

Documented chips (Table 1) were prepared by and received directly from the curatorial staff of the NASA Johnson Space Center. All subsequent sample handling, preparation, and analysis was performed by the author.

Sample selection deliberately favored large parent stones which, in general, should more faithfully preserve Antarctic weathering trends (determinable by comparing surface and interior portions) than might smaller stones (of only a few centimeters or grams in size) which could conceivably be weathered throughout. Samples from the surfaces (outer ~1–2 cm) of the respective parent stones were preferentially selected because they were expected to possess the most pronounced and, hence, most readily analyzable weathering effects. Equilibrated (petrologic type 6) ordinary chondrites were chosen because of their greater abundance relative to achondrites. Unequilibrated chondrites were avoided in order to minimize complexities in weathering behavior which would be expected for their components of glass and carbonaceous material, etc. Finally, two H- and four L-group chondrites were selected in order to facilitate a comparison of weathering effects as a function of initial metal content of the rocks.

In order to compare the weathered Antarctic chondrites with other chondrites of well-known terrestrial residence age, samples of the Holbrook, Arizona L6 chondrite were studied in parallel with the Allan Hills samples. The Holbrook shower occurred in 1912 and weathered stones were subsequently recovered in 1931 and 1968, thereby providing a natural record of weathering trends in an ordinary chondrite (Gibson and Bogard, 1978). Chips comparable to the Allan Hills specimens in size and parent stone surface proximity were selected from Holbrook-1912 (American Museum of Natural History #4351), -1931 (Arizona State University #H.52), and -1968 (E. K. Gibson, Jr., NASA Johnson Space Center) stones and were subsequently handled and analyzed by methods identical to those used for the Antarctic specimens.

ANALYTICAL METHODS

Each chondrite chip of ~5 g represented the outer ~10–20 mm (including fusion crust) portion of its respective parent stone and, after stereo-microscopic examination, was sliced using a diamond wafer-blade saw with dry (<0.2% water) isopropanol as coolant. Following reflectance spectrophotometric scans (see below) of its two sides, each slice was prepared as a diamond-polished thin-section using either dry isopropanol or light oil as the polishing lubricant. Each section was then cleaned ultrasonically in dry isopropanol. Thus, exposure of the specimens to liquid water was carefully avoided during sample handling and thin-section preparation.

A whole-rock split (~0.6–1 g) of the surface (outer 5–10 mm) portion of each chondrite chip was individually crushed in a compressed sapphire mortar and pestle and then pulverized in air to a fine powder by ten minutes of agitation in a WIG-L-BUG automatic stainless steel ball mill. The resultant powders were studied by x-ray diffractometry (XRD), visible and near-infrared (VIS/NIR) reflectance spectrophotometry, and differential thermal analysis (DTA).

XRD scans were made at 1°/min over the range $2\theta = 4$–76° using Cu-Kα radiation and a Philips/Norelco automatic diffractometer equipped with a scintillation detector, graphite crystal monochromator, 0.2° receiving slit, and solid-state single-channel analyzer and rate meter. The 2θ calibration was checked at the end of each sample scan by scanning National Bureau of Standards SRM-640 Si powder. Each sample diffraction pattern was indexed by comparison with patterns of known minerals and by reference to standard data tabulations (JCPDS, 1974; Brown, 1961).

Reflectance spectrophotometry utilized a Beckman DK-2A ratio-recording spectro-reflectometer with a tungsten source, "smoked" MgO reference and integrating sphere, and photomultiplier (VIS range, 0.35–0.70 μm) and lead sulfide (NIR range, 0.50–2.5 μm) detectors.

DTA was conducted under dynamic flow of dry, high-purity nitrogen gas using a Harrop TA-600 analyzer with direct-immersion furnaces and chromel-alumel thermocouples. Each sample (~150–200 mg, settled into block by tapping without compaction) was heated at 10°C/min from 25° to 1000°C and both heating and free-cooling curves were recorded. Pure aluminum oxide powder was used in the reference block during initial analyses although U.S. Geological Survey standard peridotite PCC-1 (Flanagan, 1973), which possesses a better thermal conductivity match to the chondrites, was found to provide less baseline drift and, hence, was used as the reference material in later analyses. Calibration of the analyzer was checked repeatedly using standards recommended by the International Confederation for Thermal Analysis (NBS material sets GM-758 and GM-760). It should be noted that sharp endotherms at ~560–580°C occurred in many of the chondrite DTA curves. Careful testing showed that these endotherms were not intrinsic to the chondrites but seemed to develop through repeated use of the same thermocouples for many analyses. It is hypothesized that reaction of the chromel (Ni–Cr)–alumel (Ni–Al) thermocouples with FeS in the chondrites formed Ni_3S_2 or Ni_7S_6, both of which are known to show structural changes at ~560–580°C (Craig and Scott, 1974). Thus, the reaction product endotherms have been excluded from Figs. 6 and 8 which display only thermal features known to be intrinsic to the samples.

Scanning electron microscopy (SEM) was performed on both natural surfaces and polished sections of selected chips using a Cambridge Stereoscan S-410 instrument equipped with an energy-dispersive x-ray analyzer. Both secondary electron images and elemental x-ray maps were obtained.

Electron microprobe analysis (EMPA) was performed on polished thin-sections using a MAC 5-SA3 instrument operated at 15 kV with 50 nA sample current and a beam spot size of ~10 μm. Well-characterized natural and synthetic minerals were used as standards and Bence-Albee corrections were made. Analyses were attempted for F, Na, Mg, Al, Si, P, S, Cl, K, Ca, Ti, Cr, Mn, Fe, and Ni.

RESULTS AND DISCUSSION

The secondary (weathering product) minerals found in meteorites recovered from regions other than Antarctica have been described previously (e.g., Buddhue, 1957; Ramdohr, 1963, 1973) so that a comprehensive review of such results will not be attempted here. For convenience, though, minerals which are germane to the following discussion (but not necessarily present in the samples studied) are listed in Table 2.

Surface morphologies and mineral deposits

The surfaces of the weathered Holbrook (1931 and 1968) samples and those of the Allan Hills Series A (ALHA) samples show clear evidence for fracturing and partial exfoliation of fusion crust. Polygonal fracture patterns are well developed on three of the ALHA samples (77155, 77270, 77272) and are composed of four- to five-sided cells of ~2–4-mm size which resemble those on the surface of Holbrook-1968. Loss of the blue-gray to black fusion crust, which in some places has turned reddish brown, reveals underlying yellow-brown zones. Degradation of the ALHA surfaces is much less extensive than that

Table 2. Secondary (weathering product) oxide and salt minerals which might occur in weathered chondrites.[a]

Mineral Name	Chemical Formula[b]
Magnetite	$Fe^{2+}Fe^{3+}_2O_4$
Hematite	$\alpha\text{-}Fe^{3+}_2O_3$
Maghemite	$\gamma\text{-}Fe^{3+}_2O_3$
Goethite	$\alpha\text{-}Fe^{3+}O(OH)$
Akaganéite	$\beta\text{-}Fe^{3+}O(OH, Cl)$
Feroxyhyte	$\delta\text{-}Fe^{3+}O(OH)$
Lepidocrocite	$\gamma\text{-}Fe^{3+}O(OH)$
Ferrihydrite	$5Fe^{3+}_2O_3 \cdot 9H_2O$
Melanterite	$Fe^{2+}SO_4 \cdot 7H_2O$
Szomolnokite	$Fe^{2+}SO_4 \cdot H_2O$
Copiapite	$Fe^{2+}Fe^{3+}_4(SO_4)_6(OH)_2 \cdot 20H_2O$
Coquimbite	$Fe^{3+}_2(SO_4)_3 \cdot 9H_2O$
Voltaite	$K_2Fe^{2+}_5Fe^{3+}_4(SO_4)_{12} \cdot 18H_2O$
Metavoltine	$K_2Na_6Fe^{2+}Fe^{3+}_6(SO_4)_{12}O_2 \cdot 18H_2O$
Slakvikite	$NaMg_2Fe^{3+}_5(SO_4)_7(OH)_6 \cdot 33H_2O$
Jarosite	$KFe^{3+}_3(SO_4)_2(OH)_6$
Epsomite	$MgSO_4 \cdot 7H_2O$
Starkeyite	$MgSO_4 \cdot 4H_2O$
Kieserite	$MgSO_4 \cdot H_2O$
Nesquehonite	$Mg(HCO_3)(OH) \cdot 2H_2O$
Hydromagnesite	$Mg_5(CO_3)_4(OH)_2 \cdot 4H_2O$
Reevesite	$Ni_6Fe^{3+}_2(CO_3)(OH)_{16} \cdot 4H_2O$
Zaratite	$Ni_3(CO_3)(OH)_4 \cdot 4H_2O$

[a] not a comprehensive list; discussed in text though not necessarily identified in present investigation
[b] as given by Fleischer (1980)

of Holbrook-1931, is comparable to that of Holbrook-1968, but is noticeably greater than that of Holbrook-1912 which is essentially untouched by weathering.

Deposits of nesquehonite were previously found to occur on the surface of the H5/6 Antarctic chondrite Yamato-74371 (Yabuki et al., 1976). Nesquehonite, hydromagnesite, and epsomite have also been identified among deposits on the surfaces of five ordinary chondrites, one carbonaceous chondrite, and one ureilite from Allan Hills (Marvin and Motylewski, 1980). Clearly, these delicate phases must have formed after the meteorites arrived on Earth and, in fact, such fluffy white deposits have apparently developed on some specimens *during* storage in the curatorial facility at the Johnson Space Center (D. D. Bogard, pers. comm., 1981), possibly by efflorescence from the interiors of these meteorites.

All of the ALHA samples examined in this study possessed remnants of fusion crust which, in places, could be interpreted as bearing trace deposits of secondary minerals. However, such deposits were most obvious and, hence, most confidently attributable to weathering, on 76006 and 77296. The 76006 deposit occurred as a single, very thin yellow-brown patch of ~ 2 mm^2 area (and $\ll 1$-mm thick) whereas the 77296 deposit occurred as a sprinkling of irregular fluffy white particles which, individually, were no more than ~ 0.1 mm in size and which resided in fractures and in the myriad of tortuous microcavities in the fusion crust. Thus, both occurrences were volumetrically much smaller than those described by Yabuki et al. (1976) and Marvin and Motylewski (1980) and could not be isolated for study in pure form as done by these other workers.

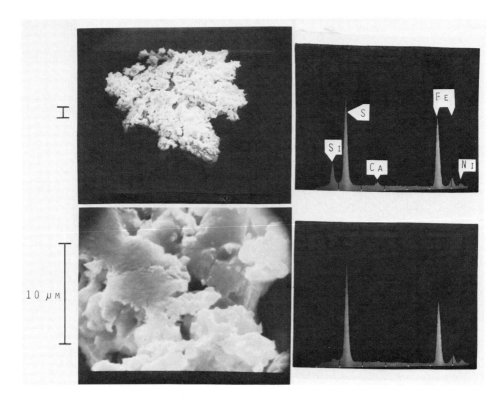

Fig. 1. Scanning electron photomicrographs and energy-dispersive x-ray spectra of a representative particle from the secondary mineral deposit on the surface of ALHA76006,3. The upper spectrum represents the central portion of the porous particle while the lower spectrum represents a spot on one of the flat blades which are aggregated together in the bulk particle. All available evidence suggests that the particle is a variety of ferric sulfate.

Consequently, particles were scraped from the surface of 76006 (which did not lend itself to fracturing at the places required to remove the total deposit) and epoxy-mounted individually for SEM study whereas a sub-sample of the 77296 chip was removed intact for SEM study.

Prior to carbon-coating, the 76006 particles were microscopically verified to be yellow-brown to orange in color and, hence, representative of the original deposit. SEM and energy-dispersive x-ray analysis showed the particles to be porous aggregates of plates and irregular blades and rich in Fe and S (Fig. 1). These particles are clearly different from the black, dense polyhedral blocks of troilite which were observed on the surface of the Holbrook-1912 reference sample and are most reasonably identified as ferric sulfates, possible coquimbite or copiapite. However, some ferrous sulfates, including szomolnokite, cannot be excluded from candidacy.

The mamillary and botryoidal habits of the 77296 surface particles are distinctly different from those of the 76006 particles although the occurrence of the 77296 particles in cavities and fractures, where electron and x-ray scattering problems may be substantial, makes the interpretation of their energy-dispersive x-ray spectra more difficult (Fig. 2). The white color of the particles and the K, Si, and Al abundances indicated by their x-ray spectra suggest low-Al zeolites such as clinoptilolite, dachiardite, stilbite, or phillipsite. However, the particle morphologies are unlike those of zeolites which are

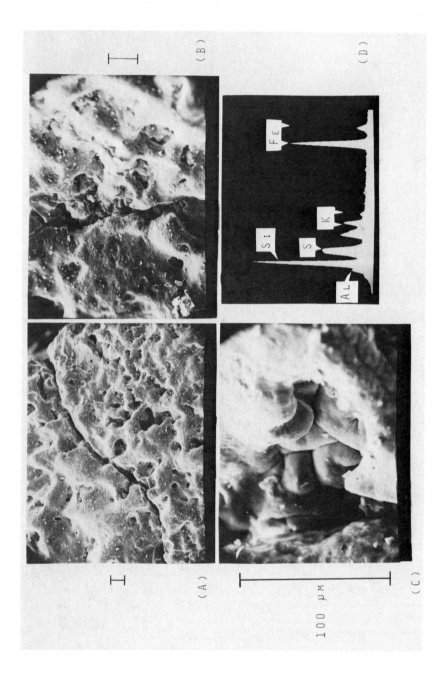

Fig. 2. Scanning electron photomicrographs of three views (A–C) of a fracture on the surface of ALHA77296,9 and an energy-dispersive x-ray spectrum (D) of a protruding surface of the botryoidal phase (C) which lines the fracture. In (D), the unlabeled peak and the low-energy shoulder of the S peak are due to the Au–Pd coating. The fracture-filling phase may be a mixed Fe–K-sulfate.

ordinarily fibrous and radiating-acicular. Furthermore, it is very difficult to understand why zeolites would form preferentially in and on magnetite-rich fusion crusts. The Fe and S peaks are probably too intense to attribute to matrix effects in the x-ray analyses so that the occurrence of Fe-sulfates within the botryoidal particles must be entertained. However, the white color of the particles is incompatible with those of most Fe-sulfates including mixed Fe-K-sulfates such as voltaite, metavoltine, or slavikite which are reported to be red-brown or green-yellow. A pale variety of melanterite, though, cannot be excluded. Still, the particle morphologies are unlike those reported for most sulfates. However, jarosite is reported to occur in massive, nodular habits although it usually is yellow-brown unless diluted with whitish alunite. In the absence of XRD data, which could not be obtained for paucity of sample material, the 76006 and 77296 surface particles, though almost certainly secondary minerals, must be considered as yet unidentified. However, if they are, in fact, Fe-sulfates, they might represent the first reported occurrence of such phases among ALHA surface deposits which are dominated by Mg-salts in other samples (Yabuki et al., 1976; Marvin and Motylewski, 1980).

No signs of appreciable alteration were found in SEM study of Holbrook-1912 surfaces. Holbrook-1931 surfaces were obviously altered but showed no mineral deposits of the types seen on ALHA surfaces. Holbrook-1968 surfaces contained millimeter-sized mounds of granules (possible quartz and calcite) which were morphologically dissimilar to the ALHA deposits and which were interpreted to be adhering residues of soil from the recovery site in Arizona.

Secondary mineral formation and weathering textures

The XRD patterns of bulk samples of all Holbrook and ALHA samples were dominated by peaks from olivine and hypersthene with minor contributions from troilite and kamacite. For any given sample, d-spacings (± 0.02 Å) of the most intense peaks were at 2.48 Å [olivine (112)], 2.53 Å [olivine (131)], or 3.20 Å [hypersthene (321; 420)]. Evidence for secondary minerals was found by searching the patterns of weathered chondrites for peaks which did not appear in the patterns of the unweathered Holbrook-1912 control sample. The only such peak which consistently and reproducibly appeared in most weathered samples was at 4.20 Å and is most reasonably attributed to the intense (110) peak of goethite. This peak occurs at a relative intensity of $\sim 10\%$ in Holbrook-1931, ALHA76006, and 76008 and at $\sim 5\%$ relative intensity in other weathered samples except for 77270 and 77296 for which it was not reproducibly detectable above background. Peaks at 4.46 Å were also found in 76008 and 77272 and might be attributable to the intense (120) or ($\bar{1}11$) peaks of starkeyite. However, no evidence was found for either epsomite of kieserite.

No convincing XRD evidence was found for the principal peaks of lepidocrocite at 6.26 Å or 3.29 Å, those of akaganéite at 7.40 Å, 3.31 Å, or 1.635 Å, or those of feroxyhyte at 1.47 Å and 2.23 Å. Unfortunately, the principal peaks of hematite, maghemite, and magnetite are all subject to severe interference from strong peaks due to olivine, hypersthene, and troilite and, thus, these potentially important Fe-oxide weathering products could not be unambiguously identified or excluded as constituents of the weathered chondrites. It is possible, though that more sophisticated peak-stripping analyses of high-resolution XRD scans, which were beyond the scope of this study, might solve the Fe-oxide identification problem. Notably, however, no convincing evidence was found for the existence of clay minerals, zeolites, zaratite, or reevesite in the weathered chondrites. Zaratite and reevesite, in particular, are known to occur in other weathered meteorites (Buddhue, 1957; Ramdohr, 1963, 1973) and should produce several strong XRD peaks free of interference from olivine and pyroxene. Thus, their *apparent* absence from the samples studied here may be due to their very low abundances. Certainly, further detailed analyses of greater sensitivity, including concentration and isolation of individual weathering products, should be attempted in future work.

Weathering textures of the chondrites, as revealed in polished thin-sections, consist

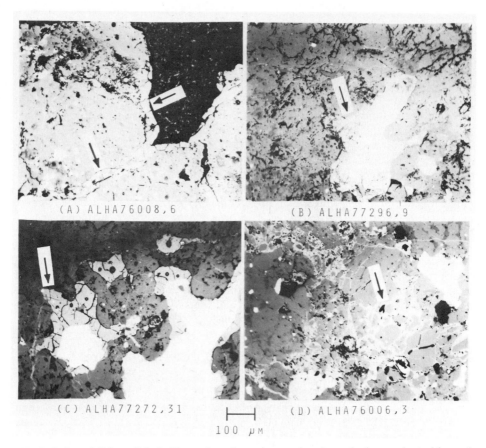

Fig. 3. Reflected light polished thin-section photomicrographs of weathering products (shown by arrows) in ALHA chondrites. (A) illustrates a type I alteration texture (rind on fusion crust) which is connected to a type II (interior vein) area. (B) and (C) show two characteristic morphologies of type III alteration (rinds on metal particles). (D) shows a typical type II area which is only incidentally associated with local metal particles.

mainly of four types of features which are
  (I) rinds, veins and fillings in fusion crust,
  (II) veins, pockets and other fillings in areas other than fusion crust,
  (III) rinds or mantles on Ni–Fe metal or sulfide particles, and
  (IV) "myrmekitic" oxide fillings within Ni–Fe metal particles.

Optical microscopy shows that at least types I–III (Fig. 3) are blood-red in transmitted light in thin areas but gray in reflected light. In reflected light, at least two different phases can be seen among the weathering products. Both appear to be of similar hardness although the first phase is less abundant, usually confined to thin rinds on metal particles, and may be bireflectant (medium gray to blue-gray against Ni–Fe metal). The second and dominant phase is light gray with only weak to nonexistent bireflectance and anisotropy. It is distinctly different from the brown-gray magnetite in the fusion crusts. Occasionally, the dominant light gray phase is actually resolvable into two phases of similar color so that a total of at least three different phases probably occurs among the weathering products illustrated in Fig. 3. However, those phases observed optically and by SEM appear to be less distinctive and diverse mineralogically than those described by Ramdohr (1963, 1973) for other weathered meteorites.

The textures of all four types of alteration clearly imply replacement of magnetite (fusion crust), troilite and Ni–Fe metal by oxidation/hydration products which can be

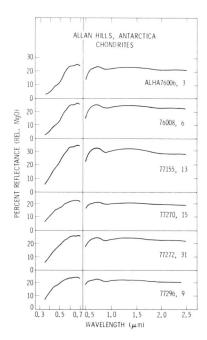

Fig. 4. Visible and near-infrared reflectance spectra of powdered samples of the outer ~5–10-mm portions of ALHA chondrites. Weathering has caused steepening of some of the spectral curves over the ~0.35–0.70 μm range and introduced absorptions, due to Fe^{3+}-oxides, at ~0.60–0.65 μm. The 0.5–2.5 μm spectra are typical of ordinary chondrites.

conveniently called "Fe-oxides." Furthermore, the colloform textures of type III, which include layering parallel to oxide/metal borders and fractures perpendicular to them (Fig. 3B, C), strongly suggest formation in the presence of abundant liquid water followed by various degrees of post-formational desiccation. Type IV textures, which probably indicate a comparatively advanced stage of weathering, are well-developed in Holbrook-1931 but show only incipient occurrences in the ALHA samples.

VIS/NIR spectra of the chondrites (Fig. 4) are similar to those of other equilibrated chondrites (Salisbury et al., 1975). No absorptions at ~1.4 and ~1.9 μm, as would be expected for samples containing significant amounts of hydrous phases, are present although such features could be masked by opaque phases in the samples (Salisbury and Hunt, 1974). In fact, spectral evidence for weathering effects in these samples seems confined to the VIS (0.35–0.70 μm) region where steepening of the curves and absorptions at ~0.60–0.65 μm occur in the spectra of the more heavily weathered samples. The 0.60–0.65 μm absorptions are attributable to Fe^{3+} although the abundances of Fe^{3+}-bearing phases are apparently too small to have affected the positions of the absorptions at ~0.9 μm which, in unweathered chondrites, are due to Fe^{2+} in olivine and pyroxene. As pointed out by Salisbury and Hunt (1974), the 0.5 μm/0.6 μm reflectance ratio can be used as an index of degree of weathering in stony meteorites. Such an approach, modified to incorporate information regarding the absolute reflectance of each sample, shows that the Holbrook and ALHA samples can, indeed, be spectrally ranked according to degree of weathering (Fig. 5). Further interpretations of this ranking will be given in the next section.

DTA results can be used to set additional limits to the types and abundances of weathering products in the chondrites. DTA curves (Fig. 6) show that the only intrinsic thermal feature which is common to all chondrites studied is a sharp endotherm at ~155–165°C which can be attributed to the inversion of troilite from the hexagonal 2C to hexagonal 1C pyrrhotite structure (Craig and Scott, 1974). Allowing for systematically high temperature readings of the analyzer which may be as great as 20°C, the observed transition temperature agrees well with the literature value of 140°C. A second sharp endotherm at ~940°C (feature E, Fig. 6) was reversible upon cooling but was observed only in ALHA76006 and 76008 and might represent a structural change in Ni–Fe metal.

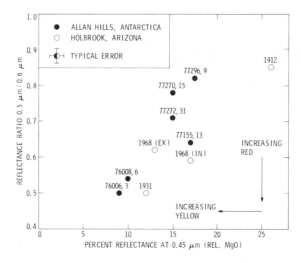

Fig. 5. Spectral reflectance characteristics of weathered ordinary chondrites. Each point is the average of three analyses of powdered surface (outer ~5–10-mm) specimens. Exterior (EX) and interior (IN; >10-mm depth) specimens of Holbrook-1968 are contrasted. Progressive weathering is accompanied by increasing orange color (as indicated by the trend from upper right to lower left) due to Fe^{3+}-oxide formation.

All other thermal features intrinsic to the samples occur either as a series of shallow endotherms over the ~200–600°C range or as sharper endotherms at ~890°C (feature D, Fig. 6) and ~960°C (feature F, Fig. 6). None of these endotherms are reversible upon cooling and are thought to represent decomposition reactions of weathering products. Unfortunately, these features could not be unambiguously identified with specific minerals although it is speculated that the low-temperature (200–600°C) endotherms represent dehydration of hydrous oxides, carbonates, and sulfates whereas the high-temperature (890, 960°C) endotherms may represent decarbonation of anhydrous carbonates. Reference analyses showed that epsomite does not provide a good match to the

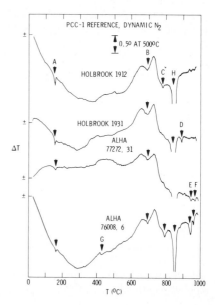

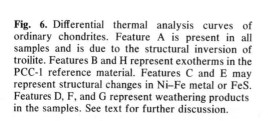

Fig. 6. Differential thermal analysis curves of ordinary chondrites. Feature A is present in all samples and is due to the structural inversion of troilite. Features B and H represent exotherms in the PCC-1 reference material. Features C and E may represent structural changes in Ni–Fe metal or FeS. Features D, F, and G represent weathering products in the samples. See text for further discussion.

chondrite weathering product endotherms although the ~300°C and ~750°C endotherms of melanterite could conceivably be hidden in the chondrite DTA curves. Thus, hydrous Fe-sulfates remain viable condidates as weathering products. It is noteworthy, though, that the broad endotherms at ~100°C and ~500°C which are characteristic of smectites, are absent from the chondrite DTA curves, thereby confirming XRD indications that clay minerals are not major weathering products in these samples. Detection limits, as estimated for goethite, are discussed below.

Despite DTA evidence for trace contents of a variety of weathering products, the dominant phases, as seen in thin-section, are the "Fe-oxides" described previously. Most meteorite petrographers acknowledge the occurrence of "limonite" veins, stains, and rust haloes in thin-section. However, "limonite" usually connotes a mixture of goethite, lepidocrocite and akaganéite (and, possibly, feroxyhyte and ferrihdrite) without specifying the relative abundances of the various phases. In the present study, XRD results indicate that goethite is the dominant hydrous ferric oxide although optical microscopy shows that the Fe-oxide weathering products are probably multiple-phase assemblages.

Electron microprobe analyses of the Fe-oxide weathering products (Table 3) demonstrate that they are complex and variable in composition. Analytical totals are usually no more than ~90% although tests made on olivine and pyroxene grains near Fe-oxide occurrences in the same sections gave excellent totals (99–100%) and appropriate structural formulas, demonstrating that analytical procedures were valid. Thus, deficits in the Fe-oxide analyses must represent the existence of H, C, and O (e.g., as H_2O and CO_2) in these phases. Most Fe-oxide spots contain ~80% Fe_2O_3 whereas stoichiometric goethite contains 89.9% Fe_2O_3. However, the chondrite Fe-oxide weathering products contain substantial amounts of Ni, S, and Cl which, in oxide form, total to ~7–9%. Thus, these weathering products could contain ~10–15% total H_2O and CO_2. Jarosewich's bulk chemical analysis of ALHA76008 (Olsen et al., 1978) found 2.67% total H_2O (with 2.20% as H_2O^+) which would correspond to ~25–30% Fe-oxide weathering products as analyzed in the present study.

Not surprisingly, Ni contents of the Fe-oxides tend to be greatest for weathering products on metal particles whereas S contents are greatest for weathering products on sulfide particles. However, the highest S concentration observed in the Fe-oxide weathering products was found in a type I alteration area on the surface of 76008 (Fig. 3A) which was not in contact with any discernible particles of sulfide. Furthermore, the highest recorded Cl contents were for types II and III alteration rather than for type I alteration which might have otherwise been expected if surface contamination (e.g., by NaCl) was the major source of Cl.

The sources and mineralogical forms of S and Cl in the Fe-oxide weathering products are of considerable interest. S was almost certainly derived by alteration of primary sulfides in the chondrites although Cl could have been derived either by the alteration of indigenous lawrencite ($FeCl_2$), as advocated for meteorites, in general, by Buddhue (1957), or by introduction of terrestrial salt solutions. Each ~1% Cl in the Fe-oxides would have required the consumption of ~1.8% lawrencite and, hence, the production of a complementary ~1.1% Fe_2O_3. Alternatively, each ~1% Cl in the Fe-oxides would be equivalent to the introduction of ~1.6% NaCl. S and Cl abundances are only weakly correlated (correlation coefficient, r = 0.36 for a linear least squares regression analysis of the data in Table 3) so that the processes by which these two elements were concentrated into the Fe-oxide weathering products need not have been strongly coupled.

Na and Mg are trace constituents of the Fe-oxide weathering products such that Na- and Mg-salts could be components of these complex assemblages. For halite (NaCl), the stoichiometric weight ratio of Cl/Na is 1.542. However, the Cl/Na ratio in the analyzed Fe-oxides is usually much greater or much less than the halite ratio so that arguments for halite in the weathering products cannot be convincingly made. Furthermore, the Na and Cl abundances do not appear to be correlated (r = −0.17 for the data of Table 3). Similarly, the stoichiometric weight ratio of SO_3/MgO is 1.590 for epsomite whereas the SO_3/MgO ratios in the Fe-oxides exhibit various values although none match those for

Table 3. Electron microprobe analyses of Fe-oxide weathering products in chondrites[a].

Chondrite:	HOLBROOK 1931			ALHA76006,3		ALHA76008,6			ALHA77272,31	ALHA77296,9		
Notes:	b,d	b,d	b,d	b,d	c,d	b,d	c,d	c,d	b,d	b,d	b,d	b,d
Wt. % SiO$_2$	2.1	1.4	1.4	0.40	0.76	0.64	0.07	0.10	0.29	1.6	0.06	4.8
Fe$_2$O$_3$[e]	82.5	79.8	82.5	75.9	79.4	79.4	80.6	78.4	78.8	77.5	74.7	75.5
NiO	5.9	6.6	6.2	5.4	1.4	10.2	3.8	5.1	3.0	4.0	0.86	3.8
MgO	bdl	bdl	0.98	0.15	0.43	1.1	1.3	1.4	0.52	0.41	bdl	1.1
CaO	0.06	0.09	0.15	0.05	bdl	0.75	0.35	0.10	0.19	0.12	bdl	0.30
Na$_2$O	0.50	0.52	1.3	bdl	0.22	0.27	0.29	0.05	0.07	0.10	0.12	0.09
P$_2$O$_5$	0.05	bdl	0.28	bdl	bdl	bdl	bdl	bdl	bdl	bdl	bdl	bdl
SO$_3$	0.66	0.58	0.50	1.2	8.4	0.88	3.9	1.5	0.91	2.2	5.3	1.2
Cl	0.23	0.24	0.46	1.0	0.36	0.47	0.27	0.65	0.14	0.16	3.8	0.21
Total[f]	92.00	89.23	93.77	84.10	90.97	93.71	90.58	87.30	83.92	86.09	84.84	87.00

[a]bdl = below detection limit; other elements below detection limit include F, Al, K, Ti, Cr, and Mn
[b]results for one spot
[c]average results for two spots
[d]each spot analyzed once for Ni, P, S, and Cl and twice for other elements
[e]all iron assumed to occur as Fe^{3+}
[f]difference from 100 interpreted to be wt. % H$_2$O + CO$_2$

Mg-sulfates and the Mg/S abundances are not correlated (r = −0.12 for data of Table 3). It seems significant, though, that >1% MgO occurs in some Fe-oxides which possess virtually no SiO_2 (e.g., column 3 for 76008 and column 1 for 77272, Table 3) such that contribution of "apparent MgO" from silicate fluorescence can be dismissed. Consequently, the occurrence of Mg-carbonates as components of the Fe-oxide assemblages cannot be excluded. In fact, pyrolysis experiments by Gibson and Andrawes (1980) indicated that release of CO_2 from ALHA chondrites at ~200–600°C could be due to carbonate minerals.

Because of the apparent dominance of hydrous Fe-oxides among the weathering products of the chondrites, a special effort was made to quantitatively estimate their goethite contents. Accordingly, a weathered chondrite simulant was prepared by carefully mixing together precisely measured quantities of U.S.G.S. Standard Peridotite PCC-1 (Flanagan, 1973) with reagent-grade FeS and natural goethite (D. J. Mineral Co., Butte, Montana) powders. The homogenized mixture containing 5% FeS and 5% goethite was analyzed by XRD and DTA under the same conditions used for the chondrites. The 5% goethite component was easily detectable both by XRD (Fig. 7) and DTA (Fig. 8). In fact, based on the count rate for the 4.20 Å goethite peak, it appears that the XRD detection limit for goethite was <1% in this study whereas the abundance of goethite in the most heavily weathered chondrite was <2% (Fig. 7). Dehydration of the goethite used in this study occurred at ~380°C (Fig. 8) although dehydration endotherms for natural goethites can occur at nearly any temperature over the range of ~200–400°C (Neumann, 1977). Thus, the position of the goethite endotherm in the chondrite DTA curves remains uncertain although it is clear that 5% goethite should be readily detectable (Fig. 8). In fact, the detection limit of the method used in this study can be estimated as ~3%. Combining the XRD and DTA sensitivity information, it is then estimated that the most heavily weathered chondrites studied here (Holbrook-1931, ALHA76006, 76008) each contain no more than 5 wt.% goethite. Further indications of DTA sensitivity stem from observed intensities of the troilite endotherms (Fig. 8) which, for average contents of ~5–6% troilite in ordinary chondrites (Keil, 1962), also suggest that thermally active phases should be readily detectable in the chondrites at abundances of ~3%. Additional tests

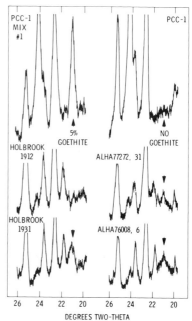

Fig. 7. Identification of goethite in the x-ray diffraction patterns of weathered chondrites. The goethite (110) peak (shown by arrows) is easily detectable in a simulated weathered chondrite (PCC-1 Mix #1) composed of peridotite with 5% FeS and 5% goethite. By comparison, the most intensely weathered chondrites (Holbrook-1931, ALHA76008) are inferred to contain much less than 5% goethite. Other peaks in all patterns are due to olivine and pyroxene.

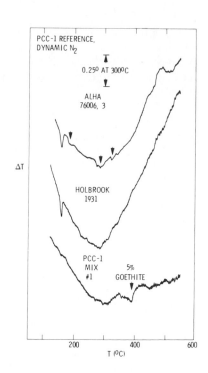

Fig. 8. Differential thermal analysis curves of weathered chondrites compared with that of a simulated weathered chondrite (PCC-1 Mix #1), composed of peridotite with 5% FeS and 5% goethite. No endotherm is observed at ~155°C in PCC-1 Mix #1 since the reagent-grade FeS used in the mixture does not possess the troilite structure. The goethite component is easily detectable and its endotherm strength suggests that the abundance of goethite (or other FeO(OH) polymorph) in the severely weathered ALHA76006 (possibly represented by one of the endotherms marked by arrows) is less than 5%.

with synthetic hydromagnesite in a PCC-1/FeS matrix suggest that the DTA detection limit for hydrous carbonates may actually be ≤1%.

As noted above, Jarosewich found 2.20% H_2O^+ in ALHA76008 (Olsen et al., 1978) and, from the data of Table 3, it is inferred that the Fe-oxide weathering products contain ~10% H_2O. The observed H_2O^+ content of the chondrite could then be explained by the presence of ~20% Fe-oxide weathering products. Similarly, Jarosewich's determination of 12.63% Fe_2O_3 in 76008 (Olsen et al., 1978) and the assumption of ~80% Fe_2O_3 in the Fe-oxides (Table 3) implies ~15% weathering products in the chondrite. However, no more than ~5% goethite is evidently present in 76008 such that the equivalent of ~10–15% Fe-oxide weathering products must exist in forms other than goethite. Since neither lepidocrocite, akaganéite, nor feroxyhyte were detected by XRD, the unidentified Fe-oxide component may exist as an amorphous or very poorly crystalline phase such as ferrihydrite (Murray, 1979). If the other chondrites carry the same weathering products as does 76008, as appears to be the case, goethite may only represent ~1/3 of the total complement of Fe-oxide weathering products.

The processes which controlled the weathering of the Antarctic chondrites remain unclear although the strong resemblance of weathering features in ALHA samples to those in the Holbrook samples show that liquid water must have played a major role in the Antarctic weathering. The general sequence of events was probably that suggested by Gibson and Andrawes (1980) with repeated exposure to glacial melt water, from multiple freeze-thaw cycles, providing the environment in which the vast majority of weathering effects developed. Although the origins of the carbonate and sulfate deposits on ALHA surfaces are not yet fully understood, Gibson and Andrawes (1980) showed that uptake of terrestrial CO_2 was almost certainly involved in carbonate formation whereas the present study strongly suggests that sulfates derived their S from the interiors of the meteorites. Thermodynamic calculations (e.g., Gooding, 1978) suggest that formation of hydrous Fe-sulfates from FeS is favorable in oxidizing environments even when the supply of water is limited to atmospheric water vapor. However, hydrous Fe-oxides form as stable products at low pressure only in the presence of liquid water.

The extensive networks of veins which run through many ALHA samples show that both inward and outward transport of elements by solutions may have affected the meteorites. Thus, the efflorescent migration of hydrous salts to meteorite surfaces with subsequent loss by eolian ablation, may have led to the net loss from these meteorites of Fe, Ni, S, and Cl as well as numerous chalcophile and siderophile trace elements. In fact, Biswas *et al.* (1980) did observe differences in trace element abundances in surface samples, relative to interior samples, of four ALHA stony meteorites. However, their two chondrites were both unequilibrated (petrologic type 3) and little weathered (weathering index A) relative to the samples examined in this study such that detailed correlations of trace element data with weathering effects are not easily performed. It is possible though, that the enrichment of Cd in the surface of ureilite ALHA77257 observed by Biswas *et al.* (1980) might have resulted from outward migration of Fe-oxide weathering products of primary sulfides, as observed in the present study. The observed surface enrichments of Rb and Cs may represent contamination by terrestrial sea salt sprays, as suggested by Biswas *et al.* (1980), or, at least for the ureilite, may represent outward migration of weathered alkali halides which occur as primary phases in the carbonaceous matrices of ureilites (Berkley *et al.*, 1978).

In any case, Antarctic weathering seems to provide ample opportunity for the net loss from ALHA chondrites of elements associated with the primary chondrite sulfide and metal phases. The extent of such losses should probably depend upon the duration of exposure to liquid water and the opportunity for outward migration and surface loss by wind ablation which each meteorite experiences. If the Fe-oxide weathering products observed in the present study are representative of those in all Antarctic ordinary chondrites, their occurrence at depths of 2 cm below chondrite surfaces suggest that studies aimed at determining pre-terrestrial geochemical properties of these specimens should utilize only samples from *at least* 2-cm depth and, preferably, from even greater depths. Ideally, each geochemical aliquot should be mineralogically characterized prior to analysis in order to assess its weathering history.

Relationships between terrestrial ages and degrees of weathering

The terrestrial residence ages of meteorites are derived from concentrations of the cosmogenic radionuclides ^{14}C, ^{26}Al, ^{53}Mn (e.g., Fireman *et al.*, 1979; Imamura *et al.*, 1979) and ^{36}Cl (Nishiizume *et al.*, 1979) measured in samples. In addition to uncertainties related to the production pathways for the nuclides, changes in the concentrations of C, Al, Mn, and Cl due to weathering are assumed to be negligible. However, a controversy has arisen over whether ^{53}Mn is (Fireman *et al.*, 1979) or is not (Imamura *et al.*, 1979) lost from meteorites during weathering, thereby yielding systematically low terrestrial ages. Although Imamura *et al.* (1979) reported 30 ppm Mn (equivalent to 0.004% MnO) in bulk metal and ~2300–2700 ppm Mn (~0.3% MnO) in non-magnetic Fe-oxides and silicates separated from ALHA76008, the present study found no evidence for Mn (all results below EMPA detection limit of 0.05% MnO) in the Fe-oxide weathering products of 76008 or other chondrites. However, olivine and pyroxene crystals were found to contain nominal concentrations of ~0.4–0.5% MnO. Thus, it seems likely that the non-magnetic sample fractions of Imamura *et al.* (1979), which were not mineralogically characterized, derived their high Mn concentrations from olivine and pyroxene components. The observation, by the same workers, that normalized count rates for ^{53}Mn are essentially the same in the 76008 metal as in separated portions of its Fe-oxide rinds seems more significant and shows that ^{53}Mn is, indeed, incorporated into the Fe-oxide weathering products of the chondrites. Given the apparent mobility of chalcophile and siderophile elements toward meteorite surfaces once they are incorporated into weathering products, the possible net loss of ^{53}Mn from weathered meteorites through time cannot be excluded. Only the magnitudes of such effects seem to be in question. However, measurement of ^{53}Mn activities in weathered surface and unweathered interior portions of various chondrites might help to place limits on the degrees and rates of ^{53}Mn migration.

Although weathering effects should be critical considerations in all terrestrial-age methods which emphasize the use of meteoritic metal fractions, it appears that the ^{53}Mn method may be less seriously disturbed than that based on ^{36}Cl. The most prolific target element for ^{36}Cl production in meteorites is Ca with lesser contributions from Fe and Ni (Nishiizumi et al., 1979) so that the Ca concentrations in the magnetic (presumably, mostly metal) separates used for ^{36}Cl analysis must be carefully monitored. The magnetic fractions analyzed by Nishiizumi et al. (1979) contained ~150–4000 ppm Ca (equivalent to ~0.02–0.6% CaO), including 920 ppm (0.13% CaO) in ALHA76008 and 3950 ppm (0.55% CaO) in 77002. The present study found similar concentrations (~0.1–0.8% CaO) in the Fe-oxide weathering products of 76008 and other chondrites even though Si and Al abundances were at or below EMPA detection limits such that the contribution of "apparent CaO" by fluorescence of plagioclase could be dismissed. In the same phases, P concentrations were usually very low (<0.05% P_2O_5) so that contribution of Ca by Ca-phosphates was probably negligible. However, Cl concentrations in the weathering products were very high, approaching or exceeding 1% in some places. As discussed in the previous section, explanation of the high Cl concentrations by invoking NaCl contamination would not be consistent with the poor correlation between Na and Cl abundances. As also discussed above, weathering of lawrencite may have contributed most of the Cl although mobilization of Fe and Ni into the Fe-oxide weathering products must have been accompanied by similar mobilization of radiogenic ^{36}Cl. Thus, the outward transport and eventual loss of these weathering products from the meteorite should ultimately lead to a net decrease in ^{36}Cl and, hence, possible systematic underestimates in terrestrial ages by the ^{36}Cl method. As for ^{53}Mn, though, the magnitude of the predicted loss remains to be checked by determination of ^{36}Cl concentrations as a function of depth in a variety of weathered stony meteorites.

The nondependence of the ^{26}Al method on metal separations and the very low Al contents (<0.05% Al_2O_3) observed for Fe-oxide weathering products in the present study suggest that, in principle, ^{26}Al age measurements for stony meteorites may be the least perturbed by terrestrial weathering effects. Unfortunately, ^{26}Al terrestrial residence ages are available for only four of the chondrites studied here (Table 1) but suggest ages increasing in the order

$$76006 < 77155 < 77272 \ll 76008.$$

In contrast, the relative degrees of weathering of the same chondrites, as measured by the effect of Fe-oxide weathering products on the chondrite spectral reflectance characteristics (Fig. 5), suggest that weathering intensifies in the order

$$77272 < 77155 \ll 76008 < 76006.$$

Thus, virtually no correlation seems to exist between ^{26}Al age and degree of weathering. Even when the effect of metal abundance is removed, by considering the H- and L-chondrites separately, the correlation does not improve. This test assumes, of course, that both the sub-samples used for ^{26}Al analysis and those used in the present study are representative of their respective parent stones in terms of bulk mineralogy and degree of weathering. However, intensities of weathering effects may vary not only with depth in a stone but also laterally on its surface if weathering occurred in intermittent puddles of water such that bottom surfaces were more severely attacked than were top surfaces. Unfortunately, despite highly commendable efforts at documentation in the field, the top/bottom orientations of individual stones recovered from the Antarctic ice cannot always be redetermined in the curatorial facility (D. D. Bogard, pers. comm., 1981). Thus, questions regarding meteorite heterogeneity and representative sampling, an old problem in meteoritics, may contribute to the difficulty in correlating degrees of weathering with terrestrial residence ages of the same meteorites.

Still, the results of this study suggest that it may be possible, through a combination of XRD, DTA, and VIS/NIR spectral analyses, to more quantitatively assess the relative degrees of weathering of chondrites of common chemical-petrologic type. For example, preliminary curatorial examinations of the four L6 ALHA chondrites studied here indicated that three of the four were weathered to about the same degree, namely the A/B category (Table 1). However, work reported here shows that these four samples are weathered to degrees which can be clearly distinguished from each other (e.g., Fig. 5) and from the various members of the Holbrook weathering sequence. Figure 5 suggests that the four L6 ALHA chondrites studied here are all much less weathered than the weathered samples of the L6 chondrite Holbrook. As also observed by Gibson and Bogard (1978), though, Holbrook-1968 seems to be less weathered than Holbrook-1931. Thus, it could be argued that the effects of weathering for a few hundred thousand years on barren Antarctic ice are less severe than 56 years, or perhaps even 19 years, of weathering in central Arizona. This observation can be understood if weathering is effectively arrested while the Antarctic meteorites are solidly frozen in glacial ice but resumes at times when thawing occurs (perhaps by radiant solar heating of the meteorites) and liquid water becomes available. The challenge then becomes determination of the timing and duration of events which expose the meteorites to water. It seems entirely possible that exposure history to water, rather than terrestrial residence age, is the decisive factor in determining the weathering effects which are recorded in the Antarctic meteorites.

CONCLUSIONS

A mineralogical survey of two H6 and four L6 chondrites from Allan Hills, Antarctica and a parallel survey of Holbrook, Arizona (L6) control samples have provided information on the processes and products of chemical weathering of ordinary chondrites in the Antarctic environment. Significant observations and inferences are that

1. The dominant weathering products in these rocks are complex, multiple-phase, hydrous ferric oxides which formed by alteration of Ni–Fe metal and sulfide particles under the influence of liquid water.
2. The Fe-oxide weathering products may comprise ~15–20 wt.% of the most intensely weathered samples although the same samples contain ≤5% goethite as the only well-crystallized hydrous ferric oxide. Thus, ~2/3 of the Fe-oxide weathering products may be amorphous or otherwise not identifiable as stoichiometric minerals.
3. The Fe-oxide weathering products contain up to ~3% S and ~4% Cl in mineralogical forms which remain obscure but which may include Fe-sulfates. However, the probable solution transport of weathering products toward meteorite surfaces may be related to the formation of sulfates and carbonates within and on fusion crust. Furthermore, the net loss of Fe, Ni, S, Cl and, to a lesser extent, Mg and chalcophile and siderophile trace elements by this weathering process seems entirely possible.
4. The effects of terrestrial weathering on the mobilization and redistribution of cosmogenic ^{53}Mn and ^{36}Cl remain uncertain and merit further study. Among the ALHA samples studied here, no obvious correlation exists between degree of weathering and terrestrial residence age although a few hundred thousand years in or on Antarctic ice apparently weathers L6 chondrites less severely than do only a few decades in the Arizona desert.

Acknowledgments—I thank the curatorial staff of the Johnson Space Center, especially Donald Bogard, for providing excellent samples. Samples of the Holbrook chondrite were generously furnished by Martin Prinz (1912 collection), Carleton Moore (1931), and Everett Gibson (1968). Barry Wawak expertly assisted with scanning electron microscopy and Arden Albee and Arthur Chodos kindly provided access to their electron microprobe facilities. Kathleen Baird and Warren Rachwitz provided priceless cooperation in organizing the laboratory required for this study. Constructive comments by H. Y. McSween, Jr., J. S. Delaney, and an anonymous reviewer were beneficial. Sue Officer graciously typed the manuscript. The research described in this paper was performed at the Jet Propulsion Laboratory, California Institute of Technology, under contract with the National Aeronautics and Space Administration.

REFERENCES

Berkeley J. L., Taylor G. J., and Keil K. (1978) Fluorescent accessory phases in the carbonaceous matrix of ureilites. *Geophys. Res. Lett.* **5**, 1075–1078.
Biswas S., Ngo H. T., and Lipschutz M. E. (1980) Trace element contents of selected Antarctic meteorites. I. Weathering effects and ALHA77005, A77257, A77278, and A77299. *Z. Naturforsch.* **35a**, 191–196.
Brown G., ed. (1961) *The X-ray Identification and Crystal Structures of Clay Minerals.* Mineral. Soc., London. 544 pp.
Buddhue J. D. (1957) *The Oxidation and Weathering of Meteorites.* Univ. of New Mexico, Albuquerque, 161 pp.
Craig J. R. and Scott S. D. (1974) Sulfide phase equilibria. In *Sulfide Mineralogy*, Vol. 1 (P. H. Ribbe, ed.), p. cs1–cs110. Mineral. Soc. Amer. Rev. in Mineralogy.
Evans J. C. and Rancitelli L. A. (1980) Terrestrial ages. In *Catalog of Antarctic Meteorites, 1977–1978* (U. B. Marvin and B. Mason, eds.), *Smithsonian Contr. Earth. Sci.* **23**, 45–46.
Fireman E. L. (1979) ^{14}C and ^{39}Ar abundances in Allan Hills meteorites. *Proc. Lunar Planet. Sci. Conf. 10th*, p. 1053–1060.
Fireman E. L., Rancitelli L. A., and Kirsten T. (1979) Terrestrial ages of four Allan Hills meteorites: consequences for Antarctic ice. *Science* **203**, 453–455.
Flanagan F. J. (1973) 1972 values for international geochemical reference samples. *Geochim. Cosmochim. Acta* **37**, 1189–1200.
Fleischer M. (1980) *Glossary of Mineral Species 1980.* Mineral. Rec., Tucson. 192 pp.
Gibson E. K. Jr. and Andrawes F. F. (1980) The Antarctic environment and its effect upon the total carbon and sulfur abundances in recovered meteorites. *Proc. Lunar Planet Sci. Conf. 11th*, p. 1223–1234.
Gibson E. K. Jr. and Bogard D. D. (1978) Chemical alterations of the Holbrook chondrite resulting from terrestrial weathering. *Meteoritics* **13**, 277–289.
Gooding J. L. (1978) Chemical weathering on Mars: thermodynamic stabilities of primary minerals (and their alteration products) from mafic igneous rocks. *Icarus* **33**, 483–513.
Imamura M., Nishiizumi K., and Honda M. (1979) Cosmogenic ^{53}Mn in Antarctic meteorites and their exposure history. *National Institute of Polar Research, Memoirs. Special Issue No. 15.* Tokyo, p. 227–242.
JCPDS (1974) *Selected Powder Diffraction Data for Minerals (Data Book)*, Publ. DBM-1-23. Joint Committee on Powder Diffraction Standards, Swarthmore, Pennsylvania.
Keil K. (1962) On the phase composition of meteorites. *J. Geophys. Res.* **67**, 4055–4061.
King T. V. V., Score R., Gabel E. M., and Mason B. (1980) Meteorite descriptions. In *Catalog of Antarctic Meteorites, 1977–1978*, (U. B. Marvin and B. Mason eds.), *Smithsonian Contr. Earth Sci.* **23**, p. 12–44.
Marvin U. B. and Motylewski K. (1980) Mg-carbonates and sulfates on Antarctic meteorites (abstract). In *Lunar and Planetary Science XI*, p. 669–670, Lunar and Planetary Institute, Houston.
Murray J. W. (1979) Iron oxides. In *Marine Minerals*, Vol. 6 (R. G. Burns ed.), p. 47–98, Mineral. Soc. Amer. Rev. in Mineralogy.
Neumann B. S. (1977) Thermal techniques. In *Physical Methods in Determinative Mineralogy* (J. Zussman, ed.), p. 605–662. Academic, N.Y.
Nishiizumi K., Arnold J. R., Elmore D., Ferraro R. D., Gove H. E., Finkel R. C., Beukens R. P., Chang K. H., and Kilius L. R. (1979) Measurements of ^{36}Cl in Antarctic meteorites and Antarctic ice using a Van de Graaff accelerator. *Earth Planet. Sci. Lett.* **45**, 285–292.
Olsen E. J., Noonan A., Fredriksson E., Jarosewich E., and Moreland G. (1978) Eleven new meteorites from Antarctica, 1976–1977. *Meteoritics* **13**, 209–225.
Ramdohr P. (1963) The opaque minerals in stony meteorites. *J. Geophys. Res.* **68**, 2011–2036.
Ramdohr P. (1973) *The Opaque Minerals in Stony Meteorites.* Elsevier, Amsterdam, 245 pp.
Salisbury J. W. and Hunt G. R. (1974) Meteorite spectra and weathering. *J. Geophys. Res.* **79**, 4439–4441.
Salisbury J. W., Hunt G. R., and Lenhoff C. J. (1975) Visible and near-infrared spectra: X. Stony meteorites. *Modern Geol.* **5**, 115–126.
Score R., Schwarz C. M., King T. V. V., Mason B., Bogard D. D., and Gabel E. M. (1981) *Antarctic Meteorite Descriptions 1976–1977–1978–1979.* Publ. 54, Lunar Curatorial Facility, NASA Johnson Space Center, Houston. 144 pp.
Yabuki H., Okada A., and Shima M. (1976) Nesquehonite found on the Yamato 74371 meteorite. *Sci. Pap. Inst. Phys. Chem. Res.* **70**, 22–29. Wako-shi, Saitama, Japan.
Zussman J. (1977) *Physical Methods of Determinative Mineralogy.* Academic, N.Y. 720 pp.

Conditions of formation of pyroxene excentroradial chondrules

Roger H. Hewins,[1] Lisa C. Klein[2] and Benjamin V. Fasano[2]

[1]Department of Geological Sciences, Rutgers University, New Brunswick, New Jersey 08903
[2]Department of Ceramics, Rutgers University, New Brunswick, New Jersey 08903

Abstract—Dynamic crystallization experiments were performed on a synthetic analog of pyroxene-rich chondrules in Manych (LL3 chondrite). Silicate glass beads held on Pt loops were cooled from the liquidus (1445°C) at different rates. The textures produced with cooling rates from 50°C/hr to 3,000°C/hr resemble those of natural chondrules. The average dendrite width for inverted protopyroxene decreases from about 200 microns to less than 10 microns as cooling rate increases from 50°C/hr to greater than 3,000°C/hr. Al is more strongly enriched in pyroxene, as the Fe content increases, in the more rapidly cooled runs. Strong Al enrichment in Manych chondrule pyroxene correlates with decreased dendrite width.

The range of pyroxene dendrite widths in the natural chondrules is greater than in the experimental runs suggesting a wide range of cooling rates. This requires the presence of a gas-chondrule medium, the varying thickness of which influences the rate of radiative cooling of the droplets. The experimental results, textures indicative of incomplete melting and isotopic heterogeneities in chondrules are compatible with an origin for chondrules by reheating of primordial material in the early solar nebula.

INTRODUCTION

Chondrites are of fundamental cosmogonical importance because of their undifferentiated chemical nature. The recent discovery that individual chondrules and inclusions in unequilibrated ordinary chondrites show isotopic heterogeneities (Mayeda *et al.*, 1980; Gooding *et al.*, 1980b) indicates the primordial character of these individual particles. Understanding the exact mechanism of origin of chondrules and the attendant physical conditions, therefore, remains a key factor in understanding the early evolution of the solar system.

There have been several attempts to synthesize particles resembling chondrules and thus indicate possible conditions for their formation. The classic laser-melting experiments of Nelson *et al.* (1972) and Blander *et al.* (1976) did much to demonstrate how droplets of silicate melt crystallize. This study, however, involved very simple bulk compositions and produced only spherulitic olivine as a crystalline phase. Further insight into the textural and chemical diversity of chondrules has been gained from more recent experimental work (Planner and Keil, 1981; Tsuchiyama *et al.*, 1980; Tsuchiyama and Nagahara, 1981; McMahon *et al.*, 1981).

An alternative approach to chondrule genesis has relied on interpretations of textural and compositional data (Reid *et al.*, 1974; Dodd, 1978 a, b; Gooding *et al.*, 1980a; Kimura and Yagi, 1980; McMahon and Haggerty, 1980; Simon and Haggerty, 1980; Rambaldi and Wasson, 1981; Nagahara, 1981). In our own previous work (Klein *et al.*, 1980a; Hewins and Klein, 1980), we have attempted to combine the experimental approach with petrographic study of the Manych LL3 chondrite (Jarosewich and Dodd, 1981) using the extensive analytical data of Dodd (1978 a, b) as a base. We concluded from the observed range of texture and crystallinity and from the range of critical cooling rates required to form glass for different bulk compositions, that Manych droplet chondrules experienced a wide range of cooling histories. By comparison of microporphyry textures with olivine morphology experiments of Donaldson (1976), we concluded that the microporphyry experienced very little supercooling. These conclusions are more consistent with an origin involving reheating of pre-existing material than condensation of the solar nebula to liquid.

One problem in interpreting chondrites is that different kinds of chondrules may have different origins. Microporphyritic chondrules are very different from droplet chondrules (e.g., Dodd, 1978a, b). Since no previous experimental work had produced objects resembling pyroxene excentroradial chondrules, we conducted dynamic crystallization studies with droplets of the appropriate composition and reproduced the textures. In this paper we interpret the range of cooling histories needed to make chondrules, along with chemical constraints, as requiring generation of chondrules, by reheating of primordial particles.

EXPERIMENTAL METHODS

Chondrule-like spherules were formed in dynamic crystallization experiments, using gas-mixing facilities at the Johnson Space Center. The starting material used in these experiments was taken from glass slabs prepared in our previous study (Klein et al., 1980a). The composition of the glass is given in Table 1. It was powdered, pressed into pellets and attached to Pt wire loops 4 mm in diameter. Runs were made at 1 atm. in a furnace with a $CO-CO_2$ gas mixture. Oxygen fugacity was measured by a Ca-stabilized zirconia cell and maintained 1/2 log unit below the iron-wustite buffer. Temperature was measured using a Pt-Rh_{10} thermocouple which was calibrated against the melting point of gold.

The liquidus was determined as 1445°C, since runs quenched from 1450°C produced a glass but a run quenched from 1435°C contained abundant fine settled crystals. The runs described in this paper were initially melted for two hours at 1455 ±5°C. Controlled cooling rates in the range 10°/hr–450°/hr were used. In addition, a cooling rate of approximately 3000°/hr (nearly linear) was achieved by turning off the electrical power used in heating the furnace. Samples cooled from the liquidus by manually pulling the holder out of the top of the furnace grew spherulitic pyroxene. Charges were quenched to a glass only when they were dropped from the hot spot of the furnace through a port into hot water. The experiments included runs cooled from just above the liquidus to about 1000°C, which form the basis of this paper, plus runs with a final temperature of 1300°C and some isothermal runs.

EXPERIMENTAL RESULTS

Runs quenched after melting for two hours produced bluish green glass. The center of the supporting Pt wire contains 1–6% Fe. The iron loss from the charge is about 0.5 wt% (Table 1). Note that this loss is largely accomodated by an increase in MgO relative to the starting composition (Table 1). No comparable increase is shown by SiO_2 and Na_2O which could possibly indicate some volatile loss. These changes from the starting composition are very slight, however, and the composition run is well within the range of Fe/Fe + Mg ratios shown by Manych chondrules (Table 1).

The principal phase encountered in run products is inverted protopyroxene identified by lamellar twinning and abundant cracks. Protopyroxene is the expected liquidus phase in this Si-rich bulk composition, based on pyroxene excentroradial chondrules (e.g., Dodd, 1978b) and pyroxene liquids with similar Fe/Mg ratios (Huebner, 1980). Very minor amounts of late nucleated more Fe- and Ca-rich pyroxene were encountered in some runs.

Droplets formed by quenching from the lowest temperature (1000°C) show the greatest resemblance to natural chondrules, because of the small amount of residual glass, and therefore are described in detail. Inverted protopyroxene occurs as elongate parallel dendritic growths, in many cases apparently nucleated on the droplet surface. Pyroxene dendrite width decreases with increased cooling rate (Fig. 1). These sizes are 200 microns average (range 50–300 microns) for 50°/hr. and 50 microns average (range 30–100 microns) for 450°/hr. Note also dendrites a few microns across in these photomicrographs (Fig. 1a, b). Dendrite width is about ten microns for a cooling rate of 3000°/hr. (Fig. 1c) while in a "quenched" specimen (the charge for which quenching was attempted by hand-pulling) the width is up to eight microns for distinctly chain-like dendrites (Fig. 1d).

Table 1. Composition of natural and synthetic chondrule material.*

	1	2	3	4
SiO_2	58.2	58.1	54.5	56.0
FeO	10.7	10.2	11.0	12.5
MgO	22.1	22.8	26.3	21.1
TiO_2	0.17	0.16	0.12	0.20
Al_2O_3	3.43	3.48	3.30	4.60
Cr_2O_3	0.76	0.79	0.74	0.80
MnO	0.47	0.51	0.46	0.50
CaO	2.37	2.41	2.44	3.30
Na_2O	0.94	0.81	1.09	2.30
Total	99.1	99.2	99.9	101.5

*These also contain about 0.2 wt.% K_2O
1. Starting glass
2. Glass run on Pt wire for two hours
3. Average Manych pyroxene-rich chondrule (Dodd, 1978b)
4. Selected Manych pyroxene-rich chondrule (Dodd, 1978b)

Figure 1 shows the profound influence of cooling rate on the crystallization of liquids cooled from just above the liquidus. The textures in Manych pyroxene-rich chondrules have been illustrated elsewhere (Dodd, 1978b; Klein et al., 1980a) and they are similar to textures in the synthetic droplets, particularly in the extreme range of grain sizes (Fig. 2). The range of dendrite widths for typical excentroradial pyroxene chondrules is approximately the same as those in the present experiments (Fig. 1, 2). However, "glassy" chondrules contain hair-like dendrites of the order of one micron wide (Fig. 2d). In addition, Manych contains granular and porphyritic pyroxene-rich chondrules with little glass, where pyroxene grains are coarser and more equant. The natural chondrules are thus compatible with a wider range of cooling rates, both higher and lower, than those used in these experiments, although some other factors could also have influenced these textures.

Systematic relationships between crystal dimensions, particularly dendrite spacings, and melt cooling rate are well known for a wide variety of phases (e.g., Blau et al., 1973; Grove and Walker, 1977). The present data are presented in the standard log-log plot of size versus cooling rate in Fig. 3. This shows the decrease in dendrite width with increasing cooling rate. Note that the more slowly cooled runs contain some dendrites similar in size to those in the more rapidly cooled runs. These dendrites are interpreted as late nucleated crystals rather than quench crystals, since they are extremely rare in charges quenched from 1300°C. This would seem to imply an increase in undercooling as the temperature falls. However, a factor to be considered is the nucleation of new phases: the extreme Fe- and Ca- contents of some of these are compatible with pigeonite and even sub-calcic augite. The cooling rate estimated for individual natural chondrules from Fig. 3 should be consistent with estimates from analytical data, since pyroxene compositions also vary systematically with cooling rate.

The Al content of low-Ca pyroxene has been shown empirically to increase with increasing cooling rate (e.g., Walker et al., 1978). This trend is displayed by the Manych data, as shown in Fig. 4. The range of Al_2O_3 concentrations shown in Fig. 4 corresponds to the minimum value and the concentration at En_{85}, which is the most Fe-rich pyroxene found in the "quenched" run. (The mean value is indicated and Al contents of late Fe-rich pyroxene in the other runs are also shown). The correlation between pyroxene Al concentration and cooling rate (Fig. 4) is clearly not related to suppression of Al-rich

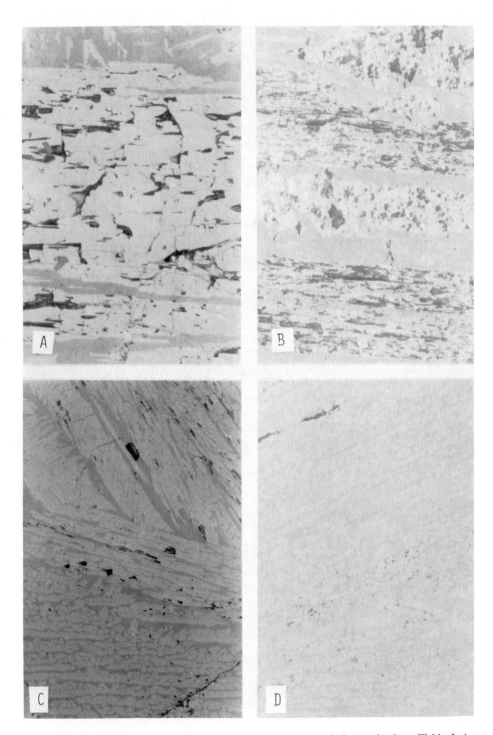

Fig. 1. Photomicrographs of run products: light grey is pyroxene, dark grey is glass. Field of view 0.5 mm long; reflected light. Cooling rates (a) 50°C/hr (b) 450°C/hr (c) 3000°C/hr (d) attempted quench.

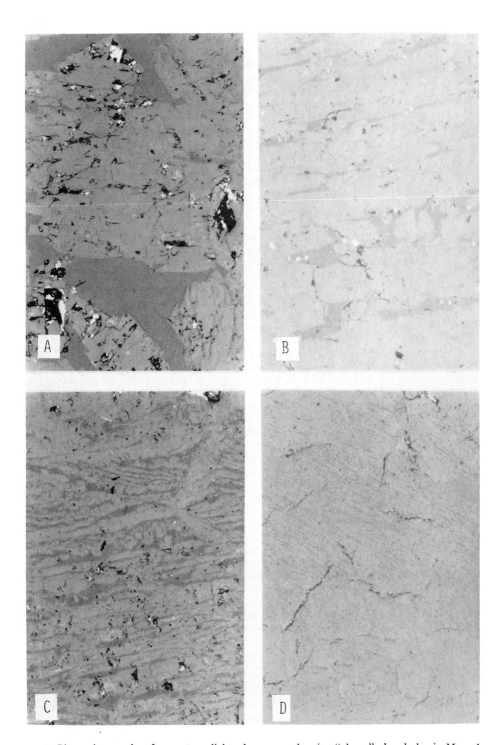

Fig. 2. Photomicrographs of excentroradial and pyroxene-bearing "glassy" chondrules in Manych. Field of view 0.5 mm long.

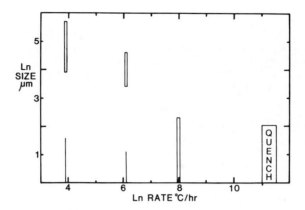

Fig. 3. Influence of cooling rate on pyroxene dendrite width. The open symbols show the size range for the large crystals; the lines show the size range of late nucleated dendrites.

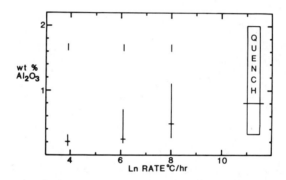

Fig. 4. Influence of cooling rate on Al content of pyroxene. The range shown is from the minimum Al to the value at En_{85}, and the mean is indicated. The Al content of late Fe-rich pyroxene is also shown.

phases, since none were encountered in any of these runs. The higher Al concentrations are most easily explained by deviations from equilibrium partitioning.

The most magnesian pyroxene encountered is $En_{93}Wo_0$, but the major element concentrations in pyroxene are, like Al, strongly influenced by cooling rate. As shown in Fig. 5, the Mg content of the initial pyroxene decreases, the extent of Fe-enrichment of pyroxene decreases and the average Ca content (or the Wo content at En_{88}) increases with increasing cooling rate. These trends are similar to relationships seen for low-Ca pyroxene in other dynamic crystallization experiments. The Mg decrease and Ca increase have been explained by changes in partitioning (Grove and Bence, 1977; Powell et al., 1980). The Mg increase in initial pyroxene in more slowly cooled charges may have been enhanced by greater Fe-loss in these longer runs. The increased Fe-enrichment is a result of diffusion over longer distances, with more impingement of gradients in the liquid, and longer amounts of time for fractional crystallization in the more slowly cooled runs. In the rapidly cooled runs, crystal growth was rapid relative to diffusion rates.

Pyroxene compositions are clearly measures of cooling rate, similar to dendrite widths. Figs. 4 and 5 could thus be used for natural chondrules to obtain estimates of cooling history. However, besides being somewhat imprecise measures, these pyroxene compositions are also dependant on bulk composition. A plot of pyroxene Al versus En content (Fig. 6) has more potential for interpreting the cooling history of individual

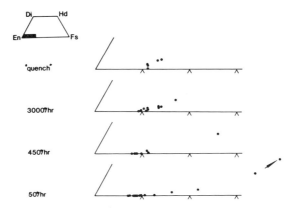

Fig. 5. Influence of cooling rate on major element composition of pyroxene. The quadrilateral segment shown is En_{100} to En_{70}.

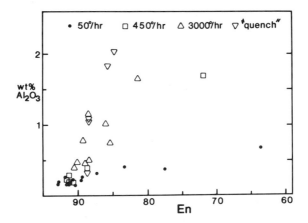

Fig. 6. Variation of Al and En contents of pyroxene as a function of cooling rate.

chondrules. The slope of fractionation trends in this diagram is very strongly influenced by cooling rate. Higher Al or Fe contents in chondrule melts probably displace data points with little change in slope. If corrections can be made for slope changes due to bulk composition, this diagram should be a tool as useful as dendrite width in estimating chondrule cooling rates.

Data for selected Manych chondrules are shown on an Al–En plot (Fig. 7). The two examples with moderate slope have dendrite widths of 50–100 microns; the two with steep slopes have widths of 10–20 microns. Both size and composition data are consistent with cooling rates of about 450°/hr. and 3000°/hr. The Fe-rich points (triangles) are for the composite chondrite shown in Fig. 8: the analytical data confirm the inference from grain sizes that the superposed partial chondrule cooled more rapidly than the original chondrule. The other chondrules are shown in Fig. 2b and c.

DISCUSSION

It must be remembered that there are a wide variety of chondrule types in chondrites which need not all have identical origins. However, pyroxene excentroradial chondrules are easily recognized and widely observed in chondrites, and the physical conditions pertaining to their origin are clearly relevant to the events which preceded chondrite

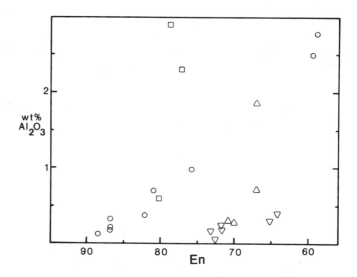

Fig. 7. Variation of Al and En contents of pyroxene in selected pyroxene chondrules from Manych. These chondrules are illustrated in Fig. 2b (circles), Fig. 2c (squares) and Fig. 8 (triangles; upright triangles are outer rim, inverted triangles are inner chondrule). Dendrite widths are 50–100 microns (circles, inverted triangles) and 10–20 microns (squares, upright triangles).

Fig. 8. Composite chondrule in Manych. Inner chondrule has pyroxene crystals about 50 microns across; outer region has dendrites about 10 microns wide.

aggregation. Reproduction of the textures in pyroxene-rich chondrules (Figs. 1, 2) shows that it is possible that chondrules formed by cooling from just above the liquidus at rates ranging from less than 50°/hr. to greater than 3000°/hr. Of course, other factors influence textures and these results do not necessarily exclude all other possible models for chondrule genesis. However, for reasons discussed below, cooling from the liquidus at varying rates appears most plausible.

One possible major influence on chondrule texture is bulk composition as discussed by, e.g., Tsuchiyama *et al.* (1980) and Simon and Haggerty (1980). This is not of major importance for pyroxene chondrules, because Dodd (1978b) showed a similar range of compositions for excentroradial and nearly glassy chondrules in Manych. Olivine-rich chondrules also show a wide variety of textures (Dodd, 1978a, b; Klein *et al.*, 1980a; Hewins and Klein, 1980).

A number of dynamic crystallization studies have shown the importance of initial temperature, i.e., superheat or supercooling and presence of nucleii, on textures (Lofgren *et al.*, 1978; Walker *et al.*, 1978; Grove and Beaty, 1980). We have not carried out a systematic study of these effects and indeed have aimed at a fixed initial temperature in the runs described here. However, there is some information bearing on this question. We also conducted some runs in which complete melting was not achieved. The charges showed small settled crystals plus many pyroxene crystals grown from the melt as stubby laths rather than elongate dendrites. Tsuchiyama *et al.* (1980) cooled a very similar composition from 1600°C, i.e., a very superheated condition, and produced radial olivine. Essentially the same textures were produced with initial temperatures close to the liquidus (Tsuchiyama and Nagahara, 1981). We have not noticed any of these textures in Manych chondrules. The abundant chondrules with elongate pyroxene may well have formed from liquids with a limited range of superheat prior to cooling.

Dendritic crystals grow as a response to supercooling of a liquid, either because of rapid cooling or as a directly imposed initial condition. The possibility that chondrules result from the second case, by condensation of solar nebula to metastable highly supercooled liquid, has long been considered (e.g., Blander *et al.*, 1976; Kimura and Yagi, 1980). Our present experiments do not rule out this possibility for pyroxene chondrules, but there is evidence that other chondrules did not form this way. The euhedral-hopper olivine in microporphyritic chondrules demonstrates little supercooling (less than 40°) from comparison with Donaldson's (1976) results (Dodd, 1978a; Klein *et al.*, 1980a; Hewins and Klein, 1980). It is not easy to extrapolate from one bulk composition to another, but the higher normative olivine content and inferred lower viscosity of the chondrule melt suggest less supercooling than for Donaldson's melt E. The difficulty of explaining these specific chondrules, and the general diversity of chondrule compositions, make the condensation model unattractive.

In addition to arguments about cooling history, other important constraints on chondrule genesis are provided by textures and chemistry. Some chondrules contain relict grains of primitive composition with reverse zoning suggesting melting (Rambaldi and Wasson, 1981; Nagahara, 1981). Chemical data for other chondrules are most easily explained by melting of pre-existing material rather than condensation from the nebula (Dodd, 1978a, b; Gooding *et al.*, 1980a). However, individual chondrules and inclusions in some ordinary chondrites are not equilibrated in oxygen isotopic compositions (Mayeda *et al.*, 1980; Gooding *et al.*, 1980b). These chondrules plot along a mixing line towards a composition rich in ^{16}O. They are therefore primitive particles which were never equilibrated in a large parent body. We can therefore reaffirm our previous conclusion that chondrules formed by reheating of pre-existing solids (Klein *et al.*, 1980a; Hewins and Klein, 1980) but consider only mechanisms that would act on small primordial particles.

The cooling rates inferred for excentroradial chondrules provide additional information on the environment and hence origin of chondrules. The excentroradial pyroxene chondrules are inferred to have cooled at a range of rates, certainly at cooling rates much lower than the $10^{5°}\,hr^{-1}$ associated with radiative cooling of small bodies in a vacuum (e.g., Klein *et al.*, 1980b; Clayton, 1980; Tsuchigama *et al.*, 1980). For bodies of chondrule size, the cooling rate cannot be much higher than about 3000°C/hr to achieve the observed textures. This indicates the presence of a medium retarding heat loss, presumably the solar nebula. The actual cooling rate for a chondrule would depend on the local gas density and abundance of particles. Chondrule formation took place over some length of time, as there are composite chondrules in which later melt cooled on a pre-existing chondrule already cooled (Fig. 8).

The presence of gas required to explain the cooling rates is also consistent with grain size frequency distributions for chondrules. Chondrules, including droplet chondrules and all particles in chondrites, are sorted by size (Dodd, 1976; King and King, 1978, 1979). Aerodynamic drag in the gas phase explains the accretion of particles of a certain size to a moving parent body, while smaller particles are swept away.

Chondrule genesis thus involves the melting of primoridal particles in the early solar nebula. Our experiments do not constrain the heating mechanism, but the isotopic data appear to eliminate any operating after the formation of large bodies. Cosmic heating mechanisms as discussed by many authors (e.g., Wood and McSween, 1977; Planner and Keil, 1981; Gooding et al., 1980a; Clayton, 1980) remain possible.

CONCLUSIONS

Textures in pyroxene-rich chondrules are well reproduced in the laboratory by cooling melts of chondrule composition from just above the liquidus. The range of pyroxene dendrite widths is slightly greater for natural chondrules than for droplets cooled from 50°C/hr to greater than 3000°C/hr. A blanketing medium of variable thickness explains the cooling rates lower than expected for radiative cooling. Experimental results, textures indicating incomplete melting, and isotopic disequilibrium are compatible with reheating of primitive material in the early solar nebula.

Acknowledgments—NASA provided access to the experimental petrology laboratory at the Johnson Space Center and financial support (NSG 7327 - RHH; NSG 9065 - LCK). The study would have been impossible without the guidance, assistance and encouragement of G. E. Lofgren and D. P. Smith. R. T. Dodd kindly provided PTS of Manych containing analyzed chondrules. Reviews of the manuscript by T. L. Grove and R. J. Kirkpatrick were helpful.

REFERENCES

Blander M., Planner H. N., Keil K., Nelson L. S., and Richardson N. L. (1976) The origin of chondrules: experimental investigation of metastable liquids in the system Mg_2SiO_4–SiO_2. *Geochim. Cosmochim. Acta* **40**, 889–896.

Blau P. J., Axon H. J., and Goldstein J. I. (1973) An investigation of the Canyon Diablo metallic spheroids and their relationship to the breakup of the Canyon Diablo meteorite. *J. Geophys. Res.* **78**, 363–374.

Clayton D. D. (1980) Chemical energy in cold-cloud aggregates: the origin of meteoritic chondrules. *Astrophys. J.* **239**, L37–41.

Dodd R. T. (1976) Accretion of the ordinary chondrites. *Earth Planet. Sci. Lett.* **30**, 281–291.

Dodd R. T. (1978a) The composition and origin of large microporphyritic chondrules in the Manych (L-3) chondrite. *Earth Planet. Sci. Lett.* **39**, 52–66.

Dodd R. T. (1978b) Compositions of droplet chondrules in the Manych (L-3) chondrite and the origin of chondrules. *Earth Planet. Sci. Lett.* **30**, 71–82.

Donaldson C. H. (1976) An experimental investigation of olivine morphology. *Contrib. Mineral. Petrol.* **57**, 187–213.

Gooding J. L., Keil K., Fukuoka T., and Schmitt R. A. (1980a) Elemental abundances in chondrules from unequilibrated chondrites: evidence for chondrule origin by melting of pre-existing materials. *Earth Planet. Sci. Lett.* **50**, 171–180.

Gooding J. L., Keil K., Mayeda T. K., Clayton R. N., Fukuoka T., and Schmitt R. A. (1980a) Oxygen isotopic compositions of petrologically characterized chondrules from unequilibrated chondrites. *Meteoritics* **15**, 295.

Grove T. L. and Beaty D. W. (1980) Classification, experimental petrology and possible volcanic histories of the Apollo 11 high-K basalts. *Proc. Lunar Planet. Sci. Conf. 11th*, p. 149–177.

Grove T. L. and Bence A. E. (1977) Experimental study of pyroxene-liquid interaction in quartz-normative basalt 15597. *Proc. Lunar Sci. Conf. 8th*, p. 1549–1579.

Grove T. L. and Walker D. (1977) Cooling histories of Apollo 15 quartz-normative basalts. *Proc. Lunar Sci. Conf. 8th*, p. 1501–1520.

Hewins R. H. and Klein L. C. (1980) Cooling histories of chondrules in the Manych (L-3) chondrite. *Meteoritics* **15**, 302.

Huebner J. S. (1980) Pyroxene phase equilibria at low pressure. *Rev. Mineral.* **7**, 213–288.

Jarosewich E. and Dodd R. T. (1981) Chemical variation among L group chondrites, II: chemical distinctions between L3 and LL3 chondrites. *Meteoritics* **16**, 83–91.

Kimura M. and Yagi K. (1980) Crystallization of chondrules in ordinary chondrites. *Geochim. Cosmochim. Acta* **44**, 589–602.

King T. V. V. and King E. A. (1978) Grain size and petrography of C2 and C3 carbonaceous chondrites. *Meteoritics* **13**, 47–72.

King T. V. V. and King E. A. (1979) Size frequency distributions of fluid drop chondrules in ordinary chondrites. *Meteoritics* **13**, 91–96.

Klein L. C., Fasano B. V., and Hewins R. H. (1980a) Flow behavior of droplet chondrules in the Manych (L-3) chondrite. *Proc. Lunar Planet. Sci. Conf. 11th*, p. 865–878.

Klein L. C., Yinnon H., and Uhlmann D. R. (1980b) Viscous flow and crystallization behavior of tektite glasses. *J. Geophys. Res.* **85**, 5485–5489.

Lofgren G. E., Smith D. P., and Brown R. W. (1978) Dynamic crystallization and kinetic melting of the lunar soil. *Proc. Lunar Planet. Sci. Conf. 9th*, p. 959–975.

Mayeda T. K., Clayton R. N., and Olsen E. J. (1980) Oxygen isotopic anomalies in an ordinary chondrite. *Meteoritics* **15**, 330–331.

McMahon B. M. and Haggerty S. E. (1980) Experimental studies bearing on the magnetite-alloy-sulfide association in the Allende meteorite: constraints on the conditions of chondrule formation. *Proc. Lunar Planet. Sci. Conf. 11th*, p. 1003–1025.

McMahon B. M., Tompkins L. A., and Haggerty S. E. (1981) Bulk compositional variations, viscosities, and the effects of oxidization and sulfide abundances on the formation of chondrules in the Allende meteorite (abstract). In *Lunar and Planetary Science XII*, p. 697–698. Lunar and Planetary Institute, Houston.

Nagahara M. (1981) Petrology of chondrules in ALHA 77015 (L3) chondrite. Sixth Antarctic Meteorite Symposium, National Institute for Polar Research, Japan, p. 31–32.

Nelson L. S., Blander M., Skaggs S. R., and Keil K. (1972) Use of a CO_2 laser to prepare chondrule-like spherules from supercooled molten oxide and silicate droplets. *Earth Planet. Sci. Lett.* **14**, 338-344.

Planner H. N. and Keil K. (1981) Evidence for the three-stage cooling history of fluid droplet chondrules. *Geochim. Cosmochim. Acta.* In press.

Powell M. A., Walker D., and Hays J. F. (1980) Controlled cooling and crystallization of a eucrite: microprobe studies. *Proc. Lunar Planet. Sci. Conf. 11th*, p. 1153–1168.

Rambaldi E. R. and Wasson J. T. (1981) Unequilibrated metals and sulfides in the Bishunpur L3 chondrite. *Geochim. Cosmochim. Acta* **45**, 1001–1015.

Reid A. M., Williams R. J., Gibson E. K. Jr., and Fredriksson, K. (1974) A refractory glass chondrule in the Vigarano chondrite. *Meteoritics* **9**, 35–45.

Simon S. B. and Haggerty S. E. (1980) Bulk compositions of chondrules in the Allende meteorite. *Proc. Lunar Planet. Sci. Conf. 11th*, p. 901–927.

Tsuchiyama A. and Nagahara H. (1981) Experimental reproduction of textures of chondrules: II-effect of residual crystals. Sixth Antarctic Meteorite Symposium, National Institute for Polar Research, Japan, p. 35–36.

Tsuchiyama A., Nagahara H., and Kushiro I. (1980) Experimental reproduction of textures of chondrules. *Earth Planet. Sci. Lett.* **48**, 155–165.

Walker D., Powell M. A., Lofgren G. E., and Hays J. F. (1978) Dynamic crystallization of a eucrite basalt. *Proc. Lunar Planet. Sci. Conf. 9th*, p. 1369–1391.

Wood J. A. and McSween H. Y. Jr. (1977) Chondrules as condensation products. In "Comets, Asteroids and Meteorites" (A. H. Delsemme, ed.), p. 365–373. Univ. Toledo, Ohio.

A revision of metallographic cooling rate curves for chondrites

John Willis and Joseph I. Goldstein

Department of Metallurgy and Materials Engineering, Lehigh University, Bethelehem, Pennsylvania 18015

Abstract—New metallographic cooling rate curves for the chondritic meteorites have been calculated. Based on these curves, estimated cooling rates for the chondrites are twice as fast as those determined using the Wood (1967a) curves. This change in the estimated rates results from the use of the most recent Fe-Ni phase diagram and the use of more accurate computational techniques. The new cooling rate curves are applicable to meteorites with P contents in the metal phase of <0.01 wt%. They should be applied with some caution to meteorites, such as the unequilibrated ordinary chondrites, in which the metal grains may not have equilibrated above ~850 K, or to metallic phases which contain P > 0.01 wt% and/or phosphides.

INTRODUCTION

Much attention has been focused recently on refining the metallographic cooling rate calculations for the iron meteorites. The effects of phosphorus have been incorporated into the cooling rate programs (Willis and Wasson, 1978; Moren and Goldstein, 1979), more sophisticated computational schemes have been developed (Moren and Goldstein, 1978), and the relevant phase diagrams have been redetermined and employed (Romig and Goldstein, 1981). Although cooling rates continue to be published for the chondrites (Scott and Rajan, 1981), no new cooling rate curves have been generated since the pioneering work of Wood (1967a). This is surprising in light of the criticisms which have been leveled against the technique and the original work.

The metallographic technique (Wood, 1967a) involves measuring the Ni content at the center of taenite grains and measuring the size of each grain. The Ni contents are plotted as a function of the apparent taenite grain size. This measured distribution of Ni content vs. taenite grain size is then compared with calculated Ni content-taenite grain size curves to determine the cooling rate. The composition of the taenite grain center is established during cooling of the metal through the kamacite-taenite two-phase field. Therefore the metallographic cooling rates apply to the temperature range from 1000 K to 600 K and the values usually reported are those at 773 K.

Four major discrepancies have been associated with metallographic cooling rates, each indicating that the cooling rates obtained are too slow by a factor of 5 or more. Cooling rates deduced from ^{40}Ar-^{39}Ar ages are four times faster for equilibrated ordinary chondrites and at least seven times faster for unequilibrated ordinary chondrites (Turner et al., 1978; Wood, 1979). Recently revised ^{244}Pu fission track cooling rates are in agreement with those indicated by the radiometric ages (Pellas and Storzer, 1981). A third problem is that many unequilibrated ordinary chondrites have silicates showing quench textures, yet their metallic minerals give some of the slower cooling rates (Bevan and Axon, 1980; Hutchison et al., 1980). Finally, it is hard to reconcile cooling rates of 0.1 K/Myr obtained for mesosiderites and some unequilibrated ordinary chondrites with their proposed origins in a planetary regolith (Powell, 1971).

It has been fifteen years since Wood developed the appropriate cooling rate curves for the chondrites. In view of the problems associated with the metallographic cooling rates and the availability of the new phase diagram and computational techniques, we have redetermined the cooling rate curves for the chondrites.

COMPUTER SIMULATION

The computer program used to generate the Wood-type plots was similar to that developed by Moren and Goldstein (1978). The Crank-Nicholson finite differences approximation to the diffusion equation was used because of the accuracy in its higher order terms and because it has no theoretical stability limit. The problem of adding or moving grid points by linear interpolation was overcome by employing the Murray-Landis variable grid transformation for the interface movement. An exponential time (t)-temperature (T) relationship of the following form

$$T = (T_0 - T_b) \exp(- CR \cdot t/(773 - T_b)) + T_b \tag{1}$$

was used. In this equation T_0 is the temperature at $t = 0$, T_b is the temperature towards which the system cools (taken as 170 K, after Wood, 1979), and CR is the cooling rate at 773 K.

The major difference between our model and that of Moren and Goldstein was the consideration of spherical rather than planar geometry. For spherical particles, Fick's diffusion equation takes the form

$$\frac{\partial C}{\partial t} = \frac{1}{r^2} \frac{\partial}{\partial r}\left(r^2 D \frac{\partial C}{\partial r}\right) \tag{2}$$

where r is the radius of the sphere, C is the Ni content, and D is the diffusivity. In comparison with the planar growth model, this geometry allows more Ni to reach the grain center. Because of the r^3 dependence of grain volume, less taenite grain shrinkage occurs to sustain kamacite growth. As a result, Ni diffusivity in taenite becomes a more dominant process and the effect of kamacite diffusivity on the cooling rate curves is reduced to the point that selection of the appropriate Ni diffusivity value is unimportant.

Kamacite was assumed to form on the outer edge of an initial sphere of taenite and the kamacite/taenite interface was assumed to move radially inward during cooling. In this model, shown in Fig. 1 as Model 1, the total metallic sphere diameter 2R remains constant. This model produces cooling rate curves nearly identical to those produced by a model in which kamacite and taenite particles exist as separate entities, either in contact or communicating through silicates. The Ni gradients in taenite are calculated as the taenite particle decreases in size. This model is the same as Model 1 for the taenite but the kamacite forms somewhere else in the system.

Besides the more accurate computational devices employed by our computer program, we employed the new phase diagram data of Romig and Goldstein (1980). This phase diagram was experimentally determined down to 573 K and shows the retrograde solubility for the $\alpha/(\alpha + \gamma)$ solvus line below 760 K. We used the Goldstein et al. (1965) diffusivity for Ni in taenite and the Borg and Lai (1963) diffusivity for Ni in kamacite. Although these diffusivities are 16 and 18 years old, respectively, no new diffusivities were available.

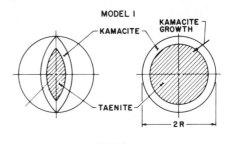

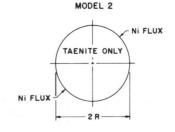

Fig. 1. The basic model assumed for the growth of kamacite is represented as Model 1. Kamacite is assumed to form on the outer edge of a taenite sphere and the kamacite/taenite interface is assumed to move radially inward during cooling. In Model 2, the size of the taenite sphere remains constant and Ni is supplied from some unspecified source.

RESULTS

Figure 2 shows the development of the M-shaped Ni concentration profile in a spherical taenite particle for three different cooling rates, 1, 10 and 100 K/Myr. For the same bulk Ni content and initial radius, slower cooling rates result in higher Ni concentrations at the taenite grain center and smaller diameter taenite grains. The results of numerous such calculations can be combined to produce a set of cooling rate curves on a plot of Ni concentration at the taenite sphere center vs. the final taenite sphere radius. Our cooling rate curves for particles with a uniform initial Ni content of 10% are shown in Fig. 3. The curves of Wood (1967a) are shown for comparison. Our curves are shifted towards higher Ni values and yield faster cooling rates by a factor of 2. Table 1 contains a list of cooling rates for the chondrites derived using our cooling rate curves. Except in a few cases the revised cooling rates were determined by plotting electron microprobe data from the literature on the new cooling rate curves. We deduced the cooling rates by drawing a line below 80% of the data. The 80% level was chosen to allow for off-center grain sectioning and for the presence of grains with planar faces. These effects will be discussed in the following section.

One criticism of Wood's work has been that his curves were derived for grains initially with a uniform bulk Ni content of 10 wt% and did not account for a variation in Ni content between metal particles or between meteorites. Taylor (1976) pointed out that the LL chondrites have much higher Ni contents. We derived cooling rate curves for 7.5, 10, 20 and 30% Ni and, as shown in Fig. 4, the initial Ni content of the metal grains has very little effect on the cooling rate curves. Significant deviations occur only when the Ni concentration at the center of taenite approaches that of the starting composition.

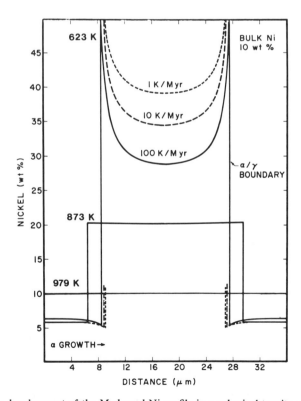

Fig. 2. The development of the M-shaped Ni profile in a spherical taenite particle with a bulk Ni content of 10 wt%. The final profiles at 623 K are shown for three different cooling rates, 1, 10 and 100 K/Myr. The intermediate profiles at 873 K and the starting profiles at 979 K are identical for these cooling rates.

Table 1. Revised cooling rates for ordinary chondritic meteorites.

Meteorite	Type	Old Cooling Rate K/Myr	Reference*	Revised Cooling Rate[+] K/Myr
Tieschitz	H3	1	1	2
Bath	H4	25	2	80
Fayetteville	H4-6	1–100	1, 3	1–100
Sena	H4	10	7	(20)
Wellman	H4	4	2	8
Weston	H3-6	10–1000	3	10–1000
Cee Vee	H5	15	2	25
Ehole	H5	2	1	4
Forest City	H5	10	1	20
Leighton	H5	100	1, 2	200
Malotas	H5	4	2	8
Pantar	H5	5	1	10
Sutton	H5	2	2	4
Mt. Browne	H6	10	2	30
Salles	H6	5	1	10
Mezö-Madaras	L3	1	1, 3	2
Bjurböle	L4	1	1, 2	2
Adelie Land	L5	1	2	2
Ausson	L5	—	1	1–100
Elenovka	L5, 6	2	2	4
Bruderheim	L6	6	5	(12)
Holbrook	L6	100	1, 2	100–200
Kandahar	L6	3	2	6
Mocs	L6	10	1, 2	20
Tillaberi	L6	130	4	400
Waconda	L6	2	2	4
Johnson City	L6	—	1	~20
Shaw	L6?	10^3–10^4	8	10^3–10^4
Bhola	LL3	0.1	3	0.2
Chainpur	LL3	0.2	1	0.5
Olivenza	LL5	4	6	(16)
Ensisheim	LL6	1	6	(4)
Lake Labyrinth	LL6	50	6	(200)
St. Severin	LL6	1	6	(2)

*References: (1) Wood (1967a). (2) Taylor and Heymann (1971). (3) Scott and Rajan (1981). (4) Michel-Levy and Malezieux (1976). (5) Wood (1967b). (6) Taylor (1976). (7) Malezieux in Pellas and Storzer (1977). (8) Scott and Rajan (1979).

[+]Cooling rates were determined by replotting electron microprobe data from the literature on the new cooling rate curves. Revised cooling rates for Sena and Bruderheim were obtained by multiplying the literature values by 2, the difference between the Wood (1967) curves and the new curves. Other revised cooling rates in parentheses were obtained by multiplying the literature values by 4. E. R. D. Scott (personal communication) pointed out that Taylor (1976) did not use the Wood (1967) curves as claimed but curves revised by Wood. These revised curves yield cooling rates about a factor of 2 slower than the Wood (1967) curves.

A slightly different criticism was voiced by Bevan and Axon (1980) upon investigation of polycrystalline taenite grains. They claimed that as a result of rapid solidification, the taenite grains are not initially homogeneous but are zoned. The initial conditions of our cooling rate calculations were altered to assume various initial profiles and new cooling rate curves were produced (Willis and Goldstein, 1981). Only for cooling rates faster than 1000 K/Myr and when grain radii exceed 20 μm do these curves differ from those computed with the models using unzoned taenite grains. Examination of calculated taenite Ni profiles at intermediate temperatures in the cooling rate computations showed that by 870 K, the Ni profiles starting with a uniform particle and those starting with zoned particles were indistinguishable.

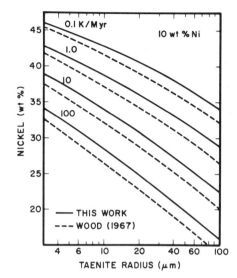

Fig. 3. Cooling rate curves calculated for particles with a uniform initial Ni content of 10 wt%. The curves of Wood (1967a) are shown for comparison.

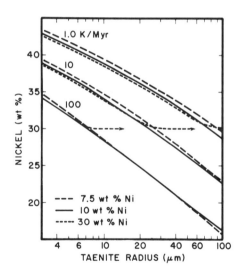

Fig. 4. Sets of cooling rate curves calculated using initial Ni contents of 7.5 wt%, 10 wt% and 30 wt%.

DISCUSSION

Previous work (Willis and Wasson, 1978; Moren and Goldstein, 1979) concluded that the cooling rate curves derived for iron meteorites are accurate to within a factor of 2 for low Ni irons and a factor of 4 for high Ni irons. The larger uncertainty in the curves for the high Ni irons is attributed to the uncertainty in the effect of P on kamacite diffusivities and solubilities. The phosphorus content of the metal in ordinary chondrites is < 0.01 wt% (Reed, 1969; Olsen et al., 1973), much less than in the metal of iron meteorites. Consequently, the cooling rate curves for chondrites are based on P-free parameters and the uncertainty in the curves will be less than that of the irons. A reasonable estimate is that the curves are accurate to within a factor of 2.

A more serious problem occurs with the uncertainties inherent in the measurement of the data points. Some typical data from Wood (1967a) and from Taylor and Heymann (1971) are plotted using the new cooling rate curves in Figs. 5 and 6. Even though Wood

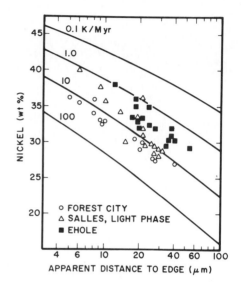

Fig. 5. Data of Wood (1967a) for the chondrites Forest City, Salles (light phase), and Ehole plotted on cooling rate curves developed in this study.

claims that his data points plot coherently (i.e., parallel to the computed curves), the Ehole data show a large amount of scatter and the Salles data show a cooling rate increasing with taenite size. The Bath and Sutton data also show a large amount of scatter similar to that of Ehole. A small portion of the scatter, similar to that seen in the Forest City data (Fig. 5), can be attributed to experimental uncertainties. Other explanations must be found to explain the complete spread of the data. Some possibilities are undercooling prior to kamacite nucleation, the geometry of the taenite grains, and the proximity of the Ni analysis point to the taenite grain center.

We tested the effect of 100 degrees of undercooling on the cooling rate curves. For grains with $>10\%$ Ni and for grains smaller than 60 μm radius, the cooling rate curves were not affected. We are in agreement with Wood (1967a) that the effects of metal particle geometry are minimal. To examine this effect we computed cooling rate curves using a model in which the size of the spherical taenite grain remains fixed while Ni is supplied from some unspecified source (Model 2 in Fig. 1). These "No Growth Model" curves are shown in Fig. 7 and are virtually identical to those computed using the "basic" model (Model 1 of Fig. 1). Also shown in Fig. 7 are curves calculated using a planar front

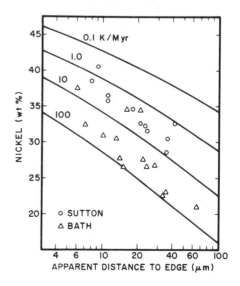

Fig. 6. Data of Taylor and Heymann (1971) for the chondrites Bath and Sutton plotted on the new cooling rate curves. Not shown in this figure are the tetrataenite data points, 14 for Bath and 7 for Sutton, which plot above 50% Ni and are not used to estimate the cooling rates of these meteorites.

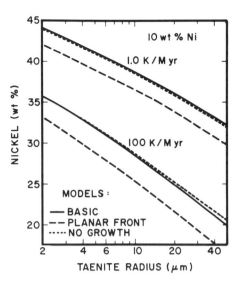

Fig. 7. Cooling rate curves for 1 K/Myr and 100 K/Myr for particles of 10 wt% Ni calculated using three different growth models. The basic model (Model 1 of Fig. 1) assumes a sphere of taenite surrounded by kamacite and a kamacite/taenite interface that moves radially inward during cooling. The planar front model assumes alternating kamacite and taenite plates with diffusion and growth in only one dimension. The no growth model (Model 2 of Fig. 1) assumes a sphere of taenite remaining at a constant diameter during cooling with Ni being supplied from an unspecified source.

growth model (Moren and Goldstein, 1978). These "planar front" curves give cooling rates about 3X slower than those of the spherical model. Thus, the actual geometry of the grain probably does not result in more than a factor of 3 error in an individual measured data point.

Uncertainties introduced by off-center grain sectioning yield slower cooling rates (Wood, 1967a). Statistically, 75% of the data points will yield cooling rates between 60% and 100% of the actual value. However, 10% of the points will yield cooling rates nearly an order of magnitude too low. Because of this effect, Wood (1967a) and Taylor and Heymann (1971) recommended defining the cooling rate of a chondrite by its fastest cooled grains. On the other hand, if a grain is non-spherical and tending to have planar faces, its apparent cooling rate will be greater than its actual cooling rate. This effect can be offset by reducing the cooling rates of the fastest cooled grains by ~50%.

Although the effects of off-center sectioning and grain geometry can account for some of the scatter in the cooling rate data for a particular meteorite, other factors must also be taken into account. One possibility is that the initial metal grains in a chondrite have different Ni contents. This will influence the apparent cooling rates of the larger grains where the central Ni contents approach those of the starting composition.

Another possibility has been proposed by Scott and Rajan (1981). They suggest that the metal grains observed in some chondrites, Fayetteville for example, cooled through 770 K at different locations before compaction into the Fayetteville parent body. After compaction, the parent body was not reheated above 650 K. If the initial locations in which the metal grains cooled were of different burial depths, then the grains will record different cooling rates.

Wood (1967a) made the suggestion that it may be possible for taenite grains to form in favorable sites, which would allow taenite grain growth independent of a diffusion mechanism. This growth mechanism would produce grains which showed no taenite Ni content-grain size relationship. However, because of the Ni content-grain size relationship shown for most taenite grains in the chondrites, Wood felt this to be an unimportant mechanism.

The work of Taylor and Heymann (1971) revealed another serious problem with employing the Wood cooling rate curves. They observed that many taenite grains do not have Ni zoning and remain "clear" when the sample is etched. Such grains were later shown to be a new phase, tetrataenite, by Scott and Clarke (1979). On the cooling rate plots, these tetrataenite grains plot at >50% Ni and show no Ni content-grain size correlation even though some grains are as large as 100 μm across. Tetrataenite Ni content-taenite grain size points have been left off Fig. 6 even though they represent 50% of the taenite grains analyzed in Bath and 33% of the grains analyzed in Sutton. Our cooling rate program and the phase diagram of Romig and Goldstein (1980) do not take into account the presence of tetrataenite. The mechanism for the formation of tetrataenite is still a subject for debate, but it is postulated to form below 590 K (Albertsen *et al.*, 1980). This temperature region is below that considered in this study for the growth of kamacite and below the minimum temperature of 623 K investigated by Romig and Goldstein for the Fe-Ni phase diagram.

Powell (1969) applied the metallographic cooling rate technique to mesosiderites and reported the lowest cooling rates of any meteorites, ~ 0.1 K/Myr. The applicability of the metallographic technique to mesosiderites is questionable for two reasons. Firstly, the P content of the metal in mesosiderites is estimated to be 0.1 to 0.5 wt% (Powell, 1971; Kulpecz and Hewins, 1978). This amount of P significantly alters the Ni diffusivities and solubilities in kamacite and taenite. Chondritic cooling rate curves were calculated using the P-free phase diagram and diffusivities and cannot be applied to meteorites with high P contents in the metal. Secondly, significant fractions of the high Ni metal in mesosiderites is tetrataenite (Clarke and Scott, 1980) which forms below the low temperature limit of our calculations. This transformation was again not modeled in our calculations. The apparent slow cooling rate of these meteorites probably reflects mild heating to ~ 570 K for a long time period.

In order for the metallographic cooling rate technique to be applicable, the metal must have had time to equilibrate at or above 870 K. For most chondrites which show evidence of having experienced thermal metamorphism at peak temperatures in excess of 1000 K, this criterion poses no problem. However, for the unequilibrated ordinary chondrites, the question of metal equilibration must be examined before the cooling rate curves of this paper can be applied.

CONCLUSIONS

This study of the metallographic cooling rate curves for chondrites has shown that the cooling rates obtained using the Wood (1967a) curves need only be increased by a factor of 2 to bring them into agreement with rates based on the most recent phase diagram and computational schemes. This study failed to predict the existence of two populations of taenite grains, the zoned and tetrataenite grains. To predict these two populations will require a new phase diagram, a new thermal history model or both.

Some precautions should also be observed in determining metallographic cooling rates of chondrites. (1) Even though off-center sectioning leads to data points yielding slower rates, the presence of planar surfaces on grains can yield faster cooling rates. Cooling rates should be selected to account for both of these effects. (2) The cooling rate curves cannot be applied to chondrites in which the metal grains have not been equilibrated at or above 850 K. (3) Because the P-free phase diagram and diffusivities have been used, these cooling rate curves are not applicable to meteorites with P contents > 0.01 wt% in their metallic phases.

Acknowledgments—We would like to thank Ed Scott (University of New Mexico) for many helpful discussions during the course of this work and Jeff Taylor (University of New Mexico) for providing us with his data. This work was supported by NASA Grant NGR 39-007-043.

REFERENCES

Albertsen J. F., Knudsen J. M., Roy-Paulsen N. O., and Vistisea L. (1980) Meteorites and thermodynamic equilibrium on FCC iron-nickel alloys (25–50% Ni). *Physica Scripta* **22**, 171–175.
Bevan A. W. R. and Axon H. J. (1980) Metallography and thermal history of the Tieschitz unequilibrated meteorite-metallic chondrules and the origin of polycrystalline taenite. *Earth Planet. Sci. Lett.* **47**, 353–360.

Borg R. J. and Lai D. Y. F. (1963) The diffusion of gold, nickel and cobalt in alpha iron: a study of the effect of ferromagnetism upon diffusion. *Acta Met.* **11**, 861–866.

Clarke R. S. Jr. and Scott E. R. D. (1980) Tetrataenite-ordered FeNi, a new mineral in meteorites. *Amer. Mineral.* **65**, 624–630.

Goldstein J. I., Hanneman R. E., and Ogilvie R. E. (1965) Diffusion in the Fe-Ni system at 1 atm and 40 kbar pressure. *Trans. Met. Soc. AIME* **233**, 812–820.

Hutchison R., Bevan A. W. R., Agrell S. O., and Ashworth J. R. (1980) Thermal history of the H-group of chondritic meteorites. *Nature* **287**, 787–790.

Kulpecz A. A. Jr. and Hewins R. H. (1978) Cooling rate based on schreibersite growth for the Emery mesosiderite. *Geochim. Cosmochim. Acta* **42**, 1495–1500.

Michel-Lévy M. C. and Malezieux J. M. (1976) La meteorite de Tillaberi. *Meteoritics* **11**, 217–224.

Moren A. E. and Goldstein J. I. (1978) Cooling rate variations of group IVA iron meteorites. *Earth Planet. Sci. Lett.* **40**, 151–161.

Moren A. E. and Goldstein J. I. (1979) Cooling rates of group IVA iron meteorites determined from a ternary Fe-Ni-P model. *Earth Planet. Sci. Lett.* **43**, 182–196.

Olsen E., Fuchs L. H., and Forbes W. C. (1973) Chromium and phosphorus enrichment in the metal of Type II (C2) carbonaceous chondrites. *Geochim. Cosmochim. Acta* **37**, 2037–2042.

Pellas P. and Storzer D. (1977) On the early thermal history of chondritic asteroids derived by ^{244}Pu fission track thermometry. In *Comets, Asteroids, Meteorites* (A. H. Delsemme, ed.), p. 355–362. Univ. Toledo, Ohio.

Pellas P. and Storzer D. (1981) ^{244}Pu fission track thermometry and its application to stony meteorites. *Proc. Roy. Soc. Lond.* **A374**, 253–270.

Powell B. N. (1969) Petrology and chemistry of mesosiderites—I. Textures and composition of nickel-iron. *Geochim. Cosmochim. Acta* **33**, 789–810.

Powell B. N. (1971) Petrology and chemistry of mesosiderites—II. Silicate textures and compositions and metal-silicate relationships. *Geochim. Cosmochim. Acta* **35**, 5–34.

Reed S. J. B. (1969) Phosphorus in meteoritic nickel-iron. In *Meteorite Research* (P. M. Millman, ed.), p. 743–762. Reidel, Dordrecht.

Romig A. D. Jr. and Goldstein J. I. (1980) Determination of the Fe-Ni and Fe-Ni-P phase diagrams at low temperatures (700 to 300°C). *Met. Trans.* **A11**, 1151–1159.

Romig A. D. Jr. and Goldstein J. I. (1981) Low temperature phase equilibria in the Fe-Ni and Fe-Ni-P systems: application to the thermal history of metallic phases in meteorites. *Geochim. Cosmochim. Acta.* **45**, 1187–1197.

Scott E. R. D. and Clarke R. S. Jr. (1979) Identification of clear taenite in meteorites as ordered FeNi. *Nature* **281**, 360–362.

Scott E. R. D. and Rajan R. S. (1979) Thermal history of the Shaw chondrite. *Proc. Lunar Planet. Sci. Conf. 10th*, p. 1031–1043.

Scott E. R. D. and Rajan R. S. (1981) Metallic minerals, thermal histories and parent bodies of some xenolithic, ordinary chondrite meteorites. *Geochim. Cosmochim. Acta* **45**, 53–67.

Taylor G. J. (1976) Cooling rates of LL-chondrites (abstract). *Meteoritics* **11**, 374–375.

Taylor G. J. and Heymann D. (1971) The formation of clear taenite in ordinary chondrites. *Geochim. Cosmochim. Acta* **35**, 175–188.

Turner G., Enright M. C., and Cadogan P. H. (1978) The early history of chondrite parent bodies inferred from ^{40}Ar-^{39}Ar ages. *Proc. Lunar Planet. Sci. Conf. 9th*, p. 989–1025.

Willis J. and Goldstein J. I. (1981) The effects of solidification zoning on the metallographic cooling rates of chondrites. *Nature*. In press.

Willis J. and Wasson J. T. (1978) Cooling rates of group IVA iron meteorites. *Earth Planet. Sci. Lett.* **40**, 141–150.

Wood J. A. (1967a) Chondrites: their metallic minerals, thermal histories and parent planets. *Icarus* **6**, 1–49.

Wood J. A. (1967b) Criticism of paper by H. E. Suess and H. Wänke, 'Metamorphosis and equilibration in chondrites.' *J. Geophys. Res.* **72**, 6379–6383.

Wood J. A. (1979) Review of the metallographic cooling rates of meteorites and a new model for the planetesimals in which they formed. In *Asteroids* (T. Gehrels and M. S. Mathews, eds.) p. 849–891. Univ. Arizona, Tucson.

Ordinary chondrite parent body: An internal heating model

Masamichi Miyamoto[1]*, Naoyuki Fujii[1], and Hiroshi Takeda[2]

[1]Department of Earth Sciences, Faculty of Science, Kobe University, Rokkodai-cho, Nada-ku, Kobe 657, Japan [2]Mineralogical Institute, Faculty of Science, University of Tokyo, Hongo, Tokyo 113, Japan

Abstract—By assuming that the decay energy of ^{26}Al was the heat source of the parent body of ordinary chondrites, we performed a model calculation by employing best available constraints including the metamorphic temperature of each petrologic type, the formation interval between type 3 and type 6, etc. The results we obtained are:

1. The radius of the parent body is 85 km for both H- and L-chondrites.
2. The initial temperature (T_0) of the L-chondrite parent body must be lower than that of the H-chondrite body.
3. The volume fraction of each petrologic type in the parent body agrees with the distribution of the ordinary chondrite falls among the petrologic types.
4. The maximum attainable temperature at the center of the L-chondrite parent body is higher than that of the H-chondrite, because of its higher bulk Al content than H's.
5. The contradiction that the cooling rate of the type 3 chondrite estimated from Fe–Ni data is slower than that of type 6 may be explained even in the case of internal heating model.
6. ^{26}Al/^{27}Al ratio at the time when the parent body was formed is estimated to be 5×10^{-6}.

INTRODUCTION

The ordinary chondrites (H, L and LL) are the most common meteorite groups, and account for about 80% of observed meteorite falls. On the basis of chemical, petrological or mineralogical studies on the ordinary chondrites, models of their parent bodies have been proposed by several workers. For example, Wasson (1972; 1974) discussed the genesis and the parent body of ordinary chondrites. Anders (1978) discussed the origin of stony meteorites from their trapped solar-wind gases and proposed a parent body model for the L chondrites. Minster and Allégre (1979) proposed a thermal evolution model of the H-chondrite parent body heated by ^{26}Al. Miyamoto (1979), Miyamoto and Fujii (1980) and Fujii *et al.* (1979) compared an internal heating model of the ordinary chondrite parent body with an external heating model and concluded that whatever the heat source may be, the size of the parent body (bodies) of the ordinary chondrites was relatively small. Oxygen isotope composition data by Clayton *et al.* (1976) point toward the existence of two parent bodies for the H and L (LL) group chondrites.

It has been a subject of controversy whether the chemical fractionations involved in the formation of chondrules and chondrites occurred in the primordial solar nebula or in parent bodies. Highly-volatile element fractionation is closely related to the evidence for (thermal) metamorphism (Wasson, 1974). We have accepted the view that the thermal metamorphism produced the entire range in petrologic types (from 3 to 6) in chondrite parent bodies. We then performed a model calculation by employing best available constraints, and estimated the radius of the parent bodies, positions and volume fraction of each petrologic type, and cooling rates. Existing internal heating models do not explain

---

*Now at Department of Pure and Applied Sciences, College of General Education, University of Tokyo, Komaba, Meguro-ku, Tokyo 153, Japan.

the observation that the L6 chondrite is the most abundant subclass of observed falls and that the cooling rate of type 3's estimated from the Fe–Ni data (Wood, 1967) is slower than that of higher petrologic types. Our model simulation answers some of these questions.

CONSTRAINTS, PARAMETERS AND A MODEL CALCULATION

Metamorphic temperatures of each petrologic type

The metamorphic temperatures for each petrologic type of chondrites have been reported by many investigators (Van Schmus and Koffman, 1967; Van Schmus and Ribbe, 1968; Dodd, 1969; Onuma et al., 1972; Bunch and Olsen, 1974; Saxena, 1976; Ishii et al., 1976; 1979; Heyse, 1978). We assume ~1150 K for the maximum metamorphic temperature of H6, and ~500 K for H3 and L3 chondrites, but no assumption was made for L6 chondrites. These values are approximately the average temperature for each type reported in the above papers. The critical temperature which distinguishes the type 5 from the type 6 chondrites was assumed to be the temperature where the slope of the maximum attainable temperature as a function of the radial distance from the center of the parent body has a steep descent. This argument will be understood when you notice that in the case of the internal heating model the parent body has the same maximum attainable temperature as far as about a half of the radius from the center, and only the outer 30 km shows a temperature gradient. The critical temperature (800–850 K) which distinguishes the type 4 from the type 5 chondrite was assumed on the basis of Mg–Fe diffusion phenomena, since Mg–Fe zoning of olivine grains in type 5 chondrites is not observed.

We assumed that the volume fraction of type 3 in the parent body was ca. 8.5% (5–10%) for both H- and L-chondrites. This value is based on the fact that the type 3 chondrite falls compose 5–10% of the H- or L-chondrites, respectively.

Formation interval between type 3 and 6

We assume that the time interval between the type 3 and type 6 formation is about 100 Ma for both H and L (LL) chondrites. Recent accurate age data for ordinary chondrites indicate that they were formed within a narrow range, around 4.55 G.y. ago. The latest Rb–Sr age data of ordinary chondrites show a subtle difference between type 3 and type 6 ordinary chondrites, though this difference is within the errors of analyses and experiments (Wasserburg et al., 1969; Gray et al., 1973; Manhes et al., 1975; Minster et al., 1976; Hamilton et al., 1979). Our assumption does not appear to be inconsistent with the available Rb–Sr data. We also assumed that the minimum temperature where the Rb–Sr system closes is about 400 K (N. Nakamura, pers. comm.).

Heat source

Three possible heat sources have been considered for the internal heating model (Wasson, 1974): (1) release of gravitational energy, (2) decay energy of long-lived radionuclides, and (3) decay energy of short-lived radionuclides. The parent body of the ordinary chondrite is considered to be so small ($< \sim 500$ km radius) that the heat sources (1) and (2) are ineffective. The Rb–Sr age data imply the heat source is (3). Since an evidence for the existence of the extinct radionuclide ^{26}Al in the early solar system has been found in the Allende meteorite (Lee et al., 1976), ^{26}Al has attracted the attention of many investigators as a heat source. We also adopted the ^{26}Al heat source. The next problem is how much ^{26}Al was in each chondrite parent body. The Al_2O_3 compositions of some H and L chondrites are given in Loverand et al. (1969), Von Michaelis et al. (1969), and Takeda et al. (1979). We note that the bulk Al_2O_3 content in the L chondrite is about

0.2 wt% higher than that in the H chondrite. It implies that heat generation (A) of L chondrite is higher than that of H chondrite, assuming that H and L parent bodies formed at the same time. We adopted the value A for the H chondrite from Herndon and Herndon (1977). It is to be noted that the heat generation, A, is expressed as a function of Al_2O_3 content and the $^{26}Al/^{27}Al$ ratio.

The model calculation

The parameters used for this model calculation are summarized in Table 1. Thermal conductivity, diffusivity and density data are evaluated from Matsui and Osako (1979).

By assuming that the decay energy of ^{26}Al was the heat source in the parent body of the ordinary chondrites and using the above parameters, we calculated temperature-time-depth relations under the conditions, the initial (uniform) temperature of the parent body T_0, and radiation at the surface into a medium at temperature T_0. The equation used for our computer calculation of an internally heated body was derived by us by modifying the equation given by Carslaw and Jaeger (1959). The equation is as follows:

$$T = T_0 + \frac{2hR^2A}{rK} \sum_{n=1}^{\infty} \frac{1}{1-\lambda/\kappa\alpha_n^2} \cdot \frac{[\exp(-\lambda t) - \exp(-\kappa\alpha_n^2)] \sin(r\alpha_n)}{\alpha_n^2[R^2\alpha_n^2 + R \cdot h(R \cdot h - 1)] \sin(R\alpha_n)}$$

where α_n, $n = 1, 2, \ldots$ are the roots of $R\alpha \cot(R\alpha) + R \cdot h - 1 = 0$. K, A, λ, κ, R, t are thermal conductivity, heat generation, decay constant, thermal diffusivity, radius of the parent body, and time, respectively; r is the radial distance from the center of the parent body, and h is a constant calculated from emissivity (E), K and T_0 (Carslaw and Jaeger, 1959, p. 21).

Table 1. Parameters used for the model calculation.

Thermal conductivity[1]	$K = 1.0$ W m^{-1} K^{-1}
Thermal diffusivity[1]	$\kappa = 5.0 \times 10^{-7}$ m^2 s^{-1}
Density[1]	$\rho = 3.2 \times 10^3$ kg m^{-3}
Heat generation[2]	$A = 11.67 \times \left(\frac{^{26}Al}{^{27}Al}\right)$ W m^{-3}
Decay constant (^{26}Al)	$\lambda = 9.63 \times 10^{-7}$ y^{-1}
Emissivity[3]	$E = 0.8$ (h = 1.0 m^{-1})

[1]Evaluated from Matsui and Osako (1979).
[2]For H chondrite, from Herndon and Herndon (1977).
[3]Evaluated from Carslaw and Jaeger (1959).

RESULTS AND INTERPRETATIONS

The temperature, T_0

The maximum metamorphic temperature of type 3 and the temperature T_0 are closely related to each other, because type 3's are located at the surface of the parent body in our model calculation. If the temperature T_0 decreases, then the volume fraction of the type 3 increases, and vice versa. Thus, the temperatures, 200 K (H) and 180 K (L) are determined by the constraints that the type 3 chondrites compose ca. 8.5% of both the H- and L-chondrites and that the maximum metamorphic temperature of the type 3 chondrites is ~500 K. If the temperature T_0 of the L-chondrite parent body is the same as that of the H chondrite, the volume fraction of the L3 chondrite is too small and is not in agreement with observation. The temperature T_0 of the L-chondrite parent body must be lower than that of the H-chondrite.

The radius

The radius, R, of the parent body is determined to be about 85 km based on the assumption that the duration of thermal metamorphism at the center of the parent body is

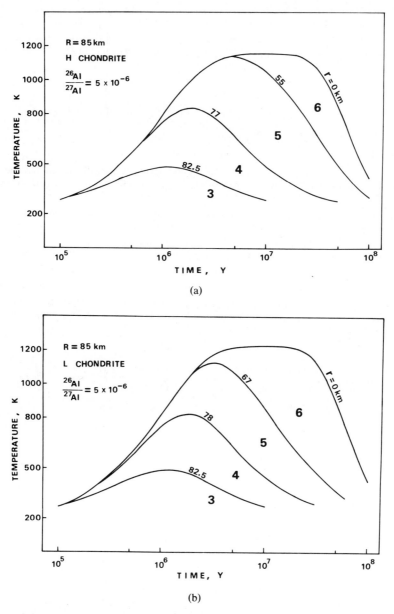

Fig. 1. Calculated temperature-time relations at four different depths. R: radius of the parent body. Numbers on curves denote the radial distance (r) from the center of the parent body. Larger numbers (3 to 6) denote the petrologic types. (a) H-chondrite parent body. (b) L-chondrite parent body.

~100 Ma. This result is also based on the assumption that the Rb–Sr system closes at the temperature, 400 K. This temperature may be a lower limit. The radius is the same between H- and L-chondrite parent bodies. Because the radius is sensitive to the thermal conductivity, the thermal conductivity is an important parameter to estimate the size of the parent body by the calculation.

Temperature-time-depth variations

The results of the temperature-time-depth variations and the internal structures of the parent bodies for H and L chondrites are shown in Figs. 1 and 2. From the profile of the temperature gradient in the parent body, we can determine the volume fraction of each petrologic type (except for the type 3, because the volume fraction of type 3 is used as a constraint for this model calculation). The results of the volume fraction of each petrologic type are tabulated in Table 2 together with the distributions of the ordinary chondrite falls among the petrologic types for comparison. For the H chondrites, close agreement was found between calculation and observation. The difference in the abundances in H and L depends mainly on its higher bulk Al content of L than that of the H chondrite. We note that the volume fraction of L6 in L chondrites is larger than that of H6 in H chondrites. Table 2 also includes the distribution of ordinary chondrites found in Antarctica. The abundance of L6 in Antarctic chondrites is lower than that of Wasson's statistics (Wasson, 1974).

The maximum temperature at the center of the L-chondrite parent body was found to be about 75 K higher than that of the H-chondrite, because of its higher bulk Al content than H's. Ishii et al. (1976; 1979) and Saxena (1976) have reported that the metamorphic temperature of L6 chondrites was higher (~10 K) than that of H6 chondrites. Although the difference may be within the limit of errors, the result of our model calculations is not inconsistent with this fact.

The cooling rate obtained by our model calculation for the type 6 chondrites is a few K/Ma. This result is consistent with that estimated from Widmanstätten structure in the Fe–Ni alloy (1–10 K/Ma at 750 K) by Wood (1967). It has been considered that the internal heating model does not satisfy the condition that the cooling rate (0.1–1 K/Ma) deduced from Fe–Ni data of the type 3 chondrites is slower than that of type 6's. The

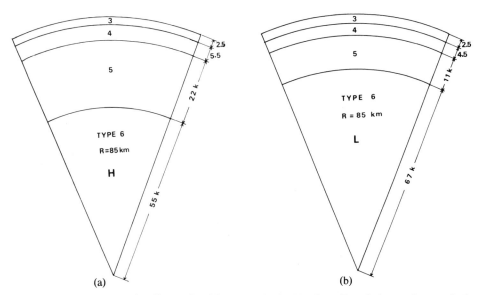

Fig. 2. Internal structure of ordinary chondrite parent body. Numbers (3 to 6) denote the petrologic types. (a) H-chondrite parent body. (b) L-chondrite parent body.

Table 2. Distribution of chondrites among the petrologic types.

			Type 3	4	5	6
H	Calc.		8.6%*	17.1%	47.2%	27.1%
	Obs.	(Wasson)[1]	5.3	20.2	46.5	28.1
		(Antarctic)[2]	5	26	36	33
L	Calc.		8.6*	14.2	28.3	49.0
	Obs.	(Wasson)[1]	5.5	6.7	17.0	70.9
		(Antarctic)[2]	9	20	15	56

[1]Ordinary chondrite falls (Wasson, 1974).
[2]Distribution of chondrites in Antarctic meteorite collection (Yanai, 1979; Takeda et al. 1979).
*Assumed.

physical meaning of the cooling rate obtained from the Fe–Ni data is not the overall cooling rate during the cooling episode but the cooling rate at the range of temperature about 500–800 K. The absolute value of the cooling rate ($\partial T/\partial t$) has a minimum at the maximum temperature which each petrologic type has experienced as was shown in Fig. 1. In our model calculation, the type 3 chondrite is assumed not to be heated above the temperature of ~600 K, although type 6 is heated up to a temperature of ~1100 K. Taking these facts into consideration, we are not in a position to exclude the internal heating model because of the contradiction for the cooling rate of each petrologic type. According to the cooling rate data derived from Pu-244 fission-track thermometry (Pellas and Storzer, 1979), the H chondrites of different petrologic types show cooling rates which significantly increase from type 6 to type 4, and a similar behavior is observed for L4 and L5 chondrites.

^{26}Al/^{27}Al ratio

The ^{26}Al/^{27}Al ratio at the time when the parent body was formed can also be estimated. In our model, it should be mentioned that the ^{26}Al/^{27}Al ratio is determined neither by the size of the parent body nor by the cooling rate data, but by the maximum thermal metamorphic temperature of the type 6 chondrites only. The thermal metamorphic temperature of the type 6 chondrites can be obtained with considerable accuracy by the pyroxene geothermometer (e.g. Ishii et al., 1976). The ^{26}Al/^{27}Al ratio, 5×10^{-6} we obtained implies the formation interval of the ordinary chondrite parent body after the Allende (CAI) formation is about 2.5 Ma, assuming the ^{26}Al/^{27}Al ratio in Allende (CAI) is about 5×10^{-5} (Lee et al., 1976). In our model calculation we neglected the duration of parent body formation, because the size of the parent body is small.

DISCUSSIONS

During the course of this study (Miyamoto, 1979), Minster and Allégre (1979) reported that the radius of the parent body of the H chondrite is about 175 km based on the internal heating model calculation, and the ^{26}Al/^{27}Al ratio at the time of accretion is $3-6 \times 10^{-6}$. The difference of the size of the parent body between their result and ours (R = 85 km) is due mainly: (1) to the fact that they used the cooling rate (2 K/Ma) in the range of 660–300 K as the constraint for their calculation, and (2) to the difference in the thermal conductivity used for the model calculation. The thermal conductivity they used is that of olivine. We selected the thermal conductivity for less weathered ordinary chondrites found in Antarctica (Matsui and Osako, 1979). Minster and Allégre (1979) do not mention the internal structure of the parent body (burial depth for each petrologic type).

At present, we cannot find a difference in the formation ages between the type 3 and type 6 chondrites on the basis of $^{40}Ar-^{39}Ar$ or Pb–Pb age data (e.g. Turner et al., 1978). Relative $^{129}I-^{129}Xe$ age data of ordinary chondrites are confusing. It is difficult to find any consistency among these data.

The temperature of the primordial solar nebula (in terms of the distance from the proto-sun), that is, T_0 in our model calculation, might have played an important role even in the internal heating model. The volume fraction of the type 3 chondrite is influenced considerably by this temperature T_0.

As was pointed out by Wasson (1972), if metamorphism occurred in internally heated bodies, some of the volatiles released would recondense in the cooler portions near the surface, and H_2O would be trapped beneath a permafrost region and would react with the metallic Fe to produce FeO. These problems are unfavourable to the internal heating model. Though the difficulties may be avoided by the high surface temperatures in the case of internal heating model, it is necessary to carry out the chemical and petrological examination in the type 3 ordinary chondrites in more detail to find any evidence either for or against the internally heated model. It is also important to search for any evidence for the presence of ^{26}Al in ordinary chondrites in order to determine whether our model is valid or not (Schramm et al., 1970).

Our model calculation is dependent mainly on the following constraints: (1) the maximum metamorphic temperature of the type 6 chondrites determines the $^{26}Al/^{27}Al$ ratio at the time when the parent body was formed. Although the bulk Al_2O_3 content is also related to this metamorphic temperature, we used the observed Al_2O_3 content, (2) the time interval between the type 3 and type 6 formation (~ 100 Ma) determines the radius (R) of the parent body, and (3) the temperature T_0 is determined by the constraints that the volume fraction of the type 3 chondrites is ca. 8.5% and that the maximum metamorphic temperature of the type 3 chondrites is ~ 500 K.

Miyamoto and Fujii (1980) examined in detail the external heating model for the ordinary chondrite parent body under the best available constraints which are similar to this internal heating model. They concluded as follows: (1) Because of the short duration of any possible source for external heating of the chondrite parent body, the radius of the parent body would have to be about 2–10 km. (2) If this is the case, the ordinary chondrites' parent body (bodies) might be the survivors of the planetesimals directly disintegrated from the dust layer in the primordial solar nebula (Goldreich and Ward, 1973).

Regardless of whether the heat source was internal or external, the size of the parent body (or bodies) of the ordinary chondrites was relatively small. This is consistent with the hypothesis that Apollo asteroids are the source of the ordinary chondrites.

Acknowledgments—We are indebted to Profs. K. Ito and N. Nakamura for discussion and to Prof. Y. Takano for taking interest in our work. We thank Profs. C. L. Chou, J. H. Jones, M. -S. Ma, and Van Schmus for critical reading of the manuscript. The calculations were made with the NEAC ACOS 700 II computer at the computer center of Kobe University.

REFERENCES

Anders E. (1978) Most stony meteorites come from the asteroid belt. In *Asteroids: An Exploration Assessment* (D. Morrison and W. C. Wells, eds.), p. 57–75. NASA-CP-2053.

Bunch T. E. and Olsen E. (1974) Restudy of pyroxene–pyroxene equilibration temperatures for ordinary chondrites. *Contrib. Mineral. Petrol.* **43**, 83–90.

Carslaw H. S. and Jaeger J. C. (1959) *Conduction of heat in solids: 2nd ed.* Oxford Univ. N.Y. 510 pp.

Clayton R. N., Onuma N., and Mayeda T. (1976) Classification of meteorites by oxygen isotope composition. *Earth Planet. Sci. Lett.* **30**, 10–18.

Dodd R. T. (1969) Metamorphism of the ordinary chondrites: A review. *Geochim. Cosmochim. Acta* **33**, 161–203.

Fujii N., Miyamoto M., and Ito K. (1979) The role of external heating and thermal metamorphism of chondritic parent body. *Planet. Sci.* **1**, 84.

Goldreich P. and Ward W. R. (1973) The formation of planetesimals. *Astrophys. J.* **183**, 1051–1061.

Gray C., Papanastassiou D. A., and Wasserburg G. J. (1973) The identification of early condensates from the solar nebula. *Icarus* **20**, 213–239.

Hamilton P. J., Evensen N. M., and O'Nions R. K. (1979) Chronology and chemistry of Parnallee (LL-3) chondrules (abstract). In *Lunar and Planetary Science X*, p. 494–496. Lunar and Planetary Institute, Houston.

Herndon J. M. and Herndon M. A. (1977) Aluminum-26 as a planetoid heat source in the early solar system. *Meteoritics* **12**, 459–465.

Heyse J. V. (1978) The metamorphic history of LL-group ordinary chondrites. *Earth Planet. Sci. Lett.* **40**, 365–381.

Ishii T., Miyamoto M., and Takeda H. (1976) Pyroxene geothermometry and crystallization-, subsolidus equilibration-temperatures of lunar and achondritic pyroxenes (abstract). In *Lunar Science VII*, p. 408–410. The Lunar Science Institute, Houston.

Ishii T., Takeda H., and Yanai K. (1979) Pyroxene geothermometry applied to a three-pyroxene achondrite from Allan Hills, Antarctica and ordinary chondrites. *Mineral. J.* **9**, 460–481.

Lee T., Papanastassiou D. A., and Wasserburg G. J. (1976) Demonstration of ^{26}Mg excess in Allende and evidence for ^{26}Al. *Geophys. Res. Lett.* **3**, 41–44.

Loverand W., Schmitt R. A., and Fisher D. E. (1969) Aluminum abundances in stony meteorites. *Geochim. Cosmochim. Acta* **33**, 375–385.

Manhes G., Minster J. F., and Allégre C. J. (1975) Lead–lead and rubidium–strontium study of the Saint-Severin LL6 chondrite. *Meteoritics* **10**, p. 451.

Matsui T. and Osako M. (1979) Thermal property measurement of Yamato meteorites. *Memoirs of the National Institute of Polar Research, Special Issue No. 15* (T. Nagata, ed.), p. 243–252. Nat'l Inst. Polar Res., Tokyo.

Minster J. F. and Allégre C. J. (1979) ^{87}Rb/^{87}Sr chronology of H chondrites: Constraint and speculations on the early evolution of their parent body. *Earth Planet. Sci. Lett.* **42**, 333–347.

Minster J. F., Birk J. L., and Allégre C. J. (1976) ^{87}Rb/^{87}Sr constraints on the primitive chronology of meteorites. *Meteoritics* **11**, 336–337.

Miyamoto M. (1979) Thermal evolution of the ordinary chondrite parent body. *Proc. 12th Lunar and Planet. Symp.*, p. 92–98, Inst. Space Aero. Sci., Univ. Tokyo.

Miyamoto M. and Fujii N. (1980) A model of the ordinary chondrite parent body: An external heating model. *Memoirs of the National Institute of Polar Research, Special Issue No. 17* (T. Nagata, ed.), p. 291–298. Nat'l Inst. Polar Res., Tokyo.

Onuma N., Clayton R. N., and Mayeda T. K. (1972) Oxygen isotope temperatures of "equilibrated" ordinary chondrites. *Geochim. Cosmochim. Acta* **36**, 157–168.

Pellas P. and Storzer D. (1979) Differences in the early cooling histories of the chondritic asteroids. *Meteoritics* **14**, p. 513–515.

Saxena S. K. (1976) Two-pyroxene geothermometer: a model with an approximate solution. *Amer. Mineral.* **61**, 643–652.

Schramm D. N., Tera F., and Wasserburg G. J. (1970) The isotopic abundance of ^{26}Mg and limits on ^{26}Al in the solar system. *Earth Planet. Sci. Lett.* **10**, 44–59.

Takeda H., Duke M. B., Ishii T., Haramura H., and Yanai K. (1979) Some unique meteorites found in Antarctica and their relation to asteroids. *Memoirs of the National Institute of Polar Research, Special Issue No. 15* (T. Nagata, ed.), p. 54–76. Nat'l. Inst. Polar Res., Tokyo.

Turner G., Enright M. C., and Cadogan P. H. (1978) The early history of chondrite parent bodies inferred from ^{40}Ar–^{39}Ar ages. *Proc. Lunar Planet. Sci. Conf. 9th*, p. 989–1025.

Van Schmus W. R. and Koffman D. M. (1967) Equilibration temperatures of iron and magnesium in chondritic meteorites. *Science* **155**, 1009–1011.

Van Schmus W. R. and Ribbe P. H. (1968) The composition and structural state of feldspar from chondritic meteorites. *Geochim. Cosmochim. Acta* **32**, 1327–1342.

Von Michaelis H., Ahrens L. H., and Willis J. P. (1969) The composition of stony meteorites II. The analytical data and an assessment of their quality. *Earth Planet. Sci. Lett.* **5**, 387–394.

Wasserburg G. J., Papanastassiou D. A., and Sanz H. G. (1969) Initial Sr for a chondrite and the determination of a metamorphism or formation interval. *Earth Planet. Sci. Lett.* **7**, 33–43.

Wasson J. T. (1972) Formation of ordinary chondrites. *Rev. Geophys. Space Phys.* **10**, 711–759.

Wasson J. T. (1974) *Meteorites: Classification and Properties*. Springer-Verlag, N.Y. 316 pp.

Wood J. A. (1967) Chondrites: Their metallic minerals, thermal histories, and parent planets. *Icarus* **6**, 1–49.

Yanai K. comp. (1979) *Catalog of Yamato Meteorites, 1st ed.* Nat'l. Inst. Polar Res., Tokyo. 188 pp. with 10 pls.

Electron microscopy of carbonaceous matter in Allende acid residues

Gregory R. Lumpkin

Department of Physics, University of California, Berkeley, California 94720

Abstract—On the basis of characteristic diffuse ring diffraction patterns, much of the carbonaceous matter in a large suite of Allende acid residues has been identified as a variety of turbostratic carbon. Crystallites of this phase contain randomly stacked sp^2 hybridized carbon layers and diffraction patterns resemble those from carbon black and glassy carbon. Carbynes are probably absent, and are certainly restricted to less than 0.5% of these acid residues. The work of Ott et al. (1981) provides a basis for the possibility that turbostratic carbon is a carrier of noble gases, but an additional component–amorphous carbon–may be necessary to explain the high release temperatures of noble gases as well as the glassy character of many of the carbonaceous particles. Carbynes are considered to be questionable as important carriers of noble gases in the Allende acid residues.

INTRODUCTION

Carbon is a minor, but common component of meteorites of various classes. It occurs at the trace level in most meteorites, attaining maximum concentrations in the carbonaceous chondrites (~5%) and the ureilites (~4%) [Nagy (1975) and references therein]. Polymorphs of elemental carbon appear to be the dominant forms in non-carbonaceous meteorites. Graphite is a common accessory mineral in iron meteorites and a few chondrites (Mason, 1972; Rubin et al., 1981; Scott et al., 1981). Carbon-rich veins in ureilites contain graphite, diamond, chaoite, and lonsdaleite (Frondel and Marvin, 1967; Vdovykin, 1969, 1970). The Fe-Ni carbides cohenite and haxonite are found in minor quantities in iron meteorites and enstatite chondrites (Brett, 1967; Scott and Agrell, 1971) and have recently been found in several unequilibrated ordinary chondrites (Taylor et al., 1981).

In the carbonaceous chondrites, carbon is present in the form of structurally and chemically complex organic matter, as opposed to polymorphs of pure carbon. Most of the carbon resides in a poorly understood solvent-insoluble carbonaceous substance, with lesser amounts in hydrocarbons and other compounds (Nagy, 1975). In the Allende C3V meteorite about 0.3% carbon is associated with the inorganic matrix (Chang et al., 1978), mainly as a thin surface film and micromounds on mineral grains (Bauman et al., 1973; Bunch and Chang, 1980). The carbonaceous matter is reportedly amorphous to x-rays (Breger et al., 1972; Bauman et al., 1973). Differential thermal analyses by Simmonds et al. (1969) and Bauman et al. (1973) indicate that the material is unlike graphite and, to some extent, resembles coal. Making the story even more interesting, Green et al. (1971) found diffraction patterns characteristic of "poorly crystalline graphite" during the course of their transmission electron microscope (TEM) study of the Allende matrix.

Following the initial excitement generated by the fall of Allende in 1969 and the discovery of its petrologically interesting inclusions, another burst of interest was generated when, in 1975, Lewis et al. (1975) made the important discovery that most of the planetary noble gases in Allende are associated with the small (~0.5 wt.%) residue produced by dissolution of the meteorite in HF and HCl. Further treatment with HNO_3 resulted in severe depletion of noble gases, despite little weight loss, while leaving behind sparse amounts of isotopically anomalous gas. The host phase model developed by the Chicago group postulated a sulfide mineral, dubbed Q, as the major carrier of the isotopically normal noble gases, with the anomalous gases distributed between Q, chromite, and carbon (Lewis et al., 1975, 1977; Gros and Anders, 1977). The chromite host was later discribed by Lewis et al. (1979) as "ferrichromite", a phase intermediate in composition between chromite and magnetite. Carbynes, the proposed triply bonded polymorphs of carbon (Whittaker, 1978), have recently been identified as the major carbonaceous phases in Allende and labeled as important carriers of noble gases (Whittaker et al., 1980; Hayatsu et al., 1980). Whittaker et al. (1980) further suggested that the identification of Q as a sulfide should be reconsidered.

Workers in our laboratory, on the other hand, have consistently argued that carbonaceous matter is the primary carrier of both the normal *and* anomalous noble gases (Frick and Chang, 1978; Reynolds et al., 1978; Ott et al., 1981). Even so, only recently have we begun to obtain a better understanding of the nature of the carbonaceous matter. Smith and Buseck (1980, 1981a) used TEM lattice imaging techniques to show that most of the carbonaceous matter in several of the Allende acid residues prepared by Ott et al. (1981) is a poorly ordered form of graphitic carbon. Preliminary work on a larger suite of these residues by the author, using electron diffraction methods, again showed that the carbonaceous matter consists predominantly of poorly ordered graphitic carbon (Lumpkin, 1981). In this paper I will describe in greater detail the characterization of the carbon-rich residues from the Allende meteorite.

EXPERIMENTAL PROCEDURES

Sample preparation

A suite of 16 acid residues[1] was kindly provided by S. Chang of NASA-Ames Research Center. The procedures used in the preparation of these samples are abstracted in Table 1. Complete details, including chemical analyses and isotopic data, can be found in Ott et al. (1981). Three major groups of samples are represented: the A-I group, subjected to *very* harsh demineralization; the A-II group, prepared by harsh demineralization; and the A-III group, prepared with milder treatment following

Table 1. Outline of sample preparation for Allende residues.

Sample	Acid treatments and physical separations
A-I-B2B-4K	Very harsh HF/HCl → 1.81–1.91 g/cc → centrifuge 4,000 rpm/1 hr.
A-I-B2B-10K	As above → centrifuge 10,000 rpm/1 hr.
A-I-B2B-14K/1	As above → centrifuge 14,000 rpm/1 hr.
A-I-B2B-14K/20	As above → centrifuge 14,000 rpm/20 hrs.
A-II-D	Harsh HF/HCl → 1.91–2.17 g/cc
A-II-D-4hrHNO$_3$	4 hours concentrated HNO$_3$ on A-II-D
A-II-D-17hrHNO$_3$	17 hours concentrated HNO$_3$ on A-II-D
A-II-D-fum HNO$_3$	1 hour fuming HNO$_3$ on A-II-D
A-II-D-HClO$_4$	1.5 hours concentrated HClO$_4$ on A-II-D
A-III-A1	Mild HF/HCl → sedimentation ≥ 20 cm/min.
A-III-A2	Further mild HF/HCl on A-III-A1
A-III-A2L	≤ 2.25 g/cc on A-III-A2
A-III-B	Mild HF/HCl → sedimentation 2–20 cm/min.
A-III-BL	≤ 2.25 g/cc on A-III-B
A-III-C	Mild HF/HCl → sedimentation ≤ 2 cm/min.
A-III-CL	≤ 2.25 g/cc on A-III-C

---

[1] Nine of the samples examined in this paper were obtained for lattice imaging studies by P. P. K. Smith and P. R. Buseck, including six A-III residues and three A-II-D residues etched in HNO$_3$. Smith and Buseck (1981a) report detailed results from A-III-BL and A-II-D-fumHNO$_3$.

the Chicago method (Lewis et al., 1975). A subgroup of A-II is constituted by the samples etched in HNO_3 and $HClO_4$. The various physical methods applied to the samples are outlined in Table 1; they involved centrifugation and sedimentation to achieve particle separations according to effective density and size.

Scanning electron microscopy

A Cambridge Stereoscan 600 scanning electron microscope (SEM) was used to obtain basic textural and mineralogical information from several of the acid residues. One or two samples from each group of residues were dispersed onto Al sample holders, the surface of which had been previously coated with colloidal graphite to eliminate interference from the Al metal during energy dispersive x-ray analysis and to provide a conducting medium for the sample material. About 200 Å of carbon was evaporated onto the samples to enhance the electrical conductivity of inorganic mineral grains. Semiquantitative analyses of spinel group minerals were obtained at 15 kV using a Kevex-ray 5000 A energy dispersive spectrometer (EDS) attached to the SEM. Qualitative EDS spectra from carbonaceous particles were recorded after counting up to 20,000 integral counts in a 20 channel window centered at 2.957 keV (ArK_α).

Transmission electron microscopy

Aliquots of the acid residues were dispersed onto a Formvar film and floated onto Cu TEM grids using distilled water. The grids were coated with Au or Al for calibration of selected area diffraction (SAD) patterns. This allows SAD patterns of the standard and unknown to be recorded simultaneously and is particularly advantageous for calibration of single crystal SAD patterns. Because the metal coating tends to mask the weak, diffuse ring SAD patterns of poorly crystalline meterials, surface areas of several of the grids were only half coated with Au or Al. Polycrystalline ring patterns were calibrated by recording the unknown and standard patterns separately, without changing any of the instrumental conditions. The accuracy attainable in d-spacings using these procedures was estimated by recording single crystal SAD patterns along [001], [011], and [012] zone axes of NaCl. For the simultaneous calibration, errors are estimated to range from ± 0.001 Å at $d = 0.5$ Å to ± 0.02 Å at $d = 4.0$ Å. When SAD patterns are calibrated separately, the errors are larger by a factor of approximately two.

Electron diffraction work was carried out on a Siemens 1A TEM operated initially at 60 kV and later at 100 kV as a precaution for detecting beam-sensitive phases, for example carbynes (Whittaker et al., 1980; Whittaker, pers. comm., 1980). Single crystal SAD patterns showing hexagonal symmetry were characterized by measuring the value of d_{max}, which lies in the range 4.3–4.7 Å for most of the patterns. I was unable to obtain additional diffraction data along other zone axes from these particles using the double-tilting stage available with the Siemens 1A, which has a tilt range of 0–30°. The results reported in this work apply to the 0.5–2.0 μm grain size population.

Particles that gave hexagonal SAD patterns in sample A-III-A1 were later analyzed qualitatively for Mg, Al, Si, K, Ca, Ti, and Fe using a Tracor-Northern energy dispersive x-ray analyzer attached to a JEOL JEMIOOC TEM. All analyses were performed using an operating voltage of 100 kV.

RESULTS

General description of the residues

The basic textural and mineralogical features of the Allende acid residues are described in this section. The main differences between the three groups of samples are the relative amounts of carbonaceous matter and inorganic mineral grains, and textural features of the carbonaceous matter. The A-I and A-II samples, subjected to very harsh and harsh HF/HCl treatments, respectively, consist of carbonaceous clumps and little or no coarse grained ($\geq 1.0 \mu$m) minerals (Figs. 1 and 2). On the other hand, coarse grained spinel, chromite, and minor hibonite are found in the A-III samples (Fig. 3), which received the mild HF/HCl treatment. Two types of carbonaceous particles are observed in the residues: soft, irregular clumps and hard, glass-like[2] particles which show conchoidal fracture. Glass-like particles are abundant in the A-I and A-II samples, but are virtually absent in the residues of the A-III group.

---

[2]The term glass-like, as used here, simply refers to the morphological appearance of the hard carbonaceous particles, not to any structural property of the material.

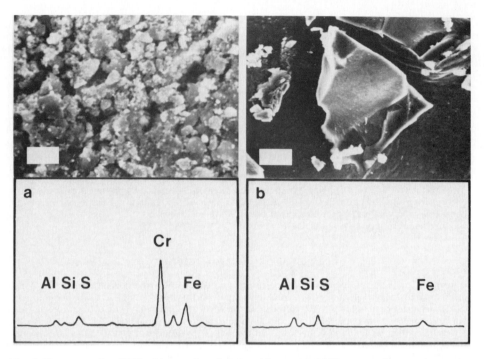

Fig. 1. Representative SEM photographs of A-I residues with EDS spectra from carbonaceous particles. (a) Soft carbonaceous clumps in A-I-B2B-4K. Scale bar = 10 μm. (b) Glass-like carbonaceous particle in A-I-B2B-14K/20. Scale bar = 40 μm. In all EDS spectra K_α and K_β lines are resolved for $Z > 16$.

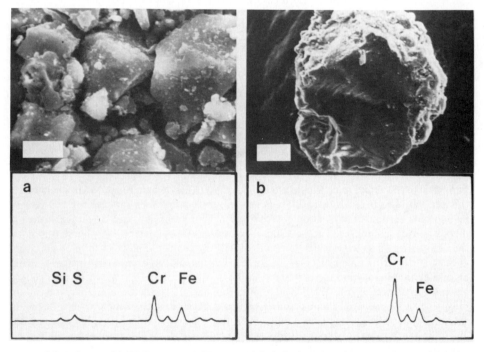

Fig. 2. SEM photos with EDS spectra, A-II group. (a) Relatively soft carbonaceous matter in A-II-D. Scale bar = 20 μm. (b) Very large, hard carbonaceous particle in A-II-D-fumHNO$_3$. Scale bar = 40 μm.

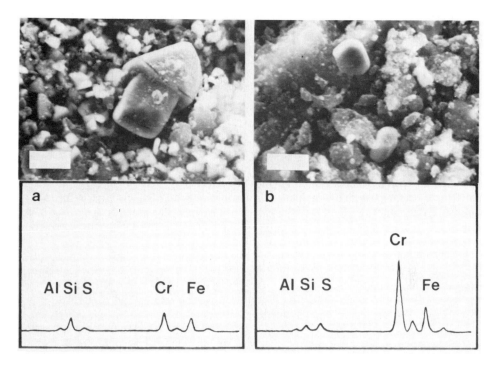

Fig. 3. SEM photos with EDS spectra, A-III group. (a) Abundant rounded spinel octahedra and small carbonaceous clumps are characteristic of A-III-A2 (shown here) and A-III-Al. (b) Carbonaceous clumps are larger and more abundant in the A-III-B (shown here) and A-III-C samples. Scale bars = 10 μm.

Most of the carbonaceous clumps and particles show Cr:Fe ratios of about 2:1 in the EDS spectra (Figs. 1a, 2, and 3), suggesting that fine grained chromite is present in the carbonaceous matter. X-ray diffraction work on sample BA, prepared by harsh HF/HCl treatment by Ott et al. (1981), shows prominent lines of a spinel group mineral with $a = 8.372(2)$ Å, consistent with chromite ($a = 8.378$ Å). Fine grained residue chromites have been observed in previous TEM studies by several authors (Fraundorf et al., 1977; Dran et al., 1979; Housely and Clarke, 1980; Smith and Buseck, 1980).

Differences are also observed between samples of the same group. Returning to Fig. 1, it is apparent that the relative amounts of soft clumps and glass-like grains are variable within samples of the A-I group. Samples A-I-B2B-4K and A-I-B2B-14K/20 are estimated to contain >90 and <25 vol.% of the soft carbonaceous clumps, respectively. The glass-like particles in Fig. 1b are interesting in that very little Cr is present above the detection limit of 0.1 wt.%, indicating a lack of fine grained chromite inclusions, as opposed to the soft clumps in Fig. 1a which show substantial Cr and Fe peaks.

General features of the A-II samples are illustrated in Fig. 2. In addition to the carbonaceous clumps shown in Fig. 2a, sample A-II-D contains at least 50 vol.% of the glassy carbonaceous particles. These particles form essentially 100% of the carbonaceous matter in the harshly etched sample shown in Fig. 2b.

Within the A-III group, both the overall abundance of coarse grained minerals and the ratio of spinel to chromite decrease with decreasing sedimentation rate from A-III-Al to A-III-B to A-III-C (Fig. 3). The light density separates A-III-A2L, A-III-BL, and A-III-CL are further depleted in the fraction of coarse grained minerals relative to their parent samples and are the most carbon-rich residues of the A-III group (Ott et al., 1981). Spinels range in composition from nearly pure *spinel* ($MgAl_2O_4$) to Fe and Cr-rich varieties with up to 45 mole% *hercynite* ($FeAl_2O_4$) and up to 25 mole% *chromite* ($FeCr_2O_4$). Ti and V are usually below 1.0 wt.% in these spinels. Coarse grained chromites generally contain less than 30% of the *hercynite* molecule and less than 10% of the *spinel* molecule in solid solution. Up to 5 wt.% of Ti + V was detected in the chromites.

Electron diffraction data

During the course of this study none of the well known polymorphs of carbon (diamond, lonsdaleite, graphite, rhombohedral graphite) were identified in the Allende acid residues. By far, most of the particles examined from all three groups of residues exhibited diffuse ring SAD patterns characteristic of poorly ordered "graphitic" carbon. Single crystal patterns were occasionally observed from fine grained chromite with $a = 8.32–8.36$ Å, but much of the chromite appears to be too fine grained ($\ll 1000$ Å) to give diffraction patterns. To distinguish the "graphitic" carbon from graphite proper, I will refer to it as turbostratic carbon throughout the remainder of this paper. The modifier *turbostratic* was originally used by Biscoe and Warren (1942) to specify materials having a structure composed of randomly stacked two dimensional layers.

Shown in Fig. 4 are representative electron micrographs and SAD patterns of carbonaceous matter in an Allende residue (A-I-B2B-14K/1), a synthetic Fischer-Tropsch sample, and a residue (anticolloid) from the Dimmitt H3 chondrite which was analyzed for noble gases by Moniot (1980). The Fischer-Tropsch sample was prepared by G. Yuen of NASA-Ames Research Center following the method of Hayatsu *et al.* (1977) and is currently being analyzed for noble gases in our laboratory. The tangled microstructure and SAD patterns of the Allende and Fischer-Tropsch samples are compared in Figs. 4a and 4b. The SAD patterns of both samples are nearly identical to those obtained from carbon black (Heidenreich *et al.*, 1968) and glassy carbon (Saxena, 1976). The ring patterns exhibit only the basal plane (00l) and asymmetric (hk) bands indicative of turbostratic carbons (Warren, 1941; Biscoe and Warren, 1942). The (00l) d-spacings are expanded by about 3% relative to graphite, whereas the values of the (hk) d-spacings are probably within error of the corresponding ($hk0$) reflections in graphite (Table 2), indicating that the C-C distances within the sp^2 layers of the turbostratic crystallites in the Allende residue and Fischer-Tropsch material are similar to the C-C distances in graphite.

To further emphasize the structural analogy between the Allende carbonaceous matter, Fischer-Tropsch material, carbon black, and glassy carbon, the observed ranges in d_{002} are given in Table 3. The minimum value of d_{002} for the turbostratic carbons is close to 3.4 Å, suggesting that no more than about 20% of the sp^2 carbon layers have the ordered stacking sequence of graphite, based on the ordering parameters of Richards (1968).

Table 2. Electron diffraction data for carbonaceous samples.

Allende			Fischer-Tropsch			Carbon black[1]			Dimmitt			Graphite[2]		
d(Å)	I	hkl	d(Å)	I	hkl	d(Å)	I	hkl	d(Å)	I	hkl	d(Å)	I	hkl
3.47	VS	002	3.46	VS	002	3.45	VS	002	3.38	M	002	3.36	VS	002
2.10	S	10	2.13	S	10	2.12	M	10	2.10	S	100	2.13	W	100
1.73	W	004	1.70	W	004	1.72	W	004				1.68	S	004
1.22	M	11	1.22	M	11	1.22	W	11	1.22	M	110	1.23	M	110
									1.15	W	112	1.16	M	112
1.16	VW	006	1.15	VW	006	1.14	VW	006				1.12	W	006
1.06	W	20	1.06	W	20	1.07	VW	20	1.05	W	200	(1.06)		200
0.81	VW	21							0.80	VW	210	(0.80)		210
0.71	VW	30							0.71	VW	300	(0.71)		300

[1]From Fig. 4 of Biscoe and Warren (1942).
[2]Natural graphite from Ceylon (ASTM card no. 23–64). Values in parentheses were calculated. Additional lines of moderate intensity are at 2.03 Å (101), 0.99 Å (114), and 0.83 Å (116). Relative intensities follow the order VS > S > M > W > VW.

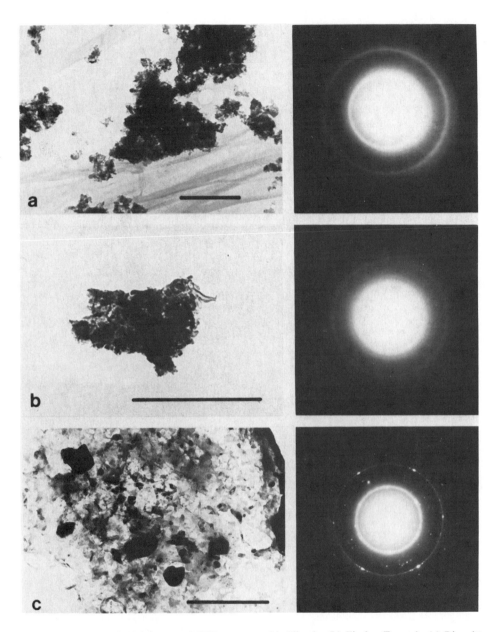

Fig. 4. TEM bright field images and SAD patterns. (a) Allende. (b) Fischer-Tropsch. (c) Dimmitt: discontinuous spotty rings in the SAD pattern are from chromite, visible as dark grains in the size range 100–3000 Å. Scale bars = 1.0 μm.

Only 11 of the 810 particles examined in the Allende residues gave hexagonal SAD patterns. These were previously (Lumpkin, 1981) thought to be consistent with carbynes, as defined by Whittaker (see Whittaker et al., 1980 and references therein). It now appears that considerable overlap exists between the d_{hk0} values of the proposed carbynes and of phyllosilicates, which invariably give hexagonal SAD patterns due to preferred orientation (Smith and Buseck, 1981b). The hexagonal flakes tentatively identified as carbynes in the A-III samples studied by Smith and Buseck (1980, 1981a) turned out to be kaolinite when EDS spectra were obtained (Smith and Buseck, 1981b). In

Table 3. Variation of d_{002} in meteorite and synthetic carbons.

d_{002}	Technique	Sample	Ref.
3.4 –3.9	Lattice imaging	Allende acid residues	1
3.4	Lattice imaging	Heated glassy carbon	2
3.39–3.51	Electron diffraction	Allende acid residues	3
3.36–3.40	Electron diffraction	Dimmitt acid residue	3
3.41–3.48	Electron diffraction	Fischer-Tropsch carbon	3
3.44–3.65	Electron diffraction	Heated glassy carbon	4
3.40–3.48	X-ray diffraction	Heated glassy carbon	4
3.45–3.55	X-ray diffraction	Heated carbon black	5
3.35–3.37	Range of values for well ordered graphites		4

1. Smith and Buseck (1980, 1981a).
2. Bose et al. (1978).
3. This work.
4. Saxena (1976).
5. Biscoe and Warren (1942).

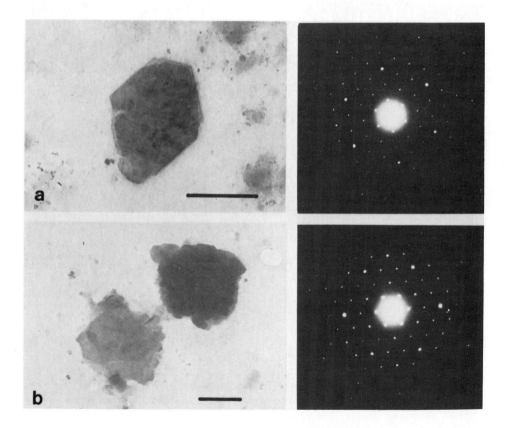

Fig. 5. TEM bright field images and SAD patterns of clays in Allende A-III-Al. (a) Kaolinite. Al and Si show up in the EDS spectrum. (b) Illite-montmorillonite (?). Mg. Al, Si, K, Ca, Ti, and Fe were detected in these particles. Intense third order (060) reflections are typical of some phyllosilicates. Scale bars = 0.5 μm.

light of their finding, I reexamined the eight particles in A-III-Al by EDS analysis and found the spectra to be consistent with montmorillonite (1 grain), kaolinite (2 grains), and what appears to be an illite-montmorillonite mixed layer clay (5 grains). Three additional grains of the last type were found during the TEM/EDS work.

Typical clays in Allende residue A-III-Al are shown in Fig. 5. Kaolinite particles are usually hexagonal flakes of uniform thickness with d_{020} values of $4.46-4.47(\pm 0.02)$Å (Fig. 5a), whereas the illite-montmorillonite particles are irregular in shape and variable in thickness and have d_{020} values in the range of $4.48-4.56(\pm 0.02 - 0.03)$Å (Fig. 5b). The electron micrographs shown here are comparable to micrographs of terrestrial clays, which display the same morphology and size range [cf., Beutelspacher and Van der Marel (1968), Figs. 19–32, 70–95, 112–122]. Furthermore, the SAD patterns shown in Fig. 5 are identical in appearance to those obtained from sheet silicates. The third order reflections (d_{060}) on the hexagonal SAD patterns are generally more intense than the other diffraction spots (Brindley and DeKimpe, 1961), a feature displayed by the SAD patterns in Fig. 5. The three unanalyzed grains resemble illite-montmorillonite in morphology and their SAD patterns show the intense (060) reflections characteristic of clays.

Carbonaceous matter in the Dimmitt residue is different in several respects from that in the Allende residues and Fischer-Tropsch material. Crystalline aggregates like the one illustrated in Fig. 4c are relatively transparent to electrons and give sharper rings in the SAD patterns than the turbostratic carbon in Allende and Fischer-Tropsch samples. Diffuse bands appropriate to turbostratic carbon were not observed in the Dimmitt sample. Measured d-spacings (Table 2) are indicative of a structure approaching that of graphite; the d_{002} reflection is reduced to 3.38 Å and the weak line at 1.15 Å is interpreted to be the three dimensional reflection (112). Diminished intensity in the (002) reflection combined with the absence of higher order (00l) reflections and normally intense (hkl) reflections like (101) and (114) is suggestive of preferred orientation of graphite crystallites in the carbonaceous aggregates. Spotty ($hk0$) rings in the SAD patterns (Fig. 4c) and the observation of several single crystal c-axis SAD patterns consistent with relatively ordered graphite support this interpretation.

Using Richards' (1968) ordering parameters, the average value of $d_{002} = 3.38$ Å can be used to predict that about 50% of the sp^2 carbon layers are ordered, although the range of measured d_{002} values (Table 3) leads to a rather broad estimate of 20–80%. Clearly, more precise measurements are necessary for definitive statements about the degree of ordering in this sample.

DISCUSSION

Carbonaceous matter in Allende acid residues

The results of this work are summarized in Table 4 and show that most of the carbonaceous matter in the Allende residues of Ott *et al.* (1981) is a variety of turbostratic carbon characterized by a random stacking of graphite-like sp^2 hybridized layers. The material should not be confused with graphite proper, in which the layers have an ordered ... ABABAB... stacking sequence. The gross diffraction structure of the residue carbon resembles that of carbon black and glassy carbon (e.g., Heidenreich *et al.*, 1968; Saxena, 1976). Previous lattice imaging by Smith and Buseck (1980, 1981a) has shown that much of the turbostratic carbon (which they call "graphitic") in the Allende residues has a tangled microstructure similar in appearance to that of glassy carbon (cf., Bose *et al.*, 1978). In the words of Biscoe and Warren (1942), turbostratic carbon can generally be described as "... a simple and definite example of an intermediate form of matter, which is distinctly different from both the crystalline and the amorphous states."

Strictly speaking, the evidence presented here does not rule out the presence of amorphous carbon, but proves that the carbonaceous matter in the Allende residues is *not completely* amorphous as suggested earlier (Breger *et al.*, 1972; Bauman *et al.*, 1973; Dran *et al.*, 1979). Models of the structure of glassy carbon generally include an appreciable

Table 4. Summary of results for Allende residues.

Sample	^{130}Xe[1]	C(%)	Diffuse rings[2]	Hexagonal[3]	Total[4]
A-I-B2B-4K	2.58	44	100	0	100
A-I-B2B-10K	4.87	75	*	0	40
A-I-B2B-14K/1	5.35	79			
A-I-B2B-14K/20	6.00	78	100	0	100
A-II-D	6.19	73	29	1	30
A-II-D-4hrHNO$_3$	1.13	68	*	0	87
A-II-D-17hrHNO$_3$	0.47	60	40	0	40
A-II-D-FumHNO$_3$		56	20	0	20
A-II-D-HClO$_4$	1.34	33	40	0	40
A-III-A1	1.36	12	*	8	33
A-III-A2	1.68	16	*	1	100
A-III-A2L		41	20	0	20
A-III-B	3.65	42	98	0	98
A-III-BL		61	*	0	30
A-III-C	4.86	62			
A-III-CL		83	70	1	72

[1] Abundance in 10^{-8} cc/g. ^{130}Xe and C data from Ott et al. (1981).
[2] Turbostratic carbon. Stars indicate samples where the Au or Al film was too thick, preventing observation of diffuse ring patterns.
[3] All 8 original plus three additional grains in A-III-A1 were relocated and identified as clays by EDS analysis.
[4] Total number of grains examined in each sample.

amount of amorphous carbon, principally sp^3 hybridized, surrounding and linking together crystallites of turbostratic carbon (Noda et al., 1969; Saxena, 1976). It is highly probable that amorphous carbon (with minor N, H) is associated with turbostratic carbon in the Allende residues, considering the similarity of the microstructure to glassy carbon.

In contrast to the recent reports by Whittaker et al. (1980) and Hayatsu et al. (1980), I have not found unambiguous evidence for carbynes. Out of 810 particles, only 11 were found to give hexagonal diffraction patterns, 8 of which have been relocated and analyzed by EDS. All 8 grains gave spectra consistent with clays, thus constraining the maximum abundance of carbynes to less than 0.5% of the 0.5–2.0 μm grain size population examined in this study. This constraint applies to carbynes as crystalline phases and does not include the possibility of triple bonding in amorphous carbon, if present. The clay particles were probably introduced as contaminants during preparation of the TEM grids as it is highly unlikely that they survived the HF/HCl treatment.

These results, in conjunction with those of Smith and Buseck (1980, 1981a, 1981b), cast a sizeable shadow of doubt on the importance of carbynes in Allende. Interestingly, the "irregular carbyne grains" shown in Fig. 2A of Whittaker et al. (1980) closely resemble the textures of many of the turbostratic carbon particles observed in this study (e.g., Fig. 4a). Even though the 5 grains they identified as carbynes by electron diffraction could be explained as silicates, their ion probe results are more difficult to rationalize. Three grains from the Allende BK residue were reported by Whittaker et al. (1980) to have carbyne contents of 80–91% based on the C_1/C_2 intensity ratio in the negative ion spectrum using graphite and carbyne reference samples. However, if turbostratic (and perhaps amorphous) carbon did not behave entirely like graphite during ion probe analysis, but rather behaved like carbynes by contributing C_2^- ions to the spectra, it could have led to overestimation of the amount of carbynes in the sample. Whittaker et al. (1980) apparently did not take the precautionary measure of analyzing other types of carbonaceous reference samples in their work. What may have been their most accurate method for the specific identification of carbynes – x-ray diffraction – was mentioned only briefly as limiting the amounts of graphitic carbon or other polymorphs to less than 10%.

Other evidence in support of the case for turbostratic carbon in Allende comes from the ten year old work of Green *et al.* (1971), who obtained diffuse ring SAD patterns from "poorly crystalline graphite" during their high voltage TEM study of the Allende matrix. Smith and Buseck's (1981a) preliminary results on carbon *in situ* in the meteorite show that it is similar to the material they imaged in the acid residues. Based on diffraction data and lattice imaging, it seems that the basic structure (turbostratic) of Allende carbonaceous matter has not been greatly altered during the acid treatments, but the textural differences observed with the SEM (Figs. 1, 2, 3) suggest that there are subtle changes in the structure which need to be accounted for.

Implications for noble gases

The Berkeley-Ames group has consistently argued that carbonaceous phases are the major carriers of planetary-type noble gases in Allende (e.g., Frick and Chang, 1978; Reynolds *et al.*, 1978), providing an interesting contrast to the pioneering work of the Chicago group in which a sulfide mineral, chromite, and amorphous carbon were regarded as the primary host phases (Lewis *et al.*, 1975, 1977; Gros and Anders, 1977). Ott *et al.* (1981) attempted to resolve the problem by obtaining an extensive body of chemical and isotopic data on a large suite of acid residues. Their data are consistent with the earlier views of the Berkeley-Ames group. Evidence presented by Ott *et al.* (1981) in support of carbonaceous host phases of both the isotopically normal and anomalous gases includes 1) concentration of both types of gas in low density and fine grained fractions, 2) a high degree of correlation between abundances of carbon and isotopically normal noble gases (see Table 4), 3) a corresponding lack of correlation between Cr, Fe, S and (normal) noble gases, and 4) the combustibility of the host phases of both types of gas.

The major contribution of the present work has been to document the identification of turbostratic carbon in the Allende residues of Ott *et al.* (1981). Characteristic SAD patterns of this phase were found in all three groups of samples (Table 4), including the daughter samples of A-II-D etched in oxidizing acids. It would be premature to assign noble gas components to particular phases at this time, but the current status of the candidates can be outlined as follows. For the isotopically normal noble gases:

1. Turbostratic carbon is a possibility, but the layer structure of this phase alone, like graphite [which is a gas poor mineral in ureilites (Göbel *et al.*, 1978)], is difficult to associate with effective trapping and retention of noble gases, which are typically released at high temperatures in the Allende residues (Ott *et al.*, 1981).
2. Amorphous carbon is a possibility, especially if the structure of glassy carbon (cf., Noda *et al.*, 1969) is an approximate model for carbonaceous matter in the Allende residues. In this scenario, amorphous sp^3 hybridized carbon atoms link together the turbostratic crystallites and could form three dimensional "cages" suitable for noble gas trapping and retention [somewhat like diamonds in ureilites, Göbel *et al.* (1978)].
3. Carbynes are not important inasmuch as they are constrained to less than 0.5% of the carbonaceous matter in the residues examined here.

For the isotopically anomalous noble gases:

1. Turbostratic carbon was found in the etched A-II-D samples (Table 4) and must again be considered as a candidate.
2. Amorphous carbon may be a stronger candidate here than for the isotopically normal gas, at least if the results of Smith and Buseck (1981a) are taken at face value. These authors report that much of the carbon in the A-II-D-fumHNO$_3$ sample appears to be amorphous.
3. Carbynes are questionable. The Chicago group claim that carbynes comprise $\geq 80\%$

of their harshly etched residue BK (Whittaker *et al.*, 1980), but Smith and Buseck (1981b) examined the sample and did not find any carbynes. In this study, 187 particles from analogous residues of the Berkeley-Ames group were analyzed and no carbynes were found (Table 4).

4. Fine grained chromite cannot be ruled out from the results of this work. For discussion of this subject the reader is referred to the papers by Frick and Chang (1978), Lewis *et al.* (1979), and Ott *et al.* (1981).

The SEM observations outlined in the results of this work generally support the arguments of Ott *et al.* (1981) as to the complexity of the acid insoluble carbonaceous matter in the Allende residues. Figures 1–3 clearly show that there are (at least) two types of carbonaceous particles in the residues: relatively soft clumps and hard, glass-like grains. The relative hardness is probably a function of the degree of cross-bonding between the turbostratic crystallites in the particles, with the hard particles having a higher degree of cross-bonding. Although there are insufficient data at the present time to draw firm conclusions, some evidence indicates that the glass-like particles may contain larger amounts of isotopically normal noble gases than the soft clumps. For instance, in sample A-I-B2B-14K/20 which has >75% glassy particles, ^{130}Xe is elevated by a factor of 2.4, but carbon is increased by a factor of only 1.8 relative to sample A-I-B2B-4K which has <10% glassy particles [Table 4, see also Ott *et al.* (1981) for He, Ne, Ar, Kr abundances]. This observation underscores the importance of understanding the subtle structural differences between the carbonaceous samples, possibly manifested as an amorphous component intimately associated with the turbostatic component.

CONCLUSIONS

Most of the 0.5–2.0 μm carbonaceous particles in a large suite of Allende acid residues give diffuse ring SAD patterns characteristic of turbostratic carbon, a phase whose crystallites consist of a random stacking sequence of sp^2 hybridized layers of carbon atoms. This study does not rule out the presence of amorphous carbon, which may in fact account for the glassy character of many of the carbonaceous particles. Only 11 out of 810 particles gave hexagonal, single crystal SAD patterns, 8 of which were relocated and shown by EDS analysis to be consistent with clay minerals. The overall abundance of carbynes is thus constrained to be less than 0.5% of the total number of grains examined.

Turbostratic carbon is considered as a carrier of both the isotopically normal and anomalous noble gases, but the layer structure of this phase alone appears to be incompatible with retention of noble gases at high temperatures. It may be necessary to invoke an amorphous component to provide suitably retentive sites for noble gases. In terms of the model structure of glassy carbon, the results presented here are not inconsistent with the presence of such a component. Carbynes are questioned as important carriers of both types of noble gases in the Allende residues, but the possibility of having triply bonded carbon in an amorphous component cannot be eliminated.

Acknowledgments—I would like to express my gratitude to the following individuals for their assistance during the course of this study. Dr. G. Thomas and Dr. H.-R. Wenk kindly provided access to their TEM facilities at the University of California, Berkeley, and A. Gronsky and B. Smith gave technical assistance in the operation of the microscopes. Dr. S. Chang provided the Allende and Fischer-Tropsch samples and Dr. B. Srinivasan provided the Dimmitt sample. The manuscript has been considerably improved through helpful discussions with Dr. S. Chang, Dr. M. Dziczkaniec, and Dr. J. H. Reynolds. Critical reviews of the manuscript were authored by Dr. P. P. K. Smith and Dr. Ian Mackinnon. Ms. Sandy Ewing provided invaluable assistance in preparation of the manuscript. This work was supported by NASA under grant NGL 05-003-409, and bears paper number 130.

REFERENCES

Bauman A. J., Devaney J. R., and Bollin E. M. (1973) Allende meteorite carbonaceous phase: intractable nature and scanning electron morphology. *Nature* **241**, 264–267.

Beutelspacher H. and Van der Marel H. W. (1968) *Atlas of Electron Microscopy of Clay Minerals and Their Admixtures.* Elsevier, Amsterdam. 333 pp.

Biscoe J. and Warren B. E. (1942) An x-ray study of carbon black. *J. Appl. Phys.* **13**, 364–371.

Bose S., Dahmen U., Bragg R. H., and Thomas G. (1978) Lattice image of LMSC glassy carbon. *J. Amer. Ceram. Soc.* **61**, 174.

Breger I. A., Zubovic P., Chandler J. C., and Clarke R. S. (1972) Occurrence and significance of formaldehyde in the Allende carbonaceous chondrite. *Nature* **236**, 155–158.

Brett R. (1967) Cohenite: its occurrence and a proposed origin. *Geochim. Cosmochim. Acta* **31**, 143–160.

Brindley G. W. and DeKimpe C. (1961) Identification of clay minerals by single crystal electron diffraction. *Amer. Mineral.* **46**, 1005–1016.

Bunch T. E. and Chang S. (1980) Carbonaceous chondrites – II. Carbonaceous chondrite phyllosilicates and light element geochemistry as indicators of parent body processes and surface conditions. *Geochim. Cosmochim. Acta* **44**, 1543–1577.

Chang S., Mack R., and Lennon K. (1978) Carbon chemistry of separated phases of Murchison and Allende meteorites (abstract). In *Lunar and Planetary Science IX*, p. 157–159. Lunar and Planetary Institute, Houston.

Dran J. C., Klossa J., and Maurette M. (1979) A preliminary microanalysis of the Berkeley gas-rich Allende residue (abstract). In *Lunar and Planetary Science X*, p. 312–314. Lunar and Planetary Institute, Houston.

Fraundorf P., Flynn G. J., Shirck J. R., and Walker R. M. (1977) Search for fission tracks from superheavy elements in Allende. *Earth Planet. Sci. Lett.* **37**, 285–295.

Frick U. and Chang S. (1978) Elimination of chromite and novel sulfides as important carriers of noble gases in carbonaceous meteorites. *Meteoritics* **13**, 465–470.

Frondel C. and Marvin U. B. (1967) Lonsdaleite, a hexagonal polymorph of diamond. *Nature* **214**, 587–589.

Göbel R., Ott U., and Begemann F. (1978) On trapped noble gases in ureilites. *J. Geophys. Res.* **83**, 855–867.

Green H. W., Radcliffe S. V. and Heuer A. H. (1971) Allende meteorite: a high voltage electron petrographic study. *Science* **172**, 936–939.

Gros J. and Anders E. (1977) Gas-rich minerals in the Allende meteorite: attempted chemical characterization. *Earth Planet. Sci. Lett.* **33**, 401–406.

Hayatsu R., Matsuoka S., Scott R. G., Studier M. H., and Anders E. (1977) Origin of organic matter in the early solar system – VII. The organic polymer in carbonaceous chondrites. *Geochim. Cosmochim. Acta* **41**, 1325–1339.

Hayatsu R., Scott R. G., Studier M. H., Lewis R. S., and Anders E. (1980) Carbynes in meteorites: detection, low-temperature origin, and implications for interstellar molecules. *Science* **209**, 1515–1518.

Heidenreich R. D., Hess W. M., and Ban L. L. (1968) A test object and criteria for high resolution electron microscopy. *J. Appl. Crystallogr.* **1**, 1–19.

Housley R. M. and Clarke D. R. (1980) XPS and STEM studies of Allende acid insoluble residues. *Proc. Lunar Planet. Sci. Conf. 11th*, p. 945–958.

Lewis R. S., Alaerts L., and Anders E. (1979) Ferrichromite: a major host phase of isotopically anomalous noble gases in primitive meteorites (abstract). In *Lunar and Planetary Science X*, p. 725–727. Lunar and Planetary Institute, Houston.

Lewis R. S., Gros J., and Anders E. (1977) Isotopic anomalies of noble gases in meteorites and their origins – 2. Separated minerals from Allende. *J. Geophys. Res.* **82**, 779–792.

Lewis R. S., Srinivasan B., and Anders E. (1975) Host phase of a strange xenon component in Allende. *Science* **190**, 1251–1262.

Lumpkin G. R. (1981) Electron microscopy of carbon in Allende acid residues (abstract). In *Lunar and Planetary Science XII*, p. 631–633. Lunar and Planetary Institute, Houston.

Mason B. (1972) The mineralogy of meteorites. *Meteoritics* **7**, 309–326.

Moniot R. K. (1980) Noble-gas-rich separates from ordinary chondrites. *Geochim. Cosmochim. Acta* **44**, 253–271.

Nagy B. (1975) *Carbonaceous Meteorites.* Elsevier, Amsterdam. 747 pp.

Noda T., Inagaki M., and Yamada S. (1969) Glass-like carbons. *J. Non-Cryst. Solids* **1**, 285–302.

Ott U., Mack R., and Chang S. (1981) Noble-gas-rich separates from the Allende meteorite. *Geochim. Cosmochim. Acta.* In press.

Reynolds J. H., Frick U., Neil J. M., and Phinney D. L. (1978) Rare-gas-rich separates from carbonaceous chondrites. *Geochim. Cosmochim. Acta* **42**, 1775–1797.

Richards B. P. (1968) Relationships between interlayer spacing, stacking order and crystallinity in carbon materials. *J. Appl. Crystallogr.* **1**, 35–48.

Rubin A. E., McKinley S., Scott E. R. D., Taylor G. J., and Keil K. (1981) A new kind of unequilibrated ordinary chondrite with a graphite-magnetite matrix (abstract). In *Lunar and Planetary Science XII*, p. 908–910. Lunar and Planetary Institute, Houston.

Saxena R. R. (1976) The structure and electrical properties of glassy carbon. Ph.D. Thesis, Univ. Calif., Berkeley. 161 pp.

Scott E. R. D. and Agrell S. O. (1971) The occurrence of carbides in iron meteorites. *Meteoritics* **6**, 312–313.

Scott E. R. D., Taylor G. J., Rubin A. E., Okada A., Keil K., Hudson B., and Hohenberg C. M. (1981) Graphite-magnetite inclusions in ordinary chondrites: An important constituent in the early solar system (abstract). In *Lunar and Planetary Science XII*, p. 955–957. Lunar and Planetary Institute, Houston.

Simmonds P. G., Bauman A. J., Bollin E. M., Gelp E., and Oró J. (1969) The unextractable organic fraction of the Pueblito de Allende meteorite: evidence for its indigenous nature. *Proc. Natl. Acad. Sci. USA* **64**, 1027–1034.

Smith P. P. K. and Buseck P. R. (1980) High resolution transmission electron microscopy of an Allende acid residue (abstract). In *Meteoritics* **15**, 368–369.

Smith P. P. K. and Buseck P. R. (1981a) Graphitic carbon in the Allende meteorite: a microstructural study. *Science* **212**, 322–324.

Smith P. P. K. and Buseck P. R. (1981b) Carbynes in carbonaceous chondrites: a cautionary comment (abstract). In *Lunar and Planetary Science XII*, p. 1017–1019. Lunar and Planetary Institute, Houston.

Taylor G. J., Okada, A., Scott E. R. D., Rubin A. E., Huss G. R., and Keil K. (1981) The occurrence and implications of carbide-magnetite assemblages in unequilibrated ordinary chondrites (abstract). In *Lunar and Planetary Science XII*, p. 1076–1078. Lunar and Planetary Institute, Houston.

Vdovykin G. P. (1969) New hexagonal modification of carbon in meteorites. *Geochem. Int.* **6**, 915–918.

Vdovykin G. P. (1970) Ureilites. *Space Sci. Rev.* **10**, 483–510.

Warren B. E. (1941) X-ray diffraction in random layer lattices. *Phys. Rev.* **59**, 693–698.

Whittaker A. G. (1978) Carbon: a new view of its high-temperature behavior. *Science* **200**, 763–764.

Whittaker A. G., Watts E. J., Lewis R. S., and Anders E. (1980) Carbynes: carriers of primordial noble gases in meteorites. *Science* **209**, 1512–1514.

Carbon in the Allende meteorite: Evidence for poorly graphitized carbon rather than carbyne

P. P. K. Smith* and Peter R. Buseck

Departments of Geology and Chemistry, Arizona State University, Tempe, Arizona 85287

Abstract—A carbon-rich acid residue from the Allende carbonaceous chondrite was examined by high-resolution transmission electron microscopy (HRTEM) and also by analytical electron microscopy. A TEM mount of this residue that has previously been reported to contain carbyne forms of carbon was shown to contain sheet silicate contaminants. These sheet silicate grains give electron diffraction patterns similar to those reported for carbynes, thus raising questions about the previous report of carbynes in this residue. Furthermore, two crystals from a glacier microspherule, which had previously been identified as carbyne VIII, were shown by microanalysis to be talc. In view of these observations it is suggested that identifications of carbyne by electron diffraction should be supported by microanalyses of the individual grains. HRTEM investigations of the allende carbon indicate that it is a poorly crystalline graphite, structurally similar to "glassy" carbon.

INTRODUCTION

Over the past year two conflicting views have been presented regarding the structural state of the carbon in the Allende C3V carbonaceous chondrite. Lewis *et al.* (1980) and Whittaker *et al.* (1980) reported that $\geq 80\%$ of the carbon in a carbon-rich acid residue from Allende is in the form of carbynes, a relatively recently discovered series of polymorphs of elemental carbon containing triply bonded carbon. Carbynes were recognized in Allende by means of electron diffraction, by their transformation behaviour in the electron microscope, and by negative-ion mass spectra obtained in an ion-probe. Carbynes were also reported to be present in the Murchison C2 chondrite. Thermal pyrolysis mass spectra obtained by Hayatsu *et al.* (1980) provided further evidence for the presence of carbynes in Allende. Smith and Buseck (1980, 1981), on the other hand, using high-resolution transmission electron microscopy (HRTEM) found the Allende carbon to be a poorly crystalline graphite. The carbon was recognized as being graphitic on the basis of the prominent crystal lattice fringes observed in HRTEM images, with d-spacings of 3.4–3.9 Å, corresponding to the (0001) layer planes of graphite. The HRTEM images are very similar to those described by Jenkins *et al.* (1972) and Ban *et al.* (1975) for poorly graphitized or "glassy" carbon; detailed studies by these authors, combining X-ray diffraction and HRTEM information, show that such carbon is not in fact a true glass; rather it is a tangled aggregate of ribbon-shaped packets of graphitic layer planes, most of the crystallites being less than ten graphitic layers thick.

Part of the interest in the carbon content of carbonaceous chondrites has stemmed from the discovery that carbonaceous phases are important carriers of isotopically-anomalous noble gases in these meteorites (e.g. Frick, 1977; Frick and Reynolds, 1977; Lewis *et al.*, 1979). These noble gases include two components that are believed to be of presolar origin: Ne-E (Black and Pepin, 1969) and s-process Xe (Srinivasan and Anders, 1978). Identification of the carrier phases of the noble gases is clearly of importance to an understanding of the mechanism(s) by which the noble gases were incorporated into the

---

*Present address: Research School of Chemistry, Australian National University, Canberra, ACT 2600, Australia.

meteorites and the phases and environment existing at the time of incorporation. For an introduction to the literature on noble-gas carriers in meteorites and for a description of the present state of knowledge the interested reader is referred to recent papers by Alaerts *et al.* (1980) and Ott *et al.* (1981).

A further cause for interest in the carbonaceous material in meteorites is the suggestion made by Webster (1979) that carbynes may be an important constituent of the interstellar dust. This idea gained indirect support (Webster, 1980) from the observation of carbynes in carbonaceous meteorites by Whittaker *et al.* (1980).

The first detailed study of carbonaceous material in Allende was that of Simmonds *et al.* (1969), who analysed whole-rock portions of the meteorite by a combined pyrolysis-gas chromatography-mass spectrometry method and also by differential thermal analysis. The yield of pyrolysis products was only about 22 ppm, compared to a total carbon content in the meteorite of 0.27%; these products consisted chiefly of aromatic and substituted aromatic hydrocarbons. Differential thermal analysis gave no indication of the presence of graphite, an observation also made in a similar study by Bauman *et al.* (1973).

A high-voltage transmission electron microscope study by Green *et al.* (1971) showed that the carbon occurs interstitially to the silicate grains, and electron diffraction indicated the carbon to be a poorly crystalline graphite. A 50 Å-resolution transmission electron microscope study by Dran *et al.* (1979) indicated the carbon to be amorphous; however, the microscope resolution was insufficient to image the kind of microcrystalline graphitic structure that was subsequently reported by Smith and Buseck (1980, 1981).

The conflicting reports described above of carbyne (Whittaker *et al.*, 1980) and graphitic carbon (Smith and Buseck, 1980, 1981) centred on carbon-rich acid residues of the kind now commonly used for noble-gas isotope studies, as pioneered by Lewis *et al.* (1975). Smith and Buseck examined residues prepared by a group at the University of California, Berkeley, including both unetched residues and residues etched in HNO_3 to remove the ordinary gases, leaving the anomalous gases in the residue (see Ott *et al.*, 1981 for details of the preparation of these residues). Whittaker *et al.* (1980) examined a HNO_3-etched residue, BK, which is nearly pure carbon.

The aim of the present paper is to throw some light on the two discordant sets of observations, with particular emphasis on the electron microscope results. In order to make possible a direct comparison between the electron diffraction results of Whittaker *et al.* (1980) and the HRTEM work of Smith and Buseck (1980, 1981) we have now examined sample BK, both on electron microscope mounts that we prepared ourselves and on mounts provided by Whittaker. As in our previous study, we used HRTEM to search for graphitic carbon and electron diffraction to locate possible carbyne grains. In addition to these methods we used analytical microscopy to obtain semi-quantitative chemical analyses of possible carbyne grains.

As part of this study we also examined electron microscope mounts prepared from carbon-rich microspheres from glacier ice. Herr *et al.* (1980) reported that these microspheres, which are believed to be of meteoritic or possibly of cometary origin, contain carbyne forms of carbon. The material that we examined (provided by Whittaker) is on locator grids, so that grains identified as carbyne on the basis of electron diffraction observations by Herr *et al.*, could be relocated in our TEM. The initial aim of this part of the study was to familiarise ourselves with the appearance of carbynes in the electron microscope. In fact these grains proved to be sheet silicates rather than carbynes, but we report the results here as they draw attention to the danger of mistaking sheet silicates for carbynes when identification is based on diffraction data alone.

SAMPLES STUDIED AND EXPERIMENTAL METHODS

The preparation of the Allende carbon-rich residue BK has been described by Whittaker *et al.* (1980). For the HRTEM observations we pipetted a few drops of a suspension of BK in ethanol onto a perforated carbon support film; images were recorded only from areas that project over holes in the support film. The high-resolution images were recorded with a JEOL JEM 100B electron microscope operating at 100 kV.

The electron microscope mount of BK that was provided by Whittaker is supported on a non-perforated amorphous carbon film. The lack of perforations together with the poor stability of this film in the electron beam made high-resolution imaging impossible for most of this sample. However, we were able to record high-resolution images from an area of sample that is supported directly by a copper grid bar, adjacent to a grid square where the carbon support film is absent.

The glacier microsphere material, also provided by Whittaker, was supplied to us on two locator TEM grids. These grids are internally labelled, so that any particular grid square can be identified and subsequently relocated. We were able to relocate two grains on these grids, both considered by Whittaker (pers. comm.) to be the carbyne carbon VIII. A third carbyne grain observed by Whittaker could not be relocated; this grain presumably fell off the support film during transport or handling.

Electron microscope microanalysis was performed using a Philips EM400 electron microscope equipped with an energy-dispersive X-ray spectrometer (EDS). The spectrometer detects the characteristic X-rays produced by the specimen in the electron beam for all elements with $Z > 10$. The instrument cannot therefore be used to analyse for carbon, but it is ideally suited to the identification of sheet silicates that might be mistaken for carbynes on diffraction information alone. We have not yet calibrated our spectrometer for quantitative analysis, but in the present study a qualitative analysis together with electron diffraction information enabled us to identify various sheet silicates unambiguously.

Electron diffraction was also carried out using the Philips EM400 electron microscope. Specimens were always brought to the eucentric position in the microscope column, at the intersection of the column axis and the specimen tilt axis. By operating in this way with the specimen at a fixed position in the microscope column, calibration errors in the diffraction patterns were reduced to a minimum. For sample BK no internal standard was used, and the estimated accuracy in the d-spacings obtained from diffraction patterns was $\pm 2\%$, equivalent to ± 0.09 Å in a d-spacing of about 4.5 Å. In the case of the glacier microsphere samples, we evaporated gold onto the specimen as an internal diffraction standard (after obtaining the chemical analyses), thereby improving the accuracy to about $\pm 0.5\%$ (± 0.02 Å).

OBSERVATIONS

A typical HRTEM image of residue BK, on a TEM mount prepared in our laboratory, is shown in Fig. 1. Since the image is very similar to those that we have previously obtained for other Allende residues (Smith and Buseck, 1980, 1981) we will not comment on it at length here. Crystal lattice fringes with spacings of 3.5 to 4.0 Å are seen throughout the specimen. Because of the similarity of such images with previously published images of glassy carbon (Jenkins *et al.*, 1972; Ban *et al.*, 1975), and considering also that BK is a nearly pure carbon residue, we consider that a poorly graphitized or glassy carbon serves as a useful model for the Allende carbon. We did not observe any grains in this TEM mount of residue BK that gave carbyne-like diffraction patterns.

In contrast to this observation, a search of the TEM mount of BK provided by Whittaker revealed 11 grains that gave hexagonal diffraction patterns with d-spacings in the range 4.4–4.6 Å. These grains are generally thin, electron-transparent flakes, with diameters up to 10 μm, although most are smaller than 1.5 μm. Some of the grains have a euhedral to subhedral hexagonal outline. The d-spacings in the hexagonal diffraction patterns of these grains would in each case be consistent with an identification as carbyne. However, X-ray microanalysis revealed that these 11 grains are all sheet silicates: 5 are kaolinite, 3 talc, 2 chlorite and 1 mica. In each case both the chemical and diffraction information are consistent with these identifications. Our search of this BK sample was not exhaustive, and it is likely that other such grains are present. All grains that give carbyne-type diffraction patterns proved to be sheet silicates. Also, high-resolution imaging demonstrated that the graphitic-type carbon is present on this TEM mount.

Since hydrated sheet silicate minerals are not generally considered to be present in Allende, and in any case would not survive the HF/HCl treatment that was used to prepare the carbon-rich residue, they were presumably introduced as contaminants after the preparation of the residue. This may have occurred during the preparation of the TEM mount. This view is supported by our failure to find any such grains on the TEM mount of BK that we prepared in our laboratory.

The two grains from the glacier microsphere samples, previously identified as the

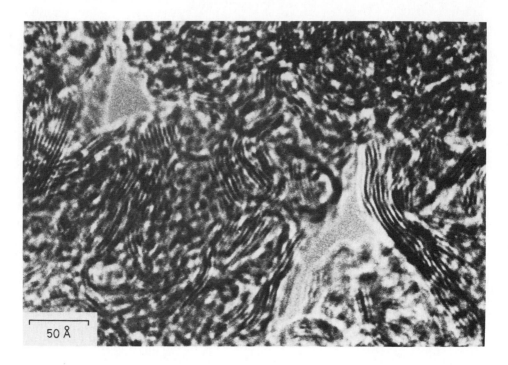

Fig. 1. High-resolution transmission electron micrograph of the carbon-rich residue BK from the Allende meteorite. The crystal lattice fringe spacings of 3.5 to 4.0 Å correspond to those of poorly crystalline graphite.

carbyne form designated carbon VIII, were relocated and shown to be talc. Electron diffraction results ($d_{020} = 4.59 \pm 0.02$ Å) and EDS spectra are both consistent with this identification. In a search of one of the two TEM grids of this material we located 22 more grains that gave hexagonal diffraction patterns. Of these grains, 14 are talc, 6 kaolinite, 1 chlorite and 1 biotite; as in the case of sample BK, all grains that give carbyne-like diffraction patterns proved to be sheet silicates. This TEM grid had been prepared from a glacier spherule that had been shown to be rather pure carbon (Whittaker, pers. comm.). It therefore seems likely that these silicates are contaminants, although they could be a minor component of the glacier spherule. It is perhaps significant that talc has been observed to be a ubiquitous component of atmospheric dusts, with mica, chlorite and kaolinite occurring in smaller amounts (Windom et al., 1967).

We previously reported a small fraction (<2%) of possible carbyne grains in an Allende residue prepared by the Berkeley group, although we were not prepared to identify these grains positively on the basis of diffraction information from a single zone axis (Smith and Buseck, 1980, 1981). X-ray microanalysis shows that these grains are in fact kaolinite. As in the case of residue BK, these must result from contamination; it is not clear whether this occurred during the preparation and handling of the residue, or during the preparation of our electron microscope mount. This observation further emphasises the need for great care not to introduce any foreign material whilst characterising acid residues by electron microscopy. Contaminants that only constitute an extremely small fraction of the sample can readily be detected, particularly if they have a distinctive feature such as the sharp diffraction patterns given by the sheet silicates.

DISCUSSION

Our observations of the glacier spherule material clearly demonstrate the danger of mistaking sheet silicates for carbynes when identification is based solely on electron diffraction—two crystals previously identified as carbyne in fact proved to be talc. Figure 2 illustrates the similarity between the c-axis electron diffraction patterns for the carbynes and the (001) orientation electron diffraction patterns obtained from sheet silicates. The carbyne polymorphs are characterized by their 110 d-spacing, which is the fundamental reflection in the hexagonal c-axis pattern. Although most sheet silicates have monoclinic crystal symmetry (kaolinite is slightly triclinic), they give rise to accurately hexagonal diffraction patterns when the electron beam is perpendicular to the (001) cleavage plane (Brindley and De Kimpe, 1961); this of course is the usual orientation for grains dispersed on a TEM support film. The d-spacing obtained from such patterns is the 020 spacing for the particular sheet silicate. Bearing in mind the likely error in the electron diffraction measurement, together with variations in d_{020} resulting from chemical substitutions in the

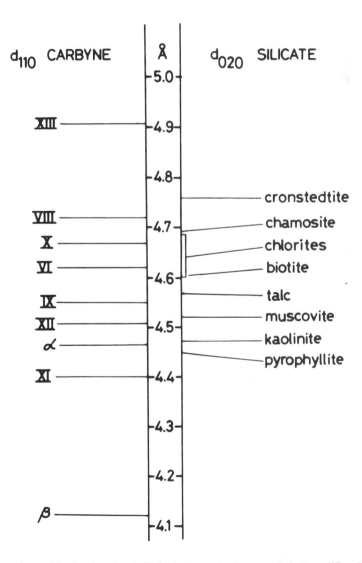

Fig. 2. Comparison table showing the similarity between the hexagonal electron diffraction patterns for the carbyne polymorphs and those for some common sheet silicate minerals. The fundamental spacing in the hexagonal pattern corresponds to the 110 d-spacing for the carbynes, or to the 020 d-spacing for the sheet silicates.

sheet silicates, it is clear that electron diffraction cannot safely be used to recognize all of the carbyne forms. Thus we feel that future identifications of carbyne should be supported by X-ray microanalysis of individual grains; this can conveniently be carried out using a TEM equipped with an EDS system.

In view of our observations of the two different TEM mounts of the Allende residue BK, it seems likely that many, and possibly all, of the carbyne diffraction patterns obtained by Whittaker et al. (1980) were from contaminants. It is perhaps significant that, whereas a carbyne content of $\geq 80\%$ was obtained from ion-probe data by these authors, only five carbyne grains were measured by electron diffraction. The electron diffraction identifications of carbyne by Whittaker et al. (1980) were supported by observations of the alteration behaviour of the crystals in the electron beam. A crystal tentatively identified by them as the carbyne CVI transformed to rhombohedral graphite in a 60 kV electron beam. As mentioned by Whittaker et al. (1980), this behaviour is not consistent with the previous report of CVI (Whittaker and Wolten, 1972) in which it was found to be stable in a 75 kV electron beam, but damaged rapidly at 100 kV. Furthermore, there have been no previous reports of the transformation of CVI to graphite; Whittaker and Wolten (1972) only described radiation damage, implying a transformation to an amorphous state. Since there are no other papers describing the electron beam damage of the carbynes, it is difficult to comment on Whittaker et al.'s (1980) identification of CXI on the basis of its radiation sensitivity.

In general, we have reservations about the use of radiation damage as a diagnostic criterion. We suggest that if identifications are to be based on radiation sensitivity, then it is important to measure the beam current density at the specimen, as well as the accelerating potential, since in our experience the beam current is an important factor in determining specimen damage rates.

Taken alone, our HRTEM observations provide clear evidence of a glassy carbon-type graphitic structure for the bulk of the Allende carbon, although the HRTEM images do not rule out the possible presence of a small fraction of amorphous carbon. Also, as discussed by Ban (1971), the interatomic distances in the distorted layer planes of glassy carbon may vary to such an extent that the perfect hexagonal arrangement of atoms is disrupted; the extent of such imperfections within the layer planes cannot yet be determined from HRTEM images. The electron diffraction results of Green et al. (1971) and Lumpkin (1981) also support a graphitic model. Lumpkin did find possible carbyne grains in 3 out of 15 Allende residues examined, but this identification has yet to be confirmed by X-ray microanalysis.

The TEM results are inconsistent with the ion-probe data of Whittaker et al. (1980), which indicate the presence of abundant carbyne in an Allende acid residue, with little if any graphitic carbon. In their study carbynes were recognized by the preponderance of even-numbered carbon species ($C_{2\bar{n}}$) in the negative-ion spectrum obtained from the sample. This feature is characteristic of carbynes and is believed to represent the tendency of a $-C \equiv C-C \equiv C-$ chain to cleave at the single bonds rather than at the triple bonds (Stuckey and Whittaker, 1971). Whittaker et al. (1980) estimated the carbyne content of their Allende sample by comparing the negative-ion spectrum from it with reference spectra obtained from graphite and carbyne samples. Since the graphitic carbon that has been recognized in Allende on the basis of HRTEM and electron diffraction observations is poorly crystalline, it would be worthwhile to investigate whether the negative-ion spectrum of graphite depends on the degree of crystallinity of the sample. Also, in view of the concerns that we express below concerning the identification of carbynes, we feel that the negative-ion spectra for the reference carbynes should be published, together with details of the characterization of these samples.

In a study parallel to that of Whittaker et al. (1980), Hayatsu et al. (1980) obtained thermal pyrolysis mass spectra from a carbon-rich residue from Allende. The positive-ion mass spectra obtained by Hayatsu et al. (1980) show a predominance of even numbered carbon species in the mass range up to C_{10}, a feature that they argued to be indicative of carbynes. It should be noted that this result only applies to a small fraction of their acid residue, since most of the sample remained unvolatilised. Clearly it is necessary to obtain thermal pyrolysis mass spectra from known carbynes before this method can be used positively to identify carbyne in unknown samples. However, we note that the mass spectra obtained by Hayatsu et al. (1980) are different from those of graphite. Further work is needed to investigate the suggestion made by Anders (1981, pers. comm.) that there may be intermediate forms of elemental carbon that incorporate structural features of both carbyne and graphitic carbon.

Although carbynes were first described in 1967 their structure remains poorly understood; further information in this area would clearly be of use in their identification. Kasatochkin et al. (1967) synthesised carbyne by polymerising acetylene; the structure of the product was considered to consist of linear chains of carbon atoms with alternating single and triple bonds. These chains are then stacked parallel to each other in a hexagonal array, the chain direction defining the c-axis of a hexagonal unit cell. Different stacking arrangements relative to the triple bond positions are then considered to account for the eight or more polymorphs of carbyne (Whittaker, 1978). Although this proposed structure has not been verified, it appears to have become accepted in the absence of any alternatives. However, there is a severe problem with this structure: it does not appear to be consistent with the previously reported morphology of carbynes, which generally consist of extremely thin (0001) plates, sometimes bounded by hexagonal outlines (Whittaker et al., 1971; Whittaker and Wolten, 1972). This morphology is more indicative of a sheet rather than of a chain structure.

Bearing in mind the similarity between the hk0 electron diffraction patterns for sheet silicates and those for carbynes, it is important to note that in none of the descriptions of carbynes published to date have the compositions of individual carbyne grains been verified by X-ray microanalysis (Kasatochkin et al., 1967, 1974; Setaka and Sekikawa, 1980; Whittaker, 1978, 1979; Whittaker and Kintner, 1969; Whittaker and Wolten, 1972; Whittaker et al., 1971). Until such data are published combining diffraction and chemical information from the same crystals, there clearly remains the possibility that the diffraction patterns of these reported carbyne examples in fact arise from silicate contaminants.

The formation conditions of carbynes also remain unclear. Whittaker (1978) reported a stability field for carbynes in the vicinity of the solid-liquid-vapor triple point of the phase diagram for elemental carbon, with β carbyne being the stable solid phase at the triple point. However, several other researchers have been able to locate a graphite-liquid-vapor triple point extremely precisely, and have not noted any other solid phases in this region of the system (e.g., Haaland, 1976; Gokcen et al., 1976). In a recent review article Bundy (1980) discussed this problem, preferring the graphite model to that involving carbynes, although the question is not yet resolved.

CONCLUSIONS

The principal conclusions of this study are as follows:

(i) High-resolution transmission electron microscopy indicates that most of the carbon in acid residues from the Allende meteorite is a poorly graphitized carbon, structurally similar to the "glassy" carbon described by Jenkins et al. (1972).

(ii) Because of the similarity between the electron diffraction patterns for sheet silicate and those reported for carbynes, identifications of carbyne by electron diffraction should be supported by microanalyses of the individual grains in question.

(iii) The carbyne forms of carbon are still poorly characterized and their crystal structure has not yet been determined. Further studies are required to determine the status of carbynes as phases in the elemental carbon system, a point brought into question recently by Bundy (1980).

Acknowledgments—We thank A. G. Whittaker and R. S. Lewis for providing samples. Reviews of the manuscript by J. F. Kerridge, R. S. Lewis and G. R. Lumpkin are gratefully acknowledged. The electron microscopy was carried out in the Facility for High Resolution Electron Microscopy, Arizona State University, established with support from the NSF regional instrumentation program (grant CHE-7916098). Research support for this project was provided by NASA grant NAG 9-4.

REFERENCES

Alaerts L., Lewis R. S., Matsuda J., and Anders E. (1980) Isotopic anomalies of noble gases in meteorites and their origins – VI. Presolar components in the Murchison C2 chondrite. *Geochim. Cosmochim. Acta* **44**, 189–209.
Ban L. L. (1971) Direct study of structural imperfections by high-resolution electron microscopy. *Sur. Defect Prop. Solids* **1**, 54–94.
Ban L. L., Crawford D., and Marsh H. (1975) Lattice-resolution electron microscopy in structural studies of non-graphitizing carbons from polyvinylidene chloride (PVDC). *J. Appl. Crystallogr.* **8**, 415–420.
Bauman A. J., Devaney J. R., and Bollin E. M. (1973) Allende meteorite carbonaceous phase: intractable nature and scanning electron morphology. *Science* **241**, 264–267.
Black D. C. and Pepin R. O. (1969) Trapped neon in meteorites – II. *Earth Planet. Sci. Lett.* **6**, 395–405.
Brindley G. W. and DeKimpe C. (1961) Identification of clay minerals by single crystal electron diffraction. *Amer. Mineral.* **46**, 1005–1016.
Bundy F. P. (1980) The P, T phase and reaction diagram for elemental carbon, 1979. *J. Geophys. Res.* **85**, 6930–6936.
Dran J. C., Klossa J., and Maurette M. (1979) A preliminary microanalysis of the Berkeley gas-rich Allende residue (abstract). In *Lunar and Planetary Science X*, p. 312–314. Lunar and Planetary Institute, Houston.
Frick U. (1977) Anomalous krypton in the Allende meteorite (abstract). In *Lunar Science VIII*, p. 310–311. The Lunar Science Institute, Houston.
Frick U. and Reynolds J. H. (1977) On the host phases for planetary noble gases in Allende (abstract). In *Lunar Science VIII*, p. 319–321. The Lunar Science Institute, Houston.
Gokcen N. A., Chang E. T., Poston T. M., and Spencer D. J. (1976) Determination of graphite/liquid/vapor triple point by laser heating. *High Temp. Sci.* **8**, 81–97.
Green H. W., Radcliffe S. V., and Heuer A. H. (1971) Allende meteorite: A high-voltage electron petrographic study. *Science* **172**, 936–939.
Haaland D. M. (1976) Determination of the solid-liquid-vapor triple point pressure of carbon. *Report SAND-76-0074*, Sandia Lab., Albuquerque, N.M. 45 pp.
Hayatsu R., Scott R. G., Studier M. H., Lewis R. S., and Anders E. (1980) Carbynes in meteorites: detection, low-temperature origin, and implications for interstellar molecules. *Science* **209**, 1515–1518.
Herr W., Englert P., Herpers U., Watts E. J., and Whittaker A. G. (1980) A contribution to the riddle about the origin of certain glassy spherules. *Meteoritics* **15**, 300.
Jenkins G. M., Kawamura K., and Ban L. L. (1972) Formation and structure of glassy carbons. *Proc. Roy. Soc. London* **A327**, 501–517.
Kasatochkin V. I., Korshak V. V., Kudryavtsev Y. P., Sladkov A. M., and Elizen V. M. (1974) Polymorphism of carbyne. *Dokl. Chem.* **214**, 84–86. In English.
Kasatochkin V. I., Sladkov A. M., Kudryavtsev Y. P., Popov N. M., and Korshak V. V. (1967) Crystalline forms of the linear modification of carbon. *Dokl. Akad. Nauk. S.S.S.R.* **177**, 358–360.
Lewis R. S., Alaerts L., Matsuda J., and Anders E. (1979) Stellar condensates in meteorites: isotopic evidence from noble gases. *Astrophys. J.* **234**, L165–L168.
Lewis R. S., Matsuda J., Whittaker A. G., Watts E. J., and Anders E. (1980) Carbynes: carriers of primordial noble gases in meteorites (abstract). In *Lunar and Planetary Science XI*, p. 624–625. Lunar and Planetary Institute, Houston.
Lewis R. S., Srinivasan B., and Anders E. (1975) Host phase of a strange xenon component in Allende. *Science* **190**, 1251–1262.
Lumpkin G. R. (1981) Electron microscopy of carbon in Allende acid residues (abstract). In *Lunar and Planetary Science XII*, p. 631–633. Lunar and Planetary Institute, Houston.
Ott U., Mack R., and Chang S. (1981) Noble-gas-rich separates from the Allende meteorite. *Geochim. Cosmochim. Acta*. In press.
Setaka N. and Sekikawa Y. (1980) Chaoite, a new allotropic form of carbon, produced by shock compression. *J. Amer. Ceram. Soc.* **63**, 238–239.
Simmonds P. G., Bauman A. J., Bollin E. M., Gelpi E., and Oró J. (1969) The unextractable organic fraction of the Pueblito de Allende meteorite: evidence for its indigenous nature. *Proc. Nat. Acad. Sci. U.S.* **64**, 1027–1034.
Smith P. P. K. and Buseck P. R. (1980) High resolution transmission electron microscopy of an Allende acid residue. *Meteoritics* **15**, 368–369.
Smith P. P. K. and Buseck P. R. (1981) Graphitic carbon in the Allende meteorite: a microstructural study. *Science* **212**, 322–324.

Srinivasan B. and Anders E. (1978) Noble gases in the Murchison meteorite: possible relics of s-process nucleosynthesis. *Science* **201**, 51–56.
Stuckey W. K. and Whittaker A. G. (1971) Paper TP-177, presented at the 10th Carbon Conference, Lehigh University, Bethelehem, PA, 27 June to 2 July, 1971.
Webster A. S. (1979) Meeting of the Royal Astronomical Society on 1978 October 13th. *Observatory* **99**, 29.
Webster A. S. (1980) Carbyne as a possible constituent of the interstellar dust. *Mon. Not. Roy. Astr. Soc.* **192**, 7P–9P.
Whittaker A. G. (1978) Carbon: a new view of its high-temperature behaviour. *Science* **200**, 763–764.
Whittaker A. G. (1979) Carbon: occurrence of carbyne forms of carbon in natural graphite. *Carbon* **17**, 21–24.
Whittaker A. G., Donnay G., and Lonsdale K. (1971) Additional data on chaoite. *Carnegie Inst. Wash. Yearb.* **69**, 311.
Whittaker A. G., and Kintner P. L. (1969) Carbon: observations on the new allotropic form. *Science* **165**, 589–591.
Whittaker A. G., Watts E. J., Lewis R. S., and Anders E. (1980) Carbynes: carriers of primordial noble gases in meteorites. *Science* **209**, 1512–1514.
Whittaker A. G. and Wolten G. M. (1972) Carbon: a suggested new hexagonal crystal form. *Science* **178**, 54–56.
Windom H., Griffin J., and Goldberg E. D. (1967) Talc in atmospheric dusts. *Environ. Sci. Technol.* **1**, 923–926.

Noble gas trapping by laboratory carbon condensates

S. Niemeyer* and K. Marti

Chemistry Department, B-017, University of California at San Diego, La Jolla, California 92093

Abstract—Trapping of noble gases by carbon-rich matter was investigated by synthesizing carbon condensates in a noble gas atmosphere. Laser evaporation of a solid carbon target yielded submicron grains which proved to be efficient noble gas trappers (Xe distribution coefficients up to 13 cm^3 STP g^{-1} atm^{-1}). The carbon condensates are better noble gas trappers than previously reported synthetic samples, except one, but coefficients inferred for meteoritic acid-residues are still orders of magnitude higher. The trapped noble gases are loosely bound and elementally strongly fractionated, but isotopic fractionations were not detected. Although this experiment does not simulate nebular conditions, the results support the evidence that carbon-rich phases in meteorites may be carriers of noble gases from early solar system reservoirs. The trapped elemental noble gas fractionations are remarkably similar to both those inferred for meteorites and those of planetary atmospheres for Earth, Mars and Venus.

INTRODUCTION

An important discovery in meteorite research was Reynold's (1961) recognition that noble gas elemental patterns in carbonaceous chondrites differ systematically from those of gas-rich brecciated meteorites. This difference is depicted in Fig. 1 where the gas-rich meteorite Pesyanoe is used for normalization, i.e., to represent the early solar system reservoir, and consequently plots on a horizontal line at zero. Patterns for bulk samples of carbonaceous chondrites, represented here by the C-I's, show strong elemental fractionations in which the heavy noble gases are enriched. We now know that the brecciated gas-rich meteorites gained their major noble gas component by solar wind implantation (Signer, 1964; Suess *et al.*, 1964; Wänke, 1965; Eberhardt *et al.*, 1965) and may be taken as an approximation of the solar system inventory of noble gases (Marti *et al.*, 1972).

Also shown in Fig. 1 are the inferred fractionation patterns of the noble gases in atmospheres of terrestrial planets. The earth shows depletions of Ne and Ar relative to Kr which are similar in magnitude to carbonaceous chondrites. The depletion of Xe may partly be due to trapping of Xe in shales (Canalas *et al.*, 1968, Fanale and Cannon, 1971). The martian noble gas pattern (Owen *et al.*, 1977) matches within the rather large uncertainties the meteoritic depletions at Ne and Ar, but, as is the case for the earth, Xe is lower. Again we may speculate that a significant portion of Mars' Xe still resides in its crust. Noble-gas data for Venus have not yet produced a consensus, and in Fig. 1 two patterns are plotted: one set using data from Hoffman *et al*'s. (1980) review paper, and a second set based on Donahue *et al*'s (1981) abundances.

Carbonaceous chondrites have experienced relatively little chemical processing. Therefore, characterization of the noble gas carriers in these meteorites may prove helpful in understanding how planets acquired their initial complement of noble gases. The discovery that trapped noble gases in the carbonaceous chondrites are located mostly in acid insoluble residues (Lewis *et al.*, 1975; Lewis *et al.*, 1977; Frick and Moniot, 1977;

---

*Present Address: University of California, Lawrence Livermore Laboratory, L-232, Livermore, California 94550.

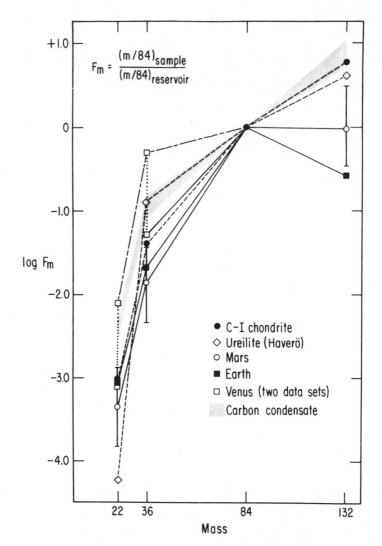

Fig. 1. A comparison of various elemental fractionation patterns. The solar-type noble gases in Pesyanoe are assumed to represent the solar system reservoir (Marti et al., 1972). The average C-I carbonaceous chondrite data are from Mazor et al. (1970), and the Haverö data are used for ureilites (Göbel et al., 1978). Data plotted for terrestrial planets represent only their atmospheric abundances. The Mars data are from Owen et al. (1977) and two sets of Venus data are from Hoffman et al. (1980) [connected by the dotted line] and Donahue et al. (1981). The stippled region shows the range of fractionations for trapped noble gases in the carbon condensates with respect to the ambient atmosphere. In light of the presumed differences in formation conditions for the meteorites, planetary atmospheres, and the carbon condensates, the degree of similarity for these fractionation patterns is striking.

Reynolds et al., 1978) permitted significant progress, but the identity of the principal carrier is still disputed. The Chicago group from the outset has argued that an inorganic phase is the carrier of the isotopically normal planetary gases, whereas the Berkeley group has consistently argued for an unspecified type of carbonaceous matter. This study was initiated partly in response to the controversy, although more recently the Chicago workers have acknowledged that carbonaceous phases, such as carbynes or some unknown phases, are also viable candidates (Alaerts et al., 1980). Our approach was to study in the laboratory the trapping of noble gases by condensates formed from a hot gas (Kothari et al., 1979; Niemeyer and Marti, 1980). Of primary interest in this study are (1) comparison of the trapping efficiencies of carbon condensates to those of other inorganic condensates and meteoritic acid residues, and (2) comparison of the elemental and isotopic noble gas fractionations of the carbon condensates to those inferred for the meteorites.

EXPERIMENTAL METHODS

The condensates are produced in the apparatus described in detail by Stephens (1979) and Stephens and Kothari (1978). A graphite target is held in a glass cell which is filled with a high-purity mixture of Ne, Ar, Kr and Xe (supplied by Matheson). Total pressures (P) are measured with a Hg manometer. The carbon vapor is produced by focussing laser pulses (~5 J/pulse) onto the target surface. Each laser pulse forms a plume volume of about 1 cm^3 at a temperature of >3500 K. The vapor nucleates into small (≤2 nm) droplets on a time scale of microseconds. The condensates cool very rapidly (10^5-10^6 K/sec) by collisions with the ambient gas and approach room temperature within tens of milliseconds. The predominant grain growth mechanism is coagulation of the droplets into chains of solid grains. Collisions with cell walls do not occur until after the grains cooled. Each run consisted of about 100 pulses, with laser pulses separated by about 1 minute. The condensates are collected for analysis by allowing them to settle onto Al foil collectors.

TEM analyses of carbon condensates (Stephens, 1979) show chains of submicron grains, which is typical of all types of condensates produced in this manner. Typical grain sizes are 100–1000 Å. Electron diffraction patterns for carbon condensates show only diffuse rings, suggesting poorly crystalline grains.

Sample weights were determined by precisely weighing the Al foils before and after collection of the condensates; uncertainties of less than 1 μg were confirmed by weighing blank foils included in some runs. Condensates produced at three different pressures were analyzed for noble gas contents. In two cases condensates were collected both near the target (A) and about 20 cm away (B). During the run at a pressure of 162 torr, an electrical discharge was introduced by placing two tesla coils just outside the glass cell, however, the discharge was strongly confined to paths between conducting

Table 1. Noble gas abundances in carbon condensates prepared in a controlled atmosphere[a].

Sample Pressure (P in torr)	Wt. (μg)	Temp-Time (°C-hrs.)	^{22}Ne $\times 10^{-5}$ cm^3 STP/g	^{40}Ar	^{84}Kr $\times 10^{-8}$ cm^3 STP/g	^{132}Xe
288.7 B	16.8	800– 1.2	1.76	2142	193.6	233.0
		1600– 1.2	0	0	0	0
161.6 B[c]	10.1	1600–1.2	1.30	1625	148.4	132.7
161.6 A[c]	79.0	125–65.5	1.03	1108	116.3	107.8
		200–20.5	1.30	527	51.2	45.8
		600– 1.2	0.76	260	30.5	21.5
		800– 1.2	0.06	<1	0.5	0.2
		1600– 1.2	0	0	0	0.4
		Total	3.15	1895	198.5	175.7
53.9 B	2.4	1600– 1.2	3.36	1030	184.4	98.0
53.9 A	7.3	125–14.5	1.01	784	94.0	35.1
		1600– 1.2	0.99	690	72.7	62.8
		Total	2.00	1474	166.7	97.9
Blanks:						
760 (in H$_2$)[d]	40	1600– 0.5	—	<13	9.0	2.0
760 (in H$_2$, N$_2$)[d]	40	1600 –0.5	—	<25	12.3	3.3

[a]Gas amounts are corrected for procedural sample system blanks: entries of zero denote no gas detected above blank levels which typically were ^{22}Ne ≤ 5 × 10^{-11} cm^3 STP, ^{40}Ar ≤ 4 × 10^{-9} cm^3 STP, ^{84}Kr ~ 5 × 10^{-13} cm^3 STP, and ^{132}Xe ≤ 1 × 10^{-13} cm^3 STP. Gas concentrations are accurate to ≤10 percent, and sample weights are uncertain by ≤1 μg. Measured noble gas proportions in the ambient noble gas atmosphere were: Ne/Ar/Kr/Xe = 0.48/ ≡ 1.0/25.2 × 10^{-6}/6.0 × 10^{-6}; all gases had terrestrial isotopic compositions.
[b]Temperatures and times refer to the extraction of noble gases from the samples by RF heating.
[c]Plasma-discharge was introduced by two external tesla coils during laser evaporation (see discussion in text).
[d]Kothari et al. (1979).

media and passed through relatively little of the region where condensation occurs. Noble gases were analyzed mass spectrometrically according to procedures described earlier (Lightner and Marti, 1974). Noble gas concentrations are given in Table 1, and isotopic compositions are discussed later in the text.

Vaporization-cell blanks, i.e., contributions from the vaporization process and the subsequent exposure to air, were determined in carbon condensates produced in H_2 and H_2 plus N_2 atmospheres (Table 1.) Because these contributions are minor for the reported condensate samples (i.e., 5–7% for Kr, 1–2% for Xe, and $\leq 1\%$ for Ar), no corrections were applied to the data in Table 1. We note that the blanks of the carbon target itself are higher, but the rare gases are effectively lost upon vaporization by the laser. It is possible that trapped noble gases might partially be lost during storage under vacuum in the spectrometer sample system. In order to evaluate diffusion losses at room temperature, we stored condensate 162A for 4 months in the sample system. However, condensate 162B, which was analysed after one month, had slightly lower noble gas concentrations.

DISTRIBUTION COEFFICIENTS

Significant concentrations (C_G) of all four noble gases were detected in every carbon condensate sample. Distribution coefficients

$$K_D(G) = C_G(\text{in cm}^3 \text{ STP g}^{-1})P_G^{-1} \text{ (in atm.)}$$

for G = Ar, Xe are listed in Table 2, along with K_D's reported for other samples which were synthesized by a condensation process. An average value is shown for each ambient total pressure P, since no systematic differences between samples A and B were observed. K_D's of the carbon condensates are substantially higher than those observed in other synthetic samples, but are still orders of magnitude lower than those inferred for acid residues of meteorites. However, since K_D's of our carbon condensates show a dependence on pressure, such comparisons with other samples require a normalization.

Different trapping mechanisms (e.g., solubility, adsorption, occlusion of gas during crystallization) are expected to yield different relations between K_D and pressure. Figure 2 shows a logarithmic plot for K_D vs. pressure for Ar, Kr and Xe. For Henry's law behavior, as observed in low pressure solubilities of gases, K_D's are independent of pressure, which is clearly not the case in Fig. 2. For the case of gas occlusion, we would expect roughly similar values of K_D for the different gases at any given P. Such a characteristic is not observed for the carbon condensates. For adsorption processes, the two most widely discussed equations for the isotherms are the Freundlich relation

Table 2. Distribution coefficients K_D for synthetic samples and meteorites.

Sample	$K_D(Ar)$	$K_D(Xe)$
	(cm^3 STP g^{-1} atm^{-1})	
Carbon condensate[1]		
289 torr	0.08	5.6
162 torr	0.12	6.7
54 torr	0.31	12.7
Propane soot[2]	0.02	1.7
Electron discharge kerogen[2]	0.035	0.46
Pyroxene condensate[3]	< 0.0006	—
Magnetite condensate[4]	2×10^{-5}	—
Meteoritic residues[5]	4×10^4	5×10^7

[1]This work.
[2]Frick et al. (1979).
[3]Kothari et al. (1979).
[4]Honda et al. (1979).
[5]Inferred from noble gas concentrations in meteoritic acid residues (e.g. Lewis et al., 1975) and assuming a gas reservoir of cosmic proportions (Cameron, 1973) at total pressures of 10^{-4}–10^{-6} atm.

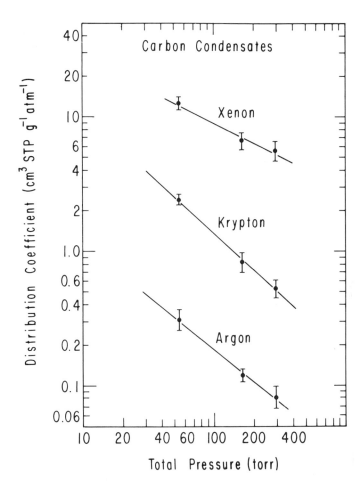

Fig. 2. Pressure dependence of the carbon condensate distribution coefficients (K_D). The 54 torr points are for the "A" sample only while the 162 torr points are averages of the "A" and "B" samples. In the investigated pressure range, each noble gas defines a straight line. (Note that to obey Henry's law, the slope would have to be zero.) The applicability of possible adsorption isotherms to the data are discussed in the text.

$V_G = kP_G^{1/n}$ and the Langmuir relation $V_G = bP_G/(1+bP_G)$, where V_G is the volume of adsorbed gas, P_G the partial pressure of ambient gas G vapor, and K, n, and b are constants (Young and Crowell, 1962). The straight lines in the Fig. 2 plot are consistent with the classical Freundlich equation, but are also compatible with the Langmuir relation. Actually, neither relation would be expected to apply strictly to our samples since theoretical derivations of these relations assume equilibrium and constant temperature. These conditions probably do not apply to the condensates which experienced extremely rapid cooling rates. These relations also assume a single gas species, and a theoretical extension to binary mixtures (Glueckauf, 1953) yielded more complex relations. Thus, empirical extrapolations of the present results to much different pressure regimes is quite uncertain, especially since we later argue that an equilibrium physical adsorption process is an unlikely trapping mechanism. Furthermore, we also anticipate that distribution coefficients depend not only on total pressure but also partial pressures of the species under consideration due to departures from equilibrium (Kovach, 1978). We note that the smooth variations observed in Fig. 2 indicate that the plasma discharge at 162 torr had no discernable effect on the amounts of trapped gases.

Since all synthetic materials listed in Table 2 were formed at similar total pressures, comparisons using the Fig. 2 trend lines for the carbon condensate values are probably meaningful. The two most efficient noble gas trappers in Frick *et al*'s. (1979) study of synthesized carbonaceous matter, namely, propane soot and electron-discharge kerogen, were made at pressures of 1 atmosphere and 400 torr respectively. Extrapolating to these pressures, we estimate that the K_D's for the carbon condensates are almost a factor of two higher than the soot for all three noble gases and 2–10 times higher than the kerogen. The magnetite condensate, formed at 100 torr, trapped about 4 orders of magnitude less Ar than carbon condensate, while the upper limit for the pyroxene condensate K_D's is 2 orders of magnitude below the carbon condensate. Overall then, the carbon condensates are somewhat better rare gas trappers than the carbonaceous phases of Frick *et al.* (1979) and are vastly superior to the two inorganic condensates studied so far.

Recently Dziczkaniec *et al.* (1981) reported results for a RF plasma discharge experiment in which carbonaceous matter was synthesized. Three of their four samples yield K_D's for Xe orders of magnitude below estimated values for our carbon condensates, but the fourth sample was extraordinarily gas-rich, corresponding to a K_D of $\sim 10^6$ cm^{3}STP g^{-1} atm^{-1}. The mechanism responsible for this very high K_D is not yet understood, but very likely is the result of the plasma discharge.

Quantitative comparison of the carbon condensates to the meteoritic residues is difficult since the residues presumably trapped their noble gases at much lower pressures. Extrapolations of the Fig. 2 trend lines point to condensate distribution coefficients which are factors of 10^3-10^5 lower than the meteoritic residues. Of course, as we discuss later, the many other differences between conditions in our laboratory experiment and the solar nebula probably play important roles, e.g., the time scale for condensation and grain growth was very much longer in nebular condensation. Furthermore, the expansion rate of the vapor plume and the dynamics of the interaction between the expanding carbon plume and the ambient gas are expected to depend on both the total and partial gas pressures (Stephens, 1979). In our initial report on this work (Kothari *et al.*, 1979), we estimated a K_D for Ar of 0.48 ± 0.24 for carbon formed in a mixture of H_2, He and Ar at one atmosphere. Despite the large uncertainty, this value lies above the trend line in Fig. 2 and hints that the availability of hydrogen may enhance the trapping efficiency of the carbon condensates.

Even though it is difficult to evaluate whether phases similar to our carbon condensates could, under nebular conditions, achieve the extremely high distribution coefficients of the meteoritic residues, the demonstration that the carbon condensates are far better noble gas trappers than the pyroxene and magnetite condensates provides valuable information regarding the identity of the principal carrier of planetary noble gases. Although sulfide condensates have not yet been studied, the recent results of the plasma synthesis experiment of Dziczkaniec *et al.* (1981) and work on the nature of noble gas-rich separates from the Allende meteorite by Ott *et al.* (1981) reinforce the implications of our study, namely, carbonaceous phases are more likely candidates for the gas-rich carrier than an inorganic phase.

ELEMENTAL AND ISOTOPIC FRACTIONATION PATTERNS

The elemental abundance patterns for the carbon condensates shown in Fig. 3 show strong fractionations favoring the heavy gases. The patterns observed at all three pressures are quite similar; however, a slight trend is suggested by the increasing fractionation of Ar relative to Kr as the pressure decreases, but the opposite trend is apparent for Xe. Since the heavier noble gases, expecially Xe, may possibly be adsorbed during exposure to air, we cannot distinguish whether these trends are due to minor air contamination or instead are indigenous to the condensates' trapping mechanism. Fractionation patterns for soot and kerogen (Frick *et al.*, 1979) are also shown in Fig. 3. The Ar value for the kerogen is only an upper limit due to meteoritic contamination in the extraction system, and the true value could be as much as an order of magnitude lower. These soot and kerogen patterns closely resemble those of our carbon condensates, a rather remarkable similarity in view of the very different formation conditions.

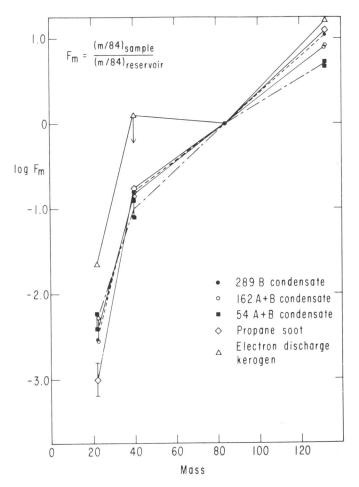

Fig. 3. Elemental noble gas fractionations for condensates of carbon. All five carbon condensate samples are plotted: single lines are shown for each pressure. The soot and kerogen samples are the two most gas-rich synthetic carbonaceous samples of Frick et al. (1979). The kerogen Ar value is only an upper limit due to meteoritic contamination, and the true value may be an order of magnitude lower.

Referring again to Fig. 1, we compare the carbon condensate fractionation patterns to the inferred patterns for the natural samples. The range of carbon condensate fractionations is shown by the stippled region. A ureilite is also plotted because in these meteorites the noble gases are concentrated in carbon-rich phases in veins rich in diamonds (Göbel et al., 1978). For Ar, Kr and Xe, these condensate patterns provide good matches to the fractionation patterns observed in carbonaceous chondrites and the ureilites. Regarding Ne, the condensates are higher and ureilites clearly lower than the C-I's, however, Ne is more susceptible to diffusive loss in natural samples. Likewise, the noble gas fractionations for the atmospheres of all three terrestrial planets are fairly similar to the carbon condensate patterns. Given the presumably different conditions prevailing during formation of the meteorites, the planetary atmospheres, and the carbon condensates, the degree of similarity in the elemental noble gas fractionations is striking.

No significant mass fractionations were detected for any of the isotopic patterns in the carbon condensates. Since isotopic fractionations are most readily detected in Xe, we plot in Fig. 4 the per mil deviations with respect to air for two of the Xe analyses. The agreement with atmospheric Xe is good, and we obtain an upper limit for mass frac-

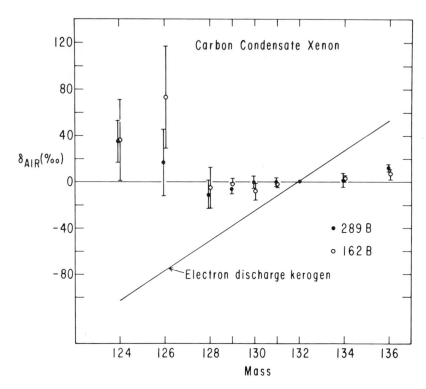

Fig. 4. Xenon isotopic compositions of carbon condensates are plotted as relative deviations $\delta_{AIR} = [((i/132)_{SAMPLE}/(i/132)_{AIR}) - 1] \times 1000$. Data for two Xe analyses are plotted (1σ uncertainties) and agree within error limits with the composition of terrestrial Xe. We obtain an upper limit of $\approx$ 2‰/amu for mass fractionation of Xe. The observed mass fractionation for the kerogen sample of Frick *et al.* (1979) is shown by the line. Since the elemental fractionations of the carbon condensates and the kerogen are similar, it appears that the mechanisms responsible for elemental and for isotopic fractionations may be decoupled.

tionation of ~ 2‰/amu. The electron discharge kerogen of Frick *et al.* (1979), however, gave a large mass dependent fractionation which is represented by the line in Fig. 4. The plasma synthesis samples of Dziczkaniec *et al.* (1981) also yield comparably large isotopic fractionations. These are the only synthetic samples known to isotopically fractionate noble gases, and the mechanism is not yet understood. An interesting feature of these samples is that, despite the rather large isotopic fractionations, the elemental fractionations are very similar to the propane soot and the carbon condensates. It appears that the mechanisms responsible for the isotopic and the elemental fractionation may be decoupled.

GAS RELEASE AND ACTIVATION ENERGY

Stepwise heating analyses of several carbon condensate samples revealed that the trapped noble gases are very loosely bound. Prolonged (65 hours) heating of sample 162A at only 125°C was sufficient to remove about 60% of the noble gases, and after further (overnight) heating at 200°C, a total of 90% had been released. The fraction of gas released in each step was nearly the same for Ar, Kr and Xe. Estimates for the activation energy and for the coefficients of volume diffusion were obtained for sample 162A by following the Fechtig and Kalbitzer (1966) procedure which considers volume diffusion of noble gases which are initially uniformly distributed in spherical grains. In Fig. 5, the D/a^2 values are plotted vs. $1/T$. The line represents an activation energy of $E_A = 9$ Kcal mole^{-1}

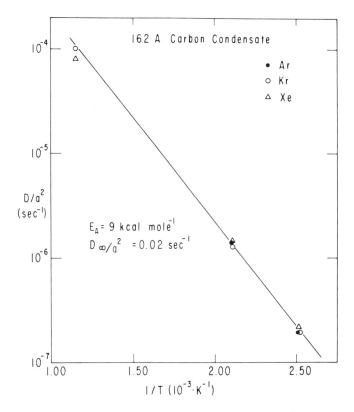

Fig. 5. Plot of "apparent" diffusion parameters vs. (temperature)$^{-1}$. "Apparent" D/a^2 values are calculated for a volume diffusion model which assumes that the noble gases are initially uniformly distributed in spherical grains. All data for Ar, Kr and Xe define a single line which corresponds to a very low activation energy of 9 Kcal/mole. However, the assumptions in this simple model calculation may not apply to the carbon condensates.

and applies to Ar, Kr and Xe, since the release patterns are similar. This activation energy is very low, at least three times lower than that of most minerals, and does not support the volume diffusion model. Instead, low and uniform E_A values are suggestive of trapping by physical adsorption. It is interesting to note that the "apparent" E_A = 9 Kcal/mole for the carbon condensates is similar to activation energies measured for terrestrial shungite (Rison, 1980). However, Rison concludes that his results are consistent with noble gas solubilities in shungite.

Physical adsorption as the trapping mechanism for noble gases in carbon-rich materials is an attractive alternative, since it provides quantitative estimates of the trapping temperature. For example, adsorption was evaluated for ureilites by Wilkening and Marti (1976) and by Göbel et al. (1978), and for synthetic samples by Frick et al. (1979). We use adsorption parameters determined for carbon black by Göbel et al. (1978) from Sams et al. (1960) data to calculate that adsorption on carbon condensates at about 210 K can account for both the amounts of trapped noble gases and the degree of elemental fractionation. Similar temperature estimates were derived by Fanale and Cannon (1971) in a study of Kr and Xe adsorption on terrestrial shales. Clearly these temperature estimates are not correct for the condensates, since the minimum temperature was at least that of the laboratory. Moreover, the absence of observable gas losses during vacuum storage is consistent with trapping well above room temperature. Therefore, equilibrium physical adsorption appears to be an unlikely trapping mechanism.

The low retentivity of the noble gases in the carbon condensates does not mimic the meteoritic acid residues which release the major portion of their noble gases at much higher (>600°C) temperature. From stepwise heatings of meteoritic acid residues, Reynolds et al. (1978) derived diffusion parameter D_∞/a^2 values of 0.1–0.9 $\sec^{-1}$ and activation energies of 23–30 Kcal $mole^{-1}$. They concluded that the assumptions of the diffusion model were probably not valid for these residues, since the very low observed diffusion coefficients appear unreasonable for the low activation energies. The carbon condensates pose a similar problem, since D_∞/a^2 values of only about 0.02 $\sec^{-1}$ are accompanied by activation energies of 9 Kcal $mole^{-1}$. Reynolds et al. (1978) proposed an alternative "chemical reaction" model whereby erosion of grain surfaces control the release of noble gases. An advantage of this model is the explanation it affords for the nearly identical temperature release patterns of all the noble gases, which is also observed for our condensates. However, there is considerable uncertainty regarding the nature of chemical reactions in the case of carbon condensates heated in an extraction system environment. A possible alternative for the condensates is that the release of gases is primarily controlled by phase transitions, recrystallization or devitrification of the partially amorphous and partially microcrystalline condensate samples.

IMPLICATIONS FOR METEORITIC GAS CARRIERS AND PLANETARY ATMOSPHERES

The many differences between conditions in the nebula and our laboratory experiment during noble gas trapping need to be taken into account in order to meaningfully compare the carbon condensates to the meteoritic acid residues. The disparate time scales for condensation and grain growth may be of greatest importance. The carbonaceous phases in meteorites may have initially condensed onto previously solidified phases, and further grain growth resulted from addition of monolayers. In contrast, in the laboratory experiment, grain growth proceeds by coagulation of nucleated droplets rather than by addition of monolayers. The entire process requires only milliseconds in the laboratory, while the duration of meteoritic grain growth was probably many orders of magnitude larger (e.g., Palme and Wlotzka, 1976). The laboratory condensation process may accordingly depart much more from equilibrium and also be far less isothermal, which would affect the K_D's. Although the conditions in the nebula are still a matter of much debate, additional, possibly significant, differences between the nebula and laboratory experiment are the presumably vastly different pressure-temperature environments, the differing noble gas partial pressures, and the presence of other gaseous species. The quantitative effects of these differences cannot yet be specified, but we would expect that, as a consequence, K_D's of condensates prepared in the laboratory may differ from values obtained under nebular conditions.

Despite our inability to quantitatively compare the carbon condensates to the meteoritic acid residues, this experimental study lends support to the recent suggestions of carbon-rich phases as the principal noble gas carrier in meteorites. With the exception of one carbonaceous sample of Dziczkaniec et al. (1981), K_D's of the carbon condensates are higher than those of other synthesized materials. Especially noteworthy is that the two inorganic condensates produced in the same manner were poorer noble gas trappers by several orders of magnitude. Laboratory carbon condensates also produce elemental noble gas fractionations which are remarkably similar to the meteoritic patterns. Although the meteoritic samples release their noble gases at much higher temperatures, this might be attributed to the different modes of grain growth or to some redistribution mechanism which enhances the retentivity. We do not mean to imply that these carbon condensates are exactly the same material as the principal noble gas carrier phase in meteorites; rather, our results enhance the plausibility of a related phase as the carrier. Actually, the relationship may be fairly close, as recent TEM studies of the meteoritic acid residues (Smith and Buseck, 1981; Lumpkin, 1981) find that poorly ordered graphite (i.e., non-amorphous) is the dominant observable phase.

This study is also relevant to theories of the origin of planetary atmospheres (see, e.g., Pollack and Black, 1979). If carbon-rich phases played an important role in planets acquiring their noble gas inventories, a suggestion consistent with noble gas studies of meteorites, then our experimental results will help to better interpret the systematics of noble gas abundances and elemental fractionations for the planets. For example, we have found that the elemental fractionations for the carbon condensates are quite similar to those determined for other carbon-rich phases which were formed under different conditions. These patterns also closely resemble those of the planets. Thus, the similar fractionations for the planets could be obtained even for fairly different trapping conditions. Our results also demonstrate that physical adsorption theories should not be used in conjunction with gas abundances to constrain temperatures during trapping, since the carbon condensates efficiently trapped noble gases at much higher temperatures than predicted for adsorption.

A longstanding puzzle in noble gas studies has been the difference in the isotopic abundances of the earth's Xe with respect to both solar and meteoritic Xe, yet with regard to elemental noble gas fractionations, the terrestrial and meteoritic patterns are similar to one another but differ from that of the sun. The experimental demonstration of an apparent decoupling of the elemental and isotopic noble gas fractionations for carbon-rich materials, although not presently understood, may provide a clue to unraveling the origin of terrestrial Xe.

SUMMARY

This study revealed the following characteristics for noble-gas trapping by carbon condensates synthesized in the laboratory:

1. The carbon condensates are efficient noble gas trappers, whereas the two inorganic condensates studied so far are vastly inferior trappers.
2. The noble gases are strongly fractionated elementally, and the patterns are similar to those inferred for meteorites and planetary atmospheres.
3. Isotopic mass fractionations were not detected.
4. Distribution coefficients increase substantially as the pressure decreases.
5. The noble gases are loosely bound and temperature release patterns are the same for Ar, Kr and Xe.

Acknowledgments—We gratefully acknowledge the tutoring of John Stephens on the use of the laser evaporation apparatus. We thank B. Srinivasan and M. Dziczkaniec for thorough and constructive reviews. Serge Regnier offered assistance to one of us (SN) in the initial stages of this work, and Irene Spellman typed the manuscript. This work was supported by NASA grant NGL 05-009-150.

REFERENCES

Alaerts L., Lewis R. S., Matsuda J., and Anders E. (1980) Isotopic anomalies of noble gases in meteorites and their origins – VI. Presolar components in the Murchison C2 chondrite. *Geochim. Cosmochim. Acta* **44**, 189–209.

Cameron A. G. W. (1973) Abundances of the elements in the solar system. *Space Sci. Rev.* **15**, 121–146.

Canalas R. A., Alexander E. C. Jr., and Manuel O. K. (1968) Terrestrial abundance of noble gases. *J. Geophys. Res.* **73**, 3331–3334.

Donahue T. M., Hoffmann J. H., and Hodges R. R. Jr. (1981) Krypton and xenon in the atmosphere of Venus. *Geophys. Res. Lett.* **8**, 513–516.

Dziczkaniec M., Lumpkin G. R., Donohoe K. and Chang S. (1981) Plasma synthesis of carbonaceous material with noble gas tracers (abstract). In *Lunar and Planetary Science XII*, p. 246–248. Lunar and Planetary Institute, Houston.

Eberhardt P., Geiss J., and Grögler N. (1965) Further evidence on the origin of trapped gases in the meteorite Khor Temiki. *J. Geophys. Res.* **70**, 4375–4378.

Fanale F. P. and Cannon W. A. (1971) Physical adsorption of rare gas on terrigenous sediments. *Earth Planet. Sci. Lett.* **11**, 362–368.

Fechtig H. and Kalbitzer S. (1966) The diffusion of argon in potassium-bearing solids. In *Potassium-Argon Dating* (O. A. Schaeffer and J. Zähringer, eds.), p. 68–107. Springer-Verlag, N.Y.

Frick U., Mack R., and Chang S. (1979) Noble gas trapping and fractionation during synthesis of carbonaceous matter. *Proc. Lunar Planet. Sci. Conf. 10th*, p. 1961–1973.

Frick U. and Moniot R. K. (1977) Planetary noble gas components in Orgueil. *Proc. Lunar Sci. Conf. 8th*, p. 229–261.

Glueckauf E. (1953) Monolayer adsorption of two species on non-uniform surfaces. *Trans. Faraday Soc.* **49**, 1066–1079.

Göbel R., Ott U., and Begemann F. (1978) On trapped noble gases in ureilites. *J. Geophys. Res.* **83**, 855–867.

Hoffman J. H., Oyama V. I., and von Zahn U. (1980) Measurements of the Venus lower atmosphere composition: A comparison of results. *J. Geophys. Res.* **85**, 7871–7881.

Honda M., Ozima M., Nakada Y., and Onaka T. (1979) Trapping of rare gases during the condensation of solids. *Earth Planet. Sci. Lett.* **43**, 197–200.

Kothari B. K., Marti K., Niemeyer S., Regnier S., and Stephens J. R. (1979) Noble gas trapping during condensation: a laboratory study (abstract). In *Lunar and Planetary Science X*, p. 682–684. Lunar and Planetary Institute, Houston.

Kovach J. L. (1978) Krypton-xenon activated carbon adsorption. In *Carbon Adsorption Handbook* (P. N. Cheremisinoff and F. Ellerbusch, eds.), p. 861. Ann Arbor Sci. Publishers, Ann Arbor, Michigan.

Lewis R. S., Srinivasan B., and Anders E. (1975) Host phase of a strange xenon component in Allende. *Science* **190**, 1251–1262.

Lewis R. S., Gros J., and Anders E. (1977) Isotopic anomalies of noble gases in meteorites and their origin -2. Separated minerals from Allende. *J. Geophys. Res.* **82**, 779–792.

Lightner B. D. and Marti K. (1974) Lunar trapped xenon. *Proc. Lunar Sci. Conf. 5th*, p. 2023–2031.

Lumpkin G. R. (1981) Electron microscopy of carbon in Allende acid residues (abstract). In *Lunar and Planetary Science XII*, p. 631–633. Lunar and Planetary Institute, Houston.

Marti K., Wilkening L. L., and Suess H. E. (1972) Solar rare gases and the abundances of the elements. *Astrophys. J.* **173**, 445–450.

Mazor E., Heymann D., and Anders E. (1970) Noble gases in carbonaceous chondrites. *Geochim. Cosmochim. Acta* **34**, 781–824.

Niemeyer S. and Marti K. (1980) "Planetary" noble gases: rare-gas trapping by laboratory carbon condesnates (abstract). In *Lunar and Planetary Science XI*, p. 812–814. Lunar and Planetary Institute, Houston.

Ott V., Mack R., and Chang S. (1981) Noble-gas-rich separates from the Allende Meteorite. *Geochim. Cosmochim. Acta.* In press.

Owen T., Biemann K., Rushneck D. R., Beller J. E., Howarth D. W., and LaFleur A. L. (1977) The composition of the atmosphere at the surface of Mars. *J. Geophys. Res.* **82**, 4635–4639.

Palme H. and Wlotzka F. (1976) A metal particle from a Ca, Al-rich inclusion from the meteorite Allende, and the condensation of refractory siderophile elements. *Earth Planet. Sci. Lett.* **33**, 45–60.

Pollack J. B. and Black D. C. (1979) Implications of the gas compositional measurements of Pioneer Venus for the origin of planetary atmospheres. *Science* **205**, 56–59.

Reynolds J. H. (1961) Xenon in stone meteorites. *Proc. Conf. Nucl. Geophys.*, Highland Park, Ill. Publ. 845, NAS-NRC, Washington, D.C.

Reynolds J. H., Frick U., Neil J. M., and Phinney D. L. (1978) Rare-gas-rich separates from carbonaceous chondrites. *Geochim. Cosmochim. Acta* **42**, 1775–1797.

Rison W. (1980) Isotopic fractionation of argon during stepwise release from shungite. *Earth Planet. Sci. Lett.* **47**, 383–390.

Sams J. R., Constabaris G., and Halsey G. D. Jr. (1960) Second virial coefficients of neon, argon, krypton and xenon with a graphitized carbon black. *J. Phys. Chem.* **64**, 1689–1696.

Signer P. (1964) Primordial rare gases in meteorites. In *The Origin and Evolution of Atmospheres and Oceans.* (P. Brancayio, ed.), p. 183–196. Wiley, N.Y.

Smith P. P. K. and Buseck P. R. (1981) Graphitic carbon in the Allende meteorite. *Science* **212**, 322–324.

Stephens J. R. (1979) Laboratory analogues to cosmic dust. Ph.D. Dissertation, Univ. California, San Diego.

Stephens J. R. and Kothari B. K. (1978) Laboratory analogues to cosmic dust. *Moon and Planets* **19**, 139–152.

Suess H. E., Wänke H., and Wlotzka F. (1964) On the origin of gasrich meteorites. *Geochim. Cosmochim. Acta* **28**, 595–607.

Wänke H. (1965) Der sonnenwind als quelle der uhedelgase in steinmeteoriten. *Z. Naturforsch.* **20a**, 946–949.

Wilkening L. L. and Marti K. (1976) Rare gases and fossil particle tracks in the Kenna ureilite. *Geochim. Cosmochim. Acta* **40**, 1465–1473.

Young D. M. and Crowell A. D. (1962) *Physical Adsorption of Gases.* Butterworths, Washington. 426 pp.

Silicon and oxygen isotopes in selected Allende inclusions†

Richard H. Becker* and Samuel Epstein

Division of Geological and Planetary Sciences, California Institute of Technology, Pasadena, California 91125

Abstract—Silicon and oxygen isotopic ratios have been measured on samples from seven Allende inclusions which have also been or are being subjected to searches for anomalies in such other elements as calcium, magnesium and titanium. Four of these samples, EGG-1, EGG-6, BG1a and 3529Yc, have oxygen and silicon isotopic ratios typical of normal Allende inclusions. Two of the samples, EGG-3 and BG10a, show clear evidence in either oxygen or silicon of being FUN inclusions. EGG-3 has a $\delta^{29}Si$ excess of about 0.5‰, a silicon fractionation of 2–3‰/amu, and its oxygen may be fractionated in the opposite direction to that seen previously in inclusions C1, EK1-4-1 and HAL. BG10a has an inferred original oxygen fractionation of 8–9‰/amu. The seventh sample, a melilite from inclusion D7, appears to be fractionated from most Allende melilites.

Density separates from Allende inclusion 3A differ by 2‰ in $\delta^{30}Si$ values, and indicate that this inclusion saw at least two isotopically distinct silicon reservoirs during its history.

INTRODUCTION

In recent years isotopic anomalies attributed to either nuclear effects or physical or chemical fractionation have been found in a number of elements in refractory inclusions from carbonaceous chondrites (Clayton *et al.*, 1973, 1978; Clayton and Mayeda, 1977; Lee *et al.*, 1978; McCulloch and Wasserburg, 1978a,b; Papanastassiou and Wasserburg, 1978; Lugmair *et al.*, 1978; Macdougall and Phinney, 1979; Niederer *et al.*, 1980; and others). The origin of these anomalies is still a subject of much discussion (Cameron and Truran, 1977; Clayton, 1977; Clayton D. D. *et al.*, 1977; Kuroda, 1979; Herbst and Rajan, 1980; Consolmagno and Cameron, 1980; Lee *et al.*, 1980; and others). It seems clear that any relationships which can be shown to exist among various anomalies, such as the observed correlated oxygen, magnesium and silicon fractionations in Allende inclusions C1 and EK1-4-1 (Clayton and Mayeda, 1977; Wasserburg *et al.*, 1977; Clayton *et al.*, 1978), would help to constrain the possible explanations for their origin. Searches for additional inclusions with correlated isotopic effects seem therefore to be in order.

Over several years analyses of silicon and oxygen isotopes in Allende inclusions have been carried out in this laboratory, using various techniques. Some results, mostly for samples picked out in this laboratory, have been reported previously (Epstein and Yeh, 1977; Yeh and Epstein, 1978). The present paper presents Si and O data for samples provided to us by G. J. Wasserburg. These Wasserburg samples were separates of pyroxene (with spinel) from EGG-1, EGG-3 and EGG-6, a melilite separate from D7, and bulk samples of BG1a, BG10a and 3529Yc. BG1a is a fine-grained aggregate, the others are coarse-grained Ca-Al rich inclusions from Allende. Mineralogic information and isotopic analyses for elements other than Si and O for some of these inclusions have been previously presented (Gray *et al.*, 1973; Lee and Papanastassiou, 1974; Esat *et al.*, 1979, 1980; Niederer and Papanastassiou, 1979; Wark and Lovering, 1980; Wark and Wasserburg, 1980). Although the data we present are in some cases not as good as in previous

---

†Contribution #3623. Division of Geological and Planetary Sciences, California Institute of Technology, Pasadena, California 91125.
*Present address: Shepherd Laboratories, University of Minnesota, Minneapolis, Minnesota 55455

work (Yeh and Epstein, 1978), where larger amounts of material were available for chemical treatment, we felt that the interest in isotopic anomalies and the lack in some cases of any more material on which to do further analyses warranted publication at this time. We also wanted to present some new silicon data for mineral separates from Allende 3A, an inclusion previously analyzed in bulk by Yeh and Epstein (1978). Lee and Papanastassiou (1974) give Mg data for Allende 3A.

ANALYTICAL PROCEDURES

O_2 was obtained by fluorinating the samples with either F_2 (Taylor and Epstein, 1970) or BrF_5 (Clayton and Mayeda, 1963). In either case, SiF_4 was held in the reaction vessel at $-196°C$ until O_2 was processed, then released using ethanol ice and transferred to nickel storage tubes. O_2 yields were determined manometrically in a calibrated Toepler pump. Silicon yields initially were not measured, because of the desire to handle SiF_4 in an all-metal system. Later, SiF_4 yields were sometimes determined after the isotope ratio measurements.

Early samples were reacted with F_2, using the total available material in one analysis. This was done to minimize errors due to analytical blanks. In particular, the F_2 used for reaction introduces 5–6 μmoles of O_2, equivalent to that obtained from about 0.5 mg of a silicate, and having a $\delta^{18}O$ value of about +1‰. Because the amount of O_2 varies from run to run, depending on the exact amount of F_2 used, and because its $\delta^{18}O$ value can vary somewhat if water vapor is present in the analytical system, correcting for blanks in small samples would be a somewhat uncertain process and was therefore avoided. Later samples were split, either to measure Si and O_2 on separate samples or to attempt to do replicate analyses. BrF_5 was preferred for the smaller samples, because of its negligible O_2 blank. However, because it can itself introduce small amounts of contaminants such as SOF_2 into SiF_4 even when carefully purified, the use of F_2 for silicon analyses was not completely halted.

A major problem in analysis of silicon isotopes is the presence of SOF_2, which gives an ion at mass 86. Since silicon isotopic ratios are determined from ratios of SiF_3^+ ions falling at masses 85, 86 and 87, SOF_2 interferes with the determination of $\delta^{29}Si$. Yeh and Epstein (1978) introduced a preparation technique to remove sulfur from samples before fluorination. However, due to the solubility of silica in wash solutions and to other losses which could not be overcome, silicate samples of the order of 2–3 mg resulted in yields of SiO_2 of only 0.1 to 0.2 mg when treated. For such small quantities of sample, SiF_4 blanks and SOF_2 from our reagent BrF_5 were sufficient to cause some errors in our analyses. SOF_2 once introduced could not be separated from SiF_4. Therefore, we decided to run untreated samples and attempt to correct the isotopic ratios obtained using mass-spectrometric background peaks to measure contamination. Although $^{34}SOF_2$ would be the best measure of $^{32}SOF_2$, since sulfur isotopic ratios should not vary by more than a few per cent from sample to sample, the actual amount of $^{34}SOF_2^+$ present was too small to measure accurately. Instead, we monitored mass 67, where the $^{32}SOF^+$ fragment occurs, and took the difference in peak heights between the sample and standard mass 67 beams, labeled the "excess mass 67", as our measure of SOF_2 contamination. This is not completely satisfactory, because although the size of the peak is easily measured, the ratio of SOF^+ to SOF_2^+ can vary as a function of mass spectrometer source conditions. From numerous analyses, we determined that the variation could be held to a reasonably narrow range, such that 1 mv of excess mass 67 corresponded to 0.005 to 0.02‰ in $\delta^{29}Si$. One aberrant set of six analyses, run on one day, was significantly outside the established range, but the normal situation was then reestablished by changing our mass spectrometer source conditions. We can thus correct our analytical results for excess mass 67 in an approximate way using our established relationship, or display our data as a function of excess mass 67 as is done in this paper, and get some measure of $\delta^{29}Si$ anomalies.

Oxygen isotopic ratios were measured against either a Rose Quartz oxygen standard or against O_2 prepared from CuO. Values were converted to SMOW using a $\delta^{18}O$ value for Rose Quartz of +8.45‰ and assuming it lies on a line through the SMOW standard with a slope of 0.50, so that its $\delta^{17}O$ value is +4.22‰. Mass spectrometer corrections were made after Craig (1957). Analytical precisions for standards are about 0.2‰ for $\delta^{18}O$ and 0.3‰ for $\delta^{17}O$ (1σ).

Silicon isotopic ratios were measured against Rose Quartz. Again, mass spectrometer corrections were made, although in most cases they are negligible because of the small δ values involved. Note that previously published results (Yeh and Epstein, 1978; Clayton et al., 1978) have not included these corrections. This is unimportant except for samples C1 and EK1-4-1. Because masses 86 and 85 are collected together when $\delta^{30}Si$ is determined, a correction is required to the $\delta^{30}Si$ values. This involves adding 4.85% of the $\delta^{29}Si$ value to $\delta^{30}Si$. For inclusion C1 (Clayton et al., 1978), this means increasing $\delta^{30}Si$ by 0.6‰, which reduces the size of the $\delta^{29}Si$ anomaly that was observed. A smaller correction is required in the case of EK1-4-1 (Yeh and Epstein, 1978). Analytical precisions for uncontaminated Rose Quartz standards are about 0.12‰ for $\delta^{30}Si$ and 0.09‰ for $\delta^{29}Si$ (1σ). Because the errors in $\delta^{30}Si$ and $\delta^{29}Si$ are correlated, the uncertainty in $\delta^{29}Si-(1/2)\delta^{30}Si$, which measures how anomalous $\delta^{29}Si$ is, is only about 0.06‰ (1σ).

RESULTS AND DISCUSSION

Oxygen

Our oxygen data are given in Table 1 and the isotopic results are displayed in a three-isotope plot in Fig. 1. Reference lines are shown for terrestrial samples, for refractory inclusions and for C2-matrix samples. The latter two lines are from data of Clayton and coworkers (Clayton R. N. et al., 1976, 1977). Samples BG1a, 3529Yc and BrF$_5$-reacted EGG-1 and EGG-6 fall essentially on the Allende inclusion line. The F$_2$-reacted samples of EGG-1 and EGG-6 are displaced towards the terrestrial line by amounts and in a direction roughly consistent with there being 6 μmoles of O$_2$ from the F$_2$ in each O$_2$ sample. They therefore can be taken to support the values gotten using BrF$_5$. These four inclusions are thus typical unfractionated Allende inclusions.

For EGG-3 both the BrF$_5$- and F$_2$-reacted samples show a displacement to the left of (above) the Allende line. If one corrects the F$_2$-reacted sample value for blank, using a blank value derived from EGG-1 and EGG-6 values, the corrected value would lie about twice as far removed vertically from the Allende line as the BrF$_5$-reacted sample value. These results seem to indicate an oxygen fractionation in EGG-3 contrary to the general trend (Lee et al., 1980), in the same way that fine-grained aggregate B29 does for magnesium (Niederer and Papanastassiou, 1979). However, since analytical errors involving contamination or fractionation would give systematic deviations from the Allende line in the same direction as the EGG-3 points, it would be wise to avoid interpreting our EGG-3 oxygen result until it can be confirmed with better analytical yield than we obtained.

Sample BG10a lies to the right of the Allende line, in an area consistent with its having been fractionated in the same fashion as FUN inclusions C1 and EK1-4-1. The O$_2$ yield is large enough so that the blank from F$_2$ will not have shifted the point significantly. The apparent original oxygen fractionation of this inclusion, obtained by passing a line from the melilite end of the Allende line through BG10a to the fractionation trend line for spinels (dashed lines in Fig. 1), following the example of Lee et al. (1980), is about 8–9‰/amu. This is comparable to the value of 10‰/amu for EK1-4-1 (Clayton and Mayeda, 1977). Based on its oxygen isotopes therefore, BG10a fits the definition of a FUN inclusion (Wasserburg et al., 1977), a conclusion confirmed by the presence of Mg fractionation as well (Lunatic Asylum, unpublished data).

Our point for D7 melilite is significantly fractionated from other Allende melilites, more or less along the C2-matrix line. This result is peculiar, for two reasons. First, in prior analyses of fractionated inclusions, melilite is one mineral which seemed to have had any original fractionation erased (Lee et al., 1980). Second, a small sample of D7 thought to contain pyroxene, spinel and some melilite was analyzed in Chicago and found to have

Table 1. Oxygen yields and isotopic compositions of Allende inclusions.

Sample	Reagent[a]	Wt. (mg)	Yield (μmoles)	δ^{18}O (‰)	δ^{17}O (‰)
D7	F$_2$	19.0	75	+8.7	+1.2
3529Yc	F$_2$	15.6	105	−6.8	−10.1
BG10a	F$_2$	12.4	70	−7.8	−14.8
BG1a	F$_2$	10.4	148	−10.2	−13.0
EGG-1	F$_2$	3.0	11.5	−6.4	−6.9
	BrF$_5$	3.5	15	−11.3	−15.4
EGG-3	F$_2$	4.1	22.8	−15.4	−13.8
	BrF$_5$	6.4	31	−20.6	−21.0
EGG-6	F$_2$	4.0	22.4	−12.7	−15.4
	BrF$_5$	7.9	42	−17.0	−20.2

[a]Samples run with F$_2$ contain 5–6 μmoles O$_2$ introduced with the F$_2$ and having δ^{18}O and δ^{17}O values of about +1.0 and +0.5‰.

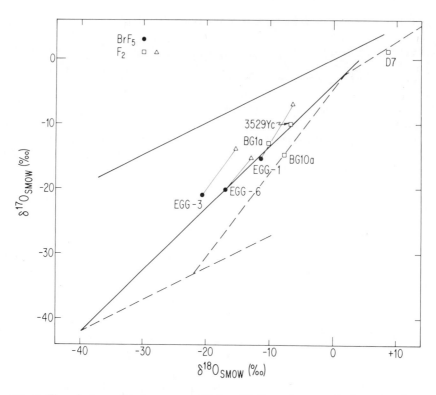

Fig. 1. Three-isotope plot for oxygen. Dark solid lines show trends for terrestrial samples, at top, and Allende inclusions. Dashed line at upper right shows trend for C2-matrix samples. Dashed line at bottom is Allende spinel fractionation trend. Dashed line through BG10a point connects melilite end of Allende line to spinel fractionation trend. Light solid lines connect duplicate analyses of EGG samples (dots and triangles).

$\delta^{18}O$ of $-5.37‰$ and $\delta^{17}O$ of $-9.10‰$ (Clayton, pers. comm.), values which put it right on the Allende line. Despite the peculiarity, we do not see any obvious source of error in our data. No contaminant other than C2-matrix material or a very high $\delta^{18}O$ chert can be envisaged that would displace an ordinary melilite to the value seen in D7, and such samples were not being analyzed at the time D7 was run. Fractionation due to incomplete yield, if such fractionation occurs, might be expected to give isotopic ratios that are too low.

Our D7 value could represent exchange of melilite with a somewhat different reservoir of oxygen than in the case of the melilites analyzed by Clayton and coworkers, or it could represent a lower temperature reaction, as for C2-matrix material. In either event, consistency with the Clayton result for D7 requires that such a melilite exchange post-dated the earlier, normal back-exchange process invoked for the typical Allende inclusions (Clayton and Mayeda, 1977; Wasserburg *et al.*, 1977; Lee *et al.*, 1980). In view of the more complicated history required to explain our D7 result in light of the Clayton result, it is unfortunate that our sample was not split for replicate analyses.

A couple of comments are in order with regard to the yields in Table 1, in view of the fact that all except for BG1a are much lower than expected. Under our reaction conditions, spinel would probably go unreacted. However, the observed yields in some cases are so low that 60–70% spinel would be required to explain them, and such high spinel contents do not fit the observations. Therefore, the major minerals melilite and pyroxene are themselves not reacting completely, for reasons that we do not understand. This could have two consequences. First, samples might appear to be fractionated. Partial reaction with BrF_5 can fractionate pyroxene by 1–2‰ in $\delta^{18}O$ (Nikolaus and Becker, unpublished data) but doesn't appear to fractionate melilite (Mayeda, pers. comm.). Partial reaction with F_2 does not appear to cause fractionation (Epstein and Taylor, 1971). The fractionation seen by Nikolaus and Becker was in the direction of yielding low $\delta^{18}O$ values. The data in Fig. 1, except possibly for EGG-3, show no evidence of such fractionation due to incomplete yield.

The second consequence of incomplete yields is that, in samples containing readily fluorinated secondary minerals, the δ values will plot higher on the Allende line than they should. This is because the product O_2 will be enriched in O_2 from the alteration minerals. This may explain why the EGG samples, which are pyroxene separates, plot so much higher up the line than do pyroxenes measured previously (Clayton R. N. et al., 1977). It is also possible, however, that these pyroxenes have simply undergone more exchange with the nebula or Allende matrix than those measured earlier.

Silicon

Results for silicon are shown in Table 2. Excesses at mass 67 are included for those samples where backgrounds were measured. Such data were unfortunately not obtained for the earliest samples run.

For D7 and BG10a the $\delta^{30}Si$ values indicate fractionation. Since these samples also show fractionation in oxygen, one would like to accept the $\delta^{30}Si$ values at face value. However, both samples, when run mass-spectrometrically, were found to be much smaller in volume than any of the other samples run at the same time. The apparent major loss of SiF_4, and the consequent fact that sample and standard isotopic ratios were determined at different beam intensities, raises doubts about the validity of these silicon isotopic ratios. Therefore, very little weight should be placed on these data.

Samples 3529Yc and BG1a, with oxygen isotopes typical of ordinary Allende inclusions (Fig. 1), have $\delta^{30}Si$ values which are also essentially within the typical range for this element (Yeh and Epstein, 1978; Clayton et al., 1978). Inclusion 3529Yc lies at the upper end of the range, among the coarse-grained type A and B inclusions measured by Clayton et al. (1978), while the fine-grained aggregate BG1a falls with the fine-grained pink inclusion that they measured. The $\delta^{29}Si$ values for both of these inclusions appear to indicate negative anomalies of about 0.5‰. However, since backgrounds were not determined, the samples were not run in duplicate, and terrestrial samples on rare occasions gave similar negative anomalies, we don't want to make too much of these values at this time.

Table 2. Silicon yields and isotopic compositions of Allende inclusions.

Sample	Reagent	Wt. (mg)	Yield (μmole)	$\delta^{30}Si$ (‰)	$\delta^{29}Si$ (‰)	Excess 67 (mv)
D7	F_2	19.0	—[a]	+8.2	+7.3	—
3529Yc	F_2	15.6	—	+3.3	+1.3	—
BG10a	F_2	12.4	—[a]	+5.7	+3.6	—
BG1a	F_2	10.4	—	−1.0	−1.2	—
EGG-1	F_2	3.0	7	+1.6	+2.4	63.0
	BrF_5	2.8	13	+3.0	+1.2	5.5
	BrF_5	3.5	9	—[b]	—[b]	—
EGG-3	F_2	4.1	7	+5.7	+3.4	3.5
	BrF_5	4.2	5	+6.9	+4.1	9.0
	BrF_5	6.4	11	+3.9	+2.3	3.0
EGG-6	F_2	4.0	14	+0.2	+0.4	13.0
	BrF_5	4.0	10	+1.6	+0.6	8.5
	BrF_5	7.9	15	+0.1	+0.1	5.0

[a] Yields were not measured but mass spectrometer beam intensities for these samples were much lower than they should have been. Isotopic ratios are questionable for these samples.
[b] Sample was lost before analysis.

For samples EGG-1 and EGG-6 the δ^{30}Si values obtained are also in the range of typical Allende samples. As with oxygen, there is no evidence for fractionation in these samples. The δ^{30}Si values for EGG-3 however, while not ideally reproducible, are beyond the range found for ordinary inclusions (Yeh and Epstein, 1978; Clayton et al., 1978). EGG-3 thus appears to be fractionated in silicon by several per mil. Although we do not know what the effect of incomplete yields would be on the δ^{30}Si of pyroxene, the fact that the yields for EGG-1 and EGG-6 are similar to that for EGG-3 suggests that the fractionation in EGG-3 is not an artifact of the procedure. EGG-3 also appears to have a δ^{29}Si anomaly (see below), confirming the FUN designation based on magnesium and calcium isotopes (Esat et al., 1979; Niederer and Papanastassiou, 1979). The silicon fractionation in EGG-3 is about 1/3 to 1/2 that in EK1-4-1 (Yeh and Epstein, 1978).

For EGG-1, EGG-3 and EGG-6, backgrounds were determined in the mass spectrometer, allowing us to check for possible δ^{29}Si anomalies. Figure 2 shows deviations of δ^{29}Si from a strict δ^{29}Si/δ^{30}Si = 1/2 relationship as a function of excess mass 67 in the sample. Broken lines indicate upper and lower bounds for the majority of the terrestrial samples run as standards and test samples. The solid line indicates the upper limit for all such samples run, and is defined by a single anomalous set of standards. Standards run interspersed with the EGG samples, which were run over a period of several days, fall within the broken lines. The EGG-1 and EGG-6 points are consistent with there being no nuclear anomaly at mass 29. All three EGG-3 values, however, indicate an excess in δ^{29}Si beyond that due to SOF_2. Although not overwhelming, the data are certainly suggestive of an excess of ^{29}Si in EGG-3 of 4–5 parts in 10^4 relative to terrestrial silicon. This is comparable to the excess seen in Cl (Clayton et al., 1978), after its δ^{30}Si value is properly corrected.

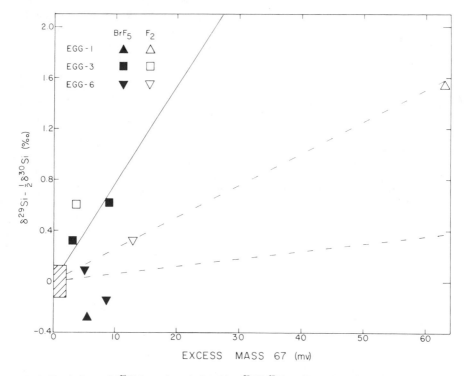

Fig. 2. Deviations of δ^{29}Si from the relationship δ^{29}Si/δ^{30}Si = 1/2 as a function of the amount of SOF_2 in the sample, measured in terms of the amount of excess mass 67 ($^{32}SOF^+$) in the sample relative to the standard (see text). Dashed lines delimit region in which most terrestrial data fall. Solid line is upper limit to all terrestrial analyses. Box at lower left shows range of variation for terrestrial samples having less than 4 mv excess mass 67, and is a measure of the analytical uncertainty.

Table 3. Silicon isotopic data for density fractions of Allende inclusion 3A.

Fraction	Reagent	Weight (mg)	Yield (μmole)	δ^{30}Si (‰)
d > 3.1	BrF$_5$	16.0	25	+0.0
d > 3.1 (HCl)[a]	BrF$_5$	2.4	7	+0.8
d < 3.1	BrF$_5$	11.2	13	+2.1
d < 3.1	BrF$_5$	10.0	13	+1.9
d < 3.1 (#2)[b]	BrF$_5$	11.2	13	+2.6
d < 3.1 (HCl)[a]	BrF$_5$	0.9	11	+3.4

[a]Residue after treatment with 1:1 HCl:H$_2$O.
[b]Two separate density separations were done on portions of inclusion 3A.

With regard to correlated isotopic effects we can make the following remarks. Inclusion BG10a has a magnesium fractionation of about 10‰/amu (Lunatic Asylum, unpublished data), similar to D7 and EGG-3 and half that for EK1-4-1 (Niederer and Papanastassiou, 1979). The oxygen fractionation for BG10a is similar to that for EK1-4-1, while that for the bulk D7 inclusion is zero (Clayton, pers. comm.) and for EGG-3 may be in the opposite direction. As noted above, silicon in EGG-3 is fractionated by 1/3 to 1/2 the amount in EK1-4-1. Inclusion HAL has a larger oxygen and smaller magnesium fractionation than EK1-4-1, by factors of more than two (Lee et al., 1980). Taken together the results indicate that the proportional fractionation effects seen for Mg, O and Si in inclusions Cl and EK1-4-1 are not a general phenomenon, at least with regard to Mg and O.

As for the new results on inclusion 3A, Table 3 shows data for two density fractions. The less dense fraction shows only gehlenite and a trace of spinel in its X-ray pattern. The denser fraction contains spinel, apparently pyroxene though its exact identification is not certain, and some minor minerals not identifiable from the X-ray pattern. Both fractions show some fine-grained alteration products under the microscope. The δ^{30}Si values of the two fractions are clearly distinguishable, and the amounts of alteration products seen in the two fractions are not enough to account for the difference (however, see below). Treatment of the two fractions with HCl, in an attempt to change the ratio of alteration product to major mineral, caused some shift in the silicon isotopic ratios but did not cause the ratios to approach one another.

The difference in the two fractions is much too large to be an equilibrium fractionation between pyroxene and melilite. This implies that the two density fractions contain silicon from isotopically different reservoirs. In the context of the scenario suggested by Clayton et al. (1978), in which nebular silicon isotope ratios vary as a function of the amount of silicon condensed, the denser fraction of inclusion 3A would have incorporated at least some of its silicon later than the less dense fraction.

One can argue that, because of our low silicon yields, the silicon isotopic ratios may be dominated by alteration products, even though the samples themselves are not. Oxygen isotopes may support this, in that the low density fraction has oxygen typical of melilite samples while the denser fraction is only about 5‰ lower on the Allende line. The less dense fraction would thus be giving a true melilite silicon isotope ratio while the other fraction is giving silicon primarily from secondary products and only 15-20% from pyroxene. In the context of the scenario mentioned above this would imply that alteration in 3A occurred before fine-grained inclusions formed. This could complicate explanations of oxygen isotopes in Allende inclusions, since fine-grained inclusions are found to have oxygen isotopic ratios lower on the Allende line than our 3A density fractions (Clayton R. N. et al., 1977). Some sort of decoupling of oxygen and silicon isotope effects would be necessary.

One might also argue that the silicon we analyzed does in fact come primarily from melilite and pyroxene, and that the oxygen data show that oxygen exchange with the nebula was significant for pyroxene. If so, and if silicon exchange occurred to some degree when oxygen exchange occurred, then our results might be taken to mean that silicon in the nebula was isotopically heavier during exchange than during formation of the inclusion. The fact that most meteorite silicon data, as well as Allende matrix silicon values, fall around zero per mil or below (Yeh and Epstein, 1978) indicates that this probably was not the case, but it doesn't completely rule out some complex scenario involving such exchange.

Sample 3A was chosen for mineral separation because it showed unexpected silicon isotopic variations when the purification techniques of Yeh and Epstein (1978) were used on it. Other samples did not show such variations and, as no other mineral separates have been run, we do not know whether 3A represents a general phenomenon or is just a peculiar sample.

CONCLUSIONS

Based on oxygen and silicon isotopes, Allende samples EGG-1, EGG-6, BG1a and 3529Yc are typical, unfractionated Allende inclusions. Where magnesium data are available, they indicate the same thing (Esat et al., 1979). The only possibly anomalous values seen were low $\delta^{29}Si$ values in BG1a and 3529Yc, and these require confirmation. The observation that pyroxene separates from EGG-1 and EGG-6 do not lie as far down the Allende line as previous pyroxene analyses do may be attributable to incomplete fluorination of pyroxene along with the presence of alteration products, or to the fact that the pyroxenes underwent some exchange with nebular oxygen.

EGG-3 is fractionated in both oxygen and silicon, as it is in magnesium (Esat et al., 1979, 1980). However, its oxygen fractionation is in the opposite direction to that previously observed for FUN inclusions (Lee et al., 1980) and may be an artifact of our analysis. The silicon fractionation is in the direction expected and is about 5‰ relative to our standard. It is accompanied by a 0.5‰ excess in $\delta^{29}Si$, presumably of nuclear origin. The silicon fractionation is somewhat erratic, possibly due to sample inhomogeneity or to variations in our analytical silicon blank.

Sample D7 melilite is fractionated in oxygen by several per mil along the C2-matrix line, away from typical Allende melilites. The observation, if real, requires an extra step in the model advanced by Lee et al. (1980) to explain the oxygen, silicon and magnesium fractionations seen in FUN inclusions. Silicon may also be fractionated in D7, but the result obtained is highly questionable.

Inclusion BG10a is a FUN inclusion with an inferred original oxygen isotopic fractionation of 8–9‰/amu relative to the Allende line. Silicon may be fractionated by about 3‰/amu, although this silicon value is also questionable. The relative sizes of the oxygen and magnesium fractionations in BG10a are different from those in C1 and EK1-4-1, something which also appears to be true for D7. Therefore, fractionations in magnesium, oxygen and perhaps silicon as well need not be related to one another in the simple way that might be inferred from C1 and EK1-4-1 data, but can vary among themselves from one inclusion to another.

Finally, silicon data for two density fractions of inclusion 3A with different mineralogies show a difference of about 2‰, which indicates that silicon, like oxygen, has been added to at least some mineral phases in Allende inclusions from a reservoir other than the one in which the inclusion first began to form.

Acknowledgments—We would like to thank G. J. Wasserburg for providing the samples used in this study. We thank G. J. Wasserburg, R. N. Clayton and T. K. Mayeda for permission to use unpublished results in this paper. We are grateful for discussions with these same persons regarding various aspects of our results. We want to acknowledge the part that H. -W. Yeh played in the early stages of this work. We thank M. H. Thiemens and an anonymous reviewer for pointing out some errors in an earlier version of this paper. Support from this work was provided by NASA grant number NGL-05-002-190.

REFERENCES

Cameron A. G. W. and Truran J. W. (1977) The supernova trigger for formation of the solar system. *Icarus* 30, 447-461.
Clayton D. D. (1977) Solar system isotopic anomalies: Supernova neighbor or presolar carriers. *Icarus* 32, 255-269.
Clayton D. D., Dwek E., and Woosley S. E. (1977) Isotopic anomalies and proton irradiation in the early solar system. *Astrophys. J.* 214, 300-315.
Clayton R. N., Grossman L., and Mayeda T. K. (1973) A component of primitive nuclear composition in carbonaceous meteorites. *Science* 182, 485-488.
Clayton R. N. and Mayeda T. K. (1963) The use of bromine pentafluoride in the extraction of oxygen from oxides and silicates for isotopic analysis. *Geochim. Cosmochim. Acta* 27, 43-52.
Clayton R. N. and Mayeda T. K. (1977) Correlated oxygen and magnesium isotope anomalies in Allende inclusions, I: Oxygen. *Geophys. Res. Lett.* 4, 295-298.
Clayton R. N., Mayeda T. K., and Epstein S. (1978) Isotopic fractionation of silicon in Allende inclusions. *Proc. Lunar Planet. Sci. Conf. 9th*, p. 1267-1278.
Clayton R. N., Onuma N., Grossman L., and Mayeda T. K. (1977) Distribution of the pre-solar component in Allende and other carbonaceous chondrites. *Earth Planet. Sci. Lett.* 34, 209-224.
Clayton R. N., Onuma N., and Mayeda T. K. (1976) A classification of meteorites based on oxygen isotopes. *Earth Planet. Sci. Lett.* 30, 10-18.
Consolmagno G. J. and Cameron A. G. W. (1980) The origin of the "FUN" anomalies and the high temperature inclusions in the Allende meteorite. *Moon and Planets* 23, 3-25.
Craig H. (1957) Isotopic standards for carbon and oxygen and correction factors for mass-spectrometric analysis of carbon dioxide. *Geochim. Cosmochim. Acta* 12, 133-149.
Epstein S. and Taylor H. P. Jr. (1971) O^{18}/O^{16}, Si^{30}/Si^{28}, D/H, and C^{13}/C^{12} ratios in lunar samples. *Proc. Lunar Sci. Conf. 2nd*, p. 1421-1441.
Epstein S. and Yeh H. -W. (1977) The $\delta^{18}O$, $\delta^{17}O$, $\delta^{30}Si$ and $\delta^{29}Si$ of oxygen and silicon in stony meteorites and Allende inclusions (abstract). In *Lunar Science VIII*, p. 287-289. The Lunar Science Institute, Houston.
Esat T. M., Papanastassiou D. A., and Wasserburg G. J. (1979) Trials and tribulations of ^{26}Al: Evidence for disturbed systems (abstract). In *Lunar and Planetary Science X*, p. 361-363. Lunar and Planetary Institute, Houston.
Esat T. M., Papanastassiou D. A., and Wasserburg G. J. (1980) The initial state of ^{26}Al and $^{26}Mg/^{24}Mg$ in the early solar system (abstract). In *Lunar and Planetary Science XI*, p. 262-264. Lunar and Planetary Institute, Houston.
Gray C. M., Papanastassiou D. A., and Wasserburg G. J. (1973) The identification of early condensates from the solar nebula. *Icarus* 20, 213-239.
Herbst W. and Rajan R. S. (1980) On the role of a supernova in the formation of the solar system. *Icarus* 42, 35-42.
Kuroda P. K. (1979) Isotopic anomalies in the early solar system. *Geochem. J.* 13, 83-90.
Lee T., Mayeda T. K., and Clayton R. N. (1980) Oxygen isotopic anomalies in Allende inclusion HAL. *Geophys. Res. Lett.* 7, 493-496.
Lee T. and Papanastassiou D. A. (1974) Mg isotopic anomalies in the Allende meteorite and correlation with O and Si effects. *Geophys. Res. Lett.* 1, 225-228.
Lee T., Papanastassiou D. A., and Wasserburg G. J. (1978) Calcium isotopic anomalies in the Allende meteorite. *Astrophys. J.* 220, L21-L25.
Lugmair G. W., Marti K., and Scheinin N. B. (1978) Incomplete mixing of products from r-, p-, and s-process nucleosynthesis: Sm-Nd systematics in Allende inclusion EK1-04-1 (abstract). In *Lunar and Planetary Science IX*, p. 672-674. Lunar and Planetary Institute, Houston.
Macdougall J. D. and Phinney D. (1979) Magnesium isotopic variations in hibonite from the Murchison meteorite: an ion microprobe study. *Geophys. Res. Lett.* 6, 215-218.
McCulloch M. T. and Wasserburg G. J. (1978a) Barium and neodymium isotopic anomalies in the Allende meteorite. *Astrophys. J.* 220, L15-L19.
McCulloch M. T. and Wasserburg G. J. (1978b) More anomalies from the Allende meteorite: Samarium. *Geophys. Res. Lett.* 5, 599-602.
Niederer F. R. and Papanastassiou D. A. (1979) Ca isotopes in Allende and Leoville inclusions (abstract). In *Lunar and Planetary Science X*, p. 913-915. Lunar and Planetary Institute, Houston.
Niederer F. R., Papanastassiou D. A., and Wasserburg G. J. (1980) Endemic isotopic anomalies in titanium. *Astrophys. J.* 240, L73-L77.
Papanastassiou D. A. and Wasserburg G. J. (1978) Strontium isotopic anomalies in the Allende meteorite. *Geophys. Res. Lett.* 5, 595-598.
Taylor H. P. Jr. and Epstein S. (1970) O^{18}/O^{16} ratios of Apollo 11 lunar rocks and minerals. *Proc. Apollo 11 Lunar Sci. Conf.*, p. 1613-1626.
Wark D. A. and Lovering J. F. (1980) More early solar system stratigraphy: Coarse-grained CAI's (abstract). In *Lunar and Planetary Science XI*, p. 1208-1210. Lunar and Planetary Institute, Houston.

Wark D. A. and Wasserburg G. J. (1980) Anomalous mineral chemistry of Allende FUN inclusions C1, EK-141 and EGG 3 (abstract). In *Lunar and Planetary Science XI*, p. 1214–1216. Lunar and Planetary Institute, Houston.

Wasserburg G. J., Lee T., and Papanastassiou D. A. (1977) Correlated O and Mg isotopic anomalies in Allende inclusions: II. Magnesium. *Geophys. Res. Lett.* **4**, 299–302.

Yeh H. -W. and Epstein S. (1978) $^{29}Si/^{28}Si$ and $^{30}Si/^{28}Si$ of meteorites and Allende inclusions (abstract). In *Lunar and Planetary Science IX*, p. 1289–1291. Lunar and Planetary Institute, Houston.

An estimate of atmospheric contamination of Allende coarse-grained inclusion 3529Z

R. Warasila, O. A. Schaeffer, and K. Frank

Department of Earth and Space Sciences, State University of New York at Stony Brook, Stony Brook, New York 11794

Abstract—An estimate of argon atmospheric contamination of Allende coarse-grained inclusion 3529Z has been obtained by exposing samples to an atmosphere of ^{38}Ar at a pressure of 0.01 atm. Samples were selected from portions of inclusion 3529Z for which laser probe and stepwise release studies have been made to investigate ages > 4.55 G.y. reported for Allende coarse-gained inclusions. The ^{38}Ar was used as a tracer in a thermal stepwise release experiment. The stepwise release data demonstrate that measurable amounts of excess ^{38}Ar persisted in the sample until it was melted. However, the amount of excess ^{38}Ar released in the temperature range 1100° to 1600°C is more than one order of magnitiude than the extra ^{40}Ar released during the thermal release study in the same temperature interval. The associated plateau yields ages > 4.55 G.y. for Allende inclusion 3529Z. We believe that for inclusion 3529Z, uptake of atmospheric argon is not the explanation for the old age.

INTRODUCTION

Several recent studies have shown that minerals can take up rare gases when exposed to the atmosphere (Zaikowski and Schaeffer, 1979; Niemeyer and Leich, 1976). As a result, K–Ar ages of meteorites may be influenced by such atmospheric contamination. Jessberger *et al.* (1980) and Herzog *et al.* (1980) considered atmospheric argon as a possible source of contamination for the Allende K–Ar ages in excess of 4.55 G.y. For terrestrial samples, atmospheric argon can usually be estimated from the amount of ^{36}Ar by assuming that the sample contains only radiogenic ^{40}Ar and argon with atmospheric isotopic composition. There are terrestrial cases, especially for rocks with a deep seated origin, in which rocks may contain some trapped argon with the same ^{36}Ar/^{38}Ar ratio as the atmosphere. If the amount of mantle argon is appreciable, the amount of atmospheric argon cannot be estimated. For meteorites, possible atmospheric argon contamination is almost always hidden as most meteorites contain two additional sources of argon: a planetary component and a cosmic ray produced component. Both of these components contain ^{36}Ar and ^{38}Ar.

The take-up of rare gases by a solid is a complicated process involving surface adsorption forming mono-layers and multi-layers as well as absorption by the bulk solid itself. The presence of capillaries and small capillary type cracks complicates the problem further. The reported study of these phenomenon is almost as old as the modern scientific literature. The first systematic study was made by T. de Saussure in 1814. Somewhat more recent studies have been carried out by A. Titoff (1910) and I. F. Homfray (1910). These early measurements clearly delineated the features and many of the thermodynamic characteristics of the overall process. In general, the adsorption is relatively rapid, of the order of hours or less for equilibrium, while the process of absorption is relatively slow and not as well studied.

In making an age determination by the K–Ar method, the sample is contained in a vacuum system before analysis, usually for the period of several days. As the processes of adsorption and its reverse, desportion, are rapid, the adsorbed gases should be removed during this period. The absorbed gases which remain may contribute significantly to the ^{40}Ar. A study of the absorption of noble gases by synthesized serpentine has been carried out by Zaikowski and Schaeffer (1979). In that experiment it has been shown that serpentine in equilibrium with the atmosphere takes up ^{40}Ar to the value of 420×10^{-8} cm^3 STP/g. It was suggested that the distribution of K–Ar ages for carbonaceous chondrites may be due to their variation in phyllosilicate content which contains atmospheric argon.

A laser microprobe study has been made for coarse-grained Ca–Al-rich inclusions from the Allende meteorite (Herzog et al., 1980). The purpose of the study was to try to elucidate the very old ages found by Jessberger et al. (1980) and Jessberger and Dominik (1979) for similar inclusions from Allende. In the course of that study, it was found that many of the coarse grained minerals contained extremely small amounts of K, ≤100 ppm, but that they contained up to 200×10^{-8} cm^3 STP/g of ^{40}Ar. It was concluded by Herzog et al. (1980) that the extra ^{40}Ar is either 1) atmospheric, 2) the result of long term, >4.55 G.y. decay of K, possibly in pre-solar grains, or 3) inherited from the matrix during a metamorphic event. As the amount of ^{40}Ar in the Allende minerals is less than the equilibrium amount found for a synthetic serpentine, we have decided to essentially repeat the experiment by using material from Allende in order to obtain a better estimate for the absorption of atmospheric argon by the minerals in Allende.

The study of the uptake of argon with synthetic serpentine was conducted as follows to ascertain that equilibrium was reached. In one study, synthetic serpentine containing 940×10^{-8} cm^3 STP/g ^{40}Ar was exposed to the atmosphere at 100°C for one week and then three weeks at 25°C. At the end of that time, the sample contained 490×10^{-8} cm^3 STP/g ^{40}Ar, i.e., the sample lost argon. The temperature release profile of the sample as synthesized and the sample after exposure to the atmosphere were different in that more of the gas was released below 400°C for the air exposed sample than for the sample as synthesized. In a second study, serpentine synthesized in an argon-free atmosphere was exposed to the atmosphere for periods of up to 18 months at 25°C. This sample took up argon to the amount of 330×10^{-8} cm^3 STP/g after six months exposure and 430×10^{-8} cm^2 STP/g after 18 months exposure. From these experiments, it would appear that the argon incorporated in the serpentine during synthesis presumably occupies both retentive and non-retentive sites, as about 1/3 is released above 1000°C. Yet, after exposure to the atmosphere, it would appear that all sites gradually equilibrate with atmospheric argon. Also, the uptake of argon from the atmosphere would appear to approach approximately a constant total argon content. Assuming that the Allende minerals behave similarly to the synthetic serpentine, the isotopic tracer, ^{38}Ar, after sufficient time, on the order of months, will attain equilibrium with both retentive and non-retentive sites.

EXPERIMENTAL PROCEDURES

The samples used in this study were unirradiated chips left over from coarse-grained Ca–Al-rich inclusion 3529Z used in our previous laser probe study of Allende; the petrography and chemistry of this inclusion are described in detail in Herzog et al. (1980). The chips were ultrasonically cleaned in acetone and then in methanol. After cleaning, the samples were crushed in a mortar and pestle to <500 μm. The crushed sample was mixed and packaged in aluminum foil in portions of approximately 20 mg. The time between crushing and placement in the ^{28}Ar vacuum system was four days.

The samples were placed in two separate removable tubes on a glass vacuum line pumped by a mercury diffusion pump. The system was pumped to less than 10^{-5} Torr prior to sample loading. After the samples were placed on the vacuum line, the tubes were evacuated overnight. Also attached to the line was a glass breakseal containing argon of isotopic composition 99.81% ^{38}Ar, 0.19% ^{40}Ar, 0.01% ^{36}Ar and with impurities of 5.6% N$_2$ and 1.4% O$_2$. By using liquid nitrogen on an activated charcoal trap connected to the system through a valve, it was possible to control the pressure in the line after the breakseal was opened. A pressure of 8 Torr, or about 0.01 atm. was chosen to simulate the partial pressure of argon in the atmosphere and was monitored by mercury manometers attached to each sample tube. The temperature of one sample tube was maintained at 20 ± 2°C for 49 days; this sample is referred to as the room temperature sample. The other sample tube was maintained at 150 ± 2°C for 71 days, and this sample is referred to as the 150°C sample. A portion of the inclusion was left uncrushed until just before loading of the samples into the extraction line. This sample was

Table 1. Argon isotope data for Allende sample 3529Z.

Sample	Temp. °C	^{38}Ar 10^{-8} cm^3 STP/g	^{40}Ar 10^{-8} cm^3 STP/g	40/36	36/38	40/38
Control	300	0.115 ± 0.008	182 ± 0.6	274 ± 5	5.77 ± 0.40	1580 ± 106
	500	0.047 ± 0.007	106 ± 0.6	410 ± 14	5.53 ± 0.82	2270 ± 320
	700	0.034 ± 0.005	344 ± 1.7	1235 ± 41	8.17 ± 2.45	10100 ± 1580
	900	0.069 ± 0.005	330 ± 1.0	875 ± 26	5.48 ± 0.46	4800 ± 370
	1100	0.400 ± 0.010	239 ± 0.7	177 ± 2	3.38 ± 0.09	597 ± 15
	1300	0.955 ± 0.022	141 ± 1.0	59.2 ± 0.8	2.49 ± 0.06	147 ± 4
	1600	1.21 ± 0.010	71.3 ± 0.6	41.6 ± 0.9	1.41 ± 0.03	58.7 ± 0.6
Totals		2.83 ± 0.030	1413 ± 5			
20°C	300	36.9 ± 0.3	32.9 ± 5.3	214 ± 20	0.005 ± 0.0005	0.893 ± 0.016
	500	5.74 ± 0.03	79.6 ± 0.6	379 ± 5.2	0.037 ± 0.002	13.9 ± 0.1
	700	1.04 ± 0.01	381 ± 2	1400 ± 50	0.263 ± 0.010	368 ± 4
	900	0.757 ± 0.013	378 ± 2	966 ± 26	0.517 ± 0.016	500 ± 10
	1100	1.34 ± 0.03	240 ± 1	193 ± 3	0.929 ± 0.022	179 ± 4
	1300	1.64 ± 0.01	152 ± 1	49.9 ± 0.4	1.86 ± 0.02	92.8 ± 1
	1600	2.06 ± 0.02	77.4 ± 1	34.5 ± 0.6	1.09 ± 0.018	37.6 ± 0.6
Totals		49.5 ± 0.4	1341 ± 7			
150°C	300	153 ± 0.2	100 ± 0.5	294 ± 12	0.002 ± 0.0001	0.655 ± 0.003
	500	18.4 ± 0.1	88.2 ± 0.4	490 ± 20	0.010 ± 0.0004	4.8 ± 0.03
	700	1.71 ± 0.01	417 ± 2	1200 ± 20	0.203 ± 0.005	243 ± 2
	900	0.898 ± 0.015	298 ± 1	872 ± 24	0.380 ± 0.013	331 ± 6
	1100	1.87 ± 0.01	248 ± 2	210 ± 2	0.632 ± 0.006	133 ± 1
	1300	2.53 ± 0.03	182 ± 1	64.1 ± 0.1	1.12 ± 0.02	71.7 ± 1
	1600	2.92 ± 0.03	70.3 ± 0.6	37.3 ± 0.3	0.65 ± 0.01	24.1 ± 0.3
Totals		181 ± 0.3	1403 ± 3			

Note: All errors represent ±1σ precision as defined by the statistics of the data. Accuracy of ^{38}Ar and ^{40}Ar abundances is estimated at approximately ±20%. Probably 10–30% of the gas has been lost during vacuum storage prior to measurement, see text.

exposed to the atmosphere for two hours after crushing prior to loading into the extraction line. This sample will be referred to as the control.

After removal from the ^{38}Ar atmosphere, the samples were placed in the sample storage section of the extraction line. The total exposure to atmosphere during the removal and transfer of the samples was approximately three hours. There was no preheating of any of the samples before the first temperature fraction at 300°C. A high vacuum valve separated the sample storage portion from the extraction system. Thus at any time the samples not being measured were isolated from the rest of the extraction system. The control, room temperature exposure to ^{38}Ar, and 150°C exposure to ^{38}Ar samples were in the vacuum line for 10 days, 12 days, and 24 days, respectively. Ordinarily, a sample was not introduced into the extraction system until a series of blank runs was completed. The 150°C sample, however, was passed through the valve before the blank run.

When the blank was run with the 150°C sample in the line, it was observed that 1.9×10^{-9} cm^3 STP g^{-1} of ^{38}Ar was degassed into the vacuim line in two hours. This compares to 181×10^{-8} cm^3 STP g^{-1} released in total from this sample during the thermal release study. The degassing rate measured over two hours after 24 days in the vacuum system for the 150°C sample is 1.3% per day. If the rate is constant, then about 30% of the sample will have been lost during vacuum storage prior to measurement. The losses for the other two samples on this basis are about half as great since they had been stored half as long in the vacuum prior to measurement.

For the thermal release experiments, the samples were heated in a type 1720 glass-walled, water-cooled furnace using an induction-heated molybdenum crucible. The extraction and cleanup procedures have been described in detail by Schaeffer and Schaeffer (1977). Thermal release steps were run at 300°C, 500°C, 700°C, 900°C, 1100°C, 1300°C, and 1600°C for one hour except for the 1600°C step which was 15 minutes. Date corrected for procedural blanks are shown in Table 1.

The blank amounts for masses 39 through 36 were the same for all temperature steps. The mass 39 and 37 background peaks corresponded to approximately 5×10^{-12} cm^3 STP, and the mass 38 and 36 backgrounds corresponded to 1×10^{-12} cm^3 STP. These background peaks were primarily due to HCl and hydrocarbons, and were found to be reproducible from blank to blank, although the 38 and 36 backgrounds are close to the sensitivity limit of the instrument and may be subject to variations of up to a factor of two in sample runs. For most measurements, however, the blanks were less than 1% of the sample. The exceptions were the ^{38}Ar blanks for the control sample fractions released below 1100°C, where they were 20%, and the ^{40}Ar blanks for the 1600°C step for all the samples The ^{40}Ar blank for the 1600°C step of all the samples was about 10%.

The measurements were made with a 20 cm radius, 60° sector, statically operated mass spectrometer. The sensitivity and discrimination at mass 40 was checked daily. The sensitivity was found to be stable to within 2%. The mass discrimination was also very stable and was 0.6%/amu. The mass peaks at 41 and 35 were monitored during every run and found to be less than 1.3×10^{-12} and 2×10^{-11} cm^3 STP equivalent argon, respectively, indicating the absence of significant hydrocarbon or chlorine contamination introduced by the samples. The data acquisition and processing system has been described by Schaeffer and Schaeffer (1977).

DISCUSSION

Homogeneity of samples

In this experiment, ^{38}Ar was used as a tracer to investigate the diffusion of argon into a coarse-grained Ca–Al-rich inclusion from Allende. It is therefore important to measure the ^{38}Ar quantities in a control sample that has not been exposed to the ^{38}Ar atmosphere. Smith *et al.* (1977) have shown that the thermal release patterns of both the amount of a given isotope and its ratio to another isotope vary considerably for samples taken from different portions of fine-grained, Ca–Al-rich inclusions. We can assume this is also true of coarse-grained inclusions; thus it is important that the homogeneity of our samples be tested.

The data available for evaluating the homogeneity of the samples are the amounts of ^{40}Ar and ^{36}Ar, since neither of these isotopes should be influenced by the exposure to ^{38}Ar, except perhaps by a random exchange of atoms. Figure 1 is a plot of the ^{40}Ar/^{36}Ar thermal release patterns of the three samples, and shows that the equivalent temperature steps are in agreement to better than 20% in all cases, and better than 10% in most cases. The most notable exception is the 300°C step of the room temperature sample which seems to have a ratio about 25% smaller than the atmospheric ratio of 295.5. As there was no prior degassing of the samples, we expect the first temperature step to be characteristic of air argon, and within the errors this expectation is realized for both the other samples. A comparison of the amounts of ^{40}Ar and ^{36}Ar released from the three samples at

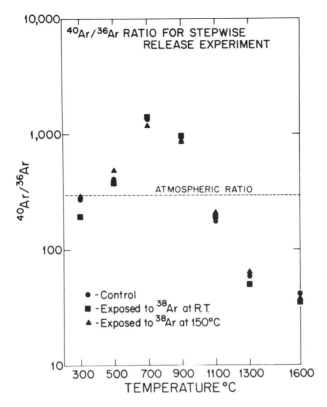

Fig. 1. ^{40}Ar/^{36}Ar ratios as a function of extraction temperature for control samples exposed to ^{38}Ar atmospheric partial pressure of Ar at room temperature and 150°C. Note that the ^{40}Ar/^{36}Ar ratio is unaffected by the ^{38}Ar exposure. See text for discussion.

300°C shows that the room temperature sample only released 18% and 33% as much ^{40}Ar as the control and 150°C samples, respectively. Thus it is possible that the 300°C step of all the samples contains both an excess air fraction and a trapped planetary argon fraction which would have a very low ^{40}Ar/^{36}Ar ratio. The interpretation is further complicated by the large uncertainty in the amount of ^{36}Ar associated with the measurement of the room temperature sample, which corresponded to approximately 2×10^{-11} cm^3 STP. The principal uncertainty is in the blank correction of 2×10^{-12} cm^3 STP attributable mostly to HCl.

In addition to examining the ^{40}Ar/^{36}Ar ratio, the amounts of ^{40}Ar and ^{36}Ar give some information about the homogeneity of the samples. As discussed above, the 300°C fractions indicate differences among the three samples. The room temperature sample shows approximately an 80% depletion of argon relative to the control sample, while the 150°C sample shows about a 50% depletion relative to the control sample. The total amounts of ^{40}Ar and ^{36}Ar summed over all the temperature fractions, however, agree to better than 5% and 7%, respectively. The principal differences appear to be the argon missing from the room temperature sample in the first thermal step and the larger amounts of ^{36}Ar in the highest temperature fractions of both the samples exposed to the ^{38}Ar atmosphere. Since one would expect the higher temperature fractions to be due principally to spallation production from Ca, we should observe correspondingly low ^{36}Ar/^{38}Ar ratios in these fractions.

Figure 2 shows the thermal release pattern for the ^{36}Ar/^{38}Ar ratio of the three samples. The first four fractions of the control sample are characterized by a ratio that is indistinguishable within the error limits from trapped planetary argon that is commonly

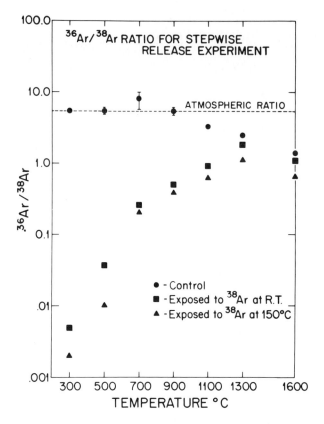

Fig. 2. $^{36}Ar/^{38}Ar$ ratios as a function of extraction temperature for samples exposed to ^{38}Ar and control sample. Compare Fig. 1. Note that the exposed samples show evidence of excess ^{38}Ar for all temperature extractions.

observed in Allende samples. The possible exception is the 700°C fraction whose $^{36}Ar/^{38}Ar$ ratio is noticeably greater than 5.35. This may not be significant as the ^{38}Ar blank correction (2×10^{-12} cm^3 STP) was somewhat uncertain for this fraction since we were working close to the limit of sensitivity of the mass spectrometer. If the ratio is accepted as being significant, however, it might be attributed to the characteristic release at 700°C of ^{36}Ar resulting from neutron produced ^{36}Cl as reported by Smith et al. (1977) in their study of fine grained inclusions. In either case, it appears that there is very little Cl-associated ^{36}Ar in these three samples.

Beginning with the 1100°C fraction, a spallation component becomes apparent by the decreasing $^{36}Ar/^{38}Ar$ ratio associated with these higher temperature steps. The total amount of spallation ^{38}Ar in the control sample released in the three high temperature fractions was 1.8×10^{-8} cm^3 STP/g. Assuming an exposure age of 5.6 m.y. (Bogard et al., 1971) and the production rate equation given by Cressy and Bogard (1976), a Ca abundance of $15 \pm 1.5\%$ was calculated for the control sample. This result agrees with the Ca content determined by the temperature release study of Herzog et al. (1980) of $15.7 \pm 0.2\%$ Ca.

Amount of ^{38}Ar diffused into samples

An examination of Fig. 2 shows that the $^{36}Ar/^{38}Ar$ ratios of the diffusion samples differ from the control most significantly at the lower temperature fractions and trend toward the control sample at the 1300°C and 1600°C fractions. Even at these high temperature fractions, the differences indicate a lower $^{36}Ar/^{38}Ar$ ratio, either an excess of ^{38}Ar or deficit of ^{36}Ar, compared to the control sample. The result of a calculation of excess ^{38}Ar amounts which assumes that the control $^{36}Ar/^{38}Ar$ is representative of the two other samples is illustrated in Table 2. Also illustrated in Table 2 is the percentage of ^{38}Ar of spallation origin in the control sample. Notice that only the 1100°C to 1600°C fractions are important sources of spallation ^{38}Ar. The result of the simple calculation shows that traces of the excess ^{38}Ar remain even in the highest temperature fractions. To be sure the amount is small, but the possibility that it may be in these fractions at all must be examined. In a study of the uptake of argon by Niemeyer and Leich (1976) of lunar anorthosite 60015, a similar effect was observed.

Table 3 shows the proportions of spallation and trapped planetary ^{38}Ar and ^{36}Ar calculated assuming $^{36}Ar/^{38}Ar_{sp} = 0.62$, $^{36}Ar/^{38}Ar_{tr} = 5.35$, and that there is no excess ^{38}Ar present. Only the total argon in the 1100°C–1600°C fractions has been used. Under these assumptions, the room temperature sample appears to contain approximately 2.5 times as much "spallation" argon as the control, and the 150°C sample approximately 4 times as much. We have previously shown, however, that the samples are homogeneous by analyzing the ^{40}Ar and ^{36}Ar; thus we conclude there must be excess ^{38}Ar present in the two samples. Table 4 shows the proportions of spallation and trapped planetary ^{38}Ar and ^{36}Ar and the amount of excess ^{38}Ar calculated on the basis of 15.7% for the Ca abundance. It is important to note that the smaller $^{36}Ar/^{38}Ar$ ratio observed for the room temperature and 150°C samples in Table 1 can only be attributed to excess ^{38}Ar amounting to about $1-5 \times 10^{-8}$ cm^3 STP/g.

Finally, we compare in Table 5 our results for ^{38}Ar taken up with the amounts of extra ^{40}Ar in the laser and temperature release study of the same inclusion, 3529Z. By extra

Table 2. % Excess ^{38}Ar relative to control sample.

Temperature	% $^{38}Ar_{sp}$ Control†	% Excess ^{38}Ar* Room Temperature Sample	150°C Sample
300°C	~0	>99	>99
500	~0	>99	>99
700	~0	97	98
900	~0	91	93
1100	43	73	81
1300	60	25	55
1600	83	23	54

†Calculated assuming $(^{36}Ar/^{38}Ar)_{sp} = 0.62$ and $(^{36}Ar/^{38}Ar)_{tr} = 5.35$.
*Calculated from % $^{38}Ar_{ex} = 100 \left[1 - \frac{(^{36}Ar/^{38}Ar) meas}{(^{36}Ar/^{38}Ar) control} \right]$.

Table 3. Spallation and trapped planetary ^{36}Ar and ^{38}Ar concentrations† in units of 10^{-8} cm^3 STP/g.

	Measured		Spallation		Trapped	
	^{38}Ar	^{36}Ar	^{38}Ar	^{36}Ar	^{38}Ar	^{36}Ar
Control	2.57 ± 0.03	5.44 ± 0.05	1.76 ± 0.05	1.09 ± 0.06	0.81 ± 0.07	4.35 ± 0.08
Room Temperature	5.04 ± 0.04	6.54 ± 0.04	4.32 ± 0.06	2.68 ± 0.06	0.72 ± 0.06	3.86 ± 0.06
150°C	7.31 ± 0.04	5.89 ± 0.03	7.02 ± 0.06	4.35 ± 0.06	0.29 ± 0.06	1.54 ± 0.06

†Calculated from the 1100°C–1600°C fractions assuming no excess ^{38}Ar is present. See text for discussion. All errors represent ±1σ precision as defined by the statistics of the data. Accuracy of ^{38}Ar and ^{36}Ar abundances is estimated at approximately ±10%.

Table 4. Spallation and trapped planetary ^{36}Ar and ^{38}Ar components† in units of 10^{-8} cm^3 STP/g.

	Spallation		Trapped		
	^{38}Ar	^{36}Ar	^{38}Ar	^{36}Ar	^{38}Ar Excess
Control	1.76 ± 0.05	1.09 ± 0.05	0.81 ± 0.07	4.35 ± 0.08	
Room Temperature	2.46 ± 0.08	1.53 ± 0.08	0.94 ± 0.10	5.01 ± 0.10	1.60 ± 0.03
150°C	2.46 ± 0.08	1.53 ± 0.08	0.81 ± 0.10	4.36 ± 0.10	4.04 ± 0.13

†Assuming 15.7% Ca abundance.

All errors represent ±1σ precision as defined by the statistics of the data. Accuracy of ^{38}Ar and ^{36}Ar abundances is estimated at approximately ±10%.

Table 5. Extra* argon in Allende inclusion 3529Z in 10^{-8} cm^3 STP/g.

		Extra ^{40}Ar† in			Acquisition of ^{38}Ar by bulk material at	
	Melilite	Anorthosite	Pyroxene	Bulk	20°C	150°C
Unheated	120	110	80	—	80	120
Temperature extractions > 600°C	40	120	60	80	2	3

*See text.
†Herzog et al. (1980).

^{40}Ar we mean the amount of ^{40}Ar needed to change the age from 4.55 G.y. to 5.9 G.y. The total K–Ar age calculated from the ^{40}Ar released above 675°C in the stepwise heating experiment. So defined, the extra ^{40}Ar in 3529Z is about 100×10^{-8} cm^3 STP/g based on either the laser or the stepwise heating data. This value is comparable to the total ^{38}Ar taken up by the bulk material in our experiment (Table 5). However, the temperature release characteristics of the extra ^{40}Ar and the tracer ^{38}Ar, are quite different. Much of the extra ^{40}Ar degasses at temperatures above 600°C, perhaps half or more; only 1 or 2% of the tracer, ^{38}Ar, survives the outgassing at 600°C.

Jessberger et al. (1980) report ages in excess of 4.55 G.y. for three samples of inclusion 3529Z. Calculating extra ^{40}Ar as above for the temperature fraction above 600°C, one obtains 57, 170, and 60×10^{-8} cm^3 STP/g^{-1}, also well in excess of the amount expected on the basis of our present absorption study.

A difference between our experiment and the ^{39}Ar–^{40}Ar work of Jessberger et al. (1980) was that our samples were not irradiated by a neutron flux. A comparison between the total ^{36}Ar released by the samples in this experiment and the Jessberger et al. (1980) work shows, however, that the largest differences are only about 40%. Thus the neutron doses used did not seem to lead to a considerably greater uptake of argon compared to the uptake in our experiment.

CONCLUSIONS

1. We believe that atmospheric contamination is not the cause for the extra argon and >4.55 G.y. ages in this Allende inclusion (Jessberger et al., 1980; and Herzog et al., 1980).
2. Absorbed atmospheric argon in these samples persists in small amounts in thermal step release experiments until the sample is melted.

Acknowledgments—We would like to acknowledge the anonymous reviewers who improved the paper and Larry Nyquist who made many helpful suggestions. This work was supported by NASA contract No. NAGW25.

REFERENCES

Bogard D. D., Clark R. S., Keith J. E., and Reynolds M. A. (1971) Noble gases and radionuclides in Lost City and other recently fallen meteorites. *J. Geophys. Res.* **76**, 4076–4083.
Cressy P. J. Jr. and Bogard D. D. (1976) On the calculations of cosmic-ray exposure ages of stone meteorites. *Geochim. Cosmochim. Acta* **40**, 749–762.
de Saussure T. (1814) *Ann. Phys.* **47**, 113–183.
Homfray I. F. (1910) Die Adsorption von Gasen durch Holzkohle. *Z. Phys. Chem.* **74**, 129–201.
Herzog G. F., Bence A. E., Bender J., Eichhorn G., Maluski H., and Schaeffer O. A. (1980) $^{39}Ar/^{40}Ar$ systematics of Allende inclusions. *Proc. Lunar Planet. Sci. Conf. 11th*, p. 959–976.
Jessberger E. K. and Dominik B. (1979) Gerontology of the Allende meteorite. *Nature* **277**, 554–555.
Jessberger E. K., Dominik B., Staudacher T., and Herzog G. F. (1980) $^{40}Ar-^{39}Ar$ ages of Allende. *Icarus* **42**, 380–405.
Niemeyer S. and Leich D. A. (1976) Atmospheric rare gases in lunar rock 60015. *Proc. Lunar Sci. Conf. 7th*, p. 587–597.
Schaeffer G. A. and Schaeffer O. A. (1977) $^{39}Ar/^{40}Ar$ ages of lunar rocks. *Proc. Lunar Sci. Conf. 8th*, p. 2253–2300.
Smith S. P., Huneke J. C., Rajon R. S., and Wasserburg, G. J. (1977) Neon and argon in the Allende meteorite. *Geochim. Cosmochim. Acta* **41**, 627–648.
Titoff A. (1910) Die Adsorption von Gasen durch Kohle. *Z. Phys. Chem.* **74**, 641–678.
Zaikowski A. and Schaeffer O. A. (1979) Solubility of noble gases in serpentine: Implications for meteoritic noble gas abundances. *Earth Planet. Sci. Lett.* **45**, 141–154.

Irradiation records of Acapulco and other small meteorites derived from ^{53}Mn and rare gas measurements

P. Englert*, U. Herpers, and W. Herr

Institut für Kernchemie der Universität zu Köln, Zülpicherstr. 47, D 5000 Köln 1, West Germany

Abstract—Cosmogenic ^{53}Mn and/or ^{26}Al has been determined in samples of eight meteorite falls with small preatmospheric masses which have previously been analyzed for cosmogenic rare gases and partially also for cosmic ray tracks. Four meteorites, some of which also show appreciable amounts of trapped rare gases, have suffered at least a two-stage irradiation: St. Germain du Pinel, Cherokee Springs, Gopalpur and Klein Wenden. ^{53}Mn-rare gas data pairs of the other small chondrites show that the linear ^{53}Mn* (saturation-activity) vs. ^{22}Ne/^{21}Ne depth (Englert and Herr, 1980a, b) and depth- and size-relations (Nishiizumi et al., 1980) may level off at high ^{22}Ne/^{21}Ne ratios, i.e., for low shielding. This suggestion is also supported by ^{53}Mn measurements on samples from Core CB4 of the St. Severin chondrite (Bhattacharya et al., 1980) which were close to the preatmospheric surface.

INTRODUCTION

Attempts to resolve the problem of the depth- and size-dependencies of the production rates of cosmogenic radionuclides within meteorites have been made recently (Englert and Herr, 1980a; Nishiizumi et al., 1980; Bhattacharya et al., 1980). Meteorites used in these studies had to fulfill at least the following requirements: a simple single-step exposure to the cosmic radiation and a high exposure age. The latter is necessary to avoid corrections for undersaturation of radionuclide activities which propagate from the still existing differences in the rare gas exposure age scales and possible recent variations of the galactic cosmic radiation (Nishiizumi et al., 1980; Müller et al., 1981). Figure 1 shows that the ^{53}Mn vs. ^{22}Ne/^{21}Ne depth relation of Core AIII of St. Severin (Englert and Herr, 1980a) fits very well into the general ^{53}Mn vs. ^{22}Ne/^{21}Ne relation of Nishiizumi et al. (1980). Recent results on cores of the Keyes chondrite (see Fig. 1) also confirm these general tendencies (Englert and Herr, 1980b). However, for the evaluation of the depth relations in Keyes and St. Severin, samples from locations very close to the preatmospheric surface were not available, so that the linearly approximated anticorrelations might not accurately represent the tendency of the relation at lower shielding, i.e., at ^{22}Ne/^{21}Ne ratios ≥ 1.15. The range of validity of the general ^{53}Mn vs. ^{22}Ne/^{21}Ne trend line of Nishiizumi et al. (1980) does not extend beyond a ^{22}Ne/^{21}Ne ratio of 1.20; the upper limit is principally determined by three ^{53}Mn–^{22}Ne/^{21}Ne data pairs only, all with rare gas ratios close to ^{22}Ne/^{21}Ne = 1.19. In order to increase the significance of the depth- and size-relations of ^{53}Mn in the region of low shielding it seemed to be necessary to analyze more meteorite samples with ^{22}Ne/^{21}Ne ratios between 1.15 and 1.25; shielding depths corresponding to ^{22}Ne/^{21}Ne ratios ≥ 1.20 are of special interest.

In order to avoid uncertainties due to long terrestrial residence times and weathering effects, samples of eight observed meteorite falls with small recovered masses were selected for ^{53}Mn and/or ^{26}Al measurements. The rare gases have been determined previously on the same aliquot. The recovered masses of these chondrites range from 0.35–8.4 kg, which corresponds to effective postatmospheric radii of 2.9–8.3 cm. The ^{3}He/^{21}Ne ratios observed in these stones range from 6.33–8.62 and the ^{22}Ne/^{21}Ne ratios

---

*Present address: University of California, San Diego, Department of Chemistry, B-017, La Jolla, California 92093, U.S.A.

from 1.132–1.248, thus generally indicating lightly shielded samples. Cosmic ray (CR) track density measurements on aliquots of the ^{53}Mn samples have been carried out for Kiel (P. Pellas, pers. comm., 1980) and Acapulco (Palme *et al.*, 1981); the track data indicate preatmospheric radii of $R_0 = 7$–10 cm and $R_0 \geq 7$ cm respectively. According to Bhandari *et al.* (1981), CR track densities in Cherokee Springs and Gopalpur also indicate preatmospheric radii on the order of $R_0 = 10$ cm, whereas for St. Germain du Pinel an $R_0 = 10$–13 cm is probable. Small preatmospheric radii are confirmed by CR track density measurements for at least five out of the eight chondrites analyzed.

The influence of size on the production of ^{3}He and the neon isotopes have been described by Eberhardt *et al.* (1966) and Nishiizumi *et al.* (1980) by means of a ^{3}He/^{21}Ne vs. ^{22}Ne/^{21}Ne correlation line. According to the interpretation given by Schultz and Signer (1976), ^{3}He/^{21}Ne–^{22}Ne/^{21}Ne data pairs of meteorites with smaller preatmospheric masses than St. Severin should lie above the ^{3}He/^{21}Ne vs. ^{22}Ne/^{21}Ne trend line for the St. Severin core AIII. The ^{3}He/^{21}Ne–^{22}Ne/^{21}Ne data pairs taken for the five small chondrites and also for San Juan Capistrano (Finkel *et al.*, 1978; Table 1) do in fact lie appreciably above the St. Severin trend line. The fact that the data pairs also follow the trend of the "Bern line" (Eberhardt *et al.*, 1966) and lie somewhat above it provides further evidence of small meteorite masses. However, Klein Wanden, Canakkale and Tromoy also have high ^{3}He/^{21}Ne and ^{22}Ne/^{21}Ne ratios; their data points occupy essentially the same area of the graph. Thus, it seems justified to assume small preatmospheric radii on the order of $R_0 = 10$ cm or even less for these three stones, from ^{3}He/^{21}Ne vs. ^{22}Ne/^{21}Ne systematics only.

EXPERIMENTAL METHODS AND RESULTS

Samples with weights between 80 and 400 mg were taken from Acapulco, Canakkale, Cherokee Springs, Gopalpur, Kiel, Klein Wenden, St. Germain du Pinel and Tromoy, where rare gas ratios have been measured previously in each case. After grinding and dissolution in HNO$_3$/HF, aliquots of the samples were analyzed for Fe, Ni, Co and Mn by neutron activation and partly also by atomic absorption. The meteoritical manganese was separated from the solution and ^{53}Mn was determined by an intense neutron activation via the ^{53}Mn(n, γ)^{54}Mn reaction, according to the method proposed by Millard (1965) and developed by Herr and coworkers (1967, 1969, 1972) and others. For the present study we have applied an improved analytical technique, which is described in more detail by Englert and Herr (1978). ^{53}Mn activities (dpm/kg Fe), corrected for the contribution from the target element Ni according to Nishiizumi (1978), are given in Table 1.

^{26}Al and ^{22}Na activities of a 14.5 g Acapulco specimen [^{22}Na $= 61.8 \pm 9.3$ dpm/kg] were determined by γ-γ coincidence spectroscopy using the counting procedures described by Heimann *et al.* (1974). ^{26}Al activities as well as rare gas ratios are also given in Table 1.

The "apparent" ^{21}Ne exposure ages (Table 1) of all chondrites are here calculated with a ^{21}Ne production rate of 0.31×10^{-8} cm^3 STP/g m.y. given in Nishiizumi *et al.* (1980) for L-chondrites. The ^{21}Ne exposure ages differ only slightly if the ^{21}Ne production rate for H-chondrites of $0.30 \pm 0.03 \times 10^{-8}$ cm^3 STP/g m.y. derived by Müller *et al.* (1981), is applied. For consistency, the first production rate was also used for Kiel and Tromoy, although their ^{22}Ne/^{21}Ne ratios exceed the range of validity of the latter calculations. The ^{21}Ne exposure ages of Canakkale, Cherokee Springs, Kiel, Klein Wenden and St. Germain du Pinel exceed 10×10^6 years; the cosmic ray exposure age of Tromoy (T $= 3.47 \times 10^6$ y) covers barely one ^{53}Mn half-life (T$_{1/2} = 3.7 \times 10^6$ y). This leads to a large uncertainty in the ^{53}Mn saturations activity (^{53}Mn*), calculated for Tromoy from the measured ^{53}Mn activity and the ^{21}Ne exposure age. ^{53}Mn* activities for all chondrites are given in Table 1. Considerable differences between these and the measured ^{53}Mn values are apparent only in the case of the Acapulco and Gopalpur samples, due to their low exposure ages.

DISCUSSION

Complex irradiation histories

The ^{53}Mn*–^{22}Ne/^{21}Ne data pairs of all chondrites are shown in Fig. 1. A first inspection of these data points of the small chondrite falls shows appreciable scattering around the established ^{53}Mn* vs. ^{22}Ne/^{21}Ne relations. Four meteorites, Gopalpur, Cherokee Springs, Klein Wenden and St. Germain du Pinel, have relatively low ^{53}Mn saturation activities, by

Table 1. Radionuclides and rare gas ratios in chondrite falls with small recovered and preatmospheric masses.

	Recovered Mass [kg]	Type	^{53}Mn [dpm/kgFe]	^{26}Al [dpm/kg]	^{3}Ne/^{21}Ne	^{22}Ne/^{21}Ne	^{21}Ne[h] exp age [10^6 y]	^{53}Mn$^{*i)}$ [dpm/kgFe]	^{53}Mn$_s^{k)}$ [dpm/kgFe]	^{53}Mn[l] exp age [10^6 y]
Acapulco	1.9	H	281 ± 23[a]	49 ± 7[a]	8.62	1.189[e]	7.47	378 ± 30	321	11.4 ± 1.1
Canakkale	4.0	L	295 ± 23[a]	49.4 ± 4.3[b]	8.21	1.169[f]	12.1	331 ± 26	344	10.7 ± 0.9
Cherokee Springs	8.4	LL5	204 ± 18[a]	54 ± 3[c]	6.59	1.161[f]	11.5	232 ± 20	354	4.7 ± 0.70
Gopalpur	1.6	H	166 ± 16[a]	29.9 ± 6.4[b]	7.05	1.182[f]	9.39	203 ± 30	329	3.86 ± 0.50
Kiel	0.73	L	317 ± 18[a]	–	8.29	1.248[g]	16.5	333 ± 19	[251]	–
Klein Wenden	3.3	H6	214 ± 15[a]	–	7.32	1.162[f]	15.4	228 ± 16	352	5.14 ± 0.80
St Germain du Pinel	4.0	H6	211 ± 15[a]	49.2 ± 1.8[d]	6.33	1.132[g]	37.1	211 ± 15	388	4.31 ± 0.70
Tromoy	0.35	H	113 ± 10[a]	–	8.21	1.28[f]	3.47	241 ± 42	[263]	3.08 ± 0.50
San Juan Capistrano	0.55	H6	310 ± 15[n]	44 ± 2[m]	7.11	1.189[m]	28.0[m]	312 ± 15	321	–

a) this work
b) Herzog and Cressy (1974)
c) Rowe and Clark (1971)
d) Cameron and Top (1975)
e) Palme et al. (1981)
f) Schultz and Kruse (1978)
g) Schultz and Signer (1974)
h) "apparent" ^{21}Ne exposure ages, calculated according to Nishiizumi et al. (1980)
i) ^{53}Mn : ^{53}Mn saturation activity, calculated from the measured ^{53}Mn activities, using the "apparent" ^{21}Ne exposure ages and a ^{53}Mn half-life of 3.7×10^6 years
k) ^{53}Mn$_s$: mean ^{53}Mn saturation activity, derived from the ^{53}Mn* vs. ^{22}Ne/^{21}Ne relation of Nishiizumi et al. (1980); see also Fig. 1
l) "apparent" ^{53}Mn exposure ages, derived from the measured ^{53}Mn activities with ^{53}Mn$_s$ as the appropriate production rate
m) Finkel et al. (1978)
n) Nishiizumi (1978)

no means even corresponding to their "apparent" shielding conditions or "apparent" ^{21}Ne exposure ages, which range from $T = 10 \times 10^6$ y to $T = 40 \times 10^6$ y; therefore, ^{53}Mn seems to be undersaturated by more than 50%. Similar discrepancies between "apparent" rare gas exposure ages and cosmogenic radionuclide contents have been observed, e.g., for Serra de Magé and Ivuna by Fuse and Anders (1969) and also for some Antarctic meteorite finds (Imamura et al., 1979; Nishiizumi and Arnold, 1981), and described in terms of a two-stage irradiation model. Unlike the falls, however, radionuclide contents of Antarctic finds might also have been affected by long terrestrial residence times, which would lead to undersaturation due to radioactive decay, or possibly to weathering losses.

A simple one-stage irradiation model does not provide a satisfactory explanation of rare gas and radionuclide data, especially if one compares the "apparent" ^{21}Ne- and ^{53}Mn-exposure ages. At least a two-stage irradiation model is necessary to explain the cosmogenic nuclide contents of the four chondrites considered. The first stage requires a long preirradiation in a large body with the "apparent" ^{21}Ne age as the minimum duration, and a second stage as a small body ($R_0 = 10$–15 cm) with the "apparent" ^{53}Mn-exposure ages (see Table 1) as the maximum duration. The major part of the cosmic ray track production might then obviously be attributed to the last irradiation step. This seems to be reasonable, as the CR-track porduction rate in larger bodies drops rapidly by several orders of magnitude (Lal et al., 1969). Our assumption would seriously affect the cosmic ray track production rate of Bhandari et al., (1981) used to derive the preatmospheric dimensions given in the introduction which have been calculated with the "apparent" ^{21}Ne exposure ages (Table 1). These considerations suggest smaller preatmospheric radii, i.e., $R_0 < 10$ cm, for Cherokee Springs, Klein Wenden and Gopalpur. In the case of St. Germain du Pinel, Bhandari et al. (1981) use a ^{21}Ne exposure age of 4.8×10^6 y, obviously

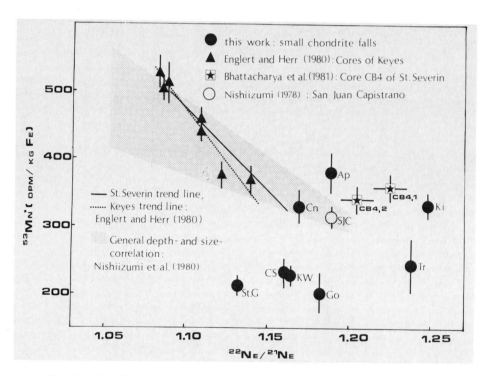

Fig. 1. ^{53}Mn* vs. ^{22}Ne/^{21}Ne in chondrites. Small chondrite falls: Ap–Acapulco, Cn–Canakkale, CS–Cherokee Springs, Go–Gopalpur, Ki–Kiel, KW–Klein Wenden, SJC–San Juan Capistrano, St. G–St. Germain du Pinel, Tr–Tromoy. CB4,1 and CB4,2 are the outermost samples of St. Severin Core CB4. The ^{53}Mn* vs. ^{22}Ne/^{21}Ne relations of St. Severin and Keyes (Englert and Herr, 1980a,b) as well as the general depth- and size-relation of Nishiizumi et al. (1980) are given for comparison.

relying on the rare gas measurements of Eberhardt *et al.* (1966). This result corresponds to our ^{53}Mn exposure age of $4.31 \pm 0.70 \times 10^6$ y, which is regarded to be the upper limit for the second irradiation step; consequently, the calculation of St. Germain du Pinel's preatmospheric radius here still holds.

^{26}Al activities have been measured for Cherokee Springs, Gopalpur and St. Germain du Pinel. As the "apparent" ^{53}Mn exposure ages of these chondrites range from 3.7–4.7×10^6 y (Table 1), the ^{26}Al activities are expected to be in saturation. With respect to shielding conditions (Nishiizumi *et al.*, 1980) this is achieved for Cherokee Springs and St. Germain du Pinel. However, for Gopalpur, the ^{26}Al activity is only 29.9 ± 6.4 dpm/kg (Herzog and Cressy, 1974), which indicates an undersaturation of at least 25%, according to Nishiizumi *et al.* (1980). Though ^{53}Mn and ^{26}Al have not been determined on aliquots of the same Gopalpur sample, the low postatmospheric mass (1.6 kg) of the latter chondrite suggests that depth effects are not the only significant factor. A multiple break up might explain the unusual cosmogenic nuclide pattern of Gopalpur.

In light of these considerations, the measured ^{22}Ne/^{21}Ne ratios of the four chondrites considered are the result of the contributions of the two components related to the respective irradiation stages. The ^{22}Ne/^{21}Ne ratio in heavily shielded bodies may converge towards ^{22}Ne/^{21}Ne $= 1.00$ due to the enhanced production of ^{21}Ne from Mg, Si and Al by low energy particles (Walton *et al.*, 1974; Herzog, 1975); in barely shielded bodies higher values are expected as, for example, in Kiel ($R_0 = 7$–10 cm) and ALHA 77081 ($R_0 = 2$ cm), where ^{22}Ne/^{21}Ne is 1.248 (Schultz and Signer, 1974) and ^{22}Ne/^{21}Ne is 1.29 (Schultz *et al.*, 1980) respectively. The range of ^{22}Ne/^{21}Ne $= 1.132$–1.182 of the four chondrites can be explained by the varying relative contributions of the two irradiation steps. St. Germain du Pinel, for example, has suffered by far the longest preirradiation and consequently shows the lowest rare gas ratios of the group. Whether the preirradiations immediately preceded the second stage cannot be inferred from the data available up to now. Cherokee Springs, Gopalpur and Klein Wenden, however, have large amounts of trapped neon (Schultz and Kruse, 1978) and therefore might be considered candidates for preirradiations in very early stages of their history (Pellas, 1973).

The ^{53}Mn* vs. ^{22}Ne/^{21}Ne trend line at low shielding depths

Out of the eight small meteorite falls analyzed, only one specimen, Canakkale, seems to follow the ^{53}Mn* vs. ^{22}Ne/^{21}Ne trendlines strictly, a behavior also found for San Juan Capistrano (Nishiizumi *et al.*, 1980; see Fig. 1). This small H6-fall had a preatmospheric radius of $R_0 = 8$ cm. The measured samples came from a preatmospheric location close to the center (Finkel *et al.* 1978). Tromoy's ^{53}Mn* activity has a very high uncertainty due to its low ^{21}Ne exposure age, as pointed out previously. In the ^{53}Mn* vs. ^{22}Ne/^{21}Ne plot (Fig. 1) however, Acapulco and Kiel show ^{53}Mn saturation activities approximately 15% and 20%, respectively, above the trendlines established for meteorites with normal irradiation histories. For both meteorites there exists some evidence for a normal one-step irradiation. We determined ^{53}Mn- and ^{26}Al-activities for Acapulco on aliquots of the same sample, so that the depth-independent ^{53}Mn/^{26}Al exposure age of 6.8×10^6 y can be calculated (Englert and Herr, 1979), in good agreement with the ^{21}Ne exposure age of 7.47×10^6 y (Table 1). The CR-track and ^{53}Mn-production rate of the Kiel specimen fit into the ranges predicted by the semi-empirical models of Kohman and Bender (1967) and Bhattacharya *et al.* (1973) for meteorites with one-step irradiation histories. Thus, the ^{53}Mn activity levels of Acapulco and Kiel can hardly be explained by an exposure to a temporarily higher flux of low energy cosmic ray particles, such as might be due to changes in orbital parameters.

The ^{53}Mn-rare gas paired sample measurements on cores of the St. Severin and Keyes chondrite do not include locations directly below the preatmospheric surface, which in the case of the latter stone is due to large ablation losses. Bhattacharya *et al.* (1980) determined ^{53}Mn in Core CB4 of the St. Severin chondrite. The preatmospheric depths of the samples, determined by CR track density measurements, were as close as 2 mm below the preatmospheric surface. Unfortunately rare gases have not been determined on aliquots of these samples, so that a direct comparison with the relations in Fig. 1 is not possible. However, for Core AIII (Lal et al., 1969; Schultz and Signer, 1976) and Core CB3 of the St. Severin chondrite there exist $^{22}Ne/^{21}Ne$–CR track density relations, which allow us to estimate appropriate $^{22}Ne/^{21}Ne$ ratios for the subsurface samples of Core CB4. The data pairs for sample CB4,1 ($^{53}Mn^* = 357 \pm 18$ dpm/kg Fe; $^{22}Ne/^{21}Ne$-estimated- = 1.22) and sampled CB4,2 ($^{53}Mn^* = 342 \pm 18$ dpm/kg Fe; $^{22}Ne/^{21}Ne$-estimated- = 1.20) are shown in Fig. 1. Indeed the $^{53}Mn^*$ activities of both outermost samples of Core CB4 also fall above the established $^{53}Mn^*$ vs. $^{22}Ne/^{21}Ne$ correlation lines.

These four data points may therefore be used to extend the $^{53}Mn^*$ vs. $^{22}Ne/^{21}Ne$ relation to larger $^{22}Ne/^{21}Ne$ ratios. Especially Kiel's data pair indicates that the general depth- and size-relation (Nishiizumi et al., 1980) might level off above $^{22}Ne/^{21}Ne = 1.19$. Additionally these results show that a simple linear approximation for the evaluation of the ^{53}Mn depth dependency in meteorite cores is only valid for that part of a chondrite where effects of low energy primary particles of the cosmic radiation on the radionuclide production can be excluded (Englert and Herr, 1980a).

Although the St. Severin-CB4, Acapulco, and Kiel data pairs (Fig. 1) show the same tendency, the "elevated" ^{53}Mn activities and the rare gas ratios are most probably due to different factors. In the case of the St. Severin Core CB3, Lal and Marti (1977) found that observed rare gas anomalies within 1 cm depth below the preatmospheric surface are consistent with the production of the rare gases by a primary low energy particle component of solar origin. So the ^{53}Mn production of sample CB4,1 (preatm. depth: 2 mm) and CB4,2 (preatm. depth: 6 mm) might also be at least partially due to the production by low energy primary solar protons.

Ablation of Acapulco and Kiel is more serious than for the surface of Core CB3 and CB4 of St. Severin. In general Kiel had an ablation loss of >80% (Bhandari et al., 1981). CR track studies on the ^{53}Mn sample (P. Pellas, pers. comm., 1980) give evidence for a preatmospheric depth of 3.5–4.0 cm in this body ($R_0 = 7$–10 cm). The minimum preatmospheric depth of the ^{53}Mn sample of Acapulco ($R_0 \geq 7$ cm) is at least considered to be >3 cm (Palme et al., 1981). Consequently ^{53}Mn cannot have been produced considerably by low-energy primary particles in either of these small chondrites. Secondary cascade particles of the galactic cosmic radiation seem to be responsible here (Kohman and Bender, 1967). However, more experimental effort, especially regarding production depth profiles of cosmogenic nuclides in small chondrites will be necessary to further clarify the trends indicated.

Acknowledgments—We are extremely thankful to Drs. H. Wänke, L. Schultz, and J. C. Lorin for making the valuable chondrite samples available to us. We also thank Dr. P. Pellas for the cosmic ray track density measurements on our Kiel sample. Helpful discussions with Dr. K. Marti and the reviewers are herewith gratefully acknowledged. This work was partially supported by the DFG (Deutsche Forschungs Gemeinschaft) and NASA grant NSG-7027.

REFERENCES

Bhandari N., Lal D., Rajan R. S., Arnold J. R., Marti K., and Moore C. B. (1980) Atmospheric ablation in meteorites: A study of cosmic ray tracks and neon isotopes. *Nucl. Tracks* **4**, 213–262.

Bhattacharya S. K., Goswami J. N., and Lal D. (1973) Semiempirical rates of formation of cosmic ray tracks in spherical objects exposed in space: Preatmospheric and postatmospheric depth profiles. *J. Geophys. Res.* **78**, 8356–8363.

Bhattacharya S. K., Imamura M., Sinha N., and Bhandari N. (1980) Depth and size dependence of ^{53}Mn activity in chondrites. *Earth Planet. Sci. Lett.* **51**, 45–57.

Cameron I. R. and Top Z. (1975) A final summary of measurements of cosmogenic ^{26}Al in stone meteorites. *Geochim. Cosmochim. Acta* **39**, 1705–1707.

Eberhardt P., Eugster O., Geiss J., and Marti K. (1966) Rare gas measurements in 30 stone meteorites. *Z. Naturforsch.* **21a**, 414–426.

Englert P., and Herr W. (1978) A study on exposure ages of chondrites based on spallogenic ^{53}Mn. *Geochim. Cosmochim. Acta* **42**, 1635–1642.

Englert P. and Herr W. (1979) Exposure- and terrestrial ages of chondrites based on ^{26}Al- and ^{53}Mn-measurements. *Meteoritics* **14**, 391–393.

Englert P. and Herr W. (1980a) On the depth dependent production of longlived spallogenic ^{53}Mn in the St. Severin chondrite. *Earth Planet. Sci. Lett.* **47**, 361–369.

Englert P. and Herr W. (1980b) Cosmogenic ^{53}Mn and ^{26}Al: Depth and size effects on the production rates in St. Severin, Keyes, Kirin and other chondrites. *Meteoritics* **15**, 288.

Finkel R. C., Kohl C. P., Marti K., Martinek B., and Rancitelli L. (1978) The cosmic ray record in the San Juan Capistrano meteorite. *Geochim. Cosmochim. Acta* **42**, 241–250.

Fuse, K. and Anders E. (1969) Aluminium-26 in meteorites-VI. Achondrites. *Geochim. Cosmochim. Acta* **33**, 653–670.

Heimann, M. Parekh, P. P. and Herr W. (1974) A comparative study on ^{26}Al and ^{53}Mn in eighteen chondrites. *Geochim. Cosmochim. Acta* **38**, 217–234.

Herpers U., Herr W., and Wölfle R. (1967) Determination of cosmic-ray produced nuclides ^{53}Mn, ^{45}Sc and ^{26}Al in meteorites by neutron activation and gamma coincidence spectroscopy. In *Radioactive Dating and Methods of Low-Level Counting*, p. 199–205. IAEA, Vienna.

Herr W., Herpers U., and Wölfle R. (1969) Die Bestimmung von ^{53}Mn, welches in meteoritischem Material durch die kosmische Strahlung erzeugt wurde, mit Hilfe der Neutronenaktivierung. *J. Radioanal. Chem.* **2**, 197–203.

Herr W., Herpers U., and Wölfle R. (1972) Study on the cosmic ray produced long lived Mn-53 in Apollo 14 samples. *Proc. Lunar Sci. Conf. 3rd*, 1763–1769.

Herzog G. F. (1975) The production of ^{21}Ne by low-energy protons in meteorites. *J. Geophys. Res.* **80**, 1109–1111.

Herzog G. F. and Cressy P. J. (1974) Variability of the ^{26}Al production rate in ordinary chondrites. *Geochim. Cosmochim. Acta* **38**, 1827–1841.

Imamura M., Nishiizumi K., and Honda M. (1979) Cosmogenic ^{53}Mn in Antarctic meteorites and their exposure history. *Memoirs of the National Institute of Polar Research, Special Issue No. 8*, (T. Nagata, ed.), p. 227–242. Nat.'l Inst. Polar Res., Tokyo.

Kohman T. P. and Bender M. L. (1967) Nuclide production by cosmic rays in meteorites and on the moon. In *High Energy Nuclear Reactions in Astrophysics*, (B.S.P. Shen, ed.), p. 169. W. A. Benjamin, N.Y.

Lal D., Lorin J. C., Pellas P., Rajan R. S., and Tamhane A. S. (1969) On the energy spectrum of iron-group nuclei as deduced from fossil-track studies in meteorite minerals. In *Meteorite Research* (P. M. Millman, ed.), p. 275–285. Reidel, Dodrecht, Holland.

Lal D. and Marti K. (1977) On the flux of low-energy particles in the solar system: The record in St. Severin meteorite. *Nuclear Track Detection* **1**, 127–130.

Millard H. T. Jr. (1965) Thermal neutron activation: Measurement of cross section for manganese-53 *Science* **147**, 503–504.

Müller O., Hampel W., Kirsten T., and Herzog G. F. (1981) Cosmic-ray constancy and cosmogenic production rates in short-lived chondrites. *Geochim. Cosmochim. Acta* **45**, 447–460.

Nishiizumi K. (1978) Cosmic ray produced ^{53}Mn in thirty-one meteorites. *Earth. Planet. Sci. Lett.* **41**, 91–100.

Nishiizumi K. and Arnold J. R. (1981) Cosmogenic nuclides in Antarctic meteorites (abstract). In *Lunar and Planetary Science XII*, p. 771–773. Lunar and Planetary Institute, Houston.

Nishiizumi K., Regnier S., and Marti K. (1980) Cosmic ray exposure ages of chondrites, pre-irradiation and constancy of the cosmic ray flux in the past. *Earth. Planet. Sci. Lett.* **50**, 156–170.

Palme H., Schultz L., Spettel B., Weber H. W., Wänke H., Michel-Levy M. C., and Lorin J. C. (1981) The Acapulco meteorite: Chemistry, mineralogy and irradiation effects. *Geochim. Cosmochim. Acta* **45**, 727–752.

Pellas P. (1973) Irradiation history of grain aggregates in ordinary chondrites. Possible clues to the advanced stages of accretion. In *From Plasma to Planet* (A. Elvius, ed.), p. 65–92. Wiley, N.Y.

Rowe M. W. and Clark R. S. (1971) Estimation of error in the determination of ^{26}Al in stone meteorites by indirect γ-ray spectrometry. *Geochim. Cosmochim. Acta* **35**, 727–730.

Schultz L. and Kruse H. (1978) Light noble gases in stony meteorites—a compilation. *Nuclear Track Detection* **2**, 65–103.

Schultz L., Palme H., Spettel B., Wänke H., and Weber H. (1980) Chemistry and noble gases of the unusual stony meteorite Allan Hills A 77081 (abstract). In *Lunar and Planetary Science XI*, p. 1003–1005. Lunar and Planetary Institute, Houston.

Schultz L. and Signer P. (1974) Helium, Neon und Argon in einigen Steinmeteoriten. In *Analyse extraterrestrischen Materials* (W. Kiesl and H. Malissa, Jr., eds.), p. 27. Springer, Wien.

Schultz L. and Signer P. (1976) Depth dependence of spallogenic helium, neon and argon in the St. Severin chondrite. *Earth. Planet. Sci. Lett.* **30**, 191–199.

Walton J. R., Yaniv A., Heyman D., Edgerlek D., and Rowe M. W. (1974) He and Ne cross-sections in natural Mg, Al, and Si targets and radionuclide cross-sections in natural Si, Ca, Ti, and Fe targets bombarded with 14- to 45-MeV protons. *J. Geophys. Res.* **78**, (1973), 6428, and *J. Geophys. Res.*, **79**, (1974), 314.

Stable NRM and mineralogy in Allende: Chondrules

P. J. Wasilewski[1] and Chhaya Saralker[2]

[1] NASA/Goddard Space Flight Center, Astrochemistry Branch, Greenbelt, Maryland 20771
[2] Computer Science Corporation, Silver Spring, Maryland 20910

Abstract—Allende chondules have NRM vectors which are stable or unstable during alternating field (AF) demagnetization. Distinctive magnetic properties and magnetic mineralogies characterize the stable and unstable chondrules. Magnetically stable chondrules exhibit thermomagnetic curves which are nearly identical to a sulfide separate (including the presence of the 150°C, 320°C, and ~600°C Curie points). Reflected light studies of the stable chondrules reveals only sulfides, as possible magnetic minerals, in a variety of textural configurations. The magnetically unstable chondrules have thermomagnetic curves without the 150°C Curie point, a barely perceptible 320° Curie point in some chondrules and curve shapes different from the stable chondrules. Thermomagnetic curves are similar to chondrule magnetic separates. Reflected light studies of the unstable chondrules identify the presence of the metal + magnetite ± sulfide aggregates found by Haggerty and McMahon (1979) to be present in 80% of the hundreds of chondrules they studied. About 90% of the NRM and 80% of the saturation remanence (SIRM) in the Allende meteorite is thermally demagnetized by 320°C. Therefore, the magnetic sulfides, present in chondrules, the matrix, and possible other inclusions, would appear to be responsible for about 90% of the observed stable NRM in Allende. The NRM in the Allende meteorite is stable during AF demagnetization, and since the unstable chondrule vectors are therefore not detected, they must be oriented randomly, and respond randomly to AF demagnetization. A sulfidation event, was probably responsible for imparting the stable NRM in the Allende meteorite.

INTRODUCTION

Carbonaceous Chondrites (CC) are generally considered to be the most primitive solar system material available for study. Allende a C3 (V) chondrite, i.e., a Vigarano subtype CC, is in a relatively high oxidation state with magnetite, Ni rich alloys and magnetic sulfide species contributing to the ferromagnetic character of the meteorite. There is some evidence that Allende may have been subject to varying degrees of post accretionary thermal recrystallization and equilibration (McSween, 1979), however, MacPherson and Grossman (1981) argue against any significant bulk metamorphism of the parent body. Allende occupies a central position in studies relating to understanding of the early solar nebula. The meteorite contains high temperature Ca Al Ti rich inclusions-condensates, a chondrule suite of lower temperature origin, and a consolidating matrix. All of these major elements contain ferromagnetic minerals which carry a record of magnetic fields present at the time of their origin. Therefore, in principle, it should be possible to reconstruct the relationship between origin of the components, assembly in the meteorite parent body, any alteration or thermochemical events, and the magnetic fields present at the time events took place. The main objective of the work described in this paper is to present the magnetic and mineralogic contrasts between chondrules which have an NRM vector which is ultra stable during alternating field (AF) demagnetization, and those which have unstable NRM vectors when subjected to the same treatment.

A summary of new magnetic results from Allende (Wasilewski, 1981a) coupled with results presented in this paper are used to argue that the stable NRM in Allende was acquired during a sulfidation event.

MAGNETIC PHASES IN ALLENDE

The dominant ferromagnetic components in the Allende meteorite have Curie points ranging from ~590°C to ~610°C (Butler, 1972; Banerjee and Hargraves, 1972; Herndon *et al.*, 1976). A FeNi alloy with about 67% Ni (Clarke *et al.*, 1970) and magnetite with several

% Al_2O_3 and Cr_2O_3 impurities (Haggerty and McMahon, 1979) are ubiquitous and their presence is consistent with the observed Curie points. The Curie point at 320°C is probably associated with pyrrhotite, (Banerjee and Hargraves, 1972), though an alternate interpretation (Butler, 1972) considers a fcc alloy with 36% Ni to be present instead of pyrrhotite. Banerjee and Hargraves (1972) observed a minor Curie point at 75°C. The presence of the 75°C Curie point has been verified (Wasilewski, 1981a) and an additional Curie point at 150°C has been confirmed (Wasilewski, 1981a). The 75°C and 150°C Curie points have not been matched with specific minerals.

In the Allende chondrules (this paper) we have identified the 150°C and 320°C Curie points (Group A-stable chondrules) and the Curie points in the vicinity of 600°C (stable and unstable groups). In the group A chondrules, the increase in I_s after heating is more or less related to the amount of material with the 150°C and 320°C Curie point, i.e., pyrrhotite, and an unknown phase.

EXPERIMENTAL PROCEDURES

Chondrules were separated by gently crushing a bulk piece of meteorite and then carefully sifting through the debris to remove those chondrules which were intact. The chondrules were then cleaned by hand to remove adherent matrix. No magnetic tools or electromagnetic separation technique were used in preparing the chondrule specimens. The REM values (ratio of NMR to SIRM) for the chondrules were less than 0.0086 except for two chondrules with values 0.026 and 0.029. This is an excellent indication that no magnetic contamination had been introduced.

Measurement of the NRM was accomplished using a superconducting rock magnetometer (Superconducting Technology Inc.) and the alternating field demagnetization was done with a GSD-1 demagnetizer (Schonstedt Instrument Co). A PAR vibrating sample magnetometer equipped with a model 151 furnace was used to obtain data on the temperature dependence of saturation magnetization. Chondrule specimens were sealed in a Boron Nitride sample holder and no control of the furnace environment was attempted since we considered it desirable to compare the chondrule data with data presented in Wasilewski (1981a) which were acquired in similar fashion. Relatively fine grained magnetite exhibits a cooling curve showing only a small change in magnetization during heating in the Boron-Nitride sample holder. Sample alteration appears to be controlled mostly by the buffering capacity within the closed sample environment.

EXPERIMENTAL RESULTS

Twenty chondrules were chosen at random for detailed magnetic studies. The NRM of each chondrule was demagnetized in steps up to the maximum field of 1000 Oe. Each chondrule was then given a SIRM in a field of 10KOe., and subsequently demagnetized. The results of these initial experiments provided the basis for separation of the chondrules into two distinct groups. The chondrule specimens are identified and the SIRM and REM values along with the MDF for SIRM are listed in Table 1. The AF demagnetization field required to demagnetize one half of the remanence is defined as MDF. The NRM vector in Group A chondrules was stable during AF demagnetization as shown in the examples (Fig. 1). The NRM demagnetization curve for the Group A chondrules always fell above the SIRM demagnetization curve, the field (MDF) required to destroy one half of the NRM was >1000 Oe, and the field (MDF) required to destroy one half the SIRM varied between 800 and 940 Oe. The NRM vector in Group B chondrules (Fig. 2) was unstable during AF demagnetization. The NRM demagnetization curve fell below, or partially below the SIRM demagnetization curve with significant changes in intensity accompanying excursions in direction. The field required to destroy one half the SIRM in Group B chondrules ranged from 300 to 525 Oe.

Demagnetization characteristics (Figs. 1 and 2) contrasting the two groups of chondrules are striking. In the Group A chondrules (Fig. 1) the directions cluster tightly at all demagnetization fields and for the Group B examples shown in Fig. 2, the directions are scattered in both hemispheres. There is additional evidence for a wide range of behavior. The NRM vectors in many other chondrules, though unstable, exhibit more organized excursions during A.F. demagnetization. Some of the unstable chondrules move progressively along a great circle path while others appear to possess two distinctive

Table 1. Chondrules from Allende divided into stable and unstable groups. Stability refers to response of NRM vector during AF (alternating field) demagnetization.

SPECIMEN	[1]SIRM(10^{-4} em/gm)	REM[1]	[1]SIRM (MDF) Oe
Stable Group A			
1	1391	0.00081	800
A2	132	0.0012	875
5	138	0.0027	940
17	118	0.0075	840
12	79	0.026	450
16	23	0.029	525
Unstable Group B			
2	2366	0.00046	380
7	1318	0.0012	375
14	1068	0.0011	400
8	1021	0.00091	300
11	999	0.00010	330
13	964	0.00093	490
A3	891	0.00186	400
6	485	0.0012	525
4	457	0.0019	490
A1	401	0.0071	350
15	319	0.0017	400
3	169	0.0013	525
10	104	0.0048	250
9	39	0.0086	>1000

[1]SIRM (Saturation isothermal remanence, $H_A = 10$ K Oe).
REM (ratio of NRM to SIRM).
SIRM (MDF)—(field at which one half the SIRM is demagnetized).

vectors with clearly different coercivity spectra. Chondrules 12 and 16 which exhibit reasonably stable vectors, but have low MDF's compared to the ultrastable chondrules have REM values of 0.026 and 0.029, respectively, compared to REM values <0.0086 for all other chondrules.

After measuring the NRM and SIRM AF demagnetization curves and then mounting and completing reflected light studies of the mounted chondrules, they were removed from their plastic mounts and thermomagnetic curves were measured. Reflected light micrographs of stable chondrules 1 and A2 and unstable chondrules 7 and 2 are shown in Fig. 3. Most of the sulfide in stable chondrules 1 and A2 is located near the edge of the chondrules. No metal or obvious magnetite was observed. Examples of the metal + magnetite ± sulfide complexes present in the interior of the unstable chondrules 7 and 2 are shown in Fig. 3(c) and (d).

Thermomagnetic curves for stable (chondrules 1 and A2) and unstable (chondrules 2 and 7) chondrules are shown in Fig. 4. The 150°C and 320°C Curie points are evident in the stable chondrule heating curves, and to a lesser extent there is some indication of the 320°C transitions in the unstable chondrule heating curves. There is no evidence of the 150°C transition in the unstable chondrules heating curves. A large increase in saturation magnetization was observed in stable chondrules during cooling from 600°C but only a moderate increase was noted for the unstable chondrules. The Curie points observed in all cooling curves was ~575°C except for chondrule 7 which also had an anomalous Curie point during the heating cycle. Much of the increase in saturation magnetization observed in the cooling curves was probably due to the production of magnetite. However, this does not explain the shift in the position of the Curie point observed in the cooling curves.

In order to gain some insight into the behavior of both the sulfide component and the metal + magnetite components we studied the properties of three separates. The first was

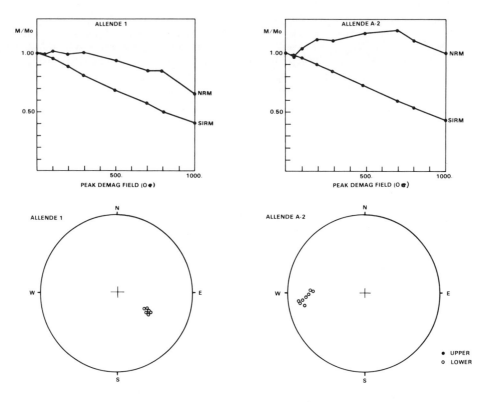

Fig. 1. Results of an alternating field (AF) demagnetization of NRM (natural remanence) and SIRM (saturation isothermal remanence) in two stable chondrules (1) and (A2).

a sulfide separate obtained by removing brassy pieces of sulfide from crushed Allende meteorite. Even after crushing these pieces to a fine powder, none of the material responded to a hand magnet. The strongly magnetic fraction was obtained by removing particles from a crushed piece of meteorite with a hand magnet. The chondrule magnetic separate was obtained by removing strongly magnetic particles with a hand magnet from crushed chondrules which had been separated from the bulk meteorite.

The SIRM demagnetization curves for the sulfide separate (Fig. 5a) is similar to the stable chondrule curves (see Fig. 1). The MDF is ~1000 Oe for this separate, while the MDF's for the stable chondrules are >800 Oe. The thermomagnetic curve for the Allende sulfide separate (Fig. 5b) is similar to the stable chondrule curves (Fig. 4). The presence of the 150°C and 320°C Curie points are evident and the increase in magnetization noted in the cooling curves is also consistent. The SIRM AF demagnetization curve for the strongly magnetic fraction (Fig. 5a) resembles the SIRM demagnetization curve for the unstable chondrules (Fig. 2). The MDF for SIRM in the strongly magnetic fraction is ~300 Oe, comparable to the 300–525 Oe MDF's for unstable chondrules. The thermomagnetic curve (Fig. 5c) for the strongly magnetic fraction is flat out to ~500°C and thereafter drops sharply to the Curie point (~590°C) with an additional tail indicating a Curie point near 615°C. This curve may be due to magnetite + NiFe alloy. The cooling curve is convex upward, more typical of ordinary ferromagnetic materials. This type of behavior has been noted in order-disorder alloys (Wasilewski, 1981b). Similar heating behavior is noted in the chondrule magnetic separate (Fig. 5d) but the cooling curve is different. There is a shift from a 590°C Curie point in the heating curve to 570°C in the cooling curve for both magnetic separates. This observation is also consistent with order-disorder behavior. From these results we suggest that the dominant ferromagnetic phases in the

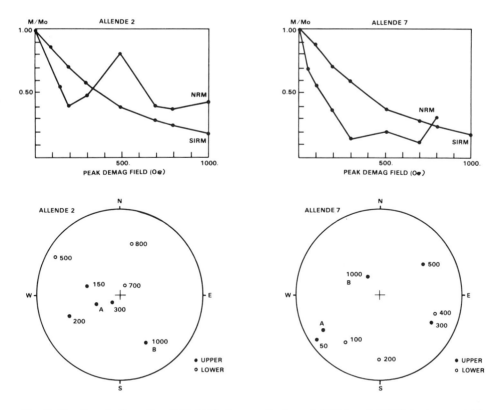

Fig. 2. Results of alternating field (AF) demagnetization of NRM and SIRM in two unstable chondrules (2) and (7).

stable chondrules are magnetic sulfides and in the unstable chondrules, magnetite and NiFe alloy which may be ordered, are the dominant ferromagnetic phases.

Interpretation of thermomagnetic curves for meteoritic sulfides requires additional experimental detail. Experiments are required in both variable heating environments and also in the presence of different magnetic fields. Shown in Fig. 6 are thermomagnetic curves for the Allende sulfide separate and for a Bruderheim troilite separate. In an applied field of 10 K Oe most of the magnetization at ambient temperature is due to the paramagnetic contribution from the sulfide. The cooling curves reflect the reaction Troilite → Magnetite + Pyrrhotite, though some structure in the cooling curves would suggest the presence of other phases or perhaps metastable configurations of both monoclinic and hexagonal pyrrhotite. Therefore the Curie point in the vicinity of 600°C need not be due to prior magnetite or FeNi alloy. The Bruderheim troilite heating and cooling curves are similar whether the sample environment has 3×10^{-4} Torr or 0.21 Torr (as in Fig. 6).

DISCUSSION

Haggerty and McMahon (1979) studied the petrography and mineral chemistry of hundreds of Allende chondrules. They found that about 80% of the chondrules contained opaque metal (NiFe) + magnetite (Mgt) ± sulfide (Tr-troilite, Pn-pentlandite) aggregates. In some chondrules only a single assemblage is present and in others all four are present. The identified assemblages are: Mgt + NiFe; Mgt + NiFe + Tr; Mgt + Tr + Pn + NiFe and Tr + Pn. The results presented in this paper indicate that chondrules which contain a stable NRM do not contain the opaque aggregates, and interestingly only four of the 20

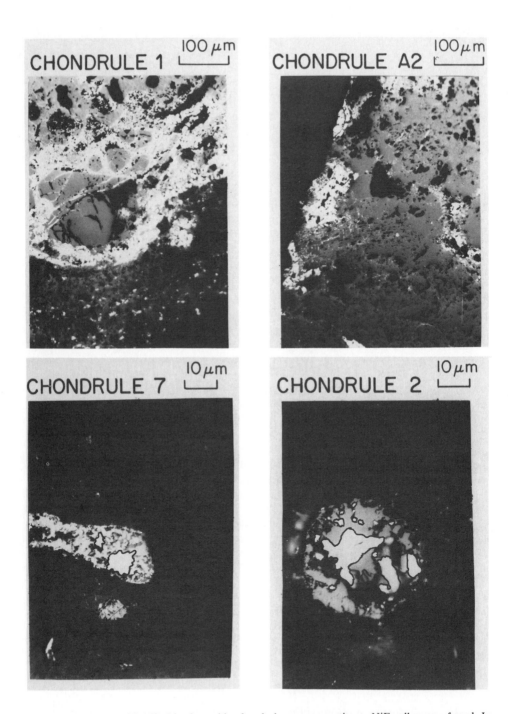

Fig. 3. Sulfides were identified in the stable chondrules, no magnetite or NiFe alloy was found. In chondrule 1 there is evidence of the sulfide cementing texture. Most of the sulfide in both stable chondrules is located near the edge of the chondrules. A metal (white) + magnetite (gray) + sulfide (light) aggregate was found in unstable chondrule (7) and a magnetite + metal aggregate was found in chondrule (2). These aggregates described by Haggerty and McMahon (1979) were found in all unstable chondrules we examined.

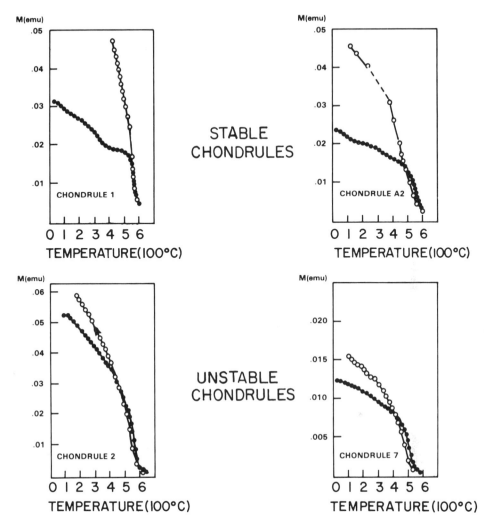

Fig. 4. Thermomagnetic curves for stable and unstable chondrules. (Black circles identify the heating curves and open circles the cooling curves). Both stable chondrules have 150°C and 320°C Curie points. The Curie points observed in both stable chondrule cooling curves are about 575°C, consistent with magnetite. During heating (see chondrule (A2)) the Curie point may be slightly higher, i.e., about 600°C. This feature is not an artifact and is observed in troilite as well (see Fig. 6). The unstable chondrules do not contain the 150°C Curie point and in Chondrule (7) there is some indication of the presence of a very minor phase with a 320°C Curie point. The heating and cooling behavior for stable and unstable chondrules is distinctively different.

studied chondrules fall in the category, i.e., roughly consistent with the estimate of Haggerty and McMahon (1979). The chondrules which contain unstable NRM's contain the metal + mgt + sulfide aggregates. There are chondrules whose NMR vectors describe very specific paths and there are two chondrules (12 and 16—see Table 1) which exhibit moderate stability and yet possess some magnetic characteristics which would place them with the unstable chondrules. Therefore this study must be considered preliminary pending more detailed study of a larger group of chondrules.

Haggerty and McMahon (1979) demonstrate that the magnetites contain moderate amounts of Al_2O_3 and Cr_2O_3 (~2 weight % for each oxide). The temperature variation of H_c and H_R has been measured for the bulk meteorite between ~300 K and 4 K (Wasilewski, 1981a). No evidence of the Verwey transition was recorded in these

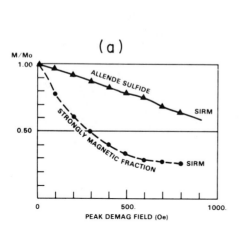
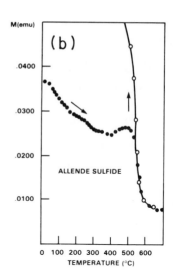
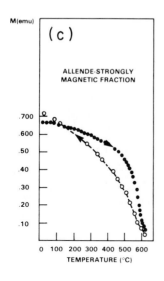
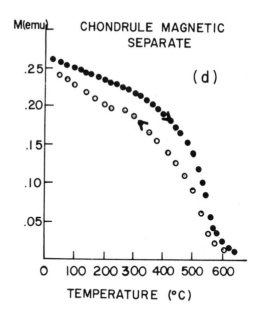

Fig. 5. Sulfide and magnetic separates were studied in order to establish the basic properties of these important magnetic components. SIRM demagnetization curves for the sulfide and strongly magnetic fraction (taken from crushed bulk meteorite) are shown in (a). The fields (MDF) required to demagnetize one half the saturation remanence are ≥ 1000 Oe for the sulfide and ~ 300 Oe for the magnetic fraction. The thermomagnetic curve for the sulfide shown in (b) is nearly identical to the thermomagnetic curve for chondrule (1) shown in Fig. 4. Thermomagnetic curves for the strongly magnetic fraction (c) and the magnetic separate from a group of chondrules (d) are distinctly different from the sulfide curve. Stable chondrule thermomagnetic curves resemble the sulfide curve and thermomagnetic curves for the unstable chondrules resemble the curves obtained from the magnetic separates. The Curie point shift recorded in the cooling curves may be associated with disordering of ordered NiFe alloy. The change in curve shape may also be associated with the disordering process.

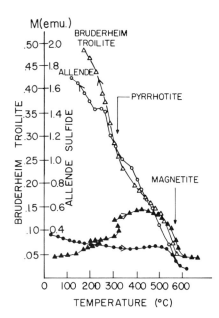

Fig. 6. Thermomagnetic curves for the Allende sulfide (same separate as in Fig. 5) and a Bruderheim troilite separate demonstrate the reaction troilite → pyrrhotite + magnetite, which is also the reason for the increase in magnetization observed in the cooling curves. The step in the vicinity of 250°C in the Allende curve may be associated with hexagonal pyrrhotite.

measurements. The absence of this transition may be a consequence of the Al_2O_3 and Cr_2O_3 impurities or else due to oxidation during cooling of the magnetite in the presence of intimately associated sulfides.

MAGNETIZATION MODEL FOR ALLENDE

The Allende meteorite has been studied in considerable detail. The major ferromagnetic phases are magnetite and NiFe alloy with about 67% Ni (Butler, 1972; Banerjee and Hargraves, 1972; Herndon et al., 1976). All estimated paleofields for the bulk meteorite are about 1 Oe (Butler, 1972; Banerjee and Hargraves, 1972; Sugiura et al., 1979). In all paleofield studies it was assumed that the NRM in Allende is of thermal origin and therefore that the Thellier-Thellier (TT) paleofield test may be correctly applied. The linear NRM-PTRM relationship exists out to ~150°C after which non-linearity develops and the TT test is not longer valid. Work on chondrules, matrix, and inclusions by the Toronto group (see Summary in Sugiura et al., 1979) indicates that the NRM in the matrix is thermally demagnetized by 300°C and a substantial part of the NRM in some chondrules (~80%) is demagnetized by about 350°C. They further demonstrate that the initial NRM directions of oriented chondrules are not random, but that the behavior during demagnetization of the chondrules is random. This indicates that the random chondrule primary vectors may have been overprinted. With the above information it remained unclear as to which phase carried the NRM, and it was only assumed that the onset of non-linear NRM-PTRM behavior marked the blocking temperature of the PTRM. The work, done by the Toronto group, on matrix, chondrules and inclusions, exposed the complex makeup of the NRM record, and indicated that certain components had random vectors that were overprinted.

Therefore a more detailed study of the Allende meteorite and in particular, of the component parts of the meteorite, was viewed as essential. The following (from Wasilewski, 1981a; and this paper) provides a refined understanding of magnetization in Allende (see Figs. 7 and 8).

The existence of magnetic phases with Curie points at 150°C and 320°C is confirmed. (Banerjee and Hargraves (1972) noted a Curie point at 75°C) and in addition a phase with Curie point near 50–75°C is observed in some specimens. During thermal demagnetization ~90% of the NRM and ~80% of the SIRM are gone by 320°C (Fig. 7), indicating that

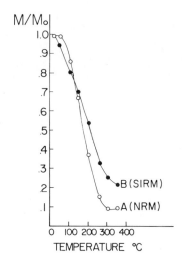

Fig. 7. Stepwise thermal demagnetization of NRM (A) and saturation isothermal remanence (SIRM)—(B) in the Allende meteorite. By 320°C fully 90% of the NRM and 80% of the SIRM are demagnetized.

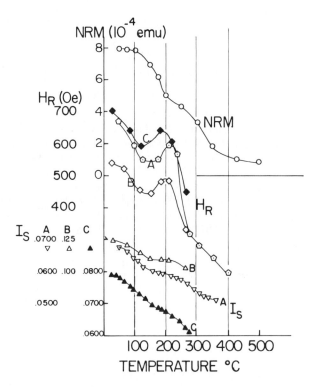

Fig. 8. Summary of magnetic results from the bulk meteorite (from Wasilewski; 1981a). Temperature dependence of H_R (remanent coercive force), I_s (saturation magnetization) and continuous thermal demagnetization of NRM are indicated. $H_R(T)$ and $I_s(T)$ data came from three separate specimens labelled A, B, and C, each specimen having a H_R (T) and I_s (T) curve as indicated. The 150°C and 320°C Curie points are indicated in the I_s (T) curves and the corresponding anamolous H_R (T) data appears related to the presence of the 150°C Curie point. The kinks in the NRM (T) curve at ~200°C coincides with the peak in the $H_R(T)$ curve.

phases with Curie points ≤320°C carry most of the NRM and most of the SIRM. These experiments lead to a conclusion that phases with Curie points ≤320°C which contribute at most 15% to the I_s value are responsible for 90% of the NRM. Though the 50–75°C and 150°C Curie points have not been ascribed to specific minerals, the ~320°C Curie point is consistent with pyrrhotite. Anomalous irreversible changes in H_R (remanent coercive force), suggest that the phase(s) responsible are thermally unstable and magnetically interacting (see Fig. 8).

The magnetically stable chondrules appear to contain only sulfides while the magnetically unstable chondrules contain the metal + oxide ± sulfide aggregates. Since the vector excursions observed in the unstable chondrules are not seen during demagnetization of the meteorite it is reasonable to assume that the unstable chondrules are randomly oriented, and behave randomly during demagnetization. However, the work of Sugiura et al. (1979) indicates that these randomly magnetized chondrules may have been overprinted, probably by the same event that was responsible for imparting the stable NRM to the meteorite.

Chondrules containing the metal + oxide ± sulfide aggregates may have acquired remanence in a magnetic field at blocking temperatures appropriate to the magnetic mineralogy. The field may have been varying or else the chondrules may have moved in a steady field. Since we have shown that chondrules which are stable or unstable have specific mineralogies, and we know that chondrules which show specific NRM vector excursions exist (though the detailed differences in mineralogy are not presently known) it is possible to document the distributed chondrule vector record which may be directly explicable in terms of the magnetic mineralogy.

The Allende chondrite NRM vector is stable during AF demagnetization, as is the case for certain chondrules, which contain sulfides only. Therefore a sulfidation process is probably responsible for the stable NRM record in Allende and must have taken place when the parent body was largely intact. The NRM would appear to have been set in the presence of a undirectional and steady magnetic field. Unstable chondrules may be encased in sulfide and in some chondrules a sulfide cementing texture is in evidence. Much of this may have taken place during the sulfidation event. If this is a correct assessment, then chondrules containing the peripheral sulfides and the cementing textures should exhibit the magnetic overprinting.

REFERENCES

Banerjee S. K. and Hargraves R. B. (1972) Natural remanent magnetizations of carbonaceous chondrites and the magnetic field in the early solar system. *Earth Planet. Sci. Lett.* **17**, 110.

Brecher A. and Arrhenius G. (1974) The paleomagnetic record in carbonaceous chondrites: Natural remance and magnetic properties. *J. Geophys. Res.* **79**, 2081.

Butler R. F. (1972) Natural remanent magnetization and thermomagnetic properties of the Allende meteorite. *Earth Planet. Sci. Lett.* **17**, 120–128.

Clarke R. S., Jarosewich E., Mason B., Nelen J., Gomez M., and Hyde J. R. (1970) The Allende Mexico, meteorite shower. *Smithson. Contrib. Earth. Sci.* **5**, 1–53.

Haggerty S. E. and McMahon B. M. (1979) Magnetite-sulfide-metal complexes in the Allende meteorite. *Proc. Lunar Sci. Conf. 10th*, p. 851–870.

Herndon J. M., Rowe M. W., Larson E. E., and Watson D. E. (1976) Thermomagnetic analysis of meteorites, Part III: C3 chondrites. *Earth Planet. Sci. Lett.* **29**, 283–290.

MacPherson G. J. and Grossman L. (1981) Clastic rims on inclusions: Clues to the accretion of the Allende parent body (abstract). In *Lunar and Planetary Science XII*, p. 646–647. Lunar and Planetary Institute, Houston.

McSween H. Y. Jr., (1979) Are carbonaceous chondrites primitive or processed? *Rev. Geophys. Space Phys.* **17**, 1059–1077.

Sugiura N., Lanoix M., and Strangway D. W. (1979) Magnetic fields of the solar nebular as recorded in chondrules from the Allende meteorite. *Phys. Earth Planet. Inter.* **20**, 342–349.

Wasilewski P. J. (1981a) New magnetic results from Allende C3(V). *Phys. Earth Planet. Inter.* In press.

Wasilewski P. J. (1981b) Magnetic properties of ordering alloys in meteorites (abstract) In *Lunar and Planetary Science XII*, p. 1159–1160. Lunar and Planetary Institute, Houston.

The composition of natural remanent magnetization of an Antarctic chondrite, ALHA 76009 (L_6)

Takesi Nagata and Minoru Funaki

National Institute of Polar Research, Tokyo 173, Japan

Abstract—The natural remanent magnetization (NRM) of 13 individual chondrules, 12 individual large grains of metal and 9 matrix parts of ALHA 76009 L_6 chondrite is separately examined. NRM of the chondrules is stable for the AF-demagnetization and the thermal demagnetization, but its direction is widely scattered within the bulk chondrite. NRM of large grains of metal is unstable and its direction is widely scattered. NRM of the matrix is stable for the AF-demagnetization below 150 Oe. peak, but it becomes unstable when the demagnetization field exceeds 150 Oe. peak. The NRM direction of matrix parts is not exactly uniform but is clustered within a hemisphere.

It appears that NRM of the chondrules was acquired as TRM in the presence of a magnetic field, but a plausible mechanism for keeping stable NRM of random orientations within the well metamorphased chondrite is not yet clarified.

1. INTRODUCTION

The natural remanent magnetization (NRM) of individual chondrules of the Allende C_3 chondrite was experimentally studied by Lanoix et al. (1978a, b), Sugiura et al. (1979) and Wasilewski and Saralkar (1981). Lanoix et al. and Sugiura et al. have shown that NRM's of individual chondrules are reasonably stable against the AF-demagnetization test and the NRM directions after the AF-demagnetization or the thermal demagnetization are widely scattered, suggesting that these chondrules acquired the stable thermoremanent magnetization (TRM) before their assembling of random orientation into the carbonaceous chondrite. The paleointensity of these chondrules, estimated on the basis of TRM-origin assumption, is reported to be 2–7 Oe. and there is an indication that these chondrules were slightly reheated, resulting in a secondary partial TRM after their assemblage (Sugiura et al., 1979). However, Wasilewski and Saralkar (1981) have shown that the Allende chondrite contains a group of chondrules which have stable NRM acquired by iron sulfide and the other group of chondrules having unstable NRM probably acquired by FeNi metal and/or magnetite. It therefore seems necessary that NRM characteristics of chondrules from other chondrites be experimentally examined and compared with those of the bulk and the matrix parts of the same chondrites.

ALHA 76009 L_6 chondrite comprises 33 fragments, the total weight of which amounts to about 407 kg (Olsen et al., 1978, Yanai, 1978, 1979). Since a considerably large piece of this chondrite is available for studying its composition of NRM in fair detail, NRM characteristics of the bulk chondrite, individual chondrules, matrix parts and metallic grains are separately examined. From a large piece of ALHA 76009 L_6 chondrite (7 kg weight), 3 pieces of bulk chondrite, 13 pieces of chondrule, 9 pieces of matrix and 12 grains of metal larger than about $\frac{1}{2}$ mm^3 in volume have been picked up, where the orientation of eached picked-up specimen relative to the mother chondrite piece is approximately determined (less than 10 decrees in angular deviation). The matrix part picked-up in the present study is so defined that it is an approximately homogeneous part between chondrules and metallic grains which are larger than $\frac{1}{2}$ mm^3 in volume. NRM's of individual specimens of the four groups have been examined with respect to their stability against the AF-demagnetization. The basic magnetic parameters such as the saturation magnetization (I_s), the saturated isothermal remanent magnetization (I_R), the coercive force (H_C) and the remanence coercive force (H_{RC}) at room temperature have also been determined for typical specimens selected from each of the four groups. As already reported (Nagata, 1979a, b), the bulk NRM of ALHA 76009 chondrite is extremely unstable against the AF-demagnetization test, whereas that of the Allende carbonaceous chondrite is fully stable. Hence, the paleointensity has been determinable for the Allende chondrite but is not determinable for the bulk of ALHA 76009 chondrite. In this regard, we must be concerned with what parts are magnetically stable and what other parts are unstable in the ALHA 76009 chondrite. On the other hand, the thermomagnetic analysis has shown that the ferromagnetic constituent in the bulk of ALHA 76009 consists of kamacite phase of 6.5 wt.% and plessite phase of 35 wt.%, on average, which indicates that this chondrite is a typical L chondrite (Nagata, 1979a).

2. OUTLINE OF PETROLOGY

In the results of a preliminary study by Olsen et al. (1978), Fa of olivine, Fs of orthopyroxene and An of plagioclase in ALHA 76009 are given by 24.1, 20.7, and 10.8 respectively. Together with observed data of metamorphic type, this chondrite is identified as L_6 chondrite (Olsen et al., 1978; Yanai, 1979). Since no other petrological data have been reported on this chondrite, preliminary results of petrological studies of a thin section of ALHA 76009, made in the National Institute of Polar Research, will be briefly summarized in the following: A small number of chondrules of 2–3 mm in diameter are still clearly observable. The chondrules consist mostly of olivine and orthopyroxene and a small amount of plogioclase and their boundaries are blurred. The principal minerals in matrix are olivine of Fa 23–25, a less amount of orthopyroxene of Fs 18–20, and a small amount of plagioclase which shows evidence of recrystallization. The presence of clinopyroxene is not clear. Fe-Ni metals of about 6% Ni less than 10% in content and approximately the same amount of troilite are observable, while the content of chromite is less than 1%. An interesting characteristic feature of the matrix is that some parts consist of large grains of olivine (1.5 mm max.) and orthopyroxene (1 mm max.), while the other parts comprise much smaller grains of olivine and orthopyroxene.

Possible weathering effects on the Rb-Sr system for this chondrite have been examined by leaching experiments by Ito et al. (1980). Results of measurements of leachibility of K, Rb and Sr with 1N-HCl in weathered and fresh specimens of this chondrite suggest that the weathering effect in the Antarctic ice was limited to the surface part less than 2 cm in thickness in regard to the Rb-Sr system. The model age of the fresh interior specimen calculated from Allende initial $^{87}SR/^{86}Sr$ ratio of 0.6988 is 4.62 ± 0.11 b.y. Ito et al. (1980) concluded, therefore, that the Rb-Sr system in this chondrite appears to be normal; namely, the effect of leaching in the Antarctic ice or of loss of Rb by late impacts in the preterrestrial history may not be so significant for this chondrite.

3. BASIC MAGNETIC PROPERTIES

The basic magnetic parameters at room temperature such as the saturation magnetization (I_s), the saturated isothermal remanent magnetization (I_R), the coercive force (H_C) and the remanence coercive force (H_{RC}) of 2 bulk specimens, 6 chondrules, 3 matrix parts and 5 metallic grains are summarized together with the intensities of their NRM (I_n) in Table 1.

For the chondrules, whose mean diameters range from 2.1 to 2.6 mm corresponding to their weight range from 0.017 to 0.032 gm, their I_s, H_C and H_{RC} values range from 1.0 to 3.7 emu/gm, from 45 to 176 Oe., and from 940 to 2650 Oe., respectively. These observed values indicate that the chondrules contain FeNi fine grains of 0.5–1.85 wt.% and these metallic grains are sufficiently small to possess comparatively large values of H_C and H_{RC}. On the contrary, the metallic grains whose mean diameters range from 0.8 to 1.7 mm corresponding to their weight range from 0.0022 to 0.022 gm are magnetically characterized by $I_s = 81–171$ emu/gm, $H_C = 0.6–13.5$ Oe. and $H_{RC} = 63–147$ Oe. Although these metallic grains mostly consist of kamacite, their observed I_s-values are considerably

Table 1. Basic magnetic properties of the bulk chondrite, chondrules, matrix parts and metallic grains of ALHA 76009 (L_6) chondrite.

	Sample	weight (gm)	I_s (emu/gm)	I_R (emu/gm)	H_C (Oe)	H_{RC} (Oe)	I_n ($\times 10^{-4}$ emu/gm)	$\dfrac{I_n}{I_R}$
Bulk	A-0	3.13	8.35	0.52	160	2100	19.9	0.0038
	A-1	2.43	9.73	0.34	92	2700	22.1	0.0065
Chondrite	D-1	0.031	3.67	0.059	115	2210	0.93	0.0016
	D-2	0.019	3.50	0.091	95	1770	0.84	0.0009
	D-3	0.017	1.30	0.016	101	1270	0.75	0.0047
	E-1	0.025	1.51	0.068	176	2650	2.68	0.0039
	E-2	0.022	2.44	0.028	48	980	0.32	0.0011
	F-2	0.032	0.99	0.012	45	940	0.57	0.0048
Matrix	C-1	0.018	5.68	0.27	132	1810	6.7	0.0025
	C-2	0.020	9.73	0.80	114	2610	35.0	0.0044
	C-3	0.015	12.75	0.35	45	1140	75.5	0.0216
Metal	B-1	0.0042	157	0.38	7.0	63	24.8	0.0065
	B-2	0.0022	171	0.28	5.0	63	26.9	0.0096
	B-3	0.0083	127	0.53	13.5	147	71.6	0.0135
	B-4	0.0114	81	0.18	0.6	135	15.2	0.0084
	B-5	0.0220	150	0.28	6.0	97	14.2	0.0051

smaller than the I_s-value of pure kamacite, 200–210 emu/gm, because these grains could not be perfectly separated from the surrounding silicate matrix. Their small H_C values may indicate that the majority of the metal is in the form of multidomain kamacite, whereas their moderate values of H_{RC} suggest that some parts of the metal are of the plessite structure. These two points have been confirmed by their thermomagnetic analysis.

The matrix parts are magnetically characterized by $I_s = 5.7$–12.8 emu/gm, $H_C = 45$–132 Oe. and $H_{RC} = 1140$–2610 Oe. The orders of magnitude of H_C and H_{RC} of the matrix are approximately the same as those of the chondrules, suggesting that the majority of metals in the matrix also are very fine grained as in the chondrules. The content of metal in the matrix, however, is estimated from its I_s-value to amount to 2.8–6 wt.%. In general, the average values of I_s, I_R, H_C and H_{RC} of the matrix parts are approximately identical to those of the bulk chondrite.

4. NATURAL REMANENT MAGNETIZATION

The histograms of NRM intensity of chondrules, matrix parts and metallic grains are separately shown in Fig. 1. Roughly speaking, the intensity of NRM (I_n) of these chondrules, matrix parts and metallic grains is approximately proportional to the saturated isothermal remanent magnetization (I_R), as indicated in Table 1.

There is a considerable difference between this L_6 chondrite and the Allende C_3 chondrite in regard to the NRM intensity of the chondrules relative to that of bulk chondrite. The I_n-values of chondrules in ALHA 76009 chondrite are definitely smaller than I_n-value of the bulk chondrite ($I_n = 21 \times 10^{-4}$ emu/gm), whereas I_n-values of chondrules in the Allende chondrite are distributed around the I_n-value of the bulk chondrite, the average value of chondrules' I_n being approximately identical to the bulk chondrite I_n (Sugiura et al., 1979). However, the absolute magnitude of chondrule I_n in ALHA 76009 ($I_n = (0.6$–$23) \times 10^{-4}$ emu/gm) is approximately in the same range as that of chondrule I_n in the Allende ($I_n = (1.6$–$150) \times 10^{-4}$ emu/gm).

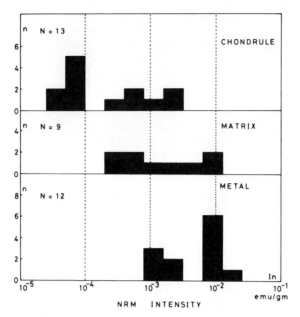

Fig. 1. Histogram of NRM intensity of chondrules, matrix parts and metallic grains in ALHA 76009 L$_6$ chondrite.

(4–1). NRM of bulk chondrite

The AF-demagnetization characteristics of 3 bulk specimens are shown in Figs. 2 and 3 for the intensity and direction, respectively. The AF-demagnetization characteristics of the total TRM acquired in a magnetic field of 0.50 Oe. for these 3 bulk specimens are also illustrated for comparison in these two figures. Although the total TRM acquired by

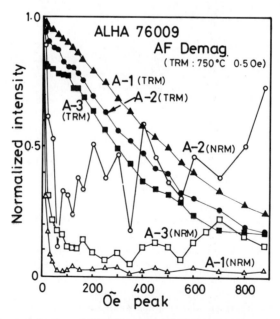

Fig. 2. AF-demagnetization of NRM intensity (hollow symbols) and total TRM intensity (full symbols) of 3 bulk specimens of ALHA 76009 chondrite.

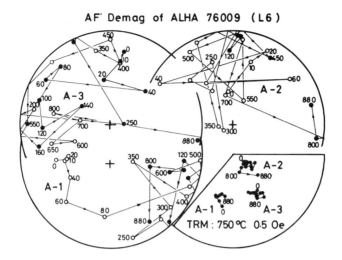

Fig. 3. AF-demagnetization of NRM direction and total TRM (Bottom-right) of 3 bulk specimens of ALHA 76009 chondrite. Full circle: lower hemisphere projection. Hollow circle: upper hemisphere projection.

heating in a vacuum of 1×10^{-5} Torr. in a 0.50 Oe. magnetic field is reasonably stable for the three bulk specimens, their NRM's are extremely unstable with respect to both intensity and direction during the AF-demagnetization test.

(4-2). NRM of chondrules

Six examples of the AF-demagnetization characteristics of chondrule NRM's are shown in Figs. 4 and 5. Although one chondrule specimen exhibits an NRM which is not stable during AF-demagnetization (the residual NRM direction changing almost over a spherical hemisphere in the course of AF-demagnetization up to 880 Oe. peak), the NRM's of the other five chondrule specimens are reasonably stable against the AF-demagnetization up to 800 Oe. peak. In Fig. 6, the distribution of the original NRM directions of individual chondrules from two separate bulk chondrite pieces is shown. In both cases, the directions of NRM of individual chondrules are widely scattered almost over the whole spherical surface within respective chondrite pieces. It may be concluded, therefore, that NRM of individual chondrules in ALHA 76009 chondrite is reasonably stable in intensity and direction, but the distribution of their NRM direction is almost at random within a bulk piece of this chondrite.

(4-3). NRM of matrix

The AF-demagnetization characteristics of 3 specimens of matrix parts are shown in Figs. 7 and 8. In these results, the intensity and direction of NRM of the C-1 matrix specimen is reasonably stable against the AF-demagnetization up to 880 Oe. peak, but the NRM of the other two specimens of matrix is stable only for the AF-demagnetization up to 140–160 Oe. peak, beyond which the direction of residual NRM becomes widely changed. It would be concluded therefore that NRM of the matrix parts is reasonably stable during AF-demagnetization up to about 150 Oe. peak, but it generally becomes unstable for the higher alternating fields.

Figure 9 shows the distribution of original directions of NRM of 9 specimens of matrix in one piece of bulk chondrite. The NRM directions of individual matrix parts are clustered within a hemisphere, but they are fairly dispersed as indicated by the radius of

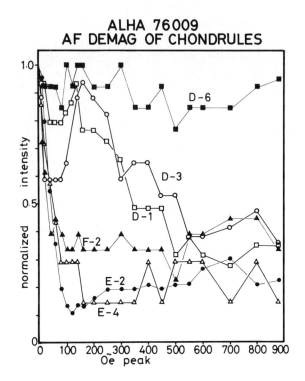

Fig. 4. AF-demagnetization of NRM intensity of 6 specimens of ALHA 76009 chondrule.

the circle of 95% confidence, $\alpha_{95} = 27°$. These observed results may suggest that NRM of the matrix is caused by a magnetic orientation mechanism in the presence of a magnetic field which is relatively weak compared with external disturbances which might be the case for formation as depositional remanent magnetization (DRM).

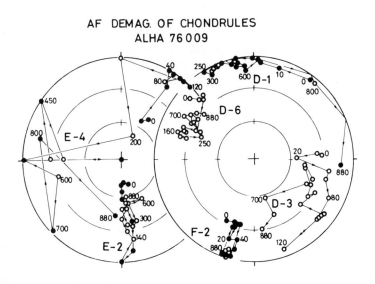

Fig. 5. AF-demagnetization of NRM direction of 6 specimens of ALHA 76009 chondrule. Full circle: lower hemisphere projection. Hollow circle: upper hemisphere projection.

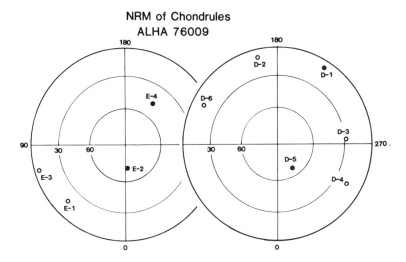

Fig. 6. Initial direction of chondrule NRM. Right: chondrules in a chondrite block D. Left: chondrules in a chondrite block E. Full circle: lower hemisphere projection. Hollow circle: upper hemisphere projection.

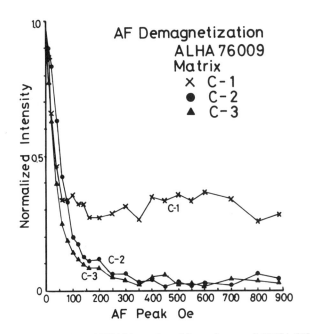

Fig. 7. AF-demagnetization of NRM intensity of 3 specimens of ALHA 76009 matrix.

(4-4). NRM of metallic grains

Figures 10 and 11 show the AF-demagnetization characteristics of the NRM for three metallic grains. The NRM of metallic grains is generally unstable against the AF-demagnetization. In particular, the direction of their NRM markedly changes even in AF-demagnetization fields less than 100 Oe. in peak intensity. As shown in Fig. 12, the direction of original NRM of metallic grains is widely scattered, being almost at random,

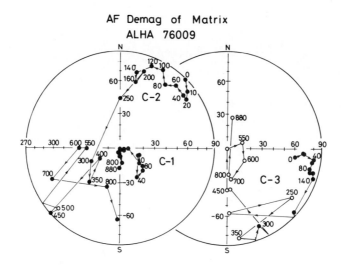

Fig. 8. AF-demagnetization of NRM direction of 3 specimens of ALHA 76009 matrix.

in a bulk specimen of the chondrite. It can thus be concluded that NRM's of metallic grains larger than $\frac{1}{2}$ mm^3 in volume are magnetically unstable and their directions are at random in this chondrite.

(4-5). Summary and comparison of NRM's of chondrule, matrix and metallic grains

Summarizing the NRM characteristics of individual chondrules, matrix parts and metallic grains in ALHA 76009 chondrite, it could be concluded that NRM's of chondrules are stable, but their directions are distributed at random, that those of matrix parts are partially stable only for weak AF-demagnetizing fields up to 150 Oe. peak, but their directions are clustered within a limited wide domain, represented by $\alpha_{95} = 27°$, and that those of metallic grains are unstable and their directions are oriented at random.

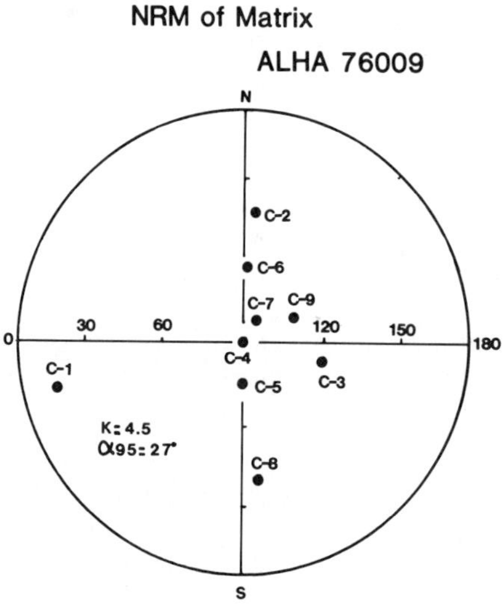

Fig. 9. Initial direction of matrix NRM.

Since the magnitudes of I_s, I_R, H_C and H_{RC} are measured for selected samples of chondrules, matrix parts and metallic grains as well as the bulk chondrite as shown in Table 1, the relative abundances of the three components of the chondrite can be approximately estimated from the observed magnetic parameters. The result of such an estimation has indicated that the relative abundances of chondrules, matrix parts and large metallic grains are 5–7%, more than 90% and less than 1% by weight, respectively. It seems likely, therefore, that NRM of the bulk chondrite can also be represented mostly by NRM of the matrix parts. Actually, the mean value of NRM intensity (I_n) of 9 matrix specimens, shown in Fig. 9, is 24×10^{-4} emu/gm, which is in approximate agreement with the average intensity of bulk NRM, 21×10^{-4} emu/gm.

Composition of natural remanent magnetization 1237

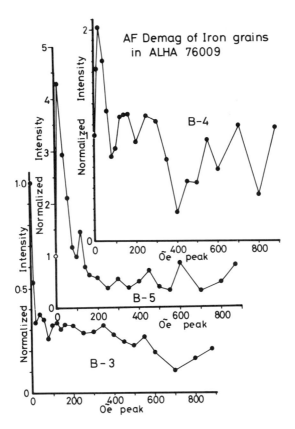

Fig. 10. AF-demagnetization of NRM intensity of 3 specimens of ALHA 76009 metallic grain.

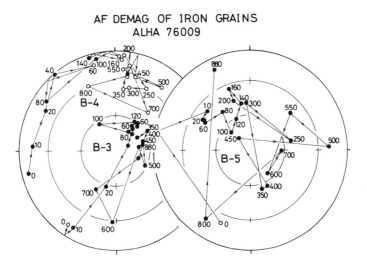

Fig. 11. AF-demagnetization of NRM direction of 3 specimens of ALHA 76009 metallic grain.

5. POSSIBLE ORIGIN OF NRM OF CHONDRULES

It seems that NRM of individual chondrules in ALHA 76009 chondrite is stable against the AF-demagnetization up to 800 Oe. peak in most cases so that the chondrule NRM's could be attributed to either TRM or chemical remanent magnetization (CRM). As the ferromagnetic constituents in the chondrules are mostly fine grains of FeNi metal, the chondrule NRM is most likely to be produced by cooling in the presence of a magnetic field either through the TRM mechanism which is based on the blocking of single domain ferromagnetic grains (kamacite grains in the present case) by their cooling or the CRM mechanism which is based on the blocking of differentiated ferromagnetic phase domains (α-phase in the present case) in the course of cooling of plessite grains.

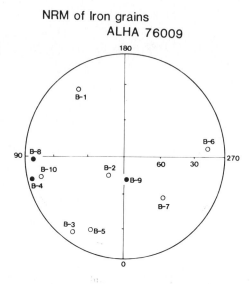

Fig. 12. Initial direction of metallic grain NRM.

Figure 13 illustrates 4 examples of changes in the residual NRM direction in the course of thermal demagnetization of the chondrule NRM. During the course of thermal demagnetization of NRM, the direction of residual NRM is confined within a reasonably small range; furthermore, the direction of residual NRM has a general tendency to make a precessional rotation either clockwise or counterclockwise. In order to emphasize the characteristic of precessional rotation, the direction of partial-thermally demagnetized component of a chondrule NRM is plotted in terms of the temperature interval of the partial-thermal demagnetization as shown in Fig. 14. It will be reasonable to presume on the basis of these thermal demagnetization data that the chondrule NRM was acquired during the process of cooling of the chondrules in the presence of a certain magnetic field. The present experimental results further suggest that individual chondrules might have made a slow precessional rotation relative to the magnetic field during their cooling process.

The Königisberger-Thellier method (KTT method) to evaluate the paleointensity is not fully applicable with L-chondrites which contain both α-phase (kamacite) and $(\alpha + \gamma)$-phase (plessite) of FeNi metallic grains, because $(\alpha + \gamma)$-phase is transformed to γ-phase only by heating above a critical temperature and a reformation of $(\alpha + \gamma)$-phase by a slow cooling of the transformed γ-phase is practically impossible within the laboratory experiment time scale. It seems still possible, however, to evaluate the paleointensity of the kamacite phase grains, if the experimental temperature range in KTT method does not exceed the $(\alpha + \gamma) \rightarrow \gamma$ transformation temperature (400–570°C in the present case). Figure 15 illustrates two examples of the NRM-remained vs. TRM-gained diagram of KTT experiment of ALHA 76009 chondrule. As shown in the figure, there is still a zig-zag relationship between NRM-remained and TRM-gained even for a low temperature range. As discussed by Pearce et al. (1976) about NRM of lunar rocks, the observed zig-zag relationship may be due to the magnetic interaction between kamacites and plessites. Ignoring the plot of temperature = 200°C in Fig. 15, the paleointensity of chondrules can be approximately evaluated as 0.9 and 0.5 Oe. for E-1 and D-4 chondrules, respectively. These paleointensity values, however, cannot be considered sufficiently reliable because of an uncertainty of the magnetic interaction mechanism.

6. CONCLUDING REMARKS

As described in the introduction, NRM of the bulk specimen of Allende C_3 chondrite is very stable and its paleointensity is estimated to be about 1 Oe. by four independent

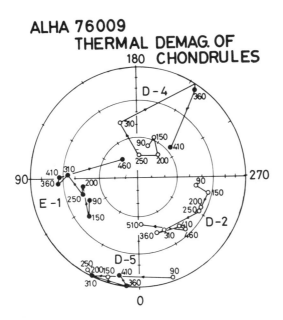

Fig. 13. Thermal demagnetization of NRM direction of 4 specimens of ALHA 76009 chondrule.

research groups, whereas NRM of the bulk specimen of ALHA 76009 L₆ chondrite is very unstable mostly owing to the presence of metallic grains of large size and smaller multi-domain metallic grains in the matrix. Nevertheless, NRM of chondrules themselves in ALHA 76009 is highly stable. In both chondrites, the direction of the stable NRM's of individual chondrules is widely scattered. This observed fact may suggest that NRM of individual chondrules was independently acquired, probably through the TRM mechanism.

In the case of the Allende C_3 chondrite, we could accept a hypothesis that individual chondrules acquired TRM before their accretion together with matrix materials including metals to form the bulk chondrite, and the TRM's thus acquired by individual chondrules

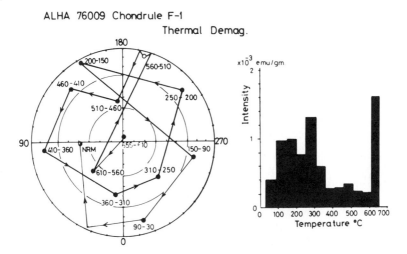

Fig. 14. Partial thermal demagnetization of an ALHA 76009 chondrule. Left: direction of partially thermal-demagnetized magnetization. Right: magnitude of partially thermal-demagnetized magnetization.

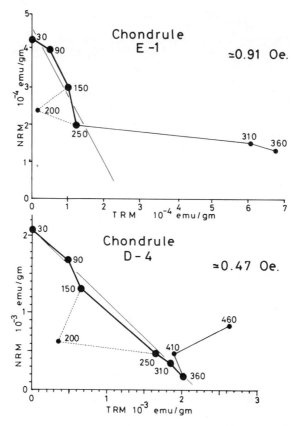

Fig. 15. NRM-remained vs. pTRM-gained diagrams of Königisberger-Thellier experiment for ALHA 76009 chondrules. Magnetic field for pTRM acquisition = 0.50 Oe.

have mostly maintained their mutually different directions through a later period of a comparatively light thermal metamorphism. In the case of the ALHA 76009 L_6 chondrite, however, it is hardly possible to assume that the original characteristics of TRM of individual chondrules have been safely kept in the course of a later heavy thermal metamorphism which resulted in metamorphic type 6 of this chondrite. Actually, the number of detectable chondrules is rare in this chondrite, and the reasonably well defined chondrules examined in the present work were picked out from a large volume of this chondrite, whereas its major parts look the typical matrix of metamorphic type 6 chondrite.

A possible interpretation for the formation of this chondrite may be a hypothesis of a re-accretion of meteoritic materials including unmetamorphased chondrules and severally metamorphased chondrules. In such a hypothesis, it is assumed that parent bodies of chondrites repeated their formation and collapse into fragments during a certain period and that meteoritic materials coming at least from a well metamorphased parent body accreted to those from a none or weakly metamorphased parent body. If a weak magnetic field was present during the final accretion process for this chondrite, fine metallic grains of single domain size possessing their spontaneous magnetization in matrix may have produced DRM, whereas NRM's of chondrules and large metallic grains may not be intense enough to produce DRM. This is, however, a possible hypothesis, which may need more observed evidence to be ascertained.

In concluding, the writers thank K. Yanai of the National Institute of Polar Research for his preliminary information of the petrological characteristics of ALHA 76009 chondrite.

REFERENCES

Ito A., Nakamura N., and Masuda A. (1980) Examination of effects of shock and weathering for the Antarctic L_6 chondrites, Yamato-74190 and Allan Hills 76009 by the Rb-Sr method. *Memoirs of the National Institute of Polar Research, Special Issue No. 17*, p. 168–176. Nat'l. Inst. Polar Res., Tokyo.

Lanoix M., Strangway D. W., and Pearce G. W. (1978a) Paleointensity determinations from Allende chondrules (abstract). In *Lunar and Planetary Science IX*, p. 630–632. Lunar and Planetary Institute, Houston.

Lanoix M., Strangway D. W., and Pearce G. W. (1978b) The primordial magnetic field preserved in chondrules of the Allende meteorite. *Geophys. Res. Lett.* **5**, 73–76.

Nagata T. (1979a) Meteorite magnetism and the early solar system magnetic field. *Phys. Earth Planet. Inter.* **20**, 324–341.

Nagata T. (1979b) Natural remanent magnetization of Antarctic meteorites. *Memoirs of the National Institute of Polar Research, Special Issue No. 12*, p. 238–249. Nat'l Inst. Polar Res., Tokyo.

Olsen E. J., Noonon A., Fredrickson K., Jarosewich E., and Moreland G. (1978) Eleven new meteorites from Antarctica. *Meteoritics* **13**, 209–225.

Pearce G. W., Hoye G. S., Strangway D. W., Walker B. M., and Taylor L. A. (1976) Some complexities in the determination of lunar paleointensities. *Proc. Lunar Sci. Conf. 7th*, p. 3271–3297.

Sugiura N., Lanoix M., and Strangway D. W. (1979) Magnetic fields of the solar nebula as recorded in chondrules from the Allende meteorite. *Phys. Earth Planet. Inter.* **20**, 342–349.

Wasilewski P. J. and Saralkar C. (1981) Stable NRM and mineralogy in Allende: Chondrules (abstract). In *Lunar and Planetary Science XII*, p. 1161–1163. Lunar and Planetary Institute, Houston.

Yanai K. (1978) First meteorites found in Victoria Land, Antarctica, December 1976 and January 1977. *Memoirs of the National Institute of Polar Research, Special Issue No. 8* (T. Nagata, ed.), p. 51–69. Nat'l Inst. Polar Res., Tokyo.

Yanai K. (1979) *Catalog of Yamato Meteorites*. Nat'l. Inst. Polar Res., Tokyo. 188 pp.

The magnetic properties of the Abee meteorite: Evidence for a strong magnetic field in the early solar system

N. Sugiura and D. W. Strangway

Department of Geology, University of Toronto, Toronto, Ontario, Canada

Abstract—The main carrier of the NRM (natural remanent magnetization) in the Abee meteorite is Fe–Ni alloy with intergrowths of cohenite (Fe_3C). Cohenite has a Curie temperature of 215°C and carries the more stable component of the NRM. This is the first report of NRM carried by cohenite in natural materials. The directions of magnetization in mutually oriented samples (two clasts and a matrix sample) are significantly different, suggesting (1) the meteorite was not reheated to temperatures much above 100°C during or after accretion (2) the clast magnetization is probably of high temperature origin and predates the formation of the meteorite (3) the NRM of the matrix was acquired at low temperature. The strength of the magnetic field before the accretion of Abee is inferred to be six oersteds using the Thellier method.

INTRODUCTION

The meteorite Abee is a type 4 enstatite chondrite with many cm-size clasts. Herndon and Rudee (1978) studied the texture of the kamacite-cohenite intergrowths and suggested that it cooled rapidly from high temperature (700°C) and was not reheated to a high temperature thereafter, although it is not possible to determine from the metallographic evidence when the rapid cooling occurred. Thus this meteorite provides us an opportunity to use the paleomagnetic conglomerate test, which consists of measurements of NRM direction of clasts and matrix (Graham, 1949), to determine if there was a magnetic field before and during the meteorite formation. Using the conglomerate test we can draw some inferences about the thermal history of the meteorite. If the NRM of all the clasts and the matrix is in the same direction, it suggests that the meteorite was reheated to >750°C (the Curie point of Kamacite) during its agglomeration or afterwards. (2) If the NRM directions of the clasts are random with respect to the matrix, it suggests that the meteorite was not reheated above the blocking temperature of the phases in the clast. (3) If the NRM directions of the clasts are scattered but are not random and have a component magnetized in the same direction as the matrix, and if the dispersion of the NRM directions increases with thermal demagnetization, it suggests that the meteorite was reheated to a moderate temperature.

This kind of conglomerate test was applied to Allende chondrules (Sugiura et al. 1979). Because of the small size of the chondrules and weak NRM intensity, the results were not conclusive, but it was inferred that the chondrules had a random magnetization with a component due to mild reheating after accretion. Since the clasts in the Abee are much larger than chondrules in Allende and since the NRM intensity is very strong, more definitive results have been obtained from the Abee meteorite.

The presence of a magnetic field is a necessary condition for the conglomerate test. Although most meteorites have relatively strong remanence, the presence of a ubiquitous magnetic field in the early solar systems has not yet been confirmed. Previous studies on the magnetic properties of meteorites have shown that (1) carbonaceous chondrites and some achondrites have a stable NRM which is probably of extra-terrestrial origin (Brecher and Arrhenius, 1974; Nagata, 1979; Brecher and Fuhrman, 1979a). (2) Ordinary chondrites have a very unstable NRM whose origin is not clear (Stacey et al., 1961;

Brecher and Ranganayaki, 1975; Sugiura, 1977; Brecher et al., 1977; Brecher and Fuhrman, 1979b). Even if a meteorite has a stable NRM, it does not necessarily mean the presence of a magnetic field in a planetary object or in the early solar system, because it could be due to exposure to an artificial magnetic field. If the NRM directions of clasts and matrix are random, however, it is strong positive evidence for the presence of a magnetic field during and/or before the accretion of the meteorite.

Sample description and experiments

Mutually oriented pieces of the Abee meteorite were studied (sample number 2,4,0; 2,5,0; 1,9,32; 1,9,34; 1,1,17; 4,1,1 of Abee consortium; coordinator Kurt Marti). Sample 2,4,0 and 2,5,0 were taken from near the fusion crust and it seems that their NRMs were disturbed by the heating during entry to the terrestrial atmosphere. Other pieces (Fig. 1) were taken more than 1 cm away from the fusion crust. Sample 1,1,17 was taken from the center of a large clast. Sample 1,9,34 were taken from just outside of clast and entirely made of matrix. Sample 4,1,1 contained both clast and matrix. These pieces were cut into smaller subsamples. The boundary between the clasts and the matrix is clear because the rim of the clasts is enriched in metallic grains.

The NRM was measured with a superconducting magnetometer. Heating experiments were done either in vacuum (thermal demagnetization) or in evacuated quartz tubes (measurements of magnetization at high temperature). A Schonstedt demagnetizer was used for alternating field (AF) demagnetization. Hysteresis parameters and magnetizations at high temperature were measured with a vibrating sample magnetometer (PAR).

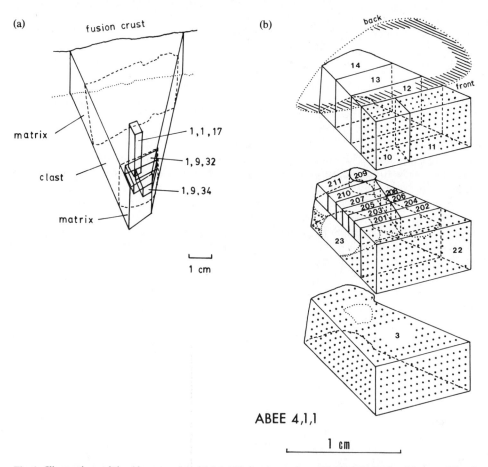

Fig. 1. Illustrations of the Abee samples. (a) 1,1,17 is in a large clast. 1,9,32 and 1,9,34 and just outside of the clast. (b) 4,1,1, spans over the boundary of a clast and the matrix. The hatched area indicates the metal rich rim of the clast on the top surface. Dotted area is the matrix. Dotted curves indicate the boundary between the clast and the matrix.

RESULTS

In Fig. 2 the saturation magnetization magnetization (Js) and the saturation remanence (Jrs) of a matrix sample are shown as a function of temperature. The component whose Curie temperature is 215°C is cohenite, which was petrologically recognized in Abee by Dawson *et al.* (1960). The thermal hysteresis of the magnetization seen during heating (730°C) and cooling (500°C) is due to the phase changes associated with kamacite. Keil (1968) showed that the kamacite contains a small amount of silicon which may affect the thermal hysteresis. The third component which contributes to the small increase of Jrs at T < 100°C during cooling is due to Fe–Ni (α_2 phase with 20–30% Ni). This could be an artificial product due to the heating experiment, because Keil (1968) did not find any metallic grains with high nickel content. During a second heating run (curves 3 and 3') a small increase of Js was observed around 530°C. This could be due to Fe–Ni which did not transform from non-magnetic phase to the ferromagnetic phase during rapid cooling. A small amount of taenite in Abee has been reported in the literature (Dawson *et al.*, 1960), but it can not be recognized in the Js(T) curve. The cohenite contributes much more to the Jrs(T) curve, suggesting it is magnetically harder than kamacite. Hysteresis parameters are shown in Table 1.

From the texture of the kamacite-cohenite grains (Herndon and Rudee, 1978), it is expected that some kind of magnetic interaction might exist between these two phases. Fig. 3 shows the results of the measurements of the magnetization at high temperature. When a pTRM < 500 < T < 750 C, H = 10 Oe) was cooled in a nonmagnetic field < H < 0.01 Oe) the remanence generally decreases to reduce the magnetostatic energy (Sugiura, 1981), but a sudden decrease of the magnetization indicates the presence of negative

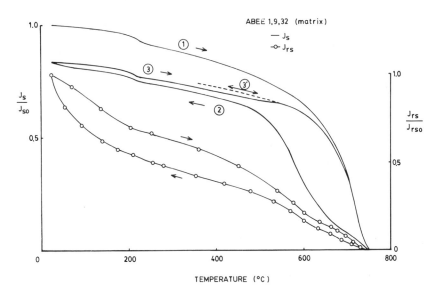

Fig. 2. Saturation and magnetization and saturation remanence as a function of temperature.

Table 1. Magnetic properties of Abee.

Js	Jrs	Ho	Hor	To	NRM
emu/s	emu/s	Oe	Oe	C	emu/s
82.4	0.596	13.6	102	215, 730	8.01E-2

Average of matrix NRM.

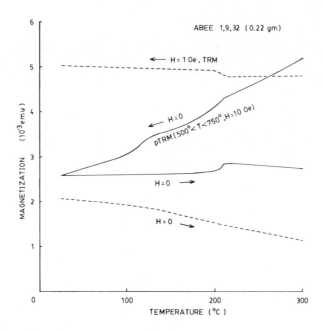

Fig. 3. Magnetization change at high temperatures. Cooling and reheating of pTRM in zero field are shown by solid curves. TRM growth in 1 Oe and its thermal demagnetization are shown by broken curves.

interactions. On the cooling curve (Fig. 3 solid curve), there are two kinks, one of which suggests negative interactions between cohenite and kamacite (210°C). On the heating curve this kink can still be seen at 210°C, suggesting that the negative interaction produced a reversed remanence in cohenite. The second kink at 120°C is probably due to α_2 phase. The kink can not be seen in the heating curve because the $\alpha \to \gamma$ transition occurs at higher temperature (600°C). The negative interaction between cohenite and kamacite does not seem to be due to an exchange interaction but to the demagnetization field produced by the remanence in kamacite; a very small external field produces positive remanence in cohenite.

In Fig. 4, the NRM direction of the two clasts and the matrix are shown. It can be seen that there are three clusters corresponding to two clasts and three matrix samples. (Because of the lamellae structure, cohenite-kamacite grains are probably magnetically anisotropic. Anisotropy of the specimen as a whole, however, is probably not significant, because the orientations of the lamellae in different grains are controlled by the crystal structure of the host kamacite, and are supposed to be random. Therefore the overall NRM directions of the clasts are not controlled by the anisotropy.) the NRM directions of clast 4,1,1, are more scattered and tend to fall into two subgroups. The larger dispersion is partly because of the smaller sample size of the 4,1,1 subsamples. But the two groupings of the NRM directions are probably significant, because all of the five samples which have lower inclinations of the NRM are interior samples (Fig. 1b) and they have relatively small NRM intensities compared with the rim samples (Fig. 5). The agreement of the NRM directions in the three separate matrix samples is good, suggesting that the matrix is homogeneously magnetized. Subdivision of sample 4,1,1 was done in our laboratory and care was taken not to expose the sample to any strong field during cutting the sample. Therefore the difference in the directions of the NRM of the matrix and the clast of the sample 4,1,1 is not artificial. Since the three matrix samples were sent to us separately and since their NRM directions agree fairly well, we may safely assume that the meteorite was not exposed to a strong artificial field before we received the samples. Thus we can conclude that each of the two clasts and the matrix of the Abee chondrite has significantly different NRM directions acquired before its encounter with the earth.

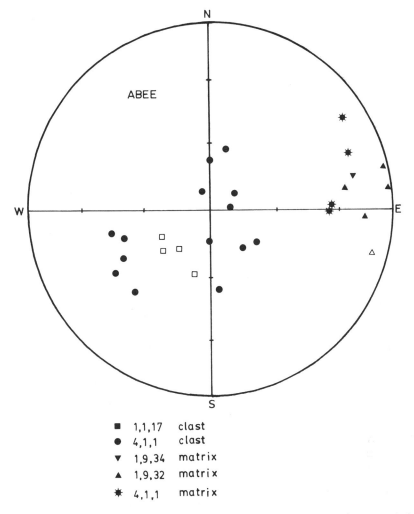

Fig. 4. NRM directions of mutually oriented samples of Abee. Solid and open symbols indicate downward and upward directions, respectively. The top of the figure is conventionally northward, but only relative orientation is significant in the case of meteorite magnetism.

Directional and intensity changes of the NRM during demagnetization are shown in Fig. 6. The NRM in the matrix sample, 1,9,32 is essentially a single component, except for a very soft small secondary component which is removed in alternating field less than 50 Oe. The NRM in clast 1,1,17 also has a small, soft secondary component. The directional change during AF demagnetization, is in the opposite sense to that seen during thermal demagnetization. This can be explained as follows. During AF demagnetization, the magnetically soft component carried by kamacite is erased first, leaving the final direction carried by the magnetically harder cohenite. During the thermal demagnetization, however, the remanence carried by the cohenite disappears first, because its Curie point (215°C) is lower. We need more detailed study to determine the origin of this secondary component. As shown in Fig. 6d, cohenite carries about half of the NRM in both the clast and the matrix. It is inferred from the thermal demagnetization curves that Abee was not reheated to temperatures much above 100°C, in the absence of magnetic field.

Fig. 6e shows the intensity changes during AF demagnetization of NRM, ARM (anhysteretic remanent magnetization) and IRM (isothermal remanent magnetization).

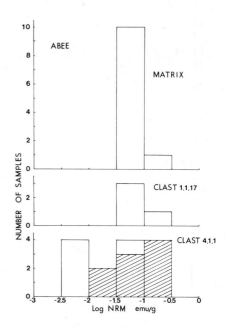

Fig. 5. Histograms of the NRM intensity of Abee. Shaded area corresponds to the metal-rich samples.

(The intensities of the ARM and IRM are about the same as that of the NRM.) The NRM is much more stable than the IRM, further suggesting the NRM is not of artificial IRM origin. The NRM is also more stable than the ARM. This is rather unusual because generally ARM and TRM (as NRM) have similar stability (Levi and Merrill, 1976). A possible interpretation is that the NRM was acquired as a TRM and was subsequently partly demagnetized.

The NRM in the clast 4,1,1 is very complex. The following features were observed.

(a) The samples from the backside of the sample 4,1,1 (14 and 209, see Fig. 1b) have only one NRM component, while others have two components (Fig. 6b). The intensity of the stable component (remaining after 100 Oe AF demagnetization) in the subsamples 14 and 209 is much larger than that in other samples (Fig. 6f).

(b) The path of the directional change during AF demagnetization is not the same for different samples. The stable component is almost antiparallel to the unstable component.

(c) All the subsamples (202, 204, 206, 211) showed little directional change of NRM during thermal demagnetization up to 400°C (Fig. 6c).

(d) The NRM of the interior samples have lower inclinations (Fig. 4) and smaller intensities (Fig. 5) than those of the rim samples.

(e) The rim samples have higher ARM intensities and lower ARM stabilities than the interior samples (Table 2).

Table 2. Remanence properties of the clast in Abee 4.1.1.

Subsample	NMR 10^{-2} emu/gm	NMR components	ARM 10^{-2} emu/gm	ARM_{100}/ARM	NRM_{100}/ARM_{100}
201	9.37	2	17	0.015	1.0
205	0.886	2	2.7	0.064	0.6
209	11.01	1	3.6	0.041	3.9

ARM was induced in a direct field of 2 Oe with a peak alternating field of 100 Oe.
ARM_{100} and NRM_{100} denote the intensity of the remanence after demagnetizing at 100 Oe.

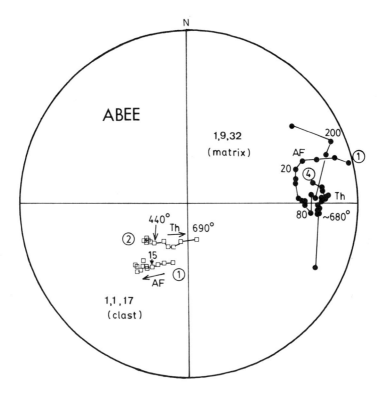

Fig. 6. (a) Directional change of NRM in 1,1,17 and 1,9,32. The numbers adjacent to some points indicate the Temperature (C) and magnetic field (Oe).

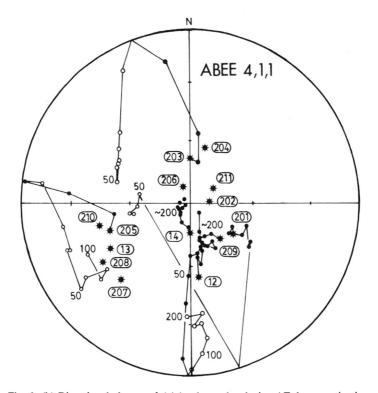

Fig. 6. (b) Directional change of 4,1,1 subsamples during AF demagnetization.

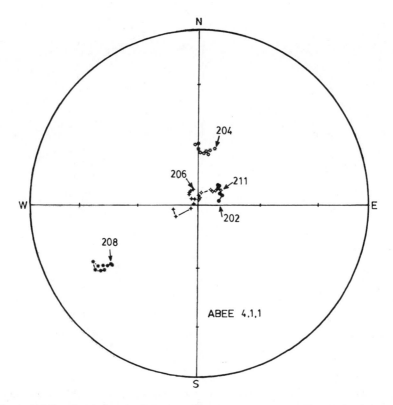

Fig. 6. (c) Directional change of 4,1,1 subsamples during thermal demagnetization. All symbols are pointing downward.

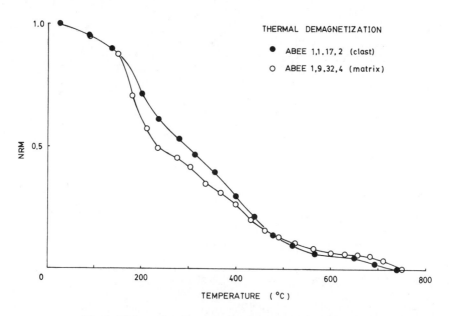

Fig. 6. (d) Intensity change during thermal demagnetization.

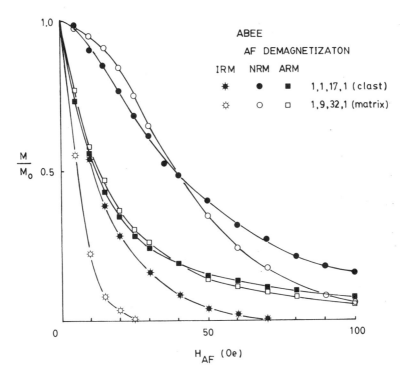

Fig. 6. (e) Intensity change of ARM, IRM and NRM during AF demagnetization are compared. IRM in 1,9,32 and 1,1,17 were given with a direct field of 30 and 120 Oe, respectively. ARM in 1,9,32 and 1,1,17 were given in a direct field 2 and 0.6 Oe, respectively, with a peak alternating field of 1000 Oe.

A possible process which created such NRM properties is: TRM acquisition by cooling from 750°C in a large field and a subsequent rapid heterogeneous reheating to >215°C from the front side of the clast in a relatively small field, keeping the temperature of the back side >215°C. As shown in Fig. 3, if a sample, which has a high temperature pTRM carried by kamacite, is cooled in zero field through the Curie point of cohenite, the cohenite acquires reversed TRM due to the demagnetization field from kamacite. Therefore the features (a), (b) and (c) are naturally explained by the above process. The feature (d) is difficult to explain unless we assume a small external field (which is pointing westward) during the reheating event. From the feature (e) it is inferred that the interior samples are enriched in cohenite relative to kamacite. Therefore they are more sensitive to low temperature ($0 > T < 215°C$) processes than the rim samples. They also quite easily show directional change because they have antiparallel components of almost equal strength. Thus it is conceivable that if a small external field which was pointing westward was present, a reheating to 215°C led to the shallow inclination of the NRM of the interior samples and left the NRM direction of the rim samples essentially unchanged. (An attempt to reproduce the directional change observed during AF demagnetization of NRM (Fig. 6b) failed. An artificial TRM was partially demagnetized by heating to 250°C in zero field, and the residual TRM was AF demagnetized. No directional change was observed during AF demagnetization, suggesting that the reversed component due to negative interaction is not strong enough to make an observable directional change.)

In Fig. 7, results of paleointensity estimates are shown. The ARM method (Stephenson and Collinson, 1974) was applied to the clast sample of 4,1,1. The ARM and the NRM have nearly equal coercivity spectrum above 10 Oe, and a paleointensity of 11.6 Oe was obtained using the conversion factor 1.34. Since a heating to 750°C produces α_2 phase and

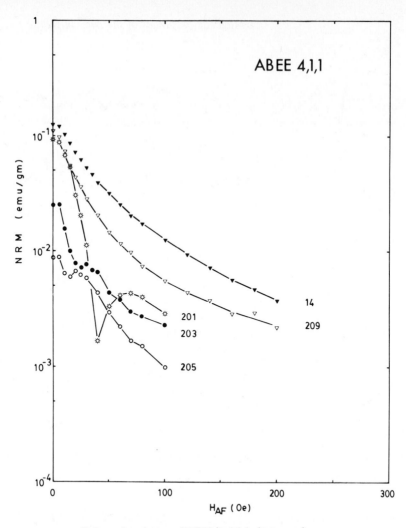

Fig. 6. (f) Intensity change of NRM in 4,1,1 clast samples.

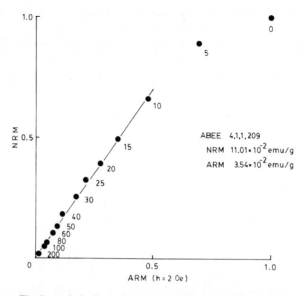

Fig. 7. (a) Paleointensity determination by ARM method.

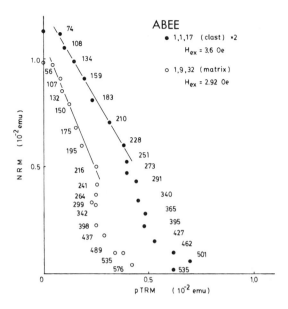

Fig. 7. (b) Paleointensity determinations by Thellier's method. Numbers adjacent to each point indicate the temperature (C) of heating.

changes the coercivity spectrum significantly, Shaw's method (Shaw, 1974) can not be applied successfully.

Thellier's method was applied to a clast and to a matrix sample. Both samples showed a linear relation between the NRM loss and the pTRM acquisition from 70°C up to the Curie point of cohenite. The paleointensities determined in this way are 6.1 and 7.0 Oe for the clast and the matrix sample, respectively. The breakdown of the linearity above the Curie point of the cohenite is either due to the negative interaction between cohenite and kamacite or due to structural and/or chemical changes of the magnetic phases due to the heating. Since the interaction between cohenite and kamacite could potentially affect the paleointensity estimates, a calibration experiment was done for a TRM produced in a known field. And it was found that the Thellier's method gives a valid estimate of the field intensity (Fig. 8). (The heated sample was later examined with an optical microscope, and it was found that the texture of the kamacite-cohenite grains had not changed by the heating except for a slight increase in the grain size of the cohenite.) Allowing for the

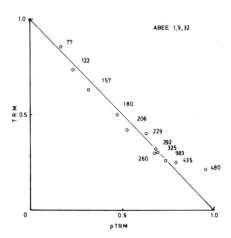

Fig. 8. Calibration of Thellier's method. Both TRM and pTRM are given in the same field. Numbers adjacent to each point indicate the temperature (C) of heating.

uncertainty (Levi and Merrill, 1976) of the ARM method, the difference of the two paleointensities for the clasts are not significant. Quantitatively, the paleointensity for the matrix sample is meaningless, because the origin of the NRM is not understood. But it does indicate that a very strong field was present at the time of agglomeration of the meteorite.

A rough estimate of the field intensity during the brief reheating of the clast in sample 4,1,1 can be made by comparing the stable component of the NRM and ARM. As shown in Table 1, the NRM_{100}/ARM_{100} for subsamples 201 and 205 are about five times smaller than that for subsample 209. Therefore, using the paleointensity of 11.6 Oe for the subsample 209, a paleointensity of about 2 Oe is obtained.

DISCUSSION AND CONCLUSIONS

The Abee meteorite has a very strong NRM (0.1 emu/gm). The stable part of the NRM seems to be carried by cohenite. This is the first report of NRM carried by cohenite in natural materials. The magnetization in cohenite-kamacite grains seems to behave normally except for a small negative interaction between the two phases. More detailed study of the cohenite-kamacite system is needed for a better understanding of the rather complex negative interaction.

Since the direction of NRM in two clasts and a matrix sample are significantly different from each other (although the number of samples is too small to make a statistically meaningful discussion) we may conclude that (1) the meteorite was not reheated to temperatures much above 100°C during or after its accretion, (2) the clast magnetization predates the formation of the meteorite, (3) the NRM of the matrix may have been acquired by one of the several processes including a DRM at low temperature, a shock remanence (SRM) or some other overprinting process. These results are consistent with the thermal history inferred from the texture of the kamacite-cohenite grains (Herdon and Rudee, 1978) and small Ar loss (Bogard, 1980).

The Abee meteorite is fairly tough and some lithification process must have been working. It is not clear whether a prolonged heating to 100°C is enough to solidify the meteorite. Ashworth and Barber (1976) argued that lithification is attributed to the passage of shock waves through porous aggregate. These shock waves could magnetize the matrix of the Abee. The disadvantages of the SRM hypothesis are: (1) There is little evidence of shock processes in the Abee meteorite (Dawson et al., 1960). (2) The SRM is generally smaller and less stable than the TRM (Cisowski and Fuller, 1978). If the NRM of the matrix was acquired as a DRM, then the lithification process must be due to a static pressure. Metallic grains which can deform quite easily and are abundant in the enstatite chondrite might have effectively cemented the meteorite under mild pressure (Dawson et al., 1960).

The primary component of the NRM in the clast samples was probably acquired during the heating (>750°C) and rapid cooling event which produced the cohenite-kamacite intergrowth texture. The enrichment of iron in the rim of the clast occurred before (or at the time of) the reheating. The clast 1,1,17 was probably never reheated to high (>100°C) temperature, while the clast in 4,1,1 was briefly heterogeneously reheated to >215°C, before the agglomeration of the Abee meteorite.

A paleointensity estimate (0.33 Oe) for Abee was previously obtained from the comparison of TRM and NRM (Brecher and Ranganayaki, 1975). There are two reasons to believe that the previous estimate is to small.

1. The estimate was obtained from the comparison of the remanence with highest coercivities but if new magnetically hard phases were produced during the heating to increase the TRM, the apparent paleointensity is lowered. Heating of Abee to 800°C produces α_2 phase which is magnetically harder than kamacite. Since the heating in Ar atmosphere usually can not prevent oxidation of Fe–Ni phases, magnetite (which

is generally magnetically harder than kamacite) was probably produced by oxidation. These α_2 phase and magnetite could make substantial contributions to the TRM.
2. The sample used in the previous study probably included matrix and clasts. The inhomogeneous magnetization directions in matrix and clasts would reduce the observed NRM intensity.

Thus we believe that the present paleointensity estimate (6 Oe), is more reliable than the previous one.

This (6 Oe) is probably the field when each clast was moving individually through interplanetary space. (Clasts and matrix must have been separated because the iron content is different.) The time required for cooling from 750°C to/or below room temperature was estimated to be about two hours (Herndon and Rudee, 1978), which seems to be too long for a transient field (generated by an impact) to survive. A steady field of 6 Oe could be either due to (1) a solar magnetic field, (2) an interplanetary field sustained by moving charged particles in a plasma, (3) a field generated by a core dynamo in a planetesimal (Levy and Sonett, 1978). The small intensity (2 Oe) of the external field during the brief heterogeneous reheating of the clast in sample 4,1,1 is puzzling. Perhaps, the clast was rotating around an axis which happened to be perpendicular to the magnetic field or perhaps this event took place in different field conditions. The time scale for this event must have been of the order of a couple of minutes.

Unless the efficiency of DRM aquisition on a planetesimal is unexpectedly large, the intensity of the magnetic field when the Abee agglomerated must have been of the order of several oersted, which is similar to the field that existed before the agglomeration.

The strong field observed in Abee has only been observed previously in Allende chondrules (Lanoix et al., 1978). Since the strong field in Allende chondrules can not be found in recent studies (Sugiura et al., 1979; Wasilewski, pers. comm.), there is a question as to whether the chondrules have been exposed to a magnet. This question has to be settled by more thorough examination of chondrules. If we ignore the chondrule data for the time being, then the next strongest paleofields (1 Oe) were found in various carbonaceous chondrites including Allende (Brecher, 1972; Butler, 1972; Banerjee and Hargraves, 1972; Nagata, 1979). Since NRM directions of chondrules in carbonaceous chondrites are probably scattered (Sugiura et al., 1979), the paleointensities might be underestimated. Since chondrules are very weakly magnetized, however, the error in the paleointensity estimates will be much less than 50%. In some models carbonaceous chondrites are considered to have condensed at the periphery of the solar nebula, while enstatite chondrites condensed at the inner edge of the nebula (Beadecker and Wasson, 1975). In this event, the field intensity would appear to decrease with increase of the distance from the sun, suggesting that the field was due to the solar magnetic field.

Ordinary chondrites and achondrites are thought to have been made in the middle of the solar nebula, but the paleointensities estimated from ordinary chondrites (0.1 Oe: Brecher and Leung, 1979) and achondrites (0.1 Oe: Nagata, 1979) do not fall between the values obtained from enstatite chondrites and carbonaceous chondrites. This is probably because these meteorites were magnetized during the metamorphic or volcanic event, which occurred long after the accretion of the meteorites. (Ureilites are exceptions. They have unique oxygen isotope composition (Clayton et al., 1976) and relatively strong (1 Oe) paleointensities (Brecher and Fuhrman, 1979).

If a strong, large scale, steady magnetic field was present in the early solar nebula, the nebula gas could be partially ionized (Freeman, 1977). Some of the anomalies of the isotopic composition observed in meteorites, may be the results of the isotopic fractionation due to the interaction of the ions and the magnetic field.

Acknowledgments—We wish to thank Dr. K. Marti for generous allocation of the Abee samples, Dr. P. Wasilewski and Dr. M. Housley for critical comments, and the Natural Sciences and Engineering Research Council of Canada for financial support.

REFERENCES

Ashworth J. R. and Barber D. J. (1976) Lithification of gas rich meteorites. *Earth Planet. Sci. Lett.* **30**, 222–233.
Banerjee S. K. and Hargraves R. E. (1972) Natural remanent magnetizations of carbonaceous chondrites and the magnetic field in the early solar system. *Earth Planet. Sci. Lett.* **17**, 110–119.
Beadecker P. A. and Wasson J. T. (1975) Elemental fractionations among enstatite chondrites. *Geochim. Cosmochim. Acta* **39**, 735–765.
Bogard D. D. (1980) 40 Ar- 39 Ar ages of Abee clasts. *Meteoritics* **15**, 267–268.
Brecher A. (1972) Memory of early magnetic fields in carbonaceous chondrites. In *On the Origin of the Solar System* (H. Reeves, ed.), p. 260–273. CNRS, Paris.
Brecher A. and Arrhenius G. (1974) The paleomagnetic record in carbonaceous chondrites: Natural remanence and magnetic properties. *J. Geophys. Res.* **79**, 2081–2106.
Brecher A. and Fuhrman M. (1979a) The magnetic effects of brecciation and shock in meteorites: II. The ureilites and evidence for strong nebular magnetic fields. *Moon and Planets* **20**, 251–263.
Brecher A. and Leung L. (1979) Ancient magnetic field determinations on selected chondritic meteorites. *Phys. Earth Planet. Inter.* **20**, 361–378.
Brecher A. and Ranganayaki R. P. (1975) Paleomagnetic systematics of ordinary chondrites. *Earth Planet. Sci. Lett.* **25**, 57–67.
Brecher A. and Fuhrman M. (1979b) Magnetism, shock and metamorphism in chondritic meteorites. *Phys. Earth Planet. Inter.* **20**, 350–360.
Brecher A., Stein J., and Fuhrman M. (1977) The magnetic effects of brecciation and shock in meteorites: I. The LL-chondrites. *The Moon* **17**, 205–216.
Butler R. F. (1972) Natural remanent magnetization and thermomagnetic properties of the Allende meteorite. *Earth Planet. Sci. Lett.* **17**, 120–128.
Cisowski S. M. and Fuller M. (1978) The effect of shock on the magnetism of terrestrial rocks. *J. Geophys. Res.* **83**, 3441–3458.
Clayton R. M., Onuma N., and Mayeda T. K. (1976) A classification of meteorites based on oxygen isotopes. *Earth Planet. Sci. Lett.* **30**, 10–18.
Dawson K. R., Maxwell J. A. and Parsons D. E. (1960) A description of the meteorite which fell near Abee, Alberta, Canada. *Geochim. Cosmochim. Acta.* **21**, 127–144.
Freeman, J. W. (1977) The magnetic field in the solar nebula. *Proc. Lunar Sci. Conf. 8th*, p. 751–755.
Graham J. W. (1949) Stability and significance of magnetism in sedimentary rocks. *J. Geophys. Res.* **54**, 132–167.
Herndon J. M. and Rudee M. L. (1978) Thermal history of the Abee enstatite chondrite. *Earth Planet. Sci. Lett.* **41**, 101–106.
Keil K. (1968) Mineralogical and chemical relationship among enstatite chondrites. *J. Geophys. Res.* **73**, 6945–6976.
Levi S. and Merrill R. T. (1976) A comparison of ARM and TRM in magnetite. *Earth Planet. Sci. Lett.* **32**, 171–184.
Levy E. H. and Sonett C. P. (1978) Meteorite magnetism and early solar system magnetic fields. In *Protostars and Planets* (T. Gehrels, ed.), p. 516–532. Univ. of Arizona, Tucson.
Lanoix M., Strangway D. W., and Pearce G. W. (1978) The primordial magnetic field preserved in chondrules of the Allende meteorite. *Geophys. Res. Lett.* **5**, 73–76.
Nagata N. (1979) Meteorite magnetism and the early solar system magnetic field. *Phys. Earth Planet. Inter.* **20**, 324–341.
Shaw J. (1974) A new method of determining the magnitude of the paleomagnetic field: Application to five historic lavas and five archaeological samples. *Geophys. J.* **39**, 133–141.
Stacey F. D., Lovering J. F., and Parry L. G. (1961) Thermomagnetic properties, natural magnetic moments and magnetic anisotropies of some chondritic meteorites. *J. Geophys. Res.* **66**, 1523–1534.
Stephenson A. and Collinson D. W. (1974) Lunar magnetic field paleointensities determined by an anhysteretic remanent magnetization method. *Earth Planet. Sci. Lett.* **23**, 220–228.
Sugiura N. (1977) Magnetic properties and remanent magnetization of stony meteorites. *J. Geomag. Geoelectr.* **29**, 519–539.
Sugiura N. (1981) A new model for the acquisition of thermoremanence by multidomain magnetite. *Can. J. Earth Sci.* **18**, 789–794.
Sugiura N., Lanoix M., and Strangway D. W. (1979) Magnetic fields of the solar nebula as recorded in chondrules from the Allende meteorite. *Phys. Earth Planet. Inter.* **20**, 342–349.

Howardites and polymict eucrites: Regolith samples from the eucrite parent body. Petrology of Bholgati, Bununu, Kapoeta, and ALHA76005

M. Fuhrman and J. J. Papike

Department of Earth and Space Sciences, State University of New York, Stony Brook, New York 11794

Abstract—Howardite meteorites are samples of the regolith of the eucrite parent body. A continuation of Labotka and Papike's (1980a) study of howardite meteorites presents modal data which confirm the following characteristics of the eucrite regolith: friable source rocks, immature soil, comminution as the major soil-forming process, and a well-mixed soil. Chemistry of single mineral clasts provides additional evidence for exchange/equilibration of Fe and Mg in pyroxenes during lithification of Bununu, Kapoeta, and ALHA 76005. This re-equilibration can take place on a scale smaller than a thin section.

Fe-rich olivines found in howardites may be the result of extreme fractionation along a zoning trend observed in pyroxene phenocrysts of ALHA 77302 (Labotka and Papike, 1980b). During annealing this olivine can react with silica grains and produce pyroxene reaction rims, as are observed in ALHA 76005.

INTRODUCTION

Howardite meteorites have long been accepted as samples of regolith from the eucrite parent body (EPB). They are polymict breccias whose clast types include lithic, fused soil (glass, breccia, and agglutinate), and single mineral clasts. Texturally, these breccias are similar to lunar soils and breccias and the petrographic methods used by Vaniman *et al.* (1979) to study lunar soils may be applied to the howardites. A petrographic study on a clast-by-clast basis such as this yields information on the source terrane, maturity, lithification, and transport processes that took place on the EPB. Labotka and Papike (1980a) used these methods to study four howardites and one polymict eucrite as representatives of the EPB regolith.

Bulk chemical analyses of howardites are intermediate between diogenites (orthopyroxenites) and eucrites (pigeonite basalts) (Moore, 1962; Mason, 1967). Mixing models that portray the bulk chemistry of howardites as two-component mixtures of diogenite and eucrite endmembers have been proposed by Jérome and Goles (1971), Fukuoka *et al.* (1977), and Dreibus *et al.* (1977).

The models based on bulk chemistry suggest a simple two-rock source for the howardites; but petrographic studies (Duke and Silver, 1967; Bunch, 1975; Dymek *et al.*, 1976) have shown that a wide range of lithic clasts occurs in howardites. These lithic clasts have diverse mineral compositions, pyroxene:plagioclase ratios, and bulk chemistries. Although these diverse clasts do not themselves fall on a diogenite-eucrite mixing line, the bulk chemistries of the meteorites do. The variations of each clast from the mixing line are canceled out by other lithic clasts. Further studies on the lithic clasts of howardites and polymict eucrites will shed more light on the petrologic relationship between lithic clasts in howardites, if any, and on the reasons why the bulk chemistries of howardites lie on a single mixing line while the individual components of howardites seem to be so complex. This is the major area of our new research on howardites.

Despite the fact that howardite lithic clasts exhibit a wide range of textures, mineralogy, and chemistry, these clasts may still be grouped into one of the two non-overlapping categories: orthopyroxenites, and plagioclase + pyroxene rocks. Labotka and Papike (1980a) carried out a petrographic study of several howardites. From their modal and mineral chemistry data, they determined relative proportions of these two major lithic types in the howardites Frankfort, Pavlovka, Yurtuk, and Malvern. These proportions agree well with diogenite:eucrite ratios determined by chemical mixing models (Dreibus et al., 1977; Fukuoka et al., 1977; McCarthy et al., 1972). Labotka and Papike (1980a) also made some preliminary conclusions about some of the EPB regolith characteristics. The present study of the meteorites Bholgati, Bununu, Kapoeta, and ALHA 76005 (polymict eucrite) is a continuation of Labotka and Papike's (1980a) study and is a further attempt to test the validity of the two-rock source model and to describe more precisely the characteristics of the EPB regolith.

METHODS

Thin sections used in this study of Bholgati, Bununu, and Kapoeta were obtained from the American Museum of Natural History (AMNH), the National Museum of Natural History (NMNH), and Arizona State University (ASU). Thin sections of ALHA 76005 were obtained from the Johnson Space Center, Antarctic Meteorite Curatorial Facility. Specimen numbers of thin sections used were: Bholgati—AMNH 4243-1, ASU; Bununu—NMNH 1571-12, 3, 1; Kapoeta—AMNH 3924-1, ASU 827D (2 sections); ALHA 76005, 9, 10, 45, 47 and 49.

Volume percent modes were obtained for all sections by optical point counting and are presented in Table 1a. In this table lithic clasts includes all lithic clasts except pyroxenite. Because pyroxenite clasts were often difficult to distinguish from fractured pyroxene grains of the same size, pyroxenite clasts are included with pyroxene mineral clasts. However, since small sections of these breccias are heterogeneous these results may not be significant. For this reason, grain count percent (number of grains/total number in size fraction × 100) were also obtained. Grain count percents for the size fractions XL (>2 mm) and L(2000–200 μm) were determined optically. Grain counts for S(200–20 μm) were done by energy dispersive electron microprobe analysis. These percentages are more meaningful than volume % modes because they help eliminate the bias created by heterogeneities in large clast populations. Grain count modes are presented in Table 1b.

Mineral compositions were determined on an ARL-EMX electron microprobe using wavelength dispersive analysis (WDS) and the correction procedure of Bence and Albee (1968). All feldspars were analyzed for Si, Al, Ca, Na, K, and Fe. All L(2000–200 μm) pyroxenes, all olivines, and some of the S(200–20 μm) pyroxenes were analyzed for Si, Al, Fe, Mg, Ca, Mn, Ti, Na, and Cr. The remaining pyroxene analyses are 4-elements (Si, Fe, Mg, and Ca). Representative analyses for each mineral are given in Appendix 1.

MODES

Modal data for Bholgati, Bununu, Kapoeta, and ALHA 76005 are given in Table 1b and are illustrated in Fig. 1. Values shown are percent of grains within a size fraction.

In the following discussion, the words 'grain' and 'clast' are used interchangeably. A lithic clast is one which is polycrystalline. A single mineral clast is a clast consisting of a single crystal (or crystal fragment) of one mineral.

Labotka and Papike (1980a) noted three major features of their modal data from four howardites and one polymict eucrite. These features also dominate these modal data. The first feature is that lithic clasts make up a small proportion of the total number of clasts in both the L and S size fractions. Dominating the modes are single-mineral clasts of pyroxene and plagioclase. Single mineral clasts occur even in the XL (>2 mm) size fraction. Coarse-grained source rocks for these single-mineral clasts must have an average grain size greater than 2 mm. In addition, finer grained source rock must be either rare or very friable to be so poorly represented as lithic fragments in either size fraction.

The second feature of the modes is the rarity of a fused soil component. Lunar soils contain at least 30 vol.% fused soil component (more typically 40–60%) (Simon et al., 1979; Taylor et al., 1979). The amount of fused soil component (glassy soil breccias and agglutinates) found in a regolith is an approximate index of the maturity of the regolith; the longer the soil is near the surface, the more fused soil component will be produced by

Table 1a. Volume % modes for Kapoeta, Bununu, Bholgati and ALHA 76005.

SIZE FRACTIONS (MICRONS)	KAPOETA			BUNUNU			BHOLGATI			ALHA 76005		
	2000	2000-200	200-20	2000	2000-200	200-20	2000	2000-200	200-20	2000	2000-200	200-20
LITHIC CLASTS	2.9	3.9	0.3	6.3	1.9	1.0	3.6	3.2	0.1	0.4	7.5	0.3
FUSED SOIL COMPONENT												
Crystallized DMB	0.6	1.0	0.1		0.7			1.3			0.5	0.2
Glassy DMB							0.3	1.0			1.0	0.1
Sulfide-Matrix Breccia		0.2						0.1			0.1	0.1
Glass and Devitrified Glass		0.2	0.1	3.2	0.4	0.2						
Recrystallized Breccia	1.2	0.2							0.1			
MINERAL CLASTS												
Pyroxene	1.6	20.9	27.9	1.0	16.7	29.7	19.4	14.8	22.9	0.7	9.6	23.2
Plagioclase		4.5	7.1		4.4	8.9		3.7	4.7		6.7	20.5
Ilmenite											0.1	0.1
Silica		0.1			0.1				0.1			0.1
Chromite		0.1	0.1			0.1			0.1			0.1
Olivine									0.1			
Metal									0.1			0.1
Sulfide		0.1	0.1			0.1						0.1
TOTAL	6.3	31.2	35.7	10.5	24.2	40.0	23.3	24.1	28.2	1.1	25.5	44.9
NUMBER OF POINTS COUNTED		3000			2316			2000			5000	
VOLUME PERCENT OF MATRIX		26.8			25.3			24.4			28.5	

Table 1b. Grain-count modes for howardites; Kapoeta, Bununu, Bholgati and ALHA 76005.

	KAPOETA			BUNUNU			BHOLGATI			ALHA 76005		
SIZE FRACTIONS (MICRONS)	2000*	2000-200	200-20	2000*	2000-200	200-20	2000*	2000-200	200-20	2000*	2000-200	200-20
LITHIC CLASTS												
Basaltic Textured Rock	3	3.3	0.6		2.4	1.1	1	3.6	1.3	1	8.2	0.6
Coarse-Grained Plagioclase + Pyroxene Rock	1	2.6			1.5		1	1.2		1	8.6	
Brecciated Plagioclase + Pyroxene Rock	1	1.0	0.2		1.2		2	3.4				
Cataclastic Rock and Granulite	1	1.2	0.8	2	4.1			4.3			2.0	
Orthopyroxenite		0.3					1	0.2				
Other	2										0.4	
FUSED SOIL COMPONENT												
Crystallized DMB	2	2.7	0.4	1	0.9			2.5			1.4	0.3
Glassy DMB	2	2.2			0.6		1	3.9			2.8	
Sulfide-Matrix Breccia		0.7			0.9			0.1			0.4	
Glass and Devitrified Glass	1	0.5		1	0.3						0.2	
MINERAL CLASTS												
Pyroxene	5	66.7	67.0	1	71.0	69.9	2	62.2	70.4		43.4	62.4
Plagioclase		18.0	24.1		17.3	24.4		18.2	22.5		32.2	34.1
Ilmenite			0.4			0.3			0.4		0.4	0.9
Silica		0.2	3.6			1.8		0.1	2.8			1.2
Chromite		0.6				0.7		0.4	0.4			0.3
Olivine			1.8			1.3			1.8			
Metal		0.3	0.2									
Sulfide		0.1	1.0			0.5			0.4			0.3
TOTAL	18	100.4	100.1	5	100.2	100.0	8	100.1	100.0	2	100.0	100.1
NUMBER OF GRAINS COUNTED	18	1574	506	5	342	614	8	1411	493	2	500	343
VOLUME PERCENT OF TOTAL ROCK	6.3	31.2	35.7	10.5	24.2	40.0	23.3	24.1	28.2	1.1	25.5	44.9

*Only the number of grains are indicated.

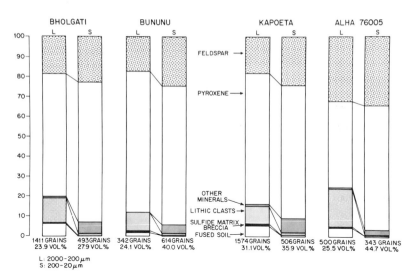

Fig. 1. Diagrammatic representation of the grain-count modes for the two size fractions 2000–200 μm and 200–20 μm.

micrometeorite impact. Matrix material (< 20 μm) in these meteorites accounts for less than 40 vol.%, as compared to lunar soils that generally contain more than 50 vol.% matrix (see for example, Taylor et al., 1977). This paucity of fused soil and of fine-grained material was also observed by Labotka and Papike (1980a) in the meteorites they studied, and indicates that the eucrite parent body regolith is less mature than most lunar soil. Exposure ages for mineral fragments and lithic clasts in Bununu and Kapoeta are also much younger than those for lunar soils (Rajan et al., 1975; 1979).

The third feature of the modes is the increase in plagioclase: pyroxene ratio from the large to small size fraction, which is accompanied by a decrease in the number of lithic clasts (Fig. 1). The contribution of single-mineral clasts from the breakdown of lithic clasts with approximately 1:1 plagioclase: pyroxene ratios can account for these observed changes; lithic clasts with this ratio are common in the large size fraction. No exotic component is needed to explain the changes in the modes between the two size fractions; comminution seems to account for all changes.

HOWARDITE CLAST TYPES

Although lithic clasts make up a small number of the total grains found in howardites, they are important because they represent the most likely source rocks of the eucrite parent body regolith. These lithic clasts are probably samples of the near-surface bedrock of the eucrite parent body and they supply material for the single-mineral clasts.

Bunch (1975) classified in detail the wide range of lithic types found as clasts in howardites. This classification, however, is too detailed for the purpose of comparing modal abundances of clast types between meteorites. The classification this study uses is the broader one of Labotka and Papike (1980a). This classification divides the lithic types into three major groups: pyroxene + plagioclase rocks, polycrystalline monomineralic clasts, and breccias. A summary is presented in Table 2. For a detailed explanation of this classification, see Labotka and Papike (1980a).

Single mineral clasts are by far the most common clasts present in howardites. Pyroxene clasts are the most abundant of the single-mineral clasts, followed by plagioclase. Also present in minor amounts are silica (tridymite), olivine, ilmenite, chromite, troilite, and Fe-Ni metal. The sources for these mineral clasts are the lithic clasts described above. Labotka and Papike (1980a) analyzed the range of pyroxene, plagio-

Table 2. Summary of lithic clast types (following Labotka and Papike, 1980a).

I. **PYROXENE + PLAGIOCLASE ROCKS:**
PLAG ≅ PYRX ± CHROMITE, ILM, TRID, TROILITE, METAL

  A. **Basaltic Textured.** Generally subophitic to intergranular; porphyritic and intersertal also occur.
  B. **Coarse-grained.** Characterized by coarse grain sizes, straight-grain boundaries and polygonal grained-shapes typical of slowly cooled plutonic rocks.
  C. **Shock-deformed.** Plagioclase replaced by maskelynite and pyroxene is brecciated.
  D. **Cataclastic and Granoblastic.** Rocks that have undergone recrystallization during and after deformation. Grain size small (~0.1 mm) and grain shapes are generally polygonal.

II. **POLYCRYSTALLINE, MONOMINERALIC ROCKS:**
Generally Rare

  A. **Pyroxenites.**
  B. **Anorthosites.** Extremely rare.

III. **BRECCIA CLASTS**

  A. **Dark Matrix Breccias (DMB) and Agglutinates.** Microcrystalline or glassy matrix; includes melt rocks in Bununu and Malvern (Bunch, 1975; Hewins and Klein, 1978; Klein and Hewins, 1979).
  B. **Sulfide Matrix Breccias.** Breccias with abundant troilite in breccia. Many annealed; textures range from fine-grained breccias to coarse-grained, olivine-orthoproxene rocks with dendritic troilite in orthopyroxene.

IV. **EXOTICA**
  A. **Carbonaceous Chondrite Clasts.** Only observed in Bholgati, but reported (Bunch, 1975; Wilkening, 1973) in Kapoeta, Jodzie, and Washougal.

clase, and olivine compositions found in the various lithic types. In the four meteorites they studied, they found that the major sources of howardite mineral clasts are orthopyroxenites and pyroxene + plagioclase rocks. The present study examines four additional meteorites and further tests the validity of this two-source model.

SINGLE MINERAL CLAST CHEMISTRY

Pyroxene

If we assume the simple two-source model of diogenites (orthopyroxenites) and pyroxene + plagioclase rocks for howardites, then we would expect a specific distribution of pyroxene compositions among the single-mineral pyroxene clasts. The large size-fraction (2000–200 μm) would consist of pyroxenes derived from the coarser-grained source rocks. Orthopyroxenite clasts, although by number of grains are not common, can be volumetrically very important. When these clasts do occur, they are usually > 2000 μm with a large average grain size (also > 2000 μm). Since these Mg-rich clasts are so important as a source for L(2000–200 μm) sized grains, in the large size fraction we would expect a relatively large number of diogenitic ($En_{70-75}Wo_{1-4}$) grains and some grains distributed throughout the compositions associated with coarse-grained pyroxene + plagioclase rocks (En_{58-35}, Wo_{5-10}). The small pyroxene grains would be a product of both comminution of the larger grains and the contribution from the finer-grained basaltic textured eucritic rocks, so this distribution should also have a large number of diogenitic grains plus a more defined peak around En_{35-40} reflecting the eucritic contribution.

The distribution of single-mineral pyroxene clast compositions in Bholgati, shown in Fig. 2, is consistent with the two-source model. The large pyroxenes peak at ~En_{75} and the small grains peak at both the diogenitic compositions and the more Fe-rich compositions.

The Kapoeta pyroxene compositions (Fig. 3) are also consistent with the two-source model, with the exception of pyroxenes found in a large recrystallized breccia clast. The

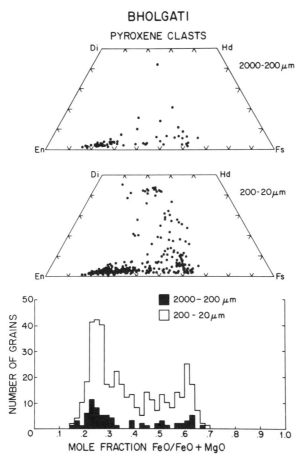

Fig. 2. Pyroxene single mineral clast compositions for Bholgati.

matrix of this clast is virtually glass-free and has a recrystallized texture. Pyroxenes in this clast are shown in the stippled pattern in Fig. 3. They tightly cluster at mole fraction FeO/(FeO + MgO) (or Fe#) equal to 0.36, though reported bulk Fe# for Kapoeta is 0.44 (Mason, 1967). The more magnesian Fe# of these equilibrated pyroxenes is probably due to partitioning of Fe into other ubiquitous Fe-bearing phases (troilite, ilmenite, chromite) during annealing.

Bununu pyroxene compositions are shown in Fig. 4. The compositions of large pyroxenes are again in the expected range. However, the small pyroxene compositions are a mixture of both the "normal" distribution and a large number of grains with an uncommon pyroxene composition of En_{62-66}. These modified pyroxenes were not found to be homogeneously distributed throughout the Bununu matrix; rather, they were found to be most common in one small (1–2 mm^2) area of a thin section. Pyroxenes elsewhere in the same section had more "normal" compositions. This is evidence for localized annealing on a scale smaller than a thin section. The only other evidence of reequilibration in the area with modified pyroxenes is a slightly lighter color to the matrix. The reported Fe# for Bununu is 0.39 (Mason, 1967); as in Kapoeta, the modified pyroxenes are slightly more magnesian than this, again probably due to Fe-partitioning into troilite, ilmenite, and chromite during annealing.

Modification of the compositions of small pyroxene grains was also observed by Labotka and Papike (1980a) along with other evidence for annealing and re-equilibration of pyroxene in the matrices of Frankfort and Yurtuk.

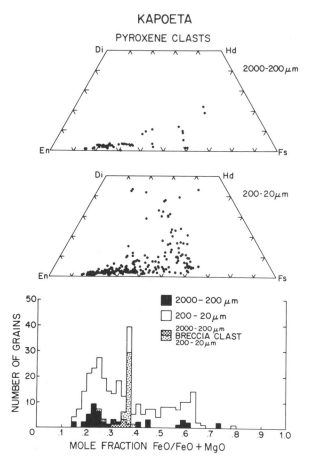

Fig. 3. Pyroxene single mineral clast compositions for Kapoeta.

ALHA 76005 is not a howardite, it is a polymict eucrite. It appears to have no diogenitic component; diogenitic pyroxenes do not occur and the bulk chemistry is eucritic (Olsen *et al.*, 1978). We would expect then, that the one pyroxene source (pyroxene + plagioclase rocks) would provide pyroxene of similar composition for both the small and large clasts. The pyroxenes found, however (Fig. 5), do not support a one-source model for pyroxenes. Labotka and Papike (1980b) observed zoning in pyroxene phenocrysts in a basaltic clast in ALHA 77302 (a similar meteorite) from cores of $En_{63}Wo_6$ to rims ranging from $En_{40}Wo_5$ to $En_{28}Wo_{37}$ (1 μm exsolution lamellae occur at rim). The 2000–200 μm grains in ALHA 76005 have relatively Mg-rich compositions, like the phenocryst cores, while the 200–20 μm grains have compositions similar to the phenocryst rims. The single-mineral pyroxene grains of ALHA 77302 also show a similar compositional difference between size fractions (Labotka and Papike, 1980a). If large pyroxenes were originally zoned like the phenocryst in ALHA 77302, the outermost edges would be broken up during comminution and contribute relatively Fe-rich pyroxene fragments to the small size fraction. The more Mg-rich cores would tend to be left as larger grains for a longer period of time as the regolith matures.

An additional feature of ALHA 76005 is the occurrence of tridymite grains with clinopyroxene reaction rims of $En_{38-40}Wo_6$. Microprobe analyses show small (~ 1 μm) areas of Ca-rich pyroxene in these rims. It is probable that this is indicative of inverted pigeonite textures. The two grains observed are shown in Figs. 6a and b. In addition, two pyroxene clasts of typical reaction rim texture without the occurrence of a silica polymorph were also observed (Figs. 6c and d). These additional clasts may either represent completely consumed silica clasts or reflect a thin section cut through the outer

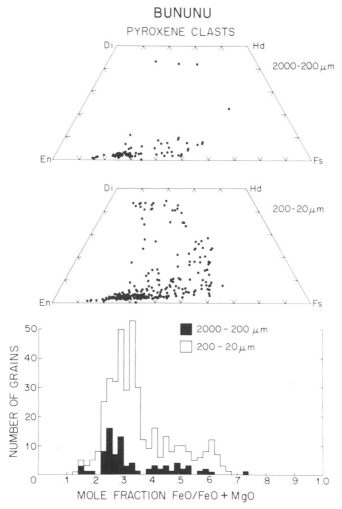

Fig. 4. Pyroxene single mineral clast compositions for Bununu.

portion of a pyroxene rim. All of the grains and rims have rounded edges and show no sign of having been through the mechanical processes of comminution and regolith formation, although all other pyroxene clasts do show such features. The implication is that these reaction rims are secondary features that are the product of chemical migration occurring *after* incorporation into the regolith. Similar rims have been observed in both polymict eucrites ALHA 78040 and ALHA 77302 (Delaney, pers. comm., 1981), and the howardite Yurtuk (Labotka and Pipike, 1980a). The rim in Yurtuk is an orthopyroxene (Wo_2En_{60}). The different composition of the rim is probably due to the difference in bulk composition between howardites and polymict eucrites. Howardites are Ca-poor and Mg-rich compared to polymict eucrites (Olsen *et al.*, 1978; Fukuoka *et al.*, 1977). Possible reactions and mechanisms to produce these rims are discussed below. However, these reaction rims are additional evidence for some kind of re-equilibration of Fe, Mg, and Si in the howardite regolith sometime during or after incorporation into the soil.

Feldspar

Feldspar clast compositions are shown in Fig. 7. Because the only feldspar-bearing lithic clasts are pyroxene + plagioclase rocks, the range in feldspar composition for mineral clasts should be the same as the range in feldspar composition for these lithic clasts, and

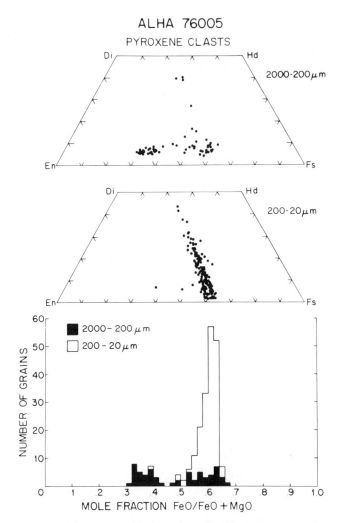

Fig. 5. Pyroxene single mineral clast compositions for ALHA 76005.

it is (An_{75-97}). However, Labotka and Papike (1980a) noted that 'basaltic' clasts (i.e., fine grained pyroxene + plagioclase rocks) have plagioclase that is more sodic than in other pyroxene + plagioclase rocks. In addition, Delaney et al. (1981) distinguished the presence of two kinds of basaltic clasts in howardites: peritectic and evolved. Their evolved basalt clasts included more sodic plagioclase compositions (An_{75-91}) than the peritectic basalts, although the plagioclase compositions overlap for most of the peritectic range. Since all basaltic clasts, in general, have a smaller average grain size than other pyroxene + plagioclase rocks (i.e., granulitic, coarse pyroxene + plagioclase) we would not be surprised if small (S) feldspars tended to be more sodic than large (L) feldspars. The feldspar compositions for mineral clasts (Fig. 7) do show more sodic compositions in the smaller size fraction. No evidence for re-equilibration of plagioclase compositions was found. This is either due to unfavorable kinetics for Ca-Na exchange or to the limited range of plagioclase compositions contributed by the one eucritic source.

Olivine

Olivine is rare in howardites and found only in single mineral clasts and sulfide matrix breccias. Delaney et al. (1980) studied eight howardites and reported a range of olivine compositions of Fo_{89}-Fo_{16} with a gap between Fo_{45} and the one Fe-rich clast found (Fo_{16}). Magnesium-rich olivine in howardites is far more common than iron-rich olivine and probably is part of a diogenitic component that represents refractory residues left after melting and extraction of more Fe-rich liquids (Delaney et al., 1980). The origin of Fe-rich olivine is more obscure; this olivine may be the result of supercooling of magma sufficient to produce metastable olivine instead of pyroxene and plagioclase that could not nucleate (Delaney et al., 1980). An alternative origin is discussed in a later section (see below).

Olivine compositions for Bholgati, Bununu, and Kapoeta are shown in Fig. 8. No olivine was found in ALHA 76005. The range seen in these three meteorites includes more Mg-rich olivine than reported by Delaney et al. (1980). An olivine as Fe-poor as Fo_{96} was found in Bholgati, and one olivine (Fo_{35}) falls within the compositional gap reported by Delaney et al. (1980). Olivine ranges reported for the howardites Macibini, Molteno, and Malvern (Desnoyer and Jérome, 1973) also extend into this gap. It is likely, then, that the compositional hiatus reported by Delaney et al. (1980) is merely a function of the difficulty in identifying olivine grains in a fine-grained howardite matrix, and not analogous to the compositional gap found in terrestrial igneous complexes, as suggested by Delaney et al. (1980).

DISCUSSION

Fe-Mg exchange during annealing

Labotka and Papike (1980a) suggested that Fe-Mg exchange took place during lithification of howardite meteorites. This study has presented evidence to further substantiate this idea. Fe-Mg exchange probably took place on a fairly local scale; different thin sections of the same meteorite show various amounts of this re-equilibration. Indeed, in the case of Bununu, the extent of re-equilibration varies drastically from one side of the thin section to the other. Re-equilibration may have taken place a number of times; an intact recrystallized breccia clast within the matrix of Kapoeta shows strong re-equilibration of pyroxene compositions – yet this clast itself is incorporated into a non-equilibrated matrix. Further evidence for mobilization of Fe-Mg after regolith formation are pyroxene reaction rims on tridymite grains in Yurtuk (Labotka and Papike, 1980a) and ALHA 76005. Prinz et al. (1980) in their modal study of several howardites and mesosiderites also reported significant amounts of pyroxene in the range common to re-equilibrated pyroxenes in this study (En_{58-65}). The howardites Kapoeta and Le Teilleul are dominated by pyroxenes of this composition, while both Bholgati and Bununu show evidence of a significant component of equilibrated pyroxenes.

Implications for source-rock modeling

Source rock modeling for howardites may be straightforward in terms of bulk chemistries; but petrologic modeling is more complex because of the wide variety of lithic clasts observed in howardites. The lithic clasts in howardites may be grouped broadly into 'orthopyroxenites' and 'pyroxene + plagioclase rocks', but attempts to subdivide the latter group fail because of overlapping pyroxene and plagioclase compositions. However, rough estimates of the contribution from each of these two broad groups can be made. Labotka and Papike (1980a) described two methods that can be used to estimate this 'orthopyroxenite': 'pyroxene + plagioclase' ratio. The first method, which assumes no overlap in pyroxene compositions between orthopyroxenites and pyroxene + plagioclase rocks, can be used for samples that exhibit little or no evidence for Fe-Mg exchange among pyroxene clasts. The second method assumes that all plagioclase is derived from pyroxene + plagioclase rocks and that all of these rocks have the same plagioclase:pyroxene ratio. These methods were applied to our data and the results are given in Table 3 (for details of this method, see Labotka and Papike, 1980a).

The average feldspar:pyroxene ratios in plagioclase + pyroxene source rocks for

Kapoeta and Bholgati were calculated by the first method. These values were then averaged with Labotka and Papike's (1980a) plagioclase:pyroxene for Pavlovka and Malvern. The average of these four values, 41:59, was used to calculate the proportions of diogenitic pyroxene in Bununu. The "dio":"euc" ratios calculated by Labotka and Papike (1980a) agreed with values calculated by bulk chemistry mixing models of Fukuoka *et al.* (1977), McCarthy *et al.* (1972), and Dreibus *et al.* (1977). The calculated

Fig. 6. (a) and (b) Silica polymorph (tridymite) grains with pyroxene ($En_{38-40}Wo_6$) reaction rims.

dio:euc ratio for Kapoeta (47:53) lies in the wide range covered by the two samples of Kapoeta modeled by Fukuoka *et al.* (1977) (20:80 and 60:40).

Mixing ratios calculated by this method have a questionable amount of significance. Although calculated ratios agree with bulk chemical model ratios, it is clear by the range of lithic types represented in howardites that the regolith samples rocks with very different chemical and thermal histories.

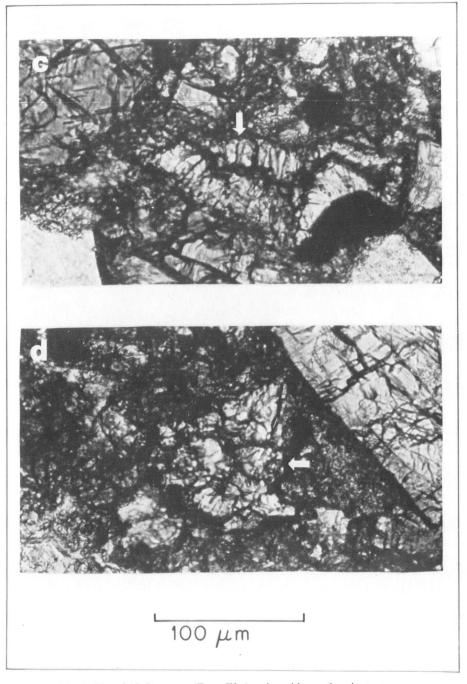

Fig. 6. (c) and (d) Pyroxene ($En_{38-40}Wo_6$) grains with reaction rim textures.

Origins of Fe-rich olivines in howardites

The occurrence of tridymite grains with pyroxene reaction rims ($En_{38}Wo_6$) in a eucrite presents a major problem. What reaction or mechanism took place to produce these rims? The typical reaction to produce such rims is SiO_2 + Olivine → Pyroxene. However, olivine has only rarely been found in eucrites. Delaney et al. (1980) cite only two known occurrences of olivine in eucrites: a metamorphic olivine in Ibitira (Steele and Smith, 1976) and olivine inclusions associated with a silica polymorph, ilmenite, and possibly Fe-rich orthopyroxene in a clinopyroxene clast ($En_{32}Wo_{38}$) found in the polymict eucrite ALHA 78040 (Delaney, unpublished data, 1980). Since the observed reaction rims seem to be post-regolith formation features, an olivine source must have been nearby to react with the silica during annealing.

Broad beam analyses (50 μm) were carried out by electron microprobe on areas of matrix (<20 μm grains only) in thin sections of ALHA 76005 (including the thin section where rims were observed). The mean compositions and 1σ deviation of 19 matrix analyses are shown in Table 4, along with the bulk analyses of ALHA 76005 reported by Olsen et al. (1978). The 1σ deviations of the broad beam analyses are understandably high, considering the difficulties of such analyses on a heterogeneously sized matrix; but in general the mean matrix composition is very close to the reported bulk composition. The matrix of ALHA 76005 is not enriched in an olivine component relative to the bulk meteorite so there must be some additional source of olivine.

If one continues to fractionate a magma along the zoning trend observed by Labotka and Papike (1980b) in a pyroxene phenocryst in ALHA 77302 (polymict eucrite), at some point of extreme fractionation the crystallizing pyroxene's composition will enter the zone of pyroxene instability (forbidden zone) (Lindsley and Munoz, 1969). A schematic diagram of this zoning trend and the forbidden zone is shown in Fig. 9a. This forbidden zone is outlined by the intersection of a volume in T-P-x space with the plane of the pyroxene quadrilateral. Inside this volume, pyroxenes in the system $MgSiO_3$-$CaSiO_3$-$FeSiO_3$ are not stable. The outline of the forbidden zone shown in Fig. 9a is the approximate boundary of the zone given by Smith (1972) for 925°C and one atmosphere pressure. Inside the forbidden zone, the stable mineral assemblage is Ca-rich pyroxene + Fa_{90-95} olivine + a silica polymorph (Smith, 1972) (see Fig. 9b). So, if a liquid continued to fractionate along the observed zoning trend, we would expect the assemblage of SiO_2 + olivine + pyroxene to be produced along the outermost crystal faces of pyroxene phenocrysts. If such a grain is then annealed, the SiO_2 + olivine would react with the more

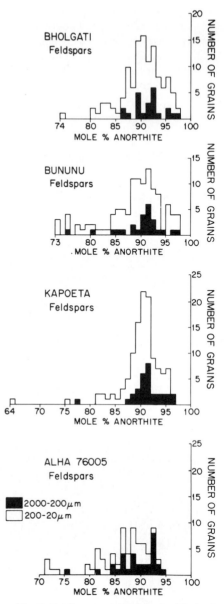

Fig. 7. Feldspar compositions for Bholgati, Bununu, Kapoeta, and ALHA 76005.

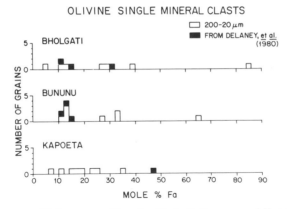

Fig. 8. Olivine compositions for Bholgati, Bununu, and Kapoeta.

Mg-rich core of the grain to produce a stable pyroxene outside of the pyroxene forbidden zone. Eucrites are silica normative and silica clasts ranging from <20 μm to ~ 200 μm make up from 1–4% vol. of howardite breccias. If pyroxene phenocrysts with outer portions consisting of the assemblage SiO_2 + olivine + pyroxene undergo comminution before annealing, the matrix of the breccia would have fine-grained olivine included in it. If annealing then occurs and silica clasts already present in the breccia react with the olivine in the matrix, then silica grains with pyroxene rims may be the result of post-regolith formation annealing. We would expect silica clasts to be available locally for reaction with matrix olivine because the most fractionated portion of a basalt would be the source for both the zoned pyroxene phenocrysts and the silica clasts. Unannealed portions and/or portions of the regolith with inefficient Fe-Mg mobilization may retain relict grains of Fe-rich olivine. Although Fe-rich olivine is uncommon, it is observed in many howardites.

Table 3. Relative proportions of sources for clasts in howardites.

	BHOLGATI			BUNUNU			KAPOETA		
	XL	L	S	XL	L	S	XL	L	S
% DIOGENITIC PYROXENE	61.6	55.9						69.8	54.2
VOLUME % OF ROCK PYROXENE	19.35	14.8	22.9	1.0	16.7	29.7	1.6	20.9	27.9
FELDSPAR		3.7	4.7		4.4	8.9		4.5	7.1
PYROXENE + PLAGIOCLASE ROCK	3.55	3.2	0.1	6.3	2.0	0.9	2.9	3.9	0.3
VOLUME % 'DIOGENITE'		41.2			28.2			31.3	
VOLUME % 'EUCRITE'		24.4			41.6			34.9	
VOLUME % 'REST'[1]		28.4			30.2			33.8	
FELDSPAR:PYROXENE RATIO[2]		35:65			41:59[3]			38:62	
'DIOGENITE':'EUCRITE' RATIO		63:37			40:60			47:53	

[1] REST = other lithic clasts + fused soil component + matrix.
[2] Ratio in eucrite.
[3] assumed ratio; average of Kapoeta, Bholgati, Pavlovka and Malvern values. (Pavlovka and Malvern from Labotka and Papike, 1980a).

Table 4. Matrix analyses of ALHA 76005.

	BULK ANALYSIS by E. JAROSEWICH (from Olsen et al., 1978)[1]	Average of 19 Broad Beam (50 μ) MICROPROBE ANALYSES of Matrix[2]	
		x	1σ
SiO_2	50.31	49.29	1.91
Al_2O_3	12.57	13.28	4.92
TiO_2	0.75	0.57	0.74
FeO	18.06	19.62	5.46
MnO	0.58	0.56	0.19
MgO	6.93	6.79	2.08
CaO	9.90	9.20	2.30
Na_2O	0.44	0.49	0.46
K_2O	0.06	0.08	0.07
Cr_2O_3	0.40	0.27	0.13
Total	100.00	100.15	

[1]Normalized to 100% excluding H_2O, P_2O_5, Fe, Ni, Co, FeS, C.
[2]Normalized to 100%.

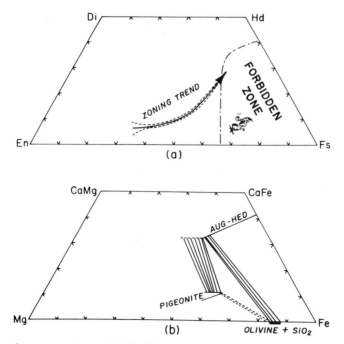

Fig. 9. Schematic representations of (a) 'forbidden zone' in relation to ALHA 77302 zoning trend (b) stable assemblage inside this zone (tie-lines do not represent real data, but are drawn to show approximate coexisting compositions).

SUMMARY AND CONCLUSIONS

This investigation of the modal petrology and mineral chemistry of Bholgati, Bununu, Kapoeta, and ALHA 76005 has allowed us to make the following observations on the characteristics and processes associated with the eucrite parent body regolith.

1. Modal data show that the characteristics of the regolith found by Labotka and Papike (1980a) hold true for these howardites also. These characteristics are: friable

source rocks, immature soil, comminution as the major soil-forming process, and a well-mixed soil.

2. Fe-Mg re-equilibration occurs locally in the regolith during annealing, perhaps during lithification of the breccia. This annealing is not necessarily a single event, but may have taken place several times. The scale of Fe-Mg re-equilibration can be smaller than a thin section.

3. Iron rich olivine ($\sim Fo_{15}$) found in howardites may be produced by extreme fractionation of a eucritic melt causing an assemblage of silica + olivine + high-Ca, Fe-rich pyroxene to be stable relative to pigeonite. If olivine produced in this manner is physically close to silica clasts during annealing in the regolith, the olivine will react with the silica to produce pyroxene.

Acknowledgments—Our meteorite research is funded by NSF grant EAR 7821857. We thank Martin Prinz of the American Museum of Natural History, Roy Clark and Brian Mason of the U.S. National Museum, Carleton Moore of Arizona State University, and the curatorial staff of Johnson Space Center for providing us with thin sections of meteorites used in this study. Constructive reviews by J. Delaney and R. Hewins are greatly appreciated. We also appreciate the help of Lois Koh, Pauline Papike, and Robin Spencer with the technical aspects of manuscript assembly.

REFERENCES

Bence A. E. and Albee A. L. (1968) Empirical correction factors for the electron microanalysis of silicates and oxides. *J. Geol.* **76**, 382–403.

Bunch T. E. (1975) Petrography and petrology of basaltic achondrite polymict breccias (howardites). *Proc. Lunar Sci. Conf. 6th*, p. 469–492.

Delaney J. S., Nehru C. E., and Prinz M. (1980) Olivine clasts from mesosiderites and howardites: Clues to the nature of achondritic parent bodies. *Proc. Lunar Planet. Sci. Conf. 11th*, p. 1073–1087.

Delaney J. S. Prinz M., Nehru C. E., and Harlow G. E. (1981) A new basalt group from howardites: Mineral chemistry and relationships with basaltic achondrites (abstract). In *Lunar and Planetary Science XII*, p. 211–213. Lunar and Planetary Institute, Houston.

Desnoyers C. and Jérome D. Y. (1973) Olivine compositions in howardites and other achondritic meteorites (abstract). *Meteoritics* **8**, 344.

Dreibus G., Kruse H., Spettel B., and Wänke H. (1977) The bulk composition of the moon and the eucrite parent body. *Proc. Lunar Sci. Conf. 8th*, p. 211–227.

Duke M. B. and Silver L. T. (1967) Petrology of eucrites, howardites and mesosiderites. *Geochim. Cosmochim. Acta* **31**, 1637–1665.

Dymek R. F., Albee A. L., Chodos A. A., and Wasserburg G. J. (1976) Petrography of isotopically-dated clasts in the Kapoeta howardite and petrologic constraints on the evolution of its parent body. *Geochim. Cosmochim. Acta* **40**, 1115–1130.

Fukuoka T., Boynton W. V., Ma M.-S., and Schmitt R. A. (1977) Genesis of howardites, diogenites, and eucrites. *Proc. Lunar Sci. Conf. 8th*, p. 187–210.

Hewins R. H. and Klein L. C. (1978) Provenance of metal and melt rock textures in the Malvern howardite. *Proc. Lunar Planet. Sci. Conf. 9th*, p. 1137–1156.

Jérome D. Y. and Goles G. G. (1971) A re-examination of relationships among pyroxene-plagioclase achondrites. In *Activation Analysis in Geochemistry and Cosmochemistry* (A. O. Brunfelt and E. Steinnes, eds.), p. 261–266. Universitetsforlaget, Oslo, Norway.

Klein L. C. and Hewins R. H. (1979) Origin of impact melt rocks in the Bununu howardite. *Proc. Lunar Planet. Sci. Conf. 10th*, p. 1127–1140.

Labotka T. C. and Papike J. J. (1980a) Howardite: Samples of the regolith of the eucrite parent-body. *Proc. Lunar Planet. Sci. Conf. 11th*, p. 1103–1130.

Labotka T. C. and Papike J. J. (1980b) Petrology of a quickly cooled pigeonite-basalt clast in the Antarctic meteorite ALHA 77302 (abstract). In *Lunar and Planetary Science XI*, p. 590–592. Lunar and Planetary Institute, Houston.

Lindsley D. H. and Munoz J. L. (1969) Subsolidus relations along the join hedenbergite-ferrosilite. *Amer. J. Sci.* **267A**, 295–324.

Mason B. (1967) The Bununu meteorite and a discussion of the pyroxene-plagioclase achondrites. *Geochim. Cosmochim. Acta* **31**, 107–115.

McCarthy T. S., Ahrens L. H., and Erland A. J. (1972) Further evidence for the mixing model for the howardite origin. *Earth Planet. Sci. Lett.* **15**, 86–93.

Moore C. B. (1962) The petrochemistry of the achondrites. In *Researches on Meteorites* (C. B. Moore, ed.), p. 164–178. Wiley and Sons, N.Y.

Olsen E. J., Noonan A., Fredriksson K., Jarosewich E., and Moreland G. (1978) Eleven new meteorites from Antarctica, 1976–1977. *Meteoritics* **13**, 209–225.

Prinz M., Nehru C. E., Delaney J. S., Harlow G. E., and Bedell R. L. (1980) Modal studies of mesosiderites and related achondrites, including the new mesosiderite ALHA 77219. *Proc. Lunar Planet. Sci. Conf. 11th*, p. 1055–1071.

Rajan R. S., Huneke J. C., Smith S. P., and Wasserburg G. J. (1975) $^{40}Ar-^{39}Ar$ chronology of isolated phases from Bununu and Malvern howardites. *Earth Planet. Sci. Lett.* **72**, 181–190.

Rajan R. S., Huneke J. C., Smith S. P., and Wasserburg G. J. (1979) Argon 40 – argon 39 chronology of lithic clasts from the Kapoeta howardite. *Geochim. Cosmochim. Acta* **43**, 957–971.

Simon S. B., Papike J. J., and Vaniman D. T. (1979) The Apollo 17 drill core. Part I: Modal petrology of the $>20~\mu m$ size fraction (abstract). In *Lunar and Planetary Science X*, p. 1122–1124. Lunar and Planetary Institute, Houston.

Smith D. (1972) Stability of iron-rich pyroxene in the system $CaSiO_3$-$FeSiO_3$-$MgSiO_3$. *Amer. Mineral.* **57**, 1413–1428.

Steele I. M. and Smith J. V. (1976) Mineralogy of the Ibitira eucrite and comparison with other eucrites and lunar samples. *Earth Planet. Sci. Lett.* **33**, 67–78.

Taylor G. J., Keil K., and Warner R. D. (1977) Petrology of Apollo 17 deep drill core: Depositional history based on modal analyses of 70009, 70008, and 70007. *Proc. Lunar Sci. Conf. 8th*, p. 3195–3222.

Taylor G. J., Warner R. D., and Keil K. K. (1979) Stratigraphy and depositional history of the Apollo 17 drill core. *Proc. Lunar Planet. Sci. Conf. 10th*, p. 1159–1184.

Vaniman D. T., Labotka T. C., Papike J. J., Simon S. B., and Laul J. C. (1979) The Apollo 17 drill core: Petrologic systematics and the identification of a possible Tycho component. *Proc. Lunar Planet. Sci. Conf. 10th*, p. 1185–1227.

Wilkening L. L. (1973) Foreign inclusions in stony meteorites–I. Chondritic xenoliths in the Kapoeta howardite. *Geochim. Cosmochim. Acta* **37**, 1985–1989.

APPENDIX

Table A1. Microprobe analyses of 2000–200 μm (L) pyroxene clasts for Bholgati and Bununu.

	BHOLGATI					BUNUNU				
	M29,1	M34,1	M58,1	M80,1	M82,1	M6,1	M8,1	M16,1	M19,1	M38,1
SiO_2	52.30	49.55	54.35	53.92	55.39	53.95	55.80	54.55	51.16	52.67
Al_2O_3	1.01	0.72	0.84	1.12	0.67	0.97	0.70	0.67	0.64	1.26
TiO_2	0.08	0.31	0.07	0.12	0.07	n.d.	n.d.	n.d.	0.28	0.13
FeO	21.58	34.65	14.66	14.54	12.37	16.97	10.30	16.10	28.50	20.22
MnO	0.63	0.96	0.55	0.54	0.42	0.66	0.35	0.44	1.04	0.74
MgO	22.52	12.87	27.81	27.32	30.20	25.14	31.08	26.48	16.79	23.24
CaO	1.05	1.13	0.94	1.17	0.78	1.69	0.48	1.26	1.04	1.50
Na_2O	n.d.	n.d.	n.d.	n.d.	n.d.	n.d.	n.d.	n.d.	n.d.	n.d.
K_2O	n.d.	n.d.	n.d.	n.d.	n.d.	n.d.	n.d.	n.d.	n.d.	n.d.
Cr_2O_3	0.35	0.55	0.50	0.82	0.24	0.59	0.72	0.48	0.17	0.58
Total	**99.52**	**100.74**	**99.72**	**99.56**	**100.14**	**99.97**	**99.43**	**99.98**	**99.62**	**100.34**
Cations/6 Oxygens										
Si	1.9590	1.9597	1.9634	1.9537	1.9677	1.9688	1.9759	1.9772	1.9813	1.9479
Al	0.0445	0.0336	0.0359	0.0480	0.0282	0.0418	0.0292	0.0288	0.0291	0.0550
Ti	0.0023	0.0092	0.0019	0.0034	0.0018	n.d.	n.d.	n.d.	0.0082	0.0037
Fe	0.6760	1.1464	0.4432	0.4406	0.3675	0.5181	0.3052	0.4882	0.9233	0.6257
Mn	0.0200	0.0323	0.0168	0.0165	0.0127	0.0205	0.0106	0.0134	0.0340	0.0232
Mg	1.2571	0.7589	1.4973	1.4755	1.5988	1.3676	1.6402	1.4305	0.9691	1.2808
Ca	0.0421	0.0480	0.0365	0.0454	0.0297	0.0659	0.0181	0.0489	0.0430	0.0593
Na	n.d.	n.d.	n.d.	n.d.	n.d.	n.d.	n.d.	n.d.	n.d.	n.d.
K	n.d.	n.d.	n.d.	n.d.	n.d.	n.d.	n.d.	n.d.	n.d.	n.d.
Cr	0.0102	0.0173	0.0142	0.0237	0.0066	0.0171	0.0202	0.0138	0.0053	0.0169
Total	**4.0112**	**4.0054**	**4.0092**	**4.0068**	**4.0130**	**3.9998**	**3.9994**	**4.0008**	**3.9933**	**4.0125**
En	63.01	38.22	75.10	74.60	79.59	69.35	83.09	72.21	49.21	64.39
Wo	2.11	2.42	1.83	2.30	1.48	3.34	0.92	2.47	2.18	2.98
Fs	34.88	59.36	23.07	23.11	18.93	27.31	16.00	25.32	48.01	32.62

n.d. = not detected.

Table A2. Microprobe analyses of 2000–200 μm (L) pyroxene clasts for Kapoeta and ALHA 76005.

	KAPOETA					ALHA 76005				
	M3, 1	M5, 1	M9, 2	M65, 1	M56, 2	M4, 1	M9, 1	M37, 1	M54, 1	M70, 1
SiO_2	54.34	53.45	55.51	52.34	49.30	49.68	52.10	50.43	51.23	49.62
Al_2O_3	1.10	1.38	0.77	0.77	0.54	0.66	1.52	1.09	1.12	1.96
TiO_2	0.14	0.21	0.06	0.14	0.15	0.22	0.11	0.59	0.19	0.99
FeO	14.13	18.60	12.79	21.96	34.86	32.73	19.57	27.31	23.40	18.66
MnO	0.49	0.54	0.46	0.64	1.11	0.96	0.60	0.84	0.71	0.51
MgO	27.87	24.24	29.65	22.18	12.85	12.93	21.60	13.79	19.24	9.42
CaO	1.38	1.60	0.65	1.27	0.78	2.19	3.44	5.95	3.29	18.67
Na_2O	n.d.	n.d.	n.d.	n.d.	n.d.	n.d.	n.d.	n.d.	n.d.	n.d.
K_2O	n.d.	n.d.	n.d.	n.d.	n.d.	n.d.	n.d.	n.d.	n.d.	n.d.
Cr_2O_3	0.60	0.48	0.73	0.31	0.08	0.18	0.64	0.61	0.56	0.29
Total	**100.05**	**100.50**	**100.62**	**99.61**	**99.67**	**99.55**	**99.58**	**100.61**	**99.74**	**100.12**
Cations/6 Oxygens										
Si	1.9546	1.9545	1.9670	1.9634	1.9722	1.9757	1.9472	1.9535	1.9483	1.9253
Al	0.0466	0.0594	0.0323	0.0340	0.0255	0.0310	0.0668	0.0500	0.0500	0.0895
Ti	0.0039	0.0057	0.0016	0.0040	0.0045	0.0065	0.0030	0.0172	0.0056	0.0290
Fe	0.4253	0.5691	0.3792	0.6891	1.1667	1.0887	0.6118	0.8850	0.7444	0.6056
Mn	0.0148	0.0168	0.0139	0.0203	0.0378	0.0322	0.0189	0.0275	0.0229	0.0169
Mg	1.4942	1.3212	1.5656	1.2399	0.7664	0.7661	1.2089	0.7959	1.0905	0.5449
Ca	0.0533	0.0627	0.0247	0.0511	0.0335	0.0934	0.1376	0.2471	0.1341	0.7761
Na	n.d.	n.d.	n.d.	n.d.	n.d.	n.d.	n.d.	n.d.	n.d.	n.d.
K	n.d.	n.d.	n.d.	n.d.	n.d.	n.d.	n.d.	n.d.	n.d.	n.d.
Cr	0.0171	0.0138	0.0206	0.0092	0.0026	0.0058	0.0188	0.0188	0.0169	0.0091
Total	**4.0098**	**4.0032**	**4.0049**	**4.0110**	**4.0092**	**3.9994**	**4.0070**	**3.9950**	**4.0127**	**3.9964**
En	75.18	67.07	78.94	61.98	38.24	38.68	61.02	40.70	54.75	28.04
Wo	2.68	3.18	1.25	2.55	1.67	4.72	6.98	12.64	6.73	39.93
Fs	22.14	29.74	19.82	35.46	60.09	56.60	32.00	46.66	38.52	32.03

n.d. = not detected.

Table A3. Microprobe analyses of 200–20 μm (S) pyroxene clasts for Bholgati.

	M010	M051	M066	M068	M072	M178	M189	M243	M372	M445
SiO_2	49.05	49.43	50.38	52.72	51.96	52.77	54.27	53.52	50.82	51.69
Al_2O_3	0.45	0.42	0.58	0.84	0.64	0.92	1.49	1.09	0.54	0.46
TiO_2	0.13	0.18	0.18	0.27	n.d.	0.21	0.06	0.13	0.23	n.d.
FeO	35.20	32.65	30.54	20.89	26.98	20.96	13.36	16.30	28.64	27.22
MnO	1.07	0.97	0.86	0.71	1.00	0.59	0.45	0.48	1.23	0.87
MgO	12.25	12.73	15.65	22.43	18.14	23.43	28.27	24.86	16.04	17.84
CaO	0.99	2.70	0.88	1.85	0.91	0.94	1.35	2.81	1.70	2.43
Na_2O	n.d.	n.d.	n.d.	n.d.	n.d.	n.d.	n.d.	n.d.	n.d.	n.d.
K_2O	n.d.	n.d.	n.d.	n.d.	n.d.	n.d.	n.d.	n.d.	n.d.	n.d.
Cr_2O_3	0.14	0.13	0.15	0.30	0.26	0.34	0.68	0.40	0.19	0.16
Total	**99.28**	**99.21**	**99.22**	**100.01**	**99.89**	**100.16**	**99.93**	**99.59**	**99.39**	**100.67**
Cations/6 Oxygens										
Si	1.9758	1.9767	1.9787	1.9625	1.9879	1.9564	1.9473	1.9611	1.9815	1.9739
Al	0.0213	0.0199	0.0269	0.0367	0.0287	0.0401	0.0632	0.0472	0.0248	0.0208
Ti	0.0041	0.0055	0.0052	0.0077	n.d.	0.0059	0.0016	0.0035	0.0068	0.0006
Fe	1.1862	1.0921	1.0032	0.6506	0.8635	0.6500	0.4010	0.4996	0.9341	0.8694
Mn	0.0367	0.0330	0.0286	0.0225	0.0323	0.0186	0.0138	0.0150	0.0405	0.0282
Mg	0.7357	0.7586	0.9158	1.2444	1.0346	1.2943	1.5118	1.3576	0.9318	1.0153
Ca	0.0429	0.1158	0.0372	0.0736	0.0371	0.0373	0.0520	0.1104	0.0708	0.0995
Na	n.d.	n.d.	n.d.	n.d.	n.d.	n.d.	n.d.	n.d.	n.d.	n.d.
K	n.d.	n.d.	n.d.	n.d.	n.d.	n.d.	n.d.	n.d.	n.d.	n.d.
Cr	0.0043	0.0041	0.0045	0.0089	0.0078	0.0100	0.0192	0.0116	0.0059	0.0049
Total	**4.0070**	**4.0057**	**4.0001**	**4.0069**	**3.9919**	**4.0126**	**4.0099**	**4.0060**	**3.9962**	**4.0126**
En	36.76	37.94	46.14	62.50	52.58	64.71	76.41	68.48	47.13	50.45
Wo	2.14	5.79	1.87	3.70	1.89	1.86	2.63	5.57	3.58	4.94
Fs	61.10	56.27	51.99	33.81	45.53	33.43	20.96	25.96	49.29	44.60

n.d. = not detected.

Table A4. Microprobe analyses of 200–20 μm (S) pyroxene clasts for Bununu.

	M005	M007	M028	M051	M092	M110	M134	M150	M302	M393
SiO_2	52.02	51.28	52.83	54.73	53.14	51.61	53.40	54.72	52.04	56.40
Al_2O_3	0.82	0.62	0.77	0.98	0.92	1.01	0.70	0.90	0.46	0.49
TiO_2	0.41	0.25	0.29	0.11	0.24	0.61	0.29	n.d.	0.13	n.d.
FeO	10.99	25.68	21.14	15.01	20.74	20.28	18.14	14.82	25.51	9.21
MnO	0.37	0.86	0.68	0.50	0.59	0.73	0.54	0.40	0.95	0.28
MgO	13.62	18.87	23.00	27.61	22.18	17.67	24.99	28.40	17.08	32.85
CaO	21.30	1.30	1.17	1.32	1.79	6.91	0.88	1.02	3.40	0.47
Na_2O	n.d.	n.d.	n.d.	n.d.	n.d.	n.d.	n.d.	n.d.	n.d.	n.d.
K_2O	n.d.	n.d.	n.d.	n.d.	n.d.	n.d.	n.d.	n.d.	n.d.	n.d.
Cr_2O_3	0.32	0.18	0.22	0.43	0.28	0.30	0.16	0.39	0.11	0.53
Total	99.85	99.04	100.10	100.69	99.88	99.12	99.10	100.65	99.68	100.23
Cations/6 Oxygens										
Si	1.9614	1.9728	1.9624	1.9614	1.9751	1.9646	1.9719	1.9575	1.9954	1.9709
Al	0.0366	0.0282	0.0339	0.0413	0.0401	0.0455	0.0303	0.0380	0.0208	0.0203
Ti	0.0117	0.0071	0.0082	0.0030	0.0068	0.0176	0.0080	n.d.	0.0038	n.d.
Fe	0.3467	0.8265	0.6570	0.4501	0.6448	0.6459	0.5604	0.4434	0.8182	0.2693
Mn	0.0118	0.0280	0.0213	0.0153	0.0187	0.0235	0.0169	0.0121	0.0309	0.0084
Mg	0.7656	1.0818	1.2736	1.4750	1.2287	1.0026	1.3753	1.5140	0.9763	1.7108
Ca	0.8604	0.0534	0.0465	0.0507	0.0714	0.2820	0.0350	0.0393	0.1398	0.0175
Na	n.d.	n.d.	n.d.	n.d.	n.d.	n.d.	n.d.	n.d.	n.d.	n.d.
K	n.d.	n.d.	n.d.	n.d.	n.d.	n.d.	n.d.	n.d.	n.d.	n.d.
Cr	0.0097	0.0054	0.0064	0.0121	0.0082	0.0089	0.0048	0.0111	0.0034	0.0145
Total	4.0039	4.0032	4.0093	4.0089	3.9938	3.9906	4.0026	4.0154	3.9886	4.0117
En	38.58	54.37	63.73	74.08	62.57	51.31	69.19	75.37	49.68	85.28
Wo	43.36	2.68	2.33	2.55	3.64	14.43	1.76	1.96	7.11	0.87
Fs	18.07	42.95	33.94	23.37	33.79	34.26	29.05	22.68	43.21	13.84

n.d. = not detected.

Table A5. Microprobe analyses of 200–20 μm (S) pyroxene clasts for Kapoeta.

	M001	M007	M012	M016	M019	M021	M170	M176	M196	M247
SiO_2	55.68	49.39	54.07	51.99	53.83	54.50	49.65	50.74	50.47	53.97
Al_2O_3	0.52	0.31	1.27	0.30	1.34	0.12	0.93	0.62	1.32	0.61
TiO_2	n.d.	0.12	0.06	0.23	n.d.	n.d.	0.06	0.24	0.35	n.d.
FeO	11.23	35.65	13.64	25.03	16.94	17.73	34.27	27.96	26.65	17.43
MnO	0.25	0.93	0.35	0.58	0.36	0.47	1.13	0.79	0.91	0.51
MgO	30.59	10.97	27.55	19.76	25.12	25.69	11.59	17.15	13.88	25.79
CaO	0.79	1.99	1.69	1.24	1.94	0.93	2.44	1.77	5.53	1.03
Na_2O	n.d.	n.d.	n.d.	n.d.	n.d.	n.d.	n.d.	n.d.	n.d.	n.d.
K_2O	n.d.	n.d.	n.d.	n.d.	n.d.	n.d.	n.d.	n.d.	n.d.	n.d.
Cr_2O_3	0.49	n.d.	0.71	0.11	0.45	0.15	0.07	0.16	0.47	0.48
Total	99.55	99.36	99.34	99.24	99.98	99.59	100.14	99.43	99.58	99.82
Cations/6 Oxygens										
Si	1.9777	1.9940	1.9556	1.9843	1.9621	1.9951	1.9776	1.9687	1.9660	1.9721
Al	0.0217	0.0148	0.0542	0.0135	0.0576	0.0051	0.0439	0.0283	0.0605	0.0263
Ti	0.0000	0.0037	0.0016	0.0066	0.0000	0.0000	0.0019	0.0071	0.0104	n.d.
Fe	0.3338	1.2042	0.4126	0.7990	0.5164	0.5428	1.1418	0.9073	0.8684	0.5329
Mn	0.0075	0.0320	0.0107	0.0188	0.0110	0.0146	0.0380	0.0260	0.0299	0.0157
Mg	1.6195	0.6600	1.4852	1.1241	1.3646	1.4018	0.6879	0.9915	0.8056	1.4041
Ca	0.0299	0.0862	0.0655	0.0509	0.0759	0.0367	0.1041	0.0737	0.2310	0.0404
Na	n.d.	n.d.	n.d.	n.d.	n.d.	n.d.	n.d.	n.d.	n.d.	n.d.
K	n.d.	n.d.	n.d.	n.d.	n.d.	n.d.	n.d.	n.d.	n.d.	n.d.
Cr	0.0139	n.d.	0.0202	0.0034	0.0130	0.0043	0.0022	0.0050	0.145	0.0140
Total	4.0040	3.9949	4.0056	4.0006	4.0006	4.0004	3.9974	4.0076	3.9863	4.0055
En	81.35	33.29	75.24	56.41	69.34	70.23	34.89	49.61	41.64	70.45
Wo	1.50	4.35	3.32	2.55	3.86	1.84	5.28	3.69	11.94	2.03
Fs	17.14	62.36	21.44	41.04	26.80	27.93	59.83	46.70	46.43	27.52

n.d. = not detected.

Table A6. Microprobe analyses of 200–20 μm (S) pyroxene clasts for ALHA 76005.

	M018	M017	M048	M152	M171	M172	M193	M196	M240	M246
SiO_2	49.55	49.23	49.56	49.82	48.11	49.00	48.74	49.41	49.80	48.22
Al_2O_3	0.98	0.58	0.82	1.62	0.97	1.83	1.72	1.49	1.34	1.78
TiO_2	0.64	0.28	0.47	0.47	0.29	0.10	0.10	0.16	0.34	0.08
FeO	28.53	35.26	29.82	23.75	35.77	35.96	32.00	29.34	24.74	35.42
MnO	0.83	1.08	0.83	0.72	0.98	1.06	0.96	0.84	0.71	1.08
MgO	10.65	11.06	10.85	9.90	10.79	11.25	11.81	15.42	11.40	12.64
CaO	8.26	1.56	7.13	12.94	1.82	0.78	3.90	2.69	11.82	1.01
Na_2O	n.d.	n.d.	n.d.	n.d.	n.d.	n.d.	n.d.	n.d.	n.d.	n.d.
K_2O	n.d.	n.d.	n.d.	n.d.	n.d.	n.d.	n.d.	n.d.	n.d.	n.d.
Cr_2O_3	0.28	0.56	0.23	0.49	0.33	0.79	0.48	0.76	0.59	0.51
Total	99.72	99.61	99.71	99.71	99.06	100.77	99.71	100.11	100.74	100.74
Cations/6 Oxygens										
Si	1.9630	1.9805	1.9692	1.9550	1.9561	1.9481	1.9416	1.9285	1.9386	1.9191
Al	0.0459	0.0276	0.0384	0.0752	0.0467	0.0860	0.0807	0.0686	0.0615	0.0834
Ti	0.0192	0.0084	0.0141	0.0139	0.0090	0.0030	0.0031	0.0048	0.0100	0.0023
Fe	0.9456	1.1865	0.9911	0.7797	1.2168	1.1961	1.0666	0.9580	0.8054	1.1791
Mn	0.0280	0.0368	0.0281	0.0238	0.0338	0.0356	0.0324	0.0277	0.0233	0.0365
Mg	0.6289	0.6631	0.6426	0.5790	0.6538	0.6665	0.7014	0.8973	0.6611	0.7494
Ca	0.3508	0.0674	0.3034	0.5443	0.0794	0.0334	0.1664	0.1126	0.4931	0.0431
Na	n.d.	n.d.	n.d.	n.d.	n.d.	n.d.	n.d.	n.d.	n.d.	n.d.
K	n.d.	n.d.	n.d.	n.d.	n.d.	n.d.	n.d.	n.d.	n.d.	n.d.
Cr	0.0088	0.0178	0.0072	0.0151	0.0107	0.0248	0.0151	0.0234	0.0183	0.0161
Total	3.9902	3.9881	3.9941	3.9860	4.0063	3.9935	4.0073	4.0209	4.0113	4.0290
En	32.20	33.94	32.70	30.05	32.96	34.51	35.66	44.96	33.34	37.32
Wo	17.96	3.45	15.44	28.25	4.00	1.73	8.46	5.64	24.87	2.15
Fs	49.84	62.61	51.86	41.70	63.04	63.77	55.88	49.39	41.79	60.53

n.d. = not detected.

Table A7. Microprobe analyses of feldspars for Bholgati and Bununu.

	BHOLGATI					BUNUNU				
	M061 S	M113 S	M137 S	M55,1 L	M49,1 L	M045 S	M049 S	M298 S	M31,1 L	M68,1 L
SiO_2	48.37	46.69	45.72	44.68	45.23	43.74	45.95	49.22	45.31	45.62
Al_2O_3	32.50	33.84	34.78	35.03	34.31	35.78	34.74	32.32	35.24	34.69
FeO	0.39	0.59	0.30	0.23	0.28	0.15	0.18	0.36	n.d.	n.d.
CaO	16.89	17.72	18.44	18.53	18.22	19.05	18.05	14.73	18.74	18.16
Na_2O	1.81	1.30	1.02	0.88	1.20	0.38	1.33	2.52	0.97	1.19
K_2O	0.25	0.07	0.04	0.03	0.09	n.d.	0.07	0.29	0.09	0.08
Total	100.21	100.21	100.30	99.38	99.33	99.10	100.32	99.44	100.35	99.74
Cations/8 Oxygens										
Si	2.2166	2.1466	2.1023	2.0753	2.1024	2.0383	2.1106	2.2588	2.0824	2.1065
Al	1.7564	1.8342	1.8857	1.9185	1.8802	1.9659	1.8817	1.7489	1.9094	1.8887
Fe	0.0148	0.0228	0.0115	0.0089	0.0108	0.0059	0.0069	0.0139	n.d.	n.d.
Ca	0.8294	0.8728	0.9086	0.9222	0.9075	0.9512	0.8882	0.7244	0.9229	0.8984
Na	0.1610	0.1156	0.0910	0.0793	0.1081	0.0346	0.1182	0.2244	0.0861	0.1065
K	0.0149	0.0041	0.0023	0.0018	0.0052	n.d.	0.0042	0.0169	0.0051	0.0047
Total	4.9931	4.9961	5.0014	5.0060	5.0142	4.9959	5.0098	4.9873	5.0059	5.0048
Ks	1.48	0.41	0.23	0.18	0.51	0.05	0.42	1.75	0.50	0.47
Ab	16.02	11.65	9.08	7.90	10.59	3.51	11.70	23.24	8.49	10.55
An	82.50	87.94	90.69	91.92	88.90	96.44	87.89	75.01	91.01	88.99

S = 200–20 μm; L = 2000–200 μm; n.d. = not detected.

Table A8. Microprobe analyses of feldspars for Kapoeta and ALHA 76005.

	KAPOETA						ALHA 76005			
	M031 S	M050 S	M143 S	M011,1 L	M45,1 L	M013 S	M061 S	M124 S	M19,1 L	M48,1 L
SiO_2	45.86	47.30	44.94	45.72	44.88	46.77	46.84	50.60	44.55	46.56
Al_2O_3	34.36	33.86	34.86	34.09	35.53	33.73	34.28	31.55	35.11	34.62
FeO	0.39	0.25	0.23	0.13	0.10	0.50	0.33	0.56	0.15	0.13
CaO	17.23	16.62	18.99	18.32	19.08	16.12	17.23	14.77	18.56	17.91
Na_2O	1.24	1.76	0.75	0.97	0.72	1.76	1.52	2.93	0.83	1.29
K_2O	0.07	0.13	0.02	0.05	0.04	0.12	0.12	0.38	0.04	0.08
Total	99.15	99.92	99.79	99.28	100.35	99.00	100.32	100.79	99.24	100.59
Cations/8 Oxygens										
Si	2.1266	2.1710	2.0804	2.1211	2.0654	2.1670	2.1459	2.2950	2.0716	2.1291
Al	1.8784	1.8327	1.9027	1.8645	1.9277	1.8426	1.8521	1.6876	1.9249	1.8665
Fe	0.0150	0.0095	0.0088	0.0052	0.0037	0.0193	0.0127	0.0211	0.0059	0.0048
Ca	0.8561	0.8172	0.9421	0.9108	0.9408	0.8002	0.8461	0.7177	0.9247	0.8774
Na	0.1119	0.1570	0.0673	0.0870	0.0642	0.1581	0.1351	0.2580	0.0750	0.1145
K	0.0043	0.0076	0.0013	0.0031	0.0021	0.0073	0.0072	0.0218	0.0026	0.0048
Total	4.9923	4.9950	5.0026	4.9917	5.0039	4.9945	4.9991	5.0012	5.0047	4.9971
Ks	0.44	0.77	0.13	0.31	0.21	0.76	0.73	2.19	0.26	0.48
Ab	11.51	15.99	6.66	8.69	6.37	16.37	13.67	25.86	7.48	11.49
An	88.05	83.23	93.21	91.00	93.42	82.87	85.60	71.95	92.26	88.03

S = 200–20 μm; L = 2000–200 μm.

Table A9. Microprobe analyses of olivine clasts[a] for Bhogati.

	M035	M104	M128	M170	M323	M383	M436
SiO_2	36.60	40.37	41.16	36.03	31.89	37.04	38.83
Al_2O_3	n.d.	n.d.	n.d.	n.d.	0.51	n.d.	n.d.
TiO_2	n.d.	n.d.	n.d.	n.d.	0.12	n.d.	n.d.
FeO	26.64	10.52	4.18	34.56	58.47	25.87	13.18
MnO	0.42	0.10	n.d.	n.d.	1.40	0.42	0.16
MgO	36.22	49.15	55.15	30.57	5.72	37.94	47.49
CaO	0.09	0.04	0.21	0.09	0.23	0.07	0.08
Na_2O	n.d.	n.d.	n.d.	n.d.	n.d.	n.d.	n.d.
K_2O	n.d.	n.d.	n.d.	n.d.	n.d.	n.d.	n.d.
Cr_2O_3	n.d.	n.d.	n.d.	n.d.	0.14	n.d.	n.d.
Total	99.97	100.18	100.70	101.25	98.49	101.34	99.74
Cations/4 Oxygens							
Si	0.9765	0.9912	0.9787	0.9829	1.0295	0.9704	0.9728
Al	n.d.	n.d.	n.d.	n.d.	0.0194	n.d.	n.d.
Ti	n.d.	n.d.	n.d.	n.d.	0.0029	n.d.	n.d.
Fe	0.5945	0.2162	0.0832	0.7887	1.5790	0.5669	0.2761
Mn	0.0096	0.0020	n.d.	n.d.	0.0383	0.0093	0.0033
Mg	1.4404	1.7984	1.9542	1.2429	0.2752	1.4812	1.7728
Ca	0.0025	0.0010	0.0053	0.0026	0.0080	0.0018	0.0022
Na	n.d.	n.d.	n.d.	n.d.	n.d.	n.d.	n.d.
K	n.d.	n.d.	n.d.	n.d.	n.d.	n.d.	n.d.
Cr	n.d.	n.d.	n.d.	n.d.	0.0036	n.d.	n.d.
Total	3.0235	3.0088	3.0214	3.0171	2.9559	3.0296	3.0272
Fo	70.45	89.18	95.92	61.18	14.54	71.99	86.39
Fa	29.55	10.82	4.08	38.82	85.46	28.01	13.61

[a] All olivines are 200–20 μm.
n.d. = not detected.

Table A10. Microprobe analyses of olivine clasts[a] for Bununu.

	M180	M240	M301	M310	M336	M337	M338	M460
SiO_2	36.70	37.17	32.69	40.64	38.38	40.49	40.65	39.69
Al_2O_3	n.d.	n.d.	1.05	n.d.	n.d.	n.d.	n.d.	n.d.
TiO_2	n.d.	n.d.	0.14	n.d.	n.d.	n.d.	n.d.	n.d.
FeO	31.25	30.53	49.67	10.51	24.74	13.12	11.96	12.75
MnO	0.40	0.44	1.12	0.11	0.39	0.13	0.17	0.15
MgO	32.47	32.57	15.07	49.76	37.89	47.52	48.35	46.44
CaO	0.08	0.07	0.08	0.06	0.08	0.12	0.06	0.10
Na_2O	n.d.	n.d.	n.d.	n.d.	n.d.	n.d.	n.d.	n.d.
K_2O	n.d.	n.d.	n.d.	n.d.	n.d.	n.d.	n.d.	n.d.
Cr_2O_3	n.d.	n.d.	0.19	n.d.	n.d.	n.d.	n.d.	n.d.
Total	**100.90**	**100.78**	**100.01**	**101.08**	**101.48**	**101.38**	**101.19**	**99.13**
Cations/4 Oxygens								
Si	0.9895	0.9988	0.9851	0.9889	0.9945	0.9936	0.9939	0.9954
Al	n.d.	n.d.	0.0373	n.d.	n.d.	n.d.	n.d.	n.d.
Ti	n.d.	n.d.	0.0032	n.d.	n.d.	n.d.	n.d.	n.d.
Fe	0.7047	0.6862	1.2522	0.2139	0.5363	0.2692	0.2447	0.2675
Mn	0.0092	0.0101	0.0286	0.0022	0.0086	0.0026	0.0036	0.0032
Mg	1.3047	1.3043	0.6769	1.8044	1.4633	1.7378	1.7622	1.7360
Ca	0.0024	0.0019	0.0026	0.0016	0.0023	0.0032	0.0017	0.0026
Na	n.d.	n.d.	n.d.	n.d.	n.d.	n.d.	n.d.	n.d.
K	n.d.	n.d.	n.d.	n.d.	n.d.	n.d.	n.d.	n.d.
Cr	n.d.	n.d.	0.0045	n.d.	n.d.	n.d.	n.d.	n.d.
Total	**3.0105**	**3.0013**	**2.9904**	**3.0110**	**3.0050**	**3.0064**	**3.0061**	**3.0047**
Fo	64.63	65.20	34.58	89.30	72.87	86.47	87.65	86.51
Fa	35.37	34.80	65.42	10.70	27.13	13.53	12.35	13.49

[a] All olivines are 200–20 μm.
n.d. = not detected.

Table A11. Microprobe analyses of olivine clasts[a] for Kapoeta.

	M092	M129	M254	M255	M256	M258	M259	M469
SiO_2	38.53	40.75	38.61	39.33	38.09	39.92	39.09	36.84
Al_2O_3	n.d.	n.d.	n.d.	n.d.	n.d.	n.d.	n.d.	n.d.
TiO_2	n.d.	n.d.	n.d.	n.d.	n.d.	n.d.	n.d.	n.d.
FeO	21.21	7.22	18.70	16.41	23.96	11.24	15.14	30.34
MnO	0.30	n.d.	0.19	0.14	0.38	0.07	0.14	0.44
MgO	39.94	54.07	43.71	45.53	38.51	49.50	46.20	33.66
CaO	0.17	0.06	0.13	0.06	0.06	0.14	0.11	0.03
Na_2O	n.d.	n.d.	n.d.	n.d.	n.d.	n.d.	n.d.	n.d.
K_2O	n.d.	n.d.	n.d.	n.d.	n.d.	n.d.	n.d.	n.d.
Cr_2O_3	n.d.	n.d.	n.d.	n.d.	n.d.	n.d.	n.d.	n.d.
Total	**100.15**	**102.10**	**101.34**	**101.47**	**101.00**	**100.87**	**100.68**	**101.31**
Cations/4 Oxygens								
Si	0.9958	0.9691	0.9755	0.9806	0.9893	0.9783	0.9777	0.9849
Al	n.d.	n.d.	n.d.	n.d.	n.d.	n.d.	n.d.	n.d.
Ti	n.d.	n.d.	n.d.	n.d.	n.d.	n.d.	n.d.	n.d.
Fe	0.4586	0.1436	0.3953	0.3424	0.5205	0.2305	0.3167	0.6784
Mn	0.0066	0.0000	0.0041	0.0030	0.0083	0.0014	0.0029	0.0099
Mg	1.5385	1.9166	1.6461	1.6919	1.4908	1.8078	1.7221	1.3411
Ca	0.0047	0.0016	0.0035	0.0016	0.0017	0.0037	0.0029	0.0008
Na	n.d.	n.d.	n.d.	n.d.	n.d.	n.d.	n.d.	n.d.
K	n.d.	n.d.	n.d.	n.d.	n.d.	n.d.	n.d.	n.d.
Cr	n.d.	n.d.	n.d.	n.d.	n.d.	n.d.	n.d.	n.d.
Total	**3.0042**	**3.0309**	**3.0245**	**3.0195**	**3.0106**	**3.0217**	**3.0223**	**3.0151**
Fo	76.78	93.03	80.47	83.05	73.82	88.63	84.35	66.08
Fa	23.22	6.97	19.53	16.95	26.18	11.37	15.65	33.92

[a] All olivines are 200–20 μm.
n.d. = not detected.

Ion probe analysis of plagioclase in three howardites and three eucrites

I. M. Steele and J. V. Smith

Department of the Geophysical Sciences, University of Chicago, Chicago, Illinois 60637

Abstract—Concentrations of Li, Na, Mg, K, Ti, Sr and Ba were measured by ion microprobe in plagioclase of lithic and mineral clasts of three howardites (Brient, Jodzie, Frankfort) and three eucrites (Juvinas, Pasamonte, Ibitira). Lithium shows a near-linear positive correlation with Na in all samples and tends to higher values in eucritic plagioclase. Similar trends were observed for lunar samples, suggesting that partial melting processes enrich Li in the melt, and that the linear trends are an indication that plagioclase grains are related petrogenetically. Potassium shows a positive correlation with Na but concentrations differ little among samples. Strontium shows a weak increase with Na, and all samples except Ibitira plot in a broad band similar to lunar data but with concentrations about half that of lunar plagioclase. Barium shows a complex relation with eucrite data grouped according to sample and clasts within samples; howardite data scatter widely. The Sr/Ba ratio of plagioclase is suggested as a possible indicator of petrogenetic processes in the eucrite parent body (bodies). Titanium values are generally lower than lunar but are grouped for each meteorite with a weak correlation with Na.

INTRODUCTION

The classical distinction (Duke and Silver, 1976; Mason *et al.*, 1979) between eucrites (unbrecciated and monomict brecciated basaltic achondrites) and howardites (polymict brecciated achondrites with both eucritic and orthopyroxenitic components), and the relationships with other meteorite classes (diogenites and mesosiderites), are being amended by recent investigations. Although many eucrites have chemical and mineralogical properties consistent with generation by crystal-liquid fractionation at low pressure from a single series of liquids, the Stannern and Ibitira eucrites do not (Stolper, 1977; Reid *et al.*, 1979). Particular attention is being paid to polymict eucrites which are distinguished from howardites by absence of an orthopyroxenite component (Miyamoto *et al.*, 1979; Simon and Haggerty, 1980; Smith and Schmitt, 1981; Takeda and Yanai, 1981). Although howardites are texturally similar to lunar breccias (Bunch, 1975), they are not simple two-component mixtures of a eucrite and an orthopyroxenite (Labotka and Papike, 1980). Dymek *et al.* (1976) concluded that the Kapoeta howardite contains a continuous series of fragments from rocks ranging from orthopyroxenites to iron-rich basalts. These were interpreted in terms of plutonic-to-volcanic differentiation of pyroxene-rich basaltic materials that were affected by thermal and impact metamorphism. Trace elements in basaltic clasts in howardites indicate a greater complexity of igneous differentiation than for eucrites (Mittlefehldt *et al.*, 1979), and complex models involving progressive melting of a single source region (Stolper, 1977) and layered parent body (Takeda, 1979) have been proposed. Whereas Ni-poor metal is associated with primary clasts, Ni-rich metal is associated with glassy melt rocks (Hewins, 1979). Of particular interest are two types of basaltic clasts in howardites (Delaney *et al.*, 1981), and a corresponding division among the basaltic eucrites.

Because achondrites are similar to lunar samples in many respects, parallel studies should be profitable. As in the case for lunar samples, minor and trace elements in silicates should prove useful for classification of individual grains and as petrogenetic indicators. We now present ion microprobe data for plagioclase from a small group of achondrites, and use lunar data (Steele *et al.*, 1980) as the basis for comparison.

SAMPLES, EXPERIMENTAL TECHNIQUE AND DATA

Three eucrites were selected for initial study because of their different properties. Ibitira crystallized at the surface of a body as indicated by spherical voids that once contained gas (Wilkening and Anders, 1975), and the coarse pyroxene lamellae require slow cooling, probably under a hot blanket of impact debris (Steele and Smith, 1976). Juvinas belongs to the main petrogenetic group of eucrites (Stolper, 1977) and has typical clouding of pyroxene and plagioclase attributable to prolonged metamorphism (Harlow and Klimentidis, 1980). Pasamonte shows sporadic clouding, and belongs to the unequilibrated group of eucrites (Reid and Barnard, 1979). The three howardites consist of the unusual Brient (Reid et al., 1979) and the well described Frankfort (Labotka and Papike, 1980) and Jodzie.

Random electron microprobe analyses of pyroxene and plagioclase in polished thin sections (unnumbered sections obtained from Field Museum of Natural History) are summarized in Figs. 1 and 2. The strongly equilibrated nature of Ibitira, and the near-equilibration of Juvinas, contrast with the wide range of pyroxene and plagioclase compositions in Pasamonte. Whereas pyroxene (Fig. 1 and data of Labotka and Papike, 1980) of the Frankfort howardite is dominated by Mg-rich and Ca-poor compositions ($En_{77-61}Wo_3$), the Brient howardite contains a large fraction of pyroxene crystals which scatter to Ca- and Fe-rich compositions. Most of the Jodzie pyroxenes show compositions near $En_{41}Wo_4$ and $En_{34}Wo_{40}$ suggesting low-temperature equilibration, while some trend towards the Mg-rich and Ca-poor groups which dominate Brient and Frankfort. Plagioclase compositions of mineral and lithic clasts (Fig. 2) overlap between 80 and 95% An, except for Ibitira whose plagioclase is restricted to 95–96%.

Whereas lithic clasts were recognized in our sections of Juvinas, Pasamonte, Brient and Jodzie, Ibitira is homogeneous and Frankfort contains only mineral clasts. Analyzed lithic clasts are illustrated in Figs. 3–6 and described in the legends.

Ion probe analyses for elements amenable to low mass resolution conditions were made for plagioclase grains from each clast, and from random grains in the equilibrated Ibitira eucrite, the Frankfort howardite and brecciated area 3 of Brient. Analytical conditions were described in Steele et al. (1980), and absolute concentrations (Table 1) were referred to Lake County plagioclase (An 68: Li 4.1, Mg 822, K 986, Ti 230, Sr 582, Ba 63 ppmw) (Meyer et al., 1974). The small scatter for Li, Mg, Sr and Ba in the Ibitira plagioclase indicates a precision better than about ±5%, and the greater scatter for Ti is considered to represent real compositional variation. Plots of trace elements (except Mg) vs. mol.% Ab (Figs. 7–11) can be compared with similar ones for lunar plagioclase (Steele et al., 1980). The range of values for any one lithic clast is generally narrow except for Mg for which scatter may result from overlap onto small Mg-rich inclusions in clouded regions (Harlow and Klimentidis, 1980). This is consistent with the low scatter of Mg for Ibitira which does not show clouding and Pasamonte plagioclases which contain inclusion-free areas. Because Ba and Sr should cohere strongly in early condensates from the solar nebula, Fig. 12 is included to show the relation between these elements for plagioclase from the eucrites and howardites because differences among samples should reflect different post-condensation processes.

DISCUSSION

At the simplest level, the trace elements in plagioclase provide a test of the chemical homogeneity of the meteorites. Ibitira is remarkably uniform in contrast to all the other specimens. Because its texture (Wilkening and Anders, 1975; Steele and Smith, 1976) is vesicular, it must have crystallized from a low-pressure liquid, and chemical zoning of the plagioclase would have been expected. The subsequent metamorphism must have been remarkably effective in causing homogenization of $CaAl_2Si_2O_8$ and $NaAlSi_3O_8$ components in view of the notorious sluggishness of Si,Al interdiffusion in crystalline plagioclase. Juvinas plagioclase is not homogeneous (Fig. 10) in spite of monotonic trends between Li, Na and K. The Ba–Sr plot and especially the Ba–Na plot show distinct groupings for the separate clasts. Pasamonte plagioclase appears to belong to one population, except with respect to Ba–Sr (Fig. 12). There is considerable scatter for the howardites, and it is obvious that a very detailed study will be needed for each one in order to delineate the full range of components; however, the plagioclase in howardites cannot be obtained from any of the three analyzed eucrites.

At this stage it is difficult to reach detailed conclusions about the petrogenetic significance of the trace element patterns. It will be necessary to analyse more eucrites before the models of partial melting of more than one parent body can be explored (Stolper, 1977; Consolmagno and Drake, 1977; Reid et al., 1979). In the meantime, some preliminary remarks are made based on the following assumptions: (a) there was no

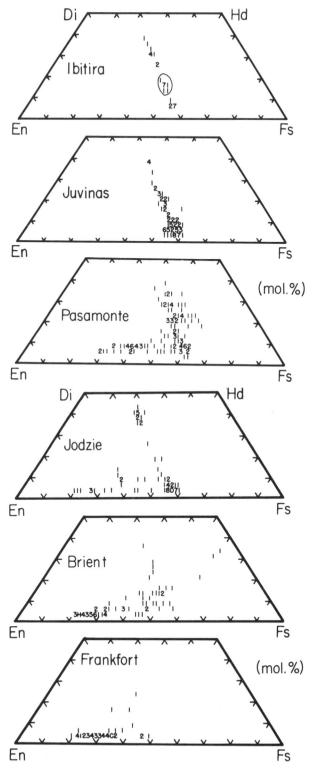

Fig. 1. Random analyses for major elements in pyroxene. Data obtained with electron probe and focused beam. Number of analyses shown for each composition (A = 10, B = 11 etc.). Circled data for Ibitira were obtained with 30 μm beam and represent bulk composition of exsolved grains.

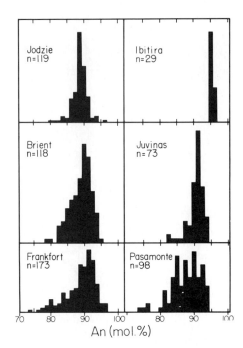

Fig. 2. Histograms of mol.% An in plagioclase. Data obtained with focused beam from random grains.

fractionation of elements condensing from the solar nebula at a similar temperature, in particular Sr and Ba in the high-temperature range, and Na and K in a medium-temperature range (cf. Morgan et al., 1978; Palme et al., 1978); (b) progressive partial melting occurred at low pressure of a source containing olivine, pigeonite, plagioclase, Cr-rich spinel and metal (Stolper, 1977); and (c) crystal-liquid fractionation occurred during crystallization near the surface of a parent body. It is assumed that K, Na and Ba would not be retained significantly in residual minerals and would ultimately end up mostly in plagioclase, and that some Sr and much Ti would go into pyroxene and oxide. The partitioning of Li is not known experimentally, but from our unpublished ion probe data similar levels occur in plagioclase, olivine and pyroxenes. The elements are now discussed in sequence, using observations on lunar plagioclase (Steele et al., 1980) as a guide.

Lithium

Lunar data indicate: (i) a positive correlation between Li and Na for plagioclase from any particular rock, and (ii) an increase of Li/Na for plagioclase from rocks with a higher content of magmatophile elements. For achondritic plagioclase, Ibitira shows a tight cluster (Fig. 7), and the Juvinas, Pasamonte, Jodzie and Frankfort specimens show single positive trends. Scatter occurs for some of the Brient data. The mean Li/Na ratio for all six specimens tends to correlate positively with the fraction of high-Ca, high Fe-pyroxene, and with the ratio of eucrite to orthopyroxenite components indicated in Fig. 1. The KREEP-rich lunar rocks are attributed to late partial melts, and their plagioclase contains higher Li than other lunar rocks, many of which must be cumulates. It will be interesting to test whether the Li/Na ratio of plagioclase will prove useful in determination of the amount of partial melting, and the extent of cumulate components in the basaltic achondrites.

Potassium

There is a strong positive correlation between K and Na (Fig. 8) for all the present data, except for two points for the Brient howardite. Although distinct trends are found for the various meteorites, the separation is much less than for the Li–Na plot. Linear trends are shown for simplicity, but a curved trend is suggested by Frankfort data with the highest Na values. These data indicate strong coherence between Na and K. The low alkali content of Ibitira was discussed by Stolper (1977).

Strontium

The Sr–Na plot (Fig. 9) shows a greater spread than for the remarkably tight trend for lunar plagioclase except for the 67667 lherzolite (Steele et al., 1980, Fig. 4). The Ibitira data are distinct, and actually fall on the lunar trend. All the other data lie scattered about the dashed trend which lies parallel to the lunar trend at about half the value of Sr. Although detailed measurements are needed of Sr in pyroxenes (and in phosphates?), the present data imply several tentative conclusions. It is necessary for the source regions of eucrites and howardites to be chemically more diverse or processes more varied than the moon. In general, the source regions of eucrites and howardites, except for Ibitira, contain a lower fraction of high-temperature Sr than low-temperature Na with respect to the moon. The Ibitira parent body has the same ratio of high/low condensates as the moon, but the similar volatile depletions in Ibitira (Stolper, 1977) and the moon might only be coincidental. In principle, the relative importance of incomplete condensation of volatile elements from the solar nebula and thermal volatilization can be tested by using various element ratios. The similarity of Na/Mn for Ibitira and the other eucrites (Palme et al., 1978) argues against differences in the extent of condensation. The constancy of K/Ba for lunar rocks was used as an argument against heterogeneous accretion and volatile loss from surface flows (e.g., Taylor, 1975), and the K/Na ratio should probably decrease during thermal volatilization. Ibitira has lower K/Ba (Fig. 10) than other eucrites, and similar but slightly higher K/Na. No firm conclusion is reached, but thermal volatilization for Ibitira appears plausible especially in view of its vesicular texture.

Barium

Figures 10 and 12 show the relations between Ba and Na, and Ba and Sr. Pasamonte shows a linear trend between Ba and Na, but there is a tendency for two groups in the Ba–Sr plot. Each clast in Juvinas has a distinct composition range (except perhaps for areas 3 and 4 for which data are sparse). The howardities scatter more than the eucrites, and the total spread in Fig. 10 is similar to that for lunar specimens.

Although the eucrite data lie about the Sr/Ba ratio for Cl meteorite (86/25; Smith, 1977), they do not lie on a single trend passing through the origin. In lunar rocks, the Sr/Ba ratio in plagioclase ranges from about 10–20 in anorthosites, ~15 in mare basalts, and only 1–4 in mg-rich plutonic KREEP-rich rocks (Steele et al., 1980). Furthermore, the Ba/Na ratios vary from low values in anorthosites and mare basalts to much higher ones in the other rock types. No simple analogy can be used for the eucrites, as is immediately seen from the large variation of Ba/Na for the eucrites and the small variations of Sr/Ba, and fractionation of Ba, Sr and Na among plagioclase, pyroxene and liquid must be considered. Thus for the Juvinas clasts, the Ba/Na ratio of the plagioclase increases with the amount of plagioclase. The small range of Ba/Sr for eucrites would be consistent with only single-stage fractionation in parent bodies which began with nearly the cosmic ratio, and the greater range of Ba/Sr for the moon would result from multi-stage fractionation.

Titanium

The upper limit of Ti (Fig. 11) is less than half of that for lunar plagioclase (Steele et al., 1980). In general, there is a weak positive correlation between Na and Ti for plagioclase from both the moon and the achondrites, indicating some systematic trend in crystal-liquid fractionation. The howardites show more complex relations than the eucrites. In detail, each eucrite is different, and only Pasamonte shows a positive trend between Ti and Na.

(a)

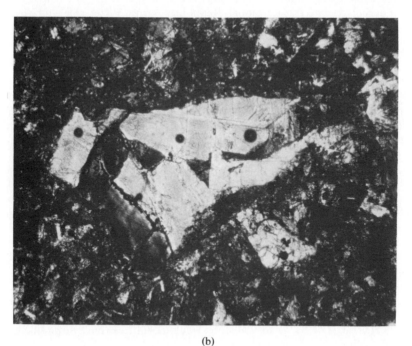

(b)

Fig. 3. Transmitted light photomicrographs of lithic clasts in Juvinas. Plagioclase is light colored, pyroxene dark; each clast is surrounded by dark matrix. Maximum dimension 1.7 mm. (a) Area 1—cluster of relatively large, blocky plagioclase with minor interstitial pyroxene. Richer in plagioclase than for other areas. (b) Area 2—small clast with distinctly more pyroxene than area 1. Plagioclase more lath shaped than in area 1. Visible pyroxene lamellae are $\sim 5\mu$m wide. (c) Area 3—lath-shaped plagioclase and fine-grained pyroxene in nearly equal amounts and minor mesostasis. (d) Area 4—one large plagioclase lath with fine pyroxene, plagioclase, ilmenite and mesostasis. All areas show clouded plagioclase with minute inclusions (Harlow and Klimentidis, 1980).

(c)

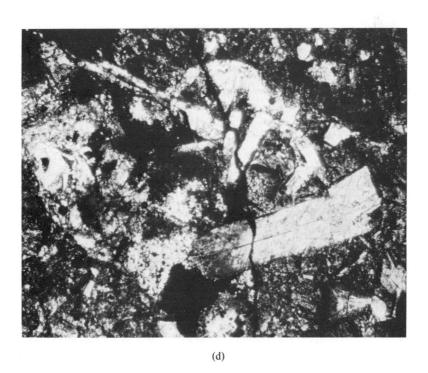

(d)

Fig. 3. (*Continued*)

(a)

(b)

Fig. 4. Transmitted light photomicrographs of lithic clasts in Pasamonte. Maximum dimension 1.7 mm. (a) Area 1—lath-shaped plagioclase with intersitial pyroxene and mesostasis (black). (b) Area 2—clast dominated by coarse, subhedral plagioclase with minor pyroxene all surrounded by dark matrix. Plagioclase shows sporadic clouding and pyroxene does not show exsolution.

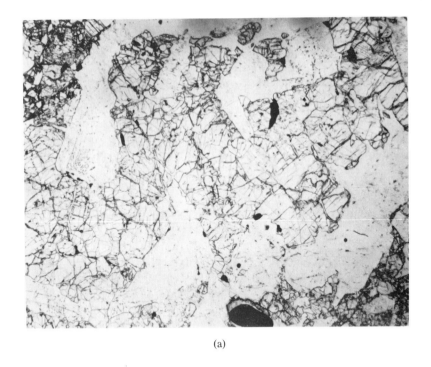

(a)

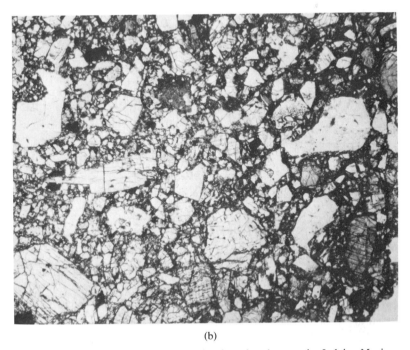

(b)

Fig. 5. Transmitted light photomicrograph of analysed areas in Jodzie. Maximum dimension 1.7 mm. (a) Area 1—lithic clast with igneous or metaigneous texture. Modal pyroxene/plagioclase ~2. Probably equivalent to cumulate ferrogabbro of Bunch *et al.* (1976). (b) Area 2—matrix with same mineralogy as area 1 except brecciated with only rare bimineralic clasts. Rare lamellae in pyroxene are up to ~3 μm wide.

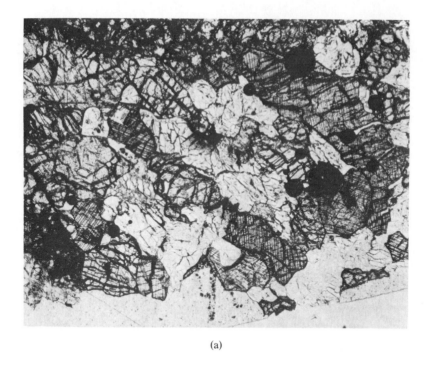

(a)

(b)

Fig. 6. Transmitted light photomicrographs of lithic clasts in Brient. Maximum dimension: (a) 1.7 mm; (b) 0.8 mm. (a) Area 1—metamorphic texture like that in Ibitira, with dominant equidimensional pyroxene whose pronounced exsolution lamellae are up to ~8 μm wide. (b) Area 2—fine-grained clast suggesting rapid cooling.

Table 1. Trace element and anorthite content of plagioclase.

			ppmw						mol.%
	area	point	Li	Mg	K	Ti	Sr	Ba	Ab
Ibitira	1	1	7.0	215	280	80	205	81	5.9
		2	7.3	220	260	60	205	81	5.9
		3	7.4	240	300	90	210	82	5.9
		4	7.3	210	270	70	205	81	5.7
		5	7.4	225	250	65	205	82	6.0
		6	7.1	225	300	80	220	84	6.0
Juvinas	1	1	8.8	2600	400	45	155	36	10.0
		2	8.0	2250	340	50	145	32	8.6
		3	9.1	2270	400	45	155	41	10.0
	2	4	8.0	1050	400	45	165	54	9.7
		5	9.4	1360	440	40	160	56	10.5
		6	7.8	650	380	35	170	56	9.5
		7	9.2	1370	390	45	175	56	10.3
	3	8	7.1	840	350	35	180	68	9.1
		9	8.7	1320	390	30	185	70	10.0
		10	9.0	530	390	35	185	70	10.7
	4	11	10.1	1210	520	50	185	74	11.8
		12	11.1	220	560	40	200	74	12.8
Pasamonte	1	1	18.7	560	800	105	205	71	16.2
		2	17.8	490	610	110	190	60	13.5
		3	18.4	560	1120	150	200	78	17.8
	2	4	15.5	200	680	115	220	66	14.4
		5	10.4	120	350	70	210	53	9.2
		6	9.6	110	320	70	200	48	8.6
		7	9.6	150	320	70	185	47	8.5
Jodzie	1	1	8.8	60	410	50	195	na	10.6
		2	12.1	1090	550	75	190	57	13.6
		3	10.6	50	470	55	200	55	12.3
		4	9.5	90	520	55	190	53	12.5
	2	5	11.9	400	560	60	185	54	13.6
		6	10.4	50	480	50	195	57	12.3
		7	11.9	100	560	60	190	59	13.8
		8	9.1	60	420	45	190	55	11.3
Brient	1	1	11.1	390	610	215	200	29	14.7
		2	5.2	410	530	220	210	26	13.7
		3	5.8	350	500	175	185	23	13.6
	2	1	5.9	2900	1260	215	230	130	14.8
	3	1	5.1	700	400	145	180	60	10.2
		2	7.9	600	1150	200	210	45	17.4
		3	6.6	760	1220	190	240	63	18.8
		4	7.1	90	410	110	150	165	11.2
		5	5.8	820	920	170	200	81	17.4
		6	4.5	750	330	85	145	36	9.7
		7	11.9	410	960	100	215	120	11.6
Frankfort	1	1	7.6	180	1740	75	190	114	20.8
		2	1.5	170	150	20	130	na	4.4
		3	4.3	230	630	140	170	na	12.8
		4	3.1	360	400	70	160	na	9.3
		5	3.5	170	460	90	160	57	10.5
		6	2.6	350	320	40	145	23	8.1
		7	2.5	700	260	100	125	34	6.5
		8	3.5	230	410	110	195	26	8.9
		9	5.2	220	850	90	215	102	15.1
		10	7.1	390	1550	150	185	100	20.4
		11	3.8	260	580	70	160	66	11.5
		12	3.7	190	470	80	155	54	10.4

na = not analysed.

1292   I. M. Steele and J. V. Smith

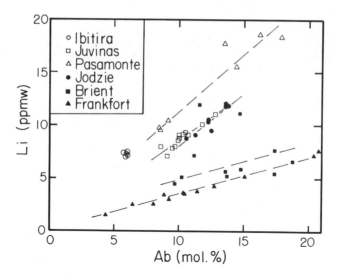

Fig. 7. Mol.% albite in plagioclase vs. Li content in ppmw. Several points for Brient fall well off any linear trend.

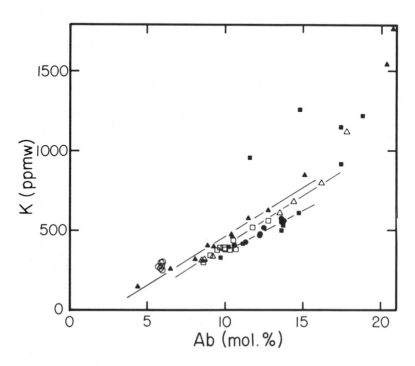

Fig. 8. Mol.% albite in plagioclase vs. K content in ppmw. Key in Fig. 7. Data for Brient show considerable scatter.

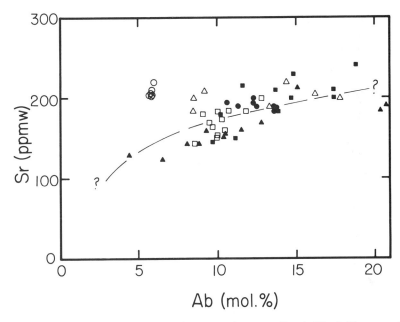

Fig. 9. Mol.% albite in plagioclase vs. Sr content in ppmw. Key in Fig. 7. The curve is parallel to the lunar trend (Steele *et al.*, 1980), and displaced to about half the Sr content of lunar plagioclase.

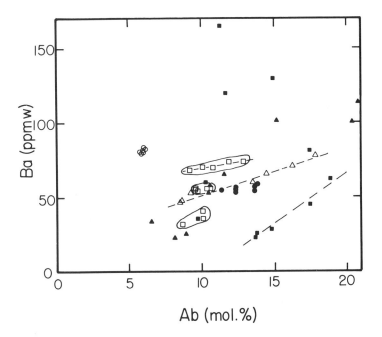

Fig. 10. Mol.% albite in plagioclase vs. Ba content in ppmw. Key in Fig. 7. The Juvinas data plot in three groups corresponding to different clasts.

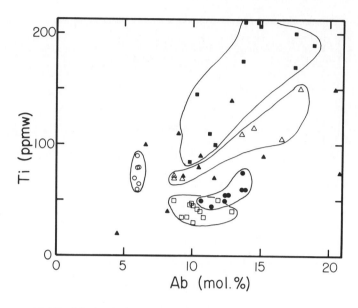

Fig. 11. Mol.% albite in plagioclase vs. Ti content in ppmw. Key given in Fig. 7.

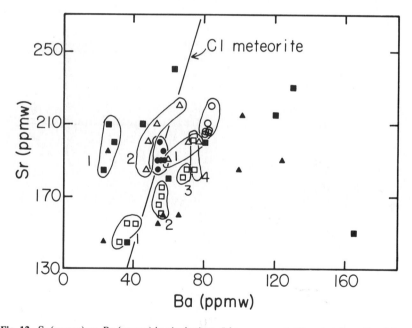

Fig. 12. Sr (ppmw) vs. Ba (ppmw) in plagioclase. Line represents Cl meteorite ratio while numbers adjacent to groupings indicate clast of Table 1. Key given in Fig. 7.

CONCLUSIONS

The above data provide encouragement in pursuing the trace element content of plagioclase in the achondrites. The trace-element signature should prove important in classifying and comparing the individual components of the polymict eucrites and howardites. Thus the simple patterns in Jodzie contrast with the more complex patterns in Frankfort and particularly Brient. To obtain full advantage from the data, it will be necessary to make further experimental measurements of the distribution of these elements among plagioclase, pyroxene and liquid before testing any partial-melting model. In particular, the existing distribution data for Ba, Sr and Ti (summarized in Table 1, Steele et al., 1981) should be extended to Li because of the indication of an important difference between eucritic and orthopyroxenitic components. The three eucrites so far measured lie close together in the main cluster of eucrites in Reid et al. (1979, Fig. 1) and systematic study of the other eucrites, especially extreme ones (e.g., Stannern, Nuevo Laredo), and the cumulate eucrites is planned. Extension to the trace elements in pyroxenes would be the next step to compare with plagioclase data for eucrites.

Acknowledgments—We thank I. Baltuska and R. Draus for technical help, and R. Hewins, J. Delaney and "another" for helpful reviews. Specimens were kindly supplied by Dr. Edward Olsen, Field Museum of Natural History, Chicago. Financial support from NASA grant 14-001-171 is appreciated.

REFERENCES

Bunch T. E. (1975) Petrography and petrology of basaltic achondrite polymict breccias (howardites). *Proc. Lunar Sci. Conf. 6th.* p. 469–492.
Bunch T. E., Chang S., Neil J. M., and Burlingame A. (1976) Unique characteristics of the Jodzie howardite. *Meteoritics* 11, 260–261.
Consolmagno G. J. and Drake M. J. (1977) Composition and evolution of the eucrite parent body: evidence from rare earth elements. *Geochim. Cosmochim. Acta* 41, 1271–1282.
Delaney J. S., Prinz M., Nehru C. E., and Harlow G. E. (1981) A new basalt group from howardites: mineral chemistry and relationships with basaltic achondrites (abstract). In *Lunar and Planetary Science XII*, p. 211–213. Lunar and Planetary Institute, Houston.
Duke M. B. and Silver L. T. (1967) Petrology of eucrites, howardites and mesosiderites. *Geochim. Cosmochim. Acta* 31, 1637–1665.
Dymek R. F., Albee A. L., Chodos A. A., and Wasserburg G. J. (1976) Petrography of isotopically-dated clasts in the Kapoeta howardite and petrologic constraints on the evolution of its parent body. *Geochim. Cosmochim. Acta* 40, 1115–1130.
Harlow G. E. and Klimentidis R. (1980) Clouding of pyroxene and plagioclase in eucrites: Implications for post-crystallization processing. *Proc. Lunar Planet. Sci. Conf. 11th*, p. 1131–1143.
Hewins R. H. (1979) The composition and origin of metal in howardites. *Geochim. Cosmochim. Acta* 43, 1663–1673.
Labotka T. C. and Papike J. J. (1980) Howardites: samples of the regolith of the eucrite parent body: petrology of Frankfort, Pavlovka, Yurtuk, Malvern and ALHA 77302. *Proc. Lunar Planet. Sci. Conf. 11th*, p. 1103–1130.
Mason B., Jarosewich E., and Nelen J. A. (1979) The pyroxene-plagioclase achondrites. *Smithsonian Contrib. Earth Sci.* # 22, p. 27–45.
Meyer C., Anderson D. H., and Bradley J. G. (1974) Ion microprobe mass analysis of plagioclase from "non-mare" lunar samples. *Proc. Lunar Sci. Conf. 5th*, p. 685–706.
Mittlefehldt D. W., Chou C.-L., and Wasson J. T. (1979) Mesosiderites and howardites: igneous formation and possible genetic relationships. *Geochim. Cosmochim. Acta* 43, 673–688.
Miyamoto M., Takeda H., and Yanai K. (1979) Eucritic polymict breccias from Allan Hills and Yamato Mountains, Antarctica (abstract). In *Lunar and Planetary Science X*, p. 847–849. Lunar and Planetary Institute, Houston.
Morgan J. W., Higuchi H., Takahashi H., and Hertogen J. (1978) A "chondritic" eucrite parent body: inference from trace elements. *Geochim. Cosmochim. Acta* 42, 27–38.
Palme H., Baddenhausen H., Blum K., Cendales M., Dreibus G., Hafmeister H., Kruse H., Palme C., Spettel B., Vilesek E., and Wänke H. (1978) New data on lunar samples and achondrites and a comparison of the least fractionated samples from the earth, the moon and the eucrite parent body. *Proc. Lunar Planet. Sci. Conf. 9th*, p. 25–57.
Reid A. M. and Barnard B. (1979) Unequilibrated and equilibrated eucrites (abstract). In *Lunar and Planetary Science X*, p. 1019–1021. Lunar and Planetary Institute, Houston.
Reid A. M., Duncan A. R., and LeRoex A. (1979) Petrogenetic models for eucrite genesis (abstract). In *Lunar and Planetary Science X*, p. 1022–1024. Lunar and Planetary Institute, Houston.
Simon S. B. and Haggerty S. E. (1980) Anarctic meteorites: a petrographic study of ALHA 77302 (abstract). In *Lunar and Planetary Science Xi*, p. 1033–1035. Lunar and Planetary Institute, Houston.
Smith J. V. (1977) Possible controls on the bulk composition of the earth: implications for the origin of the earth and moon. *Proc. Lunar Sci. Conf. 8th*, p. 333–369.
Smith M. R. and Schmitt R. A. (1981) Preliminary chemical data for some Allan Hills polymict

eucrites (abstract). In *Lunar and Planetary Science XII*, p. 1014–1016. Lunar and Planetary Institute, Houston.

Steele I. M. and Smith J. V. (1976) Mineralogy of the Ibitira eucrite and comparison with other eucrites and lunar samples. *Earth Planet. Sci. Lett.* **33**, 67–78.

Steele I. M., Hutcheon I. D., and Smith J. V. (1980) Ion microprobe analysis and petrogenetic interpretations of Li, Mg, Ti, K, Sr, Ba in lunar plagioclase. *Proc. Lunar Planet. Sci. Conf. 11th*, p. 571–590.

Steele I. M., Smith J. V., Raedeke L. D., and McCallum I. S. (1981) Ion probe analysis of Stillwater plagioclase and comparison with lunar analyses (abstract). In *Lunar and Planetary Science XII*, p. 1034–1036. Lunar and Planetary Institute, Houston.

Stolper E. (1977) Experimental petrology of eucritic meteorites. *Geochim. Cosmochim. Acta* **41**, 587–611.

Takeda H. (1979) A layered-crust model of howardite parent body. *Icarus* **40**, 455–470.

Takeda H. and Yanai K. (1981) Yamato polymict eucrites: regolith breccias excavated from a layered crust (abstract). In *Lunar and Planetary Science XII*, p. 1071–1073. Lunar and Planetary Institute, Houston.

Taylor S. R. (1975) *Lunar Science: A Post-Apollo View*. Pergamon, N.Y. 372 pp.

Wilkening L. L. and Anders E. (1975) Some studies of an unusual eurcite. *Geochim. Cosmochim. Acta* **39**, 1205–1201.

Thermal and impact histories of pyroxenes in lunar eucrite-like gabbros and eucrites

Hiroshi Takeda*[1], Hiroshi Mori[1], Teruaki Ishii[2], and Masamichi Miyamoto[3]

[1]Mineralogical Institute, Faculty of Science, University of Tokyo, Hongo, Tokyo 113, Japan
[2]Ocean Research Institute, University of Tokyo, Minamidai, Nakano-ku, Tokyo 164, Japan
[3]Department of Pure and Applied Science, College of General Education, University of Tokyo, Komaba, Meguro-ku, Tokyo 153, Japan

Abstract—Pyroxene crystals from a lunar eucrite-like gabbro (61223,47 and 61224,36) and a quartz monzodiorite (15405,148) were compared with those of three eucrites, Moore County, Juvinas and Yamato-74356. Their exsolution, inversion and deformation features were studied by analytical transmission electron microscope (ATEM) in addition to single crystal X-ray diffraction and electron microprobe techniques. The X-ray photographs of the inverted pigeonite in 61223 show augite and pigeonite reflections with (100) in common. The textures were obscured by a later shock-heating event, which might have produced minute patches of pigeonite and Mg-poor and Al-rich glass detected by the ATEM. Intragranular in situ melting due to shock effects was observed in 15405. The eucrite pigeonite does not show such extensive shock features but is characterized by stacking faults with an orthopyroxene-like structure cutting through the fine exsolution lamellae of augite on (001). The cooling histories estimated from widths and compositions of the exsolved augites, measured by the ATEM method, suggest that those of 61223 and Moore County are compatible with a plutonic origin but Juvinas and Yamato-74356 revealed much shallower origin than that previously proposed.

INTRODUCTION

To obtain a better understanding of the processes responsible for the formation of ancient planetary crusts, we have been carrying out comparative studies on pyroxene in pristine lunar crustal rocks (Warren and Wasson, 1978) and in diogenites, eucrites and howardites which are proposed to belong to the same parent body (Takeda et al., 1976). The problems with a single parent body were discussed by Delaney et al. (1981). We proposed a layered crust model for this howardite parent body (Takeda, 1979), which appears to have preserved a record of the most primitive differentiation events associated with crust formation in our solar system. We have reported a similarity in crystallization trends between KREEP basalts and quickly cooled eucrites (Takeda et al., 1980). Lunar KREEP-rich quartz-monzodiorite 15405 contains pigeonite with textures and chemical compositions similar to those of pigeonite in common eucrites (Ryder, 1976). To obtain more information on cooling histories of the two primitive crusts, we must investigate more slowly cooled pyroxenes in more deep seated rocks than the near surface basalts. However, available lunar analogues of the cumulate eucrites are very rare. With the discovery of a pristine eucritic gabbro 61224,6 from Descartes by Marvin and Warren (1980), we are able to investigate more slowly cooled and presumably deep crustal lunar rocks. Hence we have examined exsolution, inversion and deformation textures of pyroxene in samples 61224,36, 61223,47 and 15405,148 by single-crystal X-ray diffraction, electron microprobe and analytical transmission electron microscope (ATEM) techniques. For comparison we reinvestigated pyroxenes in lunar eucritic analogues, Moore County, Juvinas and Yamato-74356. The latter two eucrites are reported to contain primary augites (Reid, pers. comm., 1972; Takeda, 1979).

---

*Visiting Professor at National Institute of Polar Research.

In addition we have used quantitative ATEM techniques to determine the compositions of the host pigeonite and exsolved augite. From these ATEM data we have estimated the temperature below which exsolution no longer develops, using the pigeonite-augite geothermometer of Ishii *et al.* (1979). Also bulk chemical compositions of coexisting pigeonite-augite pairs were employed to estimate crystallization temperatures. Computer simulation of cooling histories of these exsolved pigeonites has been performed using the computer program of Miyamoto and Takeda (1977), our measurements of lamellar widths and our calculated temperature ranges. Although the models are not yet completely developed, the cooling rates thus obtained may yield some constraints on depth of crystallization of pristine nonmare rocks in the lunar crust and on the extent of subsolidus annealing events that equilibrated pigeonite in Juvinas.

SAMPLES AND EXPERIMENTAL TECHNIQUES

Thin section 61224,36 (supplied by Dr. U. Marvin) is one of three thin sections of the eucrite-like gabbro 61224,6 which was discovered in the 4 to 10 mm fraction from a white layer of the 61220 soil at Apollo 16 Station 1 (Marvin and Warren, 1980). Sample 61223,47 is a collection of small rock and pyroxene fragments similar to 61224,6 picked out by Marvin from the subsample of 61223, which is the 2–4 mm size fraction from the same sieving that produced 61224. Pyroxene crystals were separated from four rock fragments in 61223,47, which consisted of clear plagioclase and brown pyroxene surrounded by pale green glass. Pyroxene fragments 1–3 mm in diameter were also selected from lunar KREEP-rich quartz-monzodiorite 15405,148 originally studied by Ryder (1976), Moore County cumulate eucrite (Amer. Museum of Natural Hist. #4471), Juvinas (Arizona State Univ. #65a) and Yamato-74356,10 (Nat'l Inst. of Polar Res.) which is the only non-polymict eucrite from Antarctica (Takeda *et al.*, 1979b). Prior to the preparation of samples for electron microprobe and electron microscopy, pyroxene grains were examined on an X-ray precession camera to identify the orientation of any coexisting phases. $h0l$ and $0kl$ nets were taken using Zr-filtered Mo $K\alpha$ radiation. After X-ray study crystals were mounted with the b axis perpendicular to the plane of a glass slide, and polished to about 30 μm thick for electron microprobe analysis of their unmixed phases. Microprobe analyses were made of pyroxene, plagioclase and glass in thin sections. Chemical variations in the pyroxenes were recorded by measuring Ca, Mg, and Fe concentrations. The bulk compositions of the crystals were obtained by random scanning, avoiding the fractures, with a normal beam of a small area for ten seconds and by repeating this procedure about ten times to cover the entire crystal.

Crystals larger than about 0.5 mm in diameter were mounted in thick resin and sliced parallel to (010) with a 0.08 mm thick diamond blade. One slice was used for microprobe work, and the other for TEM study. For the other, smaller samples, 30 μm-thick thin sections were prepared. For direct correlation between light-optical, electron-optical and microprobe observations, some of the crystals were removed from the glass slide and glued to 3 mm molybdenum TEM grids for support. Each sample was thinned in an Edwards ion-thinning machine until perforation occurred.

Examination of the ion-thinned samples was carried out with a high-resolution TEM (JEOL-JEM 100CX) operated at 100 kV. The chemical compositions of submicroscopic exsolved phase were determined on a Hitachi analytical transmission electron microscope (H-600) equipped with a Kevex Energy Dispersive Spectrometer (EDS). For the analyses, energy dispersive spectra were acquired for 200 seconds on an approximately 60–100 nm area with a beam current of 1 nA at 100 kV. EDS data were reduced using the extended method of the Cliff-Lorimer (1975) k-factor technique. They pointed out that when a specimen is thin, the X-ray absorption and fluorescence can be neglected. For any two elements, the weight fraction ratio C_1/C_2 is proportional to the ratio of the characteristic X-ray intensities from two elements, I_1/I_2. The details of the method are also given by Lorimer and Cliff (1976), Morimoto and Kitamura (1981), and McGee *et al.* (1980).

To derive experimental k-factors, we plotted observed X-ray intensity ratios, I_X/I_{Si} for two elements X and Si against the total counts that are related to the thickness of specimens, for eight elements (X = Na, Mg, Al, Ca, Ti, Cr, Mn, and Fe) in ion-thinned pyroxene standards. The standards used in these experiments include two salites and one ferrosalite, for high-Ca pyroxenes, and two bronzites and one hypersthene for low-Ca pyroxene with uniform chemical compositions. The intensity ratios for the above elements were found to have linear relations against the increasing total counts within the range of thickness which will give sufficient total counts for a quantitative total chemical analysis including minor elements. The range of thickness for this condition may run from an area which gives rise to gradual deviation from the thin film criteria with constant k-values due to the X-ray absorption and fluorescence effects, to a thicker area which corresponds to transmission limit for the STEM mode observation. Within this thickness range for every pyroxene standard, the intensity ratios for the above elements to Si with respect to the increasing total counts can be approximated by a straight line with negative slope for elements lighter than Si, and by a positive slope for elements heavier than Si, such as Ca and Fe. An example of k-values determined from a

Table 1. Ranges and standard deviations (σ) of k-factors used for quantitative ATEM analyses.

Elements	k-factors thin → thick	σ
Mg	1.49–1.53	0.008
Al	0.98–1.05	0.05
Si	1.00–1.00	0.*
Ca	0.59–0.56	0.003
Ti	0.75–0.75	0.03
Cr	0.73–0.73	0.03
Mn	0.60–0.60	0.02
Fe	0.63–0.60	0.002

*fixed.

pyroxene standard with salite composition is given in Table 1, together with their standard deviations. The method of analysis employed in this study is justified only if we use a uniform standard whose chemical composition is close to the unknown samples. The k-values of the above elements determined from salite and bronzite, for example, have similar inclination with respect to the increasing total counts and the values are fairly similar but not identical to each other.

By employing proper k-values by the above method for given total counts, we converted the observed X-ray intensity ratios into weight fraction ratios by an equation $C_X/C_{Si} = k \cdot I_X/I_{Si}$, where C_X and C_{Si} are weight percentages of oxides for element X and Si, respectively. Then, the total weight percentage of oxides was normalized to 100%. In the analytical procedure, we did not assume cation stoichiometry for the cation sites. Therefore, good sums of cation numbers for the tetrahedral and octahedral sites per six oxygens of pyroxene can be used as an indicator of a satisfactory analysis.

The results of the quantitative analyses for the eucritic pyroxenes (Table 2) show that the total cation numbers are between 3.996 and 4.005. Good agreement was obtained between the two microprobe analyses and the ATEM data for thick augite lamellae (50–100 μm) with a uniform chemical composition in inverted pigeonites from Moore County. The difference includes minor variations in chemical composition of the augite lamellae. An example of the standard deviations, calculated from five ATEM analyses for augite lamellae (0.2 μm in width) in the Y-74356 pigeonite, is given in Table 2.

RESULTS

Lunar eucrite-like gabbro, 61223 and 61224

As was described by Marvin and Warren (1980) for thin sections 61224,6, our observation on 61223,47 shows that rock chips are coarse-grained with equal amounts of plagioclase, augite and inverted pigeonite, and minor shock-melted glass. The texture was described in detail by Marvin and Warren (1980), who reported that the rock is similar to some eucrites and is suggestive of a slowly cooled, plutonic rock. However, one of the characteristics of the lunar analogue in comparison with true cumulate eucrites is the relatively abundant occurrence of augite. The presence of augite in 61224 is due to the high proportion of normative Wo component (13.6 mol.%) in the CIPW norm calculated from the bulk chemistry of Marvin and Warren (1980). The augite coexists with 'pigeonite' in a chain of subhedral pyroxene crystals 0.5 to 3 mm in size (See Fig. 1 of Marvin and Warren, 1980).

The bulk chemical compositions of coexisting pigeonite-augite pairs in 61223,47 and 61224,36 are shown in Fig. 1 and Table 3. The host and exsolved phases in each of the coexisting pyroxenes are connected by tie-lines in Fig. 1. A representative Wo content of the inverted pigeonite is given in Table 3. The chemical compositions of 'pigeonite' and augite in 61224 are homogeneous in each bulk composition but both pyroxenes show exsolution. The pyroxene compositions from lunar quartz monzodiorite, 15405, which are more similar to those from ordinary eucrites than those of 61224, are also plotted in Fig. 1. The bulk chemical compositions of the 'pigeonite' and augite in contact with each other

Table 2. Chemical compositions of unmixed pyroxenes in pigeonites from Moore Co., Juvinas and Yamato-74356 by three different methods[†]: electron microprobe (MP), ATEM and X-ray refinement (XR).

Meteorite Pyroxene	Moore County Augite lamella			Juvinas Augite lamella				Y-74356		
Method	MP1	MP2	ATEM1	MP3	MP4	ATEM2	XR	Pig. host ATEM2	Augite lamella ATEM2	σ
SiO_2	51.28	51.3	51.45	49.68	49.68	51.11	52.54	49.30	50.90	(0.30)
TiO_2	0.57	0.60	0.64	0.25	0.19	0.23	–	0.21	0.35	(0.07)
Al_2O_3	0.82	0.79	0.80	0.79	0.41	0.92	–	0.43	0.84	(0.12)
Cr_2O_3	0.32	–	0.31	0.80	0.20	0.35	–	0.35	0.55	(0.05)
FeO	14.11	13.6	14.59	20.29	23.94	15.42	13.51	35.91	15.99	(0.10)
MnO	–	0.51	0.53	0.65	0.61	0.57	–	1.16	0.52	(0.04)
MgO	12.44	12.9	12.04	11.38	11.71	11.03	11.64	11.51	10.45	(0.18)
CaO	19.97	20.3	19.64	17.00	12.11	20.37	22.31	1.13	20.40	(0.21)
Na_2O	0.07	0.11	–	0.07	0.08	–	–	–	–	–
Total	99.58	100.11	100.00*	100.91	98.93	100.00*	100.00*	100.00*	100.00*	
				Cations per 6 oxygens						
Si	1.959	1.950	1.963	1.925	1.966	1.963	2.00	1.979	1.962	
Ti	0.016	0.017	0.018	0.007	0.006	0.007	–	0.007	0.010	
Al	0.037	0.036	0.036	0.036	0.019	0.042	–	0.020	0.038	
Cr	0.010	–	0.009	0.025	0.006	0.011	–	0.011	0.017	
Fe	0.451	0.433	0.465	0.658	0.792	0.495	0.43	1.206	0.515	
Mn	–	0.016	0.017	0.021	0.020	0.018	–	0.040	0.017	
Mg	0.708	0.732	0.685	0.657	0.690	0.631	0.66	0.689	0.600	
Ca	0.817	0.826	0.803	0.706	0.513	0.838	0.91	0.049	0.842	
Na	0.005	0.008	–	0.005	0.006	–	–	–	–	
Total	4.003	4.018	3.996	4.040	4.018	4.005	4.00	4.001	4.001	
Ca**	41.4	41.5	41.1	34.9	25.7	42.6	45.5	2.5	43.0	
Mg	35.8	36.8	35.1	32.5	34.6	32.2	33.0	35.5	30.7	
Fe	22.8	21.7	23.8	32.6	39.7	25.2	21.5	62.0	26.3	

[†]References: MP1 (Hostetler and Drake, 1978), MP2 (Ishii and Takeda, 1974), MP3 (Takeda et al., 1974), MP4 (Reid, pers. comm., 1974), XR (Takeda et al., 1974), ATEM1 (Mori and Takeda, 1981b), and ATEM2 (This study).
*Data of ATEM were normalized to 100 wt.% due to ratio-method of analysis. σ: standard deviations, for numbers of measurements greater than five.
**Atomic percentage.

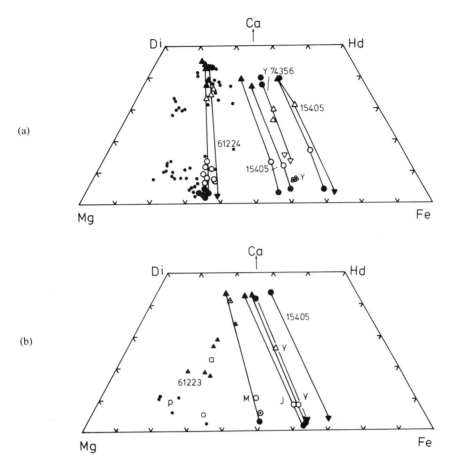

Fig. 1. (a) Pyroxene quadrilateral plot of the bulk chemical compositions of inverted pigeonite (open circle) and augite (open triangle), and exsolved pairs of host phase (solid circle) and exsolved phase (solid triangle), from lunar eucrite-like gabbro 61224,36, 61223,47 (small solid circle) and quartz monzodiorite 15405,148. Data of Yamato-74356 are plotted between the 15405 values. Small circle represents individual measurement. (b) Chemical compositions determined by ATEM, of exsolved pyroxenes from 61223 (small symbols), Moore County (M), Juvinas (J), Yamato-74356(Y), and 15405. Unmixed pair is connected by a tie-line, with their bulk chemical compositions (open symbol) in between. Compositions of a thick augite lamella (triangle with dot) and the host phases between them (circle with dot) determined by microprobe are given. Small symbols of 61223 represent augite (a), pigeonite (p) and orthopyroxene (o).

(Table 3) from 61224 give an apparent crystallization or last equilibration temperature of 1095°C (±20°C) for the pigeonite-augite geothermometer calibrated by the pigeonite geothermometer and the coexisting three pyroxene (orthopyroxene-pigeonite-augite) assemblages (Ishii *et al.*, 1979). This temperature is nearly equal to that of the minimum stability limit of pigeonite at Fe/(Fe + Mg) = 0.323, 1100°C estimated by the pigeonite geothermometer (Ishii *et al.*, 1979) or 1075°C estimated by the exsolved pair of the orthopyroxene and augite in the 61224 inverted pigeonites. The thermometer of Ishii *et al.* (1979), can be applied for both pigeonite-augite pairs and orthopyroxene-augite pairs, because it is calibrated by the three pyroxene assemblages. Other pyroxene geothermometers are also given by Ross and Huebner (1975) and Wells (1977).

Ten grains that appeared to be single crystals in 61223 were examined by the X-ray method. Three show a powder diffraction pattern with concentration of intensities at some portions of the semi-concentric powder diffraction line. Even in the single crystal diffraction photographs, powder diffraction lines and reflections from many misoriented small

Table 3. Chemical compositions (wt.% oxides) of selected pyroxenes in lunar analogues of eucrites determined by electron microprobe.

	L61224,36			L61223,47			L15405,148		
	Pig. bulk	Aug. bulk	Glass	Pig. bulk	Opx. host	Aug. bleb	Pig. bulk	Pig. host	Aug. lam.
SiO_2	51.4	50.5	49.3	52.6	52.7	51.3	50.1	50.2	49.6
Al_2O_3	1.48	2.17	15.06	1.61	0.51	3.15	0.82	0.43	1.76
TiO_2	0.55	0.98	0.44	0.49	0.46	0.78	0.69	0.40	1.09
Cr_2O_3	0.41	0.68	0.20	0.38	0.26	0.46	0.24	0.18	0.54
FeO	21.3	13.07	12.48	18.05	21.0	10.84	29.8	33.7	17.44
MnO	0.24	0.25	0.13	0.30	0.33	0.22	0.61	0.60	0.27
MgO	22.1	15.92	9.06	21.2	23.5	15.53	11.62	12.49	10.79
CaO	3.18	16.88	11.74	3.76	1.17	17.98	5.56	2.12	17.13
Na_2O	0.14	0.15	0.84	0.11	0.00	0.15	0.08	0.0	0.09
Total	100.80	100.60	99.25	98.50	99.93	100.41	99.52	100.12	98.71
Ca*	6.3	34.3	(34.4)	7.9	2.3	37.4	12.4	4.6	37.4
Mg	60.9	45.0	(37.0)	62.3	65.1	45.0	35.9	37.9	32.8
Fe	32.8	20.7	(28.6)	29.7	32.6	17.6	51.7	57.4	29.8

*Atomic percents.

crystals are observed. Of the seven crystals, four were inverted pigeonites and three were augites. The very long exposure time required for photographs with sufficient intensities may be consistent with the presence of glass in the crystals observed by ATEM. The X-ray diffraction patterns of the inverted pigeonites in 61223 show considerably misoriented spots of orthopyroxene due to mosaicing by shock effect and the presence of diffraction spots of augites with (100) in common with orthopyroxene. Faint spots of pigeonite were also observed with (100) in common with orthopyroxene. The occurrence of minor (100) pigeonite (or clinohypersthene) in orthopyroxene detected by the X-ray method is similar to that which was interpreted to have been produced by the shock-induced shear transformation from orthopyroxene found in the Johnstown diogenite (Mori and Takeda, 1981a). However, the origin may be different. Because of the poor quality of the photographs, the presence of remnant (001) augite lamellae common in inverted pigeonite could not be confirmed. The X-ray diffraction patterns of the augite crystals were also so disturbed by the shock effects that the presence of exsolved pyroxenes could not be confirmed.

Microscopic observation of both grains of 61223 in oriented thin sections revealed that the entire texture of the inverted pigeonites is broken into small slightly misoriented regions disturbed by the shock or reheating events or both, resulting in mosaicing. Many fine, complex branches of fractures appear to have been injected by Al- and Ca-rich melt (Fig. 2a). The presence of augite was confirmed by microprobe but some apparently high-Al augite analysed by microprobe may represent a mixture of pyroxene and high-Al glass. Fussy lamellae or elongated bleb-like textures found in low-Ca pyroxenes in 61223 (Fig. 2a) may not be directly correlated with the (100) augites detected by the X-ray method. The Ca content of Ca-rich clinopyroxene with low-Al contents obtained by the microprobe scan is not much higher than those of pigeonite. Although the host-orthopyroxene composition is identical to that of 61224, the primary exsolution textures are more severely obscured by the later shock events than that in 61224. In the 61224 inverted pigeonite, there are several thinner augite lamellae or blebs presumably (100) in common in the host orthopyroxene, between thick augite lamellae (3–10 μm in width) presumably on (001) of the original pigeonite. The crystals show mosaicing due to shock effects.

A combination of chemical analysis (Fig. 1b and Table 4) and selected-area electron-

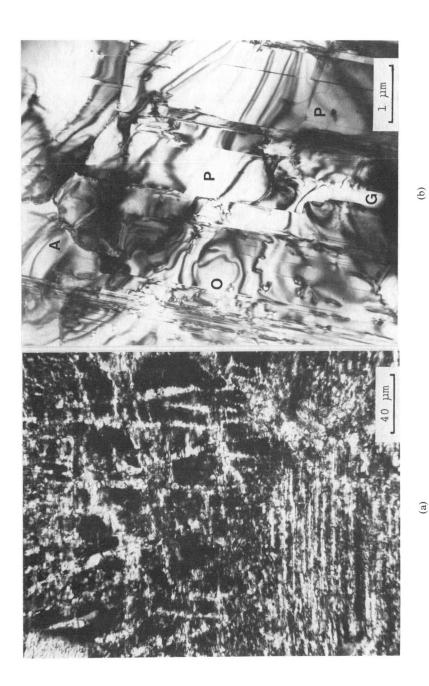

Fig. 2. (a) Photomicrograph of a part of (010) thin section of single crystal low-Ca pyroxene from 61223. Trace of the (100) plane is horizontal. Crossed nicols. (b) Electron micrograph (Bright field) of the 61223 inverted pigeonite. (100) trace is nearly vertical. Irregular intergrowth of augite (A) and pigeonite (P) is produced in orthopyroxene host (O), which show (100) stacking faults. Glass (G) may be injected melt or intragranular melt. Two micrographs were viewed along the b axis.

Table 4. Chemical compositions (wt.% oxides) of pyroxene and glass in an inverted pigeonite from lunar eucrite-like gabbro 61223,47 determined by ATEM.

	Opx. host	σ	Pig. inclusion	Aug. inclusion	Glass vein
SiO$_2$	52.43	(0.10)	55.54	52.99	53.07
Al$_2$O$_3$	0.98	(0.14)	0.58	0.64	14.08
TiO$_2$	0.36	(0.02)	0.04	0.07	3.28
Cr$_2$O$_3$	0.31	(0.05)	0.50	0.65	–
FeO	21.86	(0.11)	11.01	14.67	21.36
MnO	0.41	(0.04)	0.40	–	–
MgO	22.52	(0.12)	26.45	16.81	0.81
CaO	1.13	(0.08)	5.48	14.17	6.85
Na$_2$O	–	–	–	–	0.55
Total*	100.00		100.00	100.00	100.00
		Cations per 6 Oxygens			
Si	1.955		1.988	1.981	1.954
Al	0.043		0.025	0.028	0.611
Ti	0.010		0.001	0.002	0.091
Cr	0.009		0.013	0.019	–
Fe	0.682		0.330	0.459	0.658
Mn	0.013		0.012	–	–
Mg	1.252		1.411	0.937	0.045
Ca	0.045		0.210	0.568	0.270
Na	–		–	–	0.040
Total	4.009		3.990	3.994	3.669

*Normalized to 100% due to ratio-method of analysis.
σ: standard deviations are given for numbers of measurements greater than 5.

diffraction by ATEM confirmed the coexistence of four phases, orthopyroxene, pigeonite, augite and glass. Submicroscopic textures observed by TEM are very complicated. Irregular intergrowth of pigeonite and augite was found in the orthopyroxene host with (100) in common. Veins of Mg-poor glass rich in Fe, Ca, Ti and Al are also present in the pyroxene grain. A Fe-rich pigeonite is found along the fine veins of glass, which seem to have been injected during shock events. The augite analysed by ATEM is low in Ca and Al content. The pigeonite composition is shifted towards the Mg-rich side from the orthopyroxene-augite tie-line in the pyroxene quadrilateral [Fig. 1(b)]. One of the most representative textures is shown in Fig. 2b.

Selvage glasses produced by shock induced intergranular melting ocrur at the grain boundaries between the pyroxene and plagioclase in 61224 (Marvin and Warren, 1980). Their chemical compositions are rich in alumina, and if they are projected into the pyroxene quadrilateral, they plot more Fe-rich than the augite. A similar glass was found in 61223. It should be noted that lamella-like veins in 61223 (Fig. 2a) show high concentration of Al and Ca. Because the cation sum for six oxygens is a little low for pyroxene, the Al and Ca-rich vein may not be an augite but may be a glass produced by an injected melt, which could be supplied from intergranular melts produced between the plagioclase and pyroxene boundaries or from melts produced from the breccia matrix. Shock melted glasses with apparently Al-rich augite compositions were also found in the 15445 and 15455 norite clasts, in which primary orthopyroxene and plagioclase are the major minerals.

Lunar quartz monzodiorite, 15405

Sample 15405,148 contains both pigeonite and subcalcic augite with homogeneous bulk chemical compositions within one grain disregarding exsolution. The Fe/(Mg + Fe) ratios are identical to those in some ordinary eucrites and vary slightly from one grain to another as is shown in Fig. 1. Both pigeonite and augite show extensive unmixing and the host and exsolved phases reveal chemical differences resolvable by the electron microprobe traverse. In one grain, a pigeonite crystal with cation ratio of the bulk composition $Ca_{18}Mg_{25}Fe_{57}$ is in contact with an augite crystal $Ca_{31}Mg_{24}Fe_{45}$, both of which show extensive exsolution. The single crystal X-ray study of this grain indicated that the host and exsolved phases of each crystal have (001) in common. Such unmixed pigeonite bears a resemblance in Fe/(Mg + Fe) ratios and the width of lamellae to Fe-rich pigeonite found in 14310,90 (Takeda et al., 1974) and eucrites such as Juvinas and Sioux County, but the chief differences are the predominance of primary augite and the strongly shocked textures.

Single crystal X-ray study indicates that subcalcic augite crystals in 15405 were heavily deformed and they contained approximately equal amounts of augite and pigeonite with (001) in common, but the amount varies from grain to grain. TEM observations reveal that as a major texture they contain regularly spaced (001) pigeonite lamellae 0.7 μm to 1.5 μm in width and different areas show various deformation microstructures due to shock metamorphism (Fig. 3). One area shows extensive (100) mechanical twinning; another area shows densely distributed dislocations and some other areas show intragranular pockets of finely intergrown lamellar pyroxene. We interpret these pockets as pyroxene melt that appears to have experienced spinodal decomposition on rapid cooling when the parental mass of sample 15405 was excavated subsequent to shock heating by meteoritic impacts. This intragranular melt in pyroxene crystals is interpreted to have been formed by very heavy shock heating.

Cumulate eucrites and ordinary eucrites

The Juvinas, Moore County, and Yamato-74356 pyroxenes (Takeda, 1979) have been reexamined with ATEM to obtain accurate lamella widths and compositions. Yamato-74356 is the only known monomict eucrite among the Antarctic eucrites (Takeda et al., 1980); all others are polymict eucrites. Unlike other eucrites, Juvinas and Y-74356 are known to contain small amounts of primary augite coexisting with major pigeonite.

TEM studies of pyroxene from these three eucrites reveal no evidence of strong shock effects as were found in the above lunar pyroxenes (Mori et al., 1981). Both Moore County and Juvinas pyroxene show few deformation textures due to shock metamorphism. Some new findings include the presence of abundant stacking faults in pigeonite with (100) in common that create orthopyroxene-like structure (Fig. 4), which was predicted for Juvinas in a previous X-ray study (Takeda, 1979). Stacking of stacking faults may produce orthopyroxene slabs in the pigeonite host. Textures as in Fig. 5 was found at transitional zones between orthopyroxene inverted from pigeonite and the pigeonite host in a partially inverted pigeonite from Moore County. An electron micrograph (Fig. 5) of a part of such an area shows that many slabs of orthopyroxene cut through the thin second generation (001) augite lamellae which are present in the host between the very thick augite lamellae with (001) in common with the pigeonite (Mori et al., 1981). Finely chopped (001) augite lamellae tend to elongate and form short slabs of augite with (100) in common with the host orthopyroxene. The Ca contents in the orthopyroxene-like slabs (O in Fig. 5) between the (001) augite may be as low as those of the pigeonite host. We interpret the interface between the (001) augite and pigeonite to be semi-coherent.

To obtain accurate widths of exsolved augite lamellae in the eucrite pigeonite, measurements were made on ion-thinned pyroxene samples exactly oriented with the

Fig. 3. Electron micrographs (Bright field) of a subcalcic augite from lunar quartz monzodiorite, 15405,148. Primary exsolution texture of (001) pigeonite lamellae (P) and augite host (A) is modified by shock deformation. (a) High density of dislocations (D) and intragranular melting (M) can be seen. (b) Intragranular melts (M) whose chemical compositions are almost identical to those of surrounding pyroxene phases, have produced fine lamellar intergrowth of pigeonite-augite by rapid cooling.

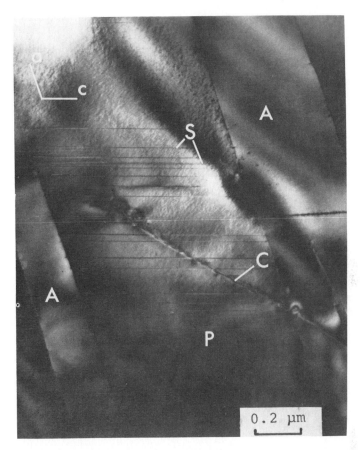

Fig. 4. Electron micrograph (Bright field) of the Juvinas pigeonite. Pigeonite host (P) has abundant (100) stacking faults (S) between (001) augite lamellae (A). C is a crack.

(010) plane perpendicular to the electron beam of the TEM. Submicroscopic augite lamellae were found in Moore County pigeonite as shown above between the thick augite (100 μm) lamellae. This finding confirms our prediction of the presence of thin second-generation (001) augite lamellae based on X-ray data (Takeda, 1979). The X-ray diffraction spots of the thicker (001) augite lamellae are slightly misoriented. However, no such two-generation lamellae were detected in Juvinas and Yamato-74356. The Juvinas pigeonite contains regularly spaced (001) augite lamellae and abundant stacking faults (Fig. 3). The mean widths in μm of these lamellae (with maximum values in parentheses) in eucrite pigeonite are: Moore County (second-generation thinner ones) 0.1 (0.3); Juvinas 0.5 (0.8); and Yamato-74356 0.2 (0.3).

The compositions of the lamellae obtained by ATEM are given in Table 2. The chemical compositions of augite lamellae in Juvinas have been studied by microprobe techniques but the resolution of the lamellae was not sufficient to give consistent values between two investigators (Table 2). The data obtained by ATEM (Table 2) are close to those estimated by X-ray crystal structure refinements, $Ca_{45.5}Mg_{33.0}Fe_{21.5}$ (Takeda et al., 1974) and the Ca content is higher than those of the microprobe data. The chemical composition $Ca_{43}Mg_{37}Fe_{20}$ [Fig. 1b] of the thinner lamellae obtained by the ATEM method for Moore County is more Mg- and Ca-rich than that of the thick one.

Yamato-74356 is a small monomict eucrite (10.0 g) partly covered by a shiny black fusion crust. Lithic clasts are abundant but some of the pyroxene crystals have become aggregates of small polygonal crystals showing a granoblastic texture. Augite in Yamato-74356 occurs as discrete grains with uniform chemical compositions of the host and

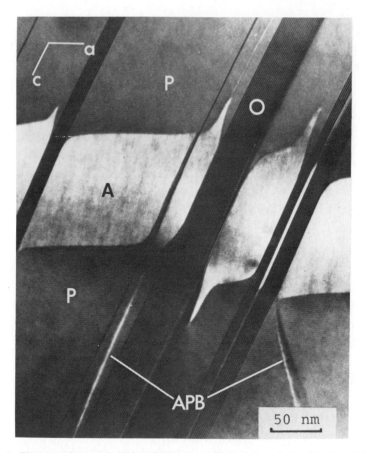

Fig. 5. Electron micrograph (Bright field) of a partially inverted pigeonite from Moore County. (001) augite lamella (A) is cut through by the (100) orthopyroxene slabs (O). Antiphase boundaries (APB) can be seen in pigeonite host (P).

exsolved phases, coexisting with major pigeonite. The pigeonite grains resemble those in Juvinas in that small chromite inclusions or clouding (Harlow and Klimentidis, 1980) are aligned along slightly curved lines. The width of pigeonite lamellae in augite is 1.5 μm on average and is up to 2.1 μm. The pyroxene compositions are given in Fig. 1(b) and Table 2. The closure temperature or that of apparent equilibration for the exsolved pair is estimated as 880°C by the pigeonite-augite geothermometer (Ishii *et al.*, 1979). We adopted 850°C for the final temperature for the development of lamellae for ordinary eucrites, since slower cooling is expected for other eucrites, and since we use the same temperature for all eucrites for comparison in our computer simulation. The temperature of crystallization obtained by the bulk chemical compositions of a coexisting pigeonite-augite grain (Takeda, 1979) is 1000°C.

ESTIMATION OF COOLING RATES

Since exsolution phenomena of pyroxene may be controlled by many complicated or possibly unknown factors, a quantitative treatment which accounts for the observed exsolution phenomena has been difficult. In an attempt to compare the cooling rates of exsolved pyroxenes in lunar rocks and meteorites, as a first approximation, cooling rates have been computed using assumptions that the most important variables in determining the exsolution textures are bulk chemistry and the width of the exsolved phase. Carrying this one step further, an attempt has been made by Miyamoto and Takeda (1977) to estimate the depth of burial of the exsolved pyroxenes in a crustal model for the eucrite parent body. In their computer simulation, a simplified model of exsolution phenomena was chosen where the growth of an exsolution lamella is controlled by the diffusion of Ca atoms perpendicular to the lamella.

In our previous studies, a temperature range from 1200 to 850°C was used in the computer simulation. In this study 1100°C, estimated for the 61223 pigeonite by the pigeonite-augite geothermometer, was taken as the initial temperature for the nucleation of augite. 1000°C was obtained for ordinary eucrite Yamato-74356 using the bulk chemistry from Takeda (1979). The final closure temperature for augite lamellae, that is, the lowest temperature of equilibration at which point equilibration was no longer able to proceed, was estimated from the chemical compositions of the host-pigeonite and exsolved augite-lamella pair obtained by ATEM for Yamato-74356. We adopted 850°C, a value slightly lower than the 880°C as estimated for Yamato-74356, for reasons given previously.

The new TEM data for the widths of exsolution lamellae were employed to calculate cooling rates and depths of burial where the exsolution was modeled by the same computer program used for our previous studies (Miyamoto and Takeda, 1977). The atomic diffusion coefficient (D) for the above simulation is $D = 4.5 \times 10^{-19} \, cm^2/sec$ at 1125°C for the ordinary eucrite compositions with activation energy of 28 kcal/mole. Justification for using these values is given in the above paper. Huebner and Nord (1981) proposed that D might change using experimental evidence and activation energy of 100 kcal/mole. Thermal diffusivity $\kappa = 0.004 \, cm^2/sec$ (Yomogida and Matsui, 1981) was used to derive the temperature-time variation within the lunar and eucrite crusts. Because the temperature-time variation curves at any depth of burial are given, the time necessary to develop lamellae can be converted into depth. A spherical parent body with radius (R) = 250 km was used for the computation on the eucrite parent body.

The results are shown in Table 5. The top value 3.44 μm for Juvinas is an old lamella half-width obtained from a microprobe scan. 0.25 μm is that obtained by TEM. For Juvinas a cooling rate of $-0.4°C/10$ years is required to produce exsolution lamellae 0.5 μm in width from 1000° to 850°C, which is equivalent to a depth of 0.5 km. The results are $-0.2°C/year$ and 200 m for the Yamato-74356. An order of magnitude larger D (4.5 ×

Table 5. Cooling histories of pigeonites from eucrites (R = 250 km) and eucrite-like lunar gabbro (1/2t = 100 km).

Half width (μm) of Aug. Lam.	Initial Ca Mol.%	Fe/(Fe + Mg)	Cooling Rates (°C/y)	Depth of Burial (km)	T°C– 850°C	Remarks
Moore County						
27.3 (thick)	10.1	49.28	$-0.16/10^2$	9.1	1200	
0.050 (thin)	5.80**	50.40	$-0.48/1$	0.15	1000	
Mt. Padbury						
3.18	15.5	60.36	$-0.49/10$	0.5	1200	
Juvinas						
3.44	8.01	60.86	$-0.40/10^2$	1.7	1200	
0.25	8.01	60.86	$-0.36/10^1$	0.50	1000	
0.25	4.55	60.50	$-0.53/10^2$	1.3	1000	
0.25	8.01	60.86	$-0.32/1$	0.16	1000	$D_A \times 10*$
Yamato-74356						
0.1	7.41	60.79	$-0.18/1$	0.22	1000	
Lunar 61224,36						
3.5	8.80	34.1	$-0.68/10^3$	4.1	1100	
5.5	10.8	35.9	$-0.65/10^3$	4.1	1100	

*$D_A = 1.064 \times 10^{-14} \, cm^2/sec$ (Miyamoto and Takeda, 1977).
R: radius of the parent body; t: thickness of the crust.
**Bulk composition of the host part which includes the thin second generation lamellae and lies between the thick ones.

10^{-18} cm^2/sec) was employed to see the minimum depth of burial for Juvinas. The use of a larger D can be justified by the fact that the diffusion of Ca will be faster for such an Fe-rich composition because the difference of their ionic radii is small. The results show that the cooling rate was reduced to $-0.3°$C/year and such cooling rates require rather shallow burial (150 m). Therefore the meteorite need not have been subjected to extended subsolidus annealing by a metamorphic event as proposed previously (Takeda *et al.*, 1978). There is evidence that there was a post-crystallization-cooling thermal history (e.g., Harlow and Klimentidis, 1980). A more complex thermal history involving an impact event can be involved but there is not strong evidence to support the events.

Simple application of the program to the lunar 61223,47 pigeonite for the thickness of the hot lunar crust of 200 km (2t in Table 5) and for a cold interior, instead of R = 250 km for the spherical howardite parent body, gave the depth of burial from 2 to 8 km (Table 5). The depth is comparable to that estimated for the cumulate eucrites, Moore County and Moama. The ATEM results for finer augite lamellae suggest a different thermal history.

DISCUSSION

In addition to the single-crystal X-ray diffraction and electron microprobe techniques, a new tool, the quantitative analytical electron microscope (ATEM) has been employed to elucidate chemical compositions of various pyroxene phases produced by the unmixing and impact processes. The detailed procedure and accuracy of the ATEM method (Morimoto and Kitamura, 1981) will be discussed in a separate paper. Using the minimum absorption thin-film criteria, quantitative analysis of pyroxene requires that the analyzed areas be less than 0.3 to 0.1 μm in thickness depending on composition (Nord, pers. comm., 1981). Although absorption has not been accounted for in the present method, experimentally determined k-factors on a pyroxene standard close to the unknown will give unbiased results. Good agreements between our ATEM and microprobe values of the Moore County augite, small standard deviations of some ATEM multiple analyses given in Tables 2 and 4, and good cation sums (3.991 to 4.012) for six oxygens for pyroxene suggest that the measurements are adequate for the present discussion. The percent mean deviations are about ±1% for major elements and ±10% for minor elements.

Some new findings on lunar pyroxenes by the ATEM technique include the presence of reinverted pigeonite and aluminum-rich glass veins in the shocked pyroxene grains of 61223. Formation of shock melts in an apparent single crystal such as in 15405 on a submicroscopic scale may be responsible for resetting of the isotopic clocks in many lunar crustal rocks. The more Mg-rich compositions of the pigeonite than the host phase, and their irregular shapes suggest that the pigeonite was caused by shock heating rather than stress-induced shear transformation in solid state. Pigeonite (clinobronzite) slabs found in Johnstown orthopyroxene were interpreted to be formed by the shear stress at high temperature during the shock event (Mori and Takeda, 1981a). Formation of the Mg-rich pigeonite is difficult to explain, but the presence of a Fe-rich glass implies that a phase separation took place during the shock heating to produce a Fe-rich melt probably under reducing conditions. However, the Al-rich composition of the glass may favor an interpretation that the original melt was injected from outside.

The low Ca contents of augite determined by ATEM indicate that the original exsolved augite was reacted with the host orthopyroxene to produce a pigeonite. The Ca ions may also be supplied from a Ca- and Al-rich intergranular melt produced by a shock event and injected into the crystal. The crystallographic orientation of the augite inclusion and the host orthopyroxene, which was presumably inverted from pigeonite, was confirmed to have (100) in common by the X-ray and electron diffraction methods. This relationship is the same as those found in Binda and other low-Ca pigeonites (Takeda, 1979), but is different from that found by Huebner *et al.* (1975) and Harlow *et al.* (1979) who reported that blebs of augite commonly show no topotactical orientation with the host.

The formation of pigeonite and melts by shock heating is a characteristic microtexture of the lunar eucrite-like pyroxenes. The true eucrite pyroxenes do not exhibit such extensive shock features. Orthopyroxene-like (100) stacking-faults are relatively common features in pigeonite from eucrites. The interpretation of formation of the orthopyroxene-like slabs produced cutting through the (001) augite in Moore County requires further study.

Comparison of lunar eucrite-like gabbros with eucrite meteorites has been made on the basis of their petrology and bulk and trace element chemistries by Marvin and Warren (1980). Our results support their conclusion that 61224 and 61223 are unique lunar rocks. One of the characteristics of the lunar analogue in comparison with true eucrites is the occurrence of abundant primary augite. The fact is compatible with the high proportions of normative Wo component (13.6 mol.%) in the CIPW norm calculated from the bulk chemistry and implied that the sources of the basaltic liquids were different. Fe-rich subcalcic augite occurs at rims of chemically-zoned pigeonite grains in the Pasamonte-type eucrites. The presence of minor primary augite was reported by Reid (pers. comm., 1972) for Juvinas. Large grains of such augite have been detected in Yamato-74356. Nehru et al. (1980) compared some features of eucrites with mesosiderite gabbros and basalts, which were initially pigeonite-augite basalts, and suggested that the latter is related by fractional crystallization rather than being quenched peritectic melts. We suggested that some eucritic melts may crystallize a low-Ca pyroxene as Mg-rich as that of some diogenites (Takeda, 1979).

Because the pristinity of a nonmare rock rests in part on the evidence for a deep-seated provenance, the depth of burial in the lunar crust estimated from the widths of exsolution lamellae of 61223 may yield some constraints on its origin. Although our present methods of computer simulation are not yet completely developed, major known factors that affect the exsolution are taken into account and the estimated depth is compatible with a plutonic origin by the same reasoning proposed for lunar rock 62236 (Takeda et al., 1979a). The depths in Table 5 may be used as a relative measure.

In interpreting the results of our computer simulation on augite lamella growth in the pigeonite host, we have to ask whether the assumptions about the development of the augite lamellae are compatible with the actual processes of the exsolution phenomenon in the eucritic pigeonites. Our diffusion model implies that the strain energy at the interface will not affect Ca diffusion, and that the lamellae are spaced wide enough not to interfere each other. Some factors which should be accounted for in a future simulation include coarsening which will increase the size of lamellae and not change their composition and nucleation density and poisoning of lamellae interfaces.

The width of the augite lamellae measured on the oriented ion-thinned section of Juvinas is much smaller than that obtained by the microprobe traverse. The new value in conjunction with a larger value for the Ca diffusion coefficient for Fe-rich pigeonite gives a shallower depth of burial for Juvinas. The slow cooling rate previously proposed is in apparent conflict with the interpretation of dynamic crystallization experiments of Walker et al. (1978), which would allow the crystallization textures to form under near-surface conditions in the rapidly-cooled portions of basaltic intrusive or extrusive bodies. In order to explain this discrepancy, extended subsolidus annealing due to a late thermal metamorphic event has been proposed. Our new result may no longer require such an extended metamorphic event and is compatible with a shallower origin. Harlow and Klimentidis (1980) believe that there have been two separate events (1) crystallization and cooling and (2) metamorphism, neither of which is inconsistent with a near surface event, but do suggest different cooling rates through different temperature regions.

In conclusion, in spite of similarities in bulk chemistries, mineral assemblages, depths of burial in the crust, and crystallization temperatures between lunar and meteoritic eucrites, there are some differences in their pyroxene assemblages and exsolution/inversion textures, which can be attributed to small but significant differences in their bulk chemistries. The largest difference is that the lunar analogues experienced a much higher degree of complex shock effects than the meteoritic eucrites. The chemistry and exsolution texture of the 15405 pigeonite are similar to those of the ordinary eucrites. However,

15405 augite shows evidence of intragranular melting produced by the shock effect. The formation of such melts on a submicroscopic scale may be responsible for resetting of the isotopic clocks in many lunar crustal rocks.

Acknowledgments—We thank Drs. Ursula Marvin, Graham Ryder, Patrick Butler, Jr., and Ms. Ruth M. Fruland who helped us in providing us with sample 61223 and 61224, Dr. M. Prinz and Professor C. B. Moore, Dr. K. Yanai (Nat'l Inst. of Polar Res.) for meteorite samples, Professors R. Sadanaga and Y. Takéuchi for their interest and equipment, Mr. O. Tachikawa, Mrs. M. Hatano, and Miss S. Yoneda for technical assistance. A part of this study has been supported by a Fund for Scientific Research of Ministry of Education and Itoh Science Foundation grant. We are indebted to Drs. U. Marvin, M. B. Duke, J. S. Delaney, and Professors P. R. Buseck, N. Morimoto, and A. M. Reid for discussion, and to Professor R. H. Hewins, and Drs. G. L. Nord, Jr. and G. E. Harlow for critical reading of the manuscript and discussion.

REFERENCES

Cliff G. and Lorimer G. W. (1975) The quantitative analysis of thin specimens. *J. Microscopy* **103**, 203–207.

Delaney J. S., Harlow G. E., Prinz M., and Nehru C. E. (1981) Metamorphism in mesosiderites: Radical chemical changes in a post-igneous silicate regolith (abstract). In *Lunar and Planetary Science XII*, p. 208–210. Lunar and Planetary Institute, Houston.

Harlow G. E. and Klimentidis R. (1980) Clouding of pyroxenes and plagioclase in eucrites: Implications for post-crystallization processing. *Proc. Lunar Planet. Sci. Conf. 11th*, p. 1131–1143.

Harlow G. E., Nehru C. E., Prinz M., Taylor G. J., and Keil K. (1979) Pyroxenes in Serra de Magé: Cooling history in comparison with Moama and Moore County. *Earth Planet. Sci. Lett.* **43**, 173–181.

Huebner J. S. and Nord G. L. Jr. (1981) Assessment of diffusion in pyroxenes: What we do and do not know (abstract). In *Lunar and Planetary Science XII*, p. 479–481. Lunar and Planetary Institute, Houston.

Huebner J. S., Ross M., and Hickling N. (1975) Significance of exsolved pyroxenes from lunar breccia 77215. *Proc. Lunar Sci. Conf. 6th*, p. 529–546.

Hostetler C. J. and Drake M. J. (1978) Quench temperatures of Moore County and other eucrites: residence time on eucrite parent body. *Geochim. Cosmochim. Acta* **42**, 517–522.

Ishii T. and Takeda H. (1974) Inversion, decomposition and exsolution phenomena of terrestrial and extraterrestrial pigeonites. *Mem. Geol. Soc. Japan*, **11**, 19–36.

Ishii T., Takeda H., and Yanai K. (1979) Pyroxene geothermometry applied to a three pyroxene achondrite from Allan Hills, Antarctica and ordinary chondrites. *Mineral J.* **9**, 460–481.

Lorimer G. W. and Cliff G. (1976) Analytical electron microscopy of minerals. In *Electron Microscopy in Mineralogy* (H.-R. Wenk, ed.), p. 506–519. Springer-Verläg, Berlin.

Marvin U. B. and Warren P. H. (1980) A pristine eucrite-like gabbro from Descartes and its exotic kindred. *Proc. Lunar Planet. Sci. Conf. 11th*, p. 507–521.

McGee J. J., Nord G. L. Jr., and Wandless M. V. (1980) Comparative thermal histories of matrix from Apollo 17 Boulder 7 fragment-laden melt rocks: An analytical transmission electron microscopy study. *Proc. Lunar Planet. Sci. Conf. 11th*, p. 611–627.

Miyamoto M. and Takeda H. (1977) Evaluation of a crust model of eucrites from the width of exsolved pyroxenes. *Geochem. J.* **11**, 161–169.

Mori H. and Takeda H. (1981a) Thermal and deformational histories of diogenites as inferred from their microtextures of orthopyroxenes. *Earth Planet. Sci. Lett.* **53**, 266–274.

Mori H. and Takeda H. (1981b) Analytical electron microscopic study of inverted pigeonite (abstract). *Mineral Soc. Japan 1981 Annual Meeting Abstr.*, p. 51. Mineral. Soc. of Japan, Tokyo.

Mori H., Takeda H., and Sadanaga R. (1981) Microstructures of pyroxenes from two eucrites and a lunar monzodiorite (abstract). In *Lunar and Planetary Science XII*, p. 717–719. Lunar and Planetary Institute, Houston.

Morimoto N. and Kitamura M. (1981) Applications of 200 kV analytical electron microscopy to the study of fine textures of minerals. *Bull. Mineral.* **104**, 241–245.

Nehru C. E., Delaney J. S., Harlow G. E. and Prinz M. (1980) Mesosiderite basalts and the eucrites. *Meteoritics* **15**, 337–338.

Ross M. and Huebner J. S. (1975) A pyroxene geothermometer based on composition-temperature relationships of naturally occurring orthopyroxene, pigeonite, and augite (abstract). *Int. Conf. Geothermometry and Geobarometry*, Pennsylvania State Univ., University Park, Pennsylvania.

Ryder G. (1976) Lunar sample 15405: Remnant of a KREEP basalt-granite differentiated pluton. *Earth Planet. Sci. Lett.* **29**, 255–268.

Takeda H. (1979) A layered crust model of a howardite parent body. *Icarus* **40**, 455–470.

Takeda H., Miyamoto M., Duke M. B., and Ishii T. (1978) Crystallization of pyroxenes in lunar KREEP basalt 15386 and meteoritic basalts. *Proc. Lunar Planet. Sci. Conf. 9th*, p. 1157–1171.

Takeda H., Miyamoto M., and Reid A. M. (1974) Crystal chemical control of element partitioning for coexisting chromite-ulvöspinel and pigeonite-augite in lunar rocks. *Proc. Lunar Sci. Conf. 5th*, p. 727–741.

Takeda H., Miyamoto M., and Ishii T. (1979a) Pyroxenes in early crustal cumulates found in achondrites and lunar highland rocks. *Proc. Lunar Planet. Sci. Conf. 10th*, p. 1095–1107.

Takeda H., Miyamoto M., and Ishii T. (1980) Comparison of basaltic clasts in lunar and eucritic polymict breccias. *Proc. Lunar Planet. Sci. Conf. 11th*, p. 135–147.

Takeda H., Miyamoto M., Ishii T., and Reid A. M. (1976) Characterization of crust formation on a parent body of achondrites and the moon by pyroxene crystallography and chemistry. *Proc. Lunar Sci. Conf. 7th*, p. 3535–3548.

Takeda H., Miyamoto M., Ishii T., Yanai K., and Matsumoto Y. (1979b) Mineralogical examination of the Yamato-75 achondrites and their layered crust model. In *Memoirs of the National Institute of Polar Research, Special Issue No. 12* (T. Nagata, ed.), p. 82–108. Nat'l. Inst. Polar Res., Tokyo.

Walker D., Powell M. A., and Hays J. F. (1978) Dynamic crystallization of a eucritic basalt (abstract). In *Lunar and Planetary Science IX*, p. 1196–1198. Lunar and Planetary Institute, Houston.

Warren P. H. and Wasson J. T. (1978) Compositional-petrographic investigation of pristine nonmare rocks. *Proc. Lunar Planet. Sci. Conf. 9th*, p. 185–217.

Wells P. R. A. (1977) Pyroxene thermometry in simple and complex systems. *Contrib. Mineral. Petrol.* **62**, 129–139.

Yomogida K. and Matsui T. (1981) Physical properties of meteorites (abstract). *Sixth Symp. on Antarctic Meteorites*, p. 83–84. Nat'l Inst. Polar Res., Tokyo.

Metamorphism in mesosiderites

Jeremy S. Delaney,[1] C. E. Nehru,[1,2] M. Prinz,[1] and G. E. Harlow[1]

[1]Department of Mineral Sciences, American Museum of Natural History, Central Park West at 79th Street, New York, New York 10024
[2]Department of Geology, Brooklyn College, City University of New York, Brooklyn, New York 11210

Abstract—High temperature metamorphism in mesosiderites has dramatically changed the chemistry and textures of the silicate clasts, the matrix and the metal fraction. The metamorphic event that produced the characteristic textures of mesosiderites was of short duration (days to months) so that, although the original polymict breccia was partly homogenized, kinetic difficulties prevented total recrystallization and chemical equilibration. Disequilibrium textures in overgrowths around orthopyroxenite clasts and in coronas around olivine are interpreted to give cooling rates of 1–100°C/day between 1150° and 900°C although experimental calibration work is needed. Mesosiderites, therefore, contain suggestions of being among the fastest *and* the slowest cooled meteorites known.

The degree of homogenization observed in individual mesosiderites is probably influenced more by kinetic factors such as the original grain size of the matrix and the degree of mechanical mixing in the matrix, than by the duration of metamorphism. The amount of modification of clasts reflects the ease of communication between reaction sites inside the clasts and the very reactive mesosiderite matrix—brecciated and small clasts are more modified than large, competent clasts.

Metamorphism of the silicate fraction was caused by mixing completely or partially molten metal (with dissolved phosphorus and sulfur) with a cool silicate regolith. The metamorphism resulted in redox exchange reactions between the silicate and metal fractions, as well as the partial homogenization of the silicate fraction. The cause of the metal-silicate mixing event is not yet known.

INTRODUCTION

Studies of mesosiderites have identified a metamorphic overprint in these meteorites (Powell, 1971; Floran, 1978) but have not explored the effects and implications of this overprint in detail. This study documents several important textural and chemical features of the mesosiderites. The silicate fraction of the mesosiderites has been metamorphosed at high temperatures (800–1200°C), and the chemistry of individual clasts may have been extensively modified by interaction with the silicate matrix and by incomplete homogenization of the silicate fraction (Delaney *et al.*, 1981a). The role of metal/silicate interaction and, in particular, the production of merrillite by redox exchange between phosphorus bearing metal and silicate minerals is discussed.

The components making up the silicate fraction of the mesosiderites are reviewed, but these are the subjects of other papers (Nehru *et al.*, 1980b,c; 1981b; Floran, 1978; Powell, 1971; Hewins, 1981a) and are not dealt with in great detail. A number of textural and chemical features seen in clasts representing these components are presented where the results were not previously discussed and they are relevant to the interpretation of the metamorphism. The unique features of the silicates in mesosiderites and the observed interaction with the metal define constraints on the evolution of the mesosiderite parent body.

ANALYTICAL TECHNIQUES AND MATERIAL STUDIED

Thin sections of all known mesosiderites were studied, and the sections used are listed in Table 1. Analyses were performed on an ARL-SEMQ electron microprobe at the American Museum of Natural History under a variety of operating conditions but usually either 15KV or 20KV and 20nA specimen current normalized on the brass standard holder. The elements Si, Ti, Al, Fe, Mg and Ca

Table 1. Mesosiderite thin sections studied.

Meteorite	P.T.S. Numbers
Allan Hills	JSC A77219,38
Barea	NMNH 1468A, 1468B
Bondoc Peninsula	AMNH 3980-4, 3983-1, -2
Budulan	AMNH 4386-1
Chinguetti	NMNH 3205
Clover Springs	AMNH 4391-1, -2, -3, -4, -5; NMNH 1633
Crab Orchard	AMNH 304-1; 902-1; -2, -3, -4
Dalgaranga	AMNH 4037-1
Dyarrl Island	NMNH 5725
Emery	AMNH 4367-1, 4441-2, -3, -4
Estherville	AMNH 327-1; 331-1; 334-1; -2
Hainholz	AMNH 316-1, -2, -3
Lowicz	AMNH 3775-1, -2, -3
Mincy	AMNH 887-1, -2, -3, -4
Morristown	AMNH 305-1, -2
Mount Padbury	AMNH 4212-1, -2
Patwar	AMNH 4112-1, -2, -3
Pinnaroo	AMNH 3950-1, -2, -3
Simondium	AMNH 4234-1, -2
Vaca Muerta	AMNH 559-1, 819-1, -2, -3, 914-1, 980-1, AMNH 986-1, -2, -3
Veramin	AMNH 313-1, -2

AMNH—American Museum of Natural History; NMNH—National Museum Natural History; JSC—Johnson Space Center.

were analyzed on fixed crystal spectrometers in most cases, while Ni, Mn, Cr, V, Na, K, P, REE, F, Cl and Ba were measured on scanning wavelength spectrometers using TAP, PET and LiF crystals. Backgrounds for elements were estimated for the six elements on fixed spectrometers using standards free of those elements. For most elements on scanning spectrometers, backgrounds were measured symmetrically on either side of the peak, although some reconnaissance analysis was done using estimated backgrounds.

COMPONENTS OF MESOSIDERITES

Much effort has been expended in recent years in identifying the igneous components that make up the stony fraction of the mesosiderites. For example, Floran (1978) suggested six components to account for the silicates, four "primary igneous" components and two secondary, "sedimentary", components. Of the four primary components, Floran distinguished between "cumulate eucrite" and "eucrite" components on the basis of textural differences. Detailed study by Nehru et al. (pers. comm.), however, suggests that these two components are part of a compositional continuum—the mesosiderite mafic suite, a series of gabbros and basalts, that are not compositionally distinct, but have a wide range of Fe/(Fe + Mg) ratios. These mafic clasts are therefore best treated as a single component of variable composition. Three primary components, therefore, are considered to make up the silicate fraction of the mesosiderites. These are: orthopyroxenites, olivine rich rocks, and the mafic suite.

(a) Orthopyroxenites

The magnesian, orthopyroxenite clasts in mesosiderites have generally been equated with the diogenites (Floran, 1978). Mittlefehldt (1979) suggests that differences in Mn and Sc between these clasts and diogenites preclude a common origin for both groups. His scandium data are, however, in conflict with those of Goles (1971) that indicate no systematic difference between mesosiderite orthopyroxene (Bondoc Peninsula and Vaca Muerta) and diogenites (Johnstown, Tatahouine). The two data sets have alsmost identical results for the diogenites. The "systematic" enrichment of scandium in mesosiderite orthopyroxene observed by Mittlefehldt (1979) probably reflects the presence of less magnesian, overgrowth pyroxene on the clasts analyzed, which was not recognized as a contaminant. Results from Göpel and Wänke (1978) suggest that iron rich pyroxene, contaminating the analyzed clasts, would increase the scandium content in the pyroxene analyses.

In addition, the manganese analyses and Fe/(Fe + Mg) variations for mesosiderites, orthopyroxenites and diogenites presented by Mittlefehldt (1979) do not show any systematic differences between the two groups. Contamination by iron-rich overgrowth pyroxene may be responsible for the greater scatter in the mesosiderite orthopyroxenite data. Scandium and manganese are positively correlated in Mittlefehldt's data. Redox exchange reactions between silicate and metal in the mesosiderites (Agosto et al., 1980; Delaney et al., 1980b; Harlow et al., 1980, 1981; Mittlefehldt et al., 1979) may also result in modification of Fe–Mn–Mg proportions in olivine and orthopyroxene.

Hewins (1979) has suggested that mesosiderite orthopyroxenites are derived from a suite of rocks similar to diogenites, although mesosiderites also sample a more magnesian source than diogenites. He argued that minor element differences exist between the two groups. In a study of 105 orthopyroxenite clasts in howardites and 97 clasts in mesosiderites, Nehru et al. (1981a) found no difference between orthopyroxene compositions in these two populations and the available diogenitic pyroxene data except that the compositional ranges sampled by mesosiderites and howardites are wider than the range for the diogenites. Element distribution data for coexisting minerals in the orthopyroxenite clasts and diogenites are, however, important for discriminating between different groups and Hewins (1981b) has shown that they yield information about the evolutionary history of each group. Such distribution studies in the mesosiderites are, however, difficult because of the combined effects of brecciation, metal invasion and metamorphism. The effect of brecciation in orthopyroxenite clasts in both howardites and mesosiderites is partly corrected by comparing element distribution patterns between orthopyroxene and minerals in cracks (which may be introduced during brecciation), with relations between orthopyroxene and minerals present as inclusions. The included minerals are more likely to be in equilibrium with the pyroxene (albeit only since the most recent recrystallization of the pyroxene). In most cases, these inclusions are probably relicts of the original igneous rock from which the orthopyroxenites were derived. It is still, however, too early to state unequivocally whether the mesosiderite orthopyroxenites sample the same source as the diogenites, but they are clearly derived by similar processes, from similar sources. Magnesian orthopyroxenes and orthopyroxenites make up more than 60% of the silicate fraction of mesosiderites (Prinz et al., 1980) and are important in the chemical homogenization resulting from metamorphism.

(b) Olivine in clasts

Olivine is a minor, but almost ubiquitous component of the mesosiderite silicate fraction. (Mean olivine content is 1.8% vol; S.D. = 1.7%: Prinz et al., 1980). Most olivine in mesosiderites occurs as mineral clasts, which may be several millimeters across, but rare lithic clasts containing subsidiary olivine have been found in Vaca Muerta and Emery. The Emery clast contains 13% olivine (Fo_{62}), 73% orthopyroxene (En_{63-69}), 6% plagioclase (An_{91}), 7% troilite and 1% chromite and is an olivine orthopyroxenite (Nehru et al., 1980c). Although the range of olivine clast compositions in mesosiderites (Fo_{90}–Fo_{55}) overlaps the Mg/(Fe + Mg) range of pyroxenite clasts great difficulty has been found in reconciling "equilibrium" Fe–Mg and Fe–Mn distribution coefficients for these minerals. (Nehru et al., 1980c; Delaney et al., 1980b; Hewins, 1981a). If an orthopyroxene clast and an olivine clast from a mesosiderite are 'paired" on the basis of experimentally determined Fe–Mg partitioning results, then in many cases Fe–Mn partitioning between the two minerals is not consistent with equilibrium results from Stolper (1977), although his results may not be directly applicable to the more magnesian, low fO_2 system (Snellenburg et al., 1979) represented by mesosiderite orthopyroxenites. The available Fe–Mn partitioning data (Watson, 1977; Longhi et al., 1978; Stolper, 1977) are insufficient to constrain likely variations caused by bulk compositional differences between the eucritic system studied by Stolper and the orthopyroxenites (Mysen and Virgo, 1980). The available evidence suggests that some olivine clasts represent samples from a suite of dunites, or perhaps harzburgites which are not simply related to the observed orthopyroxenites (Mittlefehldt, 1979; Nehru et al., 1980c, 1981; Hewins, 1979, 1981a).

It must be emphasized however, that although the most magnesian olivine in mesosiderites (Fo_{90}) probably did not coexist with any of the observed orthopyroxenites (most magnesian is En_{85}), and is probably a sample of a residual dunite horizon (Delaney et al., 1980b; Mittlefehldt, 1979) on the original parent body, the more iron rich olivines are assigned to dunitic rocks using only circumstantial evidence for the lack of an olivine-orthopyroxene source rock.

Metamorphism of the mesosiderites has produced very striking coronas around many olivines (Nehru et al., 1980c; Floran, 1978; Powell, 1971) but these are more important to this study than to recognition of the ultimate source of the olivine.

(c) Mesosiderite mafic clasts

The mesosiderite mafic clasts are discussed in detail by Nehru et al. (1981b) and reviewed briefly here. McCall (1966) drew attention to large enclaves or clasts with a variety of textures within the Mount Padbury mesosiderite. He also recognized unusual core-rim features on pyroxene in some clasts, that are considered to be products of mesosiderite metamorphism. Unfortunately, several subsequent workers have ignored McCall's important observational data in their interpretation of mafic clast chemistry. McCall presented only one chemical analysis of a very iron rich clast. This clast is indistinguishable from most basaltic eucrites (e.g., Mason et al., 1979) but does not reflect the diversity of mafic clast composition in mesosiderites (Nehru et al., pers. comm.) (Fig. 1). Mittlefehldt (1979) presented bulk analyses for six mafic clasts that illustrate some of the diversity within this suite. Unfortunately, contamination of the analyzed clasts by mesosiderite metal, and by the products of metamorphism prevents detailed use of many of Mittlefehldt's data. Electron microprobe data from Nehru et al. (pers. comm.) and Mittlefehldt (1979) are shown (Fig. 1) and illustrate the diversity of mafic clasts in mesosiderites in comparison with other basaltic achondrites.

Compositional diversity in the mafic clasts defines a trend from calcium-rich feldspar

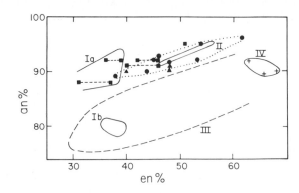

Fig. 1. Orthopyroxene (en%) vs feldspar (an%) compositions for mesosiderite mafic clasts. Symbols are: squares—data from Mittlefehldt (1979): circles—gabbroic textured clasts; triangles—basaltic textured clasts from Nehru et al. (pers. comm.). Fields for basaltic eucrites (Ia and Ib), cumulate eucrites (II), and the evolved series of basalts from howardites recognized by Delaney et al. (1981b) (III) are also shown. Field IV shows impact melt mesosiderites of Floran et al. (1978).

(An_{96}) with magnesium-rich pyroxene (En_{61}, opx) to more sodic feldspar (An_{88}) with iron rich pyroxene (En_{40}, pigeonite). This trend overlaps a similar, but tighter, trend in the cumulate eucrites (Moama, Moore Co., Nagaria, Serra de Magé) and, at its iron rich end, overlaps the monomict basaltic eucrites. Although the mafic clasts from mesosiderites have the same range of Fe/Mg ratios as the evolved basalt group from howardites (Delaney et al., 1981b) the two groups define separate but parallel trends (Fig. 1). The mesosiderite mafic clasts and similar clasts from howardites therefore, evolved independently. The mafic clasts in mesosiderites are the major contributors of plagioclase, calcic pyroxene, ilmenite and part of the tridymite found in the silicate fraction.

(d) Metal

The metal fraction of the mesosiderites form a continuous three dimensional network penetrating the silicate fraction (Powell, 1969). In general, the metal is fine grained and penetrates very fine cracks and spaces in the silicate breccia and appears to have filled all the free space in the silicate breccia. The very fine metal-filled cracks in many mesosiderite clasts indicate that the metal fraction was very mobile when it mixed with the silicate regolith. In many mesosiderites metal is concentrated in large, sub-angular nodules (Powell, 1969; Floran, 1978). Floran (1978) interpreted these nodules as metal clasts in a silicate-metal regolith. Powell (1969) pointed out, however, that these rare large metal masses have highly intricate and delicate boundaries with silicates that could not have survived had the metal and silicate been mixed as solids. Large cavities or vugs have been observed in many eucritic meteorites (Ibitira; Wilkening and Anders, 1975: Antarctic polymict eucrites; Score et al., 1981). The space filling nature of the mesosiderite metal, therefore, lends support to the suggestion by Powell (1969) that the metal of the mesosiderites was mixed with the solid silicates as a liquid rather than a solid.

THE SILICATE MATRIX OF THE MESOSIDERITES

The silicate fraction of the matrix of the mesosiderites differs from the other basaltic achondrites. It is sometimes coarse grained and often recrystallized, instead of being simply brecciated and fine grained. The matrix is the most important site for metamorphic reactions in the mesosiderites, because its originally heterogeneous composition and fine grained nature, prior to metamorphism, represent favorable kinetic conditions for equilibration. Before considering the effect of metamorphism on the mesosiderite silicates, the mineralogical composition of the matrix must be known so that comparison may be made with the entire silicate fraction.

Electron microprobe modal analyses (Prinz et al., 1980) on areas, in each of several mesosiderite thin sections, that were selected to be free of large ($200 + \mu$m) clasts, and that were typical of the general matrix in that section (Table 2) indicate that the average matrix mode and the bulk mesosiderite mode, differ for several minerals. Specifically the low-Ca pyroxene (<7% Wo in mode), olivine and chromite components are lower in matrix. These differences are also present when the matrixes of individual mesosiderites are compared to the bulk silicate fractions (Prinz et al., 1980). The minerals enriched in the matrix are those which are abundant in the mafic suite of clasts whereas the remainder, orthopyroxene and olivine, are derived from the orthopyroxenite and olivine rich suites of clasts. The mesosiderite matrix therefore contains a larger proportion of basaltic and gabbroic material than would be expected from the relative abundances of the major clast types. This difference between matrix and bulk mesosiderite silicate composition may be interpreted as a result of brecciation history of an achondritic silicate regolith that was later incorporated into the mesosiderites by mixing with metal.

A simple model of such an achondritic regolith that later evolved into the mesosiderite parent would start with development of a stratified igneous crust not unlike that suggested by Takeda (1979) for the howardite parent body (Fig. 2). Impact brecciation of the crust

Table 2. Modes of matrix areas from several mesosiderites.

Name p.t.s.	Barea NMNH 1468A	Chin-guetti NMNH 3205	Clover Springs AMNH 4391-3	Crab Orchard AMNH 902-2	Emery AMNH (all)	Esther-ville AMNH 327-1	Lowicz AMNH 3775-1	Mincy AMNH 887-1	Morris-town AMNH 345-1	Mount Padbury AMNH 4212-1	Patwar AMNH 4112-1	Vaca Muerta AMNH 986-3	Veramin AMNH 313-1	matrix mean	s.d.	metal-free matrix mean*	s.d.	metal-free whole rock mean**
olivine	—	—	0.2	—	—	—	0.7	—	—	—	—	—	1.5	0.2	0.4	0.2	0.6	2.3
low-Ca pyx	38.6	42.1	33.5	31.0	23.2	48.0	46.6	78.5	54.5	23.1	35.3	30.1	56.6	41.6	15.4	57.1	14.9	60.6
hi-Ca pyx	2.1	—	0.8	1.7	1.2	3.3	7.4	3.1	6.8	2.2	0.4	0.2	0.2	2.4	2.3	2.7	2.4	2.4
feldspar	24.6	2.9	26.2	21.1	24.8	26.6	31.6	15.4	31.3	18.1	18.4	17.3	13.6	20.9	7.9	28.8	9.7	26.9
tridymite	5.3	0.5	7.4	7.1	8.4	10.7	10.4	1.4	5.8	6.7	3.0	7.4	1.9	5.9	3.3	8.1	4.9	5.1
ilmenite	—	—	—	—	tr	—	—	—	—	—	—	—	—	tr	—	tr	—	—
rutile	—	—	—	—	0.1	—	—	—	—	—	—	—	—	tr	—	tr	—	—
chromite	—	—	0.2	0.2	0.4	0.4	1.0	—	0.5	0.1	0.4	0.1	—	0.3	0.3	0.5	0.6	0.8
merrillite	1.0	1.8	1.6	1.8	3.0	1.2	1.7	0.3	0.7	1.5	3.0	2.5	1.7	1.7	0.8	2.6	1.5	1.9
troilite	0.1	13.8	0.3	5.9	0.5	8.2	0.3	0.3	0.4	7.9	20.4	17.9	2.1	6.0	7.2	—	—	—
kamacite	26.7	38.9	28.5	29.6	35.5	1.2	—	0.7	—	38.6	17.8	23.1	19.9	20.0	14.5	—	—	—
taenite	1.6	—	1.3	1.6	2.5	0.4	—	0.3	—	1.8	1.3	1.4	2.5	1.1	0.9	—	—	—
schreibers.	—	—	—	—	0.4	—	—	—	—	—	—	—	—	tr	—	—	—	—
# points	187	215	633	493		244	297	292	285	507	230	489	477					
% silicate	72	47	70	63	62	90	<100	99	<100	52	61	58	76	73%		100%		100%

*Calculated from individually normalized matrix modes.
**Prinz et al. (1980).

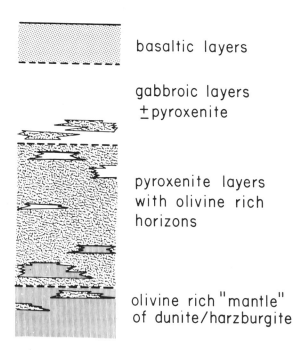

Fig. 2. Schematic diagram of a possible achondritic crust for the mesosiderite parent body before regolith formation, mixing with metal and metamorphism to form the unique features of the mesosiderites.

would comminute the higher strata (basalts and gabbros) more thoroughly than the deep horizons (pyroxenites and olivine rich rocks) because few impacts would be sufficiently energetic to penetrate to the deeper levels. A breccia that samples the crust of such a body might therefore be expected to contain large clasts from the deeper horizons with the surrounding fine grained matrix enriched in material from shallower horizons. This is exactly the nature of the mesosiderite silicate fraction.

The distribution of pyroxene compositions and the modal abundance of olivine in howardites and mesosiderites (Prinz *et al.*, 1980) suggests that the howardites have sampled pyroxene from the deeper levels of the howardite parent body less frequently than the mesosiderites have sampled pyroxene from the deep strata on their achondritic parent body. The results of Labotka and Papike (1980) also indicate that basaltic material near the surface of the howardite parent body has been more severely comminuted than the deeper pyroxenitic/diogenitic fraction. The similarities between howardites and the mesosiderite silicate fraction suggest that, prior to mesosiderite metamorphism, both suites of meteorites evolved in a similar manner with the highest levels of the crust being finely comminuted and deeper levels of the crust being brecciated, but not so severely as the surface.

It must be emphasized however, that *the matrix in mesosiderites is no longer simply a mass of brecciated clast material*. The effects of metamorphism, after the achondritic breccia was incorporated into the mesosiderite and mixed with metal, have extensively modified the texture and mineral chemistry of the matrix. In a few cases, this has led to very extensive recrystallization so that only the larger clasts, which are less reworked than the fine grained material, still show textural evidence that the silicates were originally in a breccia. Many smaller clasts, being more reactive, were totally resorbed and incorporated into the recrystallized matrix.

TEXTURAL AND CHEMICAL EVIDENCE OF METAMORPHISM

Except for the interiors of large mineral clasts and a few large lithic clasts, obvious textures produced by metamorphism are seen in all the mesosiderites but the impact melt group (Floran et al., 1978). In the latter, melting has obliterated any pre-existing metamorphic textures although Hainholz may show very minor metamorphism (Floran et al., 1978). The metamorphism of the mesosiderite silicates has produced four distinctive textural features. These are: (a) *coronas* around olivine clasts; (b) *overgrowths* of homogeneous orthopyroxene and inverted pigeonite (with compositions usually between En_{60} and En_{65}) over more magnesian orthopyroxene clasts (with compositions from En_{70} to En_{85}); (c) *rims* of similar composition to the overgrown pyroxene ($<En_{60}$) on iron rich pyroxene grains within mafic clasts, or on isolated iron rich pyroxene clasts in the matrix; (d) *poikiloblastic feldspar* in the most magnesian mesosiderites. All these features are believed to have developed as a result of subsolidus heating of the mesosiderites. A further feature of some mesosiderites, (e) *resorption* of pyroxene clasts is believed to have resulted from slight partial melting of the silicate fraction, probably at the culmination of the metamorphic event.

(a) Coronas on olivine clasts

The formation of coronas on olivine clasts has been discussed in detail by Nehru et al. (1980a). This discussion supplements Nehru et al. providing additional detail and new observations that are useful to the understanding of mesosiderite metamorphism. The coronas develop because of the instability of magnesian olivine in the presence of free silica [up to 10.7% (vol.) of the matrix of the mesosiderites (Table 2) and assumed to be tridymite]. Under most conditions silica will react with the edges of olivine clasts to form the coronas of orthopyroxene isolating the olivine core from the mesosiderite matrix. In detail, however, the coronas have a banded structure with high concentrations of merrillite (up to 20% vol.) and chromite (up to 7% vol.) in some bands. Plagioclase, calcic pyroxene and ilmenite are also present. The outer portions of the coronas often grade outward into the mesosiderite matrix, but differ from it because of their lack of tridymite and metal. In those mesosiderites with little, or no, matrix tridymite, coronas are not developed on olivine (Nehru et al., 1980a). In the impact melt mesosiderites (Floran et al., 1978), coronas are absent though tridymite is present in all three (Prinz et al., 1980). Hainholz (AMNH 316-1) contains a forsteritic olivine clast (Fo_{92}; Nehru et al., 1980a) which should have been very unstable in the presence of an SiO_2-oversaturated melt (5% vol. tridymite; Prinz et al., 1980). To prevent the olivine-SiO_2 reaction, this mesosiderite must have cooled very rapidly to below the closure temperature of this reaction. The lack of coronas in these mesosiderites (Hainholz, Simondium, Pinnaroo) that have been rapidly quenched is important evidence that the corona forming process was metamorphic rather than igneous, in agreement with Powell (1971) and Nehru et al. (1980a).

Coronas on different olivine crystals differ even within a single mesosiderite. Nehru et al. (1980a) discussed many examples and classified their inner zones into three different types of development. These are:

Stage I—Many tiny orthopyroxene crystals radiating from the central olivine with intergrown symplectic chromite (± ilmenite) (Fig. 3). Plagioclase is sometimes present but merrilite is rarely seen even though large crystals were detected in the matrix, close to metal, just outside these coronas.

Stage II—Vermicular chromite is present, but these coronas are more granular than Stage I and are probably more equilibrated. In these coronas merrillite tends to be concentrated in the outermost zone of the coronas where a little metal is sometimes present. Pyroxene is coarser than Stage I.

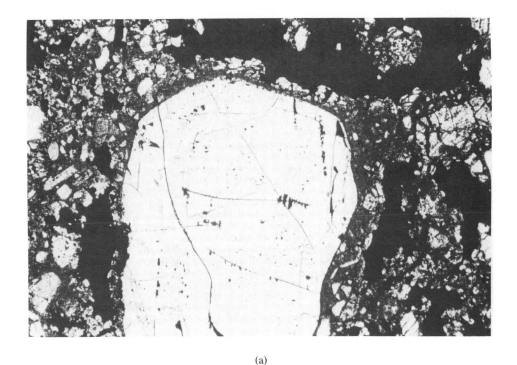

(a)

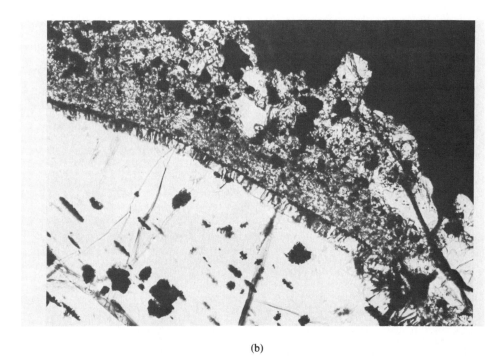

(b)

Fig. 3. (a) Large olivine clasts (1½ mm across) from Mount Padbury surrounded by very thin Stage I type of corona. Long dimension of micrograph is 3 mm. (b) Detail of margin of above corona showing fine intergrowth of chromite and orthopyroxene. Long dimension is about 450 μm.

Stage III—These are the most granular and coarse grained coronas. Chromite grains form necklaces around the outline of the olivine and are no longer vermicular. This stage (Fig. 4) probably reflects the closest approach to equilibrium in the coronas. The inner coronas are distinctive but the outer parts grade into the surrounding matrix or combine with poikiloblastic overgrowths that have modified the matrix. Unlike the Stage II coronas merrilite tends to be concentrated in the inner part of Stage III coronas, just outside the distinctive chromite rich band(s) (Fig. 4).

Chemically, the orthopyroxene formed in the inner coronas shows varying degrees of equilibration with the surrounding matrix. Experimental results for Fe–Mg distribution between olivine and orthopyroxene from Medaris (1969) provided the equilibrium constraints. A correlation is observed between the textural type of corona developed (Stage I, II or III) and the *difference* between the composition of typical groundmass orthopyroxene surrounding the corona and the 'equilibrium' orthopyroxene which should coexist with the olivine core. The least 'equilibrated' coronas (Stage I or I–II) are found when the difference between the matrix orthopyroxene and 'equilibrium' orthopyroxene is greatest. For example, an olivine corona from Morristown has an olivine core of composition Fo_{80}. This olivine should be in equilibrium with an orthopyroxene with composition En_{82-83} and indeed a thin fringe of this composition is observed. However, the surrounding matrix orthopyroxene is En_{60-65}, a contrast of 18–20% in molar composition. In contrast, olivine clasts that should equilibrate with orthopyroxene of similar composition to the surrounding matrix, tend to have coronas of the Stage III or II–III textural types. An olivine (Fo_{60-63}) from Vaca Muerta with a well developed Stage III corona has coronal orthopyroxene of composition En_{66-69} that is of essentially identical composition to the surrounding matrix orthopyroxene (En_{67-70}). Where the compositional contrast between the olivine and the matrix pyroxene was great, onset of the olivine to orthopyroxene reaction was accompanied by the formation of many pyroxene nuclei on the edges of the olivine and hence the fine grained texture of Stage I coronas. The small

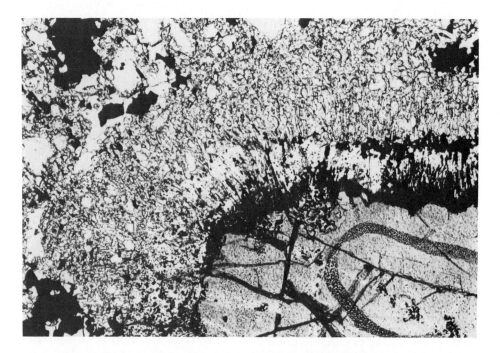

Fig. 4. Segment of a very spectacular Stage III corona from Emery showing coarsely crystallized orthopyroxene in the inner corona together with two "necklaces" of chromite. Grain size of the outer corona decreases toward the matrix (top left). Long dimension is about 2 mm.

compositional contrast in Stage III coronas resulted in less nucleation of pyroxene and hence coarser crystallization.

The most important reaction in the formation of these coronas is:

$$\underset{\text{oliv.}}{Mg_2SiO_4} + \underset{\text{trid.}}{SiO_2} \rightleftarrows \underset{\text{opx}}{2MgSiO_3} \quad (1)$$

Reaction (1) successfully accounts for the 70 to 90% of orthopyroxene in mesosiderite olivine coronas but does not explain the presence of several lesser phases (Nehru et al., 1980a). Both chromite and merrillite are characteristic of coronas but are not produced by reaction (1). Coronas on terrestrial olivines generally develop in water bearing systems of different composition from the dry polymict mesosiderite matrix (e.g., Esbensen, 1978). Nevertheless, an important class of reactions between olivine and plagioclase that is important in terrestrial coronites may be modified for application to mesosiderite coronas in addition to the reaction of Nehru et al. (1980a). Reactions (2a) and (2b) below may be used as a framework in which to consider these minor mineral phase relationships.

$$\underset{\text{forst.}}{2Mg_2SiO_4} + \underset{\text{anorth.}}{CaAl_2Si_2O_8} \rightleftarrows \underset{\text{enst.}}{2MgSiO_3} + \underset{\text{diops.}}{CaMgSi_2O_6} + \underset{\text{spinel}}{MgAl_2O_4} \quad (2a)$$

$$\underset{\text{forst.}}{2Mg_2SiO_4} + \underset{\text{anorth.}}{3CaAl_2Si_2O_8} + \underset{\text{metal}}{2P_{Fe}} + 5O \rightleftarrows \underset{\text{enst.}}{MgSiO_3} + \underset{\text{spinel}}{3MgAl_2O_4} + \underset{\text{trid.}}{7SiO_2} + \underset{\text{merril.}}{Ca_3(PO_4)_2}$$
$$(2b)$$

These reaction cells provide enstatite (in addition to that from reaction 1), spinel and minor calcic pyroxene and merrillite [approximated as the hypothetical $Ca_3(PO_4)_2$ end member]. The silica produced by reaction (2b) is unstable in the presence of olivine and would be removed by reaction (1), further increasing the orthopyroxene content of the corona. The spinel produced is obviously different than the chromite present in the coronas. The corona chromite, however, usually enriched in spinel ($MgAl_2O_4$) relative to the matrix chromite and always has a higher $Al/(Cr+Al)$ ratio. Since long range transportation of an $FeCr_2O_4$ component is unlikely, the chromite of the coronas may have formed by 'sweating' chromium out of an ($FeCrAlSiO_6$) component from earlier matrix pyroxene, together with the spinel component from the breakdown of plagioclase during the corona forming process (cf. Harlow and Klimentidis, 1980; Harlow and Delaney, 1981).

The small amount of phosphate produced by reaction (2b) may be inadequate to account for the large concentrations ($\geq 20\%$ vol.) in parts of some coronas, but this reaction may operate in addition to the clinopyroxene-metal reactions discussed by Harlow et al. (1980, 1981) that produced much of the phosphate observed in mesosiderites. The reaction mechanism of Agosto (1981) is not appropriate as the observed textures are subsolidus and they are not compatible with the silicate magma required. The variable, but localized concentration of merrillite in the olivine coronas requires a kinetic explanation in addition to being thermodynamically reasonable. A possible mechanism may be indicated by consideration of the volume relationships in reactions (1) and (2a). Both reactions involve a decrease of 10–12% in the molar volume of the solid components during reaction from left to right. The major components of the corona textures occupy a lesser volume than the original reactants.

The phosphorus required for merrillite formation must diffuse into the reaction site around the olivine and this volume decrease must produce a local pressure gradient, thereby providing a potential transport mechanism for moving a P-rich flux from metal rich areas to the coronas. The location of merrillite nucleation and growth sites, as a result of, for example, reaction (2b), would then be sensitive to the rates of nucleation and growth during reactions (1) and (2a). Coronas of the Stage III type (granular textures) appear to have resulted from slower crystal growth than coronas of the Stage I type (radiating vermicular intergrowths). If the rate of transportation of the P-bearing flux was constant, then that flux would reach the merrillite nucleation site later in the 'textural history' of the development of a rapidly growing Stage I corona, than in the history of more sluggishly growing Stage III coronas. Modal studies (Nehru *et al.*, 1980a) of different parts of Stage II and Stage III coronas reveal that merrillite is found in the inner parts of Stage III coronas, in the outer parts of Stage II coronas and is not found in Stage I coronas. This observation supports the proposed phosphorus transportation model. Studies of phosphorus volatility and transport phenomena may, therefore, give quantitative estimates of the time scale of corona development and mesosiderite high temperature metamorphism in general. If the phosphorus is transported by some gaseous diffusion mechanism, then the time for corona development is probably very short.

The coronas around olivine in mesosiderites, therefore provide important information about the response of olivine to metamorphism and may provide information about the rates of reaction, and of cooling in high temperature mesosiderite metamorphism.

(b) Overgrowths on Mg-pyroxene clasts

Powell (1971) recognized that zoning of magnesian pyroxene clasts, caused by the development of more iron rich rims, was produced after the clasts were incorporated in the mesosiderite breccia, because the rims often follow the irregular edges of the clasts. Such zoning in Patwar, from pyroxene cores of $En_{74}Wo_{1.9}$ to rims of $En_{67}Wo_{2.0}$ was ascribed to igenous fractionation by Weigand (1975), but the textural observations of Powell and others discussed below indicate that zoning profiles were developed after brecciation. Powell (1971), Floran (1978), Nehru *et al.* (1978b), Hewins (1979) and Delaney *et al.* (1980a) all recognized the presence of two generations of pyroxene. These are an early igneous clast population and a later mesosiderite overgrowth population. Powell (1971) recognized that orthopyroxene clasts might have more iron rich rims with abundant inclusions and noted the presence of inverted pigeonite in some rims. Hewins (1979) also noted the presence of inverted pigeonite mantling more magnesian orthopyroxene cores in several mesosiderites and suggested that overgrowths developed as a result of poikiloblast nucleation during pigeonite inversion. Floran (1978) used the development of overgrowths on pyroxene clasts as part of a classification scheme for the mesosiderites even though he noted the development of overgrowth textures in subgroups other than his pyroxene poikiloblastic subgroup.

The major control on the development of overgrowths on magnesian pyroxene clasts is the composition and texture of the surrounding matrix. Overgrowths may be well developed on pyroxene clasts in one part of a thin section but may be absent, or poorly developed, on clasts in another part of the same section. The mesosiderites in the pyroxene poikiloblastic subgroup of Floran (1978) undoubtedly show the most spectactular and well developed overgrowths but their presence in mesosiderites from other groups (both poorly and well recrystallized) indicates that classification of the mesosiderites on the basis of such locally controlled textural criteria may be difficult, or even misleading.

Delaney *et al.* (1980a) examined several overgrown pyroxene clasts and presented compositional profiles across four (Fig. 5). Four distinct zones were identified in well developed overgrowths. The innermost "zone" (1) (Fig. 5) is the original orthopyroxene clast or core that usually has uniform composition from its center to the inner edge of the transition zone. A few examples of zoning in these cores were seen in some clasts but there are believed to be artifacts caused by thin sectioning. These zoned clasts were most often found in mesosiderites that have overgrowth/core relationships like Lowicz (Fig. 5)—i.e. the interface between core and overgrowth is transitional rather than abrupt.

The second zone in the overgrowths is a compositional transition zone (Fig. 5, zone 2). In this zone, which may be as narrow as 50 μm or wide as 0.5 mm, the composition of the orthopyroxene changes from the magnesian core (often $>En_{70}$) to the more iron rich composition of the overgrowth, usually between En_{60} and En_{65}. Dyarrl Island is the only exception with more iron rich (En_{52}) overgrowths. Optically, this zone is either not visible

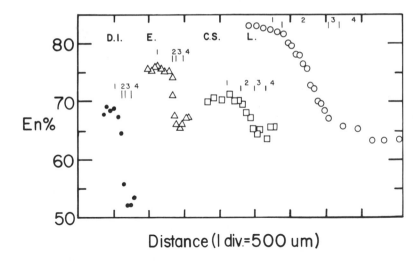

Fig. 5. Variation of orthopyroxene compositions (En%) across the core-overgrowth interface for typical overgrown clasts from Dyarrl Island (D.I.), Emery (E), Clover Springs (C.S.), and Lowicz (L). The four zones distinguished in each profile are (1) orthopyroxene core; (2) compositional transition zone; (3) two pyroxene overgrowth (not always present); and (4) overgrowth and/or poikiloblast (not always present).

because it is narrow and in the same crystallographic orientation as the core, or is seen as a band of sweeping extinction just inside the third zone. The width of this zone is of considerable importance because it is a function of the rate of diffusion of Fe and Mg between the core and the overgrowth, and the length of time during which diffusion could occur. The effect of thin sectioning a non-abrupt transition zone is always to widen the zone relative to its width in an orthogonal section (in three dimensions) through core and overgrowth. For this reason, the *narrowest* transition zones observed in each mesosiderite were taken to be the most representative, being closest to the orthogonal cut condition. The very abrupt transitions in Emery and Dyarrl Island are nearly immune to this cut-effect since they are close to 'step-function' type profiles.

The third zone of the overgrown pyroxene clasts is also often the outer-most. This is a narrow zone of homogeneous orthopyroxene usually containing between 30% and 50% (vol.) augite blebs. These blebs are often elongate but irregular and may be either radial about the central orthopyroxene clasts (Fig. 6a) or in a common orientation like augite blebs in inverted pigeonite (Fig. 6b). In many overgrown clasts, this zone is well developed and very obvious, but in a few it is very narrow and the augite blebs are less abundant (e.g. Lowicz). This zone may be the innermost part of the true overgrowth, the two inner zones being part of the original clast. The pyroxene crystallizing over the original clasts was enriched in a calcic component, probably because the matrix is enriched in material from the mafic suite. On those clasts that appear to have inverted pigeonite overgrowths rather than orthopyroxene overgrowths, this third zone may be very wide and takes the place of the fourth zone. Two-pyroxene temperatures from this zone are usually in the range 850–1000°C using the Wells (1977) two-pyroxene thermometer but these are probably closure temperatures (Ross and Huebner, 1979) and do not reflect the maximum temperature of metamorphism, or even overgrowth development. Hewins (1979) gives temperatures between 1120° and 1170°C for pigeonite inversion in overgrowths from four mesosiderites.

The fourth, and outermost, zone of the overgrowths is the poikiloblastic zone. This zone varies from nonexistent to very extensive. It is seen most spectacularly in Emery, and in Lowicz and Morristown where poikiloblastic overgrowths on magnesian pyroxene clasts fill up all the available space between clasts and include all minerals (Fig. 7). In

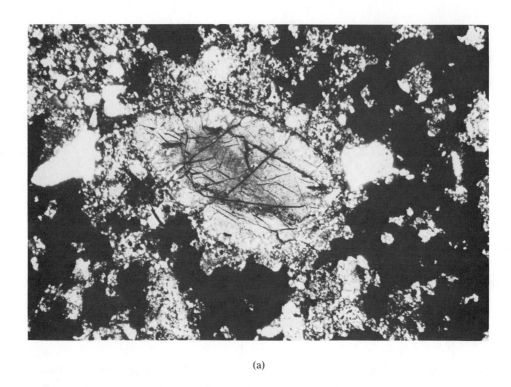

(a)

(b)

Fig. 6. (a) Pyroxene clast from Emery with narrow overgrowth containing irregular to radiating augite blebs. (b) Pyroxene clast from Morristown with broad pigeonitic overgrowth/poikiloblast. This is the same as Fig. 7 but in crossed polars to show the exsolution in the overgrowth. Long dimension in each micrograph is 3 mm.

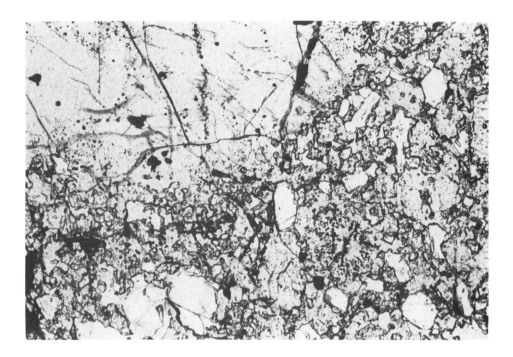

Fig. 7. Large pyroxene clast (top left) from Morristown with very extensive, space filling overgrowth filling almost entire field of view. All orthopyroxene is in same optical orientation but the overgrowth is of the inverted pigeonite type. Long dimension is 3 mm.

many mesosiderites the outer limits of the overgrowths is against metal, but in Lowicz the poikiloblastic overgrowths include the metal. In Emery also (Fig. 8), poikiloblastic overgrowth extend through and around the metal network for long distances (perhaps >2 mm).

The space filling character of the poikiloblastic overgrowths, together with the general lack of small pyroxene clasts in the matrix, or as inclusions within the poikiloblasts, suggests that the nucleation of overgrowth pyroxene was difficult and heterogeneous requiring the presence of large, preexisting pyroxene grains. This difficult nucleation event was followed by rapid crystal growth from these newly formed nuclei, accompanied by resorption of small clastic pyroxenes of variable composition to form the equilibrated and homogeneous overgrowth (Lowicz, En_{64}; Emery, En_{65}). The homogeneity of both the core pyroxene and the overgrowth pyroxene, with only a narrow transition zone between, precludes derivation of these zoning patterns by simple igneous fractionation. The occurrence of inverted pigeonite overgrowths (Hewins, 1979), however, suggests that temperatures at the time of formation of these overgrowths were high enough for partial melting to occur in the mesosiderite matrix. The preservation of irregular clastic outlines for inclusions, of minerals other than pyroxene, in the poikiloblastic overgrowths suggests that melting was not a general phenomenon in the mesosiderite matrix. Partial melting may, therefore, represent the culmination of the metamorphic event.

The average modal composition of the orthopyroxenites, the mafic clasts (Nehru et al., pers. comm.), and the average observed matrix are given in Table 3. The feldspar content of the matrix was used to calculate the proportions of orthopyroxenites (34%) and mafic clasts (66%) in the matrix, since feldspar has not participated significantly in metamorphic redox reactions. The differences between the observed and calculated matrix modes are shown in Table 3. The calculated proportions of orthopyroxenites and mafic clasts show that the

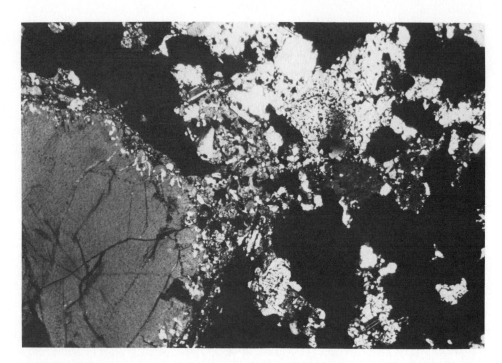

Fig. 8. Overgrowth on clast in Emery (left) that does not stop against metal but is intimately intertwined with metal network (dark). Overgrowth is in same orientation as clast and extends across half the field of view. Long dimension is 3 mm.

Table 3. Material balance calculation of mesosiderite matrix from pyroxenite (+dunite) and mafic clast components (normalized to plag.).

Mode	pyroxenite	mafic* clast	calc. matrix	obs. matrix	mat (obs.-calc)
oliv.	0.5	—	0.2	0.2	
opx	97.0	34.9	55.8	57.1	+1.5
cpx	2.5	8.9	6.7	2.7	−4.0
plag.	—	43.4	28.8	28.8	—
trid.	—	8.1	5.4	8.1	+2.7
ilm.	—	0.1	tr	tr	
cm	—	0.5	0.3	0.5	+0.2
merril.	—	*	0	2.6	+2.6
% in calc.	33.6	66.4	—	—	—

*Mafic clasts recalculated merrillite free to reduce redox effects.

matrix is enriched in the mafic clast component relative to bulk mesosiderite modes (60–75% pyroxenite, 25–40% mafic clast) described by Prinz et al. (1980). Using the 34:66 mixture to approximate the proportions of the components with typical orthopyroxenite composition between En_{70} and En_{72} and average mafic clast pyroxene between En_{45} and En_{50} the average matrix pyroxene was calculated to be En_{54-58}. This range is slightly below the typical overgrowth/poikiloblastic pyroxene compositions observed (En_{60-65}) but is consistent with those results since no allowance was made, in the calculation, for

Fe–Mg exchange between the orthopyroxenite clasts and the matrix or for removal of iron from silicates by redox reactions (Delaney et al., 1980b). A small change (~5%) in the modal proportions of the two components will also produce concordance between the modal composition of the matrix and its pyroxene composition. The compositions of overgrowths on orthopyroxene clasts and the composition of homogenized mesosiderite matrix pyroxene are close and are a further indication that overgrowth pyroxene was produced as a result of a homogenization (metamorphic) process rather than as a result of igneous fractionation.

Cooling rate estimates from overgrowths.

Delaney et al. (1980a) have used the shape of the transition zone (zone 2) between the core of the pyroxene clast and the overgrowth to estimate cooling rates during mesosiderite metamorphism. The assumption was made that overgrowth pyroxene of composition En_{60-65} (ignoring Dyarrl Island) nucleated and grew on homogeneous clasts of composition En_{70-75} because of the steep zoning profiles observed in Emery and Dyarrl Island (Fig. 5). Since the overgrowths developed at temperatures above 850°C and, in the case of some pigeonitic overgrowths, as high as 1170°C (Hewins, 1979), Fe–Mg diffusion within the pyroxene should have occurred fairly rapidly. Although experimental data for Fe–Mg exchange in pyroxene as a function of cooling rate are not available, this process has been studied in olivine (Taylor et al., 1977; Onorato et al., 1978). There are structural and chemical differences between olivine and orthopyroxene, but comparison of their anion porosities (Dowty, 1980) indicate that diffusion rates in pyroxene (and especially clinopyroxene) should be only slightly slower than in olivine. With initial compositional contrasts of 15% (mol.) forsterite across an assumed compositional step in olivine, Taylor et al. (1977) produced compositional profiles similar to those in Fig. 5 with cooling rates between 100°C/day and 1°C/day in the temperature range 900–1100°C. The steep compositional profiles of Emery and Dyarrl Island resemble those produced at cooling rates of 10–100°C/day while the gentler slopes of Clover Springs and Lowicz resemble those produced at slower rates (1–5°C/day). The close similarities between the observed compositional gradients in pyroxene from mesosiderites and the experimentally calibrated gradients in olivine indicate that the cooling conditions of the mesosiderites may have been similar to the experimental runs. Certainly rapid cooling was necessary to preserve the large compositional steps between pyroxene clasts and overgrowths.

These rapid cooling rates at high temperatures are in dramatic contrast to the slow cooling rates at low temperatures (<500°C) required by kamacite/taenite/schreibersite phase relations (Powell, 1969; Kulpecz and Hewins, 1978; Agosto et al., 1980), the presence of the ordered metal phase tetrataenite (Albertsen et al., 1978; Clarke and Scott, 1980) and fission track studies in merrillite (e.g., Crozaz and Tasker, 1980). The development of the overgrowths must have occurred during a fairly short, but hot event followed initially by a fast cooling rate that decreased asymptotically at lower temperature, perhaps because of thermal blanketing on the surface of the mesosiderite parent body (Wood, 1979). Hewins (1979) also suggested that rapid initial cooling from high temperatures occurred in mesosiderites, prior to the slow low temperature cooling recorded by the metal.

(c) Rims on iron rich pyroxene grains

Rims of more magnesian pyroxene have modified the composition of many iron rich pyroxene grains in the matrix (rare) and grains within mafic clasts. Their development mirrors formation of overgrowths on magnesian pyroxene clasts. The orthopyroxene lamellae in the cores of pyroxene from mafic clasts range in composition from En_{61} to En_{36}, and are plotted in Fig. 1 as a function of plagioclase composition. The majority of the mafic clasts have core compositions in the range En_{47} to En_{37}. In many of the mafic clasts, particularly the smaller clasts, two distinct areas are recognized. These are: (a) the clean, regularly exsolved cores, and (b) rims of lower birefringent orthopyroxene sometimes with irregular blebs of augite. The rims of orthopyroxene on inverted pigeonite were described by McCall (1966) who recognized that they were more magnesian than the core regions of the same crystals. A typical inverted pigeonite (from a mafic clast) with an orthopyroxene rim is shown in Fig. 9a. The core pigeonite has lamellae of orthopyroxene (En_{48}) but the well developed narrow rim of orthopyroxene is En_{55-60}. In Fig. 9b a similar exsolved pigeonite (En_{50}) in the matrix of Mt. Padbury is surrounded by a narrow, homogeneous rim of En_{60} orthopyroxene.

An important feature of these magnesian rims on iron rich pyroxene in mafic clasts is their spatial variability. In small clasts, rims are present on almost all the pyroxene grains and, in some cases, the cores of the grains are almost totally replaced by more magnesian pyroxene. In larger clasts the amount of rimming present on pyroxene grains decreases with increasing distance from the interface between the clast and the mesosiderite matrix. In a few cases, rimming can be seen within the clast only where mesosiderite metal has penetrated into the clast along a crack. It is not clear, however, whether the metal actively participated in the rim forming reaction. Of the clasts studied by Nehru et al. (1981b) one of the largest was an iron rich gabbro clast (Vaca Muerta 914-1) which has mineral chemistry indistinguishable from the basaltic (monomict) eucrites. This large clast shows no rimming of the pyroxene grains and contains no metal. In this respect it resembles the clast from Mount Padbury (enclave Z) analyzed by McCall (1966). A number of the large mafic clasts studied by McCall are currently under reinvestigation but initial study suggests that core-rim relationships in the pyroxene are rare. The relationship between mafic clast size and the degree of development of rims on pyroxene crystals—i.e., extensive rimming near clast edges and little rimming in clast cores—suggests that the *rim development occurred after incorporation of the clast into the mesosiderite matrix* and is therefore a metamorphic phenomenon rather than an igneous one.

The rims of orthopyroxene on iron pigeonites show the opposite zoning trend to the overgrowths on magnesian orthopyroxene. The composition of the rims, however, is very similar to that of the overgrowths. Generally, the overgrowths fall between En_{60} and En_{65}. The most magnesian rims observed are En_{62} (in several mafic clasts) with the majority falling between En_{53} and En_{61}. In common with the outer parts of the overgrowths, rims tend to be fairly homogeneous but do not show the complex zonal structure observed in many overgrowths. The interface between the cores and rims in mafic clasts is often controlled (Fig. 9) by the cleavage and exsolution lamellae of core pigeonite. In contrast to the magnesian orthopyroxene clasts and their overgrowths, there is often a distinct contrast in bulk CaO content between the pigeonite cores and their orthopyroxene rims.

Traverses along carefully selected low calcium lamellae from the cores to the rims of some pigeonites show that the $Fe/(Fe + Mg)$ ratio changes smoothly across the interface (cf. Clover Springs and Lowicz overgrowth profiles) but fine scale CaO variations prevent definitive statements at present. It is hoped that identification of several inverted pigeonites (in different mesosiderites) with broad, low calcium lamellae will permit profiles from Fe-rich cores to Mg-rich rims, complementary to Fig. 6 to be constructed. At present, no estimate of cooling rates on the basis of core-rim relations in Fe-rich pyroxene from the mafic clasts (and the rare Fe-rich pigeonites in the matrix) can be made. Because the large calcium atom does not diffuse readily through the pyroxene structure, the sharp contrast in CaO content between cores and rims does not provide useful cooling rate information.

The general similarity between the compositions of these rims (En_{55-62}) and the overgrowths (En_{60-65}) on magnesian clasts, indicates that both textures developed as a result of interaction with the hot mesosiderite matrix, and represent partial homogenization of the mesosiderites. The rims in mafic clast pyroxene formed by Fe-Mg exchange between the clast and the matrix while the overgrowths developed by recrystallization of matrix pyroxene using orthopyroxene clasts as nucleation sites.

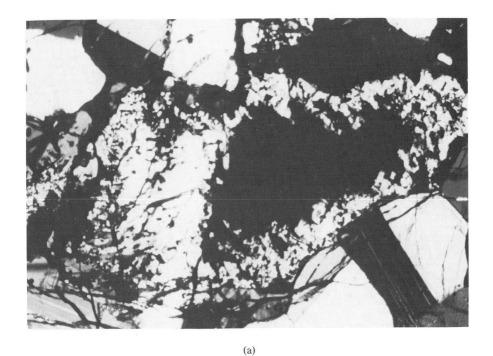

(a)

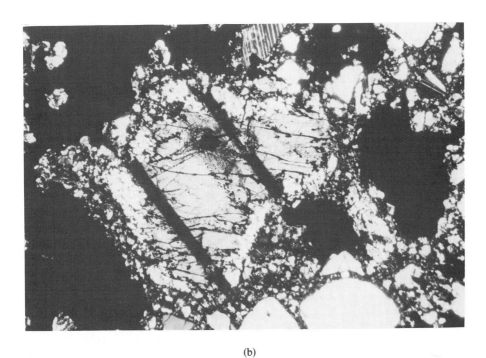

(b)

Fig. 9. (a) Iron rich pyroxene grain in gabbro clast Vaca Muerta 559 with well developed rim of more magnesian pyroxene. Core is at extinction, rim is irregular bright area. Long dimension is 2 mm (crossed polars). (b) Iron rich pigeonite pyroxene clast in Mount Padbury with more magnesian rim. Rim is brighter than the grey, regularly exsolved core (crossed polars). Long dimension is 3 mm.

(d) Poikiloblasts of plagioclase

Floran (1978) described a subgroup of the mesosiderites which is characterized by the development of poikiloblastic feldspar in the matrix. The mesosiderites in this subgroup (Bondoc, Budulan, Mincy) are dominated by the magnesian pyroxenite component (Fig. 2 in Prinz et al., 1980) with very minor amounts of the mafic clast component. Several other mesosiderites also have similar pyroxene distribution patterns (Allan Hills A77219, Barea, Veramin and perhaps Chinguetti) but do not contain large plagioclase poikiloblasts. Veramin contains poikiloblastic plagioclase but on a smaller scale than Floran's subgroup.

The lack of a significant iron rich silicate component (mafic clasts) in these mesosiderites apparently prevented the type of homogenization reactions responsible for the development of the overgrowths and rims on pyroxene described above. Minor developments of pyroxene overgrowth textures have been observed in both Mincy and Bondoc, but these are subsidiary to the poikiloblastic feldspar texture. The major feature of Bondoc, Budulan and Mincy is their thoroughly recrystallized texture (Fig. 2 in Floran, 1978). These meteorites were, therefore, subjected to metamorphic reequilibration that produced well defined pyroxene grain shapes in all three meteorites (or four including Veramin). The magnesian mesosiderites discussed in this section are probably the most nearly monomict and initially homogeneous of all the mesosiderites. Metamorphism, therefore, caused only progressive recrystallization. It is not clear if the variable degree of recrystallization in the seven magnesian mesosiderites is caused by different durations, temperatures, or kinetics during metamorphism. Recrystallization of feldspar has progressed further in these mesosiderites because it was not prevented by the development of space filling poikiloblastic pyroxene at the expense of fine grained matrix.

(e) Resorption of clasts

In the Emery mesosiderite, a few metal poor areas contain large pyroxene crystals (up to 2 mm long) which have very irregular outlines with deep embayments containing plagioclase and chromite (Fig. 10). These crystals may be extensions of poikiloblastic overgrowths. The lathy morphology of the surrounding plagioclase and the presence of deep embayments containing spinel and plagioclase suggests, however, that these clasts reacted with a melt. If this is so, then two possible sources of melt are possible. (a) Localized partial melting was induced by impact mechanisms similar to those responsible for the present textures of Hainholz, Simondium and Pinnaroo (Floran et al., 1978). (b) Partial melting in parts of the matrix at the culmination of mesosiderite metamorphism caused resorption of preexisting pyroxene grains. No criteria are yet available to distinguish between these options but if the melting is a result of metamorphism then temperatures in the mesosiderite matrix may have exceeded solidus temperatures (perhaps 1100–1200°C) for a short period of time. The presence of inverted pigeonite in overgrowth textures (Powell, 1971; Hewins, 1979) suggests that temperatures in excess of 1150°C were reached so that some melting may have accompanied metamorphism.

REDOX FORMATION OF MERRILLITE

The silicate fraction of the mesosiderites is unique among the basaltic achondrite meteorites because of its high and variable content of phosphate (mainly merrillite) (Nehru et al., 1978a). Prinz et al. (1980) show a range from 0.2% vol., (Bondoc Pen.) to 3.8% vol. (Emery) in the 21 mesosiderites. Typically the basaltic achondrites contain only a trace of either merrillite or apatite. The order of magnitude greater abundance of phosphate in the mesosiderites has been ascribed to both igneous processes and to redox exchange reactions involving the ubiquitous metal fraction (Nehru et al., 1978a, 1980b; Fuchs, 1969).

Within individual mesosiderites, the spatial distribution of merrillite is heterogeneous. Large pyroxene or pyroxenite clasts contain, at most, trace amounts, whereas individual

Fig. 10. Area in the matrix of Emery containing a large pyroxene clast showing extreme embayment with infilling of lathy plagioclase and spinel suggesting erosion or resorption by a melt. The bright irregular area in the center of this photo is the pyroxene clast and much of the surrounding melt-like material is close to extinction (crossed polars). Long dimension is 3 mm.

clasts from the mafic suite may contain up to 2.3% vol. merrillite (Nehru et al., pers. comm.). This is less than the average content of the silicate matrix (2.6%, Table 1) and considerably less than the contents in some partially melted areas in Emery that contain over 8% merrillite (Nehru et al., pers. comm.). The most extreme enrichment of phosphate, however, is found in specific zones of the olivine coronas where up to 20% merrillite has been identified.

Phosphorus should behave as an incompatible element, and Nehru et al. (pers. comm.) showed that fractionation can explain many of the features of the mesosiderite mafic suite. Phosphate minerals should, therefore, be concentrated in the most evolved basalts and gabbros of the mafic suite. No such concentration exists. The phosphate content of the mafic clasts varies irregularly and the only correlation seen is between the phosphate content of mafic clasts and the bulk phosphate of the mesosiderites in which they occur. Phosphorus does, however, behave as an incompatible element in the so-called impact melt basalts in Emery. These basalts are interpreted to be partial melts of fractions of the mesosiderite matrix. The matrix of Emery contains 3% merrillite (Table 1) but the impact melt basalts contain between 3% and 8% merrillite. This enrichment is consistent with the incompatible behavior of phosphorus in silicate melts.

The alternative source of the phosphate, by redox exchange between the metal and silicate fractions of the mesosiderites, has been investigated by several authors (Harlow et al., 1980, 1981; Agosto, 1981; Agosto et al., 1980). Agosto et al. (1980) suggested that augite is the source of the calcium needed to form merrillite from phosphorus dissolved in metal. Harlow et al. (1980, 1981) investigated several potential merrillite forming reactions and found that those involving the diopside component of pyroxene are thermodynamically the most important. An important difference between the Harlow et al., models and that of Agosto (1981) is that merrillite production is not defined in P-T-f_{O_2} space by a single univariant reaction line (Agosto, 1981) but lies in a divariant area in T-f_{O_2} space that is bounded by reactions involving different activities for the phosphorus content of the metal and the ferrosilite activity in the clinopyroxene. Within the many uncertainties of the available thermodynamic data, Harlow *et al.* (1981) defined a region in which merrillite production is very likely to occur between 950° and 1050°C and f_{O_2} between 10^{-15} and 10^{-17}. This temperature range is very similar to that inferred for mesosiderite metamorphism from two-pyroxene thermometry and from pyroxene-olivine-spinel relations (Nehru *et al.*, 1980c). The f_{O_2} range is similar to that generated by Snellenburg *et al.* (1979) for several mesosiderites using data for oxide mineral assemblages. Considerable overlap, therefore, exists between metamorphic temperatures and f_{O_2} estimates from the mesosiderites from several independent methods.

The material balance calculation done to determine the components present in the mesosiderite matrix provides further evidence supporting redox formation of phosphate. Table 3 contains the modal mineralogy of the average observed matrix (silicate fraction) of the mesosiderites and a matrix mode calculated by mixing the pyroxenite component with the mafic component in the ratio 34:66. Because feldspar is present only in the mafic component and is not actively involved in many of the allochemical metamorphic reactions, this mineral was used to determine the relative proportions of orthopyroxenite and mafic components in the matrix. The differences between the observed and calculated matrix are given in the final column. With the exception of chromite, which is in such low modal abundance that the 0.2% difference may be an artifact, the differences may be explained by redox reaction between the silicates and the metal fraction.

A reaction such as that of Agosto *et al.* (1980):

$$3CaMgSi_2O_6 + 5FeSiO_3 + 2P_{Fe} \rightleftarrows Ca_3(PO_4)_2 + 3MgSiO_3 + 8SiO_2 + 5Fe \quad (3)$$
$$\text{diops.} \qquad \text{ferrosil.} \quad \text{met.} \qquad \text{merril} \qquad \text{enst.} \qquad \text{trid.} \quad \text{met.}$$

provides a useful example of the use of redox reactions to balance the deficit and excesses between observed and calculated matrix (Table 3). The reactants of reaction 3 are removed from the matrix and show up as deficits in the mode while the products appear as excesses. Thus, the calcic pyroxene deficit in Table 3 may be explained as the loss of reactants necessary to form merrillite, tridymite and orthopyroxene which are all in excess over the calculated matrix. In detail, however, reaction 3 does not explain fully the observed modal abundances of tridymite and merrillite. The observed excess tridymite is half that predicted by reaction 3. To balance these components it is necessary to use the variable activity models of Harlow *et al.* (1980, 1981) which rely on a suite of reactions, that produce merrillite and tridymite, to account for the different proportions of these minerals.

Further evidence of the redox formation of merrillite may be revealed by electron microprobe measurements of rare earth elements (REE) in the merrillite. If merrillite formed during igneous processing the very high $D_{REE}^{merril/liquid}$ coupled with low $D_{REE}^{silicate/liquid}$ should result in levels of REE in the merrillite that are easily detectable by microprobe. Using REE data from Mittlefehldt (1979) and Mittlefehldt *et al.* (1979), estimated Ce contents for mesosiderite merrillites are of of the order 0.5–2.0%. With a detection limit (2σ) of approximately 300–500 ppm, no Ce, or any other rare earth element, was detected in any of several mesosiderite phosphates, in both mafic clasts and associated with metal in the matrix. In contrast, under the same analytical conditions fluorapatite in the polymict eucrite Allan Hills A78040 yielded up to 5% total REE. Similarities between rare earth abundances in eucrite, polymict eucrites and mesosiderite mafic clasts (Mittlefehldt, 1979; Smith and Schmitt, 1981; Wooden *et al.*, 1981) indicate that phosphates in each group should have similar high levels to ALHA 78040 apatite. Although Dowty (1976) indicates that, because of structural variations, interpretation of element distribution data for merrillite from the moon and meteorites may be difficult, at present the REE data for mesosiderite merrillite are interpreted to mean that these phosphates were never in equilibrium with igneous melts of basaltic composition. Instead they formed during metamorphism as a result of redox exchange reactions, that did not mobilize significant amounts of rare earths.

CAUSES OF METAMORPHISM

The silicates in the mesosiderites began their evolution showing many similarities to the basaltic achondrites and, in particular, to the howardites, but they were subjected to a major heating event after the formation of a silicate regolith that partially homogenized the silicate fraction. All the textural and chemical features described were produced by the process of partial chemical homogenization of the regolith and this process has produced a very different silicate fraction from any of the basaltic achondrites.

Had the metamorphic reactions gone to completion, then the mesosiderite silicate fraction would have been a granulitic orthopyroxene, clinopyroxene, plagioclase, tridymite, merrillite rock with subsidiary oxide phases coexisting with metal. Because the mesosiderites are SiO_2-normative (Powell, 1971), all the olivine would have reacted with tridymite to form orthopyroxene and all the pyroxene present would have exchanged Fe and Mg to produce a uniform orthopyroxene ($En_{58-65}Wo_{1-4}$) coexisting with diopsidic clinopyroxene. None of the metamorphic reactions did go to completion. The presence of schreibersite in most mesosiderites indicates that the redox reactions did not remove all the phosphorus in the metal. Instead sufficient P_{Fe} remained to permit exsolution of phosphide (Kulpecz and Hewins, 1978). The presence of so many incomplete reactions implies that metamorphism was sufficiently intense to trigger these reactions, but was of insufficient duration to allow the development of a homogeneous rock containing no textural evidence of its igneous forbears. The short duration of the high temperature metamorphism is confirmed by the very rapid cooling rates (1–100°C/day), between 1150° and 800°C, required to preserve the overgrowth textures and that are inferred to explain the differences in spatial concentration of phosphate in the olivine coronas.

This short duration is incompatible with heating by deeper igneous activity on the parent body which, in turn, probably eliminates metamorphic heating by short lived radioactive isotopes such as ^{26}Al. Mittlefehldt (1979) invoked an external heat source to explain supposed variations of reheating. A potential heat source is, however, still preserved within the mesosiderites. Approximately half of each mesosiderite is metal with troilite and subsidiary phosphide. In many sections the metal/silicate grain boundries show the arcuate profiles characteristic of the contacts between immiscible liquids. This, together with the pervasive invasion of metal into cracks and veins in the silicate clasts indicates that the metal (+troilite) fraction of the mesosiderites may have invaded the regolith as liquid (Powell, 1969). Floran (1978) argued that the lack of the large scale melting of the silicate fraction that would accompany invasion of silicate by liquid Fe indicates that the metal intruded as a ductile solid. The presence of significant sulfur and phosphorus in the metallic fraction, however, implies that the solidus of the metal (1100–1300°C) may have been close to the silicate solidus (1000–1200°C) (Harlow et al., 1981). The metal fraction, if introduced as a liquid, as suggested by Powell (1969), would therefore not have caused extensive melting of the silicates because of the small temperature difference between the solidi of silicate and metal fractions, and the higher enthalpies of fusion of silicates (60–80 kJ) relative to Fe metal (13.8 kJ) (Robie et al., 1978).

The mixing of a silicate regolith with a mass of phosphorus and sulfur bearing metal would provide a large pulse of heat to the silicate fraction. The insulating properties of the silicate breccia would prevent rapid dissipation of this heat, but the mixing of hot metal and cooler silicate would cause a rapid initial increase in the temperature of the silicate while quenching the metal. Although the details of this heating mechanism are not yet clear, it provides a very plausible way of reheating the mesosiderite silicate fraction to produce the unique metamorphic features of the mesosiderites at the same time as providing their obvious stony-iron characteristics.

The cause of metal-silicate mixing is also important to the history of the mesosiderites.

The metal fraction of the mesosiderites was studied by Wasson *et al.* (1974) and suggested to be unique, although it resembles metal from IIIAB, IIIE, pallasites and H-group chondrites. Harlow *et al.* (1981) have recalculated the pre-metamorphic Ni-P composition of the metal and show very close correlation with the fractionation trend of IIIAB metal. The homogeneity of the metal fraction (Wasson *et al.*, 1974) may be caused by low temperature diffusion and it, therefore, does not help identify the original composition and diversity of the metal. There is little evidence of shock in the mesosiderite regolith, but the extensive and ubiquitous metamorphic event may have obliterated much of the evidence of shock. At present, unambigous evidence of an impact origin for the mesosiderite metal is not available. Two alternative hypotheses must be considered to explain metal-silicate mixing. These are: (a) mixing a silicate regolith from the crust of a parent body with metal from the core of the same body (Chapman and Greenberg, 1981), and (b) impact mixing of a metal projectile with the regolith. Evidence supporting either of these hypotheses in preference to the other is presently lacking.

The uniform composition of the metal fraction is consistent with mixing the core and regolith of the same body but does not provide any clear evidence in support of this. Chapman and Greenberg (1981) have proposed a model for the development of mesosiderites by foundering the mesosiderite crust (as kilometer size blocks) through a liquid (olivine rich) mantle to mix with the outer layers of the core which were then quenched by the still cool regolith blocks. The Chapman and Greenberg model clearly provides for many of the features necessary in the evolution of the mesosiderites and provides an interesting dynamic framework for further studies of the metamorphism.

There are, however, several problems which must be solved before this model can be effectively applied. These are: (1) The average density of a typical mesosiderite silicate fraction is probably between 3.25 and $3.37 \text{ g} \cdot \text{cm}^{-3}$ (50–70% "diogenite" [S.G. = 3.45] + 30–50% "eucrite" [S.G. = 3.17] – McCall, 1966) whereas the density of liquid olivine mantle (arbitrarily assumed to be Fo_{90} at 1790°C) is probably close to $3.1 \text{ g} \cdot \text{cm}^{-3}$. If the parent body has a diameter of the order 10–100 km, then g should be small and the small density contrast between crust and mantle would lead to very sluggish settling unless convective downpulling also operated (cf. Walker *et al.*, 1978). (2) The molten olivine rich mantle through which the crustal blocks must settle would be extremely hot and would therefore have to overlie a superheated Fe–Ni–P–S metal core. It should cause extensive melting in the settling crustal blocks unless the sinking occurred very rapidly. (3) Assuming that the crustal blocks did survive foundering through a very hot mantle relatively unscathed and reached the core-mantle interface where bouyancy prevented further sinking, the very rapid cooling rates inferred for overgrowth and corona formation require efficient removal of excess heat. It is not clear that quenching of the metal by the silicate fraction in this environment would permit such rapid cooling. Quenching of the very fluid superheated metal in the core, by intimate mixing with cooler regolith material, may produce the initial rapid decrease in temperature required to initiate the observed disequilibrium features of the mesosiderites. The thermal mass of the hot, presumably crystallizing olivine-rich mantle above the interface zone would prevent much further cooling. To preserve the observed textures the mesosiderite metal-silicate mixture must be cooled to a temperature at which the described reactions do not proceed (probably below 800°C). To preserve the observed metamorphic textures and prevent completion of the metamorphic reactions discussed above, the parent body would have to be excavated to the core-mantle interface very rapidly after crustal foundering. In spite of the several difficulties, however, the Chapman and Greenberg model may provide a useful basis for further investigation of the mesosiderites.

The most comprehensive discussion of impact heating in mesosiderites is given by Floran (1978) and Floran *et al.* (1978) but their models assume a preexisting metal-silicate regolith and do not adequately explain the observed metal-silicate textures or how the metal fraction was initially mixed with the silicate.

In summary, the metamorphism of the mesosiderites was probably produced by mixing hot metal with a silicate regolith. The mechanism of mixing the metal and silicates is not clear but a model involving mixing of crust and core on the same body may prove useful. Elimination of the intervening mantle, however, remains difficult and cannot yet be accounted for adequately.

IMPLICATIONS OF METAMORPHISM

Metamorphism of the mesosiderite silicate fraction by interaction with the metal fraction has resulted in major mobilization of elements in both their pyroxenitic and mafic components. The partial homogenization which produced coronas, overgrowths and rims has drastically altered the chemistry of many clasts, particularly around their edges. Failure to recognize these late metamorphic effects could lead to serious misinterpretation of chemical data (e.g., Mn and Sc) obtained from many silicate clasts. It is essential that future studies of mesosiderite silicate chemistry recognize these effects and characterize them before discussing the igneous evolution of the mesosiderites.

The pervasive metamorphism described has probably influenced isotope systematics in the mesosiderites. Studies of radiogenic isotopes from *well characterized* silicate clasts in mesosiderites may provide important data about the time scale of parent body evolution. It is critical that clasts whose centers were sufficiently remote from the mesosiderite matrix and its metamorphism be identified to permit determination of igneous ages as well as metamorphic ages. If unmetamorphosed clasts can be found and compared with their metamorphosed equivalents, then useful constraints on the effects of metamorphism on differentiated parent bodies may be identified.

The interaction of the metal and silicate fractions resulted in modification of the metal chemistry. The redox exchanges discussed above may have altered the distribution of moderately siderophile elements and may be the cause of the unique metal compositions in mesosiderites. Initial results (Harlow *et al.*, 1981) suggest that mesosiderite metal has close affinities to IIIAB irons (from Ni-P relations) but more detailed chemical studies (of Ga, Ge, Ir, Au) on different metal types from mesosiderites are needed.

CONCLUSIONS

1. Metamorphism has radically changed the texture of the silicate fraction of the mesosiderites.
2. Metamorphism has changed the chemistry of both the metal fraction and the preserved igneous clasts in the silicate fraction.
3. The originally fine grained matrix of the mesosiderites was the main location for metamorphic reactions because of its high reactivity and the easy availability of all the participating reactants.
4. The overall effect of metamorphism was to chemically homogenize the polymict silicate regolith that suffered the metamorphic event. Homogenization was invariably not completed and different mesosiderites show different amounts of homogenization. All the observed metamorphic textures are reflections of this trend toward chemical homogenization. Had the metamorphism been of sufficient duration, the mesosiderites would probably be 'monomict' granulitic rocks revealing no evidence of their earlier complex igneous and brecciation histories.
5. Metamorphism of the mesosiderites occurred in a very short time (days to months) but was pervasive because of the high temperatures involved (900–1150°C). The sharp chemical contrast between magnesian pyroxene clasts and their overgrowths indicates that this texture was held at high temperatures for a very short time and cooled very rapidly (1–100°C/day). Further evidence for the short duration of the event comes from the location of merrillite rich bands in different types of coronas around olivine crystals. The different locations are interpreted to be the result of kinetic controls on the availability of phosphorus.
6. Kinetic effects have resulted in different silicate clasts suffering different amounts of metamorphism. Very small mafic clasts are often modified throughout. Larger mafic

clasts tend to have unmodified centers with metamorphic overprints clearly developed near their edges. Similarly, large clasts are more common than tiny ones as a result of both the brecciation process and resorption of smaller clasts during metamorphic homogenization.
7. The high merrillite content of the mesosiderites is the result of a suite of redox reactions involving phosphorus bearing metal and the silicate fraction, particularly the calcic pyroxene component. That these redox changes were synchronous with the metamorphism is revealed by the location of the merrillite in different parts of coronas around olivine as well as its presence in metamorphic rims around pyroxene grains and absence in their less modified igneous cores. The modification of metal chemistry resulting from these redox changes must be taken into account when comparing mesosiderite metal to other meteoritic sources of metal.
8. Metamorphism was caused by transfer of heat from partially or completely molten metal (containing dissolved phosphorus and sulfur) to a cool achondritic regolith (similar to, but distinct from, the howardite regolith) during a mixing event. The cause of metal-silicate mixing is still unclear.
9. Future isotopic studies of carefully selected silicate clasts from mesosiderites may provide useful information about the timing of the metamorphism and the earlier igneous history. Chemical studies of clasts must identify, and allow for, the presence of significant metamorphic overprints before interpreting the igneous history of the mesosiderites.

Acknowledgments—Research was supported by NASA grant #NSG 7258. Microprobe analyses by R. Klimentidis, S. Frishman and R. Bedell are acknowledged and detailed discussion with R. H. Hewins is appreciated. R. J. Floran and an anonymous reviewer provided useful comments. Typing and drafting by Gertrude Poldervaart and Carol O'Neill are gratefully acknowledged.

REFERENCES

Agosto W. N. (1981) fO_2 estimates in 14 mesosiderites from silicate-iron-oxygen equilibria of proposed parent melts (abstract). In *Lunar and Planetary Science XII*, p. 3–5. Lunar and Planetary Institute, Houston.
Agosto W. N., Hewins, R. H., and Clarke R. S. Jr. (1980) Allan Hills A77219, the first Antarctic mesosiderite. *Proc. Lunar Planet. Sci. Conf. 11th*, p. 1027–1045.
Albertsen J. F., Aydin M., and Knudsen J. M. (1978) Mössbauer effect studies of taenite lamellae of an iron meteorite Cape York (III.A). *Phys. Scr.* **17**, 467–472.
Chapman C. R. and Greenberg R. (1981). Meteorites from the asteroid belt: the stony-iron connection (abstract). In *Lunar and Planetary Science XII*, p. 129–131. Lunar and Planetary Institute, Houston.
Clarke R. S. Jr. and Scott E. R. D. (1980) Tetrataenite—ordered FeNi, a new mineral in meteorites. *Amer. Mineral.* **65**, 624–630.
Crozaz G. and Tasker D. R. (1980) Thermal history of mesosiderites. *Meteoritics* **15**, 278.
Delaney J. S., Bedell R. L., Harlow G. E., Nehru C. E., Klimentidis R., and Prinz M. (1980a) Pyroxene overgrowth textures: Evidence for rapid cooling from high temperatures in mesosiderites (abstract). In *Lunar and Planetary Science XI*, p. 204–206. Lunar and Planetary Institute, Houston.
Delaney J. S., Harlow G. E., Prinz M. and Nehru C. E. (1981a) Metamorphism in mesosiderites: Radical chemical changes in a post-igneous silicate regolith. In *Lunar and Planetary Science XII*, p. 208–210. Lunar and Planetary Institute, Houston.
Delaney J. S., Prinz M., Nehru C. E., and Harlow G. E. (1981b) A new basalt group from howardites: Mineral chemistry and relationships with basaltic achondrites. In *Lunar and Planetary Science XII*, p. 211–213. Lunar and Planetary Institute, Houston.
Delaney J. S., Nehru C. E., and Prinz M. (1980b) Olivine clasts from mesosiderites and howardites: Clues to the nature of achondritic parent bodies. *Proc. Lunar Planet. Sci. Conf. 11th*, p. 1073–1087.
Dowty E. (1976) Phosphate in Angra dos Reis: Structure and composition of $Ca_3(PO_4)_2$ minerals. *Earth Planet. Sci. Lett.* **35**, 347–351.
Dowty E. (1980) Crystal-chemical factors affecting the mobility of ions in minerals. *Amer. Mineral.* **65**, 174–182.
Esbensen K. H. (1978) Coronites from the Fongen gabbro complex, Trondheim region, Norway: Role of water in the olivine-plagioclase reaction. *N. Jb. Mineral. Abh.* **132**, 113–135.

Floran R. J. (1978) Silicate petrography, classification, and origin of the mesosiderites: Review and new observations. *Proc. Lunar Planet. Sci. Conf. 9th*, p. 1053–1081.

Floran R. J., Caulfield J. B. D., Harlow G. E., and Prinz M. (1978) Impact-melt origin for the Simondium, Pinnaroo and Hainholz mesosiderites: Implications for impact processes beyond the earth-moon system. *Proc. Lunar Planet. Sci. Conf. 9th*, p. 1083–1114.

Fuchs L. H. (1969) The phosphate mineralogy of meteorites. In *Meteorite Research* (P. M. Millman, ed.), p. 683–695. D. Reidel, Dordrecht, Holland.

Goles G. G. (1971) Scandium. In *Handbook of Elemental Abundances in Meteorites* (B. Mason, ed.), p. 175–179. Gordon and Breach Science, N.Y.

Göpel C. and Wänke H. (1978) Trace elements in single pyroxene crystals of diogenites, howardites and eucrites. *Meteoritics* 13, 477–480.

Harlow G. E. and Delaney J. S. (1981) Inclusions in minerals in howardite clasts: Indicators of processed and unprocessed regolith breccias (abstract). In *Lunar and Planetary Science XII*, p. 392–394. Lunar and Planetary Institute, Houston.

Harlow G. E., Delaney J. S., Nehru C. E., and Prinz M. (1980) The origin of abundant tridymite and phosphate in mesosiderites: Feasibility of possible reactions. *Meteoritics* 15, 297–298.

Harlow G. E., Delaney J. S., Nehru C. E. and Prinz M. (1981) Metamorphic reactions in mesosiderites: origin of abundant phosphate and silica. *Geochim. Cosmochim. Acta.* In press.

Harlow G. E. and Klimentidis R. (1980) Clouding in pyroxenes and plagioclases in eucrites: Implications for post-crystallization processing. *Proc. Lunar Planet. Sci. Conf. 11th*, p. 1131–1143.

Hewins R. H. (1979) The pyroxene chemistry of four mesosiderites. *Proc. Lunar Planet. Sci. Conf. 10th*, p. 1109–1125.

Hewins R. H. (1981a) Orthopyroxene-olivine assemblages in diogenites and mesosiderites. *Geochim. Cosmochim. Acta* 45, 123–126.

Hewins R. H. (1981b) Fractionation and equilibration in diogenites (abstract). In *Lunar and Planetary Science XII*, p. 445–447. Lunar and Planetary Institute, Houston.

Kulpecz A. A. and Hewins R. H. (1978) Cooling rate based on schreibersite growth for the Emery mesosiderite. *Geochim. Cosmochim. Acta* 42, 1495–1500.

Labotka T. C. and Papike J. J. (1980) Howardites: Samples of the regolith of the eucrite parent-body: Petrology of Frankfort, Pavlovka, Yurtuk, Malvern, and ALHA 77302. *Proc. Lunar Planet. Sci. Conf. 11th*, p. 1103–1130.

Longhi J., Walker D., and Hays J. F. (1978) The distribution of Fe and Mg between olivine and lunar basaltic liquids. *Geochim. Cosmochim. Acta* 42, 1545–1558.

Mason B., Jarosewich E., and Nelen J. A. (1979) The pyroxene-plagioclase achondrites. *Smithson. Contrib. Earth Sci.* 22, 27–45.

McCall G. J. H. (1966) The petrology of the Mount Padbury mesosiderite and its achondrite enclaves. *Mineral. Mag.* 35, 1029–1060.

Medaris L. G. Jr. (1969) Partitioning of Fe^{2+} and Mg^{2+} between coexisting synthetic olivine and orthopyroxene. *Amer. J. Sci.* 267, 945–968.

Mittlefehldt D. W. (1979) Petrographic and chemical characterization of igneous lithic clasts from mesosiderites and howardites and comparison with eucrites and diogenites. *Geochim. Cosmochim. Acta* 43, 1917–1935.

Mittlefehldt D. W., Chou C.-L., and Wasson J. T. (1979) Mesosiderites and howardites: Igneous formation and possible genetic relationships. *Geochim. Cosmochim. Acta* 43, 673–688.

Mysen B. O. and Virgo D. (1980) Trace element partitioning and melt structure: An experimental study at 1 atm pressure. *Geochim. Cosmochim. Acta*, 44, 1917–1930.

Nehru C. E., Delaney J. S., Harlow G. E., Frishman S., and Prinz M. (1981) Orthopyroxenite clasts in mesosiderites and howardites: Relationships with diogenites and orthopyroxene cumulate eucrites (abstract). In *Lunar and Planetary Science XII*, p. 765–767. Lunar and Planetary Institute, Houston.

Nehru C. E., Delaney J. S., Harlow G. E., and Prinz M. (1980a) Mesosiderite basalts and the eucrites. *Meteoritics* 15, 337–338.

Nehru C. E., Harlow G. E., Prinz M., and Hewins R. H. (1978a). The tridymite-phosphate-rich component in mesosiderites. *Meteoritics* 13, 573.

Nehru C. E., Hewins R. H., Garcia D. J., Harlow G. E., and Prinz M. (1978b) Mineralogy and petrology of the Emery mesosiderite (abstract). In *Lunar and Planetary Science IX*, p. 799–801. Lunar and Planetary Institute, Houston.

Nehru C. E., Prinz M., Delaney J. S., Harlow G. E., and Frishman S. (1980b) Gabbroic and basaltic clasts in mesosiderites: Unique achondritic tridymite-phosphate-rich, two-pyroxene rock types (abstract). In *Lunar and Planetary Science XI*, p. 803–805. Lunar and Planetary Institute, Houston.

Nehru C. E., Zucker S. M., Harlow G. E., and Prinz M. (1980c) Olivines and olivine coronas in mesosiderites. *Geochim. Cosmochim. Acta* 44, 1103–1118.

Onorato P. I. K., Uhlmann D. R., Taylor L. A., Coish R. A., and Gamble R. P. (1978) Olivine cooling speedometers. *Proc. Lunar Planet. Sci. Conf. 9th*, p. 613–628.

Powell B. N. (1969) Petrology and chemistry of mesosiderites—I. Textures and composition of nickel iron. *Geochim. Cosmochim. Acta* 33, 789–810.

Powell B. N. (1971) Petrology and chemistry of mesosiderites—II: Silicate textures and compositions and metal silicate relationships. *Geochim. Cosmochim. Acta* 35, 5–34.

Prinz M., Nehru C. E., Delaney J. S., Harlow G. E., and Bedell R. L. (1980) Modal studies of mesosiderites and related achondrites, including the new mesosiderite ALHA 77219. *Proc. Lunar Planet. Sci. Conf. 11th*, p. 1055–1071.

Robie R. A., Hemingway B. S., and Fisher J. R. (1978) Thermodynamic properties of minerals and related substances at 298.15 K and 1 bar (10^5 Pa) pressure and at higher temperatures. *U.S. Geol. Surv. Bull. 1452*, 456 pp.

Ross M. and Huebner J. S. (1979) Temperature-composition relationships between naturally occurring augite, pigeonite, and orthopyroxene at one bar pressure. *Amer. Mineral* **64**, 1133–1155.

Score R., Schwarz C. M., King T. V. V., Mason B., Bogard D. D., and Gabel E. M. (1981) Antarctic Meteorite Descriptions 1976–1977–1978–1979. Publ. 54, Lunar Curatorial Facility, NASA Johnson Space Center, Houston. 144 pp.

Smith M. R. and Schmitt R. A. (1981) Preliminary chemical data for some Allan Hills polymict eucrites (abstract). In *Lunar and Planetary Science XII*, p. 1014–1016. Lunar and Planetary Institute, Houston.

Snellenburg J. W., Nehru C. E., Caulfield J. B. D., Zucker S., and Prinz M. (1979) Petrology of temperature and oxygen fugacity indicating mineral assemblages in four low-grade mesosiderites (abstract). In *Lunar and Planetary Science X*, p. 1137–1139. Lunar and Planetary Institute, Houston.

Stolper E. (1977) Experimental petrology of eucritic meteorites. *Geochim. Cosmochim. Acta* **41**, 587–611.

Takeda H. (1979) A layered crust model of the howardite parent body. *Icarus* **40**, 455–470.

Taylor L. A., Onorato P. I. K., and Uhlmann D. R. (1977) Cooling rate estimates based on kinetic modelling of Fe-Mg diffusion in olivine. *Proc. Lunar Sci. Conf. 8th*, p. 1581–1592.

Walker D., Stolper E. M., and Hays J. F. (1978). A numerical treatment of melt/solid segregations: Size of the eucrite parent body and stability of the terrestrial low-velocity zone. *J. Geophys. Res.* **83**, 6005–6013.

Wasson J. T., Schaudy R., Bild R. W., and Chou C.-L. (1974). Mesosiderites—I. Compositions of their metallic portions and possible relationship to other metal-rich meteorite groups. *Geochim. Cosmochim. Acta* **38**, 135–149.

Watson E. B. (1977) Partitioning of manganese between forsterite and silicate liquid. *Geochim. Cosmochim. Acta* **44**, 1363–1374.

Weigand P. W. (1975) Pyroxenes in the Patwar mesosiderite. *Meteoritics* **10**, 341–351.

Wells P. R. A. (1977) Pyroxene thermometry in simple and complex systems. *Contrib. Mineral. Petrol.* **62**, 129–139.

Wilkening L. L. and Anders E. (1975) Some studies of an unusual eucrite: Ibitira. *Geochim. Cosmochim. Acta* **39**, 1205–1210.

Wood J. A. (1979) Review of the metallographic cooling rate of meteorites and a new model for the planetesimals in which they formed. In *Asteroids* (T. Gehrels, ed.), p. 849–891. Univ. Arizona, Tucson.

Wooden J., Reid A., Brown R., Bansal B., Wiesmann H. and Nyquist L. (1981) Chemical and isotopic studies of the Allan Hills polymict eucrites. In *Lunar and Planetary Science XII*, p. 1203–1205. Lunar and Planetary Institute, Houston.

Roaldite, a new nitride in iron meteorites

Hans Peter Nielsen and Vagn F. Buchwald

Department of Metallurgy, Building 204, DTH, 2800 Lyngby, Denmark

Abstract—Three nitrides have been reported previously in meteorites; TiN, osbornite; Si_2N_2O, sinoite; and CrN, carlsbergite. The present note reports a new nitride, roaldite $(Fe, Ni)_4N$, identified in iron meteorites. It forms spiky platelets in the kamacite phase and is typically 1–10 μm thick, but may be several millimeters long. Roaldite is cubic with iron atoms in f.c.c. positions and a nitrogen atom in the center of each f.c.c. cell. It has been identified in coarse octahedrites of groups IA and IIB, the type material being in Jerslev and Youndegin.

INTRODUCTION

It has been known for some time that nitrogen is present in iron meteorites; however, only in low concentrations. Buchwald (1961) reported 44 ppm in Thule; Vinogradov *et al.* (1963), 46 ppm in Sikhote-Alin. Then, in an extensive study, Gibson and Moore (1971) reported a range of 1–216 ppm N in 123 iron meteorites and also showed that the troilite is significantly enriched, relative to the metal phase. Kothari and Goel (1974) determined total nitrogen in a large number of stone meteorites and in nine iron meteorites, supporting the range and actual determinations already reported.

In all of these studies it was assumed that the nitrogen occurred in solid solution in the metal phase and in the various inclusions. Like carbon, the nitrogen atom occurs interstitially in the body-centered cubic kamacite and the face-centered cubic taenite phase, and from steels it is known that nitrogen is particularly easily soluble in the f.c.c. phase. Thus, unalloyed taenite (austenite) may contain up to 2.8 wt% N at 650°C. However, since it is also known that the low temperature b.c.c. phase will dissolve only on the order of 0.1% N at 590°C (Hansen and Anderko, 1958), it is probable that not all of the meteoritic nitrogen is in solid solution, but may be present as discrete precipitates of nitride in the kamacite. It thus became a question of where exactly the small quantity of nitrogen was located.

Part of the answer was given when Buchwald and Scott (1971) found carlsbergite CrN in many iron meteorites. The chromium nitride occurs as platelets and grain boundary precipitates and clearly represents a late exsolution of nitrogen and chromium from kamacite, which upon cooling has become super-saturated, probably at temperatures as low as 500–400°C.

Two other nitrides, TiN-osbornite, and Si_2N_2O-sinoite, have been found in meteorites, but so far have not been identified in iron meteorites (Buchwald, 1977).

ROALDITE

The new mineral is a minor mineral occurring as extended but thin platelets in the body-centered cubic iron phase, kamacite. Typically it forms planar foils a few microns thick, but many millimeters long. The observed thickness range is 0.5–15 μm.

It was seen previously by Buchwald (1975, p. 1354) and was called the X-mineral, because it could not be identified at the time. With the new Philips 301 electron microscope added to the Department in 1975–76, it became possible to work with thin film transmission electron microscopy and electron diffraction so that unambiguous results could be obtained.

The nitride plates are straight (or they may display a zig-zag growth-mode), but individual segments are planar (Figs. 1–2). The mineral is irregularly dispersed, forming "bursts" here and there in the kamacite. Because of the small width of the platelets, their composition has been established by electron microprobe analysis (Table 1). Care was taken to select the most bulky platelets for analysis, but even then the data obtained may to some extent reflect the composition of the kamacite matrix. This is very close to the bulk composition of the two meteorites which are also given.

Pure metals and a suite of synthetic iron-nickel standards were used as standards, and corrections applied for fluorescence, absorption and atomic number effects. Nitrogen could only be measured semi-quantitatively, using a BN standard, because of the low number of counts and the uncertainty in the correction procedure.

The chemical analysis corresponds to the composition $(Fe, Ni)_4N$ with additional very small amounts of cobalt. The composition of the mineral reflects the composition of the host matrix; i.e., the more nickel in the matrix, the more nickel in the mineral. Evidently the mineral forms by diffusion of nitrogen, while the substitutional atoms are rearranged only to a small degree. Electron diffraction patterns were obtained from platelets in both Jerslev and Youndegin. They have been indexed on a cubic unit cell $a = 3.79 \pm 0.04$ Å and space group O_h^1–Pm3m, L'1 (Fig. 3).

An iron nitride (without nickel) with the same morphology and structure has been known for some time as a rare occurrence in synthetic steels. This nitride usually occurs on the electron microscopic scale and is referred to as γ' (no name) (Booker *et al.*, 1957; Pitsch, 1961).

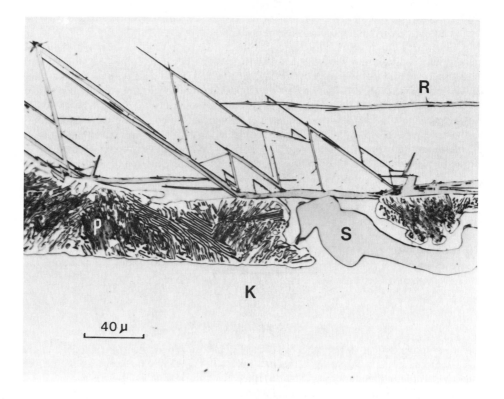

Fig. 1. The iron meteorite Youndegin, a coarse octahedrite of group IA. R = roaldite, S = schreibersite, K = kamacite, P = plessite. From Buchwald 1975, courtesy of University of California Press.

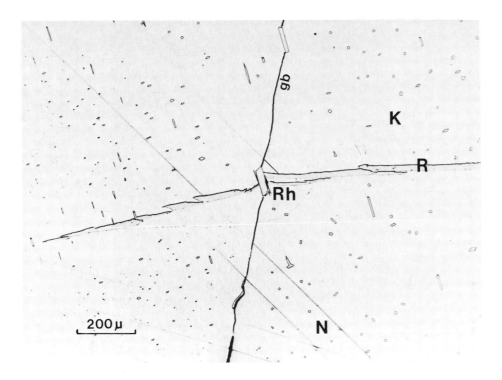

Fig. 2. Roaldite (R) and rhabdite (Rh) in two kamacite grains (K) separated by a grain boundary (gb). Neumann bands (N). Youndegin. Oil immersion photograph.

Roaldite is white in reflected light, very similar to the color of the kamacite phase in which it is imbedded, and only slightly more "bluish" then the yellowish schreibersite (Fe, Ni)$_3$P, with which it may easily be confused. However, roaldite is significantly more ductile than schreibersite, so that it will bend and twist with the metallic matrix when deformed, while schreibersite will break.

Roaldite is opaque and shows low relief relative to the kamacite in a polished section. Its Vickers hardness is estimated to be between 600 and 900. So far it has been identified in Jerslev, Sikhote-Alin, Youndegin and Toluca. The first two belong to group IIB, the others to group IA (Wasson, 1974; Kracher *et al.*, 1980). All are characterized by low volume percentages of taenite; this means that most of the nitrogen present will have been accommodated interstitially in the kamacite from which it then precipitated at low temperature.

Transmission electron microscopy reveals a characteristic internal striation, parallel to

Table 1. Analyses of roaldite (electron microprobe data) and of the meteorites Jerslev and Youndegin (bulk data).

Meteorite	Fe	Weight % Ni	Co	N	Roaldite Calculated Composition
Jerslev, roaldite	89.8	5.58	n.d.	6.3	(Fe$_{0.94}$Ni$_{0.055}$Co$_{0.005}$)$_4$N
same, bulk	93$^+$	5.84	0.49	n.d.	—
Youndegin, roaldite	88.6	6.35	0.53	7.6	(Fe$_{0.931}$Ni$_{0.064}$Co$_{0.005}$)$_4$N
same, bulk	92$^+$	6.74	0.48	26 ppm	—

$^+$Calculated by difference

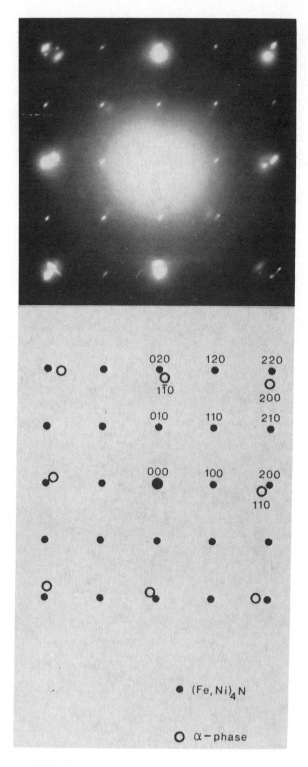

Fig. 3. Electron diffraction pattern of roaldite from the iron meteorite Youndegin. Reflections from the kamacite phase α and the roaldite $(Fe, Ni)_4N$ are shown.

Fig. 4. Transmission electron micrograph of thin film from Youndegin, showing a roaldite platelet with internal striation parallel to $(111)\gamma'$.

$(111)\gamma$. This striation has also been observed on the synthetic iron nitride (Booker, 1961), and is of diagnostic value (Fig. 4).

Roaldite is a late solid-state precipitate in the kamacite phase of iron meteorites where it occurs together with schreibersite [(Fe, Ni)$_3$P], carlsbergite (CrN), and daubreelite (FeCr$_2$S$_4$).

The mineral is named after Roald Norbach Nielsen (b. 1928), who since 1968 has been on the staff of the Department of Metallurgy, Lyngby. Mr. Nielsen is a very capable electron microprobe expert whose services have been of great help in the research and educational activities of the said department. The mineral identification and its name have been approved by the Commission on New Minerals and Mineral Names. The type material of Jerslev, a 40 kg Danish iron meteorite found in 1977, is in the Geologisk Museum of the University, Copenhagen. The iron meteorite Youndegin, a shower-producing meteorite of several tons, is widely distributed in the world's mineral and meteorite collections.

Acknowledgments—The support of a grant from Statens Teknisk-Videnskabelige Forskningsråd (516–20128) is gratefully acknowledged.

REFERENCES

Booker G. R. (1961) The growth of γ'-Fe$_4$N iron nitride precipitates in the α-iron matrix. *Acta Metall.* **9**, 590–594.

Booker G. R., Norbury J., and Sutton A. L. (1957) An investigation of nitride precipitates in pure iron and mild steels. *J. Iron Steel Inst.* **187**, 205–215.

Buchwald V. F. (1961) The iron meteorite Thule, North Greenland. *Geochim. Cosmochim. Acta* **25**, 95–98.

Buchwald V. F. (1975) *Handbook of Iron Meteorites*, Univ. Calif. Press, Los Angeles. 1400 pp.

Buchwald V. F. (1977) The mineralogy of iron meteorites. *Phil. Trans. Roy. Soc. Lond.* **A286**, 453–491.
Buchwald V. F. and Scott E. R. D. (1971) First nitride (CrN) in iron meteorites. *Nat. Phys. Sci.* **233**, 113–114.
Gibson E. K. and Moore C. B. (1971) The distribution of total nitrogen in iron meteorites. *Geochim. Cosmochim. Acta* **35**, 877–890.
Hansen M. and Anderko K. (1958) *Constitution of Binary Alloys*. McGraw-Hill, N.Y. 1305 pp.
Kothari B. K. and Goel P. S. (1974) Total nitrogen in meteorites. *Geochim. Cosmochim. Acta* **38**, 1493–1507.
Kracher A., Willis J., and Wasson J. T. (1980) Chemical classification of iron meteorites. IX. *Geochim. Cosmochim. Acta* **44**, 773–780.
Pitsch W. (1961) Die kristallographischen Eigenschaften von Eisen-nitrid-Ausscheidungen in Ferrit. *Archiv für das Eisenhüttenwesen* **32**, 493–500.
Vinogradov A. P., Florenskii K. P., and Volynets V. G. (1963) Ammonia in meteorites and igneous rocks. *Geochem.* **1963**, 905–916.
Wasson J. T. (1974) *Meteorites: Classification and properties*. Springer–Verlag, N.Y. 316 pp.

Complementary rare earth element patterns in unique achondrites, such as ALHA 77005 and shergottites, and in the earth

M.-S. Ma[1], J. C. Laul[2], and R. A. Schmitt[1]

[1]Department of Chemistry and the Radiation Center, Oregon State University, Corvallis, Oregon 97331   [2]Physical Sciences Department, Battelle Pacific Northwest Laboratory, Richland, Washington 99352

Abstract—Abundances of major, minor and trace elements were determined in the Antarctic achondrite Allan Hills (ALHA) 77005 via sequential instrumental and radiochemical neutron activation analysis. The rare earth element (REE) abundances of ALHA 77005 exhibit an unique chondritic normalized pattern, i.e., the REEs are nearly unfractionated from La to Pr at ~1.0X chondrites, monotonically increased from Pr to Gd at ~3.4X with no Eu anomaly, nearly unfractionated from Gd and Ho and monotonically decreased from Ho to Lu at ~2.2X. This unique REE pattern of ALHA 77005 can be modeled by a melting process involving the continuous melting and the progressive partial removal of melt from a light REE enriched source material. In such a model, ALHA 77005 could represent either a crystallized cumulate from such a melt or the residual source material. Calculation has shown that the parent liquids for the shergottites could also be derived from a light REE enriched source material similar to that for ALHA 77005.

The REE pattern of ALHA 77005 is similar to patterns observed in some terrestrial peridotites and is complementary to the light REE enriched patterns inferred for the parent magmas of the achondrites Angra dos Reis, nakhlites and chassignites. These observations would imply, that the source materials in the parent body or bodies of these achondrites have a great similarity to the earth's upper mantle. The similarity of complementary REE patterns of unique achondrites or their inferred parent magmas and igneous rocks from the earth would suggest that the parent body or bodies of these achondrites are large in size and that common igneous processes for the derivation of igneous rocks have been operative under similar physical and chemical conditions on various planetary bodies.

1. INTRODUCTION

ALHA 77005 is an unique achondrite recovered in the Antarctic during the field season 1977–1978 (Yanai et al., 1978). Petrological and selected trace element studies of ALHA 77005 and the shergottites (McSween et al., 1979a,b; Stolper and McSween, 1979; McSween and Stolper, 1980) have indicated significant similarities between these achondrites in terms of their mineral composition, oxidation state, shock metamorphic effects, volatile to involatile element ratios, some trace element abundances and inferred source materials. Based upon these similarities, it has been suggested (McSween et al., 1979a,b; McSween and Stolper, 1980) that ALHA 77005 may represent either a cumulate rock that crystallized prior to the shergottites but from the same or similar parent magmas, or 77005 may represent a source material with peridotite composition from which the shergottite parent magma was derived by partial melting. On the other hand, Rb–Sr and Sm–Nd isotopic data (Nyquist et al., 1979; Shih et al., 1981) indicate that these achondrites cannot be comagmatic and that their isotopic characteristics are inherited from distinct sources which were established early in the parent body's history.

We have measured the abundances of major, minor, and incompatible trace elements, particularly the group of chemically coherent REE in ALHA 77005 in order to constrain and test the hypothesis of a genetic relationship between ALHA 77005 and the shergottites. Comparisons of the observed light REE depleted patterns of ALHA 77005 with similar REE patterns in some terrestrial peridotites and the implications of this pattern to the complementary light REE enriched patterns of parent magmas calculated for other achondrites [e.g., Angra dos Reis (Ma et al., 1977), nakhlites and Chassigny (Boynton et al., 1976)] and similar patterns observed in terrestrial alkali basalts are also discussed.

2. EXPERIMENTAL RESULTS

Twenty major, minor, and trace element abundances in ALHA 77005 were measured by instrumental neutron activation analysis (INAA) at Oregon State University and 12 REE, Ba, Sr, and U were independently measured on a different 77005 aliquant by radiochemical neutron activation analysis (RNAA) at the Battelle Northwest Laboratory. Sample preparation and the detailed NAA procedure are essentially the same as reported by Laul (1979). The USGS Standard rocks BCR-1, BHVO-1, GSP-1 and PCC-1 were also analyzed for control of precision and accuracy. The results are given in Tables 1 and 2. The chondritic normalized trace element abundances are plotted in Fig. 1.

The normative mineralogy indicates that ALHA 77005 contains ~52% olivine, 26% low-Ca pyroxene, 11% high-Ca pyroxene, 10% plagioclase and 1% chromite. Comparing to shergottites, ALHA 77005 has lower TiO_2, Al_2O_3, CaO, Na_2O, and K_2O and higher MgO and Cr_2O_3 abundances (Table 1). Both $[Fe/(Fe + Mg)]_{molar}$ and $(CaO/Al_2O_3)_{wt}$ ratios of ALHA 77005 (0.30 and 1.27, respectively) are lower than those of shergottites (0.46–0.54 and 1.36–1.90, respectively). The $(Na_2O + K_2O)/(CaO + Na_2O + K_2O)$ ratio, an indicator for alkali content in plagioclase, is similar for both ALHA 77005 and the shergottites.

The REE abundances of ALHA 77005, first reported by Ma et al. (1980), exhibit an unique chondritic normalized pattern, i.e., the REE are nearly unfractionated from La to Pr at ~1.0X chondrites, monotonically increased from Pr to Gd at ~3.4X with no Eu anomaly, nearly unfractionated from Gd to Ho and monotonically decreased from Ho to Lu of ~2.2X (Fig. 1). The REE abundances are lower in ALHA 77005 and the pattern is quite different than those of the shergottites reported by Stolper and McSween (1979) but is similar to those reported by Schnetzler and Philpotts (1969) and Shih et al. (1981) (Fig. 1). Even though ALHA 77005 is a coarse-grained rock (McSween et al., 1979b), the REE abundances measured by INAA, RNAA and isotope dilution (Shih et al., 1981) on different aliquants are very similar, indicating the analyzed samples are representative of the whole rock. It is also noted that ALHA 77005 has a similar chondritic normalized medium to heavy REE (Sm–Lu), Sc, Ti, V, and Hf abundances at $2.9 \pm 0.5X$ chondrites.

Table 1. Major and minor element abundances in ALHA 77005 and shergottites.

	ALHA 77005[a] (311 mg)	Shergottites[b]					
		1	2	3	4	5	6
TiO_2 (%)	0.3 ± 0.1	0.92	0.85	0.92	0.81	0.77	0.73
Al_2O_3	3.0 ± 0.1	6.69	7.03	7.01	6.89	5.67	5.7
FeO[c]	20.2 ± 0.1	19.99	19.34	19.3	19.1	18.0	17.7
MgO	26.5 ± 0.8	9.40	9.27	9.32	9.27	11.0	11.4
CaO	3.8 ± 0.3	10.03	9.58	9.94	10.1	10.8	10.5
Na_2O	0.475 ± 0.006	1.28	—	1.34	1.37	0.99	1.2
K_2O	0.028 ± 0.006	0.16	0.18	0.17	0.16	0.14	0.1
MnO	0.435 ± 0.005	0.50	0.54	0.52	0.50	0.50	0.50
Cr_2O_3	0.83 ± 0.01	0.18	0.31	0.23	0.21	0.30	0.38

[a] Analyzed by INAA at OSU.
[b] Designated columns 1 through 4 are data for Shergotty taken respectively from Duke (1968), McCarthy et al. (1974), Smith and Hervig (1979), and Stolper and McSween (1979), and columns 5 and 6 are data for Zagami taken respectively from Stolper and McSween (1979) and Easton and Elliot (1977).
[c] All Fe as FeO.

Table 2. Trace element abundances in ALHA 77005 and shergottites.

Element	ALHA 77005 INAA (311 mg) (a)	ALHA 77005 RNAA (42.7 mg) (a)	Shergottites (d)	Shergottites (b)	Shergottites (c)	Shergottites (d)	Ziagami (b)	Ziagami (d)	Average ordinary chondritic values (e)
Sc (ppm)	22 ± 1	—	—	53	—	—	57	—	7.9
V	158 ± 7	—	—	—	—	—	—	—	71
Co	70 ± 1	—	—	35	—	—	37	—	—
Ni	320 ± 10	—	—	56	—	—	67	—	—
Sr	—	—	14.1	—	—	51.0	—	45.9	11
Ba	—	16 ± 1	4.53	—	32.0	29.4	—	25.3	3.8
La	0.33 ± 0.02	6.0 ± 0.3	0.314	2.18	—	1.50	2.07	1.60	0.329 ± 0.047
Ce	—	0.94 ± 0.05	0.74	—	5.89	3.51	—	3.75	0.863
Pr	—	0.13 ± 0.01	—	—	—	—	—	—	0.130
Nd	—	0.88 ± 0.09	0.76	—	4.96	2.60	—	2.89	0.620
Sm	0.46 ± 0.01	0.46 ± 0.01	0.45	1.36	1.89	1.01	1.42	1.17	0.206 ± 0.025
Eu	0.21 ± 0.02	0.23 ± 0.01	0.224	0.53	0.64	0.43	0.51	0.48	0.077 ± 0.007
Gd	—	0.92 ± 0.09	—	—	2.80	1.64	—	—	0.269
Tb	1.1 ± 0.06	0.18 ± 0.01	1.16	0.36	—	—	0.34	—	0.046
Dy	—	—	—	—	3.38	2.16	—	2.66	0.332
Ho	—	0.27 ± 0.01	—	—	—	—	—	—	0.080
Er	—	—	0.66	—	1.89	1.33	—	1.60	0.227
Tm	—	0.090 ± 0.010	—	—	—	—	—	—	0.038
Yb	0.53 ± 0.03	0.58 ± 0.02	0.54	1.59	1.80	1.19	1.45	1.38	0.211 ± 0.026
Lu	0.078 ± 0.005	0.080 ± 0.002	0.074	0.262	—	0.18	0.26	0.20	0.036
Hf	0.58 ± 0.05	—	—	2.0	—	—	1.9	—	0.18
Ta	—	—	—	0.27	—	—	0.22	—	0.022
Th	—	—	—	0.35	—	—	0.27	—	0.041
U	—	0.04 ± 0.01	—	—	—	—	—	—	0.011

(a) Different aliquants were analyzed by INAA at OSU (Ma and Schmitt) and by RNAA at the Battelle Northwest Laboratory (Laul).
(b) Analyzed by A. J. Irving and published in Stolper and McSween (1979).
(c) Schnetzler and Philpotts (1969).
(d) Shih et al. (1981).
(e) These normalizing chondritic values for the REE are the averages of 21 ordinary chondrites (13 CH and 8 CL) reported in the literature to 1980. Neutron activation values for Ce and Gd reported in the Handbook of Geochemistry were not included. REE values of Bruderheim and Guareña were deleted because of their exceptionally high values. Inclusion of Bruderheim data in the Handbook of Geochemistry and of Guareña would increase the average REE value by ≈ +5%. The average value for Tb of 0.055 in CH and CL chondrites has been lowered to 0.046 to conform to a smooth Gd-Tb-Dy trend as measured in C2 and C3 chondrites. Dispersions were calculated for La, Sm, Eu and Yb that are well determined analytically. The average ratio of individual REE in Leedey (CL6) by Masuda et al. (1973) to the 21 CH + CL average values above is 1.14 ± 0.03.

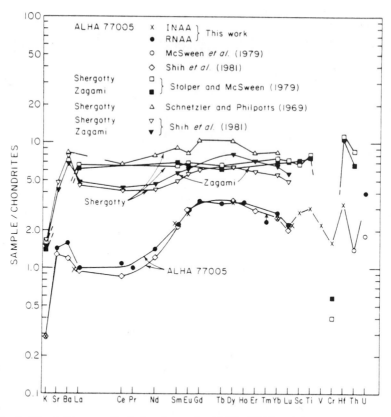

Fig. 1. Chondritic normalized abundances in ALHA 77005 and the shergottites. The "best" ordinary chondritic values (in ppm) adopted for normalization are given in Table 2, plus K: 830, Ti: 600, Cr: 3600 obtained from the literature prior to 1981.

3. DISCUSSION

3.1. Genesis of ALHA 77005

Sm–Nd isotopic data (Shih *et al.*, 1981) suggest that the source material(s) for ALHA 77005 and shergottites could have evolved in a light REE enriched environment [i.e., $\epsilon_{CHUR}(I) < 0$] for $T = 4.6-1.3$ AE and that the light REE depleted patterns of these achondrites could have been established by igneous event(s) at ~1.3 AE. Regardless of whether we model ALHA 77005 as representative of either the melt derived from partial melting or the crystallized cumulate from such a liquid or the source residue from the partial melting event, it is difficult to derive the observed REE pattern of ALHA 77005 by one step batch equilibrium partial melting of a light REE enriched source material as constrained by the isotopic data. Because the REE distribution coefficients of the mafic residual phases like olivine and pyroxenes are $D_{La}^{ol,pyx} < D_{Sm}^{ol,pyx} < D_{Lu}^{ol,pyx}$, the melt derived by one step batch equilibrium partial melting of a light REE enriched source material would also have a light REE enriched pattern. Such a pattern does not satisfy the observed light REE depleted pattern for ALHA 77005 (if we model ALHA 77005 as representative of the melt) or the equilibrium melt of ALHA 77005 (Fig. 2) (if we model ALHA 77005 as representing either a cumulate from the melt or residual source material from the partial melting event).

In order to derive a light REE depleted melt from a light REE enriched source material, continuous melting and melt removal from the source material are required. Continuous

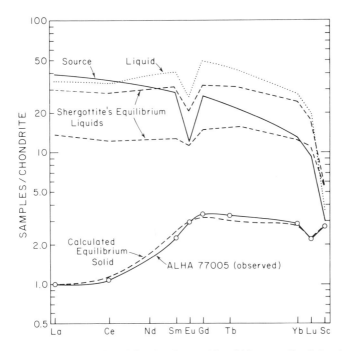

Fig. 2. The calculated REE and Sc abundances (chondritic normalized) for the source material is the solid curve labelled "source" obtained for ALHA 77005. The last 4% equilibrium melt derived by continuous melting process, as explained in the text, of such a source material is shown by the dotted curve labellel "Liquid". Cumulate or residual source material (dashed curve labelled "Calculated equilibrium solid") in equilibrium with the last 4% melt has a REE + Sc pattern that matches the observed pattern for ALHA 77005. The range of the shergottite's equilibrium liquids is calculated by using Shergotty data by Schnetzler and Philpotts (1969) and Shih *et al.* (1981) and by assuming 20–30% of trapped liquid remained in Shergotty. The distribution coefficients used for calculations are shown on Table 3 and the normative mineralogy of ALHA 77005 was used to calculate the bulk distribution coefficients for the residual source material.

melting and melt removal would progressively deplete the light REEs in successive melts because the distribution coefficients for the light REEs are smaller than those for the heavy REEs in residual phases like olivine and pyroxenes. Such a continuous melting process can be modeled mathematically as stepwise batch equilibrium partial melting with the incomplete removal of melt from the residue at each step and with additional increments of melting always retained in the residue (Langmuir *et al.*, 1977). For a continuous melting process, the light REE depleted pattern of ALHA 77005 can be modeled as either representing the source residue or the crystallized cumulate from a melt produced from a light REE enriched source material. For example, a melting process is suggested that involved 20% equilibrium melting at the first step followed by two steps of 4% incremental equilibrium partial melting. During the melt removals at each step the melt left behind in the residue is 1% of the starting matter. Such a postulated mechanism would produce a source residue exhibiting a REE + Sc pattern similar to that observed for ALHA 77005 and which is generated by continuous melting from a light REE enriched source material (Fig. 2). Alternatively, ALHA 77005 could represent a cumulate that crystallized from a melt, which represents the last 4% equilibrium melt (step 3; total fraction melted = 28%) in the previous example, and again derived from a light REE enriched source material by a continuous melting mechanism (Fig. 2). Since ALHA 77005 is an igneous rock with a cumulate texture, crystal cumulation from a light REE depleted melt is favored for its derivation. The degrees of partial melting invoked at each step in

Table 3. Distribution coefficients used in the partial melting calculation.

Element	Olivine[a]	Orthopyroxene[b]	High-Ca Clinopyroxene[c]	Plagioclase[d]
La	0.009	0.004	0.071	0.066
Ce	0.010	0.008	0.11	0.049
Sm	0.012	0.018	0.34	0.033
Eu	0.012	0.018	0.19	0.77
Gd	0.013	0.024	0.40	0.031
Tb	0.016	0.036	0.43	0.030
Yb	0.033	0.11	0.42	0.027
Lu	0.045	0.15	0.42	0.026
Sc	0.27	1.4	2.5	0

[a]Values for REE are averages from McKay and Weill (1976, 1977) and Schnetzler and Philpotts (1970) (sample HHP-66-19). Gd and Tb values are obtained by interpolation and La and Lu values are obtained by extrapolation. Sc value is from McKay and Weill (1977).
[b]Values for REE are averages from McKay and Weill (1976) at 1200°C and 1340°C. Gd and Tb values are obtained by interpolation and La and Lu values are obtained by extrapolation. Sc value is from McKay and Weill (1977).
[c]Values for REE are averages from Grutzeck et al. (1974), Schnetzler and Philpotts (1970) (sample HHP-66-19) and Arth and Hason (1975). Tb and Yb values are obtained by interpolation. $D_{Eu} = 0.33 \cdot D_{Sm} + 0.67 \cdot D_{Eu}$ at $fo_2 = \sim 10^{-12}$ (Grutzeck et al., 1974). Sc value is the average from Lindstrom (1976).
[d]Values for REE are averages from McKay and Weill (1976) at 1200°C and 1340°C, McKay and Weill (1977) at 1340°C and Shih (1972). Gd and Tb values are obtained by interpolation and La and Lu values are obtained by extrapolation. $D_{Eu} = 0.33\, D_{Sm} + 0.67\, D_{Eu}$ at $fo_2 = \sim 10^{-12}$. Sc value in plagioclase is assumed to be zero.

the above mentioned example are not unique because several parameters, like the initial REE abundances in the source, degree of partial melting at each step and the numbers of melting steps are involved in the calculation. For example, continuous melting, which is modeled as consisting of a) one step of 20% initial melting followed by three steps of 2% progressive melting, b) the melt left behind in the residue is always 1% of the matter being melted at a given stage, and c) a source material of La = 33X, Sm = 21X, Yb = 12X, Lu = 8X and Sc = 3X chondrite, would also yield a source residue with a REE + Sc pattern similar to that of ALHA 77005. The above mentioned calculations, however, illustrate that continuous melting and melt removal from a light REE enriched source material would be responsible for the derivation of ALHA 77005.

The light REE enriched source material (chondritic normalized La/Lu ~ 4) for ALHA 77005 could be derived by partial melting of material with a chondritic REE pattern, which consists primarily of olivine and pyroxenes, possibly with minor garnet. A small amount of garnet is likely in the residue because its absence requires unrealistically small degrees of partial melting (e.g., <0.1%) to produce the observed REE fractionation. The invoking of garnet in the source residue for the derivation of the light REE enriched source material for ALHA 77005 during this early melting episode (as early as 4.6 AE ago) requires a large size parent body for ALHA 77005. This is not inconsistent with the young Sm/Nd age (1.3 AE) obtained for ALHA 77005 (Shih et al., 1981).

The similar chondritic normalized heavy REE, Ti, Hf, Sc and V abundances observed in ALHA 77005 are not inconsistent with the proposed two-stage model, i.e., initial partial melting of chondritic material to produce the light REE enriched source material for ALHA 77005 followed by progressive melting to deplete the light REE in the successive melts and in the source residue. In such a model, both Ti and Hf, which are incompatible elements in mafic minerals like olivine and orthopyroxene, would be fractionated to a similar degree. On the other hand, Sc and V, which are compatible elements in orthopyroxene ($D^{opx}_{Sc,V} > 1$; McKay and Weill, 1977; Ringwood and Essene, 1970), were likely depleted (relative to REE) in the ALHA 77005 source material and remained nearly unchanged during the progressive melting episodes. Appropriate degrees of progressive melting of such source material would produce a melt from which cumulates, having comparable heavy REE, Ti, Hf, Sc and V abundances, would be derived.

3.2. Relationship with the shergottites and implication of the complementary REE patterns

Using REE data of the shergottites reported by Stolper and McSween (1979), we (Ma et al., 1980) have previously concluded the lack of a genetic relationship between ALHA 77005 and the shergottites. However, if the REE data of the shergottites by Schnetzler and Philpotts (1969) and Shih et al. (1981) are adopted for modeling, a source material having a similar REE pattern could be invoked for the derivation of ALHA 77005 and the shergottites. Because of the consistency in the REE patterns reported by Schnetzler and Philpotts (1969) and Shih et al. (1981), the samples analyzed by these investigators appear to be representative of the shergottites; therefore, their data are chosen for trace element modeling calculations.

The similar patterns between the calculated parent magmas in equilibrium with the shergottites and the ALHA 77005 (Fig. 2) indicate that the parent magmas for the shergottites could also be derived by continuous melting from a source material originally having a light REE enriched pattern similar to that for ALHA 77005. A suggestion of

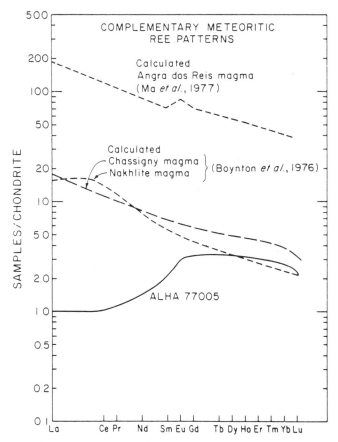

Fig. 3a. Complementary meteoritic REE patterns. A similar meteoritic magma to the calculated Chassigny magma may be inferred from another Chassigny-like achondrite, Brachina (Johnson et al., 1977).

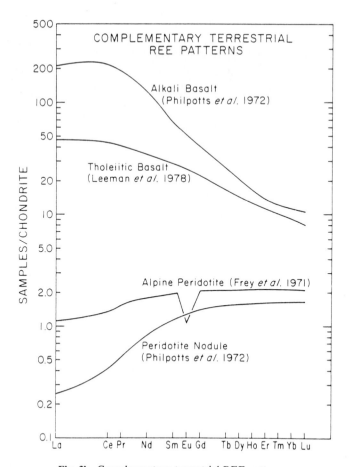

Fig. 3b. Complementary terrestrial REE patterns.

source materials in common with respect to REE pattern for ALHA 77005 and the shergottites, however, does not conflict with the isotopic constraints that ALHA 77005 and shergottites cannot be comagmatic.

The REE and other trace element pattern of ALHA 77005 shows a similarity to patterns observed in some of the terrestrial peridotites and ultramafic nodules (Figs. 3a,b). The unique light REE depleted pattern of ALHA 77005 also exhibits a complementary REE pattern to those achondrites (e.g., Angra dos Reis, nakhlites and Chassigny; Fig. 3a) which have either an observed light REE enriched pattern (nakhlites and Chassigny) or crystallized from an inferred light REE enriched magma (Ma et al., 1977; Boynton et al., 1976). The REE patterns and major element compositions of these achondrites may not allow them to be directly related by simple fractionation processes of common source material; however, these comparisons of complementary REE patterns (Figs. 3a,b) suggest that the source materials in the parent body or bodies of these achondrites have a great similarity to the earth's upper mantle. This observation confirms the conclusion reached by McSween et al. (1979a) based upon other refractory to volatile siderophile, chalcophile, and lithophile minor and trace element data of ALHA 77005. The similarity of complementary REE patterns of unique achondrites or their inferred parent magmas and igneous rocks from the earth would suggest that the parent body or bodies of these achondrites are large in size (comparable to earth) and that common igneous processes operating under similar physical-chemical conditions are responsible for the derivation of igneous rocks on various planetary bodies (Anders, 1977).

4. CONCLUSION

Summarizing the results of this study, we conclude that:

1. The unique REE pattern for ALHA 77005 can be modeled as either a source residue or cumulate that crystallized from a melt derived by continuous melting from a light REE enriched source material. The light REE enriched source material(s) could have been established early (possibly 4.6 AE ago) in the parent body of ALHA 77005 and could be similar in its REE pattern for both ALHA 77005 and the shergottites.
2. The REE pattern of ALHA 77005 is similar to those of terrestrial peridotites and is complementary to some fractionated REE patterns in achondrites like Angra dos Reis, nakhlites and Chassigny. The similarity of complementary REE patterns of unique achondrites or their inferred parent magmas and igneous rocks from the earth would imply that the parent body or bodies of these achondrites are large in size and that common igneous processes for the derivation of igneous rocks have been operative under similar physical conditions on various planetary bodies.

Acknowledgments—We are grateful to the NSF for collecting Antarctic meteorites. We acknowledge the assistance of T. V. Anderson, W. T. Carpenter, and S. L. Bennett at Oregon State University TRIGA reactor for neutron activations. This work was supported by NASA grant NGL-32-002-039 to R. A. Schmitt and NASA grant NAS-9-15357 to J. C. Laul.

REFERENCES

Anders E. (1977) Chemical compositions of the moon, earth and eucrite parent body. *Phil Trans. Roy. Soc. London* **A285**, 23–40.

Arth J. G. and Hanson G. N. (1975) Geochemistry and origin of the early Precambrian crust of northwestern Minnesota. *Geochim. Cosmochim. Acta* **39**, 325–362.

Boynton W. V., Starzyk P. M., and Schmitt R. A. (1976) Chemical evidence for the genesis of the ureilites, the achondrite Chassigny and the nakhlites. *Geochim. Cosmochim. Acta* **40**, 1439–1447.

Duke M. B. (1968) The Shergotty meteorite, magmatic and shock metamorphic features. In *Shock Metamorphism of Natural Materials* (B. M. French and N. M. Short, eds.), p. 613–621. Mono, Baltimore.

Easton A. J. and Elliott C. J. (1977) Analyses of some meteorites from the British Museum (Natural History) collection. *Meteoritics* **12**, 409–416.

Frey F. A., Haskin L. A., and Haskin M. A. (1971) Rare earth abundances in some ultramafic rocks. *J. Geophys. Res.* **76**, 2057–2070.

Grutzeck M., Kridelbaugh S., and Weill D. (1974) The distribution of Sr and REE between diopside and silicate liquid. *Geophys Res. Lett.* **1**, 273–275.

Johnson J. E., Scrymgour J. M., Jarosewich E., and Mason B. (1977) Brachina meteorite, a chassignite from South Australia. *Rec. of the S. Australian Mus.* (*Adelaide*) **17**, 309–319. Adelaide, S. A., Australia.

Langmuir C. H., Bender J. F., Bence A. E., Hanson G. N., and Taylor S. R. (1977) Petrogenesis of basalts from the FAMOUS area: Mid-Atlantic Ridge. *Earth Planet. Sci. Lett.* **36**, 133–156.

Laul J. C. (1979) Neutron activation analysis of geological materials. *Atomic Energy Review* **17**, 603–695. Int. Atomic Energy Admin.

Leeman W. P., Murali A. V., Ma M.-S., and Schmitt R. A. (1978) Mineral constitution of mantle source region for Hawaiian basalt—Rare earth evidence for mantle heterogeneity. *Proc. Chapman Conf. on Partial Melting in the Earth's Upper Mantle*, p. 169–183.

Lindstrom D. J. (1976) Experimental study of the partitioning of the transition metals between clinopyroxene and coexisting silicate liquids. Ph.D. Thesis, Univ. of Oregon, Eugene. 188 pp.

Ma M.-S., Murali A. V., and Schmitt R. A. (1977) Genesis of the Angra dos Reis and other achondritic meteorites. *Earth Planet. Sci. Lett.* **35**, 331–346.

Ma M.-S., Schmitt R. A., and Laul J. C. (1980) Genetic relationship between Allan Hills (ALHA) 77005 and shergottites—A geochemical study (abstract). *Meteoritics* **15**, 327.

Masuda A., Nakamura N., and Tanaka T. (1973) Fine structures of mutually normalized rare earth patterns in chondrites. *Geochim. Cosmochim. Acta* **37**, 239–248.

McCarthy T. S., Erlank A. G., Willis J. P., and Ahrens L. H. (1974) New chemical analyses of six achondrites and one chondrite. *Meteoritics* **9**, 215–221.

McKay G. A. and Weill D. F. (1976) Petrogenesis of KREEP. *Proc. Lunar Sci. Conf. 7th*, p. 2427–2447.

McKay G. A. and Weill D. F. (1977) KREEP petrogenesis revisited. *Proc. Lunar Sci. Conf. 8th*, p. 2339–2355.

McSween H. Y., Stolper E. M., Taylor L. A., Muntean R. A., O'Kelley G. D., Eldridge J. S., Biswas S., Ngo H. T., and Lipschutz M. E. (1979a) Petrogenetic relationship between Allan Hills 77005 and other achondrites. *Earth Planet. Sci. Lett.* **45**, 275–284.

McSween H. Y. and Stolper E. M. (1980) Basaltic meteorites. *Sci. Amer.* **242**, 54–63.

McSween H. Y., Taylor L. A., and Stolper E. M. (1979b) Allan Hills 77005: A new meteorite type found in Antarctica, *Science* **204**, 1201–1203.

Nyquist L. E., Bogard D. D., Wooden J. L., Wiesmann H., Shih C.-Y., and Bansal B. M. (1979) Early differentiation, late magmatism, and recent bombardment on the shergottite parent planet. *Meteoritics* **14**, 502.

Philpotts J. A., Schnetzler C. C., and Thomas H. H. (1972) Petrogenetic implications of some new geochemical data on eclogitic and ultrabasic inclusions. *Geochim. Cosmochim. Acta* **36**, 1131–1166.

Ringwood A. E. and Essene E. (1970) Petrogenesis of Apollo 11 basalts, internal constitution and origin of the moon. *Proc. Apollo 11 Lunar Sci. Conf.*, p. 769–799.

Schnetzler C. C. and Philpotts J. A. (1969) Genesis of the calcium-rich achondrites in light of rare-earth and barium concentrations. In *Meteorite Research* (P. M. Millman, ed.), p. 206–215. D. Reidel, Dordrecht.

Schnetzler C. C. and Philpotts J. A. (1970) Partition coefficients of rare-earth between igneous matrix material and rock forming mineral phenocrysts—II. *Geochim. Cosmochim. Acta* **34**, 331–340.

Shih C.-Y. (1972) The rare earth geochemistry of oceanic igneous rocks. Ph.D. Thesis, Columbia Univ., N.Y. 151 pp.

Shih C.-Y., Nyquist L. E., Bansal B. M., Wiesmann H., Wooden J. L., and McKay G. (1981) REE, Sr and isotopic studies on shocked achondrites—Shergotty, Zagami and ALHA 77005 (abstract). In *Lunar and Planetary Science XII*, p. 973–975. Lunar and Planetary Institute, Houston.

Smith J. V. and Hervig R. L. (1979) Shergotty meteorite mineralogy, petrography and minor elements. *Meteoritics* **14**, 121–142.

Stolper E. M. and McSween H. Y. (1979) Petrology and origin of the shergottite meteorites. *Geochim. Cosmochim. Acta* **43**, 1475–1498.

Yanai K., Cassidy W. A., Funaki M., and Glass B. P. (1978) Meteorite recoveries in Antarctica during field season 1977–1978. *Proc. Lunar Planet. Sci. Conf. 9th*, p. 977–987.

SNC meteorites: Igneous rocks from Mars?

Charles A. Wood[1] and Lewis D. Ashwal[2]

[1] SN6/Geology Branch, NASA Johnson Space Center, Houston, Texas 77058
[2] Lunar and Planetary Institute, 3303 NASA Road 1, Houston, Texas 77058

"... when you have excluded the impossible, whatever remains, however improbable, must be the truth."

The Adventure of the Beryl Coronet
Arthur Conan Doyle

Abstract—Three classes of achondrites, comprising 9 stones, are anomalous in having crystallization ages of 1.3 b.y., more than 3 b.y. younger than any other meteorite. These meteorites—shergottites, nakhlites, and chassignites (SNC)—are unbrecciated augite to olivine igneous cumulates that are volatile-rich, have complex rare earth element patterns, equilibrated near the quartz-fayalite-magnetite buffer, and formed in a magnetic field-free region of the solar system. Although SNC meteorites have been suggested as being related to eucrites, the young SNC ages and other unique characteristics make that connection exceedingly unlikely. Additionally, current understanding of the thermal evolution of asteroids is such that igneous processes as recent as 1.3 b.y. ago are difficult to envision. Mars is the only plausible parent body for SNC meteorites, as previously suggested, although the mechanism for the escape of unmelted surface material from Mars is not yet clearly understood. Assuming that SNC meteorites are Mars samples allows inferences on the history and evolution of the planet. Model ages and isotopic systematics imply that Mars formed and underwent a major chemical differentiation (core formation and/or magma ocean?) 4.6 b.y. ago. The crystallization 1.3 b.y. ago of SNC igneous rocks from four or more chemically distinct sources suggests that volcanism was vigorous late in martian history, as predicted by some crater count calibrations. Mars had only a weak or possibly no magnetic field when SNC meteorites formed, consistent with the apparent absence of a present-day field of Mars. Overall, the chemical characteristics of SNC meteorites are remarkably similar to terrestrial igneous rocks suggesting similar processes and perhaps evolution for the mantles of these two planets.

INTRODUCTION

All dated meteorites have formation ages of 4.4 to 4.6 b.y. (Sears, 1978), except for nine achondrites which crystallized 1.3 b.y. ago (Table 1). These meteorites—shergottites, nakhlites, and chassignites (SNC)—also differ chemically, petrologically, physically, and magnetically from other meteorite types, suggesting a unique origin. We propose, extending previous tentative suggestions (Wasson and Wetherill, 1979; Nyquist *et al.*, 1979a; Walker *et al.*, 1979; McSween and Stolper, 1980), that SNC meteorites are igneous rocks from Mars, and additionally, using the characteristics of SNC meteorites, we speculate on specific events in the history of Mars.

SNC METEORITES

Mineralogy and chemistry

SNC meteorites are rare members of the achondrite class of igneous meteorites, but differ substantially from eucrites, the most common (n ~ 40) achondrite type. Eucrites are composed mainly of calcic plagioclase and low-Ca pyroxene (pigeonite) that apparently originated ~4.5 b.y. ago on one of more volatile-poor bodies of asteroidal size (Duke and Silver, 1967; Stolper, 1977; Consolmagno and Drake, 1977).

Shergottites, nakhlites, and chassignites have distinguishing characteristics (see below and Table 2) but are generally uniform in their differences with eucrites and other meteorites. Compared to eucrites, SNC meteorites are enhanced in overall volatile

Table 1. SNC meteorites.

Meteorite	Type	Fall Date	Location	Crystallization Age (10^9 yrs)	Ref.	Exposure Age (10^6 yrs)	Ref.
Shergotty	S	1865	India	1.34 ± 0.06		2.1; 2.4	11; 12
Zagami	S	1962	Nigeria	Sm–Nd	1; 2	2.5	12
ALHA 77005	S	Find	Antarctica	(W.R.)		7 ± 1	13
EETA 79001	S	Find	Antarctica	?		?	
Nakhla	N	1911	Egypt	1.37 ± 0.02 Rb–Sr	3	8 ± 0.8	5
				1.24 ± 0.01 Rb–Sr	4	10.1	11
				≤ 1.3 Ar–Ar	5		
				1.3 ± 0.2 U–Pb	6		
				1.27 ± 0.06 Sm–Nd	6		
Lafayette	N	Find	Indiana	1.33 ± 0.03 Ar–Ar	5	6.5 ± 0.6 10.5; 11.2	5 11; 14
Governador Valadares	N	Find	Brazil	1.33 ± 0.01 Rb–Sr	7	8 ± 1	8
				1.32 ± 0.04 Ar–Ar	8		
Chassigny	C	1815	France	1.39 ± 0.17 K–Ar	9	8.9 ± 0.5	9
				1.27 Ar–Ar	10		
Brachina	C	Find	Australia	?		?	

S = shergottite; N = nakhlite; C = chassignite.
References: 1—Nyquist *et al.*, 1979a; 2—Shih *et al.*, 1981; 3—Papanastassiou and Wasserburg, 1974; 4—Gale *et al.*, 1975; 5—Podosek, 1973; 6—Nakamura *et al.*, 1977; 7—Wooden *et al.*, 1979; 8—Bogard and Husain, 1977; 9—Lancet and Lancet, 1971; 10—Bogard and Nyquist, 1979; 11—Fuse and Anders, 1969; 12—Heymann *et al.*, 1968; 13—Kirsten *et al.*, 1978; 14—Ganapathy and Anders, 1969.

Table 2. Comparison of eucrites and SNC meteorites.

	Eucrites	Shergottites	Nakhlites	Chassignites
Cryst. Age	~4.6 b.y.	←――――― ~1.3 b.y. ―――――→		
Exposure Age	2–40 m.y.	2.5 m.y.	8–11 m.y.	9 m.y.
Dominant Phase	Pig. + Plag.	Pig. + Aug.	Augite	Olivine
Fspar. Comp.	An_{80-95}	An_{40-60}	$An_{34} + Or_{77}$	$An_{16-37} + Or_{70}$
Oxidation	free metal	←――――― Fe–Ti oxides ―――――→		
Hydrous Phases	none	none	iddingsite	kaersutite
K/U	3,000	10,000	25,000	25,000
REE/chond.	flat	LREE depleted ←――― LREE enriched ―――→		
$\delta^{18}O/\delta^{17}O$	2.9–3.7/1.2–1.7	4.2/2.2	4.4–4.7/2.5–2.6	3.7/1.8
Shock Level	~unshocked	300 kbar	unshocked	0–~175 kbar
Texture	brecciated	←――――― unbrecciated ―――――→		
Examples	Juvinas Chervony Kut Moore County + 39 others	Shergotty Zagami ALHA 77005 EETA 79001	Nakhla Lafayette Governador Valadares	Chassigny Brachina

content, ratio of high-Ca pyroxene to low-Ca pyroxene, alkali content in feldspar, and oxidation state (Stolper *et al.*, 1979).

The shergottites Shergotty and Zagami are quite similar to each other (Table 3), containing ~70% pigeonite and augite (~1:1), ~22% maskelynite, with a few percent titanomagnetite (Stolper and McSween, 1979). Maskelynite was recognized in Shergotty

Table 3. Chemical compositions and modes of SNC meteorites.

	Shergotty[1]	Zagami[1]	ALHA 77005	EETA 79001	Nakhla[2]	Lafayette[3]	Governador Valadares[4]	Chassigny[2]	Brachina[5]
SiO_2	50.4	50.8			48.24	46.9	49.5	37.0	38.04
TiO_2	0.81	0.77			0.29	0.33	0.35	0.07	0.12
Al_2O_3	6.89	5.67			1.45	1.55	1.74	0.36	2.12
Cr_2O_3	0.21	0.30			0.42	0.18	0.21	0.83	0.58
FeO	19.1	18.0			20.64	22.7	19.74	27.44	23.69
MnO	0.50	0.50			0.54	0.79	0.67	0.53	0.34
MgO	9.27	11.0			12.47	12.9	10.9	32.83	27.27
CaO	10.1	10.8			15.08	13.4	15.8	1.99	2.10
Na_2O	1.37	0.99			0.42	0.36	0.82	0.15	0.63
K_2O	0.16	0.14			0.10	0.09	0.43	0.03	0.08
P_2O_5					0.12			0.04	0.27
H_2O									0.26
C									0.07
(Fe, Ni, Co)S									4.20
Total	98.8	99.0			100.77	99.1	100.16	101.27	99.77
			[6]		[8]	[3]	[9]	[10]	[5,11]
Pigeonite	36.3[1]	36.5[1]	26	major[7]	8				
Augite	33.5	36.5	11	minor	78.6	major	major	5.0	6
Olivine			52	+	15.5	minor	minor	91.6	79
Plagioclase	23.3*	21.7*	10*	minor*	3.7			1.7	10
Magnetite				+	1.9				
K-Feldspar					1.1				
Chromite			1			+	+	1.4	0.8
Titanomagnetite	2.25	2.1		+		+	+		
Troilite				+	+	+	+		4
Mesostasis	4.0	2.1							
Intergrowth									

Note: * = maskelynite.

References: [1]Stolper and McSween, 1979 (average of two point count estimates). [2]McCarthy et al., 1974. [3]Boctor et al., 1976. [4]Burragato et al., 1975. [5]Johnson et al., 1977. [6]Ma et al., 1981. [7]Reid, 1981. [8]Bunch and Reid, 1975. [9]Berkley et al., 1980. [10]Floran et al., 1978. [11]Nehru et al., 1979.

more than 100 years ago by G. Tschermak, and is now known to be a shock-produced alteration of plagioclase (Binns, 1967). Chemical details are not yet published for the two shergottites discovered in Antarctica (ALHA 77005 and EETA 79001). ALHA 77005 has only 37% pigeonite and augite (~2:1), 10% maskelynite, and 52% olivine, and may be similar to the peridotitic source material from which Shergotty and Zagami were derived (McSween et al., 1979; Ma et al., 1981). EETA 79001 is unique in containing two different (but related) lithologies separated by a geologic contact (Reid, 1981). About 10% of the meteorite closely resembles Shergotty but the remainder is a shocked but unbrecciated pyroxenite with pyroxene as the major phase but also containing maskelynite, Mg–Al chromite, iron sulphide, possible ilmenite, a few large olivines, and melt glasses (Reid, 1981). The main mass of EETA 79001 appears to be unique, containing maskelynite like the other shergottites, but its chromite and iron sulphide are reminiscent of chassignites.

The three nakhlites (Nakhla, Lafayette, and Governador Valadares) are fairly uniform, being nearly monomineralic (augite with minor olivine) cumulates (Bunch and Reid, 1975). Unlike shergottites, the plagioclase has not been shocked to maskelynite and occurs in both potassic and sodic phases. A hydrous alteration product of olivine, probably iddingsite, is found in nakhlites and appears to be of pre-terrestrial origin (Reid and Bunch, 1975; Berkley et al., 1980).

The two known chassignites (Chassigny and Brachina) are also nearly monomineralic cumulates with olivine being the dominant phase (80–90%) and minor augite (5–6%) (Johnson et al., 1977; Floran et al., 1978; Nehru et al., 1979). Feldspar compositions (sodic plagioclase plus minor potassic feldspar) are nearly identical to those of nakhlites, and kaersutite, a primary hydrous amphibole, occurs in melt inclusions within olivine (Floran et al., 1978).

Rare earth elements

Rare earth element (REE) patterns (Fig. 1) for most eucrites are nearly flat and cluster at 8–10 times chondrites with no Eu anomalies (Drake, 1979). A few eucrites have lower abundances with positive Eu anomalies, and others have higher abundances with negative Eu anomalies. In contrast, SNC meteorites are characterized by highly fractionated REE patterns, with no appreciable Eu anomalies (Fig. 1). Nakhla and the chassignites exhibit subparallel, strongly light REE enriched patterns, with the chassignites about 5 times lower in chondrite-normalized abundance than Nakhla (Nakamura and Masuda, 1973; Mason et al., 1975; Johnson et al., 1977; Boynton et al., 1976). A genetic relationship for the nakhlites and chassignites based on igneous differentiaton has been suggested by Mason et al. (1975) and Nakamura and Masuda (1973). Similarly, shergottites have subparallel light REE depleted patterns, with ALHA 77005 about 5 times lower in abundances than Shergotty and Zagami. Although a genetic relationship among these meteorites is possible (Ma et al., 1981), their isotopic compositions indicate that they cannot be comagmatic, as discussed subsequently.

Potassium/uranium ratios

The K/U ratios for SNC meteorites have been compared to those of the Earth, Moon, and eucrites by McSween et al. (1979). Eucrite K/U values cluster around $2-4 \times 10^3$, similar to ratios for lunar rocks, whereas shergottites and chassignites are centered at 10^4, the average value for terrestrial rocks. Nakhlites are slightly higher at $1.5-3 \times 10^4$, at the upper end of the range for terrestrial samples.

Oxidation state

Experiments (Delano and Arculus, 1980) demonstrated that the oxidation state of the Nakhla parent body was comparable to that of the Earth's upper mantle, near the wüstite-magnetite (WM) buffer. Coexisting ilmenite and Ti-magnetite analyses for Sher-

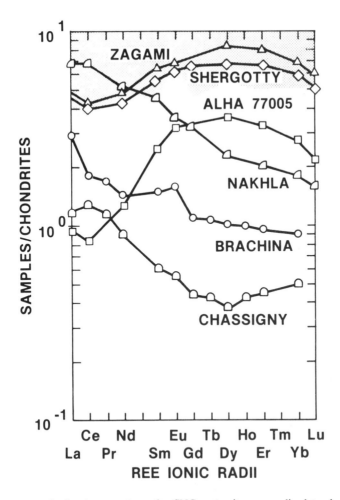

Fig. 1. Rare earth abundance patterns for SNC meteorites, normalized to chondrites. For comparison a portion of the eucrite field (Drake, 1979) is indicated by the pattern. Most eucrites are flat and 8–10 times chondrites but the range is 4–16 times chondrites, and some eucrites have pronounced Eu anomalies. Data sources: Zagami, Shergotty, and ALHA 77005-Shih *et al.*, 1981; Nakhla-Nakamura and Masuda, 1973; Brachina-Johnson *et al.*, 1977; Chassigny-Mason *et al.*, 1975.

gotty indicate an oxygen fugacity between the quartz-fayalite-magnetite (QFM) and WM buffers (Smith and Hervig, 1979). These results are consistent with the occurrence of accessory oxide phases (magnetite, ilmenite and/or Fe^{3+}-rich chromites) in various SNC meteorites suggesting equilibration near the QFM buffer (Bunch and Keil, 1971; Floran *et al.*, 1978; Stolper and McSween, 1979). Eucrites crystallized under considerably more reducing conditions (as evidenced by the presence of metallic iron) than SNC meteorites, and have oxygen fugacities similar to lunar basalts (Stolper *et al.*, 1979).

Oxygen isotopes

Meteorites have been classified by oxygen isotopes into groups that (as far as is presently understood) cannot be derived from one another by chemical processes such as fractionation (Clayton *et al.*, 1976). Eucrites and other basaltic achondrites plot in a different section of the three-isotope plot of oxygen isotopic abundances (Fig. 2) than Shergotty, Zagami, Nakhla, and Lafayette (Clayton *et al.*, 1976; Clayton, pers. comm., 1981), although Chassigny plots within the eucrite field. It should be noted that despite the high

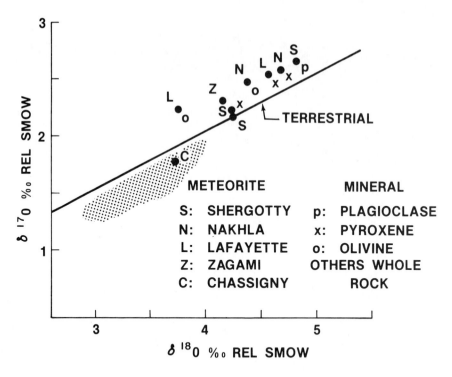

Fig. 2. Three-isotope plot (both x and y values are normalized to ^{16}O) of variations in oxygen isotopic abundances for SNC meteorites and basaltic achondrites (pattern). All data from Clayton *et al.*, 1976 and Clayton, pers. comm., 1981. Diagonal line is mass-fractionation trend defined by terrestrial and lunar materials. SMOW is standard mean ocean water.

precision of these measurements there appears to be considerable variability in replicate analyses: Clayton *et al.* (1976) give two determinations for pyroxenes from Shergotty—one plots above the terrestrial fractionation line near Zagami, but the other falls below the line, within the extreme limits of the eucrite field. Replicate analyses of Chassigny and determinations for Brachina are necessary to check whether chassignites have significantly different isotopic compositions from shergottites and nakhlites.

Ages and isotopic evolution

Ages of SNC meteorites (Table 1) average 1.31 b.y. (11 independent measurements), and have been interpreted as the time of primary crystallization. Evidence for this includes the concordancy of ages from four isotopic systems for Nakhla and from two isotopic systems for Governador Valadares and Chassigny (Table 1). Shih *et al.* (1981) have obtained a whole-rock Sm–Nd isochron of 1.34 ± 0.06 b.y. for three shergottites, but there is some uncertainty about the significance of this age because evidence exists that the samples are not comagmatic (Ma *et al.*, 1981). Rb-Sr isotopic analyses of these shergottites plot on or close to a line corresponding to an age of about 4.6 b.y. (Shih *et al.*, 1981), which may be interpreted as the time at which source regions with distinctly different Rb/Sr were produced. This event, however, cannot have produced differences in Sm/Nd among the shergottite sources, since the samples have the same initial $^{143}Nd/^{144}Nd$ ratio at the presumed crystallization age of ~1.3 b.y. Sm/Nd isotopic data for Nakhla (Nakamura *et al.*, 1977) suggest that its source had a high initial Nd ratio at 1.3 b.y. compared to a chondritic uniform reservoir (CHUR) and hence was depleted in light REE. Since the shergottite source has a low initial Nd ratio (light REE enriched) compared to CHUR at this time (Shih *et al.*, 1981), it may be argued that the nakhlite and shergottite source regions were complementary with respect to chondritic or bulk planetary evolution. This assumes, of course, that Nakhla and the shergottites were derived from the same parent body. The shergottites cannot be comagmatic with each other or with two nakhlites (Nakhla, Governador Valadares) because of the disagreement of whole-rock ages between Sm–Nd and Rb–Sr systems, and also because of the distinct differences in initial Sr ratios and/or initial Nd ratios at the presumed time of crystallization (Shih et al., 1981; Nakamura et al., 1977; Wooden et al., 1979; Papanastassiou and Wasserburg, 1974; Gale et al., 1975). In summary, the available isotopic data for SNC meteorites are consistent with a model whereby their distinctly different source regions were produced by a large-scale chemical differentiation event soon after planetary accretion. One possibility for such an event is a magma ocean, as proposed for the early evolution of the outer portions of the Moon (Wood et al., 1970).

The internal isotopic systems of the shergottites have been disturbed, presumably by a shock event between 165 and 187 m.y. ago, which transformed plagioclase into maskelynite (Shih et al., 1981; Wooden et al., 1979; Bogard et al., 1979; Nyquist et al., 1979b). This event may have ejected the meteorites from the gravitational field of their parent body. Distinct differences in initial $^{87}Sr/^{86}Sr$ ratios at this time preclude an origin for these meteorites by crystallization from a shock melt, because impact melts effectively homogenize chemical variations in the target rocks (Simonds et al., 1976).

Cosmic ray exposure ages, which measure the length of time a meteorite was exposed to space within a fragment smaller than a few meters, are less than 10 m.y. for SNC meteorites (Table 1). Shergotty and Zagami have essentially equal exposure ages of 2.5 m.y., and the other analyzed SNC meteorites cluster around 8.5 m.y. ago, suggesting two different collisions, possibly while the stones were in Earth-crossing orbits.

Magnetism

Paleomagnetic studies of meteorites have demonstrated that great variations in magnetic field strength existed when different classes of meteorites formed. Nagata (1979) has found the following average paleointensities: carbonaceous chondrites—1.02 Oe; ordinary chondrites—0.1 to 0.4 Oe; and eucrites and other basaltic achondrites—0.11 Oe. More recently a paleointensity of 0.01 Oe has been determined for the shergottite ALHA 77005 (Nagata, 1981). Cisowski (1981 and pers. comm.) found no substantial extraterrestrial magnetization associated with Shergotty, whereas Zagami's weak remanence appears to be related to a shock event that postdated the formation of maskelynite. Additionally, the thermomagnetic curves for Shergotty and Zagami are quite similar to unoxidized to slightly oxidized terrestrial basalts (Cisowski, 1981). Thus, the magnetic properties of the three SNC meteorites investigated so far suggest that they were derived from a parent body with a very small to nonexistent magnetic field, in contrast to all other meteorite types.

Shock and texture

The existence of maskelynite in shergottites demonstrates that these meteorites experienced about 300 kbar of shock (Schaal and Hörz, 1977; Nyquist et al., 1979b). Deformation of olivine, pyroxene, and feldspar indicate a shock pressure of ~200 kbar for Chassigny (Floran et al., 1978), but Brachina is unshocked (Johnson et al., 1977). No evidence of shock is found in Nakhla (Bunch and Reid, 1975) and deformed twin lamellae in Lafayette augite suggest only weak shocking, if any at all, for that meteorite (Boctor et al., 1976). None of the SNC meteorites are brecciated, indicating that they are unlikely to have been derived from an impact dominated regolith, in contrast to the eucrites and most other meteorites. This conclusion is supported by the observation that SNC fragments have not been found as clasts in other brecciated meteorites.

WHERE DO SNC METEORITES COME FROM?

There is no certainty about the specific provenance of any meteorite; all proposed sources must be plausible in terms of (a) likely chemical and petrological affinity, and (b) ease of dynamical derivation. Many of the characteristics of SNC meteorites reviewed above argue for their source being a large body, however, removal of meteorites from large bodies is dynamically difficult. Additionally, there is a strong bias against suggestions that meteorites are derived from any body other than asteroids, especially since no meteorites appear to come from the Moon. Some researchers, stressing general chemical similarities, have proposed that shergottites come from the eucrite parent body (Stolper et al., 1979; Stolper and McSween, 1979; Feierberg and Drake, 1980), although others, stressing the age differences with eucrites and other meteorites, have suggested Mars as the parent of shergottites, or even of all SNC meteorites (Wasson and Wetherill, 1979; Nyquist et al., 1979a; Walker et al., 1979; McSween and Stolper, 1980). Based on our review of SNC characteristics, and arguments presented by the above authors, we believe Mars is the most likely source of the SNC meteorites. We now briefly consider (and reject) all plausible alternative sources including the other terrestrial planets (including Earth), other asteroids, and comets. The latter type of body can presumably be ruled out, because comets appear to be very primitive (and small) bodies that should not have experienced igneous activity 1.3 b.y. ago.

Mercury and Venus

Mercury is unlikely to be the SNC parent body on both dynamical and compositional grounds. Following Consolmagno and Drake (1977), the estimate of a maximum of 6% FeO on the surface of Mercury (deduced from reflectance spectra by Adams and McCord, 1977) contrasts with measured values of ~20% FeO in SNC meteorites (Table 3). Additionally, Mercury is thought to be a volatile-poor, anhydrous planet (Consolmagno and Drake, 1977), unlike SNC meteorites. Although Mercury's relatively small gravitational field would enhance escape of crater ejecta, the planet's small mass, and location deep within the sun's gravitational well, would diminish the probability of ejecta being perturbed to the Earth. Venus is a volatile-rich planet with a minute quantity of water in its atmosphere (e.g., Oyama et al., 1980) and might be a compositionally acceptable SNC parent body, but its dense atmosphere and large gravitational field would make escape of any material very difficult. Furthermore, samples derived from the surface of Venus should show evidence of extensive alteration, due to the interaction between the hot venusian troposphere and the surface rocks (Barsukov et al., 1980).

Earth and Moon

In many respects Earth is the most likely SNC parent body. McSween and Stolper (1980, p. 56) comment that shergottites and terrestrial basalt "are so similar in composition and mineralogy that it is difficult to conceive that the shergottites could have originated elsewhere in the solar system." Furthermore the oxidation state, hydrous phases, REE patterns, K/U ratios, and thermomagnetic curves of SNC meteorites are also consistent with an origin on the Earth. But as McSween and Stolper (1980) point out, there are a number of compelling reasons for rejecting this intriguing hypothesis, the most significant being the systematic differences in oxygen isotopes between terrestrial material and SNC meteorites (Clayton et al., 1976). Additionally, the negligible magnetic paleointensity of SNC meteorites (Nagata, 1981; Cisowski, 1981) argues against their cooling within the Earth's magnetic field. Finally, the Earth's strong gravitational field makes escape of ejecta very difficult. If, however, SNC meteorites were somehow ejected from the Earth and cooled in space, it would be necessary to store them for 165–187 m.y. before they fell back to the Earth. There is no simple way to store Earth ejecta for such prolonged periods.

The Moon was seriously considered as a source of meteorites (Urey, 1959; Wänke, 1968) until the first lunar samples were demonstrated to be different isotopically and chemically from all meteorites, including SNC's.

Eucrite parent body

Because they have igneous textures and lack chondrules, SNC meteorites have been considered to be related to eucrites and other basaltic achondrites. Stolper et al. (1979) discussed the possible derivation of SNC meteorites from a protoeucrite source peridotite by addition of volatiles, but they did not suggest a mechanism to account for SNC meteorites being 3 b.y. younger than eucrites, nor for the different oxygen isotopic compositions. Drake and co-workers (Drake, 1979; Consolmagno and Drake, 1977; Feierberg and Drake, 1980) have argued that eucrites come from Vesta, the only asteroid with a reflection spectrum closely matching eucrites, and thus concluded that the shergottites probably did as well. This proposal is difficult to substantiate, and indeed, the spectral reflectance data for Vesta do not match the spectrum of the shergottites (Feierberg and Drake, 1980). Wetherill (1978) has further pointed out that Vesta is far from any resonance position in the asteroid belt and thus derivation of any meteorites— eucrites or SNC's—would be difficult. Although SNC's and eucrites are both classes of achondrites, they are sufficiently dissimilar (Table 2) such that an origin on the same parent body is highly unlikely.

Asteroids, in general, are unlikely to be parent bodies of SNC meteorites because they are too small to produce igneous activity 3 b.y. after their formation (e.g., Minear et al., 1979), and no asteroid >25 km in diameter has a spectrum that matches SNC meteorites (Drake, 1979). Impact melting on an asteroid can be rejected as a possible source of SNC meteorites. Studies of terrestrial and lunar impact melt sheets suggest that cooling of the melts occurs rapidly enough by thorough mixing with cold clastic debris that the formation of cumulates is prevented (Simonds et al., 1976). The SNC meteorites are cumulate rocks with variable initial Sr ratios at their time of crystallization, and are thus exceedingly unlikely to have been derived from any single impact melt on an asteroid. That different asteroids experienced major melt-producing impacts 1.3 b.y. ago and fed meteorites to Earth seems implausible.

Mars

To account for the young crystallization ages, various investigators (Walker et al., 1979; Nyquist et al., 1979a; Wasson and Wetherill, 1979; McSween and Stolper, 1980) have recently proposed that shergottites come from a planet large enough to have experienced igneous activity 1.3 b.y. ago. Mars is the obvious candidate because of its apparently young volcanism, and because of the close compositional similarites of Shergotty and martian soil (when corrections are made for likely weathering products; Baird and Clark, 1981). We concur that Mars is the most likely parent body for shergottites, and, because of the general chemical similarities of all SNC meteorites, and their common crystallization age, we endorse the proposal (Wasson and Wetherill, 1979) that nakhlites and chassignites also come from Mars. In addition to the arguments based on youthful volcanism, various characteristics of SNC meteorites are consistent with the present understanding of Mars. Absorption bands in reflection spectra of martian dark regions have been reinterpreted as due to augite clinopyroxenes (Singer, 1981) such as dominate (Table 2) shergottites and nakhlites. The olivine-rich chassignites have cumulate texture and most likely represent material crystallized in magma chambers, and thus would not be expected on the martian surface.

The occurrences of primary and possibly secondary hydrous phases in chassignites and nakhlites, respectively, imply that water was brought to the surface of Mars, consistent with photogeologic evidence for fluvial activity (e.g., Milton, 1973; Carr and Schaber, 1977) and the detection of atmospheric water vapor (Farmer et al., 1977). Martian lava flows are very long by terrestrial standards, suggesting dense, fluid melts. This conclusion, previously reached on the basis of estimates of the density of the martian mantle (McGetchin and Smyth, 1978), is consistent with inferred melt rheology, calculated from the chemistry of SNC meteorites (Bottinga and Weill, 1970; 1972). Assuming Brachina to have essentially the same composition as its melt (as suggested by Johnson et al., 1977), and that the range of reconstructed melt compositions for Chassigny (Boynton et al., 1976) are correct, we calculate viscosities of 2 to 45 poise, and densities of 2.8 to 3.1 gm/cm^3, for anhydrous compositions at 1150°C. For comparison, common terrestrial basalts have viscosities of 10^2 to 10^3 poise, and densities of 2.6 to 2.7 gm/cm^3 at the same temperature (Murase and McBirney, 1973). The negligible paleointensity measured for shergottites (Nagata, 1981; Cisowski, 1981) is compatible with the current weak to non-existent martian magnetic field (Russell, 1979).

Derivation of meteorites from Mars

The previous section demonstrated that Mars is the most likely parent body for SNC meteorites, based on age, chemical, and physical considerations. However, a satisfactory mechanism has not been demonstrated to explain how rocks escaped the surface of Mars without melting and evolved into Earth-crossing orbits on a time scale consistent with observed shock or exposure ages. Wasson and Wetherill (1979) have speculated that SNC meteorites may be ejecta from a cratering event on Mars. They hypothesized that martian subsurface permafrost, explosively released by an impact, could aerodynamically accelerate ejecta to escape velocity (5 km/sec). Ejecta would not be shock melted because its velocity would be similar to that of the volatiles surrounding it. The hypothesis of entrainment of ejecta within a gas is very similar to models derived by volcanologists (Wilson, 1976) to explain how terrestrial volcanic bombs can travel greater distances from the vent than predicted by ballistic calculations. On Mars, the weak gravity field and low density atmosphere would favor this process.

This model is supported by experiments in which 3 mm pyrex projectiles traveling at 2.2 km/sec impacted a target of solid wax floating on liquid wax (J. Fink, pers. comm.) In one run, in which air bubbles were accidently trapped between the solid and liquid layers, a large column of material jetted from the crater center. Other experiments, with targets lacking air bubbles, did not produce the energetic central jet, suggesting that the presence of gases in the target may dramatically change ejection characteristics. Further experiments are clearly needed to evaluate the importance of these effects.

That a significant mass of ejecta would have sufficient energy to escape from the surface of Mars is indicated by calculated relations between the mass and velocity of impacting and ejected material (O'Keefe and Ahrens; 1977). For impact velocities of ~20 km/sec (Shoemaker, 1977) martian rocks with a total mass about 0.1 times the impacting meteorite mass would achieve escape velocity (Fig. 3). The ejecta mass would equal meteorite mass for impact velocities greater than ~30 km/sec, i.e., for comet impacts. Thus, experiments and calculations support the hypothesis that significant volumes of ejecta from impact into a volatile-rich target could escape Mars' gravitational field. The orbital evolution of such material has not been investigated. Presumably much of it would be reaccreted by Mars or ejected to the outer solar system, but some fraction would be perturbed by close approaches to Mars into eccentric orbits that would finally become Earth-crossing. Based on Wetherill's (1974) estimate that Apollo asteroids (already in Earth-crossing orbits) have lifetimes of 10^6–10^7 years before colliding with Earth; the time required for a random walk evolution of a small body initially near Mars into an Earth-crossing orbit is probably ~10^8 years (R. Greenberg, pers. comm. 1981), a period commensurate with the 165–187 m.y. shock ages of shergottites.

This concordance of orbital evolution and shock age is disturbed, however, by models accounting for diffusive resetting of Rb–Sr and Ar–Ar isochrons by the 165–187 m.y. shock event (Nyquist et al., 1979b; Bogard et al., 1979). According to these analyses, the shock event buried Shergotty (and presumably the other shergottites) under a few

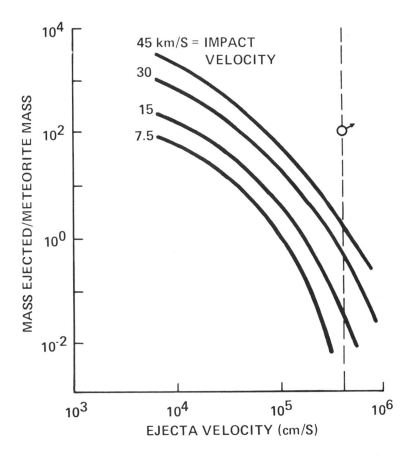

Fig. 3. Relations between velocity and mass of impacting meteorite and crater ejecta. For impact velocities greater than 15 km/sec, a significant quantity of Mars rocks will achieve escape velocity (5 km/sec). Graph from O'Keefe and Ahrens, 1977.

hundred meters of warm (300°–400°C) overburden for 10^3–40^4 years, during which time the diffusive resetting of ages occurred. Wasson and Wetherill (1979) pointed out that ejecta blocks hundreds of meters in diameter were unlikely to have escaped Mars and thus that serious difficulties beset the meteorites from Mars hypothesis. This objection is valid, but its assumptions may not be. The models quantifying ejecta dimension and isotopic resetting by shock are conceptually satisfying, but rely on parameters that are poorly known, and generally assume that severely shocked meteorite minerals behave similarly to unshocked materials for which diffusion rates, etc., are available. The models may or may not be correct, but they do not, in any case, affect the possible martian origin of SNC meteorites. We propose that the event at 165–187 m.y. did not necessarily eject material from the surface of Mars (although it may have if the diffusive models are incorrect); that may have been accomplished by a later impact, which left no shock signature. This proposal is consistent with (a) the lack of shock in nakhlites, and only minor shock recorded by chassignites, and (b) the observation that no meteorite known to us has equal shock and exposure ages. The significance of (b) is that exposure ages record the time since a meteorite fragmented from a larger body by a collision which did not engender a shock signature. For SNC's, exposure ages range between 2 and 11 m.y. (Table 1). Thus, we infer that the impact which launched igneous rocks from Mars occurred between 165 m.y. ago and 11 m.y. ago. An earlier date is more consistent with orbital evolution requirements.

Our conclusion that SNC meteorites are derived from Mars is apparently weakened by the lack of meteorites from the Moon, a much closer and gravitationally weaker world than Mars. The reason lunar meteorites are not found on Earth, however, may be related to the following observations: (1) Cratering on a large enough scale to eject material from the Moon has been statistically rare during the last few million years, and travel time to the Earth is only days to a few years (Hartung, 1981). These facts, and the observation that stony meteorites are rendered unrecognizable by erosion within a few tens to tens of thousands of years (Hartmann, 1972, p. 208) imply that lunar meteorites should rarely be found on the Earth. (2) The Moon would be less likely than Mars to yield meteorites if a volatile-rich target is required for the formation of unmelted, high-velocity ejecta.

MARTIAN EVOLUTION IMPLIED FROM SNC SAMPLES

However counter to intuition, Mars appears to be the most logical source for SNC meteorites. Assuming this to be true, SNC meteorites are Mars samples and thus offer the first detailed knowledge of that planet's history and evolution.

Model ages of ~4.6 b.y. (Shih *et al.*, 1981) for SNC samples show that by that time Mars had formed and undergone large scale differentiation that produced source regions with distinct Rb/Sr values. The early formation of Mars disagrees with gravitational accretion models of planetary formation which require more than 10^9 years for Mars to form (Weidenschilling, 1976), and also disagrees with the suggestion that Mars may have escaped early global differentiation (Hostetler and Drake, 1980). Planetary differentiation, resulting from accretional heating or core formation, may have produced a magma ocean, with subsequent formation of a low density crust—a remnant of which may have survived as the isostatically high, heavily cratered martian southern hemisphere.

The next event recorded by SNC samples is their crystallization 1.3 b.y. ago, supporting those crater-count calibrations (Hartmann, 1973; 1978; Soderblom *et al.*, 1974; Soderblom, 1977; Neukum and Hiller, 1981) which predict that volcanism extended into the last third of martian history. Initial Sr and Nd ratios at 1.3 b.y., and various chemical considerations require that four or more distinct source regions were producing magmas at 1.3 b.y., suggesting that a major episode of volcanism may have occurred at that time. Crater counts (based on the Soderblom and Hartmann chronologies) imply that the oldest lava flows on the periphery of Olympus Mons and the Tharsis shields are about 1.1 to 1.6 b.y. old (Schaber *et al.*, 1978), thus SNC meteorites may be samples of lavas from the onset of the Tharsis volcanic episode (Fig. 4). Comparison of these crater counts with a photogeologic map of volcanic units (Greeley and Spudis, 1981) suggests that no other region on Mars is young enough to be the source of SNC samples. About 165–187 m.y. ago an impact crater formed in this area of young martian volcanism, shocking the lavas sufficiently to convert plagioclase to maskelynite. This impact or perhaps a later one ejected the SNC samples off the surface of Mars.

Chemically, the most remarkable feature of Mars, as revealed by the SNC samples, is the Earth-like character of its igneous rocks. The similarities include minor and trace element abundances (McSween and Stolper, 1980), Rb/Sr and K/U ratios and REE patterns (Papanastassiou and Wasserburg, 1974), oxidation state (Delano and Arculus, 1980; Bunch and Keil, 1971), and thermomagnetic characteristics (Cisowski, 1981). These similarities must reflect similar patterns of activity and evolution for the upper mantles of the two planets, a remarkable conclusion, considering their different masses and tectonic styles. There are, however, differences which are consistent with our knowledge of Mars. Chassignites have a much higher Fe/Mg ratio than terrestrial olivine-rich rocks (Floran *et al.*, 1978), consistent with the high estimates (3.3–3.6 g/cm^3) for the zero-pressure density of the martian mantle (Arvidson *et al.*, 1980). The Fe-rich magmas produced igneous rocks that are more mafic than common on Earth, with the resulting very long lava flows.

The K/U ratio is an important quantity for understanding the thermal history of a planet (e.g., Toksöz and Hsui, 1978; Morgan and Anders, 1979). For Mars the SNC samples imply K/U values from $\sim 8 \times 10^3$ to $\sim 3 \times 10^4$ (McSween *et al.*, 1979), considerably higher than the ratio (2.2×10^3) used in Morgan and Anders' (1979) calculation of the

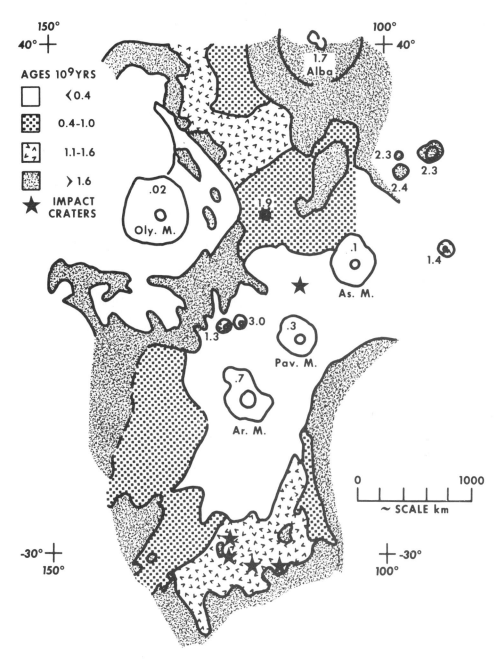

Fig. 4. Chronological map of volcanic units in the Tharsis region of Mars; compiled from crater count data by Schaber *et al.*, 1978 and Plescia and Saunders, 1979. SNC meteorites may be samples of oldest volcanic unit (1.1 to 1.6 b.y.) which is exposed on margins of region and presumably underlies younger units.

chemical composition of Mars, but comparable to values (2.5×10^4) assumed by Toksöz and Hsui (1978). Measurements of K/U by the Soviet Mars 5 spacecraft have been analyzed and calibrated repeatedly; the latest value of 6.7×10^3, with an uncertainty of about 50% (Surkov *et al.*, 1981), approaches the SNC ratio. The earth-like K/U ratio of the SNC samples implies that Mars has a volatile abundance more like the Earth than the Moon.

Currently Mars has a very weak to non-existent magnetic field (Russell, 1979), and paleo-intensity measurements on SNC meteorites (Nagata, 1981; Cisowski, 1981) provide no evidence for a field 1.3 b.y. ago. These data suggest that Mars does not have a convecting liquid core and has not had one during the last third of its history, and support thermal models such as that of Toksöz and Hsui (1978) in which core formation occurred in the first billion years of martian evolution, and not in the last billion years (e.g., Solomon, 1979).

CONCLUSIONS

The chemical and physical characteristics of SNC meteorites argue strongly against their origin in association with the eucrites, the most closely related class of meteorites. Additionally, plausible thermal models of asteroid evolution preclude igneous activity due to internal processes, and impact derived melt sheets on asteroids are not likely to yield the cumulate textures and isotopic variability of SNC meteorites. Instead, SNC meteorites were derived from an Earth-like parent body which we believe to be Mars. This conclusion, based on compilation of all available evidence, is sufficiently intriguing and important (especially since documented sample return from Mars is unlikely in the next decade) that the possible mechanisms of transport of materials from Mars to Earth should be seriously investigated. More specifically, the role of target gases and volatiles in producing unmelted ejecta from impacts on large planetary objects should be thoroughly evaluated through both theory and experiment. Calculations of the yield and travel time of Earth-crossing ejecta from Mars should also be carried out.

The prospect that we may have martian samples on Earth justifies an intensive program of SNC analysis to provide as many constraints as possible regarding the origin and evolution of Mars. Petrologic and geochemical work may define the depths of origin of the melts from which the SNC samples crystallized. Much can be learned of the SNC source regions by experimental petrological investigations at high pressures appropriate to the martian upper mantle (10–25 kbar). Further stable and radiogenic isotopic studies may clarify the relations between individual SNC meteorites, and the complex histories they seem to have experienced.

The possibility that other meteorites in terrestrial collections may have come from Mars should not be overlooked. Samples of other regions of Mars may also have been ejected into Earth-crossing orbits. Recognition and study of samples from the geologically diverse terrains of Mars are crucial to the development of a broad understanding of its evolution.

Acknowledgments—We would like to thank our colleagues at JSC for explaining isotopic systematics to us: Larry Nyquist, Joe Wooden, Don Bogard, and Chi-Yu Shih. Richard Greenberg and George Wetherill offered insights into orbital evolution, Jon Fink and Robert Clayton provided unpublished information on cratering and oxygen isotopes, respectively. Fred Hörz tutored us on shock effects, Jeff Warner was vocal in his support, and Elinor Stockton rapidly and accurately typed the manuscript when there was little time. We also thank John Wood, Richard Greenberg, and Fred Hörz for critical reviews. This paper is Lunar and Planetary Institute Contribution No. C-454. The Institute is operated by the Universities Space Research Association under Contract No. NASW-3389 with the National Aeronautics and Space Administration.

REFERENCES

Adams J. B. and McCord T. B. (1977) Mercury: Evidence for an anorthositic crust from reflectance spectra (abstract). *Bull. Amer. Astron. Soc.* **9**, 457.

Arvidson R. E., Goettel K. A., and Hohenberg C. M. (1980) A post-Viking view of martian geologic evolution. *Rev. Geophys. Space Phys.* **18**, 565–603.

Baird A. K. and Clark B. C. (1981) On the original igneous source of martian fines. *Icarus* **45**, 113–123.

Barsukov V. L., Volkov V. P., and Khodakovsky I. L. (1980) The mineral composition of Venus surface rocks: a preliminary prediction. *Proc. Lunar Planet. Sci. Conf. 11th*, 765–773.
Berkley J. L., Keil K., and Prinz M. (1980) Comparative petrology and origin of Governador Valadares and other nakhlites. *Proc. Lunar Planet. Sci. Conf. 11th*, 1089–1102.
Binns R. A. (1967) Stony meteorites bearing maskelynite. *Nature* **214**, 1111–1112.
Boctor N. W., Meyer H. O., and Kullerud G. (1976) Lafayette meteorite: Petrology and opaque mineralogy. *Earth Planet. Sci. Lett.* **32**, 69–76.
Bogard D. D. and Husain L. (1977) A new 1.3 aeon-young achondrite. *Geophys. Res. Lett.* **4**, 69–71.
Bogard D. D., Husain L., and Nyquist L. E. (1979) ^{40}Ar–^{39}Ar age of the Shergotty achondrite and implications for its post-shock thermal history. *Geochim. Cosmochim. Acta* **43**, 1047–1055.
Bogard D. D. and Nyquist L. E. (1979) Ar-39/Ar-40 chronology of related achondrites (abstract). *Meteoritics* **14**, 356.
Bottinga Y. and Weill D. F. (1970) Densities of liquid silicate systems calculated from partial molar volumes of oxide components. *Amer. J. Sci.* **269**, 169–182.
Bottinga Y. and Weill D. F. (1972) The viscosity of magmatic silicate liquids: A model for calculation. *Amer. J. Sci.* **272**, 438–475.
Boynton W. V., Starzyk P. M., and Schmitt R. A. (1976) Chemical evidence for the genesis of the ureilites, the achondrite Chassigny and the nakhlites. *Geochim. Cosmochim. Acta* **40**, 1439–1447.
Bunch T. E. and Keil K. (1971) Chromite and ilmenite in non-chondritic meteorites. *Amer. Mineral.* **56**, 146–157.
Bunch T. E. and Reid A. M. (1975) The nakhlites—I. Petrography and mineral chemistry. *Meteoritics* **10**, 303–315.
Burragato F., Cavarretta G., and Funiciello R. (1975) The new Brazilian achondrite of Governador Valadares (Minas Gerais). *Meteoritics* **10**, 374–375.
Carr M. H. and Schaber G. G. (1977) Martian permafrost features. *J. Geophys. Res.* **82**, 4039–4054.
Cisowski S. M. (1981) Magnetic properties of Shergotty and Zagami meteorites (abstract). In *Lunar and Planetary Science XII*, p. 147. Lunar and Planetary Institute, Houston.
Clayton R. N., Onuma N., and Mayeda T. K. (1976) A classification of meteorites based on oxygen isotopes. *Earth Planet. Sci. Lett.* **30**, 10–18.
Consolmagno G. and Drake M. J. (1977) Composition and evolution of the eucrite parent body: Evidence from rare earth elements. *Geochim. Cosmochim. Acta* **41**, 1271–1282.
Delano J. W. and Arculus R. J. (1980) Nakhla: Oxidation state and other constraints (abstract). In *Lunar and Planetary Science XI*, p. 219–221. Lunar and Planetary Institute, Houston.
Drake M. J. (1979) Geochemical evolution of the eucrite parent body: Possible nature and evolution of asteroid 4 Vesta? In *Asteroids* (T. Gehrels, ed.), p. 765–782. Univ. Arizona, Tucson.
Duke M. B. and Silver L. T. (1967) Petrology of eucrites, howardites, and mesosiderites. *Geochim. Cosmochim. Acta* **37**, 1637–1665.
Farmer C. B., Davies D. W., Holland A. L., LaPorte D. D., and Doms P. E. (1977) Mars: Water vapor observations from the Viking Orbiters. *J. Geophys. Res.* **82**, 4225–4248.
Feierberg M. A. and Drake M. J. (1980) The meteorite-asteroid connection: The infrared spectra of eucrites, shergottites, and Vesta. *Science* **209**, 805–807.
Floran R. J., Prinz M., Hlava P. F., Keil K., Nehru C. E., and Hinthorne J. R. (1978) The Chassigny meteorite: A cumulate dunite with hydrous amphibole-bearing melt inclusions. *Geochim. Cosmochim. Acta* **42**, 1213–1229.
Fuse K. and Anders E. (1969) Aluminum-26 in meteorites—VI. Achondrites. *Geochim. Cosmochim. Acta* **33**, 653–670.
Gale N. H., Arden J. W., and Hutchison R. (1975) The chronology of the Nakhla achondrite meteorite. *Earth Planet. Sci. Lett.* **26**, 195–206.
Ganapathy R. and Anders E. (1969) Ages of calcium-rich achondrites, II. Howardites, nakhlites and the Angra dos Reis angrite. *Geochim. Cosmochim. Acta* **33**, 775–787.
Greeley R. and Spudis, P. D. (1981) Volcanism on Mars. *Rev. Geophys. Space Phys.* **19**, 13–41.
Hartmann W. K. (1972) *Moons and Planets*. Bogden and Quigley, Tarrytown-on-Hudson. 404 pp.
Hartmann W. K. (1973) Martian cratering 4: Mariner 9 initial analysis of cratering chronology. *J. Geophys. Res.* **78**, 4096–4116.
Hartmann W. K. (1978) Martian cratering V: Toward an empirical martian chronology, and its implications. *Geophys. Res. Lett.* **5**, 450–452.
Hartung J. B. (1981) On the occurrence of Giordano Bruno ejecta on the Earth (abstract). In *Lunar and Planetary Science XII*, p. 401–403. Lunar and Planetary Institute, Houston.
Heymann D., Mazor E., and Anders E. (1968) Ages of calcium-rich achondrites—I. Eucrites. *Geochim. Cosmochim. Acta* **32**, 1241–1268.
Hostetler C. J. and Drake M. J. (1980) On the early global melting of the terrestrial planets. *Proc. Lunar Planet. Sci. Conf. 11th*, 1915–1929.
Johnson J. E., Scrymgour J. M., Jarosewich E., and Mason B. (1977) Brachina meteorite—a chassignite from South Australia. *Rec. S. Aust. Mus.* **17**, 309–319.
Kirsten T., Ries D., and Fireman E. L. (1978) Exposure and terrestrial ages of four Allan Hills Antarctic meteorites. *Meteoritics* **13**, 519–522.
Lancet M. S. and Lancet K. (1971) Cosmic-ray and gas-retention ages of the Chassigny meteorite. *Meteoritics* **6**, 81–86.

Ma M.-S., Laul J. C., and Schmitt R. A. (1981) Analogous and complementary rare earth element patterns on meteorite parent bodies and the Earth inferred from a study of the achondrite ALHA 77005, other unique achondrites, and the shergottites (abstract). In *Lunar and Planetary Science XII*, 634–636. Lunar and Planetary Institute, Houston.

Mason B., Nelen J. A., Nuir P., and Taylor S. R. (1975) The composition of the Chassigny meteorite. *Meteoritics* 11, 21–27.

McCarthy T. S., Erlank A. J., Willis J. P., and Ahrens L. H. (1974) New chemical analyses of six achondrites and one chondrite. *Meteoritics* 9, 215–222.

McGetchin T. R. and Smyth J. R. (1978) The mantle of Mars: Some possible geologic implications of its high density. *Icarus* 34, 512–536.

McSween H. Y. Jr. and Stolper E. (1980) Basaltic meteorites. *Sci. Amer.* 242, no. 6, 54–63.

McSween H. Y. Jr., Stolper E. M., Taylor L. A., Muntean R. A., O'Kelley G. D., Eldridge J. S., Swarajranjan B., Ngo H. T., and Lipschutz M. E. (1979) Petrogenetic relationship between Allan Hills 77005 and other achondrites. *Earth Planet. Sci. Lett.* 45, 275–284.

Milton D. J. (1973) Water and processes of degradation in the martian landscape. *J. Geophys. Res.* 78, 4037–4047.

Minear J. W., Clow G., and Fletcher C. R. (1979) Thermal models of asteroids (abstract). In *Lunar and Planetary Science X*, p. 842–843. Lunar and Planetary Institute, Houston.

Morgan J. W. and Anders E. (1979) Chemical composition of Mars. *Geochim. Cosmochim. Acta* 43, 1601–1610.

Murase T. and McBirney A. R. (1973) Properties of some common igneous rocks and their melts at high temperature. *Geol. Soc. Amer. Bull.* 84, 3563–3592.

Nagata T. (1979) Paleomagnetism of stony meteorites. *Memoirs of the National Institute of Polar Research, Special Issue No. 5* (T. Nagata, ed.), p. 280–293. Nat'l Inst. Polar Res., Tokyo.

Nagata T. (1981) Paleointensity of Antarctic achondrites (abstract). In *Papers Presented to the Sixth Symposium on Antarctic Meteorites*, p. 78–79. Nat'l. Inst. Polar Res., Tokyo.

Nakamura N. and Masuda A. (1973) Chondrites with peculiar rare-earth patterns. *Earth Planet. Sci. Lett.* 19, 429–437.

Nakamura N., Unruh D. M., Tatsumoto M., and Hutchison R. (1977) Nakhla: Further evidence for a young crystallization age (abstract). *Meteoritics* 12, 324–325.

Nehru E. C., Prinz M., and Zucker S. M. (1979) Brachina: Origin, melt inclusions and relationship to Chassigny (abstract). *Meteoritics* 14, 493–494.

Neukum G. and Hiller K. (1981) Martian ages. *J. Geophys. Res.* 86, 3097–3121.

Nyquist L. E., Bogard D. D., Wooden J. L., Wiesmann H., Shih C.-Y., Bansal B. M., and McKay (1979a) Early differentiation, late magmatism, and recent bombardment on the shergottite parent planet (abstract). *Meteoritics* 14, 502.

Nyquist L. E., Wooden J., Bansal B., Wiesmann H., McKay G., and Bogard D. D. (1979b) Rb–Sr age of Shergotty achondrite and implications for metamorphic resetting of isochron ages. *Geochim. Cosmochim. Acta* 43, 1057–1074.

O'Keefe J. D. and Ahrens T. J. (1977) Meteorite impact ejecta: Dependence of mass and energy lost on planetary escape velocity. *Science* 198, 1249–1251.

Oyama V. I., Carle G. C., Woeller F., Pollack J. B., Reynolds R. T., and Craig R. A. (1980) Pioneer Venus gas chromatography of the lower atmosphere of Venus. *J. Geophys. Res.* 85, 7891–7902.

Papanastassiou D. A. and Wasserburg G. J. (1974) Evidence for late formation and young metamorphism in the achondrite Nakhla. *Geophys. Res. Lett.* 1, 23–26.

Plescia J. B. and Saunders R. S. (1979) The chronology of the martian volcanoes. *Proc. Lunar Planet. Sci. Conf. 10th*, 2841–2859.

Podosek F. A. (1973) Thermal history of the nakhlites by the $^{40}Ar-^{39}Ar$ method. *Earth Planet. Sci. Lett.* 19, 135–144.

Reid A. M. (1981) EETA 79001 Petrographic description. *Antarctic Meteorite Newsletter Vol. 4, No. 1*. Code SN2, NASA Johnson Space Center, Houston.

Reid A. M. and Bunch T. E. (1975) The nakhlites—II. Where, when, and how? *Meteoritics* 10, 317–324.

Russell C. T. (1979) Planetary magnetism. *Rev. Geophys. Space Phys.* 17, 295–301.

Schaal R. B. and Hörz F. (1977) Shock metamorphism of lunar and terrestrial basalts. *Proc. Lunar Sci. Conf 8th*, 1697–1729.

Schaber G. G., Horstman K. C., and Dial A. L. Jr. (1978) Lava flow materials in the Tharsis region of Mars. *Proc. Lunar Planet. Sci. Conf. 9th*, p. 3433–3458.

Sears D. W. (1978) *The Nature and Origin of Meteorites*. Oxford Univ., London. 187 pp.

Shih C.-Y., Nyquist L. E., Bansal B., Wiesmann H., Wooden J. L., and McKay G. (1981) REE, Sr and Nd isotopic studies on shocked achondrites—Shergotty, Zagami and ALHA 77005 (abstract). In *Lunar and Planetary Science XII*, p. 973–975. Lunar and Planetary Institute, Houston.

Shoemaker E. M. (1977) Astronomically observable crater-forming projectiles. In *Impact and Explosion Cratering* (D. J. Roddy, R. O. Pepin, and R. B. Merrill, eds.), p. 617–628. Pergamon, N.Y.

Simonds C. H., Warner J. L., and Phinney W. C. (1976) Thermal regimes in cratered terrain with emphasis on the role of impact melt. *Amer. Mineral.* 61, 569–577.

Singer R. B. (1981) The composition of martian dark regions: II. Analysis of telescopically observed absorptions in near-infrared spectrophotometry. *J. Geophys. Res.* In press.

Smith J. V. and Hervig R. L. (1979) Shergotty meteorite: Mineralogy, petrography and minor elements. *Meteoritics* **14**, 121–142.

Soderblom L. A. (1977) Historical variations in the density and distribution of impacting debris in the inner solar system: Evidence from planetary imaging. In *Impact and Explosion Cratering* (D. J. Roddy, R. O. Pepin, and R. B. Merrill, eds.), p. 629–633. Pergamon, N.Y.

Soderblom L. A., Condit C. D., West R. A., Herman B. M., and Kreidler T. J. (1974) Martian planetwide crater distributions: Implications for geologic history and surface processes. *Icarus* **22**, 239–263.

Solomon S. C. (1979) Formation, history and energetics of cores in the terrestrial planets. *Phys. Earth Planet. Inter.* **19**, 168–182.

Stolper E. (1977) Experimental petrology of eucrite meteorites. *Geochim. Cosmochim. Acta* **41**, 587–611.

Stolper E. and McSween H. Y. Jr. (1979) Petrology and origin of the shergottite meteorites. *Geochim. Cosmochim. Acta* **43**, 1475–1498.

Stolper E., McSween H. Y. Jr., and Hays J. F. (1979) A petrogenetic model of the relationships among achondritic meteorites. *Geochim. Cosmochim. Acta* **43**, 589–602.

Surkov Yu. A., Moskaleva L. P., Manvelyan O. S., and Kharyukova V. P. (1981) Analysis of the gamma radiation of martian rocks based on the data of the Mars 5 spacecraft. *Cosmic Res. (Engl. Transl.)* **18**, 453–460.

Toksöz M. N. and Hsui A. T. (1978) Thermal history and evolution of Mars. *Icarus* **34**, 537–547.

Urey H. C. (1959) Primary and secondary objects. *J. Geophys. Res.* **64**, 1721–1737.

Walker D., Stolper E. M., and Hays J. F. (1979) Basaltic volcanism: The importance of planet size. *Proc. Lunar Planet. Sci. Conf. 10th*, 1995–2015.

Wänke H. (1968) Radiogenic and cosmic-ray exposure ages of meteorites, their orbits and parent bodies. In *Origin and Distribution of the Elements* (L. H. Ahrens, ed.), p. 411–421. Pergamon, Oxford.

Wasson J. T. and Wetherill G. W. (1979) Dynamical, chemical and isotopic evidence regarding the formation locations of asteroids and meteorites. In *Asteroids* (T. Gehrels, ed.), p. 926–974. Univ. Arizona, Tucson.

Weidenschilling S. J. (1976) Accretion of the terrestrial planets, II. *Icarus* **27**, 161–170.

Wetherill G. W. (1974) Solar system sources of meteorites and large meteoroids. *Ann. Rev. Earth Planet. Sci.* **2**, 303–331.

Wetherill G. W. (1978) Dynamical evidence regarding the relationship between asteroids and meteorites. In *Asteroids: An Exploration Assessment* (D. Morrison and W. Wells, eds.), p. 17–35. NASA 2053.

Wilson L. (1976) Explosive volcanic eruptions—III. Plinian eruption columns. *Geophys. J. Roy. Astron. Soc.* **45**, 543–556.

Wood J. A., Dickey J. S., Marvin U. B., and Powell B. N. (1970) Lunar anorthosites and a geophysical model of the moon. *Proc. Apollo 11 Lunar Sci. Conf.*, p. 965–988.

Wooden J. L., Nyquist L. E., Bogard D. D., Bansal B. M., Wiesmann H., Shih C.-Y., and McKay G. A. (1979) Radiometric ages for the achondrites Chervony Kut, Governador Valadares, and Allan Hills 77005 (abstract). In *Lunar and Planetary Science X*, p. 1379–1381. Lunar and Planetary Institute, Houston.

Natural radioactivity of the moon and planets

Yu. A. Surkov

Vernadsky Institute of Geochemistry and Analytical Chemistry,
Academy of Science of the USSR, V-334, Moscow

Abstract—In this report the main results of the study of natural radioactivity of the solar system bodies are considered. The radioactivity of the moon and planets was measured from orbiters and landers. The radioactivity of the returned lunar samples was studied with laboratory equipment. Analysis of the radioactivity data shows the bimodal structure of surfaces of the moon, Venus, Mars (ancient crust and young volcanic formations). Volcanic formations on all bodies, probably, consist of basaltic rocks. The compositions of ancient crusts are different (gabbro-anorthositic on the moon and maybe on Mars, granite-metamorphic on the earth and maybe on Venus).

1. INTRODUCTION

As is well known, the studies of radioactivity are of great importance for developing ideas on the moon, planets and interplanetary space.

Natural radioactivity characterizes the content of natural radioelements, U, Th and K, and thus shows the character of the rock and its origin. Radioactivity due to cosmogenic radionuclides characterizes the processes which took place on the surface of the body and in its periphery during 10 to 100 million years.

The possibility of solving a wide spectrum of scientific problems has given rise to extensive studies of the radioactivity of extraterrestrial materials both remotely from space vehicles and directly by laboratory methods through investigations of samples brought to the earth (Vinogradov et al., 1967; Trombka et al., 1975; Surkov, 1977a).

Table 1 lists space probes from which the radioactivity of certain bodies of the solar system was studied. Fifteen vehicles have explored cosmogenic and natural radioactivity of the moon and planets.

The most extensive information now available relates to the moon. The radioactivity of lunar rock was measured in situ from vehicles and in lunar samples brought to the earth.

2. EXPLORATION OF THE MOON

The first measurements of lunar radioactivity were taken from moon satellites Luna 10 and 12 in 1966 (Vinogradov et al., 1967). The satellites were entered. In elliptical orbits with perigels of 350 km (Luna 10) and 100 km (Luna 12).

Although measurements were taken from high orbits (so that in each case information was averaged for a vast area), attention was drawn to certain trends in the changes of the radioactivity between lunar maria and highlands. For instance, the area with the highest radioactivity was recorded in the vicinity of Western maria. The high radioactivity was not detected in Eastern maria. As can be seen from the results of measurements in highland regions of the moon, the content of natural radioelements similar to terrestrial granites, has not been detected anywhere. Moreover, above the highlands, the flux of gamma-quanta of natural radioelements proved to be lower than that above maria. This enabled the authors of the experiment (Vinogradov et al., 1967) to suggest that highland rocks on the moon must be slightly differentiated or not differentiated at all.

In 1971 and 1972 the moon's gamma-radiation was measured from orbital modules of Apollo 15 and 16. Measurements were performed from circular orbits at the distances of

Table 1. Space vehicles and scientific instruments used for studying the radioactivity of the moon and planets.

No	Year of Launching	Space Vehicles	Region of Measurements	Scientific Instrument
1	1966	Luna-10	Lat. ±72°	32-channel spectrometer
2	1966	Luna-12	Lat. ±20°	32-channel spectrometer
3	1971	Apollo-15	Lat. ±29°	direct transfer
4	1972	Apollo-16	Lat. ±10°	direct transfer
5	1972	Venera-8	Lat. −10° Long. 304°	64-channel spectrometer
6	1974	Mars-5	Lat. −50+20° Long. 0–150°	256-channel spectrometer
7	1975	Venera-9	Lat. 32° Long. 291°	64-channel spectrometer
8	1975	Venera-10	Lat. 16° Long. 291°	64-channel spectrometer
9	1969	Apollo-11	Mare Tranquillitatis	
10	1970	Luna-16	Mare Fecunditatis	
11	1970	Apollo-12	Oceans Procellarum	Laboratory
12	1971	Apollo-14	Crater Fra Mauro	studies
13	1971	Apollo-15	Hadley Appennines	of lunar
14	1972	Luna-20	Between Maria Fecunditatis and Crisium	samples
15	1972	Apollo-16	Cayley-Descartes	
16	1972	Apollo-17	Taurus Littrow	
17	1976	Luna-24	Mare Crisium	

100 to 120 km from the surface. The moon's equatorial region was investigated, which accounts for 20% of the entire lunar surface (Trombka et al., 1975).

The measurements of the radioactivity level in different lunar regions make it possible to define a specific structural area, especially within a region of high radioactivity.

The Eastern maria of the near side of the moon have generally shown low radioactivity, but in this area also occurs the region with the highest radioactivity. The highest radioactivity as compared to the surrounding continental area was observed in the region of Mare Crisium, the lowest in Mare Serenitatis.

The highlands on both sides of the moon have shown low radioactivity. The radioactivity maximum was recorded near the Van de Graaf crater where, as it is known, a significant magnetic anomaly has also been detected.

On the basis of uranium, thorium and potassium contents, K/U and K/Th ratios were determined; K/U turned out to be low for all regions of the lunar surface (about 2.500) as compared to the ratios known for terrestrial rocks (10.000). The K/U are somewhat higher for the Western maria than for Eastern ones.

It was noted that the radioactivity data correlate with many other properties of the lunar material which are known for different surface areas. The gamma-radiation correlates with the chemical composition data, with lunar surface profile, with magnitude of the magnetic fields and so on.

The radioactivity of lunar soil samples brought to the earth has been studied most thoroughly. At present, data for nine surface areas have been accumulated and analyzed. The comparison of the availability of radioelements in the samples of rocks, breccias and thin regolith fractions indicates a rather broad range of potassium, uranium and thorium contents for different samples brought to the Earth by Apollo and Luna missions. The bulk of investigated samples have the following concentrations of radioelements: potassium from 0.03 to 0.7% uranium from 0.1 to 5 p.p.m., thorium 0.4 to 18 p.p.m., with K/U ratios from 1.10^3 to 5.10^3 (Surkov and Fedoseyev, 1975). At the same time, in the case of isolated samples the figures lie considerably beyond these limits.

Figure 1 shows the contour map of lunar surface. The areas are marked in this map in which measurements of the rock radioactivity were carried out from orbiters or lunar samples were collected by "Apollo" and "Luna" missions. Gamma-flow from the surface of the moon can be considered (with a certain approach) proportional to concentration of natural radioelements and divided into four groups in accordance with its intensity. One can remark that in spite of the essential differences (at least one order magnitude) in the concentration of radioelements in isolated lunar samples brought to the earth, the flow density of gamma-quanta changes only factor by 1.5. It can be explained by two reasons: the first one is small variation of the average concentration of natural radioelements for different areas of surface, the second one is the fact that the main contribution in gamma-ray flow of the moon comes from cosmogenic radionuclides. (This flow depends on the composition of rocks very weakly).

One can see in this map that the highest radioactivity is observed in the Western basin of lunar maria and the lowest radioactivity on highlands of nearside and farside of the moon. Now such distribution of radioactivity became understandable especially after lunar samples were studied at the earth. It is explained by the existence of two essentially different materials of the crust: highland material of gabbro-anorthositic composition and mare material of basaltic composition.

The lunar highlands consist of gabbro-anorthositic rocks in which the predominant mineral is feldspar. It is known that these rocks occur on the ancient cratered surface. Their age is 4.0–4.5 b.y. They compose the ancient primary crust formed in the beginning of the moon's existence. This crust differentiated as a result of the heating of upper thickness of the moon during the last stage of accretion. The extraction of light refractory components to the surface and their subsequent solidification led to the formation of a feldspar crust of anorthosite-norite-tractolitic composition.

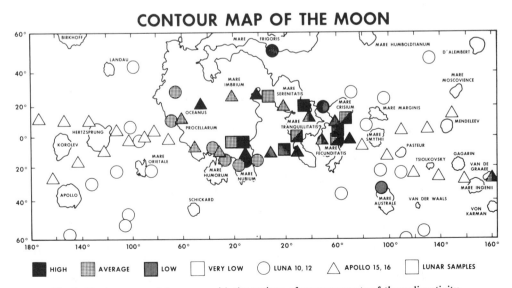

Fig. 1. Contour map of the moon with the regions of measurements of the radioactivity.

Lunar maria are covered with basaltic rocks of 3.2–3.8 b.y. old. These rocks formed later than highland rocks as a result of the radiogenic heating of the interior. Impacts of large meteorites, which perforated the thin feldspar crust, opened the way for extrusion on the surface of more differentiated basaltic rocks.

The existence of two various crusts of different nature (highland and mare crusts) on the moon is reflected quite accurately on the map of radioactivity where one can see mare regions with high radioactivity (that is, high concentration of radioelements) and highlands with low radioactivity (that is, low concentration of radioelements). The isolated regions with average level of the radioactivity probably show the mixture of the highland and mare rocks or eruption of deep highland rocks enriched in radioelements (such as Van de Graaf and another).

By comparing the amounts of radioelements on the moon and on the earth, the following points are to be noted. The bulk concentration of radioelements on the moon is somewhat higher than on the earth and the average concentration of radioelements within the lunar crust is lower than in the earth's crust. This fact indicates that the differentiation processes of the matter on the moon were not as intensive as on the earth, the result being that the lunar crust enrichment with radioelements is lower than in the case of the earth.

3. STUDIES OF THE RADIOACTIVITY OF VENUS ROCKS

In contrast to the moon, Venus has a dense atmosphere which rules out the possibility of studying the radioactivity of its rocks from orbital vehicles. That is why the radioactivity was measured by landing capsules of the Venera 8, 9 and 10 probes (Surkov, 1977b). In these experiments the presence of the Venusian atmosphere played a positive part since it absorbs cosmic rays and reduces the background of gamma-radiation induced by them.

Figure 2 presents the map of Venusian surface obtained by the Pioneer-Venus probe (Pettengill et al., 1980). The landing sites of the Soviet Venus probes are indicated on the map by numbers. Because of the low resolution of this radar-map one can speak today only about the global characteristics of the relief topography. For instance, Pettengill et al. (1980) sees three main types of provinces on the Venusian surface: 1) upland rolling plains located at the level of 0 to 2 km above an average level of the surface (they form 65% of the investigated surface), 2) highlands located at the level of 2 to 11 km above an average level of the surface (they form $\sim 8\%$) and 3) lowlands located at the level of 0 to -2 km below an average level of the surface (they form $\sim 27\%$).

Proceeding from this classification the Venera 8 landing site is located in the area of the ancient surface (upland rolling plain) covered with craters differing in size. This seems to be the ancient crust of the planet. The Venera 9 and 10 landing sites are situated on the slopes of the Beta plateau which seems to be a relatively young region of shield basalt volcanism.

Table 2 summarizes the uranium, thorium and potassium content measured from Veneras 8, 9 and 10. For the sake of comparison, data for terrestrial rocks (granite and basalt) are also given. As evident from the table, radioelement contents in Venus rocks differ from each other, but are close to the rocks of the Earth's crust. Different areas of the Venusian surface differ by no more than a factor of 4–5 in their uranium content and by a factor of 10 in their thorium and potassium contents. At the Venera 8 landing site the rock has the highest content of natural radioelements which exceeds the average level of their content in terrestrial basalts and is close to granite. Perhaps this ancient area of the Venusian surface, where Venera 8 landed, resembles in its character the continental part of the earth's crust. Data obtained from Veneras 9 and 10 permit us to classify the rocks in their landing sites as transitional between tholeitic to alkaline basalts. Probably these volcanic rocks are younger in comparison to the rock in the Venera 8 site.

The density of the rocks (2.7–2.9 g/cm^3) measured by Venera 9 and 10 also indicates that their composition is basaltic (Surkov et al., 1976a). By terrestrial analogy, this density corresponds to basalts of massive texture and low porosity. Such rocks could form during the slow cooling of basaltic lavas with low gas release.

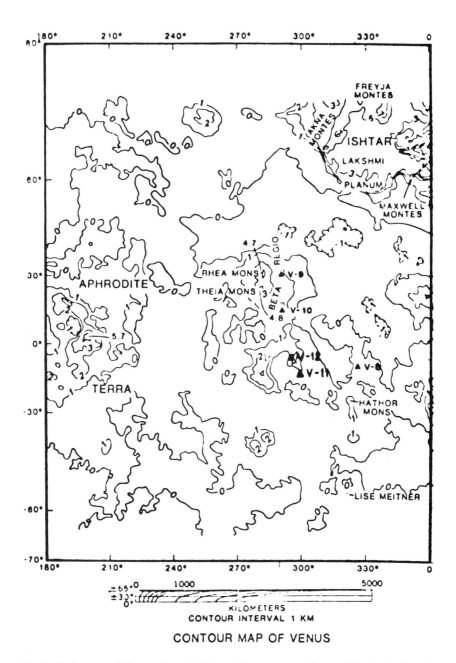

Fig. 2. Radar map of Venus taken by Pioneer-Venus spacecraft on which the landing sites of space probes "Veneras" are shown.

Thus, we see two different types of crust on Venus as well as on the moon and the earth: ancient strong cratered crust which is more like continental granite crust of the earth than gabbro-anorthositic highlands of the moon, as can be seen from the content of radioelements measured by Venera 8, and younger weakly-cratered massifs located in lowlands or depressions on rolling plains and highlands.

The K/U ratio for the Venus rocks studied is very close to the ratio typical for the bulk of the earth's magmatic rocks (about 10^4) although Venus rocks are somewhat depleted in potassium in comparison with Earth rocks. This fact indicates that if the processes which

Table 2. K, U and Th content in the rocks of Venus and the earth.

Rocks		Content		
		Potassium, %	Uranium, p.p.m.	Thorium, p.p.m.
Venus rocks:	Venera 8	4.0 ± 1.2	2.2 ± 0.7	6.5 ± 0.2
	Venera 9	0.47 ± 0.08	0.60 ± 0.16	3.65 ± 0.42
	Venera 10	0.30 ± 0.16	0.46 ± 0.26	0.70 ± 0.34
Earth's rocks:	Basalt	0.76	0.86	2.1
	Granite	3.24	9.04	21.9

differentiated the chemical composition of primary material took place on Venus (no matter what their nature), they were mainly similar to terrestrial processes.

Data on the character of Venusian rocks obtained by Veneras 8, 9 and 10 are in agreement with the concepts which follow from the results of investigating the Venusian atmosphere. Apparently the evolutionary process on Venus has been similar to that of the earth. It seems that, like the earth, in the remote past the Venusian interior was heated and differentiated. At present one can hardly say how far this process has gone, but the discovery of rocks greatly enriched with uranium, thorium and potassium on the surface of Venus implies that they were generated by melting the primary material in the planet's interior. The great amount of carbon dioxide in the Venusian atmosphere indicates that the eruption of magnetic rocks into the surface was either completed relatively recently or has been continuing.

4. STUDIES OF THE RADIOACTIVITY OF MARTIAN ROCKS

The first and so far the only measurements of Martian gamma-radiation were conducted from Mars 5 in 1974 (Surkov et al., 1976b). Measurements were taken from the distance of about 2,000 km from planetary surface. Therefore, the measured spectra are average for the vast territory of the Martian surface. This territory is shown in Fig. 3. (Gamma-radiation emitted from the surface, delineated by the contour, made up 90 per cent of the entire recorded gamma-radiation).

After processing the gamma-spectra measured, the average content of natural radio-elements and some rock-forming elements was determined. The data obtained are summarized in Table 3. For the sake of comparison the data on the composition of Martian rocks obtained from Viking 1 and 2 by the X-ray-fluorescence method (Toulmin et al., 1976) are also given. Regrettably, it is difficult to compare these data since the results of gamma-spectrometric measurements are the average characteristic of the vast region of the Martian surface, while X-ray-fluorescence data relate to individual points of the surface where Viking 1 and 2 landed. However, one conclusion can be drawn confidently. As far as the content of natural radioelements is concerned, Martian rock is close to eruptive rocks of the earth's crust of the basic composition. The data of Viking 1 and 2 on the content of the main rock-forming elements also seem to speak in favour of this similarity.

On the territory over which measurements were taken, the observed diversity of the forms of the relief and of the types of rocks can be reduced (with a certain amount of generalization) to two principal provinces: ancient terra formations and younger volcanic formations. Other formations observed on this territory, for instance, valleys of the fluvial type and fields of eolian dunes, cover a rather small part of the area, and their

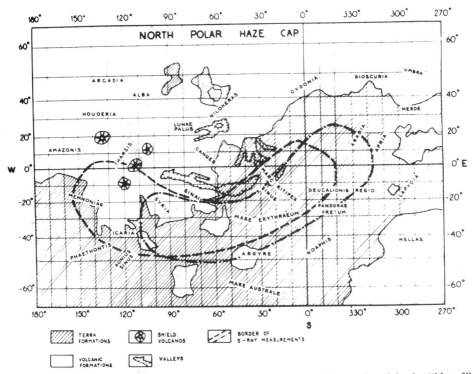

Fig. 3. Contour map of Mars with the region of measurements of the radioactivity by "Mars-5" orbiter.

contribution to recorded gamma-radiation can apparently be ignored. These two principal provinces are also shown in Fig. 3. (Basilevsky et al., 1981).

The presence of a large number of big craters is typical of Martian terra formations, and in this respect the terra formations of Mars resemble lunar highlands. Volcanic formations on the territory under investigation are represented by the Tharsis plateau, including the Arsia shield volcano, and by lowlands relating to the floors of depressions, for example, the Argyre basin floor.

The uranium and thorium concentration in the two different formations of Mars as well as of the earth, Venus and the moon are summarized in Table 4. It follows from these data that as far as the thorium and uranium concentrations are concerned volcanic rocks of the explored territories of Mars are close to basalts typical of geological formations on the earth's crust; they are also close to lunar mare basalts.

Table 3. Content of rock-forming and radioactive elements in Martian rock, %.

Elements	Mars 5	Viking 1 and 2
O	44	—
Si	17	20.9
Al	—	3.0
Fe	—	12.7
Al + Fe	19	15
K	0.3	< 0.25
U	$0.6 \cdot 10^{-4}$	—
Th	$2.1 \cdot 10^{-4}$	—

Table 4. Thorium and Uranium content (ppm) in different types of rocks of the moon and planets.

Planetary body	Types of rocks	Th	U	Th/U	Source
Moon	*Analysis of lunar samples* KREEP material Mare basalts (average) Rocks of the ANT group *Orbital gamma-spectrometry* Highland rocks of the equatorial zone Mare basalts	9.3 2.61 1.13 0.4–1.4 1.2–2.4	2.8 0.64 0.73 — —	3.4 5.8 1.55 — —	Barsukov et al. 1979[x] Trombka et al., 1975[xx]
Mars	Young volcanic rocks Ancient highland rocks	5.0 0.7	1.1 0.2	4.5 3.5	Surkov 1980[xx]
Venus	Young volcanic rocks Ancient upland plain rocks	0.7 6.5	0.5 2.2	1.4 3.0	Surkov 1976a[xx]
Earth	Ultrabasic rocks Tholeitic basalts of the oceans Alkaline basalts Granites, granodiorites	0.08 0.18 3.9 15.6	0.03 0.1 1.0 3.9	2.7 1.8 3.9 4.0	Smyslov 1974[xxx] Ronov and Yaroshevsky 1978[xxx]

[x]) Analyses of lunar samples brought to Earth; [xx]) Data of orbital gamma-spectrometry; [xxx]) Analyses of the terrestrial rocks.

The ancient formations of Mars have lower thorium and uranium concentrations than the explored Martian volcanics and differ greatly from the substance in the earth's granite continents and lunar KREEP substance of which much higher concentrations of these elements are typical. In contrast to the rocks of the anorthosite-norite-troctolite group (ANT) which are widely developed on lunar highlands, the explored Martian ancient formations have no appreciable differences in the thorium content and contain somewhat smaller amounts of uranium.

Figure 4 shows K-U-systematics of main types of the Earth crust rocks as well as of two main variants of crusts of the moon, Mars and Venus. Because of scantiness of the data on the content of radioelements on the moon and planets the average values of K/U-ratios given in the figure are approximated. One can see from Fig. 4 that K/U-ratios for the moon and planets are grouped in the vicinity of K/U-ratios typical for the most widely distributed tholeitic and alkaline basaltic rocks of the earth's crust. The difference between average concentrations of U and Th and ratios of U and Th is the lowest for the moon (as for the least differentiated body). There is somewhat more difference for Mars. On Venus there is the greatest differrence in the content and the ratio of radioelements which is close to the ratio for the earth.

Thus, one can note that one of the fundamental features of the moon and the earth's group planets is bimodal structure of their surface. Ancient upland plains (or highlands) consists of feldspar rock: gabbro-anorthositic rocks on the moon (and may be on Mars) and granite-metamorphic rocks on the earth (and may be on Venus). Young formations (mare regions on the moon and lowlands on planets) consist of basaltic rock. The formation of these two variants of crust occurred in different processes and at different times.

Primary primitive crust like lunar highlands formed about 4.5 b.y. ago as a result of heating during accretion. This primitive crust was retained only on the moon and probably on Mercury and Mars. The regions covered with basaltic lava (maria on the moon, lowlands and depressions on Venus and Mars, oceanic crust on the earth) formed much later as a result of tectonic and volcanic activity (in the period of heating of the

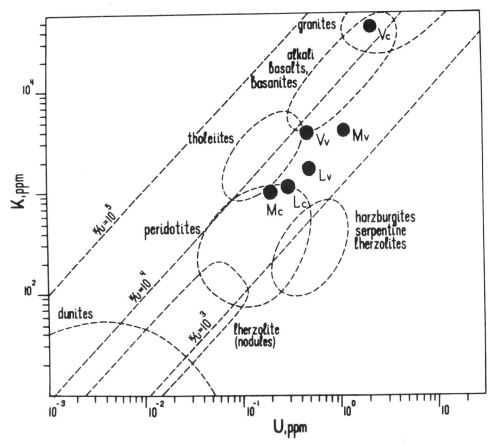

Fig. 4. K-U-diagram of the U, Th and K concentrations in the main types of the rocks of the earth, Venus, Mars and the moon. L_c—lunar continents, L_v—lunar maria, V_c—ancient crust on Venus, V_v—volcanic formation on Venus, M_c—ancient crust on Mars, M_v—volcanic formation on Mars.

upper mantle) due to accumulated radiogenic heat. Basaltic massifs are rather alike on all bodies.

Unlike the moon and other small bodies having anorthositic crust, the earth and Venus have a crust consisting of granitoid and other metamorphic rocks which are the products of later differentiation. Did an anorthositic crust exist on these bodies before? Did it submerge into upper mantle, as some investigations suppose? These problems remain unanswered for the present.

Summing up the review of the natural radioactivity of the moon and planets, it can be stated that apparently all these bodies have undergone a single geochemical process of the differentiation of matter, which divided them into envelopes, the upper of which (the crust) is predominantly composed of the rocks of the basic composition. In all cases the bimodal structure of the surface is revealed: the territories of the ancient crust and the younger territories of a volcanogenic character. On such slightly differentiated bodies as the moon and Mars the natural radioelement content on the territories of both types differs insignificantly. These territories seem to have a weakly pronounced difference in the basic rock-forming elements too. Such planets as the earth and Venus have passed through a differentiation and, therefore, their surfaces are covered with rocks which differ considerably in the content of natural and rock-forming elements. The observed difference in the character and in particular, in the content of natural radioelements, is a consequence of the difference in the degree of their differentiation which in its turn was predetermined by their primary mass and composition.

REFERENCES

Barsukov V. L., Dmitriev L. V., and Garanin A. V. (1979) The main characteristics of lunar rocks geochemistry. In *Regolith from the Highland Region of the Moon.* (V. L. Barsukov and Yu. A. Surkov, eds.), p. 18–30. Nauka, Moscow.

Basilevsky A. T., Surkov Yu. A., Moskalyova L. P., and Manvelyan O. S. (1981) Evaluations of the concentrations of Th and U in surface rock of Mars: Geochemical interpretation of gamma-radioactivity measured by "Mars-5" spacecraft. *Geokhimiya* **1**, 10–16.

Pettengill G. H., Ford P. G., Masursky H., Eliason E. (1980) The surface of Venus. *Report to 23rd COSPAR meeting* (Hungary), p. 16.

Ronov A. B. and Yaroshevsky A. A. (1978) *Tectonosphere of the Earth*, p. 379–402. Nauka, Moscow. (In Russian).

Smyslov A. A. (1974) *Uranium and Thorium in the Earth Crust*. Nedra, Lenningrad. 231 pp.

Surkov Yu. A. (1977a) Gamma-spectrometry in space explorations. *Atomizdat*, 209 pp.

Surkov Yu. A. (1977b) Geochemical studies of Venus by Venera 9 and 10 automatic interplanetary stations. *Proc. Lunar Sci. Conf. 8th*, p. 2665–2689.

Surkov Yu. A. and Fedoseyev G. A. (1975) The radioactivity of the moon, planets, and meteorites. In Cosmochemistry of the Moon and Planets (J. H. Pomeroy and N. J. Hubbard, eds.), p. 358–371. Nauka, Moscow.

Surkov Yu. A., Kirnozov F. F., Khristianov V. K., Korchuanov B. N., Glazov V. N., and Ivanov V. F. (1976a) Investigations of the Density of the Venusian surface Rocks by Venera-10, *Space Res.* **4**, 697.

Surkov Yu. A., Moskalyova L. P., Kirnozov F. F., Kharyukova V. P., Manvelyan O. S., and Shcheglov O. P. (1976b) Preliminary results of investigation of gamma-radiation on Mars from "Mars-5". *Space Res.* **16**, 993–1000.

Surkov Yu. A., Moskaleva L. P., Manvelyan O. S., Basilevsky A. T., and Kharyvkova V. P. (1980) Geochemical interpretation of the results of measuring gamma-radiation of Mars. *Proc. Lunar Planet. Sci. Conf. 11th*, p. 669–676.

Toulmin P. III, Clark B. C., Baird A. K., Keil K., Rose H. J. (1976) Preliminary results from the Viking X-ray fluorescence experiment: The first sample from Chryse Planitia, Mars. *Science* **194**, 81–84.

Trombka J. I., Arnold J. R., Metzger A. E., and Reedy R. C. (1975) Elemental composition of lunar surface according to measurements gamma- and X-ray radiation on Apollo 15 and 16. In *Cosmochemistry of the Moon and Planets* (J. H. Pomeroy and N. J. Hubbard, eds.), p. 128–152. Nauka, Moscow. (In Russian).

Vinogradov A. P., Surkov Yu. A., Chernov G. M., Kirnozov F. F., and Nazarkina G. B. (1967) Gamma-radiation and the composition of lunar rocks. *Kosm. Issled.* **5**, 974.

A possible common origin for the rare gases on Venus, Earth, and Mars

Charles J. Hostetler

Department of Planetary Sciences, University of Arizona, Tucson, Arizona 85721

Abstract—The rare gas concentrations in the terrestrial planets and the carbonaceous chondrites could have been derived from a single source—an early solar wind. The fractionation processes responsible for the varying concentration patterns in this model are differential ionization and the relative separation of neutrals from ions due to interaction with the solar magnetic field. If species dependent effects are small relative to radius dependent effects, this process can explain in detail both the relative and absolute abundances in the terrestrial planets and carbonaceous chondrites. The electron temperatures predicted by the data are consistent with astrophysical constraints previously derived for the solar nebula, and the timeline of events necessary for this model is consistent with accretion in a gas free environment. If species dependent fractionations are small, or approximately cancel each other, the Earth is found to be outgassed by an order of magnitude more efficiently than Venus or Mars.

THE RARE GAS DATA

The concentrations of the rare gases in the atmospheres of Venus and Mars have been fairly well determined by atmospheric entry and lander missions to these bodies. Although there are some disagreements between some experiments, and uncertainties in the measurements, there seems to be general agreement amongst all the data in two areas:

1. There is approximately 100 times more rare gas (i.e., nonradiogenic Ne, Ar, Kr, and Xe) in the atmosphere of Venus than Earth, and 200 times more rare gas in the atmosphere of Earth than Mars.
2. The relative concentrations on each planet, the sun, and the average carbonaceous chondrites (AVCC) (e.g., Ne/Ar, Kr/Ar, Xe/Ar) vary in a continuous and systematic manner with distance from the sun, assuming the carbonaceous chondrites formed in more distant orbits than Mars.

Evidence supporting the first statement can be found in numerous sources, e.g., Donahue *et al.* (1981), Owen *et al.* (1977). Data supporting the second statement are presented in Fig. 1. Any model for the accretion of the rare gases into the terrestrial planets must account for these characteristic concentration patterns. Traditionally, these patterns have been described as the "planetary" component of rare gases and much effort has been devoted to reconciling differences between the Earth and AVCC. In light of the Pioneer Venus results (Hoffman *et al.*, 1980a, b; Donahue *et al.*, 1981) and a reexamination of the terrestrial data (Wetherill, 1981) it seems worthwhile to consider a different model in which the abundance patterns of Venus, Earth, Mars, and the AVCC are different but related. This paper will demonstrate that the rare gases on the planetary bodies could have been derived from a single reservoir, namely an early solar wind, through processes involving differential ionization and the relative separation of neutrals from ions in the solar wind and that the different concentration patterns, varying with distance from the sun, are naturally produced by such processes.

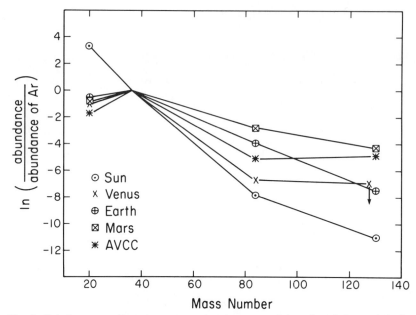

Fig. 1. Relative non-radiogenic rare gas abundances, scaled to Ar = 1 for each body. Venus Xe is an upper limit. There is a systematic variation in the slope of the curves, becoming shallower with increasing heliocentric distance. (Cameron, 1973; Hoffman et al., 1980a, b; Donahue et al., 1981; Canales et al., 1968; Owen et al., 1977; Marti, 1977; Signer and Seuss, 1963).

THE ISOTOPE PROBLEM

In addition to explaining elemental abundances, it is necessary for any successful theory to explain the isotopic abundances of the rare gases on the planetary bodies. The isotopic data for nonradiogenic Ne, Ar, Kr, and Xe on the sun, Venus, Earth, and AVCC are displayed in Fig. 2. This histogram plot is not the usual choice of rare gas workers, but the point of this plot is to show that *within the error estimates, at the few percent level, the isotope abundances are identical*, and that ninety-nine percent of the rare gas work is spent explaining one percent deviations.

This is not to suggest that these deviations are unimportant, and the deviations certainly must be explained by a successful model. However, the single source model under consideration seems to offer much more opportunity to reconcile mass variations than proposals in which the atmospheres of the terrestrial planets are accumulated from material with the "planetary" concentration pattern. For example, consider mass fractionation in Kr and Xe isotopes on the earth and carbonaceous chondrites. Marti (1967) showed that the Kr isotopes ratios (i.e., δ-values) are fractionated in the opposite sense to the Xe isotope ratios. Under the hypothesis that the terrestrial atmosphere was derived from chondrites, a mechanism is required that fractionated Kr in one sense and Xe in another. However, under the single source hypothesis, *differences* in the amounts of fractionation from solar abundances between Earth and AVCC could produce a positive relative slope for Kr and a negative relative slope for Xe. Although the isotope abundances need to be examined in further detail, at the present level of examination they do not rule out the single source hypothesis, and this hypothesis might offer some advantages in explaining isotope fractionations.

FRACTIONATION PROCESSES

Rare gas abundances are measured in the atmospheres of the planets, so we must consider the effects of processes which fractionate both elements and isotopes which

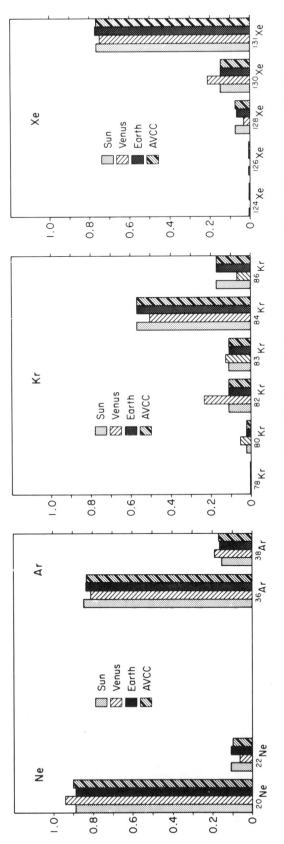

Fig. 2. Isotopic abundances on the Sun, Venus and Mars. For Venus for Kr and Xe the error in the measurements are as large as the measurement itself. (Cameron, 1973; Donahue *et al.*, 1981; Wilkening and Marti, 1976; Eugsten *et al.*, 1967).

could have operated on or around the planets from their formation to the present. These processes include entrapment of heavy rare gases in surficial reservoirs (Fanale and Cannon, 1971), atmospheric escape, planetary outgassing, differentiation of small planetesimals, and different adsorption or accretion efficiencies (e.g., Frick, 1977; Takahashi et al., 1978). We can relate the observed concentration of species i in the atmosphere of a body b, c_i^b to the nebular concentration of i, n_i in the region that went to form b vis:

$$c_i^b = f(r, i)n_i(r) \quad (1)$$

where r is the average distance from the sun of the material that b formed from and f is an unknown function. Since the effects of the fractionation process are not well known, it is difficult to proceed without some simplifying assumptions. The central assumption we make is that the unknown function f can be separated:

$$f(r, i) = R(r)I(i) \quad (2)$$

This separation is certainly valid for the accretion efficiency [the fraction of material incorporated into planet forming grains, one component of f(r, i)], which is of the form:

$$E_i^b = I_E(i)p(r)[T(r)]^{-5/2} \quad (3)$$

where p is the pressure and T the temperature of the nebula, and I_E is a function of species but not distance. Here the Langmuir adsorption process (McQuarrie, 1973) has been used as a prototype for the purpose of this discussion. The exact form of the dependence in Eq. (3) is not critical, but the separation into radial dependent and species and dependent parts is important.

We can now insert our simplified dependencies into Eq. (1):

$$\ln[c_i^b] = \ln[n_i(r)] + \ln[F^b p(r)T(r)^{-5/2}] + \ln[I_E(i)] \quad (4)$$

where F^b is the fraction of outgassing of planet b. For any given planet, the only term which is species dependent is the last term, and Eq. (3) can exhibit two general types of behavior (see Fig. 3). In the first case $\ln I(i) \ll \ln R(r)$ and the measured abundance curve is parallel to the nebular abundance curve, i.e., the slope has been preserved. I(i) can be small for two reasons, either the fractionation effects are relatively species independent, or more likely, the effects approximately cancel out. For example, even though more xenon may have been adsorbed onto grain surfaces in the solar nebula than say krypton, more xenon might also be absorbed onto planetary surficial sediments than krypton (Fanale and Cannon, 1971), the two effects working in opposite directions. In the second

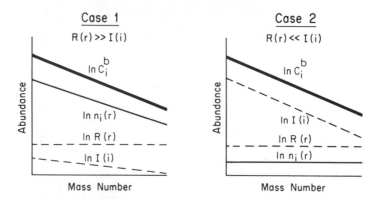

Fig. 3. Two types of Eq. (4). In the first case, the nebular abundance curve (continuous line) parallels the measured planetary data (heavy continuous line) because species dependent fractionations are less important than distance dependent fractionations. In the second case, the slope of the nebular abundance curve bears no relation to the measured planetary data.

case $\ln I(i) \gg \ln R(r)$, the measured abundances curve may have no relation to the nebular abundance curve, and all slope information will be lost. Since slope information is critical to the single source model, we will be interested in seeing how large $\ln I(i)$ could become before the single source model goes awry.

DIFFERENTIAL IONIZATION AND SEPARATION OF IONS FROM NEUTRALS

Jokipii (1964) showed that the solar magnetic field would have significantly affected the distribution of gaseous elements in a partially ionized solar wind. The relative separation of the ionized component from the neutral component of a partially ionized gas is the major process involved. The ionized component interacts directly with the magnetic field and is trapped, while the neutral component passes unaffected. The rare gas concentrations in the terrestrial atmosphere were shown to fit a model for the limiting case in which the separation of neutrals and ions was complete (Jokipii, 1964), and this work was similarly extended to Venus and Mars (Hostetler, 1980) and the ureilites (Göbel et al., 1978). In this section the mathematics of ambipolar diffusion will be considered in detail from a different viewpoint, using the equation of continuity. This theory can be used to predict rare gas abundances as a function of heliocentric distance.

Under the assumption of local ionizational equilibrium, the ratio of ionized particles of species i to neutrals of species i is given by

$$I_i(\mathbf{r}, t) = n_i^*/n_i^0 = C_i \exp(-X_i/kT_e) \quad (5)$$

where C_i is a constant which depends weakly on the species, k is Boltzmann's constant, X_i is the ionization potential of species i, and T_e is the electron temperature in the solar nebula, which varies in space and time (Elwert, 1952). The basic equation describing flow in the solar wind is the equation of continuity (Spitzer, 1978):

$$\frac{\partial}{\partial t}(n^* + n^0) = -\nabla(n^*\mathbf{V}^* + n^0\mathbf{V}^0) \quad (6)$$

where $\mathbf{V}^*$ and $\mathbf{V}^0$ are the velocities of the ionized and neutral components respectively, and are functions of space and time. Since the ions are trapped by the magnetic field $\mathbf{V}^* = 0$. We can now derive an expression for the ion concentration in the solar nebula, assuming radial symmetry

$$\frac{\partial}{\partial t}\left[n^*\left(1 + \frac{1}{I(r,t)}\right)\right] = \frac{1}{r^2}\frac{\partial}{\partial r}\left[r^2 \frac{n^*V^0(r,t)}{I(r,t)}\right]. \quad (7)$$

The concentration of ions is the concentration available for incorporation into grains, and this is equal to the nebular concentration n in Eq. (3).

The simplest case with interesting behavior is the time independent case with $V^0(r, t) = V^0$, a constant, and $I = I(r)$. In this case the solution is easily shown to be

$$n^*(r) = \frac{cI(r)}{r^2} \quad (8)$$

where c is a constant determined by the boundary conditions. As the boundary conditions are unknown, we make use of Eq. (8) by taking ratios. In particular

$$\frac{n_i^*(r)}{n_i^*(sun)} \bigg/ \frac{n_j^*(r)}{n_j^*(sun)} = \frac{I_j(sun)}{I_i(sun)} \cdot \frac{I_i(r)}{I_j(r)}. \quad (9)$$

Recalling the discussion of an earlier section, it is apparent that if the element dependent fractionation is less than the space dependent fraction (i.e., case 1), the nebular ratios are essentially equivalent to the atmospheric ratios, i.e., the relative abundances are unaffected by fractionation processes. If we assume the electron temperature near the sun is large, we can use Eqs. (8) and (4) to determine the electron temperature profile in the solar nebula. The results are in Table 1.

Table 1. Electron temperatures in the solar nebula. Errors are 1σ formal estimates.

Body	Heliocentric Distance (A.U.)	Te (K)
Sun	0	>20000
Venus	0.72	12900 ± 200
Earth	1.0	10600 ± 500
Mars	1.5	8500 ± 500
AVCC	2.7?	8700 ± 200

Furthermore, it is easily shown that

$$\frac{n_i^*(r)}{n_i^*(r^1)} = \left(\frac{r^1}{r}\right)^2 \frac{I_i(r)}{I_i(r^1)} \qquad (10)$$

which can be used to predict the nebular abundances as a function of distance. Combining this result with Eq. (4), the degree of outgassing for each planet relative to the earth is calculated (Table 2). Of course, if the nebular ratios are significantly affected by fractionation processes, the results will be drastically altered. As the electron temperature profile is changed, the outgassing ratios change as well. A straightforward error analysis shows that changing the electron temperature by changing the relative ratios as little as ±15% at Venus, Earth, or Mars would produce improbable outgassing ratios, e.g., Mars ten times more outgassed than Earth, Earth ten times more outgassed than Venus. The significance of these results will be discussed in detail below.

Table 2. Fraction of outgassing for the terrestrial planets relative to Earth.

Venus	0.14
Earth	1
Mars	0.06

DISCUSSION

The electron temperature profile

The electron temperatures derived in the previous section are between 8500 K and 13000 K. Although it is not possible to calculate a priori the electron temperature profile, temperatures of approximately 10^4 K are plausible. Jokipii (1964) showed that if the solar nebula were to exist for 10^6 to 10^8 years, the energy balance for a radiating nebula requires electron temperatures in a "stable region" near 10^4 K. Furthermore, the electron temperature derived decreases monotonically with heliocentric distance, at least out to Mars, which again is reasonable.

Chronology

It is also necessary to consider a timeline of events in the early solar system. If we take commonly accepted models of the early solar system, is there a time period in which the fractionation process previously discussed would have been effective in implanting rare gases into grains? The events we must consider are: a) the sun "turns on", b) grains condense, c) the nebula is swept away, d) grains accrete, e) the sun "turns off", and f) irradiation and implantation of rare gases into grains. Of logical necessity, a preceeds e, b preceeds c, and b preceeds d as well. Studies of solar wind penetration in the solar nebula (Housen and Wilkening, 1980) indicate that the nebular gas must be swept away before irradiation can occur. Thus the major requirement for the model is that the nebula is swept away before the grains accrete, i.e., the accretion process occurs in a gas free environment.

CONCLUSIONS

1. If gas dependent fractionation effects are small (i.e., less than 15%) relative to radius dependent effects, a model involving differential ionization and separation of ions from neutrals can explain the rare gas inventories of the terrestrial planets and chondrites. These rare gas inventories are derivable from a single source, an early solar wind.
2. Under the same assumptions as 1, the earth is found to be outgassed approximately an order of magnitude more efficiently than Venus or Mars.
3. The electron temperatures predicted by the model are consistent with astrophysical constraints. The timeline of events is also consistent with a class of accretion models – accretion in a gas free environment.
4. Gas dependent fractionation affects of up to $\pm 15\%$ can be tolerated in this model by changing the outgassing fraction. Either the radial dependent fractionations mimic the single source model fractionations within $\pm 15\%$ by coincidence, or differential ionization and separation of ions from neutrals was an effective process in distributing rare gases within a solar wind.

Acknowledgments—The author gratefully acknowledges suggestions from J. R. Jokipii, discussions with K. R. Housen and M. J. Drake, remarks from L. L. Wilkening, and criticisms from W. V. Boynton. R. E. Johnson and an anonymous reviewer provided enlightening reviews. This work was supported by NSG 7576.

REFERENCES

Cameron A. G. W. (1973) Abundances of elements in the solar system. *Space Sci. Rev.* **15**, 121–146.
Canales R. A., Alexander E. C., and Manuel O. K. (1968) Terrestrial abundances of noble gases. *J. Geophys. Res.* **73**, 3331–3334.
Donahue T. M., Hoffman J. F., and Hodges R. R. Jr. (1981) Krypton and Xenon in the atmosphere of Venus. *Geophys. Res. Lett.* In press.
Elwert G. (1952) Uber die Ionizations—und Rekombination prozesse in einem Plasma und die Ionizations formel der Sonnenkorona. *Z. Naturforsch.* **7a**, 432–439.
Eugsten O., Eberhardt P., and Geiss J. (1967) Krypton and Xenon isotopic composition in three carbonaceous chondrites. *Earth Planet. Sci. Lett.* **3**, 249–257.
Fanale F. P. and Cannon W. A. (1971) Physical absorption of rare gas on terreginous sediments. *Earth Planet. Sci. Lett.* **11**, 362–368.
Frick U. (1977) Ancient carbon and noble gas fractionation (abstract). In *Lunar Science VIII*, p. 313–315. The Lunar Science Institute, Houston.
Göbel R., Ott U., and Begemann F. (1978) On trapped noble gases in ureilites. *J. Geophys. Res.* **83**, 855–867.
Hoffman J. H., Hodges R. R. Jr., Donahue T. M., and McElroy M. B. (1980a) Composition of the Venus lower atmosphere from the Pioneer Venus Mars Spectrometer. *J. Geophys. Res.* In press.
Hoffman J. H., Oyama V. I., and von Zahn V. (1980b) Measurements of the Venus lower atmosphere composition: a comparison of results. *J. Geophys. Res.* In press.
Hostetler C. J. (1980) Ar and N systematics in Venus, Earth, and Mars. I (abstract). *Bull. Amer. Astron. Soc.* **12**, 71.
Housen K. R. and Wilkening L. L. (1980) Solar ion penetration in the early solar nebula. *Proc. Lunar Planet. Sci. Conf. 11th*, p. 1251–1269.
Jokipii J. R. (1964) The distribution of gases in the protoplanetary nebula. *Icarus* **3**, 248–252.
Marti K. (1967) isotopic composition of trapped Krypton and Xenon in chondrites. *Earth Planet. Sci. Lett.* **3**, 243–248.
McQuarrie D. A. (1973) *Statistical Thermodynamics*. Harper and Row, 343 pp.
Owen T., Biemann K., Rushnech D. R., Biller J. E., Howarth D. W., and Lafleur A. L. (1977) The composition of the atmosphere at the surface of Mars. *J. Geophys. Res.* **82**, 4635–4639.
Signer P. and Seuss H. E. (1963) Rare gases in the sun, in the atmosphere and in meteorites. In *Earth Science and Meteoritics* (J. Geiss and E. D. Goldberg, eds.), p. 241–272. Interscience, N.Y.
Spitzer L. (1978) *Physical Processes in the Interstellar Medium*. Wiley, N.Y. 318 pp.
Takashi H., Gros J., Higuchi H., Morgan J. W., and Anders E. (1978) Volatile elements in chondrites: metamorphism of nebular fractionation? *Geochim. Cosmochim Acta*, **42**, 1859–1869.
Wetherill G. W. (1981) Solar wind origin of ^{36}Ar on Venus. *Icarus* **46**, 70–80.
Wilkening L. L. and Marti K. (1976) Rare gases and fossil particle tracks in the Kenna ureilite. *Geochim. Cosmochim. Acta* **40**, 1465–1473.

Mars' global properties: Maps and applications

Hugh H. Kieffer, P. A. Davis, and L. A. Soderblom

U.S. Geological Survey, Flagstaff, Arizona 86001

Abstract—Mars data sets from many sources have been put into a common digital format and global maps at constant scale have been produced. These formatted data sets form the basis of the Mars Consortium. They currently include maps of geology, volcanic units, channels, wind features, topography, gravity, Viking approach and apoapsis color, predawn temperature residuals, thermal inertia, radiometric and solar albedo, water vapor abundance, 1.4 μm albedo, crater abundance, Earth-based radar and photographic telescope observations, and terrestrial spectral observations. The generation of these data sets is briefly described. As an example of their use, the mapped geologic units are characterized in terms of color, predawn temperature and elevation, and unusual areas identified. The Oxia Palus region is examined as an example of combining higher resolution data with the Consortium data sets. Use of the existing data sets and submission of new ones are encouraged.

INTRODUCTION

Observations of Mars have been obtained with a variety of instruments over a large range of spatial resolution. Comparisons between these observations in their original form are difficult, largely due to the varied formats of the data. The intent of the Mars Consortium is to bring together scientists of many disciplines interested in martian studies and to make readily available and useful a large amount of martian data. The major element of this effort is the Mars Consortium data base, which consists of data in a common format that enables direct comparison between measurements or analyses of diverse origin. This martian data base is modelled after that of the Lunar Consortium (Eliason and Soderblom, 1977; Frontispiece, Proceedings of the 8th Lunar Science Conference).

The major observational data sets were obtained by the Mariner 9 and Viking spacecraft, with Earth-based optical telescopes, and by Earth-based radar observations. Spatial resolution ranges from tens of meters for the low-altitude Viking imagery to hundreds of kilometers for gravity and Earth-based photometry. Data sets such as geologic maps that were derived from analyses of primary observations typically have resolutions of tens of kilometers.

In order to facilitate analysis of these data sets and correlations among them, both original and derived data sets have been transformed into a common format that can be computer processed using standard image-processing techniques. The standard representation chosen is a compromise between high resolution and storage requirements and between ease of access and scientific equity.

CONSORTIUM DATA FORMATS AND MANIPULATION

The standard format is a simple cylindrical projection, in which each element covers 1/4° latitude by 1/4° longitude. This format allows rapid and easy access to data at specific locations or to regions bounded by meridians and lines of constant latitude. It is less efficient (by a factor of $\pi/2$) than an ideal equal-area array, but the latter requires a complicated algorithm for determination of latitude and longitude. To enable agreement of direct output by conventional graphics equipment with Mars mapping conventions (north at top, 180° longitude at left and right borders), the computer storage arrangement consists of lines in order of decreasing latitude composed of samples with longitude decreasing from 180°. Line 1, sample 1 is at the upper left corner of the image.

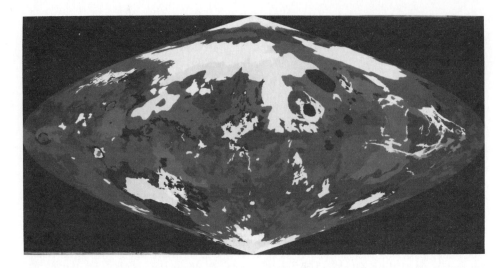

Fig. 1. Sinusoidal equal-area projection of the geologic map of Mars (Scott and Carr, 1978). In this version, 180° longitude is at the center. A histogram of this image that ignores zeros will yield the correct areal fraction of each geologic unit.

The use of 1/4° as the standard resolution element results in a complete global array having 720 lines of 1440 samples, or a total of 1,036,800 elements. Lines of integral latitude and longitude correspond to the borders of these resolution elements, not to their centers. In this way, an integral increase or decrease in resolution can be accomplished without displacement of the area represented by array elements. Array-element borders are mathematically open at the upper end (in the direction of increasing index) and closed at the lower end. Thus, the mathematical south pole is not included within the array, and 180° longitude occurs in the first sample of a line.

Where data resolutions warrant, arrays may be generated with resolution elements an

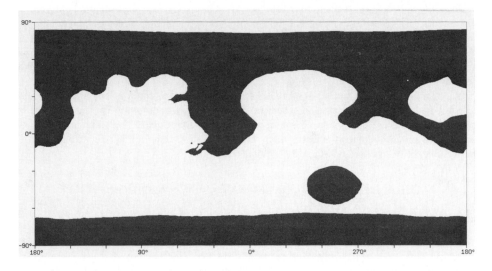

Fig. 2. Example of a stencil of topography. Normally produced as a transparency, this image would pass only areas whose elevation is greater than 1 km. High-pass and low-pass (positive and negative) transparencies can be combined to form bandpass stencils. Sets of stencils can be used in conjunction with transparencies of the Mars data sets to determine readily the distribution of materials under several constraints without requiring a computer. See also Fig. 5.

integral factor larger or smaller than the Consortium standard; linear factors of 2^n are preferred, although resolution elements of 2.5°, 5°, and 10° have been used by some investigators.

For statistical analyses, most data sets are also available in a sinusoidal, equal-area projection, where the resolution in longitude is 1/4° sin (latitude). These arrays are stored with 0° longitude at the center of each line so that they correspond to the conventional sinusoidal projection (see Fig. 1). Conversion to or from Mercator projections (normally limited to ±65° latitude) is standard. A polar stereographic projection can be used for the polar regions.

The most common format of the Mars Consortium data arrays is in 8-bit integers, with the original data scaled within the range of 1 to 255, zero representing no data. When the original data warrant and when greater precision is required, 16-bit or 32-bit files are generated.

When study of various correlations is desired without the use of a computer, overlays ("stencils") can be produced that are transparent only where a variable is within a specified range (Fig. 2). Such stencils and color or monochrome images of the Consortium data sets, all of the same scale and projection, can readily be combined in various ways to determine the spatial distribution of materials satisfying multiple constraints.

A partial list of routines available for use with the Mars data sets follows:

1. Projection changes to or from Mercator, Lambert, polar stereographic, sinusoidal equal-area, and simple cylindrical; with either averaging (for continuous variables such as topography) or replacement (for discrete variables such as geologic unit).
2. Histogram display with mean and standard deviation, with and without zero considered.
3. Two- and three-dimensional histogram and printer display.
4. Two-dimensional histogram output on film.
5. Spatial filtering with these options:
   Original data points unaltered
   No replacement of zero (no data)
   Any rectangular filter shape
   Any upper or lower bounds on data to be included in the filter
   Any minimum number of valid points required to generate a new value. Repeated application of rectangular filters can be used to approximate Gaussian or other continuous filters.
6. Filling-in of areas by expansion of point values.
7. Spatial scaling by integer or real factors independently in two dimensions.
8. Smooth, two-dimensional interpolation of contour data by successive generation of intermediate contours. The original contour lines are broadened until they meet and an intermediate contour line generated at that juncture; this process is repeated until the array is filled.

All of these computer routines are coded for rapid execution. They were developed by Eric Eliason, Kay Edwards and the authors. (See Eliason and Soderblom, 1977; Edwards and Batson, 1980).

MARS CONSORTIUM DATA SETS

Differing degrees of interpretation have gone into the Consortium data sets. Some are directly derived from observations with few assumptions and relatively simple models, e.g., the Viking approach color images. Others are derived directly from observations but require either a relatively complicated model or considerable data processing, e.g., data sets of Viking gravity residuals or thermal inertia. Other data sets are based on the inter-

Table 1. Major data sets of the Mars Consortium.

Title	Source	8-Bit Scaling	Latitude Coverage (Degrees)	Comments
Topography, M9	USGS 1:25 M map	km * 5 + 50	−90 to +90	Interpolation between 1-km contours
Geology	Scott and Carr	geologic unit	−90 to +90	
Volcanic units	Greeley and Spudis	volcanic unit	−90 to +90	
Viking 2 approach color	Soderblom et al.	Minnaert coefficient *500	−60 to +30	7 files; 3 colors, quality and 3 ratios
4-10 km crater density	Condit	craters/10^6 km^2	−65 to +45	smoothed
Gravity residuals	Sjogren		−75 to +75	Smoothing width α to s/c altitude
Predawn residual temperatures	Kieffer, Zimbelman	−100 to 350 mgal (T_{20}-model) + 50	−60 to +50	South is 1976, north is 1977-78
Thermal inertia	Palluconi, Kieffer	10 * milli-inertia	−60 to +50	From 2° bins
Radiometric albedo	Palluconi, Kieffer	albedo * 500	−60 to +60	From 2° bins
Broad-band albedo	Pleskot, Miner	albedo * 500	−60 to +60	From 1° bins
Earth-based albedo	Lowell Observatory	relative albedo	−70 to +70	5 files; 1967, 69, 71, 73, 78
Water vapor	Farmer	precipitable μm	−90 to +90	29 files at increments of 15° L_s; each is 18 × 36 (10° bins)
1.4-μm albedo	Farmer	albedo *100	−90 to +90	
Channels	Carr	channel type	−65 to +65	
Eolian features	Ward	type, wind direction	−90 to +90	3 files; type, direction, season
Radar, 1975-6 Arecibo	Simpson	Hagfors C/25.26	−12 to +24	
Wind streaks	Peterfreund	type	−65 to +65	
Thermal and Albedo Averages	Martin	Special file	−90 to +90	4 longitude regions by 2° latitude

pretation of primary data, in most instances the imaging data; examples are the geologic map and channel distribution map. Each of the available Consortium data sets is discussed briefly below. Latitude coverage and scaling within the range of 0 to 255 are listed in Table 1. In many cases, additional versions of the data are available, e.g., color ratios, unfiltered gravity, and topographic contours.

Examples of several of the data sets are shown in the Frontispiece of this volume. In the following material, the references to Plates are to those of the Frontispiece.

Geology (Plate 1)

The boundaries between units on the 1 : 25 million-scale geologic map of Mars (Scott and Carr, 1978) were digitized from the published map, which contains one Mercator and two polar sections. One array element within each area was assigned a geologic code; this point was then expanded to fill all the area within its respective boundary, and the boundaries were removed. The three sections were then transformed and combined in a

Table 2. Martian geologic units characterized by R/V ratio, albedo (A_r), predawn residual temperature (T_{20R}), and elevation. Only the highest quality data are used in the characterizations.

Geologic Unit*	Consortium Digital Number	Percent Surface Area	R/V ± σ	A_r ± σ	T_{20R} ± σ	Elevation
Plains Materials						
Apt	21	3.0	2.98 ± 0.12	0.16 ± 0.01	−16 ± 5	4–8
Aps	22	10.4	2.78 ± 0.19	0.10 ± 0.02	−2 ± 4	≤4
Apc	23	17.1	2.66 ± 0.32	0.15 ± 0.03	5 ± 5	≤5
Hpr	24	6.9	2.68 ± 0.23	0.16 ± 0.03	−6 ± 9	≤5
Hprg	26	6.2	2.55 ± 0.29	0.13 ± 0.03	−2 ± 4	3
Hpst	25	0.3	2.03 ± 0.05	0.14 ± 0.01	9 ± 1	1–2
Npm	28	5.6				
Nplc	27	20.7	2.42 ± 0.28	0.12 ± 0.03	5 ± 4	≤2
			2.64 ± 0.42	0.11 ± 0.03	−2 ± 5	≥3
Constructional Volcanic Materials						
AHvu	31	0.05				
Avy	32	0.4	2.40 ± 0.34	0.12 ± 0.01	−36 ± 7	27
AHa	35	0.6				
AHvi	33	0.7	2.82 ± 0.08	0.14 ± 0.01	−11 ± 2	0
Hvo	34	1.2	2.60 ± 0.13	0.09 ± 0.01	−3 ± 1	4
Channel and Canyon Materials						
AHcf	41	0.5	2.25 ± 0.12	0.12 ± 0.01	10 ± 3	1–3
AHct	42	0.5	2.30 ± 0.13	0.10 ± 0.01	9 ± 2	≤2
ANch	43	1.1	2.50 ± 0.09	0.15 ± 0.03	9 ± 3	3–5
Rough Terrain Materials						
HNk	51	2.9	2.38 ± 0.11	0.15 ± 0.01	0 ± 2	1
HNbr	52	1.6	2.48 ± 0.13	0.14 ± 0.01	3 ± 1	1–2
Nc	53	0.1	3.03 ± 0.09	0.11 ± 0.01	0 ± 1	1
Nm	54	0.1				
Nhc	55	18.0	2.76 ± 0.21	0.12 ± 0.01	1 ± 3	≤2
			2.63 ± 0.27	0.11 ± 0.02	−1 ± 3	≥3
Polar Materials						
Aprc	11	0.5				
Apb	12	0.9				
HNpd	13	0.5				
Regional exceptions: (boundaries approximate)						
lat 0°–30°N., long 330°–360°			2.86 ± 0.17	0.19 ± 0.02	−21 ± 3	1–3
lat 20° N.–20° S., long 150°–220°			2.78 ± 0.22	0.16 ± 0.02	−17 ± 4	≤2

*See Scott and Carr (1978) for names and descriptions of geologic units

cylindrical projection. Sinusoidal equal area projections with both 0° and 180° as the central longitude have been made for display and statistical purposes (see Fig. 1). The areal fraction of Mars represented by each unit and the encoding scheme are given in Table 2.

Volcanic units (Plate 2)

Using Viking photography, Greeley and Spudis (1981) have mapped martian volcanic units at a scale of 1 : 25 million. The unit boundaries on their original map were digitized and converted to cylindrical projection. The volcanic units were encoded and filled in following a procedure similar to that used for the geologic map.

Channels (Plate 3)

The locations of martian channel features were mapped at 1 : 5 million-scale from Viking Orbiter photos (Carr and Clow, 1981). The 28 quadrangles between latitudes 65° N. and S., in both Mercator and Lambert projections, were digitized with codes for: run-off channels, definite and uncertain runoff valley networks, escarpments bounding outflow channels, scour lines within outflow channels, and the outline of areas with no data. The digitized data from each quadrangle were then combined into a single cylindrical projection in which the presence or absence of the fluvial feature is indicated for each pixel.

Eolian features (Plate 4)

Using Viking photography, Ward (pers. comm., 1981) has mapped 10 types of eolian features on a 1 : 5 million-scale map series. A total of 4500 features were identified by type, apparent wind direction, and season of observation. These features were digitized at a scale of 1 : 5 million and combined in Consortium standard format. Arrays have been constructed both for types of features and for wind direction.

Wind streaks

A map of wind streaks compiled by Thomas and Veverka (1979) on the Mercator portion of the Mars 1 : 25 million-scale map was digitized for use with other Consortium data sets (Peterfreund, 1981). Bright and dark streaks and splotch-related streaks were coded into nine categories and converted to the Consortium standard projection.

Craters of 4- to 10-km diameter

Using Mariner 9 photography, Condit (1978) mapped the abundance of craters with diameters of 4 to 10 km. The mapping was done on Mercator and Lambert 1 : 5 million-scale maps and then combined into a single Mercator projection. The original data consisted of the number of craters per 10^6 km^2 observed in Mariner 9 wide-angle frames. These data have been converted to standard Consortium format.

Crater inventories

The extensive inventory of craters based on Mariner 9 imaging (Arvidson et al., 1974) includes their location, size, and several morphological descriptors. These data are being incorporated into the Consortium data base as vector strings for each crater and as maps of the major morphologic discriptors.

A project in progress will digitize the location of the rims of all craters with diameters larger than 1 km and their ejecta. A code will indicate the morphologic characteristics of the larger craters. This work is being done on the 1 : 2 million-scale photomosaic map series now in production. The results will be concatenated in vector format for each of the 30 1 : 5 million-scale quadrangles. A Consortium standard data base will be generated for the number of craters present in each factor of two size range in each $1/4° \times 1/4°$ area. (Two data sets may be generated, one corresponding to the location of the center of each crater, and the other indicating whether the crater covers any part of any given array element.)

Viking 2 approach color (Plates 5 and 6)

The generation of a global color map of Mars has been described in detail by Soderblom et al. (1978). Individual Viking 2 approach images in each of three colors were rectified into a Mercator projection and mosaicked. A Minnaert phase function was assumed and discontinuities at the seams between individual images were removed by filtering. The original processing was done with 4096 samples in longitude; the map was then converted into the Consortium standard cylindrical format. Images in violet ($0.45 \pm 0.03 \,\mu$m), green ($0.53 \pm 0.05 \,\mu$m), and red ($0.59 \pm 0.05 \,\mu$m) are stored as array elements of 500 times the Minnaert normal albedo. Quality of data varies across the planet due to clouds, frost, and proximity to the limb or terminator. An array indicating the quality of the observations at each location has also been generated.

Viking apoapsis color (Plates 7 and 8)

Sets of images with resolution of approximately 1 km/line pair (per pixel resolution) obtained in three colors late in the Viking mission were rectified and mosaicked into images covering about 40° latitude by 20° longitude. These mosaics are being combined into larger images with resolution of 1/16° (approximately 4 km). The data processed thus far are from Viking Orbiter 1 revolutions 500 through 700, when atmospheric conditions were relatively clear prior to the second season of global dust storms.

Predawn residual temperature (Plate 9)

Maps of 20-μm predawn residual temperatures observed by the Viking Infrared Thermal Mapper (IRTM) have been compiled by Kieffer et al. (1977) and Zimbelman and Kieffer (1979). These maps have a contour interval of 2 K and are relative to the Viking IRTM standard thermal model (Kieffer et al., 1977, appendix). The original maps have a resolution of 25–100 km; the northern portion was compiled from observations obtained one year later than those of the southern portion. The maps were digitized in a Mercator projection, interpolated by successive generation of intermediate contours, and then converted to cylindrical projection.

Solar albedo (Plate 10)

Pleskot and Miner (1981) have compiled a reference map of martian albedo based on observations by the solar band of the Viking IRTM. Their compilation excludes times of high dust opacity or transient brightenings, although clouds common in the Tharsis region probably contribute to the brightness there. The map covers the area within 60° of the equator in a cylindrical representation with 1° by 1° resolution, and presumes a Lambert reflection function for the surface. Conversion to consortium format was accomplished by a 16-fold replication of their array elements.

Thermal inertia and radiometric albedo (Plate 11)

The Viking IRTM observations for selected seasons were grouped into bins of 2° latitude $\times$ 2° longitude; the thermal inertia and radiometric albedo that provide the best least-squares fit to the data in each bin were then determined (Palluconi and Kieffer,

(1981). To first order, the thermal inertia is determined from the amplitude of the diurnal thermal variation, and the radiometric albedo is determined from the mean daily temperature. Solutions for these two parameters did not make use of the solar-band observations. The Consortium data base was constructed by 8-fold replication in both coordinates and scaling of the data into 8-bit integers.

Mars averaged thermal and albedo data

All of the Viking IRTM data has been averaged into arrays with modest resolution in latitude (2°), longitude (4 divisions), local time (1 hour), emission angle (3 ranges) and martian season (Martin, 1981). The average, standard deviation and number of points included are stored for six wavelength bands (five thermal and one reflectance), making 155,520 bins for each season. The seasonal increments are approximately 20° of L_s duration, with boundaries chosen to separate periods of different atmospheric dust content; coverage extends more than one martian year. The four divisions in longitude separate the major thermal inertia provinces (see Plate 11). There are also files which average over all longitudes, and include the temperature differences between the IRTM thermal bands.

These data are not usefully put into Consortium image format; they are documented and will be distributed in an appropriate special format.

Topography (Plate 12)

Contour lines at 1-km interval, on each of the three sections of the 1:25 million-scale topographic map (U.S. Geological Survey, 1976) were digitized, transformed into a simple cylindrical projection, and combined. The areas between contour lines were filled by successive generation of intermediate contours to produce a complete global array. The topographic information is based on Mariner 9 observations and early Earth-based radar results; the major improvements in resolution and accuracy resulting from Viking observations and later Earth-based radar observations have not yet been incorporated. The data may be in error by a few km in the south polar region.

Gravity (Plate 13)

Doppler radio tracking data from the Viking Orbiters were reduced by W. Sjogren (1979) to obtain acceleration residuals relative to a gravity field of fourth order and degree. The data for points along the subspacecraft track were submitted to the Consortium. Consortium standard arrays were made of both the spacecraft altitude and gravity residuals. The gravity data were separated into five maps based on increments of spacecraft altitude; the gravity data in each set were then broadened by application of a two-dimensional filter whose width was equal to the minimum altitude in that set. The final map was obtained by adding together each of the four filtered data sets with a weighting function inversely proportional to the square of the minimum altitude in each set. The altitude range used was 250–1700 km. In this manner, each observation was given an areal representation comparable to the part of the martian surface that it sampled, and a statistical weight appropriate to the force felt by the spacecraft.

Water-vapor abundance and 1.4-μm albedo (Plates 14 and 15)

Data from the Viking Mars Atmospheric Water Detector (MAWD) investigation were averaged into 10° square bins for 29 martian seasons of approximately equal duration. These data include the mean water abundance, in precipital micrometers, and the Lambert albedo at 1.4-μm wavelength (see Farmer et al., 1977; Farmer and Doms, 1979). The MAWD data for the entire Viking mission were reprocessed by B. Jakosky and R. Zurek and were submitted in vector format, including statistical information associated with the process of averaging into 10° bins. Because representation in the Consortium standard format would be 1600 fold redundant, these data have been put into a simple cylindrical representation with 10° resolution; each image has 18 lines of 36 samples.

Earth-based albedo

Albedo maps of Mars have been compiled from photographic observations made during each of the oppositions in 1967, 1969, 1971, 1973, and 1978 (cf. Baum *et al.*, 1970, Inge *et al.*, 1971, Baum, 1973). The original airbrush compilation for each opposition was acquired from the Planetary Patrol program at Lowell Observatory and copied by the photomechanical laboratory of the U.S.G.S., Flagstaff. The photographic transparencies were digitally scanned and converted into a computer image of approximately 1500×2500 samples. The corners of the original Mercator projections were located and least-squares transformations to simple cylindrical coordinates were performed (Fig. 3). Because at least one artistic and two photographic reproductions are involved, absolute brightness information cannot be retrieved directly from these arrays. For purposes of comparison, however, a linear stretch has been applied to each of the arrays so that they now have the same mean and variance.

Earth-based radar

Earth-based radar observations of Mars are normally compiled in vector format, including the time of flight, received peak or total power, and spectral broadening. These parameters are often then reduced to those of surface elevation, dielectric constant, root-mean-square slope, and others based on scattering models. Inventory, compilation, and reduction of the data acquired by several radar observatories over the last decade are just now being accomplished. Data acquired by Arecibo in 1975–76 at 12.6-cm wavelength (Simpson *et al.*, 1978) have been reduced to Consortium format for the Hagfors C parameter, which is a measure of surface smoothness (Fig. 4). Images of other parameters can now be routinely made, and processing of much of the available Mars radar data is planned. Because Earth-based radar observations are confined to within 26° of the equator, data sets of a restricted size are being compiled for the data obtained at each observatory during each apparition.

Earth-based spectra

Whereas radar observations pose a moderate problem for an image based system, with three to six physical parameters at each observation point, spectral observations can become a severe problem if handled only by image array techniques. Earth-based spectra photometric observations of Mars have about 40 parameters, typically covering an area hundreds of kilometers across. More extreme cases are the Mariner 9 ultraviolet spectrometer and infrared interferometric spectrometer data, both of which contain information at many wavelengths; neither has yet been converted to Consortium format. Terrestrial spectral observations of Mars (McCord and Westphal, 1971; McCord *et al.*, 1977) have been treated in a composite fashion. A standard image array was made that contains the footprint of each observation filled with an index pointing to the associated spectrum, which in turn is stored in a standard vector format. This process allows a comparison of the spectral data with other Consortium parameters without requiring extensive storage or computer operations.

DATA AVAILABILITY

All data currently in the Mars Consortium data base are available for distribution without restriction. Individuals desiring copies should mail blank standard magnetic tapes to the authors along with a statement of which files are desired. (A standard 2400-foot magnetic

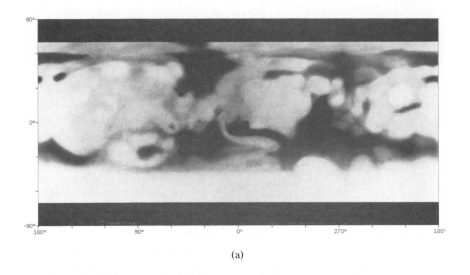

(a)

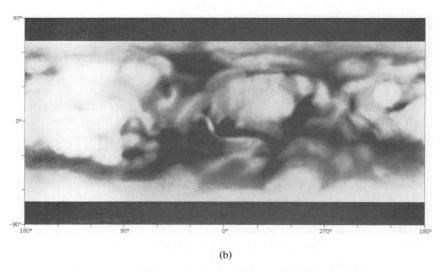

(b)

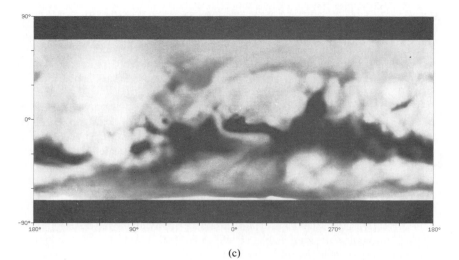

(c)

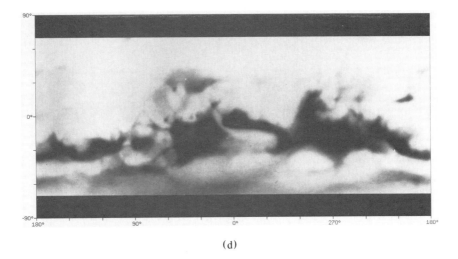

(d)

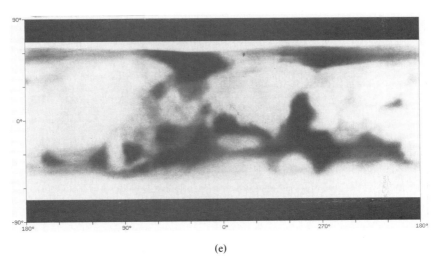

(e)

Fig. 3. Maps of albedo from Earth-based observations. For each apparition, an airbrush rendition was made from photographs obtained by Lowell Observatory and the International Planetary Patrol program (Inge et al., 1971). Photographic transparencies of the original airbrush maps were digitally scanned, then transformed to a simple cylindrical projection. The dates and martian seasons of observations are: (a) 1967 Mar. 21 to May 9, L_s 109–131 (b) 1969 May 19 to July 5, L_s 159–185 (c) 1971 July 19 to Aug 29, L_s 219–245 (d) 1973 Sept. 15 to Oct. 15, L_s 283–326 (e) 1978 Jan. 2 to Feb. 25, L_s 28–52.

tape can contain as many as 15 Consortium files at 800 bpi.) Information suitable for inclusion in the Consortium data base should be submitted in either map or digital form along with a description of the origin and content of the data and reference to any published work describing its generation. Individuals submitting data sets may impose restrictions on their distribution; such data sets will be processed with lower priority.

Reproducible transparencies and prints of each of the Mars Consortium data sets are being sent to all NASA Planetary Data Centers. These photographic products have a scale of 150 μm per array element, yielding global images of 10.9×21.7 cm.

The remainder of this paper presents examples of statistical analysis and correlations between some of the Consortium data sets, and an example of the application of higher resolution information to interpretation of the global data.

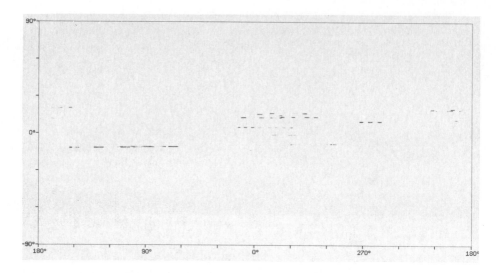

Fig. 4. Arecibo 12.6 cm radar observations of 1975–6. The radar reflection points have been broadened to 3/4° square for visibility. The plotted parameter is Hagfors C, which is a measure of the sharpness of the radar return in the frequency domain; an estimate of the rms surface slope is given by $C^{-1/2}$ radians. (See Simpson et al., 1978). In this image C ranges from 14 (15.3° rms slope, light gray) to 6518 (0.7° rms slope, black); the average value is 290 (3.4° rms slopes).

GLOBAL CORRELATIONS

The Mars Consortium data bases have been constructed in a manner that enables rapid visual or computer correlation. Every data set has been registered pixel-by-pixel so that each 1/4° pixel of the martian surface can be classified according to the geophysical and geological data contained in the Consortium data base. Although classification is rarely performed at the pixel level due to the resolution of most of the data, regional correlation studies are productive. Some examples of visual, three-dimensional cluster, and computer interactive analyses for regional characterization of the martian surface are presented here.

The most basic type of correlation study is visual comparison. This method utilizes color-coded transparencies and various black-and-white transparent stencils of each of the data sets. Stencils serve two purposes: (1) they mask out areas not to be included in visual analysis (e.g., Viking approach albedo data of low quality), and (2) they provide a means of examining relations among several variables with a variety of constraints. An example of the latter is a comparison of predawn residual temperature, Viking approach color spectral albedo (0.45, 0.53, and 0.59 μm), and topography. The predawn residual temperatures, the 0.59 μm/0.45 μm (R/N) ratio, and the 0.59-μm albedo data provide the basis for the comparison. A general correlation is observed in which the redness of the martian surface increases and the residual temperature decreases with increasing albedo (see Frontispiece Plates 6 and 9). Stenciling the three data sets to include, first, only areas of highest quality approach color data, then only areas of specific elevation, enhances this general correlation and shows it to be strongest at elevations less than 3 km above the reference geoid (Fig. 5). The correlation can be further refined by stenciling the residual temperature and Viking color-ratio data with albedo. Stencils representing 1 or 2 percent intervals in albedo show that the darkest materials (albedo $\leq 12\%$) are dominantly bluer and warmer predawn than the martian average, whereas the brightest materials (albedo $\geq 16\%$) are dominantly redder and cooler predawn. This technique has been employed by Kieffer and Soderblom (1979) and Kieffer et al. (1980).

In order to quantify observed correlations, a three-dimensional cluster analysis can be

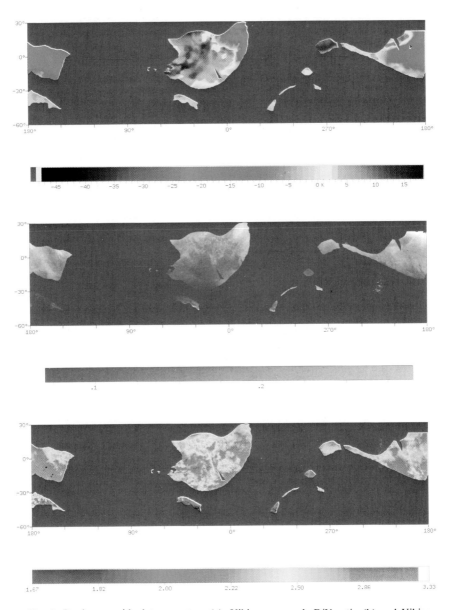

Fig. 5. Predawn residual temperature (a), Viking approach R/V ratio (b) and Viking approach red albedo (c). Each data set has been stenciled to include only areas where approach color data were of highest quality and where the elevation is less than 3 km.

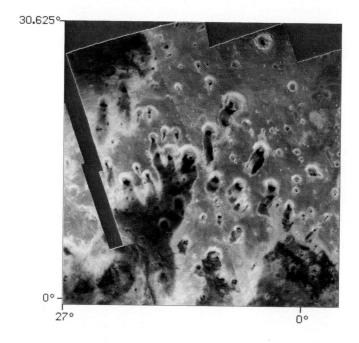

Fig. 7. Enlarged and enhanced multispectral mosaic of the Oxia Palus region (see text). This hybrid color version enables an improved discrimination among materials. A simple way to think of this version is: the red component is exaggerated "redness"; the blue component is exaggerated "blueness"; and the green component is exaggerated albedo.

employed to compare two or three data sets simultaneously. Two dimensions are represented on a page with the abundance at a particular plotted point scaled to the peak abundance. If a third dimension is desired, successive pages are used to represent successive intervals of its value. For instance, the relation between residual temperature and R/V color ratio can be shown for a progression of albedo ranges (Fig. 6). These data have been constrained to include only areas of highest quality approach color data at elevations less than 3 km. The cluster method is able to define numerically and visually materials that have distinct properties. Martian materials with an albedo $\leq 12\%$ show a linear relation between predawn residual temperature and the V/R ratio. Another material, which is redder and cooler, is associated with all higher albedos. Images must be used to determine the spatial distribution of correlations found by cluster analysis.

A highly efficient means of correlating two or more data sets is the use of an interactive display. To illustrate one way in which an interactive display can be employed in a correlation process, five data sets were selected for detailed examination. They included the geologic map, predawn residual temperature, R/V ratio, 0.59-μm albedo, and elevation. Each data set was digitally stenciled to include only areas of highest quality approach color data. The registered data sets were on individual memory planes in an interactive display. The data sets were systematically stenciled for each geologic unit, most of which occur in many separate areas. Preliminary examination of these areas showed some to have anomalous properties with respect to the average properties of the unit. Such areas were not considered in the calculation of the mean and standard deviation of each data set for each geologic unit (Table 2). The values for a particular geologic unit in Table 2 were derived by averaging the individual means and standard deviations for the different exposures of that unit, with each mean and standard deviation weighted according to the surface area of the respective exposure. The percentage of surface area covered by each geologic unit was determined from the histogram of the digitial geologic map in sinusoidal projection (Fig. 1).

The cratered plateau material (Nplc) and the hilly and cratered material (Nhc) show a wide range in properties, and have the largest areal extent of any of the units. The young volcanic material (Avy), although not abundant on the martian surface (0.4%), shows an equally wide range in R/V ratio, which increases dramatically from Arsia Mons (2.10) to Ascraeus Mons (2.86). This increase in R/V ratio may be related to a decrease in the abundance of dark patches from Arsia Mons to Ascraeus Mons. The ridged plains material (Hprg), east of Tharsis Montes, also shows a wide range in R/V ratio, which decreases from Solus Planum (2.70) northward to Lunae Planum (2.08).

The locations of the anomalous materials, along with the mean and standard deviation values of their properties, were recorded separately during the classification process. The anomalous materials occur in two large, geographically separate regions (listed at the bottom of Table 2). These materials represent extremes of the physical properties listed in Table 2 in that they have the highest albedos, very high R/V ratios, and very low predawn residual temperature.

The first region (lat 20° N.-20° S., long 150°–220°), south and southwest of Amazonis Planitia, consists of a variety of materials, including rough terrain (HNk and Nhc), volcanic material (AHvi), and plains materials (Aps, Hpr, Hpst, and Nplc). The properties of the materials in this region (red, bright, cold predawn) are very similar to those of the volcanic plains material (Apt) of the Tharsis Montes region. The only exposure of AHvi within the stenciled part of the map lies within this anomalous region.

The second region of anomalous materials is north of Schiaparelli crater (lat 0–30° N., long 330–360°) and includes the crater. This region also contains a wide range of material types and young and old geologic units, including rough terrain (Nhc and HNbr), channel (ANch), and plains materials (Apc, Hprg and Nplc). This region has properties very similar to those of the first region, (red, bright, cold predawn) and thus of the volcanic plains (Apt) of the Tharsis Montes region. The close correspondence of remotely determined properties over regions of differing inferred geologic history implies that the material being sensed is surficial and, in general, not representative of underlying geologic units.

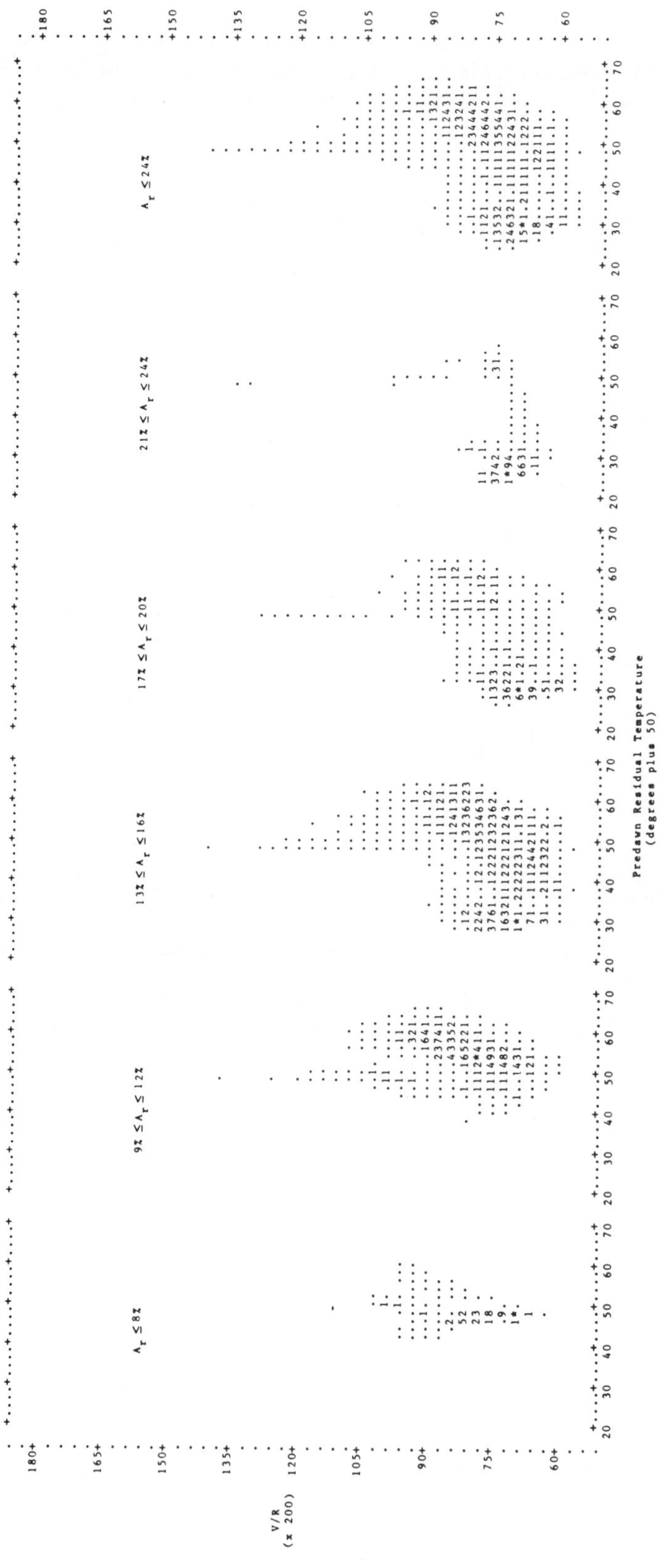

Fig. 6. Cluster plots of predawn residual temperature versus Viking approach V/R ratio. Progression from left to right represents intervals of Viking approach red albedo. Each data set is stenciled to include only areas of highest quality approach color data at elevations less than 3 km. Integer numbers within the plots reflect frequency of pixels in a plotted position in increments of 10% relative to the maximum frequency, which is shown as a star.

STUDIES OF GLOBAL SURFICIAL GEOLOGY

One area in which the Mars Consortium data sets are proving important is the study of the stratigraphy and distribution of surficial units on Mars. The following discussion is not intended as a definitive summary of the surficial geology of Mars, but as an example of how the Mars Consortium data sets can be used in conjunction with high resolution Viking images to examine this general problem.

Many of the Consortium data sets are directly related to the physical, chemical, and mineralogical properties of the surface materials. These include the Viking 2 approach color mosaic (Plates 5 and 6), the Viking 1 apoapsis color mosaics (Plates 7 and 8), thermal inertia (Plate 11), solar albedo, and the 1.4 micrometer albedo (Plate 15). Other Mars Consortium data sets contain correlations with the optical and thermal data, such as occurrence of small channels (Plate 3) and elevation (Plate 12), indicative of some connection to the nature of the surficial units.

The goal of this study is to use colorimetric, albedo and thermal properties to identify and map surfical units and use high resolution photography to understand the stratigraphy and interrelationships among these units. The Oxia Palus region shown in Fig. 7 has been chosen as an example. It must be stressed that the interpretation developed for this small region cannot be generalized to Mars as a whole. In fact, preliminary inspection of other areas suggests that the stratigraphy varies widely over the planet.

Oxia Palus was chosen because it contains a complex mixture of materials, covering the range of materials seen on Mars' surface and the materials occur in regular, repeated patterns, suggesting a reasonably simple set of relationships between the materials that might be understood. Seen in the global multispectral maps (Plates 5 and 6), Oxia Palus is generally anomalous in that the dominant material unit is an intermediate-albedo red. This plains unit surrounds the bright-rimmed dark crater tails. Elsewhere on Mars the dominant units are usually substantially brighter or darker. Oxia Palus also represents a transition from bright red plains to the north to the dark bluer equatorial regions to the south. The high abundance of small channels (Pieri, 1976) and small pristine bowl-shaped craters (Soderblom *et al.*, 1973 and Soderblom *et al.*, 1974) indicate strongly that the dark south-equatorial belt represents a stripped zone free of thick eolian debris blankets.

The intermediate albedo red materials in Oxia Palus are also thermally anomalous. As discussed above, the thermal and albedo properties of the surface materials are strongly correlated forming a homologous distribution between two end members: low thermal inertia, bright, red; and high thermal inertia, dark, blue. This relation has been interpreted as directly related to grain size and oxidation (Kieffer and Soderblom, 1979). The intermediate red materials that form the pervasive unit in Oxia Palus, however, are outside the normal distribution. These materials have higher thermal inertia than most similarly bright materials on Mars.

The features of primary interest in Oxia Palus are the bright-edged, dark plume-shaped markings (plumes) that extend southward from large craters and that are surrounded by intermediate albedo plains materials. As can be seen in Fig. 7, these features are transitional starting as isolated plumes in the north (cf. Becquerel, Rutherford, and Trouvelot, Fig. 8), coalescing to the south into much larger plumes, and finally merging with Sinus Meridiani and the dark south-equatorial belt. This association suggests that the dark central parts of the plumes and the large dark regions to the south are generically the same.

The detailed morphology of the material units recognized in Fig. 7 and the topographic character of the boundaries between these units were examined using high-resolution Viking photography.

In Oxia Palus, the darkest material occurs as small patches of uniform albedo inside the large craters. This material is very blue with red/violet ratios of about 1.5 to 2. The darkest patches have high thermal inertia (Christensen and Kieffer, 1979) and at high resolution display dune forms (Strickland, 1980). Those studies concluded that the dark patches are composed of relatively coarse (1–2 mm), less oxidized grains in the form of active dune masses.

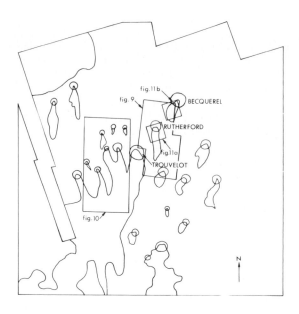

Fig. 8. Index map of the Oxia Palus region. Shown are the outline of Fig. 7 and footprints of Figs. 9, 10, and 11.

Our main concern is not the dune masses on the crater floors but the bright-rimmed plumes which extend southward from large craters. Observation of dark dune masses in the floors of the large craters have led to the suggestion that the dark blue parts of the plumes also represent deposits of dark coarse sand materials that have been blown out from the dark dune masses within large craters and deposited on top of the plains materials (cf. Strickland, 1980). This model involves a stratigraphy for the plumes in which the dark, coarse sand material and fine bright materials are superposed on the intermediate red material, which is in turn superposed on bedrock.

Soderblom *et al.* (1978) suggested that the bright-rimmed, dark plumes are primarily "windows" eroded through the intermediate red material. This suggestion was to a large extent based on the similarity in color and variegation and the continuity between the dark plume cores and the dark blue equatorial region to the south thought to be a stripped zone. They concluded further that the bright material is stratigraphically uppermost. Another interpretation, presented here, is that the main body of bright material is sandwiched between the upper intermediate red material and underlying coarser, darker, blue bedrock-derived materials seen in the central part of the plumes.

High resolution images show that the dark blue regions of the plumes display a wide variety of topographic forms including lava-flow fronts and overlapping crater ejecta blankets (Figs. 9 and 10). A mosaic of intermediate resolution images (~ 150 m/pixel) shows a cluster of craters west of the crater Trouvelot whose plumes successively merge southward (Fig. 10). At higher resolution (~ 80 m/pixel) the edges of the plumes south of Rutherford and Becquerel display progressively rougher topography across the boundaries into the interior of the plumes (Fig. 11). Many of the small craters visible in the dark plumes appear fresh and crisp-rimmed (Figure 11). By contrast, the intermediate albedo plains locally show a smoother surface, with most of the small craters appearing buried to varying degrees. These bright-rimmed dark plumes may represent windows eroded through the intermediate and bright materials, exposing coarse lags and bedrock.

The interpretation for these features as due to scour cannot be generalized to other dark crater tails, such as those described by Thomas and Veverka (1979) farther east in Sinus Meridiani. In contrast to the Oxia Palus plumes, these features display a uniform dark albedo and no bright rims. Such features are probably mobile dark sand sheets. In

Fig. 9. Mosaic of Viking images covering the crater Rutherford and plains to south (Viking Orbiter 2, revolution 212, resolution 150 m/pixel).

Fig. 10. Mosaic of Viking images of the crater Trouvelot, at upper center (Viking Orbiter 1, revolution 864, resolution 159 m/pixel).

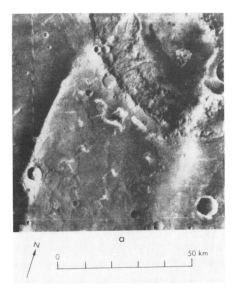 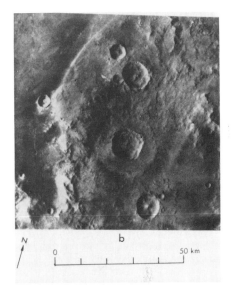

Fig. 11. High-resolution views of the west boundaries of plumes extending southward from (a) Rutherford and (b) Becquerel (Viking Orbiter 1 revolutions 212 and 209, respectively, resolution 85 m/pixel).

some regions such as Cerberus, dark regional sand sheets occur on top of the intermediate red unit. It is becoming clear that the global stratigraphy of Mars' surficial units is extremely complex.

In the model suggested here, the crater plumes and the dark blue equatorial zone (into which the dark crater tails merge to the south) are both produced by scour accounting for the greater density of craters and channels in the south equatorial zone. The intensity of the scour is probably increased by turbulence enhanced by topographic forms (cf. Greeley et al., 1974).

The relationships between the bright and intermediate red materials remains unclear. Two alternatives are offered. The first is that the bright material is locally trapped in topographic irregularities on the intermediate albedo plains (i.e., at craters and ridges) and along their contacts with the dark plumes. Alternatively, the intermediate albedo red material may lie on top of a blanket of bright fine debris and act as a caprock or duracrust. In this case bright material is "leaking" out under the intermediate red material at the plume edges. In Oxia Palus, at every transition from a dark stripped zone to the intermediate albedo plains unit, there is a zone of bright material. Commonly the transition between intermediate albedo red unit and the bright zone is sharp, for example that on the west edge of the Rutherford plume (Fig. 11a). The boundary between the dark blue part of the plume and the bright material is diffuse in all cases. This interpretation is consistent with the observation at lower resolution that intermediate albedo red plains are warmer predawn than materials of similar albedo; duracrust formation would increase the thermal inertia as the voids in the debris are filled and cemented.

A complex picture of the stratigraphy and distribution of surficial units on Mars is emerging from these studies. Grouping all dark plumes and patches together expecting a single explanation is not warranted. The details (e.g., bright-rimmed, variable albedo, surface morphology) must be carefully examined to separate the features into families and explanations for these families must be separately explored. It is clear from these studies that detailed examination of the surface morphology with high resolution data is extremely important in testing the validity of interpretations based solely on global consortium data.

Acknowledgments—The Mars Consortium is supported by the NASA Planetary Geophysics and Geochemistry Program, under the direction of William Quaide. We thank the many scientists who have submitted their data to the Mars Consortium.

Kay Edwards has made major contributions to the mapping transforms and digitizing system as well as the computer processing necessary to transform data into Consortium format. Ellen Sanchez and Pat Termain-Eliason have been instrumental in much of the computer processing of individual data sets.

REFERENCES

Arvidson R. E., Mutch T. A., and Jones K. L. (1974) Craters and associated aeolian features on Mariner 9 photographs: An automated data gathering and handling system and some preliminary results. *The Moon* **9**, 105–114.

Baum W. A., Millis R. L., Jones S. E., and Martin L. J. (1970) The International Planetary Patrol Program. *Icarus* **12**, 435–439.

Baum W. A. (1973) The International Planetary Patrol Program: An assessment of the first three years. *Planet. Space Sci.* **21**, 1511–1519.

Carr M. H. and Clow G. D. (1981) Martian channels and valleys: their characteristics, distribution and age. *Icarus*. In press.

Christensen P. R. and Kieffer H. H. (1979) Moderate resolution thermal mapping of Mars: The channel terrain around the Chryse basin. *J. Geophys. Res.* **84**, 8233–8238.

Condit C. D. (1978) Distribution and relations of 4- to 10-km-diameter craters to global geologic units of Mars. *Icarus* **34**, 465–478.

Edwards K. and Batson R. M. (1980) Preparation and presentation of digital maps in raster format. *Amer. Cartogr.* **7**, 39–49.

Eliason E. M. and Soderblom L. A. (1977) An array processing system for lunar geochemical and geophysical data. *Proc Lunar Sci. Conf. 8th*, p. 1163–1170.

Farmer C. B., Davies D. W., and Holland A. L. (1977) Mars: Water vapor observations from the Viking Orbiters. *J. Geophys. Res.* **82**, 4225–4248.

Farmer C. B. and Doms P. E. (1979) Global seasonal variation of water vapor on Mars and the implications for permafrost. *J. Geophys. Res.* **84**, 2881–2888.

Greeley R., Iversen J. D., Pollack J. B., Udovich N., and White B. (1974) Wind tunnel simulations of light and dark streaks on Mars. *Science* **183**, 847–849.

Greeley R. and Spudis P. (1981) Volcanism on Mars. *Rev. Geophys. Space Phys.* **19**, 13–41.

Inge J. L., Capen C. F., Martin L. J., Faure B. Q., and Baum W. A. (1971) A new map of Mars from Planetary Patrol photographs. *Sky and Telescope* **41**, 336–339.

Kieffer H. H., Davis P. A., and Soderblom L. A. (1980) Regional correlations of Martian remote sensing data (abstract). In *Lunar and Planetary Science XI*, p. 549–551. Lunar and Planetary Institute, Houston.

Kieffer H. H., Martin T. Z., Peterfreund A. R., Jakosky B. M., Miner E. D., and Palluconi F. D. (1977) Thermal and albedo mapping of Mars during the Viking primary mission. *J. Geophys. Res.* **82**, 4249–4291.

Kieffer H. H. and Soderblom L. A. (1979) Global variations of Martian surface materials. *Second Int. Colloq. on Mars*, p. 46, NASA Conf. Publ. 2072.

Martin T. Z. (1981) Mean thermal and albedo behavior of the Mars surface and atmosphere over a martian year. *Icarus* **45**. In press.

McCord T. B., Huguenin R. L., Mink D., and Pieters C. (1977) Spectral reflectance of martian areas during the 1973 opposition: Photoelectric filter photometry 0.33–1.10 μm. *Icarus* **31**, 25–29.

McCord T. B. and Westphal J. A. (1971) Mars: Narrow-band photometry, from 0.3 to 2.5 microns, of surface regions during the 1969 apparition. *Astrophys. J.* **168**, 141–153.

Palluconi F. D. and Kieffer H. H. (1981) Thermal inertia mapping of Mars from 60° S. to 60° N. *Icarus* **45**, 415–426.

Peterfreund A. R. (1981) Visual and Infrared observation of wind streaks on Mars. *Icarus* **45**. In press.

Pieri D. (1976) Distribution of small channels on the Martian surface. *Icarus* **27**, 25–50.

Pleskot L. and Miner E. D. (1981) A preliminary reference map of martian bolometric albedos. *Icarus*. In press.

Proceedings of 8th Lunar Science Conference (1977) Frontispiece, Global maps of lunar geochemical, geophysical and geologic variables.

Scott D. H. and Carr M. H. (1978) Geologic map of Mars. *U.S. Geol. Survey Misc. Geol. Inv. Map* I-1083.

Simpson R. A., Tyler G. L., and Campbell D. B. (1978) Arecibo radar observations of martian surface characteristics near the equator. *Icarus* **33**, 102–115.

Sjogren W. L. (1979) Mars gravity: High resolution results from Viking Orbiter 2, *Science* **203**, 1006–1010.

Soderblom L. A., Condit C. D., West R. A., Herman B. M., and Kreidler T. J. (1974) Martian planetwide crater distributions: Implications for geologic history and surface processes. *Icarus* **22**, 239–263.

Soderblom L. A., Edwards K., Eliason E. M., Sanchez E. M., and Charette M. P. (1978) Global color variations on the martian surface. *Icarus* **34**, 446–464.

Soderblom L. A., Kreidler T. J., and Masursky H. (1973) Latitudinal distribution of a debris mantle on the martan surface. *J. Geophys. Res.* **78**, 4117–4122.

Strickland E. L. (1980) Martian color/albedo units: Viking Lander 1 stratigraphy vs. Orbiter color observations (abstract). In *Lunar and Planetary Science XI*, p. 1106–1108. Lunar and Planetary Institute, Houston.

Thomas P. and Veverka J. (1979) Seasonal and secular variation of wind streaks on Mars: An analysis of Mariner 9 and Viking data. *J. Geophys. Res.* **84**, 8131–8146.

U.S. Geological Survey (1976) Topographic map of Mars. U.S. Geol. Survey Misc. Geol. Inv. Map I-961.

Zimbelman J. R. and Kieffer H. H. (1979) Thermal mapping of the northern equatorial and temperate latitudes of Mars. *J. Geophys. Res* **84**, 8239–8251.

High resolution visual, thermal, and radar observations in the northern Syrtis Major region of Mars

James R. Zimbelman and Ronald Greeley

Department of Geology, Arizona State University, Tempe, Arizona 85287

Abstract—An area in northern Syrtis Major (16°–22°N, 287°–296°W) is crossed by high resolution images, high resolution infrared data, and radar data. This provides the opportunity to assess correlations of the data in a geological context. Nighttime 20 μm surface brightness temperatures provide the following "average particle diameters" for assumed homogeneous surfaces: furrowed plains and ridged plains (0.25–0.5 mm), smooth plains (0.15–0.3 mm), fresh craters (0.5–0.6 mm), slightly degraded crater rims (0.3–0.6 mm), slightly degraded crater floor (0.2–1.5 mm), slightly degraded crater ejecta (0.2–0.4 mm). Differences between 20 μm and 11 μm brightness temperatures indicate cm-sized or larger blocks are present on all of the geologic units. Block abundances are the least on the smooth plains and the greatest on the slightly degraded crater floor material. Correlations exist between albedo and small thermal variations within the plains units; the albedo patterns are all of eolian origin. The large variation in average particle diameter for the slightly degraded crater floor material suggests very effective particle sorting occurs within craters. The smooth plains and the furrowed plains correspond to an area previously mapped as part of the oldest exposed surface on Mars; this ancient surface has been greatly modified by eolian activity. The thermal data limit possible exposed bedrock to <7 km^2. Radar scattering data (Simpson *et al.*, 1978) suggest that the ridged plains have a relatively uniform distribution of surface scattering elements (rms slope ≤2°) and the entire region is not strongly scattering (all rms slopes ≤3°). Radar resolution, however, is not sufficient to distinguish between the contributions of greatly differing geologic units in close proximity. All three data sets indicate that the uppermost surface of the portion of Syrtis Major examined in this work is dominated by eolian processes.

INTRODUCTION

The geologic investigation of Mars is dependent on remote sensing data to provide information on the present surface conditions. Observations in various portions of the electromagnetic spectrum are sensitive to the surface properties at different scales, ranging from less than one mm to more than 10 m. The spatial resolution of the data limit the effectiveness of correlating the surface properties with distinct surface units. The combination of high resolution Viking imaging and infrared data with earth-based radar data provides the opportunity to assess correlations of the data within a geological context.

The best resolution imaging of Mars gives structural and morphologic information about the surface at scales greater than ~8 m (Snyder, 1979). This information provides the basis for distinguishing between different geologic units on the martian surface. Infrared data, on the other hand, are sensitive to the size of the uppermost surface particles over the area sampled by an individual measurement. The scattering of radar signals by the martian surface provide a measure of the surface roughness at scales of from 1 to 1000 wavelengths (Simpson *et al.*, 1978) within the area sampled by an individual measurement. The combination of all three of the data types allows a more complete description of the surface to be made than is possible with any single data set. An area in the northern portion of Syrtis Major (16°–22°N, 287°–296°W) was selected for this analysis where mutual coverage exists in high resolution imaging and infrared data and earth-based radar data (see Fig. 1).

Syrtis Major is an ancient impact basin that has been modified by impact cratering and extensive volcanic flows (Meyer and Grolier, 1977). North, south, and west of Syrtis

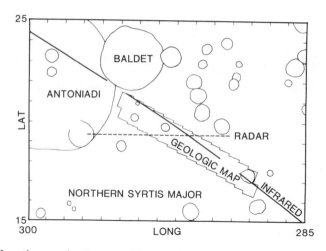

Fig. 1. Location map for data sets. Circles represent crater rims, with the larger craters Antoniadi and Baldet labeled. The imaging data used for the geologic mapping are indicated by the footprints of Viking frames 679A21-48; the geologic map is shown in Fig. 2. The ground tracks of the infrared (solid line) and radar (dashed line) data correspond to the data shown in Figs. 4 and 5, respectively. The gaps in the infrared ground track result from in-flight calibration.

Major are heavily cratered highlands considered to be some of the oldest material presently exposed on Mars (Scott and Carr, 1978). East of Syrtis Major are extensive volcanic plains that post-date both the heavily cratered highlands and the Syrtis Major volcanic plains and which are overlain by varying amounts of eolian material (Meyer and Grolier, 1977; Scott and Carr, 1978). Wind streaks and fresh-appearing dunes in the area indicate considerable recent eolian activity (Peterfreund, 1981). The Viking Infrared Thermal Mappers (IRTMs) have provided global coverage of Mars for both albedo and thermal mapping. Albedo measurements by the IRTM show portions of Syrtis Major to be very dark (albedo <0.10 at locations south of 16°N latitude); the data discussed here include surfaces with albedos of 0.13 to 0.17 (Pleskot and Miner, 1981). Global thermal mapping by the IRTM, with resolutions of 500 km^2 or greater, obtain thermal inertias of 6.0 to 7.5 for the area of mutual coverage shown in Fig. 1 (Kieffer *et al.*, 1977; Zimbelman and Kieffer, 1979). Published radar data are sparse for the martian northern hemisphere, but earth-based radar scattering measurements, from sampling areas of 10^4–10^5 km^2, indicate northern Syrtis Major (latitudes 17°N to 21°N) is "smoother" than most of the observed northern hemisphere locations (Simpson *et al.*, 1978). Viking bistatic radar data have considerably increased the amount of coverage for Syrtis Major and are consistent with the earlier results (Simpson *et al.*, 1981).

IMAGING DATA

High resolution Viking images were used to determine the geology for comparison to the other data sets. Figure 2 shows a geologic map based on images 679A21 to 48, having a resolution of approximately 80 m per line pair. The map shows the contacts between the dominant geologic units; structural features were also mapped but were not included in

Fig. 2. Geologic map of Viking images 679A21-48. The ground tracks of the infrared (solid line) and radar (dashed line) data are included. Five crater units and three plains units are mapped and their relative stratigraphic positions are indicated below the map. Resolution cells for the infrared and radar are shown at the right. The infrared ground track corresponds to the center spot of the IRTM chevron. The radar resolution is roughness dependent but corresponds approximately to a circle of radius 120 km centered on the ground track.

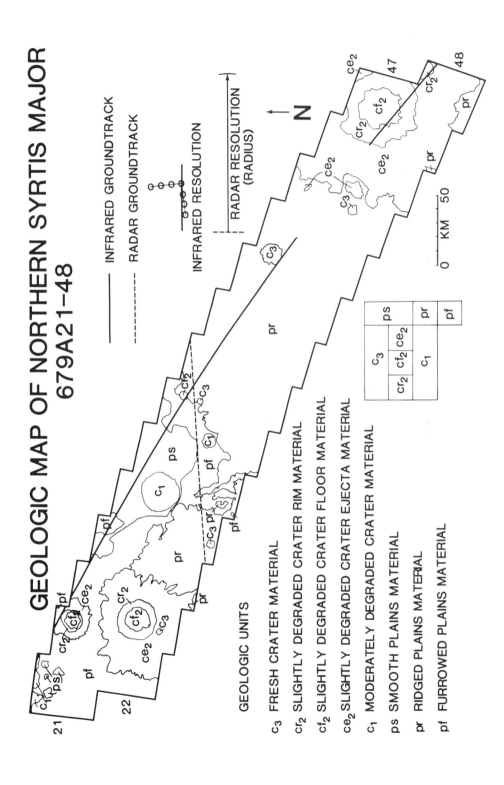

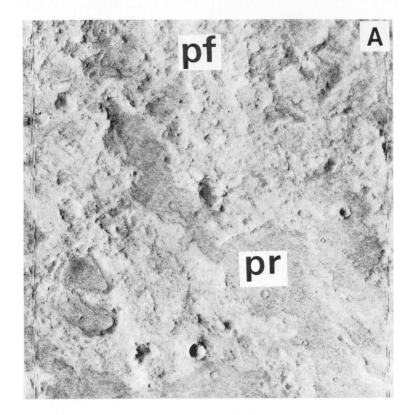

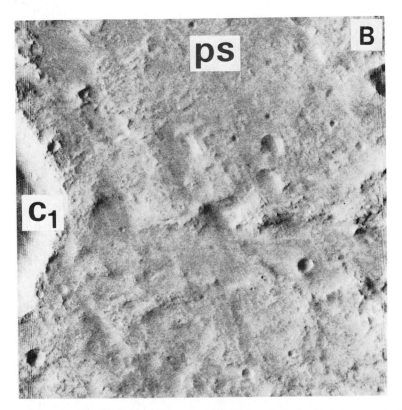

Fig. 3. Example of geologic units. Neutral gain filtered orthographic images are each 38 by 38 km, with north 15° to the left of vertical. Incidence angle is 70°. Unit designations and image locations are provided in Fig. 2. Individual images are discussed in text. (a) 679A30, (b) 679A31, (c) 679A39, (d) 679A47, (e) 679A48.

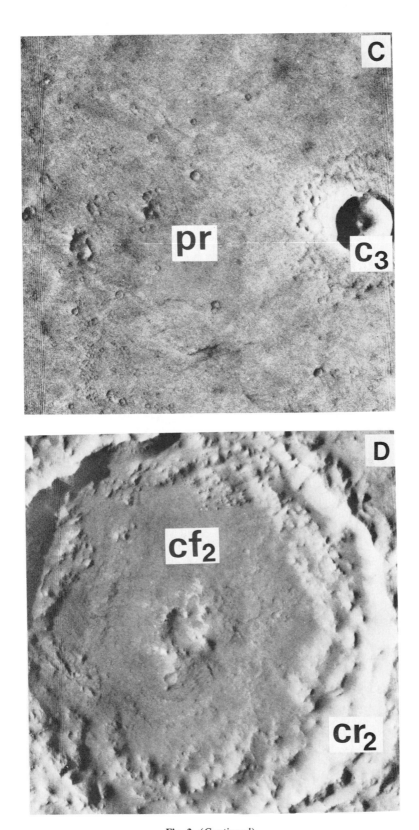

Fig. 3. (Continued).

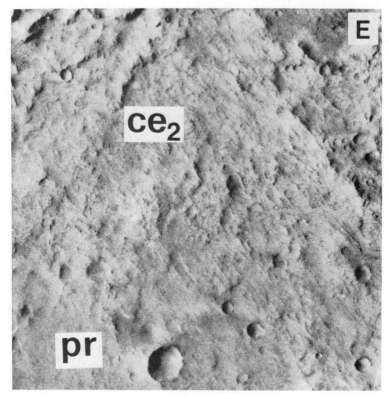

Fig. 3. (Continued).

Fig. 2 in order to emphasize the geologic relationships. Three plains units and five crater units were distinguished according to their surface morphologies.

The largest area of the map is contained in the ridged plains (pr) unit, examples of which are shown in Figs. 3a and 3c. The ridged plains are very uniform at the resolution of the images and are relatively featureless, except for numerous small, fresh craters, generally <10 km in diameter, and sinuous ridges similar to the wrinkle ridges of the lunar maria. Several of the craters on the ridged plains have associated bright streaks trending approximately 70° west of north from the craters. The ridged plains are slightly lower in albedo than the surrounding units, through the red filter used for all of the images, and they overlie the furrowed plains (pf) along the contact. The ridged plains as mapped here correspond to the ridged plains of Meyer and Grolier (1977). The furrowed plains (pf) contain numerous small knobs projecting above the general level of the plains, with some of the knobs having linear trends. Figure 3a also contains an example of the furrowed plains. These plains seem to have been furrowed or "etched" by some erosive agent that left the hummocky surface. The major proportion of small craters within the furrowed plains are slightly degraded in contrast to the fresh craters of the ridged plains. The furrowed plains mapped here are morphologically distinct from the furrowed plains of Meyer and Grolier (1977); they should be considered as separate units. The smooth plains (ps) display very subdued topography and appear to have a thick mantle of material that masks underlying topography at a scale less than several hundreds of m. Figure 3b contains an example of the smooth plains. The smooth plains grade transitionally into the furrowed plains but postdate (overlie) them. Small, fresh craters are present on the smooth plains but are less abundant than on the ridged plains. The relative stratigraphy of the three plains units is shown in Fig. 2.

The crater units range from fresh (c_3) to moderately degraded (c_1). Fresh craters are distinguished by sharp, continuous rims and range from bowl-shaped to flat-floored, some having central peaks, with increasing diameter. Many of the fresh craters have ejecta blankets with considerable surface topography (Fig. 3c); the ejecta blankets have been included with the craters in the fresh crater unit. The small size of most of the fresh craters precludes their inclusion on the map in Fig. 2; only the largest examples have been shown there. Slightly degraded craters range from 5 to 40 km in diameter and are distinguished from the fresh craters by more irregular, possibly discontinuous rims. Ejecta blankets (ce_2) are distinguishable around three slightly degraded craters greater than 20 km in diameter, as is the contact between the essentially flat floor (cf_2) and the sloping rim (cr_2). Figures 3d and 3e show these three portions of the 40 km crater at 17°N, 288°W. A central peak is present in the southernmost slightly degraded crater at 294°W, and a central pit occurs in the crater shown in Fig. 3d, but these have been included with the floor unit. Moderately degraded craters (c_1) have low and narrow rims, sometimes discontinuous, and lack central peaks. A portion of one moderately degraded crater is included in Fig. 3b. The crater units mapped here follow the divisions outlined by Meyer and Grolier (1977) except that the subscript number is reduced by one (no highly degraded crater material was included in the images). The stratigraphy of the crater units is included with the plains stratigraphy in Fig. 2.

INFRARED DATA

The infrared data were obtained around local midnight (290°W) on revolution 524 of Viking Orbiter 1. This corresponds to $L_s = 9°$, or the beginning of spring in the northern hemisphere following vernal equinox. The spacecraft was near periapsis when the data were collected, providing spatial resolution of ~ 3 km^2, but the spacecraft velocity and sample integration time combine to stretch the resolution spot along the groundtrack to ~ 13 km^2 (roughly equivalent to a rectangle 2 by 6.5 km). This represents an improvement in spatial resolution of more than an order of magnitude over previously published global data. The 20 μm surface brightness temperatures are shown in Fig. 4 as a function of longitude along the groundtrack. The standard Viking thermal model for material of thermal inertia 6.5 (10^{-3} cal cm^{-2} sec$^{-1/2}$ K^{-1}) and albedo 0.25 (Kieffer *et al.*, 1977, Appendix 1) remained essentially constant at 185 K over the longitude range covered by the geologic map. The lower albedo (0.15) of this portion of the planet increases the model temperature to 188 K for 6.5 thermal inertia material.

The nighttime temperatures provide a good indication of the thermal inertia of the surface, which can then be related to the "average particle diameter" of an ideal, homogeneous surface of uniform particle size (Kieffer *et al.*, 1977). Table 1 lists the range of average particle diameters and thermal inertias obtained for the geologic units along the groundtrack. Departure from a strictly homogeneous surface causes the average particle diameter (and the thermal inertia) to become an intermediate value between the extremes of the actual surface particle size distribution. A distribution of particle sizes, however, results in different brightness temperatures at different wavelengths. The difference between 20 μm and 11 μm temperatures for the various geologic units was used to determine the areal extent of higher thermal inertia "blocks" among a lower thermal inertia "soil" (taken here to be 5—the lowest thermal inertia observed in the area). The block abundances and thermal inertias are also listed in Table 1. The two-component surface is still an idealized situation but it provides an indication of the relative abundances of the extremes in the surface particle size distribution.

RADAR DATA

The radar data are of much poorer spatial resolution than the other data sets but they provide information about the surface at scales intermediate to the other remote sensing techniques. The scattering of radar signals by the surface of Mars is dominated by surface features with dimensions of from 1 to 1000 wavelengths of the radar signal

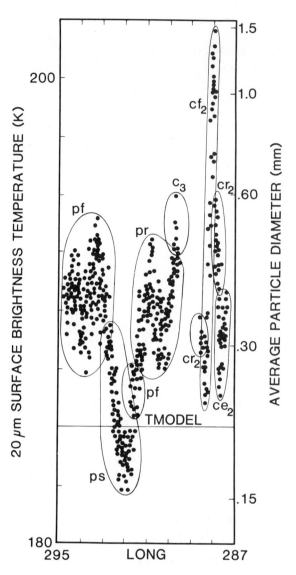

Fig. 4. Infrared data. Individual temperature measurements are shown as a function of longitude. Left vertical scale indicates brightness temperature measured in the 20 μm band of the IRTM during revolution 524 of Viking orbiter 1. Right vertical scale indicates the average particle diameter for an ideal uniform-particle-sized surface with an albedo of 0.15. The standard Viking thermal model (see text) remains constant at 185 K. Local midnight corresponds to 290°W longitude. Portions of data corresponding to geologic units in Fig. 2 are circled and labeled with the appropriate unit designation.

(Simpson et al., 1978). Figure 5 shows earth-based 12.6 cm radar scattering data from Simpson et al. (1978) for Syrtis Major; the entire region is less strongly scattering than much of the northern hemisphere. Each point in the figure corresponds to the average of from 10 to 15 individual measurements. Surface roughness is represented as an rms slope of the reflecting surfaces contributing to the signal; the size of the area contributing most strongly to the signal is directly related to the rms slope (Simpson et al., 1978). For an rms slope of 2° the sampling area is a circle of radius ~120 km centered on the groundtrack. This resolution is not sufficient to clearly isolate the contributions of the units mapped here, but the data do provide information on the regional trends. The rms roughness east

Table 1.

Geologic Unit	Average Particle Diameter (mm)	Thermal* Inertias	Block Abundances† (% surface area)	Block Thermal* Inertias
ps	0.15 to 0.30	5.0 to 6.5	0 to 5	>30
pr	0.25 to 0.50	6.0 to 8.0	14 to 25	20 to >30
pf	0.25 to 0.50	6.0 to 8.0	19 to 42	11 to >30
c₃	0.50 to 0.60	8.0 to 8.5	20 to 29	>30
cr₂	0.30 to 0.60	6.5 to 8.5	43 to 61	11 to 17
cf₂	0.20 to 1.50	5.5 to 12.0	4 to 10ª	22 to >30
			45 to 65ᵇ	21 to >30
ce₂	0.20 to 0.40	5.5 to 7.5	6 to 10	>30

*Thermal inertias in units of 10^{-3} cal cm^{-2} sec$^{-1/2}$ K^{-1} and for an albedo of 0.15. Thermal inertias of 20 and 30 correspond to average particle diameters of 10 and 90 mm, respectively.

†Block abundances obtained from differences in brightness temperatures at 20 μm and at 11 μm. A thermal inertia of 5 is assumed for the surrounding soil. ªcorresponds to 5.5 thermal inertia material, ᵇcorresponds to 12.0 thermal inertia material. Smaller abundances correspond to larger block thermal inertias.

of 292°W is quite uniform at approximately 2°. The ridged plains dominate this area and probably extend east beyond the map presented here. The ridged plains of Meyer and Grolier (1977) are the major unit south of 19°N and between 285°W and 295°W. The uniform roughness may then indicate that the ridged plains contain a relatively homogeneous distribution of scattering elements. West of 292°W there is more variation in the roughness values. Around 294–295°W the contribution from larger craters may contribute to the slightly higher values while the lower value near 293°W may indicate a contribution from the smooth plains.

DISCUSSION

The geologic units mapped here have distinctive remotely sensed surface properties. The ridged plains and furrowed plains have essentially the same average particle diameters and thermal inertias and quite similar block abundances (see Table 1). It should be noted, however, that the blocks on the furrowed plains (>1 mm) do not need to be as large as those on the ridged plains (>10 mm) to satisfy the thermal spectral differences. The stratigraphically older age of the furrowed plains would suggest that some weathering or erosive mechanism may be breaking down the blocks. There is no unique radar signature

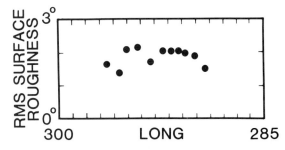

Fig. 5. Radar scattering data. Data are from Simpson *et al.*, 1978, and were collected during 10–16 June, 1976. Scattering properties of the surface are shown as an rms surface roughness. Each point corresponds to the average of from 10 to 15 individual scattering measurements.

for the furrowed plains, but the apparent uniformity of scattering surfaces on the ridged plains may also apply to the furrowed plains, in light of their very similar thermal properties. Local variations of the infrared data within both units correlate well with bright and dark eolian features visible in moderate resolution Viking images (such as 496A53 and 496A71). Darker patches have larger thermal inertias (~8) while a bright crater streak at 18.5°N, 290.5°W has a smaller thermal inertia (~6). This trend is consistent with globally observed correlations between albedo and thermal inertia (Kieffer et al., 1977).

The smooth plains have the smallest average particle diameters, thermal inertias, and block abundances of any of the mapped units. The smooth plains are brighter than the surrounding furrowed plains and ridged plains in moderate resolution imaging (496A53), again consistent with the albedo-thermal inertia correlation. If the slightly lower rms slope value of radar scattering near 293°W results from the contribution of the smooth plains, as suggested earlier, then the radar reflecting surfaces on the smooth plains would have even smaller slopes than is inferred for the ridged plains. Low slope values, small average particle diameters, and low block abundances could result from the dominance of eolian deposition and/or erosion on the smooth plains. The smooth plains occur within the unit mapped as "hilly and cratered material" and interpreted as the oldest extensively exposed surface on Mars (Scott and Carr, 1978). To the west of Syrtis Major lies a low thermal inertia (2.5 to 3.0) region that includes large portions of the same hilly and cratered unit (Zimbelman and Kieffer, 1979). It is probable that many portions of the "ancient" martian crust have eolian surfaces that mask the underlying old material to remote sensing. This is particularly important with regard to the location of potential ancient crustal material sites for future missions to Mars.

Crater units, like the plains units, have distinguishable surface characteristics in the infrared data. A fresh crater at 18.5°N, 209°W (Fig. 3c) has a larger average particle diameter, thermal inertia, and block abundance than the ridged plains on which it is located. A 40 km diameter slightly degraded crater at 18°N, 288°W (Fig. 3d) is asymmetric for the rim material, with average particle diameters of ~0.3 mm for the west rim and 0.4 to 0.6 mm for the east rim. The rim material as a whole has a large block abundance but smaller blocks than most of the other geologic units (see Table 1). The floor material of the same crater is asymmetric in the same sense as the rim material but to a much larger degree, having the largest variation in average particle diameter (0.2 to 1.5 mm) and thermal inertia (5.5 to 12.0) of the units mapped here. The largest average particle diameter material also has at least 50% of its surface composed of larger blocks while the smallest average particle diameter material has a block abundance more like that of the smooth plains. The eastern portion of the crater floor, where the larger particle size material is located, is not morphologically distinguishable from the rest of the crater floor, but there may be some patches of material on the western crater floor up against the western rim. The asymmetry is aligned with the wind direction inferred from bright streaks in the area with the finer material downwind of the coarser material. Craters have been observed to be effective traps of coarser windblown material (Christensen and Kieffer, 1979), but the coarser material here appears to lack an indentifiable morphology. The rim-floor asymmetry may reflect the variations in the flow field over a crater (see Greeley et al., 1974) with enhanced scouring around the first rim encountered. The ejecta on the east side of the crater has a finer average particle diameter (0.20 to 0.40 mm) and has a smaller block abundance than the surrounding units, similar to observations of lunar ejecta blankets (Schultz and Mendell, 1978). The radar resolution does not allow the individual crater units to be examined but it is likely that the slightly higher rms slope values around 294°W are associated with craters in that area, which would suggest that the craters increase the reflecting surface slopes.

The low rms slopes of the whole northern Syrtis Major area, relative to other northern hemisphere locations (Simpson et al., 1978), lends support to the evidence of eolian activity from the imaging and infrared data. An eolian environment should tend to reduce regional slope variations through both deposition and erosion. The lack of prominent rms slope variations indicates this activity is not confined to one particular geologic unit—all surfaces are being affected. The thermal data provide evidence of very pronounced particle size variations, such as in the crater floor material (cf_2) of Fig. 4, that could be the result of eolian redistribution. The block abundances of Table 1 also indicate that at most ~50% of the surface area within the infrared resolution cell could be composed of high thermal inertia material, limiting the potential area of exposed bedrock to <7 km^2. Thus, practically all of the surface, including craters, has been modified by the eolian activity of the area.

SUMMARY

The data examined here allow the correlation of surface properties obtained through remote sensing with the photogeologic mapping of the surface. High resolution infrared data provide an indication of the average particle diameter and the abundance of blocks on the geologic unit surfaces. The stratigraphically distinct ridged plains and furrowed plains have very similar surface properties, except for a smaller block size on the furrowed plains, while the smooth plains have a finer average particle diameter and are less blocky than either of the other plains units. Crater units are relatively blockier than the plains units with the crater ejecta material of a 40 km crater being intermediate in block abundance between the smooth plains and the ridged and furrowed plains units. Spectral differences in the thermal data limit the potential area of exposed bedrock to <7 km^2 in the mapped location. Earth-based radar scattering observations (Simpson *et al.*, 1978) indicate northern Syrtis Major has some of the lowest rms slopes for radar reflecting surfaces of the northern hemisphere and suggest the ridged plains unit is relatively homogeneous in its scattering element distribution. All three data sets indicate that eolian processes dominate the surface in northern Syrtis Major.

Acknowledgments—The comments of Stan Zisk, Alan Peterfreund, and an anonymous reviewer were very helpful during revision of the paper.
   This work was supported by the Mars Data Analysis program through NASA Grant 7583 at the University of California at Los Angeles and through NASA Grant 7548 at Arizona State University, Tempe, Arizona. From the NASA Planetary Geology Program.

REFERENCES

Christensen P. R. and Kieffer H. H. (1979) Moderate resolution thermal mapping of Mars: The channel terrain around the Chryse basin. *J. Geophys. Res.* **84**, 8233–8238.
Greeley R., Iverson J. K., Pollack J. B., Udovich N., and White B. (1974) Wind tunnel simulations of light and dark streaks on Mars. *Science* **183**, 847–849.
Kieffer H. H., Martin T. Z., Peterfreund A. R., Jakosky B. M., Miner E. D., and Palluconi F. D. (1977) Thermal and albedo mapping of Mars during the Viking primary mission. *J. Geophys. Res.* **82**, 4249–4291.
Meyer J. D. and Grolier M.J. (1977) Geologic map of the Syrtis Major Quadrangle of Mars. U.S. Geol. Survey Map I-995.
Peterfreund A. R. (1981) Visual and infrared observations of windstreaks on Mars. *Icarus* **45**, 447–467.
Pleskot L. K. and Miner E. D. (1981) Time variability of martian bolometric albedo. *Icarus*. In press.
Schultz P. H. and Mendell W. (1978) Orbital infrared observations of lunar craters and possible implications for impact ejecta emplacement. *Proc. Lunar Planet. Sci. Conf. 9th*, p. 2857–2883.
Scott D. H. and Carr M. H. (1978) Geologic map of Mars. U.S. Geol. Survey Misc. Geol. Inv. Map I-1083.
Simpson R. A., Tyler G. L., and Campbell D. B. (1978) Arecibo radar observations of Mars surface characteristics in the northern hemisphere. *Icarus* **36**, 153–173.
Simpson R. A., Tyler G. L., Harmon J. K., and Peterfreund A. R. (1981) Radar measurement of small-scale surface texture: Syrtis Major. *Icarus*. In press.
Snyder C. W. (1979) The extended mission of Viking. *J. Geophys. Res.* **84**, 7917–7933.
Zimbelman J. R. and Kieffer H. H. (1979) Thermal mapping of the northern equatorial and temperate latitudes of Mars. *J. Geophys. Res.* **84**, 8239–8251.

Late-stage summit activity of martian shield volcanoes

Peter J. Mouginis-Mark

Department of Geological Sciences, Brown University, Providence, Rhode Island 02912

Abstract—The preservation of morphologically fresh lava flows which pre-date the most recent episodes of caldera collapse at the summits of Ascraeus, Arsia and Olympus Montes indicates that explosive eruptions were not associated with this stage of Tharsis shield volcanism. The existence of resurfaced floor segments, complex wrinkle ridges, and lava terraces within the summit craters suggests that lava lakes comprised the dominant form of the intra-caldera activity. Multiple collapse episodes on Ascraeus and Olympus Montes are indicated by the nested summit craters. The most plausible cause of caldera collapse appears to be large-scale sub-terminal effusive activity, which is corroborated by the previously recognized existence of large lava flows on the flanks of these volcanoes. Due to the implied sequence of large-scale explosive (silicic) volcanism followed by effusive (basaltic) activity, it appears highly unlikely that ignimbrites or other forms of pyroclastic flows (previously proposed as possible deposits within the Olympus Mons aureole material) were ever erupted from the Tharsis Montes.

INTRODUCTION

Of fundamental importance to the current understanding of martian geological evolution was the recognition of the numerous large Tharsis and Elysium volcanoes from Mariner 9 and Viking Orbiter images (Masursky, 1973; Carr, 1973; Carr et al., 1977; Greeley and Spudis, 1981). In particular, the Tharsis volcanoes received early attention due to their size and similarity (and hence interpretable morphology) to terrestrial shield volcanoes such as Mauna Loa, Hawaii (e.g., Greeley, 1973; Carr and Greeley, 1980). Numerous lava flows were shown to characterize the summit and flanks of each Tharsis shield (Carr et al., 1977; Schaber et al., 1978), in many instances, however, the morphology and evolution of each martian volcano were treated in general, and sometimes conflicting, terms. An assortment of diverse eruptive styles were suggested by authors such as King and Riehle (1974), Hodges and Moore (1979), Morris (1980) and Scott and Tanaka (1980) in order to explain, for example, the aureole and escarpment surrounding Olympus Mons. Alternatively, analyses of volcano morphology concentrated on the regional and global significance of shield evolution for the entire geological province and/or the martian lithosphere (Carr, 1974a; Blasius and Cutts, 1976; Comer et al., 1980).

Considerable theoretical and quantitative data now exist for lunar (e.g., McGetchin and Ullrich, 1973; Head and Wilson, 1979; Wilson and Head, 1981a) and terrestrial (e.g., Wilson et al., 1980; and references contained therein) volcanic eruptions, so that reasonable extrapolations to the martian environment can be made (Blackburn, 1977; Wilson and Head, 1981b). The acquisition of high resolution (15–40 meters per pixel) Viking Orbiter images of the summit areas of Ascraeus, Arsia and Olympus Montes permits some of these theoretical predictions to be tested. In addition, the eruptive sequence and style of activity for each martian volcano can be recognized. Specifically, high resolution Viking images provide the capability to identify the spatial relationships between summit fracturing, caldera collapse and the eruption of lava flows associated with the shield volcanoes. Such attributes have been shown by Nordlie (1973) to be intimately related to different evolutionary stages of Galapagos volcanoes, and it appears reasonable to expect that similar information can be derived for the martian volcanoes. Furthermore, if plinian and sub-plinian eruptions (Walker, 1973) were to occur on Mars (Wilson and

Head, 1981b), the caldera rim may show evidence of extensive mantle deposits identifiable on Viking images: major differences in volcanic activity (effusive versus explosive) should therefore be discernible.

This paper presents observational data on three of the Tharsis Montes (Pavonis Mons is excluded due to the absence of high resolution Viking images). The caldera morphology is treated here as an indicator of the late-stage summit activity on each volcano, thereby permitting the style(s) of volcanism to be determined. Interpretations of the evolutionary sequences of terrestrial intra-plate shield volcanoes are employed to infer the individual characteristics of each martian volcano investigated here.

OBSERVATIONS

Detailed descriptions of the evolution of terrestrial intra-plate volcanoes have been given by a number of volcanologists, and it is assumed here that their observations would also be applicable to Mars were the same styles of activity to occur there. For example, Simkin and Howard (1970) described the 1968 caldera collapse of Fernandina volcano, Galapagos, and noted that rapid collapse permitted large segments of the wall to slump en masse, while slow collapse was associated with block fragmentation. Thus, on a general basis, caldera collapses on Mars could be subdivided into those produced by rapid magma chamber evacuation and those generated by a slower removal of magma. Relationships between volcano growth, fracturing, and caldera collapse have also been proposed by Nordlie (1973), Guest (1973) and Wadge (1977). Radial fractures with associated lava flows are believed to be related to volcano swelling ("tumescence") as magma rises within the volcano, while concentric fracturing and circumferential fissure eruptions are commonly associated with caldera formation (Nordlie, 1973).

In this paper the term "caldera" is applied to the entire collapse structure at the summit of a volcano, while individual pit craters are referred to as "volcanic craters". The production of a caldera in itself appears to be an indicator of the magnitude of each eruption, and the relatively shallow depth of the magma chamber (Macdonald, 1972; Williams and McBirney, 1979). To a first approximation, equivalent volumes of erupted material (lava flows, ash and pyroclastics) should initiate comparable collapse episodes. Thus, the volumetric evolution of magma chambers on Mars should be evident from the size progression of the nested craters. The presence of caldera structures in which the largest crater was the first to form would suggest that the magma chamber decreased in size as each volcano evolved. Conversely, small craters truncated by a final large crater would indicate an increase in magma chamber volume.

Wilson and Head (1981b) provide the theoretical basis for recognizing ash deposits associated with the creation of a caldera and predicting the distribution of these deposits around martian volcanoes. Explosive eruptions on terrestrial basaltic shield volcanoes typically occur when magma withdrawal permits ground water to enter the vent (Jaggar and Finch, 1924; Simkin and Howard, 1970; Tazieff, 1976–7). Because ground ice-volcano interactions have been proposed for Mars (Allen, 1979; Hodges and Moore, 1979), deposits from phreatic eruptions may also exist. In addition, martian basaltic plinian eruptions, with their associated mantle deposits around the vents, might also be observed (Wilson and Head, 1981b). This section consequently documents the characteristics of Ascraeus, Arsia and Olympus Montes to serve as primary constraints for subsequent analyses.

Ascraeus Mons: A total of eight collapse features are prominent at the summit of Ascraeus Mons (Fig. 1), and are interpreted to represent volcanic craters in various stages of preservation. Table 1 lists the individual dimensions of these craters, which range in size from about 7 to 40 km in diameter. A chronology for these collapse events can be derived from the cross-cutting sequence of the crater wall segments. Using this method, crater 8 (Fig. 1b) appears to be the oldest, while craters 6 and 7 were the next to form. Craters 5 and then 4 were produced prior to a major slump episode on the northern wall of the caldera, approximately contemporaneous with a period of circumferential graben

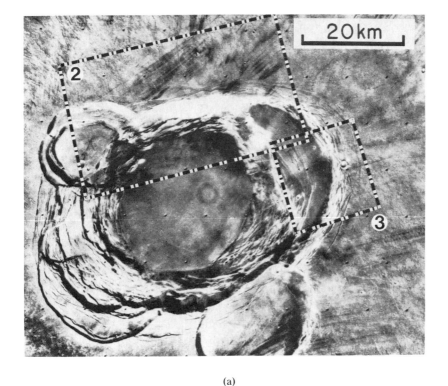

(a)

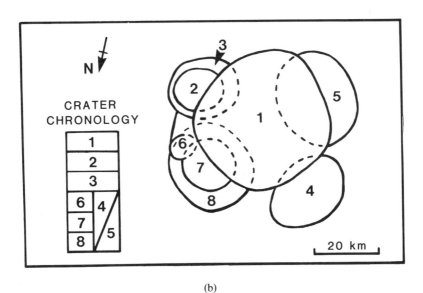

(b)

Fig. 1. (a) Medium resolution image (60 meters per pixel) of the summit of Ascraeus Mons, showing the locations of Figs. 2 and 3. Viking frame 90A50. (b) Schematic diagram of area shown in Fig. 1A, giving the crater numbers utilized in the text. The block diagram gives the inferred sequences of crater formation (1 youngest), based on cross-cutting relationships of the crater walls.

Table 1. Dimensions of Ascraeus Mons craters.

Crater No.	Diameter (km)†	Depth (km)*	Approx. Volume (km³)
1	40 × 38	3.15	3760
2	14 × 10 (14)	0.48	75
3	21 × 13 (19)	0.98	325
4	26 × 11 (17)	?	?
5	32 × 10 (27)	0.75	530
6	7 × 4 (9)	?	?
7	17 × 5 (25)	?	?
8	15 × (14)	?	?

†Dimension in parentheses gives an estimate of the original size of the crater.
*Crater depths derived in this analysis from shadow measurements.

formation (also evident in craters 6 and 7). The small craters 2 and 3 were formed on the eastern rim after this graben-forming event was complete. The most recent modification of the summit area resulted in the formation of crater 1, which now occupies the center of the caldera complex. Based purely on the size and superpositioning of each volcanic crater, it appears that several intermediate-sized (20–30 km) individual craters characterized the original summit. These craters were then modified by the formation of smaller (7–15 km) craters before the final eruption produced the single large (40 km) crater.

More detailed information on the eruption characteristics of Ascraeus Mons can be derived from Viking Orbiter images acquired on rev. 401 of Orbiter 2. Four episodes of crater formation are evident on the southern caldera wall and floor (Fig. 2). The maximum depth of the caldera at this point (from the rim to the floor of crater 1) is approximately 3.1 km (based on shadow measurements), with craters 2, 3 and 5 perched respectively 1.7 2.2 and 2.4 km above the floor of crater 1. Prominent along the entire length of the caldera rim are numerous lava flows and a few sinuous channels which flow radially away from the caldera. These lava flows are typically less than 1 km wide and 10–20 km in length, while the largest channel measures 200 meters wide and 18 km long. It is clear that the lava flows on the south rim are traceable all the way to the caldera rim crest but lack any obvious source vents. It therefore appears that each flow has been truncated by the collapse events, rather than preserving its original length. No evidence of the same flows can be identified on the lower "slump blocks", however, suggesting that each new crater floor within Ascraeus caldera experienced a resurfacing event after the collapse episode.

Numerous small wrinkle ridges upon the caldera rim are also evident (Fig. 2). These ridges appear analogous to small-scale mare ridges on the moon, and are generally orientated down the inferred maximum slope of the volcano close to the rim of crater 5. Spacing of the ridges is remarkably constant for adjacent ridges, and varies between 0.9 to 2.6 km. It is inferred here that these ridges have a tectonic origin, perhaps formed during post-tumescence contraction of the summit. Many very narrow ridges less than 3 km in length also occur on the floor of crater 1. The origin of these floor ridges is less clear, but their uniform spacing (600–800 meters) and preferential orientation (east-west) suggest that they too may be tectonic in nature, although image resolution is inadequate to preclude, for example, an eolian origin.

Sinuous channels are located on the northwest wall of crater 5 (Fig. 3). These features are interpreted as sinuous lava channels created by turbulent flow in a similar manner to lunar sinuous rilles (Hulme, 1973; Carr 1974b; Wilson and Head, 1981a). Despite the initiation of circumferential fracturing of the rim area, some of these channels can still be identified on fractured segments of the caldera rim. Apparently, at least at this locality, rim slumping was not associated with any process capable of destroying (or burying) the previously produced small surface features on the rim, implying that major explosive activity (such as plinian eruptions) can be discounted.

Arsia Mons: Arsia Mons caldera measures 113 × 128 km in diameter and has been

Fig. 2. High resolution (22 meters per pixel) mosaic of the southeastern rim of Ascraeus Mons caldera. Craters 2 and 3 (left), 1 (center) and 5 (right) are illustrated. "F" denotes the position of truncated lava flows on the caldera rim, "S" is a sinuous channel interpreted as a lava channel, and "R" marks the location of numerous wrinkle ridges: see text for discussion. Viking frames 401B17–20.

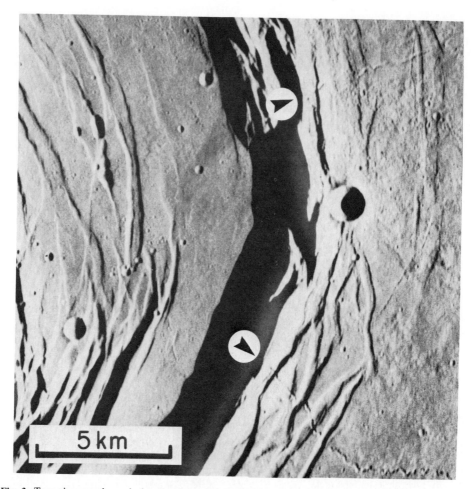

Fig. 3. Two sinuous channels (arrowed) on the southwestern rim of crater 5 on Ascraeus Mons are shown here to be cut by subsequent circumferential graben. The existence of these preserved pre-collapse features indicate that caldera enlargement was not associated with explosive eruptions that generated large air-fall deposits. Viking frame 401B17. Image resolution 22 meters per pixel.

estimated (using photogrammetry; Wu, 1980) to be approximately 900–1100 meters deep. Crumpler and Aubele (1978) interpreted the volcano to be more evolved in comparison to Ascraeus Mons. Arsia Mons differs from Ascraeus Mons in terms of the degree of subsidence of the summit, the presence of a pronounced set of concentric rim fractures, and the post-subsidence flooding of the floor by younger lava flows (Fig. 4).

Prominent on the caldera rim, and present up to distances of 60 km from the rim (Fig. 4), are numerous graben. Spacing between the major graben varies from 5–12 km, while several smaller examples are located within 1 km of each other. Transected by many of these fractures is a series of small lava flows which extend downslope toward the northwest. Several dozen individual flows can be recognized, and each measures approximately 0.5–1.3 km in width and 10–14 km in length. The thicknesses of the lava flows in this area have been estimated by Schaber et al. (1978) to be less than 10 meters. The primary source for most of these flows is a major graben located on the rim approximately 12.5 km from the edge of the current caldera floor.

Nearly all of the lava flows appear to have been emplaced prior to the fracturing of the caldera rim; they are cut by the fractures and none of them pond against any of the slump blocks (Fig. 4). One exception is a flow labelled "A" in Fig. 4b. Emanating from the same system of fissure vents which produced the other flows, this lava "fan" differs from the other eruptives in that the flow direction was toward the caldera floor and its time of emplacement was after the last episode of graben formation. A small lava delta measuring 10×16 km has been constructed by this sequence of eruptions. Image resolution (40 meters per pixel) is not, however, adequate to say confidently if the flows extend across the caldera floor, or have been buried by subsequent floor deposits. Lobate outlines on the floor suggest that individual lava flows exist within the caldera, but none can be positively identified. Several small, gently sloping elongated hills with summit craters ("B" in Fig. 4b) have been observed on the caldera floor (Carr et al., 1977) and these may be examples of martian cinder cones or hornitoes (Moore and Hodges, 1980).

Olympus Mons: Like Ascraeus Mons, the summit area of Olympus Mons (Fig. 5) is characterized by a complex nested caldera. Six coalescing volcanic craters exist, which range in diameter from 20 to 65 km (Table 2). The caldera of Olympus Mons differs from that of Ascraeus Mons in terms of the apparent sequence of the collapse events. Superposition relationships indicate that the largest crater (number 6 in Fig. 5) formed first. Successive collapse episodes subsequently produced the intermediate-sized (30–40 km) craters 3 and 4. Crater 2 represents a relatively small collapse event, and was probably associated with the flooding of the floor of crater 3. Based on the degree of preservation of craters 1 and 5, these two 20 km craters are probably the most recently formed summit craters, although the relative age of crater 5 can only be confidently placed as younger than crater 6.

The Viking Survey Mission produced extensive high resolution (~15 meters per pixel) coverage of the southeastern portion of the Olympus Mons caldera (Fig. 6). A complex sequence of wrinkle ridges exists on the floors of craters 2 and 3, while circumferential graben are located on the floor of crater 6. No obvious surface features exist on the floor of crater 1. Although Greeley and Spudis (1981) urge caution in the interpretation of the caldera ridges (Figs. 6 and 7), an analysis of lunar mare ridges (Sharpton and Head, 1980) would suggest a tectonic origin for these martian examples. These Olympus Mons ridges are much larger than ones seen within Ascraeus Mons caldera (Fig. 2): individual examples measure 0.1–3.0 km in width and ridge arches may be as much as 6 km in length. A narrow (200 meter) sinuous ridge also extends around the base of the wall within crater 3. As can be seen from Fig. 7, several very thin ridges within craters 2 and 3 possess central depressions (making them appear more like collapsed lava tubes than "mare" ridges). Other ridges possess overlapping field relationships with earlier examples, implying a multiphased mode of formation.

Olympus Mons apparently has no individual lava flows on the caldera floor; the high resolution Viking images fail to show any evidence of intra-caldera activity similar to that observed on terrestrial volcanoes (Macdonald, 1972). A semicircular depression ("V" in Fig. 6) may mark the location of a small vent at the junction of craters 1 and 2, but no mantle deposits or lava flows can be seen around this feature. Few lava flows are apparent on the volcano's near summit flanks (Fig. 8), producing a rim morphology that is appreciably different from that of Ascraeus Mons (Figs. 2 and 3). One explanation for this may be the very low relief in this area on Olympus Mons (Wu, 1981), which has precluded individual lava flows from forming and extending downslope.

Additional features of Olympus Mons caldera are also noteworthy. Within crater 6, the floor fractures appear to vary in width as a function of their location, being widest close to the crater edge and progressively narrower toward the crater center (Fig. 6). Together with the wrinkle ridges on the caldera floor, it is likely that these fractures are tectonic in origin and are probably related to the subsidence of the central portion of the caldera floor. This subsidence may have been due either to the solidification and contraction of the lava lake, or to the withdrawal of support from an underlying magma chamber. Evidently these floor fractures exerted little structural control over the evolution of the caldera, however, since the shape of crater 3 has been almost unaffected by this fracture pattern. A landslide is evident on the southern wall of crater 1 (Figs. 6 and 8), with a small lobe of material extending 2 km across the crater floor. This landslide may indicate that

Fig. 4. (a) Northwestern rim of Arsia Mons caldera. Viking frames 422A30–35, image resolution 40 meters per pixel.

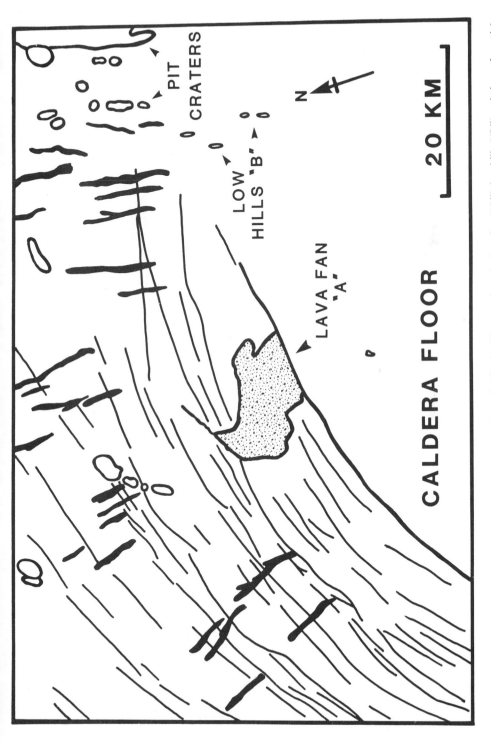

Fig. 4. (b) Schematic diagram of the area of Arsia Mons caldera shown in Fig. 4A. Illustrated are the lava fan ("A"), low hills ("B") and circumferential graben which cut the radial lava flows (solid black outlines) on the flanks of the volcano. See text for discussion.

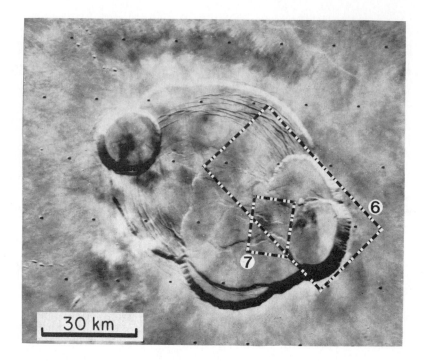

(a)

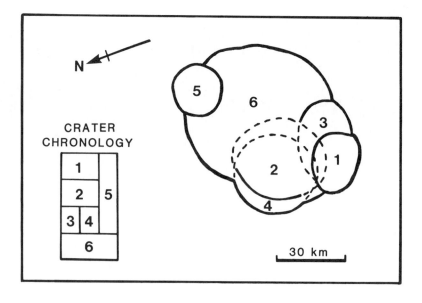

(b)

Fig. 5. (a) Medium resolution image (140 meters per pixel) of the summit caldera of Olympus Mons. The locations of Figs. 6 and 7 are also shown. Viking frame 46B31. (b) Schematic diagram of area shown in Fig. 5A, giving the crater numbers discussed in the text. Block diagram derived in the same manner as Fig. 1B.

Table 2. Dimensions of Olympus Mons craters.

Crater No.	Diameter (km)†	Depth (km)*	Approx. Volume (km³)
1	24 × 20	3.28	1245
2	36 × (30)	0.65	555
3	(40) × 25	1.37	1135
4	(38) × 6 (32)	1.42	1365
5	20 × 23	2.85	1184
6	66 × 66	1.65	5145

†Dimension in parentheses gives an estimate of the original size of the crater.
*Crater depths derived in this analysis from shadow measurements.

the wall materials of Olympus Mons are more fragmented than those of Ascraeus Mons (where no landslides are observed), and therefore more susceptible to slumping. Alternatively, the place where the landslide occurred is the closest point (within 17 km) on the caldera rim to a very recent 8 km impact crater. This impact crater post-dates the formation of crater 1, as evidenced by the secondary craters on the caldera floor. Thus, the landslide may be an anomaly induced by the subsequent disruption of the rim during the impact cratering event.

THEORETICAL CONSTRAINTS ON VOLCANIC ACTIVITY

In addition to the extrapolation of field observations of terrestrial volcanoes, numerous theoretical analyses of volcanic eruptions permit specific features of martian volcanism to be predicted and, hence, sought for on the Tharsis Montes. Interpretations of lunar and terrestrial eruptions (cf. Wilson et al., 1980; Wilson and Head, 1981a) indicate that gravity, atmospheric pressure, and volatile species are likely to be the factors which will most influence martian volcanism (Wilson and Head, 1981b). Adapting the numerical models to martian conditions should therefore permit the morphology and morphometry

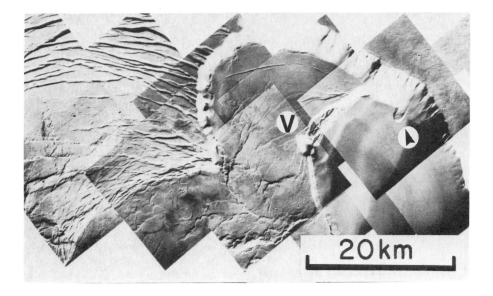

Fig. 6. High resolution mosaic (17 meters per pixel) of the southeastern floor of Olympus Mons caldera. Prominent are craters 6 (left), 3 (center) and 1 (right). A small depression, possibly a vent, is denoted by "V", while the arrow points toward a small landslide within crater 1. Viking survey mission frames from rev. 474S; part of JPL mosaic 211–5930.

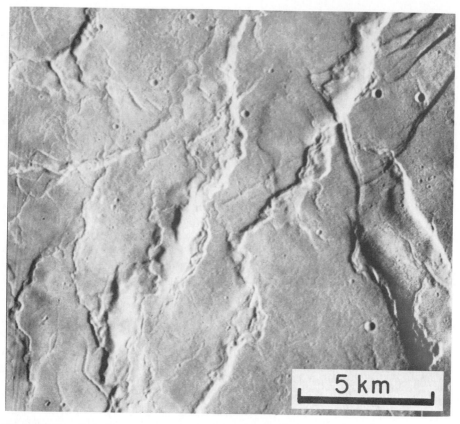

Fig. 7. High resolution image (16 meters per pixel) of the wrinkle ridges on the floor of crater 2 within Olympus Mons caldera. Numerous episodes of deformation are indicated by one set of ridges atop a second set. Several of the narrow ridges display central depressions, which may indicate that these are collapsed lava tubes. Viking frame 473S21.

of pyroclastic flows and sinuous rilles, the vent size, and the areal distribution of air-fall deposits to be predicted for different types of eruptions.

From the numerical models of Wilson and Head (1981b), sinuous rilles are predicted to form during high effusion rate martian hawaiian-style activity, while the low atmospheric pressure on Mars should favor the formation of pyroclastic flows during vulcanian eruptions due to the ease of eruption cloud collapse. Strombolian eruptions on Mars would also be influenced by the low atmospheric pressure, with fine particles ejected to a significant fraction of their vacuum ballistic range (McGetchin and Ullrich, 1973). Strombolian activity would therefore create a subdued deposit, recognizable as an albedo feature two to four kilometers in diameter around the caldera rim. Convective eruption clouds (should they occur) would also be more susceptible to collapse than their terrestrial counterparts, increasing the likelihood of lag-fall deposits close to the vent and ignimbrite sheets surrounding the volcano. Large-scale, coarse deposits would develop close to the vent during plinian eruptions, while even very low mass eruption rates for such events ($100 \text{ m}^3/\text{sec}$ dense rock equivalent) would produce vents in excess of 200 meters in diameter. Consequently, for each style of volcanic activity on Mars, the related surface features should be easily visible in Viking images with a nominal resolution of 30–100 meters per line pair.

INFERRED STYLES OF THARSIS MONTES VOLCANISM

The combination of observational data and the theoretical constraints permits the styles of volcanic activity on the three Tharsis Montes to be deduced. Of primary importance in

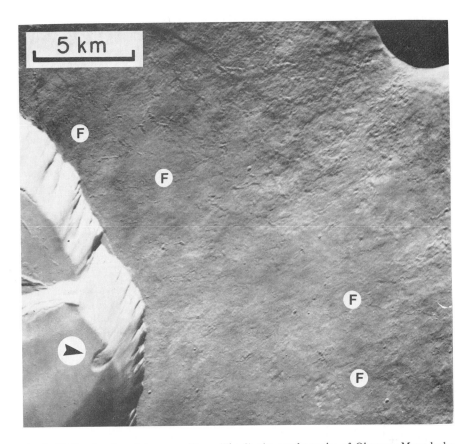

Fig. 8. Unlike the rim of Ascraeus Mons (Fig. 2), the southern rim of Olympus Mons lacks prominent lava flows. While some examples can be found (at the positions labelled "F"), this portion of the volcano has a paucity of flows and circumferential graben. The small landslide (arrowed) on the floor of crater 1 may have been initiated by the creation of the 8 km impact crater shown at top right. Viking frame 475S15, image resolution 23 meters per pixel.

distinguishing between explosive and effusive activity is the identification of the lava flows close to the caldera rim on both Ascraeus (Fig. 2) and Arsia (Fig. 4) Montes. These lava flows indicate that effusive eruptions extended to the highest points on the volcano summits, rather than being confined purely to lateral and subterminal activity. Sinuous rilles on the flanks of Ascraeus Mons (Figs. 2 and 3) imply that at least some of these eruptions involved high effusion rates of turbulent lava (Hulme, 1973; Wilson and Head, 1981a) corresponding to hawaiian-style volcanism. The absence of preserved vents (e.g., on the rim of crater 3 of Ascraeus Mons; Fig. 2) means that the areas of the summit which subsequently collapsed were no doubt similar in morphology to the preserved rim segments.

Due to the retention of these summit lava flows, a strong case can be made for saying that almost all the preserved summit activity on the three montes was non-explosive in character. Evidently lava flows on the caldera rims of Ascraeus and Arsia Montes (Figs. 3 and 4) pre-date the last collapse events (and, hence, the last major summit eruption on each volcano). These lava flows are estimated to be very thin (~10 meters; Schaber *et al.*, 1978), and so would have been quickly buried if air-fall deposits associated with plinian eruptions had been produced (Walker, 1973; Wilson, 1976; Wilson and Head, 1981b). Further evidence against air-fall deposits is that infilling of pre-existing volcanic craters by subsequent eruptions is not observed in Ascraeus Mons (Fig. 2) or Olympus Mons (Fig. 6) caldera. Because of this absence of near-rim air-fall deposits, not only can plinian eruptions be excluded for the Tharsis shields, but also the volcanic products associated with eruption cloud collapse (ignimbrites and other pyroclastic flows; Sparks and Wilson, 1976; Sparks et al., 1978) can be discounted.

Implicit in this observation (i.e., the exclusion of explosive eruptions) is the fact that volcanic materials previously identified around Olympus Mons as possible ignimbrites (King and Riehle, 1974; Scott and Tanaka, 1980) or other pyroclastic flows (Morris, 1980) could not have been produced during an eruptive phase consistent with the deduced summit activity. In order to generate either deposit around Olympus Mons, the accompanying near-vent materials (Wilson, 1976; Sparks and Wilson, 1976; Wright and Walker, 1981) would have to be formed, but these are not observed at the summit. Also absent are any pit craters within the caldera that are large enough to be the source vents for a plinian or ignimbrite-forming eruption. Although significant temporal changes in the eruption characteristics might be invoked to explain former explosive activity in the Tharsis region, the implied sequence of large-scale explosive (silicic) volcanism followed by voluminous basaltic eruptions is very unusual on Earth (Carmichael et al., 1974), and presumably this would also be true for Mars.

Several lines of evidence suggest that molten lava lakes once covered much of each caldera floor. The absence of truncated lava flows formerly on the caldera rim demonstrates that some resurfacing has occurred. The retention of fine-scale rim morphology is considered to preclude the presence of a thick ash mantle upon the floor as a result of explosive eruptions, so that floor resurfacing by lava flows/lava lakes is the most likely mechanism. Wrinkle ridges within Olympus Mons caldera (Figs. 6 and 7) are suggestive of a molten lava lake which has subsequently cooled, contracted, and been tectonically modified in a manner comparable to the mode of formation of lunar mare ridges (Lucchitta, 1976; Sharpton and Head, 1980). Perched lava terraces and the absence of prominent small lava flows on the caldera floor of each volcano also support this lava lake hypothesis. During the final summit eruption, each crater floor was totally resurfaced, rather than being modified in a piecemeal manner.

Intra-caldera activity was probably of minor importance on Mars in comparison to that observed upon terrestrial shields such as Kilauea (Carr and Greeley, 1980), the Galapagos volcanoes (Nordlie, 1973; Swanson et al., 1974), and Piton de la Fournaise, Reunion Island (Ludden, 1977). Very few individual vents within each martian caldera can be seen, with the major exceptions being the low domes within Arsia Mons caldera. More significantly, there is a total absence of any extensive constructional volcanism within the three martian calderas. Typically, central cones several hundred meters high are subsequently constructed within calderas on Earth (such as Crater Dolomieu on Piton de la Fournaise; Upton and Wadsworth, 1966), but only crater "V" within Olympus Mons caldera (Fig. 6) may be a candidate for comparable resurgent activity. Even in this instance, however, no evidence can be found for constructional growth at the summit after final caldera collapse.

MODE OF CALDERA FORMATION

Excluding the possibility of explosive activity on the Tharsis volcanoes nevertheless requires that an alternative explanation for summit collapse has to be found. There appears to be good evidence from terrestrial examples that volcano magma chambers are full at the onset of activity (Blake, 1981), so that only during (or immediately after) an eruption can any void space be created to permit caldera formation or enlargement. Thus, each martian volcanic crater must have been associated with a withdrawal of magma from the chamber either as a result of effusive activity or the intrusion of dikes within the volcanic pile. Distinguishing between the relative importance of summit versus subterminal or lateral eruptions as instigators of summit collapse does, however, depend upon the relative volumes of magma/lava involved and the timing of the collapse episodes.

Numerous very large flows in excess of 600 km in length can be seen upon the flanks of all three volcanoes (Carr et al., 1977; Schaber et al., 1978). Existing data preclude the

Table 3. Lava flows needed for crater collapse

A: Observed flow dimensions on Arsia Mons

Flow Location	Altitude of vent	Flow Length (km)†	Flow Width (km)†	Flow Thickness (m)*
Summit	>18 km	10 – 15	0.5 – 1.3	7
High Flanks	12–18 km	50 – 100	3.0 – 4.0	20
Low Flanks	<10 km	100 – 600	4.0 – 7.0	35

B: Number of flows for observed collapse events

Flow Location	Arsia Mons Caldera	Olympus Mons Crater 1	Ascraeus Mons Crater 1
Summit	$3.6 \times 10^5 - 9.2 \times 10^4$	$3.5 \times 10^4 - 9.1 \times 10^3$	$1.1 \times 10^5 - 2.7 \times 10^4$
High Flanks	$4.2 \times 10^3 - 1.6 \times 10^3$	415 – 155	$1.2 \times 10^3 - 470$
Low Flanks	896 – 85	89 – 8	268 – 26

†Data from Carr et al. (1977) and this analysis.
*Average of data given by Schaber et al. (1978).

direct correlation of individual flows with each collapse event, and even for terrestrial volcanoes this can be a very difficult task. Obviously, the number of flows required to account for the displaced caldera volume would decrease as more of the volume is accounted for by these large flank eruptions. Taking the simple case where all eruptions at a given altitude are of comparable size, the number and dimensions of flows at several elevations that are needed to initiate the observed collapse events are summarized in Table 3. Assuming that none of the magma displaced prior to collapse was distributed throughout the volcano in the form of dikes (probably an oversimplification if martian shields are similar to terrestrial examples: Simkin and Howard, 1970; Macdonald, 1972), an unrealistic number of very small summit eruptions (more than 27000 for crater 1 of Ascraeus Mons) would be needed to initiate each collapse. Plausible numbers (25–270) of very large lava flows (such as those seen at elevations of 8–12 km on Arsia Mons) could, however, account for the observed displaced volumes. Flank eruptions of this type are also consistent with activity recognized on terrestrial volcanoes, where voluminous subterminal and lateral eruptions occur due to the hydraulic fracturing of the volcano (Wadge, 1977; Ludden, 1977). Thus the existence of the calderas on Mars can be adequately explained by the observed effusive activity, with no explosive eruptions required to initiate summit collapse.

CONCLUSIONS

The variety of surface features identified within the calderas and on the summit flanks suggests that the simple interpretation of the Tharsis volcanoes as the martian equivalents of the young hawaiian volcanoes (e.g., Mauna Loa and Kilauea) is no longer adequate to describe these planetary volcanoes. The absence of lag-fall deposits within the calderas and the preservation of small lava flows on the crater rims does, however, indicate that only effusive activity characterized Tharsis Montes volcanism. This observation limits the possible terrestrial analogues to the general class of shield volcanoes, which is particularly important for the interpretation of surface materials surrounding Olympus Mons. By comparison with both field observations and theoretical models of terrestrial volcanoes, it appears that an ignimbritic (or other pyroclastic flow) origin for the aureole materials can be rejected. Were these materials to be the products of explosive volcanism, this would imply an earlier period of large-scale silicic activity, which would give the Olympus Mons magma chamber a chemical evolutionary trend that is extremely unusual for terrestrial volcanoes.

The presence of nested summit craters, wrinkle ridges on the crater floors and rims, circumferential graben, and low domes within Arsia Mons indicates that the summit of each volcano evolved over a period of time. Volcanism within each caldera was most likely restricted to the formation of lava lakes, while the large flank eruptions were the principle cause of caldera collapse. Evidently the system of conduits within each volcano permitted the magma to find repeated egress from the source chamber to both the summit area and the distal flanks. The absence of any intra-caldera eruptions after the formation of the lava lakes requires that a mechanism for terminating summit activity after each collapse has to be found. The cause of the different trends in the size-evolution of the magma chambers of Ascraeus and Olympus Montes is also unknown, but it appears that each volcano (or, more specifically, the effective volume of each magma chamber) was capable of changing in a variety of ways. In order to interpret these differences, it consequently seems that future analyses of the martian shields would benefit from additional comparisons with more evolved terrestrial volcanoes. The older hawaiian volcanoes and other shields located in, for example, Galapagos, Reunion and Iceland would appear to be prime candidates for such a comparison. Only in this manner may it be possible to interpret adequately the summit morphology of the Tharsis Montes and provide additional constraints for the interpretation of their flank deposits.

Acknowledgments—The author wishes to thank Lionel Wilson for extensive discussions during the preparation of this manuscript, and for stimulating support over the last decade in the analysis of planetary volcanoes. Richard Grieve gave valuable comments on an early version of the paper, while Chuck Wood performed admirable editorial support. Reviews were provided by Tom Simkin and an anonymous person. The talents of Sam Merrell (photography) and Sally Bosworth (typing) were appreciated during manuscript preparation. NSSDC supplied the negatives used in preparing Figs 2, 3, 7 and 8. This research was supported by NASA Grant NER 40-002-088 of the Planetary Geology Program.

REFERENCES

Allen C. C. (1979) Volcano-ice interactions on Mars. *J. Geophys. Res.* **84**, 8048–8059.
Blackburn E. A. (1977) Explosive volcanic processes on the Earth and Planets. Ph.D. thesis, Univ. Lancaster, U.K.
Blake S. (1981) Volcanism and the dynamics of open magma chambers. *Nature* **289**, 783–785.
Blasius K. R. and Cutts J. A. (1976) Shield volcanism and lithospheric structure beneath the Tharsis plateau, Mars. *Proc. Lunar Sci. Conf. 7th*, p. 3561–3573.
Carmichael I. S. E., Turner F. J., and Verhoogen J. (1974) *Igneous Petrology*, McGraw-Hill, N.Y. 739 pp.
Carr M. H. (1973) Volcanism on Mars. *J. Geophys. Res.* **78**, 4049–4062.
Carr M. H. (1974a) Tectonism and volcanism of the Tharsis region of Mars. *J. Geophys. Res.* **79**, 3943–3949.
Carr M. H. (1974b) The role of lava erosion in the formation of lunar rilles and martian channels. *Icarus* **22**, 1–23.
Carr M. H. and Greeley R. (1980) *Volcanic Features on Hawaii–A Basis for Comparison with Mars.* NASA SP–403. 211 pp.
Carr M. H., Greeley R., Blasius K. R., Guest J. E., and Murray J. B. (1977) Some volcanic features as viewed from the Viking Orbiters. *J. Geophys. Res.* **82**, 3985–4015.
Comer R. P., Solomon S. C., and Head J. W. (1980) Thickness of the martian lithosphere beneath volcanic loads: A consideration of time dependent effects (abstract). In *Lunar and Planetary Science XI*, p. 171–173. Lunar and Planetary Institute, Houston.
Crumpler L. S. and Aubele J. C. (1978) Structural evolution of Arsia Mons, Pavonis Mons and Ascraeus Mons: Tharsis region of Mars. *Icarus* **34**, 496–511.
Greeley R. (1973) Mariner 9 photographs of small volcanic structures on Mars. *Geology* **1**, 175–180.
Greeley R. and Spudis P. D. (1981) Volcanism on Mars. *Rev. Geophys. Space Phys.* **19**, 13–41.
Guest J. E. (1973) The summit of Mount Etna prior to the 1971 eruptions. *Phil. Trans. Roy. Soc. London* **A274**, 63–78.
Head J. W. and Wilson L. (1979) Alphonsus-type dark-halo craters: morphology, morphometry and eruption conditions. *Proc. Lunar Planet. Sci. Conf. 10th*, p. 2861–2897.,

Hodges C. A. and Moore H. J. (1979) The subglacial birth of Olympus Mons and its aureole. *J. Geophys. Res.* **84**, 8061–8074.
Hulme G. (1973) Turbulent lava flow and the formation of lunar sinuous rilles. *Mod. Geol.* **4**, 107–117.
King J. S. and Riehle J. R. (1974) A proposed origin of the Olympus Mons escarpment. *Icarus* **23**, 300–317.
Jaggar T. A. and Finch R. H. (1924) The explosive eruption of Kilauea, Hawaii, 1924. *Amer. J. Sci.* **V8**, 353–374.
Lucchitta B. K. (1976) Mare ridges and related highland scarps—Result of vertical tectonism? *Proc. Lunar Sci. Conf. 7th*, p. 2761–2782.
Ludden J. N. (1977) Eruptive patterns for the volcano Piton de la Fournaise, Reunion Island. *J. Vol. Geotherm. Res.* **2**, 385–395.
Macdonald G. A. (1972) *Volcanoes*. Prentice-Hall, N.J. 510 pp.
Masursky H. (1973) An overview of geologic results from Mariner 9. *J. Geophys. Res.* **70**, 4000–4030.
McGetchin T. R. and Ullrich G. W. (1973) Xenoliths in maars and diatremes with inferences for the Moon, Mars and Venus. *J. Geophys. Res.* **78**, 1833–1853.
Moore H. J. and Hodges C. A. (1980) Some martian volcanic centers with small edifices (abstract). In *Reports of Planetary Geology Program* 1980, p. 266–268. NASA TM-82385.
Morris E. C. (1980) A pyroclastic origin for the aureole deposits of Olympus Mons (abstract). In *Reports of Planetary Geology Program* 1980, p. 252–254. NASA TM-82385.
Nordlie B. E. (1973) Morphology and structure of the Western Galapagos volcanoes and a model for their origin. *Geol. Soc. Amer. Bull.* **84**, 2931–2956.
Schaber G. G., Horstman K. C., and Dial A. L. (1978) Lava flow materials in the Tharsis region of Mars. *Proc. Lunar Planet Sci. Conf. 9th*, p. 3433–3458.
Scott D. H. and Tanaka K. L. (1980) Martian ignimbrites (abstract). In *Reports of Planetary Geology Program* 1980, p. 255–257. NASA-TM 82385.
Sharpton V. L. and Head J. W. (1980) Lunar mare arches and ridges: Relation of ridge lobes to small pre-existing craters (abstract). In *Lunar and Planetary Science XI*, 1024–1026. Lunar and Planetary Institute, Houston.
Simkin T. and Howard K. A. (1970) Caldera collapse in the Galapagos Islands, 1968. *Science* **169**, 429–437.
Sparks R. S. J. and Wilson L. (1976) A model for the formation of ignimbrite by gravitational column collapse. *J. Geol. Soc. London* **132**, 441–451.
Sparks R. S. J., Wilson L., and Hulme G. (1978) Theoretical modelling of the generation, movement and emplacement of pyroclastic flows by column collapse. *J. Geophys. Res.* **83**, 1727–1739.
Swanson F. J., Baitis H. W., Lexa J., and Dymond J. (1974) Geology of Santiago, Rabida, and Pinzon Islands, Galapagos. *Geol. Soc. Amer. Bull.* **85**, 1803–1810.
Tazieff H. (1976–7) An exceptional eruption: Mt. Niragongo, Jan. 10th, 1977. *Bull. Volc.* **40-3**, 189–200.
Upton B. G. J. and Wadsworth W. J. (1966) The basalts of Reunion Island, Indian Ocean. *Bull. Volc.* **19**, 7–24.
Wadge G. (1977) The storage and release of magma on Mount Etna. *J. Volc. Geotherm. Res.* **2**, 361–384.
Walker G. P. L. (1973) Explosive volcanic eruptions—a new classification scheme. *Geol. Rundschau* **62**, 431–446.
Williams H. and McBirney A. R. (1979) *Volcanology*. Freeman, Cooper and Co., San Francisco. 397 pp.
Wilson L. (1976) Explosive volcanic eruptions—III. Plinian eruption columns. *Geophys. J. Roy. Astron. Soc.* **45**, 543–556.
Wilson L. and Head J. W. (1981a) Ascent and eruption of basaltic magma on the Earth and Moon. *J. Geophys. Res.* **86**, 2971–3001.
Wilson L. and Head J. W. (1981b) Volcanic eruption mechanisms on Mars: Some theoretical constraints (abstract). In *Lunar and Planetary Science XII*, p. 1194–1196. Lunar and Planetary Institute, Houston.
Wilson L., Sparks R. S. J., and Walker G. P. L. (1980) Explosive volcanic eruptions—IV. The control of magma properties and conduit geometry on eruption column behaviour. *Geophys. J. Roy. Astron. Soc.* **63**, 117–148.
Wright J. V. and Walker G. P. L. (1981) Eruption, transport and deposition of ignimbrite: A case study from Mexico. *J. Volc. Geotherm. Res.* **9**, 111–131.
Wu S. S. C. (1980) Special Topographic Map of Mars, Arsia Mons. U.S. Geol. Surv. Map 9/121T.
Wu S. S. C. (1981) Special Topographic Map of Mars, Olympus Mons. U.S. Geol. Surv. Map 19/134T.

Mars: A large highland volcanic province revealed by Viking images

David H. Scott and Kenneth L. Tanaka

United States Geological Survey, Flagstaff, Arizona 86001

Abstract—Regional geologic mapping from Viking images indicates that many of the mountains in the Thaumasia and Phaethontis quadrangles have morphologies characteristic of volcanoes. This postulated volcanic province lies more than 1000 km to the south and southwest of Tharsis Montes, but it is older than early lava flows from these giant shield volcanoes and its landforms do not follow the NE–SW alignments of the Tharsis structures. Some of the probable volcanoes in this province have steep sides and bold relief, suggesting that they may be composite structures formed by lava flows and interbedded pyroclastic deposits. Generally, they do not resemble other highland volcanoes such as Tyrrhena, Hadriaca, and Amphitrites Paterae which are more subdued in relief. The eastern part of the province is densely faulted and fractured in NNW to NNE directions. The mountains, provisionally mapped as volcanoes, are partly buried or embayed in places by lava flows from Arsia Mons, indicating that Tharsis volcanism occurred after an earlier period of volcanic mountain building and major tectonic activity. The central part of the province is dissected by NE trending faults and also by older sets of faults and ridges in NW directions. The volcanic structures show some preferential alignment with the older fault and ridge systems. Volcanoes in the western region commonly have oval or elongate shapes and some are aligned transverse to the major ENE fault directions of Sirenum Fossae. In places the volcanic constructs resemble fissure vents with ramparts built up along their margins; others consist of overlapping and coalescing craters along the rims of large circular structures that do not resemble impact craters and may themselves be of volcanic origin.

INTRODUCTION

Large volcanoes on Mars occur in both highland and lowland regions. These two regions, aside from elevation differences, are characterized physiographically by coarse- and fine-scale elements of relief. The highlands are composed of comparatively rough terrain with many craters and their eroded remnants; whereas the lowlands are relatively smooth, consisting of plains initially formed by lava flows. Examples of lowland volcanoes are Olympus Mons, Alba Patera, Elysium Mons, and the low relief structure and caldera within ridged plains at Syrtis Major (Scott and Carr, 1978). Highland volcanoes include Tyrrhena, Hadriaca, and Amphitrites Pataraε as well as several smaller, mostly unmapped constructs. The prominent volcanoes of Tharsis Montes and Apollinaris Patera lie on the borderland between highland and lowland provinces.

Regional geologic mapping from Viking images has revealed numerous moderate- to large-size structures and elongate fissure vents in the Thaumasia and Phaethontis quadrangles in the western highland region of Mars; these features, provisionally mapped as volcanoes, previously were interpreted to be massifs related to a huge impact basin centered at Syria Planum (Schultz and Glicken, 1979). The volcanic constructs lie 1000 km and more to the south and southwest of Tharsis Montes but they are not aligned with these giant shield volcanoes. They also appear to be considerably older morphologically, and this is substantiated in places by stratigraphic relations. Lava flows from Arsia Mons, the southernmost volcano of Tharsis Montes, have flooded large areas of this highland terrain and have overlapped or embayed many of the volcanic structures. As presently mapped, more than 20 volcanoes and suspected volcanic vents are concentrated in this region covering some 4 million km^2 between about long 90°–180°W and lat 28°S to 45°S (Fig. 1). The volcanoes range up to 100 km or more across their bases and some of

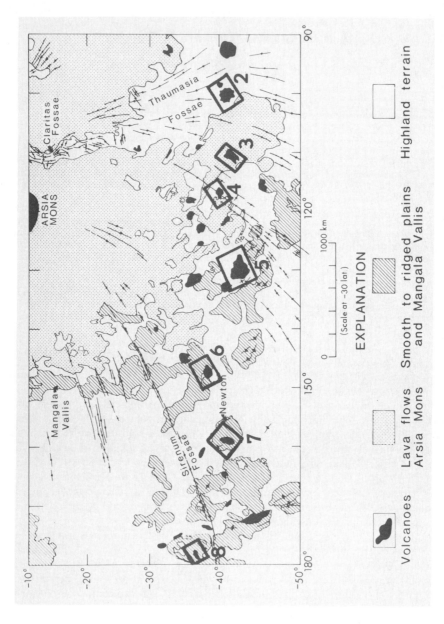

Fig. 1. Generalized geologic map of the Phaethontis–Thaumasia region of Mars. Numbered boxes show locations of Fig. 2–8 in text. Ball on line denotes graben.

them appear to have moderate- to high-relief, but no height estimates have been made. Lava flow fronts and deeply incised drainage patterns are visible in places on their flanks. Suspected calderas and summit craters are not uncommon, but generally are highly degraded and dissected by infilling, faulting, and channeling. The region is structurally complex and is transected by fault, fracture, and ridge systems having different ages and orientations. Generally, westerly to northwesterly trending faults are older than those having northeasterly directions. Many of the volcanoes and vents appear to follow the older structural systems.

OBSERVATIONS

The selected examples of volcanoes in the following discussion were chosen to give a representative sampling across the entire area. The descriptions of the figures proceed from Thaumasia Fossae on the east to Sirenum Fossae in the Phaethontis quadrangle on the west.

Within the Thaumasia quadrangle, a dense complex of intersecting faults and grabens transect the cratered highlands and fan outwards from the Tharsis Montes in SSE and SSW directions (McGill, 1978). A large part of this area has been flooded by early lava flows from Arsia Mons or surrounding fissures (Scott and Tanaka, 1981). These flows embay the highland terrain and some of the volcanoes in this region; they also bury older fault and fracture systems in the highlands, although a few younger faults transect the lava plains surface. The highland fault systems cut across the volcanic constructs in places, but are partly buried in other places by local flows from these sources. Thus it appears that highland volcanism continued after major tectonic activity, but generally

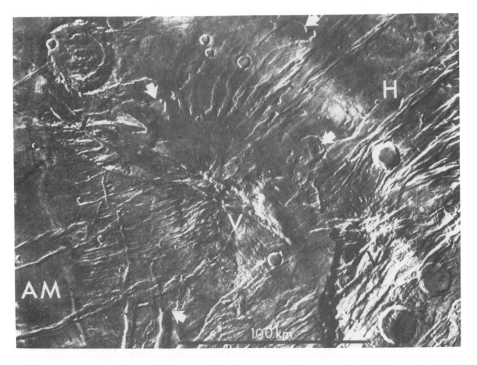

Fig. 2. These two proposed volcanoes (V) are typical for the Thaumasia highland region (H). The older volcano in the southeast corner is heavily fractured whereas the younger, central volcano post-dates much of the faulting. Note the filled fractures, partly filled crater, and eroded lava scarps (arrows). Highlands and volcanic material are embayed by Arsia Mons lava flows (AM). Viking frame 63A08 centered at −41°S, 101°W.

ceased before the voluminous, areally extensive eruptions from Tharsis Montes covered large parts of the highland terrain (Scott and Tanaka, 1980).

The two volcanoes shown in Fig. 2 are highly faulted and dissected but some of the flows on the flanks of the central structure overlap faults of the highlands and, in turn, are embayed on the west by early lava flows from Arsia Mons. This volcano also lies at the intersection of major northwest-northeast fault systems.

In Fig. 3, the elongate mountain having a broad, oval-shaped central depression is interpreted to be volcanic. It is embayed by Arsia Mons lava flows and aligned with the northeasterly trending set of faults characteristic of the Sirenum Fossae area. As in the preceeding examples, faulting is both older and younger than the volcano. The interior fault along the southeast wall of the depression or caldera is completely buried on the northeast by a large accumulation of rim material; farther to the northeast, however, its trace emerges as a linear scarp in the terra that is surrounded by Arsia Mons flows. Faults on the northwest rim of the structure transect rim material in some places but are covered or partly covered in others.

Some of the volcanoes have relatively steep sides and stubby rounded flow fronts resembling those of terrestrial composite volcanoes. The structure illustrated in Fig. 4 has a prominent central dome with a smaller peak off center from its apex. Some indication of

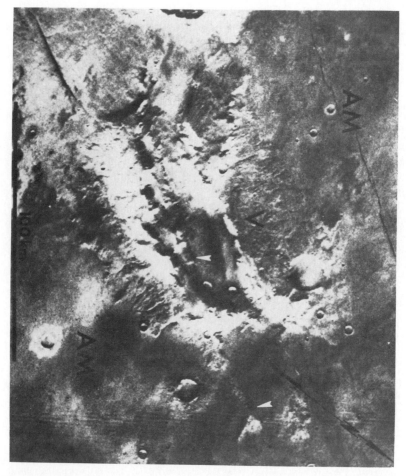

Fig. 3. Fissure vent aligned with NE–SW fractures in the Thaumasia region. Lavas (AM) from the Arsia Mons area embay the vent materials (V). Fault trace (arrows) partly buried near midpoint by bulbous mass. Viking frame 567A73 centered at $-42°S$, $111°W$.

Fig. 4. Domical volcanic center (V) in Phaethontis region from which lavas and/or volcaniclastics (D) issued. Circular deposit northwest of volcano may be derived from the channel shown by the arrow. Erosion of this deposit by fluvial activity is apparent. Plains materials (P) embay the volcanic units. Viking frame 567A70 centered at −40°S, 117°W.

caldera collapse is present around the peak. It is possible that subsidence occurred after a major eruptive period, which may have been followed by the intrusion and uplift of a nearly central core or spine; the spine-forming material appears to be younger than the faults that cut across other parts of the structure. On the southwest side of this postulated volcano an elongate collapse depression with a flat, smooth appearing floor opens into a curved channel; the channel emerges westward onto the surface of the plains where a large (40 km diameter) circular shaped deposit extends outward from its mouth. It has the appearance of an accumulation of lava or volcaniclastic flow material from the volcano. This material is also younger than some faults that transect adjacent terrain, including smooth plains which partly embay the volcano on the west.

Figure 5 is a composite of three Viking images showing a large (90 km diameter) semicircular crater with a breached eastern wall. Dark plains material within the crater appears to have been extruded through the ruptured wall, possibly leaving a bench or terrace that marks the former position of the fill material—probably basalt. On the

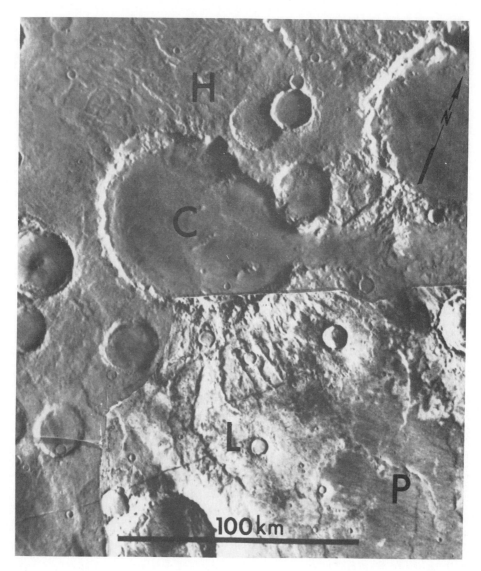

Fig. 5. The large crater (C) is a possible caldera and may be the source of the lobate flow-like deposits (L) which overlap the ridged plains (P). Note breached rim on crater's east side through which dark material has issued and partly covered highlands (H). Mosaic of Viking frames 529A78, 567A63, and 567A82 located at −41°S, 130°W.

southeast flank of the structure tongue-like, lobate patterns resemble overlapping lava flows that may have issued from fissures below its rim. The crater may be a caldera surmounting a low-relief volcano; alternatively, it may be of impact origin and subsequently filled by basalt rising through fractures in its floor. In this case, the flow-like patterns on the flank of the crater possibly could have been carved by running water.

Areas of mountainous, rugged terrain occur throughout the province that are not recognizably associated with impact crater or basin materials. Their origins are uncertain but may include either tectonic or volcanic processes. They are not mapped as volcanoes, however, unless other criteria indicative of volcanic origin such as flow fronts, calderas, etc. occur in association with them. A borderline example of this type of feature is shown in Fig. 6. The large mass of material extending across an old crater rim and onto its floor

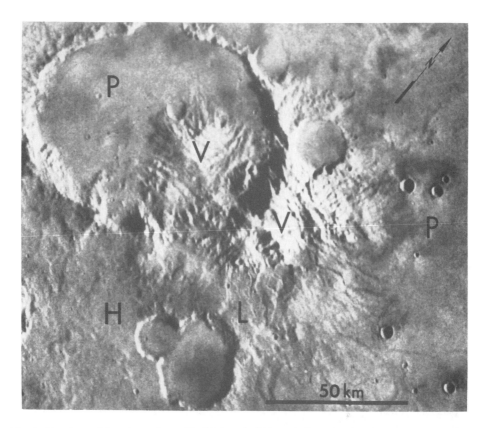

Fig. 6. The size and locations of massifs (V) found within and along the rim of the large crater imply an origin unrelated to the cratering event and thus these massifs are likely to be volcanic. Possible lavas (L) overlap highland materials (H). Smooth plains materials (P) embay the highland units (V, L, H). Viking frame 637A44 centered at −39°S, 148°W.

is much too large and in the wrong location to have been produced by the impact that formed the original crater. Several eruption centers appear to be present along the crater rim crest and within its floor. Some lava flow fronts may occur on the southern and eastern slopes of the structure but they are highly degraded and their origin and composition are conjectural.

The long, narrow volcanic structure of Fig. 7 lies within the large (250 km diameter) crater Newton whose dark floor is partly covered by ridged plains material, possibly of basaltic composition. Along the north side of the structure, low-relief flow fronts extend outward from its base and overlap ridges in the floor covering; in a few other places, however, floor material may have embayed its flanks, suggesting that formation of the volcano and extrusion of the ridged plains lava flows occurred almost concurrently. Darker patches visible along the southwest wall of the crater and across the south side of the volcano appear to be thin flows or ashfalls that terminated volcanic activity; they may have emanated from the volcano and from small vents or fissures in the crater floor. The orientation of the volcano is puzzling as the nearly north-south trend of its long axis does not conform with the regional east-northeast fault directions of Sirenum Fossae. Several other postulated volcanic structures in this northwest part of the Phaethontis quadrangle also are oriented nearly normal to the structural grain of the province. Possibly they follow older fault systems whose traces have been largely obliterated by the resurfacing of parts of the highland area.

The structure shown in Fig. 8 is a complex, highly faulted edifice whose flanks have

Fig. 7. Elongate fissure vent (V) with associated lavas and possible pyroclastics (L) located in the crater Newton. Ridged plains materials (P) filling the crater floor embay highland materials (H) and underly deposits from the vent in places. A chain of pit craters have formed along the vent's ridgecrest. Viking frame 562A09 located at −41°S, 159°W.

been partly buried by debris flows generated from adjacent areas of chaotic terrain. It resembles the construct shown in Fig. 3, and has a central low area flanked on the north and south by narrow ridges and on the east and west by larger, somewhat bulbous accumulations of material. The deposit on the east overlaps some faults interior to the structure and partly buries a prominent graben extending outward beyond its base. A series of curved, parallel lineaments along its south side may represent concentric faults dipping toward the central core of the structure or, possibly, bedding planes of layered materials in its rim. It lies along the Sirenum Fossae fault system and is thought to be of volcanotectonic origin.

Some landforms in the Phaethontis area resemble fissure vents with ramparts built up along their margins; others consist of overlapping and coalescing craters, only one of which was recognized from Mariner 9 pictures and interpreted to be volcanic (Howard, 1979). In several places these structures lie on or near the outlines of large (200 km diameter), circular, and nearly rimless features that do not clearly resemble impact

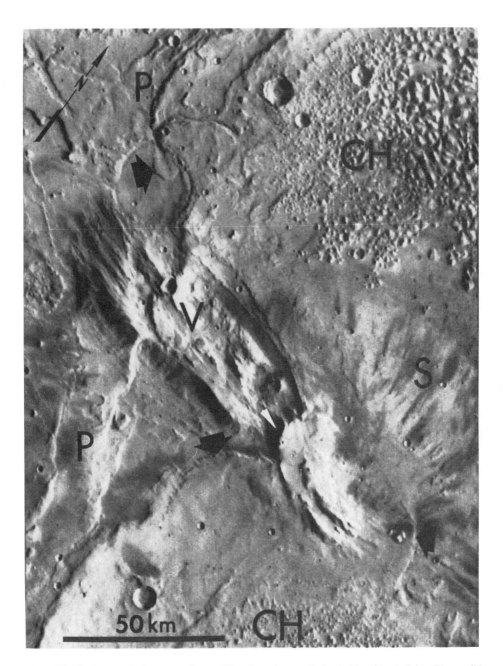

Fig. 8. Highly fractured, elongate volcano (V) embayed or partly buried by ridged plains (P), possible debris flows (black arrows), and sediments (S). Chaotic terrain (CH) may have been source of sedimentary materials. Structure is partly buried on east side by probable late stage volcano deposits (white arrows). Curved lineaments faintly visible on south flank of volcano. Viking frame 597A85 centered at −37°S, 178°W.

craters. The floors of these circular features appear to be slightly raised and consist of chaotic blocks and knobby material surrounded by dark overlapping plains having the appearance of basalt flows. Detailed studies of these structures have not yet been made; if they are of volcanic rather than impact origin, interesting possibilities exist for finding similar features in other areas of the highlands.

CONCLUSIONS

Many of the mountains in the rugged highland terrain of the Phaethontis and Thaumasia quadrangles are believed to be of volcanic origin. Those provisionally mapped as volcanoes have diagnostic characteristics such as lobate flow fronts around their bases, depressed central areas, or have massive, bulbous accumulations of material of no determinable origin other than volcanic. Most of the volcanoes are younger than materials forming the highlands but are older than early lava flows from Arsia Mons. Many are alined along older fault and ridge systems that are transverse to the more recent and prominent faults transecting the region. The older faults are generally buried by plains lava flows but their traces are visible in several places in the highlands. These faults are relatively short in length and do not appear on the map of Fig. 1.

Acknowledgments—The authors thank G. Wadge and an unknown reviewer for their helpful comments. This research was supported by NASA work order W-13,709.

REFERENCES

Howard J. H. III (1979) Geologic map of the Phaethontis quadrangle of Mars. U.S. Geol. Survey Misc. Geol. Inv. Map I-1145.
McGill G. E. (1978) Geologic map of the Thaumasia quadrangle of Mars. U.S. Geol. Survey Misc. Geol. Inv. Map I-1077.
Schultz P. H. and Glicken H. (1979) Impact crater and basin control of igneous processes on Mars. *J. Geophys. Res.* **84**, 8033–8047.
Scott D. H. and Carr M. H. (1978) Geologic map of Mars. U.S. Geol. Survey Misc. Geol. Inv. Map I-1083.
Scott D. H. and Tanaka K. L. (1980) Mars Tharsis region: Volcanotectonic events in the stratigraphic record. *Proc. Lunar Planet. Sci. Conf. 11th*, p. 2403–2421.
Scott D. H. and Tanaka K. L. (1981) Maps showing lava flows in the northwest part of the Thaumasia quadrangle and northeast part of the Phaethontis quadrangle of Mars. U.S. Geol. Survey Misc. Geol. Inv. Map I-1273, I-1281. In press.

A secondary origin for the central plateau of Hebes Chasma

Christine Peterson*

U.S. Geological Survey, 2255 North Gemini Drive, Flagstaff, Arizona 86001

Abstract—Hebes Chasma, one of the northern members of the Valles Marineris, can be divided into three physiographic provinces; chasma walls, chasma floor, and central plateau. Theories of origin of the 5-kilometer-high central plateau include (1) the plateau is an eroded remnant of the surrounding plains, and (2) the plateau is a secondary feature deposited after formation of the chasma. A secondary eolian or pyroclastic origin best explains the morphology of the plateau. The chasma probably formed by collapse of a pre-existing graben that was widened by landslides and subsequently filled with eolian or pyroclastic material. Continued mass wasting isolated the plateau from the chasma walls. The enclosed nature of Hebes Chasma may have inhibited eolian erosion and transport within the trough, so that the relatively fresh appearance of the plateau has been preserved.

INTRODUCTION

Hebes Chasma is one of the northern members of the 4500-kilometer-long Valles Marineris (Fig. 1). It is the only separate and enclosed member of the chasma system. Within it lies a 5-kilometer-high central plateau (Figs. 2, 3) composed of layered material of unknown origin. Layered deposits are present in several areas of Valles Marineris (Blasius *et al.*, 1977). Many of the deposits within the basins south and east of Hebes Chasma are relatively thin and extensively eroded, possibly as a result of intense eolian and landslide activity after breaching of the chasmata. The enclosed nature of Hebes Chasma may have inhibited eolian erosion and transport within the basin, so that the relatively fresh appearance of the central plateau has been preserved.

Layered deposits have been interpreted both as eroded remnants of the surrounding plains materials (Malin, 1976) and as secondary features deposited after formation of the troughs (Sharp, 1973; McCauley, 1978). The objectives of this study are to map the stratigraphic units within the chasma, and to suggest possible origins of the central plateau.

Mapping of the chasma was done on very high resolution Viking Orbiter 1 photographs. These pictures were enlarged to a scale of approximately 1 : 250,000 and fitted together in a mosaic. Elevation data were obtained by photogrammetry (Wu *et al.*, 1980).

STRATIGRAPHY

The ridged plains materials surrounding Hebes Chasma are thought to be composed of flood basalts similar to those of the lunar maria (McCauley, 1978). The basalts may be interlayered with volcanic ash and eolian material (McGetchin and Smith, 1978). These materials overlie intensely fractured trough-and-furrow material and the ancient cratered terrain of the southern hemisphere, both of which are exposed east and south of the study area (McCauley, 1978). Ridged plains material are overlain to the west by cratered plains materials, which consist of lava flows associated with the Tharsis volcanism that were subsequently blanketed by eolian deposits (McCauley, 1978).

---

*Presently at University of Texas, Austin, Texas.

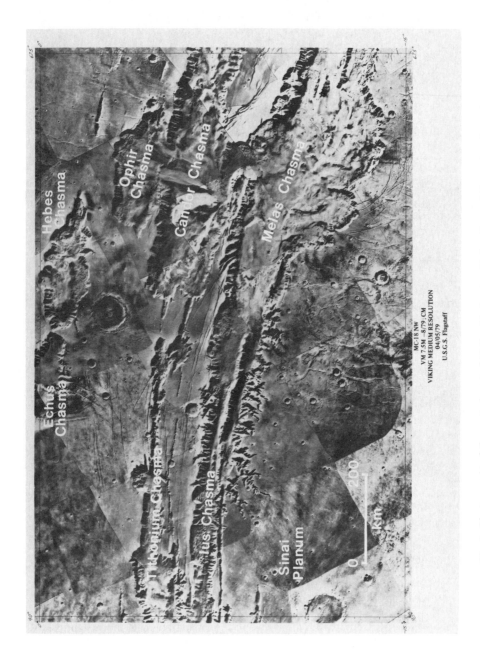

Fig. 1. Photomosaic of the central part of Valles Marineris with locations of the major chasmata.

Fig. 2. Viking 1 pictures of Hebes Chasma showing three physiographic provinces: chasma wall (cw), chasma floor (cf), and central plateau (cp). North is at top. Note normal faulting at southeast end of central plateau (hachures are on downthrown side). Local structural control is illustrated by the collapse of a tributary canyon along pre-existing grabens at "d" and by the oblique trend of spurs at "f". Layering within the plateau is especially apparent on its north slope. Broadly curved landslide scarps are shown at "g". Profile A-A' is shown in Fig. 3. (Viking Orbiter frames 682A27 and 645A60).

Hebes Chasma can be divided into three physiographic provinces, chasma wall, chasma floor, and central plateau (Fig. 2). Each province in turn can be divided into separate units according to albedo, planimetric and topographic shape, and texture.

Chasma wall material

Spur-and-gully morphology (unit cwg, Fig. 4) is the dominant morphology of the chasma walls and is expressed by deep gullies and downward-bifurcating spurs. The erosion may be gravity-induced and similar to that affecting some terrestrial scarps in desert and alpine regions (Lucchitta, 1978). Another conspicuous type of wall morphology is the landslide scarps, characterized by broadly curved to nearly straight sections of chasma wall with massive accumulations of hummocky deposits at its base (Figs. 2, 4). Exposed on upper scarps (unit cwl, Fig. 4) are apparently horizontal layers which may consist of extensive flood basalts (McCauley, 1978). These units have been eroded into vertical ribs near the top of the scarps. The lower scarps are covered by smooth talus slopes (unit cwt, Fig. 4). The formation of landslide scarps locally appears to have destroyed the spur-and-gully morphology. Spurs and gullies probably form slowly by the downhill movement of loosened material in areas unaffected by catastrophic events such as landslides (Lehman,

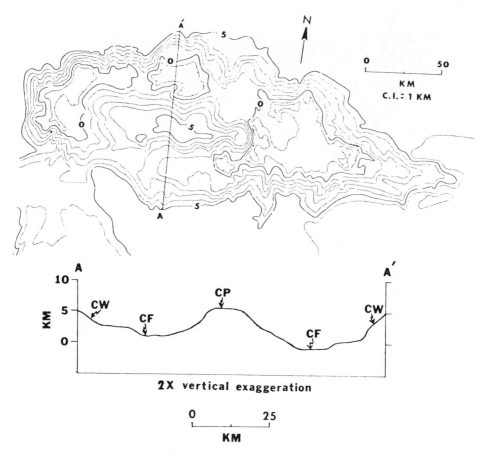

Fig. 3. Topographic map of Hebes Chasma (Wu *et al.*, 1980) and profile A-A′ through its central part. See Fig. 2 for identification of symbols.

1933; Rapp, 1960; Sharp, 1973). Thus the present landslide scarps may eventually evolve into spurs and gullies by slow mass-wasting processes (Lucchitta, 1978).

Chasma floor deposits

Most deposits on the floor of the chasma are presumably formed by landslides derived from the wall of the chasma. These floor units include longitudinally striated units (Unit fss, Fig. 4) and massive hummocky deposits (unit fsh, Fig. 4) lying at the base of the landslide scarps. Hilly deposits (unit fh, Fig. 5) and smooth material of low albedo were also mapped (unit fs, Fig. 5). Deposits resembling the fluted plateau material (described below) are found on the floor north of the central plateau (unit ff, Fig. 5).

Plateau material

The 5-kilometer-high plateau can be divided into five units. The section from the base of the plateau to the break in slope near its top is dominantly composed of massive material that is eroded in a distinctive fluted pattern (unit, pwf, Fig. 6). Less deeply eroded, laterally discontinuous, strata (unit pwr, Fig. 6) are interbedded with the fluted material. These units appear to be more resistant to erosion than the fluted plateau material.

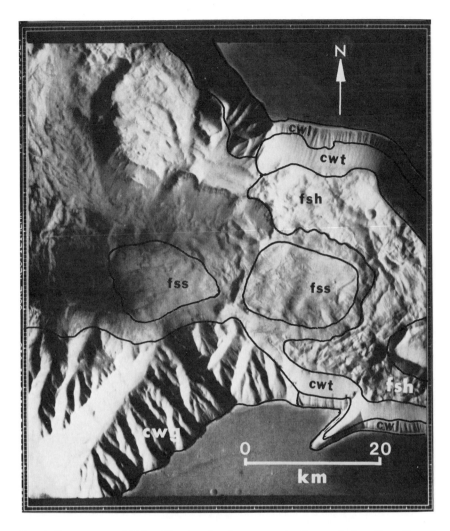

Fig. 4. Eastern part of Hebes Chasma, showing spur-and-gully morphology (in unit cwg), landslide scarps with exposed layered material near their top (unit cwl), and talus deposits (unit cwt). Also shown are longitudinally striated material (unit fss) and hummocky landslide deposits (unit fsh). (Frame 917A05).

These units are overlain by plateau-capping material (unit pcu), in which at least five broad albedo bands can be recognized on the more gently dipping north-facing slope (Fig. 6). Faint ridges and scarps less than 5 kilometers long can be seen on the western half of the plateau (Fig. 6). These ridges are very similar in appearance, although apparently smaller in scale, to those on the ridged plains surrounding the chasma. In addition, the cap unit is equal in elevation to the ridged plains. Therefore, the plateau capping material may be the same material that makes up the uppermost units of the ridged plains. Ridges on the eastern half of the plateau may be absent or mantled by a loess or ash blanket equivalent to that which covers the cratered plains west of Hebes Chasma (McCauley, 1978).

Talus deposits derived from the plateau are rare. The scarcity of recent talus may indicate that either the walls of the plateau are not presently being eroded, or the talus is being transported away by wind before large volumes can accumulate on the slopes. The latter case is favored because of the location of small talus deposits in reentrants in the plateau that may be shielded from wind action (as in Fig. 6, bottom right).

Fig. 5. Northwestern part of Hebes Chasma showing hilly floor deposit (unit fh), fluted floor deposits (unit ff), and smooth dark material (unit fs). (Frame 919A03).

STRUCTURE

The troughs of Valles Marineris probably formed in response to extensional stresses concentric to the domed Tharsis-Syria Rise (Carr, 1974). Structural control of small-scale features is exemplified within Hebes Chasma by the formation of tributary canyons along pre-existing grabens (Fig. 2). Broadly curved fractures at the west end of the chasma appear to facilitate retreat of the walls.

The trend of resistant spurs at the west end of the north wall of the chasma is further evidence of structural influence (Fig. 2). Some of these spurs trend approximately 30 degrees away from the downslope direction, reflecting the attitudes of grabens that were truncated by the chasma wall.

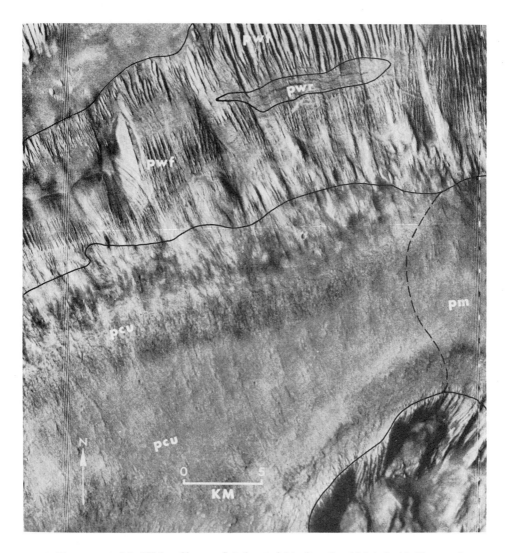

Fig. 6. Plateau material of Hebes Chasma: fluted material (unit pwf), with interbedded layers of more resistant material (unit pwr), overlain by several layers of plateau-capping material (unit pcu), and eolian mantling material (unit pm). (Frame 738A68).

ORIGIN OF THE CENTRAL PLATEAU

In this study, four theories of the origin of the plateau material are considered: 1) the plateau is an eroded remnant of the adjacent plains units, 2) the plateau is composed of lacustrine deposits, 3) the plateau units are eolian, and 4) the plateau material is pyroclastic.

The plateau as an erosional remnant of the plains (Malin, 1976)

Dissimilar weathering patterns and morphologies of the plateau and chasma walls indicate that the two physiographic provinces are not genetically related. The fluted weathering pattern of the plateau (Figs. 5 and 6) is distinctly different than the patterns of the chasma walls (Figs. 4 and 5). The flat-topped nature of the plateau itself is in clear contrast to the sharp-ridged remnants of the plains within Ius Chasma. These remnants are not eroded in

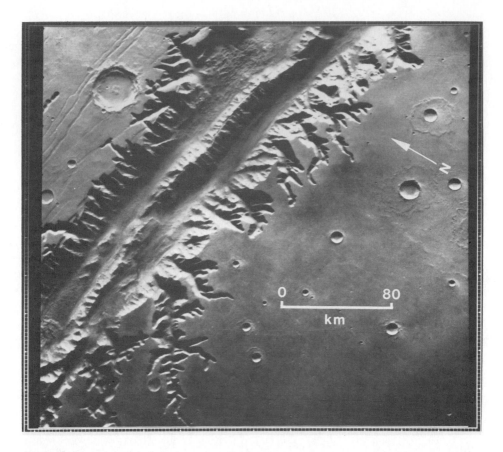

Fig. 7. Part of Ius Chasma containing sharp-ridged remnants of surrounding plains material. Compare with flat-topped, fluted material of the central plateau of Hebes Chasma (Figs. 2, 5). (Frame 645A57).

a fluted pattern, but have retained the spur-and-gully chasma wall morphology (Fig. 7). Similarly eroded ridges are located between Ophir and Candor Chasmata (Fig. 8). Furthermore, the layered strata of the upper chasma walls (unit cwl, Fig. 4) are laterally continuous, whereas the resistant layers in the plateau (unit pwr, Fig. 6) are of small areal extent. Their sparse distribution suggests that the latter occur randomly within the plateau material, rather than as an ordered succession of volcanic flows that presumably underlie the martian plains (McCauley, 1978). Thus, distinct contrasts between the morphologies and erosional characteristics of the chasma walls and the plateau material indicate that the latter was emplaced as a secondary deposit after formation of the chasma.

The plateau as lacustrine in origin

McCauley (1978) has proposed that the layered deposits throughout Valles Marineris are waterlaid sediments deposited in a low-energy environment. The deposits of the type area in Gangis Chasma are similar to the plateau material of Hebes Chasma in their erosional patterns and layered appearance. The source of the sediment in the Gangis Chasma model is material eroded from walls of the chasma, interlayered with eolian deposits derived from dust storms and with volcanic ash from the Tharsis region (McCauley, 1978).

Erosion of the sediments in this model is accomplished by catastrophic draining and exposure to wind action as a result of breaching of the chasmata walls. These processes are difficult to ascribe to the deposits within Hebes Chasma, which is an enclosed basin.

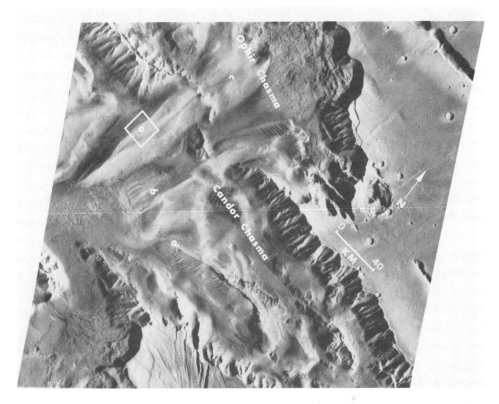

Fig. 8. Possible eolian deposits (shown at a, b, c, and d) in Ophir and Candor Chasmata. Outlined area at "a" shows approximate location of Fig. 9. (Frame 608A74).

In addition, the plateau is approximately the same elevation as the surrounding plains (Fig. 3). Deposition and subsequent re-excavation of a lacustrine sediment pile as deep as the containing basin itself is an unlikely occurrence, as pointed out by Malin (1976).

The plateau as eolian in origin

According to this model, the plateau is composed of sediments derived from planetwide or localized dust storms. Possible large-scale crossbedding of eolian transport is visible in the floor deposits in the area between Candor and Ophir Chasmata (Figs. 8, 9). These features indicate that eolian sedimentation has occurred within some areas of Valles Marineris after formation of the troughs.

The horizontal layering in the plateau of Hebes Chasma could be a result of differences in composition and water content of successive eolian depositional episodes. These differences could account for the variations in color and resistance to erosion of the layers. Isolation of the plateau from the main chasma walls might be accomplished by continued wall collapse after infilling, as outlined in Fig. 11 and discussed below.

The lack of extensive smooth eolian deposits on the floor of Hebes Chasma suggests that present eolian activity is eroding and redistributing floor material, rather than depositing large volumes of foreign material. However, Ward (1974) and Blasius *et al.* (1977) discuss the possibility that deposition of windblown material may be influenced by climatic variations due to change in Mars' orbital eccentricity and obliquity. These changes could account for the extremely regular layering in Candor and Juventae Chasmata (Blasius *et al.*, 1977) and the shift from depositional to dominantly erosional eolian activity in Hebes Chasma.

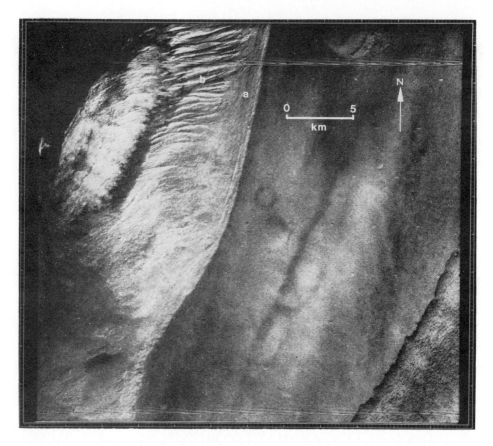

Fig. 9. Very high resolution photograph of area outlined in Fig. 8. Several crossbedded layers indicating north to south transport of eolian material are visible at "a". Note similarity of fluted weathering pattern at "b" to the fluted material of the central plateau of Hebes Chasma. (Frame 815A48).

The plateau as pyroclastic deposits

This model proposes that the central plateau consists of a thick sequence of volcanic ash, with interbedded layers of relatively resistant welded tuffs.

Early references to the existence of pyroclastics on Mars include that of King and Riehle (1974), who noted the similarity between the scarp of Olympus Mons and scarps formed in terrestrial ash-flow tuffs. West (1974) located many martian features similar in size and shape to terrestrial cinder cones such as Sunset Crater in Arizona. More recently, Morris (1980) proposed that the aureole deposits of Olympus Mons are composed of pyroclastic material. Substantial amounts of phreatic pyroclastics on Mars have also been postulated by Womer *et al.* (1979). McGetchin and Smith (1978) derived probable mineral assemblages for the dense martian mantle and concluded that partial melting of these assemblages would yield ultrabasic lavas and ferrokimberlitic ash. Thus, if the martian mantle does include abundant volatiles, pyroclastic material may be a volumetrically significant component of the martian surface.

Erosional patterns of some terrestrial ash flows are similar to the fluted pattern in the plateau of Hebes Chasma (Fig. 10). These ash flows also include resistant layers of welded tuff (Scott, 1969), which resemble the more resistant material (unit pwr) of the plateau (Fig. 6).

If the plateau materials were emplaced during a single early volcanic episode, it is

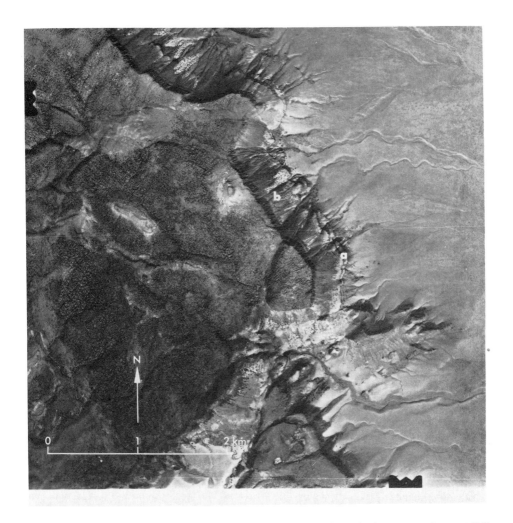

Fig. 10. Aerial photograph of ash-flow tuffs in Nevada with fluted weathering pattern shown at "b". Interbedded layers of more resistant welded tuffs shown at "a" are similar in appearance to the resistant layers in the central plateau, of Hebes Chasma; compare Fig. 6 (Southern Pancake Range, Nye County, Nevada).

reasonable to assume that the larger basins such as Hebes, Ophir, Candor, and Melas Chasmata were then only partially formed. Narrower chasmata such as Tithonium and Ius Chasmata may not have existed at all. Infilling may have occurred when the larger chasmata were isolated and still actively forming. The pyroclastics could have issued from the Tharsis volcanoes or from beneath the broad chasmata themselves. The deposits within Hebes Chasma would then be isolated from the walls of the chasma by mass wasting processes, as discussed below.

FORMATION AND EVOLUTION

Initial collapse of Hebes Chasma may have begun along a pre-existing graben that subsequently collapsed and widened to form the trough. Mechanisms for collapse include withdrawal of intruded magma and local melting of ground ice, both due to igneous activity (Sharp, 1973). Blasius et al. (1977) proposed a tectonic mechanism of trough formation by vertical adjustments of crustal blocks in response to N-S and E-W

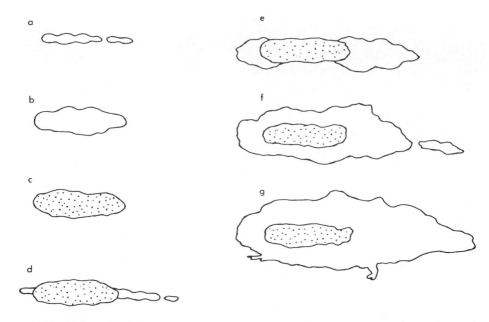

Fig. 11. Schematic diagram of proposed geologic history of Hebes Chasma. (a) Linear collapse of crater chain by evaporation of ground ice along a graben. (b) Widening of trough by landslides. (c) Infilling of trough by volcanic ash or eolian material. (d) Continued collapse along graben and lengthening of trough. (e) Widening by landslides. (f) Isolation of central plateau from main chasma walls by landslides. (g) Present configuration.

extensional stresses. Another possibility is that the broader canyons are volcano-tectonic depressions fromed by eruption of great ash flows, largely contained within their bounding scarps (Blasius, pers. comm., 1981).

The equal elevation of the plateau and ridged plains indicate that downward collapse was essentially complete by the time the chasma was filled. Thus, if volcano-tectonic activity caused most of the downward collapse of the major troughs such as Hebes Chasma, it probably became inactive before the trough was filled. Evaporation of ground ice may have performed a subsidiary role in the collapse and also may be contributing to the formation of sub-basins to the west, north, and south of the plateau (see Fig. 3). The plateau might act as a thermal blanket and thus inhibit its own subsidence.

Infilling of the chasma may have occurred when the trough was still only partially formed. Pyroclastic material could have been erupted from beneath the trough, so that little or none reached the surrounding plains. Alternatively, wind-blown material or pyroclastic debris from the Tharsis region may have been deposited over a large area, but accumulated to greater thicknesses in the chasma, which acted as a sink for the material. In addition, material within the chasma might not have been as extensively reworked by eolian activity as that deposited on the plains.

Small scarps and ridges on the surface of the plateau could indicate that the uppermost units of the plateau and the ridged plains are composed of the same material. These units may have been continuous across the chasma before continued collapse and mass wasting isolated the plateau from the chasma walls (see Fig. 11). The present location of the plateau in the western part of the chasma and the predominance of young landslide scarps near its eastern end (Fig. 3) indicate that the chasma has extended mainly eastward.

SUMMARY AND CONCLUSIONS

Hebes Chasma consists of three physiographic provinces that can be subdivided into several morphologic units.

Several possible theories of origin have been proposed for the central plateau of Hebes Chasma. These theories are summarized below:

(1) The theory of the plateau as an erosional remnant of the plains cannot adequately explain the contrasting patterns of erosion between the plateau and chasma walls, nor the flat-topped appearance of the plateau itself. Furthermore, the laterally discontinuous horizontal layering of the plateau walls is not observed in the walls of the chasma.

(2) Deposition and re-excavation of basin-filling lake sediments is a problem with a lacustrine origin. Removal of water and sediments are also difficult to resolve in the case of Hebes Chasma.

(3) The concept of the plateau as an eolian deposit proposes that the plateau deposits be composed of material derived from martian dust storms. Past climatic conditions may have been conducive to deposition of large amounts of eolian debris. Layering could be a result of variations in composition of successive depositional episodes.

(4) The pyroclastic theory of origin proposes that a thick sequence of ash-flow deposits make up the bulk of the plateau. The horizontal layers may consist of more resistant welded tuff. The source of the pyroclastics may be beneath the chasma or within the Tharsis area.

Infilling of the chasma was followed by continued collapse of the chasma walls and isolation of the central plateau. Massive landslides covered the chasma floor as widening continued. These processes are probably still active today and could eventually cause breaching of the chasma and its ultimate connection with canyons to the west and south.

Acknowledgments—Sincere thanks are extended to E. C. Morris and D. H. Scott for their review of the manuscript as well as their much appreciated contributions of ideas and suggestions. B. Lucchitta also provided important background information. This research was supported by NASA Work Order W-13,709.

REFERENCES

Blasius K. R., Cutts J. A., Guest J. E., and Masursky H. (1977) Geology of the Valles Marineris: First analysis of imaging from the Viking I Orbiter Primary Mission. *J. Geophys. Res.* **82**, 4067–4091.

Carr M. H. (1974) Tectonism and volcanism of the Tharsis region of Mars. *J. Geophys. Res.* **79**, 3943–3949.

King J. S. and Riehle J. R. (1974) A proposed origin of the Olympus Mons escarpment. *Icarus* **23**, 300–317.

Lehmann O. (1933) Morphological theory concerning the weathering of rock walls. *Vierteljahresschr. Naturforsch. Ges. Zuerich* **87**, 3–126.

Lucchitta B. K. (1978) Morphology of chasma walls, Mars. *J. Res. U.S. Geol. Surv.* **6**, 651–662.

Malin M. C. (1976) Nature and origin of intercrater plains on Mars. Ph.D. Thesis, Calif. Inst. Tech., Pasadena. 176 pp.

McCauley J. F. (1978) Geologic map of the Coprates quadrangle of Mars. *U.S. Geol. Survey Misc. Geol. Inv. Map*, I-896.

McGetchin T. and Smith J. R. (1978) The mantle of Mars: Some possible implications of its high density. *Icarus* **34**, 512–536.

Morris E. C. (1980) A pyroclastic origin for the aureole deposits of Olympus Mons (abstract). In *Reports of Planetary Geology Program 1980*, p. 252–254. NASA TM 82385.

Rapp A. (1960) Talus slopes and mountain walls at Tempelfjorden, Spitsbergen. *Norsk Polarinst. Skr.* no. 119, 93.

Scott D. H. (1969) Geology of the southern Pancake Range and lunar crater volcanic field, Nye County, Nevada. Ph.D. Thesis, Univ. Calif., Los Angeles, 128 pp.

Sharp R. P. (1973) Mars: Troughed terrain. *J. Geophys. Res.* **78**, 4063–4072.

Ward W. R. (1974) Climatic variations on Mars, astronomical theory of insolation. *J. Geophys. Res.* **79**, 3375–3386.

West M. (1974) Martian volcanism: Additional observations and evidence for pyroclastic activity. *Icarus* **21**, 1–11.

Womer M. B., Greeley R., and King J. (1979) Pyroclastic volcanism of the Snake River Plain, Idaho: Implications for Mars (abstract). In *Reports of Planetary Geology Program*, p. 265–266. NASA TM 80339.

Wu S. S. C., Schafer F. J., and Jordan R. (1980) Topographic mapping of Mars: 1:2 million contour map series (abstract). In *Reports of Planetary Geology Program*, p. 458–461. NASA TM 82385.

Spectral reflectance of weathered terrestrial and martian surfaces

Diane L. Evans, Tom G. Farr, and John B. Adams

Department of Geological Sciences, University of Washington, Seattle, Washington 98195

Abstract—Laboratory reflectance spectra of Hawaiian samples were compared to martian surface color units seen in Viking Orbiter images. Spectral characteristics of the samples are consistent with some of the martian units and with several telescopic spectra of Mars. Variations in the spectral characteristics of the Hawaiian rocks can be directly related to the age and history of the samples. Fresh samples (younger than about 1000 years old) have relatively thin, clear amorphous silica coatings that have little effect on the reflectance of the underlying rock. The reflectance spectra of more weathered samples, however, are controlled by ferric oxide-rich layers consisting of aeolian detritus that is trapped within the silica coatings. Other Hawaiian samples are characterized by hematite resulting from oxidation during eruption and/or cooling. It appears that externally derived ferric oxide-rich coatings, and in some cases primary oxidation, can explain some of the variations seen in color characteristics of martian global units observed by Viking Orbiter.

INTRODUCTION

In Evans *et al.* (1981), laboratory reflectance spectra of Hawaiian samples with known compositions were compared to Landsat multispectral images of Hawaii. We found that the extent of weathering of Hawaiian basalts could be estimated from the images in spite of the low spectral resolution provided by only four bandpasses. In the present study, which is part of a larger study by McCord *et al.* (in preparation) and described briefly in McCord *et al.* (1980), laboratory reflectance spectra of the same Hawaiian samples are compared to a global scale three color albedo map generated from Viking Orbiter images by Soderblom *et al.* (1978). The images were further studied by McCord *et al.* (1980 and in preparation), who made a red-violet cluster diagram and map of martian surface color units. The purpose of our study is to examine materials in the suite of Hawaiian samples that have spectral characteristics similar to the color units in the cluster diagram derived from the Viking Orbiter data. We discuss the weathering of these Hawaiian samples, and the implications for analogs to martian surface materials.

LABORATORY ANALYSES OF HAWAIIAN SAMPLES

Reflectance spectra of basaltic samples from semi-arid parts of the island of Hawaii show several different types of absorption features that can be attributed to the presence of ferrous and ferric iron in varying amounts, and to various degrees of hydration. Some samples are characterized by unhydrated ferric oxides such as hematite (Fe_2O_3). Our field observations indicate that these hematitic samples are not the product of weathering processes but rather the result of oxidation during emplacement and/or cooling. An example of this type of primary oxidation is sample H59, from an 80 year old basaltic spatter cone on Mauna Loa. The reflectance spectrum of this sample is shown in Fig. 1 along with the reflectance spectrum of powdered pure hematite for comparison.

Other Hawaiian samples are characterized by externally derived silica and iron-rich coatings that thicken with age on relatively unoxidized rock surfaces. From electron microprobe analyses of specially prepared thin sections, we find that rock surfaces in semi-arid areas of Hawaii exposed for as little as 60 years are coated with thin clear

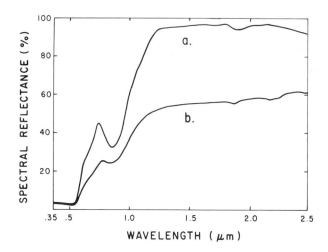

Fig. 1. Laboratory reflectance spectra of a) hematite, and b) a sample from an 80 year old spatter cone on Mauna Loa showing primary oxidation.

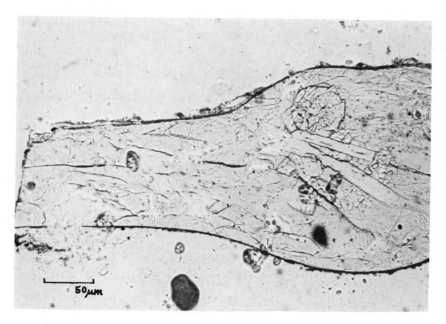

Fig. 2. Coating on H119, a sample from the 1920 Mauna Iki flow. Photo taken in plane polarized light. Note thin clear nature of the coating and its sharp contact with the underlying rock.

layers of almost pure amorphous silica gel that have sharp contacts with the underlying rock (Fig. 2). Older samples have coatings that include both silica layers and thick layers of dark ferric iron-rich material. These coatings also have sharp contacts with the underlying rock (Fig. 3). Although thickness of the coatings varies within each of the rocks sampled, maximum thickness observed on the rock surfaces increases with the known age from about 4 μm (60 years old) to 150 μm (1840 years old). The age of the older sample is based on a ^{14}C data reported by Kelley et al. (1979). We also have observed similar coatings on basalts in other semi-arid environments such as the Snake River Plain, Idaho, and in eastern Oregon.

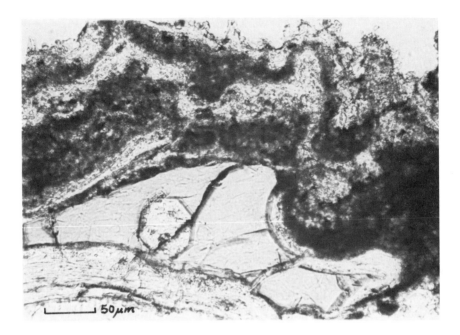

Fig. 3. Coating on H103, a prehistoric Mauna Loa pahoehoe. Photo taken in plane polarized light. This coating is 150 μm thick and contains layers of dark ferric oxides.

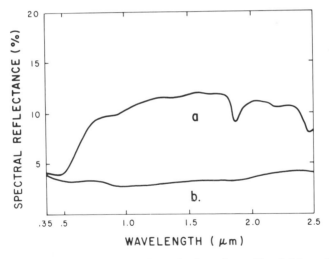

Fig. 4. Laboratory reflectance spectra of samples from a) a prehistoric Mauna Loa flow (H103) and b) the 1920 Mauna Iki flow (H119) (note expanded reflectance scale).

Laboratory reflectance specta of the coated samples show that pure silica coatings do not significantly affect the visible and near infrared reflectance of the underlying rock. However, coatings containing ferric iron have distinct spectral signatures that particularly dominate the visible reflectance spectrum of the rock surface. The spectral reflectance curve of an unweathered basalt (H119) is shown in Fig. 4. It is relatively flat and low because $Fe^{2+} - Fe^{+3}$ charge transfers in basaltic glass or finely disseminated magnetite produce a continuum optical absorption across the visible and near-infrared spectrum. When oxidized iron is present in the coating, there is a decrease in the absorption caused

Table 1. Weight Percent Oxides.

	Dark[1] Layer	Clear[1] Layer	Dark[2] Component	Basaltic[1] Glass (H103)	Weathered[3] Tephra
SiO_2	71.0	83.4	51.3	51.9	52.6
Al_2O_3	6.6	2.5	12.3	13.4	15.3
FeO*	4.2	0.86	13.8	12.0	14.0
Na_2O	0.69	0.34	2.5	2.5	2.0
MgO	2.7	0.63	5.8	6.1	5.6
K_2O	0.25	0.16	0.25	0.50	0.80
CaO	4.1	1.2	9.5	10.5	6.2
TiO_2	1.2	0.37	4.4	2.6	3.3
MnO	0.05	0	0.17	0.16	0.20
Excess Oxygen	9.2	10.47	**	**	**

[1]Electron Microprobe Analyses
[2]Calculated Composition
[3]X-ray Fluorescence Analysis
*All iron as FeO
**Renormalized to eliminate excess oxygen

by Fe^{2+}–Fe^{3+} charge transfer. The reflectance spectra of the weathered Hawaiian rocks (e.g., H103 in Fig. 4) are instead dominated by Fe^{3+} crystal field absorptions centered around 0.6 and 0.85 μm, and O^2–Fe^{3+} charge transfer around 0.34 and 0.40 μm, giving an apparent band edge around 0.7 μm (Huguenin, 1973; Singer et al., 1979).

Electron microprobe analyses of the coating on sample H103 are presented in Table 1. Table 1 shows that clear layers within the coating (Fig. 3) are nearly pure silica, while dark layers have significant contents of other oxides, including iron. This suggests that dark layers are a mixture of a silica-rich component and a dark component, possibly locally derived soil.

Field observations indicate that soil which forms at the edges of flows and in pockets on the flows in arid areas is mostly derived from physical disintegration of flow surfaces. Optical microscopy and X-ray diffraction show that a typical soil may contain slightly weathered fragments of basaltic glass, plagioclase, pyroxene, olivine and magnetite. Aeolian material supplies a variable contribution of fine grained materials such as weathered basaltic tephra described in Evans and Adams (1979, 1980). B Curtiss (pers. comm.) found that the crystalline component of rock coatings near Kilauea Crater is similar to the crystalline component of the soil surrounding the rocks. In order to determine if the chemical composition of the dark component of the dark layers in the coating on H103 is consistent with locally derived material, we assumed that the dark component had the same silica content as the surrounding soil (about 50 wt.%) and that the silica-rich component was equivalent to the adjacent clear layers. Weight percent oxides were recalculated based on these assumptions. The resulting chemical composition of the dark component is given in Table 1 along with analyses of basaltic glass from the interior of the coated sample and of a typical weathered basaltic tephra. The dark component composition is very similar to these analyses, suggesting that it may be locally derived soil material that has been deposited on the rock surface and incorporated into the silica-rich coating. However, details of the contribution of weathered tephra to the soil and to the coatings, and the processes involved in coating development are being investigated further.

COMPARISON OF HAWAIIAN SAMPLES AND MARTIAN SURFICIAL DEPOSITS

The orbital data source used in this study is a Viking Orbiter 2 three-color photometrically corrected albedo map prepared by Soderblom et al. (1978). McCord et al. (1980 and in preparation) have used the data of Soderblom et al. (1978) to prepare a unit map of Mars based on red (0.59 μm) and violet (0.45 μm) albedo values. We calculated percent red and percent violet albedos for our terrestrial samples by determining the digital numbers (DN values) that would be acquired by the red and violet channels of the Viking 2 Orbiter cameras, using the technique described in Evans and Adams (1979). Calculated red and violet albedos for the terrestrial samples were then plotted on the unit definition diagram of McCord et al. (1980 and in preparation) which displays red vs. violet albedos for all picture elements in the V.O. mosaic. The outline of the unit definition diagram developed by McCord et al. (in preparation) is shown in Fig. 5.

Singer (1981a, b) (using earth-based telescopic spectra) and Evans and Adams (1979, 1980) (using Viking Lander data) have shown that certain Hawaiian samples are reasonable analogs for martian surficial deposits. It is therefore possible to discuss the Viking Orbiter regional color units based on comporisons with the Hawaiian samples. We feel that these comparisons are valid for this data set because Soderblom et al. (1978) noted that the martian atmosphere was free of dust during the V.O. approach. Any possible effects of residual atmospheric dust are discussed further by McCord et al. (in preparation).

As previously mentioned, Fig. 5 shows the outline of the unit definition diagram developed by McCord et al. (in preparation). Our laboratory samples are also plotted on Fig. 5 based on our calculated red and violet albedos. Fresh unweathered basaltic rock surfaces plot with the lowest red and violet values along or near the R = V line since they are dark and spectrally flat. When powdered to <250 μm, the samples increase in both red and violet albedo and shift slightly toward higher red values which is consistent with the known variation in spectral reflectance of basalts with changes in particles size (Adams and Filice, 1967). Unweathered basalts do not fall within the V.O. cluster units regardless of particle size, nor do they conform to telescopic spectra. Therefore, major surface exposures of fresh basalts are unlikely on Mars on a scale of the resolution of the resolution of the orbiter data (10–20 km).

Pure samples of hematite and of geothite also do not conform to telescopic data (e.g., Adams, 1968), but are shown in Fig. 5 as reference points. The violet albedo for hematite

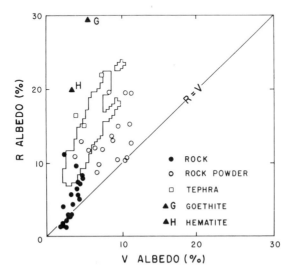

Fig. 5. Outline of unit definition diagram (McCord et al., in preparation) with laboratory samples plotted based on calculated violet and red albedos that would be measured by the Viking Orbiter cameras. Fresh basalts plot along the line R=V with fresh rock surfaces plotting at the lower left. Weathered rock surfaces and weathered tephras plot along a line joining fresh basalt surfaces and goethite.

(~3.5%) is close to that for the fresh basalts and for synthetic Fe^{3+} gel. There appears to be a limit to the violet albedo near 3% that is caused by a saturation of the Fe^{3+} absorption. Much lower values may require geometric effects such as shadowing to attenuate photons.

Goethite has higher red and violet values than does hematite. The change in the spectrum from hematite to geothite results from hydroxylation which requires changes in the size and distortion of the Fe^{3+} sites to incorporate OH into the crystal structure (Andersen and Huguenin, 1977). Points representing weathered rocks in Fig. 5 depart from the R=V line (where fresh basalts occur) and extend toward a mixing line between fresh basaltic rocks and goethite. To a first approximation, the weathered basalts (e.g., H103) plot among a mixing line between fresh basalts and weathered tephra. As previously mentioned, these weathered Hawaiian basalts are coated with layers of silica gel and locally derived soil material that includes weathered basaltic tephra. Thus, spectral data are consistent with the conclusion that these weathered basalts are in fact combinations of fresh basalt and materials similar to the weathered tephra. Weathered tephra and soils containing weathered tephra are spectrally similar. Additional work is in progress on the details of the spectral contributions of these materials to the coatings.

The suite of samples ranging from coated Hawaiian basalts to weathered tephras coincides with units on the unit diagram of McCord et al. (1980) and provides a reasonable and consistent model for the kinds of compositional differences that could be represented by these units. Some of the Hawaiian materials also provide the best matches to earth-based telescopic spectra. Singer (1981a) found that his best match to martian dark area spectra was a relatively fresh basalt with a thin coating of weathered tephra. Similar alteration products of mafic volcanic glass provide the best known spectral analogs to martian bright materials (Evans and Adams, 1979; Singer, 1981b).

Combinations of weathered tephra and coated basalts which are on a mixing line with goethite as an end member appear to be good analogs for the units of McCord et al. (1980) that are low to intermediate in brightness and are only moderately red. However, these types of materials do not appear to be good analogs for the darkest, reddest units. Analogs for the dark red units plot on a mixing line between basalt and hematite rather than goethite. In arid regions of Hawaii, hematitic volcanic rocks are related to primary oxidation (e.g., H59). These samples coincide with regional spectral units on Mars (McCord et al., 1980) that are prominent near 30°S latitude and 45° longitude, but have not yet been observed with earth-based telescopes (McCord and Singer, 1978). Our laboratory data supports the conclusion of McCord et al. (1980) that future telecopic observations of the dark red units on Mars will show a spectrum having a hematitic component.

CONCLUSIONS

Comparisons of laboratory reflectance spectra of weathered basalts in semi-arid parts of Hawaii with spectral reflectance data derived from Viking Lander and Orbiter images and with earth-based telescopic spectra support the conclusion that these terrestrial weathering products provide good analogs for surface materials on Mars. Comparisons of laboratory reflectance spectra of primary oxidized Hawaiian samples with Viking Orbiter images suggest that this type of material could be present on the surface of Mars as well.

We have found that rocks exposed to surface weathering in semi-arid areas of Hawaii develop silica and ferric iron-rich coatings that are externally derived. Other oxidized Hawaiian rocks are the result of primary oxidation during eruption. We, of course, do not know if the same weathering processes operating in semi-arid areas of Hawaii and elsewhere have also operated on Mars. However, good evidence exists on Mars for an arid environment, and for the presence of basalts, aeolian dust, and ferric iron-rich materials (e.g., Arvidson et al., 1980; Greeley and Spudis, 1981). In view of the similarities between the spectral reflectance properties of Mars and of weathered and oxidized Hawaiian basalts it appears worthwhile to investigate further whether similar processes could be at work on both planets.

Acknowledgments—We have benefitted greatly from discussions with T. B. McCord, R. B. Singer, B. R. Hawke, J. W. Head, R. L. Huguenin, C. M. Pieters, S. H. Zisk, and P. Mouginis-Mark, as part of a larger consortium project to study Viking Orbiter geologic units. We also thank them for providing information discussed here in advance of publication. We would also like to thank E. L. Strickland and R. B. Singer for their reviews, and J. A. Heidanus and F. Bardsley for their help in preparing this manuscript.

REFERENCES

Adams J. B. (1968) Lunar and martian surfaces: Petrologic significance of absorption bands in the near-infrared. *Science* **159**, 1453–1455.

Adams J. B. and Filice A. L. (1967) Spectral reflectance 0.4 to 2.0 microns of silicate rock powders. *J. Geophys. Res.* **72**, 5705–5715.

Andersen K. L. and Huguenin R. L. (1977) Photodehydration of martian dust (abstract). *Bull. Amer. Astron. Soc.* **9**, 449.

Arvidson R. E., Koettel K. A., and Hohenberg C. M. (1980) A post-Viking view of the martian geologic evolution. *Rev. Geophys. Space Phys.* **18**, 565–603.

Evans D. L. and Adams J. B. (1979) Comparison of Viking Lander multispectral images and laboratory reflectance spectra of terrestrial samples. *Proc. Lunar Planet. Sci. Conf. 10th*, p. 1829–1834.

Evans D. L. and Adams J. B. (1980) Amorphous gels as possible analogs to martian weathering products. *Proc. Lunar Planet. Sci. Conf. 11th*, p. 757–763.

Evans D. L., Farr T. G., and Adams J. B. (1981) Comparison of laboratory reflectance spectra and Landsat multispectral images of Hawaiian lava flows. *Photogrammetric Engineering and Remote Sensing*. In press.

Greeley R. and P. D. Spudis (1981) Volcanism on Mars. *Rev. Geophys. Space Phys.* **19**, 13–41.

Huguenin R. L. (1973) Photostimulated oxidation of magnetite: I. Kinetics and alteration phase identification. *J. Geophys. Res.* **78**, 848–849.

Kelley M. L., Spiker E. C., Lipman P. W., Lockwood F. P., Holcomb R. T., and Rubin M. (1979) U.S. Geological Survey, Reston, Virginia, Radiocarbon dates XV: Mauna Loa and Kilauea Volcanoes, Hawaii. *Radiocarbon* **21**, 306–320.

McCord T. B. and Singer R. B. (1978) Characterization of Mars surface units (abstract). In *Lunar and Planetary Science IX*, p. 714–716. Lunar and Planetary Institute, Houston.

McCord T. B., Singer R. B., Adams J. B., Hawke B. R., Head J. W., Huguenin R. L., Pieters C. M., Zisk S. H., and Mouginis-Mark P. (1980) Definition and characterization of Mars global surface units: Preliminary unit maps (abstract). In *Lunar and Planetary Science XI*, p. 697–699. Lunar and Planetary Institute, Houston.

Singer R. B. (1981a) The composition of the martian dark regions, I: Visible and near-infrared spectral reflectance of analog materials and interpretations of telescopically observed spectral shape. *J. Geophys. Res.* In press.

Singer R. B. (1981b) Spectral constraints on iron-rich smectites as abundant constituents of martian soils (abstract). In *Lunar and Planetary Science XII*, p. 996–998. Lunar and Planetary Institute, Houston.

Singer R. B., McCord T. B., Clark R. N., Adams J. B., and Huguenin R. L. (1979) Mars surface composition from reflectance spectroscopy: A summary. *J. Geophys. Res.* **84**, 8415–8426.

Soderblom L. A., Edwards K., Eliason E. M., Sanchez E. M., and Charette M. P. (1978) Global color variations on the martian surface. *Icarus* **34**, 446–464.

Mars weathering analogs: Secondary mineralization in Antarctic basalts

John L. Berkley

Lunar and Planetary Laboratory, University of Arizona, Tucson, Arizona 85721

Abstract—Alkalic basalt samples from Ross Island, Antarctica were evaluated as terrestrial analogs to weathered surface materials on Mars. Secondary alteration in these rocks is limited to pneumatolytic oxidation of igneous minerals and glass, rare groundmass clay and zeolite mineralization, and hydrothermal minerals coating fractures and vesicle surfaces. Hydrothermal mineral assemblages principally consist of K-feldspar, zeolites (phillipsite and chabazite), calcite, and anhydrite. Low alteration rates are attributed to cold and dry environmental factors common to both Antarctica and Mars. Mechanical weathering (aeolian abrasion) of martian equivalents to present Antarctic basalts would yield local surface fines composed of primary igneous minerals and glass, and minor hydrothermal minerals, but would produce few hydrous products such as palagonite, clay or micas. Leaching of hydrothermal vein minerals by migrating fluids and redeposition in duricrust deposits may represent an alternate process for incorporating secondary minerals of volcanic origin into martian surface fines. However, unless martian basalts are intrinsically more water-rich than the DVDP basalts in this study, the contribution of hydrothermal minerals to the martian regolith is expected to be minimal.

INTRODUCTION

Speculation regarding the nature and origin of martian surface fines analyzed by the Viking X-ray fluorescence spectrometer (Clark *et al.*, 1976) has focussed on (1) characterizing the mineralogical and chemical composition of this material (e.g., Toulmin *et al.*, 1977; Gooding and Keil, 1978; Allen *et al.*, 1980, 1981) and (2) examining possible provenance materials and weathering environments (Allen 1979; Allen *et al.*, 1980, 1981; Berkley and Drake, 1981; McGetchin and Smyth, 1978; Maderazzo and Huguenin, 1977; Huguenin, 1974; Newsom, 1980). In addition to theoretical studies, empirical evaluation of terrestrial analogs to martian materials and weathering environments has also been applied to this problem (Allen *et al.*, 1980, 1981; Gooding and Keil, 1978; Berkley and Drake, 1981). This paper summarizes the results of one such study in which core samples of alkalic basalts recovered in Antarctica by the Dry Valley Drilling Project (DVDP) were treated as terrestrial analogs to weathered martian surface materials. The term "weathering" as used here refers to any and all processes that result in the production of secondary (as opposed to primary igneous) minerals and includes deuteric reactions (hydrothermal alteration during magmatic cooling), as well as low temperature solid-gas and solid-liquid reactions.

DVDP samples were choosen for this study because they meet certain critical requirements for Mars weathering analogs, specifically: (1) the dominance of olivine-bearing, alkalic effusives (both glassy and crystalline) is consistent with some theoretical predictions regarding martian lava compositions (e.g., McGetchin and Smyth, 1978; Maderazzo and Huguenin, 1977), (2) the virtually year-round sub-freezing nature of the Antarctic climate parallels average temperatures on Mars (Kieffer, 1976; Best, 1971) and (3) low precipitation rates characteristic of desert conditions are common to both Antarctica and Mars. Furthermore, although martian surface fines have probably evolved their present characteristics through complex chemical and mechanical processes acting on multiple source materials (Toulmin *et al.*, 1977; Berkley and Drake, 1981), the observed occurrence of youthful basaltic volcanism on Mars in the form of flows and ejecta sheets suggests that weathering products of these effusives may contribute significantly to surface fine compositions. The principal tenet of this study is that volcanic material from Ross Island represents a reasonable first order analog to young martian volcanics in both composition and environment of formation and, thus, weathering products (specifically, hydrothermal minerals in this paper) in the terrestrial materials may approximate those formed on Mars (Gibson, 1981; Horowitz et al., 1972). The purpose of this report is to characterize secondary alteration products in DVDP basalts (most regarded as hydrothermal in origin) and to evaluate these as possible constituents of the present martian surface fines. The results show that secondary minerals are relatively sparse in these samples and are mainly restricted to vein and vug-filling textural modes. Principal minerals in these secondary associations include (in approximate decreasing abundance) zeolites, anhydrite, calcite, and K-feldspar with minor iron oxides. Implications for martian weathering products and processes are discussed below.

SAMPLES AND PROCEDURES

Samples were obtained from DVDP cores from site 1 and 2 located near McMurdo Station, Hut Point Peninsula, and Ross Island, Antarctic (Mudrey et al., 1973; Kyle, 1978). Most units are relatively thin (generally 1–2 m thick) and consist of interlayered pyroclastic rocks and flows belonging to a basanite–phonolite compositional suite. The igneous sequence is interrupted in some intervals by ice-cemented rock fragments or thin, clear-ice layers. A total of eight separate units from both holes were sampled and include 6 m of scoriaceous tuff and 69 m of flows. These units are between 1.15 and 1.35×10^6 years old (Kyle and Treves, 1974; Kyle, 1978).

Polished thin sections and manually separated grain mounts were analyzed by quantitative electron microprobe and scanning electron microscope methods using an ARL-SEMQ. Semiquantitative and qualitative analyses of grain mount minerals from hydrothermal vein assemblages were obtained using energy dispersive procedures. Crushed and powdered mineral separates and whole rock samples were analyzed by X-ray diffraction.

MINERALOGY–PETROLOGY

The DVDP samples in this study belong to the olivine alkali basalt petrologic group (Carmichael et al., 1974) and include basanites and Ne-hawaiites (Table 1). All gradations between glassy vitric tuffs to nearly holocrystalline flows (some sideromelane glass occurs in all) are represented. Although several distinct phenocryst assemblages occur, olivine is modally dominant in most samples (Table 1). Variable quantities and combinations of high Ti-Al augite, kaersutite amphibole, alkali feldspar, sodic plagioclase, biotite, and titanomagnetite also occur as phenocrysts or microphenocrysts. Most samples display varying degrees of trachytic or pilotaxitic crystal orientation suggesting solidification during flowage.

Basanite hyaloclastites analyzed by instrumental neutron activation display high total iron contents (av. ~10.0 wt.% as FeO), FeO/MgO (0.65–0.70), and, as expected, high total alkalis (av. 6.0 wt.% $Na_2O + K_2O$). Microprobe analyses of glass in selected hawaiites and trachybasalts suggest similar trends in these rocks as well (also see Goldich et al., 1975; Stuckless et al., 1978). Detailed primary mineralogy in DVDP basalts is described by Berkley and Drake (1981), Kyle and Treves (1974), Weiblen et al. (1974), Treves and Ali (1975), Treves (1978), and Kyle (1981).

Secondary mineralization

Primary igneous minerals and glass in DVDP samples are notable for their characteristically pristine condition in most cases (see Berkley and Drake, 1981). For instance, palagonite (hydrous alteration products of variable description; see Allen et al., 1980, 1981; Hay and Iijima, 1968) alteration of glass is rare or absent in most rocks. Minor clay mineralization occurs in certain hyaloclastites (Fig. 1; Tables 1 and 2) and is tentatively identified as a smectite clay, possibly magnesian montmorillonite. However, clay minerals are apparently quite rare and none have been identified in X-ray diffraction analyses of

Table 1. Properties of representative DVDP samples used in this study†.

Sample Number	Lithology	Core Depth* (meters)	Secondary Minerals	Occurrence	Method of Identification
1-29-101	Ne-benmoreite dike	86.5	cal-ph-ch	vein/minor gm	SEM-probe-optical
1-30-190	basanite flow	93.1	kspr-ph-cal	vein	SEM-probe-XRD
1-31-113	basanite flow	94.7	kspr-anh-cal	vein/vug	SEM-probe
1-32-116	basanite flow	96.3	kspr-anh-cal	vein	SEM-probe-XRD
2-10-39	Ne-hawaiite flow	72.2	kspr-ph-ch	vein/vug/minor gm	SEM-probe-optical
2-11A-22	Ne-hawaiite tuff	88.2	ch-cal-th	vug/minor gm	probe-optical-XRD
2-11B-30	Ne-hawaiite tuff	90.5	ch-cal-th-clay	vug/minor/gm	probe-optical-SEM-XRD
2-13-61	Ne-hawaiite flow	100.1	kspr-ph-ch-cal	vein	SEM-probe
2-13-76	basanite flow	140.6	th-cal	vien	SEM-probe
2-13-79	basanite flow	144.0	kspr-ph-cal	vein	SEM-probe
2-14-82	basanite tuff	145.7	kspr-anh-cal	vein/vug	SEM-probe-XRD

*Depth from drilling platform
Note: In addition to secondary minerals noted here, all samples contain variable degrees of red oxidation alteration.
Key to abbreviations: anh = anhydrite; cal = calcite; ch = chabazite; kspr = K-feldspar; ph = phillipsite; th = thaumasite; gm = groundmass.
†Also see Browne, 1973.

Table 2. Electron microprobe analyses (wt.%) of major secondary minerals in DVDP volcanic rocks.

	(9) K-feldspar	(16) phillipsite	(13) chabazite	(4) thaumasite	(12) smectite clay	(5) anhydrite
SiO_2	64.2	47.4	44.8	17.5	42.8	ND
TiO_2	ND	ND	ND	ND	0.10	ND
Al_2O_3	18.9	23.6	24.5	1.60	13.3	ND
FeO	0.33	ND	0.41	ND	0.09	0.07
MnO	ND	ND	ND	ND	0.11	0.05
MgO	0.08	0.55	0.31	ND	27.3	ND
CaO	0.55	2.5	4.4	28.9	2.23	42.3
Na_2O	0.72	3.3	6.3	0.76	0.17	0.23
K_2O	15.2	9.3	3.5	1.12	0.41	ND
SO_3	ND	ND	ND	14.3	ND	57.5
Total	99.98	*87.65	*84.2	#64.18	*86.51	100.15

*Low total reflect H_2O or OH^- not analyzed.
#Low total reflects CO_2, H_2O, OH^- not analyzed.
Note: Number in parentheses indicates number of individual analyses averaged.

powdered bulk samples. In spite of high bulk-rock iron contents, nontronite or other high-iron smectite clays, which have been considered as possible important martian surface fines constituents, were not identified.

Many samples, particularly hyaloclastites, display red oxidation staining of glass and silicate minerals (generally confined to margins and fractures) and coincident alteration of groundmass Fe-Ti opaque oxides (totals equal or exceed 100%) but iddingsite (Sun, 1957) or other hydrous alteration assemblages were not observed. In short, pervasive secondary alteration by replacement reaction of primary igneous minerals and glass in the present DVDP samples is limited, at best.

However, secondary minerals occur as local, thin (<1 mm thick) vein fillings and as coatings in vesicles and other vugs throughout much of the DVDP core sections. Higher

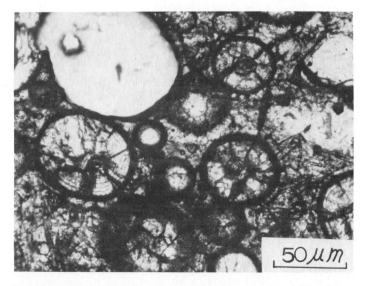

Fig. 1. Sample 2-11B-30 (basanitoid hyaloclastite). Clay minerals (magnesian smectite; Table 2) filling spherical amygdules surrounded by groundmass glass. Concentric growth rings are visible in some amygdules.

concentrations occur in pyroclastic units possibly reflecting higher permeability in these rocks compared to more compact flows. Also, this mineralization becomes more heavily concentrated with core depth as described further below.

Vein and vug minerals commonly occur in assemblages consisting of three or more minerals although monomineralic or bimineralic associations are not unknown (Table 1). Principal among these minerals are zeolites [phillipsite: $Na_2K_2Al_4Si_{12}O_{32}$ $12 H_2O$; Fig. 2a,b; Table 2; and chabazite: $(Ca, Na) (Al_2Si_4O_{12}) \cdot 6 H_2O$; Fig. 3; Table 2], calcite (Fig. 2b), anhydrite (Fig. 2c; Table 2), thaumasite [$Ca_3Si(OH)_6(CO_3)(SO_4)$ $12 H_2O$; Fig. 2d; Table 2], and K-feldspar (Fig. 3; Table 2). Transposition of coexisting mineral layers observed by SEM suggests the following common parageneses involving these minerals (early to late):

$$K\text{-feldspar} \to \text{zeolites} \pm \text{calcite}$$

or

$$K\text{-feldspar} \to \text{anhydrite} \pm \text{calcite} \to \text{thaumasite} + \text{calcite}.$$

Interestingly, zeolite and anhydrite occurrence appears to be mutually exclusive. Also, thaumasite, although invariably associated with calcite, never occurs in direct association with anhydrite. The latter relationship is readily explained since thaumasite is thought to form by reaction of carbonate-bearing solutions with anhydrite (Knill, 1960). The former relationship probably reflects differences in fluid composition of zeolite- vs. anhydrite-depositing mineralizing solutions, alkali-rich and sulfate-poor in the first case, CaO- and sulfate-poor in the second case. There is good evidence to suggest that relatively high pH favors zeolite formation over clays (Miyashiro and Shido, 1970; Hay, 1977) and fluctuations in pH may also control anhydrite vs. zeolite deposition to some extent, but this is uncertain. The occurrence of intergrown calcite with anhydrite in some samples suggests relatively alkaline conditions of formation. Calcite also occurs in intimate intergrowths with zeolites (Fig. 2b) suggesting more or less simultaneous cosaturation from the fluid phase.

Hydrothermal vein formation increases in intensity with increasing core depth. Uppermost core samples are virtually vein-free but contain efflorescent salts deposited from snow-melt solutions (Browne, 1973). The deepest core sections (Hole 2/140 m) contain irregular, pale yellow bands (0.5–3.0 mm wide) displaying andradite garnet as well as iron oxide and leucoxene alteration minerals (Browne, 1973). Above this deep core, K-feldspar–zeolite–anhydrite–calcite–thaumasite vein assemblages predominate, becoming progressively less concentrated upward. This mineralogic zonation suggests that hot hydrothermal fluids continued to circulate in the volcanic pile subsequent to extrusion, the highest temperatures being attained at depth as represented by garnet-bearing assemblages. Kyle (1981) suggested that this alteration occurred in a thick flow or lava lake and involved diffusion of H^+ away from the area with resulting oxidation (similar to what occurs in Hawaiian lava lakes). Most samples also show evidence for a previous episode of hot gaseous alteration (probably upon extrusion) as represented by anhydrous iron oxide and leucoxene alteration. Restriction of secondary mineralization to veins and vugs for the most part, however, suggests that hydrothermal circulation was relatively short lived or chemically benign. Thus primary minerals and glass largely escaped extensive replacement alteration characteristic of highly altered basic rocks (for example, subaqueous oceanic basalts).

DISCUSSION

Secondary alteration of basalt on Earth and Mars

The origin of martian surface fines has generally been attributed to aeolian mixing of chemical and mechanical weathering products derived in large part from iron-rich, basic or ultrabasic igneous rocks (e.g., Toulmin et al., 1977; Baird et al., 1976; McGetchin and

Smyth, 1978; Maderazzo and Huguenin, 1977). Although opinions differ over the precise weathering mechanisms and the relative volume of materials contributed to surface fines by diverse, globally distributed provanence areas (Berkley and Drake, 1981; Allen *et al.*, 1980, 1981; Newsom, 1980; McGetchin and Smyth, 1978; Maderazzo and Huguenin, 1977; Toulmin *et al.*, 1977; Baird *et al.*, 1976), a primarily basic or ultrabasic source material(s) modified by post-accumulation cementing agents (Gibson and Ransom, 1981; Booth and Kieffer, 1978) probably contributed much of the present material in martian surface fines. Berkley and Drake (1981) discussed four major weathering environments on Mars, two of which involve primary igneous processes, namely (1) extrusion (or intrusion) of basic

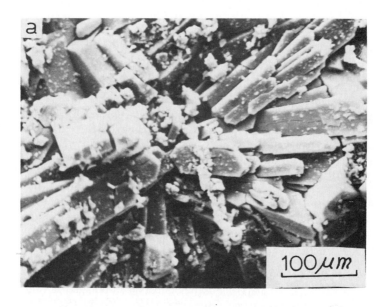

Fig. 2. Scanning electron micrographs of secondary minerals in veins and vugs. (a) Sample 2-13-79 (hawaiite flow). Radiating phillipsite crystals in hydrothermal vein. This is a characteristic habit of phillipsite. (b) Sample 2-10-39 (hawaiite flow). Intergrown calcite (long horizontal grain with chevron cleavage; tiny rhomb, lower left) and phillipsite zeolite in vein. (c) Sample 1-31-113 (hawaiite flow). Anhydrite crystals displaying perfect cleavage along {010}. Microprobe analyses of these grains consistently give anhydrite composition (Table 2); its hydrated analog, gypsum, was not encountered. (d) Sample 2-11B-30 (basanitoid hyaloclastite). Fibrous thaumasite lining vesicle (along with calcite). This mineral may represent an alteration product of anhydrite.

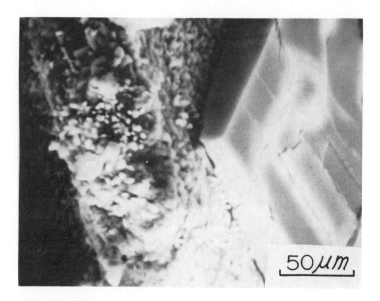

Fig. 3. Scanning electron micrograph (sample 2-10-35, hawaiite flow). Vesicle surface with chabazite rhombs (white, left of center) superimposed over fine grained K-feldspar (mostly gray) illustrating typical late paragenesis of zeolites relative to K-feldspar. Large grain at right is groundmass alkali feldspar protruding into vesicle.

magma into ground-ice (Allen, 1979; Allen *et al.*, 1980; 1981; Toulmin *et al.*, 1977) which produces considerable quantities of palagonite and other hydrous products, and (2) extrusion of basic magma subaerially above ground-ice, a process which is closely analogous to the extrusion mechanism of the present DVDP effusives. The latter environment produces little palagonite relative to the former because reaction with melted ground-ice during extrusion is minimal. A secondary effect unique to Mars and the cold desert environment of Antarctica is the possible suppression of hydrothermal circulation through unusually rapid loss of heat to the atmosphere and the absence of significant subsequent weathering by exposure to precipitation. Rainfall on Mars is probably non-existant and has only rarely been recorded in the Ross Island area (see Monthly Climate Summaries: *Antarctic Journal of U.S.*). It is likely that both volcanic eruption processes described above have occurred during martian history. In fact, any single volcanic eruption may produce significant quantities of hydrous weathering products where magma has become trapped beneath ice-saturated permafrost horizons, while simultaneously erupting nearly anhydrous (and less altered) magma directly onto the planetary surface. Assuming the validity of recent magma genesis models which hypothesize the production of very low viscosity melts on Mars (McGetchin and Smyth, 1978), eruption rates are likely to be quite high. This factor would tend to favor rapid ejection rather than near-surface ponding and sill formation, although vent roughness and diameter, the efficiency of fracture propogation and other factors also critically affect eruption rates and mechanisms. It is well known that terrestrial undersaturated alkalic basalt or ultrabasic magmas (e.g., kimberlites) are volatile-rich compared to more saturated compositions and this may be true in martian magmas (McGetchin and Smyth, 1978). Volatile-rich magmas may be capable of producing high quantities of hydrous alteration products by self-induced reaction with juvenile fluids irregardless of the eruption mechanism. However, Mars may, in fact, be depleted in volatiles relative to earth (Anders and Owen, 1977). As illustrated by the present DVDP samples, the quantity of juvenile fluids or their chemical activity in terrestrial undersaturated basalt magmas may be insufficient to cause profuse palagonite and other hydrous mineral production without the addition of meteoric water.

Secondary mineral stabilities on Mars

The distinguishing characteristics of secondary mineralization in martian basalts erupted in an analogous environment to Antarctic DVDP rocks is the following: (1) Modal palagonite/unaltered glass ratios should be quite low, (2) primary igneous minerals should remain relatively unaltered except for oxidation of ferromagnesan silicates, opaque ores, and groundmass glass, (3) phyllosilicate (clay, micas) production should be minimal, and (4) alteration products should occur prinicpally as surface coatings along fracture and vug surfaces. Specific mineral constituents in this association will vary from one locality to another; however, the minerals encountered in this study are fairly typical. These include alkali feldspar, CaO and alkali-bearing zeolites, carbonates (principally calcite), sulfates (anhydrite), and other associated phases, e.g., goethite or other hydrous iron oxides.

Low-temperature weathering of exposed surface igneous rocks on Mars (subsequent to extrusion and cooling) is probably dominated by simple gas-solid reactions (Gooding, 1978) or similar mechanisms such as ultraviolet light-stimulated oxidation (Huguenin, 1974). Gooding (1978) noted that olivine and pyroxene, in particular, are not stable in the present martian atmosphere but tend to decompose to various carbonates (magnesite–calcite–dolomite), quartz, hematite, and other compounds (ferromagnesian amphiboles may be expected to behave similarly). K- and Na-rich feldspars are thermodynamically stable compared to anorthite which decomposes to calcite, quartz, and corundum, and basaltic glass (sideromelane) should decompose to various oxides, carbonates, and sulfates (Gooding, 1978). Note that many calculated stable weathering products of primary igneous phases on Mars are represented in hydrothermal mineral assemblages in DVDP basalts, namely K-feldspar, calcite, and sulfates (anhydrite).

Although not commonly mentioned as martian weathering products, zeolites, nevertheless, are the most abundant secondary minerals in DVDP basalts. The sensitivity of H_2O content in zeolites to ambient atmospheric pH_2O is well known (Mumpton, 1977), thus any zeolites formed during hydrothermal alteration of igneous rocks on Mars should dehydrate spontaneously on exposure to the atmosphere. In addition, alkali-rich zeolites (particularly phillipsite in the present case) tend to alter to K-feldspar, either directly or indirectly with analcime ($NaAlSi_2O_6 \cdot H_2O$) as an intermediate product, during exposure to saline pore fluids (Hay, 1966; Mariner and Surdam, 1970). This suggests that zeolites may not survive physical incorporation into martian surface fines.

The mineral thaumasite is also considered unlikely to contribute significantly to martian surface fines. Although relatively common in DVDP basalts, this mineral is otherwise quite rare and its stability relations are not well defined (Knill, 1960).

Incorporation of hydrothermal minerals into martian surface fines

In addition to contributions from primary igneous silicates and glass, martian analogs to DVDP basalts could supply variable quantities of alkali feldspar, carbonates, and sulfates derived by hydrothermal alteration during and subsequent to eruption. The relative quantity of these materials available for concentration into surface fines compared to other secondary sources [e.g., atmosphere-regolith reactions (O'Conner, 1968; Toulmin et al., 1973; Huguenin, 1974; Booth and Kieffer, 1978) or aqueous precipitation during duricrust formation (Gibson and Ransom, 1981; Toulmin et al., 1977)] is not known at present. Hydrothermal mineralization in DVDP basalts displays a heterogeneous distribution throughout the core section and secondary minerals probably account for less than about 1% by volume of any given hand specimen-sized sample. The relatively higher concentration of hydrothermal minerals at depth in DVDP volcanic sequences, militates against significant surface exposure of these minerals except after considerable excavation over time. Thus, older volcanic terrains may contribute greater volumes of mechanically extracted hydrothermal minerals than younger terrains. In any event, simple mechanical weathering of hydrothermal veins is not likely to produce significant quantities of material for incorporation into soils.

An alternate means for incorporating hydrothermal minerals into the martian surface fines is by solution and redeposition by migrating pore fluids. Gibson and Ransom (1981) noted that although mechanical weathering processes dominate over chemical processes in the Antarctic Dry Valleys, chemical alteration is, nevertheless, present and plays an important role in regolith formation. They emphasized the role played by lateral fluid migration in transporting and concentrating water soluble components in near-surface deposits above permafrost horizons. Cartwright et al. (1974) also stressed the predominance of lateral fluid migration and noted that upward migration is inhibited in permafrost soil regimes.

A somewhat similar hydrologic system may occur on Mars. Clifford et al. (1979) and Clifford and Huguenin (1980) proposed a model that included an interconnected global subpermafrost groundwater system on Mars. Clifford (1981) proposed that some water from this groundwater layer might invade near-surface regolith layers by any combination of several mechanisms, including thermally driven soil moisture (vapor or liquid) transfer (Philip and deVries, 1957) and various mechanical mechanisms including impacts, earthquakes (producing permeable fault surfaces) and volcanic eruptions. These mechanical processes would act to compact fluid bearing sediments forcing water to the surface through fractures and pore channels. Under appropriate thermochemical conditions these fluids could react with hydrothermal minerals in basaltic volcanic materials (especially permeable impact-pulverized material) causing enrichment in this fluid of certain water-soluble species. For instance, fluids containing dissolved K-feldspar and zeolites would contain high K^+, Na^+, and Ca^{2+}, dissolved calcite and anhydrite would contribute Ca^{2+}, CO_3^{2-}, and SO_4^{2-}. These components would then be redeposited in surface brine deposits (as duricrust cementing material) during water evaporation and outgassing at the surface.

Note, however, that this scenario requires vertical fluid transport which is limited or absent in analogous Antarctic permafrost terrains (Cartwright et al., 1974; Kyle, pers. comm.). Also, results of this study and others (Browne, 1973) show that alkalic basalt systems intruded or extruded in Antarctic permafrost terrains are quite barren in secondary mineral content. Unless martian basalts are intrinsically more water rich or happen to interact with water or ice-laden country rock to a greater extent than in Antarctic basalts, there is no reason to infer any greater secondary mineral content in them compared to the present Antarctic basalts. Thus, unless these minerals are concentrated by repeated solution-deposition or mechanical aelian sorting mechanisms, they are not likely to add significantly to martian soil compositions.

SUMMARY AND CONCLUSIONS

Basanite and Ne-hawaiite core samples from Ross Island, Antarctica were evaluated as terrestrial analogs to weathered martian surface materials. Secondary mineralization in these rocks is mainly restricted to hydrothermal mineral assemblages coating veins and vugs typically consisting of alkali-rich zeolites, K-feldspar, calcite, anhydrite and thaumasite. Replacement alteration of primary igneous silicates and groundmass glass is negligible in most cases and palagonite, clay and other hydrous alteration products are conspicuously rare or absent. Low weathering rates in DVDP basalts is attributed to: (1) insufficient availability of chemically active, hydrous fluids during igneous cooling because of minimal incorporation of country rock-derived meteoric water, (2) suppressed hydrothermal fluid circulation, particularly in upper portions of the volcanic pile, due to perennial, subfreezing climatic conditions (high rock to atmosphere thermal gradient), and (3) deficient rainfall or active groundwater weathering. The latter two points are climate dependent parameters which may also apply to weathering processes on Mars.

Martian analogs to DVDP basalts could contribute mechanically weathered, primary igneous silicates and glass to regolith surface fines along with minor hydrothermal vein minerals. However, an alternate source of weathering materials in these rocks is provided by leaching of water soluble vein minerals in laterally migrating pore fluids with subsequent reprecipitation and concentration in duricrust deposits. DVDP hydrothermal assemblages examined here could provide readily accessible concentrations of soluble cations and ions, including K^+, Na^+, Ca^{2+}, CO_3^{2-}, and SO_4^{2-}. However, if the present DVDP basalts represent a reasonable martian analog, the relative contribution of hydrothermal minerals or their alteration products compared to any other materials is expected to be minimal. This is particularly true for the case in which basalt eruption is not accompanied by significant ground ice incorporation or other hydrous enrichment.

Acknowledgments—I am grateful to Mr. Thomas Teska for able instruction on the SEM, and Mr. Wesley Bilodeau (Department of Geology) for performing the XRD determinations. Discussions with Mr. Horton Newsom and Laurel Wilkening proved most helpful. This research was supported by NASA grant NSG 7577, Michael J. Drake, Principal Investigator.

REFERENCES

Allen C. C. (1979) Volcano-ice interactions on Mars. *J. Geophys. Res.* **84**, 8048–8059.
Allen C. C., Gooding J. L., Jercinovic M., and Keil K. (1981) Altered basaltic glass: A terrestrial analog to the soil of Mars. *Icarus* **45**, 347–369.
Allen C. C., Gooding J. L., and Keil K. (1980) Partially weathered basaltic glass – A Martian soil analog (abstract). In *Lunar and Planetary Science XI*, p. 12–14. Lunar and Planetary Institute, Houston.
Anders E. and Owen T. (1977) Mars and Earth: Origin and abundance of volatiles. *Science* **198**, 453–465.
Baird A. K., Toulmin P. III, Clark B. C., Rose H. J. Jr., Keil K., Christian R. P., and Gooding J. L. (1976) Mineralogic and petrologic implications of Viking geochemical results from Mars: Interim Report. *Science* **194**, 1288–1293.
Berkley J. L. and Drake M. J. (1981) Weathering on Mars: Antarctic analog studies. *Icarus* **45**, 231–249.
Best C. (1971) The topography of Mars. *Naturwissenschaften* **58**, 358–360.
Booth M. C. and Kieffer H. H. (1978) Carbonate formation in Mars like environments. *J. Geophys. Res.* **83**, 1809–1815.
Browne P. R. L. (1973) Secondary minerals from Dry Valley Drilling Project holes. *Antarct. J. U.S.* **8**, 159–160.
Carmichael I. S. E. (1974) *Igneous Petrology*. McGraw-Hill, N. Y. 739 pp.
Cartwright K., Harris H., and Heidari M. (1974) Hydrogeological studies in the dry valley. *Antarct. J. U.S.* **9**, 131–133.
Clark B. C., Baird A. K., Rose H. J. Jr., Toulmin P. III, Keil K., Castro A. J., Kelliher W. C., Rowe C. D., and Evans P. H. (1976) Inorganic analyses of Martian surface samples at the Viking landing sites. *Science* **194**, 1283–1288.
Clifford S. M. (1981) Mars: Ground ice replenishment from a subpermafrost ground water system (abstract). In *Lunar and Planetary Science XII*, p. 157–159. Lunar and Planetary Institute, Houston.
Clifford S. M. and Huguenin R. L. (1980) The H_2O mass balance on Mars: Implications for a global subpermafrost groundwater flow system (abstract). In *Reports of Planetary Geology Program 1979-1980*, p. 144–146. NASA TM81776.
Clifford S. M., Huguenin R. L., and Valdes J. (1979) Lifetimes of 'Oases' on Mars: Models for replenishment (abstract). *Bull. Amer. Astron. Soc.* **11**, 580.
Gibson E. K. (1981) Planetary surface analogs in the Dry Valleys, South Victoria Land, Antarctica (abstract). In *Lunar and Planetary Science XII*, p. 339–341. Lunar and Planetary Institute, Houston.
Gibson E. K. and Ransom B. (1981) Soils and weathering processes in the Dry Valleys of Antarctica: Analogs of the Martian regolith (abstract). In *Lunar and Planetary Sciences XII*, p. 342–344. Lunar and Planetary Institute, Houston.
Goldich S. S., Treves S. B., Suhr N. H., and Stuckless J. S. (1975) Geochemistry of the Cenozoic volcanic rocks of Ross Island and vicinity, Antarctica. *J. Geol.* **83**, 415–436.
Gooding J. L. (1978) Chemical weathering on Mars. *Icarus* **33**, 483–513.
Gooding J. L. and Keil K. (1978) Alteration of glass as a possible source of clay minerals on Mars. *Geophys. Res. Lett.* **5**, 727–730.
Hay R. L. (1966) Zeolites and zeolite reactions in sedimentary rocks. *Geol. Soc. Amer. Spec. Paper* **85**. 130 pp.
Hay R. L. (1977) Geology of zeolites in sedimentary rocks. *Mineral. Soc. Amer. Short Course Notes* **4**, 53–64.
Hay R. L. and Iijima A. (1968) Nature and origin of palagonite tuffs of the Honolulu group on Oahu, Hawaii. In *Studies in Volcanology: A Memoir in Honor of Howel Williams* (R. R. Coats, R. L. Hay, and C. A. Anderson, eds.), p. 331–376. Mem., Geol. Soc. Amer. 116.

Horowitz N. H., Cameron R. E., and Hubbard J. S. (1972) Microbiology of the Dry Valley of Antarctica. *Science* **46**, 242–245.
Huguenin R. L. (1974) The formation of geothite and hydrated clay minerals on Mars. *J. Geophys. Res.* **79**, 3895–3905.
Kieffer H. H. (1976) Soil and surface temperatures at the Viking landing sites. *Science* **194**, 1344–1346.
Knill D. C. (1960) Thaumasite from Co. Down, Northern Ireland. *Mineral. Mag.* **32**, 416–533.
Kyle P. R. (1978) Petrogenesis of the basanite to phonolite sequence of DVDP 2 and 3 and Observation Hill (abstract). In *Dry Valley Drilling Project Bulletin No. 8*, p. 44–45. National Institute of Polar Research, Tokyo.
Kyle P. R. (1981) Mineralogy and petrology of a basanite to phonolite sequence at Hut Point Peninsula, Antarctica, based on core from the Dry Valley Drilling Project drillholes 1, 2, and 3. *J. Petrol.* In press.
Kyle P. R. and Treves S. B. (1974) Geology of Hut Point Peninsula, Ross Island. *Antarct. J. U.S.* **9**, 232–234.
Maderazzo M. and Huguenin R. L. (1977) Petrologic interpretation of Viking XRF analyses based on reflectance spectra and the photochemical weathering model. *Bull. Amer. Astron. Soc.* **9**, 527–528.
Mariner R. H. and Surdam R. C. (1970) Alkalinity and formation of zeolites in saline alkaline lakes. *Science* **170**, 977–980.
McGetchin T. R. and Smyth J. R. (1978) The mantle of Mars: Some geological implications of its high density. *Icarus* **34**, 512–536.
Miyashiro A. and Shido F. (1970) Progressive metamorphism in zeolite assemblages. *Lithos* **3**, 251–260.
Mudrey M. G. Jr., Treves S. B., Kyle P. R., and McGinnis L. D. (1973) Frozen jigsaw puzzle: First bedrock coring in Antarctica. *Geotimes* **18**, 14–17.
Mumpton F. A. (1977) Natural zeolites. In *Mineralogy and Geology of Natural Zeolites* (F. A. Mumpton, ed.), p. 1–17. Mineral. Soc. Amer., Washington, D.C.
Newsom H. E. (1980) A model for hydrothermal alteration of impact melt sheets with implications for Mars. *Icarus* **44**, 207–216.
O'Conner J. T. (1968) Mineral stability at the Martian surface. *J. Geophys. Res.* **73**, 5301–5311.
Philip J. R. and deVries D. A. (1957) Moisture movement in porous materials under temperature gradients. *EOS (Trans. Amer. Geophys. Union)* **38**, 222–232.
Stuckless J. S., Miesch A. T., Goldich S. S., and Weiblen P. W. (1978) A petrochemical model for the genesis of the volcanic rocks from Ross Island and vicinity, Antarctica (abstract). In *Dry Valley Drilling Project Bulletin No. 8*, p. 81–89. National Institute of Polar Research, Tokyo.
Sun M-S. (1957) The nature of iddingsite in some basaltic rocks of New Mexico. *Amer. Mineral.* **42**, 525–533.
Toulmin P. III., Baird A. K., Clark B. C., Keil K., and Rose H. J. Jr. (1973) Inorganic investigation by X-ray flourescence analysis: The Viking lander. *Icarus* **20**, 153–178.
Toulmin P. III., Baird A. K., Clark B. C., Keil K., Rose H. J. Jr., Christian R. P., Evans P. H., and Kelliher W. C. (1977) Geochemical and mineralogical interpretation of the Viking inorganic chemical results. *J. Geophys. Res.* **82**, 4625–4634.
Treves S. B. (1978) DVDP 1, Hut Point Peninsula, Ross Island, Antarctica (abstract). In *Dry Valley Drilling Project Bulletin No. 8*, p. 102. National Institute of Polar Research, Tokyo.
Treves S. B. and Ali M. A. (1975) Geology and petrography of DVDP hole 1, Hut Point Peninsula, Ross Island. *Antarctic J. U.S.* **10**, 5–10.
Weiblen P. W., Stuckless J. S., Hunter W. C., Schulz K. J., and Mudrey M. G. Jr. (1974) Clinopyroxenes in alkali basalts from the Ross Island area Antarctica: Clues to stages of magma crystallization (abstract). In *Dry Valley Drilling Project Buttetin No. 8*, p. 67–71. National Institute of Polar Research, Tokyo.

Landing induced dust clouds on Venus and Mars

James B. Garvin

Department of Geological Sciences, Brown University, Providence, Rhode Island 02912

Abstract—Dust disturbances triggered by spacecraft impact on the surfaces of Venus and Mars are identifiable in images taken shortly after impact, as well as with photometers and nephelometers operating on or near the surface. Analysis of the first two Viking lander images of the martian surface reveals differences in the magnitude of dust effects at the two sites (probably a function of dust particle diameter and areal block cover). A dust cloud over Viking lander 1 shadowed the scene for ~ 28 s; a small dust disturbance at the Viking lander 2 site apparently never rose over the lander but was observed during its ~ 11 s settling period. The turbulent rise of dust from the martian surface was observed at both Viking sites to last ~ 45 s. A dust cloud shadow appears as a zone of darkening (ZD), while dust between a lander camera and the surface is recorded as a zone of bright striations (ZB) in Viking lander imagery.

Spurts of dust from the surface of Venus were recorded by Venera 9, 10, and 12 lander photometers for up to 30 s after impact. The Venera 10 panorama displays a wide zone of darkening (ZD) reminiscent of the ZD in the VL-1 image of Mars, except larger in extent. This ZD is consistent with the shadow of a dust cloud persisting for up to seven minutes in the Venera 10 camera's high resolution field of view. Dust particles on Venus are constrained through these observations to be 18–30 μm in diameter. The Pioneer Venus Day Probe nephelometer recorded a probable dust disturbance lasting up to 210 s after its impact on the venusian surface. Particles settling 1.2 cm/s with diameters in a range of 18–30 μm are consistent with the nephelometer data. Good correlation exists between dust particle sizes and settling velocities at all of the sites on Venus where dust phenomena were observed.

The percentage of surface area covered with blocks larger than a few centimeters appears to correlate with ease of landing-induced dust cloud formation. Simple computer simulation supports this hypothesis. Blocks shield surface dust from the effects of a spacecraft landing, and result in small dust spurts. A relative absence of blocks permits dust to be raised above landers on both the martian and venusian surface.

1. INTRODUCTION

Transient dust phenomena, triggered by the aerodynamic or retroengine assisted landing of a spacecraft on the surface of a planet with an atmosphere (i.e., Venus and Mars), can indirectly reveal information on the local geology of a planetary surface. The mere existence of landing-induced dust clouds indicates there are processes capable of eroding rocks into dust size particles and concentrating them in areas where they are susceptible to disturbances such as high winds or spacecraft impacts. Dust clouds, raised upon spacecraft landing, seek equilibrium by attempting to settle to the surface, but are often perturbed by winds or wake turbulence effects left over from the spacecraft. These dust effects can be detected in a variety of ways. They can appear in photographs as zones of anomalous brightening or darkening. Photometers record depressed light flux measurements in the presence of extra dust in their field of view. Particle detecting devices such as nephelometers can detect micrometer size particles in a small field of view by recording how they scatter light. In addition, optical spectrometers measure particle sizes in a variety of ranges.

Spacecraft landings on Mars and Venus provide an opportunity to seek and discover transient dust phenomena with various instruments, and to make inferences about local-scale surface modification processes. Specifically, such analyses can help resolve questions such as the size distribution of surface fines available for redistribution, the duration of dust suspension, and the modes of dust transport. Some of these questions

Table 1. Spacecraft landing parameters (Keldysh, 1979; Ragent and Blamont, 1980).

Spacecraft	Landing Date	Location* (region, planet)	IAU landing coordinates (N lat., W long.)
VL-1	July 20, 1976	Chryse Planitia, Mars	22.5°, 48°
VL-2	Sept. 3, 1976	Utopia Planitia, Mars	48° ,226°
Venera 9	Oct. 22, 1975	Beta Regio, Venus	31.7°,290.8°
Venera 10	Oct. 25, 1975	Beta Regio, Venus	16° ,291
Venera 11	Dec. 25, 1978	near Phoebe Regio, Venus	−14°,299°
Venera 12	Dec. 21, 1978	near Phoebe Regio, Venus	− 7°,294°
PV Day Probe	Dec. 9, 1978	near Hatfor Mons, Venus	−31.2 ,317°

*relative to nearest prominent feature for Venus sites.

can be answered by analyzing data from the Viking lander 1 and 2 spacecraft (VL-1 and VL-2) on Mars; the Venera 9, 10, 11, and 12 landers (V-9, V-10, V-11, V-12) on Venus; and the Pioneer Venus Day Probe on Venus. The Viking landers imaged dust disturbed by their retroengine exhausts seconds after touchdown. The Venera 9, 10 and 12 spacecraft recorded decreases in the light flux near the venusian surface. The Day Probe nephelometer operated for over an hour on the venusian surface and observed an increase in particle backscatter (dust spurt?) that lasted for over three minutes after probe impact. Since the Venera 9 and 10 spacecraft photographed the venusian surface during their short period of operation there, it is useful to examine the V-9 and V-10 panoramas for any indications of dust phenomena similar to the effects observed in VL-1 and VL-2 imagery. A zone of darkening (ZD) in the Venera 10 panorama is evaluated in this regard.

In most cases, high resolution images of planetary surfaces are not available prior to spacecraft landing, and inferences must be made about surface block cover without photographs. A possible correlation between the magnitude of a lander-created dust disturbance and the surface rock cover can be investigated by analyzing the behavior of observed dust particles, especially in the absence of photographs (i.e., at the Day Probe site on Venus.) The high quality of the photographic data returned from the surface of Mars by the Viking landers permits detailed analysis of the dust effect visible in first VL-1 and VL-2 images of Mars.

2. VIKING LANDER DUST EFFECTS

Both Viking landers initiated their first pictures of the martian surface 25 seconds after their retroengine-assisted landings (Mutch et al., 1976). Earth-based simulations indicated that considerable dust would be kicked up at the time of their impact (at $2-2.5$ m/s) provided the surface was not too rocky (Moore et al., 1977). The first images from both VL-1 and VL-2 were taken in the highest resolution mode available (0.04°/pixel or <2 mm/pixel) with camera 2 (there were 2 cameras on each lander) inclined downwards 50° from the vertical, permitting a lander footpad to be included in each first image. The area covered by the first photograph from each lander is approximately 2 m^2, and extends from 1.9 m to 2.7 m from the camera. Later it will be seen that this area corresponds well to the central portion of the V-9 and V-10 panoramas of Venus.

Figure 1 shows a portion of the initial image from Viking lander 1 in which two features are prominent. First, there is a zone of bright striations or vertical lines (ZB) at the far left of the image. These fine striations suggest fluctuations in the scene illumination on a time scale of ~0.2 s, since 0.22 s were required to scan each vertical line and there was no automatic sensitivity control system to adjust for the variations (Mutch et al., 1976; Shorthill et al., 1976; Moore et al., 1977). The ZB lasted for approximately 17 s before the scene stabilized. These lines have been attributed to the rise of a turbulent cloud of fine dust raised by the lander's retroengines during the first 42 s after touchdown (Mutch et al., 1976). The dust must have been between the camera and the section of the martian surface being photographed since this part of the scene has been brightened (Mutch et al.

Fig. 1. First image of the martian surface, initiated by Viking lander 1 (VL-1) 25 s after touchdown; the scanning direction was from left to right. The pitted and cracked rock in the upper left of this image is ~ 16 cm across. Note the bright streaks at the far left that ended ~ 47 s after touchdown. The darkened zone in the middle left lasted for ~ 30 s; it has been interpreted as a dust cloud between the lander and the sun, while the bright streaks (far left) are probably dust particles suspended between the camera (1.3 m high) and the surface. (Part of VL-1 frame 12A001.)

Fig. 2. First image of the martian surface by Viking lander 2 (VL-2) initiated 25 s after touchdown. The lowermost pitted rock in the center is ~ 20 cm across. Two zones of bright streaks appear at the left of this image, and are gone ~ 65 s after the landing. They are probably due to dust suspended between the camera (1.3 m off the surface) and the surface, as at VL-1. (Part of VL-2 frame 22A001.)

1976). Fine dust particles reflect direct sunlight and thus brighten a scene; only dense clouds of dust particles larger than 10 μm could obscure or darken the viewed surface (Moshkin et al., 1979). A similar zone of bright striations occurs in the leftmost part of the first image from VL-2 (Fig. 2). The accepted interpretation in both cases is that the fine, bright lines are due to dust rising through the lander camera's field of view and eventually out of that field of view by $\sim$ 45 s after impact (Mutch et al., 1976).

A second feature in the first picture from VL-1 is noteworthy: a zone of darkening (ZD) beginning shortly after the end of the ZB and persisting for 28 s (Fig. 1). This ZD has no counterpart in the VL-2 initial image (Fig. 2) where a second zone of bright vertical lines replaces it, but lasts for only $\sim$ 11 s. The fact that the ZD in the VL-1 frame and the second ZB in the VL-2 image begin at precisely the same time after impact ($\sim$ 60 s) is enigmatic, since the first ZB in the VL-2 scene persists at least 5 s longer than the same feature at VL-1.

The ZD in the VL-1 image is consistent with the passage of a dust cloud between the lander and the sun. The dust cloud was observed during its turbulent rise above the lander ($>$ 1.3 m off the surface) during the first 42 s after touchdown, after which it took $\sim$ 18 s for the cloud to pass between the sun and the Viking 1 lander and thus shadow the local scene. The cloud shadow lasted for only 28 s at which point it was probably blown away by the $\sim$ 2 m/s martian surface winds (Moore et al., 1977). Mutch et al. (1976) have concluded that a 100 m long dust cloud 200 m above the surface best fits the observed effects in the image.

The second ZB in the VL-2 image has not been considered previously (Fig. 2), but appears to represent dust between the camera and the surface like the first ZB. As such it could be the settling period for the dust raised during the first 60 s after VL-2 touchdown, in which case Stokes settling predicts the dust particles to be $\leq$ 52 μm in diameter (assuming minimal interference from surface winds.) Dust particles 50 μm in diameter are in general agreement with other observations of martian surface dust, but this size is far less than the 200 μm diameter predicted by Mutch et al. (1976) for particles in the cloud at VL-1. Particles 50 μm in diameter are closer to the predicted values for dust in global martian dust storms (Shorthill et al., 1976).

Since no ZD was observed in the first VL-2 image (Fig. 2), it is tempting to conclude that no large-scale dust cloud formed like the one detected at VL-1. This implies that there was some reason for the difference in magnitude of dust disturbances at the two Viking sites, even though the two landers were identical and had similar landing velocities. Particle size differences (200 μm at VL-1, 50 μm at VL-2) could be responsible, but it seems most likely that some effect due to the blocky nature of the VL-2 site relative to the VL-1 site is the reason for the discrepancy. Moore et al. (1977) have analyzed the rock populations at the two sites and found the VL-2 locality to be more than twice as blocky as the VL-1 site. Up to 20% of the visible surface at VL-2 is covered with rocks larger than 5 cm, while only 8% of the surface at VL-1 is rock-covered. Figures 3 and 4 portray general site characteristics for VL-1 and VL-2; the abundance of boulders at VL-2 is striking. In fact, it is probable that Viking 2 impacted a boulder while landing, on the basis of its tilt of over 8° (Moore et al., 1977). In order to test this hypothesis, a computer simulation of random impacts of Viking lander spacecraft on randomly generated planetary surfaces with rock coverages of 8% and 20% was carried out. The probability of landing with $>$ 50% of the spacecraft on rocks was very low for the 8% rock-covered surface ($<$ 0.02), but rather high for the 20% surface ($\sim$ 0.22). Thus, 80% of the time a Viking lander footpad would hit 1 or more rocks larger than 10 cm in diameter on a surface of which 20% is covered by rocks $>$ 5 cm in diameter. In impacting so many boulders there would be less opportunity for transferral of impact kinetic energy and retroengine exhaust to a dust surface and as a result, less dust would be raised in a dust cloud. At VL-1, regions such as "sandy flats" are essentially rock free (Fig. 1), so there is a better chance of disturbing an exposed surface dust layer with retroengine exhaust. More dust could then be entrained in the atmosphere and a large dust cloud could result.

Fig. 3. Survey panorama of VL-1 site with 2 m long boulder "Big Joe" at left. Approximately 8% of the visible surface is covered with blocks larger than ~5 cm. Note the drifts of fine materials in the far-field. (Part of frame 11A254.)

Fig. 4. Survey of the blocky VL-2 site where ~20% of the visible surface is covered with blocks larger than ~5 cm in diameter. This locality is at least twice as blocky as the VL-1 site, but not as rock covered as the Venera 9 locality on Venus. No dust cloud was recorded above the Viking 2 lander, in contrast with VL-1. The faceted block in the middle right is ~60 cm across. (Part of frame 21I110.)

3. VENERA 9 AND 10 OBSERVATIONS AND DUST PHENOMENA

In terms of Venus, we can examine the Venera 9 and 10 panoramic photography in search of features that could represent dust phenomena. In addition, analysis of light flux data by Moshkin *et al.* (1979) has identified dust spurts lasting about 25 s after the aerodynamic impact of the Venera 9 and 10 spacecraft on the venusian surface. Moshkin *et al.* (1979) and Ekonomov *et al.* (1980) examined the light flux from above the V-9 and V-10 landers and concluded that the decreased flux observed for ~20 s after impact at Venera 10 was due to a dust cloud passing overhead that was subsequently blown away by the 0.5 – 1.0 m/s venusian winds. They proposed a scenario in which wake turbulence from the Venera spacecraft after touchdown could raise dust particles over the top of the 1.5 m high spacecraft, leading to a short period of reduced light flux readings. No such effect was recorded at the rocky Venera 9 site.

An unenhanced version of the Venera 10 panorama (Fig. 5) of the venusian surface shows a ZD that stands out in the middle portion of the image (see Table 1). This zone superficially resembles the ZD in the initial VL-1 image of Mars (Fig. 1). It begins as a slight darkening and increases to a maximum; it fades gradually afterwards. No similar effect can be seen in the Venera 9 panorama (Fig. 6 and Table 1). The ZD's in both the VL-1 and V-10 images do not destroy the scene detail where they occur – this suggests that they are not surface features.

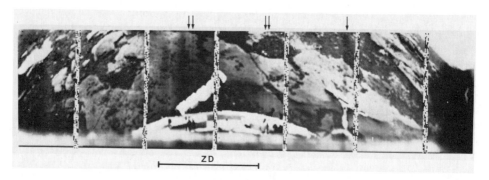

Fig. 5. Unenhanced panorama of the venusian surface at the Venera 10 (V-10) site on Venus. The heat shield cover of the camera view-port lies on the surface in the middle portion of the image and is 40 cm in length. The gamma-ray densitometer rests on the surface in the middle right. The darkened zone (ZD) in the middle section of the image extends for 39° in azimuth and may be a dust-related effect. The vertical, evenly spaced bands in the panorama are telemetry interruptions that were made every 4 minutes by the spacecraft. The arrows indicate 2–3 pixel wide dark bands.

Fig. 6. Unenhanced panorama of the venusian surface from the Venera 9 site. Telemetry cuts and data loss (at left) are visible. Note the abundance of blocks, most of which are between 5 and 80 cm in length. Over 30% of the visible surface at the V-9 site is rock-covered. The absence of a zone of darkening or any similar effects in this panorama is probably significant. See the text for details.

Along with the darkened region in the V-10 panorama, another feature can be identified. Several narrow, dark vertical lines extending from the top to the bottom of the image occur in several places, even in the middle of the ZD (Fig. 5). These 2–3 pixel-wide dark bands are not found in the V-9 panorama. Because the Venera 9 and 10 camera systems contained an automatic sensitivity control (ASC) for adjusting to brightness variations, these dark bands could be the result of the ASC's response to local variations in illumination.

What possible explanations are there for the ZD in the V-10 panorama? Keldysh (1979) and Selivanov et al. (1976) have argued that the ZD is associated with an effect of the ASC which decreases camera sensitivity due to the presence of bright details in its field of view. The presence of the camera viewport heat shield cover in the middle of the V-10 image supports this hypothesis, since the heat shield cover is saturated in the panorama (i.e., beyond the limits of the dynamic range of the imaging system). As figures 5 and 6 illustrate, a portion of the spacecraft lander ring is visible in the lower middle of both Venera panoramas. This ring is made of the same material as the heat shield cover and should saturate the image as well. If the ZD were due to the response of the ASC to saturation, then a similar effect should be seen in the V-9 panorama. Such is not the case, however (Fig. 6), and another explanation must be found.

By virtue of its superficial resemblance to the dark band (ZD) in the VL-1 image, the ZD in the Venera 10 panorama could represent a transient dust phenomenon (Garvin, 1981). On the other hand, it could also be related to parts of the spacecraft shading the near field (Figs. 7 and 8). Another possibility for the ZD involves the artificial lights and their zones of operation (Figs. 8 and 9). Lastly, the ZD could be caused by the passage of a low altitude cloud overhead. The Pioneer Venus Sounder probe, however, detected no particles or aerosols larger than 0.6 μm from 30 km to the surface (Ragent and Blamont, 1980). It appears that of any of these hypotheses, the dust cloud is the most viable and will thus be developed herein.

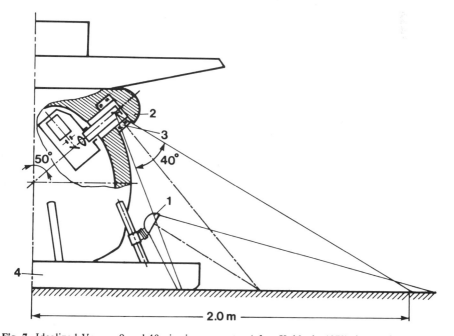

Fig. 7. Idealized Venera 9 and 10 viewing geometry (after Keldysh, 1979), i.e., a side view of a Venera lander illustrating the lights (1), lander impact ring (4), camera view port (3), and camera housing (2). The camera's vertical field of view of 40° corresponds to a ~ 1 m wide area directly in front of the spacecraft. Note that the camera is inclined 50° relative to the vertical. The V-9 and V-10 spacecraft are actually tilted relative to the horizontal on the surface of Venus.

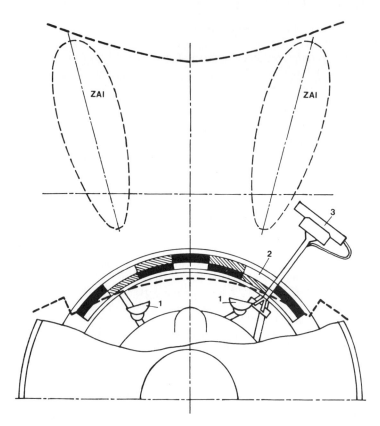

Fig. 8. Plan view of Venera lander camera viewing geometry (after Keldysh, 1979). Lights (1), gamma-ray densitometer (3), and reference brightness chart (2) are indicated in the diagram. The zones of artificial illumination (ZAI) are denoted by the oval regions; the camera's field of view is defined by the curved dashed lines at the top and lower portion of the drawing. Distortion is minimal in the ~ 1 m² area between the ZAI's, and resolution is approximately 1–2 cm.

Because of its 50° inclination and position 82.4 cm above the venusian surface, the field of view of Venera 9 and 10 cameras is rather unusual (Fig. 8). The middle section of the V-10 panorama from 65° to 110° in azimuth is approximately a 1 m² area that includes part of the lander ring with a reference brightness chart and the heat shield cover. The ZD falls entirely in this high resolution (0.35°/pixel), minimal distortion field of view (Fig. 9). The ZD extends for $\sim 39°$ in azimuth, which corresponds to a period from 13 minutes after impact to <20 minutes. Whatever is causing the ZD must persist for ~ 7 minutes, and be near enough the spacecraft to have an effect. Two possibilities emerge: first, the ZD could represent the settling of dust particles through the camera's field of view starting 13 minutes after touchdown and persisting for 7 minutes until it had completely settled; or second, it could represent a dust cloud overhead, passing between the lander and the diffuse "sunlight" at the surface of Venus. A combination of these two models is also possible. Simple Stokes settling under venusian surface conditions applied to the Venera 10 observations by Garvin (1981) indicated that 30 μm particles settling at 3.9 cm/sec were not unreasonable estimates. The size of the most easily moved dust particles on Venus is 33.4 μm (Iversen *et al.*, 1976)—a 41 m-high dust cloud would be completely settled 20 minutes after its generation if it were composed of 30 μm particles.

The dust spurt observed by Moshkin *et al.* (1979) at both Venera 9 and 10 lasted less than 25 s, and could not have extended more than 80 cm above the surface, since this is the height at which the light flux radiometer was located on Venera spacecraft. Thus,

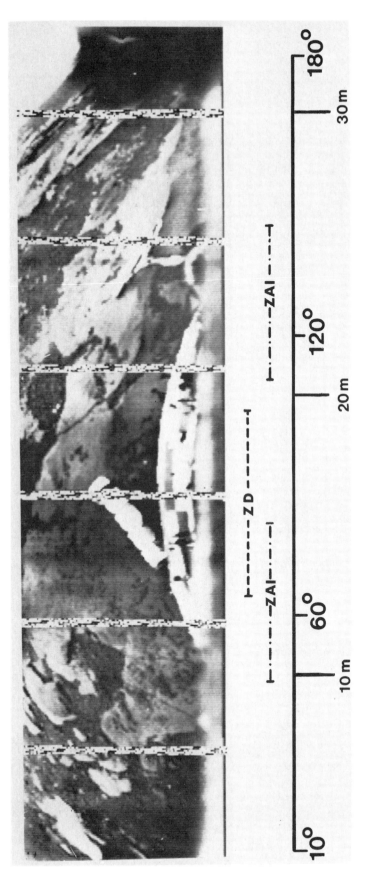

Fig. 9. Unenhanced version of the Venera 10 panorama with axes defining the time after touchdown in minutes (m) and azimuth (degrees). The zone of darkening (ZD) extends from 65° to 104° in azimuth. The two ZAI (zones of artificial illumination) each cover 33.5° in azimuth. Note that the camera scans the scene at a rate of 10 s per degree of azimuth. Imaging commenced 128 s after touchdown.

particles were settling at < 3.2 cm/s, or applying Stokes law:

$$d = \left[\frac{18\mu V_t}{(\rho - \rho_f)g_\venus}\right]^{1/2}$$

(where d is the dust particle diameter in cm, μ is the viscosity in poise, V_t is the terminal velocity in cm/s, ρ and ρ_f are the particle and fluid density in g/cm³, respectively, and $g_\venus$ is the gravitational acceleration in cm/s²). Use of $V_t = 3.2$ cm/s, $\mu = 320 \times 10^{-6}$ P, $g_\venus = 887$ cm/s², $\rho_f = 0.065$ g/cm³ (Hess, 1975), and $\rho = 2.5$ g/cm³, yields d = 29 μm. The Reynolds number for this calculation is Re = $\rho V_t d/\mu \sim 2.0$, which is not purely laminar. The height of the dust spurt can be constrained to be less than 30 cm (the height of the lights) in order that it be visible to the light flux radiometer. If we assume that the dust spurt extended only 30 cm above the surface, then $V_t = 1.2$ cm/s and d = 18 μm, for which Re < 1, and Stokes settling is valid. Thus we can constrain the size of dust particles in a settling dust cloud to be between 18 μm and 30 μm. The particles must be larger than 10 μm, otherwise they would remain suspended indefinitely (Hess, 1975).

Applying these constraints to the ZD observed in the Venera 10 panorama, a dust cloud composed of ~ 20 μm particles could settle within 80 cm of the venusian surface from a height of ~ 15 m and shadow the high resolution field of view for a period of 7 minutes, starting 13 minutes after impact. On Venus, a decrease in the overall brightness of a surface occulted by a dust cloud would create an increase in the optical depth of the atmosphere; this brightness decrease can be thought of as a "shadow" even though lighting at the surface is diffuse. Some of the narrow dark lines within the ZD (at around 15 minutes after impact; Figs. 5 and 9) could be representative of dust particles settling into the camera's field of view and causing the ASC to respond to the resulting rapid variation in illumination.

In summary, an explanation for the zone of darkening observed in the Venera 10 image of Venus that is not inconsistent with observation and theory (Moshkin et al., 1979) involves a transient dust phenomenon. On the basis of analogy with the VL-1 site on Mars and through an interpretation of photometric data by Moshkin et al. (1979), the ZD in the Venera 10 panorama could represent the shadow of a dust cloud raised on spacecraft impact—the shadow was not visible in the scene until 13 minutes after touchdown because that is precisely when the camera started to image the 1 m² area in front of the V-10 lander. It is possible that the shadow was present during part of the 13 minutes before it directly affected imaging, but Venera camera viewing geometry prevented its direct observation. A shadow effect in the venusian atmosphere would be most pronounced if it were nearby the lander and affecting the highest resolution field of view. The role of the surface wind in dissipating the dust clould would be significant except near the 1.5 m high spacecraft.

The lack of a ZD in the Venera 9 panorama could be reflective of the high areal density of blocks at the site as was the case at VL-2 on Mars. Preliminary estimates by the author (unpublished data) show that over 30% of the Venera 9 locale is covered with rocks larger than 10 cm in diameter. Computer simulation of random Venera landings on such a rocky surface indicates that the lander would come to rest entirely on rocks more than 50% of the time. Since the Veneras did not have retroengines with exhaust to disrupt surface dust, only the kinetic energy of impact coupled with wake turbulence from the spacecraft's descent through the high density venusian atmosphere could disturb surface dust. At Venera 9, over half of the landings would not be expected to affect the exposed surface directly and thus raise a dust cloud. A small dust spurt at the V-9 locality is all that could be anticipated in terms of a landing-induced dust effect.

4. PIONEER VENUS DAY PROBE NEPHELOMETER DATA

The Pioneer Venus Day Probe carried a particle-detecting nephelometer with it to the surface of Venus (Table 1). The Day probe was a 44.8 cm diameter spherical spacecraft with a window for the nephelometer located just above its equator. Ragent and Blamont

(1980) considered the possibility of dust phenomena at the Day probe site on the basis of the 67 minutes worth of backscatter data the nephelometer recorded from the surface. Figure 10 is a plot of the nephelometer data for the first 10 minutes after Day Probe impact illustrating three important events. For the first 36 s after impact, the nephelometer recorded the rise (turbulent?) of dust from the surface (large spike). During the next 13 s, the scene was relatively undisturbed as the dust settled back into the nephelometer's field of view. Finally, for 161 s the device recorded settling dust. The dust either settled or was blown away during this time period. Because the nephelometer had a narrow field of view very roughly shaped like a 2 m long cone with a 17 cm diameter base, the orientation of its viewport is critical. If the Day probe landed in its normal configuration, the nephelometer would have been able to detect dust from 54 cm above the surface to 37 cm, and out ~ 2 m from the probe. If, however, it landed with the viewport pointing upwards, the nephelometer would have been able to detect dust in a cone pointing upwards and extending from 237 cm to 45 cm above the surface. Any orientation that points the nephelometer conical field of view toward the surface is inconsistent with the returned data. If we assume the extreme case that the instrument was oriented upwards, then the dust would have risen ~ 2.5 m at ~ 7 cm/s during the first 36 s after Day Probe impact. The dust cloud would be out of the nephelometer's field of view for ~ 13 s, and settling at 1.2 cm/s. For the 161 s during which the dust was observed by the instrument, it would have been settling at 1.2 cm/s. Stokes settling would predict an optimal dust particle diameter of $\sim 18\,\mu$m with a range from 18 to 41 μm and Re < 1. This simple model correlates well with the dust spurt model on the basis of V-9 and V-10 data and serves to further constrain the size of fine particles on Venus.

A computer simulation of the Day Probe impacting a model surface with an 8% cover of rocks > 5 cm in diameter suggests that 6% of the landings would hit rocks 10 cm or

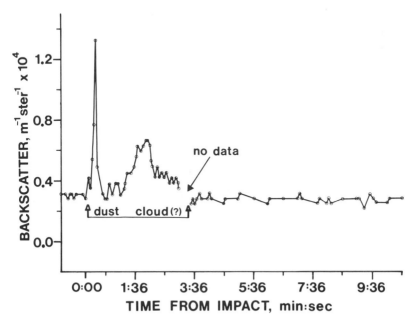

Fig. 10. Pioneer Venus Day Probe nephelometer backscatter data from just before the probe impacted the venusian surface to 10 minutes after impact. The intensity of the backscatter sensed by the particle-detecting nephelometer is an indication of particle size and concentration in the device's field of view (vertical axis). The 210 second-long disturbance in the backscatter (which began just after the Day Probe impacted the surface) could be due to a dust cloud raised by the impact, and its subsequent settling into and out of the nephelometer's field of view. (Data kindly supplied by Boris Ragent.)

larger; impacting a surface with a 30% rock cover would involve landing entirely on rocks 26% of the time.

Finally, it is worthwhile to consider the Venera 11 and 12 sites on Venus (Table 1) for any indications of dust phenomena. Ekonomov *et al.* (1979) analyzed the spectrophotometric data from the V-11 and V-12 landers just after touchdown, and found a 30 s disturbance at the V-12 site, but none whatsoever at Venera 11. The V-12 disturbance is consistent with a dust spurt triggered by lander impact—the dust would be in the 18–30 μm size range as at V-9 and V-10. The absence of dust features at V-11 is surprising and may be due to instrument error or a highly deflated surface. Ekonomov *et al.* (1979) have reported on the high albedo of the Venera 11 locality as measured from the lander; this may reflect a smooth, unmantled surface.

5. CONCLUSIONS

This study has considered possible transient dust phenomena triggered by spacecraft landings on Venus and Mars. In addition, it has attempted to relate all of the landing sites on Venus to each other in terms of dust phenomena. On Mars a difference in the magnitude of dust effects was observed between the VL-1 and VL-2 sites, which is due to either a particle size variation, or the degree of surface blockiness. In general, it seems that the more blocks per unit area, the less likely large-scale dust disturbances are to occur. It appears the blocks at VL-2 have somewhat protected the surface fines from being raised by retroengine exhaust. Evidence for the small-scale nature of the dust effect at VL-2 comes from the first photograph taken by the spacecraft in which two zones of bright striations separated by an undisturbed zone were observed. Each ZB represents dust in the camera's field of view. At VL-1, however, a zone of darkening, probably due to the shadow of a dust cloud overhead, was observed. The implication is that an insufficient volume of dust was raised at the VL-2 site to be carried overhead and thus shadow the site.

The Venera 10 panorama contains a ZD similar in appearance but greater in extent than the VL-1 zone. This zone is consistent with the shadow of a dust cloud darkening the area immediately in front of the lander, and need not be attributed to the response of the ACS to changes in scene brightness due to the presence of the heat shield cover in the field of view. Dust in the cloud at Venera 10, as at Venera 9, 11, and the Pioneer Venus Day Probe site is constrained to be 18 to 30 μm in diameter, with a maximum of 44 μm. The absence of dust effects (i.e., spurts) at Venera 11 could reflect surface characteristics different from all the other sites where at least small dust effects were observed. The Venera 11 locality could either by very rocky or extremely smooth with no dust mantle at all, unlike the Venera 10 site. Computer simulations of random spacecraft landings on surfaces with different degrees of rock cover support the conjecture that ease of dust disturbance is correlated with lack of blocks; both Venera 10 and VL-1 have large dust effects, and both are known to have low block concentrations relative to other sites; on the basis of the analysis in this paper, it is suggested the Venera 12 site is also relatively block-free.

Acknowledgments—This research was supported by NASA grant NSG-7569 of the Mars Data Analysis Program for which the author is especially grateful. The author heartily thanks Jim Head for his encouragement and valuable advice in this study. Buck Sharpton and Mark Cintala provided much needed criticism and many useful suggestions concerning the hypotheses in the paper. Lionel Wilson and Bill Patterson were most helpful on several occasions when important questions were raised. Sam Merrell must be commended for his fine photographic work. Sarah Bosworth performed miracles translating the author's handwriting into a neat manuscript. Thanks must also go to our Soviet colleagues who provided contact prints of the Venera panoramas. Boris Ragent supplied the Pioneer Venus Day Probe nephelometer data presented in Fig. 10. Helpful reviews by R. Greeley, L. Ronca, P. Mouginis-Mark, and M. Cintala (2b) were highly appreciated.

REFERENCES

Ekonomov A.P., Moshkin B.E., Golovin Yu.M., Parfent'ev N.A., and San'ko N.F. (1979) Spectrophotometric experiment on Venera 11 and Venera 12 descent modules. *Cosmic Res.* **17**, 590–600.

Ekonomov A.P., Golovin Yu.M., and Moshkin B.E. (1980) Visible radation observed near the surface of Venus: Results and their interpretation. *Icarus* **41**, 65–75.

Garvin J.B. (1981) Dust cloud observed in Venera 10 panorama of venusian surface: Inferred surface processes (abstract). In *Lunar and Planetary Science XII*, p. 324–326. Lunar and Planetary Institute, Houston.

Hess S.L. (1975) Dust on Venus. *J. Atmos. Sci.* **32**, 1076–1078.

Iversen J.D., Greeley R., and Pollack J.B. (1976) Windblown dust on Earth, Mars, and Venus. *J. Atmos. Sci.* **33**, 2425–2429.

Keldysh M.V. (ed.) (1979) *The first panoramas of the Venetian [sic] surface.* NASA TM-75706, Washington, D.C. 189 pp.

Moore H.J., Hutton R.E., Scott R.F., Spitzer C.R., and Shorthill R.W. (1977) Surface materials of the Viking landing sites. *J. Geophys. Res.* **82**, 4497–4523.

Moshkin B.E., Ekonomov A.P., and Golovin Yu.M. (1979) Dust on the surface of Venus. *Cosmic Res.* **17**, 232–237.

Mutch T.A., Binder A.B., Huck F.O., Levinthal E.C., Liebes S., Morris E.C., Patterson W.R., Pollack J.B., Sagan C., and Taylor G.R. (1976) The surface of Mars: the view from the Viking I lander. *Science* **193**, 791–801.

Ragent B. and Blamont J. (1980) The structure of the clouds of Venus: Results of the Pioneer Venus Nephelometer Experiment. *J. Geophys. Res.* **85**, 8089–8105.

Selivanov A.S., Panifilov A.S., Naraeva M.K., Chemodanov V.P., Bokhonov M.I., and Gerasimov M.A. (1976) Photometric analysis of panoramas on the surface of Venus. *Cosmic Res.* **14**, 596–600.

Shorthill R.W., Moore H. J., Scott R.F., Hutton R.E., Liebes S., and Spitzer C.R. (1976) The "soil" of Mars (Viking 1). *Science* **194**, 91–97.

Density constraints on the composition of Venus

Kenneth A. Goettel, Janet A. Shields, and Deborah A. Decker

Dept. of Earth and Planetary Sciences and McDonnell Center for the Space Sciences,
Washington University, St. Louis, Missouri 63130

Abstract—The composition of Venus is constrained most directly by the mean density of the planet, because the moment of inertia factor is unknown and seismic data do not yet exist for the interior. The density of Venus is estimated to be $1.0 \pm 0.4\%$ less dense than the density of an Earthlike Venus (i.e., a planet with the mass of Venus, but identical to Earth in composition and structure, with internal temperatures equal to temperatures in Earth at corresponding depths, except near the hot surface). Five models which have attempted to explain the density difference between Venus and Earth are reviewed: iron fractionation, basalt/eclogite ratio, oxidation state, multicomponent mixing, and equilibrium condensation. All of these models are capable of explaining the observed density difference, either with the model assumptions as published or with modest adjustments in model parameters.

Inferences about the density and composition of Venus are rather sensitive to assumptions about internal temperatures in Venus. If subduction of surface plates does not occur on Venus, then internal temperatures may be markedly increased (Turcotte et al., 1979), and the intrinsic density of Venus could be only very slightly less than the intrinsic density of Earth or even slightly higher. More definitive determination of the composition of Venus requires additional data from Venus and a better understanding of the factors governing internal temperatures in convecting planetary interiors.

INTRODUCTION

The composition of Venus is constrained most directly by the mean density of the planet because the moment of inertia factor is unknown and seismic data do not yet exist for the interior. The mean density of Venus is well determined from the observed mass and mean radius. The mass (M) of Venus is 4.869×10^{27} g, using the preferred value of the gravitational constant (G) given by Cohen and Taylor (1973) and the mean of the Mariner 5 and Mariner 10 determinations of GM of Venus (Ferrari and Bills, 1979). The mean radius of Venus is 6051.5 km, from Pioneer Orbiter radar data (Pettengill et al., 1980). These values of mass and mean radius give 5.245 g/cm^3 for the mean density of Venus. Thus, the mean density of Venus is nearly 5% less than the mean density of Earth, 5.515 g/cm^3, using GM and radius data from Moritz (1976). However, most of the difference in mean density between Venus and Earth is due to differing amounts of self compression, due to the greater mass of Earth, rather than to differences in intrinsic density. Uncertainties in equations of state introduce significant uncertainties in the densities of model planets as large as Venus or Earth. However, relative differences in density between Venus and Earth can be discussed more accurately than the densities of model planets.

One way of considering relative differences in density between Venus and Earth is to use empirical pressure-density relationships derived from seismic data for Earth's interior (e.g., Ringwood and Anderson, 1977). Ringwood and Anderson estimated that if Venus had its observed mass, but were identical to Earth in composition and structure, possessing the same ratios of core to mantle to crust, with interior temperatures the same as Earth at corresponding pressures, then Venus would have a density of 5.34 g/cm^3. With these assumptions, the density of an Earthlike Venus would be 1.8% denser than the observed mean density of Venus (Table 1). However, the assumption that temperatures in an Earthlike Venus are equal to temperatures in Earth at corresponding pressures

Table 1. Density difference between Venus and Earth mean density (g/cm^3).

Venus (observed)	5.245	
Venus (Earthlike)[a]	5.34	(1.8% denser than observed)
Venus (Earthlike)[b]	5.32	(1.4% denser than observed)
Venus (Earthlike)[c]	5.31	(1.2% denser than observed)
Venus (Earthlike)[d]	5.29 – 5.30	(0.9 – 1.05% denser than observed)

[a] Ringwood and Anderson (1977): Venus with its observed mass, but the same composition, structure, and T(P) as Earth.
[b] Present estimate: as "a", except Venus with same T(Z) as Earth.
[c] Present estimate: as "b", including thermal expansion due to high surface T on Venus.
[d] Present estimate: as "c", including suppression of basalt/eclogite transformation on Venus (basaltic crust on Venus 20% as thick as postulated by Anderson, 1980).

probably results in an overestimate of the density of an Earthlike Venus. The smaller mass of Venus implies that the depth at which a given pressure is reached is deeper in Venus than in Earth; thus, temperatures at corresponding pressures may be higher in Venus than in Earth. With the same assumptions as Ringwood and Anderson, except that temperatures at corresponding depths are equal in Venus and Earth, the density of an Earthlike Venus is estimated to be ~ 5.32 g/cm^3 (assuming an average value of $\alpha\Delta T$ in Venus of 4×10^{-3}, where α is the high temperature, high pressure thermal expansion coefficient and ΔT is the difference in temperature obtained on the assumption that temperatures in Venus and Earth are equal at corresponding depths, rather than at corresponding pressures).

The high surface temperature of Venus also affects the observed mean density of the planet. Assuming that the perturbation of the high surface temperature decays with depth, Ringwood and Anderson (1977) estimated that thermal expansion due to the high surface temperature of Venus decreases the density of an Earthlike Venus by ~ 0.01 g/cm^3. The high surface temperature of Venus also affects the mean density of the planet by suppressing the basalt/eclogite transformation in Venus (Anderson, 1980). Even if the basaltic crust of Venus is only 20% as thick as postulated by Anderson, suppression of the basalt/eclogite transformation in Venus will lower the density of Venus by 0.01 to 0.02 g/cm^3. Combining all of these effects, the density of an Earthlike Venus, with temperatures equal to temperatures in Earth at corresponding depths (except near the not surface), is estimated to be 5.29 to 5.30 g/cm^3 or 0.9 to 1.05% denser than actually observed.

Ringwood and Anderson (1977) claimed that the density of an Earthlike Venus is probably known to an accuracy of 0.01 g/cm^3. More realistically, if uncertainties in scaling internal temperatures between Venus and Earth, uncertainties about the depth to which the high surface temperature has affected internal temperatures in Venus, and uncertainties about density changes caused by suppression of the basalt/eclogite transformation are considered, the uncertainty in the density of an Earthlike Venus is probably at least twice as large or ~ 0.02 g/cm^3. Thus, an Earthlike Venus is estimated to be about $1.0 \pm 0.4\%$ denser than the actual mean density of Venus.

EXPLANATIONS FOR THE DENSITY DIFFERENCE BETWEEN VENUS AND EARTH

Viable models for the composition and internal constitution of Venus must account for the $\sim 1.0\%$ difference in intrinsic density between Venus and Earth. Five models which have been advocated to account for the density difference are listed in Table 2, with the density difference explainable by each model: iron fractionation, basalt/eclogite ratio, oxidation state, multicomponent mixing, and equilibrium condensation.

Iron fractionation, in which Venus and Earth have identical silicate compositions but differ in metal/silicate ratio, is the classical explanation for density differences among the terrestrial planets (e.g., Urey, 1952). Various mechanisms based on differences in physical

Table 2. Explanations for the density difference between Venus and Earth.

Model	Density effect	Reference
iron fractionation	several %	Urey (1952), and many others
basalt/eclogite ratio	0.8 to 1.5%	Anderson (1980)
oxidation state	1.7%	Ringwood and Anderson (1977)
multicomponent mixing	1.5%	Morgan and Anders (1980)
equilibrium condensation	1.0 to 1.4%	Lewis (1972)
Observed density difference:	$1.0 \pm 0.4\%$	

properties (strength, ductility, reflectivity, density, magnetic properties, etc.) between silicates and metal have been proposed to produce iron fractionation. Perhaps the most viable mechanism is the aerodynamic fractionation proposed by Weidenschilling (1977). Aerodynamic fractionation is capable of either increasing or decreasing the iron/silicate ratio of a planet, depending on the choice of model parameters (e.g., mean particle size and whether there is a systematic difference in size between metal and silicate grains). With an appropriate choice of parameters, aerodynamic fractionation could produce iron fractionation in the right sense to account for the density difference between Venus and Earth (i.e., to decrease the iron/silicate ratio of Venus). However, such parameters would also result in a decrease in the iron/silicate ratio of Mercury and thus would necessitate an alternative mechanism to produce the high iron/silicate ratio of Mercury. Regardless of the mechanism invoked, iron fractionation is capable, at least in principle, of producing large density differences (several percent or more) between planets, because of the large density difference between metal and silicates. Approximately 4% less iron in Venus than in Earth (i.e., an iron/silicate ratio 4% less than in Earth) would suffice to account for the observed 1% density difference between Venus and Earth.

In the basalt/eclogite model (Anderson, 1980), Venus and Earth could have identical compositions, but differ in the ratio of basalt/eclogite. Higher temperatures in the upper mantle of Venus (resulting from the high surface temperature) increase the depth of the basalt/eclogite transformation. Since basalt is substantially less dense than eclogite, suppression of the transformation lowers the density of Venus relative to Earth. Anderson estimated that this mechanism could lower the density of Venus by 0.8 to 1.5%, thus accounting for most or all of the observed density difference. However, attribution of the entire density difference to differences in basalt/eclogite ratio between Venus and Earth requires the presence of a very thick (100 to 170 km) basaltic crust on Venus as proposed by Anderson (1980).

In the oxidation state model (Ringwood and Anderson, 1977), Venus and Earth have virtually identical proportions of the major rock forming elements, but differ in the amount of Fe which is oxidized to FeO. The oxidation states of Venus and Earth were postulated to differ due to differences in the extent of reduction during planetary accretion. Ringwood and Anderson postulated that about 35% of the total Fe in Venus is oxidized to FeO, while only about 14% of the total Fe in Earth is oxidized to FeO. The model proposed by Ringwood and Anderson was estimated to account for a 1.7% density difference between Venus and Earth, a density difference larger than the present estimate of about 1.0%. However, the oxidation state of Venus which was postulated by Ringwood and Anderson was not based on a predictive, quantitative analysis of the accretion/reduction process; rather, it was chosen arbitrarily to provide a composition for Venus which matched their estimate of the density difference between Venus and Earth. Thus, their oxidation state model could be modified to match the present estimate of the density difference between Venus and Earth without violating the assumptions of the model.

In the multicomponent mixing model (Morgan and Anders, 1980), Venus and Earth are composed of different proportions of model components which are postulated to constitute the terrestrial planets. This model is based on the premise that these planets formed by exactly the same processes as the chondritic meteorites, both being condensates from the solar nebula and both experiencing the identical fractionation processes. Relative proportions of their model components in Venus and Earth were estimated from whole-planet estimates of indicator elements and ratios: Fe, U, K/U, Tl/U, and FeO/(FeO + MgO). Venus was postulated to contain slightly less Fe and S than does Earth and to have a lower percentage of FeO in mantle silicates. The detailed composition model For Venus proposed by Morgan and Anders implies a density for Venus about 1.5% less dense than Earth. This model is manifestly flexible; modest adjustments in the proportions of the components would allow this model to match the present estimate of the density difference between Venus and Earth.

In the equilibrium condensation model (Lewis, 1972), Venus and Earth differ in composition due to differences in formation temperature (i.e., the temperature at which the material constituting the planet equilibrated in the solar nebula). In the simple, end-member equilibrium condensation model, Venus and Earth have identical proportions of the major rock forming elements (except S); Venus is predicted to be virtually devoid of S and FeO. Lewis claimed that this model could account for a $\sim 1\%$ difference in mean atomic weight and a $\sim 1\%$ difference in zero pressure density between Venus and Earth. However, Ringwood and Anderson (1977) contended that Lewis' model is capable of explaining a density difference of only $\sim 0.4\%$. Detailed consideration of the density implications of the equilibrium condensation model for Venus, including effects on phase changes and effects on the thermal state of Venus, suggests that this model is capable of explaining the density difference between Venus and Earth.

The density effects arising from the composition of Venus postulated by the equilibrium condensation model are summarized in Table 3. The absence of S in this model for the composition of Venus lowers the density of Venus by about 1.1% (Ringwood and Anderson, 1977). However, the predicted absence of FeO (and corresponding increase in Fe metal) raises the density by about 0.8%, resulting in a net density decrease of only 0.3% (0.4% if changes in the depths of phase transformations are considered, according to Ringwood and Anderson, 1977). However, Ringwood and Anderson underestimated the density effects of changes in the depths of phase transformations and ignored the thermal consequences of the composition postulated by the equilibrium condensation model.

The equilibrium condensation model predicts that ferromagnesian silicates in the mantle of Venus will contain almost no FeO. Thus, the pressures required for the olivine to spinel and the spinel to perovskite phase transformations will be substantially higher than for Fe-bearing compositions. In each case, increasing the depth of the phase transformation lowers the density of Venus by about 0.1%. In the case of the olivine to spinel transformation, the higher upper mantle temperatures of Venus will further increase the depth of the transformation, lowering the density of Venus by an additional 0.1%. Since dP/dT for the spinel to perovskite transformation is probably near zero (Navrotsky, 1980), higher temperatures do not significantly affect the depth of this transformation. Thus, the composition of Venus predicted by the equilibrium conden-

Table 3. Equilibrium condensation model for Venus.

Change in Venus vs. Earth	Density decrease in Venus	Reference
no S	1.1%	Ringwood and Anderson (1977)
no FeO	−0.8%	Ringwood and Anderson (1977)
higher internal temperatures	0.4 to 0.8%	present estimate
deeper phase boundaries	0.3%	present estimate
Total	1.0 to 1.4%	

sation model results in changes in the depths of phase changes which lower the density of Venus by about 0.3%.

Three aspects of the composition of Venus predicted by the equilibrium condensation model affect the internal temperatures in Venus. First, the absence of FeO in mantle silicates implies that abundances of K, U, and Th may be about 10% higher in the silicate portion of Venus than in Earth, since both planets are predicted to have solar ratios of K, U, and Th relative to Si. The relationship between heat source density and mean temperature in a convecting planetary interior presented by Davies (1980) indicates that a 10% increase in radiogenic heat sources in the mantle of Venus would correspond to about a 1% increase in mean internal temperature or about 25°C (using 2500°C as the mean internal temperature in Venus from the thermal model of Toksöz et al., 1978).

Second, the absence of FeO in mantle silicates substantially raises the melting temperature (T_m) of the major constituents of the mantle of Venus. In a convecting planetary interior, viscosity is the major determinant of internal temperatures and the strong temperature dependence of silicate rheology dominates the viscosity (e.g., Tozer, 1965, 1972). Experimental data and theoretical considerations indicate that the subsolidus viscosity of a solid is proportional to the melting temperature of the material: "the liquidus or the solidus", according to Weertman (1970). Turcotte et al. (1979) postulated that mantle viscosities are proportional to T_m of the principal component but did not specify whether T_m meant solidus or liquidus. Following Sharpe and Peltier (1979), we assume that viscosity is proportional to the solidus temperature of the principal mantle component. The solidus temperature of olivine ($\sim Fo_{100}$) in Venus in the equilibrium condensation model is about 160°C higher than the solidus temperature of olivine ($\sim Fo_{90}$) in Earth. Assuming that convection regulates the interior of Venus at $\sim 0.8 T_m$ of the principal mantle component, temperatures in the interior of Venus may be ~ 130°C higher than in the terrestrial mantle, due to the difference in T_m.

Third, the equilibrium condensation model predicts that Venus contains virtually no water. The postulated absence of water affects the viscosity of the interior of Venus by affecting diffusion rates. Subsolidus deformation of a solid is essentially a diffusion process: whether the deformation is accommodated by surface or lattice diffusion of vacancies (Coble or Nabarro-Herring creep) or by dislocation creep, the rate limiting process is the diffusion of vacancies, either to grain boundaries or to dislocations (O'Connell, 1977). The effect of the absence of water on viscosity is complicated by the fact that depending on temperature, stress, and grain size, any one of several different deformation mechanisms may dominate. However, Blacic's (1972) results suggest that the presence of small amounts of water affects dislocation creep and lattice diffusion as well as surface diffusion, perhaps by the mechanism of hydrolytic weakening. At a given temperature, grain size, and stress, the presence of water can increase strain rates by an order of magnitude or more. Thus, the absence of water will raise viscosities in Venus by a significant amount but the effect on internal temperatures is difficult to quantify. Roughly, the absence of water may raise viscosities enough to increase temperatures in Venus by 50 to 250°C.

Combining these three compositional effects on temperature, internal temperatures in a Venus with the composition predicted by the equilibrium condensation model may be approximately 200 to 400°C higher than in Earth. Assuming a mean thermal expansion coefficient of 2×10^{-5}, this temperature increase lowers the density of Venus, relative to Earth, by 0.4 to 0.8%. Temperature increases in the outer portion of Venus, where pressures are lowest and thermal expansion coefficients are highest, probably contribute most of the density decrease due to higher internal temperatures; the outer 1250 km of Venus contains more than 50% of the volume of the planet.

The combined density effects of the compositional and thermal differences between Venus and Earth predicted by the equilibrium condensation model result in Venus being 1.0 to 1.4% less dense than Earth. Thus, contrary to the conclusion reached by Ringwood and Anderson (1977), this model is capable of explaining the observed density difference between Venus and Earth.

DISCUSSION OF MODELS

One principal conclusion which arises from these calculations is that none of the five models discussed can be excluded solely on the basis of the observed density difference between Venus and Earth. Each of these models is capable of explaining the density difference: either with the model assumptions as published or with modest adjustments in model parameters. Other data bearing on the composition of Venus exist (e.g., composition of the atmosphere, meager data on the composition of surface materials, etc.); however, application of these data to the bulk composition of Venus is tenuous at best. Therefore, evaluation of the models for Venus must be based principally on the plausibility or implausibility of their constituent assumptions and on the success or lack of success of the models in predicting the compositions of other planets for which more data exist (e.g., Earth and Mars). Arguments can be made against each of these models for the composition of Venus.

Most iron fractionation models suffer from difficulty in demonstrating that the postulated physical fractionation mechanisms suffice to produce iron/silicate fractionation to the extent required to explain the differences in density among the terrestrial planets. The aerodynamic fractionation model seems capable of explaining, at least in part, the observed density (and mass) of Mercury (Weidenschilling, 1978). However, model parameters which result in an increase in the iron/silicate ratio of Mercury also result in a slight increase in the iron/silicate ratio of Venus and thus produce iron fractionation in the sense to increase the density of Venus relative to Earth. Thus, iron fractionation by aerodynamic fractionation is probably not viable as the principal mechanism for explaining the density difference between Venus and Earth.

The basalt/eclogite model (Anderson, 1980) appears to be conceptually sound. However, explaining all or even a large fraction of the density difference between Venus and Earth by this mechanism requires the presence of an extremely thick basaltic crust on Venus. Also, uncertainties in the difference in basalt/eclogite ratio between Venus and Earth, due to uncertainties about temperature profiles and basalt compositions, may be larger than acknowledged by Anderson. More realistically, differences in basalt/eclogite ratio between Venus and Earth probably account for a fraction of the observed density difference, but probably not for all or most of the difference.

The oxidation state model (Ringwood and Anderson, 1977) has serious weaknesses, even though it is in principle capable of explaining the observed density difference. The premise that differences in accretional heating due to the 18% difference in mass between Venus and Earth have caused the postulated compositional differences has not yet been supported by a quantitative model of the accretion/reduction process. Temperature differences between Venus and Earth arising from accretion would have their most pronounced effect only in the outer few hundred kilometers of Earth. The similar K/U ratios of surface rocks on Venus and Earth do not appear to be consistent with the major differences in accretional heating required to explain the postulated difference in oxidation state, since in this model potassium abundances are determined by volatilization and loss during accretion and surface abundances are taken as representative of the whole planet abundances. The extremely hot early evolution of Earth postulated by this model does not appear to be consistent with a variety of evidence (e.g., present day degassing of noble gases from the terrestrial mantle). The complete loss of the massive CO atmosphere produced according to this model during the reduction of FeO to Fe metal has not yet been quantitatively explained. Other objections to this model for the compositions of the terrestrial planets were summarized by Goettel (1976). However, these arguments notwithstanding, the oxidation state model cannot be rejected solely on the basis of the observed density difference between Venus and Earth.

The multicomponent mixing model (Morgan and Anders, 1980) is based on the reasonable premise that the materials constituting the terrestrial planets underwent the identical fractionation processes as did the chondritic meteorites. However, fractionating components on the distance scale of meteorites or meteorite parent bodies is quite different from fractionating components on the distance scale of the accretion zones of the terrestrial planets. It is exceedingly difficult to envision the physical mechanism(s) which could have separated the postulated components (e.g., remelted and unremelted silicates) so efficiently that gravitational accretion of a planet would not have simply recombined the once-fractionated components. In addition to this fundamental objection to the basic premise of the model, the detailed, 83 element model for Venus presented by Morgan and Anders suffers from the fact the the data requisite to determine the proportions of the postulated components [Fe, U, K/U, Tl/U and FeO/(FeO + MgO) for Venus] are virtually non-existent. Thus, their model must be regarded as extremely tentative, even if the premises of the multicomponent mixing model are accepted.

The equilibrium condensation model (Lewis, 1972) can be criticized on several grounds. The limitations of this model were discussed by Goettel and Barshay (1978); one principal uncertainty is the extent to which equilibrium was attained in the solar nebula. The simple, end-member equilibrium condensation model ignores the compositional effects of possible fractionation or mixing processes in the early solar system. Objections to the equilibrium condensation model for Venus were raised by Ringwood and Anderson (1977). They objected to the model's prediction of solar (or chondritic) K/U ratios for both Venus and Earth, on the grounds that crustal rocks inhibit lower K/U ratios. However, in light of Bukowinski's (1976) prediction of drastic changes in the behavior of K at high pressure, inferences about whole-planet abundances of K which are based on crustal data or on inferences from low pressure experiments appear tenuous. Ringwood and Anderson claimed that the presence of sulfur compounds in the atmosphere of Venus violates the prediction of the equilibrium condensation model that Venus should be virtually devoid of S. However, supply of very small amounts of sulfur (an utterly negligible fraction of the planet's mass), either by incorporation of small amounts of S-bearing material into Venus by mixing materials from different heliocentric distances or by addition of small amounts of cometary or meteoritic material (Lewis, 1974) would not violate the basic premise of the equilibrium condensation model (i.e., that planetary compositions are determined primarily by condensation temperature). Ringwood and Anderson also argued that the redox state of the atmosphere of Venus is incompatible with the redox state of the interior postulated by the equilibrium condensation model. However, it appears likely that the redox state of the atmosphere of Venus, like the atmosphere of Earth, is governed largely by surface, atmospheric, and exospheric reactions (e.g., escape of hydrogen). Thus, the present redox state of the atmosphere probably provides only tenuous constraints on the redox state of the interior. These objections notwithstanding, the equilibrium condensation model is capable of explaining the density difference between Venus and Earth and cannot be rejected on the basis of present Venus data.

THERMAL STATE OF VENUS

For any of the composition models discussed, differences in thermal state between Venus and Earth may play a larger role in determining the density difference between the planets than previously recognized. Differences in composition of the major phases and differences in water content may change viscosities enough to affect internal temperatures by more than 100°C. Differences in whole planet abundances of K, U, and Th and differences in the distribution of these radiogenic heat sources between core, mantle, and crust may significantly affect internal temperatures. A difference of a factor of two in mantle heat sources corresponds to about a 7% difference in temperature (i.e., ~150°C) using the relationship between heat source density and temperature presented by Davies (1980). Differences in other heat sources, such as the rate of release of latent heat of crystallization of core materials, could also be significant.

So far we have implicitly assumed (as did Ringwood and Anderson, 1977) that the heat transfer processes in Venus are closely analogous to those in Earth. However, the presence or absence of terrestrial-style plate tectonics on Venus may have a pronounced effect on internal temperatures. The difference in mean internal temperature between a planet with a free surface (subducting plates) and a planet with a fixed surface (non-subducting plates) may be 250 to 300°C or more (Turcotte et al., 1979), because in the non-subducting case the temperature difference across the convecting layer is greatly reduced and thus internal temperatures must rise to reduce the viscosity so as to transport an equivalent heat flux. The question of whether or not Venus exhibits terrestrial-style subduction of plates is somewhat ambiguous at the resolution of the Pioneer orbiting radar data (Head et al., 1981); however, rather compelling suggestions have been made on geochemical grounds (Anderson, 1981) that subduction cannot occur on Venus. In the absence of subduction, the resulting fixed surface condition suggests that temperatures in Venus may be several hundred degrees hotter than in Earth. Combination of all of the possible compositional and dynamic effects leads to the possibility that the interior of Venus could be as much as 500 to 600°C hotter than the interior of Earth.

The sensitivity of internal temperatures in Venus to whether or not subduction occurs markedly increases the uncertainty about the intrinsic density difference between Venus and Earth. Earlier we estimated that the intrinsic density of Venus is $1.0 \pm 0.4\%$ less dense than the density of an Earthlike Venus (Table 1). However, adding the uncertainty in density which arises from the ambiguity about whether or not subduction occurs on Venus produces a somewhat startling conclusion: if subduction does not occur on Venus, the intrinsic density of Venus could be only very slightly lower than the intrinsic density of Earth, or even slightly higher. The probability of Venus being intrinsically denser than Earth would be enhanced if Venus has a thicker basaltic crust than we assumed in our calculations presented in Table 1. One intriguing possibility is that iron fractionation via aerodynamic fractionation may have slightly increased the Fe/Si ratio of Venus relative to Earth; allowing this sense of Fe/Si fractionation for Venus would allow aerodynamic fractionation to account for a significant fraction of the Fe enrichment in Mercury as suggested by Weidenschilling (1978). This increase in density for Venus relative to Earth may be compensated in part by chemical differences. For example, Venus may have substantially less S, FeO, and H_2O than does Earth, but not be as depleted in these materials as suggested by the simple, end-member equilibrium condensation model. As suggested by Anderson (1980), differences in basalt/eclogite ratio between Venus and Earth may further reduce the density of Venus relative to Earth. Differences in internal temperatures may account for the balance of the apparent density difference between Venus and Earth. Inferences about the composition of Venus are thus very sensitive to estimates of internal temperatures in Venus. Progress in understanding the relationships between composition, volatile content, viscosity, and temperature in convecting planetary interiors and progress in elucidating the role of subduction of surface plates in governing internal temperatures would significantly improve present constraints on the density and composition of Venus.

CONCLUSION

Assuming that heat transfer processes in Venus are closely analogous to those in Earth, the difference in intrinsic density between Venus and Earth is estimated to be $1.0 \pm 0.4\%$. Although objections can be raised against each of the five models for the composition of Venus which have been discussed, none can be excluded solely on the basis of the density difference between Venus and Earth. These models advocate rather disparate compositions and constitutions for Venus. The postulated FeO content of the mantle varies from ~ 0 to ~ 13 weight percent. The postulated size of the core varies from ~ 23 to ~ 32 weight percent of Venus. Postulated abundances of K, U, and Th vary by factors of more than two. These compositional differences imply correspondingly large differences in the petrologic and thermal evolution of Venus. However, the models which have been discussed in this paper are simple end-member models. Many aspects of the models are not mutually exclusive. Reality is almost certainly more complicated than any of the simple models and may incorporate aspects of several of the models.

Allowing for the possibility that the absence of subduction of surface plates on Venus markedly raises internal temperatures (Turcotte et al., 1979), the intrinsic density of Venus could be only very slightly less than the intrinsic density of Earth, or even slightly higher. In this case, substantial revision of all of the models would be necessitated. More definitive determination of the composition of Venus, and thus selection between these competing models requires a better understanding of temperatures in convecting planetary interiors and more data from Venus. Determination of the moment of inertia factor of Venus would greatly reduce the range of allowable models. Unfortunately, prospects of obtaining the moment of inertia factor are poor (Kaula, 1979), because Venus rotates slowly enough that nonhydrostatic contributions to J_2 may be large relative to the hydrostatic contribution. Definitive resolution of the present ambiguity about the composition of Venus will probably require seismic data to determine directly the core radius, elastic properties, and density of the interior of Venus. However, the near-term prospects for obtaining seismic data for the interior of Venus are certainly less than promising. Without more data from Venus, arguments about the composition of Venus will continue to be largely arguments about the plausibility of the assumptions constituting the various models.

Acknowledgments—We thank G. F. Davies, D. L. Kohlstedt, A. T. Hsui, and S. C. Solomon for helpful discussions. This work was supported in part by NASA Grant NSG 7578.

REFERENCES

Anderson D.L. (1980) Tectonics and composition of Venus. *Geophys. Res. Lett.* 7, 101–102.
Anderson D.L. (1981) Plate tectonics on Venus. *Geophys. Res. Lett.* 8, 309–311.
Blacic J.D. (1972) Effect of water on the experimental deformation of olivine. In *Flow and Fracture of Rocks* (H.C. Heard, I.Y. Borg, N.L. Carter, and C.B. Raleigh, eds.), p. 109–115. Geophysical Monograph 16, American Geophysical Union, Washington, D.C.
Bukowinski M.S.T. (1976) The effect of pressure on the physics and chemistry of potassium. *Geophys. Res. Lett.* 3, 491–494.
Cohen E.R. and Taylor B.N. (1973) The 1973 least-squares adjustment of fundamental constants. *J. Phys. Chem. Ref. Data* 2(4), 663–734.
Davies G.F. (1980) Thermal histories of convective Earth models and constraints on radiogenic heat production in the Earth. *J. Geophys. Res.* 85, 2517–2530.
Ferrari A.J. and Bills B.G. (1979) Planetary geodesy. *Rev. Geophys. Space Phys.* 17, 1663–1677.
Goettel K.A. (1976) Models for the origin and composition of the Earth and the hypothesis of potassium in the Earth's core. *Geophys. Surv.* 2, 369–397.
Goettel K.A. and Barshay S.S. (1978) The chemical equilibrium model for condensation in the solar nebula: assumptions, implications, and limitations. In *The Origin of the Solar System* (S.F. Dermott, ed.), p. 611–627. Wiley, Chichester, N.Y.
Head J.W., Yuter S.E., and Solomon S.C. (1981) Topopgraphy of Venus and Earth: a test for the presence of plate tectonics (abstract). In *Lunar and Planetary Science XII*, p. 430–432. Lunar and Planetary Institute, Houston.
Kaula W.M. (1979) The moment of inertia of Mars. *Geophys. Res. Lett.* 6, 194–196.
Lewis J.S. (1972) Metal/silicate fractionation in the solar system. *Earth Planet. Sci. Lett.* 15, 286–290.
Lewis J.S. (1974) Volatile element influx on Venus from cometary impacts. *Earth. Planet. Sci. Lett.* 22, 239–244.
Morgan J.W. and Anders E. (1980) Chemical composition of Earth, Venus and Mercury. *Proc. Natl. Acad. Sci. USA* 77, 6973–6977.
Moritz H. (1976) Fundamental geodedic constants. *Trav. Assoc. Int. Geod.* 25, 411–418.
Navrotsky A. (1980) Lower mantle phase transitions may generally have negative pressure-temperature slopes. *Geophys. Res. Lett.* 7, 709–712.
O'Connell R.J. (1977) On the scale of mantle convection. *Tectonophys.* 38, 119–136.
Pettengill G.H., Eliason E., Ford P.G., Loriot G.B., Masursky H., and McGill G.E. (1980) Pioneer Venus radar results: altimetry and surface properties. *J. Geophys. Res.* 85, 8261–8270.
Ringwood A.E. and Anderson D.L. (1977) Earth and Venus: a comparative study. *Icarus* 30, 243–253.
Sharpe H.N. and Peltier W.R. (1979) A thermal history model for the Earth with parameterized convection. *Geophys. J. Roy. Astron. Soc.* 59, 171–203.

Toksöz M.N., Hsui A.T., and Johnston D.H. (1978) Thermal evolutions of the terrestrial planets. *Moon and Planets* **18**, 281–320.
Tozer D.C. (1965) Heat transfer and convection currents. *Phil. Trans. Roy. Soc. London* **A258**, 252–271.
Tozer D.C. (1972) The present thermal state of the terrestrial planets. *Phys. Earth. Planet. Inter.* **6**, 182–197.
Turcotte D.L., Cooke F.A., and Willeman R.J. (1979) Parameterized convection within the moon and the terrestrial planets. *Proc. Lunar Planet. Sci. Conf.* 10th, p. 2375–2392.
Urey H.C. (1952) *The planets.* Yale Univ. Press, New Haven. 245 pp.
Weertman J. (1970) The creep strength of the Earth's mantle. *Rev. Geophys. Space Phys.* **8**, 145–168.
Weidenschilling S.J. (1977) Aerodynamics of solid bodies in the solar nebula. *Mon. Not. Roy. Astron. Soc.* **180**, 57–70.
Weidenschilling S.J. (1978) Iron/silicate fractionation and the origin of Mercury. *Icarus* **35**, 99–111.

Metal chloride and elemental sulfur condensates in the venusian troposphere: Are they possible?

V. L. Barsukov, I. L. Khodakovsky, V. P. Volkov, Yu. I. Sidorov, V. A. Dorofeeva and N. E. Andreeva

Vernadsky Institute of Geochemistry and Analytical Chemistry, USSR Academy of Sciences, Moscow, USSR

Abstract—Thermodynamic treatment of Venus cloud particle composition is carried out. The formation of the antimony and arsenic compounds in the cloud layer is concluded, assuming SbOCl and As_4O_6 to represent the most probable components. The subordinate elemental sulfur particles associated with dominant H_2SO_4 droplets are considered as plausible candidates.

Iron, mercury, ammonium and aluminum chlorides are ruled out. Liquid droplets of HCl solution as well as sulfonic acids are also excluded as the probable chlorine-bearing aerosols.

The theoretical implication leads to the assumption of important role of sulfur, chlorine, As, and Sb(?) in Venus meteorology similar to water in the terrestrial atmosphere.

INTRODUCTION

This report concerns advanced theoretical predictions of the chemical composition of the Venus cloud based on physicochemical treatment of atmospheric processes (Volkov et al., 1979; Barsukov et al., 1980a, b; Dorofeeva et al., 1981). A direct determination of aerosol chemical composition was obtained on the Venera 12 mission (Surkov et al., 1981). Within the 54–47 km altitude interval the mass loading of the sulfur-bearing particles is one order of magnitude less than that of the chlorine-bearing ones. These data are of special interest because of the necessity of finding agreement with the conventional sulfuric acid cloud model (Still, 1972; Young, 1973). The problem of chlorine-bearing compounds in the troposphere and Venus cloud layers is discussed in terms of our geochemical tropospheric model (Khodakovsky et al., 1979a; Barsukov et al., 1980a, b).

Gaseous chlorine compounds in the Venus troposphere

Gaseous HCl ($X_{HCl} = 6 \times 10^{-7}$) was detected in the Venus atmosphere as early as in 1967 by earth-based spectroscopy at the top of the cloud deck (Connes et al., 1967). According to Belton (1968) this value should be revised to 10^{-6}. The detailed spectral interpretation carried out by L. Young (1972) leads to the conclusion of a nonuniform HCl distribution in the Venus atmosphere. Variations of the hydrogen chloride mixing ratio at the cloud deck (about 67 km) were estimated as $(4 \text{ to } 8) \times 10^{-7}$.

The evaluation of HCl partial pressure at the Venus surface (T = 748 K, $P_{tot.}$ = 120 bar) was made in Lewis' (1970) model based on the assumption of buffering of the Venus troposphere by surface rocks. According to Lewis (1970) the HCl equilibrium partial pressure in the reaction:

$$2NaCl_{(c)} + Al_2SiO_{5(c)} + 3SiO_{2(c)} + H_2O_{(g)} = 2NaAlSi_2O_{6(c)} + 2HCl_{(g)} \tag{1}$$

halite   andalusite   quartz   jadeite

was calculated as 1.2×10^{-4} bar. Halite was assumed as one of the Venusian chlorine-bearing minerals, providing an HCl mixing ratio of 10^{-6} (Belton, 1968).

Calculations in the open multi-component system basalt-Venus troposphere (Barsukov et al., 1980a, b) allow estimation of a lower limitation the HCl partial pressure ($P_{HCl} = 1.4 \times 10^{-4}$ bar). If this limit is exceeded, the chlorine should be incorporated in marialite [$Na_8(AlSi_3O_8)_6Cl_2$], which is the most probable chlorine-bearing mineral at the Venus surface. Halite was found to be unstable in a global, open multi-component system.

The calculated value of the HCl equilibrium partial pressure corresponds to $X_{HCL} = 1.4 \times 10^{-6}$ i.e., a factor of 1.7 to 3.5 higher than the spectroscopic determinations at the cloud deck (Young, 1972). If this discrepancy in HCl mixing ratio is significant, some chlorine could be incorporated in cloud particles.

The possibility of the existence of gaseous elemental chlorine in the troposphere has been discussed in the literature. S. Kumar (1974) suggested that the HCl concentration would be lowered, and predicted a high elemental chlorine mixing ratio ($X_{Cl} \approx 10^{-4}$) above the cloud deck. It was assumed that gaseous hydrogen chlorine serves as the elemental chlorine source through photolytic decomposition of HCl (Prinn, 1973). The chlorine sink is suggested to be its conversion to HCl by reaction with H_2, as well as in the spectroscopically unidentified molecules, e.g., ClO, or even ammonium chloride (Kumar, 1974). The latter possibility was recently excluded; nevertheless, the assumption of a positive chlorine concentration gradient above the cloud deck is attractive. According to Pollack and Toon (1980) the UV-markings within the upper cloud could be explained as the result of gaseous chlorine with a mixing ratio of about 10^{-6}. Pioneer-Venus mass-spectrometric determinations give the chlorine value as "a few ppm" (Hoffman et al., 1979).

These evaluations are in agreement with the diffuse radiation spectroscopic measurements at the Venus limb by the Venera 9, 10 orbiters: the chlorine mixing ratio at the 70 to 100 km altitude level was found to be within the interval 7×10^{-7} to 10^{-5} (Krasnopolsky, 1980).

The molecular chlorine concentration within the troposphere under the cloud base is suggested to be extremely low. According to spectrophotometry in the Venera 11, 12 mission (Moroz et al., 1979, 1980) the upper limit on the Cl_2 mixing ratio below the 49 km altitude level does not exceed 10^{-8}.

A theoretical estimation of molecular chlorine concentration within the near-surface troposphere was made by Lewis (1970), assuming T = 748 K, and P_{tot} = 120 bar at the Venusian surface. The Cl_2 mixing ratio was calculated as 2×10^{-20}. Our estimation was obtained using the equilibrium calculations of the reaction:

$$2HCl_{(g)} = H_{2(g)} + Cl_{2(g)} \qquad (2)$$

using the starting values from our geochemical model:

$$X_{HCl} = 1.4 \cdot 10^{-6}; X_{H_2} = 2.4 \cdot 10^{-9} \text{ at } T = 750 \text{ K}, P_{tot.} = 97.4 \text{ bar}.$$

The Cl_2 mixing ratio was calculated as 1.5×10^{-17}. Hence the spectroscopic measurements as well as the theoretical calculations predict the hydrogen chloride as the predominant chlorine-bearing gas within the troposphere beneath the cloud base.

The most likely cloud particle candidates (review)

The detection of gaseous hydrogen chloride in the cloud deck stimulated the hypothesis of concentrated (18–26%) HCl droplets as the main aerosol component at the altitude level corresponding to temperatures near 200 K (Lewis, 1968, 1971, 1972). However, this conclusion was slightly inconsistent with the measured refractive indices as well as infrared absorption spectra of the Venus clouds, and was never applied to the main cloud layer.

Hydrogen chloride aerosol as a candidate was again evaluated in our paper (Volkov et al., 1979) in an attempt to interpret the direct chlorine determination in cloud particles carried out during the Venera 12 mission.

Consideration of the HCl-H$_2$O phase diagram leads to the conclusion that the equilibrium condensation of HCl droplets (18%, corresponding to $X_{HCl} = 10^{-6}$) is valid only in the presence of a high water vapor content ($X_{H_2O} \geq 10^{-3}$). Such a water vapor concentration was reported by the first atmospheric composition determinations on Venera 4, 5, and 6 (Vinogradov et al., 1970), as well as from Pioneer-Venus data (Oyama et al., 1980). However, H$_2$SO$_4$ and HCl aerosol coexistence for an extended period of time is highly improbable owing to the substantial difference of the water vapor pressure above the concentrated H$_2$SO$_4$ and HCl solutions. This results in the active destruction of the hydrogen chloride aerosols. Nevertheless, given a high turbulent mixing rate, the local areas of high "humidity" theoretically could arise within the cloud layer: i.e., the HCl$_{(sol)}$ stability field can be entered. Earlier we concluded (Volkov et al., 1979) that sulfuric acid droplet destruction in ascending tropospheric fluxes is favorable to HCl aerosol generation according to the reaction:

$$H_2SO_{4(sol)} + CO_{(g)} = SO_{2(g)} + H_2O_{(g)} + CO_{2(g)} \qquad (3)$$

After the publication of that paper (Volkov et al., 1979), we found a reference to relevant experimental data (Milbauer, 1918). A concentrated sulfuric acid solution (94.6%) was contained in a column (h = 9.5 cm) and subjected to a carbon monoxide flow at 250 C for 10 hours, with the flow rate of 1.2 liters per hour. The extent of conversion of H$_2$SO$_4$ to SO$_2$ was estimated as 5%. Thus, within the cloud layers at a temperature below 65°C, extensive reaction by (3) is highly improbable. The possibility of the existence of an HCl aerosol is therefore rather problematical.

Liquid condensates of oxygen-bearing sulfur and chlorine compounds (sulfonic acid, etc.) as cloud particles were also discussed in our earlier report. It was shown that such substances are easily hydrolyzed and unstable within the altitude interval of the Venera 12 aerosol composition measurements.

Ammonium chloride was discussed as a cloud particle candidate by Lewis (1968, 1969); and Kuiper (1969), but the authors suggested that equilibrium condensation of NH$_4$Cl$_{(c)}$ could be ruled out, given the predicted extremely low ammonia mixing ratios ($X_{NH_3} \leq 10^{-7}$).

At the same time the report of a direct determination of a surprisingly high ammonia concentration ($X_{NH_3} = 10^{-3}$ to 10^{-4}) in the Venus troposphere on the Venera 8 space mission (Surkov et al., 1973) resulted in a reevaluation of the ammonium salt hypothesis (Surkov and Andreichikov, 1973). Nevertheless, thermodynamic arguments (Goettel and Lewis, 1974; Florensky et al., 1976, 1978) demonstrated the impossibility of NH$_4$Cl equilibrium condensation without an extremely high water vapor concentration ($X_{H_2O} \geq 10^{-2}$). Young (1977) gave several criticisms of the extremely low reliability of Venera 8 measurements. A recent interpretation of the Venera 9, 10 mass spectra of the troposphere below the cloud base (Surkov et al., 1978) made it possible to set an upper limit on the ammonia concentration of only $X_{NH_3} < 5 \times 10^{-4}$.

A mercury and mercury halide cloud model was advanced by Lewis (1969). Complete degassing of mercury to the Venus atmosphere, followed by equilibrium condensation of liquid mercury droplets, mercury sulfides and halides, and HCl solution droplets were suggested. The total mass of mercury in the Venus troposphere was estimated as 3 g cm^{-2}, corresponding to 1.4×10^{19} g of Hg. This assumed an average mercury abundance similar to that in the Earth's crust, 8×10^{-6} wt.%.

Iron chloride (FeCl$_2 \cdot$ 2H$_2$O) was proposed as a probable cloud ingredient by Kuiper (1969). Lewis (1970) calculated that the gaseous iron chloride (FeCl$_2$) partial pressure in equilibrium with iron oxides and silicates in the Venus surface rocks (T = 748 K, P$_{tot.}$ = 120 bars) was only 10^{-8} bars. Our model of chemical equilibrium in the global system, Venus troposphere-surface rocks, is also in disagreement with the existence of FeCl$_2$ as a main cloud-forming component.

Recently, one more attempt was made to provide a chlorine-bearing cloud condensate. V. Krasnopolsky and V. Parshev (1979) suggested the formation of crystalline Al$_2$Cl$_6$ as the cloud-forming ingredient of the lower layer. The authors themselves emphasize some of the difficulties which arise from this explanation such as the possibility and amount of aluminium released and transferred from the lithosphere to the troposphere. No thermodynamic evaluation of the problem was carried out.

Elemental sulfur as the cloud component was suggested five years ago in support of the identification of dark markings in UV images. Hapke and Nelson (1975) concluded that the UV absorber could be identified as large sulfur particles ($r \approx 3$ to $10\,\mu m$) mixed with H_2SO_4 droplets, or incorporated in an aerosol polycondensate as a contaminant.

Later, Young (1979) proposed the possibility of liquid sulfur droplet condensation in the troposphere within the clouds up to 18 km altitude level, assuming a maximum sulfur-bearing gas concentration of 0.6 vol.% ($X_{\Sigma S} = 6 \times 10^{-3}$). This would result in the precipitation of liquid sulfur droplets.

The tropospheric chemical composition obtained by Venera 11, 12 (Moroz et al., 1979; Gelman et al., 1979) and Pioneer-Venus space missions (Hoffman et al., 1979; Oyama et al., 1980) proved to be in crucial disagreement with Young's hypothesis ($X_{\Sigma S} \leq 10^{-4}$) and the latter was ruled out. Atmosphere and cloud investigations in the Pioneer-Venus mission (Knollenberg et al., 1980; Pollack et al., 1979) indicate also that sulfur droplets and crystals were not the UV absorber. Consequently it became necessary to reevaluate sulfur condensate formation in light of the recent experimental results. This we present in a paper published elsewhere (Dorofeeva et al., 1981). The main conclusions of this report are summarized below.

Evaluation of a sulfur condensate in the Venus troposphere was carried out using a phase diagram plotted as the altitude versus log X_{S_n} (Fig. 1). The gaseous sulfur mixing ratio in the troposphere below the cloud base was calculated by means of our geochemical model (Khodakovsky et al., 1979a; Barsukov et al., 1980a, b) as $X_{S_2} = 1.8 \times 10^{-7}$ (calculated as sulfur dimer), the latter value being in good agreement with the experimental data obtained by Venera 11, 12 spectrophotometry (Moroz et al., 1980, San'ko, 1980). The stability field of the crystalline S_{rhomb} is found to coincide with the Venus cloud altitude interval (48 to 70 km), as is evident from Fig. 1. Liquid sulfur can exist in the cloud zone only as a metastable phase, in the form of small supercooled droplets.

An estimation of an upper altitude limit on sulfur aerosol formation was carried out by comparison of the droplet growth rate with the ascending tropospheric flux. A heterogeneous condensation mechanism was assumed, taking into account the condensation coefficient and coagulation. The growth time of a sulfur droplet from $r = 0.01\,\mu m$ up to $r = 1.0\,\mu m$) at 50 km altitude was estimated as 1 hour, assuming a super-saturation of 0.01%. Assuming an ascending tropospheric wind speed of n cm-s^{-1} (Kuz'min and Marov, 1974) the altitude level of the sulfur cloud particles is found to correspond to the saturation level calculated from the X_{S_n} value (Fig. 1).

The evaluation of a lower altitude limit for sulfur aerosol formation was carried out by comparison of the Stokes velocity of particle sedimentation with the evaporation rate. It was found that a descending spherical particle ($r = 1.0\,\mu m$) at the 50 km level would sink several centimeters over the time interval required for complete evaporation. Thus sulfur condensates in the Venus troposphere can exist at altitudes above 48 km, given the suggested value of X_{S_n}.

The maximum sulfur particle mass loading at the cloud base (48 km) was estimated as ≤ 0.6 mg-m^{-3} (calculated as elemental sulfur) given the suggested value of X_{S_n}. Hence sulfur aerosols cannot be excluded from the list of the most likely cloud particle candidates. In Dorofeeva et al., (1981), a number of possible mechanisms for gaseous sulfur formation were considered. The most plausible is the thermochemical reduction of sulfur dioxide within the near-surface tropospheric layer. Sulfur formation by photochemical or electric discharge mechanisms is extremely unlikely.

This brief review illustrates the efficiency of the physicochemical approach to the search for cloud particle candidates. These implications are useful in selecting and ruling out a number of thermodynamically "forbidden" substances, as well as in experiment planning for future space flights.

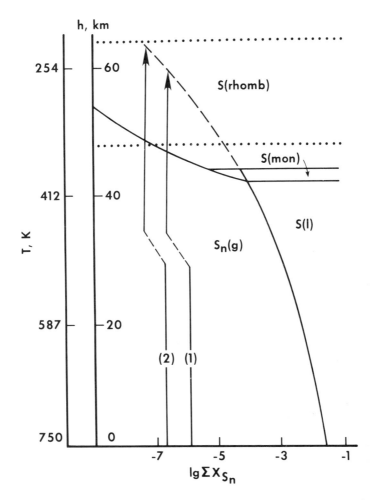

Fig. 1. The sulfur phase diagram applied to the P-T conditions of the Venus troposphere. $1 - X_{S_2} = 1.3 \times 10^{-6}$; $2 - X_{S_2} = 1.8 \times 10^{-7}$ at the datum surface level. The sulfur mixing ratios are obtained in equilibrium tropospheric calculations for different starting values depending on the error range of the experimental Venera 11, 12 values of the CO concentration. See Dorofeeva et al. (1981). The transformation of $S_{2(g)}$, dominant within the near-surface troposphere, to the octamer $S_{8(g)}$ at altitudes above 40 km, is shown by the dashed line ($X_{S_8} = 4 X_{S_2}$).

Thermodynamical consideration of cloud particle condensation stability in the Venus troposphere

A sulfuric acid aerosol has been, with remarkable assurance, concluded to be the main ingredient (in terms of numerical concentration) of the Venus clouds. Sulfuric acid (75 to 85%) is represented by average particle diameters of 0.2 to 0.3 μm (mode 1) and 2.0 to 2.5 μm (mode 2) (Knollenberg et al., 1980).

The total mass loadings of the mode 1 and 2 particles within the upper (56.5 to 70 km) and middle cloud layers (50.5 to 56.5 km) are according to Knollenberg et al. (1980), 1 and 5 mg m^{-3}, respectively. The mass loading of H$_2$SO$_4$ particles with average radii of 0.8 to 1.45 μm (recalculated as elemental sulfur) within the altitude interval of 49 to 61.5 km was evaluated using the nephelometer data from Venera 9 and 10 (Marov et al., 1978). Calculated mass loading values for different cloud layers were found to lie in the interval from 0.2 to 2.0 mg m^{-3}. Thus the lower limit on the sulfur concentration according to Venera-12 measurements of aerosol composition (0.2 mg m^{-3}) is consistent with the mass loading estimate, assuming the identification of mode 1 and 2 particles as concentrated sulfuric acid droplets. However, the maximum estimated values of particle mass loading within the middle and lower cloud layer exceed the experimental values by an order of magnitude.

The results of Pioneer-Venus experiments indicated that within the middle and lower cloud layers the mode 3 particles ($r = 4\,\mu m$) are crystals with the mass loading about 10 mg m^{-3} (Knollenberg et al., 1980). Some authors emphasized the possibility of identifying mode 3 cloud particles with the chlorine-bearing aerosol component in the Venera 12 experiment (Kransnopolsky and Parshev, 1979; Knollenberg et al., 1980). In this respect, the physico-chemical implications of chlorine-bearing cloud particles are of special interest.

Our work is based on a geochemical model of the Venus troposphere (Khodakovsky et al., 1979a; Barsukov et al., 1980a, b). The main assumption of the model is the attainment of chemical equilibrium between Venus surface rocks and the near-surface troposphere. Ascending and descending gaseous fluxes in convective tropospheric cells are assumed. The chemical compositions of the gaseous fluxes are assumed to differ with respect to minor components, especially the water vapor concentration. The variations in tropospheric chemical composition revealed by Venera 11 and 12 and by Pioneer-Venus are interpreted in terms of the model.

To first approximation, the chemical composition of the ascending tropospheric flux is assumed to correspond to a "frozen" quasiequilibrium composition in the near-surface tropospheric layer. This composition is characterized by low oxygen partial pressure ($P_{O_2} = 10^{-21}$ atm) regulated by a sulfate- sulfide buffering mineral assemblage containing pyrite, anhydrite and magnetite (Barsukov et al., 1980b). We emphasize that chemical equilibrium in the troposphere below the cloud base is not attained except in the near-surface layer.

Cloud condensate formation reactions within the ascending gaseous flux are assumed to proceed to completion. It is worthwhile to emphasize that comparative analysis of the rates of condensation and evaporation with those of turbulent mixing and of aerosol coalescence, sedimentation and transport are beyond the scope of this report.

Thermodynamic predictions of cloud condensate stability were carried out for 48 elements in the following sequence: 1) evaluation of condensed-phase stability under Venus surface P-T conditions; 2) a search for the dominant gaseous substance, e.g., elements, oxides, chlorides, oxy- and hydrochlorides, and calculation of gas mixing ratio upper limits assuming equilibrium with plausible condensed phases at the planetary surface; 3) estimation of the altitude of the condensation level using the vapor pressure curve of the dominant gases; 4) selection of the cloud particle candidates.

Three constraints were taken into account in selecting the main cloud particle candidates: 1) condensates of practical interest must be characterized by a mass loading of about 0.n to n mg m^{-3} where n = 1, 2, 3 ... This aerosol concentration was detected by a particle size spectrum determination (Knollenberg and Hunten, 1979), as well as by an X-ray florescence spectrometer (Surkov et al., 1981). Such mass loading corresponds to a gas mixing ratio of about 10^{-7} to 10^{-6}, depending on the molecular weight of the given compound and its condensation altitude. Thus condensed phases which, when fully vaporized, have mixing ratios $X_i \leq 10^{-7}$ to 10^{-6} are excluded as candidates. 2) the altitude of the condensation level must coincide with the main cloud layer. 3) the inventory of the given element resulting from complete degassing of the outer planetary shells must provide an appropriate concentration of the corresponding gas, with a mixing ratio $X_i \geq 10^{-7}$ to 10^{-6}.

An upper limit on the concentration of each element in the troposphere can be estimated by assuming complete crustal degassing and similar elemental inventories in the outer shells of Venus and Earth.

Average crustal abundances can be used directly to calculate the upper limits on the gas mixing ratios because of a fortuitous numeric value coincidence: taking the Venus crustal mass as $M\venus = 4.5 \times 10^{25}$ g (Khodakovsky et al., 1978, 1979a), the planetary surface area as $S_♀ = 4.6 \times 10^{18}$ cm^2, and atmospheric pressure at the surface as $P = 10^5$ g cm^{-2}, we obtain the simple relation:

$$X_i = \frac{P_i}{P} = \frac{M_♀ \cdot K_i}{S_♀ \cdot P \cdot 100\%}. \tag{4}$$

Here X_i is the mixing ratio of given element; P_i is the partial pressure, and K_i is the average crustal elemental abundance in weight %. After substitution of the numerical values, we find: $X_i \approx K_i$.

Hence, consideration of average crustal abundance values (Vinogradov, 1962) leads to this criterion: a chemical element will be included in the Venus cloud particle candidate list if the average crustal abundance of the element fits the requirement that $X_i = K_i \geq 10^{-7}$ to 10^{-6}.

The following part of this report deals with a detailed discussion of our thermodynamic predictions.

The set of the stable solid phases containing 10 main rock forming elements (P, Si, Al, Fe, Mn, Ti, Mg, Ca, Na, K) as well as sulfur was obtained as the result of 16-component multisystem calculations (Khodakovsky et al., 1978, 1979a, b; Barsukov et al., 1980b). The remaining 37 elements were not included in the multisystem: (Se, Te, As, Sb, Bi, Ge, Sn, Pb, B, Ga, In, Tl, Zn, Cd, Hg, Cu, Ag, Au, Co, Ni, Ru, Rh, Pd, Ir, Pt, Cr, Mo, V, Zr, Hf, Ce, Th, U, Be, Sr, Ba, Li). The stabilities of their condensed phases were estimated using equilibrium calculations which assumed the mixing ratios of CO, CO_2, SO_2, H_2O, HCl in our model (Dorofeeva et al., 1981).

About 190 gaseous and 170 solid phases were used in the calculation. Besides the mineral assemblages predicted in our previous reports, Li, Be, Zr, Th, and Sn silicates, Ba and Sr sulfates, sulfides of a number of chalcophiles (Pb, Zn, Cu, Bi, Ag, Cd), and Ni and Co are proposed as possible features of Venus mineralogy (Table 1). Oxides could be represented by Cr and Sb compounds. Gold and the elements of the platinum group are predicted to be present in the elemental form. No stable condensed phases containing As, Hg, Se or Te were revealed by our thermodynamic treatment.

The most stable condensed phases of As, Hg, Se and Te at the Venus surface ($T = 750$ K) are, respectively, liquid As_4O_6, Hg, Se and Te. Nevertheless, the saturated vapor pressures above the liquids are found to exceed the corresponding partial pressures given by the average crustal abundance constraint. Thus the inventories of As, Hg, Se and Te are probably concentrated in the Venus atmosphere, the interior being depleted in these elements.

The predominant gaseous substances are chlorides, and in some cases oxy- and hydroxychlorides. The exceptions include arsenic, antimony, selenium, tellurium, and mercury. It is evident that, for certain elements such as Si, Al, B, the dominant gas would be a fluoride or oxyfluoride.

Upper limits on the mixing ratios of the predominant gaseous substances were calculated from the corresponding mineral equilibria, assuming the "model" mixing ratios of CO_2, CO, SO_2, H_2O, HCl and the average crustal abundance constraint.

The mixing ratios of lithophile and many chalcophile chlorides do not exceed $X_i = 10^{-7}$. Consistent with our constraints, all these compounds should be excluded as cloud particle candidates. Aluminium chloride, Al_2Cl_6, is characterized with the lowest upper limit on its mixing ratio. Thus the hypothesis (Krasnopolsky and Parshev, 1979) of aluminium chloride cloud particles is contradicted by physico-chemical constraints and should probably be ruled out.

Gaseous iron chlorides also cannot be considered as the main candidates, since the calculated upper limits on their mixing ratios are only $X_{FeCl_2} \leq 10^{-11.7}$ and $X_{FeCl_3} \leq 10^{-13.9}$. This conclusion is in agreement with Lewis (1970).

The list of main ($X_i \geq 10^{-7}$ to 10^{-6}) and subordinate ($10^{-10} \leq X_i \leq 10^{-7}$) cloud particle candidates was compiled taking into account the probable errors in the thermodynamic data and probable tropospheric composition variations. Only ten chemical elements from the total 48 are left in this list: S, Se, Te, As, Sb, Bi, Pb, In, Zn, Hg (Table 1).

Table 1. The results of the thermodynamical, prediction* of the most likely Venus cloud particle condensates.

Condensable gaseous phase	Condensed phase	Gas mixing ratio upper limit within the ascending tropospheric flux [Barsukov et al., 1980a, b]	Altitude condensation level, km	Considered gases, the predominant gas is shown at the head of the sequence	Stable crystalline phases at the datum level	Considered crystalline phases (unstable at the datum level)
1	2	3	4	5	6	7
AsO_6	$\beta\text{-}As_4O_6$ (c)	$\leq 10^{-4}$	≥ 42	As_4O_6, As_4, As_2, $AsCl_3$, AsO, As	—	As, As_4O_6, $FeAsS$, $FeAs$, $FeAs_2$, Fe_2As, $MnAs$, $Ca_3(AsO_4)_2$
Sb_4O_6	$\alpha\text{-}Sb_4O_6$ (c)	$\leq 10^{-6}$	≥ 0	Sb_4O_6, $SbCl_3$, $SbCl$, Sb_4, Sb_2, Sb, SbO, $SbCl_5$	$\alpha\text{-}Sb_4O_6$ senarmontite	Sb, $Sb_2O_5Cl_2$, Sb_2S_3, $MnSb$, $FeSb_2$
$SbCl_3$	$SbOCl$ (c)	$\leq 3.2 \cdot 10^{-7}$	≥ 58			
S_8	S (c)	$\sim 1.8 \cdot 10^{-7}$	~ 48	SO_2, COS, S_2, H_2S, S_3, S, S_4	FeS_2, $CaSO_4$ pyrite anhydrite	FeS, MnS, K_2SO_4, $MgSO_4$, Na_2SO_4
S_8	S_8 (l)	$\sim 1.8 \cdot 10^{-7}$	~ 62	S_6, S_5, S_7, S_8		
Se_2	Se (l)	$\leq 10^{-5}$	≥ 18	Se_2, Se, $SeCl_2$, SeO_2, SeO	—	SeO_2, $FeSe$, $FeSe_2$, $CaSeO_3$, $Fe_2(SeO_3)_3$, $4Fe_2O_3 \cdot SeO_2$, $CaSeO_4$
Te_2	Te (c)	$\leq 10^{-7}$	≥ 18	Te_2, $TeOCl$, Te, $TeCl_2$, TeO_2, TeO, $TeCl_4$	—	TeO_2, $Te_6O_{11}Cl_2$, $FeTe_2$, $CaTeO_3$
Hg	Hg (c)	$\leq 10^{-8}$	≥ 62	Hg, $HgCl_2$, $HgCl$	—	HgO, $\alpha\text{-}HgS$, $HgCl_2$, Hg_2SO_4, Hg_2CO_3
$PbCl_2$	$PbCl_2$ (c)	$\leq 1.2 \cdot 10^{-8}$	≥ 16	$PbCl_2$, $PbCl$, Pb, PbO	PbS (galena)	PbO, PbO_2, $PbSO_4$, $PbCl_2$, $PbCl_2 \cdot 2PbO$, $PbSO_4$, $PbCO_3$, $PbSiO_3$, $PbTiO_3$
$InCl$	$InCl$ (c)	$\leq 1.6 \cdot 10^{-9}$	≥ 46	$InCl$, In, In_2O, In_2	In_2O_3,	InS, $\beta\text{-}In_2S_3$
$ZnCl_2$	$ZnCl_2$ (c)	$\leq 4 \cdot 10^{-10}$	≥ 38	$ZnCl_2$, $ZnCl$, Zn	$\alpha\text{-}ZnS$ (sphalerite)	ZnO, $ZnCO_3$, $Zn(AlO_2)_2$, $Zn(FeO_2)_2$, $Zn_2(SiO_4)$, Zn_2TiO_4
$BiCl_3$	$BiCl_3$ (c)	$\leq 4 \cdot 10^{-10}$	≥ 50	$BiCl_3$, Bi, Bi_2, $BiCl$, BiO	Bi_2S_3 (bismuthite)	$\alpha\text{-}Bi_2O_3$, $Bi_4(SiO_4)_3$

*All thermodynamical data are taken from (Glushko, 1965–1981; 1978–79)

It should be emphasized that the mixing ratios in the Table 1 represent upper limits. For example, the lack of thermodynamic data for $FeAs_2O_4$, $MgAs_2O_4$, $Ca_5(AsO_4)_3F$, $Ca_5(AsO_4)_3Cl$, $FeSb_2O_4$, $Fe_5Sb_4O_{11}$, $MgSb_2O_4$, and $CaSb_2O_4$ prevent estimation of the stability of these solid phases relative to the arsenic and antimony minerals listed in Table 1. In other words, there could exist some additional stable solid phases at the Venus surface not included in our theoretical prediction. This situation could result in a remarkable decrease in the mixing ratios of As_4O_6, Sb_4O_6, $SbCl_3$, and other arsenic- and antimony-bearing gases.

It is evident from the Table 1 that the chlorides of Pb, In, Zn, and Bi should not be considered as candidates for the main cloud particles, since their corresponding gas mixing ratios lie in the interval $4 \times 10^{-10} \leq X_i \leq 1.2 \times 10^{-8}$. Furthermore the $PbCl_2$ and $ZnCl_2$ condensation altitudes are located far below the cloud base (Table 1), and are thus inconsistent with one of our adopted constraints.

A search for mercury condensates is of special interest (Lewis, 1969). Figure 2 illustrates the predominant gaseous mercury substance at the surface of Venus to be gaseous elemental mercury Hg (g). Condensed liquid and crystalline mercury could be stable within the clouds above 52 km given a mercury mixing ratio of 10^{-5}. This value corresponds to the total mercury mass which would accumulate in the terrestrial atmosphere as a result of complete degassing of the crusts, with an average abundance of 8×10^{-6} wt.% Hg (Vinogradov, 1962).

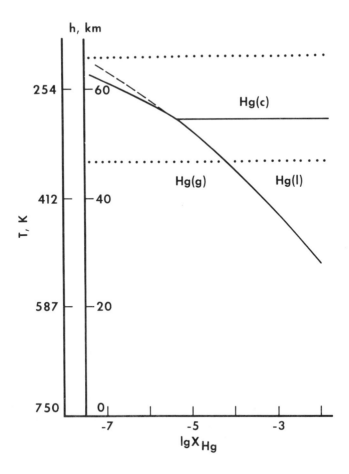

Fig. 2. The mercury phase diagram applied to the P-T conditions of the Venus troposphere. The metastable liquid mercury stability field is shown by the dashed line.

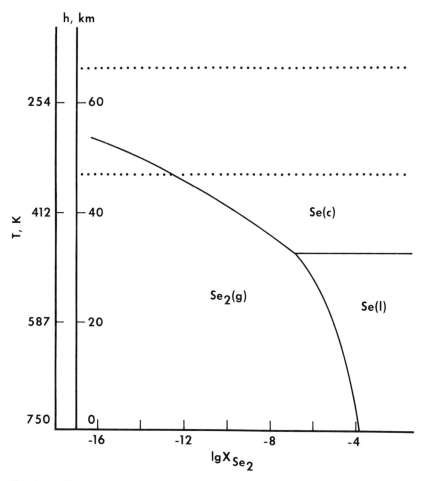

Fig. 3. (a) The selenium phase diagram applied to the P-T conditions of the Venus troposphere.

The Venera-12 measurements (Surkov et al., 1981) indicate the mercury concentration in the aerosol does not exceed 0.05 mg m^{-3}, i.e., the sensitivity limit. This value corresponds to $X_{Hg} < 10^{-8}$ at the 62 km condensation level. Thus the mercury mass in the Venus troposphere is less than 4.6×10^{15} g, or at least 3.5 orders of magnitude lower than the value (1.4×10^{19} g) obtained by assuming similar mercury inventories on Earth and Venus.

This dissimilarity could be the result of divergence in the differentiation histories of proto-Venus and proto-Earth matter, or of mercury depletion in proto-Venus, or the total effect of these factors. Variations in the average mercury abundances in terrestrial and extraterrestrial objects (lunar rocks, meteorites) are known not to exceed 2-2.5 orders of magnitude (Vinogradov, 1962; Staheev et al., 1975). Therefore we conclude that the atmospheric mercury depletion on Venus was not effected by differentiation processes in the outer shells of Venus. The alternative assumption, primary depletion from proto-Venus matter of mercury and other volatile elements (Se, Te, As), seems to us more plausible.

The radio occultation data obtained by Mariner-10 (Kliore et al., 1979) proved to be in disagreement with dispersive microwave absorption by the liquid mercury droplet clouds. This case would correspond to a fantastic aerosol mass loading of 160 kg m^{-2}. Thus direct aerosol measurements as well as theoretical estimates rule out mercury and mercury compounds as the main cloud particle candidates.

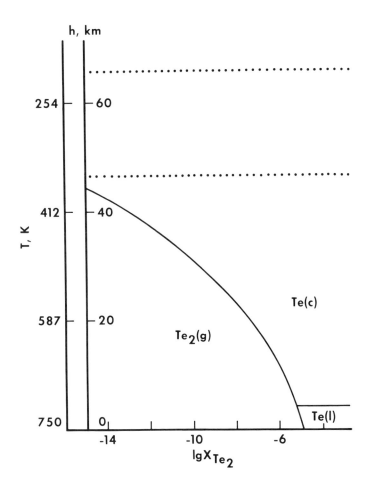

Fig. 3. (b) The tellurium phase diagram applied to the P-T conditions of the Venus troposphere.

The possible selenium and tellurium depletion of the outer layers of Venus was discussed earlier. We emphasize that the condensation altitudes of liquid selenium and crystalline tellurium (Table 1, Figs. 3a, b) are markedly below the cloud base (48 km). In contrast, elemental sulfur condensates are, as before, concluded to be one of the cloud particle candidates.

Owing to the constraints used, the total list of 48 elements as the main cloudforming components is reduced to arsenic and antimony, aside from the sulfuric acid droplets which are not the subject of this discussion (Table 2).

The most likely arsenic- and antimony-bearing condensates are α-As_4O_6 (arsenolite), β-As_4O_6 (claudetite) and crystalline SbOCl. Unfortunately, the thermodynamic properties of the arsenic oxy- and hydroxychlorides are not determined, though AsOCl is reported to have low volatility (Seppelt et al., 1978). The arsenic oxy- and hydroxychlorides are less stable than the corresponding antimony compounds We therefore suggest that crystalline arsenic oxide, As_4O_6, be considered the most probable arsenic bearing cloud particle candidate. Previously J. Lewis (1969) suggested As_2S_3 and Sb_2S_3 as probable candidates but the lack of corresponding thermodynamic values lead to the absence of the quantitative estimation.

According to the Pioneer-Venus data (Knollenberg et al., 1980) the mode 3 cloud particles within the middle and lower cloud layers (47.5–56.5 km) are tabular crystals.

Table 2. The most likely candidates to the Venus cloud particle composition.

Substance	Measured altitude interval, km	Estimated altitude condensation level, km	Aerozol particle mass loading, mg.m^{-3}	Reference
H_2SO_4 (sol.)	70.0 – 56.5		1.0	Knollenberg,
	56.5 – 50.5		5.0	Hunten, 1979
S (c)		48	0.6	Our estimate
Sulfur-bearing compounds	54.0 – 47.0		0.2	Surkov et al., 1981
Mode 3	50.5 – 47.5		10.0	Knollenberg et al., 1980
β-As_4O_6 (c)		48	~20.0	Our estimate
SbOCl (c)		58	0.2 – 2.0*	Our estimate
Chlorinebearing compounds	54.0 – 47.0		2.0	Surkov et al., 1981
Hg (c, l)	54.0 – 47.0		<0.05	Surkov et al., 1981

*Calculated to elemental chlorine

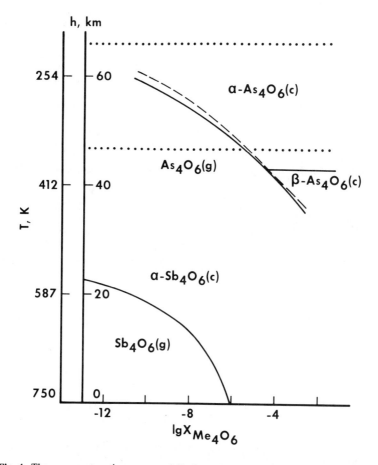

Fig. 4. The vapor saturation curves of Sb_4O_6 and As_4O_6 under the P-T conditions of the Venus troposphere. The metastable crystalline -As_4O_6 stability field is shown by the dashed line.

Is it possible to identify the mode 3 particles as arsenic oxide crystals? The α-As_4O_6 modification is reported to be stable at $T \geq 383$ K and to belong to the cubic system (White et al., 1967; Fig. 4. Thus, if mode 3 particles are identified as crystalline As_4O_6, the latter occurs in the monoclinic β-modification. This As_4O_6-form is metastable under P-T conditions of the middle and lower cloud layer. The As_4O_6 abundance at the cloud base is estimated as 20 mg m^{-3}, assuming coincidence of the gaseous As_4O_6 condensation altitude with the cloud base ($X_{As_4O_6} \leq 10^{-6}$, Fig. 4). This value is in agreement with mode 3 abundance (10 mg m^{-3}) estimates from Pioneer Venus data (Knollenberg et al., 1980). Thus the identification of mode 3 particles as crystalline β-As_4O_6 cannot be ruled out.

Antimony in the Venus troposphere is predicted to exist in two gaseous compounds: Sb_4O_6 and $SbCl_3$ (Table 1) with mixing ratios of the same order of magnitude. The condensation altitude of α-Sb_4O_6 (c) is below the cloud base (Fig. 5); thus its condensation does not satisfy one of our constraints despite the fact that the Sb_4O_6 mixing ratio upper limit ($X_{Sb_4O_6} \leq 10^{-6}$) exceeds that of $SbCl_3$ ($X_{SbCl_3} \leq 3.2 \times 10^{-7}$).

Antimony chloride, $SbCl_3$, at altitudes above 50 km is hydrolyzed according to this reaction:

$$SbCl_{3(g)} + H_2O_{(g)} = SbOCl_{(c)} + 2HCl_{(g)} \tag{5}$$

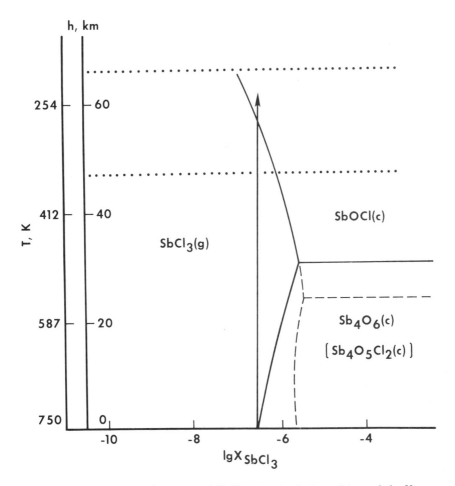

Fig. 5. The vapor saturation curve of $SbCl_3$ under the P-T conditions of the Venus troposphere. The vertical arrow refers to the upper limit on the $SbCl_3$ mixing ratio within the ascending tropospheric flux ($X_{SbCl_3} \leq 3.2 \times 10^{-7}$). The intersection of the arrow with the $SbCl_3$ vapor saturation curve corresponds to the condensation altitude of SbOCl (from $SbCl_3$ gas). The metastable $Sb_4O_5Cl_{2(c)}$ stability field is bordered with dashed lines.

Preliminary calculations indicate that antimony oxychloride, $Sb_4O_5Cl_2$, is unstable under P-T conditions in ascending tropospheric fluxes (Fig. 5). The estimated range of the $SbCl_3$ mixing ratio must be extended to $X_{SbCl_3} \leq 10^{-8}$ to 10^{-6} because thermodynamic data on $Sb_8O_{11}Cl_{2(c)}$ are lacking and data on $SbOCl_{(c)}$ and $Sb_4O_5Cl_{2(c)}$ are of low accuracy. Assuming SbOCl to be the product of $SbCl_3$ condensation, the abundance of the corresponding cloud particles within the Venus clouds is estimated as 0.2–2.0 mg m^{-3} calculated as elemental chlorine. These values are consistent with the Venera 12 aerosol measurements (Table 2).

This physico-chemical treatment of Venus cloud-forming processes takes into account only interaction of the ascending tropospheric flux with solid phases of rather simple composition, without consideration of isomorphic substitution of minerals. Thus, the complicated processes in the system troposphere-surface rocks are discussed only to the first approximation. The formation of complex sulfantiomonide compounds instead of PbS and Sb_2S_3 at the Venus surface should result in a decrease of the mixing ratios of the lead- and antimony-bearing gases in the troposphere. However, preliminary calculations of the $SbCl_3$ and $PbCl_2$ mixing ratios in the case of lead sulfantimonides (Hall, 1966; Craig and Lees, 1972; Bortnikov et al., 1978) indicates that the mixing ratio upper limit does not decrease more than a factor of 3 relative to the simple solids in reactions of PbS, Sb_2S_3, and Sb_4O_6.

CONCLUSIONS

A physico-chemical approach to the composition of the Venus cloud cover leads us to the following conclusions:

a. Formation of antimony oxychloride as a result of antimony chloride gas condensation is plausible. This conclusion is consistent with the Pioneer Venus cloud particle experiment, as well as with the detection of chlorine in the aerosol by Venera 12.
b. The mercury, arsenic, selenium and tellurium inventories probably are concentrated within the Venus atmosphere. A theoretical estimate of the total Hg, As, Se and Te masses in the troposphere indicates that the outer shells of the planet are depleted in these elements. This peculiarity could be attributed to a primary depletion of these volatiles in the material of proto-Venus.
c. We do not exclude the condensation of metastable monoclinic arsenic oxide crystals within the main cloud layer or its identification as the mode 3 particles.
d. Subordinate elemental sulfur particles associated with the dominant sulfuric acid droplets are considered to be the most likely candidates. The orthorhombic sulfur stability field coincides with the P-T conditions of the main cloud layer. Small supercooled sulfur droplets are also appropriate.
e. Iron and mercury chlorides occur in the cloud particles in negligible mass concentrations.
f. The existence of the ammonium and aluminum chlorides is excluded by thermodynamic arguments.
g. Coexistence of liquid HCl and H_2SO_4 aerosols within the cloud cover is doubtful. Suffonic acid aerosols are ruled out.

Theoretical considerations lead to the conclusion of possible important roles for sulfur, chlorine and perhaps antimony and arsenic compounds in Venus meteorology, analogous to water in the terrestrial atmosphere.

REFERENCES

Barsukov V. L., Khodakovsky I. L., Volkov V. P., and Florensky K. P. (1980a) The geochemical model of the troposphere and lithosphere of Venus based on new data. *COSPAR Space Res. XX*, 197–208.

Barsukov V. L., Volkov V. P., and Khodakovsky I. L. (1980b) The mineral composition of Venus surface rocks: a preliminary prediction. *Proc. Lunar Planet. Sci. Conf. 11th*, p. 765–773.

Belton M. (1968) Theory of the curve growth and phase effects in a cloudy atmosphere: application to Venus. *J. Atm. Sci.* **25**, 596–609.
Bortnikov N. S., Nekrasov I. Ya., and Mozgova N. N. (1978) On the phase relations in the Fe-PbSb-Zn system at 400–500°C. *Doklady Akad. Nauk SSSR* **239**, 420–423.
Connes P., Connes J., Benedict W. S., and Kaplan D. L. (1967) Traces of HCl and HF in the atmosphere of Venus. *Astrophys. J.* **147**, 1230–1237.
Craig J. R. and Lees W. R. (1972) Thermochemical data for sulfosalt ore minerals: formation from simple sulfides. *Econ. Geol.* **67**, 373–377.
Dorofeeva V. A., Andreeva N. A., Volkov V. P., and Khodakovsky I. L. (1981) On the elemental sulfur in the Venus troposphere (physico-chemical consideration). *Geokhimiya* **II**. In press.
Florensky C. P., Volkov V. P., and Nikolaeva O. V. (1976) On the geochemical model of the troposphere of Venus. *Geokhimiya* **8**, 1135–1150.
Florensky C. P., Volkov V. P., and Nikolaeva O. V. (1978) A geochemical model of the Venus troposphere. *Icarus* **33**, 537–553.
Gel'man B. G., Zolotukhin V. G., Lamonov N. L., Levcuk B. V., Lipatov A. N., Mukhin L. M., Nenarokov D. F., Rotin V. A., and Okhotnikov B. P. (1979) The gas chromatographic determination of the Venus atmospheric chemical composition at the Venera 12 space mission. *Kosm. Issled.* **XVII**, 708–713.
Glushko V. P., ed. (1965–1981) *The Thermal Constants of the Substances*, vol. 1–10. VINITI, Moscow.
Glushko V. P., ed. (1978–1979) *The Thermodynamical Properties of the Individual Substances*, vol. 1–2. Nauka, Moscow.
Goettel K. A. and Lewis J. S. (1974) Ammonia in the atmosphere of Venus. *J. Atm. Sci.* **31**, 828–830.
Hall H. T. (1966) Application of thermochemical data to problems of ore deposition. *Econ. Geol.* **61**, 622–623.
Hapke B. and Nelson R. (1975) Evidence for an elemental sulfur component of the clouds from Venus spectrophotometry. *J. Atm. Sci.* **32**, 1212–1218.
Hoffman J. H., Hodges R. R., McElroy M. B., Donahue T. M., and Kolpin M. (1979) The composition and structure of the Venus atmosphere: results from Pioneer-Venus. *Science* **205**, 49–52.
Khodakovsky I. L., Volkov V. P., Sidorov Yu I., and Borisov M. V. (1978) The preliminary prediction of the mineral composition of surface rocks and the hydration and oxidation processes of the Venus outer shell. *Geokhimiya* **12**, 1821–1835.
Khodakovsky I. L., Volkov V. P., Sidorov Yu I., Dorofeeva V. A., Borisov M. V., and Barsukov V. L. (1979a) Geochemical model of the Venus troposphere and crust according to new data. *Geokhimiya* **12**, 1747–1758.
Khodakovsky I. L., Volkov V. P., Sidorov Yu. I., and Borisov M. V. (1979b) Venus: Preliminary prediction of the mineral composition of surface rocks. *Icarus* **39**, 352–363.
Kliore A. J., Elachi C., Patel I. R., and Cimino J. B. (1979) Liquid content of the lower clouds of Venus as determined from Mariner 10 radio occultation. *Icarus* **37**, 51–72.
Knollenberg R. G. and Hunten D. M. (1979) The clouds of Venus: a preliminary assessment of microstructure. *Science* **205**, 70–74.
Knollenberg R. G., Travis L., Tomasko M., Smith P., Ragent B., Esposito L., McCleese D., Martonchik J., and Beer R. (1980) The clouds of Venus: a synthesis report. *J. Geophys. Res.* **A13**, 8059–8081.
Krasnopolsky V. A. (1980) Venera 9, 10 space mission: the diffuse radiation spectroscopy of the atmosphere above the cloud deck. *Kosm. Issled.* **XVIII**, 899–906.
Krasnopolsky V. A. and Parshev V. A. (1979) On the Venus tropospheric and cloud layer chemical composition based on the Venera 11, 12 and Pioneer-Venus space mission data. *Kosm. Issled.* **XVII**, 763–771.
Kuiper G. P. (1969) Identification of the Venus cloud layers. *Commun. Lunar and Planetary Lab. Univ. Arizona* **6**, 229–250.
Kumar S. (1974) Atomic chlorine on Venus. *Geophys. Res. Lett.* **1**, 153–156.
Kuz'min A. D., and Marov M. Ya. (1974) *The Physics of Venus*. Nauka, Moscow. 408 pp.
Lewis J. S. (1968) Composition and structure of the clouds of Venus. *Astrophys. J.* **152**, L79–L83.
Lewis J. S. (1969) Geochemistry of the volatile elements on Venus. *Icarus* **11**, 367–386.
Lewis J. S. (1970) Venus: atmospheric and lithospheric composition. *Earth Planet. Sci. Lett.* **10**, 73–80.
Lewis J. S. (1971) Composition of the Venus clouds: Refractive index of aqueous and HCl solutions. *Nature* **230**, 295–296.
Lewis J. S. (1972) Composition of the Venus cloud tops in light of recent spectroscopic data. *Astrophys. J.* **171**, L75–L79.
Marov M. Ya., Listzev V. E., and Lebedev V. N. (1978) The structure and microphysical properties of Venus clouds. Preprint of Inst. Space Research, Akad. Nauk SSSR, Pr-144.
Milbauer J. (1918) Uber die Reduction der Schwefelsäure durch Kohlenoxyd. *Chemiker-Zeitung* **42**, 313–316.
Moroz V. I., Golovin Yu. M., Ekonomov A. P., Moshkin B. E., Parfent'ev N. A., and San'ko N. F. (1980) Spectrum of the Venus day sky. *Nature* **284**, 243–244.

Moroz V. I., Parfent'ev N. A., and San'ko N. F. (1979) Spectrophotometric experiment at the Venera 11 and Venera 12 descending modules. 2. The interpretation of Venera 11 spectral data using the multilayer addition method. *Kosm. Issled.* **XVII**, 727–742.

Oyama V. I., Carle G. C., and Woeller F. (1980) Corrections in the Pioneer-Venus sounder probe gas chromatographic analysis of the lower Venus atmosphere. *Science* **208**, 399–401.

Pollack J. B., Ragent B., Boese R., Tomasko M. G., Blamont J., Knollenberg R. G., Esposito L. W., Stewart A. I., and Travis L. (1979) Nature of the UV-absorber in the Venus clouds: inferences based on Pioneer-Venus data. *Science* **205**, 76–79.

Pollack J. B., and Toon O. B. (1980) Nature of the UV-absorber in Venus atmosphere. *Bull. Amer. Astron. Soc.* **12**, 715.

Prinn R. G. (1973) Venus: composition and structure of the visible clouds *Science* **182**, 1132–1135.

San'ko N. F. (1980) Gaseous sulfur in the Venus atmosphere. *Kosm. Isled.* **XVIII**, 600–608.

Seppelt K., Lentz D., and Eysel H.-H. (1978) Oxidchloride des Arsen (V): $AsOCl_3$ und $(As_3O_3Cl_4)_n$. *Zeit. Anorg. Allgem. Chem.* **439**, 5–12.

Sill G. T. (1972) Sulfuric acid in the Venus clouds. *Commun. Lunar Planetary Lab., Univ. Arizona* **9**, 191–198.

Staheev Yu. I., Lavrukhina A. K., and Staheeva S. A. (1975) Cosmic abundance of mercury. *Geokhimiya* **9**, 1390–1398.

Surkov Yu. A. and Andreichidov B. M. (1973) Structure and composition of the Venus cloud layer. *Geokhimiya* **10**, 1435–1440.

Surkov Yu. A., Andreichikov B. M., and Kalinkina O. M. (1973) On the ammonia content in the Venus atmosphere according to Venera 8 space probe data. *Doklady Akad. Nauk SSSR* **213**, 296–298.

Surkov Yu. A., Ivanova V. F., Pudov A. N., Verkin B. I., Bagrov N. N., and Pilipenko A. P. (1978) Mass-spectra investigation of the Venus atmosphere chemical composition at the Venera 9 and Venera 10 space probes. *Geokhimiya* **4**, 506–513.

Surkov Yu. A., Kirnozov F. F., Gurjanov V. I., Glazov V. N., Dunchenko A. G., Kurochkin S. S., Rasputni V. N., Kharitonova E. G., Tatziy L. P., and Gimadov V. L. (1981) Venus cloud layer aerosol investigation at the Venera 12 space probe (preliminary data). *Geokhimiya* **1**, 3–9.

Vinogradov A. P. (1962) The average content of chemical elements in the main crustal rock types. *Geokhimiya* **7**, 555–571.

Vinogradov A. P., Surkov Yu. A., Andreichikov B. M., Kalinkina O. M., and Grechischeva I. M. (1970) Venus atmospheric chemical composition. *Kosm. Issled.* **VIII**, 578–587.

Volkov V. P., Khodakovsky I. L., Dorofeeva V. A., and Barsukov V. L. (1979) The main physico-chemical factors controlling the Venus cloud chemical composition. *Geokhimiya* **12**, 1759–1766.

White W. B., Dachille F. and Roy R. (1967) High-pressure polymorphism of As_2O_3 and Sb_2O_3. *Zeit. Kristallogr.* **125**, 450–458.

Young A. T. (1973) Are the clouds of Venus sulfuric acid? *Icarus* **18**, 564–582.

Young A. T. (1977) An improved Venus cloud model. *Icarus* **32**, 1–26.

Young A. T. (1979) The chemistry and thermodynamics of sulfur on Venus. *Geophys. Res. Lett.* **6**, 49–50.

Young L. D. G. (1972) High resolution spectra of Venus. *Icarus* **17**, 632–658.

An Io thermal model with intermittent volcanism

Guy J. Consolmagno

Department of Earth and Planetary Sciences, Massachusetts Institute of Technology, Cambridge, Massachusetts 02139

Abstract—Observations have indicated that heat flow out of Io at present may be several times greater than the heat input from enhanced tidal heating; this may indicate that the observed heat flow is intermittent, rather than continuous. A two-dimensional time-dependent thermal model of Io was constructed to examine the relationship between enhanced tidal heating, heat flow within the body, and surface volcanism. From the models one can conclude that (1) a simple homogeneous body cannot maintain enhanced tidal heating in a thin crust, but such heating is possible if the crust contains an insulating layer, as of sulfur; (2) this sulfur layer will be subject to melting by heat from the interior, resulting in hotspots which can expel heat much more rapidly than tidal heating inputs it; and (3) the location of these hotspots will vary in time and space; they can appear anywhere on the surface, even if tidal heating is concentrated at the poles, and they seem to occur roughly 10% of the time.

INTRODUCTION

In March of 1979, a theoretical paper by Peale, Cassen, and Reynolds (1979) predicted that tidal stresses inside Jupiter's innermost major moon, Io, would give rise to intense heating of Io's interior, with the possibility that "widespread and recurrent surface volcanism would occur." Within weeks after this prediction, the Voyager I flyby of Jupiter gave dramatic confirmation with the spectacular photographs of Io and its volcanic plumes (Morabito *et al.*, 1979).

Inspired by Voyager, a number of other workers (cf. Matson *et al.*, 1980, 1981; Sinton, 1981) used a number of techniques to measure the heat coming out of Io. The heat flow measured was 2000 ± 1000 ergs/cm^{2}s, a substantially larger amount than the 10–30 ergs/cm^{2}s of Earth or Moon, and much higher than the Peale *et al.* theory predicted. According to that theory, tidal heating could provide a heat flow of no more than 800 ergs/cm^{2}s under the most ideal of circumstances, and the authors felt that 400 or less was a more reasonable number (Reynolds, pers. comm.). Getting any more heat out of tidal heating would require Jupiter or Io to dissipate tidal stresses at a rate incompatible with their internal makeups, and it could rob too much energy from Io's orbit to allow the orbital configuration of the Galilean satellites as observed today to be stable over the lifetime of the solar system (Yoder, 1979a).

There is another problem with the simple tidal heating theory. The volcanic plumes seen on Io are near Io's equator, but the tidal heating theory predicts that the poles would be heated more than the equator. This may just reflect the fact that the tidal heating theory is oversimplified, compared to what is really at work inside Io; the Peale *et al.* (1979) heat flow was based on a simple homogeneous, steady-state model for Io.

It is clear, in any event, that a more sophisticated approach to heat flow is called for, to understand the distribution and evolution of heat in Io as a function of both time and location. It is certainly possible that additional energy sources, of a nature as yet unsuspected, could be responsible for the unusual nature of Io's heat; but before such a conclusion can be drawn, it behooves us to examine as thoroughly as possible those heat sources which we feel confident must operate in Io. To this end a more complex, time-dependent thermal evolution model seemed called for.

HEAT SOURCES

Four distinct processes are known which may have contributed to the heat budget of Io: the energy of accretion, the decay of radioactive isotopes, the dissipation of tidal stresses and the dissipation of electromagnetic energy arising from Io's interaction with the jovian magnetosphere.

About the first process very little is known. The formation processes which may have worked on Io are poorly understood; given the similarity in sizes between Io and the moon, one need only consult the extensive literature on the melting of the moon by accretion to appreciate the unresolved nature of the problem (cf. Ransford and Kaula, 1980).

The thermal model presented below uses the very conservative assumption that accretion occurred so slowly that no heat at all was retained, i.e., accretional heating is not considered and the starting temperature throughout the body is constant and equal to the surface temperature. Only in this way can one be confident that the evolution pattern produced by the model is not skewed by some unconstrained choice of starting conditions; but one must keep in mind that the real Io in fact almost certainly started with some remnant of accretional heat, and so the model underestimates the temperatures in the early part of Io's history.

Heat from the decay of radioactive sources can be modelled quantitatively, given reasonable assumptions for the initial abundances of the appropriate nuclides. The standard long-lived species of K, U, and Th ought to be present in chondritic abundances according to the arguments of Prinn and Fegley (1981); these are listed in Table 1.

In addition, it is interesting to note that the presence of short-lived species such as ^{26}Al and ^{60}Fe could be more important here than in the cases of the inner terrestrial planets. If Io were formed in a protojovian nebula, then it follows that its age could be the same as the age of Jupiter. The presence of Jupiter early in the solar system has been invoked to perturb planetesimals in the region where the terrestrial planets were forming (e.g., Weidenschilling, 1977, suggested that Jupiter's gravity perturbed material from the feeding zones of Mars and the asteroids, preventing the growth of larger sized planets in those regions.) Thus it follows that Io and the other Galilean satellites could be older than the terrestrial planets, the moon, or the asteroids. Even a difference of a million years in age represents several half-lives for these species, increasing the probability that Io could have incorporated significant amounts of these radionuclides before they became extinct.

Again as in the case with accretional heating, however, the models presented here are based on the most conservative assumption that only the long lived radionuclides are important, since the abundances of the short-lived species cannot be determined rigorously. And, again as in the case of accretional heating, the effect of this conservative assumption will be to underestimate the temperatures inside Io only during the earliest stages of its history.

The importance of the dissipation of tidal forces in the heat budget of Io was first pointed out by Peale et al. (1979). A tidal bulge is raised on Io by Jupiter, and there is a small eccentricity in Io's orbit imposed by its resonant interaction with Europa and Ganymede. This eccentricity means that the position and extent of the tidal bulge will vary regularly during each orbit. Since Io is not made of a perfectly elastic substance, energy will be dissipated during the motions of this bulge; Peale et al. (1979) estimated that a homogeneous, solid Io would dissipate three times as much energy as would be generated by present-day lunar levels of radioactive heating (assuming that Io has a rigidity and dissipation function similar to that of the moon). This energy dissipation would not be uniform, but concentrated towards the center of the planet and towards the poles. Profiles for the amount of heat dissipated at various points in a homogeneous spherical moon-sized body are given in Peale and Cassen (1978). In the thermal models presented below, the three-dimensional distribution of heat dissipated shown in their Fig. 2 is approximated by a two-dimensional heat distribution as presented here in Fig. 1.

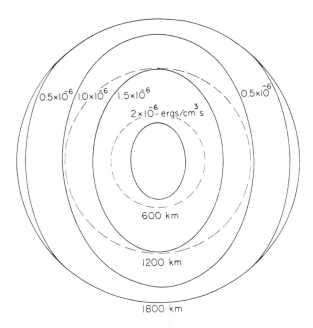

Fig. 1. The assumed heat input due only to tidal heating for Io. Solid lines are contours of constant energy input as a function of both radius and latitude. Dashed lines mark 600 km and 1200 km radius. The curves were approximated from the works of Peale and Cassen (1978) and Pearle et al. (1979), with the distribution of the heat taken from the former and the total heat input taken from the latter work.

Peale et al. (1979) pointed out that the rate of heating in a thin shell is enhanced, relative to that in a solid sphere; therefore, once melting occurs inside Io there will be an increase in the total amount of heat dissipated by the body, and this heat will be concentrated in the solid crust. The amount of enhancement depends on how thin the crust is, peaking at roughly 12 times the original heating rate for a crust whose thickness is 2.5% of the planet radius. Since the melting point of rocky material generally increases with depth, it seems likely (as Cassen et al., 1981, suggested and as is found in the thermal models below) that the structure of Io will include a thin crust with enhanced tidal heating, a liquid mantle where tidal dissipation is negligible, and a solid core where tidal heating is maintained at its original levels.

A fourth heat source which has been mentioned by several authors (cf. Gold, 1979; Drobyshevski, 1979; Yoder, 1979b; Ness et al., 1979; Colburn, 1980) is electromagnetic heating. The details of the mechanism discussed vary from author to author but the general idea is that Io's passage through the jovian magnetosphere will lead to the generation of electrical currents within Io. Since material within Io has a finite conductivity, these currents will dissipate energy which will heat the interior. The amount and location of heating is dependent upon a number of factors which are difficult to establish, including the conductivity of Io's interior and crust, the extent of Io's ionosphere, and the strength of an intrinsic or induced Io magnetic field.

Studies based on actual Voyager data for the magnetic field near Io (Ness et al., 1979) and a detailed study of the electric and magnetic wave modes within Io (Colburn, 1980) indicate that present-day conditions would supply at most two orders of magnitude less heat than tidal heating, though Colburn points out that a larger jovian magnetic field could have existed in the past. In addition, Drobyshevski (1979) calculated that even under ideal conditions, the amount of energy dissipated will be limited by the magnetic field induced by the currents themselves, to a value less than the maximum enhanced tidal heating. Given the many unanswered questions about this mechanism, we have not considered electromagnetic heating in the thermal models described below.

THERMAL MODELS OF IO

Before describing the details of the computer program used to model the thermal evolution of Io, it may be useful to remind ourselves of the nature and limitations of such computer models, since, in many instances, Io presents unique problems where these limitations become important. These limitations will directly affect the ways in which the results of the model can be used to draw conclusions about the thermal state of Io.

Numerical computer models have two fundamental limitations. First, by their nature such models consist of very long summations and these summations must converge on a single, stable solution with in a reasonable amount of time. The convergence criterion for heat flow problems demands that the quantity $(K \Delta t/C_p \rho \Delta r^2) < 1/2$, where K, C_p, and ρ are, respectively, the conductivity, heat capacity, and density of the material which makes up the planet being modelled; Δr is the spacing between the grid points; and Δt is the timestep. To increase conductivity by several orders of magnitude, as is required when one models the transport of heat by convection in a "parameterized convection" scheme, requires that the timestep be shortened by a like amount and so the computer time necessary increases by several orders of magnitude. Likewise, to make a finer grid results in increased cost in computing time, with the increase going as the square of the rate the grid size shrinks. This puts severe constraints on how precisely one can model any body.

Second, numerical models by their nature demand numerical inputs. It has been said that "you really don't understand a system until you can model it on a computer"; if this is true, it is not because the computer gives some magic answers or instant understanding, but because even before he writes the computer code which acts as the model, the modeller must already understand each step of the system well enough to assign some numerical value or algebraic function to that step. In a system like the thermal evolution of a distant planet which involves processes that often are only understood, at best, qualitatively, one can only hope that the results do not depend significantly on the guesses used as input parameters.

But given these difficulties, just what does "significant" mean? Numerical models give quantitative results; but when the input parameters are little more than guesses, one should treat with caution anything beyond a qualitative interpretation of these results. These qualitative conclusions can be very important, even profound, but they must not be overinterpreted, and the numbers on which they are based should not be taken as quantitative answers. On the other hand, the inability of these models to reproduce in specific detail the abundant display of unusual phenomena which a body such as Io exhibits should not cause us to reject the general, qualitative conclusions which can be drawn from these models.

It is in this spirit which we will proceed to discuss the thermal models for Io.

The thermal models for Io were based on the computer programs described in Consolmagno (1975), but with extensive modifications. Heat sources included radioactive nuclides and tidal heating. In the basic model, the two-dimensional tidal heating values illustrated in Fig. 1 were approximated by the equations given in Table 1. A finite difference scheme in both r and $\cos \theta$ (where θ is the colatitude) was used. The values of $\cos \theta$ ranged from 0 to 1 in steps of 0.1; 20 irregular spacings in radius from 1800 to 0 km were used, with a Δr of only 10 km for the top 100 km of the planet (5 km for the topmost layer). Changes in temperature were calculated over a timestep of 7400 years. (A longer timestep was used for the first 0.5 billion years, before the onset of solid-state convection.)

Heat transport was assumed to be by conduction until a temperature of 1300 K was reached, at which point convection was modelled simply by increasing the conductivity two orders of magnitude. Given that the onset of convection represents a sudden change in the transport of heat, it was

Table 1. Parameters used in thermal model.

Grid: 20 radius spacings (10 in upper 100 km), 0–1800 km
    10 latitude spacings, in cos θ

timestep: 25,000 years (before onset of convection)
           7,400 years (after onset of convection)

heat capacity, rock: 1.2×10^7 ergs/g K

conductivity, rock: 4.5×10^5 ergs/cm^2 s K

conductivity, sulfur-rich crust: 4.5×10^4 ergs/cm^2 s K

heat of fusion: 5×10^9 ergs/g

present day abundances of elements with radioactive isotopes:

                      K 815 ppm
                      U 0.012 ppm
                      Th 0.04 ppm

surface and initial temperature: 100 K

heat from tidal heating (ergs):

$s(r, \theta) = 0.735 \times 10^{-6}[3.28 - 2.74\, r + 2.25\, r^2 \cos^2\theta \, \exp(-1.5625\, r^2)]$

enhancement of tidal heating, E, as a function of r_m (radius of crust/melt boundary divided by total radius):

if $r_m < 0.975$ $E = 0.03285 \exp(6.421\, r_m) + 1$
if $r_m > 0.975$ $E = 727.8(1 - r_m)$

---

felt that the scheme used here would not be too unrealistic, for all its simplicity. At 1300 K, and hotter, the Nusselt number (the ratio of heat transport by convection vs. conduction) is probably greater than 100, but increasing the conductivity by any larger amount would make the timestep for convergence shrink so much that the program would take many days of computer time to run. The conductivity is, in a sense, a measure of how fast the system responds to changes in temperature; since the model only calculates changes in heat flow over a period of thousands of years, use of a higher Nusselt number would not be meaningful in any event. Below 1300 K, the Nusselt number falls rapidly, and convection should be unimportant.

Upon melting within the body, the effect of enhanced tidal heating in the crust was modelled by increasing the heating there by a factor as given in Fig. 1 of Peale et al., 1979, with a correction to concentrate their whole-body heat input into a heat per gram value applicable for the crust. Tidal heating was set to zero in any molten region; any solid core below the molten region was subject to unenhanced tidal heating.

Heat of accretion was not considered, nor were short-lived radioactive nuclides, nor was the possible affect that accretion in a sub-jovian nebula or near a contracting, luminous proto-Jupiter might lead to initial temperatures several hundreds of degrees higher than present day surface temperatures. The model starts at a temperature of 100 K everywhere. If any of these other effects were important, the effect would be to advance the timescale of thermal evolution inside Io compared to this model by roughly 0.2–0.5 billion years. Thus melting and enhanced tidal heating might have become important even earlier than these models predict. However, since tidal heating itself so quickly overwhelms all other effects, these changes would almost certainly have no long-term effects on the evolution of Io.

RESULTS OF THE THERMAL MODELS

The results of the first, simple model are shown in Fig. 2. By 0.9 billion years, the central temperature reaches 1700 K, with the dry basalt solidus of Ringwood and Essene (1970) exceeded in two partially molten pockets beneath the poles, at roughly 1600 km radius. (The dry solidus is used because Io should be thoroughly outgassed by the time melting temperatures are reached [cf. Consolmagno, 1981]; any outgassed water can escape and so is permanently lost [Pollack and Witteborn, 1980]). By 1.0 billion years, the molten region extends from 1500 km to 1775 km radius, with this mantle being somewhat thinner at the equator. This represents a time of maximum heating, as the decaying long-lived radioactive materials become less important (they provided 15% of the heat initially in Io) and the solid crust is now subjected to enhanced tidal heating. At this point the thermal gradient in the top 5 km of Io is on the order of 100 K/km; the region 5 km below the pole is 100 K hotter than that below the equator. (The surface itself is fixed at 100 K). This represents a heat flow of 450 ergs/cm^{2}s, matching the heat input into the crust.

However, the enhanced tidal heating puts the heat only into the solid crust, which obviously is always cooler than the underlying molten region. The effect of this heating is to shrink the crustal thickness, until eventually the crust is thinner than the thickness at which maximum enhancement of tidal heating occurs. At that point, tidal heating no longer can support the steep thermal gradient through the thin crust. As more heat flows out of the planet than can be replaced by the enhanced tidal heating, the molten region starts to refreeze. However, since heat is transported by convection through both the molten region and the solid core, and since the temperature increases in such a convecting region more slowly than the melting temperature, the loss of heat from the interior leads to the mantle freezing at the core, rather than at the crust. The core will grow in size as cooling continues until it meets the crust and the entire planet is refrozen.

Solid state convection will continue in this region, but tidal heating is no longer enhanced. The result is that in this model, melting never recurs; the unenhanced tidal heating can only establish a thermal gradient at the surface of roughly 30 K/km, giving a heat flow of only 130 ergs/cm^{2}s. Io would remain in this cool steady state to the present time.

Obviously this simple model fails to predict Io as we actually observe it today. The predicted heat flow is an order of magnitude lower than that observed, and there is no provision for the multiple volcanoes seen.

One deficiency of this version of the model is that it does not account for the chemical changes which certainly must have occurred in the body as it heated up. The outgassing and escape of water and carbon compounds, if they were included in sufficiently large quantities, could have absorbed and transported energy; but compared to the large amount of energy available from tidal heating, this heat loss should not change substantially the subsequent thermal evolution. If an FeS core should have formed, the heat pulse could delay, but probably not prevent, the refreezing of the planet. The refreezing as modelled here is almost certainly a geochemical oversimplification; in fact, the last material to freeze, just below the solid crust, would be some eutectic composition with a relatively lower melting point and an enhancement of incompatible elements, including the radioactive species, which could delay the refreezing of this last portion of Io (cf. Schubert et al., 1981).

But perhaps most important of all is the development of a sulfur-rich crust. Sulfur has a thermal conductivity which is roughly an order of magnitude lower than rock, when solid. However, liquid sulfur can transport heat by convection at temperatures much lower than the convective temperature of rock.

To examine the effect of a crust whose thermal properties are dominated by sulfur, a version of the thermal model was created which assumed that the thermal conductivity of the uppermost 5 km of the body, after the first 1.0 billion years, was an order of magnitude lower than that of rock. Not surprisingly, this insulating blanket served to warm up the rest of Io almost immediately, resulting in melting the uppermost regions of the body below this fixed, solid crust. Such melting would imply a crust only 5 km thick, which is clearly incompatible with the observed surface relief observed on Io (cf. Carr et al., 1979).

The difficulty, of course, is that molten rock so close to a sulfur-rich crust would melt the sulfur, which in turn could convect heat through the insulating blanket of the crust to the surface of Io. It is precisely this phenomenon—the hotspots and volcanoes on Io—which is the source of most of the heat observed being emitted from Io (Matson et al., 1981).

These volcanoes and hotspots are relatively small features, however, compared to the 5 km deep, hundreds-of-kilometers wide latitudinal stripes which are used by the two-

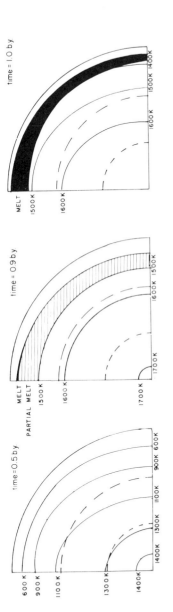
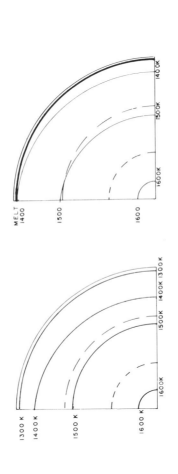

ROCK CRUST SULFUR CRUST

Fig. 2. The thermal evolution of Io with tidal heating and long-lived radioactive nuclides considered. The top three profiles show isotherms for Io over the first billion years; as melting occurs, enhanced tidal heating is confined to the crust, while convective heat transfer cools off the interior. (Dashed lines mark 600 km and 1200 km radius.) The figure on the left assumes the eventual outcomes of the heating of Io. The figure on the left assumes the crust stays essentially rocky, and shows that convection of the interior will keep temperatures just below the melting temperature everywhere; enhanced tidal heating will not occur. On the right, a low-conductivity sulfur crust 5 km thick will allow a thin melted region, enhanced tidal heating, and volcanoes; see text for details.

dimensional thermal model. In addition, the exact mechanism of these features is still a subject of current debate (cf. Reynolds *et al.*, 1980, and references therein). But it is clear that the heat flow caused by these hotspots must be accounted for, by any thermal model which hopes to portray accurately the thermal state of the interior of Io and give reasonable agreement with observed values of the heat flow from the surface.

To model this heat flow, it was assumed that the volcanism could be represented by an abrupt change in the thermal properties of the crust. When the amount of energy stored in a region below the crust reaches some critical level, a change takes place in the crust which serves to expel that energy rapidly. The way this is modelled in the program is to keep track of the temperature in the layer just below the surface. When that temperature reaches 1300 K (the critical temperature for convection in the model) the conductivity of the crust above that region is increased by two orders of magnitude (thus representing convection in that region) and the surface temperature is set at 400 K, roughly the melting point of sulfur.

It should be recalled that this model is only two dimensional, in radius and latitude, and so in three dimensions this "hotspot" corresponds to two rings of constant latitude, symmetric about the equator, where molten sulfur can be carried to the surface. Not necessarily all of the "ring" must be molten; real hotspots are probably hotter, and transport heat more efficiently, than the average heat transport of the ring as modelled here.

The eruptions in such a ring result in the expulsion of 8000 ergs/cm^{2}s through the region for the lifetime of the ring, eventually cooling the region below the ring. Thus the next ring to erupt, at some later time, will be in some other region of the planet which has not been cooled by recent volcanism.

The results of this model, as illustrated for a typical time period in Fig. 3, are quite striking. The volcanism appears roughly 10% of the time, preferentially near the poles (which is to be expected, since the most heating occurs there) but also at or near the equatorial regions, as is currently observed. The total Io-wide average heat flux during periods of volcanism is 1000 ergs/cm^{2}s; the rough agreement between this figure and the observed heat flux is likely a coincidence, however, since the crudeness of this model

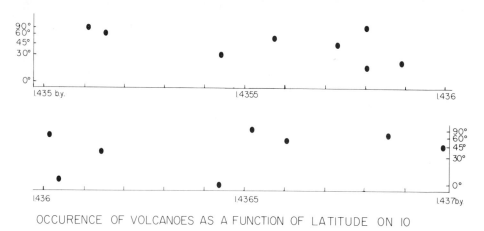

Fig. 3. In the model described in the text, volcanoes occur when the temperature in a layer immediately below the crust exceeds 1300 K. At that point the crust temperature is raised to 400 K (from 100 K) and the crust conductivity is raised by two orders of magnitude, until the region below cools to below 1300 K. This figure illustrates a typical two million year period in the model. The spots mark the occurence of volcanoes with time, with the vertical scale marking latitude (linear in sine of latitude). Volcanoes occur ten percent of the time, with roughly half of them at latitudes of 45° or below, even though tidal heating is more enhanced at the poles in our model. Note simultaneous volcanism at two latitudes at time 1.4358 billion years, and the periods of several hundred thousand years without volcanism. Heat flow during volcanism is on the same order as that observed on Io today.

gives no reason to believe that it should be capable of matching the details of the observed volcanism. During the 90% of the time when no volcanism occurs, the heat flow predicted by the model is a modest 150 ergs/cm^{2}s, and the heat flow integrated over time averages out to 180 ergs/cm^{2}s, which is consistent with heat input due to enhanced tidal heating. This model predicts a 15 km thick crust.

It must be emphasized that this model is, at best, schematic. The nature of the volcanoes of Io is almost certainly much more complicated than what has been modelled. By their natures, the individual plumes and hotspots are much smaller, and probably much shorter-lived, than any feature which could possibly be modelled by a global, time-dependent model like the one presented here. Instead, all that this model can provide is an illustration of the way heat may flow in Io, with some order-of-magnitude numbers for heat flows.

DISCUSSION

The shortcomings of this model result from the necessarily crude grid size, and the equally crude manner in which volcanism is modelled. For example, the length of time the volcanism lasts merely reflects the size of the timestep; the duration of real volcanoes is likely determined by factors, such as stress within the crust, which are not modelled here. On the other hand, the conclusion that volcanism takes place somewhere on Io approximately 10% of the time can be drawn, regardless of the length of the timestep used. This result held constant for models run with much shorter timesteps.

Likewise, the total heat output calculated from this model is biased by the large surface area which each grid space represents—10% of the total surface area of Io. However, a model with twice as many grid spaces (so that each space represented 5% of the total area) showed volcanism occurring simultaneously in two or more grid spots for 10% of the volcanic events. This was a much higher occurence rate than in the standard model, and so the total heat output given by the nominal model may not be totally misleading. Again, the extent of volcanism at any given time will certainly be determined by the structure of the crust as well as the heat flow within the crust.

All models aside, it is obvious that an Io which has an average heat input of 200 ergs/cm^{2}s (as implied by the Peale *et al.* (1979), enhanced tidal heating scheme, given the thickness of the crust demanded by the observed surface relief) but which shows a heat outflow of 2000 ergs/cm^{2}s can only maintain such heat outflow 10% of the time. Either there is some additional, unknown heat source in Io, or the volcanism is in fact intermittent. What the models do show is that such intermittent volcanism shows up very naturally, even with the crudest of assumptions; and that this volcanism will vary in space as well as in time, given the non-uniform input of tidal heating.

Another factor, not explicitly modelled here, is that the convection deep within Io may itself be periodic. Recent work on modelling the three-dimensional motions of the Earth's mantle, for example, suggests that convection may be intermittent, with a 100 million year periodicity (Arkani-Hamed *et al.*, 1981). Modelling the transport of heat by convection is clearly the weakest part of any thermal model, since it involves both the largest number of unknown parameters and the greatest theoretical uncertainty.

Finally, the refreezing phenomena described for the first model (without the insulating sulfur crust) may lead to yet a third periodicity, as suggested by Greenberg (1981). The evolution of Io's eccentricity, upon which the strength of the tidal heating depends, is also affected by the tidal heating itself. If Io refreezes, its orbit becomes more eccentric, resulting in an increase of heating in Io which may lead ultimately to remelting.

In summary, then, one can conclude that the volcanism and high heat flow observed on Io today is a transient phenomena. The source of the heat needed can be the dissipation of tidal stresses, providing heat at a rate well below the maximum rate allowed by structural and dynamical arguments; no unknown heat sources are demanded. A low conductivity crust is suggested, to prevent the refreezing of the thin liquid mantle which is necessary for enhanced tidal heating; and hotspots through the crust are necessary to keep the crust thick enough to maintain the surface relief seen on Io (as was also recently concluded, independently, by O'Reilly and Davies, 1981). Finally, the hotspots must be intermittent, to allow for periods of high heat flow, as are currently observed. The model predicts that volcanism and associated high heat flow will occur roughly 10% of the time. These hotspots can occur anywhere on the Io, even though the tidal heating may be preferentially concentrated at the poles.

Acknowledgments—This paper has been greatly helped by discussions with Sean Solomon, Nafi Toksöz, David Stevenson, and especially Ray Reynolds (none of whom, however, should be held responsible for the conclusions drawn herein). Much of this work was done under the guidance and support of A. G. W. Cameron at the Harvard/Smithsonian Center for Astrophysics; without his encouragement and his computer, the thermal models presented here could not have been developed. This work was supported by NASA grants NGR 22-007-289, NSG-7081, and NAGW-191.

REFERENCES

Arkani-Hamed J., Toksöz M. N., and Hsui A. T. (1981) Thermal evolution of the earth. *Tectonophys.* **75**, 19–30.
Carr M. H., Masursky H., Strom R. G., and Terrile R. E. (1979) Volcanic features of Io. *Nature* **280**, 729–733.
Cassen P., Peale S. J., and Reynolds R. T. (1981) Structure and thermal evolution of the Galilean satellites. In *Satellites of Jupiter* (D. Morrison, ed.) Univ. of Arizona, Tucson. In press.
Colburn D. S. (1980) Electromagnetic heating of Io. *J. Geophys. Res.* **85**, 7257–7261.
Consolmagno G. J. (1975) Thermal history models of icy satellites. MS thesis, MIT, Cambridge. 202 pp.
Consolmagno G. J. (1981) Io: thermal models and chemical evolution. *Icarus.* In press.
Drobyshevski E. M. (1979) Magnetic field of Jupiter and the volcanism and rotation of the Galilean satellites. *Nature* **282**, 811–813.
Gold T. (1979) Electrical origin of the outbursts on Io. *Science* **206**, 1071–1073.
Greenberg R. (1981) Orbital evolution of the Galilean satellites. In *Satellites of Jupiter* (D. Morrison, ed.) Univ. of Arizona, Tucson. In press.
Matson D. L., Ramsford G. A., and Johnson T. V. (1980) Heat flow from Io (JI) (abstract). In *Lunar and Planetary Science XI*, p. 686–687. Lunar and Planetary Institute, Houston.
Matson D. L., Ramsford G. A., and Johnson T. V. (1981) Heat flow from Io (J1). *J. Geophys. Res.* **86**, 1664–1672.
Morabito L. A., Synnott S. P., Kupferman P. N., and Collins S. A. (1979) Discovery of currently active extraterrestrial volcanism. *Science* **204**, 972.
Ness N. F., Acuna M. H., Lepping R. P., Burlaga L. F., Behannon K. W., and Neubauer F. M. (1979) Magnetic field studies at Jupiter by Voyager I: preliminary results. *Science* **208**, 982–986.
O'Reilly T. C., and Davies G. F. (1981) Magma transport of heat on Io: a mechanism allowing a thick lithosphere. *Geophys. Res. Lett.* **8**, 313–316.
Peale S. J. and Cassen P. (1978) Contribution of tidal dissipation to lunar thermal history. *Icarus* **36**, 245–269.
Peale S. J., Cassen P., and Reynolds R. T. (1979) Melting of Io by tidal dissipation. *Science* **203**, 892–894.
Pollack J. B. and Witteborn F. C. (1980) Evolution of Io's volatile inventory. *Icarus* **44**, 249–267.
Prinn R. N. and Fegley B. (1981) Kinetic inhibition of CO and N_2 reduction in circumplanetary nebulae: implications for satellite composition. *Astrophys. J.* In press.
Ransford G. A. and Kaula W. M. (1980) Heating of the moon by heterogeneous accretion. *J. Geophys. Res.* **85**, 6615–6627.
Reynolds R. T., Peale S. J., and Cassen P. (1980) Io: energy constraints and plume volcanism. *Icarus* **44**, 234–239.
Ringwood A. E. and Essene E. (1970) Petrogenesis of lunar basalts and the internal constitution of the moon. *Science* **167**, 607–610.
Schubert G., Stevenson D. J., and Ellsworth K. (1981) Internal structures of the Galilean satellites. *Icarus.* In press.
Sinton W. M. (1981) The thermal emission spectrum of Io and a determination of the heat flux from its hotspots. *J. Geophys. Res.* **86**, 3122–3128.
Weidenschilling S. J. (1977) The distribution of mass in the planetary system and solar nebula. *Astrophys. Space Sci.* **51**, 153–158.
Yoder C. F. (1979a) How tidal heating in Io drives the galilean oribtal resonance locks. *Nature* **279**, 767–770.
Yoder C. F. (1979b) Consequences of joule heating versus tidal heating of Io. *Bull. Amer. Astron. Soc.* **11**, 599.

Microstructure and particulate properties of the surfaces of Io and Ganymede: Comparison with other solar system bodies

Kevin D. Pang*[1], Kari Lumme[2], and Edward Bowell[2]

[1]Jet Propulsion Laboratory, California Institute of Technology, Pasadena, California 91109, and
[2]Lowell Observatory, Flagstaff, Arizona 86002

Abstract—The phase curves of Io and Ganymede between 0° and 40° solar phase angle have been compiled from Voyager photopolarimeter and groundbased observations. Modeling the data with the Lumme-Bowell theory allowed us to determine improved estimates of the surface texture and particulate properties. Accurate V-band zero-phase geometric albedos, phase integrals, and Bond albedos are also obtained. The properties of the particles on Io's surface appear to be different from those of other solar system objects whose surfaces are totally controlled by impact cratering processes. Sulfur and SO_2 frost, previously identified by spectral analysis, seem to exist in a physical form not commonly found on Earth. We postulate the agglomeration of volcanic ash into aggregates on Io's surface.

INTRODUCTION

An objective of the Voyager Photopolarimeter (PPS) experiment is to determine the microstructure and particulate properties of the surfaces of the outer satellites (Lillie et al., 1977). Preliminary analyses of PPS photometer data have been published elsewhere (Pang et al., 1979; Pang, 1980). In this paper we report our findings for the surfaces of Io and Ganymede and compare our results with those for other solar system bodies, as well as with laboratory reflectance measurements. The Voyager results are important in complementing Apollo, Mariner, Viking, and groundbased data to build up a body of information useful for studying the development and evolution of planetary and asteroidal regoliths. This work has hitherto depended almost solely on numerical simulations (Housen et al., 1979; Langevin and Maurette, 1981) and studies of meteorites believed to be former parts of an asteroidal parent body. Voyager 2 PPS observations of the Saturnian satellites are expected to add to this body of remote sensing data.

The analysis of photometric data taken in one spectral band (V) does not give direct chemical compositional information on the objects observed, nor is it our intention to do so. Our objective here is to determine the textural and particulate properties, which define the physical, but not the chemical, state of the uppermost surface layer. The chemical compositional identifications have been made using spectral data. Spectral analyses have shown that sulfur and sulfur dioxide frost are the main components of Io's surface material (Nelson and Hapke, 1978; Sagan, 1979; Hapke, 1979; Fanale et al., 1979). Modeling of Ganymede's infrared spectral reflectance yielded a water frost abundance of at least 50%, but possibly as high as 90% (Pollack et al., 1978; Clark, 1980). The remaining nonfrost component is still not identified.

In addition to the microphysical and geochemical data, some geological and geophysical background information relevant to our discussion is as follows:

Ejecta from active volcanoes are estimated to be resurfacing Io at a rate of about 1 mm

---

*Work partly done during the senior author's tenure as a National Academy of Sciences Senior Resident Research Associate.

per year (Johnson et al., 1979). The particles in the volcanic plumes have been determined to be predominantly micron- and submicron-size from the analysis of Voyager photometric data (Smith et al., 1979; Collins, 1981).

Analysis of phase curve (whole-disk brightness vs. solar phase angle) of an atmosphereless body is a useful method of determining its surface physical properties. To analyze the PPS photometric data, we make use of a generalized theory of multiple scattering in rough and porous surfaces developed by Lumme and Bowell (1981a, b). The theory treats both single and multiple scattering of light and should be applicable to particulate surfaces of any albedo, observed at any phase angle. From an analysis of photometric data for 74 atmosphereless solar system bodies, Lumme and Bowell showed that phase curves could be matched by varying a parameter termed the multiple-scattering factor (the ratio of the amount of multiply scattered light to the total scattered light at zero phase angle), and contain accessible information on the surface roughness, volume density (or porosity), and single-particle phase function (characterized by the asymmetry factor). These parameters are further defined in the table on p. 5. It was demonstrated that, for low-albedo surfaces, the effects of surface roughness and volume density are distinctly separable: The slope of the linear part of the phase curve is mainly controlled by roughness, and the opposition effect arises from porosity.

One must be particularly cautious in relating parameters defined in theoretical terms to their physical counterparts. Lumme et al. (1980) have started to undertake a series of critical laboratory experiments in order to address this problem. The good agreement between theory and experiment for several very different particulate samples has given confidence that *quantitative* surface textural information can be extracted from remote photometry of solar system objects. For example, it is gratifying that the value deduced for the volume density of the uppermost surface layers in asteroids and satellites (0.4) is matched by that of a variety of particulate samples in the laboratory. There remain, however, a number of insufficiently understood aspects of this problem that need follow-up work, and in consequence we must at this stage be wary of overinterpreting the photometric data in physical terms. For example, Pang et al. (1980) deduced from phase curve data that the surfaces of Phobos and Deimos differ only in their albedos and not in surface texture. This result apparently contradicts Viking imaging, which clearly shows that, on a scale of a meter or more the surface of Deimos is markedly smoother (Veverka and Burns, 1980). However, the photometric properties pertain to a much smaller scale than that of the Viking imaging.

DATA

The phase curves of Europa, Io and Ganymede between 0° and 40° phase angle (Fig. 1) have been compiled from Voyager photopolarimeter and groundbased observations (Millis and Thompson, 1975). The magnitudes were observed in (or have been transformed into) V, with corrections applied for longitudinal brightness variations, using Table 16.3 of Morrison and Morrison (1977). The Ganymede data are restricted to the hemisphere where 90° ≤ longitude ≤ 270°W, i.e., the hemisphere facing away from Jupiter. The Io data are restricted to longitude 0°–121° and 255°–360°W, or slightly more than the hemisphere facing Jupiter. Voyager data of other longitudes are not sufficient to define a hemispherical set. Our data agree with photometry in smaller ranges of phase angles, made by Voyager television cameras (T. V. Johnson et al., 1981, pers. comm.) and other groundbased observers (Lockwood et al., 1980). Data taken by all groundbased observers are generally consistent with one another, as reviewed by Lumme and Bowell (1981b). Comparison of the data set by Lockwood et al. (1980) with that of Millis and Thompson (1975) is depicted in Fig. 2. The phase curve of the moon (Lane and Irvine, 1973; Whitaker, 1979), and an ideal Lambert sphere are also shown in Fig. 1 (normalized to the Io and Ganymede data, respectively, at zero phase).

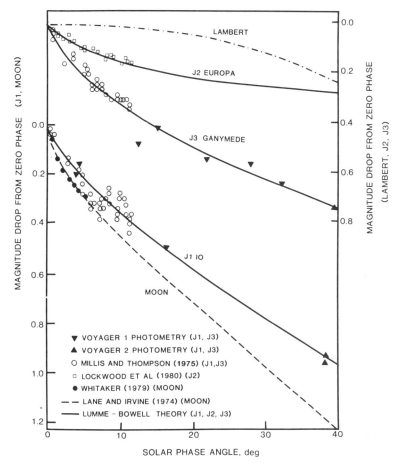

Fig. 1. Phase curves of Jovian satellites Europa, Io, and Ganymede, an ideal Lambert sphere, and the moon, normalized to zero phase angle.

PHASE CURVES AND LABORATORY DATA

We can get a good feeling for the problem by making qualitative comparisons of the satellites' phase curves with those of the moon and Europa, and with laboratory reflectance measurements of terrestrial samples (Lumme *et al.*, 1980). Although the zero-phase geometric albedo of Io ($p_V = 0.74$) is as high as that of Eurpoa, its phase curve is much steeper—more like that of a low-albedo object such as the moon (Fig. 1). In contrast, the phase curve of Europa, known to be almost completely covered with water frost (Clark, 1980), is rather flat out to at least 12° phase (Bowell and Lumme, 1980; Lockwood *et al.*, 1980), approximating that of a snow-covered sphere (Veverka, 1973). From this simple comparison we infer that Io is not covered with substances that scatter light like frost. Therefore, SO_2 frost, identified in Io's infrared reflectance spectrum (Fanale *et al.*, 1979) cannot cover much of Io's surface. This conclusion is consistent with the estimate by Nash *et al.* (1980), who found a SO_2 frost abundance ≤ 20%. The phase curve of Ganymede (Fig. 1) has a steepness intermediate between that of the moon (rocky regolith) and Europa (mostly ice regolith), which is consistent with the part-ice, part-rock regolith composition suggested by Pollack *et al.* (1978) and Clark (1980).

We can get an additional feeling for the problem by comparing the photometric properties of Io's surface to laboratory reflectance measurements. Pure ice crystals and

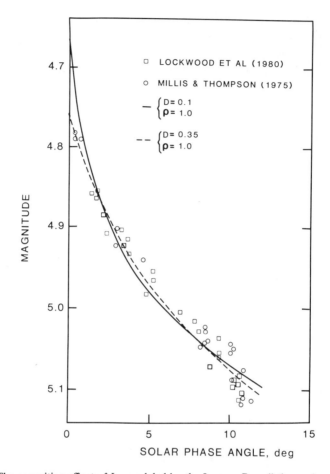

Fig. 2. The opposition effect of Io, modeled by the Lumme-Bowell theory for a highly porous (—), and a moderately porous (---) surface. The roughness of the two surfaces are the same.

orthorhombic sulfur S_8 are nearly transparent at visible wavelengths (Meyer et al., 1972), and multiple scattering of light among grains makes them good diffuse reflectors. The polar scattering diagrams of sulfur flowers and colloidal sulfur closely approximate the cosine function between 5° and 40° (Egan and Hilgeman, 1975, 1976). Indeed, the quasi-Lambertian reflecting properties of such forms of sulfur make them ideal for coating integrating spheres. Snow and frosts are also good diffuse reflectors, except when glazed, or when illuminated at nearly grazing incidence angles (Middleton and Mungal, 1952). The photometric properties of Io's surface differ markedly from those of a Lambert reflector (see Fig. 1; also Clancy and Danielson, 1981). From these comparisons we infer that sulfur exists on Io's surface in a form or forms different from the flower (large crystal) or colloidal (microcrystalline) varieties commonly found on Earth, and that neither SO_2 frost nor S_8 exist there as an optically thick deposit.

MODELING THE DATA

In order to obtain quantitative estimates of Io's and Ganymede's surface microstructure and particulate properties, we modeled the photometric observations with a generalized radiative transfer theory that has been found to work well for all atmosphereless solar system objects. A formal exposition of the theory with illustrative examples has been given by Lumme and Bowell (1981a, b). The best-fit model parameters are as follows:

	Io	Ganymede
D Volume density of surface layer (bulk density/particle specific gravity)	0.37	0.37
ρ Surface roughness (depth of typical surface irregularity/its horizontal dimensions)	0.9	1.0
Q Multiple scattering factor (proportion of multiply scattered light at zero phase)	0.35	0.48
$\tilde{\omega}_0$ Single-scattering albedo (scattering cross-section/extinction cross-section)	0.94	0.92
g Asymmetry factor of single-particle phase function (Heyney-Greenstein definition)	-0.24	-0.03

From the best-fit theoretical curves (continuous lines in Fig. 1) we obtained

p Zero-phase geometric albedo	0.74	0.50
q Phase integral	0.71	0.91
A Bond albedo	0.52	0.45

In practice, the model has only four free and not five parameters since Q, the proportion of multiply scattered light, is implicitly dependent on $\tilde{\omega}_0$ and g. The effect of perturbing D from the nominal value of 0.37 can be seen in Fig. 2. Matson and Nash (1981) found a porosity of 0.60 to 0.85 for Io's upper surface by modeling its eclipse cooling curves. Since D = 1-porosity, Matson and Nash's values correspond to values of D between 0.15 to 0.40. Their higher value is consistent with our estimate, but their lower D value is not. Using a D = 0.1 we are unable to model the opposition effect of Io well (Fig. 2). Hence, we infer that the volume density of Io's uppermost surface layer is similar to that of all atmosphereless solar system bodies for which adequate data are available. The same is true of Ganymede's surface.

Accurate photometric data over a large range of phase angles have allowed us to determine improved values of surface model parameters. Previous attempts were handicapped by both the limited range of phase angles observable from the earth (less than 12°) and the scatter in data values (see unfilled circles in Fig. 1). In particular, it is not possible to separate the effects of ρ and g unless accurate photometric data at moderate or large phase angles ($\alpha \gtrsim 30°$) are available. The scatter in groundbased observations of Io's magnitude is the worst—due to its angular proximity to Jupiter and also because of its apparent intrinsic brightness changes (Lockwood et al., 1980). The Voyager data have greatly increased our leverage and minimized the effects of uncertainty at the low phase angles. Interpretations of the derived values (compared to those of other solar system objects) are given in the next section.

INTERPRETATIONS

(1) *The inferred volume density* of Io's surface layer differs very little from that of other atmosphereless solar system objects—D = 0.37 ± 0.05. This optically modelled value is probably equivalent to the physical value. For the moon, the remotely sensed volume density agrees with direct Apollo measurements:

$$D = \frac{\text{bulk density of lunar fines}}{\text{particle specific gravity}} \simeq \frac{1.3 \text{ gm/cc}}{3.0 \text{ gm/cc}} \simeq 0.4 \text{ (Lumme and Bowell, 1981b)}.$$

The substances known to be present on Io's surface—sulfur and SO_2 frost—all have a specific gravity of about 2 gm/cc: S −2.05 to 2.09; SO_2 −1.9 (Matson and Nash, 1981). Therefore, we estimate that Io's uppermost surface layer has a bulk density of 0.7 to 0.8 gm/cc. The surface bulk density for Ganymede may be as low as 0.4 gm/cc, or as high as 0.8 gm/cc, depending on the fractional water ice content we adopt for its surface material (90% − Clark, 1980; or 50%—Pollack et al., 1978) and assuming that the non-frost component on Ganymede has the same specific gravity as lunar fines (3 gm/cc).

We point out in passing that the thickness of the surface layer probed by our optical remote sensing technique may only be a few mm deep. Thus our density values may differ from those determined by thermal infrared radiometry—for a layer a few millimeters to several centimeters deep (Morrison, 1980) or by radar backscattering—for a layer a centimeter to tens of centimeters deep (Pettengill, 1978). Needless to say, our derived values are hemispheric averages. Regional and/or local values may differ.

(2) *Inferred values of the surface roughness ρ for solar system objects are listed below in increasing order.*

	ρ	$g_0 (cm/sec^2)$	
Mars	0.6	373	Lumme et al. (1981)
Mercury	0.8	363	Lumme and Bowell (1981b)
Moon	0.8 ± 0.15	162	Lumme and Irvine (1980)
Io	0.9	179	This work
Ganymede	1.0	142	This work
Phobos, Deimos	1.1	0.5, 0.3	Pang et al. (1980)
Asteroids	0.8 − 1.2	1–40	Lumme and Bowell (1981b)

There appears to be an inverse relationship between the surface roughness of a body and the gravitational acceleration g_0 on that body's surface. A similar weak correlation is also observed to hold among C asteroids (Bowell and Lumme, 1979). The correlation is suggestive of gravitational control on regoliths. Gravitational control on planetary or asteroidal regoliths may be through seismic shaking, viz., the stronger the gravitational force on a particulate surface, the flatter the structures that can be built on that surface, and through preferential exposure of coarser grains, as the finer particles sink when the soil is shaken (Simon and Papike, 1981). Planetary activities on Io and Mars may alter the effects of simple gravitational control. Aeolian activity on Mars probably smooths out surface asperities by packing and abrasion.

(3) *The multiple-scattering factor Q is implicitly dependent on the single-scattering albedo $\tilde{\omega}_0$ and asymmetry factor g of a single particle.* The results for various solar system objects are best plotted on a "p–Q diagram" (Fig. 3), adapted from Lumme and Bowell (1981b). Lines of constant $\tilde{\omega}_0$ and g appear to demarcate regions whose surface physical processes are distinct from each other. These curves are drawn for surface roughness $\rho = 1.2$, a value that appears to pertain to a large number of atmosphereless bodies. A white body that has perfect scattering particles on its surface would lie on the $\tilde{\omega}_0 = 1.0$ line. Conversely, an ideal blackbody would be represented by the $\tilde{\omega}_0 = 0$ point (at the origin of the p, Q plane). An object with isotropically scattering surface particles would lie along the g = 0 line. When there is an abundance of smooth, forward-scattering particles on a surface, that surface would have a high positive g value, and the body would tend to be situated towards the lower right side of the p–Q diagram. On the other hand, an object with predominantly rough, backscattering surface grains would have a large negative g value and be located towards the upper left part of the diagram. Surfaces for which g = 0 and $\rho > 1.2$ occupy the upper left part of the p, Q plane, and those for which $\rho < 1.2$ lie below the g = 0 curve. Naturally, a given location in the p, Q plane can be occupied by a suite of surfaces having appropriate values of ρ and g. Only when observations are available at sufficiently large phase angles can ρ and g be separated, as noted above.

In order to understand the peculiar nature of Io's surface particles, we need first to review data on lunar and martian soils, for which we have much more intimate knowledge. *In situ* Apollo and Viking sampling measurements (McKay *et al.*, 1974; Moore *et al.*, 1979) show that an assortment, rather than a single kind, of particles is to be expected. A mixture of 40–45% agglutinates with the rest in lithic and monomineralic fragments and glass spherules may be considered to be representative of lunar soil (McKay, pers. comm., 1981). Knowledge of light scattering by irregularly shaped particles is also essential for understanding what appear to be systematic trends on the p–Q diagram.

Microwave scattering measurements performed on laboratory samples with analogous optical properties show that the shape of a particle has an important influence on its phase function. As the particle becomes more and more irregularly shaped, it scatters more and more of the incident radiation into the backward and side, rather than the forward, directions (Zerull, 1976). Algebraically, the value of g decreases for particles of equal volume in the following order: spheres, lithic fragments, agglutinates, aggregates (Pollack and Cuzzi, 1980). With this knowledge of light scattering, together with the expectancy of multimodal particles on planetary and asteroidal surfaces, we now attempt to explain the systematics found in the p–Q diagram.

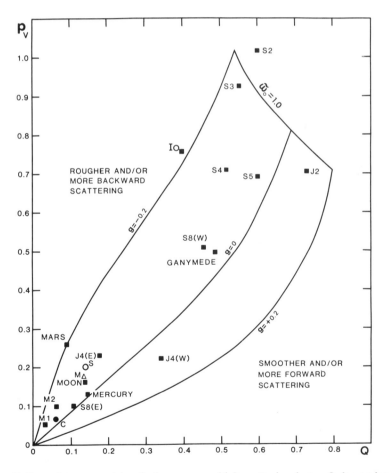

Fig. 3. Zero-phase geometric albedo p_V vs. multiple-scattering factor Q for various solar system bodies. Lines of constant single-scattering albedo $\bar{\omega}_0$ and asymmetry factor of single-particle scattering function g are drawn in for the case when the surface roughness $\rho = 1.2$. These lines seem to demarcate regions whose surface processes are distinct from each other.

The composition of a surface seems to have an important influence on the type of particles produced on that surface. Above the g = 0 line we find carbonaceous, silicate/metallic, and ice/rock objects seemingly clustered in separate groups. Grand average values for the principal taxonomic types of asteroids are denoted by a triangle (type M), and filled and unfilled circles (type C and S, respectively).

Among the silicate planets we find that the moon, whose surface has agglutinates, monomineralic and lithic fragments, and glass spherules, has a rather strongly backscattering surface. *In situ* Viking lander measurements showed that the soil on the martian surface is fine-grained, with particle sizes considerably less than 0.1 mm. All efforts to retrieve pebble-size particles failed, as apparent clumps disintegrated when shaken or sieved (Moore *et al.*, 1979). However, sand-size particles are needed to explain the formation of martian drifts and dunes. Greeley (1979) proposed that martian fines agglomerate to form sand-size aggregates, which are then deposited as sandlike features. He suggested that electrostatic bonding is the mechanism for the formation of the aggregates. Nummedal (1981) postulated cementation by salt as an alternative explanation. For our discussion we take martian soil to be a mixture of silt particles and aggregates. The highly backscattering nature of martian soil components contributes to an algebraically low value of g for Mars.

In light of this evidence, we can think of two possible explanations for Io's peculiar surface photometric function. Io's asymmetry factor, $g = -0.24$, is nearly equal to that of Mars (-0.20) (Lumme *et al.*, 1981). It was pointed out above that particles in Io's volcanic plumes are predominantly micron- and submicron-size (Smith *et al.*, 1979; Collins, 1981). The Loki plume is populated mostly by extremely small particles of radii $0.001-0.01$ μm (Collins, 1981). Dielectric particles with sizes so close to the Rayleigh limit scatter light essentially like dipoles, i.e., $g \cong 0$. Only very small metallic particles, with imaginary refractive indices (k) nearly equal to, or exceeding unity, have highly negative g values (Hansen and Travis, 1974; Lumme *et al.*, 1981). However, an explanation requiring a large k for Io's surface particles is incompatible with its known surface composition (surfur and SO_2 frost are both dielectric substances) and with its extremely low surface heat conductivity (Matson and Nash, 1981), which is decidedly nonmetallic. Clearly, another explanation is needed.

Light scattering measurements made on highly irregular "fluffy" particles and pollen show that such loose aggregates are strong backscatters (Zerull, 1976; Bickel and Stafford, 1979). Such particles are possible terrestrial analogs of the particles found on Io's surface.

In coming up with an explanation for the peculiar nature of Io's surface particles, we find our experience with martian particulates to be invaluable. Martian atmospheric aerosols are also micron-size, dielectric particles, with g values that are positive (Pollack *et al.*, 1979). However, martian surface particles are silt-size, and have large negative g values (Lumme *et al.*, 1981). To reconcile this difference in size between the aerosols and surface particles, one may postulate that fine aerosol particles agglomerate to form highly backscattering aggregates. The bonding mechanism may be electrostatic attraction between aerosols (Greeley, 1979). If this mechanism works for Mars, it should also work for Io, as the satellite is probably highly charged due to its revolution deep in Jupiter's magnetosphere, and by being constantly bombarded by trapped charged particles. The recent discovery of discharge glows from Io's volcanic plumes and polar regions supports our hypothesis (Cook *et al.*, 1981). In light of this discovery, the electrostatic charging and bonding of fine volcanic ash into aggregates in Io's plumes seem quite plausible. More theoretical calculations are needed to better define the charging and agglomeration processes involved.

The other satellites of Jupiter and Saturn, composed of mixtures of ices and rocks, occupy widely scattered positions on the p–Q diagram. Some of the scatter is due to the lack of good phase coverage (e.g., for the saturnian satellites) and to imperfectly known geometric albedos. This situation will be rectified when the Voyager 2 photopolarimeter obtains data of the saturnian satellites at large phase angles in August, 1981. Another reason for the scatter of the ice/rock objects may be that such combinations of highly volatile and refractory phases are liable to follow diverse paths in their thermal evolution (Hartmann, 1981). The proximity of a satellite to its primary, the interplay of its thermal and cratering history, its radiation environment, and contamination by débris from its neighboring satellites no doubt all contribute to making radically different regoliths. The present data base is not capable of unraveling the many complex factors involved. More observations, laboratory measurements, and theoretical calculations, e.g., of the type made by Cintala *et al.* (1980), are needed before we can understand the unique evolutionary history of the outer satellites.

SUMMARY AND DISCUSSION

Modeling our data with the Lumme-Bowell theory and assuming a sulfur and SO_2 frost composition, we find that:

1. The top surface layer of Io has a bulk density of 0.7 to 0.8 gm/cc. The corresponding value for Ganymede is 0.4 to 0.8 gm/cc.

We suggest that:

2. The extremely fine component of the ash from Io's volcanoes probably forms aggregates on Io's surface. The bonding force is likely to be electrostatic attraction among particles.

From qualitative comparison of Voyager PPS data with laboratory measurements, we conclude that:

3. Sulfur dioxide frosts are not ubiquitously present as an optically thick layer on Io's surface. The sulfur on Io's surface does not resemble the flower (large crystal) or colloidal (microcrystalline) forms commonly found on Earth. Fluffy aggregates or pollen-like structures may be considered good terrestrial analogs of the particles on Io's surface.

Our conclusions are consistent with compositional results derived by analysis of spectrophotometric data (Soderblom *et al.*, 1980; Nash *et al.*, 1980) and models of Io's surface (Hapke, 1979).

Acknowledgments—The authors thank C. W. Hord, K. E. Simmons, and A. L. Lane for their support. The research described in this paper was carried out at the Lowell Observatory under National Aeronautics and Space Administration Grant NSG-7500 and at the Jet Propulsion Laboratory, California Institute of Technology under contract with the National Aeronautics and Space Administration.

REFERENCES

Bickel W. S. and Stafford M. E. (1979) Biological particles as irregularly shaped scatterers. In *Light Scattering by Irregularly Shaped Particles* (D. W. Schuerman, ed.), p. 299–305. Plenum, N.Y.
Bowell E. and Lumme K. (1979) Colorimetry and magnitudes of asteroids. In *Asteroids* (T. Gehrels, ed.), p. 132–169. Univ. Arizona, Tucson.
Bowell E. and Lumme K. (1980) A new method for calculating the magnitudes of atmosphereless bodies (abstract). In *Lunar and Planetary Science XI*, p. 100–102. Lunar and Planetary Institute, Houston.
Cintala M. J., Head J. W., and Parmentier E. M. (1980) Impact heating of H_2O ice targets: Applications to outer planet satellites (abstract). In *Lunar and Planetary Science XI*, p. 140–142. Lunar and Planetary Institute, Houston.

Clancy R. T. and Danielson G. E. (1981) High resolution albedo measurements on Io from Voyager 1. *J. Geophys. Res.* In press.

Clark R. N. (1980) Ganymede, Europa, Callisto and Saturn's rings: Compositional analysis from reflectance spectroscopy. *Icarus* **44**, 388–409.

Collins S. A. (1981) Spatial color variations in the volcanic plume at Loki, on Io. *J. Geophys. Res.* In press.

Cook A. F., Shoemaker E. M., Smith B. A., Danielson G. E., Johnson T. V., and Synnott S. P. (1981) Volcanic origin of the eruptive plumes on Io. *Science* **211**, 1419–1422.

Egan W. G. and Hilgeman T. (1975) Development of integrating spheres for measurements between 0.185 and 12 μm. *Appl. Optics* **14**, 1137–1142.

Egan W. G. and Hilgeman T. (1976) Retroreflectance measurements of photometric standards and coatings. *Appl. Optics* **15**, 1845–1849.

Fanale F. P., Brown R. H., Cruikshank D. P., and Clark R. N. (1979) Significance of absorption features in Io's IR reflectance spectrum. *Nature* **280**, 761–763.

Greeley R. (1979) Silt-clay aggregates on Mars. *J. Geophys. Res.* **84**, 6248–6254.

Hansen J. E. and Travis L. D. (1974) Light scattering in planetary atmospheres. *Space Sci. Rev.* **16**, 527–610.

Hapke B. (1979) Io's surface and environs: A magmatic-volatile model. *Geophys. Res. Lett.* **6**, 799–802.

Hartmann W. K. (1981) Surface evolution of two-component stone/ice bodies in the Jupiter region. *Icarus* **44**, 441–453.

Housen K. R., Wilkening L. L., Chapman C. R., and Greenberg R. J. (1979) Regolith development and evolution on asteroids and the Moon. In *Asteroids* (T. Gehrels, ed.), p. 601–627. Univ. Arizona Tucson.

Johnson T. V., Cook A. F. II, Sagan C., and Soderblom L. A. (1979) Volcanic resurfacing rates and implications for volatiles on Io. *Nature* **280**, 746–750.

Lane A. P. and Irvine W. M. (1973) Monochromatic phase curves and albedos for the lunar disk. *Astron. J.* **78**, 267–277.

Langevin Y. and Maurette M. (1981) Grain size and maturity in lunar and asteroidal regoliths (abstract). In *Lunar and Planetary Science XII*, p. 595–597. Lunar and Planetary Institute, Houston.

Lillie C. F., Hord C. W., Pang K., Coffeen D. L., and Hansen J. E. (1977) The Voyager mission photopolarimeter experiment. *Space Sci. Rev.* **21**, 159–182.

Lockwood G. W., Thompson D. T., and Lumme K. (1980) A possible detection of solar variability from photometry of Io, Europa, Callisto, and Rhea, 1976–1979. *Astron. J.* **85**, 961–968.

Lumme K. and Bowell E. (1981a) Radiative transfer in the surfaces of atmosphereless bodies. I. Theory. *Astron. J.* In press.

Lumme K. and Bowell E. (1981b) Radiative transfer in the surfaces of atmosphereless bodies. II. Interpretation of phase curves. *Astron. J.* In press.

Lumme K. and Irvine W. M. (1980) Interpretation of lunar photometry using a new scattering theory. *Bull Amer. Astron. Soc.* **12**, 660.

Lumme K., Bowell E., and Zellner B. (1980) Interpretation of laboratory sample photometry by means of a generalized radiative transfer theory (abstract). In *Lunar and Planetary Science XI*, p. 637–639. Lunar and Planetary Institute, Houston.

Lumme K., Martin L. J., and Baum W. A. (1981) Theoretical interpretation of photometric properties of the martian surface and atmosphere. *Icarus* **44**, 379–397.

Matson D. L. and Nash D. B. (1981) Io's atmosphere: Pressure control by subsurface regolith coldtrapping. *J. Geophys. Res.* In press.

McKay D. S., Fruland R. M., and Heiken G. H. (1974) Grain size and evolution of lunar soils. *Proc. Lunar Sci. Conf. 5th*, p. 887–906.

Meyer B., Gouterman M., Jensen D., Oomen T. V., Spitzer K., and Stroyer-Hansen T. (1972) The spectrum of sulfur and its allotropes. *Adv. Chem. Ser.* **110**, 53–72. Amer. Chem. Soc., Washington, D.C.

Middleton W. E. K. and Mungal A. G. (1952) The luminous directional reflectance of snow. *J. Opt. Soc. Amer.* **42**, 572–579.

Millis R. L. and Thompson D. T. (1975) UBV photometry of the Galilean satellites. *Icarus* **26**, 408–419.

Moore H. J., Spitzer C. R., Bradford K. Z., Cates P. M., Hutton R. E., and Shorthill R. W. (1979) Sample fields of the Viking landers, physical properties, and aeolian processes. *J. Geophys. Res.* **84**, 8365–8378.

Morrison D. (1980) Thermal studies of planetary surfaces. In *Infrared Astronomy* (C. G. Wynn-Williams and D. P. Cruikshank, eds.), p. 89–104. Reidel, Boston.

Morrison D. and Morrison N. D. (1977) Photometry of the Galilean satellites. In *Planetary Satellites* (J. A. Burns, ed.), p. 363–378. Univ. Arizona, Tucson.

Nash D. B., Fanale F. P., and Nelson R. M. (1980) SO_2 frost: UV-visible reflectivity and Io coverage. *Geophys. Res. Lett.* **7**, 665–668.

Nelson R. M. and Hapke B. W. (1978) Spectral reflectivities of the Galilean satellites and Titan, 0.32 to 0.86 micrometers. *Icarus* **36**, 304–329.

Nummedal D. (1981) The role of salt in aggregate formation on Mars (abstract). In *Lunar and Planetary Science XII*, p. 779–781. Lunar and Planetary Institute, Houston.

Pang K. D. (1980) Composition and microstructure of Ganymede's ice surface from Voyager remote measurements. *Proc. 3rd Colloquium on Planetary Water and Polar Processes*, p. 159–167. State University of New York, Buffalo.

Pang K. D., Hord C. W., West R. A., Simmons K. E., Coffeen D. L., Bergstralh J., and Lane A. (1979) Voyager 1 photopolarimetry experiment and the phase curve and surface microstructure of Ganymede. *Nature* **280**, 804–806.

Pang K. D., Rhoads J. W., Hanover G. A., Lumme K., and Bowell E. (1980) Phase curves of Phobos and Deimos. *Bull. Amer. Astron. Soc.* **12**, 663.

Pettengill G. H. (1978) Physical properties of the planets and satellites from radar observations. *Ann. Rev. Astron. Astrophys.* **16**, 265–292.

Pollack J. B., Colburn D. F., Flasar F. M., Kahn R., Carlston C. E., and Pidek D. (1979) Properties and effects of dust particles suspended in the martian atmosphere. *J. Geophys. Res.* **84**, 2929–2946.

Pollack J. B., Witteborn F. C., Erickson E., Strecker D., Baldwin B. J., and Bunch T. E. (1978) Near-infrared spectra of the Galilean satellites: Observations and compositional implications. *Icarus* **36**, 271–303.

Pollack J. B. and Cuzzi J. N. (1980) Scattering by nonspherical particles of size comparable to a wavelength: A new semiempirical theory and its application to tropospheric aerosols. *J. Atmos. Sci.* **37**, 868–881.

Sagan C. (1979) Sulfur flows on Io. *Nature* **280**, 750–753.

Simon S. B. and Papike J. J. (1981) The lunar regolith: Comparative petrology of the Apollo and Luna soils (abstract). In *Lunar and Planetary Science XII*, p. 984–986. Lunar and Planetary Institute, Houston.

Smith B. A., Soderblom L. A., Beebe R., Boyce J., Briggs G., Carr M., Collins S. A., Cook A. F. II, Danielson G. E., Davies M. E., Hunt G. E., Ingersoll A., Johnson T. V., Masursky H., McCauley J., Morrison D., Owen T., Sagan C., Shoemaker E. M., Strom R., Suomi V. E., and Veverka J. (1979) The Galilean satellites and Jupiter: Voyager 2 imaging science results. *Science* **206**, 927–950.

Soderblom L., Johnson T., Morrison D., Danielson E., Smith B., Veverka J., Cook A., Sagan C., Kupferman P., Pieri D., Mosher J., Avis C., Gradie J., and Clancy T. (1980) Spectrophotometry of Io: Preliminary Voyager 1 results. *Geophys. Res. Lett.* **7**, 963–966.

Veverka J. (1973) The photometric properties of natural snow and snow-covered planets. *Icarus* **20**, 304–310.

Veverka J. and Burns J. A. (1980) The moons of Mars. *Ann. Rev. Earth Planet. Sci.* **8**, 527–558.

Whitaker E. A. (1979) Implications for asteroidal regolith properties from comparisons with the lunar phase relation and theoretical considerations. *Icarus* **40**, 406–417.

Zerull R. H. (1976) Scattering measurements of dielectric and absorbing nonspherical particles. *Beitr. Phys. Atmos.* **49**, 168–188.

Structures on Europa

B. K. Lucchitta, L. A. Soderblom, and H. M. Ferguson

U.S. Geological Survey, 2255 North Gemini Drive, Flagstaff, Arizona 86001

Abstract—Lineation patterns on Europa in the area covered by high resolution Voyager 2 images were divided into four classes and analyzed for their systematic trends. The classes consist of dark, wedge-shaped bands in the west-central part of the area; triple bands associated with diagonally trending, global lineations; gray, concentric, curvilinear bands in the southeastern part; and ridges near the terminator. The gray bands are probably old, the wedge-shaped bands and ridges are young, and the triple bands have developed during most of the time of formation of Europa's present crust. Most major structures on Europa's surface were apparently influenced by tidal forces centered on the axis pointing towards Jupiter, but variations in the detail of structural patterns may have been caused by internal forces. Young lineation patterns tend to rejuvenate older ones. Among the patterns most persistently reactivated is a global grid system of diagonally trending conjugate shear fractures, and possibly an old dichotomy consisting of one region dominated by straight fractures and another dominated by curvilinear fractures.

INTRODUCTION

The surface of Europa, which is thought to be composed dominantly of water ice (Pilcher *et al.*, 1972; Fanale *et al.*, 1977; Cassen *et al.*, 1979, 1980), is transected by numerous linear and curved markings. Some are of global extent, and can be recognized on Voyager 1 images (Smith *et al.*, 1979a); others appear to be restricted to local areas (Fig. 1). Pieri (1980) and Helfenstein and Parmentier (1980) analyzed the lineations and concluded that they are best explained as fractures in the planetary crust. Pieri (1980) also subdivided the patterns into distinctive sets that may reflect different origins. The four structural classes presented here partly agree with those of Pieri. This report describes our four classes and includes analyses of their trends. For a detailed description of the classes see Lucchitta and Soderblom (1981, Table 3).

The lineations are straight to curvilinear stripes of uniform albedo or light and dark bands running along side one another (Smith *et al.*, 1979b). End members of the different types of lineations are quite distinct, but individual lineations may be gradational with one another.

The various lineation types were analyzed in rose diagrams. The diagrams show the percent of the total lineations of each plot in each azimuthal bin. The plots were constructed by digitizing the trends on orthographic projections of Voyager 2 image mosaics (violet or blue filter), by summarizing trends between points spaced 30 km apart on the surface and by plotting the trends in compass segments of 20°. Individual plots include several thousand trend segments.

DARK, WEDGE-SHAPED BANDS

Figure 2 shows the dark, wedge-shaped bands to be the most conspicuous dark markings on Europa's surface. Figure 3 is a sketch map of the bands accompanied by their rose diagrams. The bands are thought to be young features because they have sharply delimited edges that cut most other lineations (Lucchitta and Soderblom, 1981). The wedge shape of several of the bands and the offset of crossing lineations suggest that rotation and expansion of the crust were involved in their formation. This possibility was explored by Schenk and Seyfert (1980), who concluded that crustal plates separated and

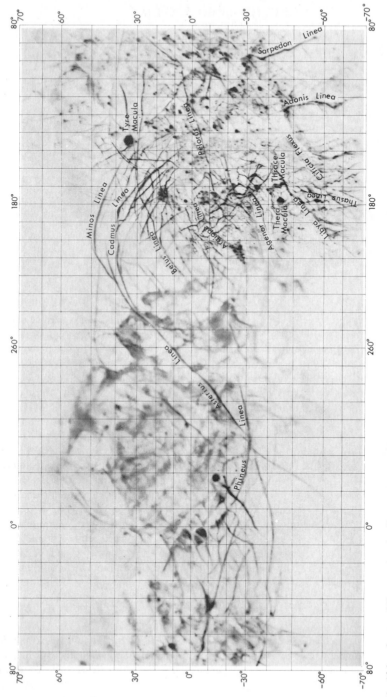

Fig. 1. Airbrush map of Europa showing linear and curved markings, some of which (Asterius Linea, Minos Linea) are of nearly global extent and form great circles. Region covered by Voyager 2 images and discussed in this report is between long 130° and 210°.

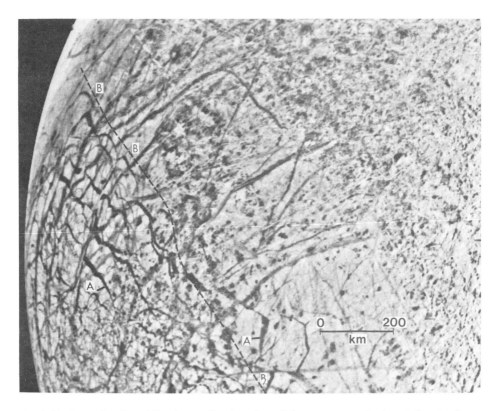

Fig. 2. Dark, wedge-shaped bands trending in subparallel rows across west-central part of area covered by Voyager 2 images. Note wedge shapes (A), as much as 25 km wide. Region to southwest of line B is dominated by curvilinear fractures; region to northeast, including some wedge-shaped bands, is dominated by straight fractures. Curvilinear shapes dominate a belt extending from about the equator at long 200° toward south pole. North toward top. (1255J2-001, clear-filter image).

rotated counterclockwise by a few degrees around a pole lying to the south, resulting in a form of incipient plate-tectonic features.

The wedge-shaped bands (lineation type 2 of Pieri, 1980) have curvilinear patterns in their western part, but are straight to the northeast (Fig. 2). The curvilinear part merges with another region to the south that is also marked by curvilinear patterns, but of a finer scale. The southern region, called fractured plains by Lucchitta and Soderblom (1981), coincides with the area of Pieri's (1980) lineation type 5, characterized by five- and six-sided polygons and interpreted by him as caused by an isotropic stress field. The straight shapes to the northeast fall into the region of Pieri's (1980) lineation type 4, characterized by three- and four-sided polygons and interpreted by him as caused by random stress fields on a sphere.

The sets of curvilinear markings in the wedge-shaped bands and in the fractured plains were thought by Pieri (1980) and Helfenstein and Parmentier (1980) to be centered on a point near the antijovian point (lat 0°, long 180°) and to be related to tidal deformation of Europa's crust. By contrast, we feel that the area of curvilinear markings, rather than being centered near the antijovian point, extends in a belt from lat 0°, long 200°, across the southern part of the Voyager 2 image area toward the south polar region (Fig. 2). This belt includes the curvilinear markings of the wedge-shaped bands and of the fractured plains, and also curvilinear markings now expressed as cycloidal ridges (Malin, 1980). We propose that the isotropic stress field that caused the five- and six-sided polygons and curvilinear markings is not centered on the antijovian point, but lies far to the south of it. Therefore, this stress pattern does not appear to reflect present-day tidal deformation. Instead, it may be related to

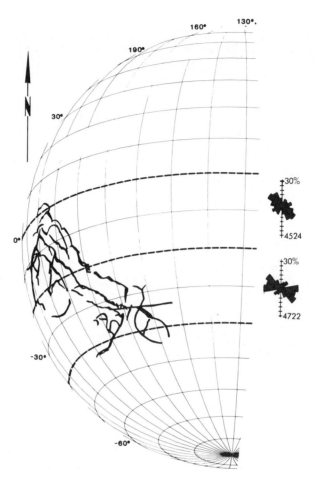

Fig. 3. Sketch map and trend diagrams of wedge-shaped bands. Because wedge-shaped dark bands and triple bands locally merge, some of these lineations are duplicated in Fig. 5. Rosette histograms are for intervals of 20° latitude, longitude 150° to 220°. Numbers indicate counted trend intervals. For discussion see text.

tidal stresses caused by a former tidal pole located elsewhere, or possibly to some kind of former volcanotectonic stresses, such as would result from doming and stretching of the crust.

Even though the isotropic stress field responsible for curvilinear patterns may not be related to present tidal deformation, the wedge-shaped bands, which are probably opened and filled fractures, do lie near the present antijovian point. The opening and filling of these fractures, a relatively young deformation, may be related to present tidal configurations. We concur with Helfenstein and Parmentier (1980) and Pieri (1980) that the fractures may be of tidal origin related to the present tidal poles, or they may be due to convective overturn locked into position by present tidal forces (Finnerty et al., 1980; Schenk and Seyfert, 1980). However, we feel that the curvilinear fracture pattern is old, and was only adopted, later, by the wedge-shaped bands. The curvilinear configuration may have resulted from an old strain pattern unrelated to the present tidal situation.

The trend diagrams of the wedge-shaped bands (Fig. 3) reflect a dominant northwesterly trend and a relatively large spread that is probably caused by the abundant five- and six-sided polygons and curved segments. The dominant northwesterly direction and subsidiary peaks in northeasterly directions suggest that the stress field associated with the late opening of the dark fractures was not isotropic. Perhaps the opening of the fractures followed not only the old isotropic strain pattern, but also developed on an additional set of lines of weakness parallel to the global lineation set discussed below.

TRIPLE BANDS

The second most conspicuous lineation type on Voyager 2 images is the triple band, consisting of a pair of dark bands running along a central, narrow, light stripe (Figs. 4 and 5). Triple bands as here defined include the orange stripe Minos Linea in the north,

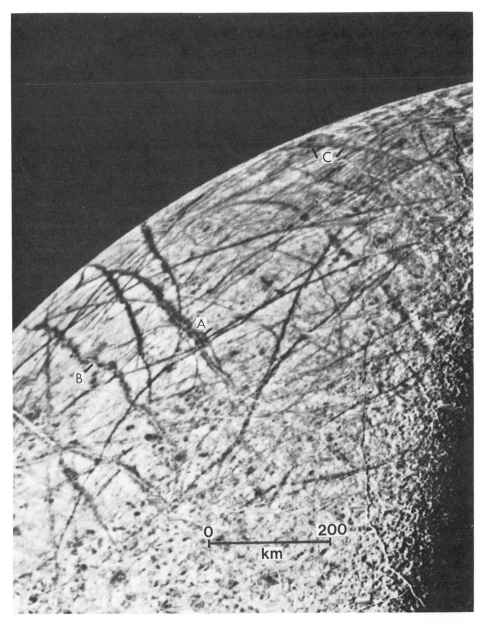

Fig. 4. Triple bands composed of dark bands or spots flanking central narrow, light-gray stripes (A). Many bands are straight and subparallel, others meander slightly (B). Orange band Minos Linea barely visible in clear image at C. Triple bands are associated with global lineations that form great circles. North toward top. (1364J2-001, clear-filter image).

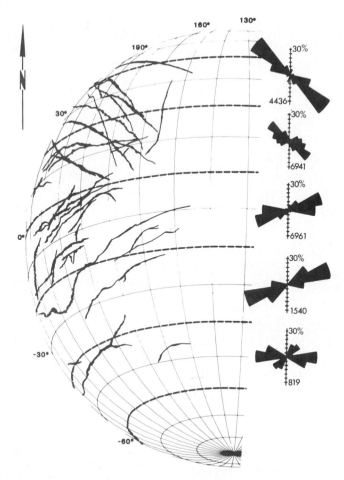

Fig. 5. Sketch map and trend diagrams of triple bands. Rosette histograms are for intervals of 20° latitude, longitude 140° to 220°. Numbers indicate counted trend intervals. For discussion see text.

identified as a triple band on clear filter images, and the white stripe Agenor Linea in the south (Fig. 1). They include Pieri's (1980) lineation types 1, 3, and 7, and many form great-circle global bands (Smith et al., 1979a).

Triple bands cut, and are cut by, most other lineations. They probably formed over a long period of time (Lucchitta and Soderblom, 1981). Malin (1980) suggested that the central stripe of triple bands is a ridge in many places. Indeed, some central stripes appear to merge with ridges in the terminator area, confirming Malin's contention. These relations suggest that some ridges in the terminator area, where ridges abound, are central stripes of triple bands whose albedo contrast cannot be seen and, conversely, that many ridges are present in high-sun areas, where their relief cannot be recognized. (The ridges in the terminator area are classified separately because of their orientations, as discussed below).

The origin of global triple bands was discussed by Finnerty et al. (1980), who suggested that expansion of the crust caused global cracking. Eventually, the cracks became dark stripes when they were intruded by breccia of ice and rock, and they acquired a light central stripe when clean ice was emplaced in a late stage of activity. Helfenstein and Parmentier (1980) concluded that the global lineations could be caused by conjugate shear fractures from cyclical tidal deformation resulting from orbital eccentricity.

Our trend analyses (Fig. 5) show that the triple bands have northwesterly trends north of the equator, and west-southwesterly trends south of the equator (except for Agenor Linea). The change in direction near the equator supports the concept that global stresses were responsible for formation of the bands. Their bisection by the equator at angles near 30° reinforces Helfenstein and Parmentier's (1980) suggestion that the global bands are conjugate shear fractures. Such a fracture pattern also agrees with the fracture pattern developed by Melosh (1980) for a triaxial body under the influence of centripetal and tidal distortion. The equatorial dividing line suggests that the tidal pole toward Jupiter was

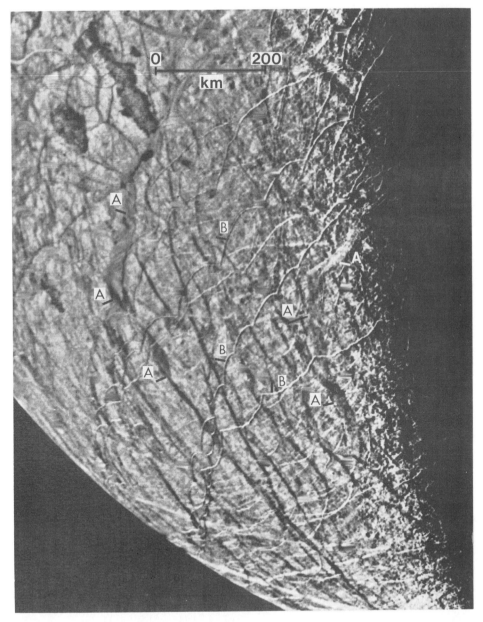

Fig. 6. Gray bands (A) concentric to a point near lat −65°, long 110°. Cycloidal ridges at B. Region is part of belt with curvilinear shapes, including curvilinear wedge-shaped bands in west center (Fig. 2) and cycloidal ridges and curvilinear gray bands in south. North toward top. (1372J2-001, clear-filter image).

located near the equator during the time of formation of the triple bands, which may well span much of the time when Europa's present crust was formed.

GRAY BANDS

Gray curvilinear bands are concentric to a point near lat −65°, long 110° (Figs. 6 and 7). Because many other lineations are superposed, they appear old (Lucchitta and Soderblom, 1981). Individual segments follow a cycloidal pattern like that seen in adjacent ridges, similar to the curvilinear pattern of the young wedge-shaped bands. Perhaps the gray bands, like the wedge-shaped bands, follow an old strain pattern that predisposed the crust in that area towards curved lineations. The origin of the gray bands could be related to a variety of mechanisms that form circular structures on planetary bodies, such as large impacts, volcanotectonic features, or patterns centered on former poles of tidal deformation.

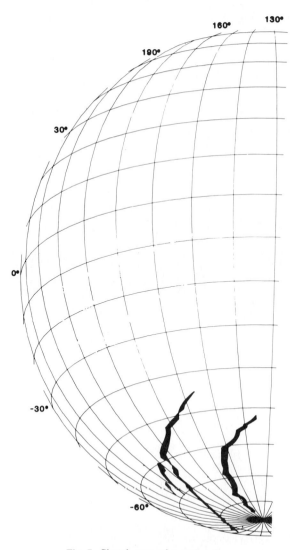

Fig. 7. Sketch map of gray bands.

RIDGES

Conspicuous on Europa's surface are numerous ridges that can be identified clearly only in the terminator area, where shadows emphasize relief (Fig. 6), but they are probably present elsewhere as well. Those recognized are sketched and diagrammed in Fig. 8. The ridges correspond to Pieri's (1980) lineation type 6. Lucchitta and Soderblom (1981) suggested that ridges are among the youngest features on Europa's surface. The ridges tend to have straight segments in the northern and central part of the area covered by Voyager 2 images, and a pronounced cycloidal pattern in the south (Malin, 1980). This dual pattern agrees with other lineations that tend to be straight in the north and center, and curvilinear in the south. The dichotomy in pattern suggests that some ridges may also follow pre-established fracture patterns, as do the wedge-shaped bands and gray bands. Perhaps the cycloidal pattern of ridges results simply from the reactivation of curvilinear cracks that were formed by an earlier isotropic stress field.

In the north, the ridges trend northwesterly; in the far south, they trend southwesterly (Fig. 8). In general the trends of the ridges are the same as those of the triple bands, supporting the interpretation that some of the ridges may be central stripes of triple bands. This similarity is especially strong in the north where the ridges, which are relatively straight and parallel, trend almost identical to the triple bands. On the other

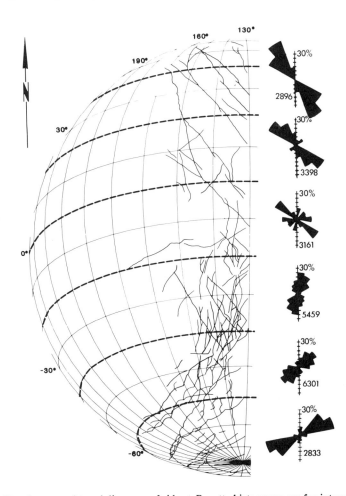

Fig. 8. Sketch map and trend diagrams of ridges. Rosette histograms are for intervals of 20° latitude, longitude 130° to 180°. Numbers indicate counted trend intervals. For discussion see text.

hand, trends of ridges and triple bands diverge substantially in the south. Also, Pieri's (1980) plot of ridges on a Mercator projection map shows that ridges form small circles centered near the antijovian point. Our plots confirm a circular arrangement, because they show a gradual change in trend direction from near north-south at the equator, to near east-west toward the south pole. Apparently, most ridges do not lie on great circles like the triple bands and, therefore, are probably a separate set of structures. However, the apparent circular arrangement may be largely an artifact of the illumination, as the subsolar point also lies near the antijovian point.

The circular arrangement of ridges on Europa around the antijovian point suggested to Finnerty et al. (1980) that the origin of the ridges is linked to tidal deformation. However, Melosh (1980) calculated that concentric patterns around tidal poles are not to be expected on tidally distorted triaxial bodies. Also, because present tidal stresses amount to only a few bars and are probably not sufficient to rupture Europa's crust, Finnerty et al. (1980) preferred a combination of tectonic and tidal forces for the origin of the concentric ridges. They proposed that second-order convection in Europa's interior became locked at the tidal axes; the currents welled up at the tidal poles, spread radially, and went down underneath the ridges, which formed by associated compression.

Alternatively, no added internal forces may have been needed to establish the concentric pattern. Perhaps the action of tidal stresses operating alone over millions of years has fatigued Europa's outer brittle ice crust sufficiently to allow the development of circular fractures. Perhaps, such fractures were later accentuated by other processes such as global expansion (Ransford et al., 1980). If the ridges did not form by compression, as suggested by Finnerty et al. (1980), their origin could be similar to the origin of the central stripes on triple bands. Thus, the ridges could be dikes intruded into cracks. Or possibly the ridges formed by the opening and closing of crustal ice segments, as in ice floes on earth that open to form leads and ram back together to form pressure ridges. The resolution of the Voyager pictures does not permit a definite assessment.

TREND COMPARISONS

A comparison of trends of the different lineation types (Fig. 9) shows that northwesterly and southwesterly trends dominate for all types in the area investigated. The apparent reversal in direction lies near the equator, which suggests that all the lineation types are influenced by a global fracture pattern; this pattern may have been formed by conjugate shear sets in the equatorial area of a tidally distorted body (Melosh, 1980; Helfenstein and Parmentier, 1980). These shear fractures may have become lines of weakness sufficiently persistent to influence other fractures that developed later, such as the wedge-shaped bands and the ridges. The possibility that northwesterly and southwesterly trending shear fractures can develop as fundamental lines of weakness in pristine crusts supports the idea that similar global grids, perhaps also formed by tidal distortion, exist on other planetary bodies.

A detailed comparison of trends lying within 20° latitude belts (Fig. 9) shows that the trends of triple bands and ridges are nearly identical north of the equator. Perhaps many of the ridges seen in that area are central stripes of triple bands. Alternatively, young concentric ridges in that area may have been diverted by lines of weakness of the global shear fracture set.

Immediately south of the equator, the ridge trends are apparently also dominated by the global shear set, as they parallel those of ridges and triple bands farther north. However, we also see minor peaks in other directions, including a peak in a direction parallel to the trend of the triple bands in the same latitude belt. These minor peaks may reflect the central ridges of triple-bands, mixed with other ridges in this area.

Only in the latitude belt from −20° to −40° can a coherent pattern be identified for wedge-shaped bands, (labelled dark bands in Fig. 9), triple bands, and ridges: the wedge-shaped bands are orthogonal to the ridges, consistent with a tensional origin for the bands, interpreted as wedge-shaped opened fractures (Schenk and Seyfert, 1980), and

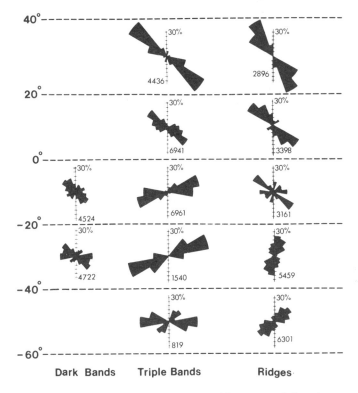

Fig. 9. Comparison of trend diagrams. Rosette histograms of lineation sets within intervals of 20° latitude. Wedge-shaped bands (labelled dark bands), long 150° to 220°. Triple bands, long 140° to 200°. Ridges, long 130° to 180°. Numbers indicate counted trend intervals. For discussion see text.

consistent with a compressional origin for the ridges (Finnerty *et al.*, 1980). The dominant trend direction of the triple bands forms an acute angle with the inferred compressional stress trajectory. The acute angle is consistent with an origin by shear (Helfenstein and Parmentier, 1980). In this region, the wedge-shaped bands, triple bands, and ridges lie relatively close to one another (within 30° longitude) so that a common stress system may indeed be responsible. On the other hand, because relations suggesting common stress systems are not consistently seen everywhere, they may be fortuitous in this latitude belt also. The minor peak in the wedge-shaped bands parallel to the major direction of the triple bands may be due to the inclusion of some triple bands in the measurements of the wedge-shaped bands (Figs. 3 and 4).

In the −40° to −60° latitude belt, the triple band Agenor Linea (see Fig. 1) trends generally east-west. This trend is oblique to the ridges but not at an angle that would suggest compressional origin for the ridges and shear origin for the triple bands. However, a possible east-west shear direction in the southern part of equatorial areas agrees with east-west shear directions proposed by Melosh (1980) for these regions on tidally distorted bodies.

HISTORY OF DEFORMATION

Shoemaker and Wolfe (1981) estimated that Europa's present crust is 30–200 million years old. Its deformation period is reflected in the lineation patterns that are now preserved on its surface. Because the crust appears to destroy old features by annealing or by rejuvenation (Lucchitta and Soderblom, 1981), no record of fracture traces due to tidal despinning is expected and none is observed (Helfenstein and Parmentier, 1980). Triple bands probably developed and formed conjugate sets throughout most of the time of formation of the present crust (Lucchitta and Soderblom, 1981). The stress field causing these shear sets appears to have been centered near the equator, on a tidal pole. This stress field apparently remained active in that area throughout the time of development of presently visible triple bands and, therefore, during most of the time of formation of the present crust. The high visibility of triple bands is probably due to expansion of the crust accentuating the fractures, as proposed by Ransford et al. (1980). This expansion, because it accompanied the formation of triple bands, probably also lasted throughout most of the time when the present crust was formed.

With the possible exception of the formation of some triple bands, the earliest recorded event on Europa's crust appears to be the development of the dichotomy between straight and curvilinear fractures: in the northern and central parts of the Voyager 2 image area, straight fractures outline three- and four-sided polygons (Pieri, 1980); in the south, curvilinear fractures define five- and six-sided polygons. All later lineations apparently adopted these old fracture traces. The pattern of curvilinear fractures in the south may have been formed by an isotropic stress field caused by stretching of the crust due to volcanotectonic activity, or due to tidal stresses centered on a pole far to the south of the present antijovian point. Existence of such a tidal pole would imply that the entire crust rotated to its present position at a later date. This supposition, however, is contradicted by the apparent centering of triple bands, which apparently developed over a long time span, on a pole near the present equator. Next, the gray concentric bands were formed near the southern part of the terminator. The responsible processes may have been external (impact), internal (volcanotectonic), or a combination of the two (tidal). Relatively recently, wedge-shaped fractures opened in the vicinity of the antijovian point, and lastly, numerous ridges were emplaced. Times of formation of these latter two features may have overlapped significantly. The wedge-shaped bands may have been opened by relatively recent internal forces resulting from incipient plate motion (Schenk and Seyfert, 1980), or they may have opened due to tidal forces near the present tidal pole (Helfenstein and Parmentier, 1980; Finnerty et al., 1980). The arrangement of the wedge-shaped bands, however, appears to have been determined to some extent by pre-existing fracture patterns, such as lines of weakness from an old isotropic stress field and from the global shear fracture set. The location of ridges forming circles centered on the antijovian point seems to be dictated by present tidal forces. The ridges may be of compressional origin, or they may be ice dikes filling circular cracks centered on the tidal pole.

In summary, the major structures on Europa's surface appear to have been formed by tidal forces centered on the axis pointing toward Jupiter, but variations in the detail of the structural patterns may have been caused by internal forces. The lineation patterns may follow a variety of stress systems, but older strain patterns appear to have reasserted their influence repeatedly by rejuvenation by younger fractures. The most persistent rejuvenated strain patterns are the global diagonal fractures, probably formed by conjugate shear, and the straight and curvilinear fractures that apparently caused an early dichotomy in fracture patterns.

Acknowledgments—We are grateful to P. Helfenstein, T. Maxwell, D. Scott, and D. Wilhelms for thoughtful reviews. We further wish to acknowledge the help of K. Edwards, who wrote the computer programs, and C. Watkins, who digitized the data. The work was performed under NASA contract W13,709.

REFERENCES

Cassen P., Peale S. J., and Reynolds R. T. (1980) Tidal dissipation in Europa: A correction. *Geophys. Res. Lett.* **7**, 963–970.

Cassen P., Reynolds R. T., and Peale S. J. (1979) Is there liquid water on Europa? *Geophys. Res. Lett.* **6**, 731–734.

Fanale F. P., Johnson T. V., and Matson D. L. (1977) Io's surface and the histories of the Galilean satellites. In *Planetary Satellites* (J. A. Burns, ed.), p. 379–405. Univ. Arizona Press, Tucson.

Finnerty A. A., Ransford G. A., Pieri D. C., and Collerson K. D. (1980) Is Europa's surface cracking due to thermal evolution? *Nature* **289**, 24–27.

Helfenstein P. and Parmentier E. M. (1980) Fractures on Europa: Possible response of an ice crust to tidal deformation. *Proc. Lunar Planet. Sci. Conf. 11th*, p. 1987–1998.

Lucchitta B. K. and Soderblom L. A. (1981) Geology of Europa. In *Satellites of Jupiter*. Univ. Arizona, Tucson. In press.

Malin M. C. (1980) Morphology of lineaments on Europa. In *The Satellites of Jupiter, IAU Colloq. 57*, May 13–16, 1980, Kailua-Kona, Hawaii, Session 7-2.

Melosh H. J. (1980) Tectonic patterns on a tidally distorted planet. *Icarus* **43**, 334–337.

Pieri D. C. (1980) Lineament and polygon patterns on Europa. *Nature* **289**, 17–21.

Pilcher C. B., Ridgway S. T., and McCord T. B. (1972) Galilean satellites: Identification of water frost. *Science* **178**, 1087–1089.

Ransford G. A., Finnerty A. A., and Collerson K. D. (1980) Europa's petrological thermal history. *Nature* **289**, 21–24.

Schenk P. M. and Seyfert C. K. (1980) Fault offsets and proposed plate motions for Europa (abstract). In *EOS Trans. Amer. Geophys. Union* **61**, 286.

Shoemaker E. M. and Wolfe R. F. (1981) Cratering timescales for the Galilean satellites. In *satellites of Jupiter*. Univ. Arizona, Tucson. In press.

Smith B. A., Soderblom L. A., Johnson T. W., Ingersoll A. P., Collins S. A., Shoemaker E. M., Hunt G. E., Masursky H., Carr M. H., Davis M. E., Cook A. F. II, Boyce J. M., Danielson G. E., Owen T., Sagan C., Beebe R. F., Veverka, J. Strom R. G., McCauley J. F., Morrison D., Briggs G. A., and Suomi V. E. (1979a) The Jupiter system through the eyes of Voyager 1. *Science* **204**, 951–972.

Smith B. A., Soderblom L. A., Johnson T. V., Ingersoll A. P., Collins S. A., Shoemaker E. M., Hunt G. E., Masursky H., Carr M. H., Davis M. E., Cook A. F., II, Boyce J. M., Danielson G. E., Owen T., Sagan C., Beebe R. F., Veverka J., Strom R. G., McCauley J. F., Morrison D., Briggs G. A., and Suomi V. E. (1979b) The Galilean satellites and Jupiter: Voyager 2 imaging science results. *Science* **206**, 927–950.

The sputter-generation of planetary coronae: Galilean satellites of Jupiter

C. C. Watson

W. K. Kellogg Radiation Laboratory, California Institute of Technology, Pasadena, California 91125

Abstract—Energetic particle fluxes which impinge upon and sputter planetary surfaces or atmospheres are effective agents for the generation or modification of atmospheric coronae. The structure of such coronae is determined by the characteristics of the sputtered particle flux, and hence of the sputtering mechanisms involved. A collisional cascade type interaction tends to produce a more extended exosphere with a greater fraction of gravitationally unbound molecules than does a thermal spike sputtering mechanism, although the total mass of the generated atmosphere might exceed an exospheric level in the latter case.

The unimpeded sputtering of Io's surface by the corotating sulphur and oxygen plasma can result in a coronal SO_2 column density of a few times 10^{14} cm^{-2}, which approaches the maximal value possible in the collision cascade regime. Under certain circumstances such a corona would be the dominant stable atmospheric component on Io. Sputtering by the higher energy constituents of the Jovian magnetosphere is probably less important on this satellite. The sputter-induced erosion of water ice on Europa may give rise to exospheric column densities in excess of a few times 10^{12} H_2O cm^{-2}. Europa's weaker gravitational interaction with sputtered molecules, when compared to Io, produces much less distinction between collisional cascade and thermal spike type coronae. The redeposition of coronal material may provide an important mass transport mechanism for volatiles on the surfaces of the Galilean satellites.

I. INTRODUCTION

Many of those bodies in the solar system whose surfaces are exposed to erosive interaction with an energetic plasma are sufficiently massive that the gravitational potential energy, U, of one of the molecules comprising their surface is of the same magnitude as the chemical binding energy of that molecule to the surface, W. The characteristic planetary (or satellitic) mass in this regard is $M = RW/Gm$, where R is the planetary radius, m is the molecular mass and G is the universal gravitational constant. A substantial fraction of the molecules which are ejected, or sputtered, from the surface of such a satellite, following impact by a fast ion, will have relatively long residence times in the immediate vicinity of their parent body. Indeed, those having kinetic energies less than $U \equiv GMm/R$ will follow ballistic trajectories reintersecting the surface, barring further collisional interactions. This halo of ejecta may be termed a sputter-induced planetary corona to distinguish it from the thermal exosphere or corona of a planetary body possessing a dense atmosphere, which has been discussed extensively by Chamberlain (1963). The two types of coronae are similar in many respects, of course. Mercury, the moon, and the Galilean satellites of Jupiter are among those objects which are massive enough to support significant sputter-induced coronae.

As a result of the Voyager missions to Jupiter the evidence is mounting that the icy Galilean satellites (Broadfoot *et al.*, 1979), and perhaps also Io (Matson and Nash, in preparation), cannot maintain atmospheres of greater than exospheric mass in thermal equilibrium with the volatiles, e.g., H_2O and SO_2, which are condensed on their surfaces. Consequently, coronal induction via sputtering may play a dominant role in determining the distribution of neutral matter in circumjovian space and in the supply of material, particularly heavy ions, to the Jovian magnetosphere. The problem of the loss rates of sputtered atomic particles from the gravitational fields of planetary bodies has been addressed by several authors in various contexts (Wehner *et al.*, 1963; Matson *et al.*, 1974;

Lanzerotti et al., 1978; Watson et al., 1980; Haff et al., 1981). In contrast, this paper is concerned primarily with characterizing the distribution of this sputtered material in space (cf. Watson et al., 1981). Before we turn to a discussion of the nature of the coronae and related effects which we expect to be associated with the Galilean satellites, we first analyze the mechanics of sputter-induced planetary coronae in a more general context.

II. SPUTTERING

As long as one is interested in atmospheres whose column density $n_c \lesssim 10^{15}$ molecules cm^{-2}, one may obtain a good approximation to their structure by considering only particles following ballistic trajectories. This neglect of intermolecular collisions is better justified for particles traveling radially than for ones on more nearly circular orbits, but in any case the accuracy of the approximation improves rapidly at lower densities. In this regime, the coronal structure is completely fixed once the energy and angular dependence of the molecular flux emanating from the surface is known. We shall assume here that this source of particles is independent of position on the surface. This assumption leads to a globally uniform exosphere. The characteristics of the sputtered particle flux depend on the mechanisms by which the energy deposited by the incident ions is transformed to the recoil motion of the target atoms. Ions whose energy is ~ 1 keV/amu interact with target atoms primarily through a screened Coulomb potential, and there is a direct transfer of nuclear kinetic energy which initiates a collisional cascade (CC) among the target particles. If the chemical binding energy of the particles to the surface, W, is much less than the incident ion's energy, E_{in}, then a good approximation for the flux of sputtered atoms with kinetic energies between E and E + dE, having velocity vectors directed into the solid angle element $d\Omega$ at an angle θ with respect to the local surface normal is (M. W. Thompson, 1968):

$$g_{cc}(E, \Omega) \, dEd\Omega = \frac{\alpha \phi E}{(E+W)^3} \frac{\cos \theta}{4\pi} \, dEd\Omega. \tag{1}$$

Here α is a constant which depends on the nuclear charges and masses of the ions and target particles, but which is independent of the incident ion flux ϕ and of the target's density. Note that W is the characteristic energy of this spectrum since the peak of the distribution occurs at E = W/2. In the analogous expression for the emitted flux of a particular species i in a target comprised of two or more components, α_i would be proportional to that species' concentration c_i, so that $\alpha_i = \alpha' c_i$, and W would be replaced by the species-specific binding energy W_i (Watson and Haff, 1980). The constant α' would now depend on the charges and masses of all species present. The total yield of species i would therefore be

$$Y_i = \frac{1}{\phi} \int \int g_i(E, \Omega) \, dEd\Omega = \frac{\alpha' c_i}{8W_i}. \tag{2}$$

Thus the sputtering yields of materials on the surface would depend on how tightly they are bound.

A planetary scientist should be cognizant of several limitations of the simplest model of collisional cascade sputtering represented by Eq. (1). For E_{in} approaching W the cascade may not involve a sufficient number of collisions to develop the indicated power law spectrum. At the other extreme, if the energy deposition (stopping power) of the impinging ions is very large (D. A. Thompson and Johar, 1980), then deviations from the spectrum of Eq. (1) may arise either because collisions between secondary recoils having energy on the order of W or greater become probable, resulting in a thermal spike type phenomenon, or perhaps due to a reduction in the effective surface binding energy itself (D. A. Thompson, 1980). The CC distribution may also exhibit a significant dependence on the incidence angle of the impinging flux as this angle approaches 90° from the normal.

A particularly interesting situation arises when one is concerned with the sputtering of a volatile molecular species adsorbed on the surface of a refractory substance, e.g., SO_2 adsorbed on elemental S, or H_2O condensed on mineral grains. In these cases the sputtering yield of the adsorbed molecule may differ considerably from that of a target comprised wholly of that species, due to differences in both the chemical binding and in the kinetic coupling of the surficial layer to the substrate. Experiments on the sputtering, by noble gas ions, of N chemisorbed on a tungsten substrate indicate yields intermediate between those characteristic of minerals and frozen volatiles (Winters and Sigmund, 1974). Finally, we should mention that the departures of the surface topography from planar geometry on a scale of a few microns or less can result in the substantial alteration of both the angular dependences of the sputtered flux and the total yield. In general, a rougher surface tends to result in a lower yield than a smoother but otherwise identical one (Hapke and Cassidy, 1978; Carey and McDonnell, 1976). Such an effect is potentially significant with regard to Io (Haff et al., 1981) since much of this satellite's surface may be covered with a fine grained, highly porous material (Matson et al., 1980; Matson and Nash, in preparation).

Despite the above qualifications it is safe to say that collisional cascade type sputtering is a universal and basically well understood phenomenon (M. W. Thompson, 1968; Sigmund, 1969), which must accompany low energy ion impact on essentially any target, whether it be metallic or dielectric, a mineral or a frozen volatile. For ion energies approaching $E_{in} = 1$ MeV/amu, however, one would expect the efficiency with which collisional cascades are generated to decrease rapidly due to the $1/E_{in}^2$ dependence of the Rutherford scattering cross section. In fact, in this regime the ion transfers most of its energy to target atoms indirectly through electronic excitation and ionization. However, in certain dielectric or insulating materials this electronic component of the energy deposition is apparently coupled efficiently to atomic recoil motion, perhaps by means of an ion explosion mechanism (Haff, 1976), with the result that sputtering yields may be enhanced by two or three orders of magnitude over what one would expect based on the nuclear component of the stopping power alone. Enhanced erosion rates associated with fast ions have been observed in such materials as UF_4 (Griffith et al., 1980), quartz (J. E. Griffith, pers. comm.), KCl (Biersack and Santner, 1976), water ice (W. L. Brown et al., 1980), SO_2 frost (R. E. Johnson, pers. comm.) and other frozen-gas films (Bøttiger et al., 1980). Yields $\sim 10^3$ H_2O molecules/ion have been measured for 5 MeV ^{19}F impact on ice (B. Cooper, pers. comm.). We must emphasize, however, that the actual existence of enhanced electronic type sputtering effects in naturally occurring materials is much more problematical than is the case for CC type sputtering. It is not clear, for example, how the presence of atomically dispersed impurities, e.g., silicates dissolved in water ice or S contamination of SO_2 frost, would affect the electronic sputtering mechanism. Nor is it known whether such materials as pure sulphur should exhibit enhanced erosion rates.

The characteristics of the flux distributions of particles sputtered by some presumably electronically mediated mechanism have not yet been well established empirically. There may in fact be more than one mechanism involved in the transmission of the deposited energy from the electron system to the lattice (Haff, in preparation). But in so far as the partitioning of this energy among atomic recoils is concerned, thermal spike pictures, which involve an approximate thermodynamic equilibration in the core region about an ion's track, have been the most extensively discussed models and have indeed met with some degree of quantitative success (Seiberling et al., 1980; Ollerhead et al., 1980). Such models are at least consistent with one's intuition that the recoil energy spectrum should be quite soft (Weller and Weller, in preparation). The most definitive experiments to date have been measurements of the differential yields of ^{235}U sputtered from UF_4 by fast ^{19}F ions (Griffith et al., 1980; Seiberling et al., 1980). Seiberling et al. (1980) have determined that the energy spectrum of the U atoms is well described by a Maxwell-Boltzmann type recoil distribution corresponding to a temperature $T = 3500°K$. The angular dependence of the flux was found to follow $\cos \theta$ closely (Griffith et al., 1980). Although the degree of universality of these results is not yet established, the potential importance of this enhanced sputtering phenomenon in astrophysical contexts, as is evidenced for instance by the concurrence of high energy particle fluxes and frozen volatiles in the Jovian system, leads us to make the assumption of an ejectile distribution appropriate to a thermal spike (TS) mechanism with a planar surface potential step,

$$g_{TS}(E, \Omega) = 4Y'\phi \frac{E}{(kT)^2} e^{-E/kT} \frac{\cos \theta}{4\pi}, \qquad (3)$$

as a useful basis for quantitative comparison with CC type sputtering. Here T is the spike temperature and Y is the total yield. The chemical work function W is incorporated in Y. An implicit assumption involved here is that the thermal spike can be well described by a single temperature (cf. Seiberling et al., 1980). Note also that the characteristic energy of this spectrum is kT and not W as for the cascade flux.

Leaving aside the question of its accuracy, the above TS spectrum derives considerable utility from its role as a limiting alternative to the CC distribution of Eq. (1). In the former case the secondary recoils are assumed to be fully equilibrated, leading to a much softer energy distribution than derives from the CC picture in which collisions between energetic secondaries are neglected. Even those workers who assert that a thermal spike picture does not provide the best model for their laboratory ice sputtering data (W. L. Brown et al., 1980; L. J. Lanzerotti, pers. comm.) have nevertheless tacitly adopted a Maxwellian energy spectrum for the emitted particle flux in their discussion of enhanced ice sputtering in astrophysical contexts (R. E. Johnson et al., 1981).

If certain of the materials exposed on a planet's surface exhibit very large sputtering yields relative to others (factors of 10^2–10^3 are not unreasonable), then one must expect the composition of the sputter-generated atmosphere to be completely dominated by the former. Thus, in the electronic sputtering regime, the stoichiometry of the corona may be quite different from that of the surface. In the case of CC type sputtering, however, the mass fractionation is governed principally by the relative values of W_i [Eq. (2)], which typically differ by a factor of 10 between rocky and icy material and by much less between the components of either type of substance. Coronae generated by this latter mechanism should therefore reflect much more closely the composition of the surface. The relative concentrations of the various species need not be uniform in altitude though, due to the role of the W_i in fixing the energy spectra of the emitted fluxes and to the action of the gravitational field in separating species of different mass.

III. CORONAE

The density of a thick, isothermal atmosphere falls off nearly exponentially with altitude over small radial ranges. But at large radii, or for atmospheres of less than an exospheric mass an exponential profile underestimates the actual column density. An accurate calculation of the exosphere's structure in these cases requires an analysis of the ballistic orbits of its constituent molecules (Chamberlain, 1963). Such an analysis may, of course, also be applied when the source flux of molecules to the exosphere does not exhibit a Maxwellian velocity spectrum. Consider a particle of mass m which is emitted from a surface located at a radius R from the center of a planet. The ballistic trajectory of such a particle in its orbital plane is completely determined by its initial kinetic energy E, and the magnitude of its angular momentum, ℓ (or angle of emission θ). Here θ is measured with respect to the local radius vector, and $\ell = (2mE)^{1/2} R \sin\theta$. If we neglect the contribution of satellite orbits, that is, orbits which do not intersect the surface and which must be populated by intermolecular collisions, and if we assume spherical symmetry, then the number density of the corona at planetocentric radii $r \geq R$, generated by the flux $g(E, \Omega)$ emanating from the surface at $r = R$ has the formal expression

$$n(r) = \left(\frac{R}{r}\right)^2 \left\{ 2 \underset{\text{bound}}{\iint} dE d\Omega + \underset{\text{escaping}}{\iint} dE d\Omega \right\} \frac{g(E, \Omega)}{\dot{r}}. \tag{4}$$

In this expression $\dot{r}$ is the radial component of the particle's velocity at radius r. We neglect any effects due to the resputtering or bouncing of emitted particles which reimpact the surface. The limits of integration are to be chosen to include all possible orbits by which a molecule may travel from radius R to radius $r > R$. The minimum kinetic energy required for a particle to reach r is $E_{min}(r) = U - GMm/r$. Molecules having kinetic energy greater than $E_{min}(r)$, but less than $E_c(r) = [r/(r+R)]U = E_{min}(r)r^2/(r^2 - R^2)$, will achieve the radius r only if $\ell^2/2mr^2 \leq E - E_{min}(r)$. Those having $E > E_c$ must attain the radius r whatever their angular momentum may be. The contribution from the escaping flux comprises all those particles having $E \geq U$.

The integrals of Eq (4) are particularly simple to evaluate when the source particle flux is spatially isotropic, so that the flux issuing from the surface at R has the form

$$g(E, \Omega) = f(E) \frac{\cos\theta}{4\pi}. \tag{5}$$

Note that both the spectra of Eqs. (1) and (3) satisfy this requirement. With these assumptions it can be shown that the angle integrated contribution to n(r) from gravitationally bound particles is

$$n_b(r) = \left(\frac{m}{2}\right)^{1/2} \left\{ \int_{E_{min}(r)}^{U} f(E)[E - E_{min}(r)]^{1/2} \frac{dE}{E} - \left(\frac{r^2 - R^2}{r^2}\right)^{1/2} \int_{E_c(r)}^{U} f(E)[E - E_c(r)]^{1/2} \frac{dE}{E} \right\}, \tag{6}$$

while that from escaping molecules is

$$n_e(r) = \frac{1}{2}\left(\frac{m}{2}\right)^{1/2} \left\{ \int_{U}^{\infty} f(E)[E - E_{min}(r)]^{1/2} \frac{dE}{E} - \left(\frac{r^2 - R^2}{r^2}\right)^{1/2} \int_{U}^{\infty} f(E)[E - E_c(r)]^{1/2} \frac{dE}{E} \right\}. \tag{7}$$

The total coronal density is then $n(r) = n_b(r) + n_e(r)$. Chamberlain (1963) developed expressions equivalent to these in his discussion of planetary coronae and atmospheric evaporation from an isothermal atmosphere. The present formulae are more general in that the energy spectrum f(E) need not be Maxwellian. Thus Eqs. (6) and (7) may be readily applied not only to TS type sputtered particle fluxes, but also to CC sputtering and those intermediate energy spectra which may be appropriate for non-thermalized ionic-explosion sputtering mechanisms.

In many cases of interest, viz., when $kT \ll U$, the TS type density profile differs only negligibly from $n_b(r) + 2n_e(r)$ since very few sputtered particles have sufficient energy to escape the planet. This latter quantity has the simple analytical form:

$$n_b(r) + 2n_e(r) = Y\phi\left(\frac{2\pi m}{kT}\right)^{1/2}\left\{\exp\left[-\frac{U}{kT}\left(\frac{r-R}{r}\right)\right] - \left(\frac{r^2-R^2}{r^2}\right)^{1/2}\exp\left[-\frac{U}{kT}\left(\frac{r}{r+R}\right)\right]\right\}. \quad (8)$$

It is this distribution which we have used in our discussion of TS induced sputtering effects on Io.

It follows from Eqs. (1), (3), (6) and (7) that $n(r)$, and thus the coronal column density n_c is proportional to the incident sputtering flux ϕ. For ϕ sufficiently large n_c will begin to approach the critical value $n_c \sim \sigma^{-1} \approx 4 \times 10^{14}$ cm^{-2}, where $\sigma \approx 25$ Å^2 is a conservative estimate for typical intermolecular elastic collision cross sections. At this point the impinging ions will begin to sputter the corona by the same CC mechanism which is operative in solids (Haff et al., 1978; Haff and Watson, 1979; Watson et al., 1980), creating a sink for exospheric mass of magnitude comparable to the flux supplied to the corona by surface sputtering. For this reason, coronae generated by sputtering in the low energy, collision cascade regime should be self-limiting at a roughly exospheric mass. This same conclusion would hold even for ions in the high energy regime, unless the surface sputtering is dominated by some enhanced electronic mechanism which is not operative in the atmosphere. In this latter scenario the rate of sputter depletion of the corona might not balance the surface source flux until a column density substantially in excess of the above value of n_c had been attained. For such more massive atmospheres effects due to the energy loss and straggling of the ions penetrating the corona may become significant.

The concept of a self-limiting atmosphere has been discussed by Yung and McElroy (1977) who argued that the increasing opacity of an O_2 atmosphere on Ganymede would serve to limit the rate of H_2O photolysis which supplies this atmosphere, thereby limiting its basal pressure to $\sim 10^{-3}$ mbar. Lanzerotti et al. (1978) subsequently proposed that the accrual of such a dense O_2 atmosphere on this satellite [which, however, is now known not to exist (Broadfoot et al., 1979)] would also limit the contribution of water ice sputtering by magnetospheric protons to the H_2O vapor reservoir. In general, though, the presence of an atmosphere might actually enhance the erosion of a satellite's surface by slowing very fast ions down to energies more favorable for sputtering (Matson et al., 1974). The determining factors, of course, are the atmosphere's density and the energy spectra of the impinging ions. In any case, we wish to emphasize here the distinction between a supply-limiting atmospheric generation mechanism, such as that proposed by Yung and McElroy, and the loss-limited corona discussed above. For when coronal generation occurs primarily through conventional CC sputtering, it is the sputtering of the atmosphere itself which places the upper limit on its bulk at an exospheric mass.

IV. IO

The sputter erosion of Io's surface has been proposed (Matson et al., 1974) as one of the most viable mechanisms for the supply of material to the neutral sodium cloud associated with Io's orbit (R. A. Brown, 1974). The same point could be made with respect to the neutral potassium (Trafton, 1975) and oxygen (R. A. Brown, 1981) clouds near Io. Recent estimates imply that the source flux required to maintain the Na cloud, globally averaged and referred to Io's surface is on the order of 2×10^9 Na atoms cm^{-2} s^{-1} (R. A. Brown, in preparation). This flux, together with those sputtered particles of lower velocity which remain gravitationally bound to Io, could in itself comprise a substantial neutral corona. However, the composition of Io's surface appears to be dominated by elemental S, SO_2 and other sulphur bearing compounds (Fanale et al., 1979; Nash and Nelson, 1979) suggesting the predominance of S and SO_2 in the total sputtered flux.

Although it is not the primary purpose of this paper to provide estimates for mass transfer rates between Io and the torus, the sputtering rates corresponding to the coronal generation processes discussed below are consistent with such estimates made by Haff et al. (1981) who showed that the sputtering of an SO_2 atmosphere on Io can supply gravitationally unbound S and O at a sufficient rate (10^{10}–10^{12} cm^{-2} s^{-1}; Kumar, 1979) to sustain the plasma torus, while the sputtering of refractory sulphur-bearing surface materials probably could not. On the other hand, the sputtering of a volatile SO_2 surface frost, which was not discussed quantitatively by Haff et al. (1981), is a viable S and O source mechanism for the torus, as we shall see here. Moreover, there may be significant secondary mass loss processes associated with the sputter-generated coronae. For example, depending upon how closely the hot ($\sim 10^5$°K; Shemansky, 1980b) torus electron plasma penetrates to Io's surface, much of the corona and not just those particles having $E > U$ could be exposed to ionization and pickup by the corotating magnetic field. In fact, a certain fraction of the sputtered particles may leave the surface in an ionized state (Haff et al., 1981). Cloutier et al. (1978) have pointed out that the Lorentz force on such ions would considerably exceed Io's gravitational attraction. A zeroth-order estimate of the ionization-entrainment loss flux is $\phi_{ion} = n_c/\tau$, where n_c is the column density of the corona and τ is the lifetime of the neutral species against ionization. For SO_2, τ may be on the order of 10^5 seconds (Kumar, 1979), while for atomic S and O, τ is somewhat smaller (Shemansky, 1980a). Thus if $n_c \sim 10^{15}$ SO_2 cm^{-2}, then $\phi_{ion} \gtrsim 10^{10}$ SO_2 cm^{-2} s^{-1}, which is similar to estimates by Ip and Axford (1979) and by Goertz (1980).

One may observe that if the dominant ionization mechanism involved here is in fact impact by the torus electron plasma, as has been widely postulated, then τ^{-1} should be proportional to the torus plasma density and hence ϕ. But we have previously noted that coronal generation through sputtering by the corotating plasma also results in a proportionality between ϕ and n_c. It therefore follows that the ionization-entrainment rate in Io's vicinity should exhibit a ϕ^2 dependence in the present scenario. Such nonlinear behavior would not lead to runaway mass loading at Io however because of the exospheric limit to the sputter-generated coronal mass in the CC regime discussed in the last section. An estimate of the critical value of ϕ involved in this picture will emerge from our quantitative analysis below.

In any model of the Io-torus interaction which involves substantial mass pick-up within a few Ionian radii (Ip and Axford, 1980; Kumar, 1979), the characteristics of the plasma flow past the satellite must be strongly tied to the distribution of the neutral coronal reservoir. The fact that the supply flux of S and O ions from Io required to energize and maintain the torus is comparable to or greater than the torus flux itself suggests that the plasma density in Io's immediate vicinity will be dominated by freshly ionized atoms and molecules. Such a mass injection rate would strongly influence the characteristics of the plasma flow past Io. This observation is consistent with the proposal of Eviatar et al. (1979) that the anomalously low gyrovelocities of torus ions (Bridge et al., 1979) may be due to the retardation of the plasma flow in the source region near Io, perhaps accompanied by mass loading rates in excess of 10^{28} ions s^{-1} ($\sim 10^{11}$ ions cm^{-2} s^{-1}; Eviatar and Siscoe, 1980; Hill, 1980). Even if Io possesses an intrinsic magnetic field (Kivelson et al., 1979), there may be substantial penetration of the energetic plasma to Io's surface (Kivelson, pers. comm.) so that the sputter-related processes discussed here are not excluded. In any event, it is important to understand the structure of that part of the corona extending above the magnetopause. The interesting question arises in this context of the interplay between magnetic field and mass loading effects in determining the dynamics of the plasma flow close to Io.

In their model for a comet-like interaction between Io and the torus, Cloutier et al. (1978) suggest that the ionospheric profile observed by Pioneer 10 (Kliore et al., 1975) could be supported by electron impact on a dynamic, sputter-induced neutral atmosphere, generated through magnetospheric ion bombardment of Io's surface. The requisite neutral densities estimated by these authors for a NH_3 atmosphere and electron energies $\gtrsim 1$ keV considerably exceed the maximal value which we have argued could be maintained through CC type sputtering. However, if one extrapolates their calculations to conditions more in line with our present knowledge of Io's environment, namely, an intense ($\gtrsim 10^{11}$ cm^{-2} s^{-1}), low-energy ($kT \sim 10^5$°K) electron flux and an atmosphere of SO_2 molecules having a much larger ionization cross section than is appropriate for NH_3, then the estimated neutral column density necessary to produce the observed ionosphere is quite consistent with the sputter-generation hypothesis. We must caution though that the interpretation of the Pioneer 10 radio occulation data in terms of an ionospheric profile is itself the subject of some controversy (Matson and Nash, 1981). Departures of Io's atmosphere from a spherical, isothermal structure due to surface temperature variations, volcanic plumes or sputtering effects could significantly influence this interpretation (D. L. Matson, pers. comm.).

The Pioneer 10 observation of Io's ionosphere (Kliore et al., 1975), the discovery of active volcanism (Morabito et al., 1979), the identification of gaseous (Pearl et al., 1979) and condensed (Fanale et al., 1979; Nash and Nelson, 1979) SO_2 on Io, together with the abundance of S and O ions in the plasma torus (Broadfoot et al., 1979) have led several authors (Pearl et al., 1979; Ip and Axford, 1980; Kumar, 1980) to postulate that Io maintains a substantial SO_2 atmosphere on its dayside ($\sim 10^{-7}$ bar $\approx 5 \times 10^{18}$ SO_2 cm^{-2} in the subsolar region), in approximate thermal equilibrium with an SO_2 surface frost. Kumar (1979) however has pointed out that such an atmosphere must be shielded from the circumjovian plasma and continuously replenished in order to be stable. Recently, Matson and Nash (in preparation) have proposed that Io's equilibrium atmosphere is controlled by the subsurface cold-trapping of SO_2 in a very porous regolith. They find that the partial pressure of SO_2 at the subsolar point in this model is only $\sim 10^{-12}$ bar $\approx 5 \times 10^{13}$ SO_2 cm^{-2}. Under these circumstances, a sputter-generated corona could be the dominant component of Io's static atmosphere (as opposed to transients such as volcanic plumes) even on the dayside. In addition, the redeposition of sputtered material would likely be an important mechanism for the transport of SO_2 from its sources toward the polar regions. More significantly, if SO_2 is indeed a major component of Io's surface material, but is present over large regions only as a monolayer adsorbed on other substances, then the sputter erosion of this "skin" could be inordinately effective in altering Io's surface constitution and hence its behavior under sputtering. An erosion rate of 10^{11} SO_2 cm^{-2} s^{-1} (see below) could remove such a monolayer in about an hour. A portion of this material would disappear into subsurface cold-traps, and if not continuously replenished would expose a skin of less volatile material to the ambient plasma. The alterations of coronal mass, composition and structure engendered would be reflected throughout the Io-torus system.

It is true, of course, that wherever volcanic resurfacing proceeds at rates $\gtrsim 5 \times 10^{-4}$ cm yr^{-1} on Io sputter erosion effects are probably negligible compared to the deposition of fresh material (cf. Haff et al., 1979). The globally averaged resurfacing rate may exceed 10^{-1} cm yr^{-1} (T. V. Johnson et al., 1979) but the local variations in this rate are potentially quite large, particularly in the absence of a dense, stable atmosphere. Volcanic plumes could transport material to large distances from the source point, but T. V. Johnson et al. (1979) have estimated a lower limit of only $\sim 3.5 \times 10^{-4}$ cm yr^{-1} for the total depositional rate from such volcanic fallout. From their model of Io's atmosphere Matson and Nash (in preparation) find that thermal diffusion of SO_2 from equatorial latitudes poleward could be responsible for a deposition rate of only $\sim 2 \times 10^9$ SO_2 cm^{-2} s^{-1} in the polar regions. As these authors point out, sputter erosion could easily dominate this deposition flux at high latitudes and perhaps help explain the observed lack of polar caps. Finally, we mention that in the thermal model of Consolmagno (1981), volcanic activity on Io at the levels we see today is a transient phenomenon occurring only 10% of the time, for periods of less than 1.6×10^4 years. Needless to say, sputter erosion processes could be much more significant during the intervening quiescent periods.

The point of view which emerges from the above discussion is that if Io's corona is the source of plasma for the torus, whose ions in turn generate the corona through their sputtering interaction on Io, as is strongly suggested by the presence of the neutral Na, K, and O clouds, then the problems of surficial composition and replenishment, sputtering, atmospheric and coronal structure, ionization-entrainment, the intrinsic magnetic field, plasma flow dynamics, and the density and stability of the torus are complexly intertwined questions which must ultimately be resolved in a self-consistent manner. In the remainder of this section we offer a discussion of only one piece of this solution, namely the characteristics of the coronae one might expect to be involved. Whether or not there exists a thick dayside atmosphere, Io's surface should be exposed to ion impact at least on the nightside, and it is this possibility which we first address.

One way in which an atmosphere might be generated is through the sputtering of a condensed SO_2 layer. The corotating S and O ions composing the Io plasma torus have energies of 16 eV/amu and should interact with the surface through the collisional cascade mechanism, giving rise to an SO_2 sputtered flux spectrum of the type approximated by Eq. (1). The solid curve in Fig. 1 corresponds to this spectrum for an incident flux of 10^{10} ions cm^{-2} s^{-1}, of which 1/3 are S and 2/3 are O. This flux is estimated from the product of the observed torus ion densities (Bridge et al., 1979) and the corotational velocity relative to Io, neglecting possible (but unknown) flow modifications in Io's vicinity. The SO_2 effective sputtering yield, $Y = 23$ molecules/ion, is derived in a manner similar to that described by Haff and Watson (1979) for an atmospheric target, in order to account for the molecular structure of the frost. The dashed curve represents a thermal spike type emission flux. A temperature of 2000°K was chosen for illustrative purposes to be consistent with that assumed by R. E. Johnson et al. (1981) for water ice sputtering. Actually, there is as yet no direct empirical support for such an emission temperature in either H_2O or SO_2 ice, but Seiberling et al. (1980) have at least observed spike temperatures of a similar magnitude in the enhanced sputtering of UF_4.

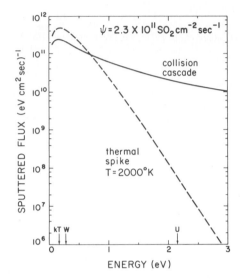

Fig. 1. Energy spectra of molecules ejected from an SO_2 frost on Io's surface, for two possible sputtering mechanisms. ψ is the total emitted flux in both cases. Along the abscissa, kT is the thermal spike energy, W is the surface binding energy and U is the gravitational potential energy of an SO_2 molecule.

The role of TS type sputtering of SO_2 on Io's surface is not clear, however, for although enhanced sputtering of an SO_2 frost has been observed, the dominant phase on Io's surface may be SO_2 adsorbed on sulphur rich grains (Matson and Nash, in preparation). How such material would respond to the impact of fast ions is unknown, but for SO_2 areal coverage much less than a monolayer it is not likely that either sputtering mechanism discussed in this paper would give a molecular yield (normalized to the areal concentration) greater than unity. Beyond this uncertainty in yields, insufficient data is available to allow detailed estimates of the composition, energies and magnitudes of the energetic ion fluxes in Io's neighborhood. Consequently we have chosen to normalize the thermal spike curve of Fig. 1 to the same total emitted SO_2 flux calculated for the collision cascade sputtering regime, i.e., $\psi = 2.3 \times 10^{11}$ cm^{-2} s^{-1}. Note that this is equivalent to a flux of $\sim 10^8$ cm^{-2} s^{-1} fast ions having $Y \approx 2 \times 10^3$. As it happens, this flux value provides a liberal estimate for fast ions in Io's vicinity, being roughly comparable to the 0.54–1.05 MeV ion flux observed at Io by the Voyager 1 LECP instrument (Krimigis *et al.*, 1979a).

The most salient feature of Fig. 1 is the dominance of the cascade spectrum for $E \gtrsim 0.7$ eV. The total flux of molecules with energies greater than the gravitational binding energy, $U = 2.2$ eV, is 5×10^{10} SO_2 cm^{-2} s^{-1} in the cascade case, compared to 1×10^7 SO_2 cm^{-2} s^{-1} for the spike mechanism. These numbers clearly imply the dominance of CC sputtering by the torus ions over TS type sputtering by higher energy magnetospheric ions as a mechanism for the direct removal of an SO_2 surface layer from Io. The mean kinetic energy of these molecules, i.e., those with $E \geq U$, less the potential energy U, is 9.3 eV for the cascade but only 0.2 eV for the thermal spike spectrum. The higher energy here, but not the lower one, is consistent with the high-velocity tails of the source distribution which have been proposed to explain the characteristics of Io's neutral Na cloud (Carlson *et al.*, 1978; Macy and Trafton, 1980). The implication is, again, that a sputter source for the Na, and thus for SO_2, must involve a substantial collision cascade component.

Upon applying Eqs. (6) through (8) to the source fluxes of Fig. 1 we find the coronal density profiles shown in Fig. 2. No corrections have been made here for the anisotropies of the impinging flux, the non-uniformities of the SO_2 distribution on the surface, or the asymmetries of Io's gravitational field. We have also neglected possible modifications due to surface roughness. Although the total emitted flux is the same for both, the column density generated by the CC type sputtering is 30% higher than that which would originate from a TS mechanism, which is a reflection of the much harder energy spectrum in the former case. Both coronae approach the critical exospheric density mentioned

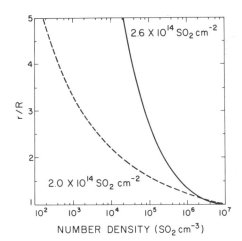

Fig. 2. Number density profiles of the coronae generated by the fluxes of Fig. 1 for the collision cascade (solid curve) and thermal spike (dashed curve) sputtering mechanisms. The profiles are labeled by the corresponding SO_2 column densities. R is the satellite's radius.

earlier. We reiterate that since only 0.2 of the sputtered flux for the CC spectrum and 5×10^{-5} of the TS flux is gravitationally unbound, most of the material represented in the profiles of Fig. 2 eventually falls back on Io's surface, even if we allow for the substantial ionization depletion rate estimated previously. Note that an SO_2 molecule having the typical emission energy of 0.5 eV will take less than $\sim 1.4 \times 10^3$ s to complete its ballistic trajectory, compared to typical lifetimes against ionization of 10^5 s. Thus sputter redeposition of SO_2 on Io may proceed at rates $\sim 10^{11}\,SO_2\,cm^{-2}\,s^{-1}$, implying substantial mass transport. The modification of surface features on the Galilean satellites due to such erosional processes has been discussed by Haff et al. (1979).

Although it is conceivable that other materials on Io's surface, such as alkali sulphides (Nash and Nelson, 1979), might exhibit enhanced sputtering under impact by the higher energy components of the Jovian magnetosphere, we shall forego further speculation about such scenarios and concentrate on the better understood low-energy sputtering processes due to the canonical $10^{10}\,cm^{-2}\,s^{-1}$ S and O torus ion flux. In particular, we consider the sputtering of pure sulphur. The most important distinction between SO_2 frost and elemental S targets is the difference in the binding energies of the atoms or molecules to the surface. For S we have assumed W = 2.9 eV (Gschneidner, 1964), whereas for SO_2 frost our estimates have been based on a value of 0.28 eV for W. This difference is evident in the sputtering yields of these materials. Averaged over the S and O components of the incident flux, $Y_{SO_2} = 23$ but $Y_S = 2.2$. The latter value was estimated on the basis of Sigmund's theory (Sigmund, 1969). Figure 3 compares the consequent S and SO_2 coronae [Eqs. (1), (6) and (7)] for the same impinging ion flux. The lower S column density is consistent with its smaller yield. Because the surface chemical binding energy of S considerably exceeds its gravitational potential energy (U = 1.1 eV) the S energy spectrum is sufficiently hard that 95% of the sputtered flux is gravitationally unbound. Obviously, mass transport due to sputter redeposition on surfaces composed of such refractory material will proceed at a considerably lower rate than it would on ones coated by condensed volatiles.

To this point we have considered only the sputtering of Io's surface. As we have mentioned, however, many workers feel that Io may possess a relatively thick, stable atmosphere in the vicinity of the subsolar point. In any case, there may be significant transient atmospheric components due to episodic volcanic venting. For an atmosphere, or plume, whose column density $\gtrsim 10^{15}\,cm^{-2}$ the exobase is the analogue of the solid's surface in the context of CC sputtering. In Fig. 4 we examine the restructuring of the exosphere due to such a collisional interaction with a torus plasma flux of 10^9 S and O ions $cm^{-2}\,s^{-1}$, for two different atmospheric models (dashed curves).

The solid curves represent the coronae generated by "sputtered", ballistic SO_2 molecules emanating from the exobase (marked by arrows). The source flux distribution

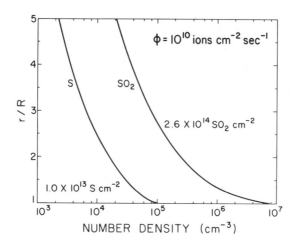

Fig. 3. Coronal profiles which could be generated through sputtering of elemental S or SO₂ frost on Io in the collision cascade regime by a globally uniform incident flux of $\phi = 10^{10}\,\text{cm}^{-2}\,\text{s}^{-1}$ low-energy S and O ions. The corresponding column densities are indicated.

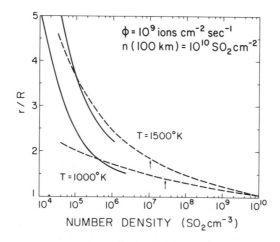

Fig. 4. The dashed curves are the number density profiles for two dense, isothermal SO₂ atmospheres on Io. The solid curves represent the exospheric modifications of these atmospheres due to the collisional interaction with an impinging S and O ion flux of $\phi = 10^9\,\text{ions}\,\text{cm}^{-2}\,\text{s}^{-1}$. The temperatures and exobases (arrows) of the isothermal components are indicated. In each case, the total atmospheric density would be approximated by the sum of the dashed and corresponding solid curves.

adopted here is the same as that used for the low-energy SO₂ frost sputtering estimate (Figs. 2 and 3), reduced by a factor of 10. In effect, we have also set $W = 0$ and cut off the energy spectrum at $E = kT$, where T is the atmospheric temperature. One may observe that since the coronal density is proportional to the incident ion flux ϕ, at least for low ϕ, the influence of the sputter corona will extend to lower altitudes as ϕ increases. For ϕ somewhere between 10^9 and 10^{10} ions cm⁻² s⁻¹, the suprathermal, sputtered particle population will completely dominate the exospheric structure in both models. At even larger ϕ values the dynamics of the energy and mass transport between the exobase region and the lower atmosphere are strongly modified, leading to a non-linear dependence of sputtered mass loss on ϕ, and a departure of the energy spectrum of coronal particles from that given in Eq. (1). The nature of such non-linear atmospheric sputtering, with emphasis on the role which it plays in mass transfer between Io and its torus, will be discussed in a separate paper. It is not accidental that ϕ values of the above magnitude are also sufficient to generate a corona of exospheric mass through the sputtering of an SO_2 frost, and remarks similar to the above apply in this context. It is perhaps coincidental but nevertheless intriguing that the observed torus plasma flux appears to be in this critical range.

V. THE ICY SATELLITES

Beyond the outer edge of the plasma torus associated with Io's orbit, the energetic magnetospheric particle fluxes observed by Voyager 1 and 2 and Pioneer 10 fall off rapidly with increasing radial distance from Jupiter (Krimigis et al., 1979a, b; Trainor et al., 1974). The inference of the details of the composition and energy spectra of these fluxes from the existing data is a delicate matter. Indeed, the data are apparently compatible with multiple interpretations (Eviatar et al., 1981). Nevertheless, provisional estimates of surface sputtering effects on the icy Galilean satellites suggested the possibility of significant erosion rates and atmospheric generation (Lanzerotti et al., 1978; R. E. Johnson et al., 1981), under the assumption that the surface ice on these satellites exhibits the enhanced sputtering yields found in the laboratory (W. L. Brown et al., 1980). Since our primary interest in this section is to compare the coronal density profiles and erosion rates one should expect from the thermal spike mechanism with the collisional cascade type sputtering of water ice, we shall not engage in extensive speculation concerning the magnitudes of the impinging particle fluxes.

The abundance of H_2O on the surfaces of the icy Galilean satellites increases as one approaches Jupiter. Curiously, very little, if any, of the H_2O observed appears to be adsorbed. The areal coverage of water frost is ~ 20–30% on Callisto, $\sim 65\%$ on Ganymede and $\sim 100\%$ on Europa (Clark and McCord, 1980). From these numbers it follows that ice should dominate over refractory surface materials as a source of coronal mass for CC type sputtering on all three satellites. A similar conclusion may or may not apply in the case of the enhanced, electronic sputtering mechanism, contingent upon the behavior of the other surficial components under fast ion bombardment. Among the three satellites, Europa with its essentially complete ice coverage, most intense impinging ion fluxes and least mass should experience the most substantial effects of sputter related processes. Consequently, we shall focus our discussion on the sputtering of water ice on this satellite.

The qualitative characteristics of the H_2O coronal generation process are identical to those described previously for SO_2 on Io. We shall assume a molecular surface binding energy of $W = 0.5$ eV in the case of H_2O ice, as opposed to the $W = 0.28$ eV appropriate for SO_2 frost, but otherwise the energy spectra of sputtered water molecules will have forms very similar to those of Fig. 1. By far the most critical difference between Europa and Io with regard to sputter induced coronae is the much lower value of the gravitational binding energy of an H_2O molecule to Europa, $U = 0.4$ eV, compared to $U = 2.2$ eV for SO_2 on Io. Thus, in the present case U is in the vicinity of the peaks of both energy spectra. One may conceive of the satellite's gravitational field as a mechanism responsible for sorting sputtered particles in the corona according to their kinetic energy, but one which is not sensitive to energies $\gtrsim U$. Therefore one must expect the CC and the TS coronae to be much more similar on Europa than was the case on Io, where the high energy tail of the CC spectrum carried considerable weight.

These remarks are confirmed by the results of the coronal density profile calculations as shown in Fig. 5. The usual assumptions of spherical symmetry have been made here, and we again adopt a sticking probability of unity for reimpacting molecules. The CC corona is appropriate to a flux 3×10^8 cm^{-2} s^{-1} S and O ions having the corotational energy at Europa of 55.3 eV/amu. Eviatar et al. (1981) have suggested that the S and O components of the Jovian magnetosphere (Krimigis et al., 1979b) which originate at Io and diffuse outward from the plasma torus would give rise to a flux of this magnitude at

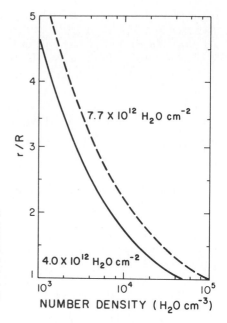

Fig. 5. Calculated H$_2$O coronal profiles on Europa resulting from the sputtering of surficial water ice in either the collision cascade (solid curve) or thermal spike (dashed curve) regime. In both cases the total H$_2$O flux emitted from the surface is assumed to be $\psi = 4.5 \times 10^9$ cm^{-2} s^{-1}, which for the collisional cascade mechanism corresponds to an isotropically incident low-energy sputtering flux of $\phi = 3 \times 10^8$ S and O ions cm^{-2} s^{-1}.

Europa's orbit. Insofar as we have neglected other low energy magnetospheric ion fluxes, the corona of Fig. 5 represents a lower bound. The effective sputtering yield of Y = 15 H$_2$O per ion utilized here was estimated according to Sigmund (1969) and averaged over a presumed 1/3 S and 2/3 O composition of the impinging flux. As before, the TS corona derives from an assumed spike temperature of T = 2000°K, and a total emitted flux of $\psi = 4.5 \times 10^9$ H$_2$O cm^{-2} s^{-1} which is equal to that estimated for the CC mechanism. This emitted flux value is about a factor of 2 lower than that estimated by R. E. Johnson *et al.* (1981) for enhanced ice sputtering due to O$^+$ impact on Europa.

In marked contrast to the situation depicted in Fig. 2, the TS type H$_2$O corona of Fig. 5 dominates over the CC profile at all altitudes. This dominance is reflected in the fact that the column density of the TS corona here is 8×10^{12} H$_2$O cm^{-2} as compared to a value of 4×10^{12} cm^{-2} for the CC case. The essence of the difference between these H$_2$O coronae and the SO$_2$ coronae on Io lies in the fact that in the former case kT ≤ U ≤ W, whereas in the latter situation we had kT ≤ U ≪ W. One obvious consequence of these relations is that a larger fraction of the sputtered H$_2$O is gravitationally unbound to Europa than was the case for SO$_2$ sputtering on Io, although the CC mechanism still dominates in this regard. We find escape fluxes of 3.4×10^9 H$_2$O cm^{-2} s^{-1} for CC and 1.6×10^9 cm^{-2} s^{-1} for the TS type sputtering. These fluxes are within the range of values which Lanzerotti *et al.* (1978) have estimated would produce significant centrifugal distortion of the ambient plasma if injected into the Jovian magnetosphere at Europa's orbit.

On the basis of this example it seems fair to assert that one need not invoke an enhanced, thermal spike type sputtering mechanism by fast magnetospheric ions in order to substantiate the importance of their collisional interaction with the H$_2$O frost on the surfaces of the icy Galilean satellites. The sputtering of this frost according to the well understood collisional cascade mechanism may in fact proceed just as effectively. Although one may certainly envision scenarios in which one or the other process dominates, a more precise, convincing analysis of sputter related processes must await more detailed knowledge of the magnetospheric particle energy spectra and fluxes, the composition of the satellite surfaces, and perhaps most importantly, a better understanding of the physical mechanisms underlying enhanced, electronic sputtering phenomena.

VI. SUMMARY

From the preceding comparison of CC and TS sputter-generated coronae on Io and Europa we have seen that their mass, structure and escape depend critically on the forms assumed for the emission spectra and on the energy scales involved, e.g., kT, W and U. Several workers, however, have simply assumed that all of the material which is sputtered from a satellite's surface directly escapes to the Jovian magnetosphere (Lanzerotti et al., 1978; Pollack and Witteborn, 1980). Without provision for a very efficient supplemental coronal depletion mechanism such an assumption leads to a substantial overestimation of the loss fluxes. Thus, for instance, the direct escape flux of SO_2 from Io proposed by Pollack and Witteborn (1980), based on the enhanced sputtering of a frost by fast heavy ions, would be reduced by a factor of $\sim 10^{-4}$ if the energy spectrum of the sputtered particles were actually of the TS nature discussed in Section IV. Particularly with regard to the SO_2 sputtering on Io, where the gravitational interaction is sufficiently strong that a power law source energy spectrum, as in Eq. (1), produces a corona quite distinct from that generated by a thermal type spectrum [Eq. (3)], it is important to base one's estimation of coronal effects on a sputtered particle energy distribution which is consistent with what one believes to be the operative sputtering mechanism.

Although the CC and TS coronal structures are much more similar on Europa, where $kT \sim U \sim W$, we should point out that our somewhat arbitrary choice of a spike temperature of $T = 2000°K$, so that $kT \sim W/2$, has coincidentally resulted in a perhaps artificially close congruence between the CC and TS energy spectra (Fig. 1). But, in general, the harder CC type spectrum tends to lead to coronae which fall off less rapidly with increasing radial distance. The sputtering of a refractory material, such as elemental S, in the low energy regime results in lower yields but an even harder energy spectrum than is obtained from a volatile frost.

In the absence of a dense, stable atmosphere on Io, it seems probable that the sputtering of surficial SO_2 by the corotating torus plasma could by itself generate a coronal density of a few times $10^{14} SO_2 cm^{-2}$. This column density approaches the maximum value obtainable by ions in the nuclear stopping energy regime. An atmosphere which is already in place due to volcanic venting or the sublimation of an SO_2 frost on Io's dayside would be subject to essentially the same sputtering interaction described here for an SO_2 frost. The unimpeded impact of the corotating torus plasma would thoroughly modify the exospheric structure of such an atmosphere.

Unless one postulates an extremely efficient mechanism for the removal of sputtered, but still gravitationally bound particles from the corona, one must conclude that much if not most of the volatile material sputtered from a satellite is redeposited on its surface. This observation raises several interesting questions concerning the distribution of volatiles on the surfaces of the Galilean satellites. We have already alluded to the role that sputter redeposition of SO_2 on Io may play in the subsurface regolith cold-trapping model of Matson and Nash (in preparation). Lane et al. (1981) have observed what appears to be SO_2 embedded in an H_2O ice matrix on Europa's trailing side. The simplest explanation is that this material derives from S and O ions, implanted from the corotating magnetospheric plasma (Eviatar et al., 1981). We have found, however, that this plasma should be an effective sputtering agent, suggesting the possibility of substantial redistribution of SO_2 to Europa's leading side which would tend to erase the observed leading edge/trailing edge asymmetry. Finally, we mention the paucity of adsorbed H_2O on the icy/rocky satellites. If an enhanced ice sputtering mechanism were the dominant erosional processes operative on these surfaces one might expect it to lead to the deposition of an adsorbed H_2O layer on the rocky material. The resolution of these problems will require a careful analysis of the mechanics of mass redeposition, incorporating not only sputtering but all significant erosional and gardening processes.

Acknowledgments—I am grateful for the critical interest which P. K. Haff has shown in this work. Several of his suggestions, particularly with regard to mass redeposition on planetary surfaces, are reflected in the text. I wish also to thank D. L. Matson and T. A. Tombrello for helpful and encouraging discussions. This work was supported in part by the National Aeronautics and Space Administration (NGR 05-002-333, NAGW-148), the National Science Foundation (PHY 79-23638) and by the Caltech President's Fund.

REFERENCES

Biersack J. P. and Santner E. (1976) Sputtering of potassium chloride by H, He, and Ar ions. *Nucl. Instrum. Meth.* **132**, 229–235.

Bøttiger J., Davis J. A., L'Ecuyer J., Matsunami N., and Ollerhead R. (1980) Erosion of frozen-gas films by MeV ions. *Rad. Effects* **49**, 119–124.

Bridge H. S., Belcher J. W., Lazarus A. J., Sullivan J. D., McNutt R. L., Bagenal F., Scudder J. D., Sittler E. C., Siscoe G. L., Vasyliunas V. M., Goertz C. K., and Yeates C. M. (1979) Plasma observations near Jupiter: initial results from Voyager 1. *Science* **204**, 987–991.

Broadfoot A. L., Belton M. J. S., Takacs P. Z., Sandel B. R., Shemansky D. E., Holdberg J. B., Ajello J. M., Atreya S. K., Donahue T. M., Moos H. W., Bertaux J. L., Blamont J. E., Strobel D. F., McConnell J. C., Dalgarno A., Goody R., and McElroy M. B. (1979) Extreme ultra violet observations from Voyager I encounter with Jupiter. *Science* **204**, 979–982.

Brown R. A. (1974) Optical line emission from Io. In *Exploration of the Planetary System: Proceedings, I.A.U. Symposium No. 65, Poland, 5–9 September 1973.* (A. Woszcyk and C. Iwaniszewska, eds.), p. 527–531. Reidel, Dordrecht-Holland.

Brown R. A. (1981) The Jupiter hot plasma torus: observed electron temperature and energy flows. *Astrophys. J.* In press.

Brown W. L., Augustyniak W. M., Lanzerotti L. J., Johnson R. E., and Evatt R. (1980) Linear and nonlinear processes in the erosion of H_2O ice by fast light ions. *Phys. Rev. Lett.* **45**, 1632–1635.

Carey W. C. and McDonnell J. A. M. (1976) Lunar surface sputter erosion: a Monte-Carlo approach to microcrater erosion and sputter redeposition. *Proc. Lunar Sci. Conf. 7th*, p. 913–926.

Carlson R. W., Matson D. L., Johnson T. V., and Bergstralh J. T. (1978) Sodium D-line emission from Io: comparison of observed and theoretical lines profiles. *Astrophys. J.* **223**, 1082–1086.

Chamberlain J. W. (1963) Planetary coronae and atmospheric evaporation. *Planet. Space Sci.* **11**, 901–960.

Clark R. N. and McCord T. B. (1980) The Galilean satellites: new near-infrared spectral reflectance measurements (0.65–2.5 μm) and a 0.325–5 μm summary. *Icarus* **41**, 323–339.

Cloutier, P. A., Daniell R. E. Jr., Dessler A. J., and Hill T. W. (1978) *Astrophys. Space Sci.* **55**, 93–112.

Consolmagno G. J. (1981) An Io thermal model with intermittent volcanism (abstract). In *Lunar and Planetary Science XII*, p. 175–177. Lunar and Planetary Institute, Houston.

Eviatar A. and Siscoe G. L. (1980) Limit on rotational energy available to excite Jovian aurora. *Geophys. Res. Lett.* **7**, 1085–1088.

Eviatar A., Siscoe G. L., Johnson T. V., and Matson D. L. (1981) Effects of Io ejecta on Europa. *Icarus.* In press.

Eviatar A., Siscoe G. L., and Mekler Y. (1979) Temperature anisotropy of the Jovian sulphur nebula. *Icarus* **39**, 450–458.

Fanale F. P., Brown R. H., Cruikshank D. P., and Clark R. N. (1979) Significance of absorption features in Io's IR reflectance spectrum. *Nature* **280**, 761–763.

Goertz C. K. (1980) Io's interaction with the plasma torus. *J. Geophys. Res.* **85**, 2949–2956.

Griffith J. E., Weller R. A., Seiberling L. E., and Tombrello T. A. (1980) Sputtering of uranium tetrafluoride in the electronic stopping regime. *Rad. Effects* **51**, 223.

Gschneidner K. A. Jr. (1964) Physical properties and interrelationships of metallic and semimetallic elements. *Solid State Phys.* **16**, 275–426.

Haff P. K. (1976) Possible new sputtering mechanism in track registering materials. *Appl. Phys. Lett.* **29**, 473–475.

Haff P. K., Switkowski Z. E., and Tombrello T. A. (1978) Solar-wind sputtering of the Martian atmosphere. *Nature* **272**, 803–804.

Haff P. K. and Watson C. C. (1979) The erosion of planetary and satellite atmospheres by energetic atomic particles. *J. Geophys. Res.* **84**, 8436–8442.

Haff P. K., Watson C. C., and Tombrello T. A. (1979) Ion erosion on the Galilean satellites of Jupiter. *Proc. Lunar Planet. Sci. Conf. 10th*, p. 1685–1699.

Haff P. K., Watson C. C., and Yung Y. L. (1981) Sputter ejection of matter from Io. *J. Geophys. Res.* **86**, 6933–6938.

Hapke B. and Cassidy W. (1978) Is the Moon really as smooth as a billiard ball? Remarks concerning recent models of sputter-fractionation on the lunar surface. *Geophys. Res. Lett.* **5**, 297–300.

Hill T. W. (1980) Corotation lag in Jupiter's magnetosphere: comparison of observation and theory. *Science* **207**, 301–302.

Ip W.-H. and Axford W. I. (1980) A weak interaction model for Io and the Jovian magnetosphere. *Nature* **283**, 180–183.

Johnson R. E., Lanzerotti L. J., Brown W. L., and Armstrong T. P. (1981) Erosion of Galilean satellite surfaces by Jovian magnetosphere particles. *Science* **212**, 1027–1030.

Johnson T. V., Cook A. F. II, Sagan C., and Soderblum L. A. (1979) Volcanic resurfacing rates and implications for volatiles on Io. *Nature* **280**, 746–750.

Kivelson M. G., Slavin J. A., and Southwood D. J. (1979) Magnetospheres of the Galilean satellites. *Science* **205**, 491–493.

Kliore A. J., Fjeldbo G., Seidel B. L., Sweetnam D. N., Sesplaukis T. T., and Woiceshyn P. M. (1975) The atmosphere of Io from Pioneer 10 radio occultation measurements. *Icarus* **24**, 407–410.

Krimigis S. M., Armstrong T. P., Axford W. I., Bostrom C. O., Fan C. Y., Gloeckler G., Lanzerotti L. J., Keath E. P., Zwickl R. D., Carbary J. F., and Hamilton D. C. (1979a) Low-energy charged particle environment at Jupiter: a first look. *Science* **204**, 998–1003.

Krimigis S. M., Armstrong T. P., Axford W. I., Bostrom C. O., Fan C. Y., Gloeckler G., Lanzerotti L. J., Keath E. P., Zwickl R. D., Carbary J. F., and Hamilton D. C. (1979b) Hot plasma environment at Jupiter: Voyager 2 results. *Science* **206**, 977–984.

Kumar S. (1979) The stability of an SO_2 atmosphere on Io. *Nature* **280**, 758–760.

Kumar S. (1980) A model of the SO_2 atmosphere and ionosphere of Io. *Geophys. Res. Lett.* **7**, 9–12.

Lane A. L., Nelson R. M., and Matson D. L. (1981) Europa's UV absorption band: evidence for sulphur implantation? *Nature* **292**, 38–39.

Lanzerotti L. J., Brown W. L., Poate J. M., and Augustyniak W. M. (1978) On the contribution of water products from Galilean satellites to the Jovian magnetosphere. *Geophys. Res. Lett.* **5**, 155–158.

Macy W. and Trafton L. (1980) The distribution of sodium in Io's cloud: implications. *Icarus* **41**, 131–141.

Matson D. L., Johnson T. V., and Fanale F. P. (1974) Sodium D-line emision from Io: sputtering and resonant scattering hypothesis. *Astrophys. J.* **192**, L43–L46.

Matson D. L., Ransford G. A., and Johnson T. V. (1980) Heat flow from Io (J1). *J. Geophys. Res.* **86**, 1664–1672.

Morabito L. A., Synnott S. P., Kupferman P. N., and Collins S. A. (1979) Discovery of currently active extraterrestrial volcanism. *Science* **204**, 972.

Nash D. B. and Nelson R. M. (1979) Spectral evidence for sublimates and adsorbates on Io. *Nature* **280**, 763–766.

Ollerhead R. W., Bøttiger J., Davies J. A., L'Ecuyer J., Haugen H. K., and Matsunami N. (1980) Evidence for a thermal spike mechanism in the erosion of frozen xenon. *Rad. Effects* **49**, 203–212.

Pearl J., Hanel R., Kunde V., Maguire W., Fox K., Gupta S., Ponnamperuma C., and Raulin F. (1979) Identification of gaseous SO_2 and new upper limits for other gases on Io. *Nature* **280**, 755–758.

Pollack J. R. and Witteborn F. C. (1980) Evolution of Io's volatile inventory. *Icarus* **44**, 249–267.

Seiberling L. E., Griffith J. E., and Tombrello T. A. (1980) A thermalized ion explosion model for high energy sputtering and track registration. *Rad. Effects* **52**, 201–210.

Shemansky D. E. (1980a) Radiative cooling efficiencies and predicted spectra of species of the Io plasma torus. *Astrophys. J.* **236**, 1043–1054.

Shemansky D. E. (1980b) Mass loading and diffusion-loss rates of the Io plasma torus. *Astrophys. J.* **242**, 1266–1277.

Sigmund P. (1969) Theory of sputtering I. Sputtering yield of amorphous and polycrystalline targets. *Phys. Rev.* **184**, 383–416.

Thompson D. A. (1980) Non-linear effects in sputtering. In *Proceedings of the Symposium on Sputtering, Perchtoldsdorf/Wien, Austria, 1980 April 28–30* (P. Varga, G. Betz, and F. P. Viehböck, eds.), p. 62–100. Technische Universität Wien, Austria.

Thompson D. A. and Johar S. S. (1980) Sputtering of silver by heavy atomic and molecular ion bombardment. *Nucl. Instrum. Meth.* **170**, 281–285.

Thompson M. W. (1968) II. The energy spectrum of ejected atoms during the high energy sputtering of gold. *Phil. Mag.* **18**, 377–414.

Trafton L. (1975) Detection of a potassium cloud near Io. *Nature* **258**, 690–692.

Trainor J. H., McDonald F. B., Teegarden B. J., Webber W. R., and Roelof E. C. (1974) Energetic particles in the Jovian magnetosphere. *J. Geophys. Res.* **79**, 3600–3613.

Watson C. C. and Haff P. K. (1980) Sputter-induced isotopic fractionation at solid surfaces. *J. Appl. Phys.* **51**, 691–699.

Watson C. C., Haff P. K., and Tombrello T. A. (1980) Solar wind sputtering effects in the atmospheres of Mars and Venus. *Proc. Lunar Planet. Sci. Conf. 11th*, p. 2479–2502.

Watson C. C., Haff P. K., and Tombrello T. A. (1981) Characteristics of a sputter-induced atmosphere on Io (abstract). In *Lunar and Planetary Science XII*, p. 1164–1166. Lunar and Planetary Institute, Houston.

Wehner G. K., Kenknight C., and Rosenburg D. L. (1963) Sputtering rates under solar wind bombardment. *Planet. Space Sci.* **11**, 885–895.

Winters H. F. and Sigmund P. (1974) Sputtering of chemisorbed gas (nitrogen on tungsten) by low-energy ions. *J. Appl. Phys.* **45**, 4760–4766.

Yung, Y. L. and McElroy M. B. (1977) Stability of an oxygen atmosphere on Ganymede. *Icarus* **30**, 97–103.

Tectonic deformation of Galileo Regio and limits to the planetary expansion of Ganymede

William B. McKinnon

Lunar and Planetary Laboratory, University of Arizona, Tucson, Arizona 85721

Abstract—Galileo Regio is the largest and most prominent unit of dark, ancient, heavily cratered terrain on Ganymede. Its major tectonic feature, an arcuate system of rimmed furrows, formed very early, as nearly the entire cratering record post-dates it. Galileo Regio has undergone very little subsequent structural alteration, as opposed to the rest of the planet (i.e., replacement by grooved and smooth terrain). Its survival as an intact lithospheric unit during the era of grooved terrain formation constrains the concomitant expansion of Ganymede, if any, to be less than *one percent* in radius. This limit is derived from an analysis which considers Galileo Regio to be a thin, freely floating, elastic spherical shell on an expanding planet. A comparison of tectonic features (both endogenic and impact produced) on Ganymede and Callisto suggests that the ultimate source that powered the creation of grooved terrain lies in the Ganymedean core.

INTRODUCTION

One of the remarkable discoveries of the Voyager project was the dichotomy of terrain type on Ganymede. Polygonal units of dark, heavily cratered terrain are separated by regions of bright grooved and smooth terrain (Smith *et al.*, 1979a, b). However, no such division is presently recognized on Callisto. While both satellites are similar in many respects (size, mean density, composition, etc.), explaining this difference remains a central goal of studies of the Jovian "planetary system".

The largest section of cratered terrain on Ganymede is Galileo Regio, shown in Fig. 1. Its major characteristics are: (1) antiquity, as determined from measurements of crater density (Smith *et al.*, 1979b); (2) vast scale, approximately 3200 km in diameter (greater than the planetary radius); (3) roughly circular shape; and (4) single dominant tectonic structure, an arcuate system of parallel to subparallel rimmed furrows (Lucchitta, 1980b). These furrows formed very early in the history of Ganymede, predating nearly all of the observed craters (Smith *et al.*, 1979b; Passey and Shoemaker 1981). They are most probably graben formed in an extensional tectonic regime (McKinnon and Melosh, 1980). They follow small circles on Galileo Regio *and* on adjacent units of cratered terrain to the south and west (e.g., Marius Regio) (Passey and Shoemaker, 1981). These traits, comparison to multiringed structures on Callisto, and theoretical analysis indicate that the source of this extension was the prompt collapse of a large impact basin (McKinnon and Melosh, 1980).

Within Galileo Regio the rimmed furrows (or ring graben) are pervasive, apparently extending to the northern limb of the planet in Fig. 1. At its southern margin (Fig. 2), Galileo Regio has been disrupted and replaced (mostly *in situ*) by grooved and smooth terrain (Lucchitta, 1980a). However, at the eastern and western boundaries, the transition is sharp and not gradational. In Fig. 3 the development of the grooved terrain has left the western margin (black arrow) unaffected. The image extends more than 1200 km into the interior, which is similarly undeformed. In general, there is a lack of furrow reactivation across superposed craters. Throughout the establishment of the cratering record, and hence development of the grooved terrain, the furrows have remained essentially inert.

The survival of Galileo Regio as an intact mechanical or lithospheric unit can be used as a test of possible thermal histories. It is widely believed that the ridges and grooves

Fig. 1. Global view of Ganymede and Galileo Regio taken by Voyager 2 at a range of 1.2 million km. Both rimmed furrow system and general circularity of outline are apparent. A frost deposit covers the northern portion of Galileo Regio. (Frame FDS 20608.11.)

that make up grooved terrain represent a form of extensional tectonics. Shoemaker and Passey (1979) once proposed that freezing of ice I in the upper mantle resulted in planetary expansion *and* grooved and smooth terrain formation. Squyres (1980b) invoked the exchange of silicates and water ice during differentiation to achieve the same effect. In this paper the amount of expansion that can be accommodated, without distorting or disrupting Galileo Regio to a greater degree than observed, is determined.

A conservative assumption is that grooved terrain normal faulting completely relieves the tensional stress, generated by planetary expansion, at the periphery of Galileo Regio. Galileo Regio can then be simply modeled as a free lithospheric shell on an expanding planet (Fig. 4). The mechanical analysis presented herein demonstrates that such expansion generates compression in the center and tension near the edge of the shell. This approach *is* conservative, as any lithospheric coupling at the periphery will result in increased tension within Galileo Regio and a *smaller* expansion limit.

The amount of edge tension can be easily estimated. As Ganymede expands, its curvature decreases, and Galileo Regio must relax to align with the new "geoid". The

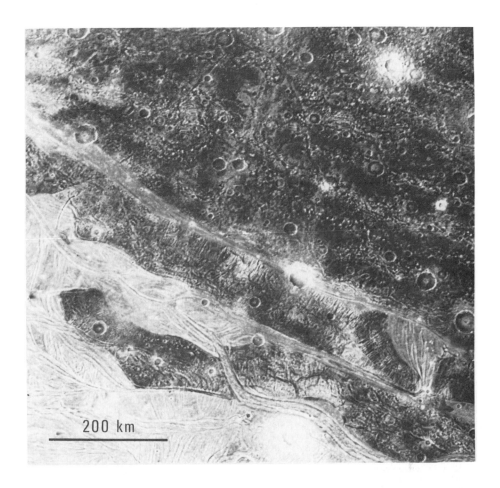

Fig. 2. Voyager 2 image of a section of the southern margin of Galileo Regio (centered near 13°N, 158°W). Bright zones and wedges of grooved and smooth terrain have penetrated the dark cratered terrain, utilizing the pre-existing lines of weakness of the furrow system. Both shear and vertical tectonics are implicated (Lucchitta, 1980a). Transection relationships between the two units are abrupt. North at top. (Frame FDS 20637.08.)

edge of the shell is free and it should deform to minimize membrane stress (Landau and Lifshitz, 1970). In the simplest approximation, the area of the shell will be conserved. As planetary radius grows, the angle Galileo Regio subtends must decrease. Overall, the circumference of the shell increases. Strain, ϵ, is given (in the limit of small change in planetary radius, ΔR) by

$$\epsilon = \left(\frac{1-\cos\theta_0}{\sin\theta_0}\right)^2 \frac{\Delta R}{R}, \tag{1}$$

where θ_0 is the co-latitude at the edge of the shell (see Fig. 4). For $\theta_0 = 35°$, appropriate to Galileo Regio, $\epsilon = 0.1\,\Delta R/R$. The corresponding (tensional) stress in the axial direction, $\sigma_{\phi\phi}$, is

$$\sigma_{\phi\phi} \simeq E\epsilon = 0.1 E\Delta R/R, \tag{2}$$

where E is Young's modulus. This may be compared to the uniform tension which results from expansion of a complete spherical shell, on the order of $E\Delta R/R$. *Both* are

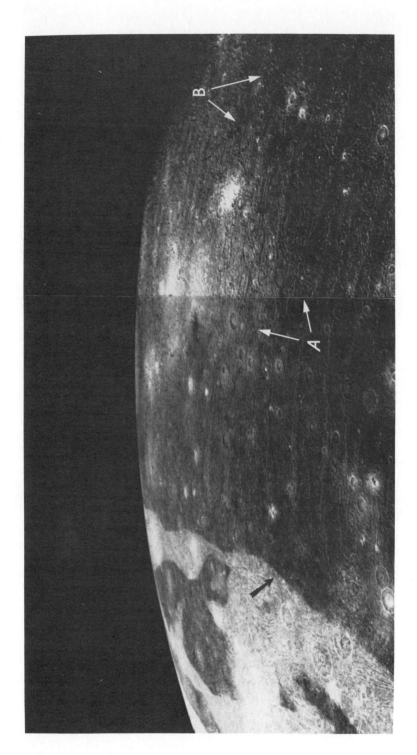

Fig. 3. Western margin (black arrow) and interior of Galileo Regio. The interior presents a remarkably uniform appearance. Only a few curvilinear lineaments (arrows A) and two "fissures" (arrows B) break the pattern of the rimmed furrows and superposed craters. See text for discussion. Increasing brightness towards top (north) is due o frost. View in the unforeshortened direction is approximately 1600 km. [Mosaic of Voyager 2 frames FDS 20636.41 and 20636.44 (centered near 58°N, 160°W)].

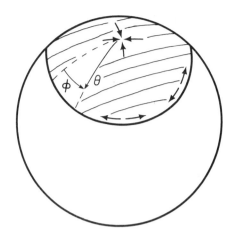

Fig. 4. Schematic Ganymede and Galileo Regio lithospheric shell. The appropriate angle variables define the spherical coordinate system. Coupled arrows denote expansion stress.

potentially large when compared to the tensile strength of an ice "lithosphere". It will be shown that the maximum possible expansion of Ganymede, since the formation of the lithosphere during late heavy bombardment, is limited to *one percent* in radius.

Removing gross planetary expansion as a mechanism for grooved terrain formation prompts examination of other possibilities. Inferences from viscous crater relaxation on Ganymede and multiringed basin structure on Ganymede and Callisto determine that global thermal conditions for generation of grooved terrain *on Callisto* were achieved within the observed cratering record. It is postulated that *regional* thermal processes associated with a silicate-core/ice-mantle boundary led to surface tectonic expression on Ganymede. The implication is that the core size determined during accretion (as examined by Schubert *et al.*, 1981) for both bodies is the discriminating variable.

THICKNESS OF THE LITHOSPHERE

Central to the analysis of this paper is the existence of a lithosphere. The rheology of ice at low (cryogenic) temperatures and moderate stress (less than a few MPa) is unmeasured, but by the theoretical work of Goodman *et al.* (1981), diffusional creep is expected to dominate in this regime. Diffusion is a physically well understood process, and measurement of appropriate material constants allows them to roughly predict the steady-state behavior of ice under these conditions. Fig. 5 is based on their calculations,

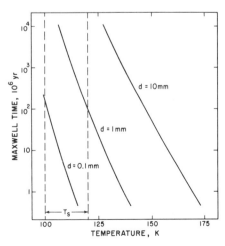

Fig. 5. Diffusional Maxwell time as a function of temperature for polycrystalline ice of three grain sizes (d). Maxwell time is defined as shear stress (normalized by the shear modulus) divided by twice the strain rate. Surface temperatures on Ganymede (T_s) are low enough that the lithosphere can remain elastic over geologic time (see text).

and gives Maxwell time as a function of absolute temperature for polycrystalline ice of various grain sizes. Growth of the lithosphere from subsolidus if not melted ice obviously favors initially large grains. The Maxwell time marks the transition from elastic deformation to solid-state creep.

T_s is a plausible range of surface temperatures. Purves and Pilcher (1980) model the maximum equatorial temperature of Ganymede to be 137 K. Squyres (1980a) estimates ≳150 K, for cratered terrain, from Voyager albedo reduction. Direct measurement by the Voyager infrared instrument (IRIS) gave 145 K for low latitudes (Hanel et al., 1979). These translate into ~100–120 K for an equatorial average, decreasing towards the poles.

It is apparent that Maxwell times are sufficiently long for a lithospheric definition to be valid. The dependence of Maxwell time on temperature is strong, however. Estimates of thermal gradients near the beginning of the preserved cratering record approach several K per km (Passey and Shoemaker, 1981). Hence the lithosphere may be quite thin, possibly only a few kilometers. Independent of theoretical arguments, the survival of furrow topography of a few hundred meters depth implies long term support for between at least 1/2 and 1 MPa (5–10 bars) of deviatoric stress. Stresses below the lithosphere will be rapidly relieved, and this region can be considered fluid on a geological timescale.

The extreme thinness of the lithosphere compared to its radius of curvature simplifies the mathematical analysis considerably. The only important stresses are membrane stresses, and the Galileo Regio spherical cap follows Ganymede's geoid as it expands. Bending (or more precisely, deviation from sphericity) will only make a small contribution within a few e-folding lengths of the cap edge. This length, ℓ, is analogous to the flexural parameter of a plane elastic plate:

$$\ell = \{EH^3/[3(1-\nu^2)(EH/R^2 + \rho g)]\}^{1/4}, \tag{3}$$

where ν is Poisson's ratio, H is lithosphere thickness, ρ is density, and g is surface gravitational acceleration. For $\nu = 0.3$, $E \sim 10^3$–10^4 MPa (10^4–10^5 bars) (Gold, 1977; Schwarz and Weeks, 1977), H = 10 km, $\rho = 1$ g/cm^3, and g = 142 cm/sec^2, $EH/R^2 \ll \rho g$ and $\ell \sim 40$ km. Bending can be safely ignored.

A THIN, ELASTIC SPHERICAL CAP ON AN EXPANDING PLANET

Consider Galileo Regio to be a circular section of a spherical elastic shell. To determine the stresses generated by planetary expansion, superpose (1) a uniform radial extension and (2) an in-plane edge compression such that the sum of the stresses normal to the boundary is zero (Fig. 6).

For uniform *radial* extension the components of the strain tensor, ϵ_{ij}, in a spherical coordinate system with Galileo Regio centered on the pole (Fig. 4), are given directly. $\epsilon_{\theta\phi}$, $\epsilon_{\theta r}$, and $\epsilon_{\phi r}$ vanish by virtue of symmetry. $\epsilon_{\theta\theta}$ and $\epsilon_{\phi\phi}$ are prescribed:

$$\epsilon_{\theta\theta} = \epsilon_{\phi\phi} = \Delta R/R. \tag{4}$$

ϵ_{rr} may be determined by using Hooke's law,

$$\sigma_{ij} = \frac{E}{1+\nu}\left(\epsilon_{ij} + \frac{\nu}{1-2\nu}\epsilon_{kk}\delta_{ij}\right), \tag{5}$$

where σ_{ij} is the stress tensor, δ_{ij} is the Kronecker delta and ϵ_{kk} is the trace of the strain tensor, and by noting that σ_{rr} can be neglected with respect to $\sigma_{\theta\theta}$ and $\sigma_{\phi\phi}$ due to the thinness of the shell:

$$\sigma_{rr} \ll \sigma_{\theta\theta}, \sigma_{\phi\phi}. \tag{6}$$

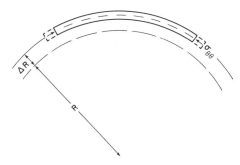

Fig. 6. Components of the elastic solution. A uniform radial extension, ΔR, is followed by an in-plane edge compression, $\sigma_{\theta\theta}$.

Reapplication of Eq. (5) gives

$$\sigma^1_{\theta\theta} = \sigma^1_{\phi\phi} = \frac{E}{1-\nu}\frac{\Delta R}{R}. \tag{7}$$

The superscript denotes that this is the radial extension solution. Note that tension is positive.

From Eq. (7) it is clear that the in-plane compression must be axially symmetric. Therefore, $\sigma_{\theta\phi}$ and $\sigma_{r\phi}$ are identically zero for this solution as well. Both σ_{rr} and $\sigma_{r\theta}$ can be neglected by thinness. Only the membrane stresses $\sigma_{\theta\theta}$ and $\sigma_{\phi\phi}$ remain.

The equilibrium equations reduce to

$$\frac{1}{r}\frac{\partial \sigma_{\theta\theta}}{\partial \theta} + \frac{1}{r}(\sigma_{\theta\theta} - \sigma_{\phi\phi})\operatorname{ctn}\theta = 0. \tag{8}$$

$\sigma_{\theta\theta}$ and $\sigma_{\phi\phi}$ are related to the ϵ_{ij} by Eq. (5), which may be expressed in terms of the displacements, u_i. However, u_ϕ is zero by axial symmetry and we require u_r, averaged over the thickness of the lithosphere, to vanish as well (i.e., no bending). Only u_θ is non-zero, and it is solely a function of θ. Thus, with the usual definitions of the ϵ_{ij} (Malvern, 1969), $\sigma_{\theta\theta}$ and $\sigma_{\phi\phi}$ can be expressed as

$$\sigma_{\theta\theta} = \frac{E}{1-\nu^2}\left(\frac{1}{r}\frac{du_\theta}{d\theta} + \nu\frac{u_\theta}{r}\operatorname{ctn}\theta\right) \tag{9a}$$

$$\sigma_{\phi\phi} = \frac{E}{1-\nu^2}\left(\frac{u_\theta}{r}\operatorname{ctn}\theta + \nu\frac{1}{r}\frac{du_\theta}{d\theta}\right). \tag{9b}$$

Substitution of Eqs. (9) into the equilibrium Eq. (8) yields

$$\frac{d^2 u_\theta}{d\theta^2} + \operatorname{ctn}\theta\,\frac{du_\theta}{d\theta} - u_\theta(\nu + \operatorname{ctn}^2\theta) = 0. \tag{10}$$

Utilizing the standard change of variables, $\cos\theta = x$, transforms the equation to

$$(1-x^2)\frac{d^2 u_\theta}{dx^2} - 2x\frac{du_\theta}{dx} + u_\theta\left[\xi(\xi+1) - \frac{1}{1-x^2}\right] = 0, \tag{11}$$

with

$$\xi = -\frac{1}{2} \pm \sqrt{5/4 - \nu}. \tag{12}$$

This is the associated Legendre equation of degree ξ and order $\sqrt{1}$. For this problem only the solutions of the first kind, $P_\xi^{\pm 1}(x)$, are needed. The solutions of the second kind, $Q_\xi^{\pm 1}(x)$, diverge at $\theta = 0$ and must be dropped. The exterior solutions *are* important, however, in models of the lithospheric fragmentation which accompanies collapse of large transient craters (see Melosh, 1981).

As $P_\xi^{-1}(x)$ is proportional to $P_\xi^1(x)$, and $P_{-\xi-1}^m(x)$ equals $P_\xi^m(x)$ (Abramowitz and Stegun, 1964), the multiplicity of $P_\xi^{\pm 1}(x)$ is reduced to a single function. Measured values of ν for fresh-water and sea ice are about 0.3 (Gold, 1977; Schwarz and Weeks, 1977). Regrettably, $P_{0.47}^1(x)$ is not very tractable, and $\nu = 1/4$ ($\xi = 1/2$) will be adopted. The expression for $P_{1/2}^1(x)$ is especially simple

$$P_{1/2}^1(\cos\theta) = \operatorname{ctn}\theta\left[K\left(\sin^2\frac{\theta}{2}\right) - \left(1 - \tan^2\frac{\theta}{2}\right)E\left(\sin^2\frac{\theta}{2}\right)\right], \tag{13}$$

where $K(m)$ and $E(m)$ are the complete elliptic integrals of the first and second kind

$$K(m) = \int_0^{\pi/2} (1 - m\sin^2\theta)^{-1/2}\,d\theta \tag{14a}$$

$$E(m) = \int_0^{\pi/2} (1 - m\sin^2\theta)^{1/2}\,d\theta. \tag{14b}$$

The stresses are given directly by Eqs. (9)

$$\sigma_{\theta\theta}^2 = -A\left[\left(2 + \csc^2\frac{\theta}{2}\right)K\left(\sin^2\frac{\theta}{2}\right) - 4(1 + \csc^2\theta)E\left(\sin^2\frac{\theta}{2}\right)\right] \tag{15a}$$

$$\sigma_{\phi\phi}^2 = A\left[\left(\csc^2\frac{\theta}{2} - 3\right)K\left(\sin^2\frac{\theta}{2}\right) + 2(3 - 2\csc^2\theta)E\left(\sin^2\frac{\theta}{2}\right)\right], \tag{15b}$$

where A is an undetermined constant and the superscript denotes the in-plane compression solution. A is specified by the boundary condition

$$\sigma_{\theta\theta}^1 + \sigma_{\theta\theta}^2 \bigg|_{\theta=\theta_0} = 0. \tag{16}$$

The total solution is then derived by summing Eqs. (7) and (15)

$$\sigma_{\theta\theta} = \frac{4}{3}\frac{E\Delta R}{R}\left[1 - \frac{\left(2 + \csc^2\frac{\theta}{2}\right)K\left(\sin^2\frac{\theta}{2}\right) - 4(1 + \csc^2\theta)E\left(\sin^2\frac{\theta}{2}\right)}{\left(2 + \csc^2\frac{\theta_0}{2}\right)K\left(\sin^2\frac{\theta_0}{2}\right) - 4(1 + \csc^2\theta_0)E\left(\sin^2\frac{\theta_0}{2}\right)}\right] \tag{17a}$$

$$\sigma_{\phi\phi} = \frac{4}{3}\frac{E\Delta R}{R}\left[1 + \frac{\left(\csc^2\frac{\theta}{2} - 3\right)K\left(\sin^2\frac{\theta}{2}\right) + 2(3 - 2\csc^2\theta)E\left(\sin^2\frac{\theta}{2}\right)}{\left(2 + \csc^2\frac{\theta_0}{2}\right)K\left(\sin^2\frac{\theta_0}{2}\right) - 4(1 + \csc^2\theta_0)E\left(\sin^2\frac{\theta_0}{2}\right)}\right]. \tag{17b}$$

This solution is in agreement (considering typographical error and mathematical convention) with that of Turcotte (1974), who solved directly for the radial pressure,

$$p_r = \frac{H}{R}(\sigma_{\theta\theta} + \sigma_{\phi\phi}), \tag{18}$$

necessary to change the curvature of a lithospheric plate.

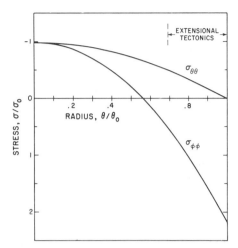

Fig. 7. Plot of normalized expansion stresses, $\sigma_{\theta\theta}$ and $\sigma_{\phi\phi}$, within Galileo Regio as a function of normalized co-latitude. Total expansion pressure is tensional within the region labeled "extensional tectonics", and may be comparable to the lithostatic pressure. (This label is informal, and serves to identify a tectonic regime where both extensional normal faults and tensional rifts are possible.) Ice invariably yields by fracture in the absence of suitable compression.

Equations (17) are plotted for $\theta_0 = 35°$ in Fig. 7. The normalization,

$$\sigma_0 = \frac{E\Delta R}{R}\frac{\theta_0^2}{8}, \tag{19}$$

is derived by expanding Eqs. (17) in powers of θ and θ_0. The colatitudinal stress, $\sigma_{\theta\theta}$, vanishes at the edge as is required and becomes negative (compressional) at the center. The axial stress, $\sigma_{\phi\phi}$, equals $\sigma_{\theta\theta}$ at the center, as it must from symmetry, but becomes positive (tensional) towards the edge. The transition to tension actually occurs at some distance from the periphery. More importantly, in the zone marked extensional tectonics, the total expansion pressure, $\sigma_{\theta\theta} + \sigma_{\phi\phi}$, is tensional. This zone comprises more than 50% of the area of the spherical cap. In the next section these results will be applied to the question of Ganymede's expansion.

DISCUSSION

Planetary expansion

The stresses generated within Galileo Regio by the expansion of Ganymede, if of sufficient magnitude, would cause brittle failure of the elastic lithosphere. Both the shearing stress intensity, $[(\sigma_{\theta\theta}^2 + \sigma_{\phi\phi}^2 - \sigma_{\theta\theta}\sigma_{\phi\phi})/3]^{1/2}$, and the maximum deviatoric stress, $\sigma_{\phi\phi} - \sigma_{\theta\theta}$, are greatest at $\theta = \theta_0$. Thus failure should first manifest itself at the edge of Galileo Regio. Strict application of the faulting criteria of Anderson (1951) predicts strike-slip faults, but it is more likely that the low magnitude of $\sigma_{\theta\theta}$ would lead to normal faults and graben propagating inward from the boundary.

Moreover, the Anderson criteria require a suitably compressive stress state. At the very least, the principal stresses must all be compressive. Lithostatic pressures in thin lithospheres on Ganymede are quite low, and this requirement may not be met. Within the "extensional tectonics" zone of Fig. 7 the potential exists for a general tensional state ($\sigma_{kk} > 0$), even in the presence of lithostatic overburden. Tension fractures would result, forming at the periphery of Galileo Regio and striking (statistically) towards the center.

From Eq. (17b), the tensional stress at the edge is

$$\sigma_{\phi\phi}|_{\theta = \theta_0} = 2.16\sigma_0 \simeq 0.1 E\Delta R/R. \tag{20}$$

Values of Young's modulus for *fresh-water* ice, in the brittle regime, cluster near 10^4 MPa (Gold, 1977). The presence of brine in *sea ice* can reduce this modulus by an order of magnitude (Schwarz and Weeks, 1977), but this is not likely to be relevant at the low temperatures found on Ganymede. The general dependence of E on temperature is probably small (Gold, 1977). Thus for an increase in planetary radius of 1%, ~10 MPa (100 bars) of stress in tension results.

The proper tensile strength to use for the Ganymedean lithosphere is more problematical. Measured brittle tensile strengths of fresh-water ice range between 1 and 2 MPa. Those for sea ice are slightly lower. Tensile (and compressive) strength is often described as increasing with decreasing temperature (i.e., Butkovich, 1959). Yet measurements of these quantities near the melting point are often complicated by dutile contributions to yield. For a given situation, cryogenic temperatures may actually lead to embrittlement and lower strengths, as is apparently the case for the impact experiments of Lange and Ahrens (1981). In addition, these tensile strengths, derived from small samples, can only be considered as upper limits for the tensile strength of a real ice lithosphere, with its attendant inhomogeneities and flaws.

There is the possibility that frozen salts or silicate contaminants may increase the strength of an ice lithosphere, in much the same manner as lithic fragments strengthen concrete. While little data exist to further explore this possibility, the conclusions of this section are unlikely to be affected, as such contamination will also act to increase E.

Even within these uncertainties, a radial expansion of 1% significantly exceeds the tensional strength of a predominantly ice lithosphere. Since the broad circumferential zone of tensional and extensional deformation (i.e., faulting) predicted is not observed, it is concluded that 1% is a reasonable and prudent upper limit for the expansion of Ganymede since the establishment of the furrow pattern. Clearly, stricter limits may be set on the expansion. For more moderate tensions due to smaller radius increases, lithostatic control on fault mechanics will be greater and compressive strength may prove to be the limiting factor. Yet even in this case, the compressive strength of an ice lithosphere probably does not exceed a few MPa.

There are several minor tectonic structures in the central region of Galileo Regio. Most are faint, wavy lineaments (Fig. 3, arrows A). As they predate the furrows they are irrelevant to the present discussion. Of greater significance are two independent fissure-like lineaments (arrows B) which formed later. Based on relative degradation and superposed craters, they appear to be of different ages. They are unrelated to the expansion stress solution and probably represent brief tensional episodes *in the interior* of Galileo Regio. Such additional tension only reinforces 1% as being an upper limit.

It could be argued that the deformation observed at the southern margin of Galileo Regio (Fig. 2) satisfies the model predictions. However, the principal uncertainties in this problem are the mechanical properties of the icy lithosphere, and defining a slightly smaller Galileo Regio to exclude the southern tectonic zone will not change the conclusions herein. Another concern is whether deformation of Galileo Regio might be accomplished on a widespread system of faults not resolved by Voyager, or by equally unresolvable movement on the furrow system. But it would be very unusual for *major* faults not to form in response to the stress levels estimated. And as evidence for furrow reactivation is meager, this hypothesis is considered very unlikely. It should be noted that tectonism is clearly identifiable elsewhere on Ganymede (i.e., in the grooved terrain).

The 1% limit, which is conservative, constrains possible thermal histories of Ganymede. The radial expansion accompanying global differentiation (~3.5%) calculated by Squyres (1980a) is ruled out. Such differentiation would have to have taken place prior to lithosphere formation, if at all. On the other hand, some expansion *can* be accommodated. Thermal models of Schubert *et al.*, (1981) predict an incompletely differentiated Ganymede. The slower differentiation of an ice-silicate lower mantle and continued core growth during the era of grooved terrain formation would result in moderate planetary expansion, more compatible with this limit.

Grooved terrain

This analysis does not imply that Ganymede underwent *zero* expansion during grooved terrain formation. Indeed, the near circularity of Galileo Regio could be an expression of global expansion, representing a minimum stress state for a lithospheric plate. However, the volume increase is surely limited, and additional mechanisms should be sought to explain grooved and smooth terrain. Faulting greatly relieves expansion stress, and if the grooved regions *are* fault sets, then their vast extent implies that planetary expansion played a diminishing role as more of the terrain formed. That is, the more faulted a lithosphere becomes, the less tension can be generated by a given increment of expansion. Nor is the *local* clustering of grooves obviously explainable in terms of *global* expansion.

Cassen et al. (1980) have suggested that when the lithosphere is thin enough, solid-state convection in a water ice mantle naturally leads to grooved terrain formation. This is difficult to accept as Galileo Regio has survived the ongoing convective activity in the mantle beneath it. Such durability would be possible if the horizontal scale of convection were large (≥ 3000 km). While such a scale cannot be totally ruled out, both the phase behavior of water ice and the high heat flows estimated for the early history of Galileo Regio from crater relaxation studies (Passey and Shoemaker, 1981) argue against it.

A comparison to Callisto also leads to trouble with this hypothesis. Interpretation of the furrows in Galileo Regio as ring graben sets a geometric lower limit of 10 km for lithosphere thickness (at the time of formation). Similar morphology for portions of the Valhalla multiringed structure (Callisto) yields a lower limit of 15–20 km (McKinnon and Melosh, 1980). While these "lithospheres" refer to the time and stress scale of basin collapse, their relative sizes should carry over to the elastic lithosphere. The relevant areas of the two ring structures are sufficiently far from their respective basin centers that thermal structure plays the dominant role in determining the thickness of the lithosphere. Thus, although the elastic lithospheres of Ganymede and Callisto were equal in thickness to within a factor of two, no grooved terrain is observed on Callisto.

Perhaps Callisto's thicker lithosphere and lower heat flow (by about a factor of two) were sufficient to prevent grooved terrain formation. However, it is important to remember that Valhalla postdates approximately 2/3 to 3/4 of the cratering record (Smith *et al.*, 1979b; Passey and Shoemaker, 1981), and the rimmed furrows predate the grooved terrain. Presumably, heat flow and lithosphere thickness, at an earlier era still preserved on Callisto, were *equal* to that of Ganymede at some point during grooved terrain formation. This is especially true if the rapid decline in heat flow on Ganymede calculated by Passey and Shoemaker is considered.

It is worth noting that as the furrow system is younger than Valhalla (McKinnon and Melosh, 1980), the estimated ratio of heat flows (≥ 1.5–2.0) exceeds that given by radionuclide mass alone (1.26). Additional energy sources for Ganymede include accretional heat and tidal dissipation (Cassen *et al.*, 1981).

Grooved terrain formation appears to be a distinct episode on Ganymede, not directly attributable to gross planetary expansion or "normal" mantle convection. While globally distributed, it is a regional process.

Thermal models and hypotheses concerning grooved terrain have progressed from mantle freezing, to global differentiation, to core/lower mantle differentiation. It is proposed that the core itself is the ultimate arbiter of grooved terrain. In Fig. 8 a highly evolved version of the Schubert *et al.* model for Ganymede is shown. The core has not only differentiated, but is convecting and indirectly causing grooved terrain formation.

The segregation of hydrous and hydrated silicates in the core allows for dehydration and eventual convective turnover. Partial melting and magma accumulation may occur depending critically on the rheology of water-rich silicates. Even in the absence of melt, violent hydrothermal activity, regionally distributed and intermittent in time, is expected at the core/mantle boundary. No details are given here concerning the transfer of heat

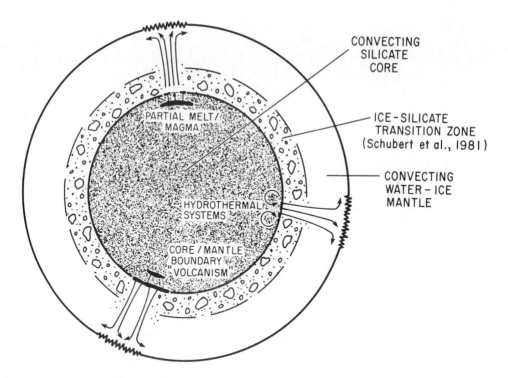

Fig. 8. Hypothetical Ganymede, based on the model of Schubert *et al.* (1981). Could grooved terrain tectonics be due to super-adiabatic mantle currents?

through the mantle (plumes, layered convection cells, etc.) or the actual mechanism of grooved terrain (cf. Fink and Fletcher, 1981). The speculative nature of this scenario is recognized, and much further work needs to be done. However, core/mantle interaction is bound to affect the surface in a profound way.

It is hypothesized that Callisto's core was never large enough to achieve the thermal state in Fig. 8. Core size is strongly "determined" by accretion (Schubert *et al.*, 1981), i.e., whether slow or fast, homogeneous or inhomogeneous, etc. and mechanisms of core growth do favor Ganymede. But it is not clear whether Callisto's lower "silicate" mass, if centrally condensed, would not have become active in the manner envisaged here. This remains a topic for future research.

The ultimate resolution of the nature of grooved terrain and tectonics on Ganymede may have to await the increased detail of images from the Galileo Orbiter. Irregardless, an observational limit on possible radius changes has been set, and must be incorporated into the theoretical framework of our understanding of the satellite.

Acknowledgments—I thank J. Spencer, whose original observations of Ganymede prompted this analysis, and P. Maloney and D. J. Stevenson for helpful discussions. This work was supported by NASA grant NSG-7146.

REFERENCES

Abramowitz M. and Stegun I. A. (1964) *Handbook of Mathematical Functions.* U.S. Government Printing Office, Washington, D.C. 1046 pp.
Anderson E. M. (1951) *The Dynamics of Faulting.* Oliver and Boyd, Edinburgh. 206 pp.
Butkovich T. R. (1959) Mechanical properties of ice. *Quart. Colo. Sch. Mines* **54**, 349–360.
Cassen P., Peale S. J., and Reynolds R. T. (1980) On the comparative evolution of Ganymede and Callisto. *Icarus* **41**, 232–239.
Cassen P. M., Peale S. J., and Reynolds R. T. (1981) Structure and thermal evolution of the Galilean satellites. In *The Satellites of Jupiter* (D. Morrison, ed.). Univ. Arizona Press, Tucson. In press.

Fink J. H. and Fletcher R. C. (1981) Variations in thickness of Ganymede's lithosphere determined by spacing of lineations (abstract). In *Lunar and Planetary Science XII*, p. 277–279. Lunar and Planetary Institute, Houston.

Gold L. W. (1977) Engineering properties of fresh-water ice. *J. Glaciol.* **19**, 197–212.

Goodman D. J., Frost H. J., and Ahsby M. F. (1981) The plasticity of polycrystalline ice. *Phil. Mag.* **A43**, 665–695.

Hanel R., Courath B., Flasar M., Herath L., Kunde V., Lowman P., Maguire W., Pearl J., Pirraglia J., Samuelson R., Gautier D., Gierasch P., Hoen L., Kuman S., and Ponnamperuma C. (1979) Infrared observations of the Jovian system from Voyager 2. *Science* **206**, 952–956.

Landau L. D. and Lifshitz E. M. (1970) *Theory of Elasticity*. Pergamon, Oxford. 165 pp.

Lange M. A. and Ahrens T. J. (1981) Impact experiments in low-temperature ice (abstract). In *Lunar and Planetary Science XII*, p. 592–594. Lunar and Planetary Institute, Houston.

Lucchitta B. K. (1980a) Grooved terrain on Ganymede. *Icarus* **44**, 481–501.

Lucchitta B. K. (1980b) Observations on Ganymede I: Cratered terrain. Paper presented at IAU Colloquium No. 57, "The Satellites of Jupiter", Kailua-Kona, Hawaii, 13–16 May 1980.

Malvern L. E. (1969) *Introduction to the Mechanics of a Continuous Medium*. Prentice-Hall, Englewood Cliffs, New Jersey. 713 pp.

McKinnon W. B. and Melosh H. J. (1980) Evolution of planetary lithospheres: Evidence from multiringed structures on Ganymede and Callisto. *Icarus* **44**, 454–471.

Melosh H. J. (1981) A simple mechanical model of Valhalla basin, Callisto. *J. Geophys. Res.* In press.

Passey Q. R. and Shoemaker E. M. (1981) Craters and basins on Ganymede and Callisto: Morphological indicators of crustal evolution. In *The Satellites of Jupiter* (D. Morrison, ed.). Univ. Arizona Press, Tucson. In press.

Purves N. G. and Pilcher C. B. (1980) Thermal migration of water on the Galilean satellites. *Icarus* **43**, 51–55.

Schubert G., Stevenson D. J., and Ellsworth K. (1981) Internal structures of the Galilean satellites. *Icarus*. In press.

Schwarz J. and Weeks W. F. (1977) Engineering properties of sea ice. *J. Glaciol.* **19**, 499–531.

Shoemaker E. M. and Passey Q. R. (1979) Tectonic history of Ganymede (abstract). *EOS (Trans Amer. Geophys. Union)*, **60**, 869.

Smith B. A., Soderblom L. A., Johnson T. V., Ingersoll A. P., Collins S. A., Shoemaker E. M., Hunt G. E., Masursky H., Carr M. H., Davies M. E., Cook A. F., Boyce J., Danielson G. E., Owen T., Sagan C., Beebe R. F., Veverka J., Strom R. G., McCauley J. F., Morrison D., Briggs G. A., and Suomi V. E. (1979a) The Jupiter system through the eyes of Voyager 1. *Science* **204**, 13–32.

Smith B. A., Soderblom L. A., Beebe R., Boyce J., Briggs G., Carr M., Collins S. A., Cook A. F., Danielson G. E., Davies M. E., Hunt G. E., Ingersoll A., Johnson T. V., Masursky H., McCauley J., Morrison D., Owen T., Sagan C., Shoemaker E. M., Strom R., Suomi V. E., and Veverka J. (1979b) The Galilean satellites and Jupiter: Voyager 2 imaging science results. *Science* **206**, 927–950.

Squyres S. W. (1980a) Surface temperatures and retention of H_2O frost on Ganymede and Callisto. *Icarus* **44**, 502–510.

Squyres S. W. (1980b) Volume changes in Ganymede and Callisto and the origin of grooved terrain. *Geophys. Res. Lett.* **7**, 593–596.

Turcotte D. L. (1974) Membrane tectonics. *Geophys. J. Roy. Astron. Soc.* **36**, 33–42.

Dark-ray craters on Ganymede

James Conca

California Institute of Technology, Pasadena, California 91125

Abstract—The distribution and ejecta ray pattern of dark-ray craters on Ganymede are found to show no systematic variation with terrain type. However, the distribution is correlated with latitude and longitude, with dark-ray craters strongly concentrated on Ganymede's trailing hemisphere increasing toward the antapex of orbital motion (antapical distribution). Ablation of H_2O from a dirty-ice layer and concentration of a surface lag deposit composed of silicate material is investigated as a model for the formation of the low-albedo deposits of dark-ray craters. Sputtering of H_2O from the surface by interaction with charged particles in Jupiter's magnetosphere is determined to be sufficient as an ablation mechanism. The non-uniform distribution of dark-ray craters may be due to the corotating magnetosphere or, more likely, to differential gardening rates. The formation of a lag deposit resulting in an albedo change can occur by sputtering on Ganymede in a relatively short time (10^5 to 10^7 years by the model presented in this paper).

INTRODUCTION

The surface of Ganymede as revealed by the Voyager images contains a great many ray craters. A significant number of these possess dark rays. The criteria for assigning a crater to the category of a dark-ray crater is the presence of a low-albedo deposit with distinct dark rays surrounding an impact crater (Fig. 1). Within the Voyager coverage of Ganymede only forty-three dark-ray craters can be identified down to the limit of resolution (1.70 km/pixel). Ten dark-ray craters reside on old cratered terrain, twenty-eight on grooved terrain, one on the boundary between old cratered and grooved terrain and the terrain type of four are uncertain because of poor resolution or because the underlying terrain is obscured by extensive ejecta deposits. However, the number of dark-ray craters with diameters greater than five kilometers is the same on both terrain types. Approximately 57% of Ganymede's surface is grooved terrain. The greater number of dark-ray craters on grooved terrain below five kilometers reflects the greater ease of identifying small, low-albedo patches on the high-albedo grooved terrain than on the lower-albedo old cratered terrain. Therefore, the distribution with respect to terrain type appears to be random. Also, the ejecta and/or ray pattern and geometric albedo of dark-ray craters do not vary systematically with terrain type.

Distribution and albedo of dark-ray craters with respect to latitude and longitude is not uniform. No dark-ray craters were observed at latitudes higher than 60° and only four are seen above 40° latitude (where the polar shroud becomes evident). Figure 2(a) shows the position of each dark-ray crater on Ganymede. In that Ganymede is in synchronous orbit around Jupiter, there are constantly leading and trailing points (apex and antapex, respectively) as shown in Fig. 2. It is clear that the dark-ray craters on Ganymede are concentrated in the trailing hemisphere. The few dark-ray craters that do occur on the leading hemisphere have higher albedos and less well-developed low-albedo deposits and rays than those occurring on the trailing hemisphere. The number of dark-ray craters per million square kilometers as a function of distance from the apex is shown in Fig. 3, and illustrates that the number of dark-ray craters increases with increasing distance from the apex. Note that the distribution around the areas of good resolution is not symmetric; even within the poor resolution areas between 120° and 180° from the apex the number of dark-ray craters continues to increase, strongly indicating that the distribution is a function of longitude and that the dark-ray craters are concentrated on the trailing

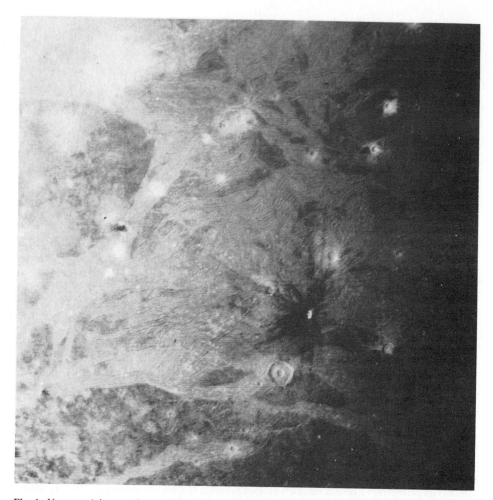

Fig. 1. Voyager 1 image of a well-developed dark-ray crater shown slightly to the right and down from center of image. Crater is located at 7° latitude and 332° longitude. Voyager picture number is 0837J1+000. Crater diameter is 21 km and the longest continuous dark ray extends over 400 km from crater rim.

hemisphere. Also, the geometric albedo decreases, while the areal extent of the low-albedo deposits surrounding dark-ray craters increases, with increasing distance from the apex. These observations suggest that the formation (or continuation) of dark-ray craters on Ganymede is controlled by a process which is longitudinally dependent.

It should be stated here that because the number of dark-ray craters is only forty-three, there is a significant probability (almost 30%) that their distribution between the hemispheres is random. However, the longitudinal variations of the albedo and areal extent of dark-ray craters coupled with their distribution suggests that the distribution is not totally random.

It is uncertain whether the low-albedo deposits of dark-ray craters are due to substrate characteristics or to characteristics of the impacting projectile. Hartmann (1980) attributes the low-albedo deposits of dark-ray craters to substrate characteristics and states that the fact that dark-ray craters are concentrated into a restricted diameter range is consistent with this hypothesis. However, the lower size limit of dark-ray craters on Ganymede probably is restricted only by the resolution of the Voyager images. Also, since the number of dark-ray craters is so few, their size range may not reflect their mechanism of formation. In several cases on Ganymede bright-ray craters are superim-

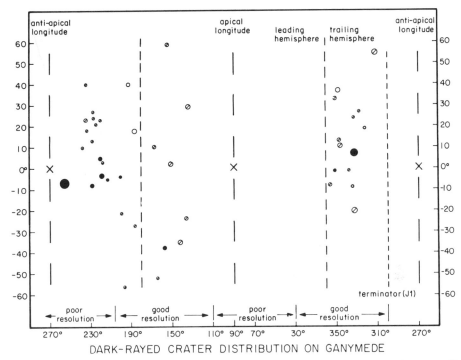

Fig. 2. (a) Distribution of dark-ray craters on Ganymede showing strong concentration on trailing hemisphere.

Fig. 2. (b) Explanations of symbols used in Fig. 2(a) showing crater size range and gross albedo range.

posed upon, underlie or are adjacent to dark-ray craters of similar sizes (e.g., Voyager image numbers 0389J2 − 001, 0392J2 − 001 and 0923J1 + 000). If both types of craters are sampling the substrate to similar depths, then it is unclear how the substrate characteristics alone could be responsible for the development of dark-ray craters. Also, dark-ray craters show no variations with apparently different substrates, i.e., between old cratered terrain and grooved terrain. These observations indicate that projectile characteristics may play a dominant role in the formation of dark-ray craters.

DISCUSSION

On icy bodies the ejecta from an impact will be a mixture of ice and projectile contamination, hereafter referred to as a dirty-ice layer. If there is ablation of H_2O from the surface of the dirty-ice layer, then the projectile contamination will be concentrated as a surface lag deposit with the projectile contaminant to ice ratio increasing through time. The effect of the dirty-ice layer on the albedo of the surface will depend on the ablation rate, projectile material and amount of contaminant.

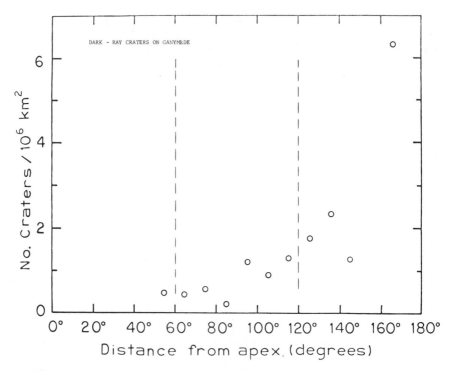

Fig. 3. The number of dark-ray craters on Ganymede per million square kilometers as a function of distance from the apex. The number of dark-ray craters increases dramatically towards the antapex of orbital motion in the trailing hemisphere, and is clearly not a resolution effect.

There are two general processes for ablation of H_2O from Ganymede's surface: interactions of H_2O with photons and with charged particles. Both processes operate on Ganymede. Purves and Pilcher (1980) discuss thermal migration of H_2O on the Galilean satellites and have determined that the rates of removal of ice from Ganymede's surface by this process can be almost six meters in 10^9 years at the equator. (See also Squyres 1980). However, according to their model thermal migration of H_2O is highly dependent on latitude. Ice removal by this process decreases sharply moving away from the equator and should not occur at latitudes higher than 20°. Since 51% of the dark-ray craters on Ganymede occur at latitudes higher than 20°, and up to 59°, other ablation processes must be operating. Significantly, the dark-ray craters occuring at latitudes below 20° have better developed dark-rays and average lower albedos than those at higher latitudes. Thus, a combination of ablation processes may be operating at lower latitudes.

The second important ablation process operating on Ganymede is sputtering. Sputtering of H_2O by charged particles is as efficient in ice removal as is thermal migration of H_2O. Sputtering of Ganymede's surface by 100 keV protons alone can remove approximately one meter of ice in 10^9 years. Sputtering of H_2O from the surface by charged particles varies with longitude because Jupiter's magnetic field is corotating with Jupiter; at Ganymede's orbit, the magnetic field is travelling 177 km/sec faster than Ganymede and in the same direction. The charged particles within the magnetosphere are, thus, overtaking Ganymede and impinging upon the trailing hemisphere at a much greater degree than on the leading hemisphere.

Although sputtering by ions with corotational energies (up to about 1 keV) will be more prevalent on Ganymede's trailing hemisphere, ions with higher energies (about 100 keV) have higher sputtering yields and uniformly sputter the surface. The slight asymmetry in sputtering due to the corotation may not be enough to cause the asymmetrical distribution of dark-ray craters. More likely is the possibility that the distribution is a result of differences in cratering rates between the apex and antapex of motion of Ganymede, resulting in a difference in gardening rates with longitude. According to Shoemaker and Wolfe (1981) the ratio of the cratering rate at the apex to the cratering rate at the antapex is 9.6. Therefore, either gardening at the apex never allows sufficient lag deposits to form, or the ray-retention ages for dark-ray craters near the apex is so short that they are not observed.

Gardening of the surface by micrometeorite impact competes with any ablation process, and the formation of a surface lag deposit, by ablation, will not take place unless the ablation rate exceeds the gardening rate. To determine the ablation rate for the sputtering process consider a dirty-ice layer formed as an ejecta blanket following the impact into an icy surface of a silicate-containing projectile. The low-albedo contaminant in the dirty-ice layer is composed of silicate particles that were not vaporized upon impact and, for simplicity, are assumed to be spherical with an average particle diameter, D. Let the fractional atomic number concentration of the silicate particles in the dirty-ice layer be given by n_m where n_m varies from zero for a pure ice layer to one for a pure silicate layer. If for a unit area of the ejecta blanket, A, $A_m(t)$ and $A_i(t)$ denote the fraction of the surface area covered by silicate particles and by ice, respectively as a function of time, then:

$$A = A_m(t) + A_i(t).$$

If area A is being bombarded by charged particles (protons in the simplest case) and R_m and R_i refer to the fluxes of sputtered silicate and ice particles, respectively, from the surface, then according to Haff (1980):

$$A_m(t) = C + Be^{-t/T} \qquad (1)$$

where

$$C = \frac{n_m}{R_m/R_i(1-n_m) + n_m} \quad \text{and} \quad B = A_m(0) - C$$

T is the steady state time, i.e., the time required to form a stable surface with respect to sputtering. The steady state time is given by:

$$T = \frac{pD}{R_i} \cdot \frac{1}{R_m/R_i(1-n_m) + n_m} \qquad (2)$$

where p is the number density of H_2O per unit volume. For ice $p = 3.46 \times 10^{22}$ molecules per cm^3. From experiment (Brown et al., 1980) and from the flux of 100 keV protons at Ganymede's orbit, $R_i \cong 10^8$ molecules $cm^{-2} sec^{-1}$ and $R_m/R_i = 0.001$.

There are no reliable constraints on the values of n_m and D, but for hypervelocity impacts reasonable values for D range from below one micron to about one hundred microns, and for n_m from 0.00001 to 0.01. A model-dependent approximation for n_m can be obtained using the geometric albedo of the surface. The albedo of the average low-albedo deposits of dark-ray craters is 0.120, and the albedo of a fresh bright-ray crater is 0.358. If the albedo of 0.120 is the result of sputtering a layer composed of a mixture of material of albedo 0.358 and carbonaceous chondrite material of albedo 0.030 as contaminant, then $A_m = 0.72$ and $n_m = 0.0026$. If we choose D to be ten microns, then the time required to form a surface layer, 72% of whose surface area is covered by contaminant, is $T = 3.05 \times 10^6$ years. Figure 4 shows a family of curves for $A_m(t)$ using selected values of D and n_m (Table 1). $A_m(t)$, the fraction of the surface area covered by contaminant, is plotted as a function of time after impact. It can be seen from Fig. 4 and Table 1 that the time, T, required to form a lag deposit by sputtering is most sensitive to the particle size, D; the smaller the particle diameter, the faster the process occurs. However, the fraction of the surface area covered is dependent primarily upon the initial contaminant concentration, n_m. Thus, for values of n_m much less than 0.0001 the surface coverage never reaches a significant amount regardless of the size of the particles or the length of time sputtered. Also, as D becomes less than one micron the ability of the surface layer to lower the albedo becomes less effective even though the layer develops more rapidly. On the other hand, as D becomes large the time required to reach the steady state coverage becomes large and other ablation or erosional processes become important. Since both particle size and initial contaminant concentration depend upon impact velocity, it may be that only those impacts within a certain velocity range (allowing n_m and A_m to be large enough to cause an albedo change and allowing D to be small enough to cause the process to occur over a reasonable time period) will ever develop into dark-ray craters by sputtering. If a rare velocity component of the impacting population is required to form a dark-ray crater, this may explain the rarity of dark-ray craters (less than 1% of all ray

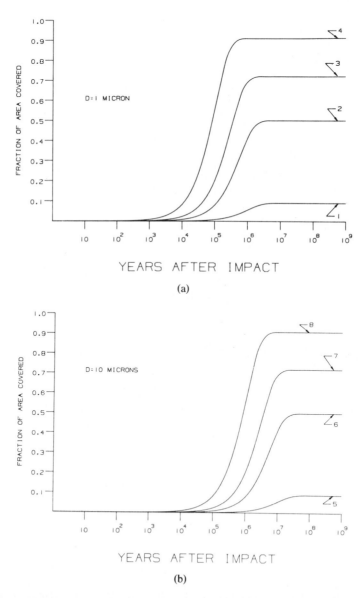

Fig. 4. (a) Evolution of the surface coverage by a sputtered lag deposit as a function of time after emplacement (impact) for an average particle size of one micron and the range of values for n_m from Table 1. (b) Same values for n_m as in (a) but using an average particle size of ten microns.

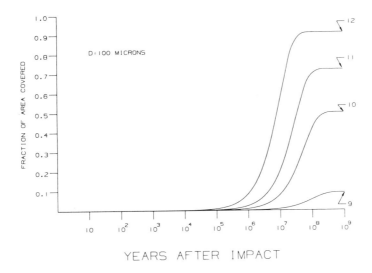

Fig. 4. (c) Same values for n_m as in (a) but using an average particle size of one hundred microns.

Table 1. Sputtering rates for different values of particle size, D, and initial contaminant concentration, n_m. Function numbers refer to the functions shown in Fig. 4.

$$A_m(t) = C + Be^{-t/T} \quad \text{where} \quad B = A_m(0) - C \cong -C$$

$$C = \frac{n_m}{R_m/R_i(1-n_m) + n_m}$$

$$T = \frac{pD}{R_i} \cdot \frac{1}{R_m/R_i(1-n_m) + n_m}$$

$R_i \cong 10^8$ molecules cm^{-2} sec^{-1} $R_m/R_i \cong 0.001$
$p = 3.46 \times 10^{22}$ molecules cm^{-3}

function number	D (microns)	n_m	C	T (years)
1	1	0.0001	0.09	1.05×10^6
2	1	0.0010	0.50	5.77×10^5
3	1	0.0026	0.72	3.05×10^5
4	1	0.0100	0.91	1.05×10^5
5	10	0.0001	0.09	1.05×10^7
6	10	0.0010	0.50	5.77×10^6
7	10	0.0026	0.72	3.05×10^6
8	10	0.0100	0.91	1.05×10^6
9	100	0.0001	0.09	1.05×10^8
10	100	0.0010	0.50	5.77×10^7
11	100	0.0026	0.72	3.05×10^7
12	100	0.0100	0.91	1.05×10^7

craters) without invoking any compositionally rare component of the impact population or of the substrate.

In general, the values for D and n_m will not be the same throughout any single ejecta deposit but will vary as a function of position within the deposit. The sputtering relationships in Eqs. (1) and (2) and in Fig. 4 must be applied separately to sufficiently small areas where the values of D and n_m are relatively constant. Thus, different areas of the same ejecta deposit will have different rates of buildup of the lag deposit. Also, the departure of the contaminant particles from perfect spheres has the same effect as increasing n_m.

Figure 4 also shows that the time required to form dark-ray craters is relatively short (10^5 to 10^7 years) compared to the rate of ray destruction by gardening, suggesting that dark-ray craters become dark early and remain dark for the rest of their history as ray craters.

The above discussion has only involved 100 keV protons for two reasons. First, in experiments with ice films by Brown *et al.* (1980), sputtering yields were seen to peak at about 100 keV for protons, and to fall off to either side of 100 keV (although significantly less sharply to the lower energy side). Therefore the above discussion has centered about the most important energies for sputtering. However, on Ganymede the sputtering rate will be the result of sputtering by all protons of energies with significant sputtering yields, i.e., between about 1 keV and 1000 keV. Sputtering will also be caused by heavier ions. But because R_i is sensitive to the ionic species, the sputtering rates due to sputtering by heavier ions will be different than for protons of the same energy. Using 0^+ ions instead of protons increases R_i by two orders of magnitude, keeping R_m/R_i approximately the same (Sigmund, 1969). Since T is inversely proportional to R_i, sputtering by 0^+ ions would decrease T by the same factor. According to Belcher *et al.* (1980) the flux of 0^+ ions is of the same order of magnitude as that of protons near Ganymede's orbit, making T much less than is shown in Fig. 4. Unfortunately, determination of the sputtering rate by integrating over all energies for all ions in Ganymede's vicinity is beyond the scope of this paper. However, the discussion of 100 keV protons is an illustrative example, and the results shown in Fig. 4 are indeed the worst case situations for sputtering on Ganymede and the values for T obtained are therefore conservative.

CONCLUSION

Sputtering of H_2O from the surface of Ganymede by interactions with charged particles in Jupiter's magnetosphere is able to concentrate a surface layer of silicate particles, i.e., a lag deposit, from a dirty-ice ejecta blanket within time periods of 10^5 to 10^7 years, and is a plausible mechanism for the formation of dark-ray craters. The emplacement of a sufficiently silicate contaminated dirty-ice layer may be due either to a compositionally rare component or an energetically rare (low velocity) component of the impacting population. Regardless of the mechanism of emplacement of the dirty-ice layer, once emplaced on the surface, lack of an atmosphere and Ganymede's location within a large energetic ion flux insures that sputtering will take place. Therefore, sputtering by charge particles may well be a rapid and significant process in the formation of dark-ray craters on Ganymede and in the evolution in general of the surface of icy satellites.

REFERENCES

Belcher J. W., Goertz C. K., and Bridge H. S. (1980) The low energy plasma in the Jovian magnetosphere. *Geophys. Res. Lett.* **7**, 17–20.

Brown W. L., Augustyniak W. M., Brody E., Cooper B., Lanzerotti L. J., Ramirez A., Evatt R., and Johnson R. E. (1980) Energy dependence on the erosion of H_2O ice films by H and He ions. *Nucl. Instrum. Methods* **170**, 321–325.

Haff P. K. (1980) A model for the formation of thin films in dirty ice targets by sputtering: Applications to the satellites of Jupiter. *Proc. of Symp. on Thin Film Interfaces and Interactions*, Electrochemical Society, **80–2**, p. 21–28.

Hartmann W. K. (1980) Surface evolution of two-component stone/ice bodies in the Jupiter region. *Icarus* **44**, 441–453.

Purves N. G. and Pilcher C. B. (1980) Thermal migration of water on the Galilean satellites. *Icarus* **43**, 51–55.

Shoemaker E. M. and Wolfe R. F. (1981) Cratering time scales for the Galilean satellites. In *The Satellites of Jupiter* (D. Morrison, ed.). Univ. Arizona Press, Tucson. In press.

Sigmund P. (1969) Theory of sputtering. I. Sputtering yield of amorphous and polycrystalline targets. *Phys. Rev.* **184**, 383–416.

Squyres S. W. (1980) Surface temperatures and retention of H_2O frost on Ganymede and Callisto. *Icarus* **44**, 502–510.

A method for estimating the initial impact conditions of terrestrial cratering events, exemplified by its application to Brent crater, Ontario*

Richard A. F. Grieve[1] and Mark J. Cintala[2]

[1]Earth Physics Branch, Department of Energy, Mines and Resources, Ottawa, Ontario, Canada K1AOY3 [2]Department of Geological Sciences, Brown University, Providence, Rhode Island 02912

Abstract—The consideration of independently-derived relationships for crater dimensions and impact melt production as functions of projectile size and impact velocity provide a method for estimating the initial conditions of impact. This procedure has been tested using the well-studied Brent crater as an example. Brent (46° 05′ N; 78° 29′ W) has a present diameter of 3.0 km and a depth to autochthonous basement of 875 m. From subsurface information and by analogy, it is estimated that the original final cavity diameter at the ground surface was 3.4 km and the depth 1.1 km, with the pre-modification transient cavity having a similar depth and a diameter of 3.1 km. The volume difference between the two cavities, 1.9 km^3, is represented by the 2.1 km^3 of slumped allochthonous breccias which partially fill the cavity. Within the breccia fill, there is a basal impact melt-zone and melt glass-bearing breccias. The volume of impact melt within the cavity is estimated at 2.2×10^{-2} km^3.

Impact melt production has been modeled for a velocity range of 15–50 km s^{-1} through a modified Gault and Heitowit treatment, with the target for Brent approximated by granite and the projectile, which was probably an L or LL chondrite, by basalt. Based on continuum mechanics analyses, it is assumed that 50% of this melt remains within the cavity. An energy scaling law of the type $D \propto KE_{EFF}^{1/3.6}$, where KE_{EFF} is the fraction of projectile kinetic energy available for cratering the target, provides a relationship for crater diameter as a function of initial conditions. When combined with the relationship for impact melt generation, the observational data gives estimates of ~109–105 m for projectile radius and corresponding impact velocities of ~30–33 km s^{-1}. With these initial conditions, the shock stress decay has been modeled and, when considered with a Z-model for the flow field, the results compare favorably with previous estimates and data for Brent. As a number of approximations and assumptions are made in the analysis, the estimated initial conditions, although yielding consistent reasonable results, are model dependent. It is suggested, however, that the method serves to differentiate in a relative sense between simple craters formed by "small, fast" and "large, slow" projectiles.

INTRODUCTION

Experimental and theoretical analyses of impact processes have produced a number of relationships between initial impact conditions and parameters such as crater diameter and volume, and the production of impact melt and vapor (Gault and Heitowit, 1963; Holsapple and Schmidt, 1981; O'Keefe and Ahrens, 1977 and 1978; Orphal et al., 1980; and others). In this contribution, we develop an analysis of the reverse problem. By considering independently derived parametric relationships in combination with observational data we attempt to estimate the initial conditions at the well-characterized Brent impact crater.

Brent (46° 05′ N; 78° 29′ W), Ontario, Canada, is the largest known terrestrial crater with a bowl-shaped form (Grieve and Robertson, 1979). The structure is 450 ± 30 m.y. old (Hartung et al., 1971; Lozej and Beales, 1975) and was excavated in an igneous-metamorphic complex on Grenvillian age. The nature of the subsurface structure and crater-fill products has been determined through extensive drilling, with approximately 5 km of core having been recovered from 12 holes. Inasmuch as Brent was formed in a crystal-

---

*Contribution from the Earth Physics Branch No. 934

line target and the subsurface character is better documented than at any other terrestrial simple crater, including Meteor Crater, Arizona (Shoemaker, 1963; Roddy, 1978), it has provided important constraints on the formation of simple craters on the terrestrial planets. Previous considerations of the cratering process at Brent have relied heavily upon structural and petrographic data to outline the dimensions of the transient cavity and its modification to the present, observed form (Dence, 1968; Dence et al., 1977). These and other semi-quantitative models of shock pressure decay and melt generation (Grieve et al., 1977; Robertson and Grieve, 1977) have been hampered, however, by the lack of information on initial conditions; that is, the relative contributions of projectile characteristics (radius, mass, composition) and impact velocity to the kinetic energy of the event.

Considerable geological, geophysical and geochemical data are available and a number of studies on specific characters of Brent have been carried out (Dence, 1964 and 1968; Robertson and Grieve, 1977; Grieve, 1978 and a listing of other studies contained therein). These studies form the basis for the derivation of parameters such as original, pre-erosional cavity and transient cavity dimensions, melt and breccia volumes, and projectile and target type. The methodology for the derivation of initial conditions has three principal components: (i) With projectile and target type constrained, the generation of impact melt in a Brent-like event is initially modeled through a modified version of the energy partitioning treatment of Gault and Heitowit (1963); (ii) An energy-crater dimension scaling relationship is derived on the basis of craters from shallow-buried nuclear explosions; and (iii) These theoretical melt and dimensional scaling relationships are combined and solved by substitution of the observational data on melt volumes and crater dimensions to estimate the initial impact conditions for the Brent event. The derived initial conditions are then used as input parameters for a model of stress decay at Brent, again based on the modified Gault and Heitowit treatment, which is combined with a Maxwell Z-model approximation of the flow field to determine internal consistency and correspondence with other information on stress decay, rim peak pressures, displacement of shock isobars in autochthonous material and relative volumes of excavation and displacement. Other energy partitioning models in combination with other proposed scaling relationships are also considered. While recognizing that there are a number of approximations and uncertainties involved, we argue that this method provides a means for determining whether a particular impact structure, for which a data base equivalent to that of Brent is available, was the result of the impact of a "small, fast" or "large, slow" body.

OBSERVATIONAL DATA

Crater dimensions

The autochthonous basement rocks at Brent define a cavity which is roughly parabolic in cross-section (Dence, 1968; Fig. 1). The present erosional diameter of the Brent crater is 3 km and the depth from the present surface to the brecciated and fractured autochthonous basement rocks in the center is 875 m (Dence, 1968; Grieve, 1978). Above 875 m depth, there is an approximately 10 m section of recrystallized monomict gneiss breccia between a basal melt-zone and the cavity floor (Fig. 1). It is not clear, however, whether this basal breccia is allochthonous or autochthonous material. On the interpretation that the 10 m brecciated gneiss section is allochthonous, Brent has a *present* d/D ratio of 0.29, where d is depth and D diameter.

A detailed analysis of the stratigraphy and the degree of compaction of the post-impact sediments within the crater by Lozej and Beales (1975) provides an estimate of the amount of erosion, with the original ground surface at the time of impact considered to be 220 m above the present erosional plane. Graphical extension of the geometry of the cavity walls to 220 m above the present surface (Fig. 2) indicates a *pre-erosional cavity* with $D_a \sim 3.4$ km, $d_a \sim 1.1$ km and $d_a/D_a \sim 0.32$, where the subscript a refers to dimensions as

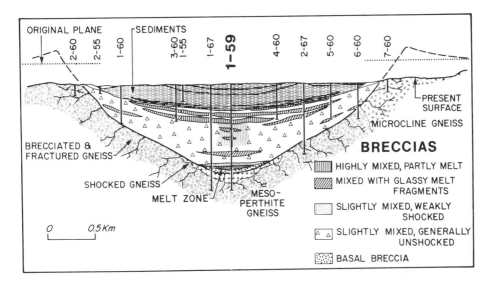

Fig. 1. Simplified cross-section of Brent, indicating major lithologies in the crater. Reprinted from Grieve (1978).

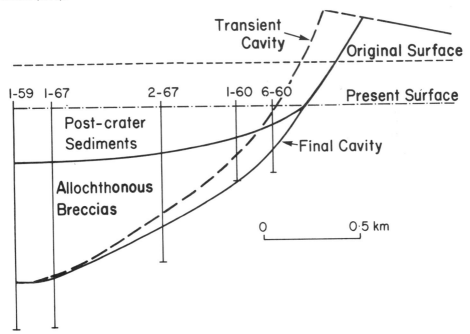

Fig. 2. Cross-section of Brent, indicating estimated dimensions of original final cavity and transient cavity. Reconstructions are based on drilling data, shown as vertical lines, and information from other terrestrial simple craters. No vertical exaggeration here or in other figures.

measured at or from the original ground surface. Thus the d_a/D_a defined by the *autochthonous basement rocks* of the cavity is close to the 1/3 ratio noted at smaller terrestrial craters with little or no breccia-fill (Dence *et al.*, 1977). Approximating cavity geometry as a paraboloid gives a volume estimate of 4.99 km^3 for the final cavity formed in the autochthonous basement and as measured at the original ground surface. The breccia and melt rocks partially filling this cavity (Fig. 1) have been preserved from erosion and have a volume of approximately 2.1 km^3.

There are no observational data at Brent pertaining to the height of the uplifted rim above the original ground surface. The pre-erosional rim crest diameter, however, has been variously estimated at 3.5–4.0 km (Innes, 1964; Dence, 1972) with the most recent estimate 3.8 km (Dence et al., 1977). Extrapolation of final cavity geometry to a diameter of 3.8 km gives a rim height of 200 m above the original ground surface (Fig. 2). This height is comparable to that obtained from scaling the rim heights at only slightly eroded terrestrial craters formed in similar quartzo-feldspathic crystalline targets; e.g., New Quebec—rim diameter 3.2 km, rim height 150 m (Pike, 1976) and Tenoumer—rim diameter 1.9 km, rim height 92 m (Fudali and Cassidy, 1972). A rim crest diameter of 3.8 km and rim height of 200 m gives 7.4 km^3 for the rim crest to rim crest volume of the pre-erosional final cavity defined by the autochthonous basement rocks.

As noted earlier, the cavity is partially filled by breccias, which are interpreted as material slumped from the wall and rim area of the transient cavity as it modified to the final cavity (Grieve et al., 1977). The transient cavity at terrestrial simple craters is considered to conform to a parabola of revolution with the radius, r, of the transient cavity at the original plane given by $r^2 = 2 p^2$, where p is the depth of the transient cavity below the original ground surface (Dence, 1973). If the estimated depth of the final cavity below the original plane is taken as a measure of transient cavity depth, i.e., p = 1.1 km, the transient cavity radius at the original surface was 1.55 km (Fig. 2).

At New Quebec crater the outer slope in the area of the final cavity rim is 10° (Currie, 1966). A similar value can be obtained from the morphometric relationships between rim height and rim flank width and diameter for "fresh" terrestrial craters in a variety of targets and over the diameter range of 0.1 to ~4 km (Pike, 1972). The radially inward extension of a 10° slope for the rim of the final cavity intersects the continuation of a parabolic transient cavity (r = 1.55 km, p = 1.1 km) at a radius of 1.64 km and height above the original plane of 250 m. This defines a transient cavity with a rim crest diameter of 3.28 km, depth of 1.35 km and volume of 5.70 km^3 (Fig. 2). Thus the difference between the original rim to rim volumes of the final cavity and the transient cavity is 1.7 km^3. This is considered to be the volume of target material slumped into the cavity to produce the bulk of the breccia-fill. The observed breccia-fill has a volume of 2.1 km^3, which corresponds to 1.9 km^3 of original target material, when a 10% bulking factor is considered in the transition from coherent crystalline rock to breccia (Innes, 1961). This relatively close correspondence between the observed volume of breccia-fill and the predicted volume of target material slumped off the transient cavity wall is further improved when it is considered that the crater-fill contains, in addition to slumped material, inclusion-rich impact melt rocks and glasses and other highly shocked, but crystalline, debris, which were not excavated from the transient cavity. The melt rocks and glasses plus their crystalline inclusions have an estimated volume of approximately 4×10^{-2} km^3 (see next section) and the volume of highly shocked crystalline debris is at least comparable to this value (Grieve, 1978).

Volume of melt rock

The impact melt retained within the final cavity at Brent occurs as altered glass clasts in the allochthonous breccias of the crater-fill and as a 34 m thick coherent melt zone at the base of the breccia lens (Fig. 1). Details of the occurrence, petrography and chemistry of the impact melt are given in Grieve (1978) and only the salient points are repeated here.

Basal melt-zone

The basal melt zone occurs at 823–857 m in the central drill hole B1-59 (Fig. 1). In B1-67, drilled 200 m to the west, it is replaced by a zone of pyroxene hornfels at equivalent depth. The pyroxene hornfels zone associated with the melt in B1-59 extends for only 10 m below the the base of the melt zone, suggesting that B1-67 came close to intersecting melt and that the maximum radial extent of the melt zone is only slightly less than 200 m.

Assuming a 190 m radius and a lenticular shape results in an estimate of 1.9×10^{-3} km^3 for the volume of material in the basal melt zone.

Not all the material in melt zone is melt. Inclusions of crystalline debris, in various stages of digestion, reaction, and alteration are abundant (Grieve, 1978). The content of microscopic lithic inclusions is variable and ranges to more than 50% at the margins of the melt zone. In addition, macroscopic inclusions of gneiss are also present. The percentage of inclusions, both macro- and microscopic, can be estimated from the available chemical data. As a general principle, the potassium content of the melt zone varies sympathetically with the number of inclusions (Fig. 9 in Grieve, 1978). Forty-eight analyses of relatively inclusion-free melt rock, melt rock with variable amounts of microscopic inclusions, and xenolithic altered gneiss inclusions are available over the 34 m section of the melt zone sampled by B1-59. On the basis that they represent a relatively complete sample, they indicate that the average K_2O content of the melt zone is 6.8%. The standard deviation of the mean is 2.1% (n = 48), indicating the high variability in K_2O between samples of different petrographic character. Essentially inclusion-free melt matrix averages 5.1% K_2O (Grieve, 1978) and analyses of discrete gneiss inclusions average 8.5% K_2O. A mass balance calculation to account for the average of 6.8% K_2O for the melt zone, as sampled in B1-59, suggest that the melt zone contains approximately 50% gneiss-derived inclusions. Thus the volume of actual melt in the melt zone is considered to be closer to 1×10^{-3} km^3.

Melt-bearing breccias

Altered, inclusion-bearing glass clasts of impact melt occur in the mixed breccias of the crater-fill. The clasts have contorted shapes and are generally in the mm-cm size range, with the largest clast encountered in the drill core having dimensions of 25 cm. The melt clasts are often associated with other highly shocked debris, such as vesiculated gneiss fragments, and are concentrated in the upper 160 m of the breccia lens (Fig. 1). Modal analyses of thin sections of core from B1-59 indicate an average glass clast content of 13% at 264–427 m, decreasing to 6–7% at 427–606 m and essentially no glass clasts at greater depths, except for an occurrence 741–763 m (Fig. 4 in Grieve, 1978). Accounting for crystalline inclusions and vesicles and integrating these glass clast values over the entire breccia lens gives an estimate of 5.5×10^{-2} km^3 for the impact melt contained within the breccia lens (Dence, 1971; Grieve et al., 1977).

This figure is probably an overestimate of the amount of melt in the mixed breccias. In particular, the breccias with 6–7% glass clasts encountered in B1-59 at 427–606 m are not typical throughout the breccia lens, being virtually absent in other holes (e.g., B1-67, Fig. 1) drilled to comparable depths. The amount of melt in the breccias has therefore been recalculated here with the volume containing 6–7% glass clasts restricted to a cylindrical central core between 427 and 606 m, with a radius of 190 m; that is slightly less than the separation of holes B1-59 and B1-67. The upper volume with 13% glass clasts was taken as parabolic in cross-section with a diameter of 2.2 km and a maximum depth of 163 m, as indicated by B1-59. On the basis of this geometry and accounting for crystalline inclusions and vesicles, the amount of melt in the mixed breccias is estimated at approximately 2.1×10^{-2} km^3. That is, the total amount of impact melt retained within the cavity at Brent is of the order of 2.2×10^{-2} km^3. This is a factor of over 2 less than previous estimates, which were based on a less detailed analysis of the melt-bearing breccias (Dence, 1971). These volume estimates, as well as the relative dimension parameters of Brent, are listed in Table 1.

Target and projectile

The geologic setting of the Brent crater is described in detail by Currie (1971) and Dence and Guy-Bray (1972). The cavity was excavated primarily in mesoperthite and microcline gneisses of granodioritic composition. Minor amphibolites and alnoite dikes are also

Table 1. Relative dimensions and volumes at Brent.

	Diameter	Depth	Volume
(i) *Crater dimensions*			
Present	3.0 km	0.875 km	3.1 km^3
Pre-erosional, at original ground surface	3.4 km	1.1 km	5.0 km^3
Pre-erosional, from rim crest	3.8 km	1.3 km	7.4 km^3
Transient cavity, at original ground surface	3.1 km	1.1 km	4.2 km^3
Transient cavity, from rim crest	3.3 km	1.35 km	5.7 km^3
(ii) *Volumes of crater fill products*			
Breccia fill	2.1 km^3		
Melt in basal melt zone	1×10^{-3} km^3		
Melt in allochthonous zone	2.1×10^{-2} km^3		
Total melt	2.2×10^{-2} km^3		

present. The regional gneissosity, which trends NW, does not appear to have influenced the crater form at the present level of exposure (Fig. 1 in Grieve, 1978).

Neutron activation analyses of the melt rock from the melt zone and glasses from the mixed breccias indicate relative enrichments in siderophile elements; e.g., up to 17.5 ppb Ir and 320 ppm Ni (Palme *et al.*, 1978 and 1981). Their relative abundance pattern indicates a chondritic projectile, with the Ni/Cr of 10 samples suggesting an L or LL chondrite for the projectile type at Brent (Palme *et al.*, 1981). Mixing model calculations suggest that the basal melt zone is chemically equivalent to a mixture of mesoperthite gneiss and 1.5% L chondrite (Grieve, 1978).

Shock deformation

With the exception of the melt glasses and associated vesiculated gneiss clasts, the bulk of the materials in the breccias of the crater-fill show few or no diagnostic indicators of shock (Grieve *et al.*, 1981). Details of the level of shock recorded in the autochthonous basement rocks of the cavity floor are given in Roberston and Grieve (1977). They note that peak pressures recorded in the cavity floor at the center are on the order of 23 GPa (230 kb). Pressure contours are roughly hemispherical in shape and an extremely steep average gradient of $P \propto r^{-20}$ exists beneath the crater floor. To account for these observations, Robertson and Grieve (1977) suggest that the section beneath the excavated cavity was permanently shortened and the shock zones thinned during displacements associated with transient cavity formation.

THEORETICAL CALCULATIONS

A modified version of the Gault and Heitowit (1963) treatment of shock pressure decay was used as a basis for deriving an estimate of the amount of melt generated on impact. Gault and Heitowit assumed a constant quantity of energy was available to a series of shells behind a hemispherically expanding shock front, centered about the impact point at the original ground surface. This assumption has been modified here to a spherical shock geometry centered on the leading edge of the projectile at its maximum depth of penetration. The depth of projectile penetration, P_p, was calculated by the relation:

$$P_p = 2(D_p/U_p)u_t$$

Table 2. Densities and parameters used in shock (U)-particle (u) velocity relationship.

	ρ, g cm^{-3}	a, km s^{-1}	b	Source
Granite	2.63	2.1	1.63	For $P \geq 35$ GPa in McQueen et al. (1967)
Basalt	2.86	2.6	1.62	Gault and Heitowit (1963)

where D_p is projectile diameter (spherical geometry), U_p is shock velocity in the projectile and u_t is particle velocity in the target. The penetration depth corresponds to the distance the original projectile-target interface moves down into the target during the time the shock wave takes to pass twice through the projectile; once in compression and once to approximate pressure release during rarefaction.

The behavior of the target and projectile under compression has been described by a linear relationship of the type:

$$U = a + bu$$

where a and b are material dependent constants. For the case of Brent, experimental shock data are unavailable for either chondrites or mesoperthite gneiss and the projectile and target have been approximated by basalt and granite, respectively. The coefficients of the shock velocity-particle velocity relationship for basalt are taken from Gault and Heitowit (1963) and for granite from McQueen et al. (1967) (Table 2). These coefficients represent average parameters and do not explicitly account for specific phase changes. As the high pressure regime induced in the target is of greatest interest, the coefficients for granite are from the ≥ 35 GPa data in McQueen et al. (1967). They differ from those derived by Kieffer and Simonds (1980), which represent averages over the entire experimental range. With respect to pressure release, the simplification made by Gault and Heitowit (1963) that the release adiabat can be approximated by the Hugoniot is retained.

Production of melt and vapor

The kinetic energy of impact is partitioned into kinetic energy and internal energy in the projectile and target. For a given projectile and target type, this partitioning is a function of impact velocity (O'Keefe and Ahrens, 1977) and, because the increase in internal energy is responsible for post-shock heating, the relative production of melt and vapor in the target varies with impact velocity. A series of computations using the modified Gault and Heitowit treatment were made for basalt, $\rho = 2.86$ g cm^{-3} ($\sim$L chondrite) impacting granite, $\rho = 2.63$ g cm^{-3} ($\sim$mesoperthite gneiss) over an impact velocity range of 15 km s^{-1} to 50 km s^{-1}. In calculating the amounts of melt and vapor, the onset of melting in the granitic target was taken to occur at an internal energy density of 1.05×10^{10} ergs g^{-1} and the onset of vaporization at 3.4×10^{10} ergs g^{-1}. These values are comparable to those for basalt (Gault and Heitowit, 1963) and gabbroic anorthosite (O'Keefe and Ahrens, 1975). The results are shown in Fig. 3 and provide the relationship:

$$M_m = (3.10 \times 10^{-5} V_i - 3.91)r^3 \qquad (1)$$

where M_m is the mass of target material elevated to temperatures above the onset of melting ($\sim 1000°C$), V_i is impact velocity and r is projectile radius, units here and in all other relationships being cgs.

The above relationship describes the total calculated mass of melt plus vapor produced upon the impact of basalt into granite at 15 to 50 km s^{-1}. It does not define the amount of impact melt remaining within the crater at the end of the excavation process. This can only be described by considering the impact-induced flow field for the target materials during the impact event. For Brent, the amount of melt plus vapor remaining within the

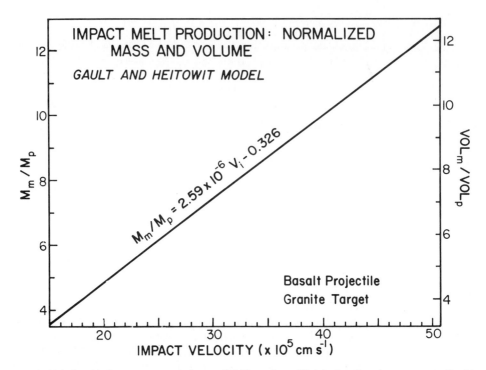

Fig. 3. Relationship between generated mass (M_m) or volume (Vol_m) of melt and vapor, normalized to that of the projectile, (M_p and Vol_p), and impact velocity, V_i, for the impact of basalt into granite at 15–50 km s^{-1}. These results are from a Gault and Heitowit treatment of energy partitioning, modified as explained in the text.

cavity is estimated at 50%, on the basis of results of the continuum mechanics calculations of Orphal *et al.* (1980) for the model impact NASA-2, which was for an iron projectile impacting gabbroic anorthosite at 15.8 km s^{-1}. The NASA-2 calculations resulted in an excavated cavity of approximately the same magnitude, diameter ~2 km, as Brent. On this basis, the volume of melt plus vapor produced by basalt impacting into granite (Fig. 3) and *remaining* within a crater with the approximate dimensions of Brent is considered to be represented by:

$$Vol_m = (5.89 \times 10^{-6}V_i - 7.43 \times 10^{-1})r^3 \qquad (2)$$

where Vol_m is the volume of melt, $\rho = 2.63$ g cm^{-3}, remaining within the cavity.

Energy considerations and scaling

The scaling of crater dimensions to projectile kinetic energy is currently the subject of considerable study. A major difficulty is determining the relevance of relationships derived from small-scale impact experiments or nuclear and HE explosion data to impact craters in the kilometer size-range (Holsapple, 1980; Schmidt, 1980). There appears, however, to be a growing consensus that the previous diameter-energy scaling relationships of the form $D \propto E^{1/3.4}$, where E is projectile kinetic energy (Cooper, 1977; Dence *et al.*, 1977; Roddy, 1978; and others), may underestimate energy requirements by as much as half an order of magnitude (Orphal, pers. comm.; Roddy *et al.*, 1980). Recent analyses of scaling relationships in various target media (Gaffney, 1978; Schmidt, 1980) suggest that a $D \propto E^{1/3.6}$ relationship is more appropriate for craters in the kilometer size-range.

Within a general scaling relationship there are specific dependences on initial conditions. These dependences are functions of the mechanical properties of the target as well as energy partitioning and its variation with projectile type, target type, and impact velocity. The relative role of these variables has been considered by Schmidt (1980), who indicates that in large impact events (energies greater than 1 kiloton TNT equivalent) target response is principally a function of the angle of internal friction in the target materials, and the size and velocity of the projectile for a given impact energy. These variables have generally not been considered in scaling laws for terrestrial craters (Dence et al., 1977; Oberbeck et al., 1975). An effort, however, has been made to assess them for the Brent event.

As an initial base for energy scaling, craters produced by shallow-buried nuclear explosive energy sources were considered. The Teapot Ess and Jangle U craters (Table 3) were selected as starting points. Both were produced by 1.2 kiloton nuclear yield explosions and therefore had energies in excess of the regime where target cohesion and tensile strength exert significant influence on terrestrial crater dimensions (Schmidt, 1980). Teapot Ess was chosen as it had a close similitude in form to Brent, with a depth/diameter ratio of ~0.31 (Table 3) compared to 0.32 for the pre-erosional apparent crater at Brent. Teapot Ess, however, was formed by a relatively deeply buried nuclear device. It was near the optimum depth of burst for cratering efficiency by a nuclear explosion, with a depth of burial to vaporization radius ratio (d/a) of 8.7 (Schmidt, 1980). This is considerably deeper than current estimates of the equivalent depth of burst for impact events, which are considered to more closely correspond to d/a of 1.5 (Holsapple, 1980). For this reason, Jangle U, which has a d/a ratio of 2.2, was also considered although its depth/diameter ratio of 0.20 differs from Brent.

Teapot Ess and Jangle U were formed in alluvium. An adjustment has therefore been made for the difference in the angle of internal friction between alluvium and granite. From Schmidt (1980), it is estimated that their radial dimensions would have been reduced by ~30% if they had been formed in granite. The energy density of a nuclear explosion is extremely high and not all the energy released is mechanically coupled to the target. An additional adjustment has therefore been made for the fact that the yield of a nuclear explosion is not as effective in producing a crater as a high energy chemical explosion of equivalent magnitude and depth of burial. It is explicitly assumed here that the cratering efficiency of a high energy explosion is comparable to that of an impact event in which an equivalent amount of kinetic energy is imparted to the target. From the data in Cooper (1977), the scaled cratering efficiencies of Teapot Ess and Jangle U relative to high energy explosions are 0.54 and 0.38, respectively (Table 3). [These adjustments were not considered fully in Grieve and Cintala (1981).] Although it has been argued that crater volume-scaling relationships are preferable (Roddy, 1978), a crater diameter relationship is used in order to avoid compounding the uncertainties in the estimates of final crater diameter and depth. The energy scaling relationship for Brent thus has the form:

$$D_a = 0.7 D_e [KE_{EFF}/cY]^{1/3.6} \tag{3}$$

where D_e is the diameter of the crater produced by the nuclear explosion of yield Y (Table 3), 0.7 and c are adjustments for target type and cratering efficiency, respectively, and KE_{EFF} is effective cratering energy of impact, as described below.

The effective cratering kinetic energy, KE_{EFF}, is the projectile kinetic energy minus that fraction which is partitioned into target heating. It is argued that it is only the fraction

Table 3. Parameters of nuclear events used as basis for scaling.

	Yield	Crater diameter	Crater depth	d/a*	c**
Teapot Ess	5.02×10^{19} ergs	89.0 m	27.4 m	8.7	0.54
Jangle U	5.02×10^{19} ergs	79.2 m	16.2 m	2.2	0.38

*Depth of burial/vaporization radius of charge, data from Schmidt (1980).
**Cratering efficiency relative to equivalent mass of high energy explosive at equivalent scaled depth (Cooper, 1977).

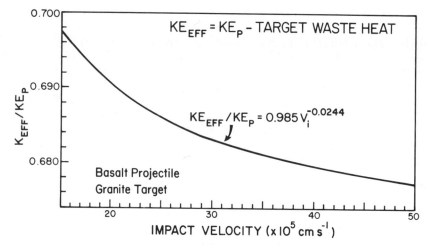

Fig. 4. Relationship between effective cratering kinetic energy (KE_{EFF}) normalized to projectile kinetic energy (KE_p) and impact velocity, for the impact of basalt into granite at 15–50 km s^{-1}. See text for the definition of KE_{EFF}.

of projectile kinetic energy which is manifested as kinetic energy in the target that contributes to cratering. Furthermore, as shock pressures are generally too low in high explosive events to produce comparable irreversible heat losses, it is believed that this correction makes a reasonable allowance for the different energy sources in impact scaling relationships based on explosive events. The projectile waste heat was not subtracted, however, since there are also significant heat losses in the detonation of explosive charges. For basalt impacting granite, the modified Gault and Heitowit treatment indicates that KE_{EFF} ranges from 0.698 to 0.677 of the total projectile kinetic energy for impact velocities of 15 to 50 km s^{-1} (Fig. 4). A least-squares fit to the data define the variation of KE_{EFF} of basalt into granite as:

$$KE_{EFF} = 5.90 r^3 V_i^{1.976} \quad (4)$$

Substituting Eq. (4) in Eq. (3) yields a scaling relationship with D_a as a function of projectile radius and impact velocity:

$$D_a = 1.146 D_e r^{0.834} V_i^{0.549} (cY)^{-0.278} \quad (5)$$

Estimate of initial conditions

The relationships derived above for energy scaling and the relative production of melt for the impact of basalt into granite can be combined with the observational data to estimate the initial conditions of the Brent event. Equation (5) can be rearranged in the form of impact velocity as a function of apparent crater diameter and projectile radius:

$$V_i = 0.780 (D_a/D_e)^{1.821} r^{-1.519} (cY)^{0.506} \quad (6)$$

This expression, when substituted in Eq. (2), gives a relation for the amount of melt remaining within the Brent crater as a function of crater diameter and projectile radius:

$$Vol_m = 4.594 \times 10^{-6} (D_a/D_e)^{1.821} r^{1.481} (cY)^{0.506} - 0.743 r^3 \quad (7)$$

The apparent diameter, D_a, and melt volume, Vol_m, at Brent have been estimated above, as 3.4 km and 2.2×10^{-2} km^3, respectively. When substituted into Eq. (7), along with D_e and cY for Teapot Ess and Jangle U (Table 3), the solutions to Eq. (7) provide estimates of the projectile radius at Brent. Insertion of these values for r into Eq. (6) then gives estimates for the impact velocity V_i. The estimated initial conditions are $r \sim 109$ m, $V_i \sim 30.4$ km s^{-1}, when Teapot Ess is used as a reference event, and $r \sim 105$ m, $V_i \sim 33.1$ km s^{-1}, when Jangle U is used.

A MODEL

The above estimates for the initial conditions have been used to model shock stress decay at Brent through the modified Gault and Heitowit treatment. The shock decay from such a model for $r \sim 105$ m and $V_i \sim 33.1$ km s^{-1}, superimposed upon a cross-section of Brent, is illustrated in Fig. 5. A model based on $r \sim 109$ m, $V_i \sim 30.4$ km s^{-1} gives essentially identical results. A number of comparisons with the observational data and previous estimates of the formational conditions at Brent are possible.

The depth of projectile penetration from the original ground surface is 241 m, giving a depth/projectile radius of 2.3 (Fig. 5). This value is comparable to 2.5 suggested by the calculations of Kieffer and Simonds (1980) for similar impact conditions. It is greater than the value of 2 to ~ 1.5 suggested by the experiments of Oberbeck (1971) and Holsapple (1980). These values, however, are for non-equivalent impact conditions and more importantly the depth of penetration (or equivalent burst) is considered to be the meteorite (explosive) center in their experiments. The definition here, depth to the leading edge of the projectile, corresponds more closely to the center of the effective flow field set up in the target by the impact (Piekutowski, 1980).

The peak shock pressure recorded in the autochthonous basement rocks of the cavity floor at Brent is 23 GPa, with the estimated pre-erosional depth from the original ground surface to the cavity floor being ~ 1.1 km. The model depth to the 23 GPa peak stress contour is 477 m (Fig. 5). It has been indicated previously, however, that the final cavity depth is determined by excavation plus a component of displacement (Stöffler *et al.*, 1975; Dence *et al.*, 1977). On the basis of pre-shot and post-shot positions in the Piledriver nuclear event in granodiorite (Borg, 1972), it has been suggested that the original, pre-displacement depth of the 23 GPa peak stress contour at Brent was at ~ 500–600 m

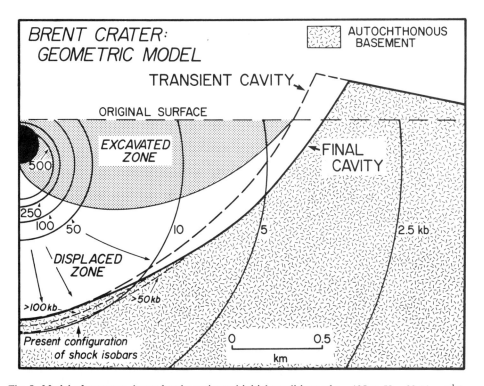

Fig. 5. Model of pressure decay for the estimated initial conditions of $r \sim 105$ m, $V_i \sim 33.1$ km s^{-1} for the Brent impact event superimposed upon an idealized cross-section of final and transient cavities at Brent. The excavated volume, ~ 2.2 km^3 is estimated from a Maxwell Z-model (Maxwell, 1977; Croft, 1980) with a Z of 2.7. See text for comparison between model and other observational results.

(Robertson and Grieve, 1977; Grieve et al., 1981). These earlier estimates compare favorably with the model depth of 477 m.

The model peak shock stress for the rim at Brent is on the order of 3.6 kb and 2.8 kb for the transient and final cavities, respectively (Fig. 5). This is less than the 10–20 kb suggested by Dence et al. (1977). Kieffer and Simonds (1980), however, argue on the basis of melt volumes and relative dimensions at a number of terrestrial craters that 10–20 kb is high and a value closer to 2 kb is more appropriate. Pressures on the order of only a few kilobars are also suggested by Cooper (1977) from explosions in hard rock. It appears, therefore, that the present estimates for rim pressures are not inconsistent with other suggestions.

The concept that the transient cavity is formed partly by excavation and partly by displacement (Dence et al., 1977; Croft, 1980) is illustrated in Fig. 5. The zone of excavation, which has its basis in the steady-flow Maxwell Z-model (Maxwell, 1977; see also Croft, 1980), is also shown (Fig. 5). The volume of material excavated is 2.2 km^2, implying that approximately 50% of the estimated transient cavity volume is due to excavation. This is in good agreement with the finite element calculations reported by Schultz et al. (1981) for the NASA-2 impact. In the absence of a more exact time-dependent Z-model (Austin et al., 1980 and 1981) or other flow field computations for flow in the zone of displacement, it is difficult to evaluate accurately the correspondence between the observed volumes of shocked "autochthonous" basement in the floor of the final cavity and those generated by the model. The present configuration of the shock zones in the cavity floor, obtained from drilling data (Robertson and Grieve, 1977; Fig. 5), yield estimates of 2.2×10^{-2} and 3.8×10^{-2} km^3 for the volumes of "autochthonous" basement with recorded shock pressures of 230–100 kb and 100–50 kb, respectively. The schematic flow field with $Z = 2.7$ and the model for the original configuration of the peak shock stress contours (Fig. 5) provide semi-quantitative estimates of 1.0×10^{-2} km^3 for the original source volume shocked to 230–100 kb and 2.0×10^{-2} km^3 for the 100–50 kb volume. The correspondence with observed volumes is within a factor of two. This comparison, however, should be viewed with caution, considering the present lack of a detailed flow field model, the uncertainties in precision of the positions of the shock zones in cores other than B1-59 (Fig. 2), and the accuracy of recorded pressure estimates based on shock-induced planar features (Grieve and Robertson, 1976; Xie and Chao, 1981).

DISCUSSION

The combination of observational data, the modified Gault and Heitowit treatment, and the scaling relationships based on considerations of craters produced by nuclear explosions results in a model for pressure decay at Brent, which compares favorably with other data and estimates for Brent and Brent-like impact events. It should be borne in mind that although an attempt has been made to minimize assumptions and approximations and to make them reasonable when necessary, there are several involved in the Gault and Heitowit approach, the scaling relationships, and the derived relationships between the various parameters at Brent. For example, the approximation of the release adiabat by the Hugoniot has been criticized by O'Keefe and Ahrens (1975, 1977), who indicate that it underestimates the amount of melt generated in an impact event. It is not the purpose of this contribution to enter into a detailed discussion of the relative merits of various models and formulations. We note, however, that given the uncertainties the present model gives reasonable initial conditions and relatively consistent results for Brent. It is possible, however, that this apparent consistency results from off-setting approximations. Whatever the reason(s), it is apparent that the present model has at least the same level of confidence with respect to relative volumes and pressure decay as other computational predictions based on continuum mechanics for observed crater dimensions (Roddy et al., 1980). Attempts to derive an estimate of the initial conditions of the Brent impact event from other formulations generally proved unsuccessful or were unrealistic. For example, various combinations of the relationships derived here, melt generation estimates based on the O'Keefe and Ahrens (1977) treatment, and scaling laws from centrifuge cratering experiments (Schmidt, 1980) yield projectile radii either in the tens of centimeter or kilometer size-range.

While further analysis is needed to determine the accuracy of the derived initial conditions and the resultant model of stress decay at Brent, it is suggested that, at a minimum, the *methodology* outlined here presents an opportunity to estimate initial conditions in impact events. As at Brent, equivalent observational data is required on crater dimensions, melt volumes, projective and target type, as is a model for pressure decay and an independently derived scaling relationship. Provided consistent assumptions and pressure decay models are used for all craters, it should be possible to differentiate between similar-sized craters produced by "big, slow" and "small, fast" projectiles.

It is apparent that the derivation of an independent energy scaling relationship for predicting the dimensions of kilometer-sized craters is still not precise (Roddy, 1978; Roddy *et al.*, 1980). The projectile kinetic energy for the estimated initial conditions in the present model is $7.1-7.7 \times 10^{25}$ ergs, depending on the reference event used, and the effective kinetic energy available for cratering is $4.8-5.2 \times 10^{25}$ ergs. These estimates can be compared to previous estimates for the projectile kinetic energy at Brent of 2×10^{24} to 8×10^{25} (Innes, 1961; Dence *et al.*, 1977). The previous preferred estimate of 1×10^{25} ergs is based on the empirical scaling relationship derived by Dence *et al.* (1977), which is in part constrained by melt volumes; this value for the projectile kinetic energy at Brent, however, is considered to be an underestimate. It assumes that all the projectile energy is available to the target and also takes a value of 1.2×10^{10} ergs g^{-1} for the energy density of the generated melt volume. This latter figure is similar to that used here for the initiation of melting but it does not consider the additional energy losses associated with the production of super-heated melt and vapor. The energy estimates generated in the present computations are consistent with the concept that previous estimates in the literature for the energy of formation of kilometer-sized craters may underestimate energy requirements (Holsapple, 1980; Orphal, pers. comm.; Roddy *et al.*, 1980; Schmidt, 1980). Although model dependent, the results for Brent are in concert with a 1/3.6 exponent for the energy term. This is in keeping with recent analyses by Gaffney (1978) and Schmidt (1980) and suggests a greater role played by gravity, than previously thought, in determining the dimensions of kilometer-sized craters (Gault *et al.*, 1975).

It would be desirable to undertake a series of finite difference calculations to derive a continuum mechanics cratering model for Brent. In the absence of such calculations, it is intended to attempt to superimpose a time-dependent Maxwell Z-model for the cratering flow field upon various model pressure decay relationships. If such an approach proves successful, it may result in a relatively rapid and inexpensive approach to first order modeling of cratering conditions in simple craters such as Brent. It may also provide at least some of the information on crater formation which is obtainable only at present through the more laborsome numerical calculations.

Acknowledgments—This work was undertaken while one author (RAFG) was on leave at Brown University from the Earth Physics Branch, Department of Energy, Mines and Resources, Ottawa, Canada. Reviews and comments by D. J. Milton and S. K. Croft are appreciated, as is an extensive critique by K. A. Holsapple. Thanks also go to Jim (APL) Garvin for leaving third base long enough to perform the Z-model calculations, while Sue Church provoked considerable admiration with her hot magenta style of drafting. The skills and patience of Sally Bosworth in manuscript preparation are gratefully acknowledged. This work was supported by NASA Grant NGR-40-002-088.

REFERENCES

Austin M. G., Thomsen J. M., Orphal D. L., Borden W. F., Larson S. A., and Schultz P. H. (1981) Gabbroic anorthosite impact cratering time dependent Z flow fields for two different impact energies (abstract). In *Lunar and Planetary Science XII*, p. 37–39. Lunar and Planetary Institute, Houston.

Austin, M. G., Thomsen J. M., Ruhl S. F., Orphal D. L., and Schultz P. H. (1980) Calculational investigation of impact cratering dynamics: Material motions during the crater growth period. *Proc. Lunar Planet. Sci. Conf. 11th*, p. 2325–2345.

Borg I. Y. (1972) Some shock effects in granodiorite to 270 kilobars at the Piledriver site. *Amer. Geophys. Union* **16**, 293–311.

Cooper H. F. Jr. (1977) A summary of explosion cratering phenomena relevant to meteor impact events. In *Impact and Explosion Cratering* (D. J. Roddy, R. O. Pepin, and R. B. Merrill, eds.), p. 11–44. Pergamon, N. Y.

Croft S. K. (1980) Cratering flow fields: Implications for the excavation and transient expansion stages of crater formation. *Proc. Lunar Planet. Sci. Conf. 11th*. p. 2347–2378.

Currie K. L. (1966) Geology of the New Quebec crater. *Geol. Surv. Canada, Bull.* **150**.

Currie K. L. (1971) A study of potash fenitization around Brent Crater, Ontario, a Paleozoic alkaline complex. *Can. J. Earth Sci.* **8**, 481–497.

Dence M. R. (1964) A comparative structural and petrographic study of probable Canadian meteorite craters. *Meteoritics* **2**, 249–270.

Dence M. R. (1968) Shock zoning at Canadian craters: Petrography and structural implications. In *Shock Metamorphism of Natural Materials* (B. M. French and N. M. Short, eds.), p. 339–362. Mono Book Corp., Baltimore.

Dence M. R. (1971) Impact melts. *J. Geophys. Res.* **76**, 5552–5565.

Dence M. R. (1972) The nature and significance of terrestrial impact structures. *International Geological Congress, XXIV Session, Sec.* 15, 77–89.

Dence M. R. (1973) Dimensional analysis of impact structures (abstract). *Meteoritics* **8**, 343–344.

Dence M. R., Grieve R. A. F., and Robertson P. B. (1977) Terrestrial impact structures: Principal characteristics and energy considerations. In *Impact and Explosion Cratering* (D. J. Roddy, R. O. Pepin, and R. B. Merrill, eds.), p. 247–276. Pergamon, N.Y.

Dence M. R. and Guy-Bray J. V. (1972) Some astroblemes, craters and crypto-volcanic structures in Ontario and Quebec. *International Geological Congress, XXIV Session, Excursion A65 Guidebook*.

Fudali R. F. and Cassidy W. A. (1972) Gravity reconnaissance of three Mauritanian craters of explosive origin. *Meteoritics* **7**, 51–70.

Gaffney E. S. (1978) Effects of gravity on explosion craters. *Proc. Lunar Planet. Sci. Conf. 9th*, p. 3831–3842.

Gault D. E., Guest J. E., Murray J. B., Dzurisin D., and Malin M. C. (1975) Some comparisons of impact craters on Mercury and the Moon. *J. Geophys. Res.* **80**, 2444–2460.

Gault D. E. and Heitowit E. D. (1963) The partitioning of energy for impact craters formed in rock. *Proc. 6th Hypervelocity Impact Symposium* **2**, 419–456.

Grieve R. A. F. (1978) The melt rocks at Brent crater, Ontario, Canada. *Proc. Lunar Planet. Sci. Conf. 9th*, p. 2579–2608.

Grieve R. A. F. and Cintala M. J. (1981) Brent crater, Ontario: Observation and theory (abstract). In *Lunar and Planetary Science XII*, p. 362–364. Lunar and Planetary Institute, Houston.

Grieve R. A. F., Dence M. R., and Robertson P. B. (1977) Cratering processes: As interpreted from the occurrence of impact melts. In *Impact and Explosion Cratering* (D. J. Roddy, R. O. Pepin and R. B. Merrill, eds.), p. 791–814. Pergamon, N.Y.

Grieve R. A. F. and Robertson P. B. (1976) Variations in shock deformation at the Slate Islands impact structure, Lake Superior. *Contrib. Mineral. Petrol.* **58**, 37–49.

Grieve R. A. F. and Robertson P. B. (1979) The terrestrial cratering record: I. Current status of observations. *Icarus* **38**, 212–229.

Grieve R. A. F., Robertson P. B., and Dence M. R. (1981) Constraints on the formation of ring impact structures, based on terrestrial data. In *Proc. Conf. Multi-Ring Basins*. Pergamon, N.Y. In press.

Hartung J. B., Dence M. R., and Adams J. A. (1971) Potassium-argon dating of shock metamorphosed rocks from the Brent impact crater, Ontario, Canada. *J. Geophys. Res.* **76**, 5437–5448.

Holsapple K. A. (1980) The equivalent depth of burst for impact cratering. *Proc. Lunar Planet. Sci. Conf. 11th*, p. 2379–2401.

Holsapple K. A. and Schmidt R. M. (1981) On the scaling of crater dimensions 2. Impact processes. *J. Geophys. Res.* In press.

Innes M. J. S. (1961) The use of gravity methods to study the underground structure and impact energy of meteorite craters. *J. Geophys. Res.* **66**, 2225–2239.

Innes M. J. S. (1964) Recent advances in meteorite crater research at the Dominion Observatory, Ottawa, Canada. *Meteoritics* **2**, 219–241.

Kieffer S. W. and Simonds C. H. (1980) The role of volatiles and lithology in the impact cratering process. *Rev. Geophys. Space Phys.* **18**, 143–181.

Lozej G. P. and Beales F. W. (1975) The unmetamorphosed sedimentary fill of the Brent meteorite crater, south-eastern Ontario. *Can. J. Earth Sci.* **12**, 606–628.

Maxwell D. E. (1977) Simple Z model of cratering, ejection and overturned flap. In *Impact and Explosion Cratering* (D. J. Roddy, R. O. Pepin, and R. B. Merrill, eds.), p. 1003–1008. Pergamon, N.Y.

McQueen R. G., Marsh S. P., and Fritz J. N. (1967) Hugoniot equation of state of twelve rocks. *J. Geophys. Res.* **72**, 4999–5036.

Oberbeck V. R. (1971) Laboratory simulation of impact cratering with high explosives. *J. Geophys. Res.* **76**, 5732–5749.
Oberbeck V. R., Hörz F., Morrison R. H., Quaide W. L., and Gault D. E. (1975) On the origin of the lunar smooth plains. *The Moon* **12**. 19–54.
O'Keefe J. D. and Ahrens T. J. (1975) Shock effects of a large impact on the moon. *Proc. Lunar. Sci. Conf. 6th*, p. 2831–2844.
O'Keefe J. D. and Ahrens T. J. (1977) Impact-induced energy partitioning, melting and vaporization on terrestrial planets. *Proc. Lunar Sci. Conf. 8th*, p. 3357–3374.
O'Keefe J. D. and Ahrens T. J. (1978) Impact flows and crater scaling on the Moon. *Phys. Earth Planet. Inter.* **16**, 341–351.
Orphal D. L., Borden W. F., Larson S. A., and Schultz P. H. (1980) Impact melt generation and transport. *Proc. Lunar Planet. Sci. Conf. 11th*, p. 2309–2323.
Palme H., Grieve R. A. F., and Wolf R. (1981) Identification of the projectile at Brent crater and further considerations of projectile types at terrestrial craters. *Geochim. Cosmochim. Acta.* In press.
Palme H., Wolf R., and Grieve R. A. F. (1978) New data on meteoritic material at terrestrial impact craters (abstract). In *Lunar and Planetary Science IX*, p. 856–858. Lunar and Planetary Institute, Houston.
Piekutowski A. J. (1980) Formation of bowl-shaped craters. *Proc. Lunar Planet. Sci. Conf. 11th*, p. 2129–2144.
Pike R. J. (1972) Geometric similitude of lunar and terrestrial craters. *International Geological Congress, XXIV Session, Sec.* **15**, 41–47.
Pike R. J. (1976) Crater dimensions from Apollo data and supplemental sources. *The Moon* **15**, 463–477.
Robertson P. B. and Grieve R. A. F. (1977) Shock attenuation at terrestrial impact structures. In *Impact and Explosion Cratering* (D. J. Roddy, R. O. Pepin, and R. B. Merrill, eds.), p. 687–706. Pergamon, N.Y.
Roddy D. J. (1978) Pre-impact geologic conditions, physical properties, energy calculations, meteorite and initial crater dimensions and orientations of joints, faults, and walls at Meteor Crater, Arizona. *Proc. Lunar Planet. Sci. Conf. 9th*, p. 3891–3930.
Roddy D. J., Schuster S. H., Kreyenhagen K. N., and Orphal D. L. (1980) Computer code simulations of the formation of Meteor Crater, Arizona: Calculations MC-1 and MC-2. *Proc. Lunar Planet. Sci. Conf. 11th*, p. 2275–2308.
Schmidt R. M. (1980) Meteor Crater: Energy of formation-implications of centrifuge scaling. *Proc. Lunar Planet. Sci. Conf. 11th*, p. 2099–2128.
Schultz P. H., Orphal D. L., Miller B., Borden W. F., and Larson S. A. (1981) Impact crater growth and ejecta characteristics: Results from computer simulations (abstract). In *Lunar and Planetary Science XII*, p. 949–951. Lunar and Planetary Institute, Houston.
Shoemaker E. M. (1963) Impact mechanics at Meteor Crater, Arizona. In *The Moon, Meteorites and Comets* (B. M. Middlehurst and G. P. Kuiper, eds.), p. 301–336. Univ. Chicago Press, Chicago.
Stöffler D., Gault, D. E., Wedekind J., and Polkowski G. (1975) Experimental hypervelocity impact in quartz sand: Distribution and shock metamorphism of ejecta. *J. Geophys. Res.* **80**, 4062–4077.
Xie X. and Chao E. C. T. (1981) Single crystal investigation of naturally shocked quartz—a restudy (abstract). In *Lunar and Planetary Science XII*, p. 1219–1220. Lunar and Planetary Institute, Houston.

Structural study of Cactus Crater

Joana Vizgirda and Thomas J. Ahrens

Seismological Laboratory, California Institute of Technology, Pasadena, California 91125

Abstract—The detailed structure of Cactus Crater, a 105 m diameter nuclear explosion crater formed in water-saturated carbonate rock of Eniwetok Atoll, is delineated using the high to low Mg calcite diagenetic transition as a stratigraphic tracer. Outside Cactus, this transition is observed as a discontinuous horizon which appears to be depressed, possibly as a result of the cratering event, near the crater. Beneath the crater, this transition occurs gradually over a 4.5 ± 0.5 m interval, leading to the following conclusions: material inside Cactus Crater underwent primarily *in situ* brecciation and mixing, the maximum depth of the excavation cavity is 20 m below sea level, and a fallback breccia lens, if it exists, has a maximum thickness of 1 m. A central uplift of 4.5 ± 0.5 m is inferred from the observation that the transition interval occurs at a 4 to 5 m greater depth at 1/2 crater radius than in the center. The excavation process, deduced from the Mg calcite transition as well as gamma well log data, involves high velocity injection of strongly shocked material to form the excavation cavity lining. The *in situ* brecciation and mixing appears to be a turbulent process, probably facilitated by fluidization of the carbonate rock. Based on the Mg calcite transition patterns beneath the crater floor, dynamic rebound is inferred as the modification mechanism for Cactus Crater. Using the Melosh Bingham model for dynamic rebound, a maximum strength of ~1 bar is inferred for the cratered carbonate medium. This strength value is representative of clays, such as those in which Snowball, a chemical explosion crater having dimensions and features similar to Cactus', was formed. Comparisons between Cactus and meteorite impact craters are also presented.

INTRODUCTION

Because of their stratified nature, sedimentary rocks, acting as target media for impact or explosion craters, offer favorable circumstances for detailed crater structure investigations. Meteorite impact craters in sedimentary rocks display deformational features such as folded and faulted central peaks (e.g., Sierra Madera; Howard *et al.*, 1972), rim deformation and ring faults (e.g., Decaturville: Offield and Pohn, 1979), and fallback and autochthonous breccia lenses (e.g., Gosses Bluff and Meteor Crater; Milton *et al.*, 1972 and Shoemaker, 1963). Explosion craters in stratified targets, such as Snowball (Roddy, 1976), provide evidence of similar cratering produced features. Laboratory scale cratering experiments using colored, layered targets, (see e.g., Gault *et al.*, 1968) corroborate the observed large scale phenomena. However, sedimentary crater (especially meteorite crater) structural observations are often sketchy and do not permit comprehensive analyses of impact induced deformation. In particular, data on a variety of possible sedimentary lithologies (with saturation and stratification as additional variables) are needed in order to allow a numerical assessment, or, at least a qualitative synthesis of the effect of target properties on crater morphologies.

The purpose of this study is two-fold. First of all, an attempt is made to present an accurate, detailed description of an explosion crater in uncemented to partially cemented carbonate sediments. As an addition to the available data base, such a description would aid in formulating a rigorous theoretical treatment of crater structural dependence on target characteristics. Secondly, detailed observations of the crater are used to infer cratering excavation and modification mechanisms, and these interpretations are then compared with theoretically and observationally based cratering models.

Cactus Crater, the object of this study, located on Runit Island in the North Pacific Eniwetok Atoll. It is one of the smallest of the Pacific Test Site nuclear craters. Insight into the details of Cactus Crater structure is provided by a unique stratigraphic tracer, the diagenetic transition boundary between high and low magnesium calcite (described in the next section). This stratigraphic marker, together with other geochemical data, also serves as the basis for interpretation of cratering mechanisms responsible for the formation of Cactus Crater.

CARBONATE MINERALOGY—BACKGROUND

It has long been observed that organically precipitated carbonate skeletal material contains relatively high concentrations (up to 30 weight %) of $MgCO_3$ (see e.g., Silliman, 1846; Bøggild, 1930). Chave (1952) showed that this magnesium is present in the mineral calcite, and not dolomite, as previously believed. In that study, $MgCO_3$ concentrations, determined by chemical analyses, were compared with X-ray diffraction spectra of the same samples. The results, shown as the solid line in Fig. 1, indicate a continuous decrease in the (104) d spacing of the calcite lattice with increasing $MgCO_3$ content.

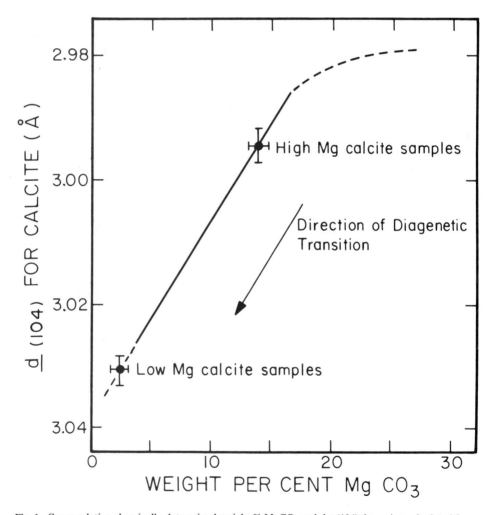

Fig. 1. Curve relating chemically determined weight % $MgCO_3$ and the (104) d spacing calculated from powder X-ray diffractometer scans of magnesian calcites; dashed portions represent considerable data scatter. (From: Chave, 1952). Points representing the range of d spacings observed for the high and low Mg calcites in Runit core samples are indicated. These variations were translated to represent compositional variations.

However, the equilibrium composition at earth's surface conditions of inorganically precipitated calcite is approximately 1 weight % $MgCO_3$ (Plummer and Mackenzie, 1974). Organically precipitated calcite with $MgCO_3$ concentrations of up to 30 weight % is, therefore, unstable at near surface conditions (excluding, of course, the biological environment in which it was produced). Thus, it is inevitable that magnesian calcite will alter to a more stable form. The method of magnesian calcite stabilization for Eniwetok samples probably involves dissolution resulting in calcite precipitation (containing less than 4 weight % $MgCO_3$) and a magnesium enriched solution, according to the following reaction (Land, 1967):

$$(Ca_{1-x}Mg_x)CO_3(calcite) + xH^+(aq) \rightleftarrows (1-x)CaCO_3(calcite) + xMg^{++}(aq) + xHCO_3^-(aq). \quad (1)$$

During sea-level lowstands relative to the atoll, typically associated with episodes of major glaciation, coral atolls are subaerially exposed. Meteoric waters percolate through the exposed coralline limestone dissolving the magnesian calcite and reprecipitating calcite with low (<4 weight %) Mg contents. As sea level rises, coral reef growth and (high) magnesian calcite deposition resumes on top of the altered low Mg calcite rock. It has been generally assumed (H. Lowenstam, 1980, pers. comm.) that the magnesian calcite diagenetic transition boundary produced under such conditions is a sharp discontinuity. Ristvet et al. (1974) first reported the presence of this diagenetic boundary at approximately 10 m depths on Eniwetok Atoll. Results on the detailed nature of this transition, as observed on Runit Island, are discussed in the following section.

OBSERVATIONS

A detailed examination of the calcite mineralogy of samples from five cores drilled outside Cactus Crater, XRU-1, 3, 4, 5 and 6, (see Fig. 2) was conducted using powder X-ray diffractometry. X-ray data was obtained with a Norelco diffractometer using Ni-filtered Cu $K\alpha$ radiation and slow scan speeds of 0.25° 2θ per minute; a minimum of 3 scans per sample were taken. Silicon powder (N.B.S. certified 99.9% pure) was used as an internal standard.

The X-ray spectra indicate the presence of two distinct calcite compositions (Fig. 1). In samples from deeper (below 15 m) levels of these cores, the strongest calcite reflection [corresponding to the (104) plane] is observed to occur between 29.43° and 29.48° 2θ, corresponding to d spacings of 3.032 Å and 3.027 Å, respectively. According to the relation between weight % $MgCO_3$ and $d_{(104)}$ spacings (hence 2θ values) determined by Chave (1952) and reproduced in Fig. 1, reflections in this range correspond to $MgCO_3$ concentrations of 2.3 ± 0.8 weight %. Judging from the considerable scatter in Chave's data at $MgCO_3$ contents less than 3%, the actual variability in $MgCO_3$ content may be somewhat larger than this; however, it almost certainly does not exceed 4 weight %. The (104) calcite reflection in samples from shallower core levels occurs between 29.81° and 29.83° 2θ, corresponding to $MgCO_3$ concentrations of 13.8 to 14.5 weight %. $MgCO_3$ calcite contents intermediate between 2.3 ± 0.8 weight % and 14.15 ± 0.35 weight % (hereafter referred to as low and high Mg calcite) were not observed. As discussed in the section on carbonate mineralogy, many organisms metastably precipitate calcite with relatively high $MgCO_3$ concentrations, which, upon exposure to meteoric waters, transforms to calcite with low $MgCO_3$ contents. This phenomenon is clearly illustrated by the XRU core samples. The high Mg calcite in the Runit samples probably represents skeletal material of foraminifera and calcareous algae. These organisms have been identified in Runit core samples (Couch et al., 1975), and the observed $MgCO_3$ concentrations in calcite correspond to those reported by Chave (1954) for skeletal material of tropical water foraminifera and calcareous algae.

The significant observation to be made in our Runit core X-ray diffraction studies is the nature of the high to low Mg calcite transition. There were no samples from any of the

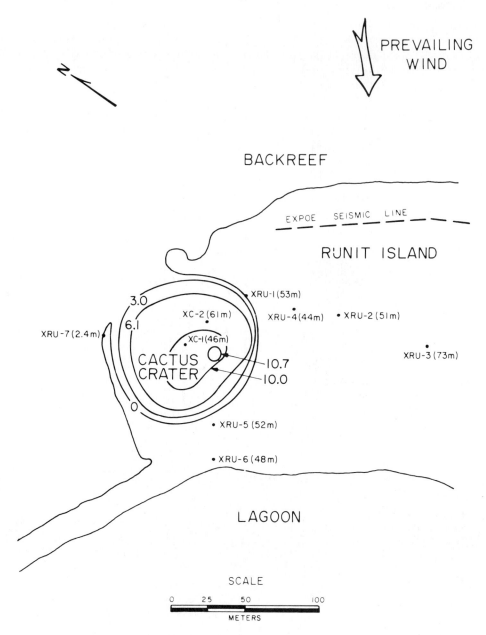

Fig. 2. Map of Runit Island, Eniwetok Atoll (11.30°N, 162.15°E) showing positions of drillholes in cratered (XC) and uncratered (XRU) portions of the island. Cactus Crater contour levels are difference contours. Dashed line indicates position of seismic refraction line.

XRU cores that contained both high and low Mg calcite. Therefore, within sampling interval limitations, magnesian calcite diagenesis on Runit Island (and, most likely, on all of Eniwetok Atoll) occurs as an abrupt transition. Depths to this transition in the XRU cores are listed in Table 1. All core sample depths are relative to the mean low water datum (Ristvet, 1981 pers. comm.). Differences in these depth levels may be a result of transectionally varying hydrologic regimes, or, may represent permanent, explosion induced displacements beneath Cactus Crater; these possibilities will be discussed in the next section.

Table 1. Runit diagenetic transition depths

Core	Depth to High-to-Low Mg[1] Calcite Transition (m)
XRU-1	13.7 ± 1.3
XRU-3	11.1 ± 0.5
XRU-4	13.6 ± 0.4
XRU-5	12.1 ± 1.0
XRU-6	13.4 ± 0.6

[1]Depths are relative to the mean low water spring datum.

The nature of the high to low Mg calcite transition, as observed in samples from the cores drilled inside Cactus Crater, XC-1 and XC-2 (see Fig. 2), markedly differs from that observed in the XRU cores taken outside the crater. Both high and low Mg calcite X-ray diffraction peaks were observed in several samples from shallow (above 20 m) levels in the XC-1 and XC-2 cores. X-ray powder diffraction spectra of the 5 shallowest available XC-1 samples are shown in Fig. 3. Peaks of both high Mg and low Mg calcite, at ~29.8° and ~29.5° 2θ, respectively, are present in the uppermost four of these samples. That of high Mg calcite, most prominent at the shallowest level, gradually decreases in intensity, while that of low Mg calcite increases in intensity with increasing depth in the core. At 15.4 m, and all levels below that, only low Mg calcite is observed. X-ray spectra of XC-2 core samples, reproduced in Fig. 4, show a similar pattern of consistently decreasing high Mg calcite and increasing low Mg calcite peak intensities. Below 20 m depth in XC-2, only the low Mg calcite peak is present.

X-ray powder diffraction spectra of XC-1 and XC-2 samples containing both high and low Mg calcite were analyzed according to the method described in the next section. (See Table 2 for Gaussian curve fit parameters.) Computer drawn observed and calculated fits to calcite peaks for four of the XC-1 samples are shown in Fig. 5. High Mg calcite concentrations were determined from integrated peak intensity ratios and the empirical calibration curve described in the next section. In Fig. 6, these concentrations are plotted against depth of sample for both XC-1 and XC-2 core material. Two noteworthy features are evident in this graph: first, the progressive nature of the two curves, and consequently, of the high to low Mg calcite transition in the XC-1 and XC-2 cores, and, secondly, the 4 to 5 m downward displacement of the XC-2 transition curve from that of XC-1. The significance of these two features will be addressed in the discussion section.

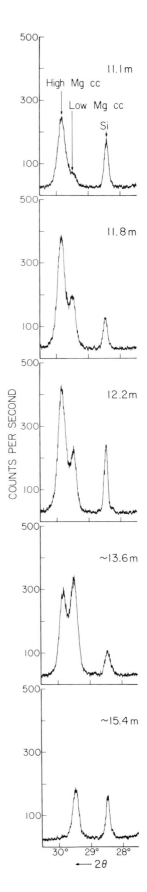

Fig. 3. Portions of powder X-ray diffractometer scans of the five shallowest XC-1 core samples. Sample depth levels are indicated in the upper right hand corner. Note the progressive variation in high and low Mg calcite (cc) peak intensities with depth in the core.

QUANTITATIVE ANALYSIS

Relative concentrations of high and low Mg calcite can be determined from X-powder diffraction spectra. The analysis involves resolution of high and low Mg calcite component peaks, calculation of integrated peak intensities (i.e., peak areas), and comparison with an empirically derived composition vs. peak intensity curve. Although the specific technique differs, the analysis presented below is analogous to that used by Neumann (1965).

Component peak resolutions and integral peak intensities were obtained in the following manner. Sections of X-ray diffractometer spectra between 28.5° and 30.5° 2θ, totally encompassing the high and low Mg calcite peaks, were digitized; additional 1° 2θ intervals on either side of these sections were used to determine a digitizing baseline. A least squares fit to the digitized data provided parameters A, B, C and D and their standard errors in the assumed Gaussian curve model:

$$I(\phi) = A \exp[-B^2(\phi-\omega)^2] + C \exp[-D^2(\phi-\omega-\Delta)^2] \quad (2)$$

where $I(\phi)$ is the peak intensity at some angle ϕ and Δ is the angular separation of the high and low Mg calcite peaks. ω is the centroid of the high Mg calcite peak:

$$\omega = \frac{\Sigma 2\theta I(2\theta)\Delta(2\theta)}{\Sigma I(2\theta)\Delta(2\theta)} \quad (3)$$

where $I(2\theta)$ is peak intensity at some angle 2θ and $\Delta 2\theta$ is the angular increment. The high Mg calcite peak was sufficiently symmetrical so that its centroid coincided with the angle of maximum amplitude, $28.82° \pm 0.01°$ 2θ. The angular separation was assumed constant at 0.37°. For the high Mg calcite peak, therefore, the integrated intensity is simply

$$\int_{-\infty}^{\infty} A \exp[-B^2(\phi-\omega)^2] d\phi = \sqrt{\pi}\frac{A}{B} \quad (4)$$

Similarly, the intensity of the low Mg calcite peak is $\sqrt{\pi}C/D$. Intensity errors are calculated from Gaussian parameter standard errors.

A calibration curve, relating weight % high Mg calcite to its fractional X-ray diffraction peak intensity, was constructed using samples of known high and low Mg calcite concentrations. The standard samples were prepared using XRU-3 core samples from 3 m (all high Mg calcite) and 58 m (all low Mg calcite) depths mixed in weight proportions of 1:9, 3:7, 5:5, 7:3 and 9:1. The weight % MgCO$_3$ was determined from this proportion and the weight % aragonite (determined

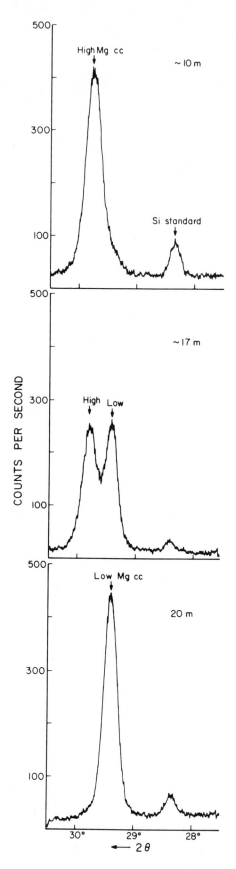

Fig. 4. Powder X-ray diffractometer spectra of XC-2 core samples. Note the progression from high to low Mg calcite with increasing depth in the core, and the similarity of this variation to that of XC-1 core samples, shown in Fig. 3.

using an X-ray calibration curve) in each sample. X-ray diffractometer spectra of these standard samples were then treated as described in the preceding paragraph and intensities of the high and low Mg calcite components were obtained. Gaussian curve fit parameters are listed in Table 2. Due to the extremely low intensity of the high Mg calcite peak in the 1:9 standard sample, a satisfactory fit to the data could not be obtained. At values of 25 weight % high Mg calcite and greater, the data lie on a straight line with a slope of 1/86.

DISCUSSION OF CACTUS CRATER STRUCTURE

The role of the Cactus cratering event in explaining the observations presented above, and what these observations, together with other geologic and geophysical data can say about the process of crater formation, are discussed below.

An initial word of caution concerning sample depth levels is warranted. Due to coring difficulties and tidal variations during drilling, there is a 1.5 meter uncertainty in the location of XC-1 and XC-2 samples; this uncertainty is supplementary to the depth level error limits indicated in Fig. 6. (See Ristvet et al., 1978, for elaboration on sampling uncertainties.) Although the arguments and conclusions presented below are not essentially altered by this additional uncertainty, quantitative values require critical assessment.

In order to evaluate possible cratering induced geologic changes, it is first necessary to establish what the undisturbed situation was. Specifically, information on the nature and positional variation of the high to low Mg calcite transition boundary is needed. As discussed in previous sections, the high to low Mg calcite transition is observed as an abrupt boundary in the XRU cores and occurs at depths between 11 and 14 m (see Table 1). Due to the role of meteoric diagenesis in this transition, a correlation between hydrologic regime and transition depth is expected. In subaerially exposed atolls such as Eniwetok, meteoric waters occur in a lens shaped configuration. It is thus anticipated that the depth to the high to low Mg calcite transition would vary along radial island transects but remain fairly

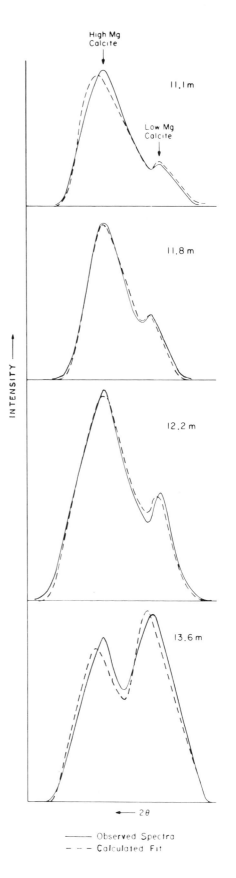

Fig. 5. Computer drawn observed and calculated X-ray diffraction peaks of high and low Mg calcite. Method used in calculating curve fits is described in the text.

Table 2. X-Ray peak Gaussian fit parameters

Sample		A	B	C	$D^{(1)}$	$\dfrac{I_{High}}{I_{High}+I_{Low}}^{(2)}$	Weight % High[3] Mg calcite
Calibration Samples	(High/Low Mg Calcite; weight %)						
	25/75	1.47 ± 0.12	3.37 ± 0.18	7.55 ± 0.06	2.92 ± 0.09	0.14 ± 0.03	
	44/56	3.67 ± 0.10	2.57 ± 0.15	7.65 ± 0.09	3.16 ± 0.11	0.37 ± 0.03	
	65/35	4.81 ± 0.10	2.46 ± 0.11	3.91 ± 0.09	3.30 ± 0.14	0.62 ± 0.03	
	88/12	6.20 ± 0.10	2.55 ± 0.15	1.58 ± 0.20	4.25 ± 0.21	0.87 ± 0.03	
XC-1 Samples	(Depth; m)						
	11.1	3.95 ± 0.35	2.51 ± 0.55	0.48 ± 0.22	1.67 ± 0.61	0.85 ± 0.11	87 ± 10
	11.8	6.31 ± 0.18	2.44 ± 0.23	2.75 ± 0.20	3.65 ± 0.31	0.77 ± 0.05	80 ± 4
	12.2	4.44 ± 0.05	2.97 ± 0.13	1.78 ± 0.06	2.99 ± 0.12	0.72 ± 0.02	75 ± 2
	13.6 ± 0.9	4.82 ± 0.13	2.73 ± 0.22	5.86 ± 0.15	3.29 ± 0.27	0.50 ± 0.05	57 ± 4
XC-2 Samples	(Depth; m)						
	15.2 ± 0.9	6.73 ± 0.38	2.26 ± 0.49	1.65 ± 0.44	5.34 ± 1.73	0.91 ± 0.07	93 ± 7
	17.2 ± 0.5	4.72 ± 0.27	2.00 ± 0.34	5.48 ± 0.17	3.71 ± 0.44	0.62 ± 0.08	68 ± 7

[1] Parameters A, B, C and D are related (but not numerically equal) to high Mg calcite peak height, width, low Mg calcite peak height and width, respectively. Gaussian fit model discussed in text.
[2] Integrated peak intensities of (high Mg calcite)/(high Mg calcite + low Mg calcite).
[3] Obtained comparing calculated intensity ratios with calibration curve.

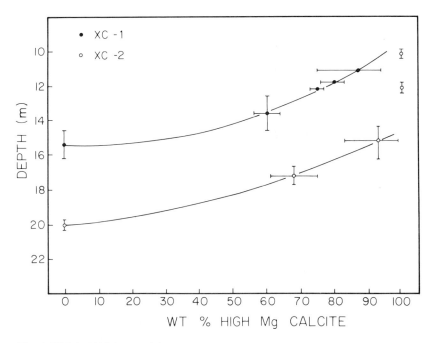

Fig. 6. Weight % high Mg calcite, as determined from powder X-ray diffraction spectra, vs. depth in the core. XC-1 and XC-2 samples are indicated by solid and open circles, respectively. Note the smooth variation in both cores of composition with depth and the 4 to 5 m displacement of the XC-2 from the XC-1 curve.

constant along any perimeter (i.e., circum-parallel to the reef crest). Ristvet et al. (1974) indicate that such lateral variations in diagenetic features of single stratigraphic units are observed at Eniwetok Atoll.

The Runit Island cores XRU-1, 4 and 3 were drilled in a line approximately parallel to both the elongated island outline and the reef crest. According to the argument just presented, depths to the high to low Mg calcite transition should be similar in these three cores. As noted in Table 1, however, the transition level in XRU-1 is depressed 1 to 3 m relative to that in XRU-3; a similar displacement is noted for XRU-4. These relations are illustrated in Fig. 7, a cross-section of Cactus Crater constructed from XRU and XC core data. The 20 and 35 m solution unconformities are based on Couch et al. (1975) data. The proximity of cores XRU-1 and 4 to Cactus Crater suggests possible cratering induced depression at these drill locations. Since the nature of the Mg calcite transition drastically changes directly beneath the crater, it is impossible to trace a sharp diagenetic transition into this area, and determining whether or not downward displacement is present here becomes difficult. Pressure attenuation considerations (Vizgirda and Ahrens, 1980) however, are consistent with stratigraphic depression occurring beneath the crater. According to results of this study, shock metamorphic levels in calcite from XC-1 require 4 to 5 m of downward displacement in order to accommodate theoretically reasonable pressure decay rates of d (depth)$^{-2}$ to d^{-3}. In addition, deeper solution unconformities (i.e., older than the first solution unconformity correlative with the Mg calcite transition) are observed to be depressed beneath the crater (see Fig. 7). Since no core samples below 2.5 m were obtained from locations to the north–west of the crater (i.e., in line with XRU-1 and XRU-3 but on the opposite side of the crater), the behavior of the solution unconformities across the crater cannot be determined. Nevertheless, available evidence from stratigraphic and pressure decay considerations suggest cratering induced depression beneath Cactus.

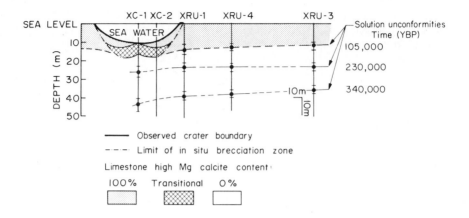

Fig. 7. Cross-section through Cactus Crater constructed on the basis of the high and low Mg calcite data and solution unconformity data of Couch et al. (1975). Note the apparent depression of the high to low Mg calcite transition boundary toward the cratered area and the inferred central peak inside the crater. The datum is mean low water level. (Two-fold vertical exaggeration.)

As discussed in previous sections, the nature of the high to low Mg calcite transition observed in samples from cores drilled inside Cactus Crater contrasts drastically with that observed in cores drilled outside the crater. The gradual change from all high to all low Mg calcite displayed by XC-1 and 2 core samples is not found anywhere else on Runit Island, nor is it expected considering the conditions under which the diagenesis occurs. Characteristics of this feature, unambiguously resulting from the cratering event, lead to several conclusions regarding the mechanics of crater formation.

First of all, the fact that the transition occurs in a consistent manner in both cores (see Figs. 2, 4 and 6), suggests that the material in this transitional interval was never ejected from the cavity, since ejection is not expected to result in such a finely stratified breccia lens. If a fallback breccia lens exists, it is limited at the center of the crater, i.e., at the XC-1 drill site, to a thickness of 1 m or less. This is the difference between the depths at which the hole was collared in and the first sample was retrieved. Instead, the bulk of material excavated during the cratering event experienced *in situ* brecciation and mixing. Mixing was probably initiated by material travelling radially downward and outward at particle velocities imparted after passage of the explosion-induced shock front. After brecciation of the originally saturated but competent coralline limestone, at least partial fluidization likely occurred, creating an inviscid, readily mixed substance. The turbulence of mixing was sufficient to completely homogenize hand specimen size samples. The extent of the gradational Mg calcite lens reflects the rate at which this turbulent mixing decreased with depth.

The observed pattern of the transitional Mg calcite layer also aids in the definition of the "excavation cavity". It the extrapolated "mixing lines" in Fig. 6 are valid, then the material experienced some degree of excavation, though not ejection, down to depths of approximately 15 and 20 m in XC-1 and XC-2 drill-core locations, respectively. Brecciation and mixing may have occurred at deeper levels within a totally low Mg calcite lithology, however, our methods would not allow its detection. The depths noted in previous sentences are, therefore, used to define the minimum limit of *in situ* brecciation sketched in Fig. 7.

Finally, the form of the transitional Mg calcite layer, sketched in Fig. 7 suggests a fundamental reclassification of Cactus Crater, which up to the present has been considered as belonging to the class of simple flat-floored craters. As depicted in Fig. 7, the transitional Mg calcite layer is 4 m thick in both cores and is depressed 4 to 5 m in XC-2 relative to that in XC-1. Neither of these observations is consistent with the structure of a simple crater. Rather, the observation that the transitional layer occurs at shallower depths in the center of the crater than at a distance of one-half crater radius outward from the center is compatible with the interpretation of Cactus as a complex crater with a central uplift. Two other possibilities, namely depth uncertainties and structural inhomogeneities, must first be investigated before any firm conclusions concerning crater structure can be drawn. As mentioned in the beginning of this section, sample depth levels in the Cactus Crater cores are not precisely known. Nevertheless, the margin of error is insufficient to alter the conclusion regarding the existence of a central uplift; its extent, however, may be under or overestimated by up to 3 m. A second situation that may produce the observed pattern is the presence of some inhomogeneity in the cratered vicinity. In this case, the depiction of the Cactus floor as symmetrical would be inaccurate, and the structural classification ambiguous. Inhomogeneities do exist in Runit Island. For example, Ristvet et al. (1978) describe a cable berm located south of borehole XC-1. Lacking good geologic controls, in particular borehole evidence in the western portion of the island, this second possibility cannot be conclusively eliminated. However, as we shall see in the next section, the observation of a complex morphology in a small crater would not be an isolated case.

Seismic investigations conducted on Runit Island during Project EXPOE (Ristvet et al., 1978) are limited to an approximately 550 m long seismic refraction profile obtained along the northeastern ocean margin (see Fig. 2). Based on this profile and geophysical logs of XRU boreholes, four seismic layers are distinguished in the upper 60 m of Runit. Apparent seismic velocities of these layers consistently vary from 0.4 km/sec to 3.1 km/sec with increasing depth, reflecting greater degrees of cementation. The shallowest layer represents unconsolidated supratidal sediments and is of uniform thickness along the extent of the seismic profile. The second layer, consisting of weakly to uncemented carbonate sediments, appears to thicken toward Cactus, and the interface (at approximately 15 m depths) between this and the third layer of moderately cemented carbonate rock appears to be depressed 3 m in the proximity of the crater. As mentioned earlier, and as shown in Fig. 7, two solution unconformities, occurring between 20 and 50 m depth, appear to be depressed beneath Cactus Crater; the extent of downward displacement is 6 m and 4 m for the first and second of these unconformities, respectively. On the basis of this, and the seismic evidence, Ristvet et al. (1978) also suggest that depression did occur beneath Cactus Crater.

Seismic velocities determined in the refraction profile may be used to further investigate the phenomenon of cratering induced brecciation. The velocities recorded for the second layer, i.e., the layer of unconsolidated to weakly cemented carbonate sediments, are observed to decrease consistently from 2.4 km/sec to 2.0 km/sec as one approaches the cratered area. Since this large a variation is not expected in a line parallel to the reef margin, it is inferred (Ristvet et al., 1978) that the velocity decrease results from shock-induced brecciation beneath the crater. The depth of this lower velocity layer at the end of the refraction line closest to the crater is about 20 m below sea level. This depth is consistent with the observed maximum excavation limit below Cactus Crater of 20 m below the datum. On the basis of the high to low Mg calcite transition evidence from cores XRU-1 and XRU-4, where the transition was observed as a sharp diagenetic break, the disturbance causing decreased velocities involves fracturing, but no mixing.

Paleontological evidence (Ristvet et al. 1978) supports speculations involving brecciation and mixing beneath the crater and the extent of the excavation cavity. Pink, unbleached shells of the encrusting foraminifera, *Homotrema rubrum*, which indicate an origin above the first unconformity, are observed in XC-1 and 2 cores down to depths of 16 ± 1.5 m below the datum. Outside Cactus, unbleached shells are found only above the first solution unconformity, correlative with the Mg calcite transition depths, listed in Table 1. This fossil evidence is unclear in that it is not noted whether bleached and unbleached foraminifera shells are observed together at any level above 16 m beneath the crater. Such an observation would be analogous to the co-existence of both high and low Mg calcite in Cactus Crater cores. Nevertheless, the *Homotrema rubrum* evidence supports mixing occurring to depths of 16 ± 1.5 m and the delineation of a minimum excavation cavity based on the Mg calcite data.

Gamma logs were obtained for most of the EXPOE boreholes. As anticipated, high radiation levels were recorded at several depths in boreholes XC-1 and XC-2; these patterns are reproduced in Fig. 8. Note the two high intensity peaks observed in both boreholes. As an accuracy check on the gamma log, radiation levels (primarily due to ^{60}Co, ^{102}Rh, ^{125}Sb and ^{155}Eu) of 24 split spoon samples were measured at the McClellan Central Laboratory, McClellan Air Force Base; this data is also presented in Fig. 8. These laboratory measurements corroborate both peaks observed in gamma logs of XC-1, and the shallower peak of XC-2. Lack of samples from 18 to 20 m depths of XC-2 may explain why the second high intensity radiation peak was not observed. Depths of the three laboratory observed high intensity peaks are consistently 2 m deeper than the well log data.

The observed distribution of gamma radiation serves to further elucidate the cratering process. Because peak penetration depths of nuclear fission products and neutrons are on the order of $10\,\mu$ and 1 m, respectively (see e.g., Fleischer et al., 1975; Burnett and Woolum, 1974), peak radiation levels are produced only in substrate materials in close proximity (within 1 m) to the nuclear device. The observed post-event radiation intensity pattern, therefore, is a result of cratering produced particle motions. The observed radiation pattern in Cactus Crater may be the result of the following sequence of events. Material exhibiting the highest measured radiation levels originated closest to the nuclear device. As the shock wave from the explosion propagated into the target, high particle velocities were induced in the same material. Some of this material travelled radially into the target and formed a lining to the excavation cavity. Other particles, engulfed by the rarefaction wave produced upon reflection of the shock wave from the free surface of the target, were ejected at high velocities from the growing crater cavity. Some of these particles fell back into the cavity and constitute a thin fallback breccia lens. It is proposed that the shallow and deep high radiation intensity peaks represent the fallback and high velocity injected material, respectively. Ristvet et al. (1978) also propose forceful injection to explain the deeper peak. This interpretation is consistent with the crater formation model derived on the basis of the Mg calcite transition data. In addition, the crater excavation limits based on the gamma well log evidence (16 m and 19 m for XC-1 and XC-2 cores, respectively) agree well with those determined from the same Mg calcite transition.

COMPARISONS WITH OTHER CRATERS AND CONCLUSIONS

In an attempt to gain a broader perspective of Cactus Crater in relation to other impact and explosion craters, the best comparison (in terms of size and substrate similarities) to be made is between Cactus and Snowball craters. The latter, described by Roddy (1976), is an experimental explosion crater, formed at the Defense Research Establishment, Suffield (Alberta) site by the detonation of 500 tons of TNT in saturated, unconsolidated layered clays, silts and silty sands. Snowball has a diameter of 107 m and a maximum depth of 9 m compared to respective dimensions of 105 m and 10.7 m for Cactus Crater. As in Cactus, water infilling occurred after the crater formed. A thin breccia lens (1 to 2 m) is observed at Snowball, corresponding to that speculated for Cactus. Major differences between the Cactus and Snowball sites are in the strength properties of the media, one being a basically competent but saturated coral rock, the other, saturated and completely unconsolidated sediments. There is a small difference in the heights of burst (HOB) of the two explosions. The 500 ton TNT hemisphere producing Snowball was detonated at ground level, i.e., zero HOB, whereas the 18 kTon TNT equivalent Cactus device had an approximately 1 m HOB. The similarity in size between the two craters despite the much greater (by a factor of 36) yield of Cactus is due to the greater cratering efficiency of high explosive, relative to nuclear explosive, sources. According to the empirical correlation curves in Cooper (1977), the cratering efficiency (defined as crater

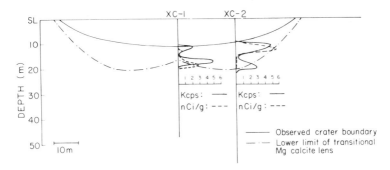

Fig. 8. Cactus Crater cross-section (no vertical exaggeration) showing gamma (γ) well log data (solid line) and laboratory measured total gamma activity (dashed line) for XC-1 and XC-2 cores; units for these two data sets are in kilocounts per second and nanocuries per gram, respectively. Figure is compiled from data presented in Ristvet et al., 1978. (Datum is mean low water level.)

volume/yield) is 50 times greater for a high explosive relative to a high energy density nuclear explosive source for craters with HOB and volume characteristics of Snowball and Cactus. A distinct central peak structure, with an uplift of 8.5 m, is evident in Snowball. Although intensely folded, the clay and silt beds in this central mount retain sufficient integrity and can be mapped. Cactus, as speculated, also has a central peak, with 4 to 5 m of uplift. The rock making up this central structure, however, appears to be intensely shattered and mixed, and, had visually distinguishable units been present, they could probably not have been mapped. Thus, significant differences exist in the size and nature of the two central peaks. Furthermore, differences in the mode of disruption are observed away from the central peak. Numbered markers used in the Snowball experiment, allowing resolution of mass movements during cratering, indicate two discrete sets of motions occurred beneath the crater floor: the top layer moved radially outward, while material below it moved toward the center and up to form the central peak. Samples from the XC-2 Cactus core suggest turbulent brecciation and mixing compared to the (relatively) ordered movements beneath Snowball. It appears, therefore, that following shock-induced brecciation, the coral rock becomes less coherent than the silt and clay sediments. Differences in shock compression properties, or in the amount or distribution of water in the two types of substrates may be responsible for the varying behavior. Despite these differences, the two craters are similar in many respects, the most important being target saturation, and the analogy of Snowball to Cactus serves as an important feasibility argument for the description of Cactus as a complex crater.

Applying the Bingham plastic model of Melosh (1981) gives some indication of the material strength of the shocked coral material. According to this model, a minimum value of 5 for the dimensionless strength parameter, $\rho g H/c$, is required before crater collapse can occur; central peak formation requires a larger value, but the limiting case is considered here. Using a water-saturated coral density, ρ, of 2.4 g/cc (from measured bulk ρ) and a transient crater depth, H, of 20 m, a maximum Bingham yield stress of ~1 bar is calculated (g is the acceleration of gravity). Interestingly, yield stresses of the same order of magnitude have been measured for a variety of water-saturated clays, comparable to those at the Snowball Crater site (Wilson, 1927, p. 58–59). The yield stress, therefore, may be the critical parameter determining crater size and shape. There is potential then, according to this model, for quantitative treatment of crater structure target dependence.

Although information on terrestrial meteorite craters is often incomplete (or nonexistant), some comparisons between these craters and Cactus can be drawn. Where depth to the bottom of the breccia lens can be ascertained, the depth to diameter ratio of simple terrestrial craters is 1:3 (Dence et al., 1977). Determining both the extent of the breccia lens and the rim position of the excavated cavity is particularly difficult for complex structures, and the depth to diameter ratio of these craters is speculative. A palinspastic reconstruction of the Sierra Madera and Gosses Bluff structures suggests a ratio of 1:4 or 1:5 (Dence et al., 1977). Using geologic observations and geophysical subsurface data, Pohl et al. (1977) speculate that the transient cavity of the Ries Crater had a similar depth to diameter ratio. The apparent 1:5 ratio for the true cavity of Cactus, therefore, is more compatible with that of complex structures.

Downward displacement has not actually been observed beneath any simple craters (Dence et al., 1977), even those which offer sufficient stratigraphic control, such as Meteor Crater (Shoemaker, 1963). On the basis of attenuation rate arguments, however, Robertson and Grieve (1977) speculate that depression and approximately 50% shortening of the rock section beneath the crater occurred at Brent, a simple crater in crystalline rock. In complex craters, rocks in the ring surrounding the central peak typically appear to have undergone downward and inward motions. At the Red Wing Creek buried structure, there is some evidence, which eventually dies out at greater depth, for depression of rocks beneath the central peak (Brenan et al., 1975). Stratigraphic depression beneath simple experimental impact craters has been clearly demonstrated by Gault et al. (1968), among others. In a laboratory scale study of explosive craters formed in loose saturated sand, Piekutowski (1977) observed downward displacement beneath a crater with a distinct central uplift. These laboratory studies, particularly those of Piekutowski which utilized saturated target materials, lend significant support to the speculated depression beneath Cactus.

In Vizgirda and Ahrens (1980), an attenuation rate of $(depth)^{-5.7}$ is calculated for Cactus on the basis of electron spin resonance detected shock metamorphism; this rate is calculated in terms of post-flow stratigraphic coordinates. According to model calculations, downward displacement of 4 to 5 m is consistent with attenuation rates of d^{-2} to d^{-3}. These rates are compatible with those inferred for simple craters (Robertson and Grieve, 1977). Considering the uncertainty in the estimate of displacement beneath Cactus, and the speculative nature of model attenuation rates, firm conclusions about either cannot be drawn. The interesting fact remains, however, that estimated amounts of displacement are consistent with reasonable attenuation rates.

The theorized thin (1 m or less) fallback breccia lens at Cactus is in agreement with general observations. Thin fallback breccia lenses, relative to the extent of autochthonous breccia, have been observed in both simple and complex structures, such as Meteor Crater (Shoemaker, 1963) and Gosses Bluff (Milton et al., 1972).

The γ ray intensity logs, discussed in the previous section, revealed two high intensity peaks, one located near the crater surface and the other at the hypothesized excavation cavity boundary. This pattern is similar to the distribution in melts or of intensely shocked and pulverized rock in simple craters in crystalline and sedimentary rock. At Brent and Lonar Lake Craters, formed in granitic gneiss and basalt, respectively, melt rocks are concentrated in the top and basal portions of the excavated cavity (Dence, 1968; Fredriksson et al., 1973). At Meteor Crater, formed entirely in sedimentary rocks, meteorite fragments and highly shocked target rock are concentrated in similar positions (Shoemaker, 1963). These two layers of highly shocked material have not been clearly identified in complex craters. The extensive melt sheets described at complex craters, for example Manicouagan, are interpreted as the material lining the excavation cavity (Grieve and Floran, 1978). Melt layers are detected at two distinct intervals beneath the ring depression of Boltysh, a 20–25 km diameter structure in pre-Cambrian crystalline rock (Yurk et al., 1975), however, the lower "unit" of melt was detected at only 1 borehole location and extended for a mere 20 m. Its lateral extent, thus, remains ambiguous, and a melt distribution analogy between Boltysh and simple craters cannot be justified. Based on the γ log evidence, an analogy between Cactus and simple craters may be warranted.

Grieve et al. (1977), on the basis of impact melt distributions in simple and complex craters, proposed a common cratering model for both types of structures. Some of the interpretations stemming from Cactus Crater observations are consistent with their model. The inferred sequence of events and material motions during the Cactus excava-

CRATER MODIFICATION MODELS

CENTRIPETAL SLIDING

REBOUND

Fig. 9. Sketches depicting two possible crater modification mechanisms. Dashed line in the rebound model represents limit of region "fluidized" during excavation process (after Melosh, 1981). Reasons for preference of the rebound model for Cactus Crater are discussed in the text.

tion stage, for example, agree closely with that described by Grieve *et al.* Evidence for the modification mechanism acting at Cactus is less clear. Centripetal sliding (depicted in Fig. 9), a commonly invoked modification means for complex craters, produces an intensely disordered central peak. As evidenced in the orderly high to low Mg calcite transition in the XC-1 core, and its close similarity to the transition trend in XC-2, such disordering does not seem to have taken place in the central uplift of Cactus Crater. This observation suggests that a rebound mechanism (Fig. 9) rather than centripetal sliding may have produced the Cactus central peak. The Melosh (1981) Bingham plastic model assumes such a hydrodynamic rebound theory to explain various morphologic features of complex craters, such as central uplifts. If rebound, and not sliding, was indeed the active mechanism in the Cactus cratering event, then the observed crater diameter closely approximates that of the transient crater; the depth to diameter ratio of 1:5 would, thus, be confirmed. The difference between this ratio and the 1:3 value commonly observed in simple craters suggests an essential difference in the transient crater geometry of simple and complex craters. This same suggestion has been proffered on the basis of terrestrial impact crater observations (see e.g., Pike, 1980) and experimental explosive crater investigations (Schmidt and Holsapple, 1981).

SUMMARY

Summarizing the key observations and interpretations, Cactus Crater can be described as a 105 m diameter complex crater with 4 to 5 m of central uplift and possible downward displacement beneath the structure. Maximum depth of the excavation cavity is 20 m, and the resultant true crater depth to diameter ratio is 1:5. The majority of the breccia infilling the excavation cavity is of autochthonous origin and a fallback breccia lens, if it exists, is limited to a maximum thickness of 1 m. The excavation process, as deduced from Mg calcite and γ log data, involves both ejection of some of the highly shocked material and high velocity injection of the remainder to form a lining to the transient crater floor. This *in situ* brecciation and mixing is envisioned as a very turbulent mechanism, facilitated by the high velocities and low viscosities of the intensely shocked and, probably, fluidized carbonate rock. Such a sequence of events is consistent with the crater excavation model of Grieve *et al.* (1977), deduced on the basis of impact melt distribution in terrestrial, particularly simple, craters. Thus, observations at Cactus Crater (specifically, the two peaks in the γ log data, inferred to be analogous to the basal and upper melt layers in simple craters) serve to support a basic contention of the Grieve *et al.* model, that is, that a single excavation model can be applicable to both simple and complex craters. Cactus

Crater modification, on the basis of the Mg calcite study, is speculated to involve a dynamic rebound mechanism rather than centripetal sliding. Using the Bingham plastic rebound model of Melosh (1981), a maximum yield stress of ~1 bar is calculated for the shock-affected carbonate medium. Clays, such as those in which Snowball (an explosion crater with many features similar to Cactus) was formed, have comparable yield stresses.

Some of the features observed at Cactus have counterparts in terrestrial impact craters and may have been produced by the same processes. Application of a cratering model based on Cactus to other impact and explosion structures, however, should proceed with caution, since the unique properties of a water-saturated carbonate target may influence cratering mechanisms as well as modulate transitions between basic morphologic categories. The detailed observations presented here will, hopefully, contribute to the quantitative evaluation of target properties on crater morphologies, and serve as useful guidelines in the formulation of theoretical cratering models.

Acknowledgments—Discussions with B. Ristvet and J. Melosh are gratefully acknowledged. This research was supported under DNA contract 001-79-C-0252. Contribution No. 3615, Division of Geological and Planetary Sciences, California Institute of Technology, Pasadena, California 91125.

REFERENCES

Bøggild O. B. (1930) The shell structure of the mollusks. *Danske Vidensk. Selsk. Skr.* **9**, pt. II, 235–326.

Brenan R. L., Peterson B. L., and Smith H. J. (1975) The origin of Red Wing Creek structure: McKenzie County, North Dakota, *Wyom. Geol. Assoc. Earth Sci. Bull.* **8**, 1–41.

Burnett D. S. and Woolum D. S. (1974) Lunar neutron capture as a tracer for regolith dynamics, *Proc. Lunar Sci. Conf. 5th*, p. 2061–2074.

Chave K. E. (1952) A solid solution between calcite and dolomite. *J. Geol.* **60**, 190–192.

Chave K. E. (1954) Aspects of the biogeochemistry of magnesium, calcareons marine organisms. *J. Geol.* **62**, 266–283.

Cooper H. F. Jr. (1977) A summary of explosion cratering phenomena relevant to meteor impact events. In *Impact and Explosion Cratering* (D. J. Roddy, R. O. Pepin, and R. B. Merrill, eds.), p. 11–44. Pergamon, N.Y.

Couch R. F., Fetzer J., Goter E., Ristvet B., Tremba E., Walter D., and Werdland V. (1975) Drilling Operations in Eniwetok Atoll During Project EXPOE. Air Force Weapons Laboratory, Rep. No. AFWL-TR-75-216. 267 pp.

Dence M. R. (1968) Shock zoning at Canadian craters: petrography and structural implications. In *Shock Metamorphism of Natural Materials* (B. M. French and N. M. Short, eds.), p. 169–184. Mono, Baltimore.

Dence M. R., Grieve R. A. F., and Robertson P. B. (1977) Terrestrial impact structures: principal characteristics and energy considerations. In *Impact and Explosion Cratering* (D. J. Roddy, R. O. Pepin, and R. B. Merrill, eds.), p. 247–275. Pergamon, N.Y.

Fleischer R. L., Price P. B., and Waler R. M. (1975) *Nuclear Tracks in Solids: Principles and Applications*. Univ. Calif., Berkeley. 605 pp.

Fredriksson K., Dube A., Milton D. J., and Balasundaram M. S. (1973) Lunar Lake, India: An impact crater in basalt. *Science* **180**, 862–864.

Gault D. E., Quaide W. L., and Oberbeck V. R. (1968) Impact cratering mechanics and structures. In *Shock Metamorphism of Natural Materials* (B. M. French and N. M. Short, eds.), p. 87–99. Mono, Baltimore.

Grieve R. A. F., Dence M. R., and Robertson P. B. (1977) Cratering processes: as interpreted from the occurrence of impact melts. In *Impact and Explosion Cratering* (D. J. Roddy, R. O. Pepin, and R. B. Merrill, eds.), p. 791–814. Pergamon, N.Y.

Grieve R. A. F. and Floran R. J. (1978) Manicouagan impact melt, Quebec Part 2. Chemical interrelations with basement and formational processes. *J. Geophys. Res.* **83**, 2761–2771.

Howard K. A., Offield T. W., and Wilshire H. G. (1972) Structure of Sierra Madera, Texas, as a guide to central peaks of lunar craters. *Geol. Soc. Amer. Bull.* **83**, 2795–2808.

Land L. S. (1967) Diagenesis of skeletal carbonates. *J. Sediment. Petrol.* **37**, 914–930.

Melosh H. J. (1981) A schematic model of crater modification by gravity. *J. Geophys. Res.* In press.

Milton D. J., Barlow B. C., Brett, R., Brown A. R., Glikson A. Y., Manwaring E. A., Moss F. J., Sedmik E. C. E., Van Son J., and Young G. A. (1972) Gosses Bluff impact structure, Australia. *Science* **175**, 1199–1207.

Neumann A. C. (1965) Processes of recent carbonate sedimentation in Harrington Sound, Bermuda. *Bull. Mar. Sci.* **15**, 987–1035.

Offield T. W. and Pohn H. A. (1979) Geology of the Decaturville impact structure, Missouri. *U.S. Geol. Survey. Prof. Paper* 1042. 48 pp.
Piekutowski A. J. (1977) Cratering mechanisms observed in laboratory-scale high-explosive experiments. In *Impact and Explosion Cratering* (D. J. Roddy, R. O. Pepin, and R. B. Merrill, eds.), p. 67–102. Pergamon, N.Y.
Pike R. J. (1980) Terrain dependence of crater morphology on Mars both yes and no (abstract). In *Lunar and Planetary Science XI*, p. 885–887. Lunar and Planetary Institute, Houston.
Plummer L. N. and Mackenzie F. T. (1974) Predicting mineral solubility from water data; application to the dissolution of magnesian calcites. *Amer. J. Sci.* **274**, 61–83.
Pohl J., Stöffler D., Gall H., and Ernstson K. (1977) The Ries impact crater. In *Impact and Explosion Cratering* (D. J. Roddy, R. O. Pepin, and R. B. Merrill, eds), p. 343–404. Pergamon, N.Y.
Ristvet B. L., Couch R. F., Fetzer J. D., Goter E. R., Tremba E. L., Walter D. R., and Wendland V. P. (1974) A Quaternary diagenetic history of Eniwetok Atoll (abstract). *Abstracts with Program, Geol. Soc. Amer.* **6**, 928–929.
Ristvet B. L., Tremba E. L., Couch R. F., Fetzer J. A., Goter E. R., Walter D. R., and Wendland V. P. (1978) Geologic and Geophysical Investigations of the Eniwetok Nuclear Craters. Air Force Weapons Laboratory, Rep. No. AFWL-TR-77-242. 330 pp.
Robertson P. B. and Grieve R. A. F. (1977) Shock attenuation at terrestrial impact structures. In *Impact and Explosion Cratering* (D. J. Roddy, R. O. Pepin, and R. B. Merrill, eds.), p. 687–702. Pergamon, N.Y.
Roddy D. J. (1976) High-explosive cratering analogs for bowl-shaped, central uplift, and multiring impact craters. *Proc. Lunar Sci. Conf. 7th*, p. 3027–3056.
Schmidt R. M. and Holsapple K. A. (1981) An experimental investigation of transient crater size (abstract). In *Lunar and Planetary Science XII*, p. 934–936. Lunar and Planetary Institute, Houston.
Shoemaker E. M. (1963) Impact mechanics at Meteor Crater, Arizona. In *The Moon, Meteorites and Comets* (D. M. Middlehurst and G. P. Kuiper, eds.), p. 301–336. Chicago, IL.
Silliman B. (1846) On the chemical composition of the calcareous corals. *Amer. J. Sci.* **1**, 2nd ser., p. 189–199.
Vizgirda J. and Ahrens T. J. (1980) Shock-induced effects in calcite from Cactus Crater. *Geochim. Cosmochim. Acta* **44**, 1059–1069.
Wilson H. (1978) *Ceramics Clay Technology.* McGraw-Hill, N.Y. 296 pp.
Yurk Y. Y., Yeremenko G. K., and Polkanov Y. A. (1975) The Boltysh depression—a fossil meteorite crater. *Int. Geol. Rev.* **18**, 196–202.

Impact accretion experiments

V. Werle[1], H. Fechtig[1], and E. Schneider[2]

[1]Max-Planck-Institut für Kernphysik, Heidelberg, W. Germany
[2]Ernst-Mach-Institut, Freiburg i.Br., W. Germany

Abstract—High velocity impact experiments have been performed using steel spheres (1.5 and 2.0 mm diameter) into low density targets (Saffile Al_2O_3, density $\rho_T = 0.18$ g/cm^3) within a velocity range between 0.5 and 8 km/sec. The morphology of these impact channels deviate considerably from impacts in low porosity silicates and metals. The amount of mass lost as ejecta was measured. As a result we conclude that mass accretion could occur during impact into very low density object for projectile velocities up to about 8 km/sec.

INTRODUCTION

Cratering on glass, silicate and metal targets shows the dependence of crater morphology on the impact velocity and mass of the projectile as well as on projectile and target material properties (Fechtig et al., 1975; Hörz et al., 1975; Brownlee, 1978). In particular it could be shown that the ratios of crater diameter (D) to crater depth (T) ratios primarily depend on the projectile density (ρ_{pr}) at impacts above a certain threshold velocity (~4 km/sec; Nagel and Fechtig, 1980). In all of the above target media experiments the target suffers considerable mass loss during hypervelocity cratering, because the ejected mass greatly exceeds the projectile mass. This means that the target would erode during impact, if the target were small enough so that the influence of its gravity could be neglected. For that reason every particle ejected from the impact crater will escape from the surface of the target (Brownlee et al., 1973; Nagel et al., 1976).

First studies of high velocity impacts using steel projectiles on low density target material have been reported by Cannon and Turner (1967) and Fechtig et al. (1980). Both papers show that craters in low density material deviate considerably in morphology from craters in low porosity silicates (Fechtig et al., 1975). We have now investigated gain or loss of target mass as a result of high velocity impacts. The investigations have been carried out as a function of target density as well as of mass and velocity of the projectile.

EXPERIMENT DESCRIPTION

Using a light gas gun at the Ernst-Mach-Institut at Freiburg i.Br. steel spheres within an average density of 7.8 g/cm^3 (1.5 and 2.0 mm diameter) have been accelerated to velocities of 0.5 to 8 km/sec. The light gas gun facility is described by Fechtig et al. (1972). As a low density target we have used a commercially available material "Saffile" with densities $\rho_T = 0.18$ and 0.38 g/cm^3. Saffile has a porous and fibrous structure with holes generally as small as 10 μm. The fibres consist of Al_2O_3 and therefore the compressed material at 0% porosity has the material properties of corundum. The noncompressed material behaviour is controlled by the strength of the forces holding together the several fibres. The target has been prepared by stacking discs 7 cm in diameter and 2 cm thick, thus allowing crater morphology to be studied. The target configuration with the important dimensions is shown in Fig. 1. We have mounted a catcher in front of the target in order to collect the emitted particles.

RESULTS

Compared to our earlier report (Fechtig et al., 1980) which describes the crater morphology for impacts of 1.5 mm steel spheres into targets with density of 0.28 g/cm^3, our

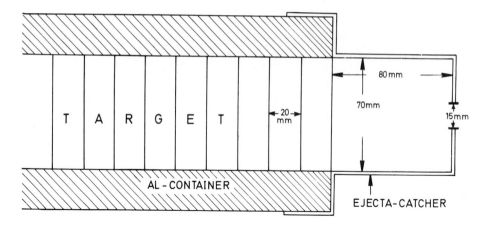

Fig. 1. Target configuration.

new results show no significant differences at different target densities ($\rho_T = 0.18$ and $\rho_T = 0.38$ g/cm^3) and the two sizes of steel spheres (1.5 and 2.0 mm diameter).

The results of crater morphology are shown in Figs. 2 and 3 for $\rho_T = 0.18$ and 0.38 g/cm^3 and the two different projectile diameters. At low velocities (<1.5 km/sec) the crater channel is rather long with a diameter slightly wider than the projectile diameter. At higher velocities an excavation occurs at about 20 mm crater depth, which increases with impact velocity. The maximum diameter D_{max} of these cavities appears in every crater at a depth of about 20 mm, independent of target density and dynamical properties of the projectile. This may be an effect of the target layering. The value of D_{max} itself depends only on the kinetic energy of the impacting projectile, not on the target density or the projectile diameter. As shown in Fig. 2b there is a variation of D_{max} of the order of a few millimeters for the two shots with the same impact velocity at 4.2 km/sec. Therefore the variation of D_{max} for comparable values of the projectile energy at different target densities of the order of a few millimeters is not significant.

The entrance widths of the crater channels at low velocities are only slightly wider than the projectile diameters. The entrance widths increase with the impact velocity but do not vary much with target density and projectile diameter. In Figs. 2 and 3 the initial masses m_i and final masses m_f of the projectiles are listed in milligrams. The final mass m_f is the mass of the projectile that is recovered at the bottom of the impact channel. The ablated mass of the projectile is the difference $m_a = m_i - m_f$. In addition, the mass lost as ejecta during impact has been determined and is indicated in Figs. 2 and 3 as m_1 in milligrams.

It was impossible to determine accurately the mass m_1, since within the collected mass not only ejected target material but also material from the sabot of the light gas gun is present. Based on its dark colour, the amount of sabot material could be estimated. It was impossible to sort out this material because it was completely intermixed with the fine grained dust of target material. It also is likely that not all of the dust could be collected. Estimates, based on the comparable densities both of target and sabot material, led us to the conclusion that the maximum error is certainly smaller than 20%. Figure 4 shows the ejected mass relative to the projectile mass as a function of impact velocity. With increasing impact velocity v_{pr} an increase in mass loss is generally observed. At velocities >3.5 km/sec the projectile sometimes splits into several pieces. The probability of splitting increases with increasing impact velocity. The split projectiles sometimes produce more ejecta mass than nonsplit projectiles. Only in four cases has mass loss been recorded. The splitting of a projectile causes a strong impulse to the ejecta and therefore gives rise to an increased amount of lost material. A least-squares fit computed for each of the two target densities leads to the conclusion that mass accretion occurs up to ~7.5 ± 2.8 km/sec for $\rho_T = 0.18$ g/cm^3 and up to ~9.2 ± 1.2 km/sec for $\rho_T = 0.38$ g/cm^3.

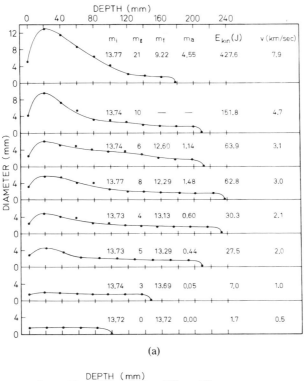

(a)

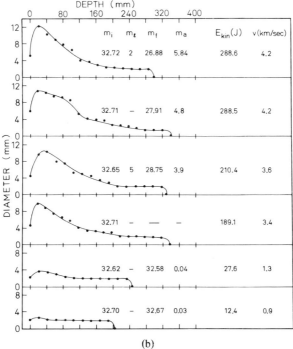

(b)

Fig. 2. Diameter D as a function of depth T for different velocities v_{pr}. Target: Saffile Al_2O_3, $\rho_T = 0.18\,g/cm^3$. Diameter and depth are given in mm; m_i = initial mass of projectile, m_f = final mass of projectile, $m_a = m_i - m_f$ = ablated mass of projectile, m_1 = lost mass = mass of ejected material (all numbers of m_i, m_f, m_a, m_1 are given in mg). (a) Porjectiles: Steel spheres, diameters 1.5 mm. (b) Projectiles: Steel spheres, diameters 2.0 mm.

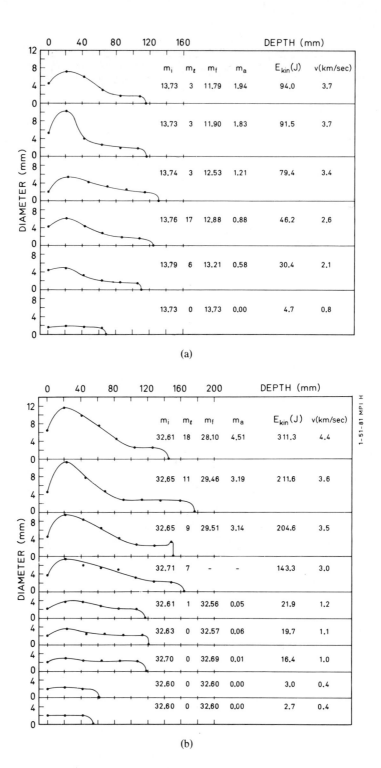

Fig. 3. Diameter D as a function of depth T for different velocities v_{pr}. Target: Saffile Al_2O_3. $\rho_T = 0.38 \text{ g/cm}^3$. m_i = initial mass of projectile, m_f = final mass of projectile, $m_a = m_i - m_f$ = ablated mass of projectile, m_l = lost mass = mass of ejected material (all numbers of m_i, m_f, m_a, m_l are given in mg). (a) Projectiles: Steel spheres, diameters 1.5 mm. (b) Projectiles: Steel spheres, diameters 2.0 mm.

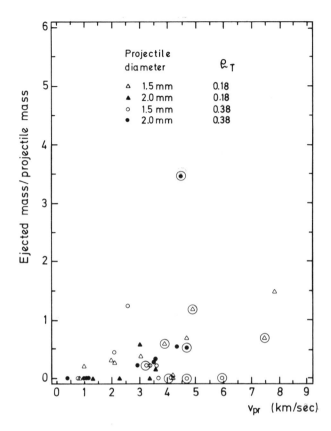

Fig. 4. Ejected mass vs. projectile mass as a function of projectile velocity v_{pr}. (Symbols in circles: split projectiles).

Figure 5 shows the crater depth T as a function of projectile velocity v_{pr}. The crater depth increases at lower velocities as a function of projectile velocity, showing a maximum at about 3–4 km/sec and decreases for higher velocities. Figure 5a shows the results of the two different target densities $\rho_{T1} = 0.18$ and $\rho_{T3} = 0.38$ g/cm³ using 1.5 mm steel spheres, in comparison with the results published earlier for $\rho_{T2} = 0.28$ g/cm³ (Fechtig et al., 1980). Figure 5b shows the results for the two target densities ρ_{T1} and ρ_{T3} using 2.0 mm steel spheres. It is clearly shown that the crater depth decreases proportional to increasing target density. The crater depth is larger for the 2.0 mm steel spheres compared to the 1.5 mm spheres for the same target density due to the larger impact energy of the bigger spheres. It was impossible to accelerate the 2.0 mm projectiles to velocities greater than 5 km/sec with our light gas gun. The shapes of the curves in Fig. 5b are therefore not as complete as they are in Fig. 5a for the 1.5 mm spheres.

CONCLUSIONS

The cratering mechanism of high velocity projectiles into low density material is clearly a superposition of two different processes: a mechanical stopping process and a high velocity cratering process that produces the cavities. It is obvious that at low velocities (<1.5 km/sec) the projectiles are stopped by displacement of the porous material along the impact path. At high velocities (>8 km/sec) an internal impact crater is found close to the target surface. Above 4 km/sec generally an ablated and split projectile is stopped in a channel below the bottom of the crater. Between 1.5 and 8 km/sec a superposition of these two effects is observed. Our main conclusion is that for low density target material,

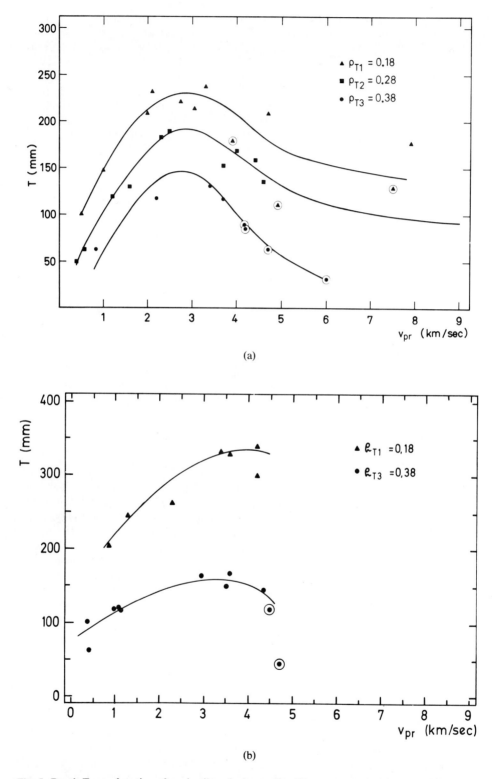

Fig. 5. Depth T as a function of projectile velocity v_{pr} for different target densities ρ_T. (Symbols in circles: split projectiles). (a) Projectiles: Steel spheres, diameter 1.5 mm. (b) Projectiles: Steel spheres, diameter 2.0 mm.

accretion occurs even at comparatively high impact velocities. For the velocities under discussion (≤ 10 km/sec) accretion processes occur for targets exceeding the produced depths; that means, for targets larger than about 0.5 m in diameter. For those small targets gravitation can be neglected. For much larger target sizes gravitational capturing of ejected material will improve the accretion process and will enlarge the velocity range. This is of considerable importance for the formation of planitesimals in the early solar system. As suggested by Völk (1981) large composite particles in the cm or even m size ranges are formed by impact accretion at low velocities. The results of the experiments described above show that high velocity impact processes that produce mass accretion with targets several meters and larger in diameter and composed of extremely low density material are possible. Such material has been measured in meteor streams (Hughes, 1978) and is assumed possible for comets.

REFERENCES

Brownlee D. E. (1978) Microparticle studies by sampling techniques. In *Cosmic Dust* (J. A. M. McDonnell, ed.) p. 295–336. Wiley, Chichester.

Brownlee D. E., Hörz F., Vedder J. F., Gault D. E., and Hartung J. B. (1973) Some physical parameters of micrometeroids. *Proc. Lunar Sci. Conf. 4th*, p. 3197–3212.

Cannon E. T. and Turner G. H. (1967) Cratering in low-density targets. NASA CR-789.

Fechtig H., Gault D. E., Neukum G., and Schneider E. (1972) Laboratory simulation of lunar craters. *Naturwissenschaften* **59**, 151–159.

Fechtig H., Gentner W., Hartung J. B., Nagel K., Neukum G., Schneider E., and Storzer D. (1975) Microcraters on lunar samples. In *Proc. Soviet-American Conference on Cosmochemistry.* NASA SP-370, p. 585–603. Washington, D.C.

Fechtig H., Nagel K., and Pailer N. (1980) Collisional processes of iron and steel projectiles on targets of different densities. In *Solid Particles in the Solar System.* Proceedings of IAU Symposium 90 held at Ottawa, Canada, 1979 (I. Halliday and B. A. McIntosh, eds.), p. 357–363. Dordrecht, Reidel.

Hörz F., Brownlee D. E., Fechtig H., Hartung J. B., Morrison D. A., Neukum G., Schneider E., Vedder J. F., and Gault D. E. (1975) Lunar microcraters: Implications for the micrometeroid complex. *Planet. Space Sci.* **23**, 151–172.

Hughes D. W. (1978) Meteors. In *Cosmic Dust* (J. A. M. McDonnell, ed.) p. 123–185. Wiley, Chichester.

Nagel K., Neukum G., Dohnanyi S. J., Fechtig H., and Gentner W. (1976) Density and chemistry of interplanetary dust particles derived from measurements of lunar microcraters. *Proc. Lunar Sci. Conf. 7th*, p. 1021–1029.

Nagel K. and Fechtig H. (1980) Diameter to depth dependence of impact craters. *Planet. Space Sci.* **28**, 567–573.

Völk H. (1981) Formation of the planetary system. *Mitt. Astron. Ges.* **51**, 63–79.

Impact cratering experiments in Bingham materials and the morphology of craters on Mars and Ganymede

Jonathan H. Fink,[1] Ronald Greeley,[1] and Donald E. Gault[2]

[1]Geology Department, Arizona State University, Tempe, Arizona, 85287
[2]Murphys Center for Planetology, P.O. Box 833, Murphys, California 95247

Abstract—Results from a series of laboratory impacts into clay slurry targets have been compared with photographs of impact craters on Mars and Ganymede. The interior and ejecta lobe morphology of rampart type craters as well as the progression of crater forms seen with increasing diameter on both Mars and Ganymede are explained qualitatively by a model for impact into Bingham materials. For increasing impact energies and constant target rheology, laboratory craters exhibit a morphologic progression from bowl shaped forms typical of dry planetary surfaces to craters with ejecta flow lobes and decreasing interior relief, characteristic of more volatile rich planets. A similar sequence is observed for uniform impact energy in slurries of decreasing yield strength. The planetary progressions are explained by assuming that volatile rich or icy planetary surfaces behave locally like Bingham materials and produce ejecta slurries with yield strengths and viscosities comparable to terrestrial debris flows. Hypothetical impacts into Mars and Ganymede are compared and it is concluded that less ejecta would be produced on Ganymede due to its lower gravitational acceleration, surface temperature, and density of surface materials. Limitations due to assumptions inherent in the model are discussed.

INTRODUCTION

Impact craters on Mars whose ejecta blankets appear to have been emplaced by surface flow have been cited as evidence for near surface volatiles in the martian crust (Head and Roth, 1976; Carr et al., 1977) (Fig. 1a). Similar crater morphology has been observed in Voyager images of Ganymede (Strom et al., 1981; Horner and Greeley, 1981) (Fig. 1b). The combined presence of flow features and steep distal escarpments on the martian ejecta lobes has led to the application of a Bingham rheological model to the ejecta material (Gault and Greeley, 1978; Greeley et al., 1980), similar to that used to describe terrestrial debris flows (Johnson, 1970), mudflows (Fink et al., 1981), and lava flows (Hulme, 1974). Alternative models for martian ejecta morphology have been proposed which consider interaction with the martian atmosphere (Schultz and Gault, 1979) or with ground water or ice (Wohletz and Sheriden, 1981) to be crucial. According to these models, much of the ejecta is emplaced as a turbulent cloud of low density, however the presence of distal scarps still implies that the flowing ejecta may be characterized as having both a viscosity and yield strength.

A second attribute of both martian and ganymedean craters is that they exhibit progressive morphologic changes with increasing diameter. In a given martian terrain, the smallest craters have bowl shapes, similar to their lunar and mercurian counterparts. With increasing diameters, craters show relatively flat floors followed by central peaks, peak rings and multiple peak rings (Wood et al., 1978; Wood, 1980). Along with these changes in interior morphology, the ejecta deposits become progressively more complex, ranging from single to double to multiple concentric ejecta lobes (Mouginis–Mark, 1979). On Ganymede, morphologic progressions are more difficult to specify due to lower resolution of the Voyager images. The principle differences from the martian sequence are that with increasing diameter, central peak craters give way to craters with central pits,

Fig. 1a. A 9-km-diameter impact crater in Chryse Planitia showing central peak, upper continuous ejecta blanket with steep scarp, and lower discontinuous ejecta deposits.

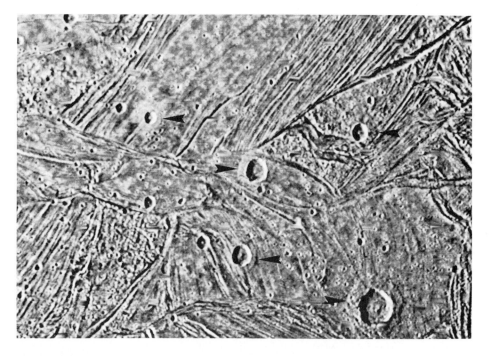

Fig. 1b. High resolution Voyager 2 filtered image showing several rampart type craters (arrows). Resolution is about 0.7 km/pixel. (FDS 20638.59). (From Horner and Greeley, 1981).

followed by large craters with little or no preserved topographic relief (palimpsests) (Passey and Shoemaker, 1981; Greeley et al., 1981).

These systematic changes in crater type have been interpreted as resulting from layering of the planetary crust. Early experimental work (Oberbeck and Quaide, 1968) demonstrated that target layering could produce a progression of crater forms with increasing diameter ranging from bowl shaped to flat floored to concentric or central pit type. These experiments used dry granular surface layers over rigid substrates for targets and their results were applied to the lunar crater population. Cintala and Mouginis-Mark (1980) measured depth to diameter ratios for 172 fresh martian craters and found that medium size craters had flatter floors than their lunar or mercurian equivalents. They interpreted this result as indicating the presence of a volatile rich subsurface layer. Whereas small craters excavate entirely within the volatile free zone above the layer and large craters incorporate great amounts of dry material from below the layer, intermediate size craters sample a large amount of volatile rich material, resulting in lower particle trajectories and hence flatter craters. Simulation of craters on Ganymede was attempted using various layered target configurations (Greeley et al., 1981). Transitions from bowl shape to central peak to flat floored to central pit to palimpsest craters with increasing diameter were achieved using dry incompetent surface layers over clay slurry substrates.

In contrast to these experiments and models employing multilayered targets, the current study is based upon impacts into homogeneous, Bingham type media which may behave as fluids under high applied stresses and as solids under lower stresses. The observed cratering processes in these targets have attributes of those in Newtonian fluids such as water, as well as those in dry cohesive targets. The final crater forms reflect the oscillation of central peaks, as in fluid targets, but also demonstrate the finite yield strength of the slurries. It is proposed that Bingham-like behavior of a volatile rich planetary crust could result within the volume of the transient crater bowl where shock stresses and temperatures are high enough to melt ice and form a slurry of rock and water. The same process could occur on both Ganymede and Mars, although the ratio of ice to rock would probably be higher on Ganymede (Lewis, 1971). Here we present a model for crater morphology resulting from impacts into materials which may exhibit Bingham properties (see also Fink et al., 1980). The model accounts for many of the features observed in both planetary and laboratory scale multilobed or rampart type impact craters.

Concurrent with our analysis of clay slurry experiments, Melosh (pers. comm.) derived a theoretical model of crater modification by gravity which relied on acoustic fluidization to induce Bingham like behavior in dry planetary surfaces during the high stresses of an impact event. According to his model, a progression from bowl shape to central peak to central pit to peak ring craters might be expected to accompany either a decrease in target rheological properties or an increase in crater diameter (or projectile kinetic energy). We will later discuss the applicability of this model to our experimental craters.

IMPACT EXPERIMENTS

Impact experiments have been conducted at the NASA Ames Vertical Gun Facility using a variety of homogeneous target materials that exhibit a range of properties, from strengthless Newtonian fluids such as water (Gault and Wedekind, 1978) and silicon oil, to dry materials with finite shear strengths such as rock and weakly bonded sand (Greeley et al., 1981). The present study is based primarily on a series of 36 experiments (Table 1) involving impacts into homogeneous, water based slurries. Spherical pyrex or aluminium projectiles of 0.32 and 0.64 cm diameter were launched at velocities up to 2.4 km s^{-1}, giving kinetic energies of impact from 0.2 to about 1000 joules. The resulting craters had diameters ranging from 17 to 60 cm. Profiles were taken across those craters which preserved measurable relief.

Rheological properties of the slurries were determined using two simple devices, a cone penetrometer and a rotational viscometer. Exact values of yield strength and viscosity are

Table 1. Impacts into water based clay slurry targets (from Greeley et al., 1981). Peak diameter, crater diameter, plume diameter, final crater depth, and final central peak height all measured after run completed. Transient central peak heights measured from high speed motion pictures.

a) Impact conditions. ρ, η and τ are target density, viscosity and yield strength. D_{pr}, Type, Mass, Veloc. and K.E. are projectile diameter, type (pyrex or aluminum), mass, velocity and kinetic energy.

Run #	Shot #	ρ kg/m^3	η Pa-s	τ Pa	D_{pr} cm	Type	Mass g	Veloc. km/s	K.E. Joules
1	800404	1515	12.8	197	0.32	PY	0.0438	1.37	41.1
2	800405	1539	14.4	197	0.32	PY	0.0446	1.59	56.4
3	800413	1534	4.5	180	0.32	PY	0.0448	1.62	58.8
4	800501	1740	43.2	424	0.32	PY	0.0445	1.50	50.1
5	800502	1740	43.2	424	0.32	PY	0.0445	1.18	31.0
6	800503	1740	40.0	487	0.32	PY	0.0444	0.60	8.0
7	800504	1740	40.0	487	0.32	PY	0.0448	1.80	72.6
8	800505	1740	48.0	487	0.32	PY	0.0448	2.10	98.8
9	800506	1740	48.0	487	0.64	PY	0.2911	0.68	67.3
10	800507	1740	46.0	487	0.64	PY	0.2906	1.10	175.8
11	800508	1740	46.0	487	0.64	PY	0.2904	1.60	371.7
12	800511	1740	43.2	423	0.64	PY	0.2997	1.57	369.4
13	800512	1740	43.2	423	0.64	PY	0.2912	1.83	487.6
14	800514	1717	51.2	486	0.64	PY	0.2913	2.22	717.8
15	800515	1661	64.0	488	0.32	PY	0.0446	0.20	0.9
16	800519	1606	8.0	152	0.32	PY	0.0446	1.49	49.5
17	800520	1606	8.0	152	0.32	PY	0.0446	1.00	22.3
18	800521	1600	8.8	152	0.32	PY	0.0445	0.57	7.2
19	800522	1600	8.8	152	0.32	PY	0.0445	0.15	0.5
20	800523	1600	8.8	152	0.32	PY	0.0445	1.78	70.5
21	800524	1630	9.6	166	0.32	PY	0.0449	1.95	85.4
22	800525	1630	9.6	166	0.32	PY	0.0445	2.15	102.9
23	800526	1630	9.6	166	0.32	PY	0.0445	2.21	108.7
24	800528	1630	9.6	166	0.64	PY	0.2917	0.79	91.0
25	800529	1630	9.6	166	0.64	PY	0.2913	1.46	310.5
26	800531	1630	9.6	166	0.64	PY	0.2997	1.85	512.9
27	800533	1600	9.9	166	0.64	PY	0.2917	2.06	618.9
28	800537	1600	130.0	3815	0.32	PY	0.0443	1.32	38.6
29	800540	1600	130.0	3815	0.64	PY	0.2995	2.24	751.4
30	800541	1600	130.0	3815	0.32	PY	0.0443	2.18	105.3
31	800542	1700	200.0	3815	0.64	PY	0.2917	1.56	354.9
32	800546	1700	200.0	3815	0.64	AL	0.3759	2.02	766.9
33	800547	1700	170.0	3815	0.64	AL	0.3759	2.22	926.3
34	800548	1700	170.0	3815	0.64	AL	0.3766	2.36	1048.8
35	800549	1720	80.0	2657	0.32	PY	0.0446	2.02	91.0
36	800550	1760	70.0	2047	0.64	PY	0.2998	2.24	752.1

b) Crater features. D_p, D_c and D_{pl} are the diameters of the central peak, crater (rim to rim), and plume deposit. d_c is the final depth and h_f is the final central peak height. h_t is the transient maximum central peak height measured in the motion pictures. P.E. is the potential energy contained in h_t. The final two columns are dimensionless strength and viscosity.

Run #	Shot #	D_p m	D_c m	D_{pl} m	d_c m	h_f m	h_t m	P.E. Joule	$\rho gd/\tau$	$\rho g^{1/2}L^{3/2}/\eta$
1	800404	0.089	0.178	0.330	0.006	—	0.046	0.031	0.45	18.08
2	800405	0.127	0.279	0.457	—	—	0.053	0.065	—	32.02
3	800413	—	0.254	0.356	—	—	0.083	0.121	—	88.24
4	800501	0.140	0.279	0.457	0.019	—	0.011	0.004	0.76	12.07
5	800502	0.127	0.254	0.330	0.019	—	0.013	0.001	0.76	10.49
6	800503	0.076	0.191	0.279	0.025	—	0.000	—	0.88	7.39
7	800504	0.140	0.305	0.406	—	0.038	0.026	0.009	1.33	14.90
8	800505	0.152	0.318	0.406	—	0.033	0.026	0.009	1.12	13.21

Table 1. (Continued).

Run #	Shot #	D_p m	D_c m	D_{pl} m	d_c m	h_f m	h_t m	P.E. Joule	$\rho g d/\tau$	$\rho g^{1/2} L^{3/2}/\eta$
9	800506	0.203	0.394	0.457	—	0.025	0.046	0.053	0.88	18.22
10	800507	0.241	0.457	0.610	—	0.013	0.061	0.173	0.46	23.75
11	800508	0.305	0.483	0.711	—	0.006	0.090	0.555	0.21	25.81
12	800511	0.241	0.508	0.660	—	—	0.086	0.490	—	29.65
13	800512	0.343	0.483	0.686	—	—	0.112	1.080	—	27.48
14	800514	0.381	0.610	0.813	—	—	0.123	1.230	—	32.48
15	800515	0.102	0.165	0.216	0.019	—	0.000	—	0.63	3.55
16	800519	—	0.305	0.445	—	—	0.118	0.374	—	68.76
17	800520	—	—	0.356	—	—	0.105	0.259	—	—
18	800521	—	—	—	—	—	0.072	0.060	—	—
19	800522	—	—	0.229	—	—	0.042	0.010	—	—
20	800523	—	—	0.508	—	—	0.114	0.350	—	—
21	800524	—	—	0.533	—	—	0.141	0.741	—	—
22	800525	—	—	0.584	—	—	0.143	0.715	—	—
23	800526	—	0.356	0.635	—	—	0.147	0.847	—	73.34
24	800528	—	0.533	—	—	—	0.178	1.770	—	134.35
25	800529	—	—	—	—	—	0.237	4.300	—	—
26	800531	—	—	—	—	—	0.255	6.010	—	—
27	800533	—	—	—	—	—	0.285	7.440	—	—
28	800537	—	0.203	—	0.089	—	0.000	—	0.37	2.30
29	800540	—	0.400	—	0.107	—	0.000	—	0.44	6.34
30	800541	—	0.224	—	0.080	—	0.000	—	0.33	2.65
31	800542	—	0.380	—	0.085	—	0.000	—	0.37	4.04
32	800546	—	0.460	—	0.115	—	0.000	—	0.50	5.40
33	800547	—	0.435	—	0.110	—	0.000	—	0.48	5.84
34	800548	—	0.470	—	0.106	—	0.000	—	0.46	6.56
35	800549	—	—	—	—	—	0.000	—	—	—
36	800550	0.178	0.483	0.559	0.035	—	0.029	0.005	0.29	17.16

difficult to measure for clay slurries due to inhomogeneous stress distributions, thixotropic effects, temporary separation of liquid from solid fractions and several other factors (van Wazer et al., 1963; Moore, 1965). As we were concerned primarily with comparing the relative consistencies of the slurries from one run to another, we used the following approximate methods.

To measure yield strength we placed a 3.5 cm diameter solid plastic cone on the surface of a sample and allowed it to sink in. Assuming that the normal stresses against the sides of the cone are negligible, then the weight of the cone is balanced by the shear stress along its sides, and the yield strength (τ) is given by (Moore, 1965, p. 48):

$$\tau = mg \cot \alpha / \pi h^2$$

where m = cone mass, g = gravity, h = depth of cone penetration and α = one half the apical angle of the cone. The assumption of negligible normal stresses applies to low apical angles. Because of the relatively high angle of our penetrometer ($\alpha = 24°$), our yield strength estimates tend to be too large. Lower limits on yield strength can also be calculated using residual crater relief (either crater depth or central peak height):

$$\tau \geq \rho g d$$

where d = relief. These calculated values are all within an order of magnitude of the values measured with the penetrometer.

Viscosity was measured with a Brookfield HBT concentric cylindrical viscometer, which records the torque required to rotate a spindle of known diameter within the fluid at a constant rate. For Newtonian fluids the viscosity can be derived from the ratio of torque to rotation rate. For non-Newtonian fluids such as clay slurries, this method only gives apparent viscosity values which decrease for increasing rotation rates (van Wazer *et al.*, 1963). These apparent values generally lie within an order of magnitude of the actual plastic viscosity values determined by other more complicated methods (Greeley *et al.*, 1980). As long as the same spindle and rotation rates are used, these apparent values may be used to compare the relative consistencies of the different slurries.

A possible source of error in the rheological measurements was that samples were not always collected from the same location relative to the center of the target bucket. Mixing of the slurry during impact and drying of the surface during chamber evacuation could cause local variations in properties (Greeley *et al.*, 1980).

Dynamics of the cratering process were observed in high speed (400 frames per second) motion pictures. In targets with low yield strengths ($\tau \gtrsim 175$ Pa) the impact process resembles that in water (Fig. 2). Following the initial impact, an ejecta plume forms which migrates outward as the crater bowl expands. After the crater attains its maximum depth, the floor begins to rise, forming a central peak. This transient peak may oscillate several

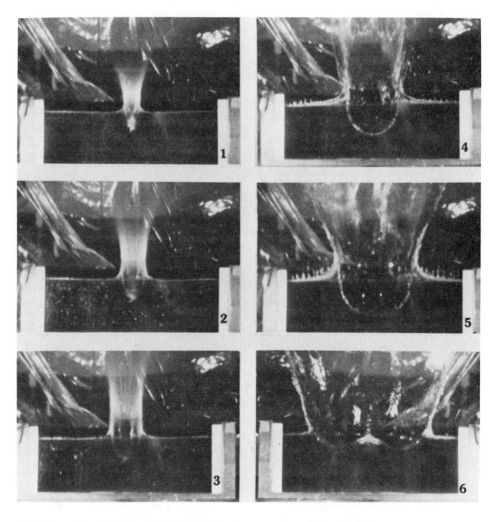

Fig. 2. Sequence of photographs showing impact into water in vacuum chamber (pressure about 250 Pa). In frames 1–5, ejecta plume moves outward as crater bowl expands. In 6, crater bowl collapses to form a central peak with material flowing in from all directions.

times before finally being damped out, and each collapse may send out a surge wave. No crater relief is preserved in water or muds of low yield strength although a circular zone of bubbles may persist after the fluid motion ceases. Heights of the initial transient central peaks were measured using the motion pictures (Table 1). These initial peaks collapsed before reflections returned from the floor or sides of the target bucket (depth = 0.3 m; diameter = 1.0 m) (see Greeley *et al.*, 1981 for discussion).

The impact process in slurries with high yield strengths ($\tau \gtrsim 1$ kPa) has attributes of the process in dry solid or granular targets. These latter materials have been used to simulate impacts into volatile depleted bodies such as the moon and Mercury (Gault and Wedekind, 1977). Such impacts initially produce ejecta plumes, but the crater floors do not rise into oscillating central peaks. The resulting bowl shaped craters have nearly constant depth to diameter ratios of 0.2 (Gault, 1973), comparable to those of small lunar and mercurian craters (Pike, 1977). Similar bowl shaped laboratory craters in sand have collapsed and slumped to form central peaks when subjected to high gravitational accelerations (Gault and Wedekind, 1977).

Impacts into targets with yield strengths between these two extremes show features of both fluid and dry targets. Comparison of crater morphology for impacts of equal kinetic energy in targets with decreasing yield strength shows the following progression (Fig. 3).

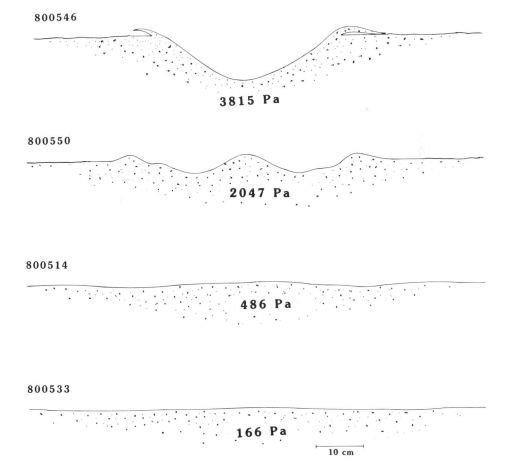

Fig. 3. Crater profiles for shots in which impact energy was maintained nearly constant (~700 joules) while target yield strength varied. Progression from bowl shaped to central peak to palimpsest type craters with decreasing strength.

In targets with high yield strength the crater form is bowl shaped. With decreasing strength, central peaks of a few centimeters height appear. As the strength decreases further, progressively less relief is preserved until in the weakest targets no relief remains at all, either in the crater interior or the ejecta lobes.

Varying the kinetic energy of impact appears to influence the morphology in a way similar to altering the target strength. Figure 4 shows a series of profiles taken across craters formed in targets all having yield strengths of between 400 and 500 Pa. For relatively low energies, bowl shaped craters form with depth to diameter ratios of about 0.2 (Fig. 5a). With increasing energy, crater diameters first become larger with no change in final morphology. At progressively higher energies (and larger crater diameters) crater floors become flatter (depth to diameter ratios decrease), then permanent central peaks form, and finally the peaks form and collapse (Fig. 5b), sending out surge waves which produce concentric overlapping ejecta lobes. At the next higher energies, residual central

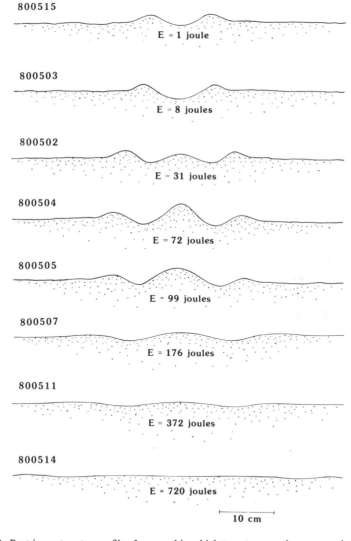

Fig. 4. Post impact crater profiles for round in which target properties were maintained constant ($\tau \simeq 450$ Pa; $\eta \simeq 45$ Pa s) while impact energies varied. Progression from bowl shaped to high relief central peak to low relief central peak to palimpsest type crater with increasing energy. Shot numbers to left (see Greeley *et al.*, 1981).

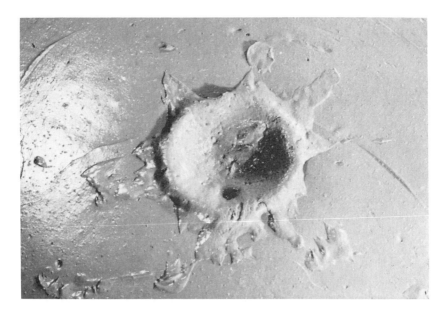

Fig. 5a. A 14-cm-diameter bowl shaped crater formed by relatively low energy (2 joules) impact into clay slurry target (Shot = 800503). Note absence of central peak. Illumination from right.

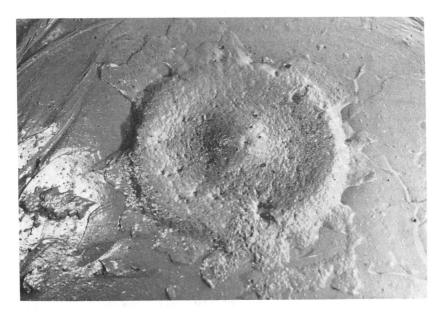

Fig. 5b. A 25-cm-diameter crater with a 1 cm high central peak formed by intermediate energy (200 joules) impact into clay clurry (Shot = 800505).

peak relief becomes less and ejecta lobe diameters increase (Fig. 5c). At the highest energies, no crater relief is preserved and the limits of the crater and the ejecta blankets can only be detected by noting textural zones within the target. The same trend is observed for targets with higher yield strengths ($\tau \simeq 4$ kPa; Fig. 6) but the morphologic transitions occur at higher energy values.

Fig. 5c. A 37-cm-diameter crater with 0.6 cm high central peak formed by high energy (700 joules) impact into clay slurry (Shot = 800507).

Sequences of transient and permanent crater features can both be delineated in these experiments. Comparing impacts into Newtonian fluids of increasing viscosities reveals a dynamic sequence similar to the morphologic sequence seen in slurries of increasing yield strengths. In low viscosity fluids like water, cratering involves several central peak oscillations. In silicon oils of increasing viscosities (Fink, Gault and Greeley, in prep.), the same impact energy causes fewer oscillations until at very high viscosities no central peaks develop at all. This sequence of transient crater forms partly mimics the preserved crater forms in the sequence accompanying increasing target strength (Fig. 3) as well as that of decreasing impact energy (Figs. 4, 6).

A MODEL FOR IMPACT INTO BINGHAM MATERIALS

Observed sequences of permanent and transient crater forms described above may be explained by considering the effect of impact on targets with Bingham rheology. During an impact, shear stresses within the target are proportional to the peak shock pressures and hence the velocity of the projectile; these stresses are maximized in a hemispherical region surrounding the point of impact. Thus for a Bingham target, shear stresses may exceed the yield strength and cause liquifaction within a region of some radius, while the surrounding material remains above yield and hence relatively rigid (Fig. 7a). Higher projectile velocities, momenta, or energies cause larger volumes to behave as a fluid and consequently produce craters less capable of sustaining relief and more influenced by hydrodynamic processes. For a given impact velocity, targets with lower yield strengths similarly produce more fluid-like morphologies than strong targets.

The particular crater form that results from a set of impact conditions and target properties reflects the ways in which energy is transferred and stress is distributed during the cratering process. The kinetic energy of an impacting projectile is partly dissipated as heat, light and sound. The resulting transient crater bowl contains potential energy corresponding to some fraction of the initial kinetic energy. The presence of this cavity also causes the target to experience gravitational stresses proportional to the crater depth. If these stresses exceed the shear strength of the target, the floor of the crater rises up as

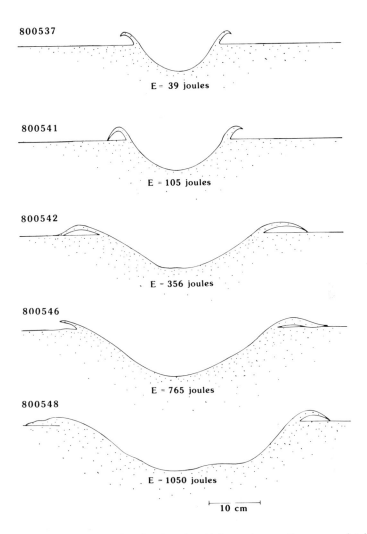

Fig. 6. Post impact crater profiles for shots in which target properties were maintained constant ($\tau \simeq 3.8$ kPa; $\eta \simeq 130\text{--}170$ Pa s) while energy was varied. Only bowl shaped craters formed within this energy range.

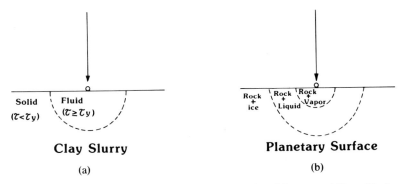

Fig. 7. (a) Schematic diagram showing 'fluidized' and 'solid' zones within a Bingham material target during impact. (b) Concentric zones around point of impact into planetary surface consisting of rock and ice. Ice is vaporized in innermost zone, melted in intermediate zone and remains solid in outer zone.

potential energy is converted back into kinetic energy. The floor continues its rise, in some cases forming a central peak, until all of its kinetic energy is expended. If gravitational stresses again exceed the target strength when this peak reaches its maximum height, it will fall, again converting potential to kinetic energy. As long as the combined gravitational and inertial stresses associated with the oscillating central peak are greater than the target strength, the motion of the target will continue (cf. Murray, 1980).

For Newtonian viscous targets which lack shear strength, central peak oscillations will continue until viscous dissipation totally converts the available energy to heat. Fluid motion may stop sooner if the inertial and gravitational stresses drop below the stress due to surface tension. Early studies of the effects of surface tension on impact cratering (Engle, 1961; Gault and Moore, 1965) used water for both projectiles and targets. These experiments indicated that surface tension influenced crater geometry only up to diameters of about 1 cm. Our preliminary experiments using pyrex and aluminum projectiles in fluids of various surface tensions show that ejecta blankets tend to become more fragmented with decreasing surface tension; however the period of central peak oscillation is not significantly reduced for craters above a few centimeters in diameter (Fink, Gault and Greeley, in prep.). Ignoring surface tension, the higher the target viscosity the fewer the number of successive central peaks and depressions that will form. For Bingham targets, motion may similarly be damped out by high plastic viscosities; however, movement will cease sooner the higher the yield strength.

The number of central peak oscillations that occur following impact may be related analytically to the yield strength and viscosity of the target material (Melosh, pers. comm.). Relating this number to the final crater morphology is more problematic however. Distinction must first be made between exterior and interior crater features. In our experiments, the morphology of concentric ejecta lobes depends upon the number of oscillations. The lowermost ejecta layer is made up of material from the ejecta plume. Successive layers either slump from the crater rim or develop as the central peak collapses (Greeley *et al.*, 1980). Collapse of more than one central peak produces additional lobes, each with decreasing diameter. Thus each oscillation may leave a record in the morphology of the ejecta deposits. In contrast, the interior morphology of our experimental craters did not reflect the number of oscillations beyond showing a transition from bowl shaped to flat floored to central peak types. Increased movement of the target surface was accompanied by decreasing crater relief, as shown in Figs. 3 and 4. Impacts with the highest kinetic energy, largest transient crater diameters, least viscous and weakest targets all resulted in practically no preserved crater morphology, even though the transient structures may have been relatively complex. This is due to the liquifaction of a large volume of target material which then could no longer sustain morphologic relief.

IMPACT GENERATION OF BINGHAM RHEOLOGY ON MARS AND GANYMEDE

In the previous section we have interpreted the dynamics and resulting morphology of impact cratering in Bingham type clay slurries. In this section we consider the applicability of the Bingham model to volatile rich planetary surfaces, and specifically to Mars and Ganymede.

In those areas where multilobed craters are common, the martian surface is most likely composed of rock with some fraction of subsurface ice or other volatiles (Carr *et al.*, 1977). This material will deform as an elastic solid before the passage of shock waves during an impact event. Within a limited cavity corresponding to the transient crater bowl, impact produces a breccia comprised of rock, ice, liquid water, and water vapor which can be mobilized as a slurry whose rheological properties may be approximated by the Bingham model. These properties will depend upon the relative amounts of liquids and solids.

The stresses and heat generated by an impact will produce several concentric zones within the target (Fig. 7b). Within the innermost zone, volatiles (and in some cases, rock) will be vaporized, and the remaining rock will be finely pulverized. The next zone will have lower temperatures and stress levels so that here rock fragments of various sizes will be mixed with melted water. This region will grade outward to one in which rock fragments and ice blocks predominate, with lesser amounts of meltwater. Beyond this zone will be a region undeformed by the impact.

A slurry consisting of fine grained particles, water, and coarse blocks will behave as a Bingham material as long as the fine particles do not interlock (Rodine and Johnson, 1976). Laboratory experiments (Moore, 1965; Rodine and Johnson, 1976) have shown that yield strength is a strong function of water content for slurries consisting of water and fine grained particles of a single grain size. For water contents greater than about 60 percent, the strength is very near that of the liquid. For water contents less than about 45 percent, these properties are near those of the dry granular solids. Over a narrow range of water contents the strength increases rapidly, and this is attributed to the interlocking of grains that are no longer sufficiently lubricated by the reduced water content. However, the presence of a variety of grain sizes, as found in ejecta from impacts (Gault et al., 1963), causes the rapid transition from fluid to solid like properties to occur at water concentrations as low as 11 percent (Rodine and Johnson, 1976, p. 231).

Terrestrial debris flows constitute a possible analog for the martian ejecta slurries. Pierson (1980) has tabulated the measured physical characteristics of fluid debris from several different sites. Debris flows with water contents as low as 10 percent have been observed with viscosities of 3000 Pa s and surface velocities of 2.5 m s^{-1} in a channel with a 9 degree slope. Flows with water contents of 30 percent may attain velocities up to 9.4 m s^{-1} on slopes as low as 8 degrees.

Thus in an impact generated slurry, a small percentage of melted ice can effectively lubricate the volume and allow it to flow. The volume of the zone in which ice melts depends upon the shock stresses and hence the projectile velocity. As velocity increases, crater size increases more slowly than the shock isobars (Schmidt, 1980). Thus, the larger the crater, the more slurry that is produced per crater volume. Within a given terrain on Mars, larger impacts should result in greater volumes of more fluid ejecta capable of developing an oscillating central peak and associated concentric ejecta flow lobes but less able to support ejecta lobe relief.

Impact processes on Ganymede may be similar in some respects to those in the frozen, volatile rich regions of Mars. Ganymede's lithosphere is believed to be composed primarily of ice with a small fraction of silicate materials derived from impacts (Fanale et al., 1977). The larger proportion of ice relative to rock on Ganymede should lead to impact generated slurries that are more fluid than those on Mars, as well as more fluid-like ejecta morphologies. However, on Ganymede the lower gravitational acceleration (1.42 versus 3.70 cm s^{-2} for Mars) and lower density of surface material (assumed roughly equal to one half that of Mars) should lead to larger craters for equivalent impact conditions. If we assume gravity scaling (Gault and Wedekind, 1977) then craters on Ganymede should be about 1.5 times larger than those on Mars. To produce craters of the same diameter on the two bodies would require an impact on Ganymede to have about one fifth the energy of its martian counterpart, which would lead to less melting and less production of impact slurry. Furthermore, ice on Ganymede is colder than in the martian subsurface, so more of the available impact energy must be partitioned into heating the ice on Ganymede before it can melt. All of these factors suggest that impact craters on Ganymede should generate smaller volumes of slurries than those on Mars.

Measurements on Voyager images of Ganymede (Horner and Greeley, pers. comm.) and Viking images of Mars (Mutch and Woronow, 1980) indicate that craters on Ganymede whose ejecta have steep scarps have total ejecta blanket areas smaller than those of comparably sized craters on Mars. Based on the above arguments, we tentatively cite this observation as evidence for the generation of proportionally lower volumes of mobile ejecta on Ganymede than on Mars, despite the presumed higher volatile content of the ganymedean crust.

The progression of crater forms on Ganymede with increasing diameter consists of bowl shapes, flat floors, central peaks, central pits and crater palimpsests. Although material within the transient crater bowl may behave as a slurry with Bingham properties, the dominantly ice surface will behave more like an elastic solid. Hence there is only limited applicability of the clay slurry experiments to this ganymedean sequence. In other experiments, we were only able to reproduce the observed transitions of crater form by using layered target media (Greeley et al., 1981) which may simulate rheological discontinuities within the ganymedean crust. More precise experimental analogs require better knowledge of the rheological behavior of ice and mixtures of ice and rock.

Another possible application of the Bingham model to cratering on Ganymede is in the case of palimpsests. These largest of Ganymede's craters show very little topographic relief. Earlier interpretations assumed that these craters originally formed with substantial relief that relaxed over time by slow viscous flow. An alternative suggested by our experiments is that very large impacts form large transient crater bowls filled with slurry that is unable to sustain any relief; the crater bowl thus collapses immediately, rather than after long periods of time.

LIMITATIONS OF THE BINGHAM MODEL

Natural materials rarely have deformational behavior that directly corresponds to idealized rheological models. This is especially true for complex materials like planetary surfaces subjected to the high stresses of an impact event, or ejecta slurries containing various material and thermal inhomogeneities. Although application of the Bingham rheological model helps explain certain features of impact crater morphology on Mars and Ganymede, detailed comparisons between the craters and the model reveal the latter's inherent limitations.

In our model we have assumed a Bingham rheology for both the crater ejecta and the interior of the crater bowl. Using a Bingham model for ejecta slurries is a reasonable rheological assumption, since terrestrial debris flows and mudflows, which are composed of similar materials to those postulated for the ejecta, have yield strengths and viscosities that have been measured. Applying the Bingham model to the crater bowl is more questionable. Although martian ejecta morphology becomes increasingly fluid-like in appearance with increasing crater diameter, the interiors of some of these craters continue to resemble those from volatile depleted planets (i.e., prominent scarps and other slump features). According to our model, formation of concentric ejecta flow lobes is usually accompanied by the development and oscillation of central peaks. However, central peaks can form by other means (e.g., Head, 1976). Based on our experiments in homogeneous clay slurry targets, crater relief should become progressively reduced with increasing diameter. Some evidence of this relationship may be seen in the low relief of large basins and palimpsests. However, the prominent relief exhibited in the interiors of large martian craters such as Arandas (Schultz and Singer, 1980) indicates that a homogeneous Bingham material cannot adequately model the martian crust for explaining the total morphology of martian craters.

One possible explanation for this discrepancy is that the martian surface may be stratified with regard to volatiles. The largest craters may then incorporate dry substrate and the smallest craters may totally sample dry surface material whereas intermediate size craters are most influenced by the volatile rich layer (Cintala and Mouginis-Mark, 1980). In our earlier experiments with simple layered targets, we found a similar relationship: small craters showed the predominant influence of the upper crustal layer, whereas larger craters showed the effect of the substrate (Greeley et al., 1981). Our Bingham model also assumes homogeneous mixing of the target material within the transient crater bowl. Alternatively, if volatiles are concentrated near the surface, the first material to be excavated could be transformed into a slurry and ejected, leaving the deeper volatile depleted material to form an interior morphology more typical of dry planetary surfaces.

According to our Bingham cratering model, there is a rapid transition from rigid to fluid behavior as volatiles and solid fractions mix. There should also be a sudden reversion to rigid behavior after the stresses drop below the effective yield strength. In practice, our clay slurries did not demonstrate this reversability. Once mixed, they did not as readily become rigid again, especially within the crater bowl. These slurries may have exhibited nonlinear behavior by which the strength and viscosity decreased at higher strain rates so that after the stresses were reduced below what had originally been the yield strength, the material still was effectively fluidized.

In his acoustic fluidization model, Melosh (pers. comm.) also assumes that a rapid transition from fluid to rigid behavior accompanies a drop in stress levels. He derives an analytic

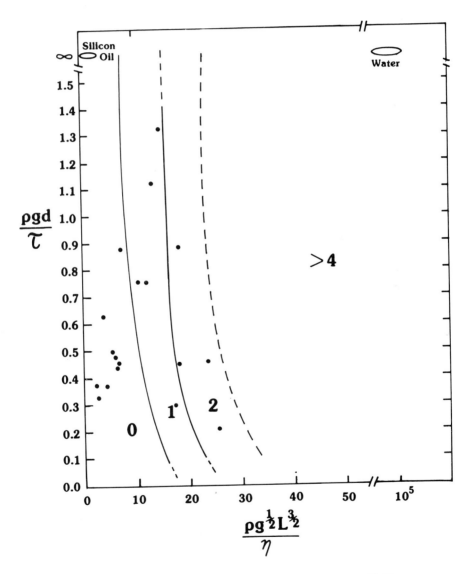

Fig. 8. Plot of dimensionless yield strength ($\rho gh/\tau$) versus viscosity ($\rho g^{1/2}L^{3/2}/\eta$). Variables defined in text. Zones on graph correspond to number of oscillations that occur before target assumes final morphology: 0 = bowl shaped crater; 1 = central peak; 2 = central peak that collapses beneath the original surface level of the target; 3 (not indicated) = oscillating peak with one full rebound; 4 = further oscillations.

relationship between the number of central peak oscillations and the target rheology and further asserts that the number of oscillations determines the interior morphology of the crater. This relationship is illustrated by locating zones in which different numbers of oscillations occur on a dimensionless plot of viscosity versus yield strength. We attempted to replicate this plot with our experimental data (Fig. 8) by calculating the dimensionless groups $\rho g d/\tau$ and $\rho g^{1/2} L^{3/2}/\eta$ for those runs in which we had sufficient information. In place of Melosh's radius of the acoustically fluidized zone (L) we used the radius of the crater ($\frac{1}{2}D_c$, in Table 1b) multiplied by a factor of 1.5. d was a measure of crater relief, either the final depth or the final height of the central peak. By comparing our data with motion pictures we were able to delineate zones on this dimensionless plot corresponding to zero, one, two or more than two oscillations. As on Melosh's plot, increased oscillation accompanied increases in crater diameter, decreases in viscosity and (to a lesser degree) decreases in yield strength. The final interior morphology of the craters reflected the number of oscillations for zero (bowl shape), one (central peak) and between zero and one (flat floored). Greater numbers of oscillations were accompanied by decreasing crater relief, as mentioned earlier. We have also plotted results for impacts into Newtonian fluids with viscosities of 10^{-3} Pa s (water) and 60 Pa s (silicon oil) along the top axis of Fig. 8, corresponding to zero yield strength. Impacts into water produced more than four oscillations, whereas those into the highly viscous oil resulted in craters with no oscillations. Ongoing experiments using targets with intermediate viscosities will allow precise determinations of the boundaries of these dynamic fields.

A remaining problem inherent to all impact experiments concerns the validity of extrapolating laboratory scale events to planetary impacts. Drawing planetary conclusions from impacts into clay slurries assumes that similar physical processes are operating across widely diverging length scales and for materials whose properties may vary by several orders of magnitude. Dimensional analysis can help reduce some of the uncertainties of scaling, but in a process as complex as cratering, exact similitude is impossible to attain. If we can demonstrate that the general processes are the same at the two scales, then simple geometric quantities such as crater volume or diameter may be related empirically to target properties and to impact conditions. A difficulty in using clay slurry target media is that their physical properties are mostly dependent upon water content. We are currently using different media to systematically evaluate and distinguish the effects of viscosity, surface tension and yield strength on the dimensions of experimental craters. Furthermore, although crater volume or diameter may be empirically related to target properties and impact conditions, there is no convenient way to relate the morphologic progressions seen in the laboratory craters to those seen on Mars and Ganymede, other than through qualitative comparisons. Theoretical models relating crater morphology to crater evolution and to target properties should assist in these extrapolations; however they may also serve to indicate discrepancies between theoretical, laboratory and planetary cratering processes.

Acknowledgments—Many thanks to the gun crew at the NASA Ames Vertical Gun Range for logistic support. Helpful reviews were provided by S. Croft, F. Horz, W. B. McKinnon, H. J. Melosh, H. J. Moore, A. R. Peterfreund and P. H. Schultz. Research supported by NASA grants NAGW-56, NAGW-132, and NSG-7548.

REFERENCES

Carr M. H., Crumpler L. S., Cutts J. A., Greeley R., Guest J. E., and Masursky H. (1977) Martian impact craters and emplacement of ejecta by surface flow. *J. Geophys. Res.* **82**, 4055–4065.
Cintala M. J. and Mouginis-Mark P. J. (1980) Martian fresh craters: More evidence for subsurface volatiles? *Geophys. Res. Lett.* **7**, 329–332.
Engle O. G. (1961) Collisions of liquid drops with liquids. *U.S Nat. Bur. Standards Tech. Note* 89.
Fanale F., Johnson T. V., and Matson D. L. (1977) Io's surface and the histories of the Galilean Satellites. In *Planetary Satellites*, (J. Burns, ed.), p. 379–405. Univ. Arizona, Tucson.

Fink J. H., Greeley R., and Gault D. E. (1980) "Fluidized" impact craters in Bingham materials and the distribution of water on Mars. In *Proc. Third Colloquium on Planetary Water and Polar Processes*, p. 64–67. SUNY, Buffalo, N.Y.

Fink J. H., Malin M. C., D'Alli R. E., and Greeley R. (1981) Rheological properties of mudflows associated with the Spring, 1980 eruptions of Mount St. Helens Volcano, Washington. *Geophys. Res. Lett.* **8**, 43–46.

Gault D. E. (1973) Displaced mass, depth, diameter and effects of oblique trajectories for impact craters formed in dense crystalline rocks. *The Moon* **6**, 32–44.

Gault D. E. and Greeley R. (1978) Exploratory experiments of impact craters formed in viscous-liquid targets: Analog for martian rampart craters? *Icarus* **34**, 486–495.

Gault D. E. and Moore H. J. (1965) Scaling relationships from microscale to megascale. *Proc. 7th Hypervelocity Impact Symp.* **6**, p. 341–351.

Gault D. E., Shoemaker E. M., and Moore H. J. (1963) Spray ejected from the lunar surface by meteoroid impact. NASA-TND-1767. NASA Ames, Calif.

Gault D. E. and Wedekind J. A. (1977) Experimental hypervelocity impact into quartz sand—II, Effects of gravitational acceleration. In *Impact and Explosion Cratering*, (D. J. Roddy, R. O. Pepin, and R. B. Merrill, eds.), p. 1231–1244. Pergamon, N.Y.

Gault D. E. and Wedekind J. A. (1978) Experimental impact "craters" formed in water: Gravity scaling realized (abstract). *EOS (Trans. Amer. Geophys. Union)* **59**, 1121.

Greeley R., Fink J. H., Gault D. E., Snyder B. D., Guest J. E., and Schultz P. E. (1980) Impact cratering in viscous targets: Laboratory experiments. *Proc. Lunar Planet. Sci. Conf. 11th*, p. 2075–2097.

Greeley R., Fink J. H., Gault D. E., and Guest J. E. (1981) Impact cratering in simulated icy satellites. In *The Satallites of Jupiter* (D. M. Morrison, ed.), Univ. Arizona, Tuscon. In press.

Head J. W. (1976) The significance of substrate characteristics in determining morphology and morphometry of lunar craters. *Proc. Lunar Sci. Conf. 7th*, p. 2913–2929.

Head J. W. and Roth R. (1976) Mars pedestal crater escarpments: Evidence for ejecta-related emplacement (abstract). In *Symposium on Planetary Cratering Mechanics*, p. 50–52. The Lunar Science Institute, Houston.

Horner V. and Greeley R. (1981) Rampart craters on Ganymede: Implications for the origin of martial rampart craters (abstract). In *Lunar and Planetary Science XII*, p. 400–402. Lunar and Planetary Institute, Houston.

Hulme G. (1974) Interpretations of lava flow morphology. *Geophys. J. Roy. Astron. Soc.* **39**, 361–383.

Johnson A. M. (1970) *Physical Processes in Geology*. Freeman and Co., San Fransisco. 577 pp.

Lewis J. S. (1971) Satellites of the outer planets: Their physical and chemical nature. *Icarus* **15**, 174–185.

Melosh H. J. (1977) Crater modification by gravity: a mechanical analysis of slumping. In *Impact and Explosion Cratering* (D. J. Roddy, R. O. Pepin, and R. B. Merrill, eds.), p. 1245–1260. Pergamon, N.Y.

Moore F. (1965) *Rheology of Ceramic Systems*. MacLaren, London. 78 pp.

Mouginis-Mark P. (1979) Martian fluidized crater morphology: Variations with crater size, latitude, altitude and target material. *J. Geophys. Res.* **84**, 8011–8022.

Murray J. B. (1980) Oscillating peak model of basin and crater formation. *Moon and Planets* **22**, 269–291.

Mutch P. and Woronow A. (1980) Martian rampart and pedestal craters' ejecta emplacement: Coprates Quadrangle. *Icarus* **41**, 259–268.

Oberbeck V. R. and Quaide W. L. (1968) Genetic implications of lunar regolith thickness variations. *Icarus* **9**, 446–465.

Passey Q. R. and Shoemaker E. M. (1981) Craters and basins on Ganymede and Callisto: Morphological indicators of crustal evolution. In *The Satellites of Jupiter* (D. Morrison, ed.), Univ. Arizona, Tuscon. In press.

Pierson T. C. (1980) Erosion and deposition by debris flows at Mt. Thomas, North Canterbury, New Zealand. *Earth Surf. Processes* **5**, 227–247.

Pike R. J. (1977) Size dependence in the shapes of fresh impact craters on the moon. In *Impact and explosion cratering* (D. J. Roddy, R. O. Pepin, and R. B. Merrill, eds.), p. 489–509. Pergamon, N.Y.

Rodine J. D. and Johnson A. M. (1976) The ability of debris, heavily freighted with coarse clastic materials, to flow on gentle slopes. *Sedimentology* **23**, 213–234.

Schmidt R. M. (1980) Meteor Crater: Energy of formation—implications of centrifuge scaling. *Proc. Lunar Planet. Sci. Conf. 11th*, p. 2099–2128.

Schultz P. H. and Gault D. E. (1979) Atmospheric effects on martian ejecta emplacement. *J. Geophys. Res.* **84**, 7669–7687.

Schultz P. H. and Singer J. (1980) A comparison of secondary craters on the moon, Mercury and Mars. *Proc. Lunar Planet. Sci. Conf. 11th*, p. 2243–2259.

Smith B. A. and the Voyager Imaging Team (1979) The Jupiter System through the eyes of Voyager I. *Science* **204**, 951–972.

Strom R., Woronow A., and Gurnis M. (1981) Crater population on Ganymede and Callistro. *J. Geophys. Res.* In press.

van Wazer J. R., Lyons J. W., Kim K. Y., and Caldwell R. E. (1963) *Viscosity and Flow Measurements.* Wiley, N.Y. 406 pp.

Wohletz K. H. and Sheridan M. F. (1981) Rampart crater ejecta: Experiments and analysis of melt-water interactions. NASA-TM-82385, p. 134–136.

Wood C. A. (1980) Martian double ring basins: New observations. *Proc. Lunar Planet. Sci. Conf. 11th,* p. 2221–2241.

Wood C. A., Head J. W., and Cintala M. J. (1978) Interior morphology of fresh martian craters: The effects of target characteristics. *Proc. Lunar Planet. Sci. Conf. 9th, p.* 3691–3709.

Fragmentation of ice by low velocity impact

Manfred A. Lange and Thomas J. Ahrens

Seismological Laboratory, California Institute of Technology, Pasadena, California 91125

Abstract—Low velocity impact experiments (0.14 to 1 km/s) carried out in polycrystalline water ice targets at 257 and 81 K resulted in interactions which can be assigned to four fragmentation classes, cratering, erosion, disruption, and total fragmentation. Specific kinetic energies for the transitions between these classes range from 1×10^5 to 7×10^5 ergs/g for 81 K ice and from 3×10^5 to $\sim 2 \times 10^6$ ergs/g for 257 K ice. These values are about one to two orders of magnitude below those for silicate rocks. The mass vs. cumulative number distribution of fragments in our experiments can be described by a simple power law, similar to that observed in fragmented rocks in both the laboratory and in nature. The logarithmic slopes of cumulative number vs. fragment weight vary between -0.9 and -1.8 decreasing with increasing projectile energy and are approximately independent of target temperature. The shapes of fragments resulting from erosion and disruption of ice targets are significantly less spherical for 257 K targets than for 81 K targets. Fragment sphericity increases with increasing projectile energy at 257 K, but no similar trend is observed for 81 K ice.

Our results support the hypothesis that the specific projectile energy is a measure for target comminution for a relatively wide range of projectile energies and target masses. We apply our results to the collisional interaction of icy planetary bodies and find that the complete destruction of a target body with radii between 50 m and 100 km range from 10^{17} to 10^{27} ergs. Energies corresponding to basaltic bodies of the same size range from 10^{18} to 10^{28} ergs. Our experiments suggest that regolith components on icy planets resemble those on rocky planetary bodies in size and shape. We predict that the initial shapes of icy particles in the Saturnian ring system were roughly spherical. The initial mass distribution of ring particles should follow a power law with a slope of ~ -1.5.

1. INTRODUCTION

An understanding of low velocity impact processes in targets of low temperature ice (~ 80 K) is important for both the accretion and surface evolution of icy planetary bodies, e.g., the Galilean and Saturnian satellites. Since impact velocities of accreting planetesimals are on the order of the escape velocity of an accreting body (Safronov, 1972), maximum impact velocities during accretion of the icy satellites did not exceed 0.15 to 2.7 km/s. The formation of a regolith layer is largely controlled by secondary impact cratering (Oberbeck, 1975). Secondary cratering on the smaller of the Saturnian satellites takes place primarily by the sweep-up of ejected debris from primary impacts. But here as in the case of the larger Galilean satellites, with escape velocities sufficient to retain primary ejecta, low velocity secondary impacts dominate the surface evolution on these bodies (Smith *et al.*, 1981).

In this paper, the following questions are addressed:

(i) What are threshold impact energies for transition between different stages of fragmentation of an icy body, as a function of temperature for decimeter sized targets?

(ii) What is the size distribution of particles resulting from impact fragmentation of an icy body as a function of temperature?

(iii) What are the shapes of these fragments, and do they change with varying target temperature?

The first problem is relevant to theoretical models describing the accretion of an icy planet and defines some of the physical parameters which define the conditions for mass accretion in contrast to mass loss during the growth of the planetary body. The latter two questions are relevant for the prediction of regolith characteristics on the surfaces of icy planets and, to a lesser degree, to the origin of icy planetary rings. The conclusions inferred from the present experiments are, in principle, testable using radar and radio reflectance data of icy regoliths and ring particles (e.g., Pollack, 1975).

The fragmentation process of silicate bodies as well as that of water ice at temperatures close to the freezing point has been experimentally studied to some degree by Gault and Wedekind (1969), Fujiwara et al. (1977) and Hartmann (1978). In order to compare impact events in which the projectile to target mass ratio is varied, the specific kinetic energy, ke, of the projectile, i.e., its kinetic energy divided by the target mass has been defined (Gault and Wedekind, 1969; Fujiwara et al., 1977). Projectile kinetic energy divided by target volume has the dimension of strength, i.e., supportable force/area and has been called impact strength (Greenberg et al., 1978). Although this quantity is conceptually different from the strength of a material in the sense of a quasi-static test, it is helpful in visualizing the magnitude of a particular impact event. In this paper we use ke to denote projectile kinetic energy divided by target mass. Critical values for ke which lead to the destruction of an impacted glass, basalt, and ice target are given as 10^7, 6×10^7, and 3×10^5 ergs/g, respectively (Gault and Wedekind, 1969; Fujiwara et al., 1977; Hartmann, 1978). However, it is expected that the physical properties of H_2O ice, controlling its behaviour under shock loading in the temperature range of interest in this study (i.e., at ~ 80 K), are different from the aforementioned materials. It should be noted here that the result for water ice, as given by Hartmann (1978) represents a different experimental approach than that of Gault and Wedekind (1969), Fujiwara et al. (1977) and the present study. Hartmann studied the destruction of ice by firing or dropping centimeter-scale ice projectiles onto flat rock targets with velocities in the range of 1 to 50 m/s. Hence, the critical energy for the destruction of ice represents kinetic projectile energy/*projectile* mass. But not only are Hartmann's experiments conceptually different, they also attempt to simulate the interaction of ice with silicate bodies, whereas we are studying primarily the interaction between icy bodies. Hence, the above given data on critical impact energies for glass, basalt, and ice will serve as a basis for comparison with our experimental results presented below.

Experimental data for the quasi-static deformation of polycrystalline ice indicate an increase in compressive strength from 20–25 bars to ~ 80 bars and an increase in tensile strength from 15 to 25 bars for temperatures between 273 and 223 K, respectively (Hobbs, 1974). However, these data are obtained at stress rates of ~ 5 bars/s (Butkovich, 1954), which are orders of magnitude below the rates of kilobars to tens of kilobars per second typical of low velocity impacts. Since it is known that the strength of ice is highly stress rate dependent (Hobbs, 1974), we will not use these data in the present study.

We performed low velocity impact experiments on polycrystalline water ice targets mainly at temperatures of 257 and 81 K. In the following, we briefly describe our experimental techniques and results. We discuss these results and how they may apply to impact and fragmentation processes of icy planetary bodies.

2. EXPERIMENTAL TECHNIQUES

The majority of our cubic target blocks (~ 19 cm sidelength) had temperatures of either 257 or 81 K. Blocks of 257 K were prepared following the technique described by Croft et al. (1979) in order to eliminate air bubbles in the sample. The container was then placed in a freezer at temperatures of ~ 257 K and remained there for several days until immediately before the experiment. A major problem in the preparation of the 81 K targets was the formation of thermoelastic induced cracks as the sample was cooled. They are caused either by volume expansion of inner, solidifying parts or by the contraction of solid outer portions of the sample which had lower temperatures due to their proximity to the coolant. A water filled plastic container was placed directly in a liquid nitrogen bath of 77 K. Crystal growth started at the bottom and along the walls of the container, and continued inward and upward. In order to avoid extensive cracking of the sample blocks, we inhibited the growth of ice crystals on the upper liquid surface. As cooling progressed, the remaining fluid was continually stirred, until the entire water volume solidified. In most cases, the surface of the freezing block was kept fluid until the temperature in the ice reached ~ 150 K. At this point, most of the volume expansion of solidifying water had occurred. For the temperature range from 273 to 150 K, the volume of ice-I decreases by $\sim 1.5\%$, while the contraction for temperatures between 150 and 81 K amounts to $\sim 0.3\%$ (Hobbs, 1974). Hence most of the total volume

Table 1. Basic experimental data.

Shot No.	Target Mass M_T, g	Target Temp. T_T, K	Impact Velocity v_i, km/s	Projectile specific kinetic energy ke, 10^5 ergs/g	Peak impact stress σ, kbar*
570	5900	81	1.05	56.83	24.03
571	5571	257	1.04	59.36	22.63
590	5458	257	0.24	4.40	5.57
591	5797	257	0.14	1.35	3.13
592	5930	81	0.23	3.70	5.91
593	6043	257	0.16	1.74	3.83
596	4657	81	0.27	6.20	6.65
597	6410	257	0.29	4.92	6.61
598	5754	257	0.35	7.13	8.01
602	6832	257	0.21	2.50	4.87
604	5882	81	0.18	2.24	4.44
605	6634	257	0.30	5.40	6.96
609	6024	257	0.51	17.70	11.49
611	5673	81	0.16	1.72	3.70

*Peak one-dimensional stresses are calculated based on Hugoniot data for ice-I by Anderson (1968) and on the elastic parameters of ice as a function of temperature given by Dantl (1969).

change could be relieved until the surface of the block closed up. The temperature in the ice was monitored by use of a thermocouple placed in the center of the target block.

The target blocks were placed inside a steel tank, under ambient pressure conditions, directly in front of the evacuated barrel of a conventional 20 mm powder gun. The blocks were impacted immediately after removal from the coolant (or from the freezer); during this interval, the increase in temperature of the blocks was insignificant (less than ~ 5 degrees). Cylindrical Lexan (a polycarbonate plastic, density = 1.2 g cm^{-3}) projectiles were used in these experiments. In order to cover a relatively wide range of specific kinetic energies, we varied the impact velocities of the projectiles from 0.14 to 1.05 km/s keeping both target and projectile mass nearly constant. Specific kinetic energies of the projectiles ranged from 1.4×10^5 to $\sim 60 \times 10^5$ ergs/g (Table 1). Immediately after the impact, the fragmented target block was photographed and a number of samples from different positions in the tank were taken for further analysis. The degree that secondary impact of fragments with the steel wall of the tank affected the fragment population is not known. In general, we tried to collect as many fragments with sizes $\geq$ cm as possible (with the exception of shots 570, 571, and 611 where a representative collection of fragments were taken). The remaining, smaller fragments were collectively weighed. The larger specimens were individually weighed and the short, intermediate, and long axes of each fragment ($=$ C, B, and A, resp.) were measured by means of a slide caliper. The error in fragment masses is not greater than ± 1 g and the lengths are accurate within ± 1 cm.

3. RESULTS

The present experiments resulted in fragmented target blocks which can be assigned to four fragmentation classes defined below. These are comparable to the destruction types given by Fujiwara *et al.* (1977) for impact experiments on basaltic targets. However, fragmentation according to their destruction type II (i.e., spallation of outer surfaces with remaining core also called core type) was not observed in our experiments. We define the following fragmentation classes based on the ratio between mass of largest fragment M_{lf} to original target mass (M_T):

I. Total destruction

The entire target block has been comminuted into fragments not exceeding sizes of ~ 10 cm. The bulk of the fragments have mean sizes between 2.5 and 9 cm. In this class, the mass of the largest fragment is less than 0.03 times the target mass (i.e., $M_{lf}/M_T \leq$

Fig. 1. Total fragmented ice target (Shot No. 570, 81 K). Initial target block was placed as shown. Impact achieved from Lexan projectile traveling from left to right. (See text and Table 1 for further details).

0.03). Figure 1 shows a typical example of class I fragmentation. The cumulative number of sampled fragments for class I experiments as a function of relative fragment mass is given in the left part of Fig. 5.

II. Disruption

This type is characterized by the target block being split into generally two to three larger fragments and a great number of small fragments. The mass of the largest fragment in this class varies from 0.03 to 0.3 times the original target mass (i.e., $0.03 < M_{lf}/M_T \leq 0.3$). Figure 2 gives a representative example of class II fragmentation and the center part of Fig. 5 shows the results for all class II experiments.

III. Erosion

This fragmentation type is intermediate between cratering and disruption of the target block. Generally, parts of the target, adjacent to the impact point, are spalled off, while most of the target block remains intact. Impacts in this class result in eroded blocks greater than 0.3 times the original target mass (i.e., $1 > M_{lf}/M_T > 0.3$). A typical example of this class is given in Fig. 3, the right part of Fig. 5 summarizes our results for class III fragmentation.

IV. Cratering

Impacts in this class result in the formation of a crater of varying dimensions and leave the target block intact (see Fig. 4; i.e., $M_{lf}/M_T \sim 1$).

Although the boundaries between the four fragmentation classes, defined as a particular ratio between the mass of the largest fragment to original target mass, seem—at this point—somewhat arbitrary, the results of our experiments confirm the utility of these definitions.

Table 2 gives the results of our experiments. The recovered target fraction (column 3) includes the collectively weighed smaller samples, as well as the individually weighed fragments. D and h (columns 7 and 8) are the diameter and depth of the craters for the type IV experiments.

In the following, the distribution of fragment masses and fragment shapes for the experiments in classes I–III are discussed. Class IV experiments are described elsewhere (Lange and Ahrens, 1981).

Fig. 2. Disrupted ice target (Shot No. 598, 257 K); configuration of initial target block as shown.

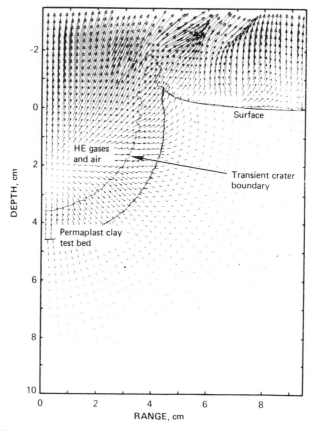

Fig. 14. Vector velocity and material boundary plot at 100 μsec for the 25-X calculation.

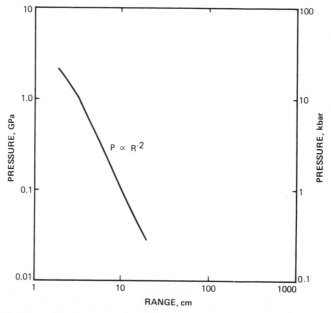

Fig. 15. Maximum shock pressure in target versus range in the downward direction from the 25-O and 25-X calculations.

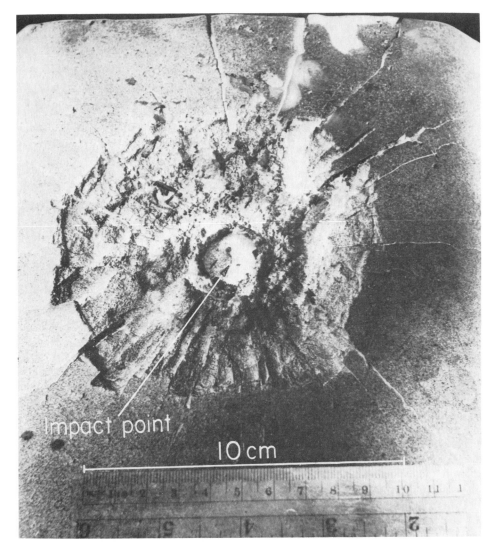

Fig. 4. Crater formation in ice (Shot No. 602, 257 K). The impact point is clearly seen as a central pit inside the crater. The target block was sprayed with black paint to enhance contrast after impact.

of fracturing (i.e., decreasing number of fragmentation class), lying at ~ 1.6, ~ 1.2, and ~ 0.9 for class I, II, and III, respectively. The dependence of b on target temperature is less obvious. The mean values of b for class I and II experiments of 81 K targets are 12% and 19% higher than for the 257 K targets. b values for class III fragmentation are essentially independent of target temperature.

Hartmann (1969) has compiled data describing the fragmentation of terrestrial and extraterrestrial rocks by a variety of natural processes. b values for terrestrial rocks range from 0.6 to 1.2. Values for the ejecta of hypervelocity impacts vary between 1.0 and 1.2. b values for (presumably impact) debris on the lunar surface lie between 0.7 and 1.1. Hence, these rock fragments from different environments yield mass-number characteristics similar to the present class II and III fragmentation.

Fujiwara *et al.* (1977) report for impact experiments in basaltic targets b values between 1.05 and 1.4 for fragment sizes from 2 mm to 1 cm. These agree well with our b values for classes I to III.

Table 2. Major experimental results.

Shot No.	ke, 10^5 ergs	Fragm. Type	Recovered Target Fraction, %	Mass of Samples <1 cm g	Mass of Samples >1 cm, g	Mass of Single Fragment, g	D*, cm	h*, cm
570	56.83	I	15	—	899	—	—	—
571	59.36	I	7	—	411	—	—	—
590	4.40	III	86	—	642	4050	—	—
591	1.35	IV	—	—	—	—	8.0	1.3
592	3.70	II	100	1790	1666	2472	—	—
593	1.74	IV	—	—	—	—	8.9	1.6
596	6.20	II	99	2153	2244	—	—	—
597	4.92	III	97	650	802	4789	—	—
598	7.13	II	100	1407	1243	3104	—	—
602	2.50	IV	—	—	—	—	10.8	1.3
604	2.24	III	98	1796	1201	2815	—	—
605	5.40	III	97	412	750	5265	—	—
609	17.70	I	96	4330	1477	—	—	—
611	1.72	III	99	1368	820	3415	—	—

*Diameter D and depth h of craters in class IV fragmentation.

The differences in impact fragmentation (for fragmentation types I to III) as a function of target temperature are shown in Fig. 8. As can be seen, the spread in fragment masses in the range from ~ 5 to ~ 100 g is significantly smaller for targets of 81 K than for 257 K targets. It can also be seen that the total number of fragments is generally smaller for the warmer targets. This does not represent incomplete sampling, as might be expected, but reflects the fact that fragments in the ~ 5 to ~ 30 g range are more numerous for targets initially at 81 K than for targets with a temperature of 257 K. Thus, while the relative distribution of fragment masses in each fragmentation class is nearly independent of temperature (i.e., the values for b do not vary greatly as a function of target temperature), the absolute number of fragments for each fragment mass is larger for an 81 K target, as compared to a 257 K target. This fact is reflected in the C values, which are larger for the 81 K targets than for the 257 K targets (see Table 3).

3.2 Fragment shapes

The ratios of intermediate to long axes (B/A) versus the ratio of short to long axes (C/A) for fragments from six experiments and mean values are shown in Fig. 9. Here B/A and C/A are called shape factors. The upper diagrams give the shape factors for targets with temperatures of 257 K, the lower ones are for 81 K targets. Roughly spheroidal or, equant, fragments will have $B/A = C/A = 1$, whereas platy fragments have $B/A \to 1$ and $C/A \to 0$, and elongated rods have $B/A \to 0$ and $C/A \to 0$. Mean values for B/A and C/A are given in Fig. 10 and Table 3. As can be seen, there is a clear distinction between the fragments of 257 K targets as compared to those of the colder targets. With the exception of class I fragmentation, the mean values of the shape factors are larger for the 81 K targets than for the 257 K targets. While the fragmentation type is of no significant influence on B/A vs. C/A for the 81 K targets, there is a transition from more bar-like fragments in class III fragmentation to more spherical fragments in class I experiments for the 257 K targets. The results for the targets with temperatures of 81 K are in agreement with the mean values of B/A and C/A obtained by Fujiwara et al. (1978) for basaltic targets. These values (coinciding with the point for our shot 604) represent the means for a total of 719 fragments greater than 4 mm from catastrophically destroyed targets with remaining cores (destruction type II of Fujiwara et al., 1977). Hartmann and Cruikshank (1978) report values of B/A for 46 fragments produced in collisional fragmentation experiments on igneous rocks at velocities between 26 and 50 cm/s and find a mean value for $B/A = 0.71$.

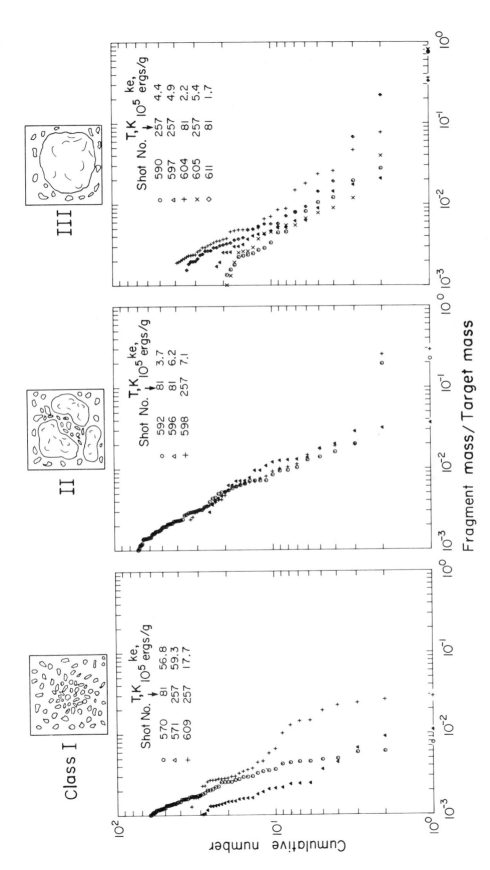

Fig. 5. Cumulative number of fragments vs. relative fragment weight.

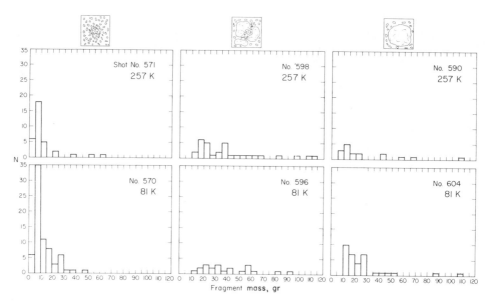

Fig. 6. Histograms of fragment number, N, vs. fragment mass for six class I to III experiments. Data for small fragments, with masses <5 g are incomplete.

In order to determine the relation between fragment mass and fragment shape in our experiments, we calculated the sphericity ψ for the resulting fragments. ψ is defined as

$$\psi = (V_p/V_s)^{1/3} \quad (2)$$

where V_p is the volume of the fragment and V_s is the volume of the smallest sphere that encloses the fragment (Friedmann and Sanders, 1978). Hörz (1969) points out that fragments originating from an experimental impact crater in granite became increasingly flattened with increasing fragment size (or fragment mass), i.e., he notes that the sphericity of his fragments decreases with increasing fragment mass. Although he defines sphericity in a slightly different way as we do, we attempted to test whether our fragments show a similar tendency. Our measurements allow the computation of ψ, since the fragment mass m and the mean density of ice ρ at ~257 K (i.e., the temperature at which the fragments were measured; $\rho = 0.019 \, g/cm^3$; Hobbs, 1974) yield the value of V_p. V_s, a sphere with diameter A, can also be calculated for our fragments. Hence, we can rewrite equation (2) as follows:

$$\psi = 1/A \, K \, m^{1/3} \quad (3)$$

where

$$K = (6/\pi \rho)^{1/3} = 1.276 \, cm \, g^{-1/3}$$

The values of the sphericity ψ as a function of fragment mass for six of our experiments are given in Fig. 11, whereas Table 3 gives the mean values of ψ. As can be seen, there is no clear correlation between ψ and m in disagreement with the observations of Hörz (1969) on fragments from an impact crater in granite. However, as already seen in Figs. 9 and 10, the sphericity of the fragments increases with decreasing temperature (with the exception of shot 571 and 609, which represent fragmentation type I for 257 K targets). The mean values of ψ are 0.73 and 0.53 for all of the 81 K and 257 K targets, respectively.

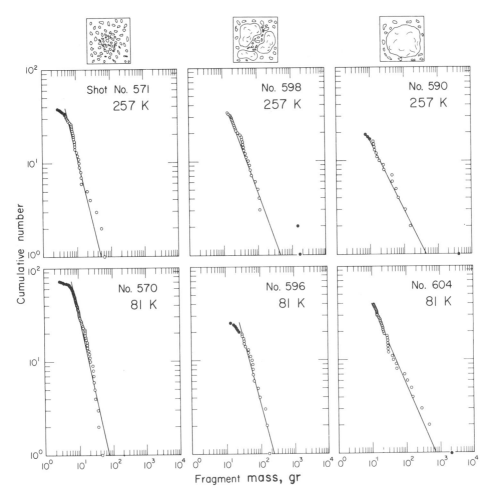

Fig. 7. Cumulative number of fragments vs. fragment mass for same data as in Fig. 6. Solid lines are regression fits for data represented by open symbols.

4. DISCUSSION AND CONCLUSION

The present experiments suggest the following conclusions. The mass ratio of the largest fragment to the target is an empirical measure representing the character of a particular impact event. It is used to assign each impact to one of the fragmentation classes as defined above, as well as to determine the transition between the fragmentation classes with respect to the required impact energies. Figure 12 gives mass ratios of largest fragment to target, M_{lf}/M_T, as a function of the specific kinetic energy ke of the projectile. The values of ke_c for the transition between the fragmentation classes (also marked in Fig. 12) are given in Table 4. As can be seen, the data points clearly fall into two distinct groups, each representing a different target temperature. The threshold specific energies ke_c for 257 K targets are ~2 to 3 times higher than for the 81 K targets. However, the curves for both target temperatures are nearly parallel, both covering a range of about 1.5 orders of magnitude in ke. Fujiwara (1980), based on the definition of destruction types by Fujiwara et al. (1977) gives explicit numbers for critical specific energies ke_c for the transition between the destruction types (Fig. 12). These energies, together with the mean values of B/A and C/A for basalt (Fujiwara et al., 1978) and data of Hartmann (1980) and Hartmann

Table 3. Details of experimental results for fragmentation types I–III.

Shot No.	Fragm. Type	T_T, K	C[+]	b[+]	B/A*	C/A*	Sphericity*
570	I	81	3.28	1.76	0.75	0.55	0.71
571	I	257	2.68	1.57	0.76	0.49	0.72
609	I	257	3.01	1.35	0.76	0.58	0.77
592	II	81	2.81	1.06	0.75	0.46	0.72
596	II	81	3.46	1.46	0.79	0.56	0.77
598	II	257	2.79	1.06	0.63	0.38	0.60
590	III	257	1.98	0.75	0.58	0.32	0.55
597	III	257	2.44	0.98	0.55	0.34	0.57
604	III	81	2.49	0.88	0.73	0.50	0.71
605	III	257	2.41	0.99	0.54	0.32	0.55
611	III	81	2.54	0.99	0.76	0.56	0.74

[+]Coefficients for fragment weight distributions in:

$$\mathrm{Log}_{10} n = C - b \, \mathrm{Log}_{10} m$$
$$n = \text{cumulative number}$$
$$m = \text{fragment mass}$$

*Mean values

and Cruikshank (1978) are given in Table 4. As can be seen, the destruction of a silicate body requires energies about one to two orders of magnitude higher than those for an icy body. The data for water ice by Hartmann (1978) indicate critical specific energies ke_c of $\sim 10^5$ ergs/g for the transition from class IV to III and $\sim 10^6$ ergs/g for the transition from III to II. Although Hartmann's data represent a different experimental approach they agree with our results for the 257 K targets (see Table 4). This indicates that the

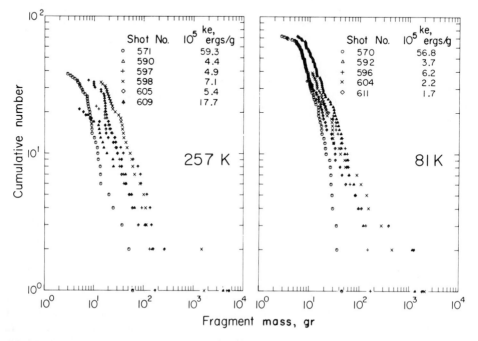

Fig. 8. Cumulative number of fragments vs. fragment mass for 257 K (left diagram) and 81 K (right diagram) target temperatures.

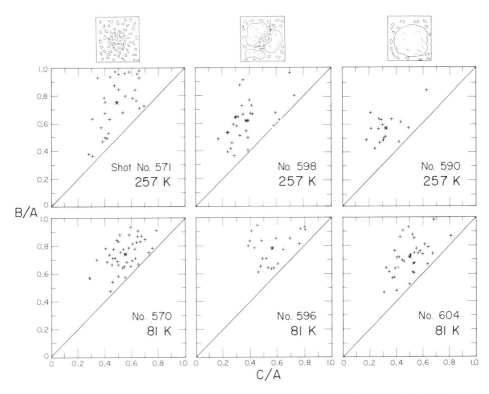

Fig. 9. Fragment shape factors (B/A and C/A) from class I, II, and III experiments. A is the long, B the intermediate, and C the short axis of each fragment. Mean values of B/A and C/A are shown as (*).

difference in the experimental techniques between Hartmann's and our study, as well as the difference in the combination of projectile and target materials (ice and rock vs. Lexan and ice) has only limited influence on the final results.

Fujiwara et al. (1977) discuss the applicability of the ke_c scaling with respect to a wider range of values of the specific kinetic energy of projectiles as found in natural impacts. They suggest that the phenomena arising from the propagation of shock waves in a target body are similar, if the imparted energies per unit mass are the same. This conclusion is based, in part, on the agreement between Fujiwara et al.'s and Gault and Wedekind's (1969) experimental results which both give threshold energies ke_c for basaltic and glass bodies, respectively, which are similar, when the difference in target strength is taken into account. While the values of ke_c for basalt are about two to three times higher than those for glass (see Fig. 10 in Fujiwara et al., 1977), the difference in quasi-static strength of glass (~1.5–2 kb; Gault and Wedekind, 1969) and typical basalt (~3 kb; Handin, 1966) also amounts to a factor of two. However, the differences in mass and velocity between the two groups of experiments is quite large (impact velocities of Gault and Wedekind's experiments were up to three times higher than those of Fujiwara et al., whereas target masses were up to 20 times larger in Fujiwara et al.'s experiments). Our results support Fujiwara et al.'s (1977) conclusion. Impact velocities in our experiments were as much as two times lower and target masses were at least twice as large as those of Fujiwara et al. However, the differences in ke_c between Fujiwara et al.'s and our experiments, as seen in Fig. 12, can be explained by the about two orders of magnitude difference in ultimate unconfined strength between basalt and water ice. Hence, the results of Gault and Wedekind (1969), Fujiwara et al. (1977), and the present study, which cover a range of ke from 10^5 to 10^8 ergs/g, seem to be consistent, when differences in the material properties of target materials are taken into account. Consequently, it appears that the processes

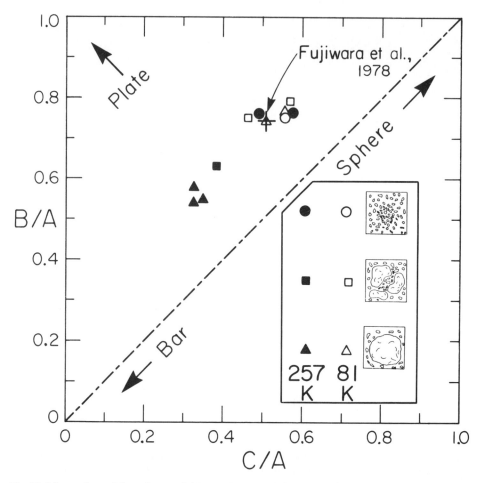

Fig. 10. Mean values of shape factors (B/A and C/A) from I to III fragmentation experiments. The cross at B/A = 0.73 and C/A = 0.50 is the mean value for 719 basaltic fragments given by Fujiwara *et al.* (1978).

related to impact fragmentation of a body can be described in terms of *ke* over a range of target masses of grams to kilograms and impact velocities in the 0.1 to 6 km/sec range.

The mass distribution of fragments resulting from the erosion, disruption, and total destruction of an icy body in our impact experiments can be described by a single power law relation [Eq. (1)], for fragment masses of up to ~ 0.1 times the target mass. The slopes b of the distributions vary between -0.9 and -1.8 and decrease with increasing specific projectile energy *ke* (or grade of fragmentation). Table 4 gives the mean values of b for our experiments with regard to fragmentation class and target temperature. It can be seen that there is no clear relation between b and target temperature. These results suggest that with increasing values of *ke* the number of small fragments in a given size range increase relative to large fragments regardless of temperature. The b values of our experiments are in good agreement with results of Fujiwara *et al.* (1977) for basalt, who find values of b of up to 1.7, and to a lesser extent with data on terrestrial and extraterrestrial rock fragments as given by Hartmann (1969) which show b values in the range from ~ 0.6 to 1.2. However, Hartmann (1980) also found b = 0.96 for an experimentally destructed artificial aggregate material.

The shapes of the fragments in our experiments have been described in terms of shape factors (i.e., ratios between intermediate and long axis and ratios between short and long axis, B/A and C/A, resp.) and the sphericity ψ of each fragment. Mean values of B/A,

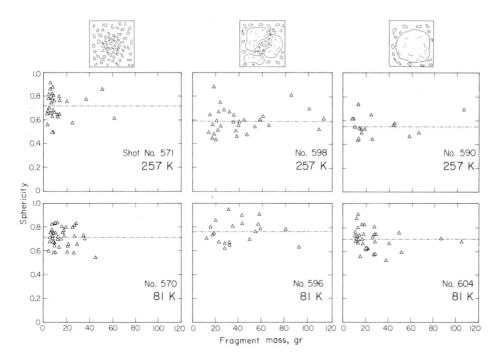

Fig. 11. Fragment sphericity, as defined in Eqs. (2) and (3), vs. fragment mass. The dash-dotted horizontal lines give the mean values for each experiment.

C/A, and ψ with respect to fragmentation class and temperature are given in Table 4. We find that fragments from destruction of 257 K targets are less spherical than those from 81 K targets. While the former have increasing values of B/A, C/A, and ψ with increasing specific projectile energy ke, the latter reveal no clear dependence of shape on fragmentation class. This is illustrated in Figs. 10 and 13. The straight lines in Fig. 13 represent linear regression fits to each set of data points. The slopes of the regression lines are given by -1.5×10^{-8} and 3.6×10^{-8} g/ergs for the 81 and 257 K targets, respectively. The shape factors of the 81 K fragments as well as those of the 257 K targets of class I impacts agree well with values for impact fragmented basaltic targets of Fujiwara *et al.* (1978) and those for experimentally destructed igneous rocks of Hartmann and Cruikshank (1978). The fragment shapes in our experiments are found to be independent on fragment mass.

In the present experiments, we believe that most of the fragmentation occurred upon nucleation of tensile failure surfaces on crack like flaws and on air bubbles in the samples. We observed that with increasing growth rate of ice crystals, the number of air bubbles in ice increases, whereas their size decreases (Hobbs, 1974). Since the growth rate in our 81 K ice reached values of up to $\sim 2 \times 10^{-4}$ m/s vs. $\sim 10^{-5}$ m/s in the 257 K targets, the colder targets contained significantly more, but smaller voids. Hence, we believe that more tensile fractures initiated and intersected in the 81 K ice than in the warmer targets. This we propose leads to a more efficient fragmentation of 81 K targets compared to 257 K targets. Secondly, an inherent difference in both target types is the different amount of volume change suffered by the target blocks (-1.8% vs. -0.3% of the total volume for 81 K and 257 K, respectively; Hobbs, 1974). Even though, a large fraction of the contraction in the colder targets could be relieved by the technique described in section 2, it is expected that the number of initial microcracks is substantially larger than in the 257 K targets. Side spallation, or spallation of outer surface layers was relatively scarce in our experiments (with the exception of class III fragmentation in

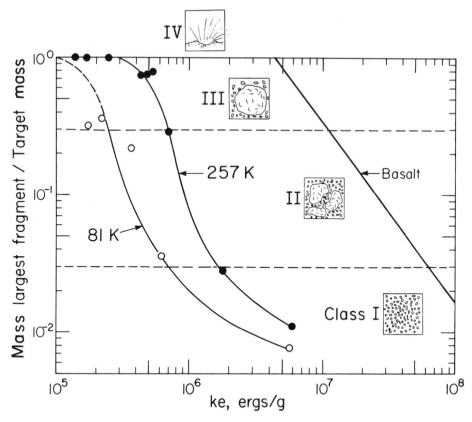

Fig. 12. Relative mass of largest fragment vs. specific kinetic energy of projectiles. Transition from class I to II fragmentation (at relative fragment weight of 0.03) and from class II to III fragmentation (at relative fragment weight of 0.3) is indicated.

257 K targets). This is mainly due to the relatively low peak stresses and the comparatively large target size. This leads to the absence of core type destruction, where the outer parts of the impacted body spall off, leaving a relatively undamaged core of the original body (Gault and Wedekind, 1969; Fujiwara et al., 1977).

An understanding of the importance of pre-cracking, as well as of the role of air bubbles in target blocks, for the fragmentation process requires a closer look at the mechanism of fracturing. Fracturing has been studied statistically under the presumption that whenever a large number of microscopic flaws becomes activated, the failure process may be statistically averaged and may yield a usable continuum description of fragmentation (e.g., Gilvarry, 1961). Microscopic fracture theories on the other hand attempt to describe explicitly the small scale rate processes which lead to failure. Curran and his co-workers have used the latter approach to derive a quantitative analytical model of the fracturing process (Curran et al., 1973; Shockey et al., 1974; Curran et al., 1977). At the present stage of our study, quantitative data which constrain the major processes in the Curran et al., models are missing. Hence, we have qualitatively applied only the basic ideas of the first step in their models, namely the activation of preexisting structural flaws, to our problem. The two principal flaws in ice crystals grown from a fluid are homogeneously dispersed, roughly spherical, air-filled voids and randomly positioned sharper, crack like flaws (Hobbs, 1974). Both flaw types have been readily observed in all of our target blocks. Shock damage will initiate primarily at the crack like flaws, because stress will be more efficiently concentrated and flaws become unstable at lower nominal stress at these points than at the bubble like voids (Curran et al., 1973).

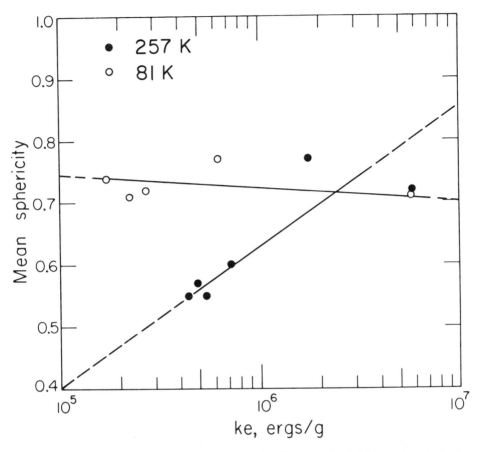

Fig. 13. Mean sphericity of fragments vs. specific projectile energy ke. Solid lines through the data represent regression lines.

Table 4. Summary of experimental results.

Target Temp., K	ke_c for transition from: I–II	II–III ergs/g	III–IV	b* I	II	III	B/A* I	II	III	C/A* I	II	III	ψ* I	II	III
257	1.7×10^6	7.1×10^5	2.9×10^5	1.5	1.1	0.9	0.8	0.6	0.5	0.5	0.4	0.3	0.8	0.6	0.5
81	7.0×10^5	2.5×10^5	1.0×10^5	1.8	1.3	0.9	0.8	0.8	0.8	0.6	0.5	0.5	0.7	0.7	0.7
Basalt$^$	1.0×10^8	1.0×10^7	4.0×10^6					0.7⁺			0.5⁺				
Igneous rocks#								0.7							
Artificial aggregate‖					1.0										

*Mean values.
$From Fujiwara, 1980.
⁺From Fujiwara et al., 1978.
#From Hartmann and Cruikshank (1978).
‖From Hartmann (1980).

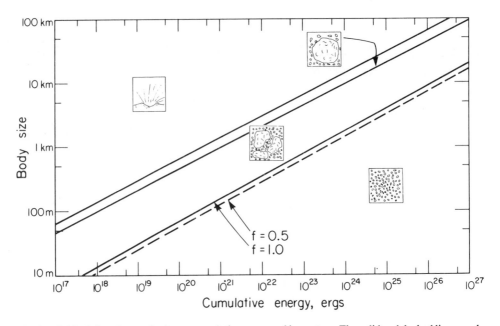

Fig. 14. Critical size of target body vs. cumulative energy of impactors. The solid and dashed lines mark the transition between different types of fragmentation suffered by the target, when hit by a number of impactors (or a single body) with a particular total energy. Each field between the solid lines in this diagram represents a type of fragmentation in the body size-energy plane depicted by the cartoons. f is the fraction of the total impact energy shared by the impacted body [Eq. (6)]. Differences in the results with regard to variation in f are identical for all of the three boundaries shown.

We therefore conclude that fracturing in our ice targets is initially, i.e., at low stress levels, controlled by the preexisting microcracks. However, at higher stresses, cracks will be initiated at the voids with equal efficiency. Since the bubbles are generally more numerous than the microcracks, the former will control crack initiation at higher stress levels (Curran et al., 1973). As outlined above, the two target types can be distinguished in that the 81 K ice includes a larger number of microcracks as well as air bubbles (of relatively small size) as compared to the 257 K ice. We believe that both of these factors lead to more extensive fragmentation of the colder targets with respect to either the stress level necessary for crack initiation, or the number of crack nucleation sites at higher stress levels.

The differences in fragment shape as a function of target temperature (see Figs. 10 and 13 and Tables 3 and 4) can also be explained in the light of the foregoing discussions. Since activation of a larger number of flaws occurs at relatively low stresses in the 81 K ice numerous fracture surfaces and thus more regularly shaped fragments are produced as compared to 257 K ice. However, at high stress levels (i.e., in class I fragmentation), the relatively large number of air bubbles controls crack initiation and allows the formation of more numerous fracture surfaces and thus more spherical fragments for both 81 K and 257 K ice, in agreement with our results.

5. APPLICATIONS

Applications of our experimental results to accretion and surface evolution of icy planetary objects is restricted to the 81 K targets. Temperatures in the region of the solar nebula where abundant water ice could have been accreted correspond to a range from ~40 to ~150 K (Lewis, 1974). Present mean surface temperatures of the Galilean and Saturnian satellites, which are the prime examples for icy planetary objects in our solar system, are 120 ± 10 K and 70 ± 10 K, respectively (e.g., Smith et al., 1979, 1981). Major obstacles in applying our small scale laboratory impact experiments to natural processes on icy planets are the differences in size and shape and the differences in internal structure and bulk composition.

Fujiwara (1980) discusses the importance of the first factors with regard to his experiments on cubic basalt targets (Fujiwara et al., 1977). He suggests that different shapes of experimental and planetary bodies are less restrictive as long as the aspect ratios for both groups of bodies is close to one. Based on scaling relations for cratering on gravity free bodies he also suggests that the dimensions of his targets (up to 10 cm sidelength) allow application to bodies of up to 100 km size. We adopt these considerations and infer that our results (obtained for cubic targets of ~ 19 cm sidelength) can be applied to accreting icy planets of $\leq \sim 100$ km radius. The size difference is less important with regard to regolith development as long as surface gravity is not a dominant parameter.

The differences in the internal structure and bulk composition of our targets and icy planets is a more severe problem. Accretion of icy planets probably involved a variety of aggregates reaching from loosely bound, highly porous "dirty snowballs" (i.e., mixtures of fine grained ice and silicates) to solid icy bodies (e.g. Hartmann, 1978). In contrast, our target blocks represent ice grown from the liquid phase thus only representing one type of icy planetesimals. Furthermore, mean densities of the icy Galilean and Saturnian satellites suggest that they contain significant fractions of silicates. This implies that accreting planetesimals consisted either of ice-silicate mixtures or that two groups of planetesimals, consisting of either ice or silicates, formed these bodies. The destruction of ice-silicate targets is a problem we have not yet addressed. Hence, the present results for pure prefractured and prestressed ice are restricted to the interaction of solid icy bodies. The crusts of the Galilean and Saturnian satellites are likely to be also multiply prefractured and prestressed. For targets of this type, Gault and Wedekind (1969) demonstrate that multiple and single impacts lead to comparable fragmentation if the cumulative energy of the multiple events equals the single impact energy. Hence, we imply in the following that energies for disruption of icy objects refer to the total energy of multiple events.

With these cautionary notes in mind, we first apply our results to the accretion of icy planetesimals. The collision of finite sized bodies requires separation of energies associated with the motion of the center-of-mass and with motions relative to it. For the latter, the impact energy, IE, is given by:

$$IE = 1/2 \, M_t M_p v^2 / (M_t + M_p) \qquad (4)$$
(Fujiwara et al., 1977)

where M_t and M_p are target and projectile mass, respectively, and v is the velocity of the projectile. We assume that a fraction f of this energy is shared by the target and a fraction (1–f) by the projectile (f depends e.g., on the ratio between projectile and target mass). Hence, we reformulate Eq. (4):

$$IE = M_t/(M_t + M_p) KE \, f \qquad (5)$$

where KE is the projectile kinetic energy. In order to determine the limiting size, R_c, of an impacted body at which damage according to one of the above defined fragmentation classes occurs, we equate IE with the product of ke_c and M_t, where ke_c is the critical transition energy as defined in the previous section. This leads to:

$$ke_c = KE \, f/(M_t + M_p)$$

which yields:

$$R_c = [3/4\pi\rho_t)(KE/ke_c \, f - M_p)]^{1/3} \qquad (6)$$

where ρ_t is the mean density of the target body (~ 1 g/cm^3). Insertion of ke_c for our 81 K targets (see Table 4) results in critical sizes (i.e., radii) for target bodies which define regions of cratering, erosion, disruption, and total fragmentation in the R_c–KE plane.

Figure 14 illustrates the results of these calculations in terms of cumulative energy vs. target size for plausible limits of f. As can be seen, variation in f has only minor effects on the critical size of the impacted body. Figure 14 allows specification of the fate of an impacted icy planetesimal of certain size, when hit by impactors with *total* energy KE. It should be noted that these results are strictly valid only for low velocity impacts (i.e., v up to ~ 1 km/s). Values for basaltic targets, corresponding to our transition from erosion to disruption of a body, lie at $\sim 10^{18}$ to 10^{28} for body sizes between 50 m and 100 km (Fujiwara, 1980). We also note that for the fragmentation of an icy planet its finite gravity has to be taken into account. Hence, even though the energy of impacting bodies might be sufficient to completely fragment the planet, separation of the fragments against the gravity field of the planet requires additional energy to be provided by the impactors. Therefore, the total energies KE in Fig. 14, defining critical sizes for the destruction of icy planetesimals have to be understood as *minimum* energies.

With regard to regolith development on icy planets (we neglect in the following the effects of silicates, mixed in the crustal layers of these bodies and assume pure water ice crusts), we predict that the shapes of regolith particles in the milimeter to centimeter range are essentially spherical (or regularly formed). The mass distribution of regolith components with masses from ~ 5 to ~ 160 g, should follow a power law with a slope of ~ -1.8 (the slope for the cumulative size distribution equals three times the slope for the mass distribution; Hartmann, 1969). Existing observational radio- and radar reflectance data of icy regoliths are as yet not conclusive (Muhleman, pers. comm.) and cannot be used to confirm this prediction. We note however that our predicted properties of icy regoliths are comparable to those of rocky planets (Hartmann, 1969).

In the case of icy planetary rings, e.g., Saturn's ring system, more complete data are available. These data suggest mean sizes of ring particles in the centimeter to meter range and can best be fit with nearly spherical bodies (Pollack, 1975). Greenberg *et al.* (1977), based on a variety of theoretical and observational results, have attempted to predict the size distribution of particles in Saturn's rings. They find that particle sizes should follow a power law [similar to our Eq. (1)], with slopes between -3.0 to -3.5, which correspond to cumulative mass distribution indices of -0.67 to -0.83. Based on our results for 81 K ice, we predict a much steeper slope for the cumulative mass distribution, namely ~ -1.5. This is the mean of b values for fragmentation classes I and II (see Table 4). Greenberg *et al.* exclude a steeper mass distribution of ring particles based on the assumption that $\sim 10\%$ of the cross section of the rings resides in particles larger than 50 m in radius whereas our result (when extrapolated to this size range) imply that 50 m particles comprise $\sim 3\%$ of the total population. We assume that Saturn's rings were formed initially by the collisional destruction of a single or several large bodies, at the present location of the rings (Pollack, 1975; Greenberg *et al.*, 1977). Hence, we strictly predict only the initial size distribution of ring particles. However, the subsequent collisional interaction of bodies in the rings would tend to produce even smaller particles and thus further steepen the mass distribution.

Acknowledgments—The technical assistance of W. Ginn, P. Gelle and M. Long is gratefully acknowledged. We thank S. K. Croft and W. K. Hartmann for constructive and helpful reviews. This work was supported under NASA grant NSG-7129. One of the authors (M. A. Lange) received a grant from the Deutsche Forschungsgemeinschaft. Contribution No. 3611, Division of Geological and Planetary Sciences, California Institute of Technology, Pasadena, California 91125.

REFERENCES

Anderson G. D. (1968) The equation of state of ice and composite frozen soil material. *CRREL Report* **257**, Hanover, N. H. 50 pp.

Butkovich T. R. (1954) The ultimate strength of ice. *SIRPE Res. Rep.* **15**, 12.

Croft S. K., Kieffer S. W., and Ahrens T. J. (1979) Low-velocity impact craters in ice and ice-saturated sand with implications for martian crater count ages. *J. Geophys. Res.* **84**, 8023–8032.

Curran D. R., Seaman L., and Shockey D. A. (1977) Dynamic failure in solids. *Physics Today* **30**, 46–55.
Curran D. R., Shockey D. A., and Seaman L. (1973) Dynamic fracture criteria for a polycarbonate. *J. Appl. Phys.* **44**, 4025–4038.
Dantl G. (1969) Elastic moduli of ice. In *Physics of Ice* (N. Riehl, B. Bullemer, and H. Engelhardt, eds.) p. 223–230, Plenum, N.Y.
Friedmann G. M. and Sanders J. E. (1978) *Principles of Sedimentology*. J. Wiley and Sons, N.Y. 792 pp.
Fujiwara A. (1980) On the mechanism of catastrophic destruction of minor planets by high-velocity impact. *Icarus* **41**, 356–364.
Fujiwara A., Kamimoto G., and Tsukamoto A. (1977) Destruction of basaltic bodies by high-velocity impact. *Icarus* **31**, 277–288.
Fujiwara A., Kamimoto G., and Tsukamoto A. (1978) Expected shape distribution of asteroids obtained from laboratory impact experiments. *Nature* **272**, 602–603.
Gault D. E. and Wedekind J. A. (1969) The destruction of tektites by micrometeoroid impact. *J. Geophys. Res.* **74**, 6780–6794.
Gilvarry J. J. (1961) Fracture of brittle solids. I. Distribution function for fragment size in single fracture (Theoretical). *J. Appl. Physics* **32**, 391–399.
Greenberg R., Davis D. R., Hartmann W. K., and Chapman C. R. (1977) Size distribution of particles in planetary rings. *Icarus* **30**, 769–779.
Greenberg R., Wacker J. F., Hartmann W. K., and Chapman C. R. (1978) Planetesimals to planets: Numerical simulation of collisional evolution. *Icarus* **35**, 1–26.
Handin J. (1966) Strength and ductility. In *Handbook of Physical Constants* (S. P. Clark, ed.), p. 223–290, *Mem. Geol. Soc. Amer.* **97**.
Hartmann W. K. (1969) Terrestrial, lunar, and interplanetary rock fragmentation. *Icarus* **10**, 201–213.
Hartmann W. K. (1978) Planet formation: Mechanism of early growth. *Icarus* **33**, 50–61.
Hartmann W. K. (1980) Continued low-velocity impact experiments at Ames Vertical Gun facility: Miscellaneous results (abstract). In *Lunar and Planetary Science XI*, p. 404–406. Lunar and Planetary Institute, Houston.
Hartmann W. K. and Cruikshank D. P. (1978) The nature of trojan asteroid 624 Hektor. *Icarus* **36**, 353–366.
Hörz F. (1969) Structural and mineralogical evaluation of an experimentally produced impact crater in granite. *Contrib. Mineral. Petrol.* **21**, 365–377.
Hobbs P. V. (1974) *Ice Physics*. Clarendon, Oxford. 837 pp.
Lewis J. S. (1974) The temperature gradient in the solar nebula. *Science* **186**, 440–443.
Oberbeck V. R. (1975) The role of ballistic erosion and sedimentation in lunar stratigraphy. *Rev. Geophys. Space Phys.* **13**, 337–362.
Pollack J. B. (1975) The rings of Saturn. *Space Sci. Rev.* **18**, 3–93.
Safranov V. S. (1972) Evolution of the protoplanetary cloud and formation of the earth and planets. NASA Technical translation F-677.
Shockey D. A., Curran D. R., Seaman L., Rosenberg J. T., and Petersen C. T. (1974) Fragmentation of rock under dynamic loads (abstract). *Int. J. Rock Mech. Sci. and Geomech.* **11**, 303–317.
Smith B. F., Soderblom L. A., Beebe R., Boyce J., Briggs G., Carr M., Collins S. A., Cook A. F., Danielson G. E., Davies M. E., Hunt G. E., Ingersoll A., Johnson T. V., Masursky H., McCauly J., Owen T., Sagan C., Shoemaker E. M., Strom S., Suomi V. E., and Veverka J. (1979) The galilean satellites and Jupiter: Voyager 2 imaging science results. *Science* **106**, 927–950.
Smith B. A., Soberblom L., Beebe R., Boyce J., Briggs G., Bunker A., Collins S. A., Hansen C. F., Johnson T. V., Mitchell J. L., Terrille R. J., Carr M., Cook A. F., Cuzzi J., Pollack J. B., Danielson G. E., Ingersoll A., Davies M. E., Hunt G. E., Masursky H., Shoemaker E., Morrison D., Owen T., Sagan C., Veverka J., Strom J., and Suomi V. E. (1981) Encounter with Saturn: Voyager I imaging science results. *Science* **212**, 163–191.

Initial energy partitioning and some excavation stage phenomenology in laboratory-scale cratering calculations in clay

Michael G. Austin, Jeffrey M. Thomsen, and Stephen F. Ruhl

Physics International Company, 2700 Merced Street, San Leandro, California 94577

Abstract—One impact and two explosive cratering calculations have been analyzed with emphasis on the *early excavation stage*. The *early excavation stage* is here defined as that part of the *excavation stage* that occurs after energy partitioning is 90% complete, but before the cratering flow field can be well described by Z-type flow fields with values of Z uniformly greater than two. Impact generated flow fields seem to have a much longer *early excavation stage* than explosion generated flow fields, due possibly to the slower momentum transfer versus energy transfer rate between projectile and target. During this time when the projectile retains a significant portion of its original momentum, Z values less than two are observed in the impact generated flow field. Z values less than two are not observed at any time in the explosion generated flow fields.

I. INTRODUCTION

In order to gain a better understanding of the *compression* and early *excavation stages* of crater growth, some previously performed calculations of laboratory-scale impact and explosion cratering in clay have been examined. In particular, comparisons of the coupling of energy to the target and the material flow immediately after passage of the shock wave generated by the initial impact or explosive detonation provide insight into some of the similarities and differences between impact and explosion cratering.

The impact and explosion calculations described here were performed for separate studies. Although the target materials are very similar, and the absolute crater dimensions are all laboratory-scale (on the order of centimeters), the depth-of-burial of the explosive was not chosen to simulate the impact, nor are the scaled energies directly comparable. Nevertheless, comparisons of the different calculations can provide a starting point for investigating some of the basic differences between impact and explosion cratering.

As described empirically by Gault *et al.* (1968), the cratering process may be divided into the *compression stage*, the *excavation stage*, and the *modification stage*. The *compression stage* begins with the initial contact between projectile and target, includes the shock compression of both, and ends when the coupling of the projectile energy and momentum to the target is essentially complete. The *excavation stage* then begins and lasts much longer. It describes the processing of the bulk of the target material by the outgoing shock wave after the shock wave has departed from the immediate vicinity of the projectile and the transient cavity. The *excavation stage* describes the modification of the initially radial motions by the continuous fan of rarefactions generated at the target surface, and the orderly ejection of material from the transient cavity. The *excavation stage* ends when all cratering motions attributable to shock waves and rarefactions have ended. The subsequent *modification stage* includes possible slumping, isostatic adjustment, and erosion.

II. INITIAL CONDITIONS OF THE CALCULATIONS

The finite difference continuum mechanics computer code PISCES 2DELK (Hancock, 1976) was used to perform two-dimensional axially symmetric calculations of laboratory-scale impact into plasticene clay and of laboratory-scale explosive detonations in permaplast clay. General considerations regarding the calculational techniques of performing these types of numerical simulations, such as zoning or Lagrangian and Eulerian reference frames, have been discussed, for example, by Thomsen et al. (1979a, b) and will not be repeated here. The target medium with lowest strength in the two calculations in Thomsen et al. (1979a) will be referenced here as the PL1 calculation. The two explosive cratering calculations reported previously by Thomsen et al. (1980) will be referred to here as the 25-O and the 25-X calculations.

The PL1 calculation is the calculation of the impact of a 6 mm diameter 2024 aluminum projectile into a plasticene target. The projectile is spherical, weights 0.3 gm, and impacts normally to the target which is assumed to be in a vacuum with terrestrial gravity present. The impact velocity is 6 km/sec so that the initial kinetic energy of the impact is 5.4 kJ. The plasticene clay target material has

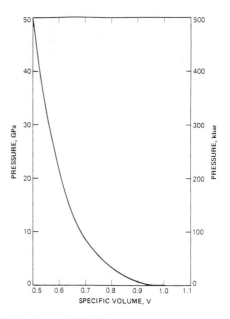

Fig. 1. Austin, Initial energy partioning, May 4, 1981.

an initial density of 1.69 Mg/m^3, a compressional wave velocity of 1.4 m/msec, a shear wave velocity of 0.475 m/msec, and a Poisson's ratio of 0.435. The Hugoniot curve for plasticene clay is given in Fig. 1. The plasticene clay has a von Mises strength failure envelope with a value of 50 kPa. The 2024 aluminium material properties and further data on the plasticene clay are given in Thomsen et al. (1979a, b).

The explosive cratering calculations 25-O and 25-X have as the explosive source a 4.08 gm PETN sphere half buried in the Permaplast clay target. Calculation 25-O includes a gravity value of 10 g and the 25-X calculation includes a gravity value of 517 g. Air is present above the target in both calculations at one atmospheric pressure initially. The initial density of the PETN is 1.73 Mg/m^3. The PETN has a Chapman-Jouguet pressure of 32 GPa (320 kbar) and a detonation velocity of 8300 m/sec (0.83 cm/μsec). A JWL equation of state is used to describe the PETN. The same Hugoniot was used for the permaplast clay as was used for the plasticene clay (Fig. 1), but the permaplast clay has a lower density of 1.53 Mg/m^3 and a different failure strength given by a Mohr-Coulomb failure envelope with cohesion equal to 11 kPa and an angle of internal friction of 1.2 degrees. Further details on the PETN, air, and permaplast clay properties are reported by Thomsen et al. (1980).

III. EARLY EXCAVATION STAGE IN THE PL1 IMPACT CALCULATION

Figures 2, 3, 4, 5, and 6 are a sequence of pressure contour and velocity vector plots from about 2.5 μ sec to 15 μ sec after the impact into the plasticene clay. At 2.5 μ sec the energy transfer from the projectile to the target material is more than 90% complete so these plots show the very beginning of the *excavation stage*. By 15.0 μsec a Z-type flow field has formed with an almost spatially uniform value of Z = 2. Before 15.0 μsec the Z-Model provides a poorer fit to the cratering flow field than it does afterwards. The projectile also has a marked effect on the Z-flow field to the extent that it has formed due to the continued overdriving by the projectile of the material in its immediate vicinity. Also, at 2.5 μsec the main shock wave is just beginning to detach itself from the area around the transient cavity, so there is not much cratering flow field in existence (the

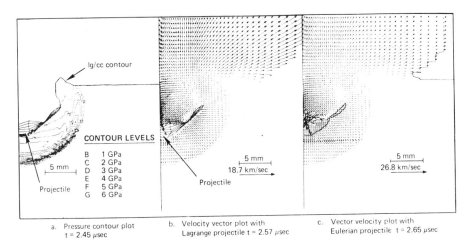

Fig. 2. Material boundary, pressure contour, and vector velocity plots at about 2.5 μsec for the PL1 calculation.

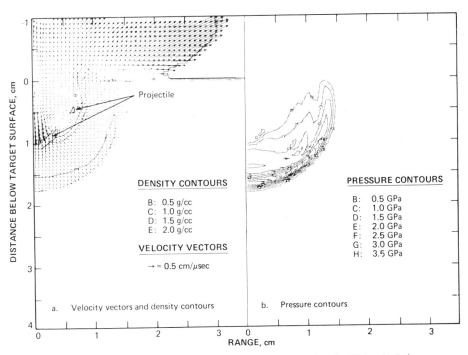

Fig. 3. Contour and velocity vector plots at 4.1 μsec for the PL1 calculation.

cratering flow field is defined as the region of material velocities behind the outgoing shock wave and beyond the transient cavity in the target). The Z-Model description of the cratering flow field is given in detail by Maxwell (1977), Orphal (1977), Thomsen et al. (1979a, b), Austin et al. (1980, 1981), and Croft (1980).

Briefly, the Z-Model observes that with respect to a fitted on-axis flow field center, the radial component $\dot{R}$ of the material flow field velocity decays spatially with the radial distance R according to $\dot{R} = \alpha R^{-Z}$, where α is a parameter which may be spatially or time dependent. For $Z = 2$, the flow is purely radial. For $Z > 2$, the material flow streamlines

1692 M. G. Austin et al.

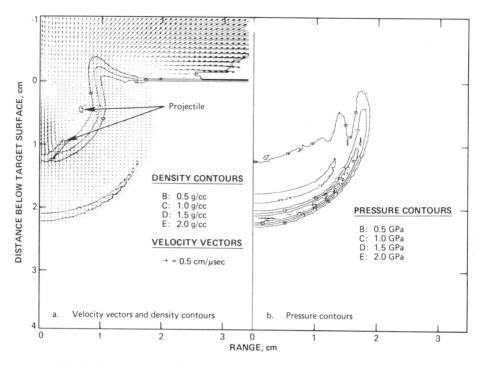

Fig. 4. Contour and velocity vector plots at 6.02 μsec for the PL1 calculation.

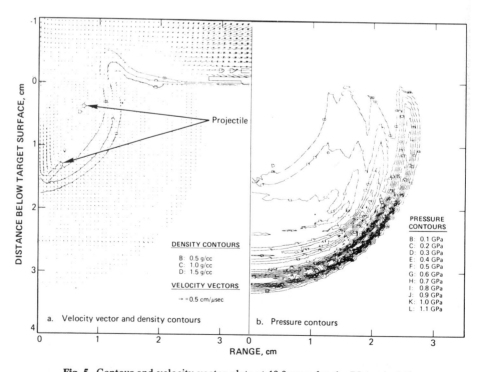

Fig. 5. Contour and velocity vector plots at 10.0 μsec for the PL1 calculation.

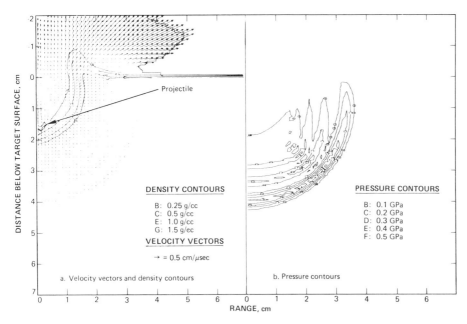

Fig. 6. Contour and velocity vector plots at 15.0 μsec for the PL1 calculation.

curve toward the free surface, and for $Z < 2$, the material flow streamlines curve downward or away from the surface.

At 4.1 μsec (Fig. 3) a lip of material is beginning to form at the rim of the transient cavity. The projectile is 1.8 projectile diameters deep and the shock wave is about 3 projectile diameters from the impact point. The projectile still retains 20% of its momentum and this overdriven condition of the target material beneath the projectile shows up in fitted Z values less than 2 with considerable variation about the fitted values.

At 6.02 μsec (Fig. 4) the cratering flow field is much better formed as the shock wave is more detached from the area of the projectile than it was at 4.1 μsec. As commented upon by Orphal *et al.* (1980), during this penetration phase of the projectile moving into the target, the depth of the transient cavity has been growing faster than its radius.

At 10.0 μsec, Fig. 5 shows the developing rarefactions from the free surface relieving the stresses in the target material and directing particle motion from a purely radial direction outward from the flow field center to motion along streamlines curving more and more toward the surface with time. Z-Model analysis of the cratering flow field at times between 10.0 and 18.0 μsec shows better fits than at earlier times with Z values of 2 or less except near the surface. These values of Z are increasing so that by 18.0 μsec, almost all Z values are greater than 2 and near the surface Z has values of about 3.

The analysis of these times for the PL1 calculation suggests the possibility of defining a part of the *excavation stage* that has distinct differences from the other times during the crater growth process. Let the *early excavation stage* be defined as the part of the *excavation stage* after energy transfer from the projectile to the target is 90% complete and before Z values are all greater than or equal to 2. For the PL1 calculation, the *early excavation stage* lasts from about 2.5 μsec to about 18 μsec.

IV. EARLY EXCAVATION STAGE IN THE 25-O AND 25-X EXPLOSIVE CALCULATIONS

Calculations 25-O and 25-X were performed to simulate two laboratory-scale explosion cratering experiments conducted on a geotechnic centrifuge by Boeing Aerospace Company (Schmidt and Holsapple, 1979), and specifically to calculate how large the transient

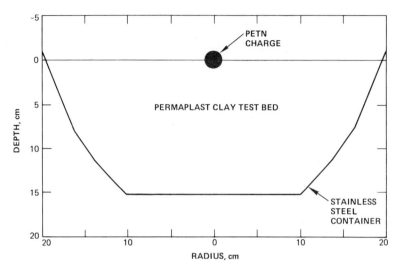

Fig. 7. Initial geometry for explosive cratering experiments in Permaplast clay performed on the centrifuge by Boeing Aerospace Company.

cavities may have been. Shot 25-O was done at 10 g's acceleration and shot 25-X was at 517 g's. Figure 7 shows the initial test bed for the experiments. The test bed boundary is about 15 cm from the charge.

Figure 8 shows the profiles of the experimentally produced craters. The 25-O (10 g) crater was nearly hemispherical and about 6 cm deep, whereas the 517 g crater was saucer-shaped, a little less than 3 cm deep, and almost 6 cm in diameter.

Figures 9, 10, 11, and 12 show the material velocity fields and transient cavity growth at times from 5.0 μsec to 16.2 μsec for the 25-O and 25-X calculations which are essentially identical at these times. At 5.0 μsec, about 90% of the energy transfer to the target is complete. This time corresponds then to 2.5 μsec in the PL1 calculation. By 16.2 μsec the kinetic energy coupling is essentially 100% complete.

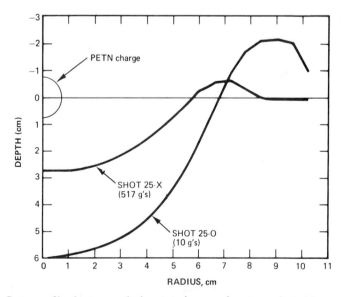

Fig. 8. Crater profiles for two explosion cratering experiments conducted in permaplast clay test beds by Boeing Aerospace Company.

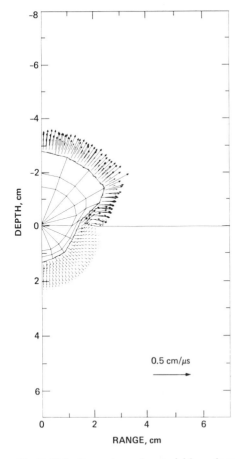 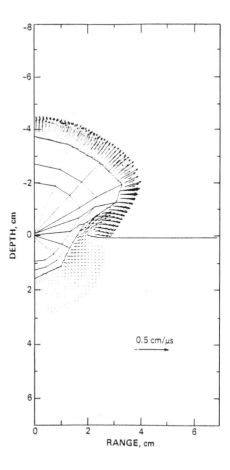

Fig. 9. Velocity vector and material boundary plot at 5.0 μsec for the 25-O and 25-X calculations.

Fig. 10. Velocity vector and material boundary plot at 7.6 μsec for the 25-O and 25-X calculations.

Preliminary Z-Model analysis of the cratering flow fields during these times which should correspond to the *early excavation stage* times of the impact calculations PL1 described in Section III shows that Z values are mostly about 2 except near the surface where they are somewhat higher and that the less than 2 values of Z found for impact cratering flow fields during the *early excavation stage* are either absent or much less frequent for closely similar explosive events.

V. LATER *EXCAVATION STAGE* OF CALCULATIONS 25-O AND 25-X

Figures 13 and 14 show the material velocity vectors at 100 μsec for calculations 25-O and 25-X, respectively. The plots are essentially identical. This indicates that at this time gravity is not yet playing a significant role in the crater formation.

Figure 15 shows the shock pressure attenuation in the target and Fig. 16 shows the time of arrival of this peak shock pressure as a function of depth in the target. By about 90 μsec, the shock velocity has decayed to sonic speed. These figures indicate that at about 85 μsec, the outgoing shock of several tenths of a kbar should reflect from the sides of the test bed container which is about 15 cm deep, and at about 170 μsec this reflected signal should come back to the vicinity of the transient cavity and possibly affect its growth. The calculations, however, used an infinite half-space for the target and so this

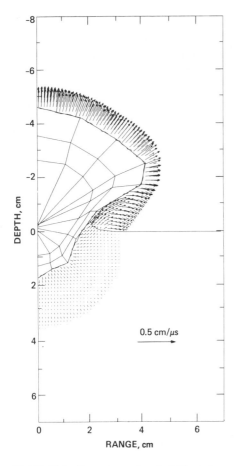 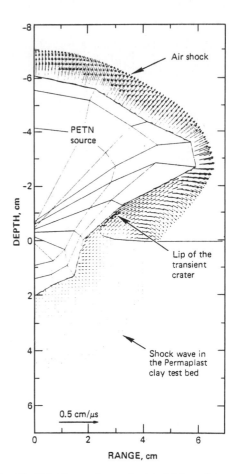

Fig. 11. Velocity vector and material boundary plot at 10.0 μsec for the 25-O and 25-X calculations.

Fig. 12. Velocity vector and material boundary plot at 16.2 μsec for the 25-O and 25-X calculations.

possible effect cannot be judged. Further calculations with the finite container boundary were begun but could not be completed within the original study.

Figure 17 shows the computed crater depth and radius versus time for the 25-O calculation and indicates continued though slowing growth at 0.5 msec. Figure 18 shows a comparison between the experimental 10 g crater profile and the calculated profile at 0.5 msec for calculation 25-O. Although the shapes are in agreement, the calculated crater is still growing at this time. The calculation was not continued further. If it had been, the crater depth and radius probably would have increased at least one centimeter. Even so, the calculations imply large transient cavities which means the experimental crater profiles are due to some rebounding motions possibly observed by Schmidt and Holsapple (1981) or to modifications after the *excavation stage* of crater growth.

VI. ENERGY PARTITIONING IN IMPACT AND EXPLOSION CRATERING

Momentum transfer in the PL1 calculation occurs at a slower rate than did the energy transfer. When the projectile has about 1% of its original kinetic energy, it still has 8% of its momentum. Figure 19 shows the projectile momentum time history. It should be noted that the curve labeled total momentum in Figure 19 is the scalar sum of the other two curves, and so is not the true total momentum. Figure 20 shows the energy in the

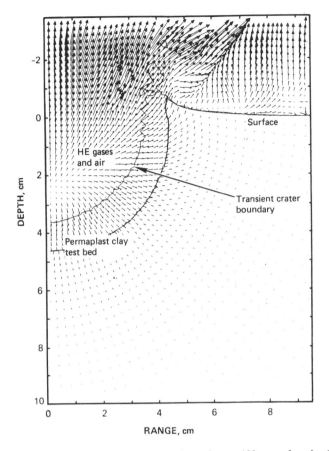

Fig. 13. Vector velocity and material boundary plot at 100 μsec for the 25-O calculation.

plasticene clay target as a function of time. This longer momentum transfer may be a reason for the existence of a substantial *early excavation stage* in the impact but not the explosive calculations.

In Fig. 20, the energy in the vaporized clay reaches a maximum at 1.6 μsec, and then decreases. 1.6 μ sec is the sime at which the shock pressure in the target has dropped below 6 GPa, the shock vaporization pressure for plasticene clay. After 1.6 μsec, the total energy in the already vaporized clay decreases as it expands.

Figure 21 shows the energy budget for the explosive calculations. Figure 22 shows the work done against gravity. It appears that it should be well over a msec before the work done against gravity in the high g calculation would even appear in Fig. 21. The gravity difference between the two calculations then does not seem to play a significant role during the time in which much crater growth occurs.

VII. DISCUSSION OF *EARLY EXCAVATION STAGE* DIFFERENCES

These calculations help provide some insight into the cratering process. Figure 23 shows a schematic diagram of the velocities associated with a particle in the cratering flow field. Proceeding outward from the center of the cratering flow field, the outgoing shock wave first accelerates the particle to a particle velocity u_p corresponding to the peak shock pressure. The fan of rarefactions trailing immediately behind the outgoing shock wave then decelerates the particle to its cratering flow field radial velocity u_c. At

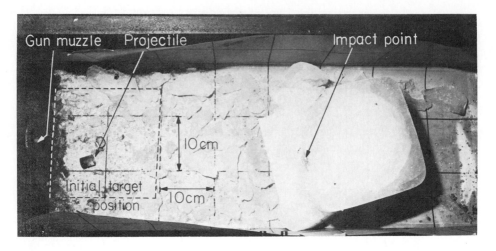

Fig. 3. Erosion of an impacted ice target (Shot No. 590, 257 K); initial target position as shown, projectile traveling from left to right.

3.1 Fragment weights

Figure 5 gives the cumulative number of fragments as a function of relative fragment masses (i.e., fragment mass divided by target mass) for all of our experiments. These diagrams served to assign each of the experiments to one of the fragmentation classes as defined above.

The mass distribution of fragments from six of our class I–III experiments are given in Figs. 6 and 7. Figure 6 gives incremental histograms, Fig. 7 cumulative distributions of fragment masses. The upper diagrams give the results for target temperatures of 257 K, the lower diagrams represent 81 K targets. As can be seen in Fig. 6, the mass distribution patterns for fragment masses of up to 120 g vary with fragmentation type and are relatively independent of target temperature. Both, class I and III experiments show a relative maximum in fragment numbers at masses between ~5 to 30 g, whereas the class II experiments reveal a relatively "flat" distribution. However, a more quantitative description of fragment mass distributions is obtained by the analysis of the cumulative distributions in Fig. 7.

It has long been noted that the distribution of rock fragments in many different geologic environments follows a power-law relation:

$$\mathrm{Log}_{10} n = C - b\, \mathrm{Log}_{10} m \tag{1}$$
(Hartmann, 1969)

where n is the cumulative number of rock fragments and m the mass of fragments. C and b are constants which define the linear relation between n and m in a log-log plot. The contribution of larger particles to the total amount of fragments within a given size range increases with decreasing value of b, whereas larger b (absolute) values reflect a higher contribution of smaller fragments. The solid lines in Fig. 7 give linear fits to the data points in our experiments which define both b and C for each fragment mass distribution. The filled symbols in each diagram are neglected from the analysis, since they represent either incomplete sampling (mainly the upper portion of the distribution, bending away from the solid line) or a different size class of fragments (i.e., the large fragments of type II and III experiments). The fits have generally been obtained for fragments with masses from ~5 to ~160 g.

In Table 3, C and b for our experiments are given. As can be seen, b clearly varies with fragmentation type. Mean values of b of our experiments increase with increasing degree

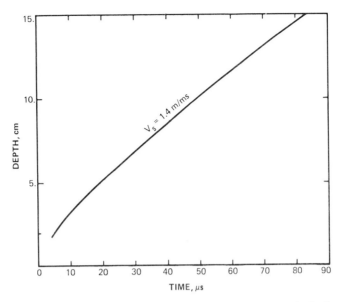

Fig. 16. Time of arrival of the maximum shock pressure at various vertically downward depths from the 25-O and 25-X calculations.

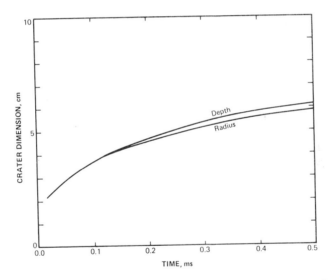

Fig. 17. Computed crater depth and radius versus time for the 25-O calculation.

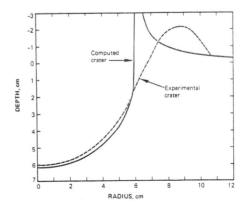

Fig. 18. Comparison between the computed crater profile at 0.5 msec (25-O calculation) and the experimental crater (Boeing Aerospace Company, Shot 25-O).

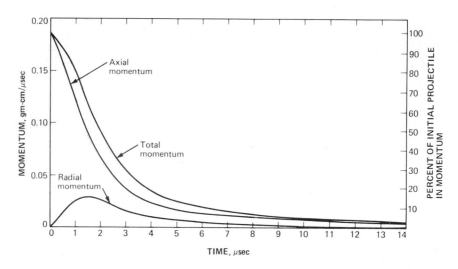

Fig. 19. Projectile momentum summary plot from the PL1 calculation.

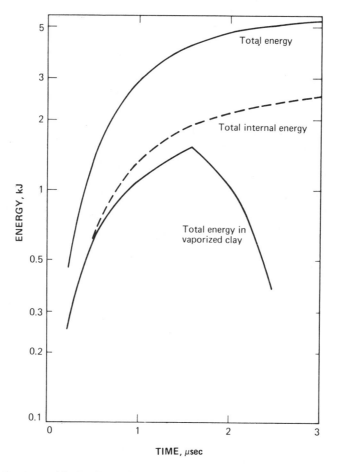

Fig. 20. Energy partitioning in the plasticene clay target over the first 3 μsec from the PL1 calculation.

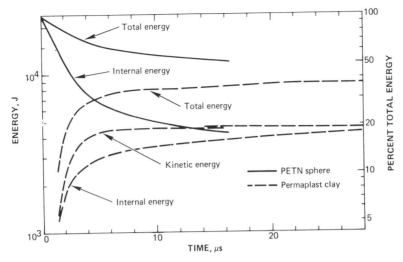

Fig. 21. Energy remaining in PETN sphere and energy coupled to the permplast clay versus time from the 25-O and 25-X calculations.

this time Z is about 2 in the explosion cases but is less than 2 during the *early excavation stage* in the impact case.

u_f is the sum of the effects due to the rarefactions behing the outgoing shock. More slowly over the remainder of the ejection process, continuous rarefactions from the free surface accelerate the particle toward the free surface by a velocity increment u_r to give the particle its final velocity u_e. During this later time, Z has been increasing to values greater than 2.

The impact and explosion calculations are for very similar target materials and for similar absolute crater dimensions. However, the data of Oberbeck (1971) indicates that zero-depth-of-burial does not provide a good overall simulation of impact. For this impact calculation corresponding depths-of-burial for explosion calculations meant to simulate it should be between 1.5 to 3.0 cm. Also, if the higher g explosion calculations are scaled to 1 g, they have higher energies than the impact calculation. However, neither of these differences should change the basic observation that for impacts Z is less than 2 during the *early excavation stage* due to the momentum still retained in the projectile, while for

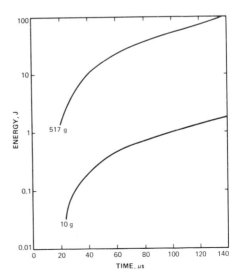

Fig. 22. Work done against gravity in the 25-O (10 g) and 25-X (517 g) calculations.

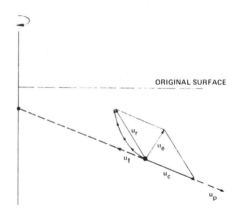

Fig. 23. Schematic vector addition diagram of the velocities associated with a particle in the cratering flow field.

explosions Z is not less than 2. A deeper depth of explosive burial may result in Z staying near 2 for a longer period of time, but it should not result in values less than 2.

Acknowledgments—This work was partially supported by the National Aeronautic and Space Administration under Contract No. NASW-3168. We appreciate the very helpful review comments of D. E. Gault and the valuable editorial assistance of S. K. Croft. The first author also deeply appreciates the special assistance given by I. J. Emerson.

REFERENCES

Austin M. G., Thomsen J. M., Ruhl S. F., Orphal D. L., Borden W. F., Larson S. A., and Schultz P. H. (1981) Z-model analysis of impact cratering: An overview. In *Multi-ring Basins, Proc. Lunar Planet. Sci.* 12A, (P. H. Schultz and R. B. Merrill, eds.), p. 197–206. Pergamon, N.Y.

Austin M. G., Thomsen J. M., Ruhl S. F., Orphal D. L., and Schultz P. H. (1980) Calculational Investigation of impact cratering dynamics: Material motions during the crater growth period. *Proc. Lunar Planet. Sci. Conf. 11th*, p. 2325–2345.

Croft S. K. (1980) Cratering flowfields: Implications for the excavation and transient expansion stages of crater formation. *Proc. Lunar Planet. Sci. Conf. 11th*, p. 2347–2378.

Gault D. E., Quaide W. L., and Oberbeck V. R. (1968) Impact cratering mechanics and structures. In *Shock Metamorphism of Natural Materials* (B. M. French and N. M. Short, eds.), p. 87–99. Mono, Baltimore.

Hancock S. L. (1976) Finite difference equations for PISCES 2DELK, a coupled Euler Lagrange continuum mechanics computer program. Report TCAM 76-2, Physics International Company, San Leandro, Calif. 174 pp.

Maxwell D. E. (1977) Simple Z-model of cratering ejection and the overturned flap. In *Impact and Explosion Cratering* (D. J. Roddy, R. O. Pepin and R. B. Merrill, eds.), p. 1003–1008. Pergamon, N.Y.

Oberbeck V. R. (1971) Laboratory simulation of impact cratering with high explosives *J. Geophys. Res.* 76, 5732–5749.

Orphal D. L. (1977) Calculations of explosion cratering—II. Cratering mechanics and phenomenology. In *Impact and Explosion Cratering* (D. J. Roddy, R. O. Pepin, and R. B. Merrill, eds.), p. 907–917. Pergamon, N.Y.

Orphal D. L., Borden W. F., Larson S. A., and Schultz P. H. (1980) Impact melt generation and transport. *Proc. Lunar Planet. Sci. Conf. 11th*, p. 2309–2323.

Schmidt R. M. and Holsapple K. A. (1979). Centrifuge crater scaling experiment II, material strength effects. Report DNA 4999. Defense Nuclear Agency, Washington, D.C.

Schmidt R. M. and Holsapple K. A. (1981) An experimental investigation of transient crater size. abstract). In *Lunar and Planetary Science XII*, p. 934–936. Lunar and Planetary Institute, Houston.

Thomsen J. M., Austin M. G., Ruhl S. F., Schultz P. H., and Orphal D. L. (1979a) Calculational investigation of impact cratering dynamics: Early time material motions. *Proc. Lunar Planet. Sci. Conf. 10th*, p. 2741–2756.

Thomsen J. M., Sauer F. M., Austin M. G., Ruhl S. F., Schultz P. H., and Orphal D. L. (1979b) Impact cratering calculations part I: Early time results. Report PIER-1220. Physics International Company, San Leandro, Calif. 76 pp.

Thomsen J. M., Ruhl S. F., and Austin M. G. (1980) Calculations of centrifuge cratering experiments: Report of an initial effort. Report PIFR-1059. Physics International Co., San Leandro, Calif. 61 pp.

Secondary cratering effects on lunar microterrain: Implications for the micrometeoroid flux

R. J. Allison and J. A. M. McDonnell

Space Sciences Laboratory, University of Kent at Canterbury, Kent, England

Abstract—We report the results of calculations on secondary and tertiary hypervelocity microcratering on the lunar surface, and consider implications for the interplanetary dust population which is deduced from such lunar records. It is shown that secondary cratering can be significant for a limited size range (about 1 order of magnitude, near 1 μm crater diameter), given favourable sample exposure geometry. The process is sensitive to the model of crater distribution chosen and the sample exposure angle. Dimensional scaling of impact parameters has only a small influence on secondary cratering effects. There are no significant tertiary contributions for crater sizes above the present lower limit of detection for lunar microcraters. Comparison of observed microcrater distributions with distributions derived from in-situ space sensor data shows a distinct discrepancy of form, even if secondary effects are taken into account. This may be due to some as yet unknown feature of the impact process which cannot yet be checked by laboratory simulation, but such evidence would also support the alternative hypothesis of dust flux variation over the past 10^6 to 10^5 years.

INTRODUCTION

The return of lunar rock samples by the Apollo expeditions has made possible laboratory studies of a variety of lunar surface effects. These include processes which contribute to lunar erosion on a submillimetre scale, e.g., micrometeoroid impact, dust accumulation, generation of accretionary particles, solar wind sputtering, solar flare and cosmic ray track formation (Crozaz *et al.*, 1970, 1971; Gault *et al.*, 1972; McDonnell *et al.*, 1974, 1976; Hartung *et al.*, 1978). These studies can also provide information about the space environment of the moon. In particular, the craters (down to sub-μm sizes) revealed by microscopic examination of lunar surface rocks are ultimately derived from hypervelocity impacts of micrometeoroids. Hence they can give information on the micrometeoroid flux, complementary to that provided by other sources such as in-situ observations by space probes (e.g., Berg and Gerloff, 1971; Hoffman *et al.*, 1975) in the vicinity of 1 A.U. heliocentric distance.

There are often severe limitations on the size, weight and power consumption of sensors, which must operate for long periods in a very extreme environment. Their results provide significant problems of noise, reliability and interpretation (see McDonnell, 1978a, ch. 6, for a review).

The interpretation of microcrater populations and their relation to dust in space is by no means straightforward. Crater studies are directly hampered by factors such as solar wind sputter erosion or accreta obscuration, which modify observable crater populations (McDonnell *et al.*, 1974; Ashworth *et al.*, 1978). Conversion to impacting particles depends on parameters such as particle speed and composition, which are themselves never known for any particular crater. Geometrical exposure factors also strongly affect absolute magnitudes of crater populations; this problem is made greater by the generally poor documentation of lunar sample exposure geometry.

It is also necessary to study the effect of craters which are not directly created by micrometeoroid impact. If particles thrown out by such direct impacts strike the surface at speeds in excess of approximately 1 km s^{-1}, they can cause secondary craters which are as yet morphologically indistinguishable from those left by primary impacts

(Schneider, 1975). It is even possible that high velocity ejecta from secondary craters can cause tertiary cratering. In this way lunar surface effects can mask the evidence which relates directly to the incoming micrometeoroids. Furthermore, the relative contributions of craters from these sources will be shown to be sensitive to exposure geometry—in general, orientations which favour secondary impacts mitigate against primary impacts.

Secondary cratering has generally been ignored in lunar microcrater studies, due to a belief that it is a negligibly small effect. Support for this view was provided by the work of Schneider (1975) who performed experiments on secondary crater formation with primary impacts of 1.58 mm sized steel spheres. Application of his results to the lunar surface led him to conclude that secondary cratering was not a significant effect. Subsequent experimental studies by Flavill and McDonnell (1977) were conducted at a smaller size scale (14 μm spall diameter); computer analysis using their results suggested a greater role for secondaries than had previously been envisaged, for micron-scale craters (Flavill et al., 1978).

The purpose of the present work is to report a very much more detailed consideration of secondary and tertiary cratering effects, based on calculations from these and other data, and taking into account the various factors discussed above.

METHOD

The procedure followed was similar to that reported in a previous paper (Flavill et al., 1978). Two types of data were used as inputs: primary crater distributions and secondary crater production functions. The latter functions show the number of secondaries produced per primary crater, as a function of the relative size of the secondary crater. A set of computer programs combined these to generate the secondary crater populations corresponding to an assumed input primary distribution. A modification of the program was then used to extract an "original" primary distribution from an observed distribution contaminated with secondaries. We considered the influence of a number of factors on the relative importance of secondaries: exposure angles, choice of model and dimensional scaling. The effect of tertiary production was also considered.

DATA

The calculations used three main primary populations and two main secondary production functions. Of the primary distributions, two were based on microcrater observations (which may, of course, contain secondaries). One was that derived from Apollo lunar sample 12054 by Morrison and Zinner (1977) and the other was the distribution presented by Le Sergeant d'Hendecourt and Lamy (1980), based on a review of microcrater observations. These two were chosen as providing recent, comprehensive and independent surveys of available crater data. The third distribution was based on in-situ space sensor data published by McDonnell (1978a). It actually consists of two distributions which diverge only at small sizes. These correspond to two different flux orientations ("solar" or β meteoroids and "apex" meteoroids); these were separately considered.

The conversion from space sensor data to a corresponding lunar crater distribution was made with the following assumptions: the average impact velocity is 20 km s^{-1}; density, 2.5 gm cm^{-3}; ratio of primary crater pit diameter to particle diameter, 2:1 for lunar rock impacts (appropriate to silicate impacts at 20 km s^{-1}—Le Sergeant d'Hendecourt and Lamy, 1980). Other assumptions are possible—Pioneer data can be interpreted to offer a range from 5 km s^{-1} to over 100 km s^{-1} (for small particles) (Berg and Grün, 1973); iron micrometeoroids would provide a higher density (3–4 gm cm^{-3}) and the "fluffy" micrometeoroid model would suggest something lower ($\sim$1 gm cm^{-3}). However, the values chosen represent a "middle way" between these extremes. To complete the conversion from a particle flux to a crater population it was also necessary to assume an exposure age for surfaces exposed to the flux. For comparability with other distributions this was initially chosen as the 1.75×10^5 years estimated for sample 12054 (Morrison and Zinner,

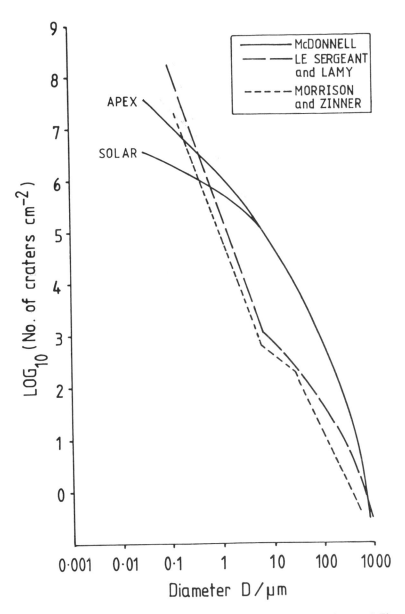

Fig. 1. Primary microcrater distributions based on data from Morrison and Zinner (1977), Le Sergeant d'Hendecourt and Lamy (1980) and McDonnell (1978b).

1977). This question of exposure ages has interesting implications which will be taken up in a later section. The three distributions are presented in Fig. 1.

The secondary production functions used were derived from experiments by Flavill and McDonnell (1977), and by Schneider (1975). The first of these was normalised to a primary spall diameter size of 14 μm and was already in suitable form. Schneider's data from impacts of a 1.58 mm steel sphere were used to derive a similar function. The two production functions used for the calculations are presented in Fig. 2.

The importance of experimental scale in determining secondary production functions was alluded to above. This raises the whole question of dimensional scaling of such functions. Any given experiment will yield a distribution of secondaries produced by

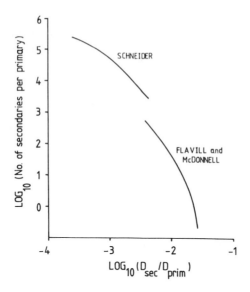

Fig. 2. Secondary crater production functions, showing numbers of secondaries (diameter D_{sec}) from a single primary (diameter D_{prim}) as a function of relative size. The curves are based on data from Schneider (1975) (1.5 mm spherical steel projectiles) and Flavill and McDonnell (1977) (micron diameter iron projectiles).

craters of the particular size used in the experiment. Can this same distribution be used for primary craters of a different size? In general, the answer to this question is 'no', as borne out by the example of the two distributions discussed above. The same result is seen with secondary distributions of low velocity ejecta (cf. Gault and Wedekind, 1969; McDonnell et al., 1976).

This problem is dealt with by scaling the distributions to convert them to curves appropriate to the particular size at which they are applied. For this study, the two secondary production functions were compared for dimensional scaling. This gave a scaling index based on the sizes at which the experiments were performed, by means of which they could be scaled to the sizes at which the various calculations were made:

$$s = \frac{\log_{10}(N_1/N_2)}{\log_{10}(D_1/D_2)}$$

where D_1 and D_2 are the sizes at which the experiments were performed, and N_1 and N_2 are secondary production values at some other size where the distributions overlap. The index itself is denoted by s. This scaling was incorporated into the calculations. s is used as an index in a power-law size scaling of secondary production:

$$N_{scaled}(\text{secondaries at size D}) = N_{measured}(\text{secondaries at size D}) \times (D/D_E)^s$$

where D_E is the dimension of the experiment from which the measurement came. From these data, s had a value of 0.408. In addition, indices of 0.2 and 0.3 were used for some further calculations to check the effect of uncertainties in the knowledge of such dimensional scaling.

RESULTS

We first consider the results obtained using the distributions of Morrison and Zinner, and of Le Sergeant and Lamy. These results, based on distributions from lunar microcrater observations, have some features in common which distinguish them from the distributions derived from space sensor data. They are presented as plots of the ratio of secondary crater number density to primary crater number density (N_s/N_p) as functions of pit diameter. This method of presentation makes clear the expected strength of secondary contributions over a range of microcrater sizes. The results appear in Figs. 3a–c. The effect of different exposure geometries has been considered by performing the calculations at a number of different values of the exposure angle, giving rise to the families of curves displayed.

The exposure angle mentioned here is the angle of elevation of the target surface, which corresponds to the solid angle over which it may receive both primary and secondary impacts. It is independent of the *ejection* angle of the secondary particles, and the surface will actually receive secondaries ejected over a range of angles coming from primary impacts in a range of distances along the lunar surface itself. Hence values used are averaged over the range of ejection angles in which ejecta are concentrated.

The most important feature of these results is the fact that all the curves show a distinct peaking in the region 1–10 μm diameter. Even the weakest peak is of the order of several per cent and this can be substantially increased by favourable geometries. The width and location of the peak are found to be model-dependent (i.e., vary for different input distributions) but is independent of the exposure angle. The height of the peaks depends on both the model and the exposure angle.

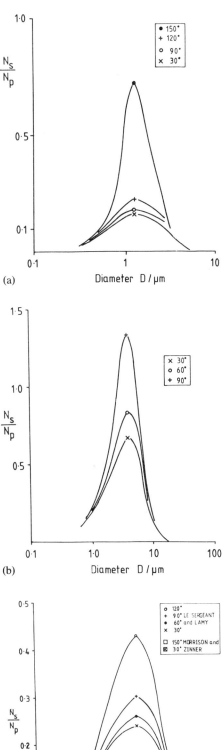

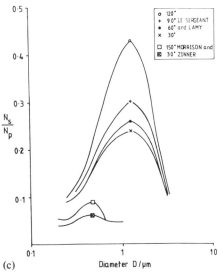

Fig. 3. (a) Numbers of secondary microcraters normalised to the primary population, as a function of size. The curves shown correspond to a range of exposure angles for the primary distribution of Morrison and Zinner (1977), using the secondary production function derived from Flavill and McDonnell (1977) at micron dimensions. (b) Numbers of secondary microcraters normalised to the primary population, as a function of size. The curves shown correspond to a range of exposure angles for the primary distribution of Le Sergeant d'Hendecourt and Lamy (1980), using the secondary production function derived from Flavill and McDonnell (1977) at micron dimensions. (c) Numbers of secondary microcraters normalised to the primary population, as a function of size. The curves shown correspond to a range of exposure angles for the primary distributions of Le Sergeant d'Hendecourt and Lamy (1980) and of Morrison and Zinner (1977), using the secondary production derived from Schneider (1975) for all cases.

Table 1.

Data Source		Position of maximum (in μm)	Peak Width ($\log_{10}(D/\mu m)$)	$(N_s/N_p)_{max}$ (at 30°)
Primary Distribution	Secondary Production			
Lamy and Le Sergeant	Schneider	1.26	1	0.24
Lamy and Le Sergeant	Flavill et al.	4	1.25	0.67
Morrison and Zinner	Schneider	0.5	0.6	0.06
Morrison and Zinner	Flavill et al.	1.3	1	0.16

Model-dependence of secondary contributions is summarised in Table 1, and the angular dependence is illustrated in Fig. 4. Clearly, the primary distribution based on Morrison and Zinner leads to weaker effects. This is probably due to a relative deficiency of this distribution at larger sizes, compared with the distribution of Le Sergeant and Lamy. Since the calculations show that secondary craters are principally contributed by primaries 2 to 3 orders of magnitude larger than themselves, it follows that such a deficiency will lead to fewer secondaries near the mid-range.

It is also clear that the secondary production function derived from Schneider's data

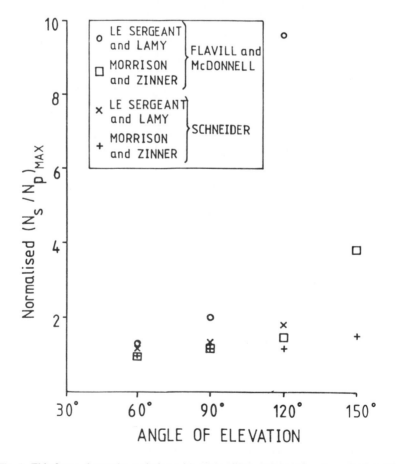

Fig. 4. This figure shows the variation with angle of the peaks in the curves depicted in Figs 3a-c. The peak heights are normalised to values for an exposure angle of 30° inclination relative to the lunar surface.

produces markedly weaker effects than that of Flavill and McDonnell. This underlines the importance of the different size scales at which such experiments are performed. The results from Flavill and McDonnell are based on micron-scale experiments and so are likely better to represent micro-terrain effects than are results from larger scale experiments such as that of Schneider.

As might be expected, increasing angle of elevation enhances secondary contamination, since it favours secondary capture while diluting the primary flux. Application of this result to observed crater distributions is made difficult by lack of information on sample exposure geometry.

We now consider the calculations of crater distributions expected using as an input the space data (see Fig. 5); by definition, craters caused by particles from space are primary. These yield characteristically different curves, with no peaking but a strong secondary contribution. This rises rapidly to dominance as size decreases below the 1–10 μm range found to be significant in the previous calculations. The ratios shown in Fig. 5 correspond to a target surface orientation of 90° to the lunar surface. Since the space data are all

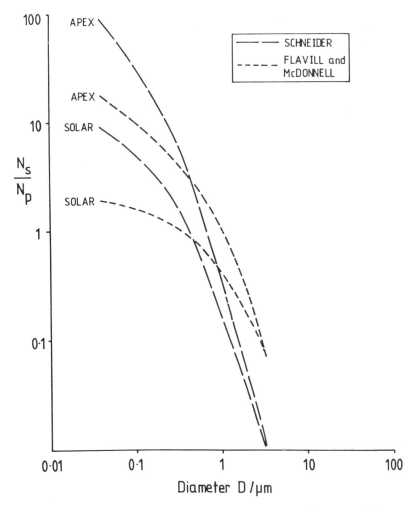

Fig. 5. Numbers of secondary microcraters normalised to the primary population, as a function of size. The curves shown result from using each of the "solar" and "apex" flux data sets of McDonnell (1978) with each of the secondary production functions, i.e., Flavill and McDonnell (1977) and Schneider (1975). The ratios given correspond to an exposure angle of 90° to the lunar surface.

primary, no computer correction of the primary population is needed (as it is for lunar data). This correction introduces angular dependence and, without it, no range of angles need be considered. As before, we find the Flavill and McDonnell curve producing a stronger effect than that from Schneider's data.

Comparing these two sets of results shows that there is a strong model dependence which distinguishes between "lunar" and "space" data. This sharp distinction between results from observed lunar crater distributions and in-situ space data has interesting implications which will be taken up in the next section.

Figure 6 shows the effect of different dimensional scaling indices on the angular variation curves for the primary distribution of Le Sergeant d'Hendecourt and Lamy, with both secondary production functions. Dimensional scaling variation has a clear effect on these curves, greater for the secondary function of Flavill and McDonnell; this function also shows increasing significance of dimensional scaling at larger angles. However, the form of the previous results is not changed and the magnitude variations are relatively small ($\ll 1$ order of magnitude). In brief, dimensional scaling is a distinct but not dominant effect in the role of secondary contributions to lunar crater populations.

It was mentioned above that most secondaries at any given size are derived mainly from observed lunar primaries of some 2 to 3 orders of magnitude larger. This emerges from inspection of the relative contributions of primaries of various sizes to secondaries of various sizes, which were provided as intermediate output by the program. These contributions tend to originate from a range of about 1 order of magnitude centred at 3 orders larger than the secondary size to which they contribute, for the Schneider data. For Flavill and McDonnell, the range is 2 orders of magnitude at about 2 to 3 orders of magnitude larger, i.e., the contributions are a little more spread out. Despite this small but discernible difference for different secondary production functions, there is virtually no difference in relative contributions for different primary distributions with the same secondary production functions. Hence it appears that the secondary production function is the dominant influence in determining which relative sizes of primaries make the main contribution to secondary craters of a given size although both this and the primary distribution are in principle responsible.

The next result to be considered deals with calculations performed using secondary crater distributions as primary input, i.e., a situation leading to the generation of tertiary populations. It was found in extensive calculations that tertiaries are never more than 1% of the total population for crater diameters $\geq 0.03~\mu$m, which is near the lower limit detectability of available data. There are no grounds for believing tertiaries to be a significant component of lunar crater populations so far observed. Comparing calculated tertiaries with extrapolated primary and secondary populations suggests that tertiaries may start to dominate by up to a factor of 3 at diameters of about 0.001 μm. This assumes the levelling-off in the crater distribution detected at the smallest sizes so far observed ($\sim 0.03~\mu$m—Morrison and Clanton, 1979). However, further studies at higher resolutions may show this to be an observational limitation, as has been found on previous occasions. Also, the characteristics of the cratering process itself are not well studied for very small sizes. The relative significance of tertiary results is shown in Figs. 7a-b, which present primary, secondary and tertiary curves for the two distributions obtained from in-situ space measurements.

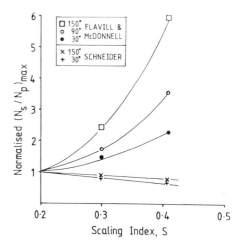

Fig. 6. Variation with dimensional scaling index of peaks in the curves of Figures 3a-c. The peak heights are normalised at s = 0.2.

For the primary crater distributions used here, the effect of secondaries does not exceed ~10% for lunar crater data. This is scarcely discernible on the log-log plots used for these distributions. However, space sensor data give results more sensitive to secondary effects, and these are considered in the next section where we discuss the implications of the calculations described above. In the light of full understanding, the derived lunar primary crater flux data and space data must fully converge. We discuss the implications presently offered at the current level of understanding of crater distribution development.

IMPLICATIONS

The region in which secondaries are found to be significant (1–10 μm) is an interesting one for lunar crater distributions. It is generally agreed (Schneider *et al.*, 1973; Brownlee *et al.*, 1975; Morrison and Zinner, 1977; Le Sergeant and Lamy, 1980 that this size range contains an inflexion of the crater size distribution; this does not, by and large, correspond to any similar feature in the micrometeoroid size distribution as revealed by in-situ sensor measurements. Many such measurements have large uncertainties which make difficult their association with a functional form. However, not even the relatively numerous (~350) measurements of Pioneers 8 and 9 plasma sensors, from which a form may be clearly defined, give support to the distinct inflexion of the lunar data (McDonnell, 1978a; Clanton *et al.*, 1980). If lunar crater data are to be understood, this discrepancy must be dealt with. In particular, we must take into account effects peculiar to the lunar surface which would mask or distort the incoming particle flux. It is precisely this point that secondary cratering studies address.

The approach chosen was to convert the space data to a lunar crater distribution, as described previously, but this time including secondary effects and choosing an exposure age such as to minimise discrepancy with observed lunar data. As Fig. 8 shows the magnitude of the distribution agrees with lunar data to within a factor of six, although not in slope, for the crater range ~30 to 1000 μm. There is a substantial discrepancy in size and slope for 3 to 30 μm and also for sizes ≤ 1 μm. In short, then, space data and lunar crater data do not agree well, even with masking effects taken into account. Moreover, such agreement as exists requires the use of an exposure age of 2.75×10^4 years, i.e., approximately six times smaller than the 12054 exposure age of 1.75×10^5 years.

The reason for this discrepancy between space and lunar data is not clear. One possibility hinges on the fact that the lunar microcrater impacts differ in dimension scaling from data based on detection of particles by other means (e.g., penetration, ionization). An exception to this rule is the observation of microcraters on Apollo-Skylab windows (Cour-Palais, 1979). These results do indeed show an inflexion similar to that found with lunar data. It should be noted, however, that this comes from only one window on one flight; the data from other flights does not suggest it. Moreover, that particular window may well have been subject to secondary effects from the surrounding shielding. Further studies by Clanton *et al.* (1980) have failed to confirm the presence of this inflexion.

Nonetheless, the possibility should be considered that the inflexion arises from the physics of the impact process rather than from any feature of the impacting particle population. In this connection it is worth remembering that the inflexion occurs near the region of transition from craters with spallation zones to spall-less pits. The resolution of this possibility unfortunately awaits further studies of the impact process, particularly experiments with micron-sized impacts, and lies beyond the scope of this study.

Another possible explanation arises from a second sharp distinction between the data sets, namely the factor of time. The lunar microcrater record represents a flux averaged over some ill-defined period of past time, while the space data constitute an essentially "instantaneous" measurement of the flux at the present time. If the flux has varied with time in the past $\sim 10^6$ years, then such a disagreement as we observe would not be unexpected. Note also that the discrepancy in exposure ages is effectively a discrepancy

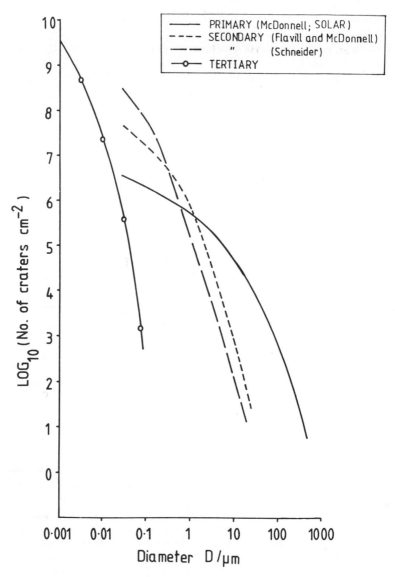

Fig. 7a. Primary, secondary and tertiary microcraters from the "solar" data of McDonnell (1978).

between the present observed flux magnitude and that implied by the microcrater record, suggesting a lower flux in time past.

Various results of lunar sample studies support the idea of the micrometeoroid flux having been lower in the past. Evidence comes from age distributions of micron-dimensioned craters, based on solar flare track, splash accreta, solid accreta and sub-micron pit studies (Hartung and Storzer, 1974; Hartung and Comstock, 1978; McDonnell, 1978b). These indicate a crater production rising ~ exponentially over the past ~10^4 years. Such an interpretation has been disputed (Hartung and Comstock, 1978) on the grounds that accreta and submicron pits are derived from the population of larger impacting particles and cannot, therefore, provide an independent "clock" for that population. Doubts also exist about the solar flare track production constancy with time (Zook et al., 1977).

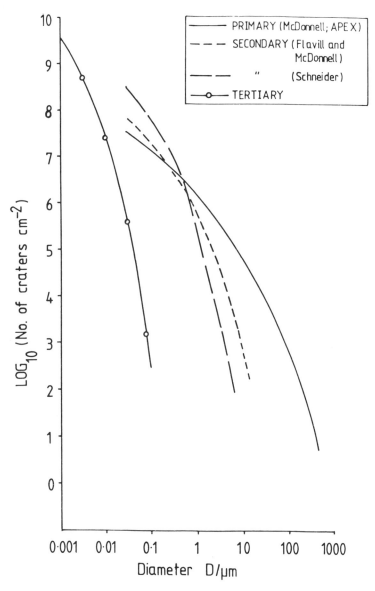

Fig. 7b. Primary, secondary and tertiary microcraters from the "apex" data of McDonnell (1978).

However, solar flare variability remains to be proven (Zook, 1979) and is, in any case, related to only one dating method. The main explanation invoked to explain an apparently lower past flux is the existence of a dust cover which slows the ageing processes. Justification for assuming such a dust layer (which had not been directly observed) comes from the above mentioned *a priori* assumption that (i) submicron pits come from particles derived in space from particles of the size which cause the larger craters for which submicron pits have been used as a clock; and (ii) splash accreta are derived from these same larger craters.

Recent studies by Le Sergeant d'Hendecourt and Lamy (1980) have strengthened the case for a two population interplanetary dust flux with transition between 1 and 10 μm crater diameter on lunar rocks, as previously suggested by various evidences (e.g.,

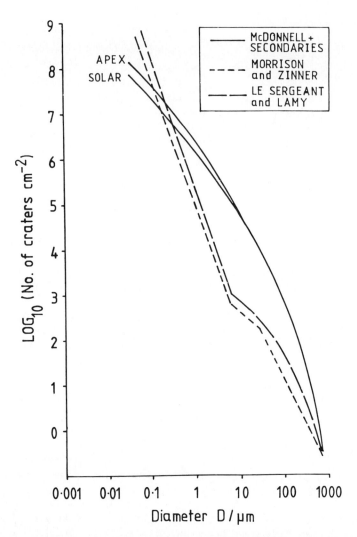

Fig. 8. Comparison of lunar and space data in terms of observed and expected microcrater populations. Lunar data are that of Morrison and Zinner (1977) and Le Sergeant d'Hendecourt and Lamy (1980). Space data are derived from McDonnell (1978), using an assumed exposure age equal to that found for Apollo 12054 by Morrison and Zinner (1977), i.e., 1.75×10^5 years.

Morrison and Zinner, 1977; Flavill et al., 1978). This would "decouple" the large and small craters, invalidating part (i) of the assumption. As for part (ii), no support has been offered in terms of a detailed study of the splash generation process. This point, then, remains largely speculative—further work, particularly microscale impact experiments, will be needed to clarify it one way or the other.

The general topic of dust effects *has* had a detailed consideration from McDonnell (1978b), who argues that the main obscuration effects come from the observed accreta populations and that much thicker dust coverings are unlikely; the work suggests that while dust effects may well distort the age distribution for craters, they do not explain the *form* of the observed distribution if the flux *is* time-constant, nor its magnitude, particularly for craters of diameters $\geq 50\ \mu m$. Moreover, Zook (1978) reports that what were previously claimed as dust effects were actually due to a rock overhang on sample 12054.

On balance, then, lunar sample cratering evidence favours a recently increasing flux of the larger, lower density particle population (causing craters $>50\,\mu$m), but a constant population of smaller particles. Cometary activity over the last 10^4 to 10^5 years would lead to this selective bombardment prior to an increase of the β meteoroid population when such particles spiral towards the near solar comminution zone (Zook and Berg, 1975). If this is so, the inflexion in the total distribution which marks the transition between populations may at present be obscured from in-situ measurements, but its existence is recorded in lunar microcrater data.

CONCLUSIONS

This study leads to a number of conclusions: 1) Secondary microcratering is a significant effect in the size range 1–10 μm crater diameter, but not outside this region, from lunar microcrater data. The effect can be enhanced by suitable sample exposure geometry. 2) The precise sizes affected are dependent on the assumed form of the crater distribution model, but are independent of the sample exposure angle. 3) The strength of the secondary cratering effect is both model-dependent and angle-dependent, increasing with increasing angle of elevation of the sample. 4) Calculations on in-situ micrometeoroid data suggest a quite different secondary crater contribution, rising sharply *below* 1–10 μm crater diameter for crater populations deriving from present day fluxes. 5) Dimensional scaling of secondary generation functions estimated from the as-yet sparse experimental data has a small but distinct influence on secondary contributions to microcrater populations. 6) Most secondary craters are derived from primaries of diameter ~ 2 to 3 orders of magnitude larger. 7) There are no indications that tertiaries are a significant part of the lunar microcrater record from consideration of either lunar data or space data. 8) Even taking secondary effects into account, there remains a distinct discrepancy between the microcrater distributions expected from the present day in-situ measurements of micrometeoroid flux. 9) If this discrepancy is associated with the fact that the lunar record represents a cratering process while in-situ data is in general derived from other effects such as plasma detection, then it may arise from as yet unexplained features of the impact process at micron dimensions. 10) If the discrepancy is associated with the different time-scales of lunar and space data, it constitutes further evidence for a lower flux of the population of larger dust particles in past times, perhaps by a factor of 6.

Acknowledgments—Thanks are due to Roger Flavill for assistance with program development, and to the SRC for support while this research was being carried out. Assistance with manuscript preparation was given by Alison Rook.

REFERENCES

Ashworth D. G., McDonnell J. A. M., and Carey W. C. (1978) The role of accretionary particles in the approach to lunar equilibrium topology. In *Space Research XVIII*, p. 439–442.
Berg O. E. and Gerloff U. (1971) More than two years of micrometeorite data from two Pioneer satellites. In *Space Research XI*, p. 225.
Berg O. E. and Grün E. (1973) Evidence of hyperbolic cosmic dust particles. In *Space Research XIII*, p. 1047.
Brownlee D. E., Hörz F., Hartung J. B., and Gault D. E. (1975) Density, chemistry and size distribution of interplanetary dust. *Proc. Lunar Sci. Conf. 6th*, p. 3409–3416.
Clanton U. S., Zook H. A., and Schultz R. A. (1980) Hypervelocity impacts on Skylab IV/Apollo windows. *Proc. Lunar Planet. Sci. Conf. 11th*, p. 2261–2273.
Cour-Palais B. G. (1979) Results of the examination of the Skylab/Apollo windows for micrometeoroid impacts. *Proc. Lunar Planet. Sci. Conf. 10th*, p. 1165–1672.
Crozaz G., Haak U., Hair M., Maurette M., Walker R., and Woolum D. (1970) Nuclear track studies of ancient solar radiations and dynamic lunar surface processes. *Proc. Apollo 11 Lunar Sci. Conf.*, p. 2051–2080.
Crozaz G., Walker R., and Woolum D. (1971) Nuclear track studies of dynamic surface processes on the moon and the constancy of solar activity. *Proc. Lunar Sci. Conf. 2nd*, p. 2543–2558.
Flavill R. P., Allison R. J., and McDonnell J. A. M. (1978) Primary, secondary and tertiary

microcrater populations on lunar rocks: Effects of hypervelocity impact ejecta on primary populations. *Proc. Lunar Planet. Sci. Conf. 9th*, p. 2539–2556.

Flavill R. P. and McDonnell J. A. M. (1977) Laboratory simulations of secondary lunar microcraters from micron scale hypervelocity impacts on lunar rock. *Meteoritics* 12, 220–225.

Gault D. A., Hörz F., and Hartung J. B. (1972) Effects of microcratering on the lunar surface. *Proc. Lunar Sci. Conf. 3rd*, p. 2713–2734.

Gault D. A. and Wedekind J. A. (1969) The destruction of tektites by micrometeoroid impact. *J. Geophys. Res.* 74, 6780–6794.

Hartung J. B. and Comstock G. M. (1978) New lunar microcrater evidence against a time-varying meteoroid flux. In *Space Research XVIII*, p. 431–434.

Hartung J. B., Hauser E. E., Hörz F., Morrison D. A., Schonfeld E., Zook H. A., Mandeville J. -C., McDonnell J. A. M., Schaal R. B., and Zinner E. (1978) Lunar surface processes: Report of the 12054 consortium. *Proc. Lunar Planet. Sci. Conf. 9th*, p. 2507–2537.

Hartung J. B. and Storzer D. (1974) Lunar microcraters and their solar flare track record. *Proc. Lunar Sci. Conf. 5th*, p. 2527–2541.

Hoffman H., Fechtig H., Grün E., and Kissel J. (1975) Temporal fluctuations and anisotropy of the micrometeoroid flux in the Earth-Moon system measured by HEOS 2. *Planet. Space Sci.* 23, 985.

McDonnell J. A. M. (1978a) Microparticle studies by space instrumentation. In *Cosmic Dust* (J. A. M. McDonnell, ed.), p. 388–389. Wiley, Chichester.

McDonnell J. A. M. (1978b) The role of accretionary particles on lunar exposure and ageing processes: lunar dust slows lunar clocks. In *Space Research XVIII*, p. 435.

McDonnell J. A. M., Flavill R. P., and Ashworth D. G. (1974) Hypervelocity impact and solar wind erosion parameters from simulated measurements on Apollo lunar samples. In *Space Research XIV*, p. 733–737.

McDonnell J. A. M., Flavill R. P., and Carey W. C. (1976) The micrometeoroid impact crater comminution distribution and accretionary populations on lunar rocks: Experimental measurements. *Proc. Lunar Sci. Conf. 7th*, p. 1055–1072.

Morrison D. A. and Clanton U. S. (1979) Properties of microcraters and cosmic dust of less than 1000 Å dimensions. *Proc. Lunar Planet. Sci. Conf. 10th*, p. 1649–1663.

Morrison D. A. and Zinner E. (1977) 12054 and 76215: New measurements of interplanetary dust and solar flare fluxes. *Proc. Lunar Sci. Conf. 8th*, p. 841–863.

Schneider E. (1975) Impact ejecta exceeding lunar escape velocity. *The Moon* 13, 173–184.

Schneider E., Storzer D., Mehl B., Hartung J. B., Fechtig H., and Gentner W. (1973) Microcraters on Apollo 15 and 16 samples and corresponding dust fluxes. *Proc. Lunar Sci. Conf. 4th*, p. 3277–3290.

Zook H. A. (1978) Dust, impact pits and accreta on lunar rock 12054. *Proc. Lunar Planet. Sci. Conf. 9th*, p. 2469–2484.

Zook H. A. (1979) Evidence from the moon for changes in solar flare activity in the past (abstract). In *Papers Presented to the Conference on the Ancient Sun*, p. 110–112. Lunar and Planetary Institute, Houston.

Zook H. A. and Berg O. E. (1975) A source for hyperbolic dust particles. *Planet. Space Sci.* 23, 183–203.

Zook H. A., Hartung J. B., and Storzer D. (1977) Solar flare activity: Evidence for large-scale changes in the past. *Icarus* 32, 106–126.

The stochastic variability of asteroidal regolith depths

Kevin R. Housen

Department of Planetary Sciences, University of Arizona, Tucson, Arizona 85721

Abstract—Modeling the depth of regolith on asteroids is approached from a statistical point of view. It is demonstrated that average values are not good descriptors of regolith depth on asteroids. Large deviations from the average can be expected to occur due to both large variations in depth over the surface of a body and to the fact that each asteroid has a unique regolith. The utility of the average depth is not significantly increased by excluding the parts of a surface which are occupied by large craters; a procedure adopted in existing regolith models. Although an asteroid's surface may be "smoothed out" by movement of debris into gravitationally low spots, the regolith depth retains its variability because of variations in topography at the bottom of the regolith layer. The large variability associated with regolith depth severely limits the power of regolith models in predicting parent body size for the brecciated meteorites.

INTRODUCTION

Asteroids are thought to be the parent bodies for many meteorites. In particular, many brecciated and gas-rich meteorites were probably derived from asteroidal regoliths (Suess *et al.*, 1964; Wilkening, 1971; Macdougall *et al.*, 1974; Rajan, 1974; Anders, 1975; see also references in Housen *et al.*, 1979b and discussions in Wasson and Wetherill, 1979). The amount of regolith that a given size body accumulates determines, to a large extent, whether or not it can yield a significant quantity of brecciated meteorites. It is important, therefore, to have estimates of asteroidal regolith depths; in principle they can be used to help pinpoint the size of parent bodies for meteorites.

Several models of regolith depth have been constructed but only recently have detailed models emerged (Housen *et al.*, 1979a,b; Langevin and Maurette, 1980). These models recognize the fact that regolith depth varies from point to point on an asteroid's surface. An analytic expression for the surficial distribution of depths is very difficult to determine and Monte Carlo calculations can become quite expensive, so generally an *average* regolith depth has been computed. Note, if the entire surface is considered then, presumably, the average becomes heavily weighted by the effects (i.e., the deep crater bowls or anomalously thick ejecta deposits) of a few large craters which are not representative of most of the surface. Therefore, Housen *et al.* (1979a, b) and, in essence, Langevin and Maurette (1980) did not average over the entire surface (see Housen, 1981a). Only the areas "saturated" by the effects of numerous small craters were considered. Over this portion of the surface it was thought that the regolith should be relatively uniform in depth and so could be adequately described by an average value.

All of the above-mentioned models are *determinate*, i.e., two asteroids of equal size, composition, etc., are considered to have identical regoliths. Actually regolith evolution is dictated by the laws of chance. For example, the number of craters produced on an asteroid cannot be predicted with certainty even though an "average" crater flux might be specified exactly. Our best estimates of the crater size-frequency distribution in the asteroid belt imply that, even though the crater flux drops sharply with increasing crater size, the largest craters on an asteroid excavate more material than the numerous small impacts (Housen *et al.*, 1979a). That is, regolith growth tends to be dominated by the larger craters. Because the large craters are also rare the number of these craters we might expect to find on an asteroid is subject to major stochastic fluctuations. Thus, we should expect regoliths to differ considerably among otherwise similar bodies.

The utility of the determinate models is of course dependent on how well simple averages can characterize regolith depth on asteroids. This, in turn, depends on how much the regolith varies over an asteroid's surface and on how much variation can be expected between bodies of equal size. The purpose of this paper is to develop a statistical framework from which the depth variation can be determined.

THE VARIANCE OF REGOLITH DEPTH

Let us consider a fictitious population of a large number of initially indistinguishable asteroids. Each body is subjected to the same impacting-projectile population and is allowed to evolve until an energetic impact causes catastrophic fragmentation. The amount of regolith developed on any given asteroid depends primarily on three factors. (1) The number of craters as a function of size. For example, a body which has been pelted by many small impacts should develop less regolith than an asteroid with a larger proportion of big craters because large craters are the ones responsible for creating new regolith via excavations into bedrock whereas small craters merely rework existing regolith. (2) The order or occurrence of craters. The regolith should be relatively deep if the largest craters form late in the evolution. If they occur early, then much of the regolith they generate will be ejected from the asteroid by numerous small impacts. (3) The relative positions of craters. The positioning of craters affects regolith depth because, as mentioned above, new regolith is generated when craters puncture through the existing debris layer and excavate bedrock. On some bodies the larger craters will, by chance, occur in the regions where the regolith layer is very thin compared to other areas on the surface. Such a body will develop more regolith than those asteroids on which the large craters preferentially formed in the deeper parts of the regolith. These three quanties are all random variables, i.e., we cannot exactly predict the number of craters produced on an asteroid (again, even though an average crater flux might be specified exactly), their order of occurrence or their relative positions. Thus, corresponding to each body is a unique surficial distribution of regolith depths, schematically illustrated in Fig. 1. In order to associate a particular distribution with a particular asteroid, we label each body with an identifying number, $I = 1, 2, \ldots$

Suppose that one asteroid is selected at random from the population. The identifying number, I, of the asteroid is a random variable. We now randomly choose a single point on the surface of this body and measure the regolith depth, X, at the point. Clearly, X is a random variable because the number of craters of given size on the asteroid, their relative positions, etc., are all random variables (which follow specific probability distributions). The probability distribution of X is just the normalized sum of all the separate depth distributions and is sketched in Fig. 1.

An expected (i.e., mean) value of X can be obtained by averaging over the surfaces of all bodies in the population. This is equivalent to averaging together the mean values for each distribution in Fig. 1. If μ denotes the expected value of X and E denotes the expectation operator, then, using standard notation (e.g., Mood et al., 1974),

$$\mu = E[E(X|I)] \qquad (1)$$

where $E(X|I)$ is the mean depth on our randomly chosen asteroid. The expectation outside the square brackets averages over all the bodies in the population.

The variance, σ^2, of X can be written as the sum of two components,

$$\sigma^2 = E[V(X|I)] + V[E(X|I)] \qquad (2)$$

where $V(X|I)$ is the variance in depth over asteroid number I and the variance operator outside the square brackets applies to the random variable I. The first component represents the variation in depth over an asteroid's surface, averaged over many bodies. This term is obtained by averaging together the squares of the standard deviations of each of the individual distributions shown in Fig. 1. The second component represents the variation in the mean depth that occurs between bodies, i.e., it indicates the degree to

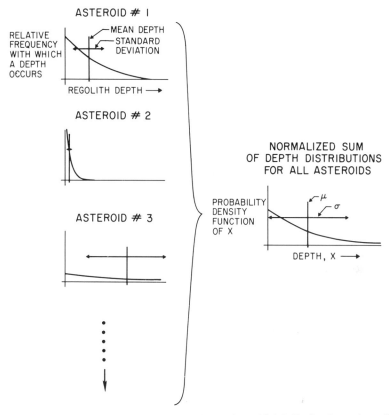

Fig. 1. On the left side of this figure are shown the surficial distributions of regolith depth for three asteroids belonging to a larger population of bodies of common size and compositional strength. Each distribution is unique because of stochastic fluctuations in the number of craters on an asteroid, their relative positions and order of occurrence. The probability density function for the random variable, X (the regolith depth measured at a randomly selected point on the surface of an asteroid chosen at random from the population), is shown on the right. The mean, μ, and standard deviation, σ, of X are also shown.

which each asteroid is unique. This term is found by computing the variance of the mean values of the distributions in the figure.

Housen (1981b) has shown that, at the time of fragmentation, X has approximately an exponential probability density function. (The proof of this result is too lengthy to warrant reproduction here.) For an exponential distribution the standard deviation is equal to the mean, i.e., $\sigma = \mu$ (e.g., Mood et al., 1974). The fact that X has such a large standard deviation is, at first thought, not too surprising because we have included the effects of large craters in computing μ and σ. Thus, the first component of variance in Eq. (2) should be large. However, as we now show, the second component in Eq. (2), which describes the differences between bodies, also contributes significantly to the variance.

A Monte Carlo computer program was written in order to simulate regolith evolution. In the program the surface of an asteroid was represented by a square grid. Craters were generated on the grid until a large impact resulted in catastrophic fragmentation. The regolith depth at each grid point was followed throughout the evolution by keeping track of the amount of debris deposited on each point and the amount excavated. For each asteroidal surface, the mean and standard deviation of regolith depth was computed. This experiment was repeated many times so that the average variation in depth over a surface and the variation in the average depth among bodies could be found.

Simulations were performed for two generic types of asteroids: "small" asteroids, on which low gravity fields result in widespread ejecta deposits, and "large" asteroids, on which ejecta are constrained to lie near craters. The actual diameter division between small and large asteroids, i.e., the division between global and nonglobal ejecta deposits, depends on the material strength of an asteroid because ejecta velocities depend on material strength (see Langevin and Maurette, 1980 for a discussion pertinent to regolith evolution). Thus, ejecta should be widespread on rocky bodies of diameter less than a couple of hundred kilometers. On weaker bodies, where ejecta velocities are lower, global ejecta deposits should occur for diameters less than a few kilometers (Housen et al., 1979a). For small asteroids we consider ejecta to be spread in a uniform layer over the surface. For large bodies three different ejecta blanket profiles were tried: (1) the thickness of ejecta decreases as the -3 power of the distance from the crater rim, (2) the thickness is a constant out to a distance of one crater radius from the rim, and zero beyond, and (3) the thickness decreases linearly from a maximum at the rim to zero at a distance of one crater radius from the rim. The results for large bodies were found to be relatively insensitive to the ejecta model used. The crater bowls were assumed to have the shape of spherical caps with depth to diameter ratio 0.2. The adopted production size-frequency distribution of craters can be found in Housen et al. (1979a).

The output for a "typical" small asteroid is shown in Fig. 2a. In the figure, the regolith depth, i.e., the thickness of the debris layer, is plotted as a function of position on the surface grid. The Monte Carlo experiments showed that, for small asteroids, the average

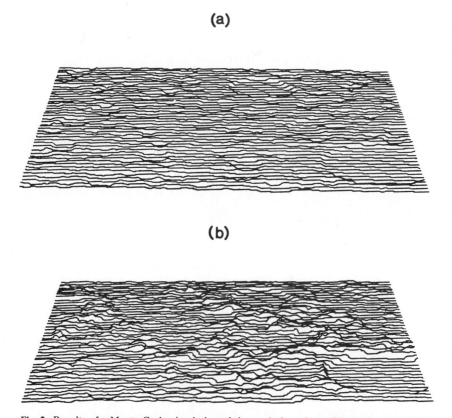

Fig. 2. Results of a Monte Carlo simulation of the evolution of regolith depth on asteroids. (a) The regolith depth as a function of position on the surface of a small body, where ejecta from a crater are globally distributed. (b) Regolith depth on a large asteroid where ejecta are constrained to lie near craters. The increased variability in depth is due to the localization of ejecta debris.

variation in depth over a surface is roughly equal to the variation in mean depth between bodies. That is, the two components of variance in Eq. (2) are about equal. A similar plot for a large asteroid is shown in Fig. 2b. [Note, in order to graphically illustrate the effects of differing ejecta morphologies on large and small bodies, the same sequence of crater sizes and positions, used in part (a) of the figure, were used in part (b).] The experiments for large bodies showed that the average surficial variation in depth [i.e., the first component in Eq. (2)] is 1.5 to 2 times larger than the variance of the mean depths [the second component in Eq. (2)]. This is due to increased roughness caused by the localization of ejecta. Thus, it is apparent that neither component vastly dominates the other.

As mentioned earlier, existing regolith models do not average over the entire surface of an asteroid. They attempt to reduce the variability of X by considering only the parts of the surface which are saturated by the effects of small craters. In order to assess the effectiveness of this procedure, the mean and standard deviation of regolith depth were recomputed, this time ignoring the same parts of an asteroid's surface that are ignored in the determinate models.

For small asteroids the large-crater bowls themselves were excluded while their widespread ejecta deposits were still included in the averaging. The procedure used in deciding which craters are small (i.e., saturated) or large (i.e., nonsaturated) is described in detail in the paper of Housen et al. (1979a). Avoiding the areas occupied by large craters, where regolith depth tends to be low, caused the mean depth (and standard deviation) to increase. The ratio σ/μ decreased slightly but was still found to be $\simeq 1$. Thus, excluding the large crater bowls does not significantly decrease the uncertainty associated with regolith depth. For large asteroids the parts of the surface occupied by the ejecta from big craters were also excluded from the averaging, following the methodology of Housen et al. (1979b). Avoiding these thick ejecta deposits caused the mean depth and standard deviation to decrease. Again, the ratio σ/μ decreased slightly but was still $\simeq 1$.

DISCUSSION

The calculations described in the previous section demonstrate that asteroidal regoliths are quite variable in depth regardless of whether or not the effects of large craters are included. We might reasonably ask if there are any processes which might tend to reduce the variability of regolith depth below that calculated. For example, if an asteroid does not conform to an equilibrium shape, then components of the gravity field which are tangent to the local surface may redistribute the regolith, thus "smoothing" the debris layer to conform to the geoid (see discussion in Cintala et al., 1979). This mechanism may have played a role in forming the smooth appearance of Deimos' surface (Cintala et al., 1979; Thomas, 1979).

It is, however, worthwhile to note that a smooth surface does not guarantee a regolith of uniform depth; variability also arises from the topography at the bottom of a regolith layer (Head, 1976). To illustrate this, we can consider the extreme case where the regolith is sufficiently mobile to redistribute itself to conform to the geoid. For the Monte Carlo calculations described earlier, the geoid is represented by a plane. In order to estimate how smoothing affects the variance of regolith depth, the regoliths generated by the afore-mentioned Monte Carlo program were allowed to "relax" so that the top of the regolith layer had a smooth, planar, surface.

After the surfaces had been smoothed, the standard deviation of the surficial depth distribution [i.e., the square root of the first component in Eq. (2)] was observed to increase by roughly a factor of 1.7. Figure 3 shows an example. Parts (a) and (b) of the figure are plots of the *surface elevation* before and after smoothing. Parts (c) and (d) show the *regolith depth* (i.e., the thickness of the regolith layer) before and after smoothing. The reason for the increased variance is clear; by removing the regolith from high surface elevations and accumulating it in the gravitationally low spots (e.g., the bottoms of large craters) we tend to add more weight to the tails of the depth distributions shown in Fig. 1.

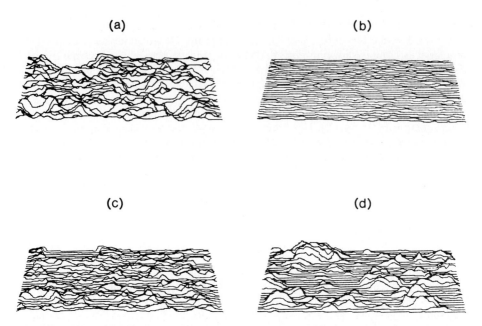

Fig. 3. This figure illustrates how the variability of regolith depth on an asteroid is affected by redistributing the debris so that the regolith has a "smooth" top surface. (This might happen on asteroids because of movement of debris into gravitationally "low spots", for instance.) (a) The elevation of the top surface of a regolith before smoothing. (b) The surface elevation on the same asteroid after the regolith has been redistributed. (c) The regolith depth (i.e., the thickness of the debris layer) corresponding to the surface shown in part a. (d) The regolith depth after the surface was smoothed. Smoothing an asteroid's surface can actually increase the variance of the regolith depth distribution because, by removing regolith from high surface elevations and accumulating it in the gravitationally low spots (e.g., the bottoms of large craters), more weight is added to the tails of the depth distributions shown in Fig. 1.

The above simulation was meant to be a simple example of the effects of smoothing an asteroid's surface. The actual process is difficult to simulate because in reality the extent to which a regolith is redistributed depends on how far the asteroid deviates from an equilibrium figure, on the cohesive properties of its regolith and on how frequently impacts "shake" the regolith (Cintala et al., 1979; Langevin and Maurette, 1981). However, the simulations do indicate that, while an asteroid (or planetary satellite) may have a smooth surface, its regolith may be quite variable in depth. We now consider a simple illustrative example of how this large variance can affect conclusions based on average depths alone.

A dominant reason for studying asteroidal regolith evolution is to shed light on the origins of brecciated and gas-rich meteorites, which are thought to be derived from regoliths (see the references cited in the first paragraph of the Introduction). For instance, we can try to constrain parent body size by comparing the observed properties of meteorites to corresponding quantities computed from a regolith model. As an example, we will consider the observed volume fraction, Q, of brecciated material within a given group of meteorites and construct a simple model in an attempt to predict the size of the parent body corresponding to a given value of Q.

For the sake of argument, let us suppose that a group of meteorites under consideration was ejected from a parent-body regolith in a single cratering event. (This assumption might be justified if the cosmic ray exposure ages in the group were clustered about a common value.) The value of Q will depend on the size of the crater which ejected the meteoritic material and on the depth of regolith at the point of impact. These are both random variables, so Q is also random. Determinate regolith models would yield only the average value of Q for a given size of asteroid. For illustration, Fig. 4 shows the average

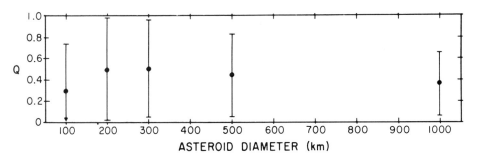

Fig. 4. The volume fraction, Q, of regolith-derived material ejected from an asteroid by a single cratering event. Q is a random variable because both the diameter of the crater ejecting the material and the regolith depth at the point of impact are random. The average value of Q is represented by the filled circles. The verticle lines represent plus and minus one standard deviation.

value of Q, plotted as filled circles, for five different sizes of rocky asteroids. The standard deviation of Q, also shown in the figure, is discussed below. For the smaller asteroids in the figure, the average increases with diameter because more ejecta are retained, resulting in thicker regoliths and, hence, larger values of Q. For larger bodies (diameters > 200–300 km) the mean value decreases again because now the size of the largest craters (which dominate regolith depth) are controlled by gravity, as opposed to smaller bodies where crater size is determined mainly by the material strength of the target. The result is that, as asteroid diameter increases, the size of the largest craters grows more slowly in proportion to asteroid diameter. But the large craters tend to dominate regolith growth, so the rate at which the average regolith depth grows also decreases, which causes the mean value of Q to decrease as asteroid size increases. As an aside, note that the values of Q shown in Fig. 4 would be increased if, for example, the flux of impacting projectiles in the asteroid belt was much higher than the modern-day value (upon which Fig. 4 was based). Also, an anonymous referee has suggested that small bodies, being fragments of larger "parent" asteroids, might inherit some regolith from their parents. Both of these effects could increase the regolith depth beyond that calculated, thus increasing the values of Q.

The standard deviation, σ_Q, of the random variable Q can be computed from the standard deviations of regolith depth and of the diameter of crater which ejected the meteorites (Housen, 1981b). Because there is a large variance associated with regolith depth, σ_Q is also large. In fact, as shown in Fig. 4, σ_Q is sufficiently large to make any size of asteroid capable of producing virtually any value of Q. The point is that, if we were to try to predict parent body size for a given group of meteorites by comparing the observed value of Q to the mean values shown in Fig. 4 (which is all a determinate model would give), then we could easily be led to erroneous conclusions.

As a final comment, it should be noted that regolith modeling is also subject to uncertainty caused by poorly-known input parameters (Langevin and Maurette, 1980). At present, this uncertainty is at least as large as the statistical variation discussed above. However, estimates of input parameters can be improved. The statistical uncertainties, which are inherent to regolith evolution and cannot be avoided, indicate that simple average values cannot be expected to precisely characterize the depth of an asteroid's regolith or any function of the regolith depth.

Acknowledgments—Special thanks are extended to Charles Hostetler, Laurel Wilkening, and Alex Woronow for useful discussions and comments. The careful reviews of Mark Cintala and an anonymous reviewer improved the manuscript. This work was supported by NASA grant NSG 7011.

REFERENCES

Anders E. (1975) Do stony meteorites come from comets? *Icarus* **24**, 363–371.
Cintala M. J., Head J. W., and Wilson L. (1979) The nature and effects of impact cratering on small bodies. In *Asteroids* (T. Gehrels, ed.), p. 579–600. Univ. Arizona Press, Tucson.
Head J. W. (1976) The significance of substrate characteristics in determining morphology and morphometry of lunar craters. *Proc. Lunar Sci. Conf. 7th*, p. 2913–2929.
Housen K. R., Wilkening L. L., Chapman C. R., and Greenberg R. (1979a) Asteroidal regoliths. *Icarus* **39**, 317–351.
Housen K. R., Wilkening L. L., Chapman C. R., and Greenberg R. (1979b) Regolith development and evolution on asteroids and the Moon. In *Asteroids* (T. Gehrels, ed.), p. 601–627. Univ. Arizona Press, Tucson.
Housen K. R. (1981a) A comparison of current asteroidal-regolith models (abstract). In *Lunar and Planetary Science XII*, p. 471–473. Lunar and Planetary Institute, Houston.
Housen K. R. (1981b) *Ph.D. Dissertation*, Univ. of Arizona, Tucson. 203 pp.
Langevin Y. and Maurette M. (1980) A model for small body regolith evolution: The critical parameters (abstract). In *Lunar and Planetary Science XI*, p. 602–604. Lunar and Planetary Institute, Houston.
Langevin Y. and Maurette M. (1981) Grain size and maturity in lunar and asteroidal regoliths (abstract). In *Lunar and Planetary Science XII*, p. 595–597.
Macdougall D., Rajan R. S., and Price P. B. (1974) Gas-rich meteorites: Possible evidence for origin on a regolith. *Science* **183**, 73–74.
Mood A. M., Graybill F. A., and Boes D. C. (1974) *Introduction to the Theory of Statistics*. McGraw-Hill, N.Y. 564 pp.
Rajan R. S. (1974) On the irradiation history and origin of gas-rich meteorites. *Geochim. Cosmochim. Acta* **38**, 777–788.
Suess H. E., Wanke H., and Wlotzka F. (1964) On the origin of gas-rich meteorites. *Geochim. Cosmochim. Acta* **28**, 595–607.
Thomas P. (1979) Surface features of Phobos and Deimos. *Icarus* **40**, 223–243.
Wasson J. T. and Wetherill G. W. (1979) Dynamical, chemical and isotopic evidence regarding the formation locations of asteroids and meteorites. In *Asteroids* (T. Gehrels, ed.), p. 926–974. Univ. Arizona Press, Tucson.
Wilkening L. L. (1971) Particle track studies and the origin of gas-rich meteorites. Center for Meteorite Studies (Contribution No. 11, Arizona State Univ., Tempe, Arizona).

Xenon diffusion following ion implantation into feldspar: Dependence on implantation dose

C. L. Melcher, D. S. Burnett, and T. A. Tombrello

Kellogg Radiation Laboratory 106-38, California Institute of Technology, Pasadena, California 91125

Abstract—The diffusion of Xe in feldspar, a major mineral in both meteorites and lunar samples, was studied. Xe ions were implanted at 200 KeV into single-crystal plagioclase targets and the depth profiles were measured by alpha particle backscattering before and after annealing for 1 hour at 900 or 1000°C. The fraction of implanted Xe retained following annealing was strongly dependent on the implantation dose. Maximum retention of 100% occurred for an implantation dose of 3×10^{15} Xe ions/cm^2. Retention was less at both lower and higher doses. The Xe retention was unaffected by two-step anneals (750 and 1000°C) or by implanting both Xe and He or Xe and Ar. Helium was implanted to a dose comparable to that expected in a typical solar wind exposure time. Three models to explain the dose-dependent diffusion properties are considered: i) epitaxial crystal regrowth during annealing which is controlled by the extent of the radiation damage produced by implantation; ii) creation of "trapping sites" by radiation damage during implantation; and iii) inhibition of recrystallization during annealing by the presence of Xe, except at low doses. Due to the complex dependence of the diffusion of Xe on implantation dose, it is difficult to extrapolate from these experiments to the much lower concentrations of solar-wind-implanted Xe in lunar materials or radiogenic Xe in meteorites.

INTRODUCTION

The diffusion properties of Xe in meteoritic minerals are important in the interpretation of ^{244}Pu/Xe and ^{129}I/^{129}Xe chronologies of the early solar system. Chronological discussions of the radiogenic Xe isotopic data are based, either explicitly or implicitly, on the assumption of a well-defined "time of Xe retention," before which Xe is quantitatively lost, and after which Xe is quantitatively retained. Improved understanding of the isotopic data in terms of more realistic thermal histories is, in principle, possible provided that the mineralogical sites and diffusion coefficients of the radiogenic Xe are known. Numerous efforts have been directed toward identification of the mineralogical sites (e.g., Lewis *et al.*, 1975; Alaerts *et al.*, 1979; Zaikowski, 1980; Rison *et al.*, 1980; Jones and Burnett, 1979). The present study deals with the diffusion properties of Xe.

The overall aim of this work is to study the diffusion of Xe implanted into terrestrial analogs of major meteoritic and lunar minerals. Results for feldspar are reported here. Jones and Burnett (1979) have proposed that in the St. Severin meteorite, and presumably in other equilibrated ordinary chondrites as well, ^{244}Pu resides on grain boundaries. For such a distribution, recoil fission Xe will be implanted to depths of a few microns in the surfaces of all major mineral grains including feldspar. We intend to use the method that we have developed to study the diffusion of Xe in other major minerals such as olivine and pyroxene, and the method can, in principle, be applied to the diffusion of radiogenic ^{129}Xe once the host minerals have been identified.

The diffusion coefficients of Xe are also important in evaluating the retention times of Xe implanted by the solar wind in lunar materials. The implantation energy used here (200 keV) is similar to the energy of Xe in the solar wind but the concentrations implanted in our experiments are much larger than implanted by the solar wind. Therefore, it becomes important to understand how Xe diffusion varies with implantation dose. In addition, it is possible that the more abundant elements in the solar wind create enough radiation damage to significantly affect the diffusion of Xe. He, in particular, seems a likely candidate in this regard. Lunar dust grains are exposed to a He dose of $\sim 8 \times$

10^{16} ions/cm^2 in a typical exposure time which apparently is sufficient to produce an amorphous layer on the grain surfaces (Borg et al., 1980). Also, Muller et al. (1976) and Ducati et al. (1973) have proposed damage trapping of He itself in lunar grains. We implanted He and Ar in addition to Xe in some samples to assess the effect of additional radiation damage on the Xe diffusion.

EXPERIMENTAL PROCEDURE

The experimental procedure consisted basically of four steps. (i) A target was first implanted with a known dose of Xe ions. (ii) The depth profile of the implanted Xe was analyzed by means of backscattering spectrometry of alpha particles (see Chu et al., 1978). (iii) The target was then annealed, typically for one hour at 900 or 1000°C, and (iv) the Xe depth profile was measured again. The basic idea is to observe broadening of the implanted profile due to thermal diffusion. The temperatures were selected to be below where Xe is readily outgassed from non-carbonaceous meteorites but high enough so that significant perturbations of the implanted profile were observed. This method has been used previously by a number of workers to study the diffusion of implanted ions primarily in Be, Al, and Si (see, for example, Sippel, 1959; Myers et al., 1974; Fontell et al., 1973; Ohkawa et al., 1974). The present study appears to be the first attempt to apply the backscattering technique to the diffusion of implanted ions in a mineral.

The Xe implantations were performed at an energy of 200 keV with a 400 keV ion accelerator. The beam current density was kept below $3\mu A/cm^2$ to avoid heating of the targets. The targets were coupled to the target holder via a thermally conducting grease and remained at room temperature or slightly below during implantation due to radiation to a cooled shield in the target chamber. The pressure in the target chamber was $\leq 10^{-7}$ torr. Values of the bombardment doses were obtained by current integration and are estimated to be accurate within ±20%. Spatially uniform implantations were obtained by scanning the beam across the target. The bombardment doses ranged from 3×10^{14} to 1×10^{17} ions/cm^2. Significant sputtering of the target surface occurs for Xe$^+$ ions incident at this energy (see below) such that for bombardment doses $\geq 2 \times 10^{16}$ ions/cm^2 the thickness of the target which is sputtered away exceeds the range of the incident Xe ions. Thus, some fraction of the implanted Xe will be lost with the sputtered target material at high doses. Consequently, for bombardment doses $\geq 2 \times 10^{16}$ ions/cm^2, the resulting implanted dose is less than the bombardment dose. Of most interest, however, are doses of $\leq 3 \times 10^{15}$ ions/cm^2 and the thickness sputtered from the target (≤ 120 Å) is much less than the range of the Xe ions (~ 650 Å). For these doses the xenon concentrations calculated from the backscattering spectra agree with the implantation doses measured by beam integration during implantation within 30%.

The Ar implantation was performed in a similar manner at an energy of 60 keV. The dose was 10^{16} ions/cm^2.

The He implantations were performed at an energy of ~ 6.4 keV with a 25 keV ion accelerator. The bombardment doses ranged from 3×10^{15} ions/cm^2 to 5×10^{16} ions/cm^2. The doses are less certain in this case but are estimated to be accurate to within 50%.

The depth profiles of the implanted Xe were determined by Rutherford backscattering of alpha particles. Alpha particles with energies usually in the range of 1 to 2 MeV are directed at the target and a small fraction of these are scattered elastically by the various nuclides in the target. Those particles backscattered at some large angle with respect to the incident beam (typically 170°) are detected and counted. The energy of a detected (backscattered) particle is a function of the mass of the nuclide from which it was scattered. Alpha particles which are scattered by heavy nuclei lose a smaller fraction of their incident energy than those particles scattered by light nuclei. The energy of a detected particle is further reduced by the energy it loses in passing through a finite thickness of the target before and after being scattered. Thus, the energy spectrum of backscattered particles can provide information on both the masses of the elements present in a target and their distribution with depth.

A typical backscattering spectrum of a feldspar target implanted with Xe is shown in Fig. 1. The major elements present in the target (O, Al, Si, Ca) produce a series of steps. The high energy edge of each step corresponds to alpha particles scattered by atoms of that element located at the surface of the target. Alpha particles scattered by atoms deeper in the target lose additional energy before reaching the detector. The strong increase in counting rate with decreasing energy for a uniformly distributed element, e.g., O in Fig. 1, reflects the increase in the scattering cross section with decreasing energy. The implanted Xe atoms, on the other hand, are not evenly distributed but have a roughly Gaussian depth distribution at a mean depth of ≈ 650Å. Therefore the Xe signal consists of an apparently Gaussian peak rather than a step. An unimplanted portion of each target was also analyzed in order to subtract the spectrum of the feldspar from that of the implanted Xe distribution. This correction, however, proved to be quite small since the abundances of heavy elements which might overlap the Xe signal are very low in these targets.

The incident alpha particle beam (1.0 or 1.5 MeV) made an angle of $\theta_1 = 0°$ or 30° with the normal to the target surface. The backscattered particles were detected by a silicon surface barrier detector

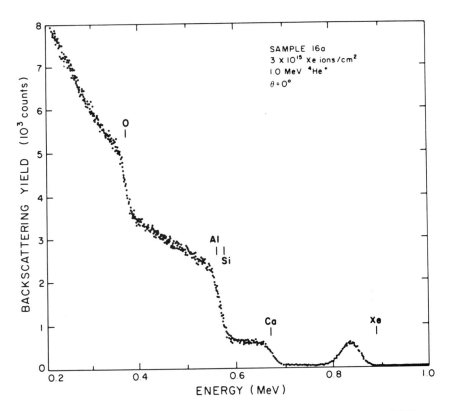

Fig. 1. Backscattering spectrum of 1.0 MeV alpha particles striking a Xe-implanted feldspar target at normal incidence. Stoichiometric elements in the target (O, Al, Si, Ca) result in step-like signals; the high energy edge of each step is indicated corresponding to scattering by an atom of that element on the surface. The implanted Xe distribution results in a Gaussian peak located ~48 keV below the energy of alpha particles scattered from Xe atoms at the surface of the target (indicated). The energy distribution of counts in the Xe peak can be directly scaled to a Xe depth distribution.

located at 170° with respect to the incident beam. The dose of alpha particles to the target during the analysis is typically $10^{14}-10^{15}$ cm^{-2}. The energy resolution of the system is ≤20 KeV (FWHM), primarily due to the energy resolution of the detector. This corresponds to a depth resolution of ≤260 Å for 1.0 MeV alpha particles incident at $\theta_1 = 0°$ and to ≤250 Å for 1.5 MeV alpha particles incident at $\theta_1 = 30°$. The positions of the peaks are reproducible to within ±40 Å.

The energy scales of the Xe spectra (Figs. 2–5) were converted to depth scales using the surface energy approximation (Chu et al., 1978, Ch. 3).

$$x = \frac{\Delta E}{N} \left[\frac{\epsilon(E_0)K}{\cos \theta_1} + \frac{\epsilon(KE_0)}{\cos \theta_2} \right]^{-1}$$

ΔE is the difference between the energy of an alpha particle which has scattered from a Xe atom at the surface of the target and the energy of one that has scattered from a Xe atom at depth x. N is the atomic density of feldspar. K is the ratio of the final to the incident energy when an alpha particle scatters from a Xe atom. $\epsilon(E_0)$ is the stopping cross section for alpha particles of incident energy E_0 in feldspar and $\epsilon(KE_0)$ is the stopping cross section evaluated at KE_0. $N\epsilon(E)$ is the stopping power, dE/dx, at energy E. θ_1 and θ_2 are the angles between the target normal and the direction of the incident beam and the direction of the scattered particle, respectively. This formula yields a depth conversion scale of 80 eV/Å for 1.5 MeV alpha particles and $\theta_1 = 30°$ and a depth conversion scale of 78 eV/Å for 1.0 MeV alpha particles and $\theta_1 = 0°$.

After measurement of the implanted depth profiles, targets were annealed in a tube furnace at a pressure of $\sim 2 \times 10^{-6}$ torr and a temperature of 900°C or 1000°C. The temperature was constant to better than ±3°C. After annealing, the depth profiles were measured again in an identical manner.

Table 1. Bombardment doses, energies, ion ranges (R_p), straggling of implanted profiles (ΔR_p), annealing temperatures (all for 1 hr.), and fraction of implanted Xe retained after anneal. Values of R_p and ΔR_p for Xe were determined from the backscattering spectra and are reproducible to within ±40Å. Values of R_p and ΔR_p for He are estimated from Ziegler (1977), noting that the targets were tilted 45° with respect to the incident beam. Values of R_p and ΔR_p for Ar are estimated from Davies *et al.* (1963).

	Dose (cm^{-2})	Energy (kev)	R_p(Å)	ΔR_p(Å)	Anneal Temperature (°C)	Xe retained after anneal (%)
Xe:	1×10^{17}‡	200	450†	330	900	36
Xe:	1×10^{16}	200	450†	240	900	78
Xe:	3×10^{15}	200	610	240	750	100
Xe:	3×10^{15}	200	610	240	1000	100
Xe:	1×10^{15}	200	650*	240	900	46
Xe:	1×10^{15}	200	650*	240	1000	45
Xe:	3×10^{14}	200	670	240	750	100
Xe:	3×10^{14}	200	670	240	1000	0
Xe:	3×10^{14}	200	670	240	750 then 1000	0
Xe:	3×10^{15}	200	610	240	750 then 1000	100
Xe:	3×10^{14}	200	720	230	1000	0
He:	3×10^{15}	6.4	600	280		
Xe:	3×10^{14}	200	720	210	1000	0
He:	5×10^{16}	6.4	600	280		
Xe:	3×10^{14}	200	680	220	1000	0**
Ar:	1×10^{16}	60	550	250		

* Post-anneal profile is located ~100 Å closer to surface.
† Post-anneal profile is significantly different in shape.
‡ Implanted dose is significantly less than bombardment dose due to sputtering of surface.
** The implanted Ar was also completely lost during the anneal.

All targets were in the form of thick slabs ($\sim 1 \times 10 \times 10$ mm) which were cut with approximately the same crystal orientation from two large single crystals of Lake County labradorite. The cut slabs were ground on 600 grit carborundum paper. The resulting surfaces were rough on a 1–10 micron scale. Consequently, some loss of resolution in the backscattering spectra was expected (Schmid and Ryssel, 1974). However, comparison of backscattering spectra of thin films deposited on these rough surfaces with the backscattering spectra of thin films deposited on highly polished sapphire surfaces showed only a slight ($\leq 25\%$) loss in resolution.

RESULTS

The obvious complication to our technique is the effect of radiation damage produced during implantation. Consequently our efforts have focused on studying the effects of varying implantation doses. Targets were implanted with Xe doses ranging from 3×10^{14} to 10^{17} ions/cm^2. In addition, two targets were implanted with both Xe and He, and one target with Xe and Ar in order to investigate the effect of additional implantation damage on the behavior of the Xe. Table 1 summarizes the implantation doses which were investigated.

High doses: Figure 2 shows Xe depth profiles for two large implantation doses. At 10^{16} ions/cm^2 the pre-anneal distribution appears nearly Gaussian but the peak occurs only 450 Å below the surface. This is ~ 200 Å nearer the surface than calculated from LSS theory (Lindhard *et al.*, 1963). The most plausible explanation is that the surface has been sputtered during the implantation. The actual distribution is then determined by simultaneous implantation and sputter erosion. The observed 200 Å shift in the peak probably implies that about 400 Å has been sputtered away. This requires a sputtering yield of ~ 30 which is in agreement with the sputtering yields of 200 kev Xe$^+$ ions on a number of metals (Sigmund, 1969). However, Bibring *et al.* (1981) report much larger yields for heavy ions incident on mica. Following a 1 hour anneal at 900°C, the shape of the Xe depth profile has changed significantly. The post-anneal profile is asymmetric with the

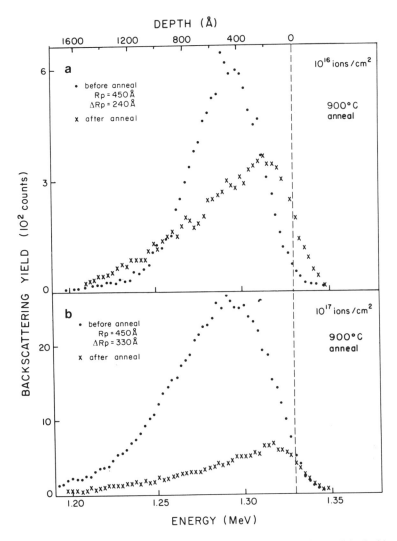

Fig. 2. Xe energy (depth) profiles from backscattering of 1.5 MeV alpha particles incident at 30° to the target normal. The spectra were taken before and after annealing for one hour at 900°C. The bombardment doses were a) 10^{16} Xe ions/cm² and b) 10^{17} Xe ions/cm². The counts at negative depths are a consequence of finite detector resolution.

peak shifted nearer the surface. Apparently Xe diffused more rapidly through the radiation-damaged surface region and more slowly, if at all, toward the interior of the crystal. About 22% of the implanted Xe has been lost, evidently by diffusion to the surface.

At a dose of 10^{17} ions/cm², the pre-anneal distribution is asymmetric and the peak occurs 450 Å below the surface. The quantity of Xe in the target is roughly a factor of four less than the bombardment dose suggesting loss of Xe by sputtering comparable to or larger than the ion range. (Actually the symmetry observed in Fig. 2a is not easily understood since an asymmetric peak would be expected given the amount of sputtering.) About 64% of the Xe is lost through the surface after one hour at 900°C. As with the 10^{16} cm^{-2} sample there appears to be no movement of Xe toward the interior of the crystal although Xe has been lost from all depths.

Intermediate doses: Figure 3 shows the depth profile of Xe implanted to a dose of

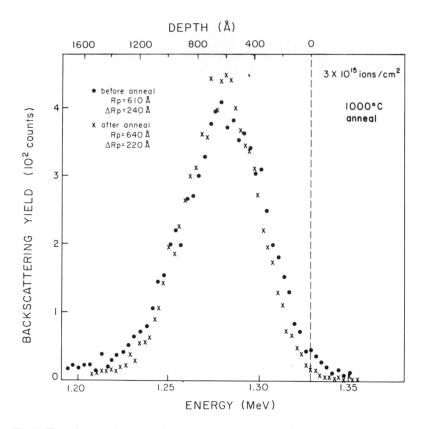

Fig. 3. Xe energy (depth) profile from backscattering of 1.5 MeV alpha particles incident at 30° to the target normal. The bombardment dose was 3×10^{15} Xe ions/cm^2. The spectra were taken before and after annealing at 1000°C for one hour.

3×10^{15} ions/cm^2. The profile is Gaussian with the peak located ~610 Å below the surface, which is somewhat less than the 650 Å predicted by theory (Lindhard et al., 1963). The difference can be accounted for by sputtering. The straggling ΔR_p (standard deviation) of the distribution is ~240 Å and the full width at half maximum is ~570 Å. After annealing for 1 hour at 1000°C, the Xe distribution is unchanged. There appears to be no movement of the Xe when implanted at this dose and annealed at 1000°C.

Figure 4 shows depth profiles of Xe implanted to a dose of 10^{15} ions/cm^2. The distributions appear nearly Gaussian with the peaks located ~650 Å below the surface in good agreement with the range predicted by theory. Sputtering of the surface is expected to be small (about 40 Å) at this bombardment dose. The straggling of the pre-anneal distributions is ~240 Å. After annealing for 1 hour at 900°C (Fig. 4a), the shape of the Xe depth profile measured after the anneal is similar to the pre-anneal profile but the number of Xe counts is smaller by a factor of 2. In addition, the peak of the profile is shifted ~100 Å nearer the surface. After annealing for 1 hour at 1000°C (Fig. 4b) about half of the Xe is lost and again the profile is shifted ~100–150 Å toward the surface.

Low doses: Figure 5 shows depth profiles of Xe implanted to a dose of 3×10^{14} ions/cm^2. The depth of the peak is ~670 Å and the straggling is ~240 Å. One sample (Fig. 5a) was annealed for 1 hour at 750°C. The post-anneal distribution of Xe is quite similar to the pre-anneal distribution, indicating no movement of the Xe at this temperature. The second sample (Fig. 5b) was annealed for 1 hour at 1000°C. In contrast to the higher dose cases, the Xe was completely lost from the sample during annealing.

Two-stage anneals: Additional targets were implanted to doses of 3×10^{14} and $3 \times$

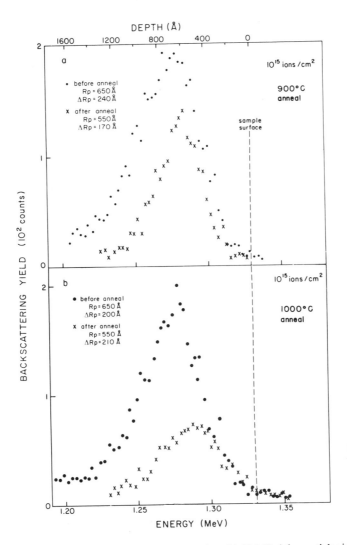

Fig. 4. Xe energy (depth) profiles from backscattering of 1.5 MeV alpha particles incident at 30° to the target normal. The dose was 10^{15} Xe ions/cm^2. The spectra were taken before and after annealing for one hour at (a) 900°C and (b) 1000°C.

10^{15} ions/cm^2 and subjected to two-stage anneals. They were first annealed at 750°C for 1 hour and then at 1000°C for 1 hour. The Xe depth profiles were measured after each step. The two-stage anneal was an attempt to anneal radiation damage (produced during implantation) at the lower temperature and to observe the effect of this on the behaviour of Xe at the higher temperature. A temperature of 750°C was chosen for the first step since Fe group cosmic ray tracks anneal in plagioclase at this temperature (Fleischer et al., 1975). In both cases, however, the end results of the two-stage anneal were identical to the single-stage anneals performed at 1000°C. The target implanted to 3×10^{14} ions/cm^2 showed total retention of Xe after the 750°C anneal step and total loss of Xe after the 1000°C anneal step. The target implanted to 3×10^{15} ions/cm^2 showed total retention of Xe after each anneal step.

Double implantations: In order to test the possibility that the retention of Xe when implanted to a dose of 3×10^{15} ions/cm^2 was due to damage trapping, targets implanted to 3×10^{14} ions/cm^2 were subjected to additional radiation damage by He or Ar ion im-

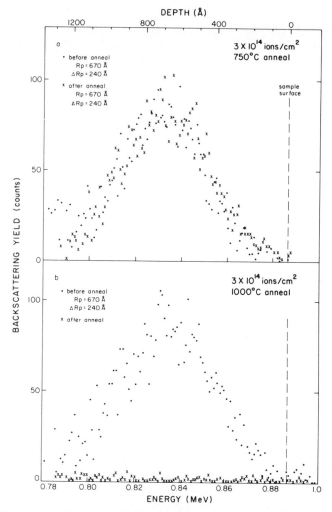

Fig. 5. Xe energy (depth) profiles from backscattering of 1.0 MeV alpha particles at normal incidence. The dose was 3×10^{14} Xe ions/cm². The spectra were taken before and after annealing for one hour at (a) 750°C and (b) 1000°C.

plantation. He doses of 3×10^{15} and 5×10^{16} ions/cm² were implanted into targets which were previously implanted with 3×10^{14} Xe ions/cm². The targets were then annealed as before (1 hour at 1000°C). The backscattering spectra taken following the anneals showed no Xe present in the targets. In other words, there was no evidence that the additional damage created by the He implantations affected the diffusion of Xe.

An Ar dose of 1×10^{16} ions/cm² (60 keV) was implanted into a target previously implanted with 3×10^{14} Xe ions/cm². The total energy deposited in the target is comparable to that deposited by the Xe-only implantation of 3×10^{15} ions/cm² (200 keV) (Fig. 3) in which the Xe profile is unchanged after annealing. Both the Ar and Xe distributions were clearly evident in the pre-anneal spectrum although a quantitative analysis of the Ar profile was difficult due to overlap of the signal with the Ca step. The range of 60 keV Ar in feldspar is estimated to be ~550 Å from the data of Davies et al. (1963). Since the straggling of both the Xe and Ar profiles is ≥220 Å, the profiles overlap considerably. The spectrum taken after a one hour anneal at 1000°C showed no Ar or Xe present in the target. As in the case of the Xe + He implantations, the additional damage created by the Ar implantation did not trap the Xe atoms.

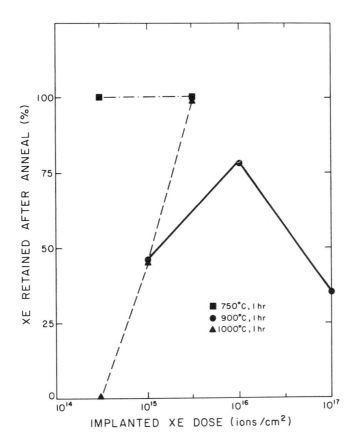

Fig. 6. Fraction of implanted Xe retained in targets as a function of dose after annealing at various temperatures for one hour.

Figure 6 summarizes the diffusion experiments in terms of the fraction of implanted Xe retained after anneals at 750°C, 900°C, and 1000°C. This fraction is seen to vary drastically with implantation dose. It reaches a maximum at a dose of 3×10^{15} ions/cm^2 and decreases for both higher and lower doses. In the next section we speculate on possible mechanisms to explain this complex behavior.

DISCUSSION

There are a number of factors which may affect the diffusion of implanted Xe atoms. The formation of Xe bubbles has been observed in studies of Xe-implanted semiconductors (see, for example Tsaur et al., 1979; Dearnaley et al., 1973). Bubbles typically form at doses of $\sim 10^{16}$ ions/cm^2 and are usually located at an interface, such as the interface between a silicon substrate and a thin metal film, or at grain boundaries (Liau and Sheng 1978; Cullis et al., 1978). The presence of bubbles characteristically results in one or more sharp spikes superimposed on the broader Xe backscattering signal. Perhaps surprisingly, the Xe profiles in Figs. 2–5 show no evidence of bubble formation, even at doses up to 10^{17} ions/cm^2.

Another factor which may affect the diffusion of Xe is the recrystallization of a radiation-damage amorphous layer during the anneal. It is probable that implantation produces an amorphous region comparable in depth to the range of the incident ions. When annealed to 900°C or 1000°C, this amorphous region may regrow epitaxially, become polycrystalline, or remain amorphous. At a given temperature, its fate probably depends on the degree of radiation damage and the amount of Xe which is present. Possibly there exists some threshold Xe concentration above which recrystallization is inhibited.

If one scales (by nuclear charge) the critical doses of H, He, and Ar ions found to produce amorphous layers in feldspar (Borg et al., 1980) to Xe, the estimated critical dose for Xe is less than 10^{14} ions/cm^2. Therefore, all of the doses studied here were probably sufficient to produce amorphous regions. In addition, radiation damage may create trapping sites for implanted ions. This was previously postulated by Leich et al. (1974) to explain solar wind H profiles in lunar samples. The existence of more than one activation energy was proposed by Frick et al. (1973) to explain complex release patterns of noble gases in lunar materials and this might be reflected in asymmetric depth profiles. Thus, at least three possibilities may be considered in trying to explain the dose-dependent diffusion properties summarized in Fig. 6.

i) Epitaxial crystal regrowth: In this model, the amorphous region produced by an implantation of 3×10^{15} ions/cm^2 regrows epitaxially, reforming a single crystal, and trapping the Xe which is present. At lower doses, the region is not completely amorphous and therefore, regrows polycrystalline. Due to the high density of grain boundaries, Xe atoms must diffuse only a short distance to reach a boundary and escape from the target. In both cases, recrystallization must be much faster than Xe diffusion. At higher doses, recrystallization is inhibited by the large concentration of Xe present and thus Xe diffuses from the still amorphous region (see Fig. 2). Although it cannot be completely ruled out, the results of the double implantations provide evidence against this model. The additional damage produced in targets containing low doses of Xe would be expected to allow epitaxial regrowth of these regions when annealed and high Xe retention similar to targets with intermediate Xe doses. However, the additional radiation damage had no effect on the Xe retention.

ii) Trapping: In this model, intermediate (3×10^{15} ions/cm^2) Xe implantation doses are required to produce sufficient "trapping sites" to retentively hold the Xe. At lower doses insufficient traps are created and the Xe is easily lost. The Xe behavior at high doses requires enhanced diffusion due to the extensive radiation damage. The double implantation results also argue against this model in that additional radiation damage introduced into low Xe targets does not trap the Xe. However, it is possible that the formation of traps has a "threshold" which is possible with Xe but not with He or Ar. In this model the shifted peak for the 10^{15} cm^{-2} dose can be explained as the Xe depth distribution reflecting the distribution of trapping sites which is expected to peak closer to the surface than the actual distribution of Xe atoms. There is no shift at 3×10^{15} ions/cm^2 since there are sufficient traps at all depths to retain the entire implanted profile.

iii) Inhibition of recrystallization by Xe: This model assumes that the feldspar crystal structure cannot accomodate the large Xe atoms. At low concentrations (3×10^{14} ions/cm^2) the amorphous region recrystallizes into a polycrystalline form, forcing Xe atoms to grain boundaries as it does so. Thus, the Xe is easily lost. At intermediate doses (3×10^{15} ions/cm^2), the Xe concentration is sufficiently high to completely inhibit recrystallization. The Xe remains in an amorphous region in which diffusion is slow. A similar dependence on the presence of noble gas atoms has been observed for the recrystallization of amorphous silicon by Wittmer et al. (1979). They found that ~0.2 atomic percent was the maximum concentration of Kr which would permit epitaxial growth of amorphous Si on crystalline Si. For comparison, 3×10^{15} Xe ions/cm^2 corresponds to an average concentration of ~0.5 atomic percent in plagioclase and might be expected to be sufficient to inhibit recrystallization. The behavior at higher ($\geq 10^{16}$) doses is not naturally explained, but the higher concentration gradients and the location of the peak of the distribution nearer to the surface will enhance diffusive loss. Also special radiation damage "structures" which promote diffusion might be created at the higher doses. In the double implantation experiments, the large doses of Ar and He probably did not inhibit recrystallization for one of the two following reasons: a) the Ar and He may have rapidly diffused out of the targets in a small fraction of the 1 hr. annealing time, thus allowing the targets to recrystallize in the remaining annealing time, or b) the small size of the He and Ar atoms was not sufficient to inhibit recrystallization.

This model can also explain the peak shift observed at a dose of 10^{15} ions/cm^2 (Fig. 4) if one assumes that only part of the amorphous region has regrown during the one hour anneal time. If the target material has recrystallized at depths greater than ~800 Å but remained amorphous at lesser depths, Xe may have been lost by lateral diffusion in the polycrystalline material at depths $\geq 800°$ Å but remained trapped in amorphous material at depths ≤ 800 Å. This regrowth behavior would be analogous to the epitaxial regrowth of ion-implanted amorphous Si in which regrowth begins at the amorphous-crystalline interface and proceeds toward the surface with increasing annealing time (reviewed by Lau, 1978).

Although none of the three models can be ruled out, the third is most consistent with the available data. The most important piece of missing information is the crystallinity of the implanted regions after annealing. We have attempted to obtain this information by channelling experiments but have not yet been successful. Low-angle X-ray diffraction measurements of target crystallinity are in progress.

Unfortunately, the applicability of these results to the diffusion of Xe in meteorites is not at all obvious. The complex dependence of the diffusion rate on dose makes it impossible to set either an upper or lower limit on the diffusion coefficient in meteorites at this point.

Similar problems are encountered in predicting the behavior of solar wind implanted Xe in lunar grains. The double implantations were intended to be a gross simulation of solar wind conditions. The implanted He dose was roughly equal to that expected in a typical solar wind exposure time. The Xe dose was, of course, much greater than a solar wind Xe dose and this may have obscured effects due to the implanted He. But, taking the data at face value, there is no evidence that radiation damage in lunar soil grains due to He and Ar implantation has *inhibited* the diffusive loss of Xe.

Acknowledgments—We are grateful to M.-A. Nicolet for use of the ion-implanter and for helpful discussions. We thank T. Banwell, M. Dumke, and R. Fernandez for performing the implantations. We also benefited greatly from discussions with D. Scott and C. Watson and from reviews from S. P. Smith, U. Frick, and an anonymous reviewer. We are indebted to Lou Ann Cordell for preparation of the manuscript. This work was supported by NASA grants NAGW-148 and NGR 05-002-333, and NSF grant PHY79-23638.

REFERENCES

Alaerts L., Lewis R. S., and Anders E. (1979) Isotopic anomalies of noble gases in meteorites and their origins-IV. C3 (Ornans) carbonaceous chondrites. *Geochim. Cosmochim. Acta* **43**, 1421–1432.

Bibring J. P., Borg J., Dran J. C., Langevin Y., Maurette M., Petit J. C., and Rocard F. (1981) Accumulation of solar wind effects in lunar dust grains: a reappraisal of solar wind simulation experiments (abstracts). In *Lunar and Planetary Science XII*, p. 65–67. Lunar and Planetary Institute, Houston.

Borg J., Chaumont J., Jouret C., Langevin Y., and Maurette M. (1980) Solar wind radiation damage in lunar dust grains and the characteristics of the ancient solar wind. In *Proc. Conf. Ancient Sun* (R. O. Pepin, J. A. Eddy, and R. B. Merrill, eds.), p. 431–461. Pergamon N.Y.

Chu W.-K., Mayer J. W., and Nicolet M.-A. (1978) *Backscattering Spectrometry*. Academic, N.Y. 384 pp.

Cullis A. G., Seidel T. E., and Meek R. L. (1978) Comparative study of annealed neon-, argon-, and krypton-, ion implantation damage in silicon. *J. Appl. Phys.* **49**, 5188–5198.

Davies J. A., Brown F., and McCargo M. (1963) Range of Xe^{133} and Ar^{41} ions of kiloelectron volt energies in aluminum. *Can. J. Phys.* **41**, 829–843.

Dearnaley C., Freeman J. H., Nelson R. S., and Stephen J. (1973) *Ion implantation*. North-Holland, Amsterdam. 802 pp.

Ducati H., Kalbitzer S., Kiko J., Kirsten T., and Muller H. W. (1973) Rare gas diffusion studies in individual lunar soil particles and in artificially implanted glasses. *The Moon* **8**, 210–227.

Fleischer R. L., Price P. B., and Walker R. M. (1975) *Nuclear tracks in solids: Principles and Applications*. Univ. Calif. Berkeley. 605 pp.

Fontell A., Arminen E., and Turunen M. (1973) Application of backscattering method for the measurement of diffusion of zinc in aluminum. *Phys. Status. Solidi* **15A**, 113–119.

Frick U., Baur H., Funk H., Phinney D., Schafer C., Schultz L., and Signer P. (1973) Diffusion properties of light noble gases in lunar fines. *Proc. Lunar Sci. Conf. 4th*, p. 1987–2002.

Jones J. H. and Burnett D. S. (1979) The distribution of U and Pu in the St. Severin chondrite. *Geochim. Cosmochim. Acta* **43**, 1895–1905.

Lau S. S. (1978) Regrowth of amorphous films. *J. Vac. Sci. Technol.* **15**, 1656–1661.

Leich D. A., Goldberg R. H., Burnett D. S., and Tombrello T. A. (1974) Hydrogen and fluorine in the surface of lunar samples. *Proc. Lunar Sci. Conf. 5th*, p. 1869–1884.

Lewis R. S., Srinivason B., and Anders E. (1975) Host phase of a strange xenon component in Allende. *Science* **190**, 1251–1262.

Liau Z. L. and Sheng T. T. (1978) Argon bubble formation in the sputtering of PtSi. *Appl. Phys. Lett.* **32**, 716–718.

Lindhard J., Scharff M., and Schiott H.E. (1963) Range concepts and heavy ion ranges. *Mat. Fys. Medd. Dan. Vid. Selsk.* **33**, 1–42.

Muller H. W., Jordan J., Kalbitzer S., Kiko J., and Kirsten T. (1976) Rare gas ion probe analysis of helium profiles in individual lunar soil particles. *Proc. Lunar Sci. Conf. 7th*, p. 937–951.

Myers S. M., Picraux S. T., and Prevender T. S. (1974) Study of Cu diffusion in Be using ion backscattering. *Phys. Rev.* **9**, 3953–3964.

Ohkawa S., Nakajima Y., Sakurai T., Nishi H., and Fukukawa Y. (1974) Diffusion profiles of arsenic in silicon observed by backscattering method and by electric measurement. *Japanese J. Appl. Phys.* **13**, 361.

Rison W., Zaikowski A., Lumpkin G. R., and Kirschbaum C. (1980) Search for ^{129}Xe bearing phases in Allende by laser microprobe. *Meteoritics* **15**, 354–355.

Schmid K. and Ryssel H. (1974) Backscattering measurements and surface roughness. *Nucl. Instrum. Meth.* **119**, 287–289.

Sigmund P. (1969) Theory of sputtering. I. Sputtering yield of amorphous and polycrystalline targets. *Phys. Rev.* **184**, 383–418.

Sippel R. F. (1959) Diffusion measurements in the system Cu-Au by elastic scattering. *Phys. Rev.* **115**, 1441.

Tsaur B. Y., Liau Z. L., and Mayer J. W. (1979) Inert-gas-bubble formation in the implanted metal/Si system. *J. Appl. Phys.* **50**, 3978–3984.

Wittmer M., Roth J., and Mayer J. W. (1979) The influence of noble gas atoms on the epitaxial growth of implanted and sputtered amorphous silicon. *J. Electrochem. Soc.*, 1247–1252.

Zaikowski A. (1980) I-Xe dating of Allende inclusions: Antiquity and fine structure. *Earth and Planet. Sci. Lett.* **47**, 211–222.

Ziegler J. F. (1977) Helium: Stopping powers and ranges in all elemental matter. *The Stopping and Range of Ions in Matter*, V. 4. Pergamon N.Y. 367 pp.

A brief note on the effect of interface bonding on seismic dissipation

B. R. Tittmann[1], M. Abdel-Gawad[1], C. Salvado[1], J. Bulau[1] L. Ahlberg[1], and T. W. Spencer[2]

[1]Earth and Planetary Sciences Group, Rockwell International Science Center, 1049 Camino Dos Rios, Thousand Oaks, California 91360 [2]Department of Geophysics, Texas A&M University, College Station, Texas 77843

Abstract—In this brief note, data on the dependence of seismic dissipation Q^{-1} on strain amplitude, confining pressure, frequency, and the presence of porefluid are presented. Dramatic differences are seen in the behavior of dry sandstones from two different formations. Boise sandstone shows little dependence of Q^{-1} on strain-amplitude, confining pressure, frequency, and water saturation compared with Berea sandstone. Since both rocks are similar in composition, porosity, permeability, grain size, and grain contact length, this behavior is explained on the basis of the nature of bonding (cementation) between grains, as perceived through petrofabric studies. The effects of bonding at interfaces on the propagation of SH waves in the field are explored in a simple theoretical model which gives Q^{-1} in terms of an effective interface bonding parameter.

BACKGROUND AND INTRODUCTION

Attenuation mechanisms based on frictional sliding between grains or crack surfaces have been proposed by numerous investigators (Knopoff and MacDonald, 1958; White, 1966; Walsh, 1966; and Johnston *et al.*, 1979). In particular, three observations led to the proposal of attenuation mechanisms based on frictional dissipation: (1) The Q's of rocks are much lower than the Q's of the individual single crystals that make up the rocks. For example, the Q of a single crystal of quartz is greater than 10,000 (Gordon and Davis, 1968) while the Q of Sioux quartzite is only 100 to 200. The composition of Sioux quartzite is greater than 99% quartz and the porosity is less than 0.5%. The larger attenuation in the rock is attributed to the presence of grain boundaries and/or cracks. (2) Q increases with confining pressure (Birch and Bancroft, 1938; Gordon and Davis, 1968; Tittmann *et al.*, 1976; Toksöz *et al.*, 1979; Winkler and Nur, 1979). (3) Q in dry rock is independent of frequency over a wide frequency band (Knopoff, 1964; Johnston *et al.*, 1979). A causal, linear process cannot have both a constant Q and a constant velocity with frequency (Knopoff, 1964) although it can have constant Q over a certain domain of frequency (Richardson and Tittmann, 1980).

Friction is a nonlinear process because there is a threshold stress below which no motion occurs. The threshold stress is $\sigma_c = k_s N$, where k_s is the coefficient of friction and N is the normal stress across the surfaces in contact. Therefore, constant Q and constant velocity with frequency would be possible.

Mavko (1979) used a generalized crack geometry instead of the elliptical shape previously used by Walsh (1966). Mavko found that frictional dissipation is inherently amplitude dependent and Q^{-1} is directly proportional to strain amplitude. Q^{-1} of a wide variety of rock types was found to be independent of strain amplitude for strain amplitudes less than 10^{-7} to 10^{-6} (Gordon and Davis, 1968; Johnston, 1978; Winkler *et al.*, 1979). The lack of dependence of Q^{-1} on strain amplitude at low strain amplitudes led Mavko (1979) to conclude that frictional dissipation is negligible at strain amplitudes less than 10^{-6}. Winkler *et al.* (1979) observed that stress/strain loops have cusped ends at strains greater than 10^{-6}. This result demonstrates the presence of hysteresis; 10^{-6} is the same strain amplitude below which Q^{-1} becomes independent of strain amplitude. On the basis of these observations, Winkler *et al.* (1979) suggested that a nonlinear mechanism, friction, is dominant at high strain amplitudes, but that friction is negligible at low strain amplitudes (less than 10^{-6}).

This brief report presents, for the first time, data on the strain amplitude dependence of the shear wave dissipation Q_s^{-1} in the 400 Hz frequency range. These measurements complement previous data obtained by us and others at higher and lower frequencies and/or by the use of other modes of vibration. Also presented are data on the dependence of Q^{-1} on confining pressure, frequency, and the presence of porefluid. Attention is focused on interface bonding which is examined by measuring Q^{-1} in Berea and Boise sandstones. These sandstones were selected because they possess similar composition, porosity and permeability but differ greatly in the degree of competence. The behavior of Q^{-1} is also shown to differ and an explanation is offered in terms of bonding or cementation across grain interfaces. The question of bonding at interfaces is further examined on a theoretical basis to arrive at a general expression relating the attenuation of shear (SH) waves to a bonding parameter.

MEASUREMENTS AND RESULTS

The measurement techniques and representative results under a variety of different conditions have been described elsewhere (see Tittmann et al., 1981).

Description of samples

Experiments were performed on Berea and Boise sandstones. Both samples were cores 1.41 cm in diameter by 13.34 cm long cut parallel to bedding. Low frequency resonance was attained by end loading. The Boise sandstone is competent, with strong, silica-bonded grains, a high porosity ($\sim 22\%$) and a high permeability (933 md). The porosity of the Berea is similar to the Boise ($\sim 20\%$), and the permeability is also fairly high (293 md). However, grains are cemented primarily with clays (kaolinite) and therefore Berea is relatively weak and friable. Median grain size, and grain contact angle are comparable. Boise grains are more angular and contain more feldspars than Berea. Measurements were first made on a brass cylinder for the purpose of evaluating apparatus losses. Rock samples were oven-dried and encapsulated in an 8 mil thick copper tube. By comparing the measurements presented here with the H_2O vapor pressure calibration curves (Clark et al., 1980) we estimate an effective water vapor pressure of 1–2% for the environment of both samples. In the later experiments the clad rock samples were fully saturated by flowing distilled water through them and raising the porefluid pressure to 2000 psi to dissolve any remnant gas bubbles.

Strain amplitude dependence of Q^{-1}

The effect of strain amplitude on shear wave loss is presented in Fig. 1. The maximum strain amplitude is calculated for the parts of the sample which are strained the greatest when the acceleration on the vibrating bar is zero (maximum strain). In the shear mode (torsion) this corresponds to all points on the sample perimeter.

As can be seen in Fig. 1, Q_s^{-1} in Berea sandstone shows a strong dependence on strain above strains of 10^{-6}, in agreement with previous data at other frequencies and in other modes of vibration. In contrast, Boise shows much weaker dependence on strain. Increases in confining pressure translate the curves toward lower Q^{-1} values and shift the onset of strain amplitude dependence to higher strain values. This effect is very strong for Berea but weak for Boise. Similar results have been obtained by us for Q_E^{-1}, i.e., Young's modulus waves. Compared to the rock samples, the brass reference sample shows negligible strain amplitude dependence demonstrating that the observed non-linearities are not caused by instrumentation. Measurements on the brass reference bar at high pressures show that the effect of confining pressure on Q_s^{-1} by way of aerodynamic drag is negligible at these frequencies.

The dramatic differences between Berea and Boise suggest that Boise has stronger bonding at grain and crack interfaces compared with Berea and that much higher dynamic stresses are necessary for the Boise interfaces to engage in the energy absorbing frictional losses mechanisms. That is, the onset of the strain amplitude dependence is

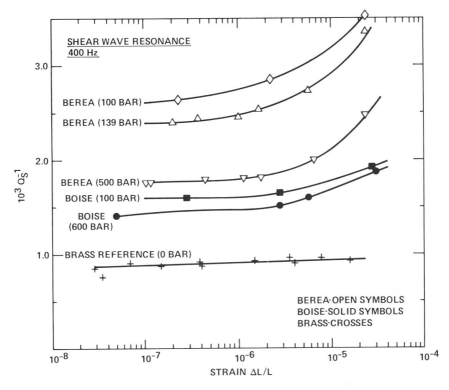

Fig. 1. Strain amplitude dependence of Q_s^{-1} for dry Berea and Boise sandstones.

shifted towards higher strains. The question of the detailed nature of the grain interface bonding was considered in extensive petrofabric and mechanical studies.

Petrofabric and mechanical studies

Table 1 gives a summary of results on the grain boundary relations as obtained from thin section photomicroscopy and from mechanical studies in uniaxial compression tests. These observations show that Berea and Boise sandstones are similar in grain size and grain boundary relations which indicate low compaction, insignificant pressure solution and deposition effects. On the other hand, they show substantially different fracture strengths. Boise has about 50% higher fracture strength than Berea implying a much stronger bonding between grains.

The difference between the grain boundaries of Berea and Boise sandstones are readily evident under the scanning electron microscope. SEM observations supplementing thin section and X-ray diffraction analysis show that the Berea sand grains are poorly cemented by clay minerals, mainly kaolinite, and by minor secondary growth of quartz, (Fig. 2a). In contrast, Boise sand grains are strongly bonded by authigenic microcystalline silica and zeolites which coat the grains and form interlocking reticulated networks and impart high rigidity to the rock frame (Fig. 2b).

Effect of porefluid on Q^{-1}

As shown previously by Winkler and Nur (1979) the strain amplitude dependence of Q^{-1} in Berea sandstone is influenced by the presence of pore fluids. As the degree of saturation was increased from dry, to partial, and to full saturation the onset of non-linearity was seen to move to lower strain amplitudes. Similar measurements by us on Boise sandstone do not show any significant change in the onset of amplitude

Table 1. Summary of petrofabric and mechanical properties.

Property	Berea	Boise
Compressive Strength (kg/cm^2)	427	614
Static Young's Modulus (10^{-10} dyne/cm^2)	2.97	3.30
No. of Grain Contracts per grain	3.6	2.4
Relative Contact Lengths (%)		
grain/grain	31	36
grain/pore	69	64
Type of Contact (%)		
Tangential	48	36.6
Longit.	43	56
Concave/Convex	9	7.3
Sutured	0	0
Grain Size Distrib.		
Median size (mm)	0.16	0.18
Sorting coefficient	1.18	1.59
Porosity (percent)	20	22
Permeability (md)	293	933

dependence. Again, the nature of the bonding between interfaces would explain the difference between Boise and Berea. An interesting question to ask is whether differences also appear in the behavior between Boise and Berea for strain amplitude well below the onset of any non-linearities.

Figure 3 compares the dependence of Q_E^{-1} on effective pressure for Berea and Boise under two conditions: dry and fully saturated. Except for the dry Boise, the curves were all obtained at about the same frequency (f = 7 kHz). The data for the dry Boise were acquired at about 500 Hz. Since extensive measurements on frequency dependence of Q^{-1} on a variety of dry sandstones by us and others (see, for example, Pandit and King, 1979) have revealed little or no frequency dependence, these data are considered approximately representative of behavior for dry Boise also at 7 kHz.

The main conclusion drawn from the results in Fig. 3 are that the changes incurred with porefluid saturation are much greater for Berea than for Boise. At first glance this is a surprising result because the losses associated with fluid flow are thought to depend primarily on porosity and permeability which are similar for Boise and Berea. However, in the mechanism proposed by O'Connell and Budianski (1977) dissipation results from the fluid flow between adjacent cracks. The flow is induced by the shear stress of the elastic waves and, therefore, the dissipation depends strongly on the crack aspect ratio. For Boise, the grain boundary cracks are strongly anchored so that their movement under the dynamic stress of the sound wave is inhibited. Fluid flow would be therefore also inhibited by comparison with Berea where grain boundary cracks are abundant and mobile. Thus, dissipation due to fluid flow would be expected to make a greater contribution to the loss in Berea than in Boise, which is in agreement with the observations.

Frequency dependence of Q^{-1}

Another manifestation of the effect of cementation between grains on Q^{-1} is shown in the frequency dependence of Q^{-1} in the saturated state. We have recently pointed to a dramatic frequency dependence of Q_E^{-1} in saturated Berea and in a similar sandstone Wingate (Tittmann et al., 1981). Q_E^{-1} was found to be substantially lower at 200 Hz than at 7 kHz, and this difference was found to decrease as the effective pressure was increased. Figure 4 shows the corresponding data for saturated Boise at 200 Hz and 7 kHz with

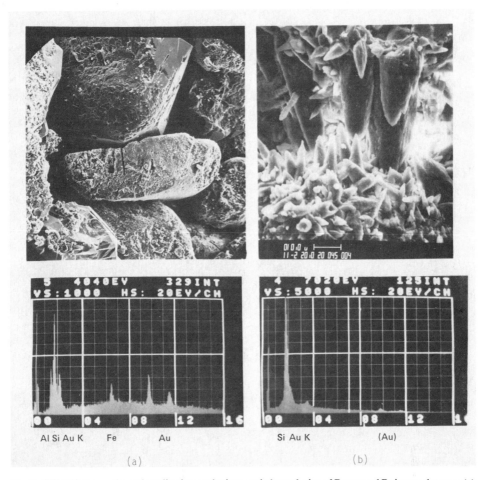

Fig. 2. SEM Photographs and qualitative analysis at grain boundaries of Berea and Boise sandstones. (a) Weakly cemented Berea grains (X240) and qualitative analysis of clay cement. (b) Reticulated networks of microcrystalline silica provide strong bonding of Boise sand grains (X1000) analysis of cementing material shows strong Si peak.

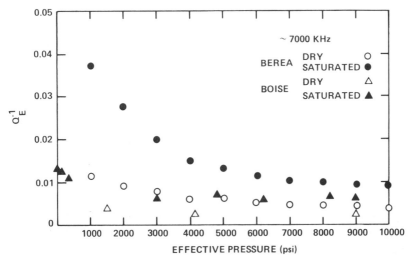

Fig. 3. Q_E^{-1} versus effective pressure for dry and saturated Berea and Boise sandstone at 7 kHz.

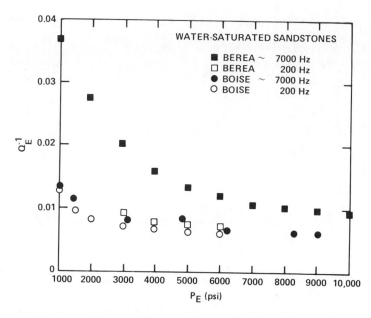

Fig. 4. Effective pressure dependence of Q_E^{-1} for saturated Boise and Berea sandstone at two frequencies.

substantially smaller differences over most of the effective pressure range. This result is in dramatic contrast with those for Berea and Wingate and suggests that because of the stronger bonding Boise acts as if already under high effective pressures. This is qualitatively confirmed by the observations which show comparable values of Q_E^{-1} for both Boise and Berea when the effective pressures are raised to about 10,000 psi.

In summary, observations are reported on the dependence of Q^{-1} on strain amplitude, confining pressure, frequency, and the presence of pore fluids. Dramatic differences are seen in these observations between two similar sandstones. These differences are explained on the basis of the nature of bonding between grains as perceived through SEM and X-ray diffraction studies of the cementation between grains. The results are complementary to those reported previously by us (Clark et al., 1980) on the effect of volatiles on Q^{-1} and velocity measured on a suite of sandstones, including Berea and Boise. The present results provide support for the mechanism proposed to explain the effect of volatiles on Q^{-1} (Tittmann et al., 1980) in terms of stress induced diffusion of water films absorbed on mobile grain boundaries and crack faces.

THEORETICAL MODEL

In general, it is assumed that no slip occurs on the interfaces until a critical value of shear stress is attained, and this critical value of stress is dependent on the normal loads across the interfaces. Unfortunately, this leads to a severely nonlinear boundary condition and unless accurate approximate techniques can be applied the problem is likely to remain in the realm of numerical tractability.

Recently, the studies of Caughey (1960), Iwan (1973), and Miller (1979) have opened the door to treating accurately and simply nonlinear boundary conditions by equivalent linearization techniques. These techniques consist in solving the problem with the aid of an "equivalent linear form" which is a superposition of an elastic and a viscous term at the boundary:

$$\tau = \tau^{(e)} + \tau^{(v)} \tag{1}$$

where

$$\tau^{(e)} = C[u]_-^+$$
$$\tau^{(v)} = K[\dot{u}]_-^+ \quad (2)$$

In Eqs. (1) and (2) τ is the shear traction, $[u]_-^+$ is the displacement jump and $[\dot{u}]_-^+$ is the velocity jump at the frictional boundary. The C and K are chosen to minimize the error between the exact nonlinear form and the "equivalent linear form." The minimization algorithm has been given generally by Iwan (1973).

However, it has been noted by Salvado' (1980) that one need not solve the problem using the complete "equivalent linear form" because there is a principle of correspondence that can be applied. The Laplace transform of Eq. (1) is given by

$$Y = [C + SK][|U|]_-^+ \quad (3)$$

where Y and U are respectively the Laplace transformed functions of τ and u, and S is the Laplace parameter. Therefore, if one has the Laplace transform of the solution of the viscous boundary condition $Y^{(v)}$ one can transform it to the solution that corresponds to the problem of the "equivalent linear form" Y by the following transformation:

$$K \rightarrow \frac{C}{S} + K \quad (4)$$

Salvado' (1980) has solved the problem of an incident SH pulse on a viscous interface separating identical halfspaces. He uses a parameterization of the viscous boundary condition in terms of bonding ϕ, $0 \leq \phi \leq 1$:

$$\tau_{YZ}(X, O, t) = \frac{\phi}{1-\phi} \frac{\mu}{\beta} [|\dot{u}_y(X, O, t)|]_-^+ \quad (5)$$

where μ is the rigidity and β is the shear wave velocity. The reflection and transmission coefficients are given by

$$R = \frac{(1-\phi) \sin \theta_S}{(1-\phi) \sin \theta_S + 2\phi}$$

$$T = \frac{2\phi}{(1-\phi) \sin \theta_S + 2\phi} \quad (6)$$

where θ_S is the angle of incidence measured from the horizontal.

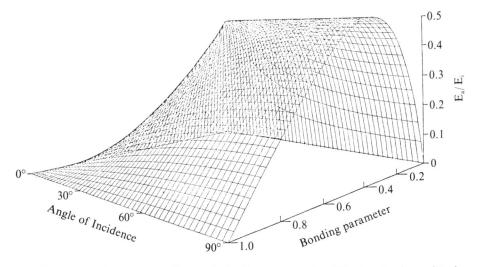

Fig. 5. The absorbed energy, normalized to the incident energy, of an SH pulse at a viscous interface as a function of bonding and angle of incidence.

Equations (5) and (6) show that, for the tractionless problem, $\phi = 0$, and reflection is perfect. For no interface, $\phi = 1$, and transmission is perfect.

The absorbed energy, E_a, normalized to the incident energy, E_i, is given by

$$2\pi Q^{-1} = \frac{E_a}{E_i} = 2\,RT. \tag{7}$$

As can be appreciated in Eq. (7) with the help of Eq. (6), for $\phi = 0$ we have $T = 0$, and for $\phi = 1$, we have $R = 0$ and therefore, no absorption occurs for these values of the bonding. This is seen in Fig. 3 where Eq. (7) is plotted as a function of bonding and angle of incidence. Maximum absorption is found to be at intermediate values of friction given by

$$\phi = \frac{\sin\theta_S}{\sin\theta_S + 2} \tag{7}$$

Derivations of corresponding expressions for Q^{-1} and ϕ have now been carried out for SV and P waves and will be published in a complete report at a later time. Currently, the calculations are being extended to the case of multiple random oriented cracks.

CONCLUSIONS

This report has presented a brief summary of experimental observations on the effect of interface bonding on dissipation Q^{-1}. Although the theory presented is for SH waves, similar results have been found for SV and P waves. Although the experiments described are for aggregate material, the theoretical model applies to rocks in general and interfaces, in general. Aside from shedding light on behavior of Q^{-1} in the laboratory, the results would seem to have also important consiquences in interpreting field data in the proximity of seismic sources and impact events where large scale cracking occurs.

Acknowledgments—This work was in part supported by a program on Rock Physical Properties sponsored by a consortium of oil companies.

REFERENCES

Birch F. and Bancroft D. (1938) The effect of pressure on the rigidity of rocks. *J. Geol.* **46**, 29–87.
Caughey T. K. (1960) Sinsoidal excitation of a system with bilinear hysteresis. *ASME J. Appl. Mech.* **44**, 652.
Clark V. A., Tittmann B. R., and Spencer T. W. (1980) Effect of volatiles on attenuation (Q^{-1}) and velocity in sedimentary rocks. *J. Geophys. Res.* **85**, 5190–5198.
Gordon R. B. and Davis L. A. (1968) Velocity and attenuation of seismic waves in imperfectly elastic rock. *J. Geophys. Res.* **73**, 3917–3935.
Iwan W. D. (1973) A generalization of the concept of equivalent linearization. *Int. J. Nonlinear Mech.* **8**, 279.
Johnston D. H., Toksöz M. N., and Timur A. (1979) Attenuation of seismic waves in dry and saturated rocks: II Mechanisms. *Geophys.* **44**, 691–711.
Johnston D. H. (1978) The Attenuation of Seismic Waves in Dry and Saturated Rocks. Ph.D. Dissertation, Massachusetts Institute of Technology, Cambridge, Mass.
Knopoff L. and MacDonald G. J. F. (1958) Attenuation of small amplitude stress waves in solids. *Rev. Mod. Phys.* **30**, 1178–1192.
Knopoff L. (1964) *Q. Rev. Geophys. Space Phys.* **2**, 625–660.
Mavko G. M. (1979) Frictional attenuation: an inherent amplitude dependence. *J. Geophys. Res.* **84**, 4769–4776.
Miller R. (1979) An estimate of the properties of Lone-type surface waves in a frictionally bonded layer. *Bull. Seismol. Soc. Amer.* **69**, 305.
O'Connell R. J. and Budianski B. (1977) Viscoelastic properties of fluid saturated cracked solids. *J. Geophys. Res.* **82**, 5719–5736.
Pandit B. I. and King M. S. (1979) The variations of elastic wave velocities and quality factor Q of a sandstone with moisture content. *Can. J. Earth Sci.* **16**, 2187–2195.
Richardson J. M. and Tittmann B. R. (1980) Phenomenological theory of attenuation and propagation velocity in rocks. *Proc. Lunar Planet. Sci. Conf. 11th*, p. 1837–1846.

Salvado' C. (1980) Scattering from a viscous plane. *Earthquake Notes* **50**, 24.
Tittmann B. R., Nadler H, Clark V. A., Ahlberg L. A., and Spencer T. W. (1981) Frequency dependence of seismic dissipation in saturated rocks. *Geophys. Res. Lett.* **8**, 36–38.
Tittmann B. R., Clark V. A., Richardson J. M., and Spencer T. W. (1980) Possible mechanism for seismic attenuation in rocks containing small amounts of volatiles. *J. Geophys. Res.* **85**, 5199–5208.
Tittmann B. R., Ahlberg L., and Curnow J. (1976) Internal friction and velocity measurements. *Proc. Lunar Sci. Conf. 7th*, p. 3123–3133.
Toksoz M. N., Johnston D. H., and Timur A. (1979) Attenuation of seismic waves in dry and saturated rocks: I. Laboratory Measurements. *Geophys.* **44**, 681–690.
Walsh J. B. (1966) Seismic wave attenuation in rock due to friction. *J. Geophys. Res.* **71**, 2591–2599.
White J. E. (1966) Static friction as a source of attenuation. *Geophys.* **31**, 333–339.
Winkler K. and Nur A. (1979) Pore fluids and seismic attenuation in rocks. *Geophys. Res. Lett.* **6**, 1–4.
Winkler K., Nur A., and Gladwin M. (1979) Friction and Seismic attenuation in rocks. *Nature* **277**, 528–531.

On the estimation of lunar paleointensities—Studies of synthetic analogues of stably magnetized samples

J. R. Dunn[1], M. Fuller[1], and D. A. Clauter[2]

[1]Department of Geological Sciences, University of California, Santa Barbara, California 93106, and
[2]Department of Earth Sciences, New Mexico State University, Las Cruces, New Mexico 88003

Abstract—Certain stably magnetized lunar samples give promise for the determination of ancient lunar field intensities, but their behaviour upon heating is anomalous, so that standard intensity methods yield unreliable results. We have prepared synthetic analogues of the carriers of natural remanent magnetization (NRM) in the stably magnetized samples, so that we can examine their behaviour during intensity determinations. Two aspects of the samples appear to explain much of the anomalous behaviour during intensity determinations. Interactions in multidomain material can yield an anomalously high field value. Magnetic viscosity at high temperature can yield anomalously low field estimates. A continuous version of the intensity determination, in which observations are made at high temperature, gave some improvement in the estimates.

1.0 INTRODUCTION

The principal quantity sought in paleomagnetic studies of lunar and meteoritic samples is the intensity of the magnetic field in which they were formed. The accurate determination of the intensity of such fields is of considerable interest as an aspect of the early history of the solar system. So far, it has not proved possible to obtain convincing estimates of these intensities. Part of the difficulty arises from the magnetic characteristics of the samples. For example, some of the lunar samples have unstable natural remanent magnetization (NRM), which is not suitable for paleointensity determinations. There are, however, certain samples whose magnetization is stable against alternating field (AF) demagnetization and thus gives some promise for paleointensity work. However, these samples tend to exhibit anomalous magnetic behaviour upon heating, so that standard intensity determinations such as the Koenigsberger-Thellier-Thellier (KTT) method (Koenigsberger, 1938; Thellier and Thellier, 1959) cannot be used.

In this paper, we describe the preparation and the study of the magnetic behaviour of synthetic analogues of the carriers of NRM in stably magnetized lunar samples. The relevant magnetic phase is iron dispersed in troilite. The synthetic samples were used to see how well the standard paleointensity method recovered known fields in which the samples had been given thermoremanent magnetization (TRM). Results of the KTT intensity determinations for these synthetic samples were anomalous. Moreover, the data they yielded were similar to those obtained from certain meteorite and lunar samples, which had been considered evidence of high magnetic fields in the early solar system. However, it now appears that at least some of the results are artifacts of the method used.

2.0 CHARACTERISTICS OF STABLY MAGNETIZED MARE BASALTS

10017, 10020, 10049, and 12022 are examples of stably magnetized basalts. They are stably magnetized in the sense that the direction of NRM does not change markedly upon demagnetization in fields of a few hundred oersted and the magnitude of remanence decreases in a systematic manner with a median destructive field (MDF) of more than 100 oe (e.g., Fuller *et al.*, 1979).

2.1 Magnetic mineralogy

Reflected light microscopy has shown that a number of the stably magnetized samples, including all those listed above, contain native iron as blebs in troilite (Haggerty et al., 1970). The samples are in general the finer grain size mare basalts and the iron and troilite occur in mesostasis. In some samples, glass occurs. However, the presence of large amounts of glass gives rise to magnetic viscosity due to the superparamagnetic iron in the glass. The occurrence of iron in troilite has been reported by Pearce et al. (1976) in other magnetically stable lunar rocks. They also noted its occurence in a synthetic analogue of 10017 prepared by R. M. Housley.

2.2 Magnetic behaviour

The analysis of the NRM of the stably magnetized mare basalts has been reported by Runcorn et al. (1971), Sugiara et al. (1978), Hoffman et al. (1978) and Fuller et al. (1979). The stability against AF demagnetization defines the group. Moreover, different subsamples from a single sample have similar directions of stable magnetization. The intensities tend, however, to vary considerably. Thermal demagnetization brings about a major reduction in NRM at temperature between 200 and 300°C.

The remanence of the samples is dominantly carried by metallic iron. The grain size of the iron varies within individual samples, so that there is magnetically hard and soft material in a single sample. The ratio of saturation remanence to saturation magnetization (J_{rs}/J_s) is generally close to 0.01, revealing that the material is not single domain. Remanent coercivities of as high as several hundred oersted have been observed, but there is much variability in both remanent coercivity (H_{rc}) and coercive force (H_c).

A particularly important series of experiments, carried out by Pearce et al. (1976), showed that stably magnetized highland samples, containing metallic iron in troilite, had anomalous pTRM acquisition. There were peaks in the cumulative curves between 200 and 300°C and close to 600 and 700°C. The lower temperature peak was interpreted as a negative interaction between the remanence of the iron and a defect ferromagnetism in the antiferromagnetic troilite. The higher peaks were related to the breakdown of troilite and of ilmenite to give additional iron. The lower temperature peak was also reported by Hoffman et al. (1978) and by Fuller et al. (1979) in stably magnetized basalts. It was demonstrated that there was no substantial increase in saturation magnetization upon heating to 850°C, so that the high temperature increases in remanent magnetization could not have been simply due to the production of additional iron.

The most puzzling feature of the magnetic behaviour of these samples is their response to heating. Both thermal demagnetization and the acquisition to pTRM are anomalous in ways yet to be interpreted. As noted in the introduction, this behaviour precludes satisfactory intensity determination by the classical methods.

3.0 SYNTHETIC ANALOGUES OF MAGNETIC CARRIERS IN STABLY MAGNETIZED MARE BASALTS

3.1 Preparation of samples

Synthetic analogues of the iron bearing troilite were prepared by heating mixtures of iron and sulphur for 24 hours in evacuated silica tubes at 850°C. The tubes containing the mixtures were enclosed in a second tube containing a Titanium getter (Taylor, 1979). The iron was electrolytic and had been previously heated 850°C for 5 hours in hydrogen to reduce oxide coatings. The samples were analyzed by electron microprobe, reflection microscopy, X-ray and thermomagnetic methods, which demonstrated that the product was indeed troilite with varying amounts of metallic iron in it. Some pyrite was also detected, but no magnetite, nor pyrrhotite. The presence of the pyrite is puzzling, but we ascribe it to local excess sulfur in the charge. Five samples were produced. Sample 1 had an initial Fe:S ratio of 1:1. Samples 2 and 3 had an excess of metallic iron to sulphur of 1.1:1, and Samples 4 and 5 of 1.2:1.

3.2 Magnetic behaviour of synthetics

The hysteresis properties of the synthetics were similar to those of the lunar samples, although none were as magnetically hard. The J_{rs}/J_s ratio was 0.14 and H_{rc}/H_c was equal to 1.8. The remaining samples clearly had multidomain characteristics, although the remanent coercivity was remarkably high.

The TRM which the samples acquired in their initial cooling after preparation was subjected to AF demagnetization. The demagnetization curves were consistent with the characteristics of the samples established by the hysteresis measurements. Sample 1 exhibited a typical curve for fine grain iron, with very little demagnetization in the weakest fields (Fig. 1a). The remainder of the samples were typically multidomain with substantial loss of remanence in the weakest fields, e.g., Sample 2 (Fig. 1b). AF demagnetization of saturation isothermal remanence (IRM_s) was compared with that of the TRM. In Sample 1, the weak field TRM was clearly harder than IRM_s. In contrast, in the other samples the IRM_s was marginally hardest. This is consistent with the TRM of all but Sample 1 being dominated by multidomain components.

Thermal demagnetization of TRM and saturation IRM was carried out using the stepwise technique. Sample 1 behaved normally; both IRM_s and TRM demagnetized monotonically although both unblocked mainly between 600 and 650°C, well below the Curie point of iron (Fig. 2a). The saturation IRM of the other samples also behaved normally. However, the demagnetization of weak field TRM in Sample 2 gave rise to a marked increase in magnetization between 550 and 700°C (Fig. 2b).

The acquisition of pTRM by Sample 1 was anomalously large between 200 and 300°C, but was otherwise normal for fine grain iron. In contrast, Sample 2 exhibited a peak in the cumulative curves between 500 and 700°C (Figs. 2a and b).

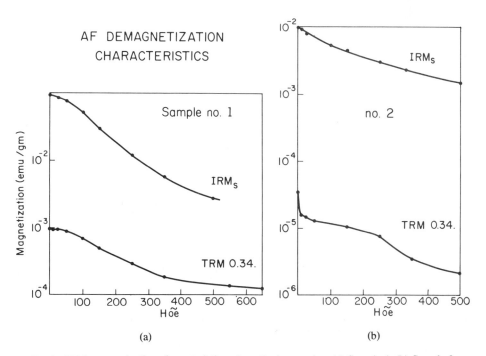

Fig. 1. AF demagnetization characteristics of synthetic samples. (a) Sample 1. (b) Sample 2.

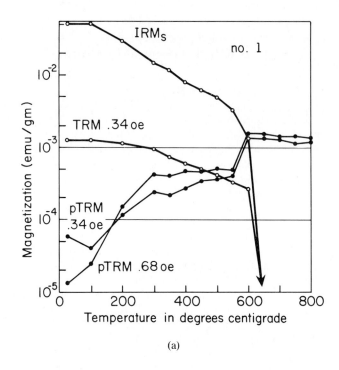

(a)

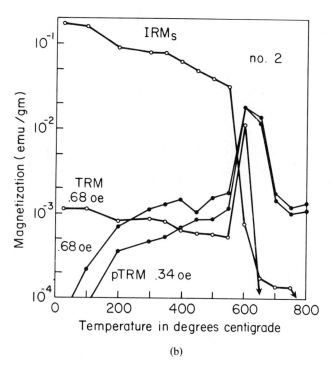

(b)

Fig. 2. Thermal demagnetization characteristics and partial thermoremanent magnetization acquisition. (a) Sample 1. (b) Sample 2.

The acquisition of total TRM by these samples in fields of 0.034, 0.1, 0.2, 0.34, 0.68 and 0.95 oersteds does not depart importantly from linearity.

KTT intensity determinations were made after the samples had been given a TRM in a 0.5 oe field. The determinations were carried out using the standard double heating method in fields of the same magnitude and opposite sign. The results are shown for Samples 1 and 2 in Figs. 3a and 3b.

Sample 1 gives an estimate for the field which is considerably larger than the laboratory field in which TRM was actually given to the sample. The discrepancy is particularly evident in the low temperature range. Note also that in this temperature range chemical effects are not likely to be important; the sample was initially heated to far higher temperature and slowly cooled. The plot is also anomalous in that it has a V-shaped departure from the linear trend representing a peak in pTRM acquisition at about 600°C. This result is quite similar to many lunar paleointensity determinations (e.g., Pearce et al., 1976).

Sample 2 exhibits very unusual behaviour. However, a field intensity estimate based on the total TRM is not as anomalous, as are those based on estimates for intermediate temperatures.

The most anomalous behaviour of the synthetic samples lies in their thermal demagnetization and in the acquisition of pTRM. It is therefore hardly surprising that the KTT method, which relies on a comparison between thermal demagnetization and pTRM acquisition, fails. The total TRM was linearly dependent upon the inducing field, however. Unfortunately, we cannot use this property to estimate the fields in which the NRM of the lunar samples was acquired because chemical alteration of the samples during heating must be checked. It is therefore important to see if the KTT experiment can be modified so that it can be used with the lunar samples. Before this is likely to be achieved, we must understand why it fails.

4.0 ANALYSIS OF KOENIGSBERGER-THELLIER-THELLIER INTENSITY METHOD

The KTT method assumes that magnetization is to be associated with temperature in a discrete manner; heating to a particular temperature demagnetizes all magnetization acquired (or blocked) below that temperature and does not affect magnetization acquired (or blocked) above that temperature. Moreover, it is assumed that the blocked and unblocked magnetizations do not interact; each magnetization increment acts independently. The NRM demagnetization in a particular temperature interval can then be compared with that acquired in the same interval by the standard KTT method, which is illustrated in Fig. 4. In this figure, the measurements of magnetization are shown on the ordinate, which represents room temperature observations. The magnetization demagnetized and that acquired as pTRM in the temperature range are obtained by the familiar sum and difference method; i.e.,

$$J(+) = J_{T_2}^{T_1}(H_L) + J_{T_c}^{T_2}(H_u),$$

$$J(-) = -J_{T_2}^{T_1}(H_L) + J_{T_c}^{T_2}(H_u),$$

$$\Sigma = 2J_{T_c}^{T_2}(H_u),$$

$$\Delta = 2J_{T_2}^{T_1}(H_L).$$

$J(+)$ and $J(-)$ are the magnetizations after the positive and negative heating cycles, assuming that the pTRM is given parallel and antiparallel to the NRM direction. T_c is the Curie point, T_2 the maximum temperature in the cycle and T_1 room temperature. H_u is the laboratory field in which "NRM" (TRM) was initially given to the samples and H_L the field used in the KTT experiment. We obtain $J_{T_c}^{T_2}(H_u)$ and the "NRM" demagnetized is

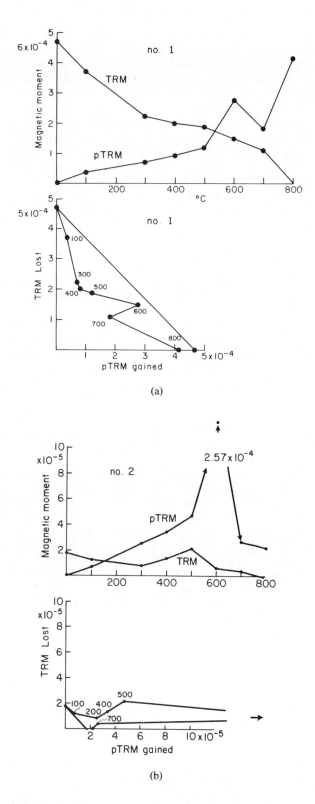

Fig. 3. Koenigsberger-Thellier-Thellier (KTT) intensity determination. (a) Sample 1. (b) Sample 2.

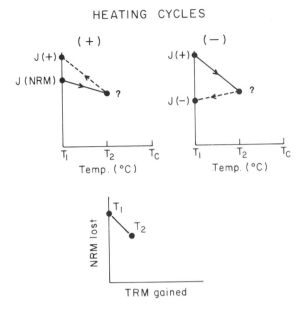

Fig. 4. Principle of KTT intensity determination.

simply

$$J_{T_2}^{T_1}(H_u) = NRM - J_{T_C}^{T_2}(H_u).$$

Thus we can compare the magnetization $J_{T_2}^{T_1}(H_L)$ acquired in the temperature cycle with the "NRM" demagnetized. Note that this is achieved without any observation at high temperature, so that the magnetization at T_2 is obtained correctly only if the sum and difference technique is justified. The results are generally plotted in terms of pNRM lost against pTRM gained (Fig. 4).

In the hands of the Thelliers the method was exploited to determine field intensities recorded by the NRM of fired pottery and other archeological artifacts which happen, for the most part, to have single domain carriers that are non-interacting. When the method is extended to rocks, the results are frequently less convincing. For example, with increasingly multidomain samples, the method becomes less and less reliable (Levi, 1977). Actually, the non-ideal behavior studied by Levi was for the pseudosingle domain grain size. The departures were less than those seen in Sample 1 (Fig. 3a), but in the same sense. In multidomain material the KTT behaviour has been reported to be chaotic (Levi 1977). Sample 2, which is more multidomain than Sample 1, departed from ideal behaviour in the opposite sense from Sample 1 and too a much greater degree (Fig. 3b). We now consider possible explanations of the non-ideal behaviour.

One of the major distinctions between single domain and multidomain magnetic materials is the difference in importance of interactions. In samples containing low concentrations of well dispersed single domain particles, each particle sees the applied field unaffected by the presence of the other magnetic material in the sample. In contrast, in a multidomain grain, the magnetization of one part of a grain is inevitably affected by the remainder of the particle. Let us now consider the behaviour during a heating cycle of a multidomain particle.

The application of a field (H_{app}) to a particle induces a magnetization. This gives rise to a demagnetizing field H_D, which can be visualized by drawing lines of force from the positive poles on one surface to the negative poles on the oppositive surface. The demagnetizing field is dependent upon sample geometry and the magnitude of magnetization. Thus the internal field (H_I) of the particle is

$$H_I = H_{app} - H_D.$$

Recently, the discussion of the nature of this demagnetizing field in multidomain grains has been reopened by Merrill (1981), who rightly pointed out that it is more complicated than the commonly used product of the magnetization and demagnetizing coefficient

$$H_D = NJ.$$

Rather, its dependence upon the actual domain configuration must be recognized. The question has also been reviewed by Halgedahl (1981), who argues that the approach of Kooy and Enz (1960) gives an adequate representation of the demagnetization effect. For our purposes, it is not important what the absolute magnitude of this quantity is, providing that it is a quantitatively important aspect of the magnetization process. We call this demagnetizing field $H_{D(app)}$.

When a sample is carrying remanence, the remanence also gives rise to a demagnetizing field. Thus at T_2 the remanence which remains blocked generates a demagnetizing field within the sample. The demagnetizing field is dependent upon the particle geometry, the domain configuration and the blocked remanent magnetization. We call this $H_D\ [J_{T_C}^{T_2}(H_u)]$.

We now assume that all magnetization processes involved are linear, so we can superpose either fields or the magnetizations they generate. To simplify matters further, we assume that the magnetization acquired in the temperature range T_2 to T_1 does not change the demagnetizing field importantly during the blocking process. Then we can write as an equivalent to the standard KTT expressions:

$$J(+) = J_{T_2}^{T_1}(H_L) - J_{T_2}^{T_1}(H_{D(app)}) - J_{T_2}^{T_1}(H_{D[J_{T_c}^{T_2}(H_u)]}) + J_{T_c}^{T_2}(H_u),$$

$$J(-) = -J_{T_2}^{T_1}(H_L) + J_{T_2}^{T_1}(H_{D(app)}) - J_{T_2}^{T_1}(H_{D[J_{T_c}^{T_2}(H_u)]}) + J_{T_c}^{T_1}(H_c),$$

$$\Sigma = 2J_{T_c}^{T_2}(H_u) - 2J_{T_2}^{T_1}(H_{D[J_{T_c}^{T_2}(H_u)]}),$$

$$\Delta = 2J_{T_2}^{T_1}(H_{app}) - 2J_{T_2}^{T_1}(H_{D(app)}).$$

The sum gives the "NRM", which has remained blocked minus magnetization aquired between T_2 and T_1 due to the demagnetizing field of this blocked "NRM". The difference gives the magnetization acquired field minus magnetization due to the demagnetizing effect of the induced moment. This is equivalent to regarding the internal field as the inducing field. Note that, because the sign of the demagnetizing field due to the unblocked magnetization does not change sign with the applied field, a fraction of the magnetization blocked in the temperature range $T_2 - T_1$ does not appear in the difference term. Hence, the difference is smaller than it would be if there were no interactions between the blocked and unblocked magnetization. The result is that the TRM blocked in the temperature range $T_2 - T_1$ is underestimated. The effect is to give a departure from ideal behaviour with values following below the ideal line—sometimes referred to as a sag-down result. This was observed by Levi (1977) and by us (Fig. 3a).

The departure from ideal behaviour exhibited by Sample 2 is indeed spectacular; it is virtually impossible to plot the result without using logarithmic graph paper to show the ideal line of pNRM lost against pTRM gained and the observed values (Fig. 3b). The form of the anomalous departure is that the pTRM acquired in each temperature increment is too large. Moreover, as temperature increases, the anomalous departure increases passing through a maximum at 600°C. This sample exhibits marked magnetic viscosity. There therefore appears to be a very simple explanation of the observed behaviour in terms of the role of a high temperature viscosity (VRM), or isothermal remanence (IRM). Such remanence invalidates the sum and difference method of separating blocked remanence from magnetization acquired in the thermal cycle. The distinction between VRM and IRM is probably not critical; either can cause the observed non-ideal behaviour. Writing this high temperature viscosity or remanence as J (H_L), the standard KTT expression becomes

$$J(+) = J_{T_2}^{T_1}(H_L) + J(H_L) + J_{T_c}^{T_2}(H_u),$$

$$J(-) = -J_{T_2}^{T_1}(H_L) - J(H_L) + J_{T_c}^{T_2}(H_u),$$

$$\sum = 2J_{T_c}^{T_2}(H_u),$$

$$\Delta = 2J_{T_2}^{T_1}(H_L) + 2J(H_L).$$

Thus the difference term now includes the remanence acquired at T_2. This magnetization contributes to the difference term because it follows the field. Yet, it should not be considered a part of the pTRM acquired between T_2 and T_1 because it is acquired at T_2 and not blocked in the normal TRM manner. It therefore appears as a false contribution to the pTRM gained and can account for the observed anomalously large values of pTRM observed. Clearly this explanation of the behaviour of Sample 2 assumes that the effect is much larger than the interaction effect described in explaining the behaviour of Sample 1. This appears reasonable in the light of the relative magnitude of the departures from ideal behaviour by the two samples.

A second effect of the high temperature viscous, or isothermal, magnetization may explain V-shaped departures from ideal behaviour in pNRM vs. pTRM plots, such as that seen in Fig. 3a and in the results of Pearce et al. (1976). The interpretation is that, at the temperature represented by the peak, the sample is particularly susceptible to the acquisition of viscous, or isothermal, remanence. This magnetization is then blocked as a contribution to the pTRM acquired as described in the explanation of the behaviour of Sample 2. Thus the V-shaped anomalies reflect enhanced viscosity, or IRM acquisition, at these temperatures.

The ideas presented in this section appear to afford relatively simple explanations of the two departures from ideal behaviour which have been observed. It remains to test their relevance. The proposed role of interactions has the merit of also explaining behaviour of nickel, when it is cooled in zero field, from high behaviour of nickel, when it is cooled in zero field, from high temperature, carrying various magnitudes of blocked remanence. Zero field cooling, when the sample carries a weak field pTRM blocked in the high temperature range, brings about a small increase in remanence consistent with the increase of saturation magnetization. However, if a strong field pTRM is carried, zero field cooling brings about strong demagnetization (Fuller, unpublished data). Clearly, the change in magnetization as the sample cools is in part determined by the demagnetizing field of the sample. The idea will be tested further, but the effect seems inevitable in multidomain material. The proposed role of the high temperature VRM or IRM in explaining the behaviour of Sample 2 is simply to recognize a well known problem of the KTT method. It too must be tested, however. Some preliminary tests have emerged from the high temperature observations to be described next.

5.0 PRELIMINARY OBSERVATIONS AT ELEVATED TEMPERATURE—A CONTINUOUS VERSION OF THE KTT EXPERIMENT

One of the experimentally attractive features of the KTT method is that all measurements of remanence are made at room temperature. The price which is paid is that one uses an indirect measurement of the blocked remanence at high temperature. As we have noted in the previous section, it is very likely that the assumptions required by this indirect determination of the blocked remanence are not justified in multidomain material. It is therefore natural to attempt to develop a version of the KTT experiment in which the high temperature measurement is made directly. Several years ago we attempted such an experiment (Day *et al.*, 1977). The technique used was to heat the sample inside a cryogenic magnetometer with a field trapped in the superconducting shield of the magnetometer. By reversing the sample, one can then carry out the equivalent of the KTT experiment in a continuous manner. The experiment failed because the measurement of remanent magnetization at high temperature, in the presence of a field, was not successful; it was not practicable to reset the field in the shield to zero zones for each measurement. We therefore redesigned the experiment so that the sample was heated outside of the magnetometer and moved into field free space for the measurement. The sample was heated with a 20 watt Korad laser. It can be moved into the magnetometer for measurements with its temperature maintained. Preliminary experiments with this technique have demonstrated two relevant results.

The possibility of a successful intensity determination for Sample 1, although it fails to give a satisfactory result using the classical KTT result, is demonstrated by the following experiment. The sample is first given a TRM over a particular temperature range, with acquisition monitored by measurements at temperatures within the interval. The sample is next thermally demagnetized, which can be monitored continuously. The sample is then given the same intensity of TRM over the temperature interval and the acquisition process again monitored. As Fig. 5 illustrates, the second acquisition curve is parallel to the first. Thus, the increment of magnetization acquired in each temperature interval is indeed the same. This magnetization can now be thermally demagnetized continuously. Although the thermal demagnetization and acquisition curves are somewhat different, the two acquisition curves have the same slope. To make a practical intensity method from these observations, one compares the two thermal demagnetization curves. Ideally, one would then change the field used in the laboratory intensity experiment until the thermal demagnetization curves had identical slopes over the lowest temperature increment studied. One would then proceed to the next temperature increment and repeat the process with the same field. Consistent results in successive temperature increments

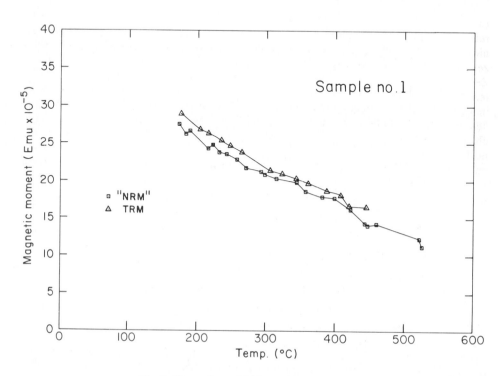

Fig. 5. Thermal acquisition curves for Sample 1.

would then provide a strong argument that the NRM was indeed a TRM and that no important degradation of the sample had taken place during the experiment. The essential difference between this method and the classical KTT experiment is that by measuring at high temperature, one eliminates the need for the sum and difference technique. This then permits one to carry out the entire experiment with pTRM acquired under identical conditions to those in which the initial NRM was required.

The capability to observe remanent magnetization at high temperature permits one to measure the high temperature viscosity of Sample 2. In this way, one can test the suggestion that it is this viscosity which causes the anomalous behaviour in the standard KTT experiment. At 500°C the magnetic viscosity of Sample 2 has increased in comparison with its room temperature value, such that it decays to half its initial value with seven minutes. Sample 1 did not show any such strong viscosity. This suggests that viscosity is indeed playing a role in the behaviour of Sample 2.

The high temperature results give one confidence in the interpretations of the anomalous results of the KTT intensity determinations for the "NRM" in the synthetic samples. They suggest that meteorite and lunar samples similar to Sample 1 may yield good intensity determinations, if the high temperature observations are made. It seems unlikely that samples as viscous as Sample 2 will yield good results using the high temperature method. Indeed, it is not clear what such viscous samples can tell us about ancient fields.

6.0 DISCUSSION

The field intensity determinations with the synthetic samples led to possible explanations of two types of anomalous KTT results. The role of interactions in multidomain material appears to account for the anomalous results in which the values fall below the ideal line in plots of pNRM vs. pTRM (the sag-down result). The V-shaped departure from linear trends in the same plots appear to be due to viscosity or isothermal remanence acquisition at the maximum temperature in the cycle.

The "sag-down" results have already been interpreted by Levi (1977) in terms of the Schmidt (1973) model of TRM in a two domain particle. This interpretation recognizes the importance of demagnetizing energy. The interpretation offered here has broad generality and is not dependent on a particular model of TRM. Another effect which may be important is thermal hysteresis, demonstrated most recently by Bol'shakov and Shcherbakova (1979). However, the high temperature observations made with Sample 1 suggest that at least in that sample it is not an important effect.

The "sag-down" results are commonly seen in intensity determinations of lunar and meteorite samples. Our work demonstrates that such results can arise, if interactions are important, even when the field in which the "NRM" was acquired is identical to that in which the KTT experiment is carried out. Before such results are cited as evidence of magnetic fields of high intensity, a second experiment should be carried out in which the inferred high field is used in the KTT experiment. If the result still "sags-down", it should probably be discarded, or, at least, treated with the utmost suspicion. If, on the other hand, the sample shows ideal behaviour with the appropriate one to one relationship between pNRM and pTRM in the KTT experiment using the high field, it can be trusted.

Results such as those exhibited by Sample 2 have not been reported in the literature, but there have been many of the V-shaped departures from linear trends. These are likely to be due to high temperature viscosity or isothermal remanence, so that the results at the temperature of the peak should be discounted. The remainder of the linear trend may be of value.

The preliminary results of the continuous version of the KTT experiment suggests that the technique may eliminate "sag-down" results. This may permit many samples, which at present are considered unsuitable for intensity work, to yield useful data. The key improvement is that the pTRM is acquired by the sample under the same condition as was the "NRM".

Despite efforts by many people, the intensity determinations of lunar and meteoritic samples remain equivocable. We hope that the approach developed in this paper will eventually bring us closer to reliable intensity determination for extraterrestrial material.

REFERENCES

Bol'shakov A. S. and Shcherbakova V. V. (1979) Thermomagnetic criterion for determining the domain structure of ferrimagnets. *Izv. Earth Phys.* **15**, 111–116.

Day R., Dunn J. R., and Fuller M. (1977) Intensity determination by continuous thermal cycling. *Phys. Earth Planet. Inter.* **13**, 301–304.

Fuller M., Meshkov E., Cisowski S. M., and Hale C. J. (1979) On the natural remanent magnetism of certain mare basalts. *Proc. Lunar Planet. Sci. Conf. 10th*, p. 2211–2233.

Haggerty S. E., Boyd F. R., Bell P. M., Finger L. W., and Bryan W. B. (1970) Opaque minerals and olivine in lunar basalts and breccias from Mare Tranquillitatis. *Proc. Apollo 11 Lunar Sci. Conf.*, p. 513–538.

Haledahl S. (1981) The dependence of domain structure upon magnetization state. PhD. Thesis, Univ. Calif., Santa Barbara. 208 pp.

Hoffman K. A., Baker J. R., and Banerjee S. K. (1978) Combining paleointensity methods; a dual valued determination on lunar sample 10017.135 (abstract). In *Papers Presented to the Conference Origins of Planetary Magnetism*, p. 45–47. Lunar and Planetary Institute, Houston.

Koenigsberger J. G. (1938) Natural residual magnetism of eruptive rocks, Parts I and II. *Terr. Mag. Atmos. Phys.* **43**, 119–127.

Kooy C. and Enz U. (1960) Experimental and theoretical studies of the domain configuration in thin layers of $Ba_{12}O_{19}$. *Philipps Res. Rept.* **15**, 7–29. Eindhoven, Netherlands.

Levi S. (1977) The effect of magnetic particle size on paleointensity determinations of the geomagnetic field. *Phys. Earth Planet. Inter.* **13**, 245–259.

Merrill R. (1981) Towards a better theory of thermal remanent magnetization. *J. Geophys. Res.* **86**, 937–949.

Pearce G. W., Hoye G. S., Strangway D. W., Walker B. M., and Taylor L. A. (1976) Some complexities in the determination of lunar paleointensities. *Proc. Lunar Sci. Conf. 7th*, p. 3271–3297.

Runcorn S. K., Collinson D. W., O'Reilly W., Stephenson A., Battey M. H., Manson A. J., and Readman P. W. (1971) Magnetic properties of Apollo 12 lunar samples. *Proc. Roy. Soc. London* **A325**, p. 157–174.

Schmidt V. A. (1973) A multidomain model of thermoremanence. *Earth Planet. Sci. Lett.* **20**, 440–446.

Sugiura N., Strangway D. W., and Pearce G. W. (1978) Heating experiments and paleointensity determinations. *Proc. Lunar Planet. Sci. Conf. 9th*, p. 3151–3163.

Taylor L. A. (1979) An effective sample preparation technique for paleointensity determinations at elevated temperature (abstract). In *Lunar and Planetary Science X*, p. 1209–1211. Lunar and Planetary Institute, Houston.

Thellier E. and Thellier O. (1959) Sur l'intensite du champ magnetique terrestre dans le passe historique et geologique. *Ann. Geophys.* **15**, 285–376.

Flow behavior of ten iron-containing silicate compositions

Lisa C. Klein, Benjamin V. Fasano, and Jenn Ming Wu

Department of Ceramics, Rutgers University, P.O. Box 909, Piscataway, New Jersey 08854

Abstract—The viscosity of ten iron-containing silicates has been measured over the temperature range 400° to 650°C. The compositions contain trace amounts up to 30 wt. pct. or 20 mole. pct. Fe_3O_4, from 10 to 20 wt. pct. or 10 to 20 mole. pct. Na_2O and the remainder is SiO_2. Samples were melted in air to form oxidized samples, in forming gas (95% N_2–5% H_2) to form mildly reduced samples, and in forming gas with carbon in the batch to form strongly reduced samples. In general, the viscosity decreased with increasing mole. pct. Na_2O and decreasing mole. pct. SiO_2 for a given temperature. In samples with increasing mole. pct. FeO_x the difference in the viscosity between the oxidized sample and the reduced sample grew in magnitude at a given temperature. In all of the compositions, the log (viscosity) vs 1/T relation shows curvature with an apparent activation energy between 90 and 100 kcal/mole.

INTRODUCTION

The viscous flow behavior of iron-containing silicates is one of the most important aspects for study in extraterrestrial materials. An evaluation of the viscosity over a wide range of temperatures and compositions is essential for understanding the physical behavior of magmas. Experimental measurements of viscosities have been performed on lunar compositions (Uhlmann and Klein, 1976), chondrule compositions (Klein *et al.*, 1980a), a tektite composition (Klein *et al.*, 1980b), plagioclases (Cukierman and Uhlmann, 1973; Uhlmann *et al.*, 1980), industrial slags (Williamson *et al.*, 1968), as well as other systems. These data are used in a kinetic analysis of the thermal history of selected natural samples (Klein and Uhlmann, 1976; Uhlmann *et al.*, 1979), and in models for sintering of breccias, compaction of ash flow tuffs and control of volcano morphology.

All of the natural systems investigated are complex silicates with more than four oxide components. In many systems iron oxides make-up 5 to 30 wt. pct. of the composition, and iron has been shown to affect the viscosity very strongly (Cukierman and Uhlmann, 1974; Bottinga and Weill, 1972; Shaw, 1972). Both the iron content and its oxidation state have an effect on the viscosity, especially at lower temperatures approaching the glass transition. For example, in lunar composition 15555 which contains over 20 wt. pct. FeO, it was reported that a change in the Fe^{2+}/total Fe ratio from 0.76 to 0.20 had a pronounced effect in increasing the viscosity by nearly 3 orders of magnitude at temperatures below 800°C (Cukierman and Uhlmann, 1974). However, the oxidized and reduced samples showed the same temperature dependence. This result was not expected, and several explanations were suggested that involved changes in configuration for Fe^{2+} and Fe^{3+}. There was also the competition between Fe^{3+} and Al^{3+} for tetrahedral sites, along with the complexity of a multicomponent system with MgO and CaO.

A simpler system was selected here for study. It is desirable to look first at the Fe–Si–O system and then add components such as Al_2O_3 and alkaline earth oxides, but the compositions are experimentally difficult to form as glasses. Therefore, third component Na_2O was selected to aid in sample preparation. The effect of Na_2O in iron-containing silicates is interesting, though it is recognized that large alkali contents are not found in extraterrestrial samples. In experiments by Mysen and Virgo (1978) on the system $NaAlSi_2O_6$–$NaFeSi_2O_6$ at atmospheric pressure, they found that composition, as well as partial pressure of oxygen, influenced the Fe^{2+}/total Fe ratio. Glasses with less than

10 mole. pct. NaFeSi$_2$O$_6$ were green and had no trivalent iron, while glasses with more that 15 mole. pct. NaFeSi$_2$O$_6$ were brown and had Fe^{2+}/total Fe ratios less than 0.70. Using Mossbauer spectroscopy this transition at about 15 mole. pct. acmite was described as dramatic. It appeared that the ratio of Fe^{2+} to total Fe was determined by the composition of the melt more strongly than by the oxidation conditions during melting. For this reason, the mole. pct. Na$_2$O in the present study has been kept fairly constant at either 10 or 20 mole. pct. to see the effect of oxidation conditions during melting on viscosity. Also no Al^{3+} is present in the melt.

In experiments by Bandyopadhyay *et al.* (1980) on Na$_2$Si$_3$O$_7$ glasses with iron additions, a similar transition in glass structure was seen at Fe^{2+}/total Fe ratio equal to about 0.70. They noted the effect of this ratio on internal friction measurements. These same changes in structure corresponding to changes in Fe^{2+}/total Fe due to Na$_2$O/SiO$_2$ ratio and oxidation condition should influence the viscosity of these glasses. Therefore, a series of glasses in the Na$_2$O–FeO$_x$–SiO$_2$ system have been studied here, and their viscosity has been measured over a range of temperatures.

EXPERIMENTAL PROCEDURE

The materials used in this investigation were synthesized from reagent grade source materials according to the compositions listed in Table 1. Glasses 1, 2, 6, 9 and 10 were batched on a wt. pct. basis, and glasses 3, 4, 5, 7 and 8 were batched on a mole. pct. basis. In all cases, appropriate amounts of reagent powders were combined and milled together for about 24 hours prior to melting. The melting was carried out in silica crucibles in air or forming gas (95% N$_2$–5% H$_2$) for 30 minutes in an induction melter. A large batch size of about 500 g was used to insure homogeneity. Glasses 4, 5, 7 and 8 were also prepared with Fe$_2$O$_3$ for the oxidized samples and with the stoichiometric amount of carbon needed to reduce Fe$_3$O$_4$ to FeO for the reduced samples. Molten samples were poured onto copper plates to form slabs of glass. These slabs were annealed between 400°C and 500°C for one hour and furnace cooled to room temperature. The resulting glasses were examined for homogeneity of color and absence of bubbles or crystals. These glasses were not chemically analyzed, but in the companion paper (Hewins *et al.*, 1981), microprobe analysis showed glass compositions were the same as batch compositions for samples prepared under the conditions outlined here.

Specimens for use in a beam-bending viscometer were cut from the annealed glass slabs. The beams are eight cm in length and between four and six mm square. The beam-bending viscometer, capable of measuring viscosities between 10^8 and 10^{15} poise, was supplied with a forming gas atmosphere for the reduced samples and air for the oxidized samples. The operation of the viscometer involved measuring the rate of viscous deformation of an end-supported beam which carried a central load connected to an alumina rod coupled to a linear variable differential transformer (LVDT). The range of viscosity measurements represented a temperature region between 400°C and 650°C. This instrument was a modification of a previous design (Cukierman *et al.*, 1972) and was calibrated with NBS standard glasses 710 and 711. Measurements of viscosity could not be made at temperatures greater than about 650°C because of crystallization of the sample during measurement.

Table 1. Iron-containing silicate compositions in mole. pct. and weight pct.

Glass	Mole pct.			Weight pct.			Tg°C $\eta = 10^{13}$ poise
	SiO$_2$	Na$_2$O	FeO$_x$	SiO$_2$	Na$_2$O	Fe$_3$O$_4$	
1	73.5	10.2	16.3	70.0	10.0	20.0	510
2	64.6	10.4	25.0	60.0	10.0	30.0	491
3	80.0	20.0	—	79.5	20.5	—	483
4	78.0	20.2	2.0	77.6	20.5	1.9	475
5	75.0	20.0	5.0	73.3	20.0	4.7	470
6	72.0	20.0	8.0	70.0	20.0	10.0	470
7	70.0	20.0	10.0	69.8	20.6	9.6	465
8	60.0	20.0	20.0	60.0	20.7	19.3	462
9	63.2	20.5	16.3	60.0	20.0	20.0	456
10	53.9	20.9	25.2	50.0	20.0	30.0	451

EXPERIMENTAL RESULTS

The variations of viscosity with temperature determined for the ten compositions under mildly reducing conditions are shown in Fig. 1. Glass 1 which has the lowest mole. pct. Na_2O has the highest viscosity overall. Its viscosity is 30 times higher than glass 3 which has only trace iron content. In general, the viscosity decreases with increasing Na_2O at a given temperature. For an approximately constant mole. pct. Na_2O, such as the series of glasses 3, 4, 5, 6, 7, 8, 9 and 10 which have 20 mole. pct. Na_2O, the viscosity decreases with increasing Fe_3O_4 at a given temperature.

The glass transition temperature ($\eta = 10^{13}$ poise) is listed in Table 1. This temperature ranges from 510°C for glass 1 to 451°C for glass 10. All of these values are lower than those measured in lunar (Uhlmann et al., 1979), howardite (Hewins and Klein, 1978; Klein and Hewins, 1979), and chondrule (Klein et al., 1980a) compositions. These compositions of extraterrestrial materials contain Al_2O_3 which raises the viscosity, and almost no Na_2O which lowers the viscosity. One of the lowest viscosity materials is Apollo 15 green glass

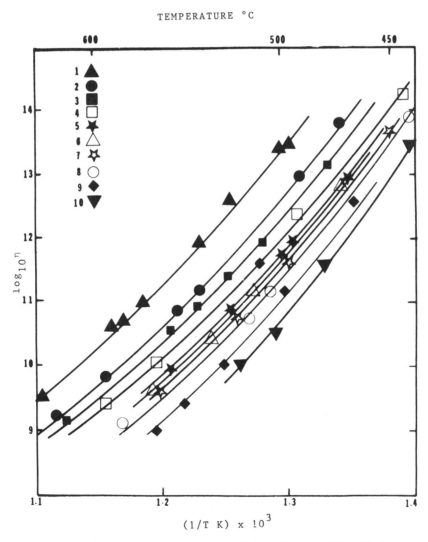

Fig. 1. Variation of viscosity with temperature for ten iron-containing silicate compositions measured in forming gas.

(Uhlmann *et al.*, 1974) which has a high FeO and MgO content, and its glass transition temperature is about 600°C.

The viscosity-temperature relation shows some curvature over the temperature range investigated. The apparent activation energy for flow in this range is between 90 and 100 kcal/mole. A value of 100 kcal/mole was reported for Apollo 15 green glass (Uhlmann *et al.*, 1974), but higher values have been reported for other lunar compositions (Uhlmann *et al.*, 1977).

A comparison of the viscosity-temperature relation for glass 4 prepared in air, in forming gas and with carbon in the batch is shown in Fig. 2. Likewise, a comparison for glass 5 is shown in Fig. 3, for glass 7 in Fig. 4, and for glass 8 in Fig. 5. In Figs. 2–5, a smooth curve is drawn to the left of the data that represents the data for glass 3 from Fig. 1. Glass 3 has iron as an impurity in the batch materials. The iron content increases on a mole. pct. FeO_x basis in the order glass 4, 5, 7, 8. In glass 4, there is hardly a difference in viscosities measured under different oxidation conditions. In glass 5, the viscosities

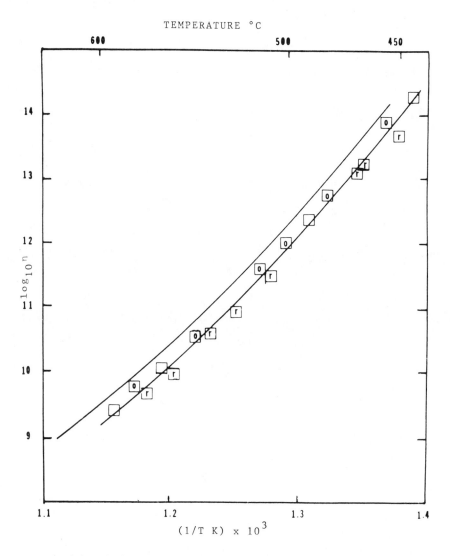

Fig. 2. Variation of viscosity with temperature for glass 4, prepared in air (0), with carbon (r) and in forming gas (open squares). The smooth curve for glass 3 is shown above.

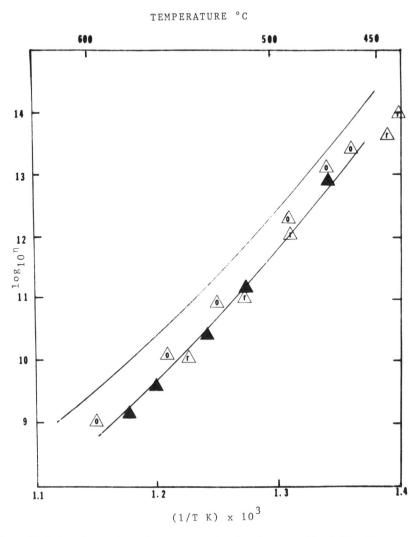

Fig. 3. Variation of viscosity with temperature for glass 5, prepared in air (0), with carbon (r) and in forming gas (open triangles). The smooth curve for glass 3 is shown above.

measured in the oxidized samples are slightly higher than in the reduced samples. In glasses 7 and 8 the effect of melting in air is to raise the viscosity and the effect of melting with carbon is to lower the viscosity. Based on a visual inspection of glass color, the glasses melted in air may have as much as 90 pct. iron in the trivalent state giving a yellow green color, and the glasses melted with carbon should have about 8 pct. iron in the divalent state giving a dark blue green color.

The magnitude of the increase or decrease in viscosity relative to the mildly reduced samples grows with increases in iron content from 10 mole. pct. in glass 7 to 20 mole. pct. in glass 8. This is expected because iron is the only component in these glasses that can be either divalent or trivalent. Also, it should be noted that glass 8 is an equimolar mixture of SiO_2 and $NaFeSi_2O_6$. As pointed out above (Mysen and Virgo, 1978), in glasses having 15 mole. pct. acmite or more, the effect of oxidation state on viscosity was very pronounced. In this study, even without Al^{3+}, there is a dramatic increase in viscosity for the oxidized sample of glass 8.

The increase in viscosity for the oxidized samples is greater than the decrease in viscosity for the reduced samples since the data from Fig. 1 represents mildly reduced

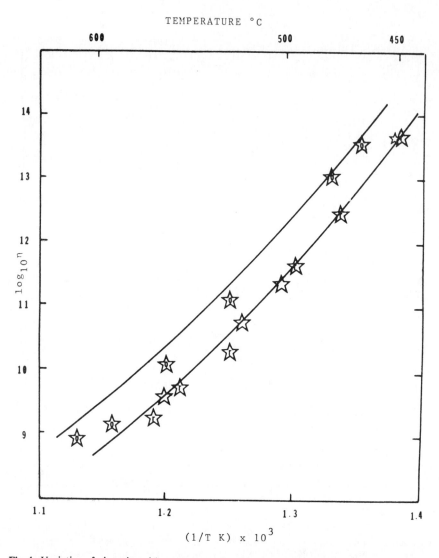

Fig. 4. Variation of viscosity with temperature for glass 7, prepared in air (0), with carbon (r) and in forming gas (open stars). The smooth curve for glass 3 is shown above.

samples. The largest increase in viscosity from strongly reduced to oxidized conditions is shown in Fig. 5 for glass 8 which contains 20 mole. pct. FeO_x. Here the viscosity increases by an order of magnitude at a given temperature.

The viscosity-temperature relation for all oxidation conditions is similar and gives an apparent activation energy between 90 and 100 kcal/mole.

DISCUSSION

There are at present no satisfactory models for describing flow behavior in the high viscosity region. For lunar compositions, the data in this region have been fit to a Vogel-Fulcher relation using $Tg(=10^{15}$ poise) as a corresponding states parameter (Uhlmann *et al.*, 1979). The Vogel-Fulcher relation used was

$$\log_{10} \eta = A + B/(T/Tg - \alpha) \tag{1}$$

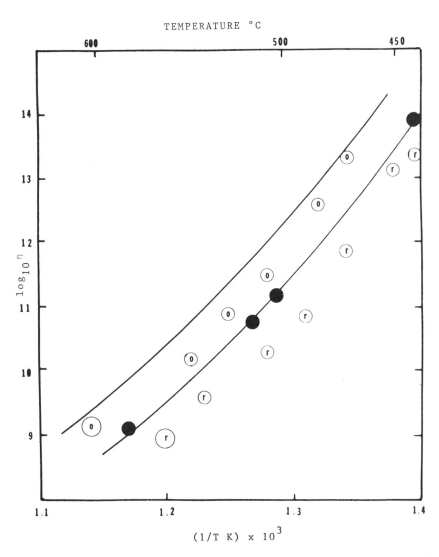

Fig. 5. Variation of viscosity with temperature for glass 8, prepared in air (0), with carbon (r) and in forming gas (open circles). The smooth curve for glass 3 is shown above.

where A, B and α are constants. A least squares fit of the data for the ten iron-containing compositions measured under mildly reducing conditions is shown in Fig. 6. The best fit to the data using $T_g (= 10^{13}$ poise) as a corresponding states parameter was

$$\log_{10} \eta = -3.0 + 7.84/(T/T_g - 0.51). \tag{2}$$

As seen in Fig. 6 the fit is within a factor of two.

For all of the compositions melted under identical conditions the trend lower viscosity with increasing Na_2O content and decreasing SiO_2 content was expected. The interesting trends are seen when viscosity data are compared for a particular glass prepared under different oxidation conditions. According to experiments by Mysen and Virgo (1978) and Bandyopadhyay et al. (1980), the glass structure of an iron-containing glass is influenced by oxidation state and chemical composition. Since the viscosity of a glass reflects its structure, this means that the effect of oxidation state on the viscosity data for glass 4 and glass 5 might be masked by the effect of chemical composition, while the effect of

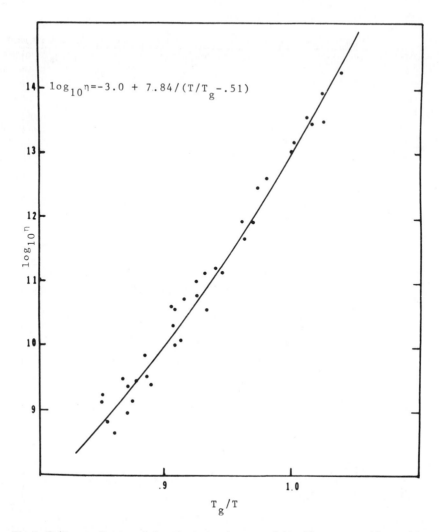

Fig. 6. Tg/T normalization of viscosity for ten iron-containing silicate compositions, with least squares Vogel-Fulcher fit.

oxidation state for glass 7 and glass 8 could account for the increase in viscosity under oxidizing conditions and the decrease in viscosity under strongly reducing conditions. The effect of oxidation state becomes apparent in glasses containing greater than 10 mole. pct. FeO_x which is in line with previous studies (Mysen and Virgo, 1978; Cukierman and Uhlmann, 1974). Above this iron concentration, it is suggested (Bandyopadhyay *et al.*, 1980) Fe^{3+} behaves like Al^{3+} and Fe^{2+} behaves like an alkaline earth. To confirm the behavior of divalent and trivalent iron, the ratio must be known with more certainty, and atomic absorption spectrometry is being used to analyze chemical compositions. When this data is available it will help in understanding the flow behavior seen here.

In the study by Cukierman and Uhlmann (1974), it was pointed out that the temperature dependence of the viscosity did not change with with change in oxidation state. No change was seen here in the apparent activation energy for oxidized and reduced samples. In fact all of the data for the ten compositions show an apparent activation energy between 90 and 100 kcal/mole. More studies are being carried out with Al_2O_3 substitutions for iron and with other transition metals to see their effect on the temperature dependence for viscous flow.

SUMMARY

For the ten iron-containing glasses investigated, the viscosity at a given temperature decreases with increasing mole. pct. FeO_x for a constant mole. pct. Na_2O, with the two glasses with 10 mole. pct. Na_2O having higher viscosities overall than the eight glasses with 20 mole. pct. Na_2O. The viscosities for the ten glasses melted under mildly reducing conditions can be represented by a Vogel-Fulcher relation using $Tg(=10^{13}$ poise) as a corresponding states parameter.

For the four glasses melted in air and with carbon in forming gas, two showed little effect of oxidation state, but the glasses with 10 and 20 mole. pct. FeO_x showed a large effect. The glass with 20 mole. pct. FeO_x showed an order of magnitude increase from the reduced to the oxidized sample. These results agree with earlier studies that showed both the effect of composition on Fe^{2+}/total Fe and the effect of partial pressure of oxygen on Fe^{2+}/total Fe. All compositions show an apparent activation energy between 90 and 100 kcal/mole for this temperature region.

Acknowledgments—The technical assistance of T. Williams with sample preparation is gratefully acknowledged, as is financial support provided by NASA grant number NSG-9065. R. J. Williams, D. R. Uhlmann and Associate Editor G. Lofgren are thanked for their comments which resulted in improvements to the manuscript.

REFERENCES

Bandyopadhyay A. K., Auric P., Phalippou J., and Zarzycki J. (1980) Spectroscopic behavior and internal friction of glasses containing iron having different redox ratios. *J. Mat. Sci.* **15**, 2081–2090.

Bottinga Y. and Weill D. F. (1972) The viscosity of magmatic silicate liquids: A model for calculation. *Amer. J. Sci.* **272**, 438–475.

Cukierman M., Tutts P. M., and Uhlmann D. R. (1972) Viscous flow behavior of lunar compositions 14259 and 14310. *Proc. Lunar Sci. Conf. 3rd*, p. 2619–2625.

Cukierman M. and Uhlmann D. R. (1973) Viscosity of liquid anorthite. *J. Geophys. Res.* **78**, 4920–4923.

Cukierman M. and Uhlmann D. R. (1974) Effects of iron oxidation state on viscosity, lunar composition 15555. *J. Geophys. Res.* **79**, 1594–1598.

Hewins R. H. and Klein L. C. (1978) Provenance of metal and melt rock textures in the Malvern howardite. *Proc. Lunar Planet. Sci. Conf. 9th*, p. 1137–1156.

Hewins R. H., Klein L. C., and Fasano B. V., (1981) Conditions of formation of pyroxene excentroradial chondrules. *Proc. Lunar Planet. Sci.*, 12B. This volume.

Klein L. C., Fasano B. V., and Hewins R. H., (1980a) Flow behavior of droplet chondrules in the Manych (L-3) chondrite. *Proc. Lunar Planet. Sci. Conf. 11th*, p. 865–878.

Klein L. C. and Hewins R. H. (1979) Origin of impact melt rocks in the Bununu howardite. *Proc. Lunar Planet. Sci. Conf. 10th*, p. 1127–1140.

Klein L. C. and Uhlmann D. R. (1976) The kinetics of lunar glass formation, revisited. *Proc. Lunar Sci. Conf. 7th*, p. 1113–1121.

Klein L. C., Yinnon H., and Uhlmann D. R., (1980b) Viscous flow and crystallization behavior of tektite glasses. *J. Geophys. Res.* **85**, 5485–5489.

Mysen B. O. and Virgo D. (1978) Influence of pressure, temperature and bulk composition on melt structures in the system $NaAlSi_2O_6$–$NaFe^{3+}Si_2O_6$. *Amer. J. Sci.* **278**, 1307–1322.

Shaw H. R. (1972) Viscosities of magmatic silicate liquids: An empirical method of prediction. *Amer. J. Sci.* **272**, 870–893.

Uhlmann D. R. and Klein L. C. (1976) Crystallization kinetics, viscous flow and thermal histories of lunar breccias 15286 and 15498. *Proc. Lunar Sci. Conf. 7th*, p. 2529–2541.

Uhlmann D. R., Klein L. C., and Handwerker C. A. (1977) Crystallization kinetics, viscous flow and thermal history of lunar breccia 67975. *Proc. Lunar Sci. Conf. 8th*, p. 2067–2078.

Unlmann D. R., Klein L., Kritchevsky G., and Hopper R. W. (1974) The formation of lunar glasses. *Proc. Lunar Sci. Conf. 5th*, p. 2317–2331.

Uhlmann D. R., Onorate P. I. K., and Scherer G. W. (1979) A simplified model for glass formation. *Proc. Lunar Planet Sci. Conf. 10th*, p. 375–381.

Uhlmann D. R., Yinnon H., and Cranmer D. (1980) Crystallization behavior of albite. In *Lunar and Planetary Science XI*, p. 1178–1180. Lunar and Planetary Institute, Houston.

Williamson J. Tipple A. J., and Rogers P. S. (1968) Influence of iron oxides on kinetics of crystal growth in CaO–MgO–Al_2O_3–SiO_2 glasses. *J. Iron Steel Inst.* **206**, 898–903.

Effects of body shape on disk-integrated spectral reflectance

Jonathan Gradie and Joseph Veverka

Laboratory for Planetary Studies, Cornell University, Ithaca,
New York 14853

Abstract—Variations in scattering geometry have been shown to affect the shapes of spectral reflectance curves for a variety of powdered materials of planetary interest. Photometric data obtained over a range of incidence angle, emission angle, and wavelength have been used to determine which of several photometric functions most accurately describes these effects. While photometric functions of the type proposed recently by Hapke and by Goguen represent our data well, we also find that a simpler photometric function which is a combination of Hapke-Irvine and Lambert terms is adequate to describe the available observations over all wavelengths.

The wavelength dependence of such functions implies that (1) there will be a difference in the shapes of the spectral reflectance curves of a flat sample and a spherical planet made of the same material, and (2) that there will be a difference between the shapes of the spectral reflectance of a spherical planet and an ellipsoidal one. The last fact can be applied to asteroids to demonstrate that elongated asteroids should appear redder at maximum light than at minimum light even if their surface material is laterally homogeneous.

INTRODUCTION

The spectrophotometric properties of various powdered substances have been shown to depend upon the scattering geometry, as well as composition and particle size. In specific studies Gradie *et al.* (1980a,b) demonstrated that variations in scattering geometry affect the shapes of the spectral reflectance curves of flat laboratory samples, sometimes quite significantly and in different ways for different materials. For example, Fig. 1 illustrates this effect for a flat sample of powdered basalt when the incidence angle and emission angle are changed while keeping the phase angle constant. The sample becomes redder with increasing emission angle.

Gradie *et al.* drew several important conclusions from their study. For instance, it was noted that the averaging of spectral data obtained over a range of phase angles may not be appropriate if high quality comparisons of laboratory data and astronomical observations are intended. These conclusions were based primarily on measurements taken from flat samples. The exact connection between the properties of samples and those of spherical planetary objects is not obvious unless one knows the appropriate photometric function of the materials involved.

In this paper we examine how well several simple photometric functions describe the photometric data measured for samples of the ordinary chondrite Bruderheim, basalt, and the carbonaceous chondrite, Allende. Some application of these results to problems of current interest in planetary science are discussed in the closing section of this paper and elsewhere by Gradie and Veverka (1981).

EXPERIMENTAL METHODS

The experimental method has been described by Gradie *et al.* (1981a). In summary, the materials studied here, a slightly oxidized basalt ($r_n = 0.22$) from Gardener, Montana, the C3 carbonaceous chondrite Allende ($r_n = 0.11$) and the ordinary chondrite Bruderheim ($r_n = 0.26$) were ground in ceramic mortars and sieved to appropriate particle sizes. (Here r_n denotes the normal reflectance and the values quoted refer to a wavelength of 0.55 μm). The basalt and ordinary chondrite were

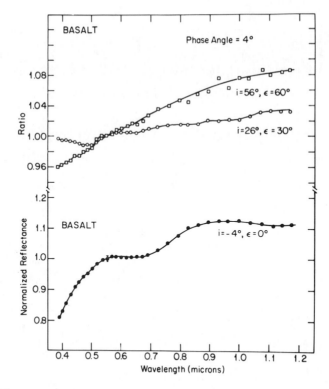

Fig. 1. The spectral reflectance of basalt (particle size 45–75 microns) normalized at $\lambda = 0.55$ ($r_n = 0.22$) microns at an incidence angle, $i = 4°$ and an emission angle, $\epsilon = 0°$, is shown in the lower half. The upper half of this figure illustrates the ratio of the basalt viewed at two other arbitrary geometries to the basalt viewed at $i = 4°$, $\epsilon = 0°$.

washed in methanol to remove the micron-sized grains and dust particles which adhere to the larger sample grains. All samples were carefully leveled but not compacted, then dusted with a thin coating of the same size fraction to simulate an unpacked regolith. The reflectance of the samples is measured relative to that of a spray-painted (Kodak Paint No. 6080) $BaSO_4$ standard. The photometric properties of the $BaSO_4$ standard have been shown by Goguen (1981) to closely approximate those of a flat Lambert surface.

All measurements were made with the Cornell Goniometer. The angles of incidence, i, and emission, ϵ, were kept coplanar with the normal to the sample but are varied from 0° to 60° to achieve a wide range of phase angles. Wavelengths for observation are obtained by varying either of two circular filter monochrometers, one in the visible from 0.4 to 0.7 microns, and the other in the infrared from 0.7 to 1.2 microns. The spectral resolution in both bands is about 0.02 microns and the central wavelength can be selected to within 0.001 microns in the visible and 0.002 microns in the infrared. Statistical errors for each measurement are less than ~1%, assuming that the errors are entirely due to photon statistics.

THE PHOTOMETRIC FUNCTION

The scattering properties of a surface can be described by the photometric function which most accurately fits the available data. This photometric function can then be used to examine any arbitrary surface in detail.

A variety of forms for the photometric functions have been suggested in the literature. Gradie *et al.* (1980b) assumed that for materials of low reflectance a Hapke-Irvine scattering law of the form

$$I_\lambda(i, \epsilon, \alpha) = \frac{\cos i}{\cos i + \cos \epsilon} A_\lambda f_\lambda(\alpha) \cdot F_\lambda \qquad (1)$$

could apply at each wavelength. Here I_λ is the intensity of the scattered light at wavelength, λ; πF_λ is the incident flux. A_λ is a constant related to the absolute reflectance; $f_\lambda(\alpha)$ is a function which depends solely on the phase angle, α, and is a product of the single particle phase function and the shadowing function derived by Hapke (1963) and Irvine (1966). Here we have assumed that $f_\lambda(\alpha)$ does not depend strongly on i and ϵ independently (cf. Goguen, 1981). We stress that Eq. (1) is *not* the Lommel–Seeliger law, which would result from the physically unrealistic assumption that $f_\lambda(\alpha) = 1$ at all α in Eq. (1).

Often the dependence of $f(\alpha)$ on α at small and moderate phase angles is approximately exponential—i.e., $f(\alpha)$ expressed in terms of *magnitudes* varies approximately linearly with α, at least between 4° and 60° phase. However, to describe accurately the measurements a more complicated form is necessary for the following reasons: (1) as noted by Gradie *et al.* (1980b), the measured dependence of $f(\alpha)$ on α is usually more complicated than a simple exponential, particularly at phase angles greater than 60°; (2) accurate measurements reveal that $I/(\cos i/\cos i + \cos \epsilon)$ is not precisely independent of i and ϵ, but does depend somewhat on the actual values of i and ϵ.

An example of this second effect is illustrated in Fig. 1 for the basalt sample where the dependence on i and ϵ of the scattering properties are shown as differences between the normalized reflectance at 4° and the reflectance at emission angles of 30° and 60°.

A photometric function which attempts to account for the small i and ϵ dependence is the following:

$$I_\lambda(i, \epsilon, \alpha) = \left[\frac{\cos i}{\cos i + \cos \epsilon} \cdot A_\lambda f_\lambda(\alpha) + B_\lambda \cos i\right] F_\lambda, \qquad (2)$$

essentially a linear combination of a Hapke-Irvine law which is known to approximate well the scattering properties of dark surfaces (such as that of the moon), and Lambert's law which should approximate the scattering behavior of very bright surfaces. Bowell and Lumme (1979) have demonstrated that Eq. (2) adequately describes the disc-integrated properties of asteroids and satellites. No α dependence is needed for A_λ since it always occurs as the product $A_\lambda f_\lambda(\alpha)$, but there may be some dependence of B_λ on α (see Fig. 2).

An alternate approach has been suggested by Goguen (1981) and by Hapke (1981). They derive the photometric function for an optically thick layer of scatters using the elements of radiative transfer theory and arrive at a function of the general form:

$$I_\lambda(i, \epsilon, \alpha) = A_\lambda \frac{\mu_0}{\mu + \mu_0} f_\lambda(\alpha)(a_\lambda + b_\lambda \mu)(a_\lambda + b_\lambda \mu_0) \cdot F_\lambda, \qquad (3)$$

where $I_\lambda(i, \epsilon, \alpha)$ is the intensity of scattered light at each wavelength, λ, $f_\lambda(\alpha)$ is the phase function, $\mu_0 = \cos i$, $\mu = \cos \epsilon$ and a_λ and b_λ are wavelength dependent parameters which are independent of i, ϵ or α.

In the most ideal model, in which the surface is assumed to consist of identical scatterers and mutual shadowing is ignored, A_λ and b_λ are determined entirely by $\tilde{\omega}_0$ the single scattering albedo of the particles making up the layer, and $f_\lambda(\alpha)$ is the scattering phase function of the particles. Actual surfaces are much more complicated, and in practice one can treat $f_\lambda(\alpha)$, A_λ and b_λ as parameters to be derived from the measurements; allowing the latter two to vary with α increases the flexibility of Eq. (3).

Often one can assume that $a_\lambda \sim 1$, simplifying Eq. (3). Goguen (1981) suggests that this modification should provide a useful fit to measurements on powdered surfaces of arbitrary albedo. In this paper we will treat Eq. (3) only in its general form. Specific versions discussed by Hapke (1981) and by Goguen (1971) will be compared with measurements in a subsequent publication.

COMPARISON OF PHOTOMETRIC FUNCTIONS

The three photometric functions described in the preceding section were compared to our measurements over a variety of i and ϵ by using a nonlinear least squares technique to determine the coefficients for $\alpha = 4°$, 30° and 60°.

Table 1.

	Mean Residuals		
	Eq. (1)*	Eq. (2)*	Eq. (3)*
Allende ($r_n = 0.11$)			
$\alpha = 4°$	0.0019	0.0015	0.0015
30°	0.0006	0.0006	0.0005
60°	0.0006	0.0006	0.0004
Basalt ($r_n = 0.22$)			
$\alpha = 4°$	0.0120	0.0015	0.0014
30°	0.0056	0.0015	0.0015
60°	0.0027	0.0010	0.0010
Bruderheim ($r_n = 0.26$)			
$\alpha = 4°$	0.0215	0.0036	0.0042
30°	0.0105	0.0051	0.0049
60°	0.0106	0.0042	0.0027

*See text.

The applicability of each photometric function was tested by comparing the mean residuals of the fit over all wavelengths. The residuals are the mean of the difference between the observed and calculated reflectances at each phase angle. Comparisons of the residuals are given in Table 1, where the mean residual over all wavelengths is expressed at each phase angle for the Allende, Bruderheim, and basalt samples.

It is obvious that Eq. (1) provides the poorest fit, especially as the normal reflectance, r_n, and the phase angles become large. This form of the photometric function has already been shown to be a good approximation for dark surfaces in general, but not for highly reflecting ones (Veverka, 1970; Veverka *et al.*, 1978). An interesting exception is provided by the surface of Ganymede. While Ganymede's surface has a reflectance in the range of 0.4, Eq. (1) provides an adequate fit to its photometric properties (Squyres and Veverka, 1981).

For the more highly reflecting surfaces, i.e., the basalt and Bruderheim, the more complex photometric functions work much better than does Eq. (1). Theory predicts that

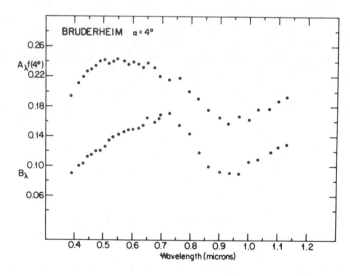

Fig. 2. The wavelength dependence of the photometric coefficients $A_\lambda f_\lambda (4°)$ and B_λ defined in Eq. (2) for a sample of Bruderheim (particle size 45–75 microns) at a phase angle of 4°.

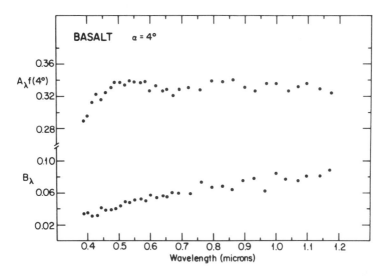

Fig. 3. The wavelength dependence of the photometric coefficients $A_\lambda f_\lambda(4°)$ and B_λ defined in Eq. (2) for a sample of basalt (particle size 45–75 microns) at a phase angle of 4°.

the values of b_λ in Eq. (2) and B_λ in Eq. (3) will increase with increasing reflectance (Goguen, 1981). This effect is best illustrated in Fig. 2, where the wavelength dependence of the quantity $A_\lambda f_\lambda(4°)$ is plotted for the Bruderheim sample. As the coefficient $A_\lambda f_\lambda(4°)$ varies over the absorption band a corresponding behavior in B_λ is observed. The effect is also evident in the basalt data (Fig. 3). Additionally, if one compares the values of B_λ at any specific wavelength for the three materials: Bruderheim (Fig. 2), basalt (Fig. 3), and Allende (Fig. 4), one finds that the value of B_λ varies directly with $A_\lambda f_\lambda(4°)$.

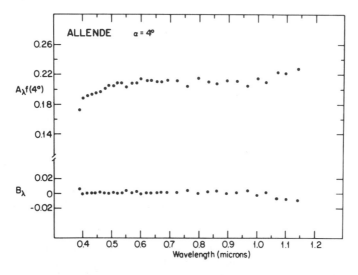

Fig. 4. The wavelength dependence of the photometric coefficients $A_\lambda f_\lambda(4°)$ and B_λ defined in Eq. (2) for a sample of Allende (particle size <75 microns) at a phase angle of 4°.

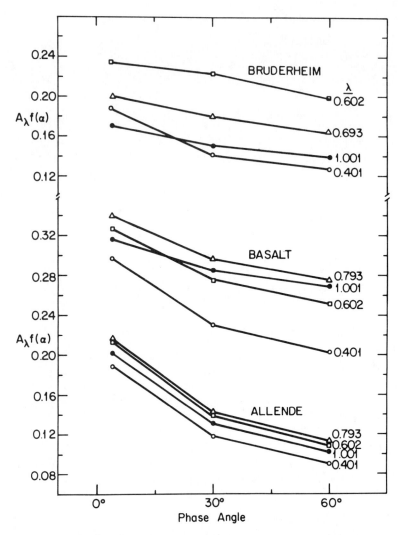

Fig. 5. The phase dependence of the $A_\lambda f(\alpha)$ coefficients of the Allende, basalt and Bruderheim samples at four wavelengths, $\lambda = 0.401$, 0.602, 0.783, and 1.001 microns.

On the basis of our empirical analysis, we conclude that Eqs. (2) and (3) are superior to Eq. (1) in describing our measurements. For the data in hand, all of which pertain to a coplanar scattering geometry, Eq. (3) does not provide a significantly better fit than the simpler Eq. (2). We note that in the analysis presented here, we have chosen to keep B_λ in Eq. (2) independent of α, as suggested by Bowell and Lumme (1981). To verify the validity of this approach we obtained fits using Eq. (3) in which B_λ was allowed to vary with α, but in all cases (except Bruderheim at $\alpha = 60°$) we found no significant α dependence.

The phase dependence of the quantities $A_\lambda f_\lambda(\alpha)$ at several wavelengths are illustrated in Fig. 5 for Allende, basalt, and Bruderheim. As the normal reflectance of the sample increases (from $r_n = 0.11$ for Allende to $r_n = 0.26$ for Bruderheim), the dependence of $A_\lambda f_\lambda(\alpha)$ on phase becomes less. This simply means that $f_\lambda(\alpha)$ depends on the normal reflectance, i.e., darker samples have steeper phase functions. The wavelength dependence of $f_\lambda(\alpha)$ is also evident in these graphs and is discussed elsewhere (Gradie and Veverka, 1981).

DISCUSSION

For the remainder of the paper we shall work in the context of Eq. (2) because of its simpler form, but we note that in both Eq. (2) and in Eq. (3) it is the wavelength dependence of the various parameters ($A_\lambda f_\lambda$, B_λ, a_λ, b_λ) which is of fundamental importance for the problem under consideration: it is precisely the wavelength dependence of these parameters that produces the changes in spectral shape with scattering geometry reported by Gradie *et al.* (1980) and others. Once the wavelength dependence of these parameters is known for a particular surface, one can predict the spectral reflectance curve for any geometry. In a recent paper Hapke (1981) demonstrates theoretically that such parameters must be wavelength dependent in the usual case in which the optical properties of the surface material change with wavelength.

As a test of the usefulness of Eq. (2), we show in Fig. 6 a comparison between the measurements and the model calculation for our Bruderheim sample at $i = 10°$, $\epsilon = 40°$ (phase angle = 30°). The difference between the observed and calculated values is shown in magnitudes. The agreement is quite good.

As an application of our measurements we briefly discuss the implication of our results for observed color variations of asteroids. There are two important implications of photometric functions such as Eq. (2) and (3). First, they imply that the spectral reflectance of a flat sample will differ from that of a spherical planet made of the same material (Gradie and Veverka, 1981). Second, they imply that there will also be differences between the shape of spectral reflectance curves of a spherical planet and an ellipsoidal one. As an example, we have studied the case of an ellipsoidal object for which the ratio of major to minor axes (b = c), a/b = 3, corresponds to a light curve amplitude of about 1.2 mag (typical of asteroid 433 Eros, for example). Figure 7 illustrates

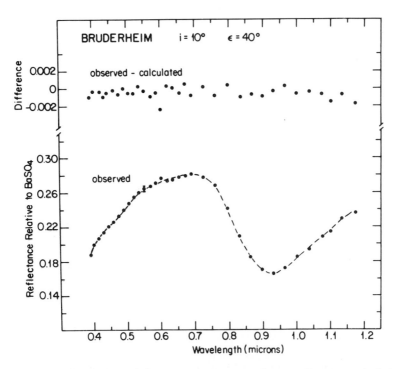

Fig. 6. A comparison between the observed spectral reflectance (lower curve) of the Bruderheim sample and the spectral reflectance calculated according to Eq. (2) using coefficients $A_\lambda f_\lambda(30°)$ and B_λ. The difference between the two spectral reflectances is shown in the upper half.

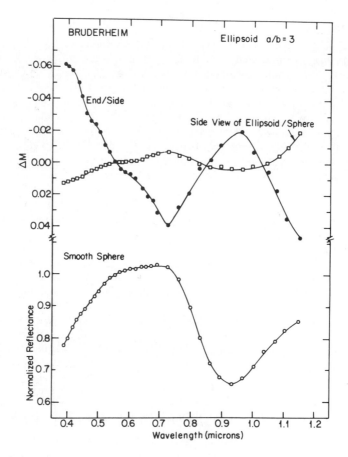

Fig. 7. The normalized spectral reflectance of a Bruderheim ($r_n = 0.26$) covered sphere (bottom) and the differences in magnitudes between the sphere and the end and side views of an ellipsoid at 4° phase.

our results for such an ellipsoid covered with Bruderheim-like material (L6 ordinary chondrite) viewed at a phase angle $\alpha = 4°$. For reference, the spectral reflectance calculated for a smooth sphere is shown at the bottom of the figure. The upper portion shows the differences (expressed in magnitudes) in the spectral reflectance for two cases: (1) the difference between the sphere and the side view of the ellipsoid and (2) the difference between the end view and side view of the ellipsoid. These differences can be thought of as the amplitudes of the color light curves with respect to the visible (0.55 microns). For Case (1) the differences are small, only a few percent over the spectral region samples (the side view of the ellipsoid showing slightly more spectral contrast than does the sphere). For Case (2), the differences amount to about ten percent. The end view of the ellipsoid shows less spectral contrast than does the side view, particularly for wavelengths below 0.7 microns. The depth of the 0.95 micron band does not change with respect to the visible part of the spectrum, but it does change by about five percent with respect to the 0.7 micron shoulder. An observer watching the rotation of this model asteroid would see the amplitude of the light curve vary as a function of wavelength by about ten percent over the spectral region 0.4 to 0.7 microns.

Figure 8 shows similar results for a sample of basalt. The lower curve is the normalized spectral reflectance of a sphere covered with such material; the upper curves show the differences for Cases (1) and (2) above. For the basalt, as for Bruderheim, the side view of the ellipsoid is slightly redder than the sphere, the difference being less than two

Effects of body shape 1777

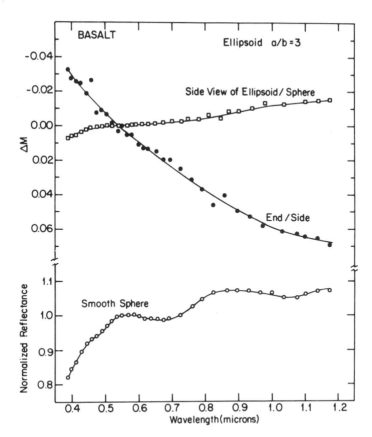

Fig. 8. The same as Fig. 7 for basalt ($r_n = 0.22$).

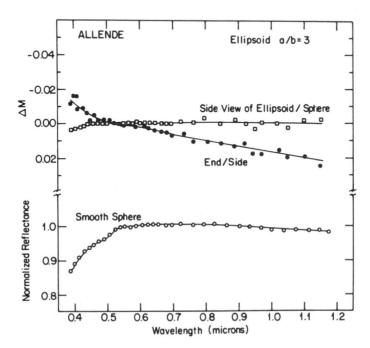

Fig. 9. The same as Fig. 7 for Allende ($r_n = 0.11$).

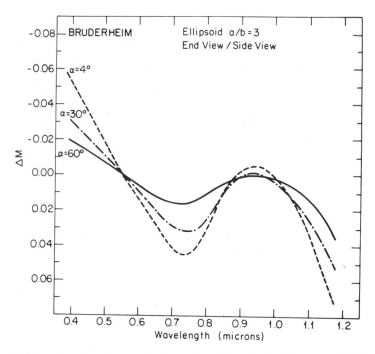

Fig. 10. The phase dependence of color effects for a Bruderheim covered ellipsoid (a/b = 3).

percent overall. However, in this case, the end view of the ellipsoid is about ten percent less reddened than the side view.

The results for the dark C3 carbonaceous chondrite Allende are shown in Fig. 9. Allende is much darker than either the basalt or Bruderheim and the above effects are subsequently smaller. There is little spectral difference between the sphere and the side view of the ellipsoid and only a three percent overall variation in slope between the end and side views.

The amplitude of these color lightcurve effects do depend on the phase angle of the observation. Figure 10 illustrates this effect for the Bruderheim ellipsoid. As the phase angle increases, the difference in spectral contrast between the end and side views of the ellipsoids changes considerably. In fact, between 30° and 60° phase, the effect vanishes for wavelengths shortward of 0.7 microns.

While most asteroids are elongated, few reach shapes as extreme as a/b ∼ 3; hence, the effects in general will be smaller. However, on the basis of our model, we predict that, in general, asteroids should appear redder at maximum light than at minimum light if the color lightcurve is synchronous with the visible lightcurves (i.e., in the absence of spots on the surface). Indeed, Bowell and Lumme (1979) have noted in their statistical analysis of the mean B-V color indices of 160 C and 127 S asteroids that these objects do appear redder by several thousandths of a magnitude when brighter than their mean magnitude. Unfortunately, on a case by case basis, relevant data are still scanty.

Our results suggest that observers of asteroid lightcurves should be wary of small color variations with amplitudes less than $0^m.03$, especially if they are synchronous with the maximum or minimum at the visible lightcurve. Such small variations probably result from the non-spherical geometry of the object, and may not indicate compositional variations (spots).

Acknowledgments—This research was supported by NASA Grant NSG-7606. We wish to thank Lorna Wong for computational assistance, and B. Hapke and M. Gaffey for helpful comments.

REFERENCES

Bowell E. and Lumme K. (1979) Colorimetry and magnitudes of asteroids. In *Asteroids* (T. Gehrels, ed.), 132–169. Univ. Arizona, Tucson.

Goguen J. (1981) A theoretical and experimental investigation of the photometric functions of particulate surfaces. Ph.D. thesis, Cornell Univ., Ithaca, NY. 209 pp.

Gradie J. and Veverka J. (1981) The effects of scattering geometry on spectral reflectance: color light curves of asteroids (abstract). In *Lunar and Planetary Science XII*, p. 359–361. Lunar and Planetary Institute, Houston.

Gradie J., Veverka J., and Buratti B. (1980a) The effects of photometric geometry on spectral reflectance (abstract). In *Lunar and Planetary Science XI*, 357–359. Lunar and Planetary Institute, Houston.

Gradie J., Veverka J., and Buratti B. (1980b) The effects of scattering geometry on the spectrophotometric properties of powdered materials. In *Proc. Lunar Planet. Sci. Conf. 11th*, p. 799–815.

Hapke B. W. (1963) A theoretical photometric function for the lunar surface. *J. Geophys. Res.* **68**, 4571–4586.

Hapke B. W. (1981) Bi-directional reflectance spectroscopy. I. Theory. *J. Geophys. Res.* **86**, 3039–3054.

Irvine W. M. (1966) The shadowing effect in diffuse reflection. *J. Geophys. Res.* **71**, 2931–2937.

Squyres S. W. and Veverka J. (1981) Voyager photometry of surface features on Ganymede and Callisto. *Icarus* **46**, 137–155.

Veverka J. (1970) Photometric and polarimetric studies of minor planets and satellites. Ph.D. thesis, Harvard Univ., Cambridge. 283 pp.

Veverka J., Goguen J., Young S., and Elliot J. (1978) Scattering of light from particulate surfaces. I. A laboratory assessment of multiple scattering effects. *Icarus* **34**, 406–414.

Some key issues in isotopic anomalies: Astrophysical history and aggregation

Donald D. Clayton

Department of Space Physics and Astronomy, Rice University, Houston, Texas 77001

Abstract—We describe ways in which astrophysical history is utilized to understand isotopic anomalies in meteorites, concentrating on those elements, oxygen, titanium, and noble gases, whose anomalies are ubiquitous in carbonaceous meteorites. Attention is given to ideas of nucleosynthesis and chemical condensation that seem capable of generating understanding. Key arguments are: (1) cosmic chemical memory, rather than an inhomogeneous injection, is the source of isotopic anomalies; (2) volatility patterns and some isotopic patterns are mapped onto a grain-size spectrum; (3) interstellar sputtering generates the FUN systematics; (4) meteoritic He and Ne are "presolar"; (5) ^{50}Ti is chemically fixed by appearing in a metallic, rather than oxidizing or sulfurous, SUNOCON environment; (6) Condensation of ^{49}Ti as its ^{49}V (330d) progenitor in SUNOCONs results in a special extinct anomaly at ^{49}Ti; (7) STARDUST condensed in s-process red-giant atmospheres provides a correlated source of ^{50}Ti excesses with ^{47}Ti deficiencies.

INTRODUCTION

In this work I will emphasize astrophysical history as the key element in understanding isotopic anomalies. The key issue is that of how isotopic anomalies get from nucleosynthesis events to meteoritic samples, and yet that issue is almost ignored (except for a series of works by this writer). It is paradoxical that while isotopic measurements are made with such great precision, and nucleosynthesis calculation made with such nuclear detail and computing power, the journey from star to sample continues to be treated with simplistic wave-of-the-hand generalities. Papers in the meteoritic literature continue to interpret data in terms of condensation within a hot gaseous solar system made somewhat inhomogeneous spatially by admixture from a late neighboring supernova. This scenario will be, as usual, discarded in this work in favor of aggregation in a cold dusty solar-system-parent cloud of largely precondensed matter (Clayton, 1978). The basic issues have been largely laid out in that work and in a lengthy review (Clayton, 1979a) of related issues.

This imbalance is worsened by casual remarks from pioneering experimental teams, who seem to expect complicated measurements and astrophysical history to be summed up by an easy model. As a specific example, Wasserburg *et al.* (1980), following a lengthy review of their measurements, urge in conclusion, "it is hoped that this report will *provoke serious scholarly inquiry* into the basic *nuclear astrophysical processes* that *must be the cause* of the isotopic effects." (Italics are this writer's.) In actual fact, serious scholarly inquiries of relevant nucleosynthesis abound, as found in the rererence list of Clayton's (1979a) review and in many papers since then. The resistance of the overall problem to solution revolves instead about the history of the matter between the time of nucleosynthesis and the formation of the meteorites. For example, Heymann and Dziczkaniec (1980), following a very lengthy and numerically detailed study of Xe and Kr isotopic profiles in a massive stellar interior, recognize this problem by stating in conclusion, "However, we do not pretend to understand the odyssey of the abnormal components from the source to the meteorites." One set of researchers who has tried to grapple with the central problem are Manuel and coworkers, whose struggle to paint a chemically realistic history has led them to the astrophysically implausible solution

(Manuel and Sabu, 1975; Ballad *et al.*, 1979) that the entire solar system condensed from the spatially inhomogeneous debris of a single supernova. The abundant good arguments in the many papers from this team have been a stimulus for the meteoritic community, but their model solution is not dynamically sound and must be rejected. Consolmagno and Cameron (1980) have also attempted to come to grips with the issue by accumulating CaAl-rich inclusions in hot gaseous protoplanets, with the isotopic effects resulting from the differential memory of chemical phases along the general lines of Clayton (1978); but it remains to be shown that such inclusions can retain the disequilibrated textures and features actually found in CAI. Unfortunately, however, most nuclear astrophysical calculations do not address the chemical history that could even conceivably lead to the actual inclusions, chondrules, and matrix as observed. Nor can we solve the problem here with much greater detail and assurance than in our previous efforts; but we will emphasize a few key points that are relevant.

What we have advocated from the beginning of our research (Clayton, 1975a, b; 1976; 1977a, b, c; 1978) into this problem is that the nucleosynthesis patterns can be and certainly *must be remembered chemically*, without being mixed with all other nucleosynthesis sources. We like to call this "cosmic chemical memory." Its largest terms result from different isotopic patterns impregnating different chemical forms condensing (SUNOCONS) during the thermal expansion of the sites of nucleosynthesis. This chemical memory resides at all times and at all places in the interstellar medium (ISM), much as the DNA genetic code resides in each cell of the living species. The chemical memory is altered by the evolution of the several phases of the ISM, just as the DNA mutates under the vicissitudes of time and environment. A study of ISM physics and chemistry is required to uncover the likely ways in which this memory is carried, altered, and later extracted. Displays of this memory are to be found in the puzzling chemical and isotopic anomalies found in diverse solar-system objects. To draw upon the biological analogy once again, our solar system is the only known species from which we can infer in detail the workings of the cosmic chemical memory. But the observed anomalous samples are not the memory any more than the dinosaur is its DNA. Our basic view is that macroscopic outer-solar-system bodies accumulated from cold ISM dust and gas, heated later only mildly in an accretion disk (Cameron, 1978). In that case, differing bulk chemical and isotopic compositions of undifferentiated primitive bodies reflect a difference in bulk of the different ISM ingredients. As simple as this idea sounds, it has been very controversial and is far reaching if correct.

ISM HOMOGENEITY IN BULK

The average expectation for the ISM is that it is well mixed and homogeneous in the following sense: in a large volume of ISM both the elemental abundances and the isotopic constitution of the elements assume their normal values. One should, to the extent allowed by the observations, take that condition to apply also to each volume element of the cold cloud whose collapse generated the solar system. The ISM is not homogeneous, however, even on average, in at least two other senses.

The first is that the ISM is not chemically homogeneous. For example, volatile elements tend toward the gas phase whereas the most refractory ones concentrate in the dust. Nor should the dust itself be homogeneous. Magnesium, to take one element as an example, is expected to exist in at least three (Clayton, 1978) chemically distinct forms of dust: (1) SUNOCONS from the carbon-burning zone thermally precipitate Mg during the expansion of the supernova interior into refractory mineral forms; (2) STARDUST ejected during mass loss from stars will have different refractory forms for Mg; (3) NEBCONS collected when gaseous Mg and other elements stick onto cold ISM grains lead to yet other chemical forms for Mg. Each of these separate forms should differ isotopically. *Thus macroscopic aggregates differing in bulk constitution can be recovered from an ISM whose bulk chemical abundances are homogeneous* simply by any scenario that can mechanically vary the mixture of the differing dust types.

The second inhomogeneity is that the ISM in bulk exists in *at least* four separate phases: (1) the hot ionized medium HIM; (2) the warm ionized medium WIM; (3) the warm neutral medium WNM; and (4) the cold clouds CC. These four media evolve continuously, one into another. The serious reader should study McKee and Ostriker (1977) for an idea of these phase transmutations of the galactic organism. I have schematically illustrated several of these features in Fig. 1. The point here is that although the solar system collapse probably occurred from the cold cloud phase, the constitution of the dust in CC is an evolutionary product of its residence times in the several phases. This astrophysical history of mutations has evolved the chemical memory to its final form before solar nebula collapse.

Of course it is clear that some spatial inhomogeneity in abundances will exist owing to different nucleosynthesis rates in different parts of the galaxy and in those supernova ejecta that are too young to have been mixed well with the ISM; but we should attempt to understand the isotopic anomalies without such inhomogeneities—except where the data call for them. Almost all of the experimental discoveries of isotopic anomalies have appealed to "an inhomogeneously admixed injection." But the *only* logical argument (in this writer's opinion) that the solar system was part of such an inhomogeneity is the belief by some that live ^{26}Al existed in the solar system at concentration levels far exceeding a reasonable average concentration in the ISM (Cameron, 1962; Lee *et al.*, 1977). And even that argument is bedeviled by the existence of very primitive inclusions whose Al-rich minerals do not retain excess ^{26}Mg, so that the inhomogeneity has to be so fine as to differ in separate parts of the solar system making separate CaAl-rich inclusions! And yet, no nucleogenetic isotopic differences have been found that correlate with the existence or non-existence of a ^{26}Mg excess, except perhaps for ^{41}K excesses (Huneke *et al.*, 1981) that were *predicted* to exist (Clayton, 1975b), along with ^{26}Mg excesses, before the latter were documented. So strongly has this writer felt about both the correctness and the predictive power of the astrophysical ideas brought to bear on these problems, that he (Clayton, 1977c) concluded a rediscussion of that prediction by this challenge: "But it will be very difficult for the SUNOCON picture of the origin of the CaAl-rich inclusions if ^{41}K excesses cannot be detected in Ca-rich mineral separates of Allende inclusions." That a

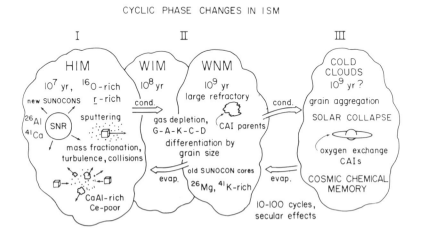

Fig. 1. The chemical evolution of matter evolves through the three main phases of the interstellar medium—hot, warm and cold (cf. McKee and Ostriker, 1977). New SUNOCONs are soon placed in the HIM by connecting supernova tunnels, and sputtered by supernova shocks. Accumulation of larger than average refractory rich particles may occur over the 10^{10} year galactic lifetime as matter cycles through these phases on indicated timescales. The solar collapse, aggregatation of macroscopic bodies and final oxygen exchange happens in an eventual cold cloud phase. The entire procedure generates cosmic chemical memory. (From Clayton, 1981a).

GRAIN-SIZE AND FRACTIONATION PATTERNS

The existence of fractionated abundance patterns according to elemental volatilities has long been attributed (Larimer and Anders, 1967) to sequential thermal condensation within slowly cooling, hot, gaseous solar nebula. Even the measurements of the distribution of the ^{16}O anomaly are generally interpreted (R. Clayton et al., 1976) in terms of a few tightly bound presolar carriers that escaped total vaporization in the otherwise hot and gaseous nebula. This writer (e.g., Clayton, 1978) has consistently eschewed a hot (> 1500 K) solar system for matter aggregating in the outer solar system into meteorite parent bodies, recommending instead a not-very-well defined historical hierarchy of segregation of dust and gas during accumulation. The history of the dust during the evolution of the phases of the ISM in Fig. 1 offers many chances for separating elements according to volatility. The opposition to such an approach has consistently centered on the belief that an aggregate of dust must contain the solar abundances of the condensed elements. This belief was explicitly met from the beginning, as expressed in the words of a thorough critical review by a referee of Clayton (1978). I quote those referee's remarks as germane to the present discussion:

"He has four components in this model: gas, NEBCONS, SUNOCONS, and STARDUST. I can understand how these components have drastically different compositions, but I do not see how they can be separated and collected in different ratios. Dust/gas separations *are easy*, of course, but how does one separate SUNOCONS from STARDUST? Solid/solid separations are very difficult to perform in an astrophysical environment. Considering that these two components both have NEBCON mantles makes it even more difficult. It is equally difficult to separate NEBCONS from the other solid components."

The italicizing of two words is my own, because there is more to separating gas and dust than meets the eye, and this same hidden feature may be the key to separating different solids according to volatility. Cameron (1973b) has provided an informative discussion in this regard of how gas and dust in a cold cloud may aggregate into larger objects. Because their thermal speeds are so much higher it is likely that condensable (stickable) gaseous atoms in a cold cloud will all accrete onto cold dust before enough dust-dust collisions can occur to aggregate larger objects. In that case even dust/gas separations are not so transparently easy, because the once gaseous atoms (an Mg, atom, say) have joined the NEBCON mantles. *But the separation still exists in the grain-size distribution.* A large number of small grains will have gathered the once-gaseous atoms much more readily than a smaller population of large grains. Clayton (1980a) has argued that gas/dust separations are in this way mapped onto the grain-size distribution. A macroscopic separation is then effected by any dynamic process (e.g., sedimentation, turbulent transport) that then separates dust particles according to size.

This same principle applies to any element that was at any time more gaseous than the most refractory dust. The entire presolar history of the ISM is a very rich and complicated one as suggested by Fig. 1. The cumulative effect of repeated volatilizations of dust, either thermally or by sputtering, may be a redistribution of the entire chemical chart according to volatility, and may map that distribution onto grain-size populations.

Clayton (1980a) showed that this happens in two distinct modes, both of which are of great interest to the chemical evolution of the ISM. The first is a previously overlooked effect in thermal condensation. When gaseous constitution B condenses thermally around condensed phase A, but does not chemically react with A, the final B/A ratio depends upon the grain size. The most interesting example is Al/Mg fractionation, which can yield much more Al-rich large SUNOCONS than small ones, provided that the initial Al SUNOCONS are characterized by a wide range in size, with the smaller being more numerous. Numerical examples are calculated by Clayton (1980a). If the independently nucleating Mg silicates form numerous *new* particles instead of mantles on the spinels, the final fractionation is greatly increased further. This geometrical aspect of condensation can provide an important first step to extracting later CaAl-rich accumulates from a cold cloud having solar abundances. The second mode is the already discussed accretion of the NEBCON component onto grains. Redistribution following extensive sputtering of the ISM dust component by supernova shock waves (Dwek and Scalo, 1980) is especially effective if some abundant population of small nonrefractory particles (probably carbonaceous, e.g., Hoyle and Wickramasinghe, pers. comm., 1981) serves as a getter for layers sputtered away from the refractory particles. Fractionation patterns according to volatility are also partly established by this sputtering history (Barlow and Silk, 1977; Clayton, 1981a).

Perhaps the simplest idea that can lead to the existence of the CaAl-rich inclusions is a special case of these ideas: when the cold solar cloud is on the verge of collapse, it contains within it a subset of larger than average dust particles that are enriched in the very refractory elements. Not only are refractory particles better able to partially withstand the 10^{10} years of evolution through the phases of Fig. 1, but energetic inelastic collisions throughout that history can perhaps have fused larger-than-average refractory particles by fracturing and vaporizing less refractory mantles and sintering refractory cores. Perhaps these particles can then be gathered as the first sediments to a disk in the solar system (Clayton, 1980a). There even larger refractory aggregates, which are the parents of the CAI as we now find them, are collected and fused into macroscopic CaAl-rich mineral fields by external heating and exothermic chemistry (Clayton, 1980b). Spallation gases from this interstellar history were probably largely removed during the solar system heating.

This principle does not lead to CAI any more than the principle of sequential thermal condensation does; but it does point to a lengthy line of thought that may yield the answer.

OXYGEN AND FUN

Three isotopic anomalies are routine in Allende and other carbonaceous chondrites: oxygen, titanium, and noble gases. The oxygen anomalies are related in some way to the nuclear anomalies in the sense that different macroscopic aggregates (the *parents* of CAI) possessed bulk oxygen compositions related to the existence of detectable nuclear anomalies. This relationship is illustrated in Fig. 2 in a form first suggested by this writer (Clayton, 1979b; 1981a) on the basis of the measurements and conclusions presented by R. N. Clayton and Mayeda (1977). I apologize to that group for plotting only schematic points in formats much larger than the observational errors (for the sake of conceptual clarity) and for changing their naming of key points in order to achieve a simpler mnemonic: E = *E*xchange oxygen, K = bulk parent of E K 1-4-1, C = bulk parent of *C* 1, A = bulk parent of most Allende inclusions, D = heavily sputtered *D*ust, and G = *G*as sputtered from grains into the ISM. Another bulk parent H = *H*AL is not shown, but falls near where I have unintentionally placed the point D (Lee *et al.*, 1980). The basic idea is that severe sputtering (Dwek and Scalo, 1980; Draine and Salpeter, 1979) of refractory interstellar dust produces a severe isotopic fractionation between the dust and gas components of interstellar O, Mg and Si, and that the remixing of that dust and gas generates mixtures along the DG line. The reaccretion of the sputtered atoms occurs in different proportions on dust of different size or of different type altogether. The usefulness of this concept is that if A K C H is a *mixing* line generated by different dust types incorporating different proportions of D and of G, it follows naturally from the cosmic-chemical-memory theory that they will also have different nuclear isotopic compositions. This is much more natural than the usual interpretation that K, C, and H are fractionated versions of A, because in that case one must find a compelling reason for mass fractionation to *cause* nuclear anomalies in Ca, Ba, Nd, Sm, etc. Interpretation of the line in terms of a mixing process between initial states prepared by astrochemical history is more economical. Note that at the time of actual aggregation of the dust, the once-sputtered atoms may have been accreted by a large population of small dust

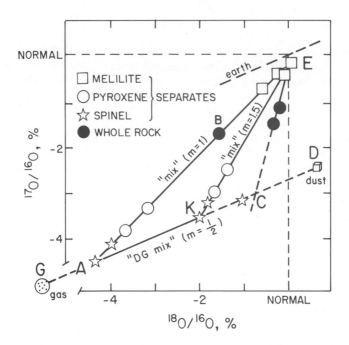

Fig. 2. The oxygen isotope systematics of Allende inclusions following R. Clayton and Mayeda (1977). Illustrative mineral separates are shown in formats larger than the actual errors in routine Allende inclusions AE, and in FUN inclusions EK1-4-1 on KE and C1 on CE. Exchange oxygen is E. Sputtered refractory interstellar dust is at D and the gas sputtered away is at G. The nomenclature is changed from the Chicago papers for a better mnemonic. A similar graph exists for Mg and Si, so that remixing along DG generates accumulates related by fractionation, but by a mixing process. For oxygen, G mixes with the bulk of interstellar O to produce E, which is then exchanged when the mineralogy of the inclusions is set by a heating event. (From Clayton, 1979b).

particles of a different type, for example a large population of small carbonaceous grains (Hoyle and Wickramasinghe, pers. comm., 1981). In this sense G of Fig. 1 refers to all atoms that *were ever* gaseous, not to only atoms that were gaseous during the final aggregation of the dust. *It is primarily through grain-size effects that the once gaseous component is remembered.*

This split into D and G seems natural enough for Mg and Si (see Clayton, 1981a), because *all* the Mg and Si atoms are accreted by dust of some type before the dust itself aggregates; but it glosses over a major complication for oxygen in that most of it remains gaseous because it is too abundant to ever have been condensed (except as ice or dry ice). The amount of oxygen in refractory dust cannot exceed about 20% of all O, so the amount of G sputtered away from refractory dust has the same upper bound. When this lesser amount of oxygen at G is mixed with the remainder of interstellar O, only a reduced shift of interstellar O toward G results. This shift could conceivably result in composition E! We are then within bounds to speculate that if K, C, and H are aggregates of increasingly sputtered (per unit mass) interstellar dust, their reincorporation of sufficient oxygen for the mineral stoichiometry generates bulk parents lying on the mixing lines AE, KE, CE and HE (not shown, but see Lee *et al.*, 1980). The exothermic sintering of these refractory dust aggregates leaves spinels near the GD line, pyroxenes somewhat displaced toward E, and melilites almost to E. The ^{16}O-richness of the original dust results from its SUNOCON component (Clayton, 1975a; 1977a).

Despite many oversimplifications of the interstellar complexity, it is illuminating to quantitatively estimate the previous conjecture in terms of a two-component model—gas and dust. Suppose oxygen in initial dust lies near A, about $^{17}\delta \simeq {}^{18}\delta \simeq -45‰$. The sputtered oxygen G could be lighter by 4% per amu (Clayton, 1981a), so that the residual D could lie as far to the right (writing isotopic shifts in ppm) as $(-45, -45) + (40, 80) \simeq (-5, 35)$, well to the right of HAL hibonite near $(-12.3, 13.7)$ but on the same line of slope 1/2 through A. The gas sputtered away with this composition, on the other hand, would be near $G \simeq (-45, -45) - (40, 80) = (-85, -125)$. Assuming that only half of the 20% of 0 in dust is sputtered away, and that it mixes with a homogeneous gaseous O_{ISM} to produce point E, we require

$$(0.9)^{-1}[0.1(-85, -125) + 0.8 \, O_{ISM}] \simeq E \simeq (-1, +4)$$

with the solution $O_{ISM} \simeq (9, 20)$. This O_{ISM}, for what it's worth, would lie rather near the right end of the terrestrial fractionation line. The point of these numbers, however, is only to illustrate characteristic magnitudes involved in this way of thinking. The initial ISM must be much more complex, and in particular must include some ^{17}O-rich component involved in the ordinary chondrites. Clayton (1980c, d) has constructed a numerical multi-carrier model of cold accretion to show how the chondrite families *might* work, suggesting that an oxidizing agent (he chose O_2) deficient in ^{16}O is involved, because a specific ^{16}O deficiency maps into ^{17}O excesses when coupled with mass fractionation. This concept of a ^{16}O-deficient oxidizer has also been utilized recently by R. N. Clayton *et al.* (1981).

It is exceedingly interesting that the carbon and oxygen isotopic ratios observed (Wilson *et al.*, 1981) in the local interstellar medium are normal except for an $^{17}O/^{18}O$ ratio almost twice the solar value. This is probably a ^{17}O excess in the CO molecule, but whether this is a nucleosynthesis effect or a chemical effect is unknown. It indicates the type of specific information needed for a full accumulation model, however.

How then can it be imagined, even in principle, that K, C, and H are aggregates of increasingly sputtered refractory dust? The most physically direct idea would seem to involve grain-size effects, so that dynamic effects responsive to grain size could provide a sorting mechanism prior to aggregation. If interstellar grains were homogeneous, the most thoroughly sputtered grains would be the smallest grains, because only a skin of 100 Å or less should be altered by the sputtering. Considering this one might suppose that the common aggregates at A are of the largest refractory particles, with K, C, and H being increasingly fine-grained refractory aggregates. The obstacle to this line of thinking seems to be the difficulty of assembling refractory rich aggregates in the first place if the refractory dust is to be finely sized. In that case the CAI would have to be the last aggregates in a sedimenting system.

Clayton (1980a) has argued instead that the early accumulation of refractory aggregates comes about because these are composed of the largest particles. Then they can be aggregated preferentially. In this view the refractories have in part amalgamated into the largest particles, both because the largest SUNOCONs are initially the most refractory-rich (Clayton, 1980a), and because high-speed interstellar collisions resulting from turbulence may have fused the refractory cores of SUNOCONs into larger refractory particles while fracturing, vaporizing and dispersing their less refractory mantles. It is the 10^{10} years of cosmic chemical evolution that have produced larger than average refractory-rich particles as a component of the ISM. These are not nearly so large as the inclusions themselves, but might be large enough to be preferentially aggregated into parents of the inclusions.

The key is that refractory particles are layered, with increasingly less refractory elements outward. This means that the sputtering destroys magnesian silicate mantles before destroying the first SUNOCON condensates, and explains why Mg and Si are so much more heavily fractionated than is Ca (in K or C). On this view of chemical processing, accompanied by collisional core buildup, the oldest ISM refractory particles are the largest. They would then also be the most heavily sputtered owing to their long residence times in the ISM and have more heavily fractionated mantles also from the inelastic collisions that fused CaAl-rich cores. It would then follow that K and C are increasingly larger-grained aggregates of ISM refractory particles.

It is necessary to take public exception to the remark by Lee *et al.* (1980): "If sputtering is the fractionation mechanism, *ad hoc* assumptions such as the layering of elements according to their volatity in the target have to be invoked (D. D. Clayton, 1981a)." This layering is not *ad hoc* at all, but was predicted from first principles (Clayton 1977b, c; 1978; 1980a) of thermal condensation chemistry—predicted before the sputtering interpretation was found (Clayton, 1979b). This layered nature for the refractory sequence is very useful in explaining why Ca and Al are so much more heavily condensed in the ISM than are Mg and Si. The Mg and Si must be sputtered away before the Ca and Al are exposed. It is perhaps also generally overlooked that atmospheric nuclear tests produce a distribution of grain sizes and that the various radionuclides are layered on those grains according to the volatilities of their progenitors at the time of condensation (e.g., Freiling, 1963; Nathans *et al.*, 1970; Heft *et al.*, 1971).

The value of a theoretical picture is that it can throw old facts into a new light and also predict new correlations. This layered sputtering model is no exception, as we illustrate with a remarkable consequence for fossil ^{26}Mg. Grant, then, the following suppositions: (1) All Al condenses in SUNOCONs and remains there in the ISM, where ^{26}Al decays to Mg; (2) The coproduced Mg condenses in a refractory mantle around the Al and in a larger number of independently condensing Mg bearing particles; (3) this grain assembly is sputtered in the ISM. A straightforward consequence is that the heavily fractionated Mg will also bear a deficit of ^{26}Mg. This anticipated ^{26}Mg deficit was named a "ghost" by Clayton (1977b), who estimated its value as $^{26}\delta = -1.45\%_o$. This is very similar to the deficits after fractionation correction found by Wasserburg *et al.* (1977) in the FUN inclusions where Mg is strongly fractionated. Thus one can perhaps understand why there is a correlation between strongly fractionated Mg and ^{26}Mg deficits—at least in the ISM. It is a part of the cosmic chemical memory. Exactly how that correlation survives in the FUN inclusions is admittedly not so easy to see, but the existence of the possibility in principle is interesting.

Although it is plausible that high-speed collisions between core-mantle particles will preferentially destroy and disperse the mantles, it is by no means obvious that they can, except in improbably favorable circumstances, stick the cores together. In either case, refractory-rich particles are produced, but only in the latter case do they secularly grow in size owing to the high-speed collisions. Even if collisions at speeds in excess of 100 km s^{-1} are only disruptive, it is entirely possible that collisions at more moderate speeds within cold clouds have fused the larger refractory-rich particles. There is ample evidence of such particle growth in cold clouds and that it is due to grain aggregation rather than to the sticking of atoms and molecules. The reader may wish to consult a thorough description of one of the best studied cold clouds, R Cr A (Vrba *et al.*, 1981). They show that grain sizes increase into the cloud, and they present very detailed arguments to show that the velocity of the grain-grain collisions is about 50 ms^{-1} and that this velocity is understandable as a drift velocity of charged grains across magnetic field lines under the external force of gravity. The gravity would collapse the cloud except for the stabilizing effect of the magnetic pressure. Because this drift velocity is proportional to the mass of the particle, the initial spectrum of masses and compositions will be altered in interesting ways during this aggregation process (which occurs near T = 50 K). Much good theoretical work remains to be done on this problem. Keep in mind that these aggregated grains in part come back out of the cloud when it is dispersed by internal energetic events. The grains are then processed again by interstellar sputtering and higher-speed grain collisions in the warmer phases of the ISM. Thereafter they again enter another cold cloud in modified form to continue the evolutionary cycle. It is this context of cosmic chemical memory that may prepare a larger-than-average component of refractory-rich grains that may be sedimented preferentially when the solar system finally does form.

Haff *et al.* (1981) raise a serious question about the applicability of sputtering to the correlated fractionations of O, Mg and Si. Their "surface flux" model of the sputtering process predicts an opposite sense to the fractionation of O from those of Mg and Si in magnesian silicates. Experiments to test this prediction are urgently needed. If it is correct, a theory built on sputtering may have to be eliminated. If an enrichment of heavy O isotopes is only obtained from sputtered mantles of ice and dry ice (where O is the heavier species), it may prove impossible for those mantles to provide the O to go with the sputtered Mg and Si dust. Certainly the simple beauty of a sputtering interpretation is challenged by their theory.

METEORITIC HE AND NE ARE "PRESOLAR"

Manuel and Sabu (1975), Frick (1977), and Sabu and Manuel (1980) have repeatedly demonstrated that meteoritic He and Ne behave like "isotopes" of anomalous Xe-X. That is, such graphs as $^{136}Xe/^{132}Xe$ versus $He/^{132}Xe$ or $^{20}Ne/^{132}Xe$ from carbonaceous meteorites show good linear correlations. I here exclude the solar gas-rich meteorites. Such "3-isotope plots" suggest strongly (Sabu and Manuel, 1980) that the noble gases exist within the meteorites primarily in two pools: (1) meteoritic He, Ne, anomalous Ar, Xe-X, and anomalous Kr; (2) normal Ar, Kr, Xe without He and Ne. Our interpretation of Xe-X (Clayton, 1975a; 1976; 1977a; Clayton and Ward, 1978) is that of a special nucleosynthetic component carried in the ISM in special dust grains, as opposed to three other views: (a) Manuel and Sabu's (1975) interpretation of the solar system as the spatially inhomogeneous debris of a supernova; (b) Lewis et al.'s (1975) interpretation as heavy-element fission plus mass fractionation in the solar system; and (c) Cameron and Truran's (1977) interpretation as an inhomogeneous admixture to the solar system from a neighboring supernova trigger. I continue to regard my interpretation to be correct. It implies that the carriers were deficient in s-process Xe (Clayton, 1975a; Clayton and Ward, 1978), contained r-process and perhaps short-lived fission Xe (Clayton, 1975a; 1977a), contains some special p-contribution (Clayton, 1975a; 1976), and is probably strongly fractionated in some or all of these components. The spirit of these nucleosynthetic connections was already evident in the pioneering work of Manuel et al. (1972), but our emphasis lies on the process of supernova condensation of carriers as a means of getting those products into the meteorites. Considerable confusion over these distinct approaches is probably the result of not knowing which is actually correct.

The implications for the correlation of meteoritic trapped He and Ne with this nucleosynthetic component are far reaching. *Our approach implies that the meteoritic He and Ne are also presolar.* Let us describe that this way. Envision the time of the cold interstellar cloud that would one day collapse to form the solar system. Among the many dust particles within that cloud are a population of small carbonaceous carriers that are loaded with isotopically anomalous Xe-X and Kr-X. They are also loaded with He and Ne, although they are the only types of grains that manage to hold those light gases tightly. This would not be so surprising, because He and Ne are probably very difficult to trap in interstellar dust, but we know that the carbonaceous population is a very good trapper of noble gas or it would not have been able to condense in the expansions of supernova interiors and trap the p and r xenon ambient there.

Is it any wonder then that meteoritic trapped Ne is unlike that in either solar wind or terrestrial atmosphere, or that widespread isotopic anomalies in Ne are ubiquitous in meteorites? On the other hand it is also not surprising that a collection of these grains from a large number of past supernovae should have a neon isotopic composition resembling that of solar, because all Ne isotopes are synthesized in supernovae. But aggregates of different grain sizes and types may isotopically differ for Ne. It is especially interesting to note that interstellar Ne is ^{21}Ne-rich. Manuel and Sabu (1975) showed that, except for some cosmic-ray produced ^{21}Ne richness in gas-poor samples, the $^{21}Ne/^{22}Ne$ ratio increases almost linearly with either He/Xe ratio or $^{136}Xe/^{132}Xe$ ratio (the monitors of the richness in the interstellar component). It should not be excluded that this ^{21}Ne richness is a result of nuclear reactions within the dust during its interstellar history (Srinivasan et al., 1977; Ott et al., pers. comm., 1981). Such a component might ultimately explain why a Cold Bokkeveld (2×10^5 yr exposure age) residue has a higher $^{21}Ne/^{22}Ne$ ratio than Murray ($3-4 \times 10^6$ yr exposure age) residue (Reynolds et al., 1978). That this dust is not still much more rich in ^{21}Ne is explained by the fact that the carbonaceous carriers have a low density of cosmic ray targets (Mg, Na, Si) and that the spallation recoil may free most ^{21}Ne products from the probably small (100–1000 Å) carriers. The spallogenic gases from the silicate particles themselves may have been more easily lost.

That ^{38}Ar/^{36}Ar also correlates with He/Xe (Manuel and Sabu, 1975; Frick, 1977) indicates that SUNOCON-carried Ar is ^{38}Ar-rich. This could easily come about despite the fact that both ^{36}Ar and ^{38}Ar are supernova products. The ^{38}Ar/^{36}Ar ratio is high in the burning shells of C, O and Ne and falls to very low values in the Si-burning zone which is synthesizing large amounts of ^{36}Ar. This clue seems to suggest that the carriers formed in or outside of the expanding C and O-burning shells, rather than from deeper layers of the star. Whether this makes sense with the isotopic pattern of Xe-X or not depends upon whether r-and-p xenon also come from those burning shells. It seems unlikely that the major r-process yield comes from these zones, but they could contain a minor r-process (Howard et al., 1972; Lee et al., 1979) and a p-process (Woosley and Howard, 1978). Heymann and Dziczkaniec (1979, 1980) have published extensive numerical studies of the modifications of Xe and Kr isotopic compositions during the explosions of these intermediate zones of massive stars. They show that highly suggestive r-and-p-rich patterns can be established there. The issue not addressed in their work, nor in almost all work on nucleosynthetic production of isotopic anomalies, is the transport of the anomalies into the meteoritic samples. It is to this very problem that we have directed emphasis, leading to the development of a theory based on a hierarchy of precondensed phases of matter in a cold solar system (Clayton, 1977a; 1978; 1979a). In that spirit we repeat: *the only viable way to implant the type -X noble gas patterns in the meteoritic samples is to condense dust (SUNOCONS) in the expanding ejecta of the many past supernovae that have synthesized the elements, and to later have that dust admix differentially during the accumulation processes leading to the meteorites.*

Although the details are not clear, it is easy to conjecture how on this picture the bulk of the planetary gases (Ar, Kr, Xe) are isotopically normal and almost devoid of the He and Ne. The bulk of the noble gases are initially in the gaseous state, and that component is isotopically normal. As the density of the interstellar cloud increases, it becomes colder owing to the large efficiency for radiating internal energy as infrared energy from the grains. A temperature of order $T = 10$–50 K is reached and maintained in the collapse (e.g., Larson, 1973). At such low temperature Ar, Kr and Xe should in part freeze out onto grains, whereas He and Ne may remain gaseous. One perhaps needs a *threshold effect* such as this for incorporating relatively unfractionated Ar, Kr and Xe without incorporating He or Ne. Grains aggregate with other grains owing to collisions derived either from the turbulence of the collapsing cloud (Cameron, 1973b) or from the gravity induced drift velocity (Vrba et al., 1981). When the temperatures of these aggregates later rise, say owing to the temperature of a viscous disk, they will then be larger aggregates having a concentrated internal atmosphere of normal Ar, Kr and Xe. Exothermic chemical rearrangement is expected to fuse new minerals at that time (Clayton, 1980b), and those minerals may trap varying amounts of this isotopically normal planetary Ar, Kr, Xe that is being degassed from the chemically active aggregates. New carriers of both anomalous gases (X-pattern) and of normal gases (Y-pattern) may be chemically fused in those first reactions in the cold aggregates, but the He and Ne was carried into the aggregates in the same first generation carriers as were the X-pattern gases.

Consider the implausabilities in other alternatives. Because even refractory dust is vaporized partially by thermal and nonthermal sputtering in the interstellar medium, it is virtually certain that the noble gases were at one time gaseous in the interstellar medium. The 50% variations in ^{136}Xe/^{132}Xe ratios can survive in a small fraction of the Xe that remains in specific dust phases, but the bulk of the xenon, which is gaseous, is surely much more homogeneous. The model of Xe-X based on fission of a superheavy element within the solar system seems hopeless to me in view of the powerful arguments that (1) light Xe would not correlate with fission Xe (Manuel et al., 1972) or with anomalous Kr (Frick, 1977) and (2) bulk He and Ne should not correlate with fission Xe (Manuel and

Sabu, 1975). The supernova model of Manuel and Sabu (1975) fails on simple dynamic grounds. Following a supernova explosion, its interior is not cooled by expansion to 2000 K until it has expanded to 10 AU (Clayton, 1979,a), at which time it has a streaming kinetic energy of at least 10^{16} erg/gm ($\sim$ 1000 km/sec radial). The entire system, except for perhaps a neutron star or black hole, is blasted away violently, and is the last place where the solar system can form. Consider then the more popular notion (Cameron and Truran, 1977) that ejecta from a neighboring supernova were admixed into the condensing solar system. The minerals that trap enough injected ^{136}Xe to raise the ^{136}Xe/^{132}Xe ratio by 50% must also trap He and Ne to preserve that correlation. By contrast, the normal solar Ar, Kr, Xe must condense without condensing He or Ne. This peculiar contrast would demand that the supernova gases are kinetically *implanted* by their high speed into solar system dust, as described below in more general terms. But high speed gases would have almost no penetration into a cold cloud, so that such grains would have to receive the implantation at the edge of the solar cloud and later mix to the region of meteorite formation. Because of the sizeable astrophysical dust transport involved, it seems more reasonable to regard this idea as a special case of supernova dust. The larger problem with the spatial inhomogeneity of a single trigger lies in the lack of correlation of the different types of isotopic anomalies. For example, why does not excess ^{129}I or excess ^{36}Cl correlate with Xe-X? Of course this does not argue against a trigger dynamically, but only argues that a spatial inhomogeneity caused by the trigger is not the source of the isotopic inhomogeneity.

Let us agree for the moment that the conclusion (Clayton, 1975a; 1976) that Xe-X and its associated type-X noble gases are trapped in SUNOCONs is correct. Two exceedingly interesting questions are raised by the carbonaceous nature of the type-X carriers (Lewis *et al.*, 1975; Frick and Reynolds, 1977; Reynolds *et al.*, 1978; Ott *et al.*, pers. comm., 1981). The first is whether the carriers in the meteorites today were the interstellar carriers, and the second is how the interstellar SUNOCONs first condensed.

To the first question we may give only an opinion, but one that is reasonably based on facts. It is likely that the carriers surviving the acid dissolution of the meteorite were not identically the same carriers that harbored the X-gas in the ISM. When the cold-cloud carbonaceous matter aggregated along with its high X concentration, it was likely to have been composed of a wide variety of chemically unequilibrated forms. Subsequent heating of that primary aggregate probably led to sudden exothermic rearrangement (Clayton, 1980b) of the carbonaceous matter into the forms that today exist in the carbonaceous meteorites. No doubt some of the X-gases were lost in this process, but because it happened quickly, the initial carriers were either not totally degassed or the temporary internal atmosphere (many orders of magnitude more dense than in the nebula) was quickly retrapped by the reorganizing carbonaceous matter. If this picture is correct it could explain why initially separate components seem irretrievably mixed. For example, the *r* (fission) isotopes and the *p*-isotopes of Xe-X could have initially resided in separate but similar interstellar SUNOCONs that fused in nearly constant proportions into the meteoritic carriers. Likewise, interstellar trapped He and Ne need only to have resided in similar carriers. So it must be clear that it is not necessary for each carbonaceous SUNOCON to trap all of the X-type gases. The chemical fusion of today's carriers in the early solar system can also explain the relative isotopic normality of the carbon (e.g., Ott *et al.*, pers. comm., 1981). Carbonaceous matter condensing in inner zones of a supernova should isotopically be ^{12}C. But if many different types of interstellar carbonaceous matter have fused together during quick exothermic chemistry in the early solar system, the bulk carbon (and some other elements) may be nearly normal. Only those elements that condensed heavily in carbonaceous SUNOCONs may ultimately still be isotopically abnormal. The special feature of the noble gases is very clear: they are so volatile that the small percentage of them carried in the carbonaceous ISM matter can later be a significant fraction *of all trapped gas*. The principle is similar to that operative for oxygen (Clayton, 1977a), where the anomalous condensed portion remains in evidence because the huge abundance of gaseous O cannot be condensed in minerals.

The second question, how the implantation of type-X noble gas into SUNOCONs initially occurred, constitutes a major problem for scientific research. For the moment we point to two major possibilities and some associated problems. The major requirement is that X-gas be implanted before it mixes with the interstellar medium, which would make it isotopically normal. One approach is to condense SUNOCONs, which, according to their definition (Clayton, 1978), precipitate in the supernova interior during its expansion. The second approach is to implant the rapidly expanding noble gases in circumsupernova carbonaceous matter that condensed during the earlier loss of a carbon-star envelope. These two modes are illustrated as Figs. 3 and 4.

Mode A in Fig. 3 has several problems, any one of which could prove fatal in a complete analysis. The first is that almost the entirety of the massive-star interior has more oxygen than carbon. If $O > C$, thermodynamic equilibrium would lock up the C in CO molecules, after which it would be virtually impossible to gather those CO molecules into carbonaceous dust. Hoyle and Wickramasinghe (1977) have already suggested one possible way out of this constraint. The radiative combination of C and O atoms as a two-body radiative process has a small cross section, so that equilibrium may not be maintained. Nature may then find a way of growing carbon particles before the carbon is consumed into CO. That the ambient Xe and Kr can have the X signature has been illustrated by calculations by Heymann and Dziczkaniec (1979, 1980).

Carbon actually exceeds oxygen only in those parts of a massive star where He has not yet been totally consumed. Sudden expansion of partially burned He could then have $C > O$ and not face the CO trap. Another advantage of this zone is that it has He still available for trapping, because in the interior SUNOCONs He should be absent by virtue

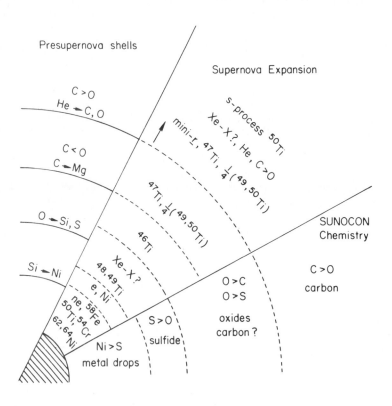

Fig. 3. Relevant isotopic and chemical effects in supernova. Presupernova structure is on the left; key isotopic effects in the expanding shells are in the center; SUNOCON chemistry is in the lower wedge. Illustrated is the relationship between the chemical form of SUNOCONs and their isotopic content.

of its absence from the ambient gas. Thus the association of He with Xe-X in C-SUNOCONs implies that either the He-burning zone is where Xe-X is made, or that turbulent mixing of the supernova interior has been extensive prior to the fall of the temperature to the point where SUNOCONs can condense. It is tempting to attribute the neutron-rich isotopes of Xe-X to the mini-r-process that can accompany the shock heating of the He zone (Truran *et al.*, 1978; Thielemann *et al.*, 1979; Blake *et al.*, 1981).

Mode B in Fig. 4 illustrates a second approach to the problem. The anomalous gases and He and Ne could be implanted in circumsupernova carbonaceous dust by the high speeds of the ejection, which range from several hundred to 10^4 km/sec. The approach illustrated suggests that the massive star ejected a carbon shell prior to the explosion of the remaining core. The carbon-rich shell could conceivably have been prepared by the same processes of carbon nucleosynthesis and mixing that result in the observed carbon stars (Iben, pers. comm., 1981). The loss of such a shell could result in the condensation of carbonanceous matter. This condensation could have trapped He directly, or the He could have been implanted when the dust is later overtaken by the fast moving supernova ejecta. In either case the anomalous heavy gases are simply driven into the dust. Such a process could not have very high efficiency, but it does not need to. If only one part in 10^5 of the supernova Xe-X is implanted in this way, that would suffice to account for the Xe-X concentration in carbonaceous chondrites which later accumulate from the dust. The Kr-X (Frick, 1977) must of course be a closely related process.

In another variant of Mode B, the dust simply preexists in the neighborhood of the supernovae, perhaps by virtue of the supernovae exploding within cold molecular clouds. The carbonaceous matter is later polymerized by the ultraviolet action of starlight on icy mantles (Greenberg, 1976) with the noble gases trapped during the polymerization process. Such a particle is also shown in Fig. 4. In this variant the preference for

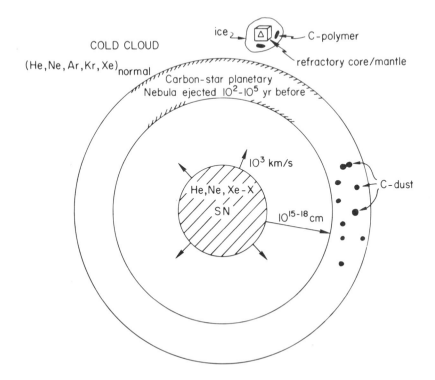

Fig. 4. A model for implanting He, Ne and Xe-X in circumstellar carbon dust. The carbon dust formed in a prior loss of a carbon-rich shell, and the noble gases are implanted by high speed combardment by the later supernova ejecta before these gases can mix with the ISM.

anomalous gases is explained by their being the fast moving gases. The normal ambient noble gases may move too slowly for the initial implantation.

The abundance ratio He/Ne ≈ 170 found in the type-X gas is about a factor of four smaller than the solar ratio. This is conventionally attributed to abundance fractionation in the trapping process. Ott et al., pers. comm., (1981) present an interesting expression for this fractionation. However it could also reflect in part the He/Ne ratio present in the above scenarios. I can see at present no easy way to decide between these alternatives, but it will ultimately be a useful fact.

How in these pictures do I interpret the failure of the Xe-X component to correlate with excess ^{129}Xe? Certainly any carbonaceous matter trapping Xe should equally well trap I. Lewis and Anders (1981) have reemphasized this argument as evidence in favor of their view that fission fragments are the origin of the heavy-isotope excesses in Xe-X. This explanation remains a possibility in this picture too; in fact, the initial suggestion (Clayton, 1975a) that Xe-X was carried in SUNOCONs also made the suggestion that shortlived ($\tau > 1$ month) fissioning progenitors at the source may have been trapped, rather than being required to survive until the formation of the solar system. I still see this as a possibility. But suppose it is anomalous Xe-X and Kr-X itself being trapped. The absence of ^{129}Xe excess then demands an absence of ^{129}I in the gas being trapped. One possibility would be that Xe-X condenses in SUNOCONs from a very specific zone that has not synthesized ^{129}I. Heymann (1981) has made this argument in these proceedings. I am skeptical of this argument, not because the nuclear astrophysics is inadequate, but rather because I cannot see how the condensation of the carbonaceous SUNOCONs could have been restricted to only this special zone. Neighboring zones are not greatly different in chemical composition and thermal history. But what other possibility for Xe-X exists other than SUNOCON transport for the journey from nucleosynthesis to meteorite parent? A more plausible resolution of this problem is that iodine is totally condensed before the anomalous noble gases are trapped. It is not impossible that all elements emerge from the interiors of the supernovae in condensed forms. My calculations show that if the sticking coefficients are taken to be unity, there is time to condense all of the interior matter in the outward flow (assuming that abundant nucleation has yielded a large number of small, 0.01–0.1 μ, particles rather than a few large ones). In that case the gaseous matter in supernova remnants consists of the interaction of its outer layers with the ISM and the subsequent sputtering of the high velocity dust after ejection. What is then indicated by the absence of fresh ^{129}I in Xe-X carriers is that the gaseous Xe-X is either trapped or implanted in carbonaceous SUNOCONs or circumstellar grains *after* the epoch of thermal condensation of the iodine. Halide condensation is not well studied, especially in environments lacking hydrogen and, in many cases, oxygen. But it is a relatively low-T condensate, so that either the carbonaceous SUNOCONs must be capable of condensing (or reforming by aggregation) at still lower temperatures or the implantation scenario of Fig. 4 is the applicable one. Yet another possibility is that the ^{129}I *is* trapped in part with Xe-X, but, because it is iodine, it rearranges into other chemical forms *before* decay and is therefore removed from the carbonaceous matter that will retain the noble gas. The SUNOCONs are, after all, subjected to some severe chemical treatment during their slowing down interactions with the surrounding ISM. It *is* a definite clue, as Lewis and Anders (1981) reemphasize, that Xe-X contains no ^{129}Xe excess, but it poses a greater problem for an injection from a supernova trigger than it does for a SUNOCON carried component. On the other hand, their fission-*in-situ*-plus-mass-fractionation model is of little credibility until it explains naturally the other gases associated with Xe-X, as Manuel and coworkers have repeatedly argued.

UBIQUITOUS TITANIUM ISOTOPIC ANOMALIES

The discoverers of isotopic anomalies in meteoritic titanium (Heydegger et al., 1979; Niederer et al., 1980; Niemeyer and Lugmair, 1981) have demonstrated that its occurrence is not restricted to FUN inclusions. Their presence is indeed ubiquitous in carbonaceous-chondrite matter, including the inclusions of oxygen-isotopic-signature A in Fig. 2 and matrix material. Clayton (1981b) has added to the discoverers' remarks on nucleosynthesis complexity his own view of how that complexity actually leads to the ubiquitous titanium anomalies. The key is the idea of cosmic chemical memory. As Clayton (1981b) said: "The clue is to be found in the nucleosynthesis of the Ti isotopes; however it is not the nucleosynthesis itself that provides the answer, but rather *the chemical circumstances within which the freshly synthesized nuclei find themselves.*"

Understanding of ^{50}Ti nucleosynthesis and the associated excesses to be expected in Cr, Fe and Ni isotopes is to be found in the concept of nuclear statistical equilibrium. Whenever the peak stellar temperatures exceed 4×10^9 K, the rates of all nuclear reactions come into equilibrium with their inverse reactions (Clayton, 1968). The nucleosynthesis in such a thermal equilibrium is called the *e*-process, following the designations of Burbidge *et al.* (1957), even though their concept of how it works has largely been replaced by the idea that the equilibrium shells eject proton-rich ^{54}Fe, ^{56,57,58,60}Ni instead of neutron-rich ^{56,57,58}Fe, ^{60}Ni. Hainebach *et al.* (1974) have discussed these ideas thoroughly, and in their work one can see clearly the basic equilibrium issues.

In a nuclear equilibrium, the isotopic composition of the elements depends upon the neutron/proton ratio of the matter. This parameter is normally described in terms of the *excess* number of neutrons per total number of nuclei: $\eta \equiv (N_n - N_p)/(N_n + N_p)$, where the numbers of neutrons and protons include both free nucleons and those bound into nuclei. It is stellar evolution that determines the evolution of η as the star evolves. It remains slightly above the value of zero that it would have were the evolution from *pure* He through *pure* ^{16}O burning only because of the other initial nuclei present in the He, roughly $\eta = 0.002$ to $\eta = 0.004$ during most of stellar evolution indicated in Fig. 3. But as ^{28}Si burns to ^{56}Ni (Bodansky *et al.*, 1968), the ^{56}Ni can capture electrons in the densest inner parts of the presupernova as those portions attempt to contract onto a collapsing core. This leads to a neutron-rich equilibrium, labeled ne in Fig. 3, near the matter attempting to follow the collapse to a neutron star (or black hole). In those inner zones, which can be dynamically ejected by shock heating, the number of excess neutrons per nucleon can easily approach $\eta = 0.1$ or more. Figure 5 shows, in four panels from calculations of Hainebach *et al.* (1974), the fraction by mass in the equilibrated matter of the progenitors of: ^{48,49,50}Ti and ^{50,51}V; (b) ^{50,52,53,54}Cr and ^{55}Mn; (c) ^{54,56,57,58}Fe and ^{59}Co; and (d) ^{58,60,61,62}Ni. The thing to notice is that as η approaches 0.08 excess neutrons per nucleon, the ^{50}Ti concentration rises dramatically, reaching 0.08 grms/gm at $\eta = 0.1$. Under the same transition, the other panels show a similar rise in ^{54}Cr, ^{58}Fe and ^{62}Ni. Because these four neutron-rich isotopes are underproduced in an otherwise impressive main line of nucleosynthesis, their origin has been ascribed to these *ne* zones, which can be thought of as a small fraction of the neutronized matter that almost fell onto the neutron star.

Figure 6 displays the best fit (in a least-squares sense) that can be obtained to the iron peak from two equilibrium zones that have been expanded and cooled. The low-η zone calculations are solid dots (the one for ^{50}Cr falling awkwardly into the triangle representing the natural abundance of ^{50}Ti). The high-η zone, represented by solid squares, rather naturally accounts for ^{50}Ti (low by a factor 2 for this single $\eta = 0.0769$), ^{54}Cr, ^{58}Fe and ^{62}Ni, which are otherwise not synthesized at all by the low-η matter. The ^{50}Ti production is still greater in admixtures of still higher η, as Fig. 5 shows. The inset shows the relative fractions of the two zones in the mixture. This is an impressive fit to the natural abundances in the iron abundance peak, suggesting that it is not far removed from the truth. It can be improved still further by going to three η zones (Hainebach *et al.*, 1974).

As far as titanium is concerned, one also notes that the ^{48}Ti/^{49}Ti ratio is produced about right in the low-η zone, and this production occurs in silicon burning as well as in the *e*-process (Bodansky *et al.*, 1968; Woosley *et al.*, 1973). I have shown this zone separation of Ti isotopes also in Fig. 3. That is, the lion's share of ^{48,49}Ti are coproduced in silicon-burning zones and in their transition to the low-η *e*-process. These calculations also showed that the nucleosynthesis in these zones is of their proton-rich progenitors,

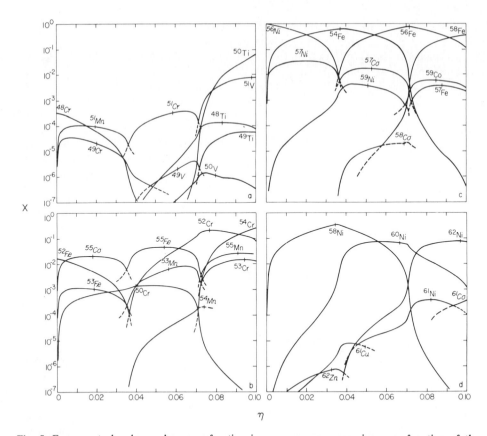

Fig. 5. Freeze-out abundances by mass fraction in an e-process expansion as a function of the number of excess neutrons per nucleon from Hainebach et al. (1974). Panels a, b, c, d show respectively the progenitors of (a) ^{48,49,50}Ti and ^{50,51}V; (b) ^{50,52,53,54}Cr and ^{55}Mn; (c) ^{54,56,57,58}Fe and ^{59}Co; and (d) ^{58,60,61,62}Ni. Correlated synthesis of ^{50}Ti, ^{54}Cr, ^{58}Fe, and ^{62}Ni occurs for matter more neutron-rich than $\eta = 0.08$. In low-η zones ^{48,49}Ti are synthesized as radioactive ^{48,49}Cr. A mixture of a large low-η mass and a small high-η mass is shown in Fig. 6.

^{48}Cr and ^{49}Cr (see also Fig. 5). During SUNOCON condensation some 1–6 months later, portions of these isotopes will condense as the progenitors ^{48}V(16d) and ^{49}V(330d). This condensation as V will clearly be much more severe at A = 49, so that a special extinct anomaly may be detectable in ^{49}Ti.

The ^{46}Ti is produced almost monoisotopically in the explosive ejection of oxygen-burning matter (Woosley et al., 1973). That is indicated as a ^{46}Ti zone in Fig. 3.

Howard et al. (1972) showed that ^{47}Ti and about one quarter of ^{49}Ti and ^{50}Ti can be produced by short explosive bursts of neutrons, such as in explosive carbon burning or helium burning. We await more modern recalculations of the titanium yield in such a mini-r process; but in Fig. 3, I show them as accounting for ^{47}Ti $+ \frac{1}{4}(^{49}$Ti $+ {}^{50}$Ti$)$ from the combined zones. This notation is repeated in both mini-r zones of Fig. 3 as a reminder that both contribute to this yield.

The relevance of this discussion, even with its many uncertainties and simplifications, is that it allows the construction of a chemical SUNOCON table for Ti. This approach has been begun by this writer (Clayton, 1978), but the ne zone was omitted from that discussion. It is here that one comes to a key element of cosmic chemical memory. The Ti SUNOCONs from different shells condense in different chemical forms. There is a natural transition inward in a massive star from carbonaceous SUNOCON, oxide

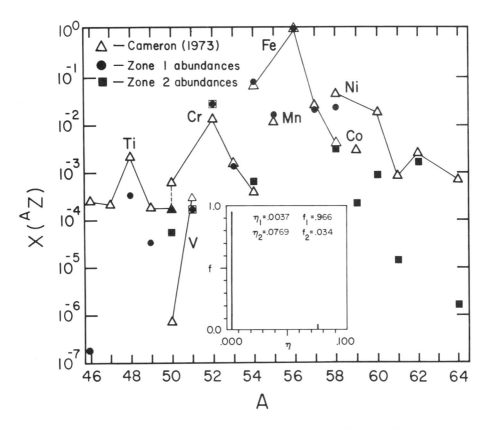

Fig. 6. Nucleosynthesis yields from the sum of a low-η zone and a high-η zone from Hainebach et al. (1974). Solid dots are the low-η yields and solid squares the high-η yields. Open triangles are natural solar abundances as tabulated by Cameron (1973a). Note that low-η ^{50}Cr yield has accidentally fallen in the ^{50}Ti abundance, which is itself produced only in the high-η zone. Inset gives values of η and mass fractions f of the two zones. The special correlated synthesis of ^{50}Ti, ^{54}Cr, ^{58}Fe, and ^{62}Ni is evident.

SUNOCON, sulfide SUNOCON, to metallic droplet SUNOCON. This chemical transition was first explicitly discussed by Clayton and Ramadurai (1977), but it was implied by Clayton (1975b). What this separation produces is the ejection of different Ti isotopic patterns in different chemical forms. It is this chemical separation that Clayton (1981b) argued to be the cause of ubiquitous isotopic anomalies in Ti. The transition in SUNOCON chemistry is labeled along the bottom of Fig. 3.

The reason for ^{50}Ti anomalies to be much the largest in Allende inclusions seems to relate naturally to its condensation in a metallic environment, without oxygen or sulfur. One is led to suspect small metal droplets in the ISM that are enriched in ^{50}Ti, ^{54}Cr, ^{58}Fe, and ^{62}Ni. One is also led to the realization that the condensation efficiency for ^{50}Ti will differ from that of the other Ti isotopes. For example, if the extreme conditions near the mass cut should prevent efficient condensation of the metal there, the ^{50}Ti could initially appear more gaseous in the ISM. It will then later plate out on the surfaces of abundant small particles. This separation by grain size provides the chemical fixer. Of course, it could be just the reverse—the metal could condense more efficiently than the sulfurous and oxidized Ti. At present, not enough is known to decide even the sign of that differential condensation effect.

Assuming 100% condensation of all the Ti, however, one can construct a SUNOCON table to replace the previous Ti table of Clayton (1978). This is done in Table 1. It is intended as a first approach to a theoretical interpretation of how the cosmos remembers

Table 1. Idealized Ti SUNOCONs[a].

(SUNOCON)$_{Zone}$	% sll SN Ti	$Ti(\Sigma\ Ti_j = 100)$
$^{(CaTiO_3)_{II}+(MgTiO_3)}III$	10%	(0, 74, 0, 13, 13)
$^{(CaTi(O,S)_3+TiS_2)}IV$	8%	(100, 0, 0, 0, 0,)
$^{(CaTiS_3+TiS_2)}V$	78%	(0, 0, 0.95, 5[b], 0)
$^{(FeNiCrTi)}Vne$	4%	(0, 0, 0, 0, 100)

[a]Format of Clayton (1978) to replace the Ti entry of that table. A zero in the idealized isotopic composition is to indicate that that yield is small so that it would not likely show up as an excess.
[b]Should condense as $^{49}V(330d)$ progenitor.

the titanium isotopes. The format of the table is identical to Clayton (1978), where it is explained further.

At the present time it seems adequate to simply identify the plausibility of such SUNOCON features. In what follows I turn from the explosive production of Ti isotopes to those that can have been synthesized in the prior hydrostatic s-process, both in the same presupernova stars and in other stars altogether.

The s-process of stellar neutron capture (Clayton *et al.*, 1961; Clayton, 1968) provides another potential source of ^{50}Ti excesses. Peters *et al.* (1972) analyzed this source while studying the idea of a weak neutron fluence (weak s-process) as the primary origins of ^{54}Cr, ^{58}Fe, and $^{61,62}Ni$. They observed that such a weak s-process also produces ^{50}Ti-rich titanium, but they discounted this origin as *the major source of all* ^{50}Ti because its overabundance factor was about a factor of ten smaller than that of ^{58}Fe. Furthermore, since the ^{54}Cr overabundance, which is the other significant one in the metal peak, was also 8–10 times smaller than that of ^{58}Fe, Peters *et al.* rightly concluded that a neutron-rich e-process rather than a weak s-process was the major origin of ^{50}Ti, ^{54}Cr, ^{58}Fe, and $^{62,64}Ni$, as discussed above. No more than about 5% of all ^{50}Ti can have been synthesized by this process.

Nonetheless, the s-process does occur and does create red-giant stellar atmospheres enriched in its products. During the loss of those atmospheres, s-rich STARDUST no doubt condenses to some degree, especially for refractory Ti. These should be oxide STARDUST. In fact, the bands of TiO are prominent in red-giant atmospheres, although large ^{50}Ti excesses cannot be seen in the relative band strengths in those stars that have been studied at high dispersion (Clegg *et al.*, 1979). More recently, Truran and Iben (1977) and Woosley and Weaver (1981) have also calculated the s-overabundances in giant atmospheres and confirm the significant excesses to be expected at ^{50}Ti, ^{54}Cr, and ^{58}Fe (in increasing overabundance). At first approximation these are almost monoisotopic sources.

Although this nucleosynthesis origin can be discounted as the primary producer of ^{50}Ti, these ^{50}Ti-rich STARDUST could conceivably be the cause of the small but ubiquitous isotopic anomalies in the meteorites. Their special chemical compositions and grain sizes could be preferentially taken up or rejected in specific macroscopic accumulates. Curiously enough, they will be accompanied by the same excesses (^{54}Cr, ^{58}Fe, $^{61,62}Ni$) suggested by the metallic SUNOCON picture. Note, however, that the excess ^{61}Ni isotope differs in the two pictures. Of course, these other metals need not be in the same oxide STARDUST, depending on the condensation chemistry. Another interesting point in my own unpublished calculations is that the ^{50}Ti excess from the s-process should be accompanied by a smaller ^{47}Ti deficit. This pattern is so suggestive of that found in Allende inclusions (Niederer *et al.*, 1980; Niemeyer and Lugmair, 1981) that I am eager to publish details; however, the lack of measured cross sections for the Ti isotopes holds up the quantitative meaningfulness of that calculation. At present, therefore, I am content to reveal the existence of this correlation. Perhaps s-process STARDUST is playing a major role in these Ti anomalies.

The inverted Ti pattern of the FUN inclusion Cl in comparison with normal inclusions suggests on this s-process interpretation that Cl is missing this s-process component. The biggest exception to this interpretation is the FUN inclusion EK1-4-1, which seems to be the only inclusion studied in which the $^{47}\delta$ and $^{50}\delta$ do not have opposite signs. So perhaps there is a component with opposite $^{47}\delta$ and $^{50}\delta$ that is masked in EK1-4-1 by another aggregation effect that masks the s-pattern. For example, it could contain extra SUNO-CONs from the mini-r-processes of He and C-burning zones of Fig. 3, which are estimated to (combined) produce all of ^{47}Ti but only $\frac{1}{4}$ of ^{49,50}Ti. In any case, the impression remains clear that Cl is missing several nuclei that are present in most Allende inclusions. It has been speculated (Clayton, 1980a; 1981a, b) that this deficiency has resulted from the more severe sputtering history of the Cl grains and the preferential accumulation of the largest refractory-rich grains into its parent accumulate. As incomplete as these lines of thinking are, they do point the way toward an approach to understanding based on the ways in which astrophysical history and aggregation prepare the cosmic chemical memory.

Acknowledgments—I thank the large numbers of meteoriticists who have encouraged me to continue to display my point of view, including Doug Phinney, who urged me to submit this paper when I was reluctant to do so, and Sid Niemeyer and Gunter Lugmair, who allowed me to discuss their Ti data prior to its publication, and Oliver Manuel and Urs Frick, who provided helpful reviews of this manuscript. But my great debt is to Ahmed El Goresy and Till Kirsten for four years of help and encouragement and to Al Cameron for inspiring many of the relevant ideas. This research was supported by NASA grant NSG-7361.

REFERENCES

Ballad R. V., Oliver L. L., Downing R. G., and Manuel O. K. (1979) Isotopes of Te, Xe and Kr in Allende meteorite retain record of nucleosynthesis. *Nature* **277**, 615–620.
Barlow M. J. and Silk J. (1977) Sputtering in interstellar shocks: a model for heavy element depletion. *Astrophys. J. Lett.* **211**, L83–L88.
Blake J. B., Woosley S. E., Weaver T. A., and Schramn D. N. (1981) Nucleosynthesis of nuetron-rich heavy nuclei during explosive helium burning in massive stars. *Astrophys. J.* **248**, 315–320.
Bodansky D., Clayton D. D., and Fowler W. A. (1968) Nuclear quasiequilibrium during silicon burning. *Astrophys. J. Suppl. Ser.* **16**, 299–371.
Burbidge E. M., Burbidge G. R., Fowler W. A., and Hoyle F. (1957) Synthesis of the elements in stars. *Rev. Mod. Phys.* **29**, 547–650.
Cameron A. G. W. (1962) Formation of the sun and planets. *Icarus* **1**, 13–69.
Cameron A. G. W. (1973a) Abundances of the elements in the solar system. *Space Sci. Rev.* **15**, 121–146.
Cameron A. G. W. (1973b) Accumulation processes in the primitive solar nebula. *Icarus* **18**, 407–450.
Cameron A. G. W. (1978) The primitive solar accretion disk and the formation of the planets. In *The Origin of the Solar System* (S. F. Dermott, ed.), p. 49–74 Wiley, N.Y.
Cameron A. G. W. and Truran J. W. (1977) The supernova trigger for the formation of the solar system. *Icarus* **30**, 447–461.
Clayton D. D. (1968) *Principles of Stellar Evolution and Nucleosynthesis.* McGraw-Hill, N.Y. 612 pp.
Clayton D. D. (1975a) Extinct radioactivities: trapped residuals of presolar grains. *Astrophys. J.* **199**, 765–769.
Clayton D. D. (1975b) ^{22}Na, Ne-E, extinct radioactive anomalies, and unsupported ^{40}Ar. *Nature* **257**, 36–37.
Clayton D. D. (1976) Spectrum of carbonaceous chondrite fission xenon. *Geochim. Cosmochim. Acta* **40**, 563–565.
Clayton D. D. (1977a) Solar system isotopic anomalies: supernova neighbor or presolar carriers. *Icarus* **32**, 255–269.
Clayton D. D. (1977b) Cosmoradiogenic ghosts and the origin of CaAl-rich inclusions. *Earth Planet. Sci. Lett.* **35**, 398–410.
Clayton D. D. (1977c) Interstellar potassium and argon. *Earth Planet. Sci. Lett.* **36**, 381–390.
Clayton D. D. (1978) Precondensed matter: key to the early solar system. *Moon and Planets* **19**, 109–137.
Clayton D. D. (1979a) Supernovae and the origin of the solar system. *Space Sci. Rev.* **24**, 147–226.
Clayton D. D. (1979b) Sputtering-produced FUN anomalies in the interstellar medium. *Meteoritics* **14**, 368–371.
Clayton D. D. (1980a) Chemical and isotopic fractionation by grain size separates. *Earth Planet. Sci. Lett.* **47**, 199–210.
Clayton D. D. (1980b) Chemical energy in cold-cloud aggregates: the origin of meteoritic chondrules. *Astrophys. J. Lett.* **239**, L37–L41.

Clayton D. D. (1980c) Internal chemical energy: heating of parent bodies and oxygen isotope anomalies. *Nukleonika* **25**, 1477–1490.
Clayton D. D. (1980d) A cold-accumulation model for oxygen isotopes. *Meteoritics* **15**, 275.
Clayton D. D. (1981a) Origin of Ca Al-rich inclusions II. Sputtering and collisions in the three-phase interstellar medium. *Astrophys. J.* **250**. In press.
Clayton D. D. (1981b) Ubiquitous titanium isotopic anomalies (abstract). In *Lunar and Planetary Science XII*, p. 151–153. Lunar and Planetary Institute, Houston.
Clayton D. D., Fowler W. A., Hull T. E., and Zimmerman B. A. (1961) Neutron capture chains in heavy element synthesis. *Ann. Phys.* **12**, 331–408.
Clayton D. D. and Remadurai S. (1977) On presolar meteoritic sulfides. *Nature* **265**, 427–428.
Clayton D. D. and Ward R. A. (1978) s-process studies: xenon and krypton isotopic abundances. *Astrophys. J.* **224**, 1000–1006.
Clayton R. N. and Mayeda T. K. (1977) Correlated oxygen and magnesium isotopic anomalies in Allende inclusions I. oxygen. *Geophys. Res. Lett.* **4**, 295–298.
Clayton R. N., Mayeda T. K., Gooding J. L., Keil K., and Olson E. J. (1981) Redox processes in chondrules and chondrites (abstract). In *Lunar and Planetary Science XII*, 154–156. Lunar and Planetary Institute, Houston.
Clayton R. N., Onuma N., and Mayeda T. K. (1976) A classification of meteorites based on oxygen isotopes. *Earth Planet. Sci. Lett.* **30**, 10–18.
Clegg R. E. S., Lambert D. L., and Bell R. A. (1979) Isotopes of Ti in cool stars. *Astrophys. J.* **234**, 188–199.
Consolmagno G. J. and Cameron A. G. W. (1980) The origin of the FUN anomalies and the high temperature inclusions in the Allende meteorite. *Moon and Planets* **23**, 3–25.
Draine B. T. and Salpeter E. E. (1979) On the physics of dust grains in a hot gas. *Astrophys. J.* **231**, 77–94.
Dwek E. and Scalo J. M. (1980) The evolution of refractory interstellar grains in the solar neighborhood. *Astrophys. J.* **239**, 193–211.
Freiling E. C. (1963) Theoretical basis for logarithmic correlations of fractionated radionuclide compositions. *Science* **139**, 1058–1059.
Frick U. (1977) Anomalous krypton in the Allende meteorite. *Proc. Lunar Sci. Conf. 8th*, p. 273–292.
Frick U. and Reynolds J. H. (1977) On the host phases of planetary noble gases in Allende (abstract). In *Lunar Science VIII*, 319–321. The Lunar Science Institute, Houston.
Greenberg J. M. (1976) Radical formation, chemical processing, and explosion of interstellar grains. *Astrophys. Space Sci.* **39**, 9–18.
Haff P. K., Watson C. C., and Tombrello T. A. (1981) Prediction of isotopic fractionation effects in sputtered minerals (abstract). In *Lunar and Planetary Science XII*, p. 380–382. Lunar and Planetary Institute, Houston.
Hainebach K. L., Clayton D. D., Arnett W. D., and Woosley S. E. (1974) On the e-process: its components and their neutron excesses. *Astrophys. J.* **193**, 157–168.
Heft R. E., Phillips W. and Steele W. (1971) Radionuclide distribution in the particle population produced by the Schooner cratering detonation. *Nucl. Technol.* **11**, 403–443.
Heydegger H. R., Foster J. J., and Compston W. (1979) Evidence of a new isotopic anomaly from titanium isotopic ratios in meteoritic minerals. *Nature* **278**, 704–707.
Heymann D. (1981) Comments on: "^{129}Xe and the origin of CCF xenon in meteorites" by R. S. Lewis and E. Anders in *Lunar and Planetary Science XII*. *Proc. Lunar Planet. Sci.*, 12B. This volume.
Heymann D. and Dziczkaniec M. (1979) Isotopic composition of Xe from intermediate zones of supernova. *Proc. Lunar Planet. Sci. Conf. 10th*, p. 1943–1960.
Heymann D. and Dziczkaniec D. (1980) A first roadmap of kryptology. *Proc. Lunar Planet Sci. Conf. 11th*, p. 1179–1213.
Howard W. M., Arnett W. D., Clayton D. D., and Woosley S. E. (1972) Nucleosynthesis of rare nuclei from seed nuclei in explosive carbon burning. *Astrophys. J.* **175**, 201–216.
Hoyle F. and Wickramasinghe N. C. (1977) Origin and nature of carbonaceous material in the galaxy. *Nature* **270**, 701–703.
Huneke J. C., Armstrong J. T., and Wasserburg G. J. (1981) ^{41}K and ^{26}Mg in Allende inclusions and a hint of ^{41}Ca in the early solar system (abstract). In *Lunar and Planetary Science XII*, p. 482–484. Lunar and Planetary Institute, Houston.
Larimer J. W. and Anders E. (1967) Chemical fractionation in meteorites–II. Abundance patterns and their interpretation. *Geochim. Cosmochim. Acta* **31**, 1239–1270.
Larson R. B. (1973) Processes in collapsing interstellar clouds. *Ann. Rev. Astron. Astrophys.* **11**, 219–238.
Lee T., Mayeda T. K., and Clayton R. N. (1980) Oxygen isotopic anomalies in Allende inclusion HAL. *Geophys. Res. Lett.* **7**, 493–496.
Lee T., Papanastassiou D. A., and Wasserburg G. J. (1977) ^{26}Al in the early solar system: fossil or fuel? *Astrophys. J. Lett.* **211**, L107–L110.
Lee T., Schramm D. N., Wefel J., and Blake J. B. (1979) Isotopic anomalies from neutron reactions during explosive carbon burning. *Astrophys. J.* **232**, 854–862.

Lewis R. S. and Anders E. (1981) ^{129}Xe and the origin of CCF xenon in meteorites (abstract). In *Lunar and Planetary Science XII*, p. 616–681. Lunar and Planetary Institute, Houston.

Lewis R. S., Srinivasan B., and Anders E. (1975) Host phase of a strange xenon component in Allende. *Science* **190**, 1251–1262.

Manuel O. K., Hennecke E. W., and Sabu D. D. (1972) Xenon in carbonaceous chondrites. *Nature* **240**, 99–101.

Manuel O. K. and Sabu D. D. (1975) Elemental and isotopic inhomogenities in noble gases: the case for local synthesis of the chemical elements. *Trans. Mo. Acad. Sci.* **9**, 104–122.

McKee C. F. and Ostriker J. P. (1977) A theory of the interstellar medium: three components regulated by supernova explosions in an inhomogeneous substrate. *Astrophys. J.* **218**, 148–169.

Nathans M. W., Thews R., Holland W. D., and Benson P. A. (1970) Particle size distributions in clouds from nuclear airbursts. *J. Geophys. Res.* **75**, 7559–7572.

Niederer F. R., Papanastassiou D. A., and Wasserburg G. J. (1980) Endemic isotopic anomalies in Ti. *Astrophys. J. Lett.* **240**, L73–L77.

Niemeyer S. and Lugmair G. W. (1981) Ubiquitous isotopic anomalies in Ti from normal Allende inclusions. *Earth Planet. Sci. Lett.* **53**, 211–225.

Peters J. G., Fowler W. A., and Clayton D. D. (1972) Weak s-process irradiations. *Astrophys. J.* **173**, 637–648.

Reynolds J. H., Frick U., Neil J. M., and Phinney D. L. (1978) Rare-gas-rich separates from carbonaceous chondrites. *Geochim. Cosmochim. Acta* **42**, 1775–1797.

Sabu D. D. and Manuel O. K. (1980) Noble gas anomalies and synthesis of the chemical elements. *Meteoritics* **15**, 117–138.

Srinivasan B., Gros J., and Anders E. (1977) Noble gases in separated meteoritic minerals: Murchison (C2), Ornans (C3), Karoonda (C5) and Abee (E4). *J. Geophys. Res.* **82**, 762–778.

Thielemann F., Arnould M., and Hillebrandt W. (1979) Meteoritic anomalies and explosive neutron processing of He-burning shells. *Astron. Astrophys.* **74**, 175–185.

Truran J. W., Cowan J. J., and Cameron A. G. W. (1978) The He-driven r-process in supernovae. *Astrophys. J. Lett.* **222**, L63–L68.

Truran J. W. and Iben I. (1977) On s-process nucleoynthesis in thermally pulsing stars. *Astrophys. J.* **216**, 797–810.

Vrba F. J., Coyne G. V., and Tapia S. (1981) Observations grain and magnetic field properties of the R Coronae Australis dark cloud. *Astrophys. J.* **243**, 489–511.

Wasserburg G. J., Lee T., and Papanastassiou D. A. (1977) Correlated O and Mg isotopic anomalies in Allende inclusions II. Mg. *Geophys. Res. Lett.* **4**, 299–302.

Wasserburg G. J., Papanastassiou D. A., and Lee T. (1980) Isotopic heterogeneities in the solar system. In *Early Solar Processes and the Present Solar System*, p. 144–191. Soc. Italiana di Fisica, Bologna.

Wilson R. W., Langer W. D., and Goldsmith P. F. (1981) A determination of the carbon and oxygen isotopic ratios in the local interstellar medium. *Astrophys. J.* **243**, L47–L52.

Woosley S. E., Arnett W. D., and Clayton D. D. (1973) The explosive burning of oxygen and silicon. *Astrophys. J. Suppl. Ser.* **26**, 231–312.

Woosley S. E. and Howard W. M. (1978) The p-process in supernovae. *Astrophys. J. Suppl. Ser.* **36**, 285–304.

Comments on: "Xe^{129} and the origin of CCF xenon in meteorites" by R. S. Lewis and E. Anders

D. Heymann

Department of Geology, Department of Space Physics and Astronomy, Rice University, Houston, Texas 77001

Abstract—The case of Lewis and Anders (1981a, b) against "supernova" theories for the origin of CCF-Xe would have been a welcome constraint for these theories if the case were firmly acceptable, which it is not. There are locations in supernovae where the ^{129}I/^{136}Xe yield ratio of a "mini r-process" could be as small as, or smaller than 2.4×10^{-4}.

There are two major models for the origin of CCF-Xe (= Carbonaceous Chondrite Fission Xenon*) in primitive meteorites. One model posits that superheavy elements have been formed by an r-process of nucleosynthesis (Anders and Heyman, 1969), that these elements have decayed quantitatively either in interstellar media (Clayton, 1975), or else that comparatively volatile superheavy elements have decayed *in situ* (= in the meteorites of interest) after their condensation from the solar nebula (Lewis et al., 1975; Anders et al., 1975). The cosmochemical properties of the superheavy elements for the "*in situ* decay variant*" were loosely deduced as volatile, perhaps as volatile as Xe itself, by Anders and Heymann (1969). The work by Anders et al. (1975) is the first in a series of attempts to pin down these properties better. In this variant, the abundance of CCF-Xe in meteorites has the potential of becoming an ingredient of a cosmochronometer and/or solar system chronometer. The other ingredients are the cosmochemical properties of the CCF-progenitor and its half-life. If the half-life of the progenitor is close to, or longer than, that of ^{244}Pu, then CCF-Xe might be a relic from general, continuous nucleosynthesis. If, however, its half-life is close to, or shorter than, those of ^{129}I and ^{247}Cm, then CCF-Xe is likely to be the relic from r-process nucleosynthesis in a nearby source with late stage "pollution" of the solar system by the superheavy elements. I call the general variant with *in situ* decay the "superheavy" model.

The other model, called here the "supernova" model (Manuel et al., 1972; Heymann and Dziczkaniec, 1979, 1980a) proposes the direct formation of CCF-Xe by an r-process in a supernova without the intermediary of fissile progenitors. Heymann and Dziczkaniec (1979, 1980a) have studied the nature of an r-process which satisfies the principal boundary condition that the ^{136}Xe/^{134}Xe yield must be 1.4. Although the "CCF-Xe" of this variant does not have to be brought from its source to the solar system comparatively quickly because no unstable nuclei are involved, a comparatively fast transfer from a nearby source to the solar-system, first proposed by Cameron and Truran (1977), has been adopted in our work (Heymann and Dziczkaniec, 1979, 1980a). In our "supernova" model, CCF-Xe must be trapped in supernova condensates soon after its formation, without excessive dilution by unprocessed Xe from other zones of the massive star (see Heymann and Dziczkaniec, 1979, 1980a).

Lewis and Anders (1981a, b) have published a test for "supernova" models along the

---
*CCF-Xe is a misnomer for those "supernova" models which do not assume that fission is involved in its formation. Heymann and Dziczkaniec (1980a) therefore use the term H-Xe, coined by Pepin and Phinney (1981).

1803

following lines:

1. ^{129}I is formed primarily by *r*-process nucleosynthesis in comparable amounts to ^{136}Xe (Heymann and Dziczkaniec have pointed out in their 1979 paper that significant ^{129}I/^{136}Xe yield ratios are expected from locations of the supernova where the peak explosion temperature is 2.0×10^9 °K, i.e., $T_{9m} = 2.0$).
2. Sample 3CS5 of the Allende meteorite contains bromine, hence probably iodine in near-cosmic abundance (but normalized to Fe $\equiv 1.0$), whereas the abundances of ^{129}Xe and ^{130}Xe are only 10^{-4}. The Br abundance plus those of other elements in 3CS5 define a "condensation sequence", calculated for the solar nebula, but posited to be not too different in a post-supernova environment.
3. *Ergo*: where is the I-derived radiogenic ^{129}Xe in 3CS5 as "predicted" by the "supernova" models?

This is, in principle, a valid test which touches not only on the origin of CCF-Xe (an *r*-process of nucleosynthesis lies at the core of both models!), but also on known problems of ^{129}I-^{129}Xe systematics in meteorites (see Wasserburg and Huneke, 1979; Villa *et al.*, 1981). Let us first remember that the apparent absence of radiogenic ^{129}Xe in 3CS5 with so much Br (= I?) around can be potentially just as embarrassing for the "super-heavy" model if the half-life of the CCF-progenitor is comparable to, or shorter than, that of ^{129}I. Let us also keep in mind that the test is far from conclusive from terrestrially acid-treated and chemically etched samples such as 3CS5 unless all of the following are correct:

1. That the Br-carrying phase or phases in 3CS5 (in the following I will use the singular only) has not been formed by metamorphism on the Allende parent body without live ^{129}I present,
2. That the Br-carrying phase in 3CS5 has not been formed by the chemical treatments in the Chicago laboratory,
3. That the Br-carrying phase in 3CS5 is not a genuine pre-terrestrial phase which has picked up Br during the chemical treatments without detectable changes of its internal atomic arrangements.

Live ^{129}I *has* occurred in the Allende meteorite (see Villa *et al.*, 1981; Armstrong and Wasserburg, 1981). Very little, if any of it is associated with the acid-resistant, etched residues as Lewis and Anders have pointed out, but its earlier presence is clearly evident in other parts of this meteorite, e.g., the Pink Angel inclusion. In the context of the theory of Lewis and Anders this would seem to require that the metamorphism of the interior of Pink Angel predates the formation of the Br-carrying phase in the residues. In my thinking, the observation dovetails with the idea that the pollution of the solar-system with explosive products from a single, nearby supernova has been grossly nonuniform, i.e., greatly localized.

Let me finish with an addendum to the "supernova" theory as developed by Heymann and Dziczkaniec (1979, 1980a). These investigators have reported on the formation of ^{129}Xe and ^{129}Te by neutron capture in an *s*-enhanced seed during a "mini *r*-process" at locations of a supernova where the peak explosion temperature is 2.0×10^9 °K ($T_{9m} = 2.0$), but they have not reported on the formation of ^{129}Sb and ^{129}Sn, nor on the probability of survival of A = 129 species at $T_{9m} = 2.0$ or locations of higher explosion temperatures deeper into the O, Ne-rich zone of a massive star. Table 1 reports the results of calculations (Dziczkaniec, 1980) for the formation of nuclei of interest here by rapid neutron capture in an *s*-enhanced seed at $T_9 = 2.0$. The calculation is parametrized with neutron dose Φ_n in units of mole cm^{-3} s, the primary parameter because the deduced neutron doses in supernovae are variable and somewhat uncertain. Note how "waves" of nuclei travel through all points Z, A, except 54,136 (= ^{136}Xe) where the neutron capture chain in Xe is arrested for the less than one second duration of this "mini *r*-process".

Table 1. Abundance of $A = 129$ and $A = 130$ species of interest and of ^{136}Xe, for neutron capture in an s-enhanced seed at $T_9 = 2.0$. The parameter Φ_n is neutron dose (i.e., integral of neutron density and time) in mole cm^{-3} s. All abundances are normalized to that of ^{132}Xe in the untransformed, s-enhanced seed. Numbers in parentheses are factors of ten. Data from Dziczkaniec (1980).

Species	From Seed	Φ_n(mole cm^{-3} s)							
		1(−8)	5(−8)	1(−7)	2(−7)	4(−7)	6(−7)	8(−7)	(−6)
^{136}Xe	Xe	9.4(−6)	4.2(−3)	4.4(−2)	3.1(−1)	1.1(0)	1.5(0)	1.7(0)	1.7(0)
^{129}Xe	Xe	4.6(−2)	1.1(−2)	1.3(−3)	2.0(−5)	1.9(−6)	9.2(−13)	2.0(−16)	4.3(−20)
^{130}I	I	2.0(−3)	8.6(−3)	1.6(−3)	2.1(−5)	2.8(−9)	8.6(−13)	4.7(−17)	5.4(−19)
^{129}I	I	1.4(−2)	1.4(−2)	1.9(−3)	2.3(−5)	3.0(−9)	3.9(−13)	5.1(−17)	6.6(−21)
^{129}Te	Te	1.5(−3)	8.0(−2)	2.3(−1)	2.6(−1)	4.4(−2)	4.0(−3)	3.1(−4)	2.4(−5)
^{130}Sb	Sb	1.3(−9)	1.9(−4)	4.6(−3)	7.6(−3)	2.1(−4)	2.5(−6)	2.8(−8)	3.2(−10)
^{129}Sb	Sb	5.0(−8)	1.4(−3)	1.5(−2)	1.2(−2)	2.1(−4)	2.5(−6)	2.8(−8)	3.2(−10)
^{129}Sn	Sn	<(−50)	1.5(−10)	5.3(−8)	1.4(−5)	2.1(−3)	2.4(−2)	1.0(−1)	2.4(−1)

Because neutron capture rates at constant neutron density ρ_n vary less than 5% with temperature near 2.0×10^9 °K (see Heymann and Dziczkaniec, 1979), the results of Table 1 are essentially valid for locations in the supernova where $1.8 \leq T_{9m} \leq 2.2$.

The best-established, hence primary boundary condition for this "mini r-process" is that the ^{136}Xe/^{134}Xe yield ratio must be 1.4. Heymann and Dziczkaniec (1979) have shown that this condition is satisfied for $\Phi_n \approx 3 \times 10^{-7}$ mole cm^{-3} s for locations where $T_{9m} = 2.0$, and is $\Phi_n \approx 6 \times 10^{-7}$ for locations where $T_{9m} = 2.2$. The necessary increase of Φ_n from $T_{9m} = 2.0$ to $T_{9m} = 2.2$ is due to the post-neutron capture transformation of ^{135}Xe nuclei to ^{134}Xe nuclei by photoneutron emission in a still very hot environment. Note from Table 1 that the ^{129}Xe/^{136}Xe yield ratio is very small in the range $3 \times 10^{-7} < \Phi_n \leq 6 \times 10^{-7}$ mole cm^{-3} s, that the ^{130}I/^{136}Xe yield ratio is small, that the ^{129}I/^{136}Xe yield ratio is small, that the ^{130}Sb/^{136}Xe yield ratio is small, and that the ^{129}Sb/^{136}Xe yield ratio is small. The potentially "dangerous" suppliers of ^{129}I in the post-explosion matter from locations $T_{9m} = 2.2$ are ^{129}Te and ^{129}Sn which decay to ^{129}I. At the end of the neutron capture phase, with $\Phi_n = 6 \times 10^{-7}$ mole cm^{-3} s, the ^{129}I/^{136}Xe, ^{129}Xe/^{136}Xe, ^{129}Te/^{136}Xe, ^{129}Sb/^{136}Xe, and ^{129}Sn/^{136}Xe yield ratios are 2.5×10^{-13}, 2.7×10^{-3}, 1.7×10^{-6}, and 1.6×10^{-2} respectively, but $\Phi_n = 6 \times 10^{-7}$ mole cm^{-3} s implies $T_{9m} = 2.2$.

Table 2 presents probabilities of survival of nuclear species of interest in an environment which contains no free neutrons, for exponential cooling with a time-parameter of one second for $T_{9m} = 1.8$, 2.0, or 2.2. Note that ^{136}Xe, ^{130}Xe, ^{129}I, ^{130}Te, and ^{129}Sb survive near-quantitatively, once formed by neutron capture, at all locations of the supernova where $1.8 \leq T_{9m} \leq 2.2$, but that ^{130}I, ^{129}Te, and ^{129}Sn can be substantially transformed to ^{129}I, ^{128}Te, and ^{128}Sn at locations with increasing T_{9m}. The transformation of ^{130}I to ^{129}I is not a very salient process here because of the small abundance of ^{130}I for $3 \times 10^{-7} \leq \Phi_n \leq 6 \times 10^{-7}$ mole cm^{-3} s (analogous arguments show that ^{130}Sb and ^{130}Sn can contribute only little to A = 129 by their "photoerosion"). However, the vulnerabilities of ^{129}Te and ^{129}Sn to photoneutron emission are salient because they mean that the *corrected* ^{129}Te/^{136}Xe and ^{129}Sn/^{136}Xe could be at least two orders of magnitude smaller than those mentioned earlier, i.e., $\leq 4 \times 10^{-5}$ and $\leq 2.4 \times 10^{-4}$ respectively.

The "mini r-process" treated by Heymann and Dziczkaniec (1979, 1980a) does not operate at locations in supernovae where $T_{9m} \gtrsim 2.3$ (see also Woosley and Howard, 1978) because it is masked there by the formation of p-nuclei by photodisintegration. The process is unlikely to be significant at locations $T_{9m} \lesssim 1.8$ because of a dearth of free

Table 2. Probabilities of survival of nuclear species of interest for exponential cooling in an environment which contains no free neutrons from $T_{9m} = 1.8$, 2.0, or 2.2 with an assumed time-parameter of one second. The probability of quantitative survival is 1.00.... Data are either from Dziczkaniec (1980) or are calculated with the method outlined in Dziczkaniec and Heymann (1980b). The latter are for ^{129}Sb and ^{129}Sn. For time-parameters longer than one second, survival probabilities decrease and *vice versa*.

Species	Survival Probability		
	$T_{9m} = 1.8$	$T_{9m} = 2.0$	$T_{9m} = 2.2$
^{136}Xe	> 0.99	> 0.99	0.98
^{130}Xe	> 0.99	> 0.99	> 0.99
^{129}I	> 0.99	> 0.99	> 0.99
^{130}Te	> 0.99	> 0.99	0.99
^{129}Sb	> 0.99	> 0.99	> 0.99
^{130}I	0.99	0.36	≤ 0.01
^{129}Te	0.98	0.31	≤ 0.01
^{129}Sn	0.99	0.37	≤ 0.01

neutrons (see also Weaver and Woosley, 1980). The point which is to be distilled from Tables 1 and 2 is that the $A = 129/^{136}$Xe yield ratio of matter accompanying CCF-Xe cannot be deduced for our model from what is predicted by the "classical" variant of the r-process, that the yield ratio could be as large as 0.1 (Heymann and Dziczkaniec, 1979) or as small as, or smaller than, 2.4×10^{-4} depending on the *exact* conditions in the O, Ne-rich zone where $1.8 \leq T_{9m} \leq 2.2$, and depending on whether CCF-Xe in the "supernova" model represents matter from all of these locations, or only from a limited selection of them.

Acknowledgments—I want to thank Drs. D. D. Clayton and P. Pellas for their valuable comments. The work is supported by NASA grant NGL 44-006-127.

REFERENCES

Anders E. and Heymann D. (1969) Elements 112 to 119: Were they present in meteorites? *Science* **164**, 821–823.

Anders E., Higuchi H., Takahashi H., and Morgan J. W. (1975) Extinct superheavy element in the Allende meteorite. *Science* **190**, 1262–1271.

Armstrong J. T. and Wasserburg G. J. (1981) The Allende Pink Angel: Its mineralogy, petrology, and the constraints of its genesis (abstract). In *Lunar and Planetary Science XII*, 25–27. Lunar and Planetary Institute, Houston.

Cameron A. G. W. and Truran J. W. (1977) The supernova trigger for formation of the solar system. *Icarus* **30**, 447–461.

Clayton D. D. (1975) Extinct radioactivities: trapped residuals of presolar grains. *Astrophys. J.* **199**, 765–769.

Dziczkaniec M. (1980) Isotopic anomalies from nuclear processes in stars. Ph.D. Thesis, Rice University, Houston. 176 pp.

Heymann D. and Dziczkaniec M. (1979) Xenon from intermediate zones of supernovae. *Proc. Lunar and Planet. Sci. Conf. 10th*, 1943–1959.

Heymann D. and Dziczkaniec M. (1980a) A first roadmap for kryptology. *Proc. Lunar Planet. Sci. Conf. 11th*, 1179–1213.

Heymann D. and Dziczkaniec M. (1980b) Xenon, osmium, and lead formed in O-shells and C-shells of massive stars. *Meteoritics* **15**, 1–14.

Lewis R. S., Srinivasan B., and Anders E. (1975) Host phases of a strange Xe component in Allende. *Science* **190**, 1251–1262.

Lewis R. S. and Anders E. (1981a) Xe^{129} and the origin of CCF xenon in meteorites (abstract). In *Lunar and Planetary Science XII*, p. 616–618. Lunar and Planetary Institute, Houston.

Lewis R. S. and Anders E. (1981b) Isotopically anomalous xenon in meteorites: A new clue to its origin. *Astrophys. J.* In press.

Manuel O. K., Hennecke E. W., and Sabu D. D. (1972) Xenon in carbonaceous chondrites. *Nature* **240**, 99–101.

Pepin R. O. and Phinney D. (1981) Components of xenon in the solar system. *Moon and Planets*. In press.

Villa I. M., Huneke J. C., Papanastassiou D. A., and Wasserburg G. J. (1981) The Allende Pink Angel: Chronological constraints from Xe, Ar, and Mg (abstract). In *Lunar and Planetary Science XII*, p. 1115–1117. Lunar and Planetary Institute, Houston.

Wasserburg G. J. and Huneke J. C. (1979) I-Xe dating of I-bearing phases in Allende (abstract). In *Lunar and Planetary Science X*, p. 1307–1309. Lunar and Planetary Institute, Houston.

Weaver T. A. and Woosley S. E. (1980) Evolution and explosion of massive stars. *Ann. N. Y. Acad. Sci.* **336**, 335–357.

Woosley S. E. and Howard W. M. (1978) The p-process in supernovae. *Astrophys. J. Suppl.* **36**, 285–304.

Cosmic-ray-produced stable nuclides: Various production rates and their implications

Robert C. Reedy

Nuclear Chemistry Group, Mail Stop 514, University of California, Los Alamos National Laboratory, Los Alamos, New Mexico 87545

Abstract—The rates for a number of reactions producing certain stable nuclides, such as ^{3}He and ^{4}He, and fission in the moon are calculated for galactic-cosmic-ray particles and for solar protons. Solar-proton-induced reactions with bromine usually are not an important source of cosmogenic Kr isotopes. The ^{130}Ba(n, p) reaction cannot account for the undercalculation of ^{130}Xe production rates. Calculated production rates of ^{15}N, ^{13}C, and ^{2}H agree fairly well with rates inferred from measured excesses of these isotopes in samples with long exposure ages. Cosmic-ray-induced fission of U and Th can produce significant amounts of fission tracks and of ^{86}Kr, ^{134}Xe, and ^{136}Xe, especially in samples with long exposures to cosmic-ray particles.

INTRODUCTION

Cosmic-ray-induced reactions in extraterrestrial matter are important sources of certain stable nuclides, such as the noble gases (Hohenberg et al., 1978) and ^{15}N (Becker et al., 1976). Cosmic-ray-induced reactions with uranium and thorium could produce a significant fraction of the fissions observed in certain samples (Woolum and Burnett, 1974; Damm et al., 1978), and cosmogenic fissions might be an important source of certain nuclides, such as ^{86}Kr and ^{136}Xe (Regnier, 1977). These stable products of cosmic-ray reactions in extraterrestrial matter are important both for studies of the exposure of samples to cosmic rays and for corrections which must be applied to raw data to deduce other "components" (e.g., "trapped" noble gases).

Calculated production rates for neon, argon, krypton, and xenon were reported and compared with observed results by Hohenberg et al. (1978). Improved krypton calculations and comparisons were given by Regnier et al. (1979). Calculated production rates for several noble-gas nuclides (e.g., ^{130}Xe) did not agree well with rates inferred from observed data and certain target elements (bromine, thorium, and uranium) were not included. The reaction ^{130}Ba(n, p) is studied here to see if it could improve the agreement of calculated and observed ^{130}Xe concentrations. The possible importance of solar-proton reactions with bromine in producing Kr isotopes is examined. Cosmogenic fission is considered as sources of ^{86}Kr, ^{134}Xe, and ^{136}Xe. Rates for a number of cosmogenic reactions have not been calculated (e.g., ^{13}C) or reported (e.g., He) previously. Reported here are the rates for producing ^{3}He, ^{4}He, ^{15}O (which decays to ^{15}N), ^{13}C, and ^{2}H and for the cosmogenic fission of thorium and uranium by energetic cosmic-ray particles (including the production of ^{86}Kr, ^{134}Xe, and ^{136}Xe).

Reactions induced by both galactic-cosmic-ray (GCR) particles and solar protons are considered for most products. The lunar particle fluxes of Reedy and Arnold (1972) were used. Two spectral parameters for solar protons (R_0 = 100 and 200 MV) were used with the exponential-rigidity expression, $dJ/dR = K \exp(-R/R_0)$, for proton flux J. In lunar samples, most radionuclides have been produced by solar protons which had a spectral shape with R_0 = 100 MV (e.g., ^{26}Al and ^{53}Mn; Kohl et al., 1978). However, the spectral shape of solar protons producing 2.1×10^5-year ^{81}Kr appears to have been much harder (Reedy, 1980), hence the inclusion of rates for a spectral parameter of R_0 = 200 MV. An omnidirectional (4π) flux of 100 protons/cm^2 s for protons with E > 10 MeV was adopted.

The actual fluxes of solar protons during various time periods have varied considerably about this adopted flux (Reedy, 1980). Wherever possible, experimental excitation functions—cross sections as a function of energy—were used; otherwise, excitation functions were estimated from cross sections for analogous reactions or by theoretical systematics. Production rates for making ^{22}Ne from Mg calculated with the new excitation functions of Reedy et al. (1979) agreed within 1% with those reported earlier by Hohenberg et al. (1978).

HELIUM

Rates for the GCR and solar-proton production of ^{3}He and ^{4}He were not reported in Hohenberg et al. (1978) because there were no good observed He data with which to make comparisons. There appear to have been significant ($\sim$30%) diffusion losses of Ne in lunar samples (Hohenberg et al., 1978), and diffusion losses of He are expected to be much larger. Yaniv and Kirsten (1981) reported retention of only $\cong$1% of cosmogenic ^{3}He in the top centimeter of lunar rock 65315. They also noted that an observed excess of ^{3}He above that expected from cosmic-ray-induced reactions in the top millimeter was evidence for solar-flare-implanted ^{3}He. Rao and Venkatesan (1980) and Venkatesan et al. (1980) observed that the ratios of cosmogenic ^{3}He to cosmogenic ^{21}Ne varied over narrow ranges in several samples of rocks 61016 and 66435 and that the solar-proton exposure ages for ^{3}He agreed with the ^{21}Ne and ^{38}Ar exposure ages, and concluded that there was no significant depth-dependent diffusion in their samples. Cosmogenic helium generally is retained in meteorites, and ^{3}He is often used in studies of cosmic-ray exposure ages (see, e.g., Cressy and Bogard, 1976). Meteorite data will be used below in comparing observed and calculated ^{3}He production rates.

Excitation functions were compiled for the production of ^{3}He and ^{4}He from the major target elements (oxygen, magnesium, aluminum, silicon, and iron). The excitation functions in Reedy and Arnold (1972) for the production of ^{3}H (which decays to ^{3}He) were used. For the direct production of ^{3}He from oxygen by both GCR particles and protons, the excitation function was assumed to have the shape of that for producing ^{3}H from oxygen and was normalized to the experimental cross sections at 600 and 3000 MeV of Kruger and Heymann (1973). For the direct production of ^{3}He from Mg, Al, and Si by protons, the measured cross sections of Walton et al. (1976a), Goebel et al. (1964), and Kruger and Heymann (1973) were used. The excitation functions for the production of ^{3}He by GCR-produced neutrons were the same as for the protons, but with the energies below $\cong$30 MeV raised by 2 or 3 MeV for Al or lowered by 3 or 4 MeV for Si to correct for differences in threshold energies for the neutron- and proton-induced reactions. For the GCR-particle and proton production of ^{3}He from Fe, the experimental data above 150 MeV that were compiled by Goebel et al. (1964) were used; estimated cross sections were used at lower energies. The adopted cross sections near 30 MeV are similar to the Ni(p, x)^{3}He cross sections measured by Pulfer (1979).

For the proton production of ^{4}He from these elements, experimental cross sections from the same references cited above for ^{3}He were used. For iron and protons with $E < 20$ MeV, the measured (p, α) cross sections of Kumabe et al. (1963) and Sherr and Brady (1961) and the shape of the (p, α) excitation function of Fulmer and Goodman (1960) were used. For the production of ^{4}He by GCR particles with oxygen, the threshold energy was lowered from 8 MeV to 4 MeV. For GCR production of ^{4}He from Mg, Al, and Si for $E < 10$ MeV, the ^{27}Al(n, α)^{24}Na excitation function (see compilation of Garber and Kinsey, 1976) was used. Similarly, the ^{54}Fe(n, α)^{51}Cr excitation function was used for ^{4}He production from Fe by GCR secondary neutrons with $E < 10$ MeV.

The calculated production rates of ^{3}He (including ^{3}H decay) and ^{4}He are given in Table 1 for GCR particles and in Tables 2 and 3 for solar protons with spectral shapes having $R_0 = 100$ and 200 MV, respectively. Solar-proton production rates also were calculated using spectrum A of Walton et al. (1976b), which has a shape of $dJ/dE = K E^{-2.55}$. For depths within 1 g/cm^2 of the surface, the calculated results generally agreed with those of Walton et al. (1976b) within 10%. At greater depths, the calculated rates of Walton et al. (1976b) were much lower than those calculated here (by a factor of $\sim$2 at 20 g/cm^2); the disagreement was due mainly to the fact that Walton et al. (1976b) ignored protons with $E > 200$ MeV, whereas this work considered all energies with $E < 3668$ MeV.

Of the total GCR production of ^{3}He, slightly less than 0.5 was calculated as being due to the production and decay of ^{3}H. For solar-proton-induced reactions, approximately 0.4

Table 1. GCR production rates (per minute per kg-element) versus depth for particles with E ≥ 0.2 MeV.

Product	Target	Depths (g/cm²) 0	5	10	20	40	65	100	150	225	350	500
²H	O	1949	1984	2016	2037	1925	1700	1358	991	634	294	91
²H	Mg	1005	995	987	964	875	749	582	410	251	110	34
²H	Al	900	890	884	864	784	672	523	370	227	100	31
²H	Si	870	861	854	835	757	648	504	355	217	96	29
²H	Fe	409	395	383	363	316	262	198	134	78	32	9.6
³He	O	438	441	444	442	409	354	278	198	122	54	16.3
³He	Mg	314	317	320	321	302	266	213	155	100	47	14.4
³He	Al	292	294	297	298	280	246	196	143	91	43	13.2
³He	Si	301	304	306	307	287	252	201	146	93	43	13.3
³He	Fe	154	150	147	141	126	106	81	57	34	14.6	4.4
⁴He	O	2698	2928	3148	3447	3600	3434	2940	2331	1659	873	281
⁴He	Mg	1756	1938	2114	2357	2513	2432	2108	1694	1224	655	212
⁴He	Al	1606	1746	1880	2063	2159	2063	1769	1405	1002	529	170
⁴He	Si	1530	1658	1782	1949	2032	1937	1657	1312	933	490	158
⁴He	Fe	825	861	896	940	936	863	718	552	379	192	61
¹³C	O	360	442	521	640	755	782	716	610	471	271	89
¹⁵O	O	369	407	442	491	519	500	431	343	246	130	41.9
Fission	Th	523	589	653	742	807	791	693	563	412	223	72.4
Fission	U	1000	1158	1311	1534	1721	1725	1539	1276	957	533	174
⁸⁶Kr	Th	17.0	19.8	22.5	26.5	30.0	30.2	27.1	22.6	17.1	9.6	3.1
⁸⁶Kr	U	14.0	16.2	18.3	21.5	24.2	24.2	21.7	18.0	13.5	7.5	2.5
¹³⁴Xe	Th	11.4	13.7	16.0	19.4	22.5	23.1	21.0	17.7	13.6	7.7	2.5
¹³⁴Xe	U	32.0	40.4	48.6	60.9	73.1	76.4	70.4	60.3	46.8	27.0	8.9
¹³⁶Xe	Th	9.4	11.5	13.6	16.7	19.7	20.3	18.6	15.8	12.2	7.0	2.3
¹³⁶Xe	U	22.4	28.9	35.2	44.7	54.3	57.2	52.9	45.6	35.7	20.7	6.9

of the ^{3}He is made by way of ^{3}H. In lunar samples, Reedy (1977) found that the measured ^{3}H activities at deep depths were about 1.4 times the calculated GCR production rates. There are no good experimental data for comparing observed and calculated production rates of ^{3}H in meteorites or of ^{3}He in lunar samples. For the Keyes and St. Séverin chondrites, the calculated ^{3}He exposure ages were 1.3–1.4 times those calculated using ^{21}Ne and ^{38}Ar data (Reedy et al., 1978; Reedy et al., 1979). Because the calculated ^{21}Ne production rates of Reedy et al. (1979) agree fairly well with those determined from ^{53}Mn exposure ages (Nishiizumi et al., 1980; Müller et al., 1981), this factor of 1.3–1.4 probably applies also to the observed/calculated ^{3}He production rates. Thus, both lunar ^{3}H data and meteoritic noble-gas data imply that the calculated GCR production rates for ^{3}He should be multiplied by ≅1.4 to get actual ^{3}He production rates. Because oxygen produces ≅0.6 of the total ^{3}He and because production cross sections of ^{3}He from oxygen are not known well, this observed/calculated ratio of ≅1.4 probably results from the O(p, x)^{3}He cross sections used here being too low.

Most of the ^{4}He in extraterrestrial samples is radiogenic, being produced by the alpha decay of U and Th and their daughter isotopes. Using the meteoritic particle fluxes of Reedy et al. (1979), a ^{4}He/^{3}He production ratio of about 7 was calculated. Using the ^{3}He normalization factor of 1.4 given above, the predicted cosmogenic ^{4}He/^{3}He production ratio is about 5, the same number that Schultz and Signer (1976) used to correct measured ^{4}He concentrations for cosmogenic ^{4}He. In iron meteorites, almost all of the observed ^{4}He is produced by cosmic-ray-induced reactions with Fe. The measured ^{3}He/^{4}He ratios in iron meteorites are usually 0.22–0.28 (Signer and Nier, 1962), much larger than the production ratios calculated here for Fe (0.07–0.19).

Table 2. Solar-proton production rates (per minute per kg-element) vs. depth using $R_0 = 100$ MV and $J(4\pi, E > 10$ MeV$) = 100$ protons/cm²s.

Product	Target	Depths (g/cm²) 0	0.07	0.15	0.325	0.70	1.50	3.20	7.00	15.0	30.0
^{2}H	O	3976	3454	3108	2623	2038	1430	894	483	242	149
^{2}H	Mg	1058	954	878	765	618	453	295	166	85	53.3
^{2}H	Al	1091	965	878	753	597	429	275	153	78	48.6
^{2}H	Si	938	841	772	670	540	394	256	144	74	46.2
^{2}H	Fe	286	245	219	185	145	104	68	38.5	20.3	12.9
^{3}He	O	495	450	417	367	300	223	148	84.8	44.5	28.0
^{3}He	Mg	641	569	518	443	348	246	154	82.5	40.7	24.8
^{3}He	Al	610	538	487	415	325	229	143	76.8	38.0	23.2
^{3}He	Si	550	492	449	387	307	220	139	75.5	37.6	23.0
^{3}He	Fe	144	127	116	99	79	57	36.6	20.4	10.5	6.6
^{4}He	O	24378	18876	15904	12295	8600	5378	2985	1433	648	378
^{4}He	Mg	17463	13844	11802	9244	6539	4110	2276	1082	482	279
^{4}He	Al	14463	10932	9173	7085	4970	3123	1743	841	381	223
^{4}He	Si	10984	8914	7697	6131	4424	2845	1616	789	360	211
^{4}He	Fe	2684	2078	1767	1394	1008	662	391	201	97	59
^{13}C	O	1462	1136	962	752	538	349	203	103	48	28.5
^{15}O	O	3138	2712	2418	2002	1499	989	566	273	122	70.1
^{78}Kr	Br	4369	3144	2500	1755	1062	547	241	89.7	32.2	16.7
^{80}Kr	Br	5910	4213	3349	2363	1453	773	359	144	55.7	30.2
^{81}Kr	Br	3885	1643	1050	581	283	123	48.5	16.7	5.8	2.9
Fission	Th	4974	3950	3369	2640	1869	1177	655	313	141	82
Fission	U	7839	6205	5282	4130	2915	1831	1016	485	217	126
^{86}Kr	Th	141	111	94	73	50	31	16.8	7.8	3.4	1.9
^{86}Kr	U	53	45	40	33	25	17	10.2	5.2	2.4	1.4
^{134}Xe	Th	168	126	104	77	51	29	14.8	6.5	2.7	1.5
^{134}Xe	U	263	207	176	138	97	61	33.5	15.7	6.8	3.9
^{136}Xe	Th	185	136	111	80	52	29	14.2	6.0	2.4	1.3
^{136}Xe	U	196	150	125	96	67	41	22.1	10.2	4.3	2.5

SOLAR-PROTON PRODUCTION OF KRYPTON ISOTOPES FROM BROMINE

Production rates of krypton isotopes by cosmic-ray particles have been calculated previously (Regnier et al., 1979) for the target elements rubidium, strontium, yttrium, and zirconium, and for neutron-capture reactions with bromine. Because of the recent use of ^{81}Kr to study solar-proton fluxes over the last $\simeq 3 \times 10^5$ years (Reedy, 1980), some production-rate calculations were made to see whether reactions with bromine, like ^{81}Br(p, n)^{81}Kr, could be important in producing Kr isotopes in the surface layers of lunar rocks. Excitation functions for the ^{81}Br(p, n)^{81}Kr, ^{81}Br(p, 2n)^{80}Kr, ^{81}Br(p, pn)^{80}Br, ^{81}Br(p, 4n)^{78}Kr, and ^{79}Br(p, 2n)^{78}Kr reactions were constructed using analogous reactions with ^{89}Y (Birattari et al., 1973), ^{79}Br (Collé and Kishore, 1974), and other odd-even nuclei in this $A \sim 80$ mass region. All but the (p, 4n) cross sections are probably well estimated because most analogous reactions had similar excitation functions after correcting for differences in threshold energies. Tables 2 and 3 give the calculated ^{81}Kr, ^{80}Kr, and ^{78}Kr production rates from bromine for several depths using solar protons with $R_0 = 100$ and 200 MV spectral shapes, respectively.

For ^{81}Kr, solar-proton-induced reactions with bromine contributed very little, being most important for the surface layers ($\simeq 5\%$ for the top 0.5 mm of lunar rock 68815). This calculated contribution is probably an upper limit as most of the ^{82}Kr and ^{80}Kr produced by neutron-capture reactions with Br are not retained in lunar rocks (Regnier et al., 1979). Bromine was most important for ^{78}Kr at the surface, where $\simeq 40\%$ of the calculated solar-proton production was from Br. However, most of ^{78}Kr, even at the surface, is

Table 3. Solar-proton production rates (per minute per kg-element) vs. depth using $R_0 = 200$ MV and $J(4\pi, E > 10$ MeV$) = 100$ protons/cm^{2}s.

Depths (g/cm²) Product	Target	0	0.07	0.15	0.325	0.70	1.50	3.20	7.00	15.0	30.0
^{2}H	O	7856	7428	7109	6620	5946	5106	4163	3179	2338	1860
^{2}H	Mg	2593	2490	2409	2277	2084	1828	1523	1187	888	713
^{2}H	Al	2460	2347	2261	2126	1934	1686	1398	1087	813	653
^{2}H	Si	2270	2177	2105	1988	1818	1594	1329	1037	777	625
^{2}H	Fe	677	644	621	586	537	475	403	323	249	204
^{3}He	O	1329	1283	1245	1183	1090	965	813	641	484	391
^{3}He	Mg	1276	1212	1162	1082	971	828	668	502	362	285
^{3}He	Al	1202	1139	1091	1015	910	777	628	473	342	270
^{3}He	Si	1161	1107	1064	996	898	773	628	476	347	274
^{3}He	Fe	338	325	312	294	269	236	197	155	117	95
^{4}He	O	28426	25004	22913	20097	16768	13241	9880	6886	4670	3551
^{4}He	Mg	20776	18436	16951	14902	12423	9753	7197	4938	3294	2479
^{4}He	Al	16671	14564	13342	11717	9802	7765	5811	4060	2758	2098
^{4}He	Si	14334	12933	12009	10703	9076	7272	5487	3855	2628	2002
^{4}He	Fe	3742	3367	3142	2835	2462	2046	1619	1204	867	682
^{13}C	O	1877	1673	1549	1381	1180	960	739	530	367	281
^{15}O	O	4484	4143	3889	3504	2990	2386	1772	1211	800	597
^{78}Kr	Br	3363	2661	2261	1763	1248	801	473	256	139	94
^{80}Kr	Br	4746	3786	3250	2587	1899	1290	822	490	293	208
^{81}Kr	Br	1980	1019	728	473	285	162	88	45.7	24.2	16.2
Fission	Th	5992	5328	4904	4319	3611	2847	2113	1460	981	741
Fission	U	9355	8301	7632	6712	5602	4410	3269	2257	1515	1145
^{86}Kr	Th	158	139	127	110	90	70	57	34	22.6	16.9
^{86}Kr	U	83	77	73	66	58	48	37	26	18.1	13.8
^{134}Xe	Th	160	135	120	101	79	58	40	26	16.8	12.4
^{134}Xe	U	300	265	243	212	175	136	98	65	42.3	31.2
^{136}Xe	Th	164	135	119	98	75	53	36	22	13.8	10.0
^{136}Xe	U	207	179	162	140	114	87	62	41	25.8	18.8

made by galactic-cosmic-ray particles. Therefore production of Kr isotopes from Br should be unimportant unless a sample has a very high abundance of Br or, for ^{81}Kr, unless there were many more protons with energies between 2 and 10 MeV than observed recently via the ^{56}Fe(p, n)^{56}Co reaction. (Protons with energies below 10 MeV cannot produce ^{26}Al, ^{53}Mn, and most other long-lived radionuclides, so there is no way to detect their presence in the past by these cosmogenic radionuclides.)

PRODUCTION OF ^{130}Xe BY THE ^{130}Ba(n, p) REACTION

Kaiser (1977) measured the proton-induced cross sections for the production of Xe isotopes from natural barium. Production rates calculated using his cross sections have matched most measured Xe isotopic ratios. However the ^{130}Xe/^{126}Xe ratio is undercalculated by $\cong 40\%$ (Hohenberg et al., 1978). Kaiser (1977) and Hohenberg et al. (1978) felt that the deficiency in calculating ^{130}Xe production rates may have been the omission of neutron-induced reactions, specifically ^{130}Ba(n, p). Neutron-deficient isotopes of an element have larger (n, p) cross sections than the isotopes with higher masses, so a large ^{130}Ba(n, p) cross section is possible even though ^{136}Ba and ^{138}Ba have (n, p) cross sections of only a few millibarns (Garber and Kinsey, 1976).

To see quantitatively how important the ^{130}Ba(n, p) reaction could be in calculating ^{130}Xe production rates from barium, an excitation function for that reaction was constructed using a shape obtained from other (n, p) reactions in the A ~130 mass region and assuming a peak cross section of 0.5 barns at 20 MeV. [Kaiser (1977) has estimated that the ^{130}Ba(n, p) reaction might have a peak cross section of ~0.55 b at an energy of

25 MeV.] Calculated production rates of ^{130}Xe via the ^{130}Ba(n, p) reaction were 0.002 to 0.02 of those calculated via the Ba(p, x)^{130}Xe excitation function of Kaiser (1977). The reason that the ^{130}Ba(n, p) contributes so little is that only 0.11% of the isotopes of natural barium are ^{130}Ba, so a 0.5-barn cross section for ^{130}Ba is only about 0.5 mb for natural barium. Kaiser (1977) measured a 0.39 mb cross section for 38-MeV protons producing ^{130}Xe and a peak Ba(p, x)^{130}Xe cross section of 31.5 mb at 300 MeV.

These calculated ^{130}Ba(n, p)/Ba(p, x)^{130}Xe production ratios show that the majority of ^{130}Xe production from Ba via galactic-cosmic-ray particles is by spallation reactions with protons having energies above $\sim$100 MeV. The ^{130}Ba(n, p) reaction should not have a cross section which is considerably greater than 0.5 b because the total nonelastic cross section is 2 b (Garber and Kinsey, 1976), so this reaction should not be an important source of cosmogenic ^{130}Xe. Kaiser and Berman (1972) irradiated natural barium with neutrons having a continuum of energies below about 30 MeV and did not see any ^{130}Xe above their blank levels, while they saw large amounts of ^{131}Xe made by the ^{130}Ba(n, γ) reaction. Eberhardt et al. (1971) didn't observe any ^{130}Xe in a sample of Ba-feldspar irradiated in a reactor with epithermal and fast neutrons. Thus the cause of the undercalculation of ^{130}Xe production rates by Hohenberg et al. (1978) and Kaiser (1977) is not known.

NITROGEN-15 (OXYGEN-15)

In a number of lunar samples, there is a ^{15}N-rich component of nitrogen which is released at high temperatures and which is made by cosmic-ray-induced spallation reactions (Becker et al., 1976). To calculate production-rate-versus-depth profiles for ^{15}N, the ^{16}O(p, pn)^{15}O cross sections compiled by Audouze et al. (1967) were adopted. The ^{15}O-production cross sections were used because there are very few cross sections for the direct production of ^{15}N, and both nuclides are produced by similar reactions. The calculated ^{15}O production rates are given in Tables 1, 2, and 3.

The only measured ^{16}O(n, 2n)^{15}O cross sections (Brill' et al., 1961) have a maximum of 16 mb ($\pm$30%) at 31 MeV, while the ^{16}O(p, pn)^{15}O cross sections used here were 30 mb at 30 MeV and 74 mb at 55 MeV. An excitation function for the production of ^{15}O by GCR particles was constructed using these (n, 2n) cross sections at low energies (mainly E < 100 MeV) and the (p, pn) cross sections at high energies. The production rates calculated with the (p, pn) cross-section set were from 1.9 (at the lunar surface) to 3.15 (for depths >500 g/cm^2) of those calculated with the GCR cross-section set. The shape of an excitation function determines the production-rate profile as a function of depth (Reedy and Arnold, 1972). The excitation function which really should be used for ^{15}N production from ^{16}O should include the sum of the (n, 2n), (n, d), (n, np), (n, pn), and (n, 2p) cross sections for neutron-induced reactions and the sum of the (p, pn), (p, np), and (p, 2p) cross sections for proton-induced reactions. (All of these reactions with ^{16}O produce mass-15 nuclides, including ^{15}O and ^{15}C which rapidly decay to ^{15}N.) The choice of which of the two excitations functions to use for ^{15}N production was based on which shape was most similar to the shape of the excitation function which really should be used. Because the ^{16}O(p, pn)^{15}O excitation function has a shape which probably is the better approximation of the desired shape, it was used here.

In their studies of cosmogenic ^{15}N in lunar samples with known exposure ages, Becker et al. (1976) empirically determined a ^{15}N production rate of 3.6 ($\pm$0.8) pg ^{15}N/g/m.y., which is equivalent to 275 atoms/min/kg. This production rate is 1.55 times greater than the calculated rate in Table 1 for ^{15}O at 5 g/cm^2, assuming an oxygen abundance of 0.435. This factor of 0.55 for direct ^{15}N production relative to ^{15}O production seems low, but is not an unreasonable value, given both experimental and calculational uncertainties. The empirical production rate of Becker et al. (1976) could be low because some cosmogenic ^{15}N might be released at low temperatures. As noted above, there also are uncertainties in the production cross sections for ^{15}O and ^{15}N. The excitation function which used ^{16}O(n, 2n)^{15}O cross sections at low energies gave a calculated production rate at a depth of 5 g/cm^2 which was 0.50 of that given in Table 1, so the empirical/calculated production-rate ratio of 1.55 would become 3.1.

In the atmospheres of Mars and Earth, the ^{14}N/^{15}N ratios are 165 and 277, respectively (Owens et al., 1977). Assuming that nitrogen in the Earth's atmosphere is typical of solar-system nitrogen, the martian atmosphere has a ^{15}N excess of $\cong 6 \times 10^{19}$ atoms/cm^2. This enrichment in ^{15}N usually is considered to have resulted from isotopic fractionation when large amounts of atmosphere escaped from Mars. An excess of ^{15}N also could be produced by cosmic-ray-induced reactions with oxygen atoms in the martian atmosphere or surface. When the lunar ^{15}O production rates are integrated over all depths, an oxygen abundance of 45% is assumed, and the factor of 1.55 is used to convert to total ^{15}N production, the total GCR and $R_0 = 100$ MV solar-proton production rates are 89 and 7 atoms/min/cm^2, respectively. Over 4.5×10^9 years, the production of cosmogenic ^{15}N is 2.3×10^{17} atoms/cm^2, slightly less than 0.5% of the observed excess of ^{15}N in the martian atmosphere. Yanagita and Imamura (1978), using simpler calculations, also have concluded that contemporary cosmic-ray fluxes could not produce this excess ^{15}N in the martian atmosphere.

CARBON-13

Like ^{15}N, ^{13}C also is enriched in the high-temperature fractions from lunar samples and the amount of the enrichment is proportional to the sample's exposure age (Des Marais, 1980). Using the Monte Carlo calculations of Armstrong and Alsmiller (1971) for the production rates of ^{15}N and ^{13}C, Des Marais (1978) estimated a ^{13}C/^{15}N production ratio of approximately 0.32. If all the production rates of mass-13 and mass-15 isobars in Fig. 2 of Armstrong and Alsmiller (1971) are summed, then a net ^{13}C/^{15}N production ratio of about 0.30 is obtained. Shapiro and Silberberg (1970) give ^{13}C/^{15}N production ratios from oxygen of 0.40 to 0.48 for protons with $E = 150$ and ≥ 2300 MeV, respectively. To check the ^{13}C/^{15}N production ratio, excitation functions for the cumulative production of ^{13}C (including ^{13}N, which decays to ^{13}C) were prepared.

For protons with $E < 50$ MeV, the experimental cross sections for the ^{16}O(p, x)^{13}N reaction of Albouy et al. (1962) were used. The peak ^{16}O(p, α)^{13}N cross section is 19 mb for $E = 19$–25 MeV. For neutrons with $E < 20$ MeV, the ^{16}O(n, α)^{13}C cross sections evaluated by Foster and Young (1972) were adopted. Peak cross sections for the ^{16}O(n, α)^{13}C reactions were 120 mb for $E \cong 5$ MeV and $\cong 200$ mb near 10 MeV. Above 50 MeV, there are only data for ^{13}N production, the cross sections being $\cong 5$ mb from $E \cong 80$ MeV to $E = 5.7$ GeV (Audouze et al., 1967). Armstrong and Alsmiller (1971) calculated that the ^{13}C/^{13}N production ratio was about 5, which would imply a ^{13}C + ^{13}N production cross section of $\cong 30$ mb. Shapiro and Silberberg (1970) calculated total ^{13}C production cross sections from oxygen of 20–25 mb for $E > 150$ MeV. For $E > 75$ MeV, the average of these cross sections, 25 mb, was adopted. The ^{16}O(p, pn)^{15}O cross sections are greater than those for ^{13}C production from 25 to $\cong 2500$ MeV, but lower than the ^{16}O(p, α)^{13}N or ^{16}O(n, α)^{13}C cross sections below 25 MeV. The production rates for ^{13}C from oxygen are given in Table 1 for GCR reactions and Tables 2 and 3 for solar protons.

The ^{13}C/^{15}O production ratios calculated for GCR particles ranged from 0.98 at the surface to 2.13 below 500 g/cm^2. Assuming a ^{15}N/^{15}O production ratio of 1.55 (see above), the ^{13}C/^{15}N GCR production ratios would range from 0.63 to 1.38. The ^{13}C/^{15}N production ratio is calculated to be 1.0 at 65 g/cm^2. The ^{16}O(n, α)^{13}C reaction, which has large cross sections, is the main reason that the ^{13}C/^{15}N production ratio is much higher than predicted by Shapiro and Silberberg (1970) for reactions by high-energy protons. Because Armstrong and Alsmiller (1971) included low-energy neutrons in their calculations, it is not clear why their ^{13}C/^{15}N production ratio is a factor of ~3 below those calculated here.

The largest ^{13}C enrichment observed by Des Marais (1980) was for lunar rock 15499. Assuming this rock's excess ^{13}C corresponded to a δ^{13}C of 75 per mil, that its exposure age (in 10^6 years) divided by its C abundance (in μg/g) was 480, and its oxygen abundance was 0.435, the observed production rate of ^{13}C in 15499 was $\cong 380$ atoms/min/kg(O). This ^{13}C production rate is comparable to, but somewhat smaller than, those given in Table 1. As with ^{15}O, there are uncertainties in both the observed and calculated ^{13}C production rate. About 65% of the ^{13}C is calculated to have been produced by particles with $E > 30$ MeV and, as noted above, there are considerable uncertainties in the cross sections for ^{13}C production from oxygen at such energies.

The amount of excess ^{13}C observed by Des Marais (1980) is consistent with the production rates calculated here. Spallogenic ^{13}C usually will be harder to detect than spallogenic ^{15}N because (1) the average ^{13}C/^{15}N production rate is <1 for most samples, (2) the carbon content of most samples is much higher than their nitrogen content, and (3) ^{13}C is 1.11% of natural carbon, while ^{15}N is only 0.37% of normal nitrogen.

DEUTERIUM

The heavy isotope of hydrogen, deuterium (^{2}H), like ^{13}C and ^{15}N, is found normally in low abundances in most lunar materials. Because the ^{2}H/H ratio in the solar wind is expected to be very low (see, e.g., Epstein and Taylor, 1971), the relatively high ^{2}H contents of lunar samples were recognized by Epstein and Taylor (1971) and Friedman et al. (1971) to be due to deuterium production by spallation reactions. Merlivat et al. (1974) made a rough estimation of the production rate of deuterium on the lunar surface. Merlivat et al. (1976) measured the deuterium content of eight samples in 70215, including five with a distribution of known depths, and determined a production rate of 0.46×10^{-10} mole ^{2}H$_2$/g/10^8 years (1050 atoms/min/kg). They didn't see any variation in spallogenic ^{2}H content with sample depth, a trend consistent with the rock's long, complex exposure history. Merlivat et al. (1976) reported unpublished deuterium production rates calculated by Yokoyama and Tobailem of about 1030 and 1010 atoms/min/kg at 4 and 8 cm, respectively, for an erosion rate of 0.5 mm/10^6 years, and noted that their inferred ^{2}H production rate was about 4.6 times the measured tritium (^{3}H) activities in other lunar samples.

There are very few measured cross sections for the production of deuterium, so the excitation functions used here in calculating production rates have considerable uncertainties. For the major target element oxygen, the evaluated ^{16}O(n, d) cross sections of Foster and Young (1972) were used for GCR particles with E < 15 MeV. For proton-induced reactions, the energies for these cross sections were raised by 4 MeV to account for the different threshold energies. At 39 and 62 MeV, the experimental ^{16}O(p, x)^{2}H cross sections of Bertrand and Peelle (1973) were used. At 300 and 600 MeV, the tritium cross sections of Reedy and Arnold (1972) times the ^{2}H/^{3}H production ratio of 9.1 reported by Badhwar and Daniel (1963) for 300-MeV protons reacting with oxygen were adopted. Badhwar and Daniel (1963) also reported that 190-MeV protons reacting with Al and Ni had ^{2}H/^{3}H production ratios of 5.6 and 4.7, respectively. Similar ^{2}H/^{3}H production ratios were measured by Bertrand and Peelle (1973) for 62 MeV protons reacting with Al and Fe. For the cross sections for ^{2}H production from Mg, Al, Si, and Fe by both GCR particles and protons, the tritium cross sections of Reedy and Arnold (1972) were multiplied by 6, 6, 6, and 5, respectively, and the energies for the reaction thresholds were lowered. The calculated deuterium production rates from these five target elements are given in Table 1 for GCR particles and in Tables 2 and 3 for solar protons.

In the top 40 g/cm^2 of a lunar rock, the GCR production rate of deuterium would be about 1200 atoms/min/kg, in quite good agreement with the inferrred and calculated values reported in Merlivat et al. (1976). In lunar samples, ≅70–80% of the ^{2}H is calculated as having been produced from reactions with oxygen. Within 100 g/cm^2 of the moon's surface ≅80–95% of the GCR production of ^{2}H is induced by particles with E > 100 MeV. The calculated production rates for ^{2}H are 8.0–9.5 times those for tritium. Because ^{3}H production rates are undercalculated by a factor of 1.4 (Reedy, 1977), the calculated ^{2}H production rates would have been expected to be lower than the observed rates. Perhaps the ^{2}H/^{3}H production ratios adopted here are too high, or the excitation functions for producing ^{2}H and ^{3}H have different shapes.

COSMOGENIC FISSION OF URANIUM AND THORIUM

Cosmic-ray particles, especially secondary neutrons, can induce the fission of ^{232}Th, ^{235}U, and ^{238}U. The resulting fission products are sources both of nuclides, like ^{136}Xe, and of fossil charged-particle tracks. The spontaneous fission of ^{238}U (which has a partial half-life of 8.2×10^{15} years) is the source of most fission in lunar samples, occurring at a rate of 404 fissions/min/kg-U. Because the cross sections for cosmic-ray-induced fission of Th and U isotopes are high (~1 barn), rates for cosmogenic fission are similar to those for spontaneous fission. In samples with long exposures to the cosmic rays (e.g., lunar cores), cosmogenic fission could be a significant fraction of the fissions which have occurred (Damm *et al.*, 1978). As noted by K. Marti (pers. comm., 1980), certain nuclides, like ^{136}Xe and ^{86}Kr, are made at relatively low rates by spallation reactions (Hohenberg *et al.*, 1978; Regnier *et al.*, 1979), but in high yields by fission reactions; so cosmogenic fission could be an important source of these nuclides in lunar samples. To study quantitatively the importance of cosmogenic fission in lunar samples, fission cross sections were evaluated and used to calculate rates for the fission of Th and U isotopes by GCR particles and by solar protons. Damm *et al.* (1978) calculated rates for the fission of U, Th, Bi, Pb, and Au by GCR particles in the moon, but used (p, f) cross sections. Their results showed that the fission of Bi, Pb, and Au are relatively unimportant in lunar samples compared to the fission of U and Th.

The cross sections for the fission of ^{232}Th, ^{235}U, and ^{238}U by galactic-cosmic-ray particles were adopted or estimated from a number of sources. For neutrons with E < 40 MeV, the fission cross sections compiled in BNL-235 (Garber and Kinsey, 1976) were used. For protons with E > 100 MeV, the cross sections summarized in de Carvalho *et al.* (1963) and Friedlander (1965) were used. In the preliminary calculations reported in Reedy (1981), the fission cross sections were assumed to remain constant for E > 400 MeV. Above ~400 MeV, the product nuclides with 50 < A < 170, which can be made only by fission reactions at lower energies, can be produced by other types of reactions. Although the total production cross sections for nuclides with 50 < A < 170 remains constant for E > 200 MeV, the cross sections for binary fission decrease for energies above ~100 MeV (Friedlander, 1965). The shape of the excitation functions for proton-induced fission for 40 < E < 100 MeV was estimated from the cross sections for several fission products made by the ^{238}U(p, f) reaction (Diksic *et al.*, 1974). There is an appreciable flux of neutrons for E ~ 50 MeV, so the fission cross sections from 40 to about 60 MeV were increased above the proton-induced values because, at 30 MeV, cross sections for neutron-induced fission are higher than those for proton-induced fission. Because of the scarcity of measured total fission cross sections for 30 < E < 100 MeV, especially with neutrons, there are considerable uncertainties in the cross sections adopted for this energy region. In particlar, the U(n, f) cross sections for 30 to 40 MeV (Garber and Kinsey, 1976), which are used here, seem too high.

The excitation functions for proton-induced fission of natural uranium and thorium at low energies were based on the cross sections measured by Boyce *et al.* (1974) for U(p, f) with E_p < 30 MeV and by Choppin *et al.* (1963) and Eaker and Choppin (1976) for Th(p, f) with E_p < 16 MeV. Cross sections summarized in de Carvalho *et al.* (1963) were used to connect these low-energy data with the adopted GCR cross sections at 100 MeV. These (p, f) excitation functions are very similar to those of Damm *et al.* (1978). When these (p, f) cross sections were used to calculate fission rates of U and Th by GCR particles, the results agreed within about 2% or less with the calculated production rates of Damm *et al.* (1978). The cross sections for (p, f) reactions are lower than those for (n, f) reactions with Th and U for energies below about 18 and 100 MeV, respectively. For the lunar GCR-particle spectrum, the GCR cross-section set for U gave fission rates higher than those calculated with the U(p, f) excitation function by factors of 1.87 and 1.39 for 10 < E < 30 MeV and 30 < E < 100 MeV, respectively.

The excitation functions for the production of ^{86}Kr, ^{134}Xe, and ^{136}Xe from ^{232}Th, ^{235}U, and ^{238}U by GCR particles were constructed from evaluated fission yields and the adopted fission cross sections for E < 20 MeV and from measured cross sections for producing these nuclides at energies above 60 MeV. Cross sections for other fission products were not constructed because such nuclides are made in high yields by other cosmogenic nuclear reactions (e.g., ^{131}Xe) or are naturally present in samples (e.g., barium isotopes). All radioactive fission products with these masses which decay to these three nuclides (e.g., ^{136}Te and ^{136}I for ^{136}Xe) were included, and fission products which decayed to other stable isotopes (e.g., ^{136}Cs, which decays to ^{136}Ba) were excluded. The evaluated fission yields of Rider and Meek (1978) were used for fission-spectrum (~1 MeV) and 14-MeV neutrons and linearly interpolated for 1 < E < 14 MeV. For E > 60 MeV, cross sections for the fission products which decay to these three nuclides were determined from a large number of cumulative or independent cross sections. The cross sections for the production of noble-gas nuclides as measured by Regnier (1977) for Th(p, f) reactions at 0.15 and 1.05 GeV, by Hudis *et al.* (1970) for U(p, f) reactions at 3 and 29 GeV, and by Yu *et al.* (1973) for the U(p, f) reaction at 11.5 GeV were used. For the production of ^{134}Xe by ^{238}U(p, f) reactions, the total cross sections for producing the mass-134 chain, less the independent cross sections for making isotopes which decayed to ^{134}Ba (such as ^{134}Cs and ^{134}La), were used. Smooth trends in cross sections or isotope ratios (e.g., ^{134}Cs/^{134}Xe) versus energy were used when no experimental data were available. Cross sections for the other eight target-product combinations were determined similarly. For 15 < E < 60 MeV, product yields were interpolated from the measured 14-MeV yields and yields inferred from the cross sections for these products and for fission at E ≅ 60 MeV. Separate excitation functions for each target-product combination had to be determined because the yields of most fission products per fission vary with energy (see, e.g., Friedlander, 1965). For example, the cumulative yield of ^{136}Xe per fission of ^{238}U varies from 6.85% for ~1-MeV neutrons to 0.8% for 1-GeV protons. These excitation functions are different than those used in the preliminary calculations reported by Reedy (1981).

The excitation functions used for the production of ^{134}Xe and ^{136}Xe by (p, f) reactions below 100 MeV were based mainly on measured cross sections. Estimated yields were used at many energies to get ^{86}Kr production cross sections, so the calculated solar-proton production rates for ^{86}Kr have greater uncertainties than those for other fission reactions. For the production of all three of these nuclides by the Th(p, f) reaction at 15.6 MeV, the cross sections of Eaker and Choppin (1976) were used. The excitation functions for the production of ^{134}Xe and ^{136}Xe by Th(p, f) reactions for $25 < E < 87$ MeV were based on the measured cross sections of Pate et al. (1958). The Th(p, f)^{86}Kr cross sections for energies between 15.6 and 150 MeV and for $E < 15.6$ MeV were estimated by assuming that the yield of ^{86}Kr per fission varied slowly with energy. The cross sections for the production of ^{134}Xe and ^{136}Xe by the U(p, f) reaction were based mainly on the measurements of Diksic et al. (1974). The U(p, f)^{86}Kr excitation function below 100 MeV was constructed by assuming fission yields of about 1.25% for $E > 50$ MeV and 0.3–0.4% for $E < 40$ MeV.

The rates for the fission or the production of ^{86}Kr, ^{134}Xe, and ^{136}Xe from Th and natural uranium were calculated using the lunar GCR fluxes of Reedy and Arnold (1972) and are given in Table 1. Only particles with $E \geq 0.2$ MeV were considered, as the Reedy–Arnold GCR-particle-flux model is good only for $E > 0.5$ MeV. Relatively few fission reactions occur for particles having energies between 0.2 and 0.5 MeV. Only ^{235}U can be fissioned by particles with $E < 0.2$ MeV, and the rates for the ^{235}U(n, f) reaction as a function of depth were determined from the Lunar Neutron Probe Experiment (LNPE) by Woolum and Burnett (1974). Because natural uranium consists of only 0.72% ^{235}U, almost all ($\cong 99\%$) of the fissions induced in U by particles with $E \geq 0.2$ MeV are from ^{238}U. The average yields of ^{86}Kr, ^{134}Xe, and ^{136}Xe per fission usually increased with depth because of the increased ratio of secondary neutrons to high-energy particles. The ratios for ^{134}Xe/^{136}Xe production rates decreased less than 10% with depth, but the ^{86}Kr/^{136}Xe production ratios decreased with depth by factors of 1.33 and 1.73 for Th and U, respectively.

The total fission rates calculated here are higher than those calculated by Damm et al. (1978) because of the higher cross sections used here for energies below 18 and 100 MeV for the fission of Th and U, respectively (see above in paragraph on proton-induced fission cross sections). For natural uranium, the fission rates given in Table 1 for the lunar surface and a depth of 225 g/cm^2 are higher than those of Damm et al. (1978) by factors of 1.29 and 1.85, respectively. Because 40% (at the lunar surface) to 84% (depths below 500 g/cm^2) of the cosmogenic fissions of ^{238}U are induced by particles with $E < 100$ MeV, the differences in the GCR and (p, f) excitation functions can account for these differences in calculated production rates. At depths of 0 and 225 g/cm^2, the rates calculated here for the fission of Th are 1.04 and 1.23 times those of Damm et al. (1978). This difference is due mainly to fission of Th induced by particles with $E < 15$ MeV.

The cosmogenic fission rates calculated here or determined from the LNPE data by Woolum and Burnett (1974) are compared with the rate for the spontaneous fission of ^{238}U in Fig. 1, assuming the lunar Th/U ratio of 3.8 and the present ^{235}U isotopic abundance of 0.72%. The ^{235}U(n, f) rates are from the smooth curve in Fig. 4 of Woolum and Burnett (1974). Their results should be divided by a factor of 1.21 to convert the rates measured by the LNPE during the Apollo-17 mission to those averaged over a solar cycle (Woolum and Burnett, 1974). However, the Apollo 17 soil into which the LNPE was placed had a high macroscopic cross section, Σ, for neutron capture (0.00936 cm^2/g) because of its high concentrations of Fe and Ti. Rates for neutron-capture reactions vary with Σ, and the rate for the ^{235}U(n, f) reaction would increase by a factor of 1.21 for a more typical Σ of 0.0075 cm^2/g (Reedy, 1978). Thus, the uncorrected LNPE rates would apply for a solar-cycle averaged rate in a typical lunar soil. At these depths, and other lunar depths above 200 g/cm^2, the total rates for the cosmogenic fission of ^{232}Th, ^{235}U, and ^{238}U are $\cong 12$ times the rate for the spontaneous fission of ^{238}U. A similar ratio for cosmogenic to spontaneous fission rates was calculated by Damm et al. (1978). Thus cosmic-ray-induced reactions can produce a significant fraction of the total number of fissions that have occurred in lunar samples with long ($\sim 10^8$ years or more) exposure ages, so cosmogenic fission must be considered in studies involving fission-track dating, ^{244}Pu fission, and fission products.

In discussing the origins of fossil charged-particle tracks in meteorites, Fleischer et al.

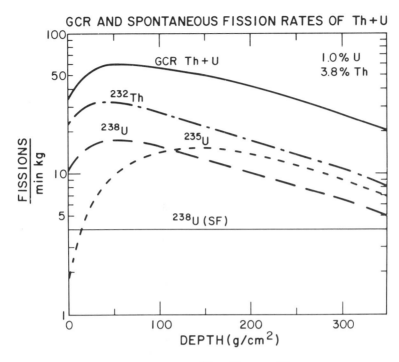

Fig. 1. Calculated rates for the fission of ^{232}Th, ^{235}U, and ^{238}U by GCR particles as a function of depth are compared with the rate for the spontaneous fission of ^{238}U. The rates for the fission of ^{232}Th and ^{238}U are from Table 1; those for ^{235}U(n, f) are from Woolum and Burnett (1974). The lunar Th/U ratio of 3.8 and the contemporary ^{235}U abundance of 0.72% were assumed.

(1967) noted that there are two types of cosmogenic fission tracks. Fission induced by high-energy ("energies greater than ~100 MeV") particles produces V-shaped tracks because of the large momentum of the incident projectile. Fission induced by low-energy particles produces linear tracks like those made by spontaneous fission. The fractions of fissions induced by particles with E > 100 MeV range from 70% to 24% for Th and from 60% to 16% for ^{238}U. Almost all of the fission of ^{235}U is induced by low-energy neutrons. At lunar depths of 40 and 225 g/cm^2, 38% and 18% of the total fissions of ^{232}Th, ^{235}U, and ^{238}U are induced by particles with E > 100 MeV; and 4.6% and 1.0%, respectively, of the fissions at these two depths are induced by particles with E > 1 GeV. As noted by Fleischer et al. (1967), the densities of cosmogenic fission tracks vary with depth, and the V-shaped tracks decrease in density more rapidly with greater depths than do the other cosmogenic fission tracks.

Because the half-lives of these three fissile nuclides are about the same as the age of the moon but are different (^{232}Th = 1.40 × 10^{10} years, ^{235}U = 7.04 × 10^8 years, and ^{238}U = 4.47 × 10^9 years), the ratios of their fission rates were different during the early history of the moon. In particular, the rates for the ^{235}U(n, f) reaction >10^9 years ago were relatively much greater than those for the fission of ^{232}Th or ^{238}U. Eugster et al. (1979) observed excess ^{136}Xe in the soils of the 74001/2 cores which they concluded was produced by the ^{235}U(n, f) reaction about 3.8 × 10^9 years ago. As noted above, the rates for the ^{235}U(n, f) reaction also change with neutron macroscopic cross section.

The yields of ^{136}Xe, ^{134}Xe, and ^{86}Kr per cosmogenic fission varied with depth and were different from their yields by the spontaneous fission of ^{238}U. The yields of ^{136}Xe, ^{134}Xe, and ^{86}Kr per cosmogenic fission of ^{232}Th and ^{238}U were ~0.4, ~0.6, and ~2-3, respectively, of their yields for the spontaneous fission of ^{238}U (6.4, 5.3, and 0.9%, respectively). Thus, as with total fission, cosmogenic fission could produce significant amounts of these three fission products (especially ^{86}Kr). The production of other Kr and Xe isotopes by cosmic-ray-induced fission also could be relatively important because, like ^{86}Kr, these nuclides are produced in greater yields by cosmogenic fission than they are by the spontaneous fission of ^{238}U.

These three noble-gas nuclides are made in relatively low yields by spallation reactions and ^{86}Kr and ^{134}Xe are more abundant in lunar rocks than calculated for spallation reactions (Hohenberg et al., 1978; Regnier et al., 1979). The average lunar composition of Reedy (1978) was used to compare the production rates for these three nuclides by spallation reactions and by cosmogenic fission of ^{232}Th and ^{238}U. At the lunar surface and below 500 g/cm^2, the fission/spallation production ratios were 0.12–0.38 for ^{86}Kr, 0.13–1.15 for ^{134}Xe, and 0.9–7.1 for ^{136}Xe. The ^{235}U(n, f) reaction was not included, but would increase these ratios, especially at depths below ~100 g/cm^2. Therefore, cosmogenic fission cannot account for the undercalculation of cosmogenic ^{86}Kr and ^{134}Xe in lunar rocks, but is an important source of cosmic-ray-produced ^{136}Xe. (The undercalculation of ^{86}Kr and ^{134}Xe production rates for spallation reactions probably is caused by low estimated cross sections for their production from Zr and rare earth elements.)

DISCUSSION

The production rates reported here were for a variety of reactions, and most reactions have not had their production rates published elsewhere. Production rates for ^{3}He, ^{15}N, ^{13}C, and ^{2}H previously had been inferred from their measured concentrations and experimentally determined exposure ages. The production rates calculated here for these nuclides agreed with the inferred values within factors of $\cong 1.5$, quite good agreement considering the uncertainties in both the experimental results and/or in the excitation functions used in these calculations. For the reactions studied here, the production-rate-versus-depth profiles probably are calculated very well. Erosion of rock surfaces or gardening of the lunar soil will modify these production profiles, especially for the solar-proton-induced reactions and for samples with long (~10^7 years or more) exposure ages.

The ^{3}He production rates calculated here, multiplied by 1.4 for GCR-induced reactions, can be used to predict the concentrations of cosmogenic ^{3}He in studies of ^{3}He in lunar samples. Solar-proton-induced reactions with bromine should produce only a small fraction of the ^{78}Kr, ^{80}Kr, and ^{81}Kr observed in lunar samples. The ^{130}Ba(n, p) reaction produces ~1% of the ^{130}Xe made from barium; thus the reason that the ^{130}Xe production rate is undercalculated in lunar samples by Hohenberg et al. (1978) is not known. The calculated production rates of ^{15}N (actually about 1.5 times the ^{15}O values), ^{13}C, and ^{2}H agree approximately with the measured data for these isotopes in lunar samples with long exposure ages, confirming that cosmic-ray-induced reactions can account for most of the excesses of these isotopes observed in such lunar samples. The contemporary fluxes of cosmic-ray particles would produce $\cong 0.5\%$ of the excess ^{15}N observed in the atmosphere of Mars.

The rates for the fission of U and Th by cosmic-ray particles within 200 g/cm^2 of the lunar surface are $\cong 12$ times the rate for the spontaneous fission of ^{238}U. The productions of ^{86}Kr and ^{134}Xe by cosmogenic fission are important sources of these nuclides, and cosmogenic fission produces more ^{136}Xe than do spallation reactions with Ba and rare earth elements. Rates for the ^{235}U(n, f) reaction vary with the chemical composition, and would be relatively more important, compared to fission of ^{232}Th and ^{238}U, more than ~10^9 years ago. Cosmogenic fission of Th and U can be important sources of fission tracks or fission-product nuclides, especially in samples with long exposure ages.

Acknowledgments—This work was supported by NASA under work order W-14,084 and done under the auspices of the U.S. Department of Energy. The author thanks K. Marti for suggesting some of the reactions studied in this work and L. M. Mitchell, C. Gallegos, and M. L. Ennis for their help in preparing the manuscript.

REFERENCES

Albouy G., Cohen J.-P., Gusakow M., Poffe N., Sergolle H., and Valentin L. (1962) Spallation de l'oxygene par des protons de 20 à 150 MeV. *Phys. Lett.* **2**, 306–307.
Armstrong T. W. and Alsmiller R. G. (1971) Calculation of cosmogenic radionuclides in the Moon and comparison with Apollo measurements. *Proc. Lunar Sci. Conf. 2nd*, p. 1729–1745.
Audouze J., Epherre M., and Reeves H. (1967) Survey of experimental cross sections for proton-induced spallation reactions in He^4, C^{12}, N^{14}, and O^{16}. In *High-Energy Nuclear Reactions in Astrophysics* (B. S. P. Shen, ed.), p. 255–271. W. A. Benjamin, New York.
Badhwar G. D. and Daniel R. R. (1963) Some remarks concerning the energy dependence of the intensities of nuclei of helium-3, lithium, beryllium, and boron in the galactic cosmic radiation. *Progr. Theor. Phys.* **30**, 615–626.
Becker R. H., Clayton R. N., and Mayeda T. K. (1976) Characterization of lunar nitrogen components. *Proc. Lunar Sci. Conf. 7th*, p. 441–458.
Bertrand F. E. and Peelle R. W. (1973) Complete hydrogen and helium particle spectra from 30- to 60-MeV proton bombardment of nuclei with A = 12 to 209 and comparison with the intranuclear cascade model. *Phys. Rev. C* **8**, 1045–1064.
Birattari C., Gadioli E., Gadioli Erda E., Grassi Strini A. M., Strini G., and Tagliaferri G. (1973) Pre-equilibrium processes in (p, n) reactions. *Nucl. Phys.* **A201**, 579–592.
Boyce J. R., Hayward T. D., Bass R., Newson H. W., Bilpuch E. G., Purser F. O., and Schmitt H. W. (1974) Absolute cross sections for proton-induced fission of the uranium isotopes. *Phys. Rev. C* **10**, 231–244.
Brill' O. D., Vlasov N. A., Kalinin S. P., and Sokolov L. S. (1961) Cross section of the (n, 2n) reaction in C^{12}, N^{14}, O^{16}, and F^{19} in the energy interval 10–37 MeV. *Dokl. Akad. Nauk SSSR* **136**, 55–57; also in *Sov. Phys. Dokl.* **6**, 24–26. In English.
Choppin G. R., Meriwether J. R., and Fox J. D. (1963) Low-energy charged-particle-induced fission. *Phys. Rev.* **131**, 2149–2152.
Collé R. and Kishore R. (1974) Excitation functions for (p, n) reactions on ^{79}Br and ^{127}I. *Phys. Rev. C* **9**, 2166–2170.
Cressy P. J. and Bogard D. D. (1976) On the calculation of cosmic-ray exposure ages of stone meteorites. *Geochim. Cosmochim. Acta* **40**, 749–762.
Damm G., Thiel K., and Herr W. (1978) Cosmic-ray-induced fission of heavy nuclides: Possible influence on apparent ^{238}U-fission track ages of extraterrestrial samples. *Earth Planet. Sci. Lett.* **40**, 439–444.
de Carvalho H. G., Cortini G., Muchnik M., Potenza G., and Rinzivillo R. (1963) Fission of uranium, thorium and bismuth by 20 GeV protons. *Nuovo Cimento* **27**, 468–474.
Des Marais D. J. (1978) Carbon, nitrogen, and sulfur in Apollo 15, 16, and 17 rocks. *Proc. Lunar Planet. Sci. Conf. 9th*, p. 2451–2467.
Des Marais D. J. (1980) Six lunar rocks have little carbon and nitrogen and some rocks have detectible spallogenic ^{13}C (abstract). In *Lunar and Planetary Science XI*, p. 228–230. Lunar and Planetary Institute, Houston.
Diksic M., McMillan D. K., and Yaffe L. (1974) Nuclear charge dispersion in mass chains 130–135 from the fission of ^{238}U by medium-energy protons. *J. Inorg. Nucl. Chem.* **36**, 7–16.
Eaker R. W. and Choppin G. R. (1976) Mass and charge distribution in the fission of ^{232}Th. *J. Inorg. Nucl. Chem.* **38**, 31–36.
Eberhardt P., Geiss J., and Graf H. (1971) On the origin of excess ^{131}Xe in lunar rocks. *Earth Planet. Sci. Lett.* **12**, 260–262.
Epstein S. and Taylor H. P. (1971) O^{18}/O^{16}, Si^{30}/Si^{28}, D/H, and C^{13}/C^{12} ratios in lunar samples. *Proc. Lunar Sci. Conf. 2nd*, p. 1421–1441.
Eugster O., Grögler N., Eberhardt P., and Geiss J. (1979) Double drive tube 74001/2: History of the black and orange glass; determination of a pre-exposure 3.7 AE ago by ^{136}Xe/^{235}U dating. *Proc. Lunar Planet. Sci. Conf. 10th*, p. 1351–1379.
Fleischer R. L., Price P. B., Walker R. M., and Maurette M. (1967) Origins of fossil charged-particle tracks in meteorites. *J. Geophys. Res.* **72**, 331–353.
Foster D. G. and Young P. G. (1972) A preliminary evaluation of the neutron and photon-production cross sections of oxygen. Atomic Energy Commission report LA-4780. 30 pp.
Friedlander G. (1965) Fission of heavy elements by high-energy protons. In *Physics and Chemistry of Fission*, Vol. II, p. 265–282. IAEA, Vienna.
Friedman I., O'Neil J. R., Gleason J. D., and Hardcastle K. (1971) The carbon and hydrogen content and isotopic composition of some Apollo 12 materials. *Proc. Lunar Sci. Conf. 2nd*, p. 1407–1415.
Fulmer C. B. and Goodman C. D. (1960) (p, α) reactions induced by protons in the energy range of 9.5–23 MeV. *Phys. Rev.* **117**, 1339–1344.
Garber D. I. and Kinsey R. R. (1976) Neutron cross sections volume II, curves. ERDA report BNL-325, 3rd edition, Vol. II, 491 pp.
Goebel K., Schultes H., and Zaehringer J. (1964) Production cross sections of tritium and rare gases in various target elements. CERN (European Commission for Nuclear Research) Rep. CERN-64-12. 78 pp.

Hohenberg C. M., Marti K., Podosek F. A., Reedy R. C., and Shirck J. R. (1978) Comparisons between observed and predicted cosmogenic noble gases in lunar samples. *Proc. Lunar Planet. Sci. Conf. 9th*, p. 2311–2344.

Hudis J., Kirsten T., Stoenner R. W., and Schaeffer O. A. (1970) Yields of stable and radioactive rare-gas isotopes formed by 3- and 29-GeV proton bombardment of Cu, Ag, Au, and U. *Phys. Rev. C* **1**, 2019–2030.

Kaiser W. A. (1977) The excitation function of $Ba(p, X)^M Xe$ (M = 124–136) in the energy range 38–600 MeV; the use of 'cosmogenic' xenon for estimating 'burial' depths and 'real' exposure ages. *Phil. Trans. Roy Soc. London* **A285**, 337–362.

Kaiser W. A. and Berman B. L. (1972) The average $^{130}Ba(n, \gamma)$ cross section and the origin of ^{131}Xe on the moon. *Earth Planet. Sci. Lett.* **15**, 320–324.

Kohl C. P., Murrell M. T., Russ G. P., and Arnold J. R. (1978) Evidence for the constancy of the solar cosmic ray flux over the past ten million years: ^{53}Mn and ^{26}Al measurements. *Proc. Lunar Planet. Sci. Conf. 9th*, p. 2299–2310.

Kruger S. T. and Heymann D. (1973) High-energy proton production of ^3H, ^3He, and ^4He in light targets. *Phys. Rev. C* **7**, 2179–2187.

Kumabe I., Ogata H., Komatuzaki T., Inoue N., Tomita S., Yamada Y., Yamaki T., and Matsumoto S. (1963) (p, α) reactions on the even nuclei Ni^{58}, Ni^{60} and Fe^{56}. *Nucl. Phys.* **46**, 437–453.

Merlivat L., Lelu M., Nief G., and Roth E. (1974) Deuterium, hydrogen, and water content of lunar material. *Proc. Lunar Sci. Conf. 5th*, p. 1885–1895.

Merlivat L., Lelu M., Nief G., and Roth E. (1976) Spallation deuterium in rock 70215. *Proc. Lunar Sci. Conf. 7th*, p. 649–658.

Müller O., Hampel W., Kirsten T., and Herzog G. F. (1981) Cosmic-ray constancy and cosmogenic production rates in short-lived chondrites. *Geochim. Cosmochim. Acta* **45**, 447–460.

Nishiizumi K., Regnier S., and Marti K. (1980) Cosmic ray exposure ages of chondrites, pre-irradiation and constancy of cosmic ray flux in the past. *Earth Planet. Sci. Lett.* **50**, 156–170.

Owens T., Biemann K., Rushneck D. R., Biller J. E., Howarth D. W., and Lafleur A. L. (1977) The composition of the atmosphere at the surface of Mars. *J. Geophys. Res.* **82**, 4635–4639.

Pate B. D., Foster J. S., and Yaffe L. (1958) Distribution of nuclear charge in the proton-induced fission of Th^{232}. *Can. J. Chem.* **36**, 1691–1706.

Pulfer P. (1979) Bestimmung von absoluten produktionsquerschnitten fuer protoneninduzierte reaktionen im energiebereich zwischen 15 MeV und 72 MeV und bei 1820 MeV. Modellrechnungen fuer die produktion von Ne20, Ne21 und Ne22 an der mondoberflaeche durch die bestrahlung mit solaren protonen. Ph.D. Thesis, Univ. Bern. 150 pp.

Rao M. N. and Venkatesan T. R. (1980) Solar-flare produced ^3He in lunar samples. *Nature* **286**, 788–790.

Reedy R. C. (1977) Solar proton fluxes since 1956. *Proc. Lunar Sci. Conf. 8th*, p. 825–839.

Reedy R. C. (1978) Planetary gamma-ray spectroscopy. *Proc. Lunar Planet. Sci. Conf. 9th*, p. 2961–2984.

Reedy R. C. (1980) Lunar radionuclide records of average solar-cosmic-ray fluxes over the last ten million years. In *Proc. Conf. Ancient Sun* (R. O. Pepin, J. A. Eddy, and R. B. Merrill, eds.), p. 365–386. Pergamon, N.Y.

Reedy R. C. (1981) Cosmic-ray-produced stable nuclides: Various production rates and their implications (abstract). In *Lunar and Planetary Science XII*, p. 871–873. Lunar and Planetary Institute, Houston.

Reedy R. C. and Arnold J. R. (1972) Interaction of solar and galactic cosmic-ray particles with the moon. *J. Geophys. Res.* **77**, 537–555.

Reedy R. C., Herzog G. F., and Jessberger E. K. (1978) Depth variations of spallogenic nuclides in meteorites (abstract). In *Lunar and Planetary Science IX*, p. 940–942. Lunar and Planetary Institute, Houston.

Reedy R. C., Herzog G. F., and Jessberger E. K. (1979) The reaction $Mg(n, \alpha)Ne$ at 14.1 and 14.7 MeV: Cross sections and implications for meteorites. *Earth Planet. Sci. Lett.* **44**, 341–348.

Regnier S. (1977) Production des gaz rares Ne, Ar, Kr et Xe au cours des réactions nucléaires induites par des protons de haute énergie dans des cibles allant de ^{27}Al à ^{232}Th. Ph.D. Thesis, Univ. de Bordeaux I. 106 pp.

Regnier S., Hohenberg C. M., Marti K., and Reedy R. C. (1979) Predicted versus observed cosmic-ray-produced noble gases in lunar samples: Improved Kr production ratios. *Proc. Lunar Planet. Sci. Conf. 10th*, p. 1565–1586.

Rider B. F. and Meek M. E. (1978) Compilation of fission product yields. General Electric Co., Vallecitos Nuclear Center report NEDO-12154-2(E).

Schultz L. and Signer P. (1976) Depth dependence of spallogenic helium, neon, and argon in the St. Severin chondrite. *Earth Planet. Sci. Lett.* **30**, 191–199.

Shapiro M. M. and Silberberg R. (1970) Heavy cosmic ray nuclei. *Ann. Rev. Nucl. Sci.* **20**, 323–392.

Sherr R. and Brady F. P. (1961) Spectra of (p, α) and (p, p') reactions and the evaporation model. *Phys. Rev.* **124**, 1928–1943.

Signer P. and Nier A. O. C. (1962) The measurement and interpretation of rare gas concentrations in iron meteorites. In *Researches on Meteorites* (C. B. Moore, ed.), p. 7–35. John Wiley & Sons, N.Y.

Venkatesan T. R., Nautiyal C. M., Padia J. T., and Rao M. N. (1980) Solar (flare) cosmic ray proton fluxes in the recent past. *Proc. Lunar Planet. Sci. Conf. 11th*, p. 1271–1284.

Walton J. R., Heymann D., Yaniv A., Edgerley D., and Rowe M. W. (1976a) Cross sections for He and Ne isotopes in natural Mg, Al, and Si, He isotopes in CaF_2, Ar isotopes in natural Ca, and radionuclides in natural Al, Si, Ti, Cr, and stainless steel induced by 12- to 45-MeV protons. *J. Geophys. Res.* **81**, 5689–5699.

Walton J. R., Heymann D., and Yaniv A. (1976b) Production of He, Ne, and Ar isotopes and U^{236} in lunar materials by solar cosmic ray protons—production rate calculations. *J. Geophys. Res.* **81**, 5701–5710.

Woolum D. S. and Burnett D. S. (1974) In-situ measurement of the rate of ^{235}U fission induced by lunar neutrons. *Earth Planet. Sci. Lett.* **21**, 153–164.

Yanagita S. and Imamura M. (1978) Excess ^{15}N in the martian atmosphere and cosmic rays in the early solar system. *Nature* **274**, 234–235.

Yaniv A. and Kirsten T. (1981) Isotopic ratios of rare gases and the role of nuclear reactions in solar flares (abstract). In *Lunar and Planetary Science XII*, p. 1224–1226. Lunar and Planetary Institute, Houston.

Yu Y. W., Porile N. T., Warasila R., and Schaeffer O. A. (1973) Cross sections and recoil properties of xenon isotopes formed in 11.5-GeV proton bombardment of ^{238}U. *Phys. Rev. C* **8**, 1091–1098.

Errata

NOTE TO FUTURE CONTRIBUTORS: As you detect *important* typographical misprints, drafting errors, miscalculations, fallacious logic, or other mistakes in papers published in current or past *Proceedings* volumes, please send corrections to the following address:

Publications Office
Lunar and Planetary Institute
3303 NASA Road One
Houston, Texas 77058

Please keep notices of *errata* brief. Contributions greater in length than half of a printed page will be assessed page charges at the then current rate.

Proceedings of the Eleventh Lunar and Planetary Science Conference
Geochimica et Cosmochimica Acta, Supplement 14

VOLUME 2

Agosto W. N., Hewins R. H., and *Clarke R. S. Jr.* Allen Hills A77219, the first Antarctic mesosiderite, p. 1027–1045.

Page 1039: 8 lines up from bottom of page, change the word *increase* to *decrease*.

Page 1042: 3 lines up from bottom of page, change 0.43 wt% to 0.49 wt%.

Page 1044: first line on page, change *poikilitic* to *poikiloblastic*.

Wieler R., Etique Ph., Signer P., and *Poupeau G.* Record of the solar corpuscular radiation in minerals from lunar soils: A comparative study of noble gases and tracks, p. 1369–1393.

Page 1373: In Table 2, some $^{20}Ne/^{22}Ne$- and $^{22}Ne/^{21}Ne$-ratios for plagioclase should be changed as follows:

Nr.	soil	$^{20}Ne/^{22}Ne$	$^{22}Ne/^{21}Ne$
13	64421 clean	11.62	10.69
15	67601 clean	12.26	22.68
16	67601 dirty	12.65	26.73
21	67960 clean	12.48	23.95
23	72261 clean	12.67	23.22

Page 1382: Figure 3, upper part: The correct unit value on the ordinate is 10'000 ($\times 10^{-8}$ cm^3 ^{4}He/g).

Lunar Sample Index

Pages   1–1018: Section 1, The Moon
Pages 1019–1823: Section 2, Planets, Asteroids and Satellites

Index entries were compiled from information supplied by the authors and refer only to opening pages of articles.

10009	10074	12054
607	607	1703
10010	10075	12057
607	607	339
10017	10082	12070
1747	607	339
10018	10084	12073
289, 339	339, 371, 409, 421,	21
10019	607, 627, 891	14003
339, 607	10093	965
10020	607	14078
1747	10094	1
10021	607	14143
339	12001	1
10023	339, 371, 409, 421	14148
607	12008	627, 965
10044	339	14156
323	12013	965
10047	117, 173	14160
421	12015	21
10048	339	14163
339, 607	12022	371, 409, 421, 559
10049	1747	14172
1747	12023	21
10056	339	14179
339, 607	12029	21
10059	339	14240
339	12032	339
10060	339	14259
281, 339	12033	339, 371, 965
10061	21, 339, 371, 409, 421	14276
339	12037	1027
10064	339	14301
607	12038	99, 409
10065	339	14303
607	12040	21
10070	323	14305
607	12042	21
10073	339	
607		

14310
  67, 323, 421, 791, 903, 965
14316
  21
14321
  21, 117
15003
  421, 485
15007
  451, 463
15008
  451, 463
15010
  421, 433, 475, 485, 567
15011
  421, 433, 463, 475, 485, 567
15021
  433, 509
15028
  281
15031
  433, 509
15041
  433, 509
15065
  451
15071
  433, 509
15076
  339, 577
15081
  509
15086
  281
15091
  433, 509
15101
  281, 409, 509
15211
  451, 509
15221
  371, 409, 421, 433, 485, 509, 567
15231
  509, 567
15251
  509
15261
  509
15271
  371, 409, 421, 433, 485, 509

15286
  281
15291
  509
15301
  281, 433, 509
15318
  339, 915
15362
  21, 727
15382
  903, 965
15386
  421, 509, 903
15401
  451
15405
  117, 1297
15411
  509
15415
  21, 451, 509, 727, 1001
15425
  339, 915
15426
  339, 509, 915, 935
15427
  339, 915
15431
  509
15434
  421
15445
  67, 727, 1297
15455
  67, 99, 727
15465
  21
15471
  433, 509
15498
  281
15499
  281, 1809
15501
  433, 509
15511
  509
15515
  509
15531
  433, 475

15545
  421
15555
  1759
15565
  21
15601
  433, 475, 509
21000
  371, 389
22001
  339, 371, 389
24077
  1
24114
  1
24170
  1
24999
  371, 389
26261
  567
60001
  577
60002
  577
60003
  339
60007
  577
60009
  577
60010
  577
60015
  253, 305
60016
  253
60017
  185, 253, 305
60018
  253, 305
60019
  235, 253
60025
  209, 253
60035
  253
60051
  577
60055
  253

60056	61156	62237
253	209, 253, 577	253, 305
60075	61161	62241
235, 253	529, 577, 965	529, 577, 965
60095	61175	62255
253	235, 253, 305	253
60115	61181	62275
253	577	253
60135	61195	62281
253	253	577, 965
60215	61220	62295
253	1297	253, 509, 577
60235	61221	63321
253	529, 577, 627	577
60255	61223	63335
253, 281	1297	185, 253, 305
60275	61224	63341
253	1297	577
60315	61241	63355
209, 253	339, 577	185, 253
60335	61281	63500
253, 421	577, 965	209
60501	61295	63501
577, 965	235, 253, 339	529, 577
60503	61500	63503
339	209	185, 209
60525	61501	63505
253	577	185, 253
60601	61503	63506
577, 965	209, 339	185, 253
60615	61536	63507
253	253	185
60618	61546	63508
253	253	185
60625	61547	63509
253	253	185
60639	61548	63515
253	253	185
60665	61558	63525
253	253	185, 253
60666	61568	63526
253	253	185
61015	61569	63527
209, 253	253, 577	185, 253
61016	61575	63528
209, 253, 295, 627, 1809	253	185
61135	62231	63529
235, 253	577	185, 253
61141	62235	63535
577	209, 253, 577, 791	185, 253
61155	62236	63536
253	305	185

63537
  185, 253
63538
  185, 253
63539
  185
63545
  185
63546
  185, 253
63547
  185
63548
  185
63549
  185, 253, 577
63555
  185
63556
  185, 577, 791
63557
  185
63558
  185
63559
  185, 253
63565
  185
63566
  185
63567
  185
63568
  185
63569
  185
63575
  185
63576
  185
63577
  185, 253
63578
  185
63579
  185, 253
63585
  185, 253
63586
  185
63587
  185, 253

63588
  185
63589
  185
63595
  185
63596
  185, 253
63597
  185, 253
63598
  185, 253
64421
  529, 577, 965
64435
  253, 305, 627, 1809
64455
  253, 305
64475
  253, 577
64476
  253
64478
  253
64501
  371, 409, 421, 529, 577, 965
64503
  339
64535
  253
64536
  253
64537
  253
64538
  253
64548
  577
64559
  253
64567
  253, 577
64576
  253
64579
  253
64585
  253
64586
  253
64801
  577

64810
  577
64811
  529
64815
  253
64816
  253
64818
  253
65015
  209, 253, 577, 791
65016
  253, 281
65035
  253
65055
  209, 253, 577
65056
  253
65075
  253
65095
  253
65315
  21, 253, 1809
65325
  253
65326
  253
65359
  253
65501
  529, 577
65511
  577
65701
  529, 577, 965
65715
  253
65759
  253
65766
  253
65777
  577, 791
65779
  253
65785
  577
65901
  577

66031
  529, 577
66035
  253
66036
  253
66037
  253
66041
  577, 965
66043
  209
66055
  253, 305
66075
  117, 235, 253
66081
  577, 965
66095
  209, 253, 261, 295, 323,
  529, 577
67015
  185, 209, 235, 253, 305
67016
  185, 209, 235, 253, 305
67025
  185, 253
67035
  185, 209, 235, 253, 305, 577
67055
  185, 253
67075
  21, 185, 253, 305
67095
  185, 253, 305, 577
67115
  185, 209, 253, 305
67215
  185, 253
67235
  185, 253
67415
  185, 235, 253, 305
67435
  185, 253, 305
67455
  185, 209, 235, 253, 261,
  305, 451
67461
  371, 409, 421, 577
67475
  185, 235, 253, 305

67480
  209
67481
  577
67483
  185, 209
67485
  185
67486
  185
67487
  185
67488
  185
67489
  185
67495
  185
67511
  577
67515
  185
67516
  185
67517
  185
67518
  185
67519
  185
67525
  185
67526
  185
67527
  185
67528
  185
67529
  185
67535
  185
67536
  185
67537
  185
67538
  185
67539
  185
67545
  185

67546
  185
67547
  185
67548
  185
67549
  185
67555
  185
67556
  185
67557
  185
67558
  185
67559
  185, 577
67565
  185, 253
67566
  185
67567
  185
67568
  185
67569
  185
67575
  185
67576
  185
67600
  209
67601
  529, 577
67603
  185, 209
67605
  253
67615
  185
67616
  185
67617
  185
67618
  185
67619
  185
67625
  185

67626	67685	67736
185	185	185
67627	67686	67737
185	185	185
67628	67687	67738
185	185	185
67629	67688	67739
185	185	185
67635	67695	67745
185	185	185
67636	67696	67746
185	185	185
67637	67697	67747
185	185	185
67638	67700	67748
185, 253	209	185
67639	67701	67749
185	577	185
67645	67703	67755
185	185, 209	185
67646	67705	67756
185	185, 253	185
67647	67706	67757
185	185	185
67648	67707	67758
185	185	185
67649	67708	67759
185	185	185
67655	67711	67765
185	529, 577	185
67656	67715	67766
185	185	185
67657	67716	67767
185	185	185
67658	67717	67768
185	185	185
67659	67718	67769
185	185	185
67665	67719	67775
185	185	185
67666	67725	67776
185	185	185
67667	67726	67915
577	185	185, 253, 305
67668	67727	67935
185	185	185, 253
67669	67728	67936
185, 253	185	185, 253
67675	67729	67937
185	185, 253	185, 253
67676	67735	67941
185	185, 253	577

67945	68821	73141
185	577	339
67946	68841	73215
185	577, 965	173
67947	69921	73224
185	567, 577, 627, 965	21
67948	69935	73244
185	253, 567	21
67955	69941	74001
185, 209, 253, 305	577, 965	541, 1809
67956	69945	74002
185	253	541, 1809
67957	69955	74220
185	253, 567	935
67961	69961	75061
577	567, 577, 965	339
67975	69963	75081
253, 305	339	529
68035	70001	76001
253	409, 519	371, 463
68115	70002	76241
253	409, 519	567
68121	70003	76501
529, 577	409, 519	371, 389
68415	70004	76535
209, 253, 577	409, 519	99, 371, 891
68416	70005	77017
209, 253, 577	409, 519	281
68500	70006	77075
209	409, 421, 519	577
68501	70007	78221
577	409, 519	371, 389
68503	70008	78235
209 339	409, 519	577
68505	70009	78236
253	371, 409, 519	67
68515	70164	79035
253	21	289
68517	70181	79135
253	371	289
68519	70215	79155
253	1809	281
68525	71501	79215
253	891	627
68526	72501	Luna 24087
253	371, 389	627
68529		
253		
68815		
253, 1809		

Heavenly Body Index

Pages 1–1018: Section 1, The Moon
Pages 1019–1823: Section 2, Planets, Asteroids and Satellites
Index entries were compiled from information supplied by the authors and refer only to opening pages of articles.

Asteroids, 1543
  Apollo, 1145

Earth, 1001, 1177, 1473, 1507, 1543
  Antarctic dry valleys, 1481
  Brent Crater, 1607
  Cactus Crater, 1623
  Snowball Crater, 1623

Jupiter, 1543, 1569, 1599, 1737

Mars, 837, 1177, 1359, 1387, 1449, 1473, 1481, 1543, 1649, 1737, 1809
  Antoniad, 1419
  Arsia Mons, 1449
  Baldet, 1419
  Candor Chasma, 1459
  Echus Chasma, 1459
  Gangis Chasma, 1459
  Ganymede, 1649
  Hebes Chasma, 1459
  Ius Chasma, 1459
  Melas Chasma, 1459
  Newton Crater, 1449
  Olympus Mons, 1459
  Ophir Chasma, 1459
  Sirenum Fossae, 1449
  Syrtis Major, 1419
  Tharsis, 1359, 1459
  Tharsis Montes, 1449
  Thaumasia Fossae, 1449
  Tithonium Chasma, 1459
  Valles Marineris, 1459
  Viking Lander 1 site, 1493
  Viking Lander 2 site, 1493

Mercury, 695, 837, 1543, 1569
  ALHA 77005, 1359
  Abee, 1243
  Acapulco, 1209
  Achondrite 77005, 1349
  Allan Hills 76005, 1019, 1257
  Allan Hills 76006, 1019, 1105
  Allan Hills 76007, 1019
  Allan Hills 76008, 1019, 1105
  Allan Hills 76009, 1229
  Allan Hills 77003, 1019
  Allan Hills 77004, 1019
  Allan Hills 77005, 1359
  Allan Hills 77011, 1039
  Allan Hills 77015, 1039
  Allan Hills 77033, 1039
  Allan Hills 77043, 1039
  Allan Hills 77140, 1039
  Allan Hills 77155, 1105
  Allan Hills 77160, 1039
  Allan Hills 77164, 1039
  Allan Hills 77165, 1039
  Allan Hills 77167, 1039
  Allan Hills 77214, 1019, 1039
  Allan Hills 77249, 1039
  Allan Hills 77256, 1019
  Allan Hills 77260, 1039
  Allan Hills 77270, 1105
  Allan Hills 77272, 1105
  Allan Hills 77282, 1019
  Allan Hills 77294, 1019
  Allan Hills 77296, 1105
  Allan Hills 77297, 1019
  Allan Hills 77302, 1257
  Allan Hills 78038, 1039
  Allan Hills 78188, 1039
  Allan Hills 79001, 1039
  Allan Hills 79045, 1039

Allan Hills A77219, 1315
Allan Hills A78040, 1315
Allende, 1069, 1079, 1145, 1153, 1167,
    1177, 1189, 1199, 1229, 1769, 1781, 1803
Angra Dos Reis, 1349
Antarctic, 1145
Barea, 1315
Bath, 1135
Bholgati, 1257
Bishunpur, 1039
Bondoc Peninsula, 1315
Bouvante, 949
Brachina, 1349, 1359
Brient, 1281
Bruderheim, 541, 559, 1349, 1769
Budulan, 1315
Bununu, 1257
Chassigny, 1349, 1359
Chinguetti, 1315
Clover Springs 1315
Colomera, 1027
Crab Orchard, 1315
Dalgarauga, 1315
Dimmitt H3, 1153
Dyarrl Island, 1315
EETA 79001, 1359
Ehloe, 1135
Elga, 1049
Emery, 1315
Estherville, 1315
Forest City, 1135
Frankfort, 1281
Governador Valadares, 1359
Guarena, 1349
Hainholz Lowicz, 1315
Holbrook, Arizona, 1105
Ibitira, 949
Jersley, 1343
Jodzie, 1281
Johnstown, 1297
Juvinas, 949, 1281, 1297
Kapoeta, 1257
Keyes, 1809
Lafayette, 1359
Leedey, 1349
Leighton, 1093
Manych, 1123
Melrose-b, 1349
Mincy, 1315
Mocs, 541
Moore County, 1297
Morristown, 1315

Mount Padbury, 1315
Murchison, 1027
Nakhla, 1349, 1359
Pasamonte, 949, 1281
Patwar, 1315
Pesyanoe, 1177
Pinnarroo, 1315
Salles, 1135
Sharps, 1039
Shergotty, 1349, 1359
Sikhote-Alin, 1343
Simondium, 1315
Sioux County, 949
St. Severin, 1809
Stannern, 949
Sutton, 1135
Toluca, 1343
Ureilites, 1177
Vaca Muerta, 1315
Veramin, 1315
Yamato, 1145
Yamato 74356, 1297
Youndegin, 1343
Zagami, 1349, 1359

Moon, 173, 281, 323, 389, 607, 639, 651, 695,
    831, 837, 903, 979, 1001, 1543, 1569, 1703,
    1737, 1809
Andel Crater, 791
Apennine Front, 451, 485, 727
Apollo 11, 409, 703
Apollo 11 landing site, 21, 607
Apollo 12, 409
Apollo 12 landing site, 21
Apollo 14, 409
Apollo 14 landing site, 21
Apollo 15, 409, 451
Apollo 15 landing site, 21
Apollo 16, 409
Apollo 16 landing site, 21, 209, 235, 781
Apollo 17, 409
Apollo 17 landing site, 21, 781
Aristillus, 339
Audel Crater, 781
Autolycus Crater, 339
Balmer Crater, 715
Camelot, 519
Cavalerius Crater, 679
Cayley Plains, 185, 209, 305, 577, 751, 767,
    809
Central Cluster, 519

Central Highlands, 751, 767
Copernicus Crater, 117, 665, 949
Copernicus H Crater, 665
Crisium Basin, 781
Descartes Crater, 781
Descartes Formation, 185, 209, 751, 767, 1297
Descartes Highlands, 185, 305
Descartes Mountains, 235, 577, 767
Dollond Crater, 185
Far-eastern landing sites, 21
Fra Mauro, 791, 965
Gator Crater, 185
Gerasimovich Crater, 817
Hadley Apennine region, 727
Hadley Rille, 475, 485
House Rock, 577
Imbrium Backslope, 727
Imbrium Basin, 185
Inghirami W Crater, 665
Kant Plateau, 185, 235, 577, 751, 767, 781
Kiva Crater, 185
Langrenus, 715
Light Mantle, 519
Luna 16 landing site, 21, 409
Luna 20 landing site, 21, 409
Luna 24 landing site, 21, 409
Mare Crisium, 1, 371, 809
Mare Fecunditatis, 371
Mare Imbrium, 339, 727, 915, 935, 965
Mare Nectaris, 339, 767, 781
Mare Procellarum, 117
Mare Serenitatis, 727, 935
Mare Tranquillitatis, 339, 607, 809
Montes Apennines, 727
Montes Archimedes, 727
Near-eastern landing sites, 21
Nectaris Basin, 185, 209, 767, 781
Noggerath F Crater, 665
North Massif, 371, 519
North Ray Crater, 185, 209, 235, 305, 529, 577, 767
Oceanus Procellarum, 339, 679, 965
Olbers A Crater, 679
Orientale Basin, 185, 665, 781
Orientale Crater, 791
Palmetto Crater, 185
Palus Putredinis, 433, 475, 727
Plum Crater, 577
Ptolemaeus Crater, 767, 781
Ravine Crater, 185
Reiner Crater, 679
Reiner Gamma Formation, 679, 817
Schickard Crater, 665
Schickard R Crater, 665
Schiller Crater, 665
Schiller Plains, 665
Schiller-Zucchius Basin, 665
Sculptured Hills, 371
Serenitatis Basin, 781
Shergotty, 67
Shorty Crater, 541
Smoky Mountain, 185
South Massif, 371, 519
South Ray Crater, 185, 209, 305, 577
St. George Crater, 463
Stubby Crater, 185
Sulpicius Gallus, 727, 809
Taurus Littrow Valley, 389, 519
Taurus-Littrow, 371, 781
Theophilus Crater, 767, 781
Trap Crater, 185
Tycho, 703
Unnamed crater B, 185
Van Serg Crater, 627
Van de Graaf Crater, 791
Western landing sites, 21
Wreck Crater, 185

Moons
  Callisto, 1543, 1569, 1585
  Deimos, 1543, 1717
  Dione, 1543
  Europa, 1543, 1555, 1569
  Ganymede, 1543, 1585, 1599, 1649
  Iapetus, 1543
  Io, 1533, 1543, 1569
  Phobos, 1543
  Rhea, 1543

Saturn, 1543, 1737

Sun, 627

Venus, 1177, 1387, 1507
  Pioneer Venus Day Probe impact site, 1493
  Venera landing sites, 1493

Subject Index

Pages   1–1018: Section 1, The Moon
Pages 1019–1823: Section 2, Planets, Asteroids and Satellites

Index entries were compiled from information supplied by the authors, and refer only to opening pages of articles.

ANT rocks, 21
Ablation, 1599
Accretion, 979, 1641
Acid residues, 1153
Adsorption, 891, 1177, 1199
Agglutinate, 409, 421, 475, 519, 529
Aggregation, 1781
Air composition, 1019
Akaganeite, 253, 261, 323
Albedo, 1599
Aluminous basalts, 607
Aluminum, 567, 727, 767, 1145
Ambipolar diffusion, 1387
Analytical electron microscope, 1297
Anomalies, 1189
Antarctic, 1105, 1481
    ice, 1019
    meteorites, 1019, 1039, 1145
Apatite, 295, 323
    green glass, 695
    mare basalt, 915
Apollo 16, 235, 529, 577
    landing site, 767
Apparent depth, 639
Argon, 1199
    dating, 209
Asteroids, 1641, 1717
    Apollo, 1145
Asymmetry, 979
Atmosphere, 891, 1569
Avalanche, 519

Basalt, 117, 651, 935
    genesis, 1
Basin, 639, 651, 715, 781
    chronology, 209
    volcanism, 651
Black soil, 541
Boulders, 567

Breccias, 1, 117, 173, 253, 261, 289, 305, 1093
    crystalline melt, 185
    dimict, 209
    evolution, 295
    feldspathic fragmental, 209, 235
    fragmental, 185
    welded, 607
Bromine, 295
Byrd core ice, 1019

Cadmium, 529
Carbon, 289, 559, 621, 1019, 1153, 1167, 1781, 1809
Carbonate targets, 1623
Carbynes, 1153, 1167
Carlsbergite, 1343
Chasma, 1459
Chlorine, 253, 295, 323
Chlorine-phosphorus ratio, 295
Chondrules, 1069, 1123, 1229
Chromite, 1069
Chromium, 1069
Chromium nitride, 1343
Chronology, 173
Clasts, 305
Clay, 1689
Cohenite, 1243
Comets, 1641
Condensates, 1177
Condensation, 1123
Convection, 991
Cooling, 1123
    rate, 281, 1135, 1145, 1297, 1315
Core, 409, 463, 519, 567, 831
    formation, 949
Coronae, 1315, 1569
Cosmogenic noble gases, 541
Cosmogenic nuclides, 1809

Cratering, 1689
  dynamics, 1689, 639, 1641
  bottom, 463
  dark-halo, 665
  dark-ray, 1599
  excavation, 1623
  explosion, 1623
  modification, 1623
  morphology, 639
  morphometry, 639
  multi-lobed, 1649
  rampart, 1649
  simple, 1607
Crust, 1001
  cooling, 67
Crystallization, 281, 903
Cumulate sources, 915, 1349

Dark mantle, 809
Deconvolution, 867
Degassing, 891
Dendrite, 1123
Deuterium, 1809
Differentiation, 979
Diffusion, 1079, 1725
Direct loading, 1027
Dissipation factor, 1737
Distribution coefficients, 1001
Drill core, 433, 475
Dust, 1493, 1641, 1703

Earth, 1387
  mantle, 1349
  sediments, 433
Ejecta, 463, 519, 781, 1641, 1689
Electron diffraction, 1343
Energy partitioning, 1607
Equilibration, 1093
Equilibrium, 1001
Erosion, 1641
Europium, 21
Excavated cavity, 1607
Exosphere, 1569
Exsolution, 1297
Extrusion, 965

Faults, 1449
Feldspar, 1725
Felsite, 117, 173, 979
Fines, 389

Fission, 979, 1809
Flexure, 837
Flow behavior, 1759
Fluid flow, 1737
Fractionation, 935, 991, 1189, 1781
Fracture patterns, 1555
Fragment-laden melt, 117
Fusion, 409

Galactic cosmic ray, 627
Gamma-ray spectrometer, 751
Geochemical provinces, 781
Geochemistry, 727
Geochronology, 529
Geology, 1449
Glass, 1759
  coating, 305
  formation, 281
  spectra, 695
Glasses, 281, 339
  brown, 915
  green, 915, 935
Global grid, 1555
Grain size, 1781
Grains, 621
Granite, 117, 173
Granulite, 235, 305
Graphite, 1039, 1153, 1167
Graphite-magnetite, 1039
Gravity, 639, 837
Grooved terrain, 1585

Halogens, 295
Helium, 1809
High temperature spectra, 695
High-K basalts, 607
Highlands, 173, 235, 253, 261, 639, 727, 767, 781, 791, 891, 1449
  chemistry, 751
  enrichment gradient, 509
  rocks, 185
Huss matrix, 1039
Hydrogen, 559

Ice, 1599
  ablation, 1599
  crust, 1555
  deformation, 1585
  erosion, 1599
Icy planets, 1667

formation of, 1667
Icy satellites, 1649
Igneous fractionation, 915, 1349
Imbrium ejecta, 185
Immiscibility, 117
Impact, 817, 1641
Impact melts, 253, 261, 305, 1607
  conditions, 1607
  craters, 665, 1649, 1689
  fragmentation, 1667
  glass, 485
Inclusions, 1079, 1189, 1199
Incompatible elements, 21, 949
Internal heating, 1145
Interstellar medium, 1781
Inverted pigeonite, 1297
Iodine, 1803
Ion microbe, 1281
Iron, 727
Iron/argon, 67
Isotopes, 99, 289, 1189
  rubidium, 173
  strontium, 173
Isotopic anomalies, 1781

KREEP, 1, 117, 295, 371, 389, 529, 577, 727, 791, 903, 965, 979
KREEP volcanism, 767
Kamacite, 1243
Komatiite, 1001
Krypton, 1809

LKFM, 727
Laboratory impacts, 1649
Lava, 651, 1449
Leaching, 295
Lead, 529
Light plains, 665, 715
Lineations, 1555
Liquid separation, 991
Liquidus, 1001
Lithospheres, 1585
Longitudinal fractionations, 21
Low-K basalts, 607
Lumme-Bowell, 1543
Luna 24 brown glass, 695
Lunar chronology, 99
Lunar lithosphere, 853
Lunar primitive matter, 915
Lunar swirls, 679

Mafic clasts, 1315
Magma ocean, 903, 915
Magnesian basalt, 1
Magnesium, 727, 767
Magnesium/aluminum ratios, 809
Magnetic anomalies, 679
Magnetic fields, 1243
Magnetism, 831, 1243, 1747
Magnetite, 1039
Magnetization, 817
Major elements, 305, 607
Manganese-53, 567
Mantle, 1001
Mare, 639, 651
  basalts, 607, 979
Mars analog materials, 1473
Martian weathering, 1473
Mascons, 853
Maturity, 409, 433, 475, 529
Megaregolith, 639
Melt rocks, 117, 209
Merrillite, 1315
Mesostasis, 389, 421
Metabasalt, 1
Metal, 1039, 1049, 1135
Metal-silicate partitioning, 949
Metallic grains, 1229
Metallography, 1135
Metamorphism, 323
Meteorites, 463, 1049, 1093, 1105, 1135, 1257, 1359, 1717, 1725
  abundance, 1145
  achondrites, 1257, 1281, 1349, 1359
  alkali-rich, 21
  anorthosites, 305, 577, 767, 781, 791, 979
  chassignites, 1359
  chondrites, 1093, 1105, 1123, 1135, 1145, 1229
    classification, 1039
    matrix, 1039
    unequilibrated ordinary, 1039, 1123
  eucrites, 949, 1281, 1297, 1359
  ferroan, 21
  iron, 1343
Methane, 621
Microcraters, 1703
Micrometeoroids, 1703
Microstructures, 1543
Mineralogy, 1049
Mixing, 173, 389, 577, 607, 791
  diagrams, 607
  models, 371, 509, 781, 791, 915
Mobile elements, 1093

Modal analysis, 433, 475
Modal petrology, 371
Model ages, 173
Models, 1533
Monzonite, 979
Mössbauer effect, 1069
Multi-ring basins, 715
Multispectral imaging, 451, 1473
    cores, 451

NRM, 1229, 1747
Nakhlites, 1359
Nebula, 1123
Nephelometer, 1493
Neutron produced krypton, 541
Nitrogen, 289, 1809
    in meteorites, 1343
Noble gases, 99, 627, 1153, 1167, 1781
Nonmare crust, 21
Nuclear reactions, 1809
Nucleosynthesis, 1803

Olivine, 1069
Olivine clasts, 1315
Optical microscopy, 1069
Orange glass, 935
Orange soil, 541
Orbital geochemistry, 715, 781
Orthopyroxenite clasts, 1315
Overgrowths, 1315
Overturning, 463
Oxygen, 1189
Oxygen fugacity, 935

Paleointensity, 1229, 1243, 1747
Parent body, 1145
Parent magmas, 1349
Partial melting, 915, 935, 949, 1001, 1349
Petrography, 1049
Petrology, 781, 1049
Phase equilibria, 1001
Phase functions, 1769
Phosphorus, 253, 295, 323
Photogeology, 651
Photometric functions, 1769
Pioneer Venus Day Probe, 1493
Plagioclase analyses, 1281
Planetary differentiation, 949
Planetary expansion, 1585
Planetary interiors, 935

Planetesimals, 1641
Post-accretion, 1229
Potassium-argon age, 1199
Potassium-sodium feldspar, 1049
Pre-accretion, 1229
Pre-exposure, 541
Pristine rocks, 21, 253, 781, 1001
Production rates, 541
Protopyroxene, 1123
Provenance, 371, 475
Pyroclastics, 727, 809, 1459
Pyroxene, 1297, 1315

Quartz monzodiorite, 1297

Radar, 1419
Radiation, 1725
Radionuclides, 567
Rake samples, 185
Rare earth elements (REE) 117, 903, 915, 1349
Rare gas trapping, 1177
Rare gases, 1387, 1725
Reflectance spectra, 451, 665, 695
Regolith, 339, 371, 389, 409, 433, 475, 529, 567, 1093, 1257, 1481, 1717
Remote sensing, 665, 679, 695, 751, 767, 791
Resolution, 1419
Reststrahlen, 703
Reworking, 519
Rheology, 1649
Rims, 1079
Roaldite, 1343
Rubidium/strontium, 67, 1027
Rust, 253, 261, 323

SUNOCON, 1781
Samarium, 21
Saturnian ring system, 1667
Scanning electron microscopy (SEM), 1069
Scattering geometry, 1769
Schreibersite, 253, 261, 323
Sedimentology, 1623
Seismic attenuation, 1737
Shear fractures, 1555
Shock, 409
    decay, 1607
    effect, 1297
    metamorphism, 67, 1049, 1607
Siderophile elements, 21, 949

Silicates, 703
Size fraction, 389
Soil, 409, 475, 607, 621, 791
  chemistry, 1
  composition, 577
  evolution, 433
  grain size fraction, 577
  maturity, 433
  mixing, 577
  size fraction, 433
Solar cosmic ray, 627
Solar flare, 627
Solar nebula, 1387
Solar neon, 627
Solar wind, 99, 541, 559, 627, 1725
Soret effect, 991
Spallation reactions, 1809
Spectral reflectance, 679, 1473, 1769
Spectrophotometry, 1769
Sphalerite, 261
Spinel, 1069
Sputtering, 1569, 1599, 1781
Stratigraphy, 451, 463
Stress, 837
  state, 853
  systems, 1555
Substrate, 639
Suevite, 235
Sulfur, 935
Superheavy elements, 1803
Supernova, 1803
Surface, 1543, 1599
  erosion, 1599
  exposure age, 627
  soils, 371

TRM, 1229
Taenite, 1135, 1243
Target properties, 1623
Tectonics, 1585
Tertiaries, 1703
Tetrataenite, 1135
Thermal, 1419
  demagnetization, 1229
  diffusion, 991

histories, 281, 1145
history, 1145
release profiles, 529
Tholeiite, 1001
Thorium, 727, 751
Tidal forces, 1555
Titanium, 727
Trace elements, 305, 607, 1349
Transient cavity, 1607
Transmission electron microscopy, 1297
Trapped noble gases, 541
Tritium, 559
Trivalent iron, 1049
Turbostratic carbon, 1153

VLT basalt, 1
Vaporization, 339
Velocity structure, 867
Venera 9, 1493
Venera 10, 1493
Viking Lander 1, 1493
Viking Lander 2, 1493
Viscosity, 1759
Vogel-Fulcher relation, 1759
Volatile elements, 1093
  immobility, 295
Volatile transport, 99
Volatiles, 253, 261, 323, 1649
Volcanism, 651, 727, 935, 1533
  ancient, 665
  early, 767
  highland, 809
Volcanoes, 1449

Weathering processes, 1105, 1481

X-ray fluorescence data, 715, 767
Xenon, 99, 891, 1725, 1803, 1809

Z-model, 1607, 1689
Zinc, 529

Author Index

Pages 1–1018: Section 1, The Moon
Pages 1019–1823: Section 2, Planets, Asteroids and Satellites

Abdel-Gawad, M., 1737
Adams, J. B., 1473
Ahlberg, L., 1737
Ahrens, T. J., 1623, 1667
Albee, A. L., 117, 173
Allen, J. M., 1079
Allison, R. J., 1703
Andre, C. G., 715, 767
Andreeva, N. E., 1517
Ashwal, L. D., 1359
Austin, M. G., 1689

Bansal, B.M., 67
Barsukov, V. L., 1507
Baryshnikova, G. V., 1049
Basu, A., 433, 475
Becker, R. H., 289, 1189
Beckett, J. R., 1079
Bell, J. F., 665, 679
Berkley, J. L., 1481
Bernatowicz, T. J., 891
Biswas, S., 1093
Blanchard, D. P., 607
Bogard, D. D., 67
Borchardt, R., 185
Bowell, E., 1543
Buchwald, V. F., 1343
Bulau, J., 1737
Burnett, D. S., 1725
Burns, R. G., 695
Buseck, P. R., 1167

Caffee, M., 99
Cintala, M. J., 1607
Cirlin, E. H., 529
Clark, P. E., 727
Clauter, D., 1747
Clayton, D. D., 1781
Coleman, P. J. Jr., 831
Conca, J., 1599
Consolmagno, G. J., 1533

Davis, P. A., 1395
De Hon, R., 639
Decker, D. A., 1507
Delaney, J. S., 1315
Delano, J. W., 339
Dorofeeva, V. A., 1517
Dunn, J. R., 1747
Dyar, M. D., 695
Eberhardt, P., 541
El-Baz, F., 767
Englert, P., 1209
Epstein, S., 289, 1189
Etchegaray-Ramirez, M. I., 751
Eugster, O., 541
Evans, D. L., 1473
Evans, J. C., 567

Fang, C.-Y., 281
Farr, T. G., 1473
Fasano, B. V., 1123, 1759
Fechtig, H., 1641
Ferguson, H. M., 1555
Fink, J. H., 1649
Fireman, E. L., 559, 1019
Frank, K., 1199
Fruchter, J. S., 567
Fuhrman, M., 1257
Fujii, N., 1145
Fuller, M., 1747
Funaki, M., 1229

Garvin, J. B., 1493
Gault, D. E., 1649
Geiss, J., 541
Goettel, K. A., 1507
Goldstein, B. E., 831
Goldstein, J. I., 1135
Gooding, J. L., 1105
Gradie, J., 1769
Greeley, R., 651, 1419, 1649

Grieve, R. A. F., 1607
Griffiths, S. A., 433, 475
Grogler, N., 541
Grömet, L. P., 903
Grossman, L., 1079
Grove, T. L., 935

Haines, E. L., 751
Harlow, G. E., 1315
Haskin, L. A., 41, 791
Hawke, B. R., 451, 665, 679, 727, 781
Hays, J. F., 991
Herpers, U., 1209
Herr, W., 1209
Hess, P. C., 903
Hewins, R. H., 1123
Heymann, D., 1803
Hohenberg, C. M., 99
Hood, L. L., 817
Horvath, P., 867
Hostetler, C. J., 1387
Housen, K. R., 1717
Housley, R. M., 529, 1069
Hudson, B., 99
Hunter, R. H., 253, 261, 323, 337

Ishii, T., 1297

James, O. B., 117, 173, 209
Jovanovic, S., 295, **333**

Keil, K., 21, 1039
Khodakovsky, I. L., 1517
Kieffer, H. H., 1395
Kiesl, W., 541
Klein, L. C., 1123, 1759
Korotev, R. L., 577, 791
Kracher, A., 1

Kramer, F. E., 891
Kurat, G., 1

Lambeck, K., 853
Lange, M. A., 1667
Laul, J. C., 371, 389, 409, 1349
Lesher, C. E., 991
Lindsley, D. H., 339
Lindstrom, D. J., 41
Lindstrom, M. M., 41, 305
Lipschutz, M. E., 1093
Liu, Y.-G., 915
Longhi, J., 1001
Lucchitta, B. K., 1555
Lumme, K., 1543
Lumpkin, G. R., 1153

Ma, M.-S., 915, 1349
MacPherson, G. J., 1079
Malley, J., 185
Marshall, C., 21
Marti, K., 1177
Maxwell, T. A., 715
McDonnell, J. A. M., 1703
McKay, D. S., 433, 475
McKinley, S. G., 1039
McKinnon, W. B., 1585
McSween, H. Y. Jr., 1093
Melcher, C. L., 1725
Meloy, A., 451
Metzger, A. E., 751
Miyamoto, M., 1145, 1297
Morgan, T. H., 703
Mori, H., 1297
Mouginis-Mark, P. J., 1431

Nace, G., 433, 475
Nagata, T., 1229
Nagle, J. S., 451, 463, 519
Nautiyal, C. M., 627
Nehru, C. E., 1315
Nielsen, H. P., 1343
Niemeyer, S., 1177
Norman, M. D., 235
Norris, T., 1019
Novikov, G. V., 1049
Nyquist, L. E., 67

Osadchii, Eu. G., 1049

Ostertag, R., 185

Padia, J. T., 627
Palme, H., 949
Pang, K. D., 1543
Papanastassiou, D. A., 1027
Papike, J. J., 371, 389, 409, 421, 485, 509, 1257
Perkins, R. W., 567
Peterson, C., 1459
Pieters, C. M., 451
Podosek, F. A., 891
Potter, A. E., Jr., 703
Prinz, M., 1315
Pullan, S., 853

Quick, J. E., 117, 173

Rammensee, W., 949
Rao, M. N., 627
Reed, G. W. Jr., 295, 333
Reedy, R. C., 1809
Reeves, J. H., 567
Rehfeldt, A., 185
Reimold, W. U., 67, 185
Rhodes, J. M., 607
Rudowski, R., 339
Ruhl, S. F., 1689
Russell, C. T., 831
Rutherford, M. J., 903

Salpas, P. A., 41, 305
Salvado, C., 1737
Saralkar, C., 1217
Schaeffer, O. A., 1199
Schmitt, R. A., 915, 1349
Schneider, E., 1641
Schonfeld, E., 809
Scott, D. H., 1449
Scott, E. R. D., 1039
Shields, J. A., 1507
Shih, C.-Y., 67
Shirley, D. N., 965
Sidorov, Yu. I., 1517
Simon, S. B., 371, 389, 409
Smith, J. V., 979, 1281
Smith, P. P. K., 1167
Soderblom, L. A. 1395, 1555
Spencer, T. W., 1737

Spudis, P. D., 781
Steele, I. M., 1281
Stoenner, R. W., 559
Stöffler, D., 185
Strangway, D. W., 1243
Sugiura, N., 1243
Surkov, Yu. A., 1377

Takeda, H., 1145, 1297
Tamhane, A. S., 621
Tanaka, K. L., 1449
Taylor, G. J., 21, 1039
Taylor, L. A., 253, 261, 323, 337
Thomsen, J. M., 1689
Tittmann, B. R., 1737
Tombrello, T. A., 1725
Turcotte, D. L., 837

Uhlmann, D. R., 281

Venkatesan, T. R., 627
Veverka, J., 1769
Vizgirda, J., 1623
Volkov, V. P., 1517

Walker, D., 991
Walker, R. J., 421, 485, 509
Warasila, R., 1199
Warren, P. H., 21
Wasilewski, P., 1217
Wasserburg, G. J., 1027
Wasson, J. T., 21, 965
Watson, C. C., 1569
Werle, V., 1641
White, C. C., 409
Wiesmann, H., 67
Willeman, R. J., 837
Willis, J., 1135
Womer, M. B., 651
Wood, C. A., 1359
Wooden, J. L., 67
Wu, J. M., 1759

Yinnon, H., 281

Zimbelman, J. R., 1419